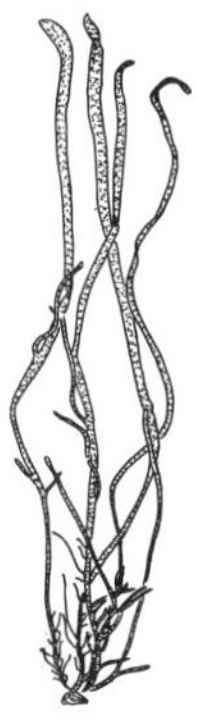

The Seaweeds of Florida

University Press of Florida

Florida A&M University, Tallahassee
Florida Atlantic University, Boca Raton
Florida Gulf Coast University, Ft. Myers
Florida International University, Miami
Florida State University, Tallahassee
New College of Florida, Sarasota
University of Central Florida, Orlando
University of Florida, Gainesville
University of North Florida, Jacksonville
University of South Florida, Tampa
University of West Florida, Pensacola

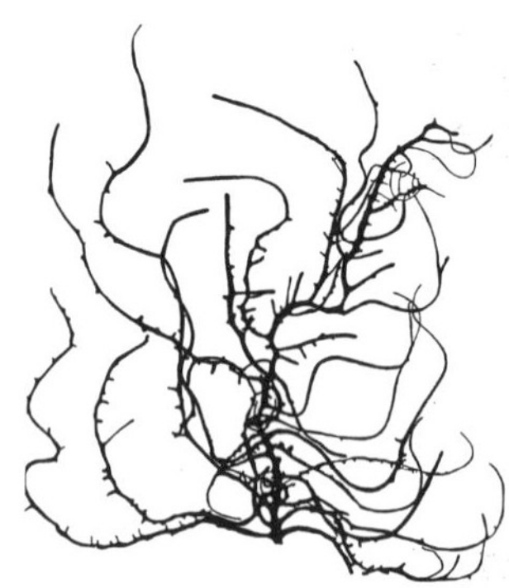

University Press of Florida

Gainesville Tallahassee Tampa Boca Raton Pensacola Orlando Miami Jacksonville Ft. Myers Sarasota

The Seaweeds

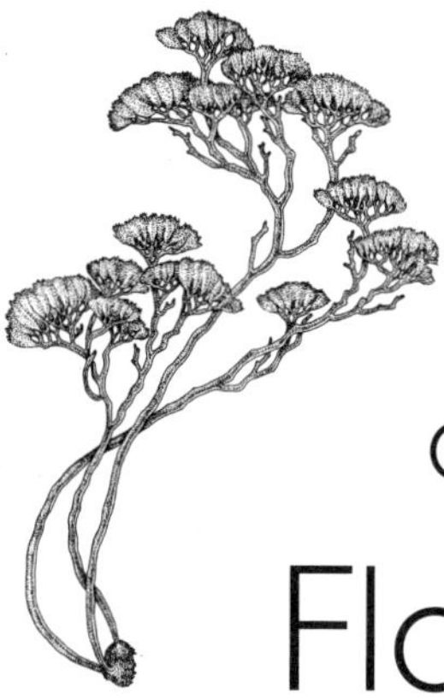

# of Florida

Clinton J. Dawes and Arthur C. Mathieson

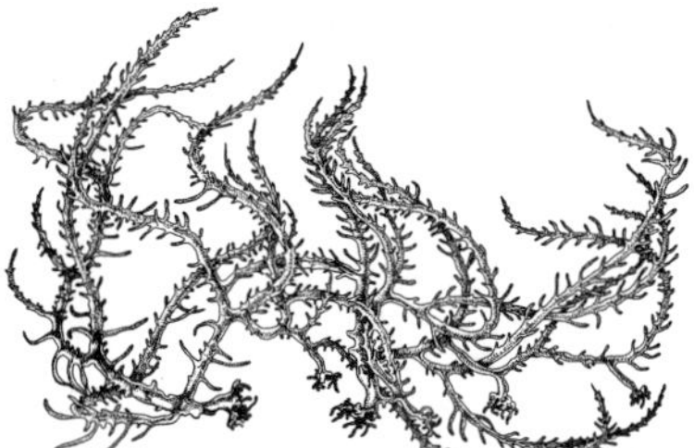

13 12 11 10 09 08 6 5 4 3 2 1

Library of Congress Cataloging-in-Publication Data
Dawes, Clinton J.
The seaweeds of Florida / Clinton J. Dawes and Arthur C. Mathieson.
p. cm.
Includes bibliographical references and index.
ISBN 978-0-8130-3148-4 (alk. paper)
1. Marine algae—Florida—Identification. I. Mathieson, Arthur C. II. Title.
QK571.5.F6D385 2007
579.8'809759—dc22 2007027918

The University Press of Florida is the scholarly publishing agency for the State University System of Florida, comprising Florida A&M University, Florida Atlantic University, Florida Gulf Coast University, Florida International University, Florida State University, New College of Florida, University of Central Florida, University of Florida, University of North Florida, University of South Florida, and University of West Florida.

University Press of Florida
15 Northwest 15th Street
Gainesville, FL 32611-2079
http://www.upf.com

Contents

Figures

Preface

Knowledge of Florida seaweeds has increased significantly since the publication of William Randolph Taylor's classic text in 1960 on the tropical marine algae of the Western Atlantic. Recent data include descriptions of new species, particularly based upon molecular data, new distributional records, and detailed studies of seaweed ecology. During earlier evaluations of seaweeds from Tampa Bay (Dawes, 1967) and the west coast of Florida (Dawes, 1974) it became apparent that a comprehensive text dealing with Florida's total coastline was needed. However, this effort has been delayed until the present collaboration was possible and adequate time available.

In preparing this text we have included ecological and detailed taxonomic and descriptive information for each genus and species as well as taxonomic references, to make the book useful to experts, students, and novices. In so doing we have included one or more figures for almost all of the 693 taxa presented herein and give the derivations of generic and species names.

Many people have helped with the development of the text, especially students of the senior author since his arrival at the University of South Florida in 1964. We are indebted to several professionals who provided invaluable help and advice. In particular, Michael Wynne (University of Michigan) checked many critical references, determined the derivation of several generic and species names, identified many type locations, and helped us obtain several herbarium specimens at the University of Michigan. His checklists of tropical and subtropical western Atlantic seaweeds (Wynne, 1986, 1998a, 2005a) were invaluable to us. Bill Woelkerling (La Trobe University) advised us on the taxonomy of several members of the Corallinales. Donald Pfister (Farlow Herbarium, Harvard University) provided access to several of W. G. Farlow's reprints at Harvard University. Karen Fagerberg (Interlibrary Loan Office, Reference Department, University of New Hampshire) helped with many searches for historical articles and texts. Heather Talbot (Jackson Estuarine Lab, University of New Hampshire) helped create the index and with manuscript preparation and Kathleen Hotchkiss (Department of Biology, University of South Florida) helped modify the plates. Robert Sims and James Norris provided a listing of Florida marine algae deposited at the Smithsonian Institution. Willem Prud'homme van Reine (Rijksherbarium, Leiden) sent several key references, including the early studies of Sprach (1841), Wight (1839), and Krauss (1846). Some of the Latin and French descriptions were translated at the University of South Florida by Professors Richard Wunderlin and John Lawrence, respectively. Generic names were reviewed using *Index Nominum Genericorum (Plantarum)* (http://rathbun.sci.edu/botany/ing) and generic and species names using Algaebase (www.algaebase.org). We thank the scientists who produced these invaluable resources.

Many of the figures in this guide were redrawn by us from a variety of published sources, as named in the figure descriptions. Figures were also drawn from living algae by Linda Baumhart, Jessie Futch, Frank Hoff, Ana Lisa Paul, and Robin Holbrook, all former students at the University of South Florida. Some were scanned from publications and modified by Gary Landtrip of the Department of Biology at USF. Copies of the final line drawings (plates and figures) were scanned and digitized by Bill Mitchell of Printing Services at the University of New Hampshire.

The guide is dedicated to several of our former teachers (I. A. Abbott, E. Y. Dawson, A. W. Haupt, G. J. Hollenberg, and R. F. Scagel) and to our various students (graduate and undergraduate) who have worked with us at our respective institutions for more than 30 years. We acknowledge our respect and admiration for the late Michael Neushul, who introduced each of us to the study of seaweeds while we were graduate students together at the University of California in Los Angeles.

1

Introduction

Historical Review

The study of Florida seaweeds can be divided into three phases, similar to those outlined by Ducker (1981) and Silva et al. (1996) for Australia and the Indian Ocean, respectively. In the first phase, naturalists made collections during scientific cruises and species were described in general floras. General Caribbean collections were made in Jamaica and the West Indies during the late seventeenth and early eighteenth centuries by Sloane (1696, 1707–1725) and Browne (1756, 1789). In the second stage, collections were made by visitors and residents and sent to European experts for analysis and publication, including studies by Harvey (1852, 1853, 1858), Jordan (1874), Eaton (1875), Melvill (1875), and Murray (1888, 1889). Detailed early collections were made by Floretta A. Curtiss, Mrs. G. A. Hall, and Mary S. Snyder (Taylor, 1960) in the Florida Keys and on the east coast, with some of these specimens included in *Phycotheca boreali-americana* (Collins et al., 1895–1919) and *Algae exsiccatae Americae borealis* (Farlow et al., 1877–1889). The third stage overlapped the second and was marked by the collections and publications of resident phycologists, including Hooper (1850), Farlow (1876), Howe (1904), Collins (1909b), Hoyt (1920), and Taylor (1928). Until the 1950s, most Florida collections were made by "northern" visitors during the winter, which contrasts with the primarily summer collections made by Anderson, Collins, Eaton, Farlow, Setchell, Taylor, and others in New England (Taylor 1937a, 1957).

The seaweeds of Florida's west coast and panhandle were essentially ignored because of extensive salt marshes (Earle, 1969). General collections included those from the eastern Gulf of Mexico (Taylor, 1936, 1954a,b; Humm, 1963), Apalachee Bay (Humm, 1952b), Crystal River Bay (Phillips, 1960b; Steidinger and van Breedveld, 1971), the Anclote River Estuary (Hamm and Humm, 1976), and Tampa Bay (Phillips, 1960a; Dawes, 1985). Evaluation of deep water seaweeds included studies from 10–20 m off Pinellas County (Phillips and Springer, 1960), 20–60 m off Tampa and Sarasota Bays (Dawes and van Breedveld, 1969), 25–60 m on the Florida Middle Grounds (Cheney and Dyer, 1974), 20 m off Anclote Key (Dawes and Lawrence, 1990), and the north Florida Reef tract (Lirman and Biber, 2000). Detailed seasonal collections included synopses of green algae from northwestern Florida (Humm and Taylor, 1961), the seaweeds of Tampa Bay (Dawes, 1967), the brown algae from the eastern Gulf of Mexico (Earle, 1969), the intertidal flora at Point of Rocks on Siesta Key (Eiseman, 1975), the flora of the central west coast (Dawes, 1974), and phenological patterns in the Florida Keys and Anclote Key (Mathieson and Dawes 1975).

Floristic evaluations of east coast seaweeds included studies of deep water plants (Eiseman, 1979; Hanisak and Blair, 1988), the intertidal flora of Marineland (Humm, 1952a), the flora of worm reefs off Vero Beach (Juett et al., 1976), and the nearshore ecology at Hutchinson Island

(Moffler and van Breedveld, 1979). Estuarine studies were also conducted in the Indian River Lagoon and Tampa Bay (Phillips, 1961; Benz et al., 1979; Virnstein and Carbonara, 1985).

Although the Florida Keys have been the subject of several ecological studies, few publications have focused on their overall seaweed floras, with most emphasizing descriptions of new species or distributional records. Two early detailed studies included an evaluation of Dry Tortugas flora (Taylor, 1928) and intertidal zonation at Vaca, Crawl, Plantation, and West Summerland Keys (Stephenson and Stephenson, 1972). In addition, a four-year seasonal study of seaweeds from Content Keys (Croley and Dawes, 1970) was followed by a second set of multiple collections that showed a decline in species richness after thirty years (Dawes et al., 1999b). Bucher et al. (1990) also conducted floristic studies at Molasses Key, describing several new records from Florida and new species of seaweeds.

Anthropogenic effects have been occurring in Florida for a very long time. Thus historian Michael Gannon (Ewel, 1990) stated, "When the pilgrims landed at Plymouth Rock, St. Augustine was up for urban renewal." Coastline modification and human impacts on coastal waters have accelerated since the publication of Taylor's (1960) *Marine Algae of the Eastern Tropical and Subtropical Coasts of the Americas*, which provided a baseline for future studies. In 2000, the population of Florida exceeded 15 million; thus it had more than doubled since the publication of Taylor's (1960) text. The loss of seagrass beds (Livingston, 1990; Dawes et al., 2004), salt marshes (Montague and Wiegert, 1990), and mangrove forests (Odum and McIvor, 1990) has been extensive, resulting from coastal development, canal building, agriculture, and pollution. Since 1970, coastal development has limited shore access, damaged tidal marshes, and polluted coastal waters. Data from the Florida Keys Sanctuary (est. 1990) indicate that the Florida coral reef tract is under severe stress because of macroalgal growth, sedimentation, turbidity, and pollution (Jaap and Hallock, 1990; Lapointe, 1997, 1999; Lapointe and Barile, 2001; Lapointe et al., 1994, 2004, 2005a).

Public and governmental agencies are now aware that habitat restoration and preservation are critical for marine communities. Protection of sensitive marine habitats in the Keys includes efforts by Everglades National Park, Florida Keys Sanctuary, Pennycamp State Park, and Looe Key Preserve. On the east coast preservation efforts include Biscayne Bay National Park, the South Florida Water Management District (protection of Florida Bay and restoration of the Everglades), and the Indian River Estuary Program. On the west coast restoration and protection efforts include the South and Southwest Water Management Districts, and National Estuary Programs in Charlotte Harbor, Tampa Bay, Suwannee River, and Pensacola Bay, as well as local preserves (for example, Hillsborough County's Cockroach Bay Aquatic Preserve). However, because of increasing population, more restrictions on coastal development and protection of submerged habitats are needed. Hopefully restoration of habitats and improvement of coastal water quality in Florida will continue and allow the recovery of many seaweed habitats.

Florida's Environment

Geology

The Florida peninsula is a tectonically stable carbonate platform that was sutured onto the North American continent about 220 million YBP (Webb, 1990). The peninsula is covered by Quaternary quartz sands, being relatively thick on the east coast and less so on the west coast. Florida's coastline has changed significantly within the past 100,000 years. For example, about 18,000 YBP, Florida was two to three times larger than today and five to six times larger than it was 110,000 YBP (fig.

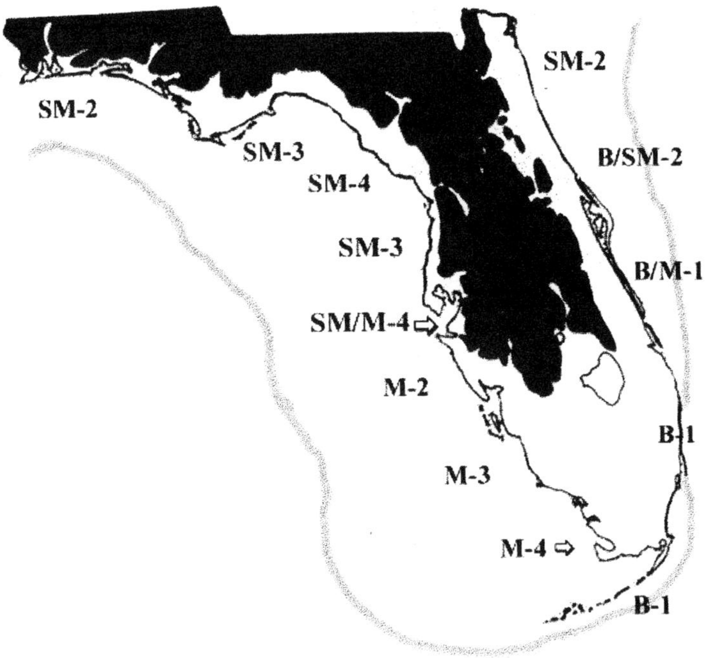

Figure 1. An outline of Florida's historical and present coastlines (after Webb, 1990), coastal communities, and levels of wave energy (after Tanner, 1960). *Stippled line*: Surface of Florida about 18,000 YBP, during the last major glacial period. *Solid line*: Present shoreline. *Black area*: Surface of Florida about 110,000 years ago, during the last interglacial period. Coastal communities include salt marshes (SM), mangrove forests (M), and barrier islands (B). Average coastal wave energy ranges from greater than 50 cm (1), to 10–50 cm (2), 0–10 cm (3), and 0 cm (4).

1). The Florida peninsular coast can be divided into five geomorphic elements: (1) an east coast barrier system; (2) a southern limestone island arc that comprises the Florida Keys; (3) the mangrove coast of southwest Florida; (4) the central Gulf barrier island system; and (5) the marsh coast of the Big Bend area (Davis et al., 1992). In addition, a sixth component, the northeastern "panhandle," is characterized by a chain of barrier islands. The carbonate platform was active, accumulating carbonate sediments from its inception during the Mesozoic until the late Paleocene, when the Suwannee Straits seaway (Gulf Trough) was filled. Until then, the seaway prevented transportation of silica-based sediments from the southern Appalachian Mountains to the carbonate-producing environments of Florida. The continental shelf of Florida's northeastern coast has a narrow, steep sloping shelf that becomes further constricted in the south. From Cape Canaveral to the Florida Keys, the shelf is narrow (<1 km wide) and steep. Wave energy along the east coast is also high (fig. 1), with mean annual wave heights of 70 cm (Tanner, 1960). By contrast, the west coast, with its broad shelf (ca. 200 km wide) and limited fetch from the Gulf of Mexico, experiences low mean annual wave heights of 10–25 cm (Tanner, 1960; Davis et al., 1992), while the panhandle experiences moderate wave energy with average breaker heights to 50 cm (fig. 1).

Coastal Features

Tidal ranges throughout Florida are microtidal (<2 m), except for the northeast section, where they may be mesotidal (>2 m) during spring tides. While the east coast and Florida Keys are wave dominated, the southwest mangrove communities and coastal northwest tidal marshes are tidally dominated, with spring tides of about 1.3 m and low wave energy. The northeastern part of Florida is a mixed-energy coast where tidal ranges are maximal (to 1.7 m), whereas the west central barrier island system has tides that reach 1.0 m. Severe erosion occurs on the lower east coast (for example, West Palm Beach) where the continental shelf is limited and slopes are steep. Similar patterns occur on the west central Gulf barrier system, where there is a limited supply of sediment. Thus, on the east coast, mangals, salt marshes, and seagrass beds occur behind barrier islands like those of the Indian River Lagoon (Littler and Littler, 2003).

On the west coast, mangrove forests border the southwest area (for example, Ten Thousand Islands) and occur behind barrier islands and in bays on the west central coast (for example, Longboat Key, Sarasota Bay, Tampa Bay). Salt marshes fringe the northwest coasts, including Apalachicola Bay. As a result of high wave energy on the northern coast (i.e., Florida Panhandle), salt marshes again occur behind barrier islands (fig. 1). The intertidal and shallow hard bottom communities of the Florida Keys support the most extensive seaweed communities. By contrast, intertidal natural hard substratum (i.e., limestone) is limited to beach rock outcroppings on Siesta Key on the west coast, as well as Marineland, Blowing Rocks, and West Palm Beach on the east coast. In addition, there are many manmade structures, including forts, piers, pilings, seawalls, breakwaters, groins, and jetties. In the south, the low wave energy coasts of Florida Bay and its islands support mangrove forests, while the southern, more exposed coast of the Florida Keys consists of limestone with a diverse seaweed flora.

Climatic Features

Florida's high seaweed diversity (693 total taxa) probably reflects the state's latitudinal span (6.5°) and length (~700 km) that includes warm temperate to tropical habitats (fig. 2). Lying between 24.5° and 31.0° north latitudes, the state is strongly influenced by the western side of the Bermuda/Azores high. It has a humid subtropical climate as a result of maritime influences of the Caribbean Sea and Gulf of Mexico, with a progressive shift from temperate to tropical waters below 26.5° north latitude, as shown by the north-to-south increase in annual temperatures (Chen and Gerber, 1990). Water temperatures range from 18 to 24 °C in winter to 27 to 30 °C in summer. The state's coastal waters are influenced by the strong climatic cycle on land; the cool, dry air of winter lowers water temperatures and raises salinities. In summer, the warm rainy weather increases water temperatures and lowers salinities. In northern Florida, climatic events include winter and spring cold fronts that produce freezing temperatures and thus limit the northern expansion of mangroves (Chapman 1977a; Dawes 1998). In the warm, wet summers and falls, tropical storms and hurricanes can cause severe coastal damage that results in the production of new tidal channels, destruction of mangrove forests (Dawes et al., 1995), and the removal of seagrass and macroalgal beds (Fourqurean and Rutten, 2004).

Seaweeds

Characteristics

Seaweeds are macroscopic marine algae (singular = alga) or "photosynthetic, nonvascular plants that contain chlorophyll *a* and have simple reproductive structures" (Trainor, 1978). Algae lack true roots, stems, leaves, and vascular tissue (i.e., xylem and phloem), in contrast to flowering

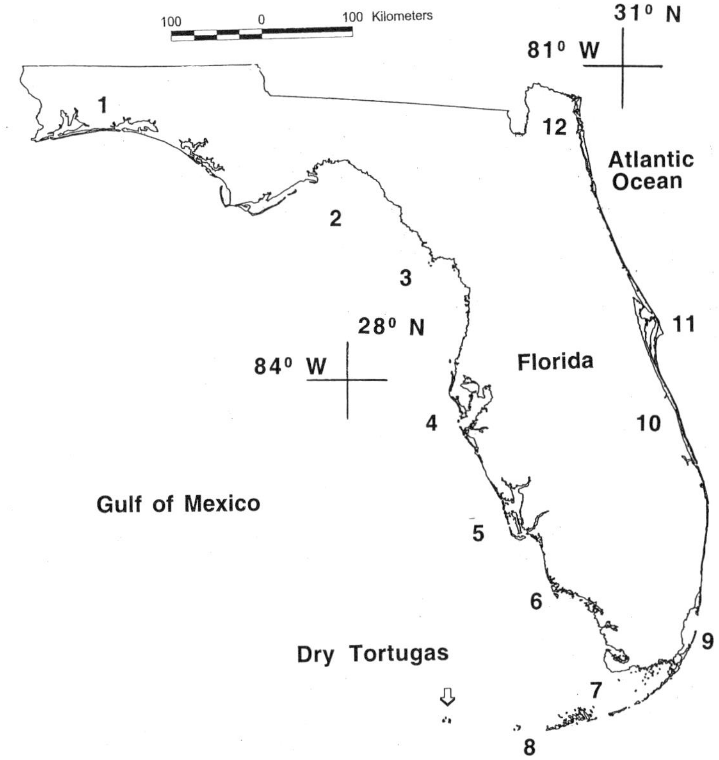

Figure 2. Selected Florida coastal sites. 1. Florida Panhandle, Pensacola Bay. 2. Apalachicola Bay. 3. Cedar Key. 4. Tampa Bay. 5. Sanibel Island, Charlotte Harbor. 6. Cape Sable. 7. Cape San Blas at Florida Bay. 8. Dry Tortugas (*arrow*) and Florida Keys. 9. Biscayne Bay. 10. Indian River Lagoon. 11. Cape Canaveral. 12. Amelia Island.

plants. Further, unlike mosses and vascular plants, algae have sex organs that are usually one celled or, if multicellular, all cells function in reproduction. According to Baldauf (2003) the phyla Chlorophyta (green algae) and Rhodophyta (red algae) are placed in the Kingdom Plantae, while the brown seaweeds (class Phaeophyceae) are placed in the phylum Heterokontophyta and kingdom Protista. There are about 900 species of green algae, 997 brown algae, and 2,540 red algae worldwide (Dring, 1982).

A variety of morphological and anatomical features are useful in identifying seaweeds, including types of pigments and reserve foods, cell wall chemistry, thallus (body) construction, branching patterns, presence of specialized cells, and modes of growth (apical, intercalary, or diffuse). The basic morphologies of seaweeds include: (1) uniseriate or multiseriate filaments (with one or more rows of cells) that are either unbranched (*Chaetomorpha* spp.; plates II and III) or branched (*Cladophora* spp.; plate III); (2) axes of intertwined filaments that are siphonous (*Codium* spp.; plate V) or filamentous/hollow in construction (*Cladosiphon occidentalis*; plate XVII, fig. 1, 2); (3) axes that are solid (*Hypnea cornuta*; plate XLII, fig. 21, 22) or hollow (*Rosenvingea intricata*; plate XVIII, fig. 18, 19); and (4) flattened, foliose, or bladelike plants (*Ulva lactuca*; plate I, fig. 20). Cylindrical or bladed species may have a dense central filamentous core or medulla (*Eucheuma isiforme* var. *denudatum*; plate XLIV, fig. 10), a pseudoparenchymatous core with spherical or cubelike cells (*Hummia onusta*; plate XVII, fig. 7), or be mostly hollow with a few medullary filaments (*Solieria filiformis*; plate XLIV, fig. 17). Axes can also be parenchymatous, with closely compacted isodiametric cells that lack filaments (*Dictyota menstrualis*; plate XIII, fig. 14).

Apical growth in some seaweeds includes division of a single cell, as in the brown alga *Dictyota crispata* (Plate XIII, fig. 10), or division by a group of apical cells (meristem), as in the red seaweed *Wurdemannia miniata* (plate XLV, fig. 3). By contrast, growth by cell division may occur throughout the plant (diffuse meristem) or be restricted to specific areas (intercalary meristem), as in the filamentous brown alga *Bachelotia antillarum* (plate XVII, fig. 20). Trichothallic growth is another form of intercalary growth where cell divisions produce apical hairs or filaments in addition to new branch tissue, as seen in the brown seaweed *Sporochnus pedunculatus* (plate XIX, fig. 10). Trichothallic hairs may be ephemeral or long lasting.

Classification

Algae are placed in various phyla, like those of the plant (Plantae) and animal (Animalia) kingdoms. Each phylum can be subdivided into one or more classes and orders. Families make up the next subgroup and within each family various genera and their species occur. Species are named according to the binomial system of nomenclature established by Linnaeus (1753) in his first edition of *Species Plantarum*. In this volume he listed 56 seaweeds with their previous polynomial names plus one new species. The 57 species were placed in three genera: 27 species in *Fucus* (cartilaginous thalli), 9 in *Ulva* (bladelike thalli), and 21 in *Conferva* (filamentous thalli). Examples of Florida seaweeds named by Linnaeus include the green alga *Ulva lactuca* Linnaeus, the brown *Sargassum natans* (Linnaeus) Gaillon (originally named *Fucus natans*), and the red *Acanthophora muscoides* (Linnaeus) Bory (originally named *Fucus muscoides* Linnaeus). A scientific name is a binomial generic and specific designation (in Latin), followed by the name(s) of the author(s) who initially described it or transferred it to another taxon. Generic and specific epithets (i.e., names) are descriptive terms denoting a feature of the plant, a name of a person, or a location.

One example of a classification is the green alga *Ulva lactuca* Linnaeus, which is commonly called "sea lettuce." *Ulva* is the generic name, based upon the Latin term for marsh plant. Its specific epithet *lactuca* is a descriptive term meaning "lettuce." The alga is classified as follows:

Taxonomic level	Names	Common endings of names
Phylum	Chlorophyta	-phyta
Class	Ulvophyceae	-phyceae
Order	Ulvales	-ales
Family	Ulvaceae	-aceae
Name (genus and species)	*Ulva lactuca* Linneaus	

The alga's name is italicized (or underlined) because it is in Latin, and it includes the author. If a species is transferred to another genus, the original author's name for the species is retained and the person(s) who made the transfer is also cited. For example, the green seaweed *Caulerpa prolifera* (Forsskål) Lamouroux indicates that Forsskål is the original author of the species *Fucus prolifera* (i.e., the basionym) and that Lamouroux transferred it to *Caulerpa*.

Floristics

The present guide to Florida seaweeds contains 693 taxa, including 224 green (Chlorophyta), 102 brown (Heterokontophyta), and 367 red (Rhodophyta) algae (table 1). The Heterokontophyta contain 96 brown (Phaeophyceae), 3 yellow green (Xanthophyceae), 1 golden brown (Chrysophyceae), and 1 colonial cryptomonad-like species (*Sedis incertae*). The numbers of taxa given in the present account are higher than those listed for Florida by Taylor (1960) and Littler and Littler (2000). As shown in table 1, the numbers of algal taxa described for Florida exceed those reported for Bermuda (Schneider, 2003), the Malaysia-Indonesia area (Weber-van Bosse 1901, 1904a,b,c, 1913, 1914, 1921, 1923, 1928) and tropical west Africa (Lawson and John 1987), while they are significantly lower than those reported for the subtropical and tropical coasts of the western Atlantic, where they range from 752 taxa to 1,442 taxa (Taylor, 1960; Diaz-Piferrer, 1969; Wynne, 2005a). The high diversity in the Indian Ocean (2,991 taxa) probably reflects the large area and variety of habitats in this geography (Silva et al., 1996).

The two floristic indexes outlined in table 1 show the complexity and tropical nature of Florida's seaweed flora. That is, the R:B Index of Feldmann (1937) compares the number of red algal taxa divided by the number of browns and the R+G:B Index of Cheney (1977) represents the total number of red and green algae divided by the number of browns. In both ratios, the seaweed flora of Florida shows strong tropical affinities (3.6 and 5.9), similar to tropical West Africa (4.6 and 6.0)

Table 1. Comparison of Tropical Seaweed Floras by Class

Area	Total taxa[a] (species, varieties, forms)				Ratios[b]		Source
	Green	Brown	Red	Total	R:B	R+G:B	
Florida							
	116	56	226	398	4.0	6.1	Taylor 1960[c,d]
	79	29	150	258	5.2	7.9	Croley & Dawes 1970[e]
	83	42	136	261	3.2	5.2	Dawes 1974[e]
	134	60	186	380	3.1	5.3	Littler & Littler 2000[c,e,f]
	224	102	367	693	3.6	5.9	present text[e]
Gulf of Mexico							
	61	38	85	184	2.2	3.8	Waters et al. 2001[d]
Subtropical and Tropical Western Atlantic							
	233	98	421	752	4.3	6.7	Taylor 1960[d]
	221	103	496	820	4.8	7.0	Diaz-Piferrer 1969[d]
	415	200	827	1442	4.1	6.2	Wynne 2005a[e]
	187	64	266	517	4.2	7.1	Littler & Littler 2000[e,f]
Bermuda							
	118	66	240	431	3.6	5.4	Schneider 2003[d]
Canary Islands							
	73	59	97	229	1.6	3.0	Børgesen 1925, 1926, 1927, 1929, 1930a[e]
Malaysia and Indonesia							
	139	94	396	602	4.2	5.7	Weber-van Bosse 1901, 1904a,b,c, 1913, 1914, 1921, 1923, 1928[e]
Tropical W. Africa							
	63	45	208	316	4.6	6.0	Lawson & Johns 1987[e]
Indian Ocean							
	585	596	1810	2991	3.0	4.0	Silva et al. 1996[3]

a. green = Chlorophyceae; brown = Phaeophyceae; red = Rhodophyceae
b. R:B = red to brown; R+G:B = red + green to brown
c. Listed for Florida without specific sites.
d. Species level only.
e. Includes subspecies, varieties, and forms.
f. Seaweeds from reef areas of Caribbean.

and the subtropical to tropical western Atlantic (4.1 and 6.2). The ratios for Florida are similar to those of Bermuda (3.6 and 5.4) and much higher than those of the Canary Islands (1.6 and 3.0), which experience cooler Atlantic water. Florida's ratio is also lower than that of the entire Caribbean (table 1.1: Taylor, 1960; Diaz- Piferrer, 1969; Wynne, 1998a; Littler and Littler, 2000), which reflects the inclusion of temperate brown algae on Florida's northeastern and western coasts. The temperate aspect of the northern Gulf of Mexico is also demonstrated by the low ratios recorded for seaweeds on oil rigs within the northern Gulf of Mexico (2.2 and 3.8, respectively). None of the floras outlined in table 1 have low R+G:B ratios that are typical of arctic and cold temperate waters (Mathieson et al., 1991) that range from 1.1 (Greenland) to 3.1 (Gulf of Maine).

Seaweeds play major roles in primary productivity and food chains of marine and estuarine communities, including salt marshes, mangrove forests, and seagrass beds, hard substrata habitats (for example, "live bottoms"), and coral reefs. Deep water (ca. 20–200 m) seaweed communities off the east and west coasts of Florida and the Dry Tortugas have been described from dredge, SCUBA, and submersible samples; they include new species records and range extensions from the Caribbean. Additional records of new species and enhanced distributional records will no doubt occur based upon future studies in the Florida Keys Sanctuary, on both coasts, and from detailed molecular studies.

Salt Marshes

Florida's salt marshes form prairielike intertidal communities (fig. 3) on the northern, eastern, western, and panhandle coasts (fig. 1). Estimates of their coverage range from 155,000 ha (Draft highlights, 1984) to 189,000 ha (Lewis et al., 1985a). Salt marshes occur on the open coast southward to Tampa Bay, as well as behind mangals farther south or where frost occurs on the west coast. On the east coast, salt marshes extend southward to Merritt Island, as well as behind fringing mangals along the Indian River Lagoon (fig. 1; also see Littler and Littler, 2003).

About 60 percent of Florida's salt marshes are on the northwest and Panhandle coasts, 20 percent on the northeastern (fig. 4b), and 10 percent each in the Indian River Lagoon and on the south Florida coasts (Montague and Wiegert, 1990). West coast marshes north of Tampa Bay and along the Panhandle have irregular tidal cycles and are dominated by *Juncus roemerianus* Scheele (fig. 3, 4a), with *Spartina alterniflora* Loiseleur-Deslongchamps usually forming a border on the tidal channels or river edges (Kurz and Wagner, 1957; Kruczynski, 1982). By contrast, east coast salt marshes from Marineland through Georgia have similar vegetation and regular diel tidal flooding (fig. 4b), as well as hydrology, climate, and extensive coverage by *S. alterniflora* (Teal, 1962; De la Cruz, 1981). Salt marshes on the Indian River are mostly found along mosquito control channels and usually above mean high water (MHW); they are flooded by windblown tides or seasonal rises in sea level (Provost, 1973). Salt marshes on both coasts of south Florida compete with mangroves and are restricted to seaward and landward areas of mangals (Lewis et al., 1985a); they are most extensive where freezing temperatures are more common. High marshes are formed landward of fringing mangrove forests in areas of infrequent frosts (for example, northern Tampa Bay and Indian River).

Vascular Plants

As outlined by Kurz and Wagner (1957), the two dominant species in Florida's salt marshes are black needle rush (*Juncus roemerianus*) and smooth cordgrass (*Spartina alterniflora*). Populations

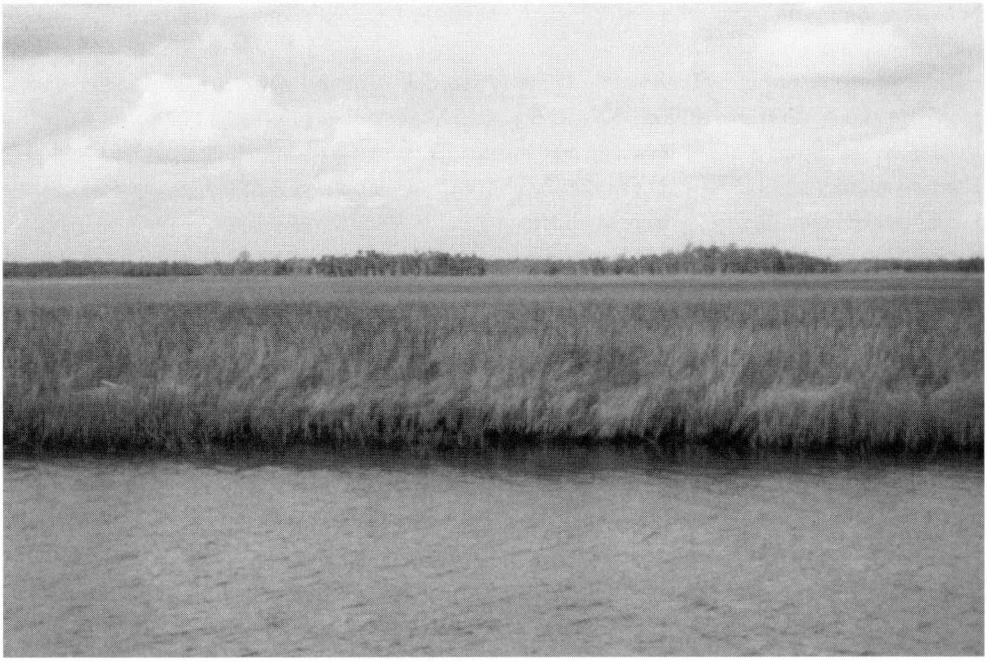

Figure 3. The salt marsh at the Weeki Wachee River (west central Florida) that is dominated by the black rush, *Juncus roemerianus*.

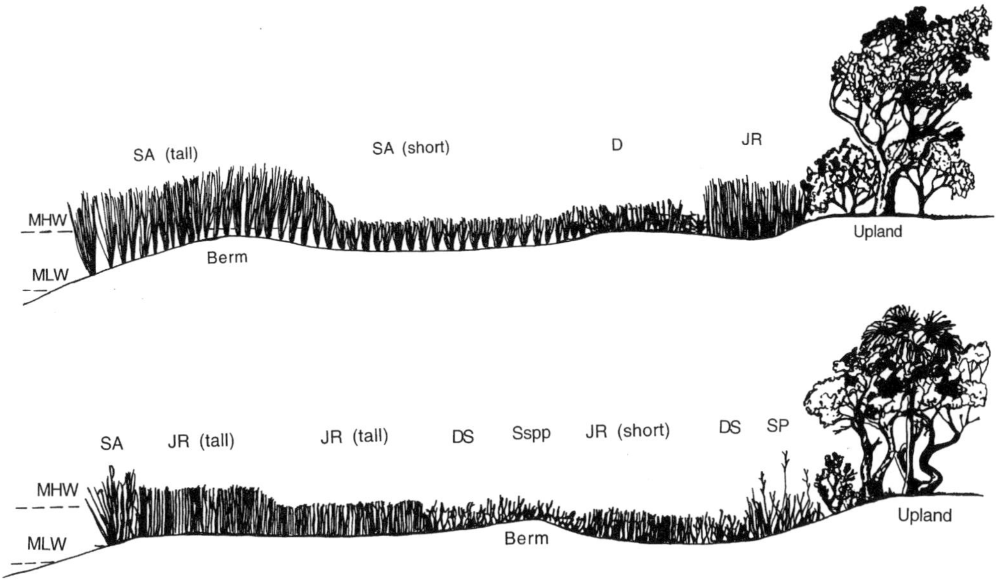

Figure 4. Florida salt marsh zonation diagrams (after Montague and Wiegert, 1990), including Apalachicola Bay, northwestern Florida (*above*) and the west side of Amelia Island, northeastern Florida (*below*). D = Succulent and woody dicots (e.g., *Batis maritima* and *Borrichia frutescens*), DS = *Distichilis spicata*, JR = *Juncus romerianus*, MHW = mean high water, MLW = mean low water, SA = *Spartina alterniflora*, SP = *Spartina patens*, Sspp = *Salicornia* spp., Upland = Coastal trees and shrubs.

of *S. alterniflora* are usually limited to narrow bands along tidal channels or rivers, while *J. roemerianus* commonly forms extensive, prairielike fields. Montague and Wiegert (1990) list 23 common salt marsh plants for Florida, including succulent dicots [for example, *Batis maritima* Linneaus, *Suaeda linearis* (Elliott) Mog.], dry xerophytic monocots [i.e., spike grass *Distichlis spicata* (Linnaeus) Greene], and woody dicots [i.e., *Borrichia frutescens* (Linnaeus) De Candolle]. *Spartina patens* (Aiton) Muhl., or salt meadow cordgrass, that is common in high marshes of New England (Chapman, 1977a), forms small fringing patches shoreward of the primary marsh. Salt barrens (salt pans) usually occur on the landward side of salt marshes and are fringed by succulents such as *Sesuvium portulacastrum* (Linnaeus) Linnaeus and *Salicornia perennis* Mill. (Hoffman and Dawes, 1997).

Zonation and Ecology

Abrupt changes from monospecific stands of *Spartina alterniflora* to *Juncus roemerianus* usually occur within a few cm of vertical relief. If this elevational change is rapid, the two species may grow together (Stout, 1984). In general, Florida salt marshes are limited to the upper and mid intertidal zones, with a bare zone of sand or mud in the low intertidal area (fig. 4a,b). Plant zonation in tidal marshes is influenced by physical factors, including wave energy, tidal regime, coastal topography, sediment type and rate of deposition, as well as biological factors including competition and grazing (Adam, 1990; Chapman, 1977a; Dawes, 1998). Sediments in Florida marshes usually contain less silt and clay and more sand than Georgia marshes. The zonation in coastal tidelands frequently consists of overlapping vertical ranges of individual species (Gray, 1992). Eight plant zones have been identified in northwest Florida (Kurz and Wagner, 1957), two on the east coast (Provost, 1973), and three on the west coast (Stout, 1984).

In west coast salt marshes, the bare mud or sandy surfaces of the lower intertidal area are colonized by edaphic microalgae (greens and diatoms) and cyanobacteria. Above this, *Spartina alterniflora* may form a narrow (fringing) lower border (fig. 4a), while *Juncus roemerianus* is the dominant meadow plant in the lower slope (Dawes, 1974). The upper slope is usually dominated by a combination of *Distichlis spicata* and *S. patens* that grades into an upper border covered by succulents such as *Salicornia* spp., as well as halophytic herbs or scrubby bushes [*Iva imbricata* Linnaeus and *Fimbristylis spadicea* (Linnaeus) Vahl]. By contrast, salt marshes of northeastern Florida usually contain large expanses of tall and short *Spartina alterniflora* (fig. 4b) and are flooded and drained twice daily, similar to Georgia marshes (De la Cruz, 1981).

Salt barrens or salterns are common in many salt marshes throughout Florida, occurring between upland communities and salt marshes or mangrove forests (fig. 6b; Hoffman and Dawes, 1997; Dawes, 1998). Salterns are flat expanses of firm sediment with fringing halophytes comprising mostly succulents such as *Batis maritima*, *Sesuvium portulacastrum*, and *Salicornia perennis*. Areas of bare sediment contain cyanobacteria and often have a pink hue resulting from sulfur bacteria. Sediment salinities in Florida mangrove and salterns range from 40 to 170 ppt during dry periods (Hoffman and Dawes, 1997). The high salinities are due to evaporation after spring tides flood the upper regions of marshes. Based on elevational surveys, Florida salterns result from shallow berms 5–10 cm high that prevent complete flushing and cause standing seawater. Salterns have three major ecological roles: storage and release of salts, including orthophosphate (Ridd et al., 1988), provision of transitional areas for plant succession (Adams, 1963), and habitat for fish and wildlife based on an abundance of brine shrimp and open areas for wading birds (Anonymous, 1986; Dawes, 1998).

Salt marshes have at least five ecological roles in coastal ecosystems, including primary production, a direct source of food, habitat, stabilization of sediments, and filtration of land runoff (Dawes, 1998). Primary production varies with the marsh and site, ranging from 130 to 2,500 g dry matter m^{-2} y^{-1}. Kruczynski et al. (1978) estimated that aboveground production in the St. Marks National Wildlife Refuge (on Apalachee Bay; fig. 2, site 2) ranged from 250 to 950 g dry matter m^{-2} y^{-1}. Similar estimates are known for the Crystal River Bay north of Tampa Bay (fig. 2, between sites 3 and 4; Knight and Coggins, 1982). On the east coast, *Spartina bakeri* Merrill was found to be the dominant species in a restored site on Merritt Island, with aboveground net production of 2,000 to 2,500 g dry matter m^{-2} y^{-1} (Chynoweth, 1975). Usually only 10 percent of the primary production is used directly as food, with most plant biomass decomposing and forming detritus in the marsh (Montague et al., 1987). With their high productivity and dense stands of plants, salt marshes function as habitats and nurseries for a variety of resident and transient animals, especially in tidal creeks and pools. By contrast, the diversity of insects and spiders in salt marshes is low compared to terrestrial habitats, presumably reflecting the lower diversity of plants (Pfeiffer and Wiegert, 1981).

Micro- and Macroalgae

Diatoms account for almost 70 percent of the microalgae on sediments in salt marshes (Montague and Wiegert, 1990) and they are similar to those recorded in Mississippi (Stout, 1984) and Georgia (Pomeroy et al., 1981). About 90 percent of salt marsh diatom species belong to four genera: *Cylindrotheca*, *Gyrosigma*, *Navicula*, and *Nitzschia* (Williams, 1962). Common cyanobacteria include species of *Anabaena*, *Microcoleus*, and *Schizothrix*.

The diversity of macroalgae in Florida salt marshes exceeds that reported for Georgia (Pomeroy et al., 1981); these marshes are most diverse and abundant on the west coast north of Tampa Bay (Dawes, 1974). The above pattern may reflect the effect of warm coastal waters and spring-fed rivers. Macroalgae may be attached to driftwood, shells, and culms of salt marsh plants; they may also form a turf, occur in drift, or grow directly on sediments.

Mangrove Forests

Mangrove communities (mangals) form impressive intertidal communities on the central and southern coasts of Florida where wave action is low (fig. 5). All three Caribbean mangrove species are present (fig. 6a,b), including *Rhizophora mangle* Linnaeus (red mangrove), *Avicennia germinans* (Linnaeus) Linnaeus (black mangrove), and *Laguncularia racemosa* (Linnaeus) C. F. Gaertner (white mangrove). *Conocarpus erectus* Linnaeus (buttonwood) is a common upland plant or "mangrove associate" that does not have any degree of root modification for saturated saline soils nor does it reproduce by vivipary (Odum and McIvor, 1990).

Mangrove forests in Florida cover between 189,725 ha (Anonymous, 1974) and 272,873 ha (Lewis et al., 1985a), with 90 percent of the forests occurring in the four southern counties of Lee, Collier, Monroe, and Dade (fig. 1, M-3, M-4). Mangrove forests have been extensively damaged by anthropogenic impacts on the east (for example, Biscyane Bay) and west coasts (Tampa and Sarasota Bays), plus the Florida Keys, but not on the southwest coast (Ten Thousand Islands). Being tropical trees, mangroves do not survive where the annual average temperature falls below 19 °C (Waisel, 1972). Of the three species, the red mangrove is the most sensitive to frost. During the past fifty years, fluctuations in Florida's climate have allowed mangroves to extend north of Tampa Bay on the west coast to Crystal River Bay (fig. 1, SM/M-4) and to Ponce de Leon Inlet on the east coast (fig. 1, B/M-1).

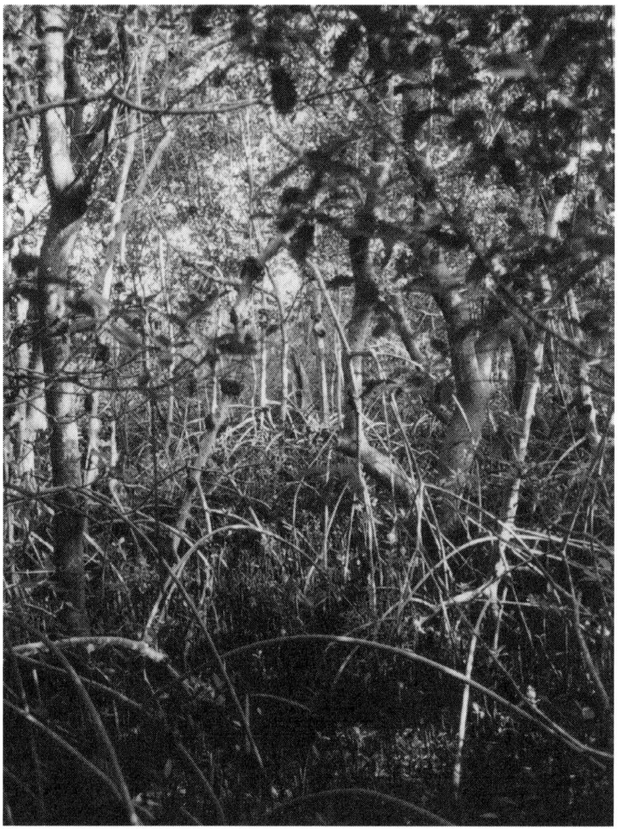

Figure 5. The interior of a fringing mangrove forest in Tampa Bay. The prop and drop roots of the red mangrove (*Rhizophora mangle*) and pneumatophores of the black mangrove (*Avicennia germinans*) form a dense thicket.

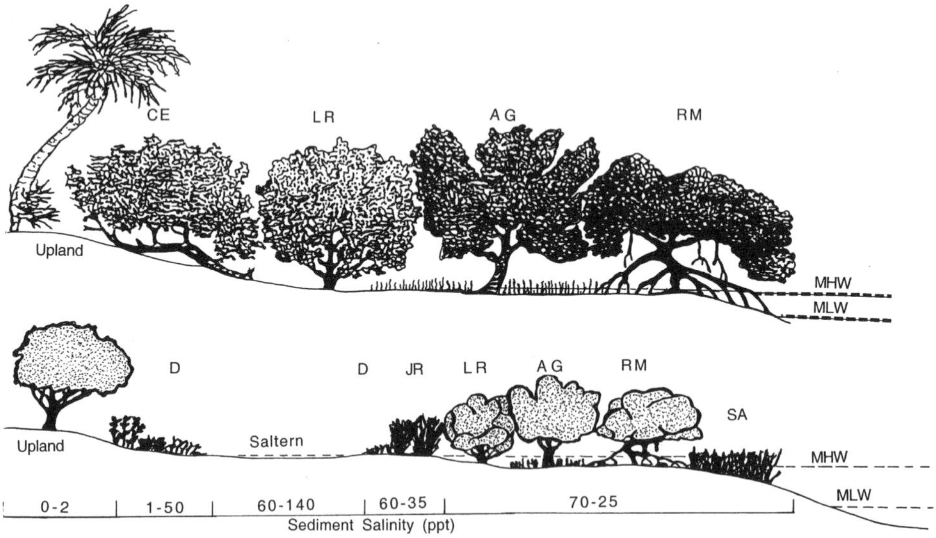

Figure 6. Zonation diagrams of Florida mangals (with permission of R. R. Lewis III), including a fringing mangal in south Florida (*above*) and in Tampa Bay with a saltern and sediment salinity profile (*below*). AG = *Avicennia germinans*, CE = *Conocarpus erectus*, D = succulent and woody dicots (e.g., *Batis maritima*, *Borrichia frutescens*, and *Salicornia* spp.), DS = *Distichlis spicata*, JR = *Juncus roemerianus*, LR = *Laguncularia racemosa*, MHW = mean high water, MLW = mean low water, RM = *Rhizophora mangle*, SA = *Spartina alterniflora*, Upland = coastal trees and shrubs.

The distribution of mangroves and salt marsh species is linked to several major coastal features: climate, coastline protection, shallow depths, tidal currents, salinity, tidal amplitude, and substrata (Chapman, 1977a; Odum et al., 1982; Robertson and Alongi, 1992; Smith, 1992; Dawes, 1998). Mangals also form forests on oceanic islands, such as the limestone reefs of the Florida Keys where freshwater and sediment are limited (Croley and Dawes, 1970). As outlined by several authors (Lugo and Snedaker, 1974; Cintròn et al., 1985; Dawes, 1998), Florida mangals can be classified as six major types: fringing (coastal), overwash islands, riverine (along rivers), basin (behind berms), hammock (mouths of rivers), and dwarf communities.

Associated Vascular Plants

In contrast to salt marshes, mangals contain few associated flowering plants or understory species, except for macroalgae (Janzen, 1985; Lugo, 1986; Snedaker and Lahmann, 1988). Most associated flowering plants occur on the fringes of mangals; for example, *Spartina alterniflora* may grow on the water's edge and *Juncus roemerianus* on the landward edge (fig. 6b). Large and long-term canopy disruptions will allow the development of salt marsh plants, including succulent (*Batis maritima*) and woody dicots (*Myrica* spp.), grasses (*Distichlis spicata*), and the leather fern *Acrostichum aureum* Linnaeus. Snedaker and Lahmann (1988) proposed that the high metabolic cost associated with living in intertidal areas prevents the survival of understory halophytes.

Zonation and Ecology

Red, black, and white mangroves often show vertical zonation (fig. 6a,b). In fringing and riverine sites, *Rhizophora mangle* is usually the species growing nearest mean low water (MLW). *Avicennia germinans* typically dominates near mean high water (MHW), while *Laguncularia racemosa* occurs near mean higher high water (MHHW). The fringing red mangroves may be replaced by *Spartina alterniflora* (fig. 6b) as a result of low temperatures or erosion by waves from boat wakes or tidal currents. Monotypic forests of red mangroves form overwash islands or occur in basins that lack a high intertidal zone. If tidal flushing is limited because of loss of tidal channels, blockage by causeways, or storm berms, then basin sediments can become hypersaline as a result of evaporation. Often only dwarfed black mangroves grow under these conditions because of their tolerance to high sediment salinities.

Similar to salt marshes, salterns or salt barrens occur near the upper intertidal zone (fig. 6b) and form a continuum between upland (terrestrial) communities and mangals (Hoffman and Dawes, 1997). Trees near salterns show dwarfing (scrub morphology), sclerophylly (thickened, reduced leaves), and usually gradation into monotypic stands of black mangroves. Mangroves may be replaced by succulent halophytes, such as species of *Sesuvium*, *Batis*, and *Salicornia* that fringe salt barrens where pore water salinities are higher than 70 ppt during dry seasons.

Succession in terms of predictable shifts of species has been demonstrated in mangals disturbed by hurricanes (Davis, 1940; Smith, 1992; Dawes et al., 1995), although recovery is slow. For example, Hurricane Donna destroyed or severely damaged mangrove communities near Lostman's River in south Florida during 1960; Hurricane Andrew destroyed them again in 1992. Thirty-two years after Hurricane Donna, tree height indicated that recovery was still in progress, and it was estimated that recovery from Hurricane Andrew would take forty years (Dawes et al., 1995). The extent of damage to mangals in southwest Florida by Hurricane Charlie (September 2004) is still unknown.

Mangals have the same ecological roles as salt marshes, including filtering land runoff, stabilizing sediments, and serving as habitat for a variety of wildlife. In south Florida, primary production of a fringing mangal is high (about 2,442 g C m^{-2} y^{-1}; Mitsch and Gosselink, 1993) versus that in Puerto Rico (572 g C m^{-2} y^{-1}) or the east coast of Mexico (1,607 to 2,468 g C m^{-2} y^{-1}). Litter fall is also high in Florida mangals, with northern mangals of Tampa Bay producing a biomass of 3.3 g dwt m^{-2} d^{-1} (Dawes et al., 1999a) versus 2–3 g dwt m^{-2} d^{-1} for a southern fringing forest at Rookery Bay near Cape Sable (fig. 2, site 6; Odum et al., 1982). Detrital export has been estimated to be 0.5–0.7 g C m^{-2} d^{-1} for fringing mangals, 1.2–2.7 for riverine forests, and 0.15–0.18 for basin forests (Odum and McIvor, 1990). Based on stable isotope ratios, mangrove detritus plays a significant role in adjacent benthic communities (Dawes, 1998).

Mangals serve as nurseries and habitats for more than 100 species of fishes, many being economically important (Day et al., 1989). As noted in several preservation and protection studies of these intertidal forests, many higher vertebrates, such as birds, mammals, reptiles, and amphibians use mangals. Odum et al. (1982) reported that a Florida mangrove supported 22 species of fishes plus 24 reptiles and amphibians (crocodiles, sea turtles, snakes, lizards, and frogs). In addition, the ecosystem might have up to 18 mammals (raccoon, mink, river otter, black bear, striped skunk, key deer, manatee, and Florida panther) plus 181 species of birds, including the bald eagle, osprey, American peregrine, and brown pelican.

Micro- and Macroalgae

Unlike salt marsh communities, diatomaceous microalgae probably do not play a significant role in mangals. While cyanobacteria are common, especially filamentous species, seaweeds are the most obvious algal component, presumably because of the shaded, humid conditions in these habitats versus salt marshes (Dawes, 1974; 1998). Macroalgal communities can be divided into subtidal forms that grow in tidal channels (Almodovar and Pagan, 1971; Rodriguez and Stoner, 1990), plus intertidal root epiphytes and epipelic turf communities under forest canopies (Dawes, 1974, 1996; Dawes et al., 1999a). A diverse macroalgal flora occurs in tidal channels of Florida Key mangals, including several green (*Caulerpa* spp., *Ulva* spp., *Cladophoropsis membranacea*), brown (*Dictyota* and *Padina* spp.), and red algae (*Acanthophora spicifera, Laurencia obtusa, Polysiphonia* spp.). Shade-loving macroalgae found under forest canopies are usually dominated by species of *Bostrychia* and *Caloglossa* (i.e., red algae)that grow on pneumatophores of black mangroves and prop roots of red mangroves, as well as turf-forming filamentous green (*Boodleopsis, Chaetomorpha, Cladophora, Rhizoclonium*) and red algae (*Herposiphonia* and *Lophosiphonia*). Primary productivity is high for all three macroalgal floras, with rates ranging from 1.5–2.2 mg C g dwt^{-1} h^{-1} for channel epiphytes (Burkholder and Almodovar, 1973), to 15.0 for epiphytes, and 21.9 mg C g dry wt^{-1} h^{-1} for canopy algae (Dawes, 1996).

Seagrasses

Seagrasses are the only submerged marine Angiosperms in Florida (fig. 7) and they are usually placed in five related families within the class Helobiae (Tomlinson, 1982; Larkum et al., 1989). Thus, they are not true grasses (i.e., members of the Poaceae). Seagrasses are adapted to submerged aquatic and saline environments and have a variety of morphological and anatomical features similar to freshwater hydrophytes (Dawes, 1998). Seagrasses can grow in nutrient-poor water because of the high absorption efficiency of their blades and roots; thus, they are effective colonizers of many shallow-water habitats in Florida, where they form vast meadows (fig. 7). Conversely, eutrophica-

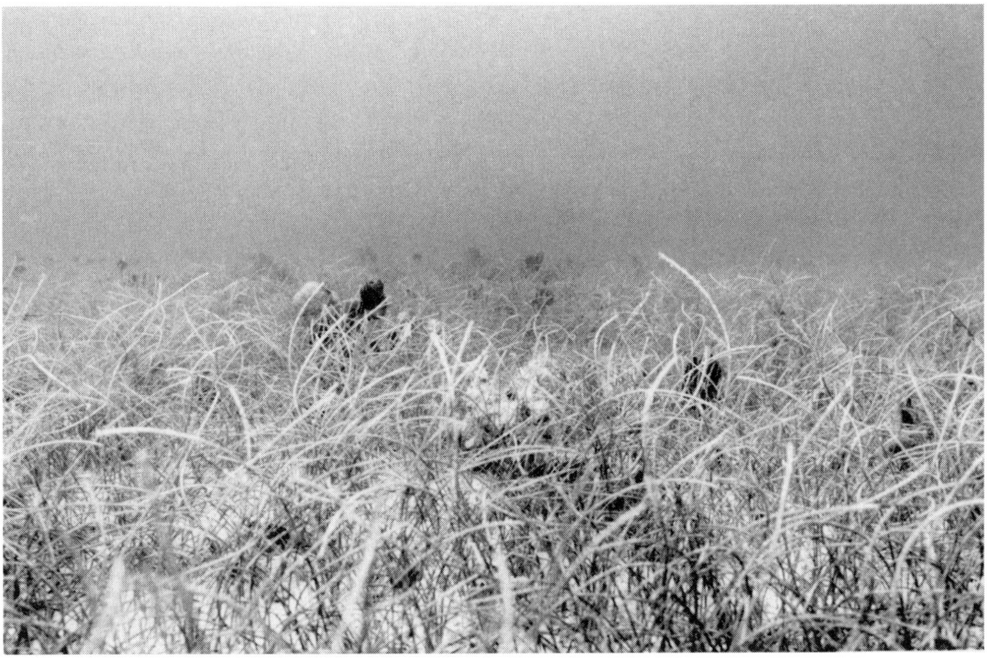

Figure 7. Submarine view of a mixed seagrass bed in the Florida Keys containing turtle grass, *Thalassia testudinum* (flat blades), and manatee grass, *Syringodium filiforme* (cylindrical blades). Black pipe sponges and *Porities* sp. coral heads are visible.

tion will support the growth of epiphytic and planktonic algae that reduce submarine illumination and can result in seagrass diebacks as shown in the Florida Keys (Lapointe et al., 1994, 1997, 1999, 2004; Lapointe and Barile, 1999, 2004; Lapointe and Matzie, 2002; Livingston, 2006), Sarasota Bay (Tomasko et al., 1996), and Tampa Bay (Lewis et al., 1985b). In Florida, seagrass beds have been severely damaged in urbanized estuaries (Lewis et al., 1985b; Durako, 1988). Reviews of seagrass communities include those found in the Indian River (Dawes et al., 1995), south Florida (Zieman, 1982), and the west coast of Florida (Zieman and Zieman, 1989; Dawes et al., 2004).

Types and Ecology

Of the approximately 58 species of seagrasses known worldwide, seven occur in Florida (Dawes et al., 2004), where they form more than 919,962 ha of nearshore beds, of which 39,991 ha are continuous and 520,056 ha patchy (F. Sargent, pers. comm.). The seagrass beds in Florida, which are dominated by *Thalassia testudinum* Banks and Solander *ex* König (turtle grass), extend to about 10 m on all coasts and to about 3 m in estuaries. Four of the seven Florida species are placed in the family Hydrocharitaceae (*T. testudinum*; *Halophila decipiens* Ostenfeld, paddle grass; *H. engelmanii* Ascherson, star grass; *H. johnsonii* Eiseman, Johnson's grass), two in the Cymodoceaceae (*Syringodium filiforme* Kützing, manatee grass; *Halodule wrightii* Ascherson, shoal grass), and one in the Ruppiaceae (*Ruppia maritima* Linnaeus, widgeon grass). Seagrasses have high primary productivity, serve as direct food sources for sea urchins and fish, and are a major component of detritus and dissolved organic carbon. Seagrass beds also serve as nurseries and habitats for diverse fauna, including the Atlantic green sea turtle [*Chelonia mydas* (L.)], manatee (*Trichechus manatus* Linnaeus), bottlenose dolphin [*Tursiops truncatus* (Montagu)], and various wading and diving birds. For example, a single bed of *H. wrightii* in Apalachicola Bay may contain 58 species of infaunal and epifaunal species, which serve as prey for larger invertebrates and fish, including drum, sea bass, porgies, grunts, and snappers; the same bed may also contain shrimp, blue crabs, and spiny lobsters (Sheridan and Livingston, 1983). The pink shrimp, which is one of the most important fisheries in Florida, also relies on seagrass beds for its life cycle (Jaap and Hallock, 1990). The occurrence of introduced species of *Caulerpa* (cf. Lapointe et al., 2005b) raises concerns regarding their potential impact on native seagrass populations (cf. Ceccherelli and Cinelli, 1999; York et al., 2006), while physical disturbance may also cause enhanced spatial competition between native species of *Caulerpa* (i.e., *C. prolifera*) and *Halodule wrightii* (Stafford and Bell, 2006).

Macroalgae

The seaweeds that are common to seagrass beds include epiphytes, drift, and psammophytic species such as *Halimeda* and *Penicillus*. The macroalgae of these beds can compete for space, light, and nutrients and play significant roles as primary producers, food sources, and habitats (Fourqurean and Rutten, 2004).

Competition for space in shallow subtidal environments is intense, as shown by the variety and abundance of epiphytes on seagrass blades. Humm (1964) recorded 113 species of macroalgal epiphytes on turtle grass blades in Florida, while Hall (1988) found 25 species at Egmont Key at the mouth of Tampa Bay (fig. 2, site 4). Ballantine and Humm (1975) identified 66 algal species on four different seagrass taxa within the Anclote River estuary, while Hall and Eiseman (1981) reported 41 taxa of algal epiphytes on three different seagrass taxa within the Indian River estuary. Common algal epiphytes of seagrass blades include multiple species of *Cladophora* and *Ulva* (i.e., greens), different species of *Ectocarpus* and *Hincksia*, plus *Hummia onusta* and *Cladosiphon occidentalis* (all brown algae). Red algal epiphytes include species of *Acrochaetium*, *Colaconema*, and *Polysiphonia*, crustose corallines (*Hydrolithon farinosum*), and fleshy forms (*Hypnea musciformis*, *Chondria* spp.). Epiphytes serve as primary producers, habitats for small invertebrates, and direct food sources (Dawes, 1998). The levels of primary production by epiphytic algae are high, with those on the seagrass *Thalassia testudinum* contributing 25–33 percent of the fixed carbon in that community (Borowitzka and Lethbridge, 1989).

Drift algae also provide food and refuge for the diverse fauna in seagrass beds (Hamm and Humm, 1976; Benz et al., 1979; Virnstein and Carbonara, 1985); they also exhibit high levels of primary production (Brown, 2001) and contribute to dissolved organic carbon through decomposition (Zieman, 1975). Most species of drift seaweeds in Florida originate from attached plants, including some epiphytes of seagrass blades. Benz et al. (1979) identified 63 species of drift algae from Indian River seagrass beds, while Hamm and Humm (1976) found 65 species from the Anclote River estuary, and Kelly (1995) 13 species from Hillsborough Bay in Tampa Bay. Brown (2001) recorded 25 species of drift macroalgae during a year-long study at three sites on the west coast. Drift algal biomass can be high; for example, the spring biomass in the Indian River Lagoon averaged 164 g dry wt m^{-2} (Virnstein and Carbonara, 1985) and by early fall the biomass was higher than the aboveground standing stock of the seagrasses. The dominant drift seaweeds on Florida's coasts include various species of green (*Anadyomene*, *Chaetomorpha*, *Cladophora*, *Ulva*), brown (*Dictyota*, *Hincksia*, *Rosenvingea*), and red algae (*Acanthophora*, *Chondria*, *Chondrophycus*, *Digenea*, *Gracilaria*, *Hypnea*, *Jania*, *Laurencia*, *Solieria*, and *Spyridia*), with mean total biomass ranging from 1.8 to 240 g dwt m^{-2} (Brown, 2001).

Although most seaweeds (ca. 98%) grow attached to hard substrata, several tropical species are psammophytic and grow on unconsolidated sediments. In a 15-month study of seagrass beds in Apalachee Bay (fig. 2, site 2), 20 species of attached macroalgae were recorded (Zimmerman and Livingston, 1976), including psammophytic species of the green algal genera *Caulerpa*, *Halimeda*, *Penicillus*, and *Udotea*. In addition, several lithophytic algae can grow on shells and coral rubble in seagrass beds, including species of green (*Acetabularia*, *Dasycladus*, *Neomeris*), brown (*Dictyota*, *Padina*, *Sargassum*), and red algae (*Digenea*, *Eucheuma*, *Gracilaria*, *Halymenia*, *Jania*, *Laurencia*, *Polysiphonia*, *Spyridia*). The calcareous green algae *Halimeda*, *Penicillus*, *Rhipocephalus* and *Udotea* are also important primary producers and contribute carbonate (aragonite)to the sediments.

Intertidal Algal Communities

As noted earlier, most seaweeds are lithophytes that grow attached to hard substrata and form some of the most productive communities in the world (Dawes, 1998). The natural hard bottom in Florida is composed of limestone, which is characteristic of warm climates and low wave energy; it is easily shaped by chemical and biological activity.

The key factors controlling zonation of intertidal algae in Florida and the Caribbean are tides, exposure, substrata, and biotic interactions (Dawes et al., 1991; Dawes, 1998), all of which play significant roles in structuring the distribution of intertidal seaweeds. Limestone is white when bleached and thus reflects light; it is also more soluble, porous, and retains water. Thus, while some macroalgae grow directly on intertidal limestone, a variety of species occur in crevices, ledges, or solution holes (pool-like structures formed by dissolution from water). For example, the green algae *Dictyosphaeria cavernosa* and *D. ocellata* grow under projecting limestone ridges in the Florida Keys.

Tides are of primary importance in structuring intertidal zonation with tidal amplitudes of approximately 1 m and 2 m on the west and east coasts, respectively. Further, exposure to sun and high temperatures will result in desiccation. The combination of intense sun, high temperatures, and low tides limits the number of intertidal species. For example, the biweekly spring low tides, during daylight hours in May and June, cause bleaching and kill intertidal seaweeds such as *Palisada papillosa* and *Padina sanctae-crucis* in the Content Keys. Tough, densely branched, turflike morphologies tend to persist within intertidal tropical communities, as seen on wave-exposed worm and shell reef habitats off Vero Beach on the east coast and coquina shell hash at Point of Rocks on the west coast.

With the exception of the Florida Keys, most of the east and west coasts and the Panhandle have little natural hard intertidal substrata because of a covering of Quaternary sediments (for example, silica sand). Thus, natural intertidal hard bottoms are limited to coquina shell deposits, such as those at Marineland (Humm, 1952a), Point of Rocks (Eiseman, 1975), and the shell and worm reefs off Vero Beach (Juett et al., 1976). Manmade structures often have highly diverse floras and include a sunken bunker at Mullet Key (Dawes, 1974), piers and pilings at Bayport near the mouth of the Homosassa River, breakwaters at East Point near Apalachicola, jetties at Destin near Ft. Walton Beach (fig. 2), seawalls near Jupiter Inlet at Jupiter (fig. 2), and groins at Vero Beach (fig. 1).

The Stephensons (1972) studied intertidal zonation at Summerland Key (fig. 8), while Croley and Dawes (1970) conducted a three-year study of seaweed zonation on a north-facing limestone reef at Content Keys (fig. 9), reporting 258 seaweed taxa. The maritime and spray zones in the

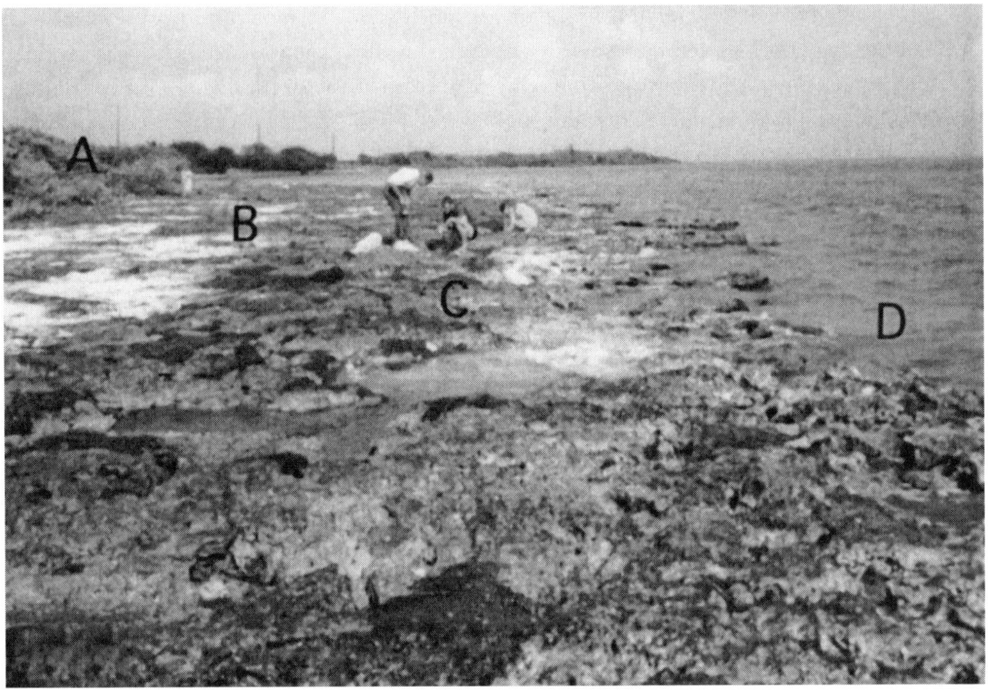

Figure 8. Intertidal zonation of the limestone shore at West Summerland Key (Florida Keys) at low tide. A = maritime, spray zones, B = gray-black zone, C = yellow zone.

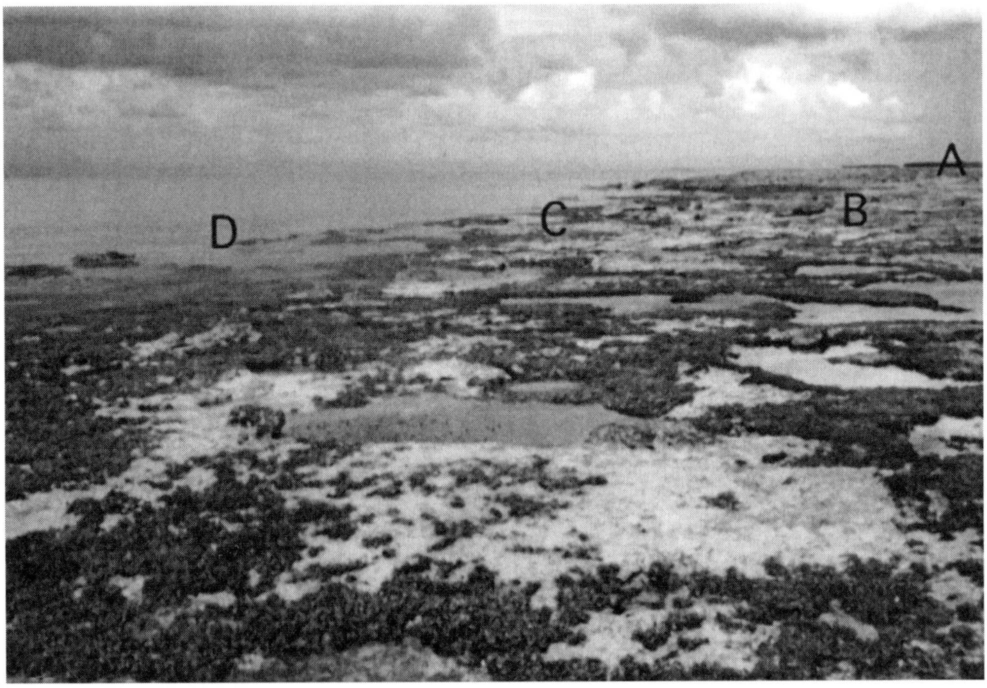

Figure 9. Intertidal zonation of the limestone shoreline on the north side of the Content Keys (Florida Keys) at low tide. A = maritime, spray zones, B = gray-black zone, C = yellow zone, D = low intertidal zone.

Keys are affected by wave splash and spray (fig. 8–10, Zone A), with the area vegetated by coastal flowering plants, including *Ipomoea imperati* (Vahl) Grisels., *Opuntia* spp., *Sesuvium portulacastrum*, *Batis maritima*, and *Panicum* spp. Scattered and stunted black mangroves extend into the gray-black zone of the upper intertidal, with their roots penetrating the fragmented limestone. The orange lichen, *Pyrenula cerina* Eschw., grows on mangrove trunks, while species of blue-green algae or Cyanobacteria (for example, *Merismopedia* and *Oscillatoria*) grow on shells and coral rubble. The gray-black color of the limestone in the upper intertidal area between 45 and 65 cm above mean low water (MLW; fig. 8–10; Zone B) is caused mostly by Cyanobacteria (Stephenson and Stephenson, 1972; Hoek et al., 1972). Storm waves and spring tides flood the epiphytes of black mangrove roots, which include *Catenella caespitosa*, various species of *Bostrychia*, and *Gayralia oxysperma*. There are also several species of lithophytic cyanobacteria (i.e., *Calothrix, Lyngbya*, and *Rivularia*) and macroalgae growing in depressions, including various green (i.e., *Caulerpa* and *Cladophora*) and red (i.e., *Amphiroa, Chondrophycus, Laurencia*, and *Lithothamnion*) algal taxa. Snails are also common in the lower maritime zone, including members of the genera *Littorina, Nerita*, and *Tectarius*.

The pale yellow mid intertidal zone lies between 35 and 45 cm above MLW (fig. 8–10; Zone C) and supports barnacles (*Chthamalus* sp.), cyanobacteria, and macroalgae. Stunted species of the green algae *Caulerpa, Halimeda, Penicillus*, and *Udotea* are partially buried in sand-filled solution holes, while *Dasycladus vermicularis* and *Neomeris annulata* grow with the brown alga *Padina sanctae-crucis* under rock ledges. *Palisada papillosa, Gelidiella acerosa*, and species of *Gelidium* often form a tall turf (~2 cm) in depressions that are impregnated with sand.

The low intertidal zone (fig. 8–10; Zone D) extends from 10 to 30 cm above MLW and is yellow to gold in color. In wave-exposed habitats there are turf or mosslike growths of several genera of red (*Centroceras, Herposiphonia, Hypnea*, and *Laurencia*) and green algae (*Avrainvillea, Cladophoropsis*, and *Halimeda*). The brown alga *Cystoseira myrica* forms erect clumps in protected basins. Cyanobacteria also penetrate limestone surfaces, while mussels (*Mytilus* sp.) occur in crevices.

The subtidal fringe, which occurs from 10 cm below to 20 cm above MLW, often lacks seaweeds because of waves. However, solution holes and tide pools contain seagrasses (often *Thalassia testudinum*) and species of several coenocytic green algal genera (*Halimeda, Penicillus*, and *Udotea*). Other macroalgae (*Palisada papillosa* and *Padina gymnospora*) grow on the surface,

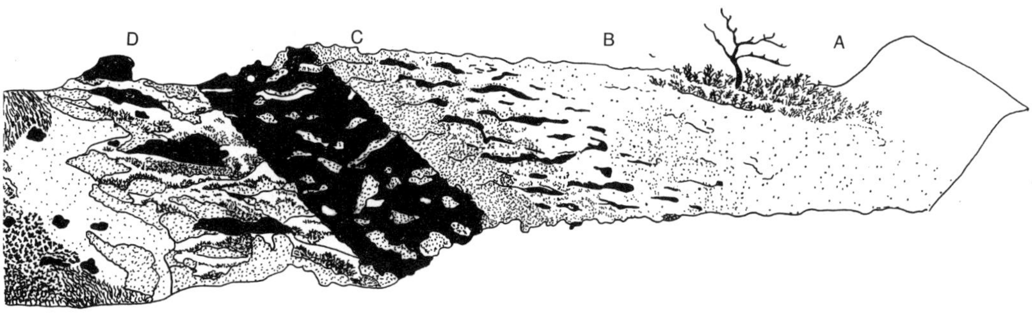

Figure 10. Intertidal zonation diagram of the limestone shoreline at low tide in the Florida Keys (after Stephenson and Stephenson, 1972). A = maritime, spray zones, B = gray-black zone, C = yellow zone, D = low intertidal zone.

where dense populations of *Gelidiella acerosa*, *Halimeda opuntia*, and *Palisada poiteaui* can trap sand in tide pools.

Subtidal Algal Communities

Offshore water transparency on both coasts is often high, with visibility of 10 m or more resulting from low concentrations of dissolved nutrients and limited discharge of sediments from rivers and estuaries. If hard substrata are available, lithophytic seaweeds may extend to 100 m or deeper. Subtidal hard bottom communities also include manmade structures such as artificial reefs, seawalls, and bridge pilings that have been placed ubiquitously on Florida's coasts. By contrast, estuarine floras are limited to shallower than 2 m (for example, Tampa Bay, Indian River Lagoon) because of turbidity and limited hard substrata.

Shallow water hard-bottom seaweed communities that extend to about 20 m on the east and west coasts are heavily impacted by wave activity and coastal processes; thus, they support euryhaline, temperate to subtropical species (Dawes, 1974). Shallow subtidal habitats have been studied off Anclote Key in 9 m (Mathieson and Dawes, 1975) and 20 m (Dawes and Lawrence, 1990), near Vero Beach in 3–8 m (Juett et al., 1976), west of the mouth of Florida Bay (Dawes et al., 1967), and in the Keys (Croley and Dawes, 1970). The conspicuous seaweeds found in shallow subtidal communities are often cartilaginous and firm, including *Eucheuma isiforme* and various species of *Gracilaria* and *Sargassum* that grow with soft corals. Variations in species richness between sites with similar hydrographic conditions usually reflect differences in substrata and light. On the west coast, where the 50 m contour line can be more than 100 km offshore, Pleistocene coral reefs that are 0.5–2 m high are common and support a diverse seaweed flora (Dawes and van Breedvelt, 1969). The Florida Middle Grounds, located northwest of Tampa Bay, provide a unique example of hard-bottom ("live bottom") communities at depths ranging 18–40 m and dominated by a tropical seaweed flora (Cheney and Dyer, 1974). Similar hard-bottom communities occur off the east coast, especially south of Cape Canaveral (Moe 1963; Jones et al., 1986). Limestone outcroppings at 5–15 m (for example, off Jupiter Island near North Palm Beach; fig. 2, between sites 9 and 10) also support a diverse seaweed flora.

Deep water seaweed communities (i.e., 30 m or deeper) are less influenced by wave activity and coastal processes and are more tropical, with this pattern documented on both the west (Dawes and van Breedveld, 1969) and east coasts (Hanisak and Blair, 1988). Extensive use of the Johnson Sea-Link Submersible by Harbor Branch Oceanographic Institution revealed a highly diverse and rich seaweed flora off the southeast coast of Florida (fig. 2, site 10; Hanisak and Blair, 1988), including a large number of new species (Eiseman, 1979; Eiseman and Moe, 1981; Eiseman and J. N. Norris, 1981). Hanisak and Blair's study on the east coast was based upon 170 submersible dives; they reported 208 algal taxa, with 105 species at 30 m, 141 at 50 m, 49 species at 70 m, and fewer than 10 at 98 m. Dawes and van Breedveld identified 157 species of seaweeds off Tampa and Sarasota based on 28 monthly dredge collections. Most species (81 taxa) occurred at 30 m or less, with 53 species at 50 m and 38 at 70 m. Because of reduced submarine illumination at deep depths (30 m or greater), subtidal seaweeds in these zones are more sparsely distributed and tend to have flattened morphologies (peltate fronds, broad blades, or crusts) that aid in capturing light (cf. Littler and Littler, 1980). Recent studies of deep water habitats (>27 m) in the Bahamas also document patterns of high species richness and many new depth and geographical records (Aponte and Ballantine, 2001; Ballantine and Aponte, 2003, 2005), which can probably be duplicated in other Florida and Caribbean sites.

Biotic Reefs

Florida is endowed with a variety of biotic reefs that can form substrata for macroalgae, including ancient Pleistocene coral reefs (Dragastan et al., 2003) and present-day worm tube and shell reefs, coral reefs, and coral banks (Jaap and Hallock, 1990). Live worm reefs occur from Cape Canaveral to Key Biscayne on the east coast and extend from the intertidal to 10 m. The worm-tube reefs are aggregations of *Phragmatopoma lapidosa* Kinberg that cement sand grains together. The vermetid reefs off the Ten Thousand Islands are no longer actively growing and were produced by a mollusc, *Petaloconchus* sp. The coral *Oculina varicosa* Le Sueur produces banks on the east coast that extend from Jacksonville to the St. Lucie Inlet and cover hectares of sea bottom.

Reef-building, hermatypic corals are unique to the continental shelf of North America and similar in species composition to Caribbean and Bahamian tropical reefs; they extend along the southern edge of the Florida Keys and northward into the lower part of Biscayne Bay (Jaap and Hallock, 1990). The coral reefs in the Keys include bank and patch formations, while those in the Dry Tortugas are more irregular and dominated by staghorn coral (*Acropora cervicornis* Lamarck). Bank reefs develop parallel to the south coast in the Florida Keys (fig. 11a), as well as behind the larger islands that protect them from storm waves; the largest reefs are off Key Largo and between Big Pine Key and Key West. Bank reefs include two major habitats: an upper reef flat that is barren and composed of coral rubble, and lower spur and groove formations (fig. 11b). Spurs are elongated reef limestone structures that support living corals, while grooves are valleys containing unconsolidated limestone rubble and sediment.

Depth distributions of spur and groove bank reefs are usually between 1 and 10 m below MLW, with the possibility of a fore reef to 35 m in the Florida Keys. Patch reefs occur throughout the Florida reef tract and are most important between northern Elliott Key and Key Largo; they occur 2–9 m below MLW and are circular, ranging from 30–700 m in diameter (Jaap and Hallock, 1990).

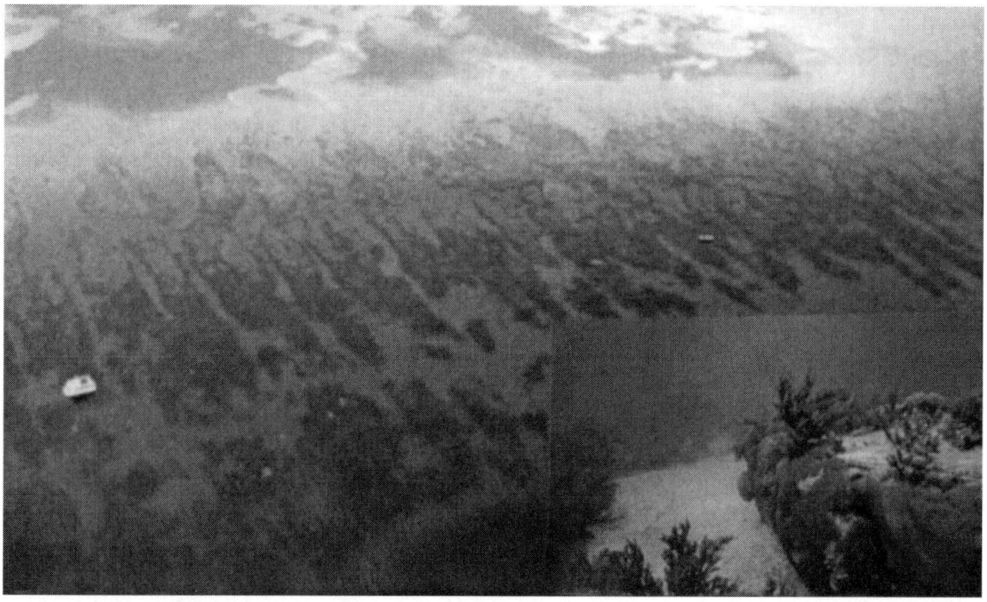

Figure 11. Coral reefs of Florida. 11a. Aerial view of the spur and groove bank reef at Looe Key (Florida Keys). 11b. Submarine view of a coral spur and sandy groove (courtesy of W. Jaap, Florida Marine Research Institute, St. Petersburg).

The seaweeds of the Florida Keys coral reefs are similar to those described for the Content Keys, the Northern Florida Reef Tract south of Biscayne Bay (Lirman and Biber, 2000), and Molasses Reef off Key Largo (Bucher et al., 1990). Algal biomass is often low because of extensive grazing by fish and sea urchins.

Processing Seaweeds

Field collections

Shallow subtidal and intertidal communities can be visited using snorkel, mask, and wading shoes, while deeper subtidal samplings will require SCUBA. A less suitable method for collecting deep water algae is dredging, as it damages *in situ* bottom communities and samples. Collecting pails and dive bags should include a scraper to remove algal holdfasts, plastic bags to keep collections separate, water-resistant labels, and a field book. Field measurements should include salinity and temperature. Physical traits such as topography, wave action, and type of substratum should be noted in order to characterize the seaweed's habitat. In Florida, seaweeds usually begin to decompose within a few hours after collection; hence, they should be stored in ice coolers, with the ice kept in separate bags, or in a refrigerator prior to processing. Initial sorting, identification, and pressing (see below) of plants should be carried out within 24 hours. If there is a delay, the plants can be frozen (but thawed specimens are difficult to work with) or preserved in 5% formalin in seawater (see details below) and kept in the dark until processing. Formalin preservation is not appropriate if molecular evaluations are to be conducted; if such evaluations are needed small portions of plant material should be preserved in silica gel (cf. Klein et al., 2003).

Specimen Preparation

Two types of specimens should be made when studying seaweed floras: herbarium (pressed on paper) and wet-stack (stored in liquid in vials). Herbarium specimens require a plant press, paper with high rag content such as herbarium paper (standard size: 42.5 x 25.5 cm) or heavy drawing paper, and a dryer with a fan blowing on a rack that may be heated by light bulbs. All supplies are available from biological supply firms. Initially the seaweed is floated out in a large tray with tap water (if preserved) or seawater (if fresh) and the pressing paper is slid under it after providing an identifying number in pencil. The alga and paper are then carefully removed from the tray and placed in a plant press. If the plant is delicate, a small brush can be used to spread the branches. Pressing is like constructing a sandwich. First a sheet of cardboard (with fluting to allow air flow) is placed on the press board, followed by a blotter (sheet of absorbent paper), an open newspaper (1–2 pages), and finally the herbarium paper with the alga. The specimen is then covered with a sheet of thin muslin, cheesecloth, or wax paper to prevent sticking. The newspaper is folded over, followed by a blotter, and then a corrugated cardboard is added; the process is then repeated. Each specimen should be labeled, including its species name and relevant field data (for example, date, latitude, longitude, site name, depth, type of substratum, wave action, currents, etc.).

Preserved or wet-stack specimens should also be made if identifications are to follow pressing. The most inexpensive and readily available preservative is the noxious reagent formalin, which unlike alcohol retains seaweed color and structure. A 5% solution of formalin in seawater should be stored in a fume hood. A portion of the alga can be stored in small (ca. 30–50 ml) vials in the dark until needed for further study.

Tools for Identification

Successful identification of seaweeds requires sufficient material, including the holdfast and young branches. Some seaweeds can be identified with a millimeter ruler, while others require evaluation under a dissecting (to 50 X) or compound microscope (usually to 400 X). Both microscopes should have calibrated ocular micrometers. Cross sections may be necessary to determine the internal anatomy and construction (filamentous, parenchymatous, polysiphonous, hollow), while whole mounts of apical portions should be made to characterize modes of growth and thallus construction. Most seaweeds can be sectioned by hand using a sharp single-edge razor blade. That is, hold the alga with your index finger on a slide under a dissecting microscope and mince the extended portion of the alga using your finger as a guide. Often this is best done over an illuminated white background.

Brown algae can appear green and red algae can look green or brown because the plant's pigments may be affected by nutrient or light conditions. To determine the division to which an alga belongs, first examine a part or section of the plant with a dissecting microscope. Green algae are typically grass green in color (chlorophylls *a* and *b*), brown seaweeds olive green or brown, and reds usually rose red to deep purple red. If the material is fresh (not preserved), a red alga will become green when placed in boiling water for a few minutes because the red phycoerythrin pigment is water soluble. In contrast, brown algae will not change color. If placed in heated 70% isopropyl alcohol, a brown alga will turn green, while a red alga will not because the xanthophylls, especially fucoxanthin, are soluble in alcohol. A starch test can be used to differentiate green from red or brown algae. A drop of potassium iodide (I_2KI) solution added to a section or fragment of a live green alga will indicate the presence of starch by staining it dark blue to black. The potassium iodide solution is made with 2 g KI plus 1 g I dissolved in 25 ml of water. A drop of a 1% aqueous solution of HCl (hydrochloric acid) will cause bubbling (release of hydrogen) if the algal is calcified (presence of $CaCO_3$).

Molecular evaluations have become increasingly important in differentiating many cryptic taxa (cf. Zuccarello and West, 2002; Klein et al., 2003; West et al., 2005; Verbruggen et al., 2006), including the use of nuclear, mitochondrial, and plastid DNA markers. As noted previously, representative samples should be preserved in silica gel rather than a formalin solution.

Algal Keys

The keys to the genera of algae (chapter 5) use vegetative features whenever possible and thus are "artificial." The keys emphasize external and internal vegetative structures, while reproductive features, often thought to express evolutionary relationships, have been avoided where possible. This is because production of reproductive structures may be seasonal and they are often absent or difficult to identify. All keys are dichotomous and each "leg" has a pair of opposing alternatives. Both alternatives should be examined before moving to the next pair, as there are often a number of characters presented; the glossary at the end of the book should be used for specialized terms.

Text Organization

Keys to the species and descriptions for all taxonomic levels are given in chapters 2 (Chlorophyta), 3 (Heterokontophyta), and 4 (Rhodophyta), while generic keys are in chapter 5 and a glossary is included at the end of the text. Generic and other levels of taxonomy are organized as follows: genus, subgenus (subg.), species (sp.), subspecies (subsp.), variety (var.), and form (f.).

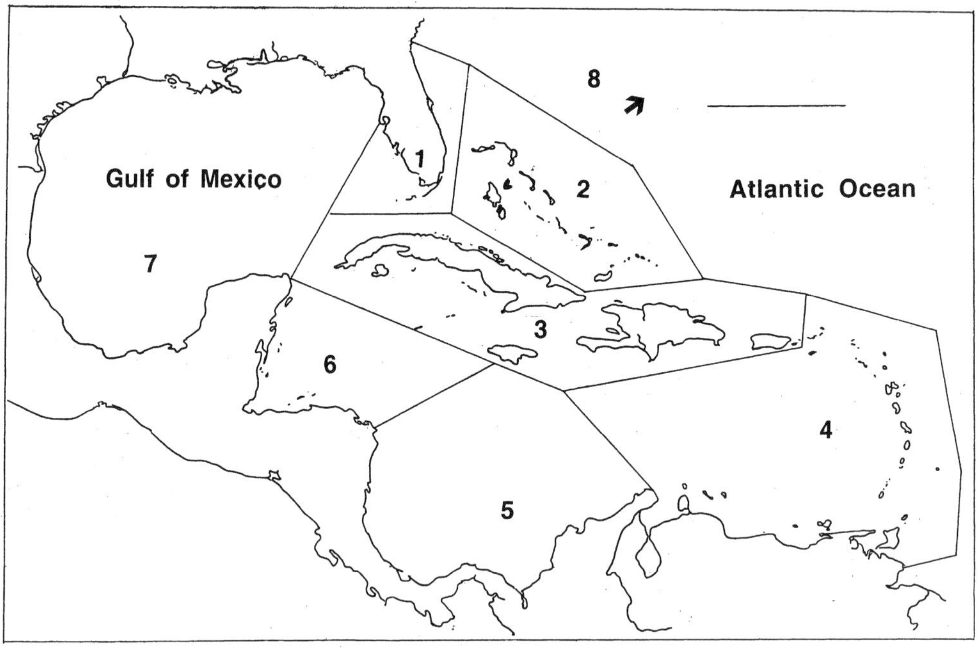

Figure 12. Geographic areas of the western tropical Atlantic Ocean (after Littler and Littler, 2000) including, (1) Florida, (2) the Bahamas, (3) Greater Antilles, (4) Lesser Antilles, (5) southern Caribbean, (6) western Caribbean, (7) Gulf of Mexico, and (8) Bermuda (*arrow*).

The first element of the description is the alga's name, authorship, and reference, followed by the source of the taxa's name (L. = Latin; Gr. = Greek) and a listing of the plate and figure numbers where illustrations are provided in the present volume. A listing of the basionym and any synonyms relevant to the Florida and Caribbean then follows, along with the taxonomic reference. Occasionally, an explanatory paragraph on taxonomic features will be included. All generic names were checked using *Index Nominum Genericorum (Plantarum)* (http://rathbun.sci.edu/botany/ing). Generic, specific, and subspecific name were also assessed, plus various synonyms, using Algaebase (Guiry and Nic Dhonncha, 2006; www.algaebase.org) and a variety of papers.

The first paragraph is followed by an initial statement of its overall morphology, then details on sizes, shapes, and reproductive features. The third element is a description of the frequency (rare, uncommon, common), ecology, and distribution of the taxon in Florida, plus its general distribution in the northwest Atlantic from the Canadian Arctic to Brazil (Uruguay, Argentina) and Gulf of Mexico (figure 12, after Littler and Littler, 2000) and worldwide based on published floras and descriptions (see appendix I for a detailed listing). In order to avoid specific geographic details regarding each country, we've used the following general geographic terms: eastern North Atlantic (Arctic through northern Europe), Mediterranean, eastern tropical Atlantic (west coast of Africa and islands), and Indo-Pacific (the Indian and Pacific Oceans and minor seas between the two in the general area of Indonesia). {Figure 1.12 near here}

Each description concludes with specific references and pagination for floras that describe Florida species, starting with Taylor (1960) and extending to the present day.

2

Phylum Chlorophyta

The green algae are dominated by unicellular freshwater species. Of the 16,800 known species (Norton et al., 1996) about 10 percent are marine. Chlorophylls *a* and *b*, which also occur in higher plants, provide the typical "green plant" coloration found in most green algae. In addition, green algae contain ß-carotene and the xanthophylls lutein and zeaxanthin, which also occur in higher plants. Some members of the Chlorophyta may contain ß-carotene, plus violaxanthin, neoxanthin, siphonein, and siphonoxanthin (Hoek et al., 1995). Siphonoxanthin plays a role in the acclimation of deep water species to the blue-green spectral quality of submarine illumination. The subtropical to tropical waters of Florida contain diverse green seaweeds, which range from uninucleate (Ulvales), to multinucleated multicellular plants (Cladophorales, Dasycladales), to true coenocytes that are multinucleate and lack transverse walls (Bryopsidales). Chloroplast structure varies from cup-shaped, to discoid, reticulate, and laminate. Pyrenoids are amalyase-containing protein bodies in the chloroplasts of some green algae (for example, Cladophorales). Reserve food occurs as starch (amylose and amylopectin) that is chemically similar to that found in flowering plants. Green seaweeds are of economic interest, with species of *Caulerpa*, *Gayralia*, and *Ulva* farmed as "sea vegetables," while others like *Ulva* are associated with fouling and excessive growth under eutrophic conditions (Morand and Briand, 1996). The shape of crystalline cell inclusions (calcium oxalate, calcium carbonate, proteins) has recently been used to help distinguish various taxa of the Ulvophyceae (Leliaert and Coppejans, 2004).

The taxonomy of the green algae has been the subject of several reviews (Bold and Wynne, 1985; Kumar, 1989; Hoek et al., 1995; Graham and Wilcox, 2000) that vary in grouping them in three to five classes. The marine Chlorophyta described in this text are assigned to three classes, the Charophyceae, Chlorophyceae, and Ulvophyceae.

CHAROPHYCEAE

The Charophyceae include the evolutionary line of algae that led to flowering plants. Motile cells of the more advanced members are similar in structure to those in the Bryophyta (mosses, liverworts) and vascular cryptogams and are asymmetrical with two subapical flagella. As in flowering plants, the mitotic spindle is persistent during cytokinesis and cell division is by means of a phragmoplast.

KLEBSORMIDIALES

The order contains unbranched filaments that lack holdfasts and plasmodesmata. Motile cells lack a cell wall (naked) and escape through a pore in the filament wall.

Stichococcaceae

Lokhorst (1991) placed the family in the Klebsormidiales of the Charophyceae.
Filaments are short, with cells loosely arranged in a gelatinous to tubular sheath and often dissociated. Asexual reproduction is by fragmentation and zoospores.

Stichococcus Nägeli 1849: p. 77 (Gr. *stichos*: a row, series + *coccus*: berry-shaped).

Plants are filamentous or composed of solitary cells; filaments are cylindrical and attached by a basal cell. Cells are uninucleate; they have a single, platelike plastid that covers about one-half of the cell and contain a single pyrenoid. Asexual reproduction is by aplanospores or zoospores, with one spore produced per sporangial cell.

Stichococcus marinus **(Wille) Hazen 1902: p. 161, pl. 21, fig. 8, 9 (L. *marinus*: growing by or in the sea). Plate I, fig. 1.**

Basionym. *Ulothrix variabilis* var. *marina* Wille 1885: p. 87, pl. 13, fig. 8.

Burrows (1991) placed *Stichococcus marinus* in synonymy with *S. bacillaris* Nägeli (also Guiry and Nic Dhonncha, 2006), although the former tolerated high salinities (Haywood, 1974), while Nägeli (1849) stated that the species never occurred in typically marine habitats. Others (Hazen, 1902; Collins, 1909b) separated the taxa and designated the Florida and Caribbean plants as *S. marinus* (Humm and Taylor, 1961; Wynne, 2005a). Sears (2002) noted that *S. marinus* was difficult to distinguish from *Klebsormidium* and *Ulothrix*.

Filaments are unbranched, pale to yellow green, 5–6 µm in diam., and attached by an unmodified basal cell having thick walls. Filaments are 20–75 µm tall by 4.5–7.5 µm in diam., have 3–6 cells 1–3 diam. 1ong, and a parietal plastid covering more than half the cell with a single pyrenoid. Asexual reproduction is by biflagellate zoospores that lack an eyespot and are formed singly in a cell.

Rare, a year-round, aseasonal annual, occurring within intertidal estuarine sites on the northwest coast; also known from Northumberland Strait (Canada) to Connecticut, the British Isles, France, the Arctic Ocean (type location: Novaya Zemlya, northeastern coast of Russia; Wille, 1885), plus western North and Central America.

Humm and Taylor 1961: p. 335.

CHLOROPHYCEAE

The Chlorophyceae are dominated by motile or nonmotile freshwater species that are unicellular, colonial, or filamentous. Motile cells have a cell wall, which is a feature of this class.

Order TETRASPORALES

Algae form colonies; cells are embedded in a gelatinous matrix, are uninucleate, and have a single plastid with a central pyrenoid.

Palmellopsidaceae

Cells occur in a matrix at the periphery of a colony.

Pseudotetraspora Wille 1906: p. 20 (Gr. *pseudo*: false + *tetra*: four + *spora*: spore).

Colonies are gelatinous; cells are in groups of 2–4; plastids are parietal, lobed or radiating, and have a central pyrenoid.

***Pseudotetraspora marina* Wille 1906: p. 20, pl. 1, fig. 32–36 (L. *marina*: growing by or in the sea). Plate I, fig. 2.**

Synonym. *Pseudotetraspora antillarum* M. Howe 1920a: p. 597

Plants form small globular or flattened yellowish brown gelatinous masses, 2–20 mm in diam., and are attached to other algae or various substrata. Cells usually occur in a matrix; they are mostly 3–7 µm in diam., globose to elliptical, and often occur in clusters of four.

Common, on seagrasses and macroalgae on all coasts; also known from 47–61 m in the Bahamas (Ballantine and Aponte, 2003), Caribbean, and Gulf of Mexico.

Taylor 1960: p. 45; Humm and Taylor 1961: p. 342; Dawes 1974: p. 64.

Verdigellas D. L. Ballantine and J. N. Norris 1994: p. 368 (L. *verdi*: green + *gellas*: jelly).

Plants are soft, gelatinous, green, irregular to uniformly compressed, erect, and either on a short stalk or spreading over the surface. The gelatinous matrix contains irregularly arranged small spherical cells, which are uninucleate and contain one cup-shaped plastid that lacks pyrenoids.

***Verdigellas peltata* D. L. Ballantine and J. N. Norris 1994: p. 362, fig. 5–7 (L. *peltata*: a small shield). Plate I, fig. 3, 4.**

Plants are prostrate to peltate gelatinous discs, to 6 cm in diam., 2 mm thick, and bright green in color. Cells are 6–9 µm, scattered in the matrix, and have one cup-shaped plastid.

Rare, a deep water species reported from 80 m off Pulley Ridge reef near the Dry Tortugas (see http://soundwaves.usgs.gov/2001/08/); also known from the Bahamas and Puerto Rico (type location: Mona Island).

Littler and Littler 2000: p. 298.

ULVOPHYCEAE

See Hoek et al. (1995) and Graham and Wilcox (2000) for details of this class. Molecular data suggested that the Ulotrichales and Ulvales were monophyletic (Hayden and Waaland, 2002) and should be placed in the class Ulvophyceae (Floyd and O'Kelly, 1990).

Motile cells have a persistent interzonal spindle during mitosis and lack a cell wall (Hoek et al. 1995). Florida species occur in all six orders of this class (Bryopsidales, Cladophorales, Dasycladales, Phaeophilales, Ulotrichales, and Ulvales).

ULOTRICHALES

Plants are filamentous or bladelike, simple or branched, and often attached by a differentiated basal cell. Their cells have visible walls as well as one parietal plastid with or without one or more pyrenoids, and are uninucleate.

Ulotrichaceae

Plants are filamentous and either have or lack a basal attachment cell; cell walls are thin and gelatinous; plastids are band- or platelike, and with or without pyrenoids.

Ulothrix Kützing 1833: p. 517 (Gr. *ulo*: shaggy + *thrix*: hair).

Filaments are uniseriate, unbranched, and initially attached by a single basal cell; later they are often detached. Cell walls are thin and gelatinous; cells have a single parietal bandlike plastid with one or more pyrenoids. Two to many quadriflagellate zoospores are formed in each sporangium.

Ulothrix subflaccida **Wille 1901: p. 27, pl. 3, Figs. 90–100 (L. *sub*: slightly, nearly + *flaccida*: weak, soft). Plate I, fig. 5.**

There are varying interpretations of *Ulothrix subflaccida* versus *U. implexa* (Kützing) Kützing (1849). Some authors consider the former to be a synonym of the latter (Hazen, 1902; Setchell and Gardner, 1920; Burrows, 1991) and others (Hamel, 1930) include *U. implexa* in *U. subflaccida* along with *U. cutleriae* Thuret in Le Jolis (1863). Field and culture studies of *U. implexa* (Lokhorst and Vroman, 1974) have shown a broad tolerance to salinity and stable morphological features, suggesting a single taxon (Burrows, 1991). Lokhorst (1978) segregated *U. implexa* and *U. subflaccida* as distinct species, as do Silva et al. (1996) and Guiry and Nic Dhonncha (2006).

Filaments form yellow-green tufts to 3 cm long; they are uniseriate, unbranched, 6–12(18) µm in diam., uniform in width or slightly constricted at cross walls, and attached by a tapering basal rhizoidal cell. The plants cells are 0.5–1.0 diam. long and have plastids that either encircle or partially cover the side walls.

Rare, from 11 m offshore Alligator Point and in the adjacent harbor; also known from Labrador to Virginia, and Cuba, as well as the eastern North Atlantic (type location: Goes, Netherlands; Burrows, 1991), Mediterranean, Morocco, Pacific Ocean, Australia, and Antarctica.

Humm and Taylor 1961: p. 333.

ULVALES

Plants range from branched, uniseriate filaments to monostromatic or distromatic blades or tubes. Cells are uninucleate, their plastids are laminate, parietal, and have one or more pyrenoids. Zoospores have four flagella, while gametes have two; flagella are apically inserted.

Gayraliaceae

Molecular studies (Hayden and Waaland, 2002) suggested that members were more closely allied with the Ulotrichales.

Plants are foliaceous, membranous, and monostromatic; life histories are variable, with most being asexual or heteromorphic.

Gayralia K. L. Vinogradova 1969: p. 1351, 1354 (named for the French phycologist Mme P. Gayral, who conducted field and culture studies of the Ulvales).

The genus has been assigned to several different genera using life histories and germination patterns of reproductive cells (Abbott and Huisman, 2004). *Ulva oxysperma* Kützing, the basionym of *Gayralia oxysperma*, was assigned to *Ulvaria* Ruprecht (1851) and *Monostroma* Thuret (1854), but then considered to be an independent genus (*Gayralia*) by Vinogradova (1969). Others (Golden and Garbary, 1984; Burrows, 1991) believe that the separation is unnecessary. However, this separation is supported by molecular studies (Hayden and Waaland, 2002).

Plants are initially saccate and subsequently split into flattened blades or strips. Blades are monostromatic, have thin-walled cells (except *G. oxysperma*), and usually have undulate margins; their fronds are attached by rhizoids.

Gayralia oxysperma (Kützing) K. L. Vinogradova _ex_ Scagel et al. 1989: p. 72 (Gr. _oxy_: sharp, pointed, or sour + _spermum_: a seed or semen). Plate I, fig. 8–10.

Basionym. *Ulva oxysperma* Kützing 1843: p. 296.
Synonyms. *Ulvaria oxysperma* (Kützing) Bliding 1968: p. 585, fig. 31–33.
Monostroma oxyspermum (Kützing) Doty 1974: p. 12.

Plants are initially saccate and then split open. Blades are monostromatic, soft, and light green, to 10(60) cm long, 20–40 µm thick, and attached by rhizoids from basal cells. Their cell walls are thick, while cells are rounded, angular to ovoid, 7–18 µm in diam., and either irregularly arranged or in groups of 2–4.

Common, on mangrove roots, oyster shells, and limestone, as well as in drift, on all coasts; also known from Labrador to Brazil, Bermuda, eastern North and tropical Atlantic (type location: Germany; Abbott and Huisman, 2004), the Mediterranean, Black Sea, Indo-Pacific, and Australia.

Taylor 1960: p. 64; Dawes 1974: p. 69; Schneider and Searles 1991: p. 33; Littler and Littler 2000: p. 298.

Gomontiaceae

Gabrielson et al. (2004) and Wynne, (2005a) note that the name Gomontiaceae De Toni (1889) has priority over Monostromataceae Kuneida (1934) or Collinsiellaceae Chihara (1967).

Filaments are irregularly branched; their cells are variously shaped, they have one to several nuclei, and plastids are parietal, lobed or reticulate.

Gomontia Bornet and Flahault 1888: p. 163, 164; also 1889: p. CLII, pl. VI-VIII (named for M. A. Gomont, a French phycologist).

Filaments are irregularly branched and penetrate shells.

Gomontia polyrhiza (Lagerheim) Bornet and Flahault 1888: p 163. (L. _polyrhiza_: many-rooted, with numerous roots). Plate I, fig. 6, 7.

Basionym. *Codiolum polyrhizum* Lagerheim 1885: p. 21, pl. XXVIII.

See O'Kelly et al. (2004) regarding molecular evaluations of several shell-boring *Codiolum*-like phases. *Gomontia polyrhiza* may be confused with these heteromorphic life history stages that include *Entocladia perforans* (Huber) Levring and *Monostroma grevillei* (Thuret) Wittrock (Kornmann 1962; Kuckuck, 1962; Wilkinson and Burrows 1972; Schneider and Searles, 1991; O'Kelly et al. 2004).

Filaments are irregularly branched and penetrate shells, turning them green. Cells are 4–8 µu}m in diam. and 2–6 times as long; sporangia are 30–40 µm by 150–250 µm.

Common, in shells within the intertidal and shallow subtidal zones on all coasts; also known from the Canadian Arctic to the Caribbean and Bermuda, as well as from the eastern North and tropical Atlantic (type location: Kristineberg, Sweden; Burrows, 1991), Mediterranean, Black Sea, Indo-Pacific, and Australia.

Taylor 1960: p. 53; Humm and Taylor 1961: p. 342; Dawes 1974: p. 64; Schneider and Searles 1991: p. 32.

Ulvaceae

Synonym. Percursariaceae

Plants are biseriate filaments that arise from a parenchymatous disc; they exhibit a heterotrichous type of growth and form monostromatic tubes, distromatic blades, or biseriate filaments, and attach by rhizoids.

Percursaria Bory 1823a (1822–1831): p. 393 (L. *percurso*: to ramble over or about).

Plants are bright green, slender filaments that are mostly biseriate aggregations of cells. Growth is heterotrichous, with erect filaments arising from a parenchymatous disc.

Percursaria percursa (C. Agardh) Rosenvinge 1893: p. 963 (L. *percursa*: to run through or over). Plate I, fig. 11, 12.

Basionym. *Conferva percursa* C. Agardh 1817: p. 87.

Synonyms. *Enteromorpha percursa* (C. Agardh) J. Agardh 1842: p. 15.

In contrast to the above authorship, Nielsen and Gunnarsson (2001) have recorded the plant as *Percursaria percursa* (C. Agardh) Bory.

Plants have a parenchymatous disc attached by rhizoids and produce 1–4 erect filaments that are 1–2 cm tall, typically biseriate or with 1–4 cell rows. Cells are evenly placed, rectangular, 28 μm long by 10–15 μm wide, and have thick walls.

Uncommon, in drift, entangled with other algae, or attached to intertidal limestone on the west coast and in the Keys; also known from Labrador to New Jersey, the Gulf of Mexico, and Venezuela, as well as eastern North and tropical Atlantic (type location: Hofmansgave, Denmark; Womersley, 1984), the Mediterranean, Black Sea, Indo-Pacific, Australia, and Antarctica.

Taylor 1960: p. 54.

Ulva Linnaeus 1753: p. 1163, *nom. cons.* (Celtic and L. *ulva*: a sedge, marsh plant).

Synonym. *Enteromorpha* Link in Nees 1820: p. 5

See Hayden et al. (2003) and Shimada et al. (2003) regarding the merger of *Enteromorpha* into *Ulva*. Culture studies have shown that tubular (i.e., "*Enteromorpha*") and bladelike forms (i.e.,"*Ulva*") can exist within the same species (Gayral, 1959; Foyn, 1960, 1961; Gayral, 1967; Bonneau, 1977) and that plants in axenic cultures can form morphologies that include *Enteromorpha* and *Ulva* (Provasoli, 1965; Berglund, 1969; Kapraun, 1970; Fries, 1975; Provasoli and Pintner, 1980). Molecular data also support combining *Enteromorpha* with *Ulva* (Blomster et al. 1998, 1999; Malta et al. 1999; Tan et al. 1999; Hayden et al. 2003).

Plants are linear to spreading distromatic blades or monostromatic tubes that are often branched. Fronds are attached by rhizoids growing from basal cells. Cells are typically uninucleate, except for some multinucleate rhizoids; plastids are cup- to plate-shaped, usually occur on the outer cell face, and have 1 to many pyrenoids.

Key to *Ulva* species (13), including tubular "*Enteromorpha*-like" plants

1. Blades foliose 2
1. Blades tubular in part or total; if solid then 1–3 cells in cross section 5
    2. Plants either having simple blades or being broadly lobed 3
    2. Plants divided into narrow segments or strips . . . . *U. fasciata*
3. Blades not naturally perforated or with a regular pattern of holes. 4
3. Blades regularly perforated with various sized holes . . . . *U. profunda*
    4. Blades irregularly lobed and with microscopic marginal teeth; in cross section the cells of older portions are taller than broad . . . . *U. rigida*
    4. Blades often regularly lobed, with margins lacking teeth; in cross section the cells of older parts are nearly square or slightly taller than broad . . . . *U. lactuca*

5. Plants tubular, if compressed or flattened then with hollow margins 6
5. Axes mostly 1–3 cells in cross section and not tubular . . . . *U. chaetomorphoides*
6. Plants locally or continuously tubular or compressed. 7
6. Blades linear or strap-shaped, hollow along margins, tapering to base. . . . . *U. linza*
7. Tubular axes branched from base or also from upper parts. 8
7. Not as above. 9
8. Branching mostly from bases of axes; pyrenoids 1(2) per cell. *U. compressa*
8. Branching from base and upper parts of axes; pyrenoids 1(2)–5 per cell . . . . *U. flexuosa.*
9. Axes tubular, profusely branched; branchlets uniseriate . . . . *U. flexuosa* subsp. *paradoxa*
9. Axes compressed. 10
10. Cells in longitudinal rows, at least within packets in some parts of axes 11
10. Cells not arranged in rows or in packets . . . . *U. intestinalis*
11. Tubular branches not expanding above; cells only in clear rows near tips. 12
11. Tubular branches expanding above; cells in clear rows in all axes and not only at tips . . *Ulva flexuosa* subsp. *flexuosa*
12. Branching abundant or not; with short and long spinelike branchlets; 2–6 pyrenoids per cell . . .*U. clathrata.*
12. Branching abundant; spiny branchlets absent; 1–3 pyrenoids per cell . . . . *U. prolifera.*

Ulva chaetomorphoides (Børgesen) Hayden, Blomster, Maggs, Silva, Stanhope, and Waaland 2003: p. 288 (Gr. *chaeto*: hair + *morpha*: form + *oides*: like; thus, similar to the green alga *Chaetomorpha*). Plate I, fig. 13.

Basionym. *Enteromorpha chaetomorphoides* Børgesen 1911: p. 149, fig. 12.

Filaments are bright green, simple or abundantly branched, cylindrical, to 60 µm in diam., and attached by rhizoids. Branches terminate in a single series of cells; basal regions are 2–3 cells in cross section and 45 µm in diam. Cells are square to rectangular, 16–18 µm in diam., and in longitudinal rows; each cell has 2–3 pyrenoids.

Common, on salt marsh plants, mangrove roots, coarse algae, and shells on all coasts; also known from Bermuda, Jamaica, Puerto Rico, the Virgin Islands (type location: St. Thomas; Schneider and Lane, 2005), Brazil, and West Africa.

Taylor 1960: p. 57; Dawes 1974: p. 67.

Ulva clathrata (Roth) C. Agardh 1811 (1810–1812): p. 23 (Gr. *clathratus*: lattice). Plate I, fig. 24–26.

Basionym: *Conferva clathrata* Roth 1806: p. 175–178.
Synonyms: *Ulva muscoides* Clemente y Rubio 1807: p. 320.
Enteromorpha clathrata (Roth) Greville 1830: lxvi, p. 181.
Enteromorpha muscoides (Clemente y Rubio) Cremades in Cremades and Pérez-Cirera 1990: p. 489.

Molecular studies by Blomster et al. (1999) and Hayden et al. (2003) showed infraspecific variation within this species.

Plants are erect, tubular, 10–30(40) cm tall, and strap-shaped to filiform; they may form stiff mats and are attached by rhizoidal filaments. Fronds have abundant short and long alternating spinelike branchlets that end in a single cell or a uniseriate tip of 5 or more cells. Cells are polygonal, quadrate, or rounded, 10–30 µm wide, to 40 µm long, and unordered in older parts; cells in younger parts are rectangular, 15 by 20 µm, and in longitudinal rows. Plastids lie on outer walls and have 2–6(–10) pyrenoids per cell.

Common during winter, on seagrasses, as mats on stones and shells at wave exposed sites on all coasts; also present from the Canadian Arctic to Georgia, Bermuda, Puerto Rico, and Brazil, as well as the eastern North and tropical Atlantic [type locations: Algeciras, Spain (Cremades and Pérez-Cirera, 1909) or Baltic Sea (Abbott and Huisman, 2004)], Indo-Pacific, and Australia.

Taylor 1960: p. 58, 60; Humm and Taylor 1961: p. 343, 348; Dawes 1974: p. 67, 69; Schneider and Searles 1991: p. 43, 49.

Ulva compressa Linnaeus 1753: p. 1163; also 1755: p. 433, no. 1155 (L. *compressa*: pressed together). Plate I, fig. 14, 15.

Synonyms. *Enteromorpha compressa* (Linnaeus) Greville 1830: p. 180.
Enteromorpha clathrata (Roth) C. Agardh f. *linkiana* (Greville) V. J. Chapman 1937: p. 232.

Burrows (1991) proposed that plants designated as *Enteromorpha compressa* (branched, lower intertidal) and *E. intestinalis* (unbranched, upper intertidal) were subspecies of *E. intestinalis*. H. Kim and Lee (1994) found different tubular morphologies, with rapidly growing forms in eutrophic habitats being highly branched. By contrast, Blomster et al. (2000) reported that the species are separated by low levels of DNA sequence variations.

Plants are usually gregarious and to 35 cm long; fronds are morphologically variable, being partially tubular, compressed, or collapsed with hollow margins, and attached by rhizoids. Main axis is to 0.5 cm wide, tapering below and attenuate at tips. Branching is mostly basal and with branches to 0.3 cm wide. Cells form packets of well-ordered rows in the midregion; basal cells are rounded and unordered, while others are rectangular and 15 μm wide by 18 μm long. Most cells have one pyrenoid per cell. *Ulva compressa* and *U. linza* have flattened and expanded foliar tubes with hollow or 1-layer margins (Abbott and Huisman, 2004). When fresh, *U. compressa* is coarse and not slippery, while *U. linza* is delicate and slippery.

Common, during winter in Tampa Bay and year-round in the Keys; growing on limestone, pilings, and shells on all coasts; also known from Labrador to Georgia, Bermuda, the Caribbean, Brazil, and Uruguay, as well as eastern North and tropical Atlantic (likely type location: Bangor, England; Hayden et al. 2003), the Mediterranean, Indo-Pacific, and Australia.

Taylor 1960: p. 60; Dawes 1974: p. 67; Schneider and Searles 1991: p. 43.

Ulva fasciata Delile 1813 (1813–1826): p. 297, pl. 58, fig. 5 (L. *fasciata*: bundled, bound together, striped, or banded). Plate I, fig. 16.

Blades are distromatic, thin, 10–50 (100) cm long, bright green, and with shield-shaped bases; stalks are inconspicuous or absent and attached by basal rhizoids. Fronds are irregularly or palmately divided into linear (0.5–3.0 cm wide) segments and with margins either entire and smooth or ruffled. Cells are 8–20 μm wide, 14–40 μm long, with the center of blades thicker and cells 16–20 μm tall; plastids are parietal and have 1–2 pyrenoids.

Common, usually in winter, on wave-exposed intertidal habitats; also known from North Carolina, Georgia, Bermuda, the Caribbean and Gulf of Mexico to Uruguay, as well as France to Portugal, the Mediterranean (type location: Alexandria Egypt; Silva et al. 1996), eastern tropical Atlantic, Red Sea, and Australia.

Taylor 1960: p. 66; Humm and Taylor 1961: p. 351; Dawes 1974: p. 70; Schneider and Searles 1991: p. 53; Littler and Littler 2000: p. 304.

Ulva flexuosa Wulfen 1803: XXII, p. 1 (L. *flexuosa*: winding, wavy, tortuous, zig-zag). Plate I, fig. 17, 18.

Basionym. *Conferva flexuosa* Roth 1800b: p. 188–190, *nom. illeg.*

Plants are flaccid, gregarious, to 25 cm tall, and attached by a pad of rhizoids; branches filiform and hollow or partially compressed. Cells are rectangular, 10–28 μm wide, 8.5–28 μm long, in longitudinal or transverse rows in midregions and rounded within basal areas; plastids are parietal, cylindrical, and with 1–2 pyrenoids.

subsp. *flexuosa* (L. *flexuosa*: wavy). Plate I, fig. 36.

Synonyms. *Enteromorpha flexuosa* (Wulfen) J. Agardh 1883: p. 126–128.
Enteromorpha prolifera var. *flexuosa* (Wulfen) Doty 1948: p. 259.

According to Bliding (1963) the green alga designated as *Enteromorpha lingulata* J. Agardh in the Agardh Herbarium (Lund) is conspecific with or closely related to *E. flexuosa* subsp. *flexuosa*. Hence Wynne (2005a) treats *E. lingulata* as a synonym of *Ulva flexuosa* subsp. *flexuosa*. However, Hayden et al. (2003) notes that *E. lingulata* J. Agardh (1883: p. 143–144) cannot be transferred to *Ulva* because of the prior existence of *U. lingulata* A. P. de Candolle in Lamarck and de Candolle (1805: p. 14), a name occupied by a red alga.

Filaments are tubular, bright green, in entangled tufts, 3–7(15) cm tall, and have rhizoidal holdfasts. Axes are sparsely to abundantly branched, hollow, and 1–2 mm in diam.; branchlets are longer above and terminate in rows of single cells. The plant's cells are rectangular or square, 9–12 μm wide by 9–28 μm long, occur in longitudinal rows, and have thin lateral walls.

subsp. *paradoxa* (C. Agardh) M. J. Wynne 2005a: p. 62 (L. *paradoxa*: unusual). Plate I, fig. 27, 28.

Basionym. *Ulva paradoxa* C. Agardh 1817: p. 12.
Synonyms. *Enteromorpha erecta* (Lyngbye) Carmichael in W. J. Hooker 1833: p. 314.
Enteromorpha plumosa Kützing 1843: p. 300, pl. 20, fig. 1.
Enteromorpha paradoxa (C. Agardh) Kützing 1845: p. 247.
Enteromorpha flexuosa (C. Agardh) Kützing subsp. *paradoxa* (C. Agardh) Bliding 1963: p. 79, fig. 42–45.

Axes are small bright green tubes 8(–20) cm long, abundantly branched, and attached by a dense pad of fine rhizoids. Branches are covered with uniseriate branchlets that taper at the tips to 8 μm in diam. Tube cells are rectangular or polygonal, to 25 μm wide by 20–50 μm long; plastids appear as dentate discs with 2–5 pyrenoids; cells occur in longitudinal and transverse rows throughout.

Common during winter and spring on shells, jetties, and mangrove roots in shallow estuarine habitats on all coasts; also known from the Canadian Arctic to North Carolina, Bermuda, the Bahamas, Caribbean to Uruguay, and Gulf of Mexico (syntype locations of subsp. *flexuosa*: North Atlantic, Gulf of Mexico, Tasmania, New Zealand; Silva et al. 1996), as well as the eastern North and tropical Atlantic (paratype locations of subsp. *paradoxa*: Bangor, Wales and Brighton, England; Abbott and Huisman, 2004), Mediterranean (type location of species: Duino, Italy; Burrows, 1991), and Indo-Pacific.

Taylor 1960: p. 58, 60, 61; Humm and Taylor 1961: p. 345, 347; Dawes 1974: p. 67, 68; Schneider and Searles 1991: p. 44, 45; Littler and Littler 2000: p. 300, 302.

Ulva intestinalis Linnaeus 1753: p. 1163 (L. *intestinalis*: internal or shaped like intestines). Plate I, fig. 19.

Synonym. *Enteromorpha intestinalis* (Linnaeus) Nees 1820: Index 2.

Link (1820: p. 5, fig. 8) first named *Enteromorpha intestinalis* and Nees placed it in the index of a collective work in which Link's article was published (cf. Burrows, 1991). See also *Ulva compressa* and Burrows' (1991) statement regarding its designation as a subspecies of *E. intestinalis*.

Plants are gregarious, often unbranched, and 30(–50) cm long by 4(–10) cm wide; axes are tubular to compressed, contorted, with weakly adherent walls, hollow margins, and attached by a rhizoidal pad. Cells are 7–15 μm in diam. and do not occur in linear rows; basal cells are 18–22 μm long and rounded; plastids are parietal and have two pyrenoids.

Common, in winter-spring, in estuarine habitats on shells, rock jetties, limestone (to 5 m), and as an epiphyte on all coasts; also known from the Canadian Arctic to the Caribbean, Gulf of Mexico, and Brazil, as well as the eastern North and tropical Atlantic (type location: Lapland; Burrows, 1991), Mediterranean, Indo-Pacific, and Australia.

Taylor 1960: p. 62; Humm and Taylor 1961: p. 345; Dawes 1974: p. 67; Schneider and Searles 1991: p. 45; Littler and Litter 2000: p. 302.

Ulva lactuca Linnaeus 1753: p. 1163 (lettuce, the genus *Lactuca*). Plate I, fig. 20.

Blades are solitary or clustered, orbicular to irregular, to 50 cm long, lack a stalk, and are attached by rhizoids from a disc-shaped holdfast. Cells are rounded to quadrate, in pairs or short rows, 8–12 μm wide by 12–14 μm long, have a single parietal plastid with 1–2 pyrenoids and thick outer walls.

Common, year-round, on all coasts, and to 10 m; also known throughout the world (type location: the west coast of Sweden; Womersley, 1984) and often associated with high nutrient conditions (Cotton, 1910; Sawyer, 1965; Menesguen, 1992; Morand and Briand, 1996).

Taylor 1960: p. 65; Humm and Taylor 1961: p. 351; Dawes 1974: p. 70; Littler and Littler 2000: p 306.

Ulva linza Linnaeus 1753: p. 1163 (L. *linza*: a name used by Roman Imperators to describe the plant's strap-shape appearance). Plate I, fig. 21–23.

Synonym. *Enteromorpha linza* (Linnaeus) J. Agardh 1883: p. 134, pl. 4, fig. 110–112

See comments above about morphological separation of *Ulva compressa* and *U. linza*.

Blades are mostly unbranched and 5–40 cm tall; main axes are 3–10 cm wide, tapering from base, and attached by a rhizoidal pad. Fronds are hollow or adhering distally, strap-shaped, lanceolate to linear, and with ruffled margins. Cells in mid to upper regions are quadrangular to rectangular, often polygonal, 10–20 μm in diam., and in longitudinal rows; cells of lower areas are unordered and irregularly polygonal, with plastids having 1–2 pyrenoids per cell.

Common, mostly during winter, on shells and mangrove roots, at the bases of salt marsh plants, and attached to limestone on all coasts; also known from the Canadian Arctic to Georgia, Bermuda, the Caribbean, Gulf of Mexico, and Brazil, as well as the eastern North and tropical Atlantic (type location: probably England; Scagel, 1996), the Mediterranean, Indo-Pacific, Australia, and Antarctic.

Taylor 1960: p. 63; Humm and Taylor 1961: p. 347; Dawes 1974: p. 69; Schneider and Searles 1991: p. 47.

Ulva profunda W. R. Taylor, 1928: p. 57, pl. 3, fig. 2, 3, 16, 17 (L. *profunda*: deep). Plate I, fig. 29, 30.

Blades are oval to broad, irregular, to 40 cm long, with regular pores 0.5–1.5 cm in diam., on a short stalk (1 mm long), and attached by a shieldlike rhizoidal base. Cells are angular and 18–36 μm in diam.; each cell has one parietal plastid with 1–2 pyrenoids.

Rare, from 15–67 m at the Dry Tortugas (type location); also known from Mauritius and India.

Taylor 1960: p. 67.

Ulva prolifera O. F. Müller 1778: p. 7, pl. 763, fig. 1 (L. *prolifera*: proliferous, with many offshoots) plate I, fig. 31, 32.

Synonyms. *Enteromorpha salina* Kützing 1845: p. 247.
Enteromorpha salina var. *polyclados* Kützing 1845: p. 248.
Enteromorpha prolifera (O. F. Müller) J. Agardh 1883: p. 129, pl. 4, fig. 103, 104.
Enteromorpha torta (Mertens) Reinbold 1893: p. 196, 199, 205.

Primary axes are tubular, distinct, to 60 cm tall, and attached by a rhizoidal pad. Branching is often abundant along entire axes; branches are 1.0–1.5 cm wide; branchlets are mostly less than 500 µm long. Cells are 9–27 µm in diam., irregular to rectangular in older axes, and in rows on younger axes; plastids are parietal and have 1–3 pyrenoids.

Common during winter and spring, forming mats or clumps on shells and limestone within the intertidal and to 2 m on all coasts; also known from the Canadian Arctic (James Bay) to Georgia, Bermuda, the Caribbean, Gulf of Mexico, and Brazil, as well as the eastern North and tropical Atlantic (type location: Denmark; Abbott and Huisman, 2004), Mediterranean, Indo-Pacific, and Australia.

Taylor 1960: p. 56, 57, 59; Humm and Taylor 1961: p. 348; Dawes 1974: p. 69; Schneider and Searles 1991: p. 48, 50; Littler and Littler 2000: p. 304.

Ulva rigida C. Agardh 1823 (1820–1828): p. 410, 411 (L. *rigida*: stiff). Plate I, fig. 33–35.

Synonyms. *Ulva lactuca* L. var. *rigida* (C. Agardh) Le Jolis 1863: p. 38.
Ulva lactuca var. *lacinulata* (Kützing) W. R. Taylor 1960: p. 65.

Blades are bright green, solitary or tufted, 10(–100) cm long, strap-shaped to lobed, and attached by basal rhizoids. Blades are usually 32–40 µm thick and to 200 µm thick at bases. Margins are lobed, smooth to ruffled, and with microscopic projections. Cells are rectangular to rounded, 11–18 µm wide, 15–22 µm long in surface view, and in rows in upper parts. Cells are 29 by 16 µm in young blades, taller than broad (21 µm) at margins, to 70 µm tall near base, and have one parietal plastid with 2(–8) pyrenoids.

Uncommon, on limestone, jetties, and mangrove roots, as well as adrift in the Keys; also known from the Canadian Arctic to Georgia, Bermuda, and the Caribbean, as well as the eastern North and tropical Atlantic, Mediterranean (type location: Cádiz, Spain; Papenfuss, 1960), Indo-Pacific, Australia, Antarctic, and sub-Antarctic islands.

Taylor 1960: p. 65; Dawes 1974: p. 70; Schneider and Searles 1991: p. 54; Littler and Littler 2000: p. 306.

Ulvellaceae

See Silva et al. (1996) regarding the placement of the family in the Ctenocladales rather than the Ulvales.

Plants are small epiphytes or endophytes with branched uniseriate filaments; some erect filaments may be pseudoparenchymatous.

Entocladia Reinke 1879: p. 476 (Gr. *ento*: inside + L. *cladia*: a small branch, shoot).

Synonyms. *Entoderma* Wille 1890: p. 94.
Ectochaete (Huber) Wille 1909 (1909–1910): p. 79.

Based on studies of *Entocladia viridis* (type species) by Nielson (1979), the generic placement is unclear (Schneider, 2003).

Plants are epiphytic, form green disclike or pseudoparenchymatous patches that are thicker in the center than on the margins, and have rhizoids that may penetrate the outer walls of various hosts. The patches consist of filaments that radiate outward toward the margin.

Key to the species (2) of *Entocladia*

1. Common, epiphytic on various algae and seagrasses . . . . *E. viridis*
1. Rare, endophytic in the cell walls of brown and red algae . . . . *E. wittrockii*

Entocladia viridis Reinke 1879: p. 476–478, pl. 6, fig. 6–9 (L. *viridis*: green). Plate II, fig. 4.

Synonyms. *Phaeophila viridis* (Reinke) Burrows in Parke and Dixon 1976: p. 568.
Acrochaete viridis (Reinke) Nielsen 1979: p. 442.

Plants are microscopic with branched uniseriate filaments forming a central membrane that appears pseudoparenchymatous and has margins of radiating filaments. Cells are irregular, polygonal, swollen, or elongate, and 3.5–7.0(12) µm in diam. by 1–4(6) diam. long; plastids are parietal and have 1(–4) pyrenoids.

Common, a cosmopolitan and often overlooked epiphyte on seaweeds, seagrasses, and bryozoans on all coasts; also known from the Canadian Arctic to Bermuda, the Caribbean, Gulf of Mexico, and Brazil, as well as the eastern North and tropical Atlantic, Mediterranean (type location: Naples Italy), and Indo-Pacific.

Taylor 1960: p. 50; Humm and Taylor 1961: p. 338; Dawes 1974: p. 65; Schneider and Searles 1991: p. 36.

Entocladia wittrockii Wille 1880: p. 1–4, pl. 1 (named for V. B. Wittrock, a Swedish phycologist). Plate II, fig. 2, 3.

Synonyms. *Endoderma wittrockii* (Wille) Lagerheim 1883: p. 75.
Endoderma wittrockii (Wille) De Toni 1889: p. 209.
Ectochaete wittrockii (Wille) Kylin 1938b: p. 72.
Phaeophila wittrockii (Wille) Nielsen 1972: p. 264.
Acrochaete wittrockii (Wille) Nielsen 1983: p. 692.

Filaments are microscopic, dark to light green, irregularly branched, 5–10 µm in diam., and often in gelatinous coatings on various hosts. Cells are cylindrical, 1.0–1.5 diam. long, and have one parietal plastid with a single pyrenoid.

Rare, endophytic in cell walls of subtidal brown and red algae on the northwest coast; also known from Newfoundland to Florida and the Gulf of Mexico, as well as the eastern North Atlantic (type location: Oslo, Norway; Burrows, 1991), Mediterranean, and Indo-Pacific.

Taylor 1957: p. 54; Humm and Taylor 1961: p. 338.

Epicladia Reinke 1889b: 31 (L. *epi*: on, above + *clados*: small branch, shoot).

Filaments are uniseriate, branched, and form either tufts or pseudoparenchymatous pads. Hairs are absent and cells have a parietal chloroplast with one or more pyrenoids. Asexual propagation is by pyriform bi- or quadriflagellate zoospores.

Epicladia flustrae Reinke 1889b: p. 31, pl. XXIV, fig. 5–9; also 1889a: p. 86 (L. *flustra*: small piece; also the bryozoan genus *Flustra*). Plate II, fig. 1.

Synonym. *Entocladia flustrae* (Reinke) W. R. Taylor 1937a: p. 54.

Plants are small with irregular, densely branched filaments; mature plants have a central membrane, a filamentous margin, and are attached by rhizoids. Plastids are parietal and have one pyrenoid; central cells are irregularly polygonal and 7–12 μm in diam. or larger.

Rare, on bryozoans and sponges on the northeast coast in shallow water; also known from Labrador to Long Island Sound, Texas, as well as the eastern North Atlantic (type location: Germany; Burrows, 1991), Mediterranean, Madeira Archipelago, and British Columbia.

Humm and Taylor 1961: p. 338.

Pringsheimiella Höhnel 1920: p. 97 (named for N. Pringsheim, a German botanist and phycologist + L. *ella*: lesser, a diminutive ending).

Plants have marginal growth and form monostromatic discs with filaments closely joined. Central cells are taller than marginal ones that are radially elongate.

Pringsheimiella scutata **(Reinke) Höhnel *ex* Marchewianka 1925: p. 42, as '*scuttata*' (L. *scutata*: shield-shaped, armed with a shield). Plate II, fig. 5.**

Basionym. *Pringsheimiella scutata* Reinke 1888: p. 241; also 1889a: p. 81; 1889b: p. 33, pl. 25.
Synonyms. *Pringsheimiella udoteae* (Børgesen) W. R. Taylor 1960: p. 51.

Plants form bright green epiphytic discs or crusts, to 2.0 mm in diam. and 10–12 μm thick. Cells are 5–25 μm wide by 10–40 μm long and vary in shape. Marginal cells are radiating, elongate, and laterally branched. Central cells are more columnar (to 12 μm) than marginal ones and extend into vertical hairs, 8–10 μm wide and to 140 μm tall. Sporangia are ovoid to pear-shaped, 15–22 μm in diam. by 28–38 μm long, and have an apical pore.

Common, on algae and seagrasses on all coasts; also known from the Canadian Arctic to North Carolina, Bermuda, the Caribbean, Gulf of Mexico, and Brazil, as well as the eastern North and tropical Atlantic (type location: Kieler Förde, Baltic Sea; Burrows, 1991), Mediterranean, and Indo-Pacific.

Taylor 1960: p. 51; Humm and Taylor 1961: p. 339; Dawes 1974: p. 66; Schneider and Searles 1991: p. 38; Littler and Littler 2000: p. 306.

Pseudendoclonium Wille 1901: p. 29, pl. 3, fig. 101–134 (Gr. *pseud*: false + *endo*: inside + *clonium*: a branch).

Plants produce crustose pseudoparenchymatous patches on wood or rubble; filaments are irregularly branched, prostate and erect, and are attached by 1-celled rhizoids.

Pseudendoclonium submarinum **Wille 1901: p. 29, pl. 3, fig. 101–134 (L. *submarinum*: under the sea). Plate II, fig. 6.**

Crusts are pseudoparenchymatous, with prostrate (15–20 cells) and erect (3–7 cells) irregularly branched filaments and a few rhizoids. Young filaments are 3–7 cells long; most cells are 6–9 μm in diam. by 7–14 μm long, while older ones may be short and broad.

Common, on wood, shells, barnacles, and crab carapaces; often forming a green band at high tide on jetties on all coasts. Also known from Newfoundland to Florida, Bermuda, the Caribbean, and Gulf of Mexico, as well as the eastern North and tropical Atlantic (type location: O. Drøbak, Norway; Burrows, 1991), Mediterranean, and Australia.

Taylor 1960: p. 52; Humm and Taylor 1961: p. 341; Dawes 1974: p. 66; Schneider and Searles 1991: p. 51.

Ulvella P. Crouan and H. Crouan 1859: p. 288. (L. *ulva*: a sedge, marsh plant + *ella*: lesser, a diminutive ending; thus like a small *Ulva*).

Synonym. *Pseudulvella* Willie 1909 (1909–1910): p. 90.

Crusts are epiphytic, endophytic, or lithophytic, dark green, irregular, and have a central multicellular region. Margins consist of radiating filaments with forked apical cells; plastids lack pyrenoids.

Ulvella lens P. Crouan and H. Crouan 1859: p. 288–289, pl. 22, fig. E (L. _lens_: a lentil, a leguminous fruit). Plate II, fig. 7.

Crusts are bright green, thin, circular, 0.1–5 mm in diam. by 8–25 µm thick, and attached by basal rhizoids. Its central area is compact and composed of 2–3 layers. Marginal apical cells radiate and fork; surface hairs are absent. Central cells are elongate to rectangular and 5–12(20) µm long , while marginal cells are elongate, 3–4(15) µm in diam., and 15–30 µm long. Plastids are parietal discs that fill an entire cell face. Sporangia are formed by central cells and produce 4–16 biflagellate zooids.

Common but inconspicuous, growing on coarse algae, seagrass blades, ascidians, barnacles, shells, and limestone on all coasts. Also known from the Canadian Arctic to South Carolina, Bermuda, the Caribbean, and Gulf of Mexico, as well as the eastern North and tropical Atlantic (type location: Brest, France; Burrows, 1991), Mediterranean, and Indo-Pacific.

Taylor 1960: p. 52; Humm and Taylor 1961: p. 341; Dawes 1974: p. 66; Schneider and Searles 1991: p. 38; Littler and Littler 2000: p. 308.

PHAEOPHILALES

Plants are epiphytic, endophytic, or may penetrate shells. Filaments are branched, uniseriate, and have 1–3 colorless hairlike extensions that lack basal cross walls. The order is monotypic and based upon the occurrence of zoospores that are produced by simultaneous cleavage of multinucleate initials; zooids have simple, convex, terminal caps on their anterior-most pair of flagellar basal bodies (Chappell et al. 1990).

Phaeophilaceae

Chappell et al. (1990) transferred *Phaeophila* from the Ulvellaceae to its own family and order based on the plant's zoosporangial structure and development (O'Kelly and Yarish, 1981 [1980]), as well as zoospore flagellar apparatus.

The family is monotypic with the same characteristics as the order.

Phaeophila Hauck 1876: p. 56 (Gr. *phaeo*: dark, brown + *phila*: love, affinity for).

Plants are epiphytic, endophytic, or penetrate shells; filaments are branched, uniseriate, and with 1–3 colorless hairlike extensions that lack basal cross walls. Cells have lobed parietal plastids and multiple pyrenoids. Sporangial mother cells are multinucleate and produce quadriflagellate zoospores.

Phaeophila dendroides (P. Crouan and H. Crouan) Batters 1902: p. 13 (Gr. _dendroides_: resembling a tree, treelike). Plate II, fig. 9.

Basionym. *Ochlochaete dendroides* P. Crouan and H. Crouan 1852: no. 346; 1867: p. 128, pl. 18.

See O'Kelley and Yarish (1980), Correa et al. (1988), and Kitayama and Yoshida (1992) for details on morphology. O'Kelley et al. (2004b) summarized the plant's phylogenetic position.

Plants are microscopic with many irregularly branched prostrate filaments. Hairs or setae are upright, 1–3 per cell, colorless, often twisted, and lack swollen bases and cross walls. Cells are 9–40 μm in diam. and 15–50(80) μm long; plastids are parietal, lobed, and have several pyrenoids. Zoosporangia are 16–40 μm in diam. by 30–85 μm long.

Common, on/in algae and seagrass blades, in limestone and shells, and on all coasts; also known from Newfoundland to North Carolina, Bermuda, the Caribbean, Gulf of Mexico, and Brazil, as well as from the eastern North and tropical Atlantic (type location: Brest, France; Burrows, 1991), Mediterranean, Red Sea, Indo-Pacific, and Australia.

Taylor 1960: p. 48; Humm and Taylor 1961: p. 341; Dawes 1974: p. 65; Schneider and Searles 1991: p. 37.

CLADOPHORALES

Based on molecular studies, Leliaert et al. (2003) considered the order to be paraphyletic. By contrast, Bakker et al. (1994) considered the Siphonocladales to be monophyletic and distinct.

Plants are branched or unbranched uniseriate filaments, with cells formed by cross walls or internal divisions (segregative division) that produce branches externally. Cells are usually large and multinucleate, and have laminated walls of crystalline, cellulose microfibrils. Chloroplasts are numerous, discoid to angular, and often connected to form a reticulum; they have one or more pyrenoids per chloroplast.

Anadyomenaceae

Plants are delicate bright green blades formed by multinucleated, branching filaments; blades have an open or closed network and a stalk attached by rhizoids.

Anadyomene J.V. Lamouroux 1812: p. 187, *nom. cons.* as '*Anadyomena*' (in Greek mythology *Anadyomene* is the goddess Aphrodite when arising from the sea).

Synonym. *Cystodictyon* J. E. Gray 1866: p. 72.

See Littler and Littler (1991) for a review of Caribbean species.

Plants form loose to dense clusters of blades that are yellow green and crisp; they arise from an inconspicuous stalk and are attached by rhizoids from the stalk and lower rib cells. Blades arise from radiating filaments of larger cells that become "veins," with spaces between the ribs filled with cells, giving a netlike appearance.

Key to the species (4) of *Anadyomene*

1. Blades lacking a regular pattern of pores (nonperforate) 2
1. Blades having a regular pattern of pores (perforated) . . . . *A. pavonina*
   2. Primary filaments (veins) produced in a single uniseriate rows of cells 3
   2. Primary veins composed of several parallel rows of cells . . . . *A. menziesii*
3. Filler (interstitial) cells random and oval in section . . . . *A. saldanhae*
3. Filler (interstitial) cells mostly parallel and H-shaped in section . . . . *A. stellata.*

Anadyomene menziesii (J. E. Gray) J. Agardh 1887: p. 128 (named for Archibald Menzies, a surgeon and collector on HMS _Discovery_ that surveyed the west coast of North America in 1793). Plate II, fig. 10, 11.

Basionym. *Grayemma menziesii* J. E. Gray 1866: p. 49–51.

Synonym. *Anadyomene flabellata* Lamouroux 1816: p. 365, pl. 14, fig. 3 *sensu* Harvey 1858: p. 49–50.

Guiry and Nic Dhonncha (2006) cite the plant as *Anadyomene menziesii* Harvey 1858: p. 50.

Blades are single or clustered, crisp, flat or ruffled, to 20 cm in diam., and attached by massive rhizoidal pads. Fronds are monostromatic and have "veins" of 2–6 parallel filaments, with the remaining space filled with oval to elongate branchlets.

Common, in 20–30+ m off the west coast and the eastern Gulf of Mexico (type location: dredged from 60 m; Harvey, 1858); videos taken by submersibles show hectares of large leafy plants in 30 m off the Dry Tortugas.

Taylor 1960: p. 125; Dawes 1974: p. 90.

Anadyomene pavonina (J. Agardh) Wille 1910 (1909–1910): p. 114 (L. *pavonina*: shining, peacock-like, fan-shaped). Plate II, fig. 12, 13.

Basionym. *Cystodictyon pavonium* J. Agardh 1894: p. 109 (as '*pavoniuum*').

Blades are delicate, bright green, to 14 cm in diam., and with pores 0.1–10 mm in diam.; plants are sessile and attached by rhizoids from lower cells of the main axis. Monostromatic blades have veinlike filaments radiating from the centers of various parts of each blade; cells are 180–300 μm in diam. and 2–10 diam. long.

Uncommon, from 20–50 m on all coasts (type location: Florida); also known from Madagascar.

Taylor 1960: p. 124; Dawes 1974: p. 92.

Anadyomene saldanhae A. B. Joly and E. C. Oliveira 1969 ('1968'): p. 30, fig. 1–3 (named for the Brazilian research vessel *Almirante Saldanha*). Plate II, fig. 14, 15.

Plants are erect or decumbent; blades are densely packed and attached by loosely entangled rhizoids that arise from a wide area, including extensions of lower vein cells. Blades are monostromatic, crisp, fan-shaped, to 10 cm tall, lack perforations, and have branching filaments with faint veins. Main veins are uniseriate, radiating outward, 250–350 μm in diam., and polychotomous. Filler cells occur in random patterns between veins.

Rare, on mangrove roots and in >20 m in the Keys; also known from North Carolina (adrift and offshore in Onslow Bay), Bermuda, the Caribbean, Brazil (type location: Banco Dogareza), and Canary Islands.

Schneider and Searles 1991: p. 78; Littler and Littler 2000: p. 310.

Anadyomene stellata (Wulfen in Jacquin) C. Agardh 1823 (1820–1828): p. 400. fig. 44–56 (L. *stellata*: starry, like the rays of a star). Plate II, fig. 16, 17.

Basionym. *Ulva stellata* Wulfen in Jacquin 1787 ('1786'): p. 351–352.

Synonym. *Anadyomene stellata* var. *floridana* J. E. Gray 1866.

Blades are monostromatic, dark green, crisp, 2–6 cm tall, clustered, ovate to kidney-shaped, with ruffled edges, and attached by loosely entangled rhizoids from basal cells. "Veins" are single, polychotomous, and with fanlike branching; filaments are barely visible by eye; filler cells between veins are ovoid to elongate and in parallel rows.

Common, on all coasts, on pilings, limestone, mangrove roots, shells, and drift in seagrass beds; also known from Bermuda, the Caribbean, Gulf of Mexico, and Brazil, the Mediterranean (type location: Adriatic Sea; Womersley, 1984), the eastern tropical Atlantic, Indian Ocean, and Australia.

Taylor 1960: p. 125; Humm and Taylor 1961: p. 372; Dawes 1974: p. 91; Littler and Littler 2000: p. 310.

Microdictyon Decaisne 1841: p. 115 (Gr. *micro*: small, microscopic + *dictyon*: a net).

Leliaert (2004) transferred *Microdictyon curtissisiae* to *Phyllodictyon plucherrimum* as a taxonomic synonym.

Blades form a reticulate network of filaments with pores and are attached by unicellular (mostly unbranched) rhizoids of basal cells. The frond net is formed by fusion of unmodified vegetative cells or by cells with annular or crenulate pads near their tips.

***Microdictyon boergesenii* Setchell 1925: p. 106 (named for the Danish phycologist F. Børgesen, who published on the seaweeds of the Danish West Indies and Canary Islands). Plate II, fig. 18, 19.**

Fronds are pale green, delicate, 3–6 cm in diam., with an open filamentous network having edges fringed by short projecting branchlets, and attached by short rhizoids arising from obscure filamentous cells. Meshes of fronds are trapezoidal to rectangular and 0.3–0.4 mm in diam.; main cells are 145–200 µm in diam., 400–500 µm long, slightly swollen, and with walls 2–3 µm thick. Branchlet cells are 45–110 µm in diam. and have fusions formed without specialized attachments.

Locally common, growing on seaweeds or limestone in 2–10 m in the Keys and 20–160 m off the east and west coasts; also known from North Carolina, Bermuda, the Caribbean (type location: St. John, Virgin Islands; Silva et al. 1991), Gulf of Mexico, and Brazil, as well as the eastern tropical Atlantic, Madagascar, and Central Pacific Islands.

Taylor 1960: p. 120; Schneider and Searles 1991: p. 79; Littler and Littler 2000: p. 312.

Valoniopsis Børgesen 1934: p. 10 (Italian: *valonia*: the dried acorn cups of the eastern Mediterranean oak tree *Quercus aegilops* and the generic name for the green alga *Valonia* + *opsis*: resembling; thus, like *Valonia*).

Plants are cushionlike, with filaments repent, entangled, and attached by lobed to branched haptera formed at cross walls. Erect filaments are stiff, opposite to irregularly branched, and attached secondarily by their tips.

***Valoniopsis pachynema* (G. Martens) Børgesen 1934: p. 10–16, fig. 1, 2 (Gr. *pachy*: thick, stout + *nema*: thread). Plate II, fig. 8.**

Basionym. *Bryopsis pachynema* G. Martens 1868: p. 24, 62–63, pl. 4, fig. 2.

Plants are pad- to cushionlike, to 2–5 cm thick, 5–10 cm in diam., dark green, and attached by either short rhizoids or lateral branches. Erect filaments are cylindrical, bent, spreading, and 0.3–1.0 mm in diam. by 2–10 diam. long; branchlets are cylindrical, have blunt tips, and grow from the upper parts of main filaments.

Uncommon, forming stiff clumps in protected shallow waters (to 5 m) in the Keys; also recorded from Bermuda, the Bahamas and Caribbean, as well as the Indo-Pacific (syntype localities: Benkulen and Pulau Tikus; Silva et al. 1996), and Lord Howe Island.

Taylor 1960: p. 113; Littler and Littler 2000: p. 314.

Boodleaceae

Blades are monostromatic, stalked, and reticulate, with most filaments in one plane.

Phyllodictyon J. E. Gray 1866: p. 70 (Gr. *phyllo*: relating to leaves + *dictyon*: a net).

Blades are oblong, free, and netlike, and arise from slender filaments or stalks that are elongate, simple, or forked; each fork ends in a blade with a midrib; stalks are attached by rhizoidal holdfasts. Blades are solitary or in clumps and composed of a network of first- and higher-order branches growing in a single plane. Branches are formed by nonsynchronous centripetal cross wall growths that divide parent walls into approximately equal halves; thus, branch cross walls are formed similar to those resulting from cell divisions in *Cladophora* rather than by segregative divisions as in *Struvea*. Blade filaments terminate in tenacular cells that fuse with other branches to form a network that is mostly in one plane.

Key to the species (2) of *Phyllodictyon*

1. Central vein dominant, with 4–8 pairs of branchlets. . . . .*P. anastomosans*
1. With one or more central veins and 8–12 or more branchlet pairs . . . .*P. pulcherrimum*

Phyllodictyon anastomosans **(Harvey) Kraft and M. J. Wynne 1996: p. 139, fig. 16–25 (L. *anastomosans*: to fuse together). Plate II, fig. 20.**

Basionym. *Cladophora anastomosans* Harvey 1859: pl. CI (no page numbers).
Synonyms. *Struvea anastomosans* (Harvey) Piccone and Grunow in Piccone 1884: p. 20, fig. 1, 2.
Struvea anastomosans var. *caracasana* M. J. Wynne 1998a: p. 65.

Plants are erect, in sparse clusters, and 3–5 cm tall; blades are delicate, netlike, either sessile or on uniseriate stalks that are annulated in upper parts, and attached by short, branched rhizoids. Stalks may branch to form secondary fronds; their central stalks are 200–900 μm in diam. Blades arise from 4–6 pairs of opposite branchlets that divide more than once; branchlets are intensely tangled and decrease in diam. with each division; tips are 100–140 μm in diam. Main axial filaments are obscure, with opposite branching, and have cells 320–600 μm in diam. by 0.6–2.9 mm long. Branchlet cells are 250–340 μm in diam. and 2–3 diam. long; branchlets are fused by special hapteral cells, 43–114 μm in diam. by 30–80 μm wide.

Rare to uncommon, in crevices on wave-exposed limestone (to 20 m), mangrove roots, and coarse algae in the Keys; also known from the Bahamas and Caribbean, as well as the eastern tropical Atlantic, Indo-Pacific, and Australia (type location: Fremantle; Lipkin and Silva, 2002).

Taylor 1960: p. 121, 122; Littler and Littler 2000: p. 312, 328.

Phyllodictyon pulcherrimum **J. E. Gray 1866: p. 70 (as *pulcherrima*) (L. *pulcherrimum*: very pretty, most beautiful). Plate II, fig. 21–25.**

Synonyms. *Struvea ramosa* Dickie 1874c: p. 316.
Struvea pulcherrima (J. E. Gray) G. Murray and Boodle 1888: p. 281.
Microdictyon curtissiae W. R. Taylor 1955: p. 69, pl. 1, fig. 1–8; pl. 3, fig. 2.
Leliaert (2004) placed *Microdictyon curtissiae* in synonymy with *Phyllodictyon plucherrimum*.

Blades are solitary or clustered, 3–30 cm tall, bright green, and stalked; stalks are uniseriate, have 2–3 blades, and a lobed basal cell for attachment. Stalks and blade filaments are 0.2–1.5 mm in diam.; blades are reticulate, to 20 cm wide, with midrib distinct, and filaments oppositely branched. Branches are progressively smaller at each fork, with branchlet tips attached to adjacent cells by specialized hapteral cells.

Uncommon, in 20–80 m off the west coast and in 5 m at the Keys; also known from North Carolina, Bermuda, the Bahamas, Puerto Rico, Lesser Antilles, Gulf of Mexico (type location), Madeira Archipelago, and Canary Islands.

Taylor 1960: p. 123; Dawes 1974: p. 92; Schneider and Searles 1991: p. 80; Littler and Littler 2000: p. 328.

Struvea Sonder 1845: p. 49, *nom. cons.* (probably named for Friedrich Georg Wilheim von Struve, a German-born Russian astronomer who first measured stellar parallax).

Plants have one or more blades, each of which is composed of a filamentous net that is supported by a simple or branched uniseriate stalk attached by multicellular, branched rhizoids. Filaments have terminal tenacular cells that fuse with each other to form a network that is in one plane. Cell division is segregative rather than by the formation of a centripetal cross wall as in *Phyllodictyon* and *Cladophora*.

Struvea elegans Børgesen 1912: p. 264: fig. 13, 14 (L. _elegans_: graceful, tasteful). Plate II, fig. 26, 27.

Plants are erect, solitary stalked netlike blades, 10–12 cm tall; stalks are attached by creeping, branched rhizoids with fingerlike pads. Stalks are to 4.5 cm long with a basal cell 0.8–2.0 mm in diam. by 2–3 cm long; upper cells are shorter, 1–3 cm long and basally have slight ringlike swellings (annulations). Blades are to 3 cm wide, have an obvious midrib, and opposing lateral branches. The branches arise from the apex of a supporting cell; they branch 2–4 times and are attached to adjacent cells by terminal haptera cells.

Rare, in 40 m at the Dry Tortugas; also known from the Caribbean (type location: Virgin Islands), Réunion and Seychelles Islands, Fiji, and Papua New Guinea.

Taylor 1960: p. 123; Littler and Littler 2000: p. 328.

Cladophoraceae

Plants are branched or unbranched uniseriate filaments that do not form blades.

Chaetomorpha Kützing 1845: p. 203, *nom. cons.* (Gr. *chaeto*: hairlike + *morpha*: form).

Filaments are uniseriate, unbranched, and attached by a basal cell; some populations may form drifting or entangled populations. Cells are multinucleate and often have stratified walls due to layers of highly crystalline cellulose; plastids are angular, have numerous pyrenoids, and are interconnected to form reticulate nets or broken into small discoid units.

Key to the species (9) of *Chaetomorpha*

1. Filaments usually erect and basally attached; basal attachment sometimes rare. 2
1. Filaments detached and entangled, lacking basal attachment 6
    2. Filaments epiphytic or not, more than 30 μm in diameter 3
    2. Filaments epiphytic, 10–20 (40) μm in diameter. . . . . *C. minima*
3. Filaments less than 100 μm in diameter. 4
3. Filaments 100 μm or more in diameter 5
    4. Basal cell elongate (to 450 μm long); other cells 0.5–1.5 diameters long . . . . *C. nodosa*
    4. Basal cells not as elongate; other cells 2–4 diameters long . . . . *C. gracilis*
5. Filaments 150–350 μm in diam.; usually attached. . . . . *C. aerea*
5. Filaments 200–400 μm in diam.; usually detached and with *C. linum*. . . . . *C. picquotiana*
    6. Filaments 80 μm or less in diameter. 7
    6. Filaments usually more than 100 μm in diameter. 8

7. Filaments 10–20(40) µm in diameter . . . . *C. minima*
7. Filaments 30–100 µm in diameter . . . .*C. gracilis*
   8. Filaments soft, flexible; cells mostly as long as broad 9
   8. Filaments stiff, inflexible; cells 1–2 (5) diameters long . . . .*C. linum*
9. Filaments 200 µm or more in diameter . . . . 10.
9. Filaments 80–175 µm in diameter . . . .*C. brachygona.*
   10. Filaments 200–400 µm in diameter . . . . *C. piquotiana.*
   10. Filaments 300–700 (1000) µm in diameter. . . . . *C. crassa.*

Chaetomorpha aerea (Dillwyn) Kützing 1849: p. 379 (L. *aerea*: bronze, copper colored). Plate II, fig. 28.

Basionym. *Conferva aerea* Dilllwyn 1806 (1802–1809): pl. 80.

Synonym. *Chaetomorpha linum* (O. F. Müller) Kützing f. *aerea* (Dillwyn) Collins.

Christensen (1957), Abbott and Hollenberg (1976), Burrows (1991), and John et al. (2004) include *Chaetomorpha aerea* as a synonym for *C. linum*, while South and Cardinal (1970) place *C. linum* in *C. aerea*. Kornmann (1972), Silva et al. (1996), and Lipkin and Silva (2002) recognize both species. Based on isozyme studies Blair et al. (1982) suggested that *C. linum* was an ecad of *C. aerea*, while Kornmann and Sahling (1977) kept both taxa based on life histories. Chromosome studies by Hinson and Kapraun (1991) indicate that western Atlantic species of *Chaetomorpha*, including *C. aerea*, represent a polyploid series. In studying the phylogeny of the Cladophorales with 28S rDNA sequences Boedeker et al. (2005) found that *C. area* and *C. linum* represent a single lineage, consisting of both attached forms and unattached mats (i.e., green tides).

Filaments are uniseriate, unbranched, single or in clusters, 10–15(30) cm long, stiff and straight above, and attached by an elongate holdfast cell (2–4 times as long as wide) and by rhizoids from lower cells. Cells are multinucleate, 125–175(500) µm in diam., 0.5–2 diam. long, and distally swollen; fertile cells are swollen to spherical.

Common, on wave-exposed intertidal to shallow subtidal limestone on all coasts; also known from the Gulf of Saint Lawrence to North Carolina, Bermuda, the Caribbean, Gulf of Mexico, and Brazil, as well as the eastern North and tropical Atlantic (type location: Cromer, England; Womersley, 1984), Mediterranean, Indo-Pacific, and Australia.

Taylor 1960: p. 72; Humm and Taylor 1961: p. 352; Dawes 1974: p. 87; Schneider and Searles 1991: p. 57; Littler and Littler 2000: p. 316.

Chaetomorpha brachygona Harvey 1858: p. 87–88, pl. XLVI A, fig. 1, 2 (Gr. *brachy*: short + *gon*: reproductive structure, or origination). Plate II, fig. 29.

Synonyms. *Chaetomorpha tortuosa sensu* Mazé and Schramm 1878 (1870–1877): p. 47; non *Chaetomorpha torulosum* (Dillwyn) Kützing 1845: p. 204.

Rhizoclonium capillare Vickers 1905: p. 45–66.

Collins (1909b), Blair (1983), and Silva et al. (1996) compared *C. brachygona* with *Rhizoclonium tortuosum* (Dillwyn) Kützing, which has longer cells and more slender, entangled filaments.

Filaments are uniseriate, unbranched, 80–175 µm in diam., soft, entangled, and yellow green to dark green; cells are multinucleate, 60–175(400) µm long, and slightly swollen.

Common, in drift, as loose or entangled mats in salt marshes, mangrove, and seagrass communities on all coasts; also known from the Northumberland Strait (Canada) to North Carolina, Bermuda, the Caribbean, Gulf of Mexico, and Brazil (syntype localities: Key West, Florida and mouth of Rio Bravo, Texas; Silva et al. 1991), as well as the eastern North and tropical Atlantic, Mediterranean, Indo-Pacific, and Australia.

Taylor 1960: p. 70; Humm and Taylor 1961: p. 352; Dawes 1974: p. 87; Schneider and Searles 1991: p. 58.

Chaetomorpha crassa (C. Agardh) Kützing 1845: p. 204; 1853: pl. 59 (L. _crassa_: stout, fleshy, thick). Plate II, fig. 30.

Basionym. *Conferva crassa* C. Agardh 1824: p. 99.

According to Price (1967) and Parke and Dixon (1976) the species may be synonymous with *Chaetomorpha linum*.

Filaments are uniseriate, unbranched, and often form entangled dark green masses; cells are 500–550(1,000) µm in diam. and 1–2 diam. long.

Uncommon, mixed with other seaweeds, in mats in protected coves, and to 50 m in the Keys; also known from North Carolina (24–50 m), Bermuda, the Caribbean, as well as the Mediterranean (syntype localities: Trieste and Venice, Italy, and England; Silva et al. 1991), eastern tropical Atlantic, and Indo-Pacific.

Taylor 1960: p. 72; Schneider and Searles 1991: p. 58; Littler and Littler 2000: p. 318.

Chaetomorpha gracilis Kützing 1845: p. 203 (L. _gracilis_: thin, graceful). Plate II, fig. 31.

Synonyms. *Rhizoclonium tortuosum sensu auct.*, non (Dillwyn) Kützing 1845: p. 205.

Lola gracilis (Kützing) V. J. Chapman 1961: p. 73.

John et al. (2004) cite the authority as *C. gracilis* (Kützing) Kützing.

Filaments are uniseriate, unbranched, pale green to yellow green, in drift and entangled, or attached by a few rhizoidal projections from a basal cell. Attached filament cells are cylindrical, 32–100(120) µm in diam., and 1–2 diam. long; basal cells 125–360 µm long. Cells of drift filaments are often longer (i.e., 2–4 diam.) than those of attached plants.

Common, on mangrove roots, shells, adrift on all coasts, and forming spring blooms in the Keys (Lapointe et al. 2004); also known from North Carolina, Bermuda, the Caribbean, Gulf of Mexico, and Brazil, as well as the Mediterranean (type location: Trieste, Italy; Silva et al. 1991), eastern tropical Atlantic, Indo-Pacific, and Lord Howe Islands.

Taylor 1960: p. 70; Schneider and Searles 1991: p. 59; Littler and Littler 2000: p. 318.

Chaetomorpha linum (O. F. Müller) Kützing 1845: p. 204; 1849: p. 378–379 (L. _linum_: thread, net; also the generic name for flax, _Linum_). Plate II, fig. 32; plate III, fig. 2.

Basionym. *Conferva linum* O. F. Müller 1778: p. 7, pl. 771(2).

See comment above about varying interpretations regarding the separation of *Chaetomorpha aerea* and *C. linum*.

Mature filaments are uniseriate, unbranched, stiff, mostly unattached, loosely curled, and entangled. Cells may be slightly swollen with modest constrictions at cross walls, 80–300(400) µm in diam. by 1–2(5) diam. long, and with thick walls.

Common, as dark green mats on mangrove roots and in salt marshes, often in areas of high nutrients (Mathieson and Dawes, 2002), and entangled with other seaweeds or seagrasses, on all coasts; as well as from Labrador to the Caribbean, Bermuda, Gulf of Mexico, and Brazil, as well as the eastern North and tropical Atlantic (type location: Nakskov Fjord, Denmark; Womersley, 1984), Mediterranean, Red Sea, Indo-Pacific, and Australia.

Taylor 1960: p. 71; Humm and Taylor 1961: p. 354; Dawes 1974: p. 87; Schneider and Searles 1991: p. 60; Littler and Littler 2000: p. 318.

Chaetomorpha minima Collins and Hervey 1917: p. 41–42, pl. I, fig. 5–7 (L. _minima_: small). Plate III, fig. 1.

Filaments are uniseriate, unbranched, forming a green fuzz to 5 mm tall, and attached by discoid basal cells. Upper cells of filaments are 10–20(40) µm in diam. by 1.5–3 diam. long, slightly constricted at cross walls, and have strongly stratified walls; cells of unattached filaments are often to 5 diam. long.

Uncommon, on seagrass blades, adrift, or entangled on salt marsh culms and mangrove roots on all coasts; also known from Maine to North Carolina, and Bermuda (type location; Silva et al. 1996), as well as Kenya and the Maldives.

Taylor 1960: p. 72; Dawes 1974: p. 88; Schneider and Searles 1991: p. 61.

Chaetomorpha nodosa Kützing 1849: p. 376 (L. *nodos*: knotted). Plate III, fig. 3.

Filaments are in erect tufts, uniseriate, unbranched, stiff, straight, dull green, to 12 cm long, and attached by basal cells. Basal cells are long (to 450 µm), to 40 µm in diam. near attachment point, and to 80 µm in diam. at upper end. Vegetative cells are 75–95 µm in diam., 0.5–1.5 times as long as broad, and may be slightly swollen.

Rare, known in Florida only from Smithsonian Institution herbarium specimens from West Summerland and Big Pine Keys; also recorded from the Caribbean, French Guiana, as well as the eastern tropical Atlantic, Indo-Pacific, and the Antarctic.

Chaetomorpha picquotiana Montagne *ex* Kützing 1849: p. 379 (named for M. Lamare-Piquot, a French explorer of Newfoundland and Labrador). Plate III, fig. 2.

Synonyms. *Conferva picquotiana* Montagne 1849: p. 66.
Chaetomorpha melagonium f. *typica sensu* Collins 1909b: p. 323.
Chaetomorpha atrovirens Taylor 1937b: p. 227.

Filaments are usually detached and in entangled mats, filaments are curved, twisted, or curled; cells are 200–400 µm in diam. and 3(2–5) times as long as wide.

A northern species but thought to be in Florida (Blair, 1983); also known from Labrador (type location) to Long Island Sound, Bermuda, and Alaska to Oregon.

Taylor 1957: p. 79.

Cladophora Kützing 1843: p. 262, *nom. cons.* (Gr. *clados*: branch + *phora*: to bear).

Synonyms. *Aegagropila* Kützing 1843: p. 272.
Willeella Børgesen 1930b: p. 155–158.
Rama V. J. Chapman 1952: p. 55–56.

In contrast to the above synonymy, Guiry and Nic Dhonncha (2006) recognize *Willeella* as a distinct genus.

Filaments are uniseriate, sparingly to profusely branched, and adrift or attached by rhizoids or a basal cell. Plastids are parietal, angular or fused into a net, and usually have many pyrenoids. Detached plants are often difficult to identify. Species descriptions are modified from Hoek (1982) based on his earlier studies (1960b, 1963).

Key to the species (16) of *Cladophora*

Bloom populations of *Cladophora crispata* (Roth) J. Agardh and *C. crispula* Vickers have been reported from the Florida Keys (Lapointe et al. 2004) but are not included as the identifications need confirmation.

1. Growth acropetal, solely or almost only by division of apical cells 2
1. Growth intercalary or due to a combination of apical and intercalary divisions 6
   2. Apical cells usually 150 µm or more in diameter 3
   2. Apical cells usually less than 140 µm in diameter 5
3. Main axial cells less than 220 µm in diam., base lacking annular thickenings 4
3. Main axial cells 240–350 µm in diam., with basal thickenings . . . . *C. prolifera*

4. Branching angle near tips to 45°; main axes indistinct . . . . *C. blomquistii*
4. Branching angle near tips 45–90°; main axes distinct. . . . . *C. longicellulata*
5. Apical cells 55–140 µm in diam.; branching irregular, with a >45° angle. . . . . *C. coelothrix*
5. Apical cells 30–60 µm in diam.; branching regular, with a 45–90° angle . . . . *C. vadorum*
6. Growth resulting only from intercalary cell divisions 7
6. Growth resulting from a combination of apical and intercalary cell divisions 10
7. Main axis distinct, branching angle 45° or more; apical cells <8 diam. long 8
7. Axes indistinct, branching angle mostly 35–45°; apical cells 8–15 diam. long. . . . . *C. liniformis*
8. Branching mostly pseudodichotomous; branchlets straight or slightly curved 9
8. Branching mostly irregular; branchlets strongly curved to sickle-shaped . . . . *C. albida*
9. Apical cells 13–40 µm in diam.; parts of axes lacking branches. . . . . *C. montagneana*
9. Apical cells 24–65 µm in diam.; branching throughout all axes. . . . . *C. ruchingeri*
10. Main axes indistinct or prostrate and entangled. 11
10. Main axes distinct, usually erect and not entangled 12
11. Main axes erect; apical cells 50–95 µm in diameter. . . . . *C. pellucidoidea*
11. Main axes prostrate and entangled; apical cells 24–54 µm in diameter . . . . *C. socialis*
12. Branchlets curved, thornlike, or sickle-shaped 13
12. Branchlets not as above . . . . *C. sericea*
13. Plants soft and flexible; apical cells less than 150 µm in diameter 14
13. Plants stiff; apical cells 160–450 µm in diameter. . . . . *C. catenata*
14. Plants >1 cm tall; branchlets curved to sickle-shaped, not thornlike. 15
14. Plants minute, to 2 mm tall; branchlets curved and thornlike. *C. jongiorum*
15. Branchlets sickle-shaped and often in fascicles; apical cells tapering . . . . *C. vagabunda*
15. Branchlets curved and clustered; apical cells cylindrical . . . . *C. dalmatica*

Cladophora albida (Nees) Kützing 1843: p. 267 (Gr. *albida*: white). Plate III, fig. 4, 5.

Basionym. *Annulina albida* Nees 1820: Index 1.
Synonyms. *Cladophora glaucescens* (Griffiths *ex* Harvey) Kützing 1849: p. 403.
Cladophora scitula (Suhr) Kützing 1849: p. 399.

Filaments are branched, uniseriate, dark to pale green, and attached by short rhizoids from basal cells. Plants are often in spongy masses (1–7 cm in diam.) in high energy sites and in strands to 10 cm long in protected coves. Branching is irregular to dichotomous below, unilateral above; branchlets are curved, bent, or sickle-shaped; lateral branches are formed distally; cell divisions are mostly intercalary and not apical. Cells are cylindrical, 12–40(80) µm in diam., and 1–7 diam. long; apical cells are 2–7 diam. long and taper to 10–40 µm in diam.

Common, forming pale green clumps in estuaries or bright green mounds on wave-exposed sites; growing on coquina on all coasts; a spring-summer epiphyte on seagrasses. Known from Newfoundland to Georgia, the Bahamas, Antilles, and Gulf of Mexico, as well as eastern North and tropical Atlantic (type location: Selsey, England; Hoek, 1963), Black Sea, Mediterranean, Bosporus, Indo-Pacific, and Australia.

Taylor 1960: p. 86, 90; Humm and Taylor 1961: p. 357; Dawes 1974: p. 88, 90; Hoek 1982: p. 100; Schneider and Searles 1991: p. 63; Littler and Littler 2000, p. 320.

Cladophora blomquistii Hoek 1982: p. 170–172, fig. 328–331 (named for H. L. Blomquist, an American phycologist who studied seaweeds of the southeastern United States). Plate III, fig. 6.

Filaments are branched, uniseriate, in loose clumps 2–5 cm in diam., and attached by rhizoidal clusters of lower cells. Filaments have a distinct acropetal organization with dividing apical cells. Branching is dichotomous, trichotomous, or irregular; the angle of branching in young branches is about 45°; branchlets are incurved. Apical cells are 150–320 μm in diam. by 4.5–14 diam. long; basal cells are 150–205 μm in diam. by 7–23 diam. long.

Rare, in unattached clumps at Alligator Harbor, Franklin County (type location) and Cedar Key, Florida.

Hoek 1982: p. 170.

Cladophora catenata (Linneaus) Kützing 1843: p. 271; emend. Hoek, 1963: p. 123 (L. _catenata_: bound, chainlike). Plate III, fig. 7.

Basionym. *Conferva catenata* Linnaeus 1753: p. 1166.
Synonyms. *Cladophora montagnei* Kützing 1849: p. 415.
Aegagrophila montagneana (Kützing) Kützing 1854: p. 14, pl. 65.

Filaments are branched, 2–15 cm tall, and form crisp, dense, bright to dark green tufts, cushions, or balls in wave-exposed areas or in loose strands in protected sites and attached by rhizoids. Filaments are stiff and often curved and intertwined; branching is dichotomous to alternate below and usually with one branch per fork. Growth is by conspicuous apical cells and some intercalary divisions near basal regions. Main axial cells are cylindrical, to 450 μm in diam., and 3–10 diam. long; apical cells are 160–450 μm in diam. by 6–26 diam. long; cells are larger with age; walls stratified and 5–13 μm thick. The Hypomycete fungus *Blodgettiomyces* sp. often grows in the cell wall in a lichen-like relationship (Hoek, 1982).

Common, forming dark cushions, mats, or aggregated balls in tide pools, to 55 m on wave-exposed shores on all coasts, and forming blooms in the Keys (Lapointe et al. 2004); also known from Bermuda, the Caribbean (type location: Havana, Cuba; Hoek, 1982), Mediterranean [as *Cladophora lehmanniana* (Lindenberg) Kützing], Indo-Pacific, and Australia.

Taylor 1960: p. 83; Humm and Taylor 1961: p. 357; Dawes 1974: p. 88; Hoek 1982: p. 59; Littler and Littler 2000: p. 320.

Cladophora coelothrix Kützing 1843: p. 272 (Gr. _coelos_: hollow + _thrix_: hair). Plate III, fig. 8.

Synonyms. *Cladophora repens* Harvey 1849 (1846–1851): pl. CCXXXVI.
Cladophora glebifera (Kützing) Kützing var. *occidentalis* P. Crouan and H. Crouan in Mazé and Schramm 1878 (1870–1877): p. 66, *nom. nudum.*

Filaments are branched, uniseriate, in mosslike dark to dull green tufts to 5 cm or loose mats, and attached by branching rhizoids of basal cells. Branching is acropetal and acute (45° or less) in high energy sites or irregular and wider angled (45–90°) in more sheltered areas. Growth is by conspicuous apical cells with little elongation after cell division. Branch cells are 55–140 μm in diam.; tip cells are 2.5–20 in diam. and 5–16 diam long; main axial cells are 60–220 μm in diam. and largest in older plants.

Common, under ledges, as a turf in mangroves, as tufts in wave-exposed sites, and adrift on all coasts; also known from Bermuda, the Bahamas, and the Caribbean, as well as eastern North and tropical Atlantic, the Mediterranean (type location: Livorno, or the Gulf of Genoa, Italy; Hoek, 1963), Black Sea, Indo-Pacific, and Australia.

Taylor 1960: p. 82; Dawes 1974: p. 90; Hoek 1982: p. 47.

Cladophora dalmatica Kützing 1843: p. 268–269 (L. _dalmatica_: an outer vestment or robe; also the Dalmatian region of Yugoslavia on the Adriatic). Plate III, fig. 9.

Synonyms. *Cladophora luteola* Harvey 1858: p. 81–82.
Cladophora caespitosa sensu P. Crouan and H. Crouan in Schramm and Mazé 1865: p. 39, non

Cladophora caespitosa Harvey 1858: p 47.
Cladophora crassicaulis P. Crouan and H. Crouan in Schramm and Mazé 1865: p. 36; P. Crouan and H. Crouan in Mazé and Schramm 1878 (1870–1877): p. 62.
Cladophora gracillima P. Crouan and H. Crouan in Schramm and Mazé 1865: p. 38; P. Crouan and H. Crouan in Mazé and Schramm 1878 (1870–1877): p. 62 (as *C. gracilima*).
Cladophora zostericola P. Crouan and H. Crouan in Schramm and Mazé 1865: p. 38; P. Crouan and H. Crouan in Mazé and Schramm 1878 (1870–1877): p. 64.

Filaments are branched, uniseriate, in grass or pale green tufts 10 to 20 cm tall, and attached by fine rhizoids. Main axes have pseudodichotomous, trichotomous, or irregular branching; upper branches are clustered and show acropetal organization. Branchlet tips are curved and growth is by apical and intercalary divisions of the 4th to 8th cells behind the branch tip. Branch cells are 14–32 μm in diam., apical cells 4.5–13 diam. long by 25–40 μm in diam., and main axial cells 8–10 diam. long or longer in older plants. Cell walls are 0.5–1.0 thick near tips and 1–4 μm in main axial cells.

Uncommon, in tide pools, as a turf at wave-exposed sites, and on seagrasses in the Keys; also known from Massachusetts to Georgia, Bermuda, the Caribbean, and Gulf of Mexico, as well as the eastern North and tropical Atlantic, Mediterranean (type location: Split, Croatia; Hoek and Womersley, 1984), Black Sea, Bahrain, and Australia.

Taylor 1960: p. 88; Hoek 1982: p. 150; Schneider and Searles 1991: p. 64.

Cladophora jongiorum Hoek 1969a: p. 131, fig. 1–6 (named for the Dutch botanists C. de Jong and son, who collected seaweeds in Curaçao). Plate III, fig. 10.

Filaments are branched, uniseriate, in minute (to 2 mm tall) dark green spongy mats attached by branching rhizoids. Branching is pseudodichotomous; main axes are densely covered by thornlike slightly curved branchlets with 1–6 short cells. Growth is by apical and intercalary cell divisions. Apical cells are acute, 2–3.5 μm in diam., and 12–29 μm long; most main axial cells are 28–52(92) μm in diam. and 1.5–3.5 diam. long; larger cells may occur on older plants.

Rare, in shallow water near St. Theresa on the northwest coast; also known from the Greater Antilles (type location: Curaçao; Hoek, 1969a) and southern Caribbean.

Cladophora liniformis Kützing 1849: p. 405 (Gr. *liniformis*: threadlike). Plate III, fig. 11.

Filaments are branched, uniseriate, in pale to dark green spongy mats (to 10 cm tall), and attached by fine rhizoids. Branching is pseudodichotomous below, with limited unilateral branching above, and one branch per node at an angle of 35–45°. Growth is by apical and intercalary cell divisions. Branchlets are straight or curved; young branches are laterally inserted, with a steeply inclined cross wall at parent cell. Cells are 15–65(80) μm in diam. and 3.5–20 diam. long; apical cells are 20–30(45) μm in diam. and 8–15 diam. long.

Uncommon, as mats in mangrove and salt marsh lagoons on the west coast; also known from Quebec to North Carolina, Bermuda, the Antilles, and Gulf of Mexico, as well as the eastern North and tropical Atlantic, Mediterranean (type location: Venice Lagoon, Italy; Hoek, 1963), Black Sea, and Pacific Ocean.

Hoek 1982: p. 74; Schneider and Searles 1991: p. 67; Littler and Littler 2000: p. 322.

Cladophora longicellulata Hoek 1969a: p. 131–132, fig. 7–11 (L. *longi*: long + *cellula*: a cell, small room). Plate III, fig. 12, 13.

Synonyms. *Cladophora catenifera* Kützing *sensu* Collins and Hervey 1917: p. 48.
Cladophora catenifera Kützing *sensu* W. R. Taylor 1960: p. 92–93; non *Cladophora catenifera* Kützing 1849: p. 390.

Filaments are branched, uniseriate, in stiff grass-green tufts 2.5–10 (20) cm tall, and attached by fine rhizoids. Main axes are obvious and have thick-walled cells; primary branching is alternate and wide angled (45–90°). Growth is by apical cells with obvious acropetal development; apical cells are 170–320 µm in diam. by 2–7 diam. long; main axial cells are 140–420 µm wide by 9–20 in diam. long; older plants have largest cells.

Rare, reported in 20 m off Crystal River on the west coast and Bermuda (type location).

Taylor 1960: p. 92; Dawes 1974: p. 89; Hoek 1982: p. 178.

Cladophora montagneana Kützing 1847a: p. 166; 1849: p. 415 (named for the French phycologist J. F. C. Montagne, a friend of Harvey and J. W. Bailey who published on Caribbean algae in the mid-1800s). Plate III, fig. 14.

Synonyms. *Conferva brachyclados* Montagne 1837 (1837–1838): p. 349; 1842a: p. 13.
Cladophora delicatula Montagne 1850: p. 302.
Cladophora polyacantha Montagne 1850: p. 302.
Cladophora stricta P. Crouan and H. Crouan in Schramm and Mazé 1866: p. 83 *nom. illeg.*; P. Crouan and H. Crouan in Mazé and Schramm 1878 (1870–1877): p. 57; non Kützing 1853: p. 28, pl. 80.

Filaments are delicately branched, uniseriate, dull to grass green, forming hairlike tufts to 30 cm tall, and attached by basal rhizoids. Growth is irregular, nonacropetal, and by intercalary cell divisions; main axes are distinct with one or more well-defined primary laterals. Branching is pseudodichotomous and axes may not branch for some length. Cells are mostly cylindrical, often club-shaped, and thick walled in smaller, stiff plants. Apical cells are 13–30(40) µm diam. and 1.5–5.0 diam. long; main axial cells are 36–50(135) µm diam. and 0.8–4 diam. long; older plants have the largest cells.

Common, in long tufts on seagrasses in protected sites on all coasts and as summer blooms in the Keys (Lapointe et al. 2004); also known from North Carolina, Georgia, Bermuda, the Caribbean (type location: Cuba; Hoek, 1982), and Gulf of Mexico, as well as West Africa, Angola, the Pacific Coast of Panama, and Australia.

Taylor 1960: p. 86, 87; Humm and Taylor 1961: p. 354; Dawes 1974: p. 88; Hoek 1982: p. 105; Schneider and Searles 1991: p. 67.

Cladophora pellucidoidea Hoek 1982: p. 179–182, fig. 358–362, 365 (Gr. *pellucid*: clear, translucent + *oidea*: similar to, like). Plate III, fig. 15.

Filaments are branched and uniseriate, form erect, pale green tufts 1.5–6 cm tall, and are attached by rhizoids from basal holdfast cells. Growth is acropetal by obvious apical cells, with some intercalary divisions near branch bases. Main axes are indistinct, upper branching abundant, pseudodichotomous or trichotomous, and at an angle of 30°; lower branching is at an angle up to 45°. Apical cells are 50–95 µm diam. and 9–23 diam. long; main axial cells are 70–260 µm in diam. and 7–48 diam. long; cell walls in young apical cells are 1.5–4.0 µm thick versus 8–16 µm in older axial cells.

Rare, collected offshore from Apalachee and Tampa Bays; also known from the Carolinas (type location in 16–20 m off Bogue Beach, North Carolina), Georgia, the Caribbean, and Curaçao (30–65 m).

Hoek: 1982: p. 179; Schneider and Searles 1991: p. 69.

Cladophora prolifera (Roth) Kützing 1843: p. 271; 1845: p. 207 (L. *prolifera*: bearing offshoots, proliferating). Plate III, fig. 16, 17.

Basionym. *Conferva prolifera* Roth 1797: p. 182–183, pl. 3, fig. 2; 1800b: p. 213.

Filaments are branched, uniseriate, coarse, dark green, and to 25 cm tall; basal cells have annular thickenings and are attached by rhizoids. Detached plants may form balls to 10 cm in diam. Branches are clustered, have 1–3 (4) forks, with branching angles of 45° or less, stiff, and straight to slightly bent; cells are cylindrical to club-shaped. Growth is acropetal with apical cell divisions; each new cell cuts off a branch apically. Apical cells are 95–240 µm in diam. by 2–5 diam. long; main axial cells are 240–345(800) µm in diam. by 7–9(16) diam. long; walls are 2–10 µm thick above and 15–40 µm in old cells.

Uncommon, on shells at Vero Beach; also known from Bermuda, where it forms blooms in eutrophic embayments (Bach and Josselyn, 1978), as well as from the Carolinas (16–45 m), the Caribbean, and Brazil, North and tropical Atlantic, Mediterranean (neotype locality: Miramare, Italy; Hoek, 1963), and the Indo-Pacific.

Taylor 1960: p. 91; Hoek 1982: p. 166; Schneider and Searles 1991: p. 70; Littler and Littler 2000: p. 322.

Cladophora ruchingeri (C. Agardh) Kützing 1845: p. 211 (named for Giuseppe Ruchinger, who wrote the "Flora dei Lidi Veneti" in 1818). Plate III, fig. 18.

Basionym. *Conferva ruchingeri* C. Agardh 1824: p. 112.

Synonyms. *Cladophora gracilis* (Griffiths *ex* Mackay) Kützing var. *tenuis* Thuret *ex* Farlow 1881 ('1882') p. 55.

Cladophora gracilis (Griffiths *ex* Mackay) Kützing f. *elongata* Collins 1902: p. 122.

Filaments are branched, uniseriate, coarse, bright to dark green hairlike strands (10–100 cm long) attached by basal rhizoids. Growth is by irregular intercalary cell divisions. Main axes are distinct with a few main laterals at pseudodichotomous forks and a short, straight branchlet at each node with an angle of <45°. Apical cells are 24–65 µm in diam. and 3–8 diam. long; main axial cells are 50–110 µm in diam. and 2–8 diam. long; older axes have larger cells. Apical walls are 2–4 µm thick, while main axial walls are 2–6 µm thick.

Rare, reported by Schneider and Searles (1991) from Florida; also known from the Canadian Maritime Provinces to Georgia, and Gulf of Mexico, as well as from Iceland, Netherlands, the Mediterranean (type location: Venice; Hoek, 1963), and Ghana.

Hoek 1982: p. 89; Schneider and Searles 1991: p. 72.

Cladophora sericea (Hudson) Kützing 1843: p. 264 (L. *sericea*: silky). Plate III, fig. 19, 20.

Basionym. *Conferva sericea* Hudson 1762: p. 485.

Synonyms. *Conferva glaucescens* Griffiths ex Harvey 1841: p. 139, *pro parte*.

Conferva nitida (Kützing) Rabenhorst 1847: p. 111.

Filaments are branched, uniseriate, in pale to grass green tufts, to 30 cm tall, and attached by short basal rhizoids. Growth is apical and intercalary; primary axes are distinct with several main laterals and pseudodichotomous forks. Branch angles are small (30–45°); branches are irregular to disorganized, resulting from a mix of cell divisions. Apical cells are 20–70 µm in diam. and 3–9 diam. long, while main axial cells are 50–170 µm in diam. and 3.5–8 diam. long. Branchlets cells are 20–85 μm in diam. and 3–10 diam. long; the largest cells occur on older plants. Apical walls are 0.5–2 μm thick, while older axial walls are 2–6 μm.

Common, on seagrasses, in estuaries on all coasts, and to 50 m; also known from Labrador to North Carolina (15–50 m) and Georgia, as well as the eastern North and tropical Atlantic (lectotype location: Isle of Sheppey, England; Hoek, 1963), Mediterranean, Black and White Seas, Indo-Pacific, and Australia.

Taylor 1960: p. 86, 88; Humm and Taylor 1961: p. 357; Dawes 1974: p. 88, 89; Schneider and Searles 1991: p. 73.

Cladophora socialis Kützing 1849: p. 416 (L. *socialis*: allied to, in colonies). Plate III, fig. 21.

Synonym. *Cladophora constricta sensu auct.*, non Collins 1909a: p. 19, pl. 78, fig. 4, 5.

Filaments are branched, uniseriate, in dark to dull green spongy mats of entangled filaments, to 2 cm tall; prostrate axes are attached by fine rhizoids. Growth is irregular, with both apical and intercalary cell divisions. Lateral branches arise from the 4th to 8th cells below the tips and have branching angles of 45–90°. Tip cells are 24–54 μm in diam. and 7–18 diam. long; main axial cells are 44–80 μm in diam. and 2–20 diam. long. Apical cell walls are <1 μm thick, while those of main axes are 2–4 μm thick.

Uncommon, in mats on wave-exposed sites or loosely interwoven patches in mangals and seagrass beds in the Keys and at the Dry Tortugas; also known from Bermuda, the Bahamas, and the Caribbean, as well as the Mediterranean, Madeira Archipelago, Liberia, and Indo-Pacific (type location: Tahiti; Silva et al. 1996).

Hoek 1982: p. 52.

Cladophora vadorum (Areschoug) Kützing 1849: p. 402 (L. *vador*: to bind over; or *vadum*: a shoal in a river). Plate III, fig. 22.

Basionym. *Conferva vadorum* Areschoug 1843: p. 269.

Filaments are branched, uniseriate, flexuous, grass or pale green, bushy, to 30 cm tall, and attached by fine rhizoids. Main axis is pseudodichotomously branched and has apical cell divisions. Morphology is irregular, branching occurs throughout axes and is wide angled (>45°). Apical cells are long, round to tapered, and 30–60 μm in diam.; branchlets are 30–90 μm in diam. Cells of young axes are 75–120 μm in diam., while older cells are 70–133 μm wide and 2.5 to 10 diam long.

Common, on pebbles or pilings, or as mats in estuaries (Indian River) and bays (Andrews Bay) on all coasts; also known from North Carolina, Georgia, Bermuda, the Bahamas, and Antilles, as well as Norway to France (type location: Göteborg, Sweden; Hoek and Womersley, 1984), the Mediterranean, Morocco, Gabon, and Australia.

Hoek 1982: p. 156; Schneider and Searles 1991: p. 74.

Cladophora vagabunda (Linnaeus) Hoek 1963: p. 144, fig. 434, 436–439, 470–503, 505–514 (L. *vagabunda*: to wander). Plate III, fig. 23–25.

Basionym. *Conferva vagabunda* Linnaeus 1753: p. 1167.

Synonyms. *Cladophora fascicularis* (Mertens *ex* C. Agardh) Kützing 1843: p. 268; 1849: p. 393.
Cladophora expansa (Mertens) Kützing 1843: p. 265; 1853: p. 27, pl. 99, fig. 1.
Cladophora brachyclona Montagne *ex* Kützing 1849: p. 394.
Cladophora mauritiana Kützing 1849: p. 399.
Cladophora sertularina (Montagne) Kützing 1849: p. 396.

Cladophora crucigera Grunow 1867: p. 38.

See Wynne (2005a) for synonymy. Based on molecular studies, Leliaert and Coppejans (2003) stated that *Cladophora vagabunda* included four or more lineages (cf. Bot et al. 1990; Bakker et al. 1995; Hoek and Chihara, 2000). Hence, it could be interpreted as a single species or complex of cryptic taxa.

Filaments are branched and uniseriate and form loose, spongy grass green to gray green tufts (0.5–4 cm tall) on wave-exposed sites, and attach by haptera-like rhizoids. Axes are distinct, to 30 cm tall, and with terminal fascicles of branches in protected sites. Axes branch 1–4 times at forks, with branch angles of erect axes 25–45°. Branchlets are recurved or sickle-shaped and at angles of 45–60°. Growth is apical with some intercalary and basal cell divisions. Apical cells are cylindrical, tapered to conical, and 24–60 µm in diam.; branch cells are 48–75(135) µm in diam., while older axial cells are 120–350 µm in diam. Young axial cells are 4–9 diam. long and 3–10 times as long as wide; older ones are 1.5–4.0 diam. long with length-to-width ratios of 2–6. Apical cell walls are 0.5–1 µm thick versus older axial cells that are 1–7 µm.

Common and highly polymorphic, occurring on exposed and protected shores of all coasts; also known from Newfoundland to Georgia (16–60 m in North Carolina), Bermuda, the Bahamas, Antilles, Gulf of Mexico, Brazil, and Uruguay, as well as the eastern North and tropical Atlantic (type location: Selsey, England; Hoek, 1963), Mediterranean, Red Sea, and Indo-Pacific.

Taylor 1960: p. 87, 91; Humm and Taylor 1961: p. 354, 355; Dawes 1974: p. 88; Hoek 1982: p. 137; Schneider and Searles 1991: p.74; Littler and Littler 2000: p. 324.

Rhizoclonium Kützing 1843: p. 261 (Gr. *rhizo*: root, rootlike + *clonium*: a branch).

Blair et al. (1982) reported that isozymes for *Rhizoclonium riparium* and *R. tortuosum* (Dillwyn) Kützing were distinct from each other and from seven species of *Chaetomorpha*.

Filaments are uniseriate, unbranched or with a few short branches, and usually either unattached or with a few short rhizoids. Growth is by intercalary cell divisions. Cells are less than 100 µm in diam. and have netlike plastids with many pyrenoids.

Key to the species (2) of *Rhizoclonium*

1. Filaments stiff; cells 40–80 (100) µm wide, 2–4 diam. long . . . . *R. africanum*
1. Filaments flexuous; cells 10–20(30) µm wide, 1–6 diam. long . . . . *R. riparium*

Rhizoclonium africanum **Kützing 1853: p. 21, pl. 67, fig. II (L. *africanun*: African, from Africa). Plate III, fig. 26, 27.**

Synonym. *Rhizoclonium hookeri sensu auct.*, non Kützing 1849: p. 383.

Womersley and Bailey (1970) considered *Rhizoclonium africanum* and *R. hookeri* to be morphologically distinct, with the latter being a sub-Antarctic species.

Filaments are uniseriate, rarely branched, stiff, entangled, and may have forked, multicellular, colorless rhizoids. Cells are irregular in shape, 40–80 (100) µm wide by 2–4 diam. long, and have walls 4–10 µm thick. Filament dimensions of *Rhizoclonium africanum* may overlap those of *R. riparium* so that identifications are difficult unless rhizoidal branches are present (Abbott and Huisman, 2004).

Common, on all coasts, in tide pools with low salinities; also epiphytic on culms of salt marsh plants and mangrove roots, and adrift in seagrass beds. The plant is also known from the Caribbean, Gulf of Mexico, and Brazil, as well as the eastern tropical Atlantic (type location: Senegambia, Senegal or Gambia; Skelton and South, 2004), Indo-Pacific, and Australia.

Taylor 1960: p. 77; Dawes, 1974: p. 90.

Rhizoclonium riparium (Roth) Kützing _ex_ Harvey 1849 (1846–1851), pl. 238 (L. _riparium_: growing on river banks, belonging to the shore). Plate III, fig. 28, 29.

Basionym. *Conferva riparia* Roth 1806: p. 216–217.

Synonyms. *Rhizoclonium implexum* (Dillwyn) Kützing 1845: p. 206.

Chaetomorpha lanosa P. Crouan and H. Crouan in Schramm and Mazé 1865: p. 35; P. Crouan and H. Crouan in Mazé and Schramm 1878 (1870–1877): p. 51.

Rhizoclonium kerneri Stockmayer 1890: p. 582.

Schneider and Searles (1991) noted similar cell sizes for *Rhizoclonium riparium*, *R. kerneri*, and *Chaetomorpha minima*. Burrows (1991) included *R. tortuosum* in *R. riparium*, while Blair (1983) and Sears (2002) considered them to be distinct taxa.

Filaments are uniseriate, green to yellow green, seldom branched, variable in diam., and often curled or entangled; rhizoids, if present, are short and inconspicuous. Apical cells may be swollen; other cells are terete, 10–20(30) µm in diam., and 1–6 diam. long.

Common, in unattached mats in salt marshes, entangled with other algae in tide pools, and on pilings on all coasts; also known from Labrador to Georgia, Bermuda, the Caribbean, Gulf of Mexico, and Brazil, as well as the eastern North and tropical Atlantic (type location: East Frisian Islands, Germany; Silva et al. 1996), Mediterranean, and Indo-Pacific.

Taylor 1960: p. 75, 76; Humm and Taylor 1961: p. 358, 359; Dawes 1974: p. 90; Schneider and Searles 1991: p. 76; Littler and Littler 2000: p. 324.

Siphonocladaceae

Leliaert et al. (2003) found that siphonocladalean algae were monophyletic and distinct from the Cladophoraceae.

Plants consist of loosely branched, irregularly shaped cells or filaments of an elongate cell that may form a stalk with fronds and rhizoids. Branches arise via segregative cell division via internal cleavage of protoplasts into various-sized cell bodies separated by a cell wall. Internal cells form an endogenous pseudoparenchymatous tissue or extend outward through the cell wall as exogenous branches. Cells are multinucleate with lobed or reticulate chloroplasts that lack pyrenoids.

Chamaedoris Montagne 1842b: p. 261 (Gr. *chamai*: on earth, a dwarf + *dory*: a stem).

Erect plants have elongate uniseriate stalks with annular thickenings, a brushlike cap, and rarely have superficial calcification. The cap is an apical tuft of dichotomously branched filaments that lack basal walls, but have cross walls above each fork.

Chamaedoris peniculum (J. Ellis and Solander) Kuntze 1898: p. 400 (L. _peniculum_: a brush). Plate IV, fig. 1, 2.

Basionym. *Corallina peniculum* J. Ellis and Solander 1768: p. 127, pl. 7, fig. 5–8; pl. 25, fig. 1.

Plants are 10–20 cm tall, bright to dark green, and stalked with a peltate cap; stalks are usually unbranched, 0.5–2.0 mm in diam., to 10 cm tall, slightly calcified at annular constrictions, and attached by branching rhizoids. Cap is filamentous, cup-shaped to flattened, spongy, oval, and 3–10 cm in diam. Filaments are coarse and 80–250 µm in diam. Branching is pseudodichotomous below and with unevenly distributed cross walls.

Uncommon, in the Keys (2 m) and Dry Tortugas (55 m); also known from the Bahamas (type location: "American seas, —," particularly near the Bahama Islands; Ellis and Solander, 1768), Antilles, Panama, Venezuela, and Brazil.

Taylor 1960: p. 115; Littler and Littler 2000: p. 330.

Cladophoropsis Børgesen 1905: p. 288, *nom. cons.* (Gr. *clados*: branch, shoot + *phora*: to bear, + *opsis*: related to; thus, like the branching green algal genus *Cladophora*).

See the reassessment of *Cladophoropsis* by Leliaert and Coppejans (2006).

Filaments are multinucleate, septate, either free or in a yellow to dark green turf, and attached by multicellular, colorless rhizoids with fingerlike pads (haptera). Septa occur above forks, with cytoplasm continuous into branches; plastids have many pyrenoids; initially the plastids are netlike and then become fragmented and discoid.

Key to the species (2) of *Cladophoropsis*

1. Main axes more than 375 μm in diameter . . . . *C. macromeres*
1. Main axes less than 270 μm in diameter . . . . *C. membranacea*

***Cladophoropsis macromeres* W. R. Taylor 1928: p. 64, pl. 4, fig. 15, 16 (Gr. *macro*-: great, large, or long + *mere*: a part or segment). Plate IV, fig. 3, 4.**

Filaments are bright green, coarse, stiff, to 15 cm tall, and form a turf attached by multicellular rhizoids with haptera-like ends. Branching is irregular below and secund or unilateral above; main filaments are 375–460 μm in diam., while lateral filaments are 210–295 μm in diam.; they arise at the upper ends of parent cells, and lack basal cross walls.

Common, in cushionlike clumps in protected sites in the Keys to 5 m (type location: Dry Tortugas); also known from Bermuda (29–31 m), the Bahamas, Caribbean, Brazil, and Canary Islands.

Taylor 1960: p. 118; Littler and Littler 2000: p. 330.

***Cladophoropsis membranacea* (C. Agardh) Børgesen 1905: p. 289, fig. 8–13; 1913: p. 47–48, fig. 26–33 (L. *membranacea*: a thin texture). Plate IV, fig. 5.**

Basionym. *Conferva membranacea* C. Agardh 1824: p. 120–121.

Van der Strate et al. (2002) found that this widespread "species" represents a complex of cryptic species.

Filaments are uniseriate, to 10 cm tall, and form light green cushions that are attached by colorless rhizoids ending in hapteral pads. Branching is alternate below and unilateral above. Branchlets are formed at the upper ends of parent cells, lack basal septa, and are 100–175 μm in diam. Main axial cells are 170–270 μm in diam. and irregular in length.

Common, forming intertidal cushions on mangrove roots and wave-exposed limestone (20–40 m) on all coasts; also known from Bermuda, the Caribbean (type location: St. Croix, Virgin Islands; Abbott and Huisman, 2004), Gulf of Mexico, and Brazil, as well as the Mediterranean, eastern tropical Atlantic, Red Sea, and Indo-Pacific.

Taylor 1960: p. 117; Dawes 1974: p. 92: Littler and Littler 2000: p. 332.

Dictyosphaeria Decaisne *ex* Endlicher 1843: p. 18 (Gr. *dictyo*: net + *sphairi*: ball).

Young plants form solid or hollow bright green vesicles, while older ones rupture and form a thick spreading membrane. The membrane consists of multinucleated cells with thick cell walls formed by internal segregative cell divisions. Cells attach to one another along contact lines by minute hapteral cells.

Key to the species (2) of *Dictyosphaeria*

1. Cells of membrane 0.1–3.0 mm in diam.; plants hollow or saclike . . . . *D. cavernosa*
1. Cells of membrane 2–4 mm in diam.; plants forming firm crusts or clusters of irregular filaments . . *D. ocellata*

Dictyosphaeria cavernosa (Forsskål) Børgesen 1932b: p. 2, pl. 1, fig. 1 (L. *cavernosa*: with hollows or cavities). Plate IV, fig. 6–8.

Basionym. *Ulva cavernosa* Forsskål 1775: p. 187.

Plants are saclike, with one or more spherical or irregular vesicles that rupture with age; vesicles are 2–12(30) cm in diam. and attached by short, unbranched rhizoids. The membrane consists of a layer of polygonal to angular cells, 0.1 –1.0(3) mm in diam., and appear honeycomb-like. The large membrane cells adhere by minute hapteral ones arranged in continuous rows along lines of contact.

Common, in shallow protected sites or wave-exposed subtidal areas (to 55 m) on all coasts; also known from Bermuda, the Caribbean, and Brazil, as well as the Cape Verde Islands, Indo-Pacific (syntype locations: Saudi Arabia and Yemen; Skelton and South, 2004), and Australia.

Dictyosphaeria ocellata (M. Howe) Olsen-Stojkovich 1985: p. 62 (L. *ocellata*: eyelike). Plate IV, fig. 9.

Olsen-Stojkovich (1985) transferred *Valonia ocellata* to *Dictyosphaeria* based upon immunological data.

Plants form firm crusts or clusters of irregular filaments, 10–20 cm in diam., to 3 cm thick, and attached by elongate, rhizoidlike cells. Main cells are 2–4 mm in diam., irregular to angular, and adhere to one another along angles of contact by minute lens-shaped hapteral cells. Hapteral cells are 40–60 µm in diam.

Common, in crevices and on intertidal limestone and mangrove roots in the Keys, and to 30 m on the west coast; also known from Bermuda, the Bahamas (type location: San Salvador Island; Howe, 1920a), Caribbean, Gulf of Mexico, Canary Islands, and West Africa.

Taylor 1960: p. 111; Littler and Littler 2000: p. 334.

Siphonocladus F. Schmitz 1879: p 18 (Gr. *siphonos*: a tube, pipe + *klados*: a branch).

Plants arise from an elongated cell and are attached by multicellular rhizoids; older protoplasts subdivide endogenously to form irregular cells that extend outwardly to form irregular to club-shaped branches.

Key to the species (2) of *Siphonocladus*

1. Plants stiff, in tufts; stalk cell short, indistinct; branching irregular . . . . *S. rigidus*
1. Plants soft, erect; stalk cell long, distinct; branches radiating, abundant . . . . *S. tropicus*

Siphonocladus rigidus M. Howe 1905a: p. 244, pl. 13, fig. 1, pl. 14, fig. 1–11 (L. *rigidus*: rigid, stiff). Plate IV, fig. 10, 11.

Plants form coarse green tufts or cushions to 5 cm tall; initial stalk cell is short, indistinct, and attached by thick rhizoids. Branching is initially subdichotomous, later either irregular or unilateral;

older branchlet tips may fuse with branches by short hapteral extensions. Main axes are 350–1150 µm in diam. and 1–3 cells wide; final axes are 350–900 µm in diam. Axes often have internal cells and stratified cell walls, 15–70 µm thick.

Common locally, inconspicuous, free or on limestone and mangrove roots in the Keys (2–3 m; type location Key West, Silva et al. 1996) and off Tampa Bay (30–50 m); also known from Bermuda, the Bahamas, Caribbean, Gulf of Mexico, and Seychelles.

Taylor 1960: p. 114; Littler and Littler 2000: p. 336.

Siphonocladus tropicus **(P. Crouan and H. Crouan in Schramm and Mazé) J. Agardh 1887: p. 105 (L. *tropicus*: of or from the tropics). Plate IV, fig. 12, 13.**

Basionym. *Apjohnia tropica* P. Crouan and H. Crouan in Schramm and Mazé 1865: p. 47.

Axes are pale green, coarse, brushlike, to 10 cm tall, and attached by thick-walled rhizoids. Primary axis or stalk is 1–3 cells wide, to 5 cm long, 0.4–1 cm in diam., and often with annular constrictions. Lateral branches are abundant and usually 1-celled; they arise from internal cells, radiate laterally, have blunt tips, and are 300–500 µm in diam.

Uncommon, on algae, limestone, adrift, and to 18 m in the Keys; also known from Bermuda, the Caribbean (syntype localities: Guadeloupe and the French West Indies; Silva et al. 1996), as well as the eastern tropical Atlantic, Indo-Pacific, and Australia.

Taylor 1960: p. 114; Littler and Littler 2000: p. 336.

Ventricaria J. L. Olsen and J. A. West 1988: p. 104 (L. from *ventricosa*: distended).

Plants are solitary, large, multinucleated cells attached by several, minute rhizoidal cells that arise endogenously at the contracted basal region.

Ventricaria ventricosa **(J. Agardh) J. L. Olsen and J. A. West 1988: p. 104, fig. 1–4, 11 (L. *ventricosa*: swollen, distended, expanded on one side). Plate IV, fig. 14.**

Basionym. *Valonia ventricosa* J. Agardh 1887: p. 96.

Cells are glossy, dark green, solitary or with a few cells that are spherical to oblong, 1.5–3.0(10) cm in diam., and attached by minute basal hapteral cells. Large cells are easily punctured because of their thin cell wall.

Common, in intertidal limestone crevices and on mangrove roots in the Keys and from 20–40 m off the east and west coasts; also known from Bermuda, the Caribbean (type location: Guadeloupe; Olsen and West, 1988), and Gulf of Mexico, as well as the Mediterranean, Indo-Pacific, and Australia.

Taylor 1960: p. 110; Dawes 1974: p. 92; Littler and Littler 2000: p. 336.

Valoniaceae

Plants form aggregations of vesicles with internal lens-shaped cells that produce external hapteral cells on their basal surface and lack a central filamentous axis. Cells are multinucleate and in some species >1.0 cm in diam.

Ernodesmis Børgesen 1912: p. 259 (Gr. *ernodea*: branched + *desme*: a bundle, tuft).

The initial cell is solitary, erect, and club-shaped; branches eventually produce terminal clusters of similar club-shaped cells.

Ernodesmis verticillata (Kützing) Børgesen 1912: p. 259, fig. 10–12; 1913: p. 66, fig. 51–54 (L. *verticillata*: whorled). Plate IV, fig. 15.

Basionym. *Valonia verticillata* Kützing 1847a: p. 165; 1849: p. 508; 1856: pl. 88.

Plants are bushy, crisp, 5–10 cm tall, and attached by small rhizoids. The initial cell is 10–20 mm long by 1.5–2.5 mm in diam. and forms a cluster of small, similar cells at its tip; branch cells are club-shaped or swollen.

Uncommon, inconspicuous, often covered by sediment, to 6 m in protected areas of the Keys and to 45 m on the west coast; also known from Bermuda, the Caribbean (type location: St. Croix, Virgin Islands; Silva et al. 1996), Gulf of Mexico, as well as the eastern tropical Atlantic and Indo-Pacific.

Taylor 1960: p. 113; Littler and Littler 2000: p. 338.

Petrosiphon M. Howe 1905a: p. 248 (Gr. *petro*: rock + *siphon*: tubular, pipe).

Plants form disclike mats of decumbent filaments and are attached by rhizoids. Disc margins have 1–2 layers of uniseriate, multinucleated filaments with cross walls; the central region is multilayered and may have erect filaments.

Petrosiphon adhaerens M. Howe 1905a: p. 248, pl. XV (L. *adhaerens*: adhering, clinging closely). Plate IV, fig. 16, 17.

Plants are crustose, circular or with lobed discs, 2–6 cm in diam., pale green, lightly calcified, to 5 mm thick in older plants, and attached by rhizoids that penetrate various substrata. Marginal filaments are 300–850 µm in diam., 0.5–20 diam. long, linear or bent, radiating, and dichotomously branched.

Rare, on the underside of rubble at wave-exposed, shallow sites at the Dry Tortugas; also known from Bermuda (type location), the Bahamas, Jamaica, and Cuba.

Taylor 1960: p. 117.

Valonia C. Agardh 1823 (1820–1828): 428–430 (L. *valonia*: dried acorn cups of the Mediterranean oak tree *Quercus aegilops* L.).

Plants are large, multinucleated, usually in clusters, and attached by small haptera or rhizoids; daughter vesicles develop externally from parent cells.

Key to the species (3) of *Valonia*

1. Plants often form large masses; cells 5–15 mm in diameter . . . . *V. macrophysa*
1. Plants not forming large masses; cells less than 5 mm in diameter 2
    2. Branches interwoven and irregular; cells 1–3 mm in diameter . . . . *V. aegagropila*
    2. Branches swollen and irregular; cells 1.0–2.5 mm in diameter . . . . *V. utricularis*

Valonia aegagropila C. Agardh 1823 (1820–1828): p. 429–430. (L. *aegagropila*: ball-like, clustered). Plate IV, fig. 18–20.

Plants are bright to dark green balls or cushions, 4–20 cm diam., and attached by minute rhizoidal cells. Branches are clustered, tightly packed, and formed at the apical ends of parent cells. Surface cells are angular and honeycomb-like. Branch cells are subcylindrical to curved, 1–3 mm in diam., and 5–10 (-20) mm long.

Uncommon, loosely attached to limestone and mangrove roots or as free spheres in the Keys (to 7 m); also known from the Caribbean, Brazil, Mediterranean (type location: Venice, Italy; Egerod, 1952), Canary Islands, Indo-Pacific, and Australia.

Taylor 1960: p. 111; Littler and Littler 2000: p. 338.

Valonia macrophysa Kützing 1843: p. 307; 1849: p. 507 (Gr. *macro*: large + *physa*: a bladder, to swell up). Plate IV, fig. 21.

Plants are spherical clusters of glossy olive green cells 2–5 cm in diam. or indeterminate in size and shape, creeping, and attached by rhizoids that extend from basal lens-shaped cells. Cells are macroscopic, ovoid to irregular, 0.5–1.5 cm in diam., 2–3.5 cm long, and usually crowded. Branches arise from bases or exposed areas of parent cells.

Common, in crevices or under ledges near low tide in the Keys, sometimes covered by crustose coralline algae or other seaweeds; also known from Bermuda, the Caribbean, and Gulf of Mexico, as well as the Mediterranean (type location: Lessina, Croatia; Silva et al. 1996), eastern tropical Atlantic, Indo-Pacific, and Australia.

Taylor 1960: p.110; Littler and Littler 2000: p. 340.

Valonia utricularis (Roth) C. Agardh 1823 (1820–1828): p. 431 (L. *utricularis*: bladderlike, inflated). Plate IV, fig. 22, 23.

Basionym. *Conferva utricularis* Roth 1797: p. 160, pl. 1, fig. 1.

Physiological studies by Eggert et al. (2003) showed ecotypic variation between populations of this plant.

Plants are bright green and irregularly shaped; axes branched, to 5 cm tall, and attached by small hapteral rhizoids. Axes are randomly arranged and often arise from swollen ends of parent cells. Filaments are short to large, cylindrical to club-shaped, and curved or arched. Cells vary in shape and are 1.0–2.5 mm in diam. by 5–20 mm long.

Common, in crevices and under ledges in protected intertidal shores (to 3 m) in the Keys; also known from North Carolina, Bermuda, the Caribbean, Gulf of Mexico, and Brazil, as well as the eastern North and tropical Atlantic, Mediterranean (type location; Silva et al. 1996), Indo-Pacific, and Australia.

Taylor 1960: p. 112; Schneider and Searles 1991: p. 77; Littler and Littler 2000: p. 340.

BRYOPSIDALES

The order corresponds to Schneider and Searles' Caulerpales (1991) and Wynne's (2005a) Bryopsidales that include heteromorphic species of the Bryopsidaceae.

Plants are true coenocytes with multinucleate cells that lack septa, except in reproductive structures. Thalli are uniaxial, multiaxial, or have interwoven to fused coenocytic siphons (siphonaceous). Cell walls contain polymers of xylan and mannan and usually lack cellulose microfibrils; some species have deposits of aragonite crystals of calcium carbonate (calcified). Plastids are numerous, discoid, and with or without pyrenoids. Many species are psammophytic with rhizoids that penetrate sediments and may hold water and modify substrata (Bedinger and Bell, 2006).

Bryopsidaceae

Plants are true coenocytes with branched filaments or simple vesicles that lack septa, except in reproductive branches. Filaments are not interwoven to form complex structures: plants are macroscopic, generally erect, and attached by rhizoids. Plastids are numerous and discoid or elliptical; cell walls of gametophytes have cellulose and xylan and sporophytes have mannan. The life histories of local species are uncertain.

Bryopsis J.V. Lamouroux 1809b: p. 133; 1809c p. 333 (Gr. *bryo*: moss, relating to mosses + *opsis*: resembling; thus similar to a moss).

Based on morphological and molecular studies of western North Atlantic populations of *Bryopsis*, Krellwitz et al. (2001) found no cryptic taxa.

Plants are erect, tufted or spreading, and usually have a visible main axis; branchlets are pinnate, radial, or unilateral. Plastids are numerous, discoid, and with one to several pyrenoids; reproductive branchlets are separated from main axes by cross walls.

Key to the species (5) of *Bryopsis*

1. Branchlets usually abundant, in 1–2 rows 2
1. Branchlets abundant or not and radially or irregularly arranged 3
  2. Erect axis featherlike; branchlets similar in length and in 1–2 rows . . . . *B. pennata*
  2. Erect axes plumelike; branchlets in 2 rows forming triangular fronds . . . . *B. plumosa*
3. Branchlets often abundant, usually regularly and radially arranged 4
3. Branchlets sparse, irregularly scattered in pairs or solitary . . . . *B. ramulosa*
  4. Branchlets abundant, hairlike, distinctly smaller than the main axis . . . . *B. halliae*
  4. Branchlets sparse to abundant, similar in size to the main axis . . . . *B. hypnoides*

***Bryopsis halliae* W. R. Taylor 1962d: p. 26, fig. 1 (named for Mrs. G. A. Hall, an American naturalist who collected the type specimen from Key West, Florida, during the late 1890s). Plate IV, fig. 24, 25.**

Synonym. *Bryopsis duchassaingii sensu auct.*, non J. Agardh 1854: p. 107; 1887: p. 31.

See Taylor (1962d) and Wynne (1998a, 2005a) for a discussion of the species designated as *Trichosolen duchassaingii* (J. Agardh) W. R. Taylor.

Plants are pale to yellow green, bushy, 10–20 cm tall, with several unbranched main axes, and attached by rhizoidal holdfasts. Branching is limited; main axes are 0.75–1.25 mm in diam., coarse, and with numerous short slender branches that are covered by delicate branchlets, 1.0–2.5 mm long, and 20–50(75) μm in diam.

var. *filicina* (Collins and Hervey) W. R. Taylor 1962d: p. 27 (L. *filicina*: fernlike, resembling a fern).

Fronds are fernlike, pinnately branched, broadly to narrowly lanceolate, and attached by rhizoidal holdfasts. Main axes are unbranched, with branchlets densely set on the main axis from base to tip, often opposite, fine, and of equal length below tip.

Rare, on mangrove roots; the variety from the Keys and the species at the Dry Tortugas; the species is also known from Bermuda, the Bahamas, Antilles, and Venezuela (type locations for species and variety: Key West, Florida and Bermuda, respectively; Taylor, 1962d).

Taylor 1960: p. 133.

Bryopsis hypnoides J. V. Lamouroux 1809c, p. 333; 1809b: p. 135, pl. 1, fig. 2a,b (L. _hypno_: the moss genus _Hypnum_ + _oides_: like; thus, like the genus _Hypnum_). Plate IV, fig. 26.

Synonym. *Bryopsis hypnoides* f. *prolongata* J. Agardh (cf. Wynne, 2005a).

The species is similar to *Bryopsis plumosa* and recognized by some (Silva et al. 1996) and not others (Schneider and Searles, 1991). Forma *prolongata*, which has loose branching and branches similar in size to the main axes, is probably a phenotypic variant (Wynne, 2005c).

Plants are dark green, erect, to 10 cm, and attached by fibrous tightly woven rhizoids. Branches are the same diam. as main axes or may become successively smaller with each division. Branching is irregular; main axes are 80–140 μm in diam.; branchlets are thinner, 40–80 μm in diam., constricted at bases, and with rounded tips.

Common, on mangrove roots, seagrasses and wave-exposed jetties on all coasts; also known from Labrador to the Caribbean, Bermuda, Gulf of Mexico, and Uruguay, as well as the eastern North and tropical Atlantic (type location: Cette, France; Silva et al. 1996), the Mediterranean, Indo-Pacific, and Australia.

Taylor 1960: p. 130; Dawes 1974: p. 72; Littler and Littler 2000: p. 342.

Bryopsis pennata J. V. Lamouroux 1809c: p. 333; 1809b: p. 134, pl. III, fig. 1 (L. _pennata_: pinnate, feathered). Plate IV, fig. 27–30.

Synonym. *Bryopsis plumosa* (Hudson) C. Agardh var. *pennata* (J. V. Lamouroux) Børgesen 1911: p. 147.

Fronds are dark green, glossy, featherlike, linear to lanceolate or oblong, 5–8 mm wide by 2–5(10) cm tall, and attached by a compact mass of rhizoids. Main axes are sparingly divided and naked below. Branchlets are in two opposing rows above, taper to tips, and have constricted bases. Main axial diameter is 240–360 μm, and branchlets are 65–150 μm in diam.

var. _leprieurii_ (Kützing) Collins and Hervey 1917: p. 62 (named for M. Leprieur, a French naturalist who collected in French Guiana).

Basionym. *Bryopsis leprieurii* Kützing 1849: p. 490; 1856: p. 27, pl. LXXV, fig. 2.

Branching is irregular; fronds have a few short branchlets in unilateral or discontinuous series; parts of main axes may be naked.

var. _secunda_ (Harvey) Collins and Hervey 1917: p. 62 (L. _secunda_: secund, with branches turned to same side).

Basionym. *Bryopsis plumosa* (Hudson) C. Agardh var. *secunda* Harvey 1858: p. 31, pl. 45A, fig. 1–3.

Synonym ?*Bryopsis harveyana* J. Agardh 1887: p. 22–23.

Fronds are 2–9 mm wide, sparingly branched, and with branchlets usually arising from one side of main axes to produce a unilateral or secund frond.

Common (species and varieties), on mangrove roots and limestone at protected sites (to 5 m) in the Keys. The species is known from North Carolina (34 m), Georgia (17–35 m), Bermuda, the Caribbean (type location: the Lesser Antilles; Abbott and Huisman, 2004), Gulf of Mexico, and Brazil, as well as from the eastern North and tropical Atlantic, Mediterranean, and Indo- Pacific. The variety *secunda* is known from Florida (syntype locations: Key West and Sand Key; Abbott and Huisman, 2004) and the Caribbean, as well as from the Mediterranean, West Africa, Indo-Pacific, and Australia.

Taylor 1960: p. 132; Humm and Taylor 1961: p. 360; Dawes 1974: p. 72; Littler and Littler 2000: p. 342.

Bryopsis plumosa (Hudson) C. Agardh 1823: (1820–1828): p. 448 (L. _plumosa_: feathery, full of feathers). Plate IV, fig. 31, 32.

Basionym. *Ulva plumosa* Hudson 1778: p. 571.

Plants are erect, in light green clumps 10(12) cm tall, and attached by a compact mass of rhizoids. Fronds are feather- or plumelike, rounded, triangular or pyramidal, and 1.0–1.5(3) mm broad at base. Main axes are to 200 µm in diam., naked below, and sparingly to abundantly branched above. Branchlets are distinct from main axes, basally constricted, with obtuse tips, unbranched or sparingly forked, 65–100 µm in diam., and grading evenly with longest branches at base and shortest at tips. Thallus regeneration occurs via extruded cytoplasm (Kim et al. 2001; Kobayashi and Kanaizuka, 1985; Pak et al. 1991) and involves a lectin "bryohealin" (Grossman, 2006; Kim et al. 2006).

Common, in protected to partially exposed subtidal shores on all coasts; also known from Newfoundland to Georgia, Bermuda, the Caribbean, Gulf of Mexico, Brazil and Uruguay, as well as the eastern North and tropical Atlantic (type location: Exmouth, England; Womersley, 1984), Mediterranean, Indo-Pacific, and Australia.

Taylor 1960: p. 131; Dawes 1974: p. 73; Schneider and Searles 1991: p. 95; Littler and Littler 2000: p. 344.

Bryopsis ramulosa Montagne 1842c: p. 16, pl. 3, fig. 2 (L. _ramulosa_: much branched). Plate IV, fig. 33, 34.

Plants form dense dark green tufts 2–10 cm tall; erect axes are coarse, to 300 µm in diam., sparingly branched, and attached by rhizoidal bases. Fronds are to 8 mm long and naked below; branchlets are short, 65–120 µm in diam., constricted at bases, scattered in unilateral groups or pairs, and mostly in the midregion of axes. Frond tips are round.

Rare, on mangrove roots or limestone in shallow waters at the Dry Tortugas; also known from the Bahamas, Caribbean (type location: Havana, Cuba; Silva et al. 1996), Brazil, South Africa, Mozambique, and India.

Taylor 1960: p. 131; Littler and Littler 2000: p. 344.

Derbesia Solier 1846: p. 452; 1847: p. 157–158 (named for Professor Alphonse A. Derbés of Marseilles, a French phycologist).

Plants are true coenocytes, with most having heteromorphic life histories. Sporophytes are sparingly branched and filamentous. Gametophytes are spherical to subspherical, and attached by rhizoids. Cells have numerous spherical to elliptical plastids with or without pyrenoids; the cell walls contain xylan.

Key to the species (3) of *Derbesia*

1. Plants are dichotomously branched filaments 2
1. Plants are shiny balloonlike green cells, to 3 cm in diameter . . . . *D. osterhoutii*
    2. Plants brushlike; main filaments 25–50 µm in diameter . . . . *D. vaucheriaeformis*
    2. Plants are fine tufts; main filaments 50–75 (100) µm in diameter . . . . *D. fastigiata*

Derbesia fastigiata W. R. Taylor 1928: p. 94, pl. 11, fig. 1–3 (L. _fastigiata_: branches erect and close together). Plate V, fig. 1.

Sporophytes are filamentous, delicate, dark green tufts 0.5–2.0 cm tall and attached by inconspicuous basal rhizoids. Branching is dichotomous 4–7 times. Filaments are stiff, straight, or swollen at tips or forks, 50–75 (100) µm in diam., and rarely constricted. Branches decrease in diam. with each forking. Tips are 7–10 µm in diam. No spherical gametophytic stage (i.e., *Halicystis*) is reported.

Uncommon, on seagrasses, seaweeds, jetties, and limestone near low tide at the Dry Tortugas (type location); also known from the Caribbean and Pacific Ocean.

Taylor 1960: p 129 (cited as an uncertain record); Littler and Littler 2000: p. 346.

Derbesia osterhoutii (L. R. Blinks and A. C. H. Blinks) J. Z. Page 1970: p. 375, fig. 1–6 (named for Dr. W. J. V. Osterhout, an American algal physiologist). Plate V, fig. 2, 3.

Basionym. *Halicystis osterhoutii* L. R. Blinks and A. H. Blinks 1931: p. 389, pl. 22, fig. 1–8; pl. 23, fig. 9–12, text fig. 18 (the gametophytic stage of *D. osterhoutii*).

Gametophytes are solitary or in clusters, shiny green cells to 3 cm in diam., and attached by penetrating rhizoids with blunt tips. Cells are ovoid to spherical with lens-shaped plastids having many pyrenoids; walls are thin but tough. No filamentous sporophytes have been recorded from Florida.

Uncommon, inconspicuous, on crustose coralline algae in shaded crevices (to 18 m) in the Keys; also known from Bermuda (type location; Blinks and Blinks, 1931), the Caribbean, and Gulf of Mexico.

Taylor 1960: p. 109; Littler and Littler 2000: p. 348.

Derbesia vaucheriaeformis (Harvey) J. Agardh 1887: p. 34 (L. _Vaucheria_: an algal genus in the Xanthophyceae named for Jean Pierre Etienne Vaucher, a Swiss clergyman and botanist + _formis_: shaped like; thus, similar to _Vaucheria_). Plate V, fig. 4, 5.

Basionym. *Chlorodesmis vaucheriaeformis* Harvey 1858: p. 30, pl. XL.D.

Sporophytes are filamentous, in dense brushlike tufts, 0.5–5.0 cm tall, and attached by rhizoids. Filaments are 25–50 µm in diam., dichotomously branched, and have a few cross walls above the forks. Lateral sporangia are ovoid to pyriform, 100–130 µm in diam., 190–300 µm long, on stalks 50–100 µm long and 15 µm in diam., with two cross walls near their bases, and sporangia that may replace branches. No spherical gametophytic (i.e., *Halicystis*) stages have been reported for Florida.

Common, on jetties and limestone near low tide in the Keys (type location: Key West; Harvey, 1858) and on *Codium* in the Dry Tortugas; also known from Bermuda, North Carolina, the Bahamas, and Venezuela, plus widespread in other tropical seas.

Taylor 1960: p. 128.

Trichosolen Montagne 1861 ('1860'): p. 171 (Gr. *tricho*: hairy + *solen*: a tube, pipe).

The genus is variously interpreted; some authors list *Pseudobryopsis* Berthold *ex* Oltmanns (1904) as a synonym (Taylor, 1962c; Diaz-Piferrer, 1965), others cite the former as a synonym of the latter (Chihara et al. 1980; Christensen, 1987b; Silva et al. 1996), and others designate both taxa as distinct (Henne and Schnetter, 1999; Guiry and Nic Dhonncha, 2006). We follow Henne and Schnetter (1999), who state that chloroplasts in *Trichosolen* are relatively large, have an obvious pyrenoid, and lack plugs that separate gametangia from supporting branches. In contrast, chloroplasts in *Pseudobryopsis* are smaller; they lack pyrenoids, and their gametangia are separated from supporting branches by plugs (Wynne, 2005a).

Plants are true coenocytes, with erect simple to infrequently branched axes, and attached by numerous short rhizoids. Plastids are ovoid to spindle-shaped and have a single pyrenoid. Gametangia are ovoid to elongate, stalked or sessile, laterally and adaxially near branch bases, and lack a plug. *Trichosolen* and *Bryopsis* are similar morphologically (Bucher et al. 1990); the former produces "specialized" gametangia as distinct structures, while entire vegetative branches function as gametangia in the latter (Taylor, 1962c).

Key to the species (2) of *Trichosolen*

1. Main axes 0.7–1.2 mm in diam., dichotomously branched . . . . *T. duchassaingii*
1. Main axes to 370 μm in diam., little divided . . . . *T. molassensis*

Trichosolen duchassaingii (J. Agardh) W. R. Taylor 1962c: p. 62 (named for Dr. Placide D. de Fontressin Duchassaing, who collected invertebrates and plants and studied the geology of Guadeloupe, Panama, and St. Thomas). Plate V, fig. 6.

Basionym. *Bryopsis duchassaingii* J. Agardh 1854: p. 107.
Synonyms. *Bryopsis antillarum* J. Agardh 1854: p. 107; 1887: p. 31.
Trichosolen antillarum Montagne 1861 ('1860'): p. 171, pl. 11C.

Taylor (1962c) placed *Pseudobryopsis longipedicellata* [= *Trichosolen longipedicellata* (Blomquist and Díaz-Piferrer) D. M. John] as a synonym of *T. duchassaingii*, while Stegenga and Vroman (1988) and Wynne (2005a) disagreed and retained them as separate taxa.

Plants are bright green and 10–20 cm tall; main axes are 0.75–1.25 mm in diam. and attached by rhizoids. Branching is dichotomous; branchlets are short, delicate, abundant, and cover the main axes. Gametangia are lateral, 105–150 μm in diam., 160–345 μm long, with or without stalklike bases, and replace simple branchlets.

Rare, collected at Sand Key (J. Hopper, 1876); also known from Bermuda, Trinidad, and Guadeloupe (type location: J. Agardh, 1854).

Trichosolen molassensis Bucher and J. N. Norris in Bucher et al. 1990: p. 297–300, fig. 5–9 (named for Molasses Reef in the Florida Keys, the type locality). Plate V, fig. 7.

Plants are erect, small (to 17 mm tall), and attached by a cluster of yellow green rhizoids. Axes are sparingly branched, to 370 μm in diam., and taper to 30 μm in diam. near obtuse tips. Branchlets densely cover upper axes, are mostly simple, have constricted bases, grow to 2 mm long, 40 μm in diam., and taper to 12 μm in diam. at tips. Plastids are ovoid, 5–10 by 10–20 μm, and have a single pyrenoid. Gametangia are ovoid to club-shaped, 63 by 185(260) μm, and have 'release' papillae with rounded apices.

Rare, in 5–6 m on Molasses Reef, in the Florida Keys (type location).

Caulerpaceae

A monotypic family composed of true coenocytes, with multinucleated cytoplasm and lacking cross walls. Species have conspicuous or inconspicuous rhizomes (stolons) with descending branches that divide progressively into slender rhizoids that penetrate soft substrata and erect photosynthetic fronds (assimilators) that may be subdivided into branchlets. The cell wall is composed of xylan and has internal projections (trabeculae) characteristic of the genus. The rhizome and fronds contain chloroplasts and amyloplasts; the plastids contain pyrenoids in some species.

Caulerpa J.V. Lamouroux 1809c: p. 332; 1809b: p. 332 (L. *caulis*: a stem or stalk of a plant + *erpo*: to creep).

See Silva (2003) for the nomenclatural history of the genus *Caulerpa*. Svedelius (1906), Børgesen (1907), and Calvert (1976) discussed the evolution of radial versus bilateral symmetry in the genus, while Fama et al. (2002) and Stam et al. (2006) summarized molecular features of *Caulerpa*.

The genus occurs primarily in tropical to subtropical seas (Zaleski and Murray, 2006), has the same features as the family, and contains approximately 80 recognized species worldwide that are separated mostly by frond morphology (Hodgson et al. 2004). The large number of varieties and forms of some species (for example, *Caulerpa cupressoides*) reflects the morphological plasticity and ability to change with culture conditions (Barilotti, 1970; Peterson, 1972; Calvert, 1976), especially under varying light (Calvert, 1976; Senerpont Domis et al. 2003). Some species (*C. racemosa*, *C. taxifolia*) are invasive taxa that have caused serious biological problems (Ceccherelli and Piazzi, 2001; Piazzi et al. 2001, 2005; Zaleski and Muray, 2006; Stam et al., 2006), including reduced abundance of seagrass populations (Ceccherelli and Cinelli, 1997, 1999) and altered fish assemblages in these important nursery habitats (York et al. 2006). Stam et al. (2006) recommend that sales of *Caulerpa* spp. be prohibited to avoid spread of invasive taxa.

Key to the species (17) of *Caulerpa* (modified from Taylor, 1960)

1. Plants have distinctly different rhizomes and erect fronds 2
1. Plants appear filamentous with rhizomes and erect fronds similar in form . . . . *C. fastigiata*
   2. Erect fronds flat and bladelike, not divided into various types of branchlets 3
   2. Erect fronds flat or otherwise and divided into various types of branchlets 5
3. Blade margins often toothed and with an apical notch . . . . *C. brachypus* f. *parvifolia*
3. Blade margins entire and without an apical notch 4
   4. Blades greater than 4.0 mm wide and more than 4.0 cm long, tips rounded . . . . *C. prolifera*
   4. Blades 0.8–3.0 mm wide, 4.0–4.6 cm long, tips rounded to triangular . . . . *C. ollivieri*
5. Branchlets or lobes of erect fronds tipped with a minute spine (i.e., apiculate) 6
5. Branchlets or lobes of erect fronds lacking spines, smooth, rounded, or flat 14
   6. Branchlets regularly divided 7
   6. Branchlets simple and not regularly divided, with some having lobes or teeth 9
7. Erect axes bushy and large; basal stalk 1–3 mm in diam.; branchlets pinnately divided in upper portions only . . . . *C. paspaloides*
7. Erect axes delicate, stalks not showing conspicuous differentiation 8
   8. Fronds to 15 mm tall; branchlets to 250 μm long and not whorled . . . . *C. webbiana*
   8. Fronds to 3 cm tall; branchlets 0.5–2.0 mm long, in delicate whorls . . . . *C. verticillata*
9. Branchlets in 2 or more rows or randomly and radially arranged 10
9. Branchlets or teeth in 1–2 rows 11
   10. Rhizomes smooth, naked, lacking filamentous branchlets . . . . *C. cupressoides* (in part)
   10. Rhizomes covered with filamentous branchlets . . . . *C. lanuginosa*
11. Branchlets cylindrical or compressed, with blades not flat 12
11. Branchlets in two opposite rows, consisting of flat bladelets . . . . *C. mexicana*
   12. Branchlets mostly long and constricted at bases of erect axes 13
   12. Branchlets or teeth short and not constricted at bases . . . . *C. cupressoides* (in part).
13. Branchlets 0.3–0.5 mm in diam., almost terete, and bluntly pointed . . . . *C. sertularioides*
13. Branchlets 1.5–3.0 mm in diam., compressed; expanding from a narrow base to tips that are rounded and with a spine . . . . *C. floridana*

14. Branchlets mostly cylindrical, usually in 2 rows, like the pinnae of feathers 15
14. Branchlets spherical, club-shaped, or enlarged at tip; if cylindrical not in 2 rows 16
15. Plants minute, 1–2 cm tall; blades flat, with cylindrical branchlets in two rows; erect axes repeatedly divided and branchlets forked . . . . *C. vickersiae*
15. Plants large, more than 5 cm tall; branchlets swollen at tips, not forked . . . . *C. ashmeadii*
16. Branchlets subsessile, on very short stalks, spherical to berrylike 17
16. Branchlets on obvious stalks, tips vary from strongly swollen to cylindrical, club-shaped, or flat on top . . . . *C. racemosa*
17. Branchlets on short stalks; spherical, constricted at base of sphere; pigmentation uniform in branchlets . . . . *C. microphysa*
17. Branchlets not constricted at stalk-sphere interface; spheres often with a molted or patchy pigmentation . . . . *C. macrophysa*

Caulerpa ashmeadii Harvey 1858: p. 18, pl. XXXVIII, fig. A (named for Samuel Ashmead, an American phycologist). Plate V, fig. 22, 23.

Rhizomes are wide spreading, to 2 m long by 2.0–2.5 mm in diam., and attached by stout descending rhizoidal clusters. Fronds are erect, 10–30 cm tall, mostly simple below, with stalks 1.5–2.0 mm in diam. by 5–30 mm long; above producing oppositely pinnate branchlets, 5–15(30) cm tall and 6–25 mm wide. Branchlets are cylindrical to slightly club-shaped, 7–18 mm long by 0.50–1.25 µm in diam, curved upward, and close together or separated; tips truncated and rounded. Under low light, fronds are often elongate and lack pinnate branching (for example, in seagrass beds).

Common, in seagrass beds and to 40 m on all coasts (type location: Key West; Silva et al. 1996); also known from Bermuda, the Bahamas, Antilles, Gulf of Mexico, Brazil, and India.

Taylor 1960: p. 142; Humm and Taylor 1961: p. 360; Dawes 1974: p. 73; Littler and Littler 2000: p. 358.

Caulerpa brachypus Harvey 1860a: p. 333 f. *parvifolia* (Harvey) A. B. Cribb 1958: p. 209 (L. *parvi*: small + *folia*: leaf). Plate V, fig. 24, 25.

Basionym. *Caulerpa parvifolia* Harvey 1860b: pl. CLXXII.

See Cribb (1958b), South and N'Yeurt (1993), and Silva et al. (1996) for synonymy.

Blades are grass green, undivided, flat, thin, elliptic or oval to lanceolate, 0.5–4.0 mm wide, and 0.5–4 (8) mm long; margins are smooth in older blades and have teethlike serrations in younger ones; apices are indented or notched. Blades are sessile or on stalks, 300–400 µm in diam. and 1–3 mm long. Rhizomes (stolons) have branched rhizoids to 100 µm in diam. Blades of the form are smaller and more linear and rhizomes thinner than in the species (Cribb, 1958b; Taylor, 1967; Coppejans and Beeckman, 1990).

Rare, apparently introduced, and overgrowing reefs in 20–60 m off southwest Florida (Jacoby et al. 2004; Lapointe et al. 2005a); also known from the eastern tropical Atlantic, Indo-Pacific and Australia (type location: New South Wales; Silva et al. 1996).

Littler and Littler 2000: p. 358.

Caulerpa cupressoides (H. West in Vahl) C. Agardh 1817: p. 23 (L. *cupressus*: resembling the conifer genus *Cupressus* or cypress + *oides*: like; thus, resembling cypress leaves). Plate V, fig. 26–30.

Basionym. *Fucus cupressoides* H. West in Vahl 1802: p. 38.

Synonyms. *Chauvinia indica* Sonder *ex* Kützing 1857: p. 5, pl. 10, fig. 1.

Caulerpa triangularis Mazé and Schramm 1878 (1870–1877): p. 83, *pro parte*.

Caulerpa juniperoides J. Agardh 1883: p. 15.

Nielsen and Price (2001) reported that H. West collected material at St. Croix and sent it to Vahl with a handwritten manuscript dated 1795. Because no holotype of *Fucus cupressoides* could be found in Vahl's collections, they designated specimen C31688 at the Copenhagen Herbarium as the type (A-826). Morphological and molecular studies (Senerpont Domis et al. 2003) showed that the species is distinct from *Caulerpa racemosa* and *C. sertularioides*.

Rhizomes are branched, form extensive colonies, and are 20–30 cm in length by 1.5–2.5 mm in diam.; erect fronds form every 1–3 cm on rhizomes and are attached by stout rhizoidal clusters at 2–5 cm intervals. Fronds are 2.5–7.0 cm tall, usually branched, naked the first 5–15 mm, and have branchlets above. Branchlets are simple, 1–2 mm long, to 0.4 mm in diam., closely clustered, knobby, and conical; they curve upward, taper to terminal spines, and lie in 3–(4) parallel vertical rows along axes.

var. *ericifolia* (Turner) Weber-van Bosse 1898: p. 335 (L. *Erica*: the generic name for heather or heath, a fine-leaved shrub + *folia*: leaved; thus with leaves like *Erica*).

Basionym. *Fucus ericifolius* Turner 1807–1808: p. 124–125, pl. 56.
Synonym. *Caulerpa ericifolia* (Turner) C. Agardh 1817: XXIII.

Fronds are erect, to 10 cm tall, and surrounded by rows of compact cylindrical branchlets pressed against the axis; branchlets equal in length to the axial diam.

var. *flabellata* Børgesen 1907: p. 368, fig. 18, 19 (L. *flabellata:* fan-shaped, fanlike).

Fronds are erect, stiff, and to 20 cm tall. Axes naked the first 3–8 cm above rhizome; also flat, 1–2 mm wide, and branched 3–5 times. Branchlets are marginal, 0.8–1.0 mm in diam., 0.5–1.5 mm long; tips are pointed, curved upward, and usually as long as broad or reduced to serrate points.

var. *lycopodium* Weber-van Bosse 1898: p. 335, pl. 27, fig. 8–13; pl. 28, fig. 10–12, 14 (Gr. *lyco*: wolf + *podium*: foot; also the generic name for the club moss, *Lycopodium*).

Synonym. *Caulerpa lycopodium* J. Agardh 1847: p. 6, *nom. illeg.*

See Weber-van Bosse (1898) and South and N'Yeurt (1993) for synonymy. By contrast, the name *Caulerpa lycopodium* C. Agardh 1817: p. XXIII is regarded as a synonym of *Digenea simplex* (Wulfen) C. Agardh by Guiry and Nic Dhonncha (2006).

Fronds are to 20 cm tall, sparingly branched, and densely covered with rows of branchlets, which are curved upward, 0.8–1.2 mm in diam., in several ranks, cylindrical, 2–6 times as long as axial diam., and with rounded or pointed tips.

f. *alternifolia* (P. Crouan and H. Crouan) Weber-van Bosse 1898: p. 336 (L. *alterni*: alternate + *folia*: leaved).

Basionym. *Caulerpa alternifolia* P. Crouan and H. Crouan in Schramm and Mazé 1865: p. 41.

Fronds are dichotomously forked and not stalked; branchlets cover most of the axis and are in 3 (distichous) or more ranks (multiseriate).

f. *disticha* Weber-van Bosse 1898: p. 327, 338, pl. XXVII, fig. 14 (L. *disticha:* two-ranked, in two opposed ranks).

Synonym. *Caulerpa cupressoides* var. *disticha* Weber-van Bosse 1898: p. 327, 338, pl. xxvii, fig. 14.

Fronds are stalked and much branched above; branchlets are terete, 2 diam. long, opposite, and usually in two vertical rows or sometimes in multiple rows at top.

f. *elegans* (P. Crouan and H. Crouan) Weber-van Bosse 1898: p. 336, pl. 27, fig. 8, 9 (L. *elegans*: graceful, elegant, slender).

Basionym. *Caulerpa plumaris* (Forsskål) C. Agardh var. *elegans* P. Crouan and H. Crouan in Schramm and Mazé 1865: p. 39; also Mazé and Schramm 1878 (1870–1877): p. 79.

See South and N'Yeurt (1993) and Silva et al. (1996) for synonymy.

Fronds are equally forked; branchlets terete, opposite, and as long as their diam.

f. *intermedia* Weber-van Bosse 1898: p. 337 pl. 27, fig. 10 (L. *intermedia*: between or in between).

Fronds are to 30 cm tall and their stalks naked at bases; branchlets are in two or more ranks, except for the lowest cylindrical ones.

var. *mamillosa* (Montagne) Weber-van Bosse 1898: p. 332–333, pl. XXVIII, fig. 2–7, as f. *typica* (L. *mamillosa*: with small nipples).

Basionym. *Caulerpa mamillosa* Montagne 1842a: p. 13, pl. 6, fig. 3.

Synonym. *Chauvinia mamillosa* (Montagne) Kützing 1849: p. 498.

Fronds are bushy, erect, to 7 cm tall, forked 2–3 times above stalks, and densely covered with 3–5 or more spiral ranks of branchlets. Branchlets are knobby, oblong, or spinelike, to 1.0 mm in diam., and narrower than the main axis.

f. *nuda* Weber-van Bosse 1898: p. 332 (L. *nuda*: naked).

Fronds are naked below and branchlets few or absent on lower portions.

var. *serrata* (Kützing) Weber-van Bosse 1898: p. 331, pl. 27, fig. 5, 6 (L. *serrata*: sawlike, saw-toothed, or beset with teeth).

Fronds are slender, irregularly branched, and naked below; upper branchlets small, fan-shaped, in 2–3 ranks, and shorter than the diam. of the main axis.

var *turneri* Weber-van Bosse 1898: p. 330, pl. XXVII, fig. 4 (named for the British phycologist D. Turner).

Fronds are slender, conical to fanlike, and flat; branchlets are in 3–4 ranks and shorter than their diam.

Common, the species and varieties grow on soft sediments, especially in seagrass beds and on hard substrata of all coasts (to 10 m); also known from Bermuda (type location of var. *ericifolia*), Virgin Islands (type location of species: St. Croix; Abbott and Huisman, 2004; syntype localities of var. *flabellata*: St John and St. Thomas; Silva et al. 1996), Guadeloupe (type location of f. *elegans*: syntype localities of f. *alternifolia*, f. *disticha*, and f. *nuda*: Silva et al. 1996), Gulf of Mexico, and Brazil, as well as the eastern tropical Atlantic, Indo-Pacific, and Australia.

Taylor 1960: p. 146–148; Humm and Taylor 1961: p. 362; Dawes 1974: p. 74; Littler and Littler 2000: p. 358–361.

Caulerpa fastigiata Montagne 1837(1837–1838): p. 353–354 (L. *fastigiata*: branches erect and close together). Plate VI, fig. 1, 2.

The var. *confervoides* P. Crouan and H. Crouan *ex* Weber-van Bosse is not recognized due to phenotypic variability (Taylor, 1960).

Plants are filamentous, form dark green mats or bushes 1.5–3 cm tall, and show little distinction between blades, rhizomes, and rhizoids. Rhizomes are 150–210 µm in diam.; rhizoids are abundant, hyaline, hairlike, and 20–60 µm in diam. Erect filaments have dichotomous, irregular, or opposite branching, 40–120 µm in diam.

Common, inconspicuous, in tufts or clumps on mangrove roots and on sandy to muddy substrata (to 1 m) on all coasts; also known from Bermuda, the Caribbean (type location: Cuba; Silva et al. 1996), Gulf of Mexico, Brazil, Sierra Leone, the Red Sea, and Pacific Ocean.

Taylor 1960: p. 136; Dawes 1974: p. 74; Littler and Littler 2000: p. 362.

Caulerpa floridana W. R. Taylor 1960: p. 143–144, 630, pl. 11, fig. 5, 6; pl. 18, fig. 10 (L. *floridana*: of or from the state of Florida). Plate VI, fig. 3, 4.

Fronds are erect, compressed, and with opposite, pinnate branchlets; rhizomes are to 70 cm long, 2–2.5 mm in diam., and attached by rhizoidal clusters every 2–4 cm. Erect fronds arise on the rhizome every 3–5 cm; stalks are naked for 1–3 cm above the rhizome and then produce branchlets for 3–13 cm. Branchlets are 1.7–2.7 cm wide, contracted and terete toward the bases; upper branchlets are compressed, 1.5–2 mm wide, with tips rounded and ending in a spine. The species resembles *C. ashmeadii*, but is distinguished by the slightly compressed, gradually tapering branchlets that are contracted at the base and have a pointed tip.

Rare, one record from 18 to 109 m off the Dry Tortugas (type locality).

Taylor 1960: p. 143.

Caulerpa lanuginosa J. Agardh 1873: p. 28 (L. *lanuginosa:* woolly). Plate VI, fig. 5.

Synonym. *Caulerpa lycopodium* Harvey 1858: p. 19, pl. XXXVII.B, *nom. illeg.*

Rhizome and erect axes are covered by densely placed branchlets; rhizomes are 0.4–1 mm in diam., 10–30 cm long, and attached by short descending branches that end in clusters of rhizoids. Fronds are 3–13 cm tall, simple or branched 1–4 times, and densely covered with slightly upcurved branchlets that are filiform, cylindrical, 3–5 mm long by 60–180 µm in diam., and have pointed tips.

Common, firmly attached in sediments of seagrass beds, on hard substrata on all coasts (type location: Key West; Silva et al. 1996), and to 110 m off the Dry Tortugas; also known from the Caribbean, Gulf of Mexico, and Indian Ocean.

Taylor 1960: p. 145; Dawes 1974: p. 74; Littler and Littler 2000: p. 362.

Caulerpa macrophysa (Sonder *ex* Kützing) G. Murray 1887: p. 38 (Gr. *makros*: large + *physa*: inflated, bladderlike). Plate VI, fig. 6.

Basionym. *Chauvinia macrophysa* Sonder *ex* Kützing 1857: p. 6, pl. 15, fig. II.

Synonyms. *Caulerpa racemosa* var. *macrophysa* (Sonder *ex* Kützing) W. R. Taylor 1928: p. 101, pl. 12, fig. 3; pl. 13, fig. 9.

Taylor (1928; 1960) placed this plant as a variety of *Caulerpa racemosa*. Calvert (1976) showed that *C. racemosa* var. *macrophysa* resembled *C. pasapaloides* when cultured under dim light. Littler and Littler (2000) separated *C. macrophysa* by branchlet morphology and spotted pigmentation that is visible even when mixed with *C. racemosa*.

Fronds are to 6 cm tall; branchlets are beadlike and clustered; rhizomes are spreading, 3–4 mm in diam., and attached by descending branches that progressively branch to form yellow rhizoids. Branchlets often have pigment spots, are spherical to hemispherical, and 3–7 mm in diam. Stalks, if present, are 1–2 mm in diam. and to 2 mm long.

Uncommon, in compact mats on limestone in areas with moderate wave activity and to 20 m in the Keys; also known from the Caribbean, Gulf of Mexico, and Brazil, as well as the Indo-Pacific.

Type location cited as Central America (Hodgson et al. 2004) and Indonesia (Abbott and Huisman, 2004).

Taylor 1960: p. 153; Littler and Littler 2000: p. 362.

Caulerpa mexicana Sonder *ex* Kützing 1849: p. 496 (L. *mexicana*: of or from Mexico or the Gulf of Mexico). Plate VI, fig. 7, 8.

Synonyms. *Caulerpa harveyana* Kützing 1857: p. 3, pl. 5, fig. iii.
Caulerpa crassifolia (C. Agardh) J. Agardh 1873: p. 13–14.
Caulerpa crassifolia f. *mexicana* (Sonder *ex* Kützing) J. Agardh 1873: p. 14.
Caulerpa pinnata C. Agardh f. *mexicana* (Sonder *ex* Kützing) Weber-van Bosse 1898: p. 290–291, pl. XXIV, fig. 2.

Coppejans (1992), Coppejans and Prud'homme van Reine (1992), and Chisholm et al. (1995) consider *Caulerpa mexicana* to be an ecad of *C. taxifolia* (Vahl) C. Agardh, while others regard them as distinct species (Littler and Littler, 2000; Hodgson et al. 2004). Hodgson et al. (2004) note that *C. mexicana* is flatter than *C. taxifolia* and has a slight but distinctive narrowing at the base of each branchlet. Caribbean plants also have a more flattened erect axis, especially at the base, while *C. taxifolia* is clearly terete.

Fronds are erect, flat, and dissected; rhizomes are 0.5–1.25 mm in diam., 10–30 cm long, and attached by slender and descending rhizoid-bearing branches every 1–3 mm. Fronds are stalked, 15–25 cm tall, 5–25 mm wide, and arise from rhizomes every 1–4 cm. Axes are simple or forked, flat, and 1–3 mm wide; branchlets are pinnately arranged, 2–10 mm long, opposite, flat to oval or oblong, contracted at bases, and with pointed tips.

f. *laxior* (Weber-van Bosse) W. R. Taylor 1960: p. 141, pl. 12, fig. 2, 3 (L. *laxus*: loose; *laxi*: spreading).

Basionym. *Caulerpa pinnata* f. *laxior* Weber-van Bosse 1898: p. 291.

Plants are small, with fronds narrow, linear, or lanceolate.

f. *pectinata* (Kützing) W. R. Taylor 1960: p. 141 (L. *pectinata*: comblike).

Basionym. *Caulerpa pectinata* Kützing 1849: p. 495.
Synonyms. *Caulerpa pinnata* Kützing f. *pectinata* (Kützing) Weber-van Bosse 1898: p. 291.
Caulerpa crassifolia (C. Agardh) J. Agardh f. *pectinata* (Kützing) Collins 1909b: p. 414.

Fronds are narrow and branchlets distinctly separated and constricted at bases.

Common and highly polymorphic, growing on mangrove roots and limestone in tide pools, in seagrass beds, and to 15 m on all coasts. Also known from Georgia (17–21 m), Bermuda (type location of f. *laxior*; Silva et al. 1996), the Caribbean, Gulf of Mexico (type location of species, Mexico; Silva et al. 1996), Venezuela (type location of f. *pectinata*, La Guaira; Silva et al. 1996), and Brazil, as well as the Mediterranean (introduced into Israel), Canary Islands, Mauritania, Red Sea, Indo-Pacific, and Australia.

Taylor 1960: p. 141; Humm and Taylor 1961: p. 362; Dawes 1974: p. 74; Schneider and Searles 1991: p. 92; Littler and Littler 2000: p. 364.

Caulerpa microphysa (Weber-van Bosse) Feldmann 1955: p. 430 (Gr. *micro*: little + *physa*: inflated, bladderlike). Plate VI, fig. 9, 10.

Basionym. *Caulerpa racemosa* (Forsskål) J. Agardh (var. *clavifera*) f. *microphysa* Weber-van Bosse 1898: p. 361, 362, pl. 33, fig. 5.
Synonyms. *Caulerpa racemosa* var. *microphysa* (Weber-van Bosse) Reinke 1900: p. 38.
Caulerpa lentillifera J. Agardh f. *parvula* Børgesen 1949a: p. 15–18, fig. 5–7.

Plants are delicate and to 30 mm tall; axes are beadlike, from slender rhizomes, and attached by many highly branched, tapering rhizoids. Branchlets are berrylike, 0.9 to 2.4 mm in diam., clustered or solitary, constricted at bases, and nearly sessile or on short stalks (to 0.5 mm long). The species can be distinguished from *Caulerpa racemosa* by the presence of pyrenoids in the plastids (Abbott and Huisman, 2004).

Uncommon, on limestone and mangrove roots within shaded sites and from 9 m in the Keys; also known from Bermuda, the Caribbean (to 110 m), and Gulf of Mexico, as well as the eastern tropical Atlantic and Indo-Pacific (type location: Makassar, Indonesia; Abbott and Huisman, 2004).

Taylor 1960: p. 155; Littler and Littler 2000: p. 364.

Caulerpa ollivieri Dostál 1929: p. 2, fig. 3 (named for G. Ollivier, who wrote the flora of the Alpes Maritimes and gave Dostál the plants that he cultured). Plate VI, fig. 11, 12.

Blades are pale to bright green, narrow, elongate, slightly twisted, and often 20 times longer than wide. Rhizomes are fine, 0.2–0.4 mm in diam., little branched, and attached by clusters of fine rhizoids. Blades are 0.8–3.0 mm wide, 4.0–4.6 mm long, with rounded to triangular tips, and simple to rarely proliferous. Blades are morphologically similar to *Caulerpa prolifera* f. *zosterifolia* Børgesen.

Rare, in 35 m off Tampa Bay and 60 m off Dry Tortugas (Hines and Humm, 1971) and in silt in the Florida Keys; also known from the Mediterranean (type location: Villefrance), and Turkey. Lapointe et al. (2005b) reported the species as introduced to eutrophic areas in the Bahamas.

Caulerpa paspaloides (Bory) Greville 1830: p. 64 (Gr. *paspalum*: a type of grain or millet + *oides*: resembling; thus similar to millet). Plate VI, fig. 13–17.

Basionym. *Chauvinia paspaloides* Bory 1829 (1826–1829): p. 206, pl. 23, fig. 1.

Calvert (1976) reported a shift from radial to bilateral symmetry in fronds of *Caulerpa paspaloides* cultured under low light as also seen with *C. racemosa* var. *macrophysa* and *C. sertularioides* f. *farlowii* (see below).

Fronds are erect, to 25 cm tall, bushy, and palmlike; rhizomes robust, to 4 mm in diam., with stout descending axes that subdivide into numerous, yellow rhizoids. Stalks of fronds are obvious and naked for 1–5 cm; upper axes are dichotomously, irregularly, or palmately branched 1–4 times. Fronds are divided 2–3 times, with subdivisions 5–7 cm long, terete, and densely covered with branchlets (pinnae). Branchlets are cylindrical, 2–3 mm long, in 3–4 ranks, simple to forked, with pointed tips, and often have keel-like flat sides.

f. *flabellata* Weber-van Bosse 1898: p. 353 (L. *flabellata:* fanlike).

Plants are large and to 11 cm long; fronds are compressed and flat while pinnae are in 2 rows, closely placed, and turned to one side.

var. *compressa* (Weber-van Bosse) M. Howe 1920a: p. 609 (L. *compressa*: flattened sideways or pressed together).

Basionym. *Caulerpa paspaloides* f. *compressa* Weber-van Bosse 1898: p. 353, pl. 30, fig. 3, 4.

Synonym. *Caulerpa paspaloides* var. *typica* f. *compressa* Weber-van Bosse 1898: p. 353, pl. 30, fig. 3, 4.

Plants are small with fronds to 8 cm tall; stalks are 4 mm in diam., 1–2 cm long, and palmately forked. Secondary axes are 2–4 cm long, terete, and covered by pinnae 80–100 μm in diam. that are small, crowded, pinnately arranged, but not in vertical rows.

var. *laxa* Weber-van Bosse 1898: p. 353, fig. 5 (L. *laxa*: loose, lax, or weak).

Frond stalks are erect and to 18 cm long; fronds are to 28 cm long and terete or compressed. Branchlets are limited to upper third of stalks, short, subverticillate or 3-ranked, and unbranched.

var. *minima* Humm and S.E. Taylor 1961: p. 364, fig. 9J, 10A,E (L. *minima*: very small).

Plants are small, to 1 cm tall, and in entangled masses on rubble; branchlets are in 2 or 3 ranks, with two rows in close approximation and the third more or less opposite the other two.

var. *phleoides* (Bory) Weber-van Bosse 1898: p. 353 (Gr. *phleõs*: name for marsh reed; also *phleõ*: with abundance + *oides*: like; thus similar to a marsh reed).

Basionym. *Chauvinia phleoides* Bory 1829 (1826–1829): p. 205, pl. 23, fig. 2.

Axes are cylindrical and branchlets crowded and in 2 pinnate ranks.

var. *wurdemannii* Weber-van Bosse 1898: pl. 30, fig. 6 (named for Dr. Wurdemann, an American naturalist who collected in the Keys).

Frond stalks are to 8 cm long and produce branches 10 cm or longer; branchlets are pinnate, biseriate, subopposite to scattered, and with simple forked pinnae that bend to one side.

f. *phyllaphlaston* (G. Murray) Weber-van Bosse 1898: p. 353 (Gr. *phyllon*: leaf, blade + *phlastos*: plastic, formed, arranged).

Basionym. *Caulerpa phyllaphlaston* G. Murray 1891: p 208.

Plants are similar to the variety but have uniseriate pinnae.

Common, species and taxa in seagrass beds on all coasts, often in protected sites, and to 20 m (type location of var. *minima*: Cedar Key); also known from the Bahamas (type location of species; Bory, 1829), Gulf of Mexico, and Brazil.

Taylor 1960: p. 149; Humm and Taylor 1961: p. 363; Dawes 1974: p. 75; Littler and Littler 2000: p. 366–368.

Caulerpa prolifera (Forsskål) J. V. Lamouroux 1809c: p. 332 (L. *prolifera*: proliferous, producing offshoots). Plate VI, fig. 18, 19.

Basionym. *Fucus prolifera* Forsskål 1775: p. 193 (as '*prolifer*').

Taylor (1960) and Dawes (1974) recognized the robust f. *obovata* J. Agardh (1872: p. 11) and delicate f. *zosterifolia* Børgesen (1907: p. 359, fig. 9; 1913: p. 127, fig. 101). Neither of the forms are recognized here for the following reasons: (1) f. *obovata* resembles f. *zosterifolia* grown under low light (Barilotti, 1970); (2) *in situ* populations often include the delicate form (Collado-Vides, 2002); (3) gravity and light regimes may influence blade and rhizoid morphology (Collado-Vides, 2002; Senerpont Domis et al. 2003); and (4) morphology can change with nutrient availability (Malta et al. 2005).

Blades are erect, flat, thin, delicate to leathery, and oval or elongate; rhizomes are often forked, slender, and with rhizoid-bearing branches below every 0.5–2 cm. Blades are simple to proliferous, stalked for 0.5 to 1 cm, oval, linear or lanceolate, 3–13(25) mm wide, 3–15 cm long, and have a narrow to obtuse apex.

The plant's clonal growth is similar to that of seagrasses (Hemminga and Durate, 2000); its fronds remove nutrients from the water while the rhizomes and rhizoids utilize sources from the sediment (Williams and Fischer, 1984; Larned, 1988). The species may be common in eutrophic habitats, but is also nitrophilic and forages for nutrients via extensive rhizome growth (Malta et al. 2005).

Common, perhaps the most common species of *Caulerpa* in Florida, forming monotypic beds in sandy or muddy lagoons, sometimes overgrowing seagrass beds, and to 10 m on all coasts; also known from North Carolina, Bermuda, the Caribbean, Gulf of Mexico (including Lower Laguna Madre), and Brazil, as well as Spain, Portugal, the Mediterranean (type location: Egypt; Silva et al. 1996), eastern tropical Atlantic, Red Sea, and Indo-Pacific.

Taylor 1960: p. 140; Humm and Taylor 1961: p. 364; Dawes 1974: p. 77; Littler and Littler 2000: p. 368.

Caulerpa racemosa (Forsskål) J. Agardh 1873: p. 35–36 (L. *racemosa*: clustered). Plate VI, fig. 20–25.

Basionym. *Fucus racemosus* Forsskål 1775: p. 191.

Synonyms. *Caulerpa racemosa* var. *uvifera* (C. Agardh) J. Agardh 1873: p. 35.

Caulerpa racemosa var. *clavifera* (Turner) Weber-van Bosse 1898: p. 361–362, pl. 33, fig. 1–3 (including f. *typica*).

Caulerpa racemosa var. *macrophysa* (Sonder *ex* Kützing) *sensu* W. R. Taylor 1928: p. 101, pl. 12, fig. 3; pl. 13, fig. 9.

Wynne (2005a) included var. *clavifera* (Turner) Weber-van Bosse and var. *uvifera* (Turner) Weber-van Bosse under *Caulerpa racemosa* var. *racemosa* (Taylor, 1960). Lipkin and Silva (2002) reviewed the various morphologies of *C. racemosa* in different habitats in the southern Red Sea and Ohba and Enomoto (1987) found that many varieties result from varying light intensities. Morphological and molecular studies by Senerpont Domis et al. (2003) demonstrated that *C. racemosa* was distinct from *C. cupressoides* and *C. sertularioides*. As noted above *C. microphysa* differs from *C. racemosa* because of the presence of pyrenoids that are absent in the latter.

Rhizomes form colonies to 2 m in diam., have axes 1–5 cm tall and are usually covered by beadlike bright to dark green branchlets similar to clusters of grapes. They are attached by stout, descending branches that produce numerous slender, yellowish forked rhizoids. Erect axes are branched or not and often crowded on rhizomes, which are coarse, 2–3 mm in diam., often branched, and densely entangled. Branchlets are spherical, oval, or slightly club-shaped, 2–4 mm in diam. and not constricted at bases; stalks are 1–2 mm in diam. and 2–5 mm long. Vegetative fragments can enhance dispersal and colonization (Ceccherelli and Piazzi, 2001).

f. *reducta* Børgesen 1907: p. 378, 382, 386, fig. 27 (L. *reducta*: dwarf, reduced).

Plants are small, densely compacted, and to 10 mm tall; erect branches are forked with a few irregularly placed cylindrical to turbinate branchlets.

var. *lamourouxii* (Turner) Weber-van Bosse 1898: p. 369, pl. 32, fig. 1–4, as f. *typica* (named for J. V. Lamouroux, a French phycologist who published on macroalgae).

Basionym. *Fucus lamourouxii* Turner 1811–1819: p. 79–80, pl. 229.

Fronds are 10–16 cm tall, with branchlets alternately opposite, usually in 2–4 ranks, widely scattered (few), 2–3 mm in diam., to 1 cm long, and club-shaped or spherical. Axes are more compressed when few branchlets are present.

var. *occidentalis* (J. Agardh) Børgesen 1907: p. 379, fig. 28, 29 (L. *occidentalis*: west).

Basionym. *Caulerpa chemnitzia* (Esper) J. V. Lamouroux var. *occidentalis* J. Agardh 1873: p. 37.

Fronds are 2–10 cm tall; branchlets are in radial or distichous patterns; stalks are 1.0–2.5 cm long; tips are expanded or subspherical and 1.5–2.5 mm in diam.

var. *peltata* (J. V. Lamouroux) Eubank in Stephenson 1944: p. 349 (L. *peltata*: having a small shield, shield-shaped).

Basionym. *Caulerpa peltata* J. V. Lamouroux 1809c: p. 332; 1809d: p. 145, pl. 3, fig. 2.

Synonym. *Caulerpa racemosa* var. *laetevirens* (Montagne) Weber-van Bosse 1898: p. 366, 367, pl. XXXIII, fig. 16, 20 (as f. *typica*).

Several authors (Womersley and Bailey, 1970; South and N'Yeurt, 1993; Silva et al. 1996; Hodgson et al. 2004; Guiry and Nic Dhonncha, 2006) consider *Caulerpa peltata* to be distinct from *C. racemosa*, while others (Weber-van Bosse, 1898; Svedelius, 1906; Børgesen, 1925) report transitional forms. Ohba and Enomoto (1987) and Ohba et al. (1992) also report that var. *peltata* and var. *laetevirens* change morphology under high or low light regimes, as also shown by Calvert (1976). Thus, Wynne (2005a) supported Eubank's (1946) assignment of *C. peltata* as a variety. The var. *laetevirens* is also a synonym because *Fucus chemnitzia* var. *peltatus* (J. V. Lamouroux) Turner 1811–1819 has priority (Ohba and Enomoto, 1987). Using molecular techniques (i.e., rDNA), Yeh and Chen (2004) found that var. *peltata* and var. *laetevirens* were variants of a single species (Wynne, 2005a).

Plants are robust, 12–30 cm tall. Branchlets are crowded to well spaced, club-shaped, with rounded to disklike tips, 3–4 mm in diam. Stalks are 1 mm in diam. by 1–2 mm long.

var. *turbinata* (J. Agardh) Eubank 1946: p. 420–421, fig. 2o–q (L. *turbinata*: like a top).

Basionym. *Caulerpa clavifera* (Turner) C. Agardh var. *turbinata* J. Agardh 1837: p. 173.

Synonyms. *Caulerpa chemnitzia* (Esper) J. V. Lamouroux 1809c: p. 332.

Caulerpa racemosa var. *chemnitzia* (Esper) Weber-van Bosse 1898: p. 370–373, pl. XXXI, fig. 5–7.

Coppejans and Beeckman (1989) and Ohba et al. (1992) considered var. *turbinata* to be an ecad of *Caulerpa racemosa* var. *laetevirens* (see above). Peterson (1972) and Calvert (1976) found that reduced light or wave action could influence frond morphology. Yeh and Chen (2004), using rDNA analyses, reported that var. *turbinata* is closely related to var. *peltata*.

Fronds are slender, with branchlets overlapping, varying in shape, and often cylindrical near the base and truncate near the tips.

Common, species and varieties occur on all coasts (to 55 m); compact forms in wave-exposed shallow sites and elongate forms usually in protected sites. Species may form humps or knoll-like colonies that overgrow reef habitats (Lapointe, 1999; Lapointe and Matzie, 2002). The species and varieties are also known from North Carolina (37 m), Bermuda, the Caribbean (type location of f. *reducta*: St. Thomas; Silva et al. 1996; and var. *peltata*, the West Indies; Huisman, 2000), Gulf of Mexico (type locations of var. *occidentalis*, upper Gulf of Mexico to Recife, Brazil; South and N'Yeurt, 1993), Brazil, as well as the Mediterranean (type location of species and var. *turbinata*: Suez, Egypt; Abbott and Huisman, 2004), eastern tropical Atlantic, Red Sea (type location of var. *lamourouxii*; Silva et al. 1996), Indo-Pacific, and Australia.

Taylor 1960: p. 151–153; Humm and Taylor 1961: p. 364–365; Dawes 1974: p. 77; Schneider and Searles 1991: p. 93; Littler and Littler 2000: p. 370, 372.

Caulerpa sertularioides (S. G. Gmelin) M. Howe 1905b: p. 576 (L. _sertula_: to entwine; also _Sertularia_: a genus of hydroids + _oides_: like; thus similar to _Sertularia_). Plate VI, fig. 26–29.

Basionym. *Fucus sertularioides* S. G. Gmelin 1768: p. 151, pl. 15, fig. 4.

Morphological and molecular studies by Senerpont Domis et al. (2003) indicate that *Caulerpa sertularioides* is distinct from *C. racemosa* and *C. cupressoides*.

Erect fronds are featherlike, pale to dark green, and to 20 cm tall; rhizomes are robust, to 2 m long by 2.2–5.0 mm in diam., and attached by branching rhizoids growing from stout (2 mm in diam.) descending branches. Fronds are 13–25 mm wide; its central stalks are naked below, 1.0–1.5 mm in diam., and simple or forked. Branchlets (pinnae) are opposite, cylindrical to needle-shaped, slightly curved upward, regularly and closely pinnate, 180–330 μm in diam., 3–11 mm long, and with rounded and bluntly pointed tips.

f. *brevipes* (J. Agardh) Svedelius 1906: p. 114–115, fig. 7, 8 (L. *brevipes*: short-footed, short-stalked).

Basionym. *Caulerpa plumaris* (Forsskål) C. Agardh var. *brevipes* J. Agardh 1873: p. 15.

Plants are small, 1–5 cm tall, and with sessile, rarely forked fronds.

f. *corymbosa* W. R. Taylor 1942: p. 31, pl. 6, fig. 1 (L. *corymbosa*: clustered).

Fronds are 2–8 cm tall, with short stalks, densely branched to 2nd or 3rd degree, and with 10–20 fronds per stalk; pinnae are opposite, 7 mm long below, and 3 mm above.

f. *farlowii* (Weber-van Bosse) Børgesen 1907: p. 365–366, fig. 11 (named for W. G. Farlow, an American cryptogamic botanist who studied the seaweeds of the eastern United States between 1870 and 1900).

Basionym. *Caulerpa plumaris* (Forsskäl) C. Agardh f. *farlowii* Weber-van Bosse 1898: p. 295, pl. 24, fig. 5, 6 (as '*farlowi*').

Calvert (1976) reported that the fronds of *Caulerpa sertularioides* changed from radial to bilateral symmetry under low light.

Fronds are erect, 5–10 cm tall, and densely covered by closely set pinnae, which are similar to the axes, radially disposed, curved upward, cylindrical, 500–700 μm in diam. by 2–5 mm long, and with bluntly pointed tips.

f. *longiseta* (Bory) Svedelius 1906: p. 114–115, fig. 10 (L. *longi-*: long + *seta*: bristle or stiff hair).

Basionym. *Caulerpa plumaris* (Forsskäl) C. Agardh var. *longiseta* Bory 1828 (1826–1829): p. 194, pl. 22, fig. 4.

Plants form extensive colonies, with rhizomes to 1 m in length. Fronds are erect to 20 cm tall and covered in upper parts with stalked featherlike branchlets (pinnae) to 10 mm wide. Frond stalks are cylindrical, naked, and 1–3 cm long; upper axes flattened, 1.0–1.2 mm in diam., naked below and with branchlets above. Pinnae are in opposing or 3 rows, cylindrical, needle-shaped, 300–500 μm in diam., and bluntly pointed.

Common, species and various forms occur on all coasts in open sand and seagrass beds and to 60 m offshore from Tampa Bay. The f. *longiseta* occurs in seagrass beds between 8 m and 10 m where light and wave activity are low, while f. *brevipes* is often found in areas of moderate wave activity. Species and forms are also known from Bermuda, the Caribbean (syntype locations for f. *brevipes*: West Indies, Pacific Ocean, Sri Lanka, and the Red Sea; Silva et al. 1996), Gulf of Mexico, and Brazil, as well as the Mediterranean, Canary Islands, West Africa, Red Sea (type locality of species; Abbott and Huisman, 2004), Pacific Ocean, and Australia.

Taylor 1960: p. 144; Humm and Taylor 1961: p. 365; Dawes 1974: p. 77; Littler and Littler 2000: p. 374–375.

Caulerpa verticillata J. Agardh 1847: p. 6, as '*verticillatum*' (L. *verticillata*: whorled). Plate VII, fig. 1, 2.

Synonyms. *Caulerpa pusilla* (Kützing) J. Agardh 1872: p. 6; *sensu* Mazé and Schramm 1878 (1870–1877): p. 82.

Caulerpa verticillata J. Agardh var. *pusilla sensu* Vickers 1905: p. 57.

Plants appear as filaments and form delicate dark green tufts or felted colonies 1–3(10) cm high; rhizomes are slender, 300–560 µm in diam., frequently branched, and attached by a few rhizoids. Fronds have delicate whorls of branchlets (pinnae), are 5–8 mm in diam., and branch dichotomously 5–7 times. Pinnae are 100–210 µm in diam. by 5–10 diam. long, with ultimate divisions 30–40 µm in diam., and pointed tips.

f. *charoides* Weber-van Bosse 1898: p. 267–268 (Gr. *chairis*: delight, grace; also the generic name of a green alga, *Chara* + *oides*: like; thus, resembling *Chara*).

Synonym. *Caulerpa charoides* (Weber-van Bosse) Thivy and Visalakshmi 1963: p. 101–104, fig. 1–10, pl. I.

Fronds are erect and densely covered with pinnae in alternate, opposite, or scattered patterns, but not whorled.

Common, the species produces tufts or feltlike colonies on mangrove roots or shaded limestone in the Keys and has overgrown reefs (20–30 m) in southeast Florida (Lapointe, 1999; Raloff, 2000; Lapointe and Matzie, 2002); also known from Bermuda, the Caribbean (type location of species: West Indies; Abbott and Huisman, 2004), and Gulf of Mexico, as well as the Indo-Pacific (lectotype locality of f. *charoides* Tonga; Thivy and Visalakshmi, 1963), and Australia.

Taylor 1960: p.138; Littler and Littler 2000: p. 376.

Caulerpa vickersiae Børgesen 1911: p. 129–132, fig. 2; 1913: p. 119, fig. 94, as '*vickersii*' (named for A. Vickers, a French phycologist who collected and described seaweeds in Barbados). Plate VII, fig. 3–5.

Several authors (Eubank, 1946; Egerod, 1975; Prud'homme van Reine and Lokhorst, 1992; Schneider and Lane, 2005) suggest that *Caulerpa ambigua* Okamura and *C. vickersiae* are conspecific, while others (Silva et al. 2996; Guiry and Nic Dhonncha, 2006; Wynne, 2005a) cite the plant as *Caulerpella ambigua* (Okamura) Prud'homme van Reine (1992) and list *C. vickersiae* as a taxonomic synonym. By contrast, Børgesen (1949a), Dawson (1956) and Littler and Littler (2000) recognize both as species of *Caulerpa*. The Littlers stressed that *C. vickersiae* is distinct from *C. ambigua* based on field work in the Pacific and Atlantic Oceans. *Caulerpella ambigua* was proposed by Prud'homme van Reine and Lokhorst (1992) based on its reproductive whorls of gametophores rather than the holocarpic pattern in *Caulerpa* where gametes form throughout the plant. They also pointed out that *C. ambigua* has trabeculae like *Caulerpa*. Fama et al. (2002), using chloroplast tufA sequences of 23 taxa of *Caulerpa*, suggested that *Caulerpella ambigua* should be recognized as a distinct taxon (Hodgson et al. 2004). Until *C. vickersiae* is shown to be a synonym, we follow Littler and Littler's (2000) designations.

Plants are small, delicate, and form thin mats; fronds are 1(–2) cm tall; rhizomes are inconspicuous, 200–250 μm in diam., and have few branched rhizoids. Fronds are 1–2 mm wide, featherlike, forking, and to 150 μm in diam.; their bases are naked or not. Axes are covered with pinnately and oppositely arranged branchlets (pinnae) to 100 μm in diam. 5 diam. long that occasionally fork and have blunt tips.

var. *furcifolia* W. R. Taylor 1933: p. 396 (L. *furca*: forked, a fork + *folia*: leaved).

Fronds are 60–65 μm in diam., 4–7 mm tall, 1–4 times forked, and covered with pinnate branchlets above; branchlets (pinnae) are irregular to whorled, subdichotomously forked 1–3 times, and 50–85 by 170–300 μm. Rhizomes are 125–215 μm in diam., while rhizoids are sparse and branched.

var. *luxurians* W. R. Taylor 1928: p. 104, pl. 12, fig. 20; pl. 13, fig. 12 (L. *luxurians*: luxuriant, abundant).

Rhizomes are inconspicuous; fronds are solitary or clustered, sessile or stalked, 5–15 mm tall, forked 1–5 times, and pinnately branched. Pinnae are 60–85 μm in diam. and 0.5–1.0 mm long.

Uncommon, the species and varieties are easily overlooked as they often occur with other algae on mangrove roots and soft corals in the Keys (to 20 m); also known from the Caribbean (syntype locations of species: St. John and Barbados; Silva et al. 1996), Gulf of Mexico, Costa Rica, Brazil, and Indo-Pacific.

Taylor 1960, p. 137; Littler and Littler 2000: p. 378–379.

Caulerpa webbiana Montagne 1837: p. 354 (named for A. Weber-van Bosse, a Dutch phycologist who went on the Siboga Expedition, where she described species from Indonesia and Malaysia). Plate VII, fig. 6.

Plants are small and form mats to 20 cm in diam.; erect fronds are to 15 mm tall and covered with indistinct whorled branchlets; rhizomes branch and produce many padlike rhizoidal branches. Fronds have many whorls of 2–6 branchlets (pinnae) that are 250 μm long, basally terete, and flat above; branchlets fork 3–5 times, decrease in diam. with each forking, and have one or more points at tips.

f. *disticha* Weber-van Bosse 1898: p. 270, fig. 1 (L. *disticha:* two-ranked).

Wynne (2005a) cites the authorship as f. *disticha* Vickers.

Fronds are flat; pinnae are also flat, in one plane, and opposite, and fork 2–3 times.

f. *tomentella* (Harvey *ex* J. Agardh) Weber-van Bosse 1898: p. 270–271, pl. XXI, fig. 4 (L. *tomentella*: covered with minute hairs).

Basionym. *Caulerpa tomentella* Harvey ex J. Agardh 1873: p. 8.

Fronds are repent, densely covered with pinnae, and little compressed; rhizomes are elongate. Rhizoids are filamentous, branched, and arise from prostrate and erect axes.

Rare, easily overlooked, on sand-covered rubble in reef channels (to 50 m) in the Keys; also known from the Caribbean to Brazil, the eastern tropical Atlantic (type location of species: Arrecife, Canary Islands; Silva et al. 1996), Indo-Pacific (type locations for f. *tomentella*, Tonga; Silva et al. 1996; and Canary Islands; Hodgson et al. 2004), and Australia.

Taylor 1960: p. 139–140; Littler and Littler 2000: p. 378–380.

Ostreobiaceae

Cladistic and molecular studies of *Ostreobium* by Vroom et al. (1998) and Woolcott et al. (2000) have supported the creation of its own family, which is distinct from the Bryopsidaceae and Codiaceae.

Plants are endophytic or endozoic coenocytes with large ovoid or branched cells. Branches lack cross walls and have continuous cytoplasm with many nuclei. Plastids are numerous, discoid, and lack pyrenoids.

Ostreobium Bornet and Flahault 1889: p. 161 (Gr. *Ostreon*: an oyster + *bios*: living in).

Filaments are coenocytic, much branched, and irregular in shape and diam. Asexual reproduction is by aplanospores from swollen branch tips and by quadriflagellate zoospores.

Ostreobium quekettii Bornet and Flahault 1889: p. 161, pl. IX, fig. 5–8, as '_queketti_' (named for John Thomas Quekett, a popular British microscopist and lecturer who published two volumes on "Lectures on Histology"). Plate VII, fig. 7.

Ostreobium has been placed in the Bryopsidaceae (Burrows, 1991), Phyllosiphonaceae (Collins, 1909b), or Ostreobiaceae (Schneider and Searles, 1991; Wynne, 2005a). Data from life history studies are variable, with Mayhoub (1974a,b) reporting *Ostreobium*-like "protonema" in *Pseudobryopsis myura* (J. Ag.) Berthold and Kornmann and Sahling (1980) indicating that the "prothalli" are not identical to those of European *Ostreobium*. The absence of *P. myura* from the tropical and subtropical western Atlantic should be noted, while *P. blomquistii* Diaz-Piferrer, *P. oahuensis* Egerod, and *P. venezoleana* (W. R. Taylor) Henne and Schnetter are recorded (Wynne 1998a).

Plants are endophytic or endozoic coenocytes that grow in shells and urchin tests. Filaments are slender, much branched, twisted, irregular, sometimes inflated to 40 µm in diam., mostly 4–10 µm, and tapering to 2µm at tips. Plastids are abundant, small, elongate, spindle-shaped, and lack pyrenoids. Reproduction is by aplanospores or quadriflagellate zoospores from irregularly swollen branch tips cut off by a wall; zoosporangia have short or long hyaline exit tubes and zoospores are 9–12 µm long.

Common, inconspicuous, with few published reports for Florida probably because of its habit in shells or bryozoan tests that require decalcification for identification; often associated with the shell-boring green alga *Gomontia polyrhiza*. Also known from the Canadian Arctic through the Caribbean, as well as the eastern North and tropical Atlantic (lectotype location: Le Croisic, France; Lukas, 1974), the Mediterranean, Indo-Pacific, Australia, and Fiji (in barnacles and octocorals; Feussner et al. 2004).

Taylor 1960: p. 189; Humm and Taylor 1961: p. 373; Dawes 1974: p. 87; Schneider and Searles 1991: p. 98.

Codiaceae

Plants are spongy, erect or prostrate, and have interwoven colorless coenocytic siphons. Septa occur only at the base of reproductive branches or when axes are damaged. Siphons form swollen tips (utricles) at the surface that fuse along the lateral walls, contain discoid plastids that lack pyrenoids, and have mannan in their cell walls.

Codium Stackhouse 1797 (1795–1801): p. xvi, xxiv (Gr. *codium*: fleece, skin).

With the same features as the family.

Key to the species (6) of *Codium*

1. Plants or main axes either prostrate or lobed 2
1. Plants erect and with basal holdfasts. 4
 2. Plants irregularly to dichotomously branched 3
 2. Plants lobed, growing over objects; utricles 70–110µm in diameter . . . . *C. intertexum*
3. Axes 4–11 mm in diam.; utricle tip wall 8–10 µm thick . . . . *C. carolinianum*
3. Axes 1–5 mm in diam.; utricle tip wall 4–6 µm thick . . . . *C. repens.*
 4. Branches often 4 mm or larger in diameter 5
 4. Branches cylindrical, 2.5–4.0 mm in diameter . . . . *C. isthmocladum*
5. Utricles mostly 150–850 µm in diam., tip walls 4–8 µm thick . . . . *C. decorticatum*
5. Utricles mostly 55–260 µm in diam., tip walls 23 µm thick . . . . *C. taylorii.*

Codium carolinianum Searles 1972: p. 19, fig. 1a, 2 (named for the state of North Carolina). Plate V, fig. 8–10.

Plants are dark green, prostrate, 20–30(60) cm wide, with irregular branching; axes may fuse and reattach to substrata via ventral fibrous tufts and pads of rhizoids. Branches are 4–11 mm in diam, spongy, and terete to slightly flattened with forks more conspicuously flattened. Utricles are club-shaped to cylindrical, 75–210 µm in diam. by 730–1510 µm long, with lateral walls 2 µm thick and have truncate to rounded tips; tip walls are 8–10 µm thick and often have a central spine. Hairs or hair scars are 140–175 µm below utricle apex and rare or to 5 per utricle. Gametangia are 195–260 µm long, 1–3 per utricle, 240–350 µm below the apex of utricles, and on short stalks (14 µm long).

Uncommon, growing on limestone off the west coast in 15–30 m; also known from North Carolina, South Carolina (15–35 m type location), the Florida Gulf Coast, Puerto Rico, and Canary Islands.

Schneider and Searles 1991: p. 83; Littler and Littler 2000: p. 350.

Codium decorticatum (Woodward) M. Howe 1911: p. 494–495 (L. *decorticatum*: deprived of bark, lacking a covering). Plate V, fig. 11, 12.

Basionym. *Ulva decorticata* Woodward 1797: p. 55–58.

Plants are dark green, erect, 12–100 cm tall, often in clumps, and attached by fibrous padlike holdfasts. Main axes are few, but produce many dichotomous branches that are spongy, terete, 6–25 mm in diam., 6–9 cm long, compressed at forks only or throughout, and 6(–9) mm wide. Utricles are cylindrical to club-shaped, 150–850 µm in diam. by 1100–1750(2000) µm long, and with tips rounded to depressed; lateral walls are 1.5–2 µm thick and apical walls 4–8 µm thick. Hairs and scars occur in a band 145–330 µm below tips, up to 12 per utricle. Gametangia are ovoid, 1–7 per utricle, on stalks, 430–650 µm below utricle tips, 60–125 µm in diam., 150–400 µm long, and 10–15 µm in diam.

Common, on hard substrata to 20 m on all coasts; sometimes overgrowing coral reefs in southeastern Florida (Hanisak, 2001); also known from the Carolinas (14–25 m), Bermuda, Gulf of Mexico, Caribbean to Argentina, as well as the eastern North and tropical Atlantic, Mediterranean (type location; Abbott and Huisman, 2004), Indo-Pacific, and Antarctica.

Taylor 1960: p.188; Humm and Taylor 1961: p. 367; Dawes 1974: p. 80; Schneider and Searles 1991: p. 84; Littler and Littler 2000: p. 350.

Codium intertextum Collins and Hervey 1917: p. 54 (L. *intertextum*: interwoven, intertwined). Plate V, fig. 13, 14.

Plants are prostrate, solitary, or in colonies; lobes are dull green, smooth to slippery, overlapping, 2–5 mm wide, and attached by many rhizoids. Utricles are cylindrical, constricted below, and (40)70–110(215) µm in diam. by (400)500–700(880) µm long; utricle apices are rounded to truncate, sometimes indented, and with walls 3–8(20) µm thick. Hairs arise from older utricles in a band 60–145 µm below tips. Gametangia are stalked, fusiform to ovoid, 50–110 µm in diam., and 220–330 µm below utricle tips.

Common, forming a distinct band at the low tide mark on all coasts as well as extending to 20 m; also known from Bermuda (type location), the Caribbean, Gulf of Mexico, and Brazil, as well as the Salvage, Canary, and Cape Verde Islands.

Taylor 1960: p. 185; Littler and Littler 2000: p. 352.

Codium isthmocladum Vickers 1905: p. 57 (Gr. *isthmos*: a strip of land connecting two larger land masses + *cladum*: a branch). Plate V, fig. 15, 16.

Plants are erect, to 20 cm tall, attached by crustose bases, and mostly clumped. Branching is usually dichotomous above to 12 orders and proliferous in older plants. Branches are soft, terete, 2–6 mm in diam., compressed in mid to lower regions, and may be constricted above dichotomies. Utricles are cylindrical, club or pear-shaped, 120–350(475) µm in diam. by 440–850 µm long, and often constricted 130–260 µm below tips. Utricle tips are rounded or truncate, lateral walls 1.5 µm thick, and end walls 18–56(110) µm thick. Hair and scars are few to several per utricle and 40–120 µm below apices. Two to several gametangia occur on a utricle; they are ovoid, 50–130 µm in diam., 180–280 µm long, and on short (5–7 µm) stalks. Prince and Trowbridge (2004) report hermaphroditic and separate gametophytic plants, as well as variable morphologies.

var. *clavatum* (Collins and Hervey) P. C. Silva 1960: p. 500, 508, pl. 116 (L. *clavatum*: club-shaped).

Basionym. *Codium decorticatum* var. *clavatum* Collins and Hervey 1917: p. 56.

Branches are 5–10 mm in diam., with lower branches often flat; utricle tips have thinner walls than the species.

Common (both species and variety), from 2–30 m (to 73 m) on all coasts; the variety often in areas of high wave activity; the species has overgrown coral reefs in southeastern Florida (Lapointe, 1997, 1999; Hanisak, 2001). The species is also known from the Carolinas (11–45 m), Georgia, Bermuda, the Caribbean (type location: Barbados), Gulf of Mexico, and Brazil, as well as Israel, and eastern tropical Atlantic. The variety is also known from Bermuda (type location), the Caribbean, and Brazil.

Taylor 1960: p. 186, 187; Humm and Taylor 1961: p. 367; Dawes 1974: p. 80; Schneider and Searles 1991: p. 86; Littler and Littler 2000: p. 352.

Codium repens Vickers 1905: p. 56–57 (L. *repens*: creeping, prostrate, rooting). Plate V, fig. 17, 18.

Synonym. *Codium tomentosum* Stackhouse var. *subsimplex* P. Crouan and H. Crouan in Schramm and Mazé 1865: p. 47; P. Crouan and H. Crouan in Mazé and Schramm 1878 (1870–1877): p. 106.

Plants form prostrate mats to 40 cm in diam.; branching is irregular to dichotomous, with wide forks; axes attached by tufts of rhizoids. Branches are tough, entangled, terete to slightly flat, 1–5 mm in diam., and can possibly become reattached. Utricles are subcylindrical to club-shaped, slender, 100–275 µm in diam., and 330–550(875) µm long; utricle tips are rounded with walls 4–6(16) µm thick. Hairs and scars are common, 1–2 per utricle and 65–130 µm below the apex. Gametangia are rare, 1–3 per utricle, stalked, ovoid to spindle-shaped, 65–90 µm in diam., 200–225 µm long, and 290–350 µm below utricle tips.

Uncommon, under ledges at wave-exposed low intertidal sites in the Keys, at Point of Rocks, and to 3–55 m off the Dry Tortugas; also known from Bermuda, the Caribbean (type location: Vieu Fort, Guadeloupe; Silva et al. 1996), and Brazil, as well as the Canary and Cape Verde Islands, Red Sea, and Indo-Pacific.

Taylor 1960: p. 186; Silva et al. 1966: p. 861; Littler and Littler 2000: p. 354.

Codium taylorii P. C. Silva 1960: p. 510–513, pl. 112, 118b, 119, 120a,b (named for William Randolph Taylor, an American phycologist who published the seaweed floras of the northeastern and tropical Atlantic Americas). Plate V, fig. 19–21.

Plants are bushy to compact dark green clumps or erect to 15 cm tall and attached by crustose holdfasts. Branching is unequal, often cervicorn below, at wide angles, and to 7–9 orders; branches are firm, spongy, 4–25 mm wide by 3–4 mm thick, and with blunt tips. Utricles are cylindrical to club-shaped, 55–380 µm in diam., and 550–1450 µm long; utricle tips are rounded to pointed and with end walls to 23 µm thick. Hairs and scars are common and 50–105 µm below utricle tips. Gametangia are ovoid to cylindrical, 45–86 by 200–350 µm, 1–2 per utricle, on stalks to 8 µm long, and 275–430 µm below utricle tips.

Common, on mangrove roots and limestone (2–30 m) on all coasts (type location: Pass-a-Grill jetty, St Petersburg Beach); also known from North Carolina, Bermuda, the Caribbean, Gulf of Mexico (including Lower Laguna Madre, Texas; De Yoe and Hockaday, 2001), Brazil, and Uruguay, as well as Israel (introduced), the eastern tropical Atlantic, and Indo-Pacific.

Taylor 1960: p. 188; Dawes 1974: p. 80; Schneider and Searles 1991: p. 87; Littler and Littler 2000: p. 354.

Halimedaceae

The molecular studies of Lam and Zechman (2006) showed that *Halimeda* should be placed in a family distinct from the Udoteaceae and that *Pseudocodium* was most closely related to that group. However, its familial affinity was not resolved, thus, we place *Pseudocodium* in this family in contrast to its own (Wynne, 2005a).

Plants are erect and with a surface of tightly compacted utricles containing both chloroplasts and amyloplasts (heteroplasty).

Halimeda J.V. Lamouroux 1812: p. 186, *nom. cons.* (*Halimeda*: Greek mythology: a sea nymph, a daughter of Halimedon, King of the sea; also Gr. *halimos*: maritime).

On the basis of molecular and anatomical studies, Hillis et al. (1998) reported three lineages in the genus. Using similar data and historical information, Verbruggen and Kooistra (2004) divided *Halimeda* into five sections: *Halimeda, Micronesicae, Opuntia, Pseudo-opuntia*, and *Rhipsalis*. The eleven taxa recorded from Florida occur in four of these sections. Verbruggen et al. (2005a,b,c,d) emphasized that species-level taxonomy of Bryopsidalean genera (for example, *Halimeda*) is best defined on quantifiable morphological, anatomical, and molecular data.

Plants are true coenocytes, usually erect and attached by rhizoidal holdfasts. Axes are branched with flexible, noncalcified joints (geniculae). Segments are moniliform, cylindrical or discoid, simple or lobed, and weakly to strongly calcified by aragonite calcium carbonate. Axial siphons are parallel, branched, and fused at joints. The cortex arises from lateral fascicles of siphons and consists of a continuous surface of swollen tips (utricles). Gametangia are dark green, on branched stipes, clustered, and arise from cortical siphons at the surface of segments. Gametophore length appears to be one of the most informative phylogenetic characters and accounts for more than 75 percent of different lineages of *Halimeda* (Vroom and Smith, 2003). Species of the genus may account for 83 percent of the carbonate production on coral reefs (Hillis, 1997) and, with *Penicillus* spp., are often major components of psammophytic back-reef habitats (Fourqurean and Rutten, 2004). The presence of epiphytic populations of *Dictyota* on *Halimeda* species results in reduced growth and enhanced respiration (Beach et al. 2003).

Key to the species (11) of *Halimeda* (modified from Littler and Littler, 2000)

1. Segments cylindrical or flattened, ridges may be present (for example, trilobed) 2
1. Segments oval to spherical. . . . . *H. lacrimosa*
    2. Segments large, usually more than 5 mm wide. 4
    2. Segments small, less than 6 mm wide. 3
3. Plants to 2.5 cm tall, branching irregular; segments to 3.5 mm wide. . . . . *H. hummii*
3. Plants to 13 cm tall, branching regular; segments to 2–6 mm wide . . . . *H. goreauii*
    4. Segments of upper parts of plant flattened 5
    4. Segments of upper parts of plant cylindrical. . . . . *H. monile*
5. Axes and branching primarily in one plane; plants appearing flat and with one attachment site 6
5. Axes and branching in many planes; plants clumped and with more than one attachment site . . . . . *H. opuntia*
    6. Surface of segments smooth; utricles with flat or rounded tips. 7
    6. Surface of segments rough; utricles with a center spine . . . . *H. scabra*
7. Segment margins not thickened (ribbed) 8
7. Segment margins thickened because of swollen subsurface utricles . . . . *H. gracilis*
    8. Segments with distinct to weakly evident ribs, often trilobed. 9
    8. Segments rounded, not lobed or ribbed 10
9. Segments ribbed; surface utricles angular after decalcification. . . . . *H. incrassata*
9. Segments weakly ribbed; surface utricles spherical after decalcification. . . . . *H. simulans*
    10. Segments 2–4 cm wide; basal segments not stipelike . . . . *H. discoidea*
    10. Segments to 1.2 cm wide; basal segments stipelike. . . . . *H. tuna*

Section *Halimeda*

Species have medullary siphons that often divide below the joints and become entangled with one another.

Halimeda discoidea Decaisne 1842b: p. 102 (L. *discoidea*: disclike). Plate VIII, fig. 13, 14.

Synonym. *Halimeda discoidea* var. *platyloba* Børgesen 1911: p. 134, fig. 86.

Molecular studies suggest that cryptic species may occur in *Halimeda discoidea* (Kooistra et al. 2002; Verbruggen et al. 2005c,d).

Plants are bright green to gray white, erect, to 20 cm tall, solitary or in clumps, sparsely dichotomously branched, and attached by small rhizoidal masses. Segments are lightly calcified; basal segments are subterete; upper ones are smooth (not ribbed), flat, to 4 cm in diam., and 3 cm long. Older segments are oval to kidney-shaped, with entire or slightly truncate margins. Internal siphons are entangled and fused in groups of 2s or 3s at joints; subcortical utricles are inflated, 50–250 µm in diam., 100–350 µm long, and with up to 14 surface utricles that are flattened, 30–90 µm in diam., and 35–210 µm long.

Common, on all coasts, in seagrass beds, from 2 to 50 m in the Keys, and 8–30 m off the west coast; vegetative expansion can occur via fragments (Walters and Smith, 1994). This plant is also known from the Bahamas (29–61 m), Caribbean, Gulf of Mexico, and Brazil, as well as the Canary Islands and Indo-Pacific (type location: Kamtshatka, Russia; Hillis, 1959).

Taylor 1960: p. 179; Humm and Taylor 1961: p. 369; Dawes 1974: p. 81; Littler and Littler 2000: p. 400.

Halimeda hummii D. L. Ballantine 1982: p. 87, fig. 1–6 (named for Harold J. Humm, an American phycologist who studied seaweeds of Virginia, North Carolina, Florida, and the Caribbean). Plate VIII, fig. 19, 20.

Plants are minute (to 2.5 cm tall), loosely organized, moderately calcified, and attached by a small rhizoidal holdfast. Branching is irregular, with 1–4 branches arising from an enlarged segment 0.2–3.5 mm wide and 0.8–3.2 mm long. Segments are irregular, cylindrical or flat, with entire or eroded margins. Utricles are in 1–3 series; subcortical utricles are trichotomously branched, basally constricted, and enlarged above; surface utricles are 30–63 µm in diam. and sunken to form a pitted surface. Medullary filaments are fused or free at joints and 30–80 µm in diam. Gametangia form clusters on distal blade siphons (Wysor and Kooistra, 2003).

Rare, from 12 m in the Dry Tortugas (type location); also known from Puerto Rico (Ballantine and Aponte, 1997a) and Panama were it is common in shallow, exposed shores (Wysor and Kooistra, 2003).

Halimeda scabra M. Howe 1905a: p. 241, pl. 11, 12 (L. *scabra*: rough, scurfy). Plate IX, fig. 1, 2.

See comments below regarding the separation of *Halimeda scabra* and *H. tuna*.

Plants are whitish green, erect or lax, to 25 cm tall, tufted, branched in one plane, and attached by rhizoidal tufts. Segments are lightly calcified with the lower 2–3 fused to form a stipe; other segments are flat, not ribbed, disc or wedge-shaped, with rounded and entire margins, 0.2–2 cm wide by 0.1–1.5 cm long. Internal siphons are fused and entangled at joints in 2s or 3s. Utricles are in 2–4 layers; subsurface utricles are 25–60 µm in diam., to 70 µm long, and have 2–8 surface utricles that are 25–70 µm in diam. and 70 µm long. Surface utricles have a prominent central spine and are slightly fused with adjacent ones. Gametangia are often at segment margins; they are pyriform, on simple to forked stipes, and 160–320 µm in diam.

Common, on rubble or limestone in tide pools, in seagrass beds in the Keys, and in 10–30 m off the west coast; also known from Bermuda (type location), the Caribbean, and Gulf of Mexico.

Taylor 1960: p. 180; Littler and Littler 2000: p. 406.

Halimeda tuna (J. Ellis and Solander) J. V. Lamouroux 1816: p. 309 (L. from Spanish *taino*: the name for a flat, jointed prickly cactus of the genus *Opuntia*). Plate IX, fig. 5–7.

Basionym. *Corallina tuna* J. Ellis and Solander 1786: p. 111, pl. 20, fig. e.

Plants form dark green dense clusters to 10–25 cm tall; primary branching is in one plane, and fronds are attached by an inconspicuous rhizoidal tuft. Stipes are distinct, with lower 1–2 segments thick, subterete, and fused; upper segments are flat and disclike, oval, kidney-shaped, or triangular, not ribbed, to 2 cm wide by 1.3 cm long, and with entire margins. Internal siphons are entangled and fused in clusters of 2s or 3s at joints. Utricles are in 2–4 layers; subsurface ones are club-shaped, 20–110 µm in diam., 40–120 µm long, and have 2–7 surface utricles. Surface utricles are 20–125 µm in diam., 40–130 µm long, and laterally fused only 0.04–0.1 of their length.

f. *platydisca* (Decaisne) E. S. Barton 1901: p. 14, pl. 1, fig. 2 (Gr. *platy-*: flat, broad, wide + L. *disca*: disc).

Basionym. *Halimeda platydisca* Decaisne 1842b: p. 102.

Synonym. *Halimeda tuna* var. *platydisca* (Decaisne) Børgesen 1911: p. 134.

Silva et al. (1996) and Guiry and Nic Dhonncha (2006) did not differentiate this form and included it within *Halimeda tuna*.

Plants form compact bright green masses to 15 cm tall, have primary branching in one plane, and are attached by inconspicuous rhizoidal tufts. Segments are large, lightly calcified, disc-shaped, 2–4 cm in diam., flexible when alive, without ribs, and with smooth, rounded margins. The form is similar to *Halimeda scabra*, but with a smoother texture and surface utricles that lack a central spine.

Common, but inconspicuous, both the species and form grow on hard substrata to 40 m, on all coasts; small fragments may reattach themselves vegetatively (Walters et al. 2002). The species is also known from Bermuda, the Bahamas (31–61 m), Caribbean, Gulf of Mexico, Mediterranean (type location of species; Silva et al. 1996), Morocco, the Canary Islands (type location for form; Silva et al. 1996), and Indo-Pacific.

Taylor 1960: p. 178; Littler and Littler 2000: p. 408.

Section *Opuntia*

Species have internal siphons that fuse over a short distance at joints, but retain their identity.

Halimeda goreaui W. R. Taylor (= *goreauii*) 1962b: p. 173, fig. 1–7 (L., named for T. F. Goreau, a coral reef biologist who collected the plant in Jamaica). Plate VIII, fig. 15, 16.

Plants are bright green, pendent, 7–13 cm long, with branched clusters to 4 cm, and attached by small rhizoidal masses. Segments are 2.5–6.0 mm wide, 3–4 mm long, 300–500 µm thick, brittle, heavily calcified, and with faint ribs. Basal segments are cylindrical to wedge-shaped, while upper ones are trilobed or deltoid. Siphons are entangled and paired at joints. Utricles are slender and occur in three or more layers; subsurface utricles are 10–30 µm in diam. by 20–60 µm long, and produce 2–6 surface utricles that are 15–37 µm in diam. and 15–40 µm long.

Rare, locally abundant, in 20–80 m off the northern reef track in the Biscayne National Park (Lirman and Biber, 2000); also known from the Bahamas, Panama, and Caribbean (type location, Jamaica).

Littler and Littler 2000: p. 402.

Halimeda opuntia (Linnaeus) J. V. Lamouroux 1816: p. 308 (Gr. *opuntia*: a plant of Opuntis in ancient Greece; also the genus *Opuntia*, a cactus with large flat segments). Plate VIII, fig. 29–31.

Basionym. *Corallina opuntia* Linnaeus 1758: p. 805.

Based on phylogeographic studies, Kooistra and Verbruggen (2005) reported that *Halimeda opuntia* spread from the Indo-Pacific into the Atlantic during the last millennium. Presently it is probably the most common species of *Halimeda* in all warm seas (Abbott and Huisman, 2004).

Plants form white-yellow-dark green dense to diffuse colonies to 20 cm tall; they are densely branched in various planes, lack a primary base, are attached by rhizoidal masses, and can become reattached by rhizoids from segments. Segments are strongly calcified, flat to contorted, often ribbed, oval to tear-shaped, 7–20 mm wide, 4–12 mm long, and 0.5–1.2 mm thick. Medullary siphons are paired but not fused at joints. Utricles are in 4–5 layers; subsurface utricles are 11–50 µm in diam., 30–50 µm wide, and 3–5 surface utricles are 15–60 µm in diam. and 15–30(70) µm long. Plants can be reduced and discolored when next to mesenterial filaments of scleractinian corals (Nugues et al. 2004).

f. *triloba* (Decaisne) J. Agardh 1887: p. 84–85 (L. *triloba*: three-lobed).

Basionym. *Halimeda triloba* Decaisne 1842b: p. 102.

Some authors place the form in synonymy with *Halimeda opuntia* (Hillis, 1959; Ortega et al. 2001; Guiry and Nic Dhonncha, 2006), while others recognize it because of its large, long, and loose aspect (Littler and Littler, 2000; Wynne, 2005a).

Axes are highly irregularly to randomly branched and attached by diffuse, randomly developed rhizoids. Segments are heavily calcified, subterete below and flat to contorted and strongly ribbed above. Segments are trilobed, to 11 mm wide, 7 mm long, and usually placed at right angles to sister segments. Utricles are similar to those of the species.

Common, both the species and the form grow in seagrass beds on all coasts, often as large calcified cushions behind reefs in the Keys; also present offshore from Tampa Bay (12–25 m), at the Dry Tortugas, the Caribbean (type location of species: Jamaica; Hillis, 1959), Gulf of Mexico, and Brazil, as well as the Mediterranean, Indo-Pacific, and Australia.

Taylor 1960: p. 176; Dawes 1974: p. 82; Littler and Littler 2000: p. 406.

Section *Pseudo-opuntia*

Species have club-shaped, subperipheral cortical utricles.

Halimeda gracilis Harvey *ex* J. Agardh 1887: p. 82 (L. *gracilis*: thin, graceful). Plate VIII, fig. 17, 18.

Plants are light green, erect or decumbent, to 40 cm tall, and attached by basal rhizoidal masses and rhizoids from prostrate segments. Branches are irregular with up to 5 per segment. Segments are heavily calcified, brittle, flat, not ribbed, oval to wedge-shaped, to 15 mm wide, 9 mm long, 0.5–2.0 mm thick, and with thickened undulate margins. Siphons are fused in 2s or 3s at joints. Utricles are in 2–3 layers; subsurface utricles are club-shaped, swollen or with flat tips, 20–125 µm in diam. and 100–400 µm long; surface utricles are 20–70 µm in diam. by 40–110 µm long.

Uncommon, on sand plains at 60 m and mangrove peat in 5 m in the Keys; known from the Caribbean, the Indo-Pacific (type location, Sri Lanka; Abbott and Huisman, 2004), and Australia.

Taylor 1960: p. 177; Littler and Littler 2000: p. 402.

Halimeda lacrimosa M. Howe 1909: p. 93–94, pl. 4, fig. 1, pl. 6, fig. 3–12 (L. *lacrimosa*: tear drop). Plate VIII, fig. 23–26.

Plants are gray green, pendent or 2–5 cm tall, heavily calcified, lacking a distinct stipe, and attached by rhizoidal bases and rhizoids of prostrate branches. Branching is sparse, irregular, and mostly in a single plane; segments are beadlike, oval, pyriform, or tear-shaped, 1–5 mm wide by long, and with a smooth surface. Siphons of joints are fused in groups of 2–4 and not calcified. Utricles are in 2–3 layers; subcortical ones are club-shaped, 65–110 µm in diam., 105–340 µm long, and with 6–18 surface utricles. Surface utricles are conical, 30–42 µm in diam. by 40–110 µm long, with flat tips, thick outer walls, and are hexagonal in appearance in surface view; the lateral walls are in contact 0.03–0.1 of their length.

var. *globosa* Dawes and Humm *ex* Dawes 1980: p. 142 (L. *globosa*: spherical, round).

See Dawes and Humm (1980) for a detailed description of the variety.

Plants are gray green to white, fragile, to 10–25 cm long, in clumps, and with sparse to irregular branching. Segments are brittle, strongly calcified, spherical to tear-shaped, glazed, and 5–10 mm in diam. Surface utricles are 30–45 μm in diam., with their outer walls 7–9 μm; joint siphons are fused in 2s to 4s.

Uncommon, the species extends to 2 m on limestone at the Content Keys and the variety from 40 m off Tampa Bay (type location); the species is also known from the Bahamas (type location: Mariguana; Silva et al. 1996), Greater Antilles, Cuba, and Madagascar, while the variety is also known from the Bahamas and Greater Antilles.

Taylor 1960: p. 179; Littler and Littler 2000: p. 404.

Section *Rhipsalis*

Species have all or most medullary siphons fused into a single unit at their joints. Most species in this section grow on sandy or muddy substrata in tropical lagoons and mangrove forests; the large, bulbous holdfast is modified for attachment.

Halimeda incrassata (J. Ellis) J. V. Lamouroux 1816: p. 307 (L. *incrassata*: thickened, thick). Plate VIII, fig. 21, 22.

Basionym. *Corallina incrassata* J. Ellis 1768: p. 408, pl. 17, fig. 20–27; also J. Ellis and Solander 1786: p. 111, tab 20, fig. d1–d3, D1–D6.

Synonyms. *Corallina tridens* J. Ellis and Solander 1786: p. 109, pl. 20, fig. a.

Halimeda tridens (J. Ellis and Solander) J. V. Lamouroux 1812: p. 186.

Molecular studies have indicated that Atlantic and Indo-Pacific populations designated as *Halimeda incrassata* represent cryptically different species (Kooistra et al. 2002; Verbruggen et al. 2005b,c,d, 2006), which have been recently designated as *H. heteromorpha* N'Yeurt and *H. kanaloana* Vroom.

Plants are bright to dull green, erect, to 24 cm tall, often solitary, and attached by bulbous rhizoidal bases. Branching is pseudodichotomous; their main axes arise in the lower half of thallus, and segments are strongly calcified and hard. Basal segments are fused, terete to slightly trilobed, and to 8 mm wide; upper segments are flat to compressed, shield-shaped to oval to cuneate, to 14 mm wide, and often have 3–5 vertical ribs. Cortex has 2–3 utricle layers; subcortical utricles are oval or subglobose, 20–90 μm in diam. and 30–115 μm long, while 2–4 surface utricles are 43–64 μm in diam., 55–90 μm long, and fused for 0.05–0.12 of their length. Internal siphons are fused into one bundle at joints and rarely free. Gametangia are oval to pyriform, 200–380 μm in diam., and on stipes that are long, dichotomously forked, and marginally attached to utricles.

f. *gracilis* Børgesen 1913: p. 111, fig. 89 (L. *gracilis*: thin, slender, graceful).

Segments are small, 2–4 mm wide, thick, circular to broadly oval, slightly ribbed or lobed, flexible, weakly calcified, and with utricles 40–70 μm in diam.

f. *tripartita* E. S. Barton 1901: p. 27; pl. IV, fig. 43 (L. *tripartita*: three-parted).

The mid to upper segments are divided into 3 lobes that are almost cylindrical in cross section; lower segments are mostly cylindrical and not lobed.

Common, the species and both forms occur in seagrass beds and sheltered sand flats (to 60 m) on all coasts; the species is also known from Bermuda, the Bahamas (29–61 m), Caribbean (type location: West Indies; Abbott and Huisman, 2004), and Gulf of Mexico.

Taylor 1960: p. 181; Humm and Taylor 1961: p. 369; Dawes 1974: p. 81; Littler and Littler 2000: p. 402.

Halimeda monile (J. Ellis and Solander) J. V. Lamouroux 1816: p. 306 (L. *monile*: a necklace, collar). Plate VIII, fig. 27, 28.

Basionym. *Corallina monile* J. Ellis and Solander 1786: p. 110, pl. 20, fig. c.

Halimeda monile and *H. cylindracea* Decaisne are morphologically similar and may represent cryptic species (Verbruggen et al. 2005c).

Plants are dark green, erect, 10–25 cm tall, firmly calcified, bushy, and with branches clustered above and basal parts attached by masses of rhizoids. Upper segments are cylindrical, trilobed at forks, 1–5 mm in diam., and 3–8 mm long; lower segments are flat and fused to form stipes. Utricles are in 3–5 layers; subsurface utricles are oval to clavate, 35–90 μm in diam., 23–130 μm long, and produce 2–4 surface utricles. Utricles are 25–80 μm in diam., 45–115 μm long, and in contact 0.1–0.3 of their length. Siphons are fused at joints in multiples of 2s, 3s, or more.

f. *cylindrica* (Børgesen) Collins and Hervey 1917: p. 60 (L. *cylindrica*: cylinder).

Basionym. *Halimeda incrassata* f. *cylindrica* Børgesen 1913: p. 113, fig. 90.

Synonym. *Halimeda tridens* (Ellis and Solander) Lamouroux f. *cylindrica* Collins and Hervey 1917: p. 60.

Axes are slender and sparingly branched; lower and upper segments are cylindrical and rarely fused.

Common, the species and form grow in seagrass beds and sand flats on all coasts. The species is also known from Bermuda, the Caribbean, Gulf of Mexico, and the Red Sea, while the form is known throughout the Caribbean (type location: the Virgin Islands; Børgesen, 1913).

Taylor 1960: p. 182; Dawes 1974: p. 82; Littler and Littler 2000: p. 404.

Halimeda simulans M. Howe 1907: p. 503–504, pl. 29. (L. *simulans*: similar to, imitating). Plate IX, fig. 3, 4.

Synonym. *Halimeda incrassata* (J. Ellis) Lamouroux var. *simulans* (M. Howe) Børgesen 1911: p. 144.

Species boundaries between *Halimeda simulans* and *H. borneensis* W. R. Taylor 1975 contradict genetic patterns for the couplet (Verbruggen et al. 2005c). *Halimeda simulans* was reported from the Caribbean and the Indo-Pacific (Hillis-Colinvaux 1980), while *H. borneensis* W. R. Taylor was thought to be limited to the southwest Pacific. Verbruggen et al. (2005a) found that the Indo-Pacific taxon *H. simulans* did not belong to the clade of Atlantic *H. simulans* and clustered with *H. borneensis*.

Plants are light green, erect, 10–15 cm tall, heavily calcified, and with bulbous rhizoid masses. Branching is pseudodichotomous and in one plane, with the lower 1–3 segments fused into stipes. Segments are terete at the stipe; upper segments are flat, kidney-shaped to oval or trilobed, to 15 mm wide by 2–10 mm long, and distinctly ribbed or lobed. Utricles are in 3–5 layers; subsurface utricles are swollen, 30–70 μm in diam. by 30–115 μm long, and produce 2–4 surface utricles. Utricles are 25–60 μm in diam., 25–90 μm long, apically rounded, and triangular in surface view. Internal siphons are fused into one group at joints.

Common, in seagrass beds, on peat in mangrove lagoons in the Keys, and from 70 m at the Dry Tortugas; also known from Bermuda, the Caribbean (type location, Isla de Culebra, Puerto Rico; Silva et al. 1996), and Gulf of Mexico, as well as the Indo-Pacific and Australia.

Taylor 1960: p. 180; Littler and Littler 2000: p. 408.

Pseudocodium Weber-van Bosse 1896: p. 209 (Gr. *pseudes*: false + *kodion*: fleece, sheepskin; also the green algal genus *Codium*; thus, a false *Codium*).

Wynne (2005a) places this genus in its own family.

Plants are erect, pale green coenocytes with apical growth; axes cylindrical to slightly compressed, irregularly to subdichotomously branched, and attached by rhizoidal tufts. Medullary siphons are parallel to one another and radially produce a single cortical layer of tightly adherent utricles; plastids are found in siphons and not in utricles. The genus differs from *Codium* because of the absence of cellulose plugs in central filaments, the occurrence of plastids in the cortical layer, and the presence of amyloplasts.

***Pseudocodium floridanum* Dawes and Mathieson 1972: p. 273–277, fig. 1–5 (L. *floridanum*: of or from the state of Florida). Plate IX, fig. 19–21.**

Plants are erect, dark to pale green, lax, soft, flexible, and to 2–10 cm tall; branching is dichotomous to subdichotomous, branches are cylindrical, and fronds are attached by rhizoidal clusters. The main axis has 5–20 parallel internal siphons with plastids. Branch tips have 1–6 siphons that are irregularly swollen or moniliform, 40–90 in diam., and with constrictions 0–50 μm in diam. Utricles form a compact cortex and lack plastids; they are polygonal to hexagonal in surface view and 140–340 μm in diam. Gametangia (?) occur on lateral branches, arise from siphons, are spherical to oval, 80–120 μm wide by 120–260 μm long, and have plastids.

Uncommon, in 30–60 m off the west coast (type location: 20–60 m off Tampa Bay) and 90–140 m at the Dry Tortugas (Wynne, pers. comm.); also known from the Bahamas (27–61 m; Ballantine and Aponte, 2003; Norris and Olsen, 1991), Brazil (Bravin et al. 1999), Indonesia (Verheij and Prud'homme van Reine, 1993), and Papua, New Guinea (Coppejans et al. 2001a).

Dawes 1974: p. 84.

Udoteaceae

Based on *rbcL* gene sequences, Lam and Zechman (2006) find that the families Bryopsidaceae, Caulerpaceae, Codiaceae, Derbesiaceae, and Halimediaceae are monophyletic, while the Udoteaceae is not. Molecular studies (Kooistra, 2002) showed that *Udotea* and *Penicillus* are polyphyletic, with the brushlike caps in the latter representing uncorticated blades in the former. However, siphon morphologies of vegetative plants did not suggest natural groups and further studies were recommended.

Plants are true coenocytes with siphons or filaments that lack cross walls. The siphons may or may not terminate in utricles (for example, Codiaceae) and are separate or laterally united. Cell walls are composed of xylan, with some species calcified and having an aragonite form of calcium carbonate; chloroplasts and amyloplasts are present. Mass spawning of *in situ* populations of *Penicillus* spp. and *Udotea* spp. has been described (Clifton, 1997; Clifton and Clifton, 1999).

Avrainvillea Decaisne 1842b: p. 108 (named for M. Avrainville, a French naturalist who collected in Guadeloupe, the French West Indies).

See Littler and Littler (1992) for a review of Caribbean species of *Avrainvillea*, plus Littler et al. (2004) for an account of phenotypic variability of some species (i.e., single bases versus mound-building forms) under varying habitat conditions.

Plants are noncalcified, with coenocytic siphons that interweave, form cup- or bladelike fronds, and attach to soft sediments by rhizoidal masses. Medullary (interior) siphons are yellow to brown, dichotomously branched, constricted above the forks, and often moniliform (beadlike) and tortuous (with bends). Surface siphons lack such differentiation. Erect axes may be loosely organized or unconsolidated and plumose.

Key to the species (10) of *Avrainvillea* (modified from Littler and Littler, 1992)

1. Fronds flattened and paddle-shaped 2
1. Fronds thickened, peltate, or knoblike . . . . *A. rawsonii*
 2. Siphons of blades uniformly moniliform 3
 2. Siphons of blades not uniformly moniliform 5
3. Siphons of blades with little or no tapering at surface of frond . . . . *A. nigricans*
3. Siphons of blades abruptly tapered at surface of frond 4
 4. Blade siphons gradually tapering to form a loose cortex; interior siphons of blades 60–140 µm in diameter; fronds 2–5 mm thick . . . . *A. fulva*
 4. Blade siphons tapering abruptly to form a tightly knit cortex; interior siphons of blades 45–80 µm in diameter; fronds 1–2 mm thick . . . . *A. silvana*
5. Siphons of blade cylindrical throughout, regular in shape 6
5. Siphons of blade partly moniliform and/or irregular (tortuous) in shape 7
 6. Young siphons pigmented, 30–40 µm in diam.; older ones 40–80 µm . . . . *A. mazei*
 6. Young siphons not pigmented and all siphons about the same size. . . . . *A. sylvearleae*
7. Blades thickness <2 mm 8
7. Blades thickness >2 mm . . . . *A. longicaulis*
 8. Distal part of blade having a compact cortex 9
 8. Distal part of blade lacking a compact cortex . . . . *A. geppiorum*
9. Blade base truncate but not cordate; siphons not tapering at surface. . . . . *A. elliotii*
9. Blade base cordate; siphons tapering at surface 10
 10. Blades large, to 24 cm tall, and zonately banded; bases deeply cordate, with cortical siphons 12–20 µm in diameter . . . . *A. asarifolia*
 10. Blades small, to 6 cm tall, and not zonately banded; bases weakly cordate, cortical siphons 8–12 µm in diameter . . . . *A. levis*

Avrainvillea asarifolia Børgesen 1909a: p. 34, text-fig. 4, pl. 3 (Gr. *asaron*: a low shrub; *Asarum*: genus of ginger + *folia*: leaf; thus, gingerlike leaves). Plate VII, fig. 8–10.

Blades are dull green, 10–30 cm tall, and solitary or in clusters of 3–5; stipes are 3–15(20) cm long, 6–12 mm in diam., terete below, and attached by bulbous rhizoidal masses. Medullary siphons are 35–50 µm in diam. and slightly contorted or moniliform. Blades are compressed at the base, broadly kidney-shaped, have rounded to heart-shaped bases, and range to 19 cm wide, 11 cm long, and less than 2 mm thick; surface is smooth with distinct zonal bands; interior siphons are 20–50 µm in diam. and slightly moniliform; surface siphons are tapered, contorted, and 12–20 µm in diam.

f. *olivacea* D. Littler and M. Littler 1992: p. 379, fig. 2 (L. *olivacea*: olive-green).

Cabrera and Martínez-Daranas (2005) do not recognize f. *olivacea* as it falls within the typical form (Wynne, 2005a).

Blades are olive green, to 12 cm wide and 8 cm long, thin, and with smooth to ragged margins. Multiple stipes (8–12) can arise from prostrate axes and sometimes form moundlike colonies with massive holdfasts; stipes are long, slender, and to 1 cm in diam. Interior siphons are 35–50 µm in diam. and cylindrical or slightly moniliform; surface siphons are 8–12 µm in diam. and irregular to slightly moniliform. Blade siphons are similar to those of the species.

Common, f. *olivacea* occurs in seagrass beds and protected muddy lagoons, while the species grows in sandy areas near coral reefs, and to 20 m in the Keys. The species is also known from the Bahamas, Caribbean (type location of species: Virgin Islands and for f. *olivacea*: Belize; South and Skelton, 2003), Gulf of Mexico, and Fiji.

Taylor 1960: p. 161; Dawes 1974: p. 78; Littler and Littler 2000: p. 382.

Avrainvillea elliotii A. Gepp and E. Gepp 1911: p. 35–36, 138, pl. 11, fig. 99; pl. 12, fig. 99a, 100 (named for W. R. Elliott, a British phycologist who collected seaweeds in Granada). Plate VII, fig. 11, 12.

See comments below on possible synonymy with *Avrainvillea geppiorum*.

Plants are usually solitary and to 13 cm tall; blades are bright green and on stipes attached by rhizoidal masses. Fronds are fan-shaped, tapered to the stipe, to 8 cm long, 10 cm wide, >4 mm thick, loosely woven, weakly zonate, and with ragged to smooth margins. Interior frond siphons are 20–35 µm in diam. and mostly cylindrical; surface siphons 18–30 µm in diam. and slightly moniliform to irregular. Stipes are to 1 cm in diam., 4 cm long, and cylindrical to slightly compressed; interior siphons are 30–45 µm in diam. and cylindrical; surface siphons are 20–30 µm in diam. and slightly moniliform to irregular.

Uncommon, in sandy or gravely areas (to 10 m) in the Keys; also known from the Caribbean (type location: Grenada; Silva et al. 1996) and Seychelles.

Taylor 1960: p. 162; Littler and Littler 2000: p. 384.

Avrainvillea fulva (M. Howe) D. Littler and M. Littler 1992: p. 387, fig. 7 (L. *fulva*: tawny, yellowish brown, or orange-gray yellow). Plate VII, fig. 13, 14.

Basionym. *Avrainvillea nigricans* f. *fulva* M. Howe in Collins, Holden and Setchell 1908 (1895–1919): ex. no. 1480.

Plants are gray green, to 18 cm tall, and solitary or in clusters of 2–5; stipes are to 6 cm long and attached by a massive rhizoidal system. Blades are to 12 cm long, 14 cm wide, 2–5 mm thick, loosely woven, and without zonal bands; margins are smooth in young blades and eroded in older ones. Stipes are cylindrical, compressed near blade base, and to 4.5 cm in diam. Siphons are similar in both stipes and blades, with interior siphons moniliform and 60–120 µm in diam. and surface siphons branched, 20–40 µm in diam., and moniliform to irregular.

Uncommon, on soft sediments in lagoons and to 10 m in the Keys; also known from the Bahamas (type location: Castle Island), Bermuda, Caribbean, and Gulf of Mexico.

Taylor 1960: p. 160; Littler and Littler 2000: p. 384.

Avrainvillea geppiorum Børgesen 1913: p. 87–89, fig. 71, 72, as '*geppii*' (Silva et al. 1996) (named for A. and E. Gepp, British phycologists who described several species of green algae from the Siboga Expedition to Indonesia). Plate VII, fig. 15–17.

Olsen-Stojkovich (1985), Silva et al. (1996), and Guiry and Nic Dhonncha (2006) considered this plant to be synonymous with *Avrainvillea elliotii*, while Littler and Littler (1992) recognized both taxa.

Plants are brownish green, 12–20 cm tall, and with 1–3 stipes arising from a dense rhizoidal mass. Blades are oval to wedge-shaped; they have a loose texture, are thin (1–2 mm thick), lack zonal bands, are 5–10 cm long by 8–13 cm wide, and have lacerate to deeply lobed margins. Interior siphons of blades are 40–50 µm in diam. and moniliform or cylindrical; surface siphons taper to the surface, are 20–40 µm in diam., and with slightly swollen tips. Stipes are 3–5 mm thick, to 10 cm long, tapered, and flat at base of blade; interior siphons are like blade siphons; surface siphons are smaller and 18–25 µm in diam.

Uncommon, on sandy sediments in seagrass beds and mangrove lagoons in the Keys and to 16 m in the Dry Tortugas; also known from the Antilles (type location: Maho Bay, Virgin Islands; Silva et al. 1996), and possibly the Indian Ocean.

Taylor 1960: p. 161.

Avrainvillea levis M. Howe 1905b: p. 565, pl. 23, fig. 1, pl. 26, fig. 8–10 (L. *levis:* polished, smooth, free of hairs). Plate VII, fig. 18, 19.

Synonym. *Avrainvillea lacerata* Harvey *ex* J. Agardh 1887: p. 54 auct.

Littler and Littler (1992, 2000) recognized the Pacific *Avrainvillea lacerata* and Atlantic *A. levis* as distinct species.

Plants are gray to olive green, 6–11 cm tall, solitary, and with stipes attached by massive conical bases. Blades are pyriform, 4–6 cm long, 7 cm broad, thin (~1 mm thick), tightly woven, and entire or slightly lobed; young blade margins that are smooth and with or without zonal lines. Interior siphons of blades are 25–35(40) µm in diam., and slightly moniliform to cylindrical; surface siphons are compacted, tapering, 6–12(24) µm in diam., and mostly moniliform. Stipes are cylindrical to compressed above, 0.5–5 cm long, and 2–6 mm in diam.; interior siphons are 30–40 µm in diam.; surface siphons are 8–12 µm and cylindrical to slightly moniliform.

Common, in seagrass beds in the northern Gulf of Mexico, in protected lagoons in the Keys, and from 90 m at the Dry Tortugas; also known from the Bahamas (type location: Cave Cays), and the Caribbean.

Taylor 1960: p. 162; Humm and Taylor 1961: p. 365; Dawes 1974: p. 79; Littler and Littler 2000: p. 386.

Avrainvillea longicaulis (Kützing) G. Murray and Boodle 1889: p. 70 (L. *longi*: long + Gr. *kaulos*: stem of a plant). Plate VII, fig. 20–23.

Synonym. *Rhipilia longicaulis* Kützing 1858: p. 13, pl. 28, fig. II(c,d).

Blades are 10–22 cm tall, solitary or clustered, and on stipes attached by a large rhizoidal base. Blades are oval to elongate, with rounded bases, to 6 cm long, 5–10 cm wide, 2–3 mm thick, loosely woven, faintly zonate, and hairy when dry. Interior blade siphons are mostly cylindrical, 28–70 µm in diam., and little constricted at forks; surface siphons are slightly moniliform (ca. 10 percent), taper to 15–25 µm in diam., and have many forks. Stipes are mostly terete, to 12 cm long, and 13 mm in diam.; interior siphons are terete and 38–44 µm in diam., surface siphons are 8–12 µm in diam. and moniliform or irregular.

f. *laxa* D. Littler and M. Littler 1992: p. 397, fig. 13 (L. *laxa*: loose, weak).

Plants are to 30 cm tall and often in large clumps; stipes are elongate, thin, to 26 cm long, and 1 cm in diam. Stipes are attached by rhizoidal masses and old horizontal stipes that may form moundlike colonies with holdfasts in sheltered mangrove habitats (Littler et al. 2004). Blades are elongate, to 7 cm wide, 19 cm long, less than 3 mm thick, spongy, loosely woven, and with shieldlike bases. Blade siphons are mostly cylindrical and moniliform or irregular in the lower third; interior siphons are 40–60 µm in diam. with some surface ones tapering to 8 µm in diam.

Common, the species and form grow to 4 m on soft substrata in the Keys, with the form also in mangrove lagoons. The species is also known from the Carolinas (15–32 m), Bermuda, Caribbean (type location: Antilles; Silva et al. 1996), Gulf of Mexico, the Central Pacific Islands, and Indonesia, while the form is recorded from Bermuda, Cuba, Puerto Rico, Antigua, and Belize (type location: Tobacco Range).

Taylor 1960: p. 160; Schneider and Searles 1991: p. 88; Littler and Littler 2000: p. 388.

Avrainvillea mazei G. Murray and Boodle 1889: p. 70 (named for the French phycologist H. Mazé, who with A. Schramm published "Algues de la Guadeloupe"). Plate VII, fig. 24, 25.

Plants are dark to bright green and to 15 cm tall; blades are solitary or clustered and on stipes with massive rhizoidal bases. Blades are wedgelike, to 7 cm tall, 9 cm wide, <2 mm thick, loosely woven, coarse, not zonate, and with smooth to ragged margins. Interior siphons are 40–80 µm in diam., cylindrical, and constricted above forks; surface siphons are similar but 30–40 µm in diam. Stipes are terete to slightly compressed, unbranched, to 8 cm long, and 1.2 cm in diam.; medullary siphons are 55–70 µm in diam. and cylindrical or irregular; surface siphons are 12–18 µm in diam. and slightly moniliform.

Uncommon, in protected lagoons near mangroves and to 5 m in the Keys; also known from Bermuda, the Caribbean (type location: Guadeloupe; Silva et al. 1996), Gulf of Mexico, and India.

Littler and Littler 2000: p. 388.

Avrainvillea nigricans Decaisne 1842b: p. 108 (L. *nigricans*: black). Plate VII, fig. 26–28.

Plants are often solitary, to 30 cm tall, with dark green to brownish black blades that are sessile or have a stipe; bases are bulbous with rhizoids 40–50 µm in diam. that taper to 7–12 µm at the tips. Blades are weakly zonate, wedge-shaped to oval, 6–15 cm long, to 23 cm wide, less than 2 mm thick, loosely woven, and with smooth or ragged margins. Interior siphons of blades are moniliform and 30–40(50) µm in diam.; surface siphons are tapered, 20–30 µm wide, and strongly moniliform. Stipes are cylindrical, to 8 cm long, and 1–4 mm in diam.; interior siphons are 30–60 µm in diam. and slightly moniliform; surface siphons are 10–20 µm in diam. and strongly moniliform.

f. *floridana* D. Littler and M. Littler 1992: p. 402, fig. 16 (L. *floridana*: of or from the state of Florida).

Blades are highly variable, coarse, more than 2 mm thick, lack zonal banding, and have a distinct cortex. Blade siphons are 30–40 µm in diam., strongly moniliform, and not tapered. Stipes are cylindrical, to 2 cm in diam., and 5 cm long; internal siphons are 30–60 µm in diam. and slightly moniliform; surface siphons are 10–20 µm in diam. and strongly moniliform.

Common, the species and form occur in mangrove lagoons, seagrass beds, sand flats, wave-exposed tide pools, and to 30 m in the Keys (type location of form: Content Keys). The species is also known from the Caribbean (type location: Guadeloupe; Silva et al. 1996), Brazil, and Indo-Pacific, while the form is also known from the Bahamas and Martinique.

Taylor 1960: p. 160; Dawes 1974: p. 79; Littler and Littler 2000: p. 390.

Avrainvillea rawsonii (Dickie) M. Howe 1907: p. 510, pl. 30 (named for Sir William Rawson, the British governor of the Barbados, who published a review on *Filicum* for Southern Africa). Plate VIII, fig. 1, 2.

Basionym. *Rhipilia rawsonii* Dickie 1874a: p. 151, pl. 11, fig. 1, 2 (as '*rawsoni*').

Plants are gregarious, dark brown to black, 4–12 cm tall, lack distinct blades, and are attached by large mats of rhizoids. Erect axes are flat, knob- or fingerlike, irregular, with cleft lobes, 4–12 cm long, 0.5–5 cm in diam., loosely woven, spongy, and with tops flattened to lopsided in deep water. Interior and surface filaments near tips of erect axes are 50–80 µm in diam., slightly moniliform, and with blunt to swollen tips; lower siphons are slightly moniliform and 35–75 µm in diam.

Common, in the intertidal and shallow subtidal of mangrove lagoons and limestone tidal pools; peltate forms extend to 25 m in the Keys. The species is also known from Bermuda, the Caribbean (type location: Barbados; Silva et al. 1996), Gulf of Mexico, and Madagascar.

Taylor 1960: p. 159; Littler and Littler 2000: p. 392.

Avrainvillea silvana D. Littler and M. Littler 1992: p. 408, fig. 20 (L. *silvana*: named for P. C. Silva, an American algal taxonomist who published on *Codium* and the seaweeds of the Indian Ocean). Plate VIII, fig. 3–5.

Plants are olive green, erect, to 17 cm tall, usually solitary, and attached by small fibrous rhizoidal masses. Blades are to 16 cm wide, 11 cm long, thin (1–2 cm thick), smooth, lack bandings, and have ragged to eroded margins. Lower margins of deep water plants are flat or slightly lobed and wedge-shaped in shallow water. Blade medullary siphons are moniliform and 45–80 µm in diam. Blade surface siphons are moniliform, tapered to 20 µm in diam., with rounded, hooked, or tortuous apices, and form a tight cortex. Stipes are terete, rarely branched, to 7 mm in diam., and 7 cm long; surface siphons are 15–20 µm in diam., moniliform to tortuous, and often not pigmented.

Uncommon, known in Florida only from herbarium records of the Smithsonian Institution collected from Saint Teresa (type location) and Wilson's Beach; also known on sand plains (to 117 m) from the Bahamas and Caribbean.

Littler and Littler 2000: p. 392.

Avrainvillea sylvearleae D. Littler and M. Littler 1992: p. 410, fig. 21 (named for the American phycologist Sylvia A. Earle, who studied the green and brown seaweeds of the west coast of Florida). Plate VIII, fig. 6, 7.

Plants are dark green, erect, to 6 cm tall, and solitary or in clusters; stipes are attached by small dense rhizoidal masses. Blades are 5 x 5 cm, 1 mm thick, loosely woven, oval to oblong, and faintly zonate; interior siphons are 32–50 µm in diam., smooth, and constricted at fork; surface siphons are terete and 40–48 µm in diam. Stipes are cylindrical, 1–2 cm long and 2–5 mm in diam.; internal and surface siphons of upper part of stipes are 40–50 µm in diam. and smooth to slightly moniliform; surface siphons of lower stipe portion are smaller, 8–12 µm in diam., and slightly moniliform or tortuous.

Rare, in shallow (2 m) seagrass beds at Wilson's Pier, Alligator Harbor, St. Theresa, Florida (type location); also known from offshore North Carolina (29 m) and Bermuda (Schneider and Searles, 1998).

Boodleopsis A. Gepp and E. Gepp 1911: p. 64, fig. 147–152 (L. *Boodlea*: a green algal genus named for the British phycologist L. A. Boodle + L. *opsis*: like; thus, resembling the green algal genus *Boodlea*).

Filaments are coenocytic and form loosely woven mats or cushions; branching is common, wide angled, dichotomous to irregular, and often constricted at forks.

Boodleopsis pusilla (Collins) W. R. Taylor, A. B. Joly, and Bernatowitz 1953: p. 105–106, pl. I-III (L. _pusilla_: minute, small). Plate VIII, fig. 8–10.

Basionym. *Dichotomosiphon pusillus* Collins 1909b: p. 431–432.

Cassano et al. (2004) suggest that geographic records should be reviewed and specimens compared with *Boodleopsis vaucherioidea* Calderón-Sáenz and Schnetter, which lacks regular constrictions above the forks.

Filaments are coenocytic, form bright green mats, and have irregular to dichotomous branching. Filaments are cylindrical to slightly moniliform, often strongly constricted at forks, lack septa, to 90 μm in diam., and have tips that taper to 30–45 μm in diam. Subspherical sporangial structures are rare, 60–150 μm in diam., and 83–200 μm long. Akinetes are reported from cultured material. Irregular (corolloid) masses of filaments may occur in upper parts of dense mats.

Common, forming green turflike intertidal mats in mangals on all coasts and cushions on shaded intertidal limestone in the Keys; also known from North Carolina (jetties, to 35 m), Georgia, Bermuda, the Caribbean (type location: West Indies; Silva et al. 1996), Gulf of Mexico, Brazil, Indo-Pacific, and Australia.

Taylor 1960: p. 157; Dawes 1974: p. 79; Schneider and Searles 1991: p. 89; Littler and Littler 2000: p. 394.

Cladocephalus M. Howe 1905b: p. 569 (Gr. *clado*: a branch + *cephalus*: headed).

Plants are on stalks, with or without distinct blades, noncalcified, and attached by rhizoidal mats. Interior siphons are pigmented, smooth, parallel to one another, and sparingly dichotomously branched. The cortex is distinct from the medulla; inner siphons have wide-angled dichotomous forks and outer branches form a dense cortex.

Cladocephalus luteofuscus (P. Crouan and H. Crouan) Børgesen 1909a: p. 44 (L. _luteus_: golden yellow + _fuscus_: brown, dark, dusky). Plate VIII, fig. 11, 12.

Basionym. *Flabellaria luteofusca* P. Crouan and H. Crouan in Mazé and Schramm 1878 (1870–1877): p. 88.

Synonym. *Udotea luteo-fusca* (P. Crouan and H. Crouan in Mazé and Schramm) M. Howe 1907: p. 513.

Blades are yellow green, to 12 cm tall, on stalks, and attached by large rhizoidal masses. Blades are oval, rounded, or wedge-shaped, 4.5 cm long by 7 cm wide, thin, smooth, and slightly zonate. Interior siphons of blades are pigmented and 50–80 μm in diam.; surface siphons are colorless, 4–10 μm in diam., densely interwoven, and form a distinct cortex. Stipes are often branched, 2–6 mm in diam., to 9 cm long, and flat at blade juncture; the cortex consists of interwoven, fingerlike transparent siphons and green internal siphons (these are absent in *Avrainvillea*).

Uncommon, in 2–5 m at the Content Keys; also the Antilles, western Caribbean (type location: Guadeloupe; Silva et al. 1996), Gulf of Mexico, and Seychelles.

Taylor 1960: p. 163; Littler and Littler 2000: p. 396.

Penicillus Lamarck 1813 (1813–1815): p. 297 (L. *penicillus*: a small tail, a brush).

Dragastan et al. (2003) included the fossil genus *Ovulites* J. B. Lamarck *ex* J. V. F. Lamouroux with *Penicillus*. Others (Munier-Chalmas, 1879; Johnson, 1961) noted similarities of *Ovulites* to an uncorticated *Udotea* blade with free siphons (Kooistra, 2002).

Plants are erect, superficially calcified, and have a stipe and cap of dichotomously branched free siphons that are often constricted at dichotomies, but not elsewhere. Its cap is bright green when young and whitish with age. Stipe siphons are parallel to axes, while subcortical siphons are radially disposed to form a cortex. Vegetative production of new plants can occur by swelling of rhizoidal tips ("primordia") that produce ascending and descending rhizoids, which develop into typical plants (Friedmann and Roth, 1977). The *Espera* state consists of free-standing cap filaments that grow above the sediment from rhizoids (Friedmann and Roth, 1977). *Penicillus* and species of *Halimeda* may dominate psammophytic back-reef communities (Fourqurean and Rutten, 2004).

Key to the species (4) of *Penicillus* (after Taylor, 1960 and Littler and Littler, 2000)

1. Cortical branchlets of stipe siphons ending in long pointed tips; stipe rough. 2
1. Cortical branchlets of stipe siphons ending in short flat tips; stipe smooth 3
   2. Cap filaments entangled and branching at wide angles; cap flat-topped, heavily calcified, and with siphons 150–250 μm in diameter . . . . *P. pyriformis*
   2. Cap filaments not entangled nor with branching at wide angles; caps elongate, oval, lightly calcified, and with siphons 400–800 μm in diameter. . . . . *P. dumetosus*
3. Plants heavily calcified, with cap siphons 300–500 μm in diameter . . . . *P. lamourouxii*
3. Plants lightly calcified, with cap siphons 100–300 μm in diameter . . . . *P. capitatus*.

Penicillus capitatus Lamarck 1813 (1813–1815): p. 299 (L. *capitatus*: headed, a dense head; common name, mermaid's shaving brush). Plate IX, fig. 8–10.

Plants are erect, 2–25 cm tall, dull gray green, calcified, and with a brushlike cap on a stipe attached by large rhizoidal tufts. Stipes are 2–10 cm tall, to 3 mm in diam. below and 7 mm above. Stipe surface is smooth and corticated by closely set surface siphons that are dichotomously branched 4–5 times and with blunt, flat, or rounded tips. Its cap is bushy, rounded to flat-topped, and 2–6 cm in diam. Cap siphons are slender, compact, 75–300 μm in diam. by 2–3 cm long, constricted above dichotomies, and not tapering to tips.

f. *elongatus* (Decaisne) A. Gepp and E. Gepp 1911: p. 83, fig. 166, 167 (L. *elongatus*: long, lengthened).

Basionym. *Penicillus elongatus* Decaisne 1842b: p. 97 (reprint).

Plants are tall, with stipes to 15 cm long that penetrate the caps. Caps are bushy, narrow to oblong, and composed of siphons to 300 μm in diam.

f. *laxus* Børgesen 1913: p. 98, fig. 80 (L. *laxus*: loose, lax, weak).

Stipes are to 10 cm long; caps are loosely compacted and with lightly calcified siphons that are rarely over 150 μm in diam.

Common, the species and forms occur in seagrass beds, on soft sediments in mangrove lagoons, and to 10 m in the Keys. The f. *elongatus* occurs in 2–5 m on the open coast, and f. *laxus* grows in shaded sites in the Keys. The species is also known from Bermuda, the Caribbean, Gulf of Mexico, Brazil, Mediterranean, the Madeira Archipelago, and southwest Asia (Gallardo et al. 1993), while Linnaeus (1758) lists the type location as "Habi. In Asia." The various forms occur in the Caribbean (type locality of f. *laxus*: Virgin Island lagoons) and Brazil.

Taylor 1960: p. 171; Humm and Taylor 1961: p. 371; Dawes 1974: p. 83; Littler and Littler 2000: p. 410.

Penicillus dumetosus (J. V. Lamouroux) Blainville 1834: p. 553 (L. *dumetosus*: bushy, shrubby). Plate IX, fig. 11–13.

Basionym. *Nesaea dumetosa* J. V. Lamouroux 1816: p. 259, pl. 8, fig. 3a,b.
Synonyms. *Penicillus longiarticulatus* P. Crouan and H. Crouan in Schramm and Mazé 1865: p. 44; P. Crouan and H. Crouan in Mazé and Schramm 1878 (1870–1877): p. 91.
Penicillus clavatus P. Crouan and H. Crouan in Mazé and Schramm 1878 (1870–1877): p. 91.
Dragastan et al. (2003) described fossil plants from late Pleistocene limestone in the Keys.

Plants are gray green, large (to 30 cm tall), slightly calcified, with a stipe and brushlike cap, and attached by a rhizoidal mass. Stipes are often branched, stout, to 8 cm long, 15–25 mm in diam., tapered slightly to the base, and flat on top. Stipe surfaces are smooth to spongy and have loose, dichotomously branched siphons that taper abruptly to blunt tips. Caps are coarse and longer (10–15 cm) than wide; cap siphons are 3.5–8.0 cm long, 400–800 µm in diam., coarse, and constricted at dichotomies.

f. *expansus* Børgesen 1913: p. 101, as *expansa* (L. *expansus*: wide, broad).

Stipes are short (2–4 cm long), with tall (to 16 cm) spreading caps, 14–17 cm in diam. Cap siphons are loose to entangled.

Common, the species and form occur in shallow seagrass beds and protected soft sediments on all coasts, with the form extending to 30 m. The species is also known from Bermuda, the Caribbean (type location: Antilles; Lamouroux 1816), and Gulf of Mexico.

Taylor 1960: p. 172; Dawes 1974: p. 84; Littler and Littler 2000: p. 410.

Penicillus lamourouxii Decaisne 1842b: p. 97 (named for the French phycologist J. V. Lamouroux, who published on seaweeds in the 1800s). Plate IX, fig. 14–16.

Plants are erect, strongly calcified, and to 7 cm tall; stipes have a shaggy cap and are attached by rhizoidal masses. Stipes are smooth, 3–4 cm long, 5–8 mm in diam., terete to compressed, and do not penetrate caps. Surface siphons of stipe are dichotomously branched and terminate in compact fingerlike tips. Caps are spherical to oval, 2–5 cm in diam., and 2–5 cm long; siphons are spreading or matted, 300–500 µm in diam., constricted at forks, and irregularly moniliform.

var. *gracilis* A. Gepp and E. Gepp 1905: p. 2, pl. 468, fig. 2 (L. *gracilis*: thin, graceful).

Caps of the variety are more heavily calcified, larger, and denser than the species. Its siphons are also more compact, less entangled, and twice as many as in the species, plus 300–400 µm in diam.

Common, the species and variety occur in seagrass beds (to 20 m) on all coasts mixed with *Penicillus capitatus* and *P. lamourouxii*; also known from the Bahamas (type location of species), Caribbean, and Gulf of Mexico.

Taylor 1960: p. 172; Humm and Taylor 1961: p. 371; Dawes 1974: p. 84.

Penicillus pyriformis A. Gepp and E. Gepp 1905: p. 1, pl. 468, fig. 1 (L. *pyriformis*: pear-shaped, broader at base than apex). Plate IX, fig. 17, 18.

Plants are erect, to 12 cm tall, heavily calcified, and attached by rhizoidal masses. Stipes are cylindrical, spongy, 3–10 cm tall, and 5–7 mm in diam.; cortical siphons branch loosely 2–4 times, taper to blunt points, and have thick cell walls. Caps are gray green, compact, firm, cone-shaped, often flat to indented at the top, 5–7 cm long, and 3–4.5 cm wide. Cap siphons are 2–3 cm long, 150–250 µm in diam., entangled, and constricted at wide-angled forks.

Uncommon, in sandy habitats to 30 m in the Keys; also known from Bermuda, the Caribbean, and Gulf of Mexico. Syntype locations: Bahamas, Bermuda, and Key West Florida (Gepp and Gepp, 1905).

Taylor 1960: p. 170; Littler and Littler 2000: p. 412.

Rhipidosiphon Montagne 1842a: p. 14 (Gr. *rhipis*: a fan + *siphon*: tubular, pipe).

After being merged into *Udotea* (Gepp and Gepp, 1904), the genus was resurrected by Farghaly (1980) and recognized by Littler and Littler (1990b). Farghaly (1980) distinguished the genus by its lack of lateral branching, while Littler and Littler (1990b) limited it to species with a monosiphonous partially calcified stipe, plus a monostromatic blade with free siphons.

Plants are mostly calcified, except for the monosiphonous stipe that is attached by rhizoids. Blades are fan-shaped, monostromatic, and have shieldlike bases. Blade siphons are in parallel fused calcium-carbonate sheaths constricted at their forks.

Rhipidosiphon floridensis D. Littler and M. Littler 1990b: p. 34, fig. 1–5 (L. *floridensis*: of or from the state of Florida). Plate IX, fig. 31.

Plants in Florida are to 10 mm tall and have one calcified monostromatic blade on a monosiphonous stipe that is attached by many fine hyaline rhizoids. Stipes are unbranched, noncorticated, to 1 mm long, simple, erect, smooth, and 30–220 μm in diam. Blades are shield-shaped at their bases and consist of parallel siphons that are laterally fused by calcium carbonate deposits, dichotomously branched, and constricted at their forks.

Rare, in 2–6 m at Molasses Reef in the Keys and the Dry Tortugas (type location); larger plants (3 times those from Florida) are known from the Greater Antilles and Panama (Wysor, 2004; Wysor and Kooistra, 2003).

Littler and Littler 2000: p. 420.

Rhipilia Kützing 1858: p. 12, pl. 28 (Gr. *rhipis* or *rhipidos*: fan or rush).

See Millar and Kraft (2001) for a monograph of the genus.

Plants are erect, green, noncalcified, usually with a stipe, and a peltate, funnel-like to fan-shaped blade. Blades are slightly porous, have laxly interwoven siphons, and lack a distinct cortex; siphons branch at right angles and may end in hapteral connections.

Rhipilia tomentosa Kützing 1858 p. 12, pl. 28, fig. 1 (L. *tomentosa*: woolly, covered with dense, matted hairs). Plate IX, fig. 22, 23.

Plants are solitary, with spongy blades, and attached by prostrate axes (to 180 μm in diam.) or rhizoidal tufts. Stipes are 5–10 mm long and 3–5 mm in diam. Blades are peltate when young, shield-shaped to round when older, and have faint bands and lobed or eroded margins. Blade and stipe siphons are similar, 30–70 μm in diam., thin walled, loosely interwoven, and equally constricted at dichotomous forks, but unequally elsewhere. Lateral siphons are common, 50–200 μm long, often septate at the bases of forks, and terminate in 2–3 haptera or spinelike projections.

f. *zonata* A. Gepp and E. Gepp 1911: p. 55, fig. 129 (L. *zonata*: banded).

Stipes are more slender than the species, and blades are round to kidney-shaped, very thin, and often zonate.

Common, the species and form occur in tide pools on wave-exposed sandy reefs and to 30 m in the Keys; the species is also known from the Bahamas, Antilles (type location; Silva et al. 1996), southern Caribbean, Brazil, Canary Islands, and Seychelles.

Taylor 1960: p. 162; Littler and Littler 2000: p. 412.

Rhipocephalus Kützing 1843: p. 311 (Gr. *rhipis*: a fan + *cephalus*: headed).

Plants are calcified, erect, have a stipe, and are attached by a loose mass of rhizoids. Stipes have a cap of fused or free siphons arranged in the same plane; if fused they appear as small blades.

Key to the species (2) of *Rhipocephalus*

1. Cap filaments not laterally fused to form small blades . . . . *R. oblongus*
1. Cap filaments laterally fused to form small blades. . . . . *R. phoenix*

Rhipocephalus oblongus **(Decaisne) Kützing 1849: p. 506 (L. *oblongus*: elliptical with blunt ends). Plate IX, fig. 24–26.**

Basionym. *Penicillus oblongus* Decaisne 1842b: p. 10.

Plants are calcified, erect to 13 cm tall, and dark to white green; stipes have a flat to tufted cap of siphons and a rhizoidal mass. Stipes are stout, terete, to 6 mm in diam., 4–7 cm long, and have a smooth porous surface. Internal siphons are dichotomously divided 5–7 times, while surface siphons end in a loose cortex of short, fingerlike tips with thin cell walls. Caps are filamentous, conelike to elongate, 5–15 cm long, 1–3 cm in diam., and with rounded tops. Cap siphons are calcified and do not fuse or form blades. Siphons are 200–350 μm in diam. at the base, 160–200 μm wide above, 7–15 mm long, dichotomously branched with constrictions at forks, and all in one plane. The species can be distinguished from *Penicillus capitatus* by its clusters of free cap siphons aligned in the same plane, the V-shaped rather than U-shaped cap base, and the thin tip walls of surface siphons.

Uncommon, in seagrass beds, on limestone and from 40 m in the Keys; also known from the Bahamas (type location; Decaisne 1842b) and the Caribbean.

Taylor 1960: p. 173; Littler and Littler 2000: p. 416.

Rhipocephalus phoenix **(J. Ellis and Solander) Kützing 1843: p. 311; 1849: p. 506 (L. *Phoenix*: the genus for date palm that has spreading leaves). Plate IX, fig. 27–30.**

Basionym. *Corallina phoenix* J. Ellis and Solander 1786: p. 126, pl. 25, fig. 2, 3.

Synonyms. *Penicillus phoenix* Harvey 1858: p. 46, pl. XLIII.C.

Udotea plumula P. Crouan and H. Crouan in Schramm and Mazé 1865: p. 43 (as *Udotea plumula* var. *plumula*).

Udotea plumula var. *curta* P. Crouan and H. Crouan in Schramm and Mazé 1865: p. 4.

Udotea phoenix (J. V. Lamouroux) *sensu* P. Crouan and H. Crouan in Mazé and Schramm 1878 (1870–1877): p. 87.

Plants are pale to dark green, erect, 7–12 cm tall, light to moderately calcified, and anchored by a rhizoidal mass. Stipes are 2–10 cm long, 3–8 mm in diam., cylindrical, smooth, and have a cap of small blades. Surface siphons are compact, radially arranged, with thick cell walls and blunt tips, and appear fingerlike. Caps are oval and 3–7 cm long; blades are minute, 2–5 cm long, flattened, and monostromatic. Blade siphons are laterally fused, parallel, and 200–250 μm in diam. below; siphons are 50–100 μm in diam. above and mostly constricted at forks after the first 1–2 dichotomies.

f. *brevifolius* A. Gepp and E. Gepp 1905: p. 4; 1911 p. 95, pl. 31, fig. 184–186 (L. *brevi*: short, abbreviated + *folius*: blade, foliage).

Caps are cylindrical, 1.5 cm in diam., 5 cm long, and have small, closely packed, shinglelike blades 0.6–2 cm long. Siphons are cylindrical and laterally fused.

f. *longifolius* A. Gepp and E. Gepp 1905: p. 4, pl. CCCCLXVIII, fig. 3; 1911: p. 95, pl. 31, fig. 187, 188 (L. *longi*: long, elongated + *folius*: blade, foliage).

Caps are 3–4 cm long and wide and consist of small, loosely diverging blades 10–15 mm long; siphons are cylindrical and laterally fused.

Common, the species and forms occur in seagrass beds, on sand, to 10 m in the Keys, and in 20–60 m off the west coast; also known from the Bahamas (type location; J. Ellis and Solander 1786), Caribbean, and Gulf of Mexico.

Taylor 1960: p. 174; Dawes 1974: p. 84; Littler and Littler 2000: p. 418.

Udotea J.V. Lamouroux 1812: p. 186; 1813: p. 27 (Gr. *Udotea*: named for an ocean nymph from Greek mythology).

Farghaly (1980) divided *Udotea* into four genera, an existing genus *Rhipidosiphon* and three others: *Decaisnella* Fabre, *Geppina* Farghaly, and *Howella* Farghaly. Silva et al. (1996) considered the new names invalid without Latin diagnoses. Littler and Littler (1990a) retained *Rhipidosiphon*. See previous comments regarding the similarities of the fossil genus *Ovulites* to an uncorticated *Udotea* blade with free siphons (Kooistra, 2002).

Plants are erect, lightly to heavily calcified, with a fan- or funnel-shaped blade and an unbranched stipe attached by rhizoids. Stipes and blades have dichotomously branched siphons constricted at the forks and elsewhere; stipe and blade siphons have lateral branches or projections that form a distinct cortex.

Key to the species (12) of *Udotea* (modified from Littler and Littler, 2000)

1. Blades have many layers of siphons 2
1. Blades have one layer of siphons . . . . *U. unistratea*
    2. Blade siphons lack lateral appendages 3
    2. Blade siphons have lateral appendages 8
3. Appendages of stipe surface siphons have pointed tips 4
3. Appendages of stipe surface siphons have blunt, flattened, or swollen tips 5
    4. Blades thin and delicate; margin has one layer of siphons . . . . *U. looensis*
    4. Blades thick and leathery; margin with many layers of siphons . . . . *U. luna*
5. Blades typically flat and fan-shaped 6
5. Blades cup, funnel, or scoop-shaped . . . . *U. cyathiformis*
    6. Siphons of blades <70 μm in diam. and firmly adherent; blade not fibrous 7
    6. Siphons >85 μm in diam. and loosely adherent; blade fibrous . . . . *U. abbottiorum*.
7. Blades distinctly zonate; siphons have many irregular forks . . . . *U. conglutinata*
7. Blades weakly zonate; siphons straight, subparallel, and with few forks . . . . *U. caribaea*
    8. Lateral appendages of blade siphons not forming a continuous cortex 9
    8. Lateral appendages of blade siphons forming a continuous cortex 10
9. Blades often numerous and radiating from a central axis; lateral appendages of siphons terminate in rounded or flattened tips . . . . *U. wilsonii*
9. Blades single or overlapping in one plane; all surface appendages of blade siphons are pointed . . . . . *U. spinulosa*

10. Lateral appendages of blade numerous, often opposite and closely aligned 11
10. Lateral appendages of blade random, either sparse or scattered . . . . *U. flabellum*
11. Lateral appendages of blade siphons unbranched; tips with swollen lobes that have 6–12 small knobs and concave tips. . . . . *U. occidentalis*
11. Lateral appendages of blade siphons branching widely; tips not swollen but often rounded .*U. dixonii*

Udotea abbottiorum D. Littler and M. Littler 1990a: p. 210, fig. 1 (named for Isabella and Donald Abbott; the former an American specialist on red algal taxonomy and author of seaweed floras of California and Hawaii; the latter a zoologist who studied tunicates). Plate IX, fig. 32, 33.

Plants are erect and 10–15 cm tall; blades are yellow green and fan-shaped, and on narrow stipes attached by a bulbous mass of rhizoids. Stipes are cylindrical, 4–7 cm long and 2–3 mm in diam.; interior siphons are 70–100 µm in diam. and constricted above forks; surface siphon tips are swollen in upper parts and truncated below. Blades are calcified, fan-shaped, to 8 cm long, 5 cm wide, uncorticated, 1–3 mm thick, and spongy. Blade siphons are parallel, 90–110 µm in diam., lack appendages, have wide-angled forks, and are covered with a porous sheath of calcium carbonate. The species is similar to *Udotea conglutinata*, except that blade siphons are larger and stipe siphons are blunt and swollen or flat.

Rare, to 5 m on sand or limestone at Content Keys (type location); also known from Bermuda, the Bahamas, and Caribbean.

Littler and Littler 2000: p. 422.

Udotea caribaea D. Littler and M. Littler 1990a: p. 211, fig. 2. (L. *caribaea*: of or from the Caribbean Sea). Plate X, fig. 1, 2.

Plants are erect and to 9 cm tall; blades are wide, fan-shaped and heavily calcified; stipes are 2–5 mm in diam., to 2.5 cm long, extend into the blade, and are attached by a tangled rhizoidal mass. Interior stipe siphons are 60–80 µm in diam.; surface siphons are stubby and fingerlike. Blades are to 10 cm wide, 7 cm long, 0.5–1.0 mm thick, and lack a cortex; blade siphons are 30–50 µm in diam. and tightly compact.

Rare, in protected areas near mangrove islands in the Keys and to 7 m; also known from the Caribbean (type location: Tobacco Range mangrove islands, Belize) and India.

Littler and Littler 2000: p. 422.

Udotea conglutinata (J. Ellis and Solander) J. V. Lamouroux 1816: p. 312 (L. *conglutinata*: glued together, united firmly). Plate X, fig. 3, 4.

Basionym. *Corallina conglutinata* J. Ellis and Solander 1786: p. 125, pl. 25, fig. 7.

Previously, all *Udotea* plants with fan-shaped blades and siphons that lacked lateral appendages were referred to *U. conglutinata*.

Plants are erect, to 14 cm tall, and heavily calcified; stipes produce one blade and are attached by a bulbous rhizoidal tuft. Stipes are smooth, firm, expanded and flat near blades, to 2.5 cm long and 5 mm in diam. near base; interior siphons are 50–70 µm in diam. and constricted above forks; surface siphon tips are fingerlike. Blades are to 12 cm long, 9 cm wide, 1–2 mm thick, and feltlike. Blade siphons are heavily calcified, 40–60 µm in diam., repeatedly divided, and contorted.

Littler and Littler (1990a) noted that *Udotea conglutinata* resembled *U. caribaea*, but the former had a thicker blade (1.2–2.0 mm) than the latter (0.5–1.0 mm). However, Florida specimens we have examined show overlap with regard to siphon diameter and morphology. Common, growing in seagrass beds and to 60 m on all coasts; also known from Bermuda, the Bahamas (type locality; Dixon, 1960), the Caribbean, and Brazil.

Taylor 1960: p. 165; Dawes 1974: p. 84; Littler and Littler 2000: p. 422.

Udotea cyathiformis Decaisne 1842b: p. 106 (L. *cyath*: cup + *formis*: shaped). Plate X, fig. 5–7.

Plants are erect, calcified, and to 8 cm tall. Blades are usually funnel-shaped and with stipes attached by a bulbous rhizoidal mass. Stipes are cylindrical, to 2 cm long, and 4 mm in diam.; internal siphons are 50–120 µm in diam. with unequal constrictions above forks; surface siphons branch irregularly and have swollen to flat tips. Blades are delicate funnels, split or peltate, fibrous to papery, and 0.5–1.0 mm thick; siphons are 30–70 µm in diam., parallel to woven, and have equal constrictions above their forks.

f. *infundibulum* (J. Agardh) D. Littler and M. Littler 1990a: p. 216 (L. *infundibulum*: a funnel).

Basionym. *Udotea infundibulum* J. Agardh 1887: p. 71.

Blades are thin, 0.5–1.0 mm thick, papery, funnel-shaped, to 8 cm tall, and with monostromatic margins. Stipes are 1–4 mm in diam. and 0.3–2.0 cm long; siphons are parallel, laterally joined at edges, and 30–70 µm in diam.

f. *sublittoralis* (W. R. Taylor) D. Littler and M. Littler 1990a: p. 216, fig. 6. (L. *sublittoralis*: under the sea).

Basionym. *Udotea sublittoralis* W. R. Taylor 1928: p. 91, pl. 8, fig. 16; pl. 9, fig. 8, 9.

Udotea sublittoralis represents a continuum with *U. cyathiformis* and is reduced to a form.

Blades are tough and 1–2 mm thick; siphons are 60–80 µm in diam., interwoven, not parallel, and have wide forks

var. *flabellifolia* D. Littler and M. Littler 1990a: p. 220, fig. 7 (L. *flabellum*: a small fan + *folia*: blade).

Blades are fan-shaped, not cup- or funnel-like, with siphons 30–70 µm in diam. Stipe siphons are 50–120 µm in diam.; surface siphon tips are swollen to flat.

Common, the species and subspecific taxa occur in seagrass beds and mangrove lagoons in the Keys (type location for f. *sublittoralis*: Dry Tortugas) and from 20–30 m on the west and east coasts; also known from North Carolina, Bermuda, the Caribbean (type location of species: Iles des Saintes, Guadeloupe; of var. *flabellifolia*: San Blas Islands, Panama; of form *infundibulum*: West Indies), the Gulf of Mexico, and Brazil, as well as the Indo-Pacific.

Taylor 1960: p. 166; Dawes 1974: p. 86; Schneider and Searles 1991: p. 90; Littler and Littler 2000: p. 424.

Udotea dixonii D. Littler and M. Littler 1990a: p. 220, fig. 8 (named for the British-American phycologist Peter Dixon, who studied red seaweeds). Plate X, fig. 8–10.

Plants are erect, to 21 cm tall, and solitary or in clusters; stipes have a flat blade and are attached by rhizoidal masses. Stipes are 2–3 mm in diam. and 2–6 cm long; internal siphons are 30–60 µm in diam. and not constricted at forks; surface siphons end in dichotomously divided, fingerlike tips. Blades are longer than wide, 6–12 cm long, 4–10 cm wide, 0.2–0.8 mm thick, flat and lobed, with distinct concentric bands, a smooth surface, and 2–3 layers. Blade siphons are 20–35 μm in diam., cylindrical to slightly moniliform, and unequally constricted above dichotomies; surface siphon tips are swollen, knoblike, and interlocked to form a smooth compact cortex.

Rare, reported from 38 m off the east coast (Hanisak and Blair, 1988); also known from the Caribbean (type location: Curlew Cay, Belize) and Gulf of Mexico.

Littler and Littler 2000: p. 426.

Udotea flabellum (J. Ellis and Solander) M. Howe 1904: p. 94 (L. *flabellum*: fanlike). Plate X, fig. 11–13.

Basionym. *Corallina flabellum* J. Ellis and Solander 1786: p. 124–125, pl. 24.

Plants are erect and to 30 cm tall; blades are heavily calcified and gray green in color; stipes are attached by massive rhizoidal bases. Stipes are 2–4 cm long, 5–7 mm in diam., and cylindrical below; internal siphons are 20–80 μm in diam. and not constricted above forks; surface siphons are irregular, compact, and end in clusters of rounded tips. Blades are multilayered, tough, leathery, smooth, calcified, and with distinct zonal bands; older blades are highly dissected with overlapping lobes. Blade siphons are 30–50 μm in diam. and lack constrictions; surface siphons are crowded and terminate in short, rounded, dichotomously branched tips.

Common, to 10 m on all coasts; taller plants occur in areas of low light; also known from North Carolina, Bermuda, the Bahamas (29–31 m), Caribbean (type location: West Indies; Silva et al. 1996), Gulf of Mexico, Brazil, Indo-Pacific, and Australia.

Taylor 1960: p. 168; Humm and Taylor 1961: p. 372; Dawes 1974: p. 86; Schneider and Searles 1991: p. 90; Littler and Littler 2000: p. 428.

Udotea looensis D. Littler and M. Littler 1990a: p. 232, fig. 15 (named for the type locality, Looe Key in the Florida Keys). Plate X, fig. 14, 15.

Plants are erect and to 7 cm tall; blades are simple to fan-shaped, dark green, and lightly calcified; stipes are attached by small rhizoidal masses. Stipes are to 2 mm in diam., 8 mm long and widest at base of blade; internal siphons are 45–90 μm in diam.; surface siphons divide repeatedly and have long, tapering tips with equal constrictions above forks. Blades are monostromatic, 5–7 cm wide, and 3–4 cm long; margins are often torn or ragged; siphons are similar to those of the stipe; surface appendages are absent. The species resembles *Udotea spinulosa*, except that its blade siphons lack lateral appendages.

Rare, from 5–30 m on sand and coral rubble at Looe Key (type location) and Molasses Reef in the Keys; also known from the Bahamas (29–31 m) and Greater Antilles as well as the Caribbean, including Mexico (Mendoza-Gonzáles et al. 2000).

Littler and Littler 2000: p. 428.

Udotea luna D. Littler and M. Littler 1990a: p. 232, fig. 16 (L. *luna*: the moon). Plate X, fig. 16, 17.

Plants are erect, to 7 cm tall, and lightly calcified; blades are dark to light green and on stipes with tangled rhizoidal masses. Stipes are 1–3 mm in diam., to 1 cm long, and very fibrous near blade base; interior siphons are 60–90 μm in diam., smooth, and unequally constricted above forks; surface siphons end in acute tips. Blades are crescent-shaped, broader than long, to 6 cm wide and 4 cm long, to 1 mm thick, with faint bandings, and ragged, often irregular margins. Blade siphons are similar to those of the stipe and have more acute tips near the base. The species is similar to *U. looensis* except that it has many layers of siphons near the margin.

Rare, on sand-covered limestone at Content Keys (type location); also known from the Bahamas (23 m) and Caribbean.

Littler and Littler 2000: p. 428.

Udotea occidentalis A. Gepp and E. Gepp 1911: p. 127, 144, pl. 2, fig. 18, 22a,b; pl. 7, fig. 63–65. (L. *occidentale*: west, western, or occidental). Plate X, fig. 18–20.

Plants are to 10 cm tall; blades are gray green, heavily calcified, and on stipes attached by small rhizoidal masses. Stipes are simple, cylindrical, 1–2 cm long, and 2 mm in diam.; interior siphons are 40–70 μm in diam. and unequally constricted above forks; surface siphons are dichotomously branched and end in stubby, rounded projections. Blades are to 8 cm long and wide, oval to fan-shaped, papery, with shield-shaped bases and often marginal and surface proliferations. Internal blade siphons are 20–50 μm in diam., unequally constricted above with few forks; surface siphons are compact and end in knobs or lobes that form a puzzlelike surface. All surfaces are covered with minute spines visible with scanning electron microscopy. The species is similar to *Udotea flabellum*, but has thinner blades, shield-shaped bases, and siphons with minute spines.

Rare, in seagrass beds at the Content Keys; also known from Bermuda, the Caribbean (type location: St. Thomas, Virgin Islands; Littler and Littler, 1990a), Gulf of Mexico, Brazil, and the Philippines.

Taylor 1960: p. 169; Littler and Littler 2000: p. 430.

Udotea spinulosa M. Howe 1909: p. 97, pl. 4, fig. 2, pl. 8, fig. 1–7 (L. *spinulosa*: minutely, somewhat, or weakly spiny). Plate X, fig. 21–24.

Plants are erect, to 8 cm tall; blades are green to white green, heavily calcified, and on stipes that are 1–2 cm long, 1–2 mm in diam., cylindrical below, flat at blade base, and attached by a rhizoidal mass. Internal stipe siphons are 80–120 μm in diam. with constrictions equally distributed above the forks; surface siphons branch and end in long, pointed projections. Blades are 4–6 cm in diam., less than 0.8 mm thick, incompletely corticated, and often have weak vertical striations running from the stipe to its margins. Internal blade siphons are 60–90 μm in diam. and equally constricted at forks; surface siphons are branched, 50–150 μm in diam., and end in 1–8 spinelike surface appendages.

f. *palmettoidea* A. Gepp and E. Gepp 1911: p. 125, pl. 6, fig. 56 (L. palma: the palm tree + *oidea*: like; thus, palmlike).

Blades are 4–6 cm wide, 5–6 cm long, with faint vertical striations running to the stipe; surface siphons are few, 50–150 μm in diam., cylindrical, and topped by 1–8 prongs. Stipes are 1–2 cm long; surface siphons are repeatedly divided into long, pointed, fingerlike tips. Unlike the species, blades of the form have few and poorly formed siphon appendages.

Uncommon, in seagrass beds and from 9–90 m in the Keys; also known from the Bahamas (type location of species: Bimini Harbor; Silva et al. 1996), Antilles (type location of f. *palmettoidea*: St. Thomas, Virgin Islands; Littler and Littler, 1990a), Gulf of Mexico, Brazil, and the Seychelles.

Taylor 1960: p. 167; Littler and Littler 2000: p. 430, 432.

Udotea unistratea **D. Littler and M. Littler 1990a: p. 240, fig. 21 (L. *uni*: one + *strata*: layer). Plate X, fig. 25, 26.**

Plants are to 7 cm tall; blades olive green, sessile or with a stipe, and attached by fine fibrous rhizoids. Stipes are to 1 mm in diam. and 1–5 cm long; interior siphons are 80–130 μm in diam. and unequally constricted above forks; surface siphons branch dichotomously and end in short, simple or lobed, blunt tips. Blades are monostromatic, lack a cortex, are 0.1–0.4 mm thick, curved or flat, and have faint concentric lines; siphons are 100–240 μm in diam. and fused laterally.

Rare, reported from Florida with no specific location (Littler and Littler, 2000); also known from the Bahamas, Caribbean, Mexico (Mendoza-Gonzáles et al. 2000), and Belize (24–46 m; type location Carrie Bow Cay).

Littler and Littler 2000: p. 432.

Udotea wilsonii **A. Gepp, E. Gepp and M. Howe in A. Gepp and E. Gepp 1911: p. 130–131, 144–145, pl. VII, fig. 66, pl. VIII, fig. 67a, 68a: as '*wilsoni*' (named for the British phycologist P. Wilson). Plate X, fig. 27–29.**

Plants are erect, 4–13 cm tall; blades are dark to gray green, lightly calcified, with many fronds radiating from central axes, and on a stipe that is attached by fibrous masses of rhizoids. Stipes are 1–2 mm in diam., 1–4 cm long, and widest near the blade base; interior siphons are 50–100 μm in diam. and equally constricted above forks; surface siphons end in short, dichotomous, fingerlike branchlets. Blades are fan-shaped, 10 cm wide, 8 cm long, 1–2 mm thick, and multilayered except at margins; interior siphons are 40–80 μm in diam. and have a few unequal constrictions; surface siphons are numerous, short (25–90 μm long), stubby, blunt, and simple or forked. The species has many blades that radiate from a central axis.

Rare, on sand plains (18 m) in the Keys; also known from the Bahamas (type location: North End, Anguilla Cays; Silva et al. 1996), Caribbean, Gulf of Mexico, and Mozambique.

Taylor 1960: p. 167; Littler and Littler 2000: p. 434.

Sedis incertae: Smithsoniella Sears and Brawley 1982: p. 1451–1452 (named for the Smithsonian Institution in Washington, D. C., where it was found in a coral reef microcosm).

Burrows (1991) included *Smithsoniella earleae* (as *Pilinia earliae* Gallagher and Humm) in her description of *Pilinia*. See also Leonardi et al. (2002) for synonymy.

Plants are heterotrichous, uniseriate filaments with apical growth; the prostrate filaments produce erect, sparsely branched axes to 5 mm in length. Cells of erect and prostrate axes are irregularly cylindrical, longer than wide, uninucleate, have a parietal, perforated plastid with pyrenoids, and a refractive plug in the cross wall. Sporangia are ovoid, have a basal plug, and occur in secund series on erect filaments.

Smithsoniella earleae **(Gallagher and Humm) Sears and Brawley 1982: p. 1451–1461, fig. 1–21; 1983: p. 962 (named for the American phycologist Sylvia A. Earle, who studied the green and brown seaweeds of the west coast of Florida). Plate X, fig. 30–32.**

Basionym. *Pilinia earleae* Gallagher and Humm 1980: p. 532–534, fig. 1–15.

Filaments are short, 1–3(5) mm tall, and uniseriate; prostrate filaments are attached by occasional rhizoids and form a pseudoparenchymatous layer. Erect filaments are cylindrical, not constricted at cross walls, and have many ovoid, sessile sporangia, 25 µm in diam. by 40 µm long. Cells are 10–14 µm in diam., 45–75 µm long, and uninucleate; a single parietal plastid covers the cell and contains 3–12 pyrenoids.

Uncommon and easily overlooked, collected from a canal in Boca Ciega Bay on the west coast (type location: Treasure Island) and on tunicates, mollusks, and coarse seaweeds in Tampa Bay, also known from the Lesser Antilles and Coral Reef Microcosm of the Smithsonian Institution.

DASYCLADALES

See Berger and Kaever (1992) and Zechman (2003) for overviews of their phylogeny.

Plants initiate growth with a long uninucleate or multinucleated axial cell attached by rhizoidal outgrowths. The initial cell produces whorls of simple or forked branchlets that have limited growth. Plants are radially symmetrical, superficially calcified or not, and produce aplanospores on their branchlets or rays.

Dasycladaceae

The family has the same features as the order.

Batophora J. Agardh 1854: p. 107–108 (Gr. *bato*: a spine, prickly bush + L. *phora*: to carry).

See Gómez-Poot et al. (2002) for a review of vegetative and taxonomic features of the two Gulf of Mexico species.

Plants are uninucleate, with an axial cell to 10 cm long, noncalcified when vegetative, and attached by basal rhizoidal outgrowths. Erect axial cells have whorls of branches that divide 1–7 times. Gametophores produce several gametangia at each node near their branch bases, with these being oblong and dark green to black.

Key to the species (2) of *Batophora*

1. Plants to 6 cm tall; branchlets whorled every 1 mm or less; whorls to 400 µm in diam.; gametangia spherical. . . . . *B. occidentalis*
1. Plants to 14 cm tall; branchlets whorled every 1–2 mm, whorls to 750 µm in diam.; gametangia ovoid . . . . *B. oerstedii*

***Batophora occidentalis* (Harvey) Berger and Kaever *ex* M. J. Wynne 1998a: p. 108 (L. *occidentalis*: west, western, or occidental). Plate XI, fig. 1, 2.**

Basionym. *Dasycladus occidentalis* Harvey 1858: p. 38.

Synonyms. *Batophora oerstedi* J. Agardh 1854: p. 108 var. *occidentalis* (Harvey) M. Howe 1905b: p. 579.

Coccocladus occidentalis (Harvey) C. Cramer var. *laxus* M. Howe 1904: p. 95, pl. VI, fig. 1, 2.

Plants are bright green soft cylinders, 2–4 mm in diam., 2–4(6) cm tall, usually clustered, and attached by rhizoidal outgrowths. Axial cell are elongate, naked below, and covered above with whorls of di- to trichotomous branchlets that divide 1–7 times, have lobed outgrowths at their base, and end in deciduous hairlike divisions. Gametophores are on the inner 1–4 forks of branchlets, ovoid to elliptical, 325–450 by 500–1,000 µm, and with several per node. Gametangia are spherical and 40–70 µm in diam.

var. *largoensis* (J. Prince and S. Baker) Berger and Kaever *ex* M. J. Wynne 1998a: p. 108 (L. *largoensis*: of or from Key Largo in the Florida Keys).

Basionym. *Batophora largoensis* J. Prince and S. Baker 1984: p. 323, fig. 1–10.

Gómez-Poot et al. (2002) did not recognize this variety.

Plants are 3–6 cm tall. The axial cells have loose whorls of 8–13 branchlets that are dichotomously forked 6–8 times, with the final 3–4 divisions hairlike and deciduous. Gametophores are reddish brown, spherical, 500–800 µm in diam., and occur in clusters of 3–4 near the bases of branchlets. Gametangia are ovoid to oblong and 40–60 µm in diam.

Common, the species grows on shells and limestone in lagoons and open coastal habitats throughout the state; the variety occurs on loggerhead sponges and shells on the east side of Key Largo (type locality). The species is also known from Bermuda, the Caribbean (syntype locations: Key Biscayne and Key West, Florida; Harvey, 1858), Gulf of Mexico, and Canary Islands.

Taylor 1960: p. 98; Littler and Littler 2000: p. 434.

***Batophora oerstedii* J. Agardh 1854: p. 108 (named for Anders S. Ørsted, a Danish botanist). Plate XI, fig. 3–6.**

Synonyms. *Dasycladus conquerantii* P. Crouan and H. Crouan in Schramm and Mazé 1865: p. 47; P. Crouan and H. Crouan in Mazé and Schramm 1878 (1870–1877): p. 108.

Coccocladus occidentalis (Harvey) C. Cramer var. *conquerantii* (P. Crouan and H. Crouan in Schramm and Mazé) M. Howe 1904: p. 96.

Plants often grow in bright green clusters that leak a bright yellow stain if crushed; axes are erect, soft, 4–6(14) cm tall, 4–10 mm in diam., and attached by lobed basal growths. Axial cells are naked below; above they produce whorls of 8–10 monosiphonous branches every 1–2 mm. Initial branches are to 750 µm in diam., dichotomously forked, with the last two divisions hairlike and with forked tips. Gametophores are spherical, 330–700 µm in diam., 550–1000 µm long, bright green yellow, and on the inner 1–4 forks with several at each node. Gametangia are ovoid to elliptical and 40–75 µm in diam.

Common, on shells, limestone, mangrove roots, and algae (to 8 m) on all coasts; sometimes mixed with the freshwater green alga *Spirogyra* in roadside ditches with spring-fed waters from the Weeki Watchee River; also known from Bermuda, the Caribbean (type location: Virgin Islands), and Gulf of Mexico.

Taylor 1960: p. 98; Humm and Taylor 1961; p. 375; Dawes 1974: p. 94; Littler and Littler 2000: p. 436.

Cymopolia J.V. Lamouroux 1816: p. 292 (named for *Cymopolia*, the daughter of Neptune in Roman myths; L. *cyma*: surf, a wave, or sprout + *polos*: gray).

Plants are erect, bushy, calcified below, and attached by a rhizoidal holdfast. Tips are noncalcified and terminate in bright green tufts of delicate hairs. The multinucleate axial cell is constricted at regular intervals and has whorls of cortical branches. Basal cells of branches produce clusters of 4–8 distal club-shaped cells that adhere and form a smooth, calcified surface and terminal sporangia.

***Cymopolia barbata* (Linnaeus) J. V. Lamouroux 1816: p. 293. (L. *barbata*: bearded, with tufts of long hairs). Plate XI, fig. 7–9.**

Basionym. *Corallina barbata* Linnaeus 1758: p. 806.

Plants are prostrate or 10–30 cm tall; axes are terete, multinucleate, segmented, heavily calcified except at joints, and attached by dense clusters of rhizoids from lower segments. Axial tips produce whorls of bright green apical tufts of noncalcified filaments. Filaments are forked 5–6 times, with the last 2–3 forks hairlike, colorless, and deciduous; surface cells are 115–140 μm in diam. Segments are rigid, 2–4 mm in diam., 3–12 mm long, and whitened with carbonate. Gametangia are solitary, spherical to pear-shaped, 150–200 μm in diam., and in terminal forks of branches at surface.

Although uncommon, a "bloom" of the species was noted in July 2006 in 5 m off the Content Keys (Walt Avery, pers. comm.); it grows on limestone in tide pools and to 5 m in the Keys; also known from Bermuda, the Caribbean, Gulf of Mexico, and Canary Islands, with its type location listed as "Hab. Oc. Americana" (Linnaeus, 1758). Dragastan et al. (2003) reported fossil specimens from late Pleistocene limestone in the Florida Keys.

Taylor 1960: p. 102; Littler and Littler 2000: p. 436.

Dasycladus C. Agardh 1828 (1820–1828): p. 15 (L. *dasycladus*: shaggy-branched).

Plants are small, gregarious, spongy, noncalcified, olive to dark green, and attached by lobed basal outgrowths. Axial cells are elongate, initially uninucleate, and have closely packed whorls of compound determinate branches. Sporangia are solitary, on lower portions of branches, and blacken when mature.

Dasycladus vermicularis **(Scopoli) Krasser in Beck and Zahlbruckner 1898: p. 459, fig. 8 (L.** ***vermicularis*****: wormy, worm-shaped). Plate XI, fig. 10, 11.**

Basionym. *Spongia vermicularis* Scopoli 1771: p. 412, pl. 64, fig. 1454.

Plants usually are "furry" and grow in soft, green clusters; axial cells are 2–6 cm tall, 3–6 mm in diam., and attached by rhizoidal outgrowths. Axial cells are naked below; above, they produce 10–15 densely packed whorls of branches spaced 100–150 μm between whorls. Branches arise from basal cells and fork dichotomously 3–4 times, with final forks ending in 3 spinelike terminal cells. Sporangia are single, in the axils of branches, spherical, 500–800 μm in diam., and bright yellow green. Wound repair involves an initial rapid gelling of a lectin-carbohydrate ligand system followed by an oxidative burst and catechol localization in wound plugs (Ross et al. 2005a,b).

Common, on limestone and coral rubble in tide pools of wave-exposed shores and to 55 m in the Keys; also known from Bermuda, the Bahamas, Caribbean, Gulf of Mexico, Brazil, the Mediterranean, eastern tropical Atlantic, and Pacific Ocean. The type location is cited by Scopoli (1772) as "Habitat ibidem super lapides et conches" with no further details.

Taylor 1960: p. 99; Humm and Taylor 1961: p. 376; Littler and Littler 2000: p. 436.

Neomeris J.V. Lamouroux 1816: p. 241. (Gr. *neos*: new + *meris*: part; *meros*: apart).

Plants are small, gregarious, calcified, cylindrical to club-shaped, and their axial cell is attached by rhizoidal outgrowths. Axial cell tips have bright green filaments or tufts; lower parts are covered by dense whorls of branches arising from basal cells that end in two club-shaped branchlets with deciduous hairs. Sporangia are calcified and on branches.

Neomeris annulata **Dickie 1874b: p. 198. (L.** ***annulata*****: ringed, marked with rings). Plate XI, fig. 12, 13.**

Plants are small, 1–3 cm tall, 1–2 mm in diam., solitary or in clusters, heavily calcified, with a bright green tuft at the tips, and attached by lobed or padlike rhizoids. Axial cells are covered by densely packed whorls of branchlets that end in two swollen surface cells that bracket a gametangium. Surface cells are polyhedral, 80–135 μm in diam., and in annular rows that cover the axes. Gametangia are elongate, stalked, 45–80 μm in diam., 115–170 μm long, and fused into calcified sets of 5–12.

The species is grazer resistant as a result of calcification and chemical defenses (Bitar et al. 2005), including sesquiterpenes (Paul et al. 1993; Pennings et al. 1996). Also, the vegetative phase can regenerate from rhizoids where the primary nucleus resides; it can also survive low temperatures by retracting cytoplasm into the rhizoidal part in the substratum (Berger and Kaever, 1992).

Common, on limestone and shells in sandy tide pools and to 50 m in the Keys; also known from Bermuda, the Caribbean, Mexico, inner Mediterranean (Bitar et al. 2005), and the Indo-Pacific (type location: Mauritius; Silva et al. 1996). Dragastan et al. (2003) report fossil specimens from late Pleistocene limestone in the Florida Keys.

Taylor 1960: p. 101; Littler and Littler 2000: p. 438.

Polyphysaceae

See Olsen et al. (1994), Silva et al. (1996), and Berger et al. (2003) for reviews.

Plants are usually gregarious. Axial cells are uninucleate, attached by lobed basal outgrowths, have a cap or disk with one or more whorls of branchlets (rays) with calcified aplanospores, and produce gametes.

Acetabularia J.V. Lamouroux 1812: p. 185; 1816: p. 244, *nom. cons.* (L. *Acetabulum*: a cup-shaped vessel, a little cup for vinegar + *aria*: derived from).

Synonyms. *Polyphysa* Lamarck 1816: p. 151.
Acicularia D'Archiac 1843: p. 386.

Molecular studies (Olsen et al., 1994) showed that *Polyphysa penicula* (Turner) Solms-Laubach is not distinct from *Acetabularia* in contrast to other species of *Polyphysa*. Berger et al. (2003) noted that transferring *P. penicula* (as '*peniculus*') to *Acetabularia* leaves four other species and they proposed establishment of *Parvocaulis*. *Polyphysa* was separated from *Acetabularia* by its unfused rays and presence or absence of an inferior corona (Valet, 1969; Womersley, 1984; Sosa, 1985; Berger and Kaever, 1992; Wynne, 1998a; Littler and Littler, 2000; Berger et al., 2003).

Plants are calcified; axial cells are unicellular, slender, cylindrical, with or without annular wall thickenings from residual scars of deciduous hairs, and attached by rhizoidal outgrowths. Axial cell tips produce up to 3 whorls of branchlets, and an upper ring (corona superior) of small cells, a middle ring of long rays that are usually fused laterally into a disk or cup, and a lower ring (corona inferior) of small cells. The mature rays contain calcified aplanospores.

Key to the species of *Acetabularia* (4) and *Parvocaulis* (1)

1. Plants evenly calcified 2
1. Calcification concentrated between rays 4
   2. Disc with 22–30 rays; to 7 mm in diameter .... *A. calyculus*
   2. Disc with 30 or more rays; 5–20 mm in diameter 3
3. Stipe 200–300 μm in diam.; disc with 30–80 rays; cysts 55–80 μm in diam. and with up to 500 per ray .... *A. crenulata*
3. Stipe 450–600 μm in diam.; disc with 30–60 rays; cysts 60–100 μm in diam. and with up to 300 per ray .... *A. schenckii*
   4. Plants larger, to 2 cm tall; disc 4–7 mm in diam. and with 20–30 rays.......... *A. farlowii*
   4. Plants minute, to 3 mm tall; disc 1–2.5 mm in diam. and with 6–18 rays .... *P. pusilla*

Acetabularia calyculus J. V. Lamouroux in Quoy and Gaimard 1824: p. 621, pl. 90, fig. 6–7, as '_caliculus_,' also cf. Berger et al., 2003 (L. _calyculus_: cuplike; botanical name: a whorl of bracts below the calyx in flowers). Plate XI, fig. 14–17.

Plants are solitary or clustered, erect, 1–3 cm tall, bright green, calcified and with the axial cell attached by blunt, fingerlike rhizoids arising at its base. The axial cell (stipe) is slender, 200–250 μm in diam. below and 350–400 μm above, heavily calcified, spindlelike as a result of nodal swellings, and with a goblet-shaped cap or disc. When young, axial tips have whorls of branching filaments (10–16) that terminate in 1–5 branchlets, branch repeatedly 7 times with successive branchlets smaller in diam., dehisce, and leave scars on older axes. Caps or discs are lightly calcified, cup-shaped, 4–7 mm in diam., and have 22–30 rays with a central notch on the outer edge. Each ray becomes a gametangium with about 80 spherical bright green cysts, 80–160 μm in diam. The corona superior is above the ray bases and has short, blunt projections on the upper side. The corona inferior is below the ray bases and has small projections on its lower side.

Uncommon and inconspicuous, on mangrove roots and shells to 5 m in the Keys, Indian River, and Jupiter Sound (Eiseman et al., 1978); also known from the Caribbean and Brazil, the Mediterranean (introduced into Egypt and Morocco), Canary Islands, Red Sea, Pacific Ocean, and Australia (type location: Shark Bay; Womersley, 1984).

Taylor 1960: p. 105; Littler and Littler 2000: p. 442.

Acetabularia crenulata J. V. Lamouroux 1816: p. 249, pl. 8, fig. 1 (L. _crenulata_: indented, notched). Plate XI, fig. 18–21.

Synonyms. _Acetabulum caraibicum_ (Kützing) M. Howe 1901: p. 331–333.
Acetabularia caraibica Vickers in Vickers and Shaw 1908: pl. 49.

Berger et al. (2003) and Bailey et al. (1976) have commented on the plant's possible conspecificity with _Acetabularia farlowii_.

Plants are solitary or clustered, calcified, and to 7 cm tall; axial (stipe) cells are slender, have a cap or disc, and are attached by lobed basal outgrowths. Stipes are 200–300 μm in diam., heavily calcified, and may have whorls of deciduous filaments that end in 1–5 branchlets that divide 5–7 times and become smaller in diam. with each forking. Caps are concave, 6–20 mm in diam., and with 30–80 fused rays; each ray has rounded to truncate tips and a central apical spine. At maturity, rays become gametangia containing up to 500 cysts, each 65–80 μm in diam. The corona superior has elongate lobes crowded on the upper side at ray bases. Each lobe has two faint hair scars and rounded to lobed tips. The corona inferior has fingerlike tips crowded on the lower side of ray bases.

Common, growing as colonies on shells and rubble in seagrass beds, on jetties at moderately exposed sites, and to 3 m on all coasts; also known from Bermuda, the Caribbean (type location; Guiry and Nic Dhonncha, 2006), Gulf of Mexico, Sri Lanka, the Andaman Islands, Philippines, and Australia.

Taylor 1960: p. 105; Humm and Taylor 1961: p. 373; Dawes 1974: p. 92; Littler and Littler 2000: p. 442.

Acetabularia farlowii Solms-Laubach 1895: p. 27, pl. III, fig. 1 (named for W. G. Farlow, an American cryptogamic botanist who studied the seaweeds of the eastern U.S. between 1870 and 1900). Plate XI, fig. 22, 23.

Synonyms. _Acetabularia antillana sensu_ Eiseman 1970: p. 45–46, fig. 1–3.
Acetabularia crenulata sensu Bailey et al. 1976: p. 21, 23, fig. 1–12.

Eiseman (1970) suggested conspecificity of *Acetabularia farlowii* and *A.* (*Chalmasia*) *antillana*, while Bailey et al. (1976) suggested this for *A. farlowii* and *A. crenulata*. Others (Valet, 1969; Berger and Kaever, 1992; Sosa, 1996; Berger et al., 2003; Wynne, 2005a) have considered *A. farlowii* to be a distinct entity.

Plants are calcified and 1–2 cm tall; axial cells (stipes) have a cap with rays and are attached by lobed basal outgrowths. Axial cells are slender, flexible, lightly calcified, 200–400 µm in diam., and produce caps 4–7 mm in diam. with 20–30 rays. Rays are lightly calcified, weakly coherent or laterally free, and with rounded blunt tips. At maturity rays become gametangia; each ray contains 40–120 cysts. The corona superior is crowded, lobed, has two hair scars, and is above the bases of rays; a corona inferior is also present.

Common, inconspicuous, on shells and limestone in the Keys (type location: Key West), often mixed with *Acetabularia crenulata*; also known from Mexico and Martinique.

Taylor 1960: p. 105; Humm and Taylor 1961: p. 374; Dawes 1974: p. 93.

Acetabularia schenckii K. Möbius 1889: p. 318–320, pl. 10, fig. 8–12 (named for H. Schenck, a German phycologist). Plate XI, fig. 26–30.

Basionym. *Acicularia schenckii* (K. Möbius) Solms-Laubach 1895: p. 33–34, pl. 3, fig. 4, 9, 11, 12, 14, 15.

Based on molecular data, Berger et al. (2003) found that *Acicularia* (*A. schenckii*) should be included within the genus *Acetabularia*.

Plants are usually clustered and 3–8 cm tall; axial cells (stipes) are heavily calcified, have a cap of rays, and are attached by basal lobed outgrowths. Axial cells are 420–600 µm in diam. and may have whorls of green deciduous filaments that produce 1–6 branchlets, fork 5–7 times, and become smaller with each forking. Axial cells have one or more flat to cup-shaped caps, each with a whorl of 30–50 rays. Rays are wedge-shaped, 3–6(8) mm long, 60–80 µm in diam., and have a central spine at tips and crenulate lateral margins that fuse with heavy calcification. At maturity, the rays become gametangia, each with 100–200 cysts that are 60–100 µm in diam. and embedded in lime. Corona superior lobes are on the upper side at ray bases; each lobe has two faint hair scars and a deep cleft at the tip. Corona inferior lobes are on the lower side of ray bases as minor projections with forked tips.

Common, but inconspicuous; often growing with *Acetabularia crenulata*, on shells and coral rubble on all coasts, to 73 m; also known from Bermuda, the Caribbean, Gulf of Mexico, and Brazil (type location: Cabo Frio), plus the Pacific coasts of Panama and Columbia.

Taylor 1960: p. 107; Humm and Taylor 1961: p. 375; Dawes 1974: p. 93; Littler and Littler 2000: p. 444.

Chalmasia Solms-Laubach 1895: p. 32 (named for Munier-Chalmas, a French phycologist who studied *Acicularia*).

Plants are not calcified; axial cells (stipes) have funnel-shaped whorls of rays (branchlets) at their tip and are attached by lobed basal outgrowths. Axial cells are thick walled with annular thickenings; branchlets become sporangial rays that contain free, calcified aplanospores. A corona superior is present, while a corona inferior is absent.

Chalmasia antillana Solms-Laubach 1895: p. 12, pl. 3, fig. 2, 3, 5 (L. *antillana*: of or from the Antilles or West Indies). Plate XI, fig. 31–33.

Synonyms. *Acetabularia antillana* (Solms-Laubach) Egerod 1952: p. 410.

Acetabularia antillana (Solms-Laubach) Egerod *orth mut*. In Valet 1969: p. 625–626, pl. 44, fig. 10, 11.

Polyphysa antillana (Solms-Laubach) M. J. Wynne 1986: p. 2265.

See Berger et al. (2003) for descriptive details. Eiseman (1970) and Bailey et al. (1976) regarded *Acetabularia farlowii* as conspecific with *A.* (= *C.*) *antillana* and *A. crenulata.*

Plants are 0.2–3.0 cm tall; axial cells (stipes) have caps with 15–30 rays and are attached by short, lobed rhizoids. Axial cells are 300–800 µm in diam., 1–9 mm tall, and taper toward the base. Caps (discs) are to 6 mm in diam. Rays are fused laterally by dense carbonate deposits, 0.5–8.0 mm in diam., club to wedge-shaped, and have blunt tips. At maturity, rays become gametangia that contain internal cysts that are 100–150 µm in diam. The corona superior consists of lobes, which occur on the upper side near the bases of rays, are laterally free, and have 2–3 hair scars. A corona inferior is absent.

Uncommon, in 2–4 m on limestone in the Keys and back reef turfs of the Dry Tortugas (to 20 m); also known from the Caribbean (type location: West Indies) and Gulf of Mexico.

Taylor 1960: p. 103; Littler and Littler 2000: p. 440.

Parvocaulis Berger, Fettweiss, Gleissberg et al. 2003: p. 559 (L. *parvus*: short + *caulis*: stalk of plant; refers to a short corrugated stalk).

Plants are calcified; axial cells (stalks) are unicellular, slender, cylindrical, have annular wall thickenings, and are attached by rhizoidal outgrowths. Annular thickenings (corrugations) are formed during cap formation when the axial cell shortens. The young cap is covered by a velum (mucuslike layer). Stalks produce whorled branchlets; the upper ring (corona superior) consists of small cells, the middle ring has club to ovate rays that are free, and the lower ring (corona inferior) has small cells and is inconspicuous. The mature rays contain calcified aplanospores.

Parvocaulis pusilla **(M. Howe) Berger, Fettweiss, Gleissberg, et al. 2003: p. 560, fig. 15, 29, 35 (L. *pusilla*: dwarf, minute, small). Plate XI, fig. 24, 25.**

Basionym. *Acetabulum pusillum* M. Howe 1909: p. 89–91, pl. 6, fig. 13–15, pl. 7, fig. 1–4.

Synonyms. *Acetabularia pusilla* (M. Howe) Collins 1909b: p. 379–380.

Polyphysa pusilla (M. Howe) S. Berger and Kaever 1992: p. 166, fig. 3.78–3.80 (*nom. inval.*).

Plants are small and solitary; axial cells (stalks) are 1–3 mm tall, 130–300 µm in diam., have a cap of free rays, and are attached by lobed basal outgrowths. Developing caps have a transparent velum or mucuslike layer. Caps are 1.0–2.5 mm in diam. and have 6–18 rays. Rays are mostly free, 0.4–0.7 mm in diam., to 1 mm long, ovoid to club-shaped, with blunt to tapered tips, and form gametangia with 15–60 cysts. The corona superior is above the ray bases and small and rudimentary; each lobe is cylindrical, 22–35 µm in diam., and with 2–3 dorsal hairs. A corona inferior is absent.

Common, inconspicuous, in protected sites on rubble and within crevices in the Keys (to 5 m); also known from the Caribbean (type location: Montego Bay, Jamaica; Silva et al., 1996), Gulf of Mexico, Liberia, Seychelles and Aldabra Islands.

Taylor 1960: p. 104; Littler and Littler 2000: p. 444.

Fossil records of calcified green algae in Florida

Dragastan et al. (2003) found fossils in the Key Largo and Miami late Pleistocene limestone formations in the Florida Keys that included broken siphon tubes of *Penicillus dumetosus*, fragments of two Dasycladacean green algae (*Neomeris annulata* and *Cymopolia barbata*), and 12 species of *Halimeda*. Nine of the *Halimeda* species are recent (*H. copiosa*, *H. discoidea*, *H. gracilis*, *H. incrassata*, *H. monile*, *H. opuntia*, *H. scabra*, *H. simulans*, and *H. tuna*), while the other three are new. Eight of the nine fossil species are present in Florida and *H. copiosa* occurs in other Caribbean sites (Littler and Littler, 2000). Three currently recognized Florida species (*H. goreauii*, *H. hummii*, and *H. lacrimosa*) were absent. In discussing species richness for the *Halimeda* fossils, Dragastan et al. (2003) noted that all species originated earlier than the Pleistocene, during an interval of high sea levels that corresponded with a favorable period for speciation and diversification of new species. Even so, the fossil record of 12 *Halimeda* species is lower than the total of 19 recent species reported from the western Atlantic.

3

Phylum Heterokontophyta

Synonym. Ochrophyta (Wynne, 2005a)

Based on pigmentation, reserve foods, and flagellar structure, the phylum Heterokontophyta contains a diverse group of micro- and macroalgae that include marine species in the classes Chrysophyceae, Xanthophyceae, and Phaeophyceae. Although we use the term "plant" in describing brown algae, others (for example, Baldauf, 2003) do not place them in the Kingdom Plantae. Motile cells usually have two flagella, a longer, forward-directed one with two rows of stiff, three-parted hairs, and a shorter, smooth one that often has a basal granule that can sense light. Heterokont and haptophytic algae have tubular mitochondrial cristae, a storage product of short ß-1–3 linked glucan polymers, plastids with groups of three evenly adpressed thylakoids and two external endoplasmic reticulum membranes, chlorophylls *a* and *c*, and tripartite tubular hairs on immature flagella (Andersen, 2004).

CHRYSOPHYCEAE

Plants in this class are mostly unicellular, with a few colonial or multicellular species. Cells are golden brown and have a thick-walled siliceous cyst (stomatocyst).

CHRYSOMERIDALES

Marine members of the order are pluriseriate erect branched filaments with heterotrichous growth. Zoospores lack a cell wall, have violaxanthin, are filamentous, and have flagellar features of the phylum Heterokontophyta (Hoffmann et al. 2000).

Chrysomeridaceae

The family has the same features as the order (Graham and Wilcox, 2000; Schneider, 2003).

Chrysonephos W. R. Taylor 1952: p. 79 (L. *chryso*: golden, yellow + *nephos*: a cloud).

Synonym. *Chrysophaeum sensu* W. R. Taylor 1951: p. 129; non *Chrysophaeum* I. F. Lewis and H. F. Bryan 1941: p. 347

Chrysonephos was tentatively placed in the order Chrysomeridales by O'Kelly (1989). See Kristiansen and Preisig (2001) for further details on the plant's taxonomy.

Plants are filamentous, dichotomously divided, gelatinous, multicellular, and attached by rhizoids.

Cells are irregularly arranged and have 2–4 yellow-brown plastids. Pyriform zoospores develop from vegetative cells that swell or round up and have gelatinized cell walls; zooids have 2 unequal flagella and 2 plastids.

Chrysonephos lewisii (W. R. Taylor) W. R. Taylor 1952: p. 79 (named for I. F. Lewis, an American phycologist who studied brown algae, including the periodicity of sexual reproduction in _Dictyota_). Plate XII, fig. 1–3.

Synonym. *Chrysophaeum lewisii* W. R. Taylor 1951: p. 129–130; fig. 1–36.

Plants are filamentous, whitish or pale yellow, to 4 cm tall, in delicate tufts or colonies, and attached by inconspicuous holdfasts. Filaments are 20–38 µm wide and uniseriate or locally multiseriate. Cells are 8–18 µm long, 8–11 µm in diam., and occur in a hyaline, nonstratified sheath. Each cell has one yellow parietal plastid with 2–4 lobes or that is bandlike and with stalked pyrenoids. Zoospores are biflagellate, pyriform, 7–8 µm in diam., and 8–9 µm long.

Rare, as cloudy streamers over sand and seagrasses in protected sites in the Dry Tortugas; also known from Bermuda (type location: Port Royal Bay, Hunt Island), Jamaica, Puerto Rico, and the Mediterranean.

Taylor 1960: p. 195.

XANTHOPHYCEAE

Members of the class contain flagellates, coccoid colonies, filaments, and true coenocytes. Cells often have cellulose walls and lipids as reserve foods. Flagella are inserted apically in pairs and unequal in length; photoreception includes a swelling at the base of the smooth flagella and an eyespot in the chloroplast. Asexual reproduction is by zoospores and nonmotile reproductive structures (i.e., aplanospores, autospores, or cysts).

VAUCHERIALES

The order contains filamentous coenocytes or spherical cells that are attached by rhizoids. Plastids are small, with or without pyrenoids, and parietal. Food reserves are oil or starch. Asexual reproduction is by zooids or aplanospores, while sexual reproduction may be isogamous, anisogamous, or oogamous.

Vaucheriaceae

Plants in the monotypic family are filamentous coenocytes that lack cross walls except at the site of reproductive cells. Branching is irregular to dichotomous and growth is apical. Asexual zoospores are solitary, from terminal sporangia, multinucleate, and multiflagellate with multiple pairs of flagella spread over the cell. Sexual reproduction is oogamous and species are dioecious or monoecious. Antheridia and oogonia are stalked. Antheridia are curved, cylindrical, and produce many small biflagellate spermatozoids, while oogonia are single or clustered and contain one egg. Oospores contain oil and have thickened walls.

Vaucheria De Candolle 1801: p. 20 (named for Jean Pierre Etienne Vaucher of Geneva, a Swiss clergyman and botanist).

The genus has the same features as the family. It appears to have been poorly studied in Florida, with only three species reported (Gallagher and Humm, 1981). Plants occur between late October or November to March and reproduce during January to March. The few taxa recorded contrast with the 13 known from the northwest Atlantic (Sears, 2002), 10–11 from North Carolina (Ott and Hommersand, 1974; Schneider and Searles, 1991), and 24 from the British Isles (Christensen, 1987a).

Key to the species (3) of *Vaucheria*

1. Oogonia and antheridia on separate (i.e., dioecious) coenocytic filaments. 2
1. Oogonia and antheridia on the same coenocytic filaments . . . . *V. velutina*
   2. Filaments 70–95(110) µm in diam.; antheridia 500–700 µm long . . . . *V. litorea*
   2. Filaments 25–50(60) µm in diam.; antheridia 275–400 µm long . . . . *V. longicaulis*

Vaucheria litorea C. Agardh 1823 (1820–1828): p. 463 (L. *litorea*: of the shore or the littoral zone). Plate XII, fig. 4, 5.

Plants form dark green mats; filaments are long, little branched, and 70–95(110) µm in diam. Plants are dioecious and have terminal sexual organs that are usually separated by an empty cell (i.e., double cross wall). Antheridia are elongate, cylindrical, sometimes recurved, 500–700 µm long, 65–75 µm in diam., and with a terminal discharge pore and 2–6 lateral ones. Oogonia are club-shaped, terminal on a bent branch, and separated by a septate cell having a distinct mass of protoplasm. Oospores are 175–220 µm long, 162–214 µm wide, and lack conspicuous oil droplets.

Uncommon, forming feltlike carpets or irregular tufted stands in brackish ponds on the Anclote River; also known from Northumberland Strait (Canada) to North Carolina and Louisiana, as well as from the northeast Atlantic (type location: Netherlands; Christensen, 1987a), Washington State, and Australia.

Schneider and Searles 1991: p. 105.

Vaucheria longicaulis Hoppaugh 1930: p. 332–334, fig. 1, 2 (L. *longi*: elongate + *caulis*: a stem, a stalk). Plate XII, Fig. 6, 7.

Filaments are 25–50 (60) µm in diam. and form dark green mats. Plants are dioecious, with oogonia and antheridia terminal on straight branches and separated by an empty cell (i.e., double cross wall). Antheridia are 275–400 µm long with 1–4 lateral discharge pores. Oogonia are pear-shaped, 290–420 µm long, 90–135 µm wide, and have a terminal pore; oospores are spherical and do not adhere to the oogonial wall.

Common, forming mats in the low intertidal at Tarpon Springs (Anclote River estuary) and Treasure and Pine Islands on the west coast (as *V. bermudensis* W. R. Taylor and Bernatowicz; Hamm and Humm, 1976); also known from Northumberland Strait (Canada) to North Carolina, Brazil, the eastern North Atlantic, and Indo-Pacific (type location: Elkhorn Slough, California; Womersley, 1987), and Australia.

Schneider and Searles 1991: p. 106.

Vaucheria velutina C. Agardh 1824: p. 312 (L. *velutina*: like velvet, densely covered with fine short hairs). Plate XII, fig. 8.

Synonym. *Vaucheria thuretii* Woronin 1869, p 157, pl. 2, fig. 30–32.

Filaments are 50–80 µm in diam. and form dark green mats. Chloroplasts are broad and have a pyrenoid along a single edge. Antheridia are sessile or on short pedicels, in clusters of 2–4(7), sack-shaped, bent or upright, and have one apical pore. Oogonia are usually sessile, pear-shaped, and solitary or in clusters of 3 on a filament, usually with one close to an antheridium. Oospores are spherical, to 160 µm in diam., and have smooth, colorless cell walls 3 µm thick.

Common and perhaps the most frequent species in the Tampa Bay area (Dawes, 1974; Gallagher and Humm, 1981); also known from the Bay of Fundy to North Carolina, the Gulf of Mexico, eastern North Atlantic (type location: Gråen, Sweden; Christensen, 1987a), Indo-Pacific, and Australia.

Humm and Taylor 1961: p. 376; Dawes 1974: p. 95; Schneider and Searles 1991: p. 108.

Sedis incertae: Chrysophaeum I. F. Lewis and H. F. Bryan 1941: p. 343–347 (L. *chryso*: golden, yellow + *phaeum*: dusky, dark colored, dark gray).

Synonym. non *Chrysophaeum sensu* W. R. Taylor 1951: p. 129.

The placement of the genus is uncertain; it may belong to the Cryptophyta.

Colonies are highly branched, yellow, and form clumps of cells, which are subcylindrical, covered with a pellicle, have many plastids, lack flagella and eyespots, and are on long, delicate intertwined stalks. Reproduction is by "aplanospores" that produce zoospores with two equal, laterally attached flagella, one active and the other trailing.

Chrysophaeum taylorii I. F. Lewis and H. F. Bryan 1941: p. 343 (named for William R. Taylor, an American phycologist who published extensively on the seaweed floras of the northeastern and tropical Atlantic Americas). Plate XII, fig. 9, 10.

Colonies are gelatinous tufts or streamers, 1–6 cm tall, and with a sulfur mustard (yellow) coloration. Cells are 49–118 μm long, 24–38 μm wide, ovoid to cylindrical, and narrowing to the base. The cell's anterior end is indented, gulletlike, and has many irregularly ovoid plastids. Reproduction is by "aplanospores" (to 25 μm in diam.), with each releasing 16–32 zoospores that are to 9 μm long.

Rare, the yellowish mucous streamers grow on coral rubble in protected waters of the Dry Tortugas (type location); also known from Bermuda.

Taylor 1960: p. 194.

PHAEOPHYCEAE

The class of brown algae contains about 265 genera and 1,500–2,000 species (Norton et al. 1996); more than 95 percent are marine. Brown algae are more prevalent in cold (arctic or boreal) to temperate waters and range from filaments (for example, *Ectocarpus*) to large complex seaweeds with blades, stipes, and floats (for example, *Sargassum*). Some tropical brown taxa such as *Lobophora variegata* can extend to depths of 100 m, while most have a more limited vertical distribution. Typically, brown algae are lithophytes that grow attached to stable, hard substrata, while several smaller filamentous species are epiphytes or endophytes. By contrast, *Sargassum fluitans* and *S. natans* occur only as drift populations in the Sargasso Sea and are usually vegetative (Moreira and Suárez, 2002d)). Brown algae contain chlorophylls *a* and *c*, ß-carotene, fucoxanthin, neofucoxanthin, and other cartenoid pigments. Their cells are usually uninucleate and have chloroplasts with thylakoids that are grouped into bands of three, plus a girdling lamella inside the double plastid membrane. Plastids of species of the Ectocarpales and Dictyotales often have stalked or protruding pyrenoids. The life histories of less complex species usually involve plurilocular (multicellular) sporangia on both haploid (gametophytic) and diploid (sporophytic) plants. The presence of unilocular sporangia usually indicates a diploid plant. The zooids of most species have a heterokont morphology with the two flagella laterally or subapically inserted. Zoospores produced from plurilocular structures on gametophytes usually function as gametes that fuse to form a diploid sporophyte, while zoospores from plurilocular structures on sporophytes recycle the diploid generation via asexual reproduction. Zoospores from unilocular sporangia are usually produced via meiosis, are haploid, and grow into gametophytes. There are many variations of this basic life history; some taxa (for example, Dictyotales) are isomorphic and produce oogonia and antheridia, while members of the Fucales lack a multicellular haploid generation and produce oogonia and antheridia.

Hoek et al. (1995) list 16 orders, 9 of which contain species found in Florida, while Wynne (2005a) lists 10 orders, 7 of which have species in Florida. The present classification follows Wynne (2005a) and is based upon several molecular studies dealing with brown algae (de Reviers and Rousseau, 1999; Rousseau and de Reviers, 1999a,b; Rousseau et al. 2001; Draisma et al. 2001, 2003; Cho et al. 2004; Kawai and Sasaki, 2004).

DICTYOTALES

Plants have flat blades attached by rhizoidal filaments and grow by a single or a marginal row of apical cells. Blades are parenchymatous with large colorless medullary cells. The cortex has one or more layers of small pigmented cells with many discoid plastids. Hyaline hairs arise from surface cells and form bands or tufts. Sporophytes and gametophytes are isomorphic. Oogonia are unicellular and produce one egg, while the multicellular antheridia yield many sperm; both reproductive structures occur in sori. Unilocular sporangia produce nonmotile meiospores.

Dictyotaceae

Molecular studies suggest that present generic separations, based mostly on anatomical and growth features, are untenable (De Clerck et al. 2005b) and may not correspond to natural phylogenetic relationships (Lee and Bae, 2002).

The family has the same features as the order.

Dictyopteris J.V. Lamouroux 1809c: p. 332; 1809b p. 129, *nom. cons.* (Gr. *dictyo*: net, reticulate + *pteris*: a fern, wing).

Synonyms. *Neurocarpus* Weber and Mohr 1805: p. 300.
Haliseris De Toni 1895: p. 253.

Plants are bushy, strap-shaped, dichotomously branched, and attached by rhizoidal holdfasts. Growth is by a group of apical cells. Branches have a prominent midrib and sometimes lateral veins. Hairs occur in scattered clusters over the surface or along the midrib. Unilocular sporangia produce nonmotile meiospores (tetraspores) in dense sori.

Key to the species (5) of *Dictyopteris*

1. Veins and pinnately arranged veinlets present 2
1. Veins present and veinlets absent. 3
 2. Blade margins entire, fronds 1 cell thick between veinlets; branching irregularly alternate. . . . . . *D. plagiogramma*
 2. Blade margins dentate to ciliate, teeth to 1 mm long; fronds 2–4 cells thick between veinlets; branching dichotomous . . . . *D. hoytii*
3. Blades are thin, somewhat transparent, and less than 1.5 cm wide 4
3. Blades are thick, with 4 cell layers, and 5–8.0 cm wide. . . . . *D. justii*
 4. Plants 2–8 cm tall or decumbent; fronds 0.5–5.0 mm wide . . . . *D. delicatula*
 4. Plants 10–30 cm tall and erect; fronds 5–15 mm wide. . . . . *D. polypodioides*

Dictyopteris delicatula J. V. Lamouroux 1809b: p. 332, pl. 6, fig. 2b (L. *delicatula*: very delicate or thin). Plate XII, fig. 11, 12.

Synonym. *Polyzonia ?divaricata* P. Crouan and H. Crouan in Schramm and Mazé 1865: p. 27.

Wysor and De Clerck (2003) compared *Dictyopteris delicatula* and the morphologically similar *D. repens* (Okamura) Børgesen from the western Pacific. The taxonomic value of submarginal veins has been questioned by Tsuda (1972), Allender and Kraft (1983), Phillips (2000), and N'Yeurt (2001).

Plants are prostrate, spreading or erect, and 2–8 cm tall, with erect stalks attached by basal holdfasts. Blades are flat, light brown, and reattach to substrata and each other; fronds are also strap-shaped, 0.5–5.0 mm wide, irregularly or dichotomously branched, and 2 or more cells thick along their margins and midrib. Hairs occur in clusters on one surface of the blade, while sporangia are in dense sori in single rows along the midrib.

Common, on limestone and large algae in the Keys and to 20 m on the west coast; also known from the Carolinas (35–40 m), Bermuda, the Caribbean (type location: Lesser Antilles), Gulf of Mexico, and Brazil, as well as the eastern tropical Atlantic, Red Sea, Indo-Pacific, and Australia.

Taylor 1960: p. 227; Earle 1969: p. 149; Dawes 1974: p. 100; Schneider and Searles 1991: p.155; Littler and Littler 2000: p. 254.

Dictyopteris hoytii W. R. Taylor 1960: p. 229–230, 631–632, 726–727, pl. 32, fig. 1 (named for the American phycologist W. D. Hoyt, who studied the seaweeds of the southeastern U.S.A.). Plate XII, fig. 13.

Plants are bushy, yellow brown, to 30(–45) cm tall, and attached by a small pad of rhizoids. Lower axes are often denuded for 1–6 cm; above they are dichotomously branched 2–6 times and have narrow forks. Blades are strap-shaped, 1.5–3.0 cm wide, 90–150 µm thick, and undulate; margins are crenulate or smooth with dentations 1 mm long at intervals of 1–2 mm. The medulla has 2–4 layers of large colorless cells covered by small, pigmented surface cells. Lateral veins are delicate, with thick-walled cells in irregular tiers of 5–6, at intervals of 2–3 mm, and extend obliquely from midrib to margins; midribs are to 185 µm thick. Hairs are in clusters scattered over the blade's surface. Sporangia occur in sori aligned parallel to lateral veins. Oogonia are 65–130 µm in diam. and solitary or in groups of 2–3, except near the margins.

Rare, reported from 56–59 m off the southeast coast of Florida (Hanisak and Blair, 1988); also known from North Carolina in 14–47 m (type location: Beaufort, North Carolina; Silva et al. 1996), Georgia, the Caribbean, and Kenya.

Taylor 1960: p. 229; Schneider and Searles 1991: p. 155.

Dictyopteris justii J. V. Lamouroux 1809c: p. 332, pl. 6, fig. 2a (named for Saint Just Lacuée, a recipient of the Legion of Honor). Plate XII, fig. 14, 15.

Plants are erect and to 40 cm high; blades are strap-shaped, dichotomously branched, and their stalks attached by fibrous holdfasts. Blades are 1.5–8.0 cm wide, to 200 µm thick, entire, ruffled, and with obtuse or indented tips. The medulla has 2 layers of large colorless cells and small, pigmented surface cells. Midribs are obvious and have 8–14 cell layers; marginal veins are absent. Sori are scattered over the upper parts of blades with many hairs. Oogonia are ovoid and lack sterile indusial cells. Plants are often epiphytized by coralline algae and turflike species (Wysor and De Clerck, 2003).

Uncommon, on intertidal rubble at moderately exposed sites in the Keys and on the west coast (at 40 m); also known from Bermuda, the Caribbean (type location: St. Dominique, Lesser Antilles; Littler and Littler, 2000), Gulf of Mexico, and Brazil.

Taylor 1960: p. 226; Earle 1969: p. 149; Littler and Littler 2000: p. 256.

Dictyopteris plagiogramma (Montagne) Vickers 1905: p. 58 (Gr. *plagio*: oblique + *gramma*: a line). Plate XII, fig. 16, 17.

Basionym. *Haliseris plagiogramma* Montagne 1837: p. 356.

Plants are erect, to 25 cm tall, and profusely, alternately to irregularly branched; stipes arise from eroded midribs and are attached by a fibrous holdfast. Blades are flat, 3–10 mm wide, and one cell thick (35–50 µm). Midribs are thicker and more pronounced than foliar parts of blades. Pinnate veinlets are weakly developed. Blade margins are entire, smooth to undulating, and lack teeth. Hairs occur in scattered clusters or in irregular rows along the midribs. Sporangia are 80–120 µm in diam. and near the midribs.

Uncommon, on limestone and shells in the Dry Tortugas (8–55 m); also known from Bermuda, the Caribbean (type location: Havana, Cuba; Silva et al. 1996), Gulf of Mexico, and Brazil, as well as the Canary Islands, Indo-Pacific, and Australia.

Taylor 1960: p. 229; Earle 1969: p. 151; Littler and Littler 2000: p. 256.

Dictyopteris polypodioides (DeCandolle in J. V. Lamouroux and DeCandolle) J. V. Lamouroux 1809c: p. 332 (L. *poly*: many + *pod*: foot + *oides:* similar to; also *Polypodium*: the generic name for the oak-fern; thus like the fern *Polypodium*). Plate XII, fig. 18, 19.

Basionym. *Ulva polypodioides* De Candolle in Lamarck and DeCandolle 1805: p. 15.

Synonyms. *Dictyopteris membranacea* (Stackhouse) Batters 1902: p. 54.

Dictyopteris ambigua (Clemente) Cremades in Cremades and Pérez-Cirera 1990: p. 489.

As noted by Silva et al. (1996) there is some doubt regarding the conspecificity of northeast Atlantic and Mediterranean populations of this plant.

Plants are bushy, erect, 10–30 cm tall, and the stipes are attached by padlike holdfasts. Blades are irregularly dichotomously branched every 1–5(10) cm and stipes result from denudation of lower portions of fronds. Blades are strap-shaped, entire, 5–15 mm wide, and 175–200 µm in cross section; tips are rounded and 2 cells thick, except at margins. Blades arise from prominent denuded midribs and lack lateral veins. Surface hairs are 15–22 µm in diam. and in tufts opposite one another on both blade surfaces. Sporangia are located in bandlike sori near the midribs. Oogonia and antheridia are in large diffuse clusters on the surface.

Uncommon, on rubble at 18 m on the west coast and in shallow waters in the Keys; also known from the Carolinas (14–40 m), Caribbean, Gulf of Mexico, and Brazil, as well as eastern North and tropical Atlantic, the Mediterranean (type location: "Mare Numidico"; Silva et al. 1996), and the Indo-Pacific.

Taylor 1960: p. 227; Schneider and Searles 1991: p. 156; Littler and Littler 2000: p. 258.

Dictyota J.V. Lamouroux 1809c: p. 331; 1809a: p. 38, *nom. cons.* (Gr. *dictyota*: latticed).

Synonym. *Dilophus* J. Agardh 1882: p. 106.

Hörnig et al. (1992b) noted that the "multilayered medulla" used by J. Agardh (1882) to distinguish *Dilophus* from *Dictyota* was not effective and proposed merging it with the latter. De Clerck et al. (2006) moved *Dictyota cervicornis* and *D. crispata* to a new genus, *Canistrocar-*

pus based on molecular studies of Philippine plants. We retain the species in *Dictyota* until Caribbean plants have been examined.

Plants are erect, usually dichotomously branched, and attached by small fibrous mats of rhizoids. Blades are strap- to ribbonlike, with each frond ending in a single large apical cell. The medulla usually has one layer of large colorless cells and is covered by small, pigmented surface cells. Colorless hairs (paraphyses) occur in clusters on both blade surfaces. Plants are dioecious. Antheridia are in patches, on both surfaces, and have marginal sterile indusial cells. Oogonia occur in small clusters and lack an indusium. Unilocular sporangia are scattered or in sori, with each sporangium having 4 nonmotile meiospores (tetraspores). Species may dominate reef tracks in the Florida Keys and can negatively impact the growth of *Halimeda* as an epiphyte (Beach et al., 2003).

Key to the species (11) of *Dictyota*

1. Medulla of fronds usually two cells thick 2
1. Medulla of fronds usually 1 cell thick 3
  2. Fronds 2–4(5) mm wide; medulla 2–5 cells thick at margin . . . . *D. pinnatifida*
  2. Fronds 0.3–4.0 mm wide; medulla 2–4 cells thick; margins 1-celled . . . . *D. guineensis*
3. Branching near frond tips alternate 4
3. Branching near frond tips not alternate 5
  4. Primary axes distinct; lateral branches twisted; 80–120 µm thick . . . . *D. mertensii*
  4. Primary axes indistinct; branches little twisted; 180–350 µm thick . . . . *D. crispata*
5. Margins toothed 6
5. Margins smooth 7
  6. Blades to 7(12) mm wide, marginal teeth small but obvious . . . . *D. ciliolata*
  6. Blades to 3 mm wide, with marginal teeth minute and not obvious . . . . *D. crenulata*
7. Fronds not narrowing with each successive division 8
7. Fronds wider near base, each division narrowing; 0.1–0.2 mm at tips . . . . *D. pulchella*
  8. Fronds more than 3 mm wide 9
  8. Fronds less than 3 mm wide 10
9. Fronds with wide bands, 3–5 mm wide, and 100–140 µm thick . . . . *D. bartayresiana*
9. Fronds lacking wide bands, 4–15 mm wide, and 150–250 µm thick . . . . *D. menstrualis*
  10. Fronds 0.3–1.5 mm wide, 200–300 µm thick; sori lacking paraphyses . . . . *D. caribaea*
  10. Fronds 1.0–2.5 mm wide, 180–220 µm thick; sori with paraphyses . . . . *D. cervicornis*

Dictyota bartayresiana J. V. Lamouroux 1809a: p. 43 (name source not cited; probably Prof. Antoine Bartayrès, a French mathematician and naturalist). Plate XIII, fig. 1.

Synonyms. *Dictyota bartayresii* J. V. Lamouroux 1809c: p. 331, pl. 6, fig. 2A–E, *nom. illeg.*
Dictyota patens J. Agardh 1882: p. 93.
Dilophus alternans f. *acutissimus* W. R. Taylor 1969: p. 157, fig. 15, 16.
Dictyota neglecta Hörnig and Schnetter in Hörnig et al. 1992b: p. 56, fig. 5.

The confusion between *Dictyota bartayresiana*, *D. bartayresii*, *D. crispata*, and *D. neglecta* was resolved by De Clerck and Coppejans (1997). Specimens with pointed tips, surface growths, and involucrate sporangia, often incorrectly named *D. bartayresiana* (Vickers, 1908; Hörnig et al., 1992a,b), were referred to *D. crispata*. Tsuda (2004) described compact clumps, loose mats, and twisted thalli in Micronesia that had been called *D. cervicornis* and *D. friabilis* Setchell. See De Clerck (2003) for details of *D. bartayresiana*.

Plants are erect, 5–20 cm tall, clustered, and delicate; blades are dichotomously branched and attached by small matlike holdfasts. Blades are strap-shaped, 3–7 mm wide, 100–140 µm thick, and with branching angles of 45–90°; tips are broadly rounded or acute. The medulla has 1 layer of large cells that are rectangular, 45–110 µm in diam., and covered by one layer of small pigmented cells. Sporangia are spherical, 60–110 µm in diam., and in sori lacking sterile hairs. Oogonia occur in scattered sori.

Common, in dense clumps on hard substrata, on seagrasses in the Keys, and from 33 m in the Florida Middle Grounds (Cheney and Dyer, 1974); also known from Bermuda, the Bahamas, Caribbean (type location: Antilles; Lipkin and Silva, 2002), and Brazil, as well as eastern North and tropical Atlantic, the tropical Pacific, and Australia.

Taylor 1960: p. 219; Earle 1969: p. 151; Littler and Littler 2000: p. 260.

Dictyota caribaea Hörnig and Schnetter in Hörnig et al. 1992b: p. 58 (L. *caribaea*: of or from the Caribbean). Plate XIII, fig. 2.

Synonym. *Dictyota volubilis* Kützing 1849: p. 554 *sensu* Vickers 1908.

? *Dictyota indica* f. *torta* P. Crouan and H. Crouan in Mazé and Schramm 1878 (1870–1877): p. 121.

Dictyota cervicornis f. *pseudobartayresii* W. R. Taylor 1928: p. 118.

The species is easily confused with *Dictyota cervicornis* Kützing but lacks involucral cells near sporangia and has cervicorn branching (Wysor and De Clerck, 2003).

Plants are tough, bushy or entangled, to 20 cm tall, with dichotomous to cervicorn branching, and attached by small fibrous rhizoidal mats. Blades are strap-shaped, often twisted, 0.3–1.5 mm wide, and 200–300 µm thick; internodes are 1.5–4.0 cm long with pointed tips. The medulla has one layer of rectangular cells 180–280 µm tall, and is covered with a layer of small pigmented cells, 10–20 µm tall. Deciduous hairs are produced in scattered tufts. Unilocular sporangia are solitary, scattered or near margins, and lack sterile hairs.

Common, on limestone, rubble, and mangrove roots and to 27 m in the Keys; also known from Bermuda, the Caribbean (type location: Colombia; Hörnig et al. 1992b), and Gulf of Mexico.

Taylor 1960: p. 221; Earle 1969: p. 161; Littler and Littler 2000: p. 260.

Dictyota cervicornis Kützing 1859: p. 11, pl. 24, fig. II (L. *cervicornis*: cervicorn, with branches on one side, like the horns of a deer). Plate XIII, fig. 3.

Synonyms. *Dictyota indica* Sonder *ex* Kützing 1859: p. 8, pl. 17, fig. 1.

Dictyota pardalis Kützing 1859: p. 16–17, pl. 39, fig. II.

Dictyota dichotoma var. *curvula* P. Crouan and H. Crouan in Mazë and Schramm 1878 (1870–1877): p. 119.

? *Dictyota indica* f. *torta* P. Crouan and H. Crouan in Mazé and Schramm 1878 (1870–1877): p. 121.

Dictyota cervicornis f. *pseudodichotoma* W. R. Taylor 1928: p. 118.

Dictyota cervicornis f. *spiralis* W. R. Taylor 1928: p. 118.

De-Paula et al. (2001) presented chemical evidence for the conspecificity of *Dictyota pardalis* and *D. cervicornis*, while De Clerk (2003) gave a detailed description of *D. cervicornis*.

Plants are erect, bushy, to 20 cm tall, and attached by small rhizoidal mats. Branching is wide angled, subdichotomous and asymmetrical to cervicorn, and with a spur or short fork. Branches are flat, slender, 1–2.5 mm wide, 180–220 µm thick, and twisted or spiraled; internodes are 1.0–3.5.0 cm long, margins proliferous, and tips acute. The medulla has one layer; cells are rectangular, 140–180 µm tall, and covered by small rectangular pigmented cells 15–20 µm tall. Hairs are clustered along the central axes. Unilocular sporangia are scattered and with sterile hairs. Oogonia occur in scattered sori; antheridia are multichambered and in sori with sterile hairs.

Common, in sheltered areas at the Keys (type location: Key West; Silva et al. 1996) and to 20 m off Tampa Bay; also known from North Carolina, Bermuda, the Caribbean, Gulf of Mexico, and Brazil, as well as eastern North and tropical Atlantic, Red Sea, Indo-Pacific, and Australia.

Taylor 1960: p. 222; Earle 1969: p. 153; Dawes 1974: p. 100; Schneider and Searles 1991: p. 157; Littler and Littler 2000: p. 260.

Dictyota ciliolata Sonder *ex* Kützing 1859: p. 12, pl. 27, fig. I (L. *ciliolata*: having fine marginal hairs). Plate XIII, fig. 4.

Synonym. *Dictyota ciliata* J. Agardh 1841: p. 5, *nom. illeg.*

See De Clerk (2003) for a detailed description of the species.

Plants are erect, bushy, and to 15 cm tall; blades are irregularly to dichotomously branched, have narrow forks, and are attached by small fibrous rhizoidal mats. Branches are flat, spiraled or twisted, to 12 mm wide below forks, 7 mm or less above forks, and to 150 µm thick. Branch internodes are 1–5 cm long; tips are tapered and blunt; margins usually have acute, teethlike projections and/or proliferations. The medulla has one layer of rectangular cells (to 125 µm tall) and is covered by a layer of small pigmented rectangular cells to 25 µm tall. Surface hairs are scattered or in tufts and to 120 µm tall. Unilocular sporangia are in sori without hairs, ovoid, and to 25 per sori. Oogonia are to 200 µm in diam. and in scattered sori.

Common, on limestone and rubble in protected sites in the Keys (to 24 m); also known from North Carolina (15–20 m), Bermuda, the Caribbean, (type location: Guayra, Venezuela; Silva et al. 1996), Gulf of Mexico, and Brazil, as well as the eastern tropical Atlantic, Red Sea, Indo-Pacific, and Australia.

Taylor 1960: p. 223; Earle 1969: p. 155; Dawes 1974: p. 100; Schneider and Searles 1991: p. 157; Littler and Littler 2000: p. 262.

Dictyota crenulata J. Agardh 1847 ('1848'): p. 7 (L. *crenulata*: notched or somewhat scalloped). Plate XIII, fig. 5–8.

Synonyms. *Dictyota ciliolata* var. *bermudensis* Taylor 1960: p. 630, pl. 59, fig. S.
Dictyota jamaicensis Taylor 1960: p. 630–631, pl. 32, fig. 4, 5.
Dilophus crenulatus (J. Agardh) Nizamuddin and Gerloff 1980: p. 867.

Wysor and De Clerck (2003) stated that *Dictyota ciliolata* var. *bermudensis* was a less dentate form of *D. crenulata*.

Plants are erect, bushy, fluffy, 5–20 cm tall, yellow brown, and attached by a fibrous mass of filaments. Marginal rhizoids are common on blades; short terete flagellar growths may arise from the holdfast. Fronds are usually widely dichotomously branched near bases, while their tips have parallel narrow dichotomies. Blades are strap-shaped, 1–2 mm wide, 100–130 µm thick, often twisted, with tips blunt, and teeth 50–150 µm. The medulla has 1 layer of large rectangular cells (75–100 µm in diam.) and is surrounded by one layer of pigmented surface cells. Surface hairs are in small clusters and to 120 µm long. Sterile hairs are absent; sporangia are spherical and 80–120 µm in diam. Oogonia are to 500 µm long, 250 µm in diam., and evenly dispersed. Antheridia are 75–200 µm in diam. and 60–120 µm high.

Uncommon, to 15 m on hard substrata, and known in Florida only from Smithsonian Institution herbarium specimens (Cay Sal Bank in the Strait of Florida); also known from the Caribbean and Gulf of Mexico, as well as the eastern tropical Atlantic and Indo-Pacific.

Taylor 1960: p. 223; Littler and Littler 2000: p 262.

Dictyota crispata J. V. Lamouroux 1809a: p. 44 (L. _crispata_: crisped, curled, finely waved). Plate XIII, fig. 9, 10.

Synonyms. *Dictyota bartayresii* J.V. Lamouroux 1809a *sensu* Vickers 1908: p. 38, pl. 12.
Dictyota cuspidata Kützing 1859: p. 9, pl. 20, fig. 2.
Dictyota bartayresiana J. V. Lamouroux *sensu* Vickers 1908.
Dilophus alternans J. Agardh f. *acutissima* Taylor 1969: p. 157, fig. 15, 16.

The taxonomic confusion between *Dictyota bartayresiana*, *D. crispata*, and *D. neglecta* Hörnig and Schnetter was recently summarized by De Clerk and Coppejans (1997) and Wysor and De Clerck (2003). De Clerk (2003) gave a detailed description of *D. crispata*.

Plants are bushy, erect, and to 20 cm tall; blades mostly alternately branched, with many short spurlike branchlets 1–2 mm long, and attached by small rhizoidal mats. Branches are narrow, to 3 mm wide, 180–340 µm thick, and with pointed tips. The medulla has one layer of rectangular cells (to 280 µm tall) covered by small pigmented cells, 15–30 µm in diam. Deciduous hairs are tufted and scattered. Unilocular sporangia are to 150 µm in diam. and scattered among sterile involucrate hairs.

Uncommon, on wave-exposed limestone in the Keys; also recorded from the Caribbean (type location: Antilles; De Clerck, 2003), tropical Pacific, and Australia.

Littler and Littler 2000: p. 262.

Dictyota guineensis (Kützing) P. Crouan and H. Crouan in Mazé and Schramm 1878 (1870–1877): p. 126, as '_guineense_' (L. _guineensis_: of or from Guinea or tropical West Africa). Plate XIII, fig. 11, 12.

Basionym. *Spatoglossum guineense* Kützing 1843: p. 339.
Synonyms. *Dictyota bipinnata* P. Crouan and H. Crouan in Mazé and Schramm 1878: p. 124.
Dilophus guineensis (Kützing) J. Agardh 1882: p. 108.

Plants are erect, 5–12 cm tall, and bushy; blades usually occur in one plane; they have dense dichotomous or rarely alternate branching, and are attached by fibrous rhizoids. Branches are strap-shaped, 0.5–4.0 mm wide, with smooth margins and blunt to acute tips. The medulla usually has 2–(4) layers, except at margins where it is one cell thick; cells are rectangular, to 80 µm in diam., and covered by a layer of small pigmented cells to 20 µm tall. Hairs are in scattered tufts. Unilocular sporangia, antheridia, and oogonia are scattered on both surfaces near the tips of blades. Unilocular sporangia are to 100 µm in diam.

Uncommon, easily overlooked, on limestone at MLW in the Keys; also known from Bermuda, the Caribbean, Gulf of Mexico (10 m), and Brazil, as well as the North and eastern tropical Atlantic.

Taylor 1960: p. 216; Earle 1969: 163; Littler and Littler 2000: p. 264.

Dictyota menstrualis (Hoyt) Schnetter, Hörnig, and Weber-Peukert 1987: p. 195, fig. 5, 6 (L. _menstrualis_: monthly, every month). Plate XIII, fig. 13, 14.

Basionym. *Dictyota dichotoma* (Hudson) J. V. Lamouroux var. *menstrualis* Hoyt 1927: p. 616.
Synonym. *Dictyota dichotoma* var. *latifrons* P. Crouan and H. Crouan in Mazé and Schramm 1878: p. 119.

The species is polymorphic, but can be separated from *Dictyota dichotoma*, a Mediterranean species, by chromosome counts and interbreeding (Schnetter et al., 1987).

Plants are bushy, erect, and to 15–35 cm tall; blades are regularly dichotomously branched, have angles of 15–45°, and are attached by rhizoids from blade margins and bases. Branches are straplike, sometimes twisted, 150–250 μm thick, 2–15 mm wide, 1.0–2.5 cm long, narrowed above, and with blunt to slightly tapering tips. The medulla in young parts is one layer thick but multilayered in older parts; cells are rectangular and 120–180μm tall. Surface cells are pigmented and 20–30 μm in diam. Unilocular sporangia are scattered and sterile hairs are absent.

Common, on seagrasses, algae, and limestone on all coasts, and to 30 m on the west coast, including the Florida Middle Grounds (Cheney and Dyer, 1974); also known from Virginia to Georgia (type location: North Carolina; Schneider and Searles, 1991), the Caribbean, Gulf of Mexico, Brazil, and eastern tropical Atlantic.

Taylor 1960: p. 218; Earle 1969: p. 157; Dawes 1974: p. 101; Schneider and Searles 1991: p. 158; Littler and Littler 2000: p. 266.

Dictyota mertensii (Martius) Kützing 1859: p. 15, pl. 36, fig. 1 (named for Carl Mertens, a German naturalist who studied Brazilian seaweeds in the 1700s). Plate XIII, fig. 15.

Basionym. *Ulva mertensii* Martius 1828 (1828–1834): p. 5, pl. 1; 1833: p. 21.

Synonym. *Dictyota dentata* J. V. Lamouroux 1809c: p. 331.

Plants are robust, bushy, erect, to 20 cm tall, and with obvious main axes; blades are alternately branched and attached by small fibrous matlike holdfasts. Primary blades have an iridescent blue-green sheen when alive, are to 6 mm wide, 80–120 μm thick, and branch every 1–2 cm; branchlets are spurlike, alternate, 1–2 mm long, and occur every 3–7 mm. Young tips are rounded, while older ones are pointed. The medulla has one layer of rectangular cells that are to 100 μm tall and covered by a layer of pigmented, rectangular cells 15–20 μm tall. Unilocular sporangia are scattered and solitary or in clusters of 4–8. Oogonia are 80–100 μm in diam. and solitary or in clusters of 2–3. Vegetative reproduction by small plantlets may be extensive (Quan-Young et al. 2004).

Common, in shallow waters in the Keys and to 15 m on the west coast; also known from Bermuda, the Bahamas, Caribbean, and Brazil (type location: Salvador; Guiry and Nic Dhonncha, 2006), as well as the eastern tropical Atlantic and Pacific.

Taylor 1960: p. 224; Earle 1969: p. 156; Dawes 1974: p. 100; Littler and Littler 2000: p. 266.

Dictyota pinnatifida Kützing 1859: p. 16, pl. 39, fig. 1 (L. *pinnatifida*: pinnately cut or divided). Plate XIII, fig. 16, 17.

Synonyms. *Dilophus alternans* J. Agardh 1882: p. 108.

Dictyota alternans (J. Agardh) Hörnig, Schnetter and Prud'homme van Reine 1993: p. 169.

Plants are bushy, erect, and 10–15 cm tall; blades alternately to irregularly branched, and attached by small fibrous holdfasts. Blades are strap-shaped, (0.5)2–4 mm wide, with branching angles of 15–45° above; margins smooth. Tips of blades are dichotomously branched and blunt, obtuse, or indented. The medulla has 2–5 layers at edges and one in the center; cells are rectangular, 50–80 μm thick, and covered by pigmented, rectangular cells 10–20 μm tall. Hairs occur in scattered tufts. Unilocular sporangia are scattered and solitary.

Common, on limestone in the Keys and in 12–26 m off Tampa Bay; also known from Bermuda, the Caribbean (type location: Antigua; Silva et al. 1996), Gulf of Mexico, Brazil, the eastern tropical Atlantic, and Indo-Pacific.

Taylor 1960: p. 216; Earle 1969: p. 163; Dawes 1974: p. 102; Littler and Littler 2000: p. 268.

Dictyota pulchella Hörnig and Schnetter 1988: p. 287–288, fig. 7–9 (L. *pulchella*: a diminutive of beautiful, a little beauty). Plate XIV, fig. 1–4.

Synonyms. *Dictyota divaricata sensu auct.*, non. J. V. Lamouroux 1809a: p. 43.

Dictyota linearis sensu auct., non (C. Agardh) Greville 1830: xliii.

Dictyota pulchella was described after hybridization studies showed that Caribbean and European specimens of *D. dichotoma* var. *intricata* (basionym = *D. divaricata* J. V. Lamouroux) did not interbreed with western Atlantic plants.

Plants are erect, in bushy to entangled clumps, 3–12 cm tall, and brown to green brown; blades are regularly dichotomously branched with wide angles of 90–120° and attached by small matlike fibrous holdfasts. Branches are 0.5–3 mm wide below and narrower above; tips are blunt, rounded to almost filamentous, and 0.1–0.2 mm wide. Medullary cells are in one layer, rectangular, 120–200 µm tall, and covered by small pigmented rectangular cells, 20–30 µm tall. Hairs are common, tufted, and scattered on midportion of blades. Unilocular sporangia occur singly or in compact clusters near blade centers. Oogonia are in scattered sori with a clear indusium, while antheridia are formed in scattered sori.

Common, on algae, seagrasses, and limestone in shallow waters in the Keys and from 10–55 m off Tampa Bay; also known from North Carolina, Bermuda, the Caribbean (type location: Punta La Loma, Columbia; De Clerk, 2003), Gulf of Mexico, and Brazil, as well as the eastern tropical Atlantic and Indo-Pacific.

Taylor 1960: p. 219, 221; Earle 1969: p. 160, 161; Dawes 1974: p. 101, 102; Schneider and Searles 1991: p. 160; Littler and Littler 2000: p. 268.

Lobophora J. Agardh 1894: p. 21 (Gr. *lobos*: a lobe, capsule, pod + *phora*: to carry).

Synonym. *Pocockiella* Papenfuss 1943a: 463–467.

Blades are flat, fan-shaped, often clustered, have a row of apical cells, and are attached by dense rhizoidal mats. Fronds have a medulla, cortex, and surface cells. Unilocular sporangia occur in scattered sori on both surfaces; they lack sterile hairs and produce 4–8 nonmotile meiospores (tetraspores).

Lobophora variegata (J. V. Lamouroux) Womersley *ex* E. C. Oliveira 1977: p. 217 (L. *variegata*: irregularly colored, with different patterns). Plate XIV, fig. 5, 6.

Basionym. *Dictyota variegata* J. V. Lamouroux 1809a: p. 40; 1809c: p. 331.

Synonym. *Pocockiella variegata* (J. V. Lamouroux) Papenfuss 1943a: p. 463–467, fig. 1–15.

Plants are morphologically variable, being crustose, prostrate, or erect. Erect fronds are crisp, to 8 cm tall, deltoid to fan-shaped, and attached by rhizoids at base or arise from prostrate blades; rhizoids may be moniliform and matted. Blades are 100–300 µm thick. The medulla has one layer; cells are colorless, rectangular, and 40–100 µm tall. The cortex has 2–3 layers, with cells pigmented and in vertical rows; surface cells 10–20 µm in diam. Hairs are 15–25 µm in diam. and scattered on surface. Unilocular sporangia are 50–100 µm in diam., 80–150 µm long, and in scattered sori. Oogonia are ovoid, in scattered sori, and covered by an indusium. Extensive vegetative reproduction occurs by means of small fragments (Quan-Young et al., 2004).

The morphology corresponds to varying levels of light and grazing (Littler and Littler, 2000). Erect plants occur in shallow water, prostrate ones in deeper water or shaded areas, and crusts in areas of intense grazing. Prostrate populations of *Lobophora variegata* may overgrow some scleractinian corals, *Agaricia agaricites* (Linnaeus), causing tissue mortality (i.e., degradation) and competitive replacement.

Common in shallow waters on all coasts, to 60 m off the west and to 90 m off the east coasts; also known from the Bahamas (31–107 m), Carolinas (14–48 m), Caribbean (type location: West Indies; Lipkin and Silva, 2002), Gulf of Mexico, and Brazil, as well as the Mediterranean, eastern tropical Atlantic, Red Sea, Indo-Pacific, and Australia.

Taylor 1960: p. 231; Earle 1969: p. 173; Dawes 1974: p. 102; Schneider and Searles 1991: p. 160; Littler and Littler 2000: p. 268, 270.

Padina Adanson 1763: p. 13, 586, *nom. cons.* (Gr. *padinos*: flat, a plain).

Blades are fan-shaped, clustered, and attached by compact rhizoidal mats arising from stipes. Apical edge of blade is enrolled and covers a row of apical cells. Blades are two to several cells thick, often split lengthwise into wedge-shaped segments, have zonate concentric rows of hairs, and are sometimes slightly calcified. Unilocular sporangia are scattered between bands of hairs or in sori; plurilocular antheridia and unilocular oogonia occur in sori, with oogonia covered by an indusial membrane.

Key to the species (8) of *Padina* plus the "*Dictyerpa*" stage

1. Axes with one or more large fan-shaped blades; apical margins enrolled 2
1. Axes compressed, entangled; tips upturned; blades minute . . . . *Dictyerpa jamaicensis*
    2. Blades more than 2 cells thick near base and lack calcification. 5
    2. Blades 2–3 cells thick throughout and lightly to heavily calcified 3
3. Plants from shallow (to 15 m) water; blade surfaces slightly calcified 4
3. Plants from 30–60 m; blades weakly calcified on lower surface. . . . . *P. profunda*
    4. Blades lightly calcified on upper surface, moderately calcified below, and with some uncalcified bands at 1–2 mm intervals . . . . *P. perindusiata.*
    4. Blades heavily calcified on upper surface, weakly calcified below. . . . . *P. sanctae-crucis.*
5. Blades mostly 4 cells thick throughout, 6–8 cells thick at base 6
5. Blades mostly 2 cells thick throughout, except at base that is 3–4 cells thick. 7
    6. Blades 4 cells thick throughout, except at base that is 4–9 cells thick. 8
    6. Blades 3–4 cells thick throughout. . . . .*P. glabra*
7. Blade bases 3 cells thick; unilocular sporangia having an indusium. . . . .*P. pavonica*
7. Blade bases 4 cells thick; unilocular sporangia lacking an indusium . . . . *P. boergesenii*
    8. Blades 150–250 µm thick in center; 6–9 cells thick at bases. . . . . *P. gymnospora*
    8. Blades 30–70 µm thick; 6 cells thick at bases . . . . *P. antillarum*

***Dictyerpa jamaicensis* Collins 1901: p. 251, a growth stage of *Padina* (Gr. *dictyo*: net + *erpa*: to creep). Plate XIV, fig. 7.**

Synonym. *Vaughaniella rupicola* Børgesen 1950a: 3–10, fig. 1–8; 1951: p. 11–14, fig. 4.

Cribb (1951) summarized the invalidation of the genus *Vaughaniella*. Littler and Littler (2000) and Silva et al. (1996) considered it to be a growth stage for several *Padina* species.

Plants grow in mats or turfs to 3 cm thick; they are noncalcified, have irregular, opposite, or dichotomous branching, and are attached by rhizoids from prostrate axes. Axes are strap-shaped to compressed, 0.7–1.0 mm wide, 2–3 cm long, and to 500 µm thick. Axial tips are turned up, hooked, or form minute fans. The medulla is irregular and to 6 cells thick, with cells 40–60 µm wide by 60–120 µm long; surface cells are pigmented, 20–30 µm in diam., and 30–50 µm long.

Uncommon, inconspicuous, in delicate mats on rubble and coral fragments in the Keys (to 10 m); also known from the Caribbean (type location: Jamaica), Gulf of Mexico (deep water), and Indian Ocean.

Littler and Littler 2000: p. 272.

Padina antillarum (Kützing) Piccone 1886 p. 36 (L. *antillarum*: of or from the Antilles, West Indies). Plate XIV, fig. 8.

Basionym. *Zonaria antillarum* Kützing 1859: p. 29, pl. 72, fig. 11
Synonym. *Padina tetrastromatica* Hauck 1887 (1886–1889): p. 43–44

Plants are erect, to 10.5 cm tall, and attached by rhizoidal mats. Blades are fan-shaped, noncalcified, to 11.5 cm wide, with an enrolled apical margin, often split into 0.5–2.0 cm strips, and with four layers of bricklike cells, 30–70 µm thick or more near the base. Hairs are in concentric lines on the side opposite the enrolled edge. Unilocular sporangia and oogonia are in irregular banded sori on the underside of fronds and not associated with hairs. Oogonia, but not sporangia, are covered by an indusium.

Rare, reported as drift near Cocoa Beach (Wynne and De Clerck, 1999a); also known from the Caribbean (type location: Trinidad; Wynne, 1998b), eastern tropical Atlantic, Indo-Pacific, and Australia.

Padina boergesenii Allender and Kraft 1983: p. 87–89, fig. 6C,H,I, 7C,D (named for the Danish phycologist F. Børgesen, who studied the seaweeds of the Danish West Indies and Canary Islands). Plate XIV, fig. 9.

Synonyms. *Padina australis sensu auct.*, non Hauck 1887 (1886–1889): p. 44.
Padina gymnospora (Kützing) Vickers 1905: p. 58 *sensu* W. R. Taylor 1960: p. 237; non *Zonaria gymnospora* Kützing 1859: p. 29, pl. 71, fig. II.

The type specimen of *Padina boergesenii* is four cells thick in midfrond and six to eight at the base. Caribbean *P. gymnospora* (= *Zonaria gymnospora*) is mostly tristromatic (Allender and Kraft, 1983), as well as dioecious, and has a "*Vaughaniella*" (*Dictyerpa*) stage (Thivy, 1959). Records of *P. gymnospora* from India (Thivy, 1959; Misra, 1966), the Philippines (Taylor, 1966), Vietnam (Pham, 1969), and Queensland (Cribb, 1966) may be referable to *P. boergesenii* (Allender and Kraft, 1983). The same confusion may exist for Atlantic records of Børgesen, (1914b, 1926), Taylor (1960, 1969), and Earle (1969), who record tristromatic fronds. Wynne (1998a,b, 2005a) listed *P. australis* as *sensu auct.* and a synonym for *P. boergesenii*. However, *P. australis* is distromatic, monoecious (Allender and Kraft, 1983), without a *Vaughaniella* stage, and considered a distinct species by John et al. (2004).

Plants are erect, 5–10 cm tall, in leaflike clusters, and attached by rhizoidal mats. Blades are fan-shaped, 5–20 cm wide, with enrolled apical margins and moderate calcification on upper sides and less so on lower sides. Mature blades are 3 layered and 75–125 µm thick in center; the margins are 2 layered and 50–70 µm thick, while the bases are 4 layered and 110–130 µm thick. Hairs are 20–25 µm in diam. and in concentric zones that alternate between surfaces in noncalcified bands. Sporangia are ovoid, 90–125 µm in diam., and alternate between hair bands. Oogonia are spherical and to 80 µm in diam.

Uncommon, on rubble (in wave-exposed areas), mangrove roots, and to 14 m in the Keys; also known from the Caribbean (type location: Virgin Islands; Silva et al. 1996), Brazil, the Mediterranean (introduced), Red Sea, Indo-Pacific, and Australia.

Taylor 1960: p. 237; Earle 1969: p. 165.

Padina glabra Gaillard 1966: p. 226, fig. 1–3 (L. *glaba*: smooth, not hairy). Plate XIV, fig. 10–12.

Blades are fan-shaped, noncalcified, to 10–12 cm long, and attached by basal rhizoids. Blades lack hairs, have a row of apical cells, and enrolled apical margins. Older fronds may split into 2–3 cm segments. Blades are 3–4 layered, with surface cells 24–30 µm wide, 25–27 µm high, and 48–62 µm long; medullary cells are 25 µm high if 4 layered or 50 µm high if 3 layered. Unilocular sporangia are 90–120 µm in diam., usually on the upper (enrolled) side of blades, and in irregular sori or concentric bands. Minute plants may be formed by *in situ* germination of meiospores in unilocular sporangia.

Rare, in drift near Cocoa Beach (Wynne and De Clerck, 1999a), Anastasia Island, and St. Augustine; also known from West Africa (type location: Dakar, Senegal; Gaillard, 1966) and southern India.

Padina gymnospora (Kützing) Sonder 1871: p. 47 (Gr. *gymnos*: naked + *spora*: seed, spore). Plate XIV, fig. 13, 14.

Basionym. *Zonaria gymnospora* Kützing 1859: p. 29, pl. 71, fig. II.
Synonyms. *Padina variegata* (J. V. Lamouroux) Hauck 1887 (1886–1889): p. 42.
Padina variegata (J. V. Lamouroux) Vickers 1905: p. 58.
Padina variegata (J. V. Lamouroux) Vickers *sensu* Børgesen 1914b: p. 205(49).
Padina howeana Børgesen 1920: p. 442.
Padina vickersiae Hoyt in M. Howe 1920a: p. 595.

See previous comments regarding confusion with *Padina boergesenii*, plus Silva et al. (1996) for a taxonomic clarification and recognition of *P. crassa* Yamada as a distinct species.

Blades are fan-shaped, clustered, 4–22 cm tall by 5–37 cm wide, sometimes lightly calcified, with enrolled apical margins, and on short stalks attached by rhizoidal mats; older blades often split into striplike segments. Blades have concentric bands of hairs 1.5–8 mm apart, and vary in thickness; margins are 2–3 layered with cells 50–150 µm in diam.; midregions are 4 layered with cells 150–250 µm in diam. and basal areas, with 6–8 layers of cells, are 150–220 µm in diam. Unilocular sporangia are in 1–2 bands between hair lines on both surfaces, ovoid, and 90–125 µm in diam. Antheridia are in bands, 200 µm wide. Oogonia are 30–65 µm in diam.; both oogonia and unilocular sporangia are covered by indusial cells.

Common, to 14 m and within tide pools of protected to moderately exposed sites on all coasts; also known from North Carolina southward, Bermuda, the Caribbean (type location: St. Thomas, Virgin Islands; Abbott and Huisman, 2004), Gulf of Mexico, and Brazil, as well as the eastern tropical Atlantic, Indo-Pacific, and Australia.

Taylor 1960: p. 236; Earle 1969: p. 172; Dawes 1974: p. 103; Schneider and Searles 1991: p. 161; Littler and Littler 2000: p. 272.

Padina pavonica (Linnaeus) Thivy in W. R. Taylor 1960: p. 234–235 (L. *pavonica*: peacock-like; resembling *Pavonica*, the peacock genus). Plate XIV, fig. 15.

Basionym. *Fucus pavonicus* Linnaeus 1753: p. 1162.

Silva et al. (1996) and Wynne (1998a,b) considered *Fucus pavonius* Linnaeus to be an illegitimate substitute for *F. pavonicus*.

Blades are clustered, fan-shaped, with concentric bands, to 22 cm tall by 12 cm wide, and attached by rhizoidal mats. Blades are moderately calcified above, less so below, and have enrolled apical margins. Marginal areas are 2 layered and 50–65 µm thick; older parts are 3–4 layered and 80–130 µm thick. Hairs are 15–25 µm in diam. and in concentric bands, 1.5–6 mm apart. Unilocular sporangia are in sori covered by a thin indusium. Oogonia are spherical and 40–50 µm in diam.; antheridia alternate with oogonia near hair bands.

Uncommon, growing on limestone and mangrove roots below MLW and to 20 m in the Keys;

also known from Bermuda, the Caribbean, and Gulf of Mexico, as well as the eastern North and tropical Atlantic, Mediterranean, Red Sea, and Indo-Pacific. Type location is "In Mari Europae australis" (Silva et al. 1996).

Taylor 1960: p. 234; Littler and Littler 2000: p. 274.

Padina perindusiata Thivy in W. R. Taylor 1960: p. 235, 236, 632, 812, 813, pl. 75, fig. 2 (Gr. _peri_: encircling, around + L. _indusiata_: having a tunic or cover). Plate XV, fig. 1.

Blades are fan-shaped, to 15 cm tall, 9 cm wide, with enrolled apical margins, and attached by rhizoidal mats. Blades have concentric bands of calcification, are 1–2 mm wide on the lower side and lightly so on upper surfaces. Blades are distromatic, 70– 90 µm in the center, and to 105 µm thick near the base. Upper cells are rectangular, 60–70 µm long, and 45–50 µm tall; lower cells are 30–60 µm long and 30–35 µm tall. Hairs are in concentric rows that alternate on opposite sides of the blade, and are 15–25 µm in diam. Unilocular sporangia are ovoid, to 170 µm in diam., in bands (0.7–2 mm in diam.), and covered by an indusial membrane.

Uncommon, on limestone and coral rubble on reef flats and to 14 m in the Keys (type location: Dry Tortugas); also known from the Greater Antilles.

Taylor 1960: p. 235; Earle 1969: p. 165; Littler and Littler 2000: p. 274.

Padina profunda S. Earle 1969: p. 167, fig. 62–68 (L. _profunda_: deep, from a great depth). Plate XV, fig. 2–4.

Blades are erect, 10–25 cm tall, clustered, fan-shaped, golden brown, lightly calcified, and attached by rhizoidal mats. Blade margins are entire or split, have enrolled apical margins, are 2–3 layered, and 80–150 µm thick; the midregion is 3–4 layered and 125–500 µm thick, while the basal area is 3–6 layered and 700µm thick. In surface view, outer cortical cells are 55–75 µm wide and 55–130 µm long; in cross section they are 27–37 µm tall. Hairs are 23–28 µm in diam., in faint concentric lines 1.8–2.5 mm apart, and alternate on the two blade surfaces.

Rare, from 60 m at the Dry Tortugas (type location); also known from North Carolina (45 m), the Bahamas, and Greater Antilles.

Earle 1969: p. 167; Schneider and Searles 1991: p. 162; Littler and Littler 2000: p. 276.

Padina sanctae-crucis Børgesen 1914b: p. 45, 46, fig. 27, 28, as '_sanctae crucis_'; continuous pagination = p. 201–202, fig. 153, 154 (L. _sanctae_: holy + _crucis_: a cross; probably named for St. Croix in the Virgin Islands). Plate XV, fig. 5, 6.

Synonym. *Padina jamaicensis* (Collins) Papenfuss 1977: p. 272.

Blades are fan-shaped, 5–15 cm tall, 5–9 (25) cm wide, clustered, chalky white, heavily calcified on upper surfaces, and attached by rhizoidal mats. Blades are split or strap-shaped, have concentric zones, and enrolled apical margins; they are also 2 layered with rectangular cells; the upper ones are 30–60 µm wide and 45–75 µm tall and lower ones 24–35 µm wide and 35–55 µm tall. Hairs are 15–25 µm in diam., in bands that alternate on the two surfaces, and 1–3 mm apart. Unilocular sporangia are ovoid, 60–120 µm in diam., 100–170 µm long, and in bands above each hair line. Oogonia are spherical and 30–50 µm in diam. Antheridia occur in broken bands. A *Dictyerpa* stage may be present (Abbott and Huisman 2004; Wynne 2005a).

Common, on limestone and seagrass blades (Humm, 1964) in the Keys to 4 m; also known from Bermuda, the Caribbean (type location: St. Croix, Virgin Islands: Abbott and Huisman, 2004), Gulf of Mexico, Brazil, Indo-Pacific, and Australia.

Taylor 1960: p. 237; Earle 1969: p. 171; Littler and Littler 2000: p. 276.

Spatoglossum Kützing 1843: p. 339; non *Spatoglossum* DeToni 1895: p. 246 (Gr. *spat*: a spoon + *glossum*: tongue-shaped).

Plants form both erect and flat blades that are irregularly palmate to marginally lobed and have a marginal row of apical cells. Blades have stalklike bases that are attached by rhizoidal holdfasts. Blades are several cells thick with medullary cells larger than the cortical cells. Unilocular sporangia are scattered over the blade surface and produce nonmotile meiospores. Plants are dioecious; female plants have scattered oogonia, while male plants have antheridia in small, scattered, and slightly raised sori.

Spatoglossum schroederi **(C. Agardh) Kützing 1859: p. 21, pl. 51, fig. 1 (source of name unknown, cited by C. Agardh "Mertens in lit. E. Brasilia"). Plate XV, fig. 7, 8.**

Basionym. *Zonaria schroederi* C. Agardh 1824: p. 265–266 (as '*schröderi*').

Synonym. *Spatoglossum areschougii* J. Agardh 1894: p. 39.

Blades are flat, iridescent, erect, to 30 cm tall, usually divided into alternate, palmate, or dichotomous lobes, and with short stalks attached by inconspicuous rhizoidal mats. Blades are 0.5–4.0 cm wide, to 180 µm thick, with rounded tips and a marginal row of apical cells. Margins are proliferous, undulate, irregularly dentate, and usually have pointed teeth. The medulla usually has 4 layers of rectangular cells to 50 µm in diam.; surface cells are pigmented, rectangular, and to 30 µm in diam. Hairs occur in scattered tufts. Unilocular sporangia are on both surfaces. Oogonia and antheridia are scattered, with the latter are in raised sori.

Common, in shallow waters in the Keys, in 10–20 m off the west coast, and at the Dry Tortugas; also known from North Carolina (17–46 m) southward, Bermuda, the Caribbean, Gulf of Mexico, Brazil (type location; Silva et al. 1996), Mediterranean, eastern tropical Atlantic, and Indo-Pacific.

Taylor 1960: p. 225; Earle 1969: p. 176; Dawes 1974: p. 104; Schneider and Searles 1991: p.163; Littler and Littler 2000: p. 276.

Stypopodium Kützing 1843: p. 341 (Gr. *stype*: the coarse part of flax + *podium*: foot).

Plants are erect, fan-shaped bushy fronds that split into narrow segments and are attached by massive rhizoidal holdfasts. Growth is by a marginal row of apical cells. Hairs occur in concentric bands. The medulla has large cells and is covered on both sides by smaller surface cells. Unilocular sporangia are in sori along lines of sterile hairs.

Stypopodium zonale **(J. V. Lamouroux) Papenfuss 1940: p. 205–206 (L. *zonale*: banded, zoned). Plate XV, fig. 9–11.**

Basionym. *Fucus zonalis* J. V. Lamouroux 1805: p. 38–40, pl. XXV, fig. 1.

Synonym. *Zonaria zonalis* (J. V. Lamouroux) M. Howe 1918: p. 507; 1920: p. 594.

Blades are erect, clustered, to 36 cm tall, flat, fan-shaped or straplike, 1–10 (15) cm wide, zonate at irregular intervals, with a marginal row of apical cells, stalked below, and attached by massive rhizoidal holdfasts. The medulla is 2 layered in young parts and 4 layered in older parts; cells are rectangular, 60–120 µm long and 40–100 µm tall. Surface cells are pigmented, small, 30–50 µm long, 10–20 µm tall, and cover the apical margin. Hairs are in transverse rows that alternate on each side. Unilocular sporangia are 100–145 µm in diam. and in sori along hair bands.

According to Pereira et al. (2004) different populations of this species vary in their levels of atomaric acid and stypoldione compounds that can deter herbivory. Uncommon, on hard substrata in 1–10 m at the Dry Tortugas and from 55 m in the Keys and the west coast; also known from Bermuda, the Caribbean (type location: Dominican Republic; Silva et al. 1996), Gulf of Mexico (deep water), Brazil, the eastern tropical Atlantic, and Indo-Pacific.

Taylor 1960: p. 232; Earle 1969: p. 177; Littler and Littler 2000: p. 278.

Taonia J. Agardh 1848 (1848–1901): p. 101 (G. *taon*: like a peacock).

Fronds are erect, flat, lack a midrib, have a marginal row of apical cells, and are attached by matted, branched rhizoids. Blades are often deeply and irregularly divided; the medulla consists of 4–5 layers of cells in an irregular arrangement and has a covering layer of quadrate pigmented cells mostly in longitudinal series. Blades are 2 cells thick near the apex and often 4–7 cells thick near the base. Sporangial and gametangial sori are scattered or in concentric zones on both surfaces. Reproductive structures are partially embedded. Sexual thalli are rare, with oogonia and antheridia usually on different individuals; however, there are some records of other *Taonia* species with antheridia and sporangia on the same thallus (Mathieson, 1966).

Taonia abbottiana D. Littler and M. Littler 2004: p. 420, figs. 1–7 (named for Isabella Abbott, an American specialist on red algal taxonomy and the author of the seaweed floras of California and Hawaii). Plate XV, fig. 12–15.

Plants are erect, to 20 cm tall, 15 cm wide, and 180–210 µm thick; blades are clustered, iridescent yellow brown, fan-shaped when young, and attached by matted, branched rhizoids. Older blades are often split and ragged to strap-shaped with smooth or irregular margins. Blades have distinct light-colored concentric bands (lines) of sterile hairs 1–3 mm apart near margins, 6–8 mm apart in older parts, and common on the dorsal surface. The medulla of young tissue is 2- or 4-celled in older parts; cells are 50–125 µm in length and 30–60 µm in thickness. Surface cells are irregularly rectangular, 15 by 20 µm, and absent at apical margins. Sporangia occur near to and parallel with hair bands; they are most commonly found on ventral surfaces, are spherical to ovoid, 80–100 µm, and on short two-celled stalks to 30 µm in diam. Oogonia and antheridia are unknown. The species is similar to *Stypopodium zonale* but has darker concentric lines. It was first identified as "*Taonia* sp." (Littler & Littler, 2000).

Uncommon, in <15 m on limestone at Sombrero Key in the Keys; also known from the Bahamas, Jamaica (type location: Discovery Bay), Puerto Rico, and Mexico.

SPHACELARIALES

Based on molecular studies, Draisma et al. (2001) first suggested that the order was paraphyletic and had three clades: (1) Sphacelariales *sensu strictu*; (2) *Onslowia endophytica* and *Verosphacela ebrachia*; and (3) *Choristocarpus tenellus*. However, after analysis of 27 species of Sphacelariales, Draisma et al. (2002) found strong evolutionary convergence (homoplasy) and a mostly unresolved phylogenetic tree. See Prud'homme van Reine (1982) for a monographic treatment of the order.

Filaments are multiseriate and attached by discs or rhizoids; erect axes are branched and have large apical cells with subapical cells that divide longitudinally to produce bi- or triseriate parenchymatous axes. Cells have many discoid plastids that lack pyrenoids, except in zoospores. The life histories are typically isomorphic and include the production of both unilocular and plurilocular sporangia. Vegetative reproduction is by deciduous propagules.

Sphacelariaceae

Plants are erect, with apical cells dividing transversely and subapical cells dividing longitudinally to form multiseriate axes. Lateral branches develop from subapical cells and hairs are produced from lens-shaped cells cut off laterally from apical or subapical cells. The cell walls blacken when treated with bleaching solution.

Sphacelaria Lyngbye 1819: p. xxxi, 103, *nom. cons.* (Gr. *sphakelos*: gangrene, decay; i.e., referring to the plant's dark tips).

Filaments are multiseriate and form erect tufts or mats; axes are cylindrical, with lower portions ecorticate or corticated by descending rhizoids. Asexual propagules have 2–3 arms with a large apical cell. Sexual reproduction is isogamous or anisogamous.

Key to the species (4) of *Sphacelaria*

1. Propagules slender, with arms elongate and longer than broad 2
1. Propagules thick, with arms short, broader than long or not extended at all. 3
   2. Axes 60–80 µm in diam.; propagules with three straight arms. . . . .*S. fusca*
   2. Axes 16–45 µm in diam.; propagules with 2–3 tapering arms . . . .*S. rigidula*
3. Axes 45–80 µm in diam.; propagule arms to 450 µm long . . . . *S. novae-hollandiae*
3. Axes 25–60 µm in diam.; propagule arms thick and to 165 µm long. . . . . *S. tribuloides*

Sphacelaria fusca (Hudson) S. F. Gray 1821: p. 333 (L. *fusca*: a somber brown, dusky). Plate XV, fig. 16, 17.

Basionym. *Conferva fusca* Hudson 1762: p. 486; 1778: p. 602.

Filaments are multiseriate, with 1–4 longitudinal walls, 1–3 cm tall, 60–80 µm in diam., and attached by monostromatic discs. Propagules are stalked, 225–280 µm tall; they increase in diam. from base to tips and have 3 arms (triradiate) that extend 100–300 µm, with each having a large apical cell.

Rare, on shells (to 2 m) at Bayport on the west coast; also known from Newfoundland to Virginia, Bermuda, and the Gulf of Mexico, as well as the northeast Atlantic (type location: Sidmouth or York, England; Womersley, 1987), Mediterranean, eastern tropical Atlantic, Indo-Pacific, and Australia.

Taylor 1960: p. 210; Earle 1969: p. 146.

Sphacelaria novae-hollandiae Sonder 1845: p. 50, fig. 10, as '*novae hollandiae*' (L. *novae*: new + *hollandiae*: Holland; Nova Hollandia, a Dutch name for Australia). Plate XV, fig. 18.

Filaments are multiseriate, have 1–4 longitudinal walls, grow in dense tufts, and are 5–20 mm tall; erect axes are straight, cylindrical, and 40–80 µm in diam., while prostrate ones are fibrous and attached by filamentous rhizoids. Axial segments are 1–2 diam. long; branching is irregular or clustered at tips. Hairs are common, deciduous, and 15–20 µm in diam. Propagules are triradiate with arms longer than broad, 20–30 µm in diam. by 150–450 µm long, and each with a large apical cell.

Uncommon, inconspicuous, on algae and limestone in the Keys (Bucher et al. 1990); also known from Bermuda and the Caribbean, as well as the Cape Verde Islands, Indo-Pacific, and Australia (type location: near Fremantle; Silva et al. 1996).

Taylor 1960: p. 211; Littler and Littler 2000: p. 252.

Sphacelaria rigidula Kützing 1843: p. 292; 1849: p. 463 (L. *rigidula*: little to very rigid). Plate XVI, fig. 1.

Synonym. *Sphacelaria furcigera* Kützing 1855: p. 27. pl. 90, fig. II.

Filaments are multiseriate with 1–4 longitudinal walls; they are scattered or in tufts, 0.5–3.0 cm tall, and from prostrate filaments that form a disc or creeping holdfast. Filaments are terete, 30–60 μm in diam., and with segments 1–2 diam. long. Deciduous hairs are abundant and 10–20 μm in diam. Propagules have 2–3 slender arms, and are 20–30 μm in diam.; their stalk and arms are to 450 μm tall and with a large apical cell on each arm. Unilocular sporangia are spherical and 50–70 μm in diam. Plurilocular sporangia are terete, 25–30 μm in diam., 45–65 μm long, sometimes irregular, and 30–40 μm in diam. by 30–60 μm long.

Common, epiphytic or endophytic on seagrasses and algae, as well as attached to limestone (to 3 m) on all coasts; also recorded from Newfoundland to North Carolina, Bermuda, the Caribbean, Gulf of Mexico, and Brazil, as well as the eastern North and tropical Atlantic, Mediterranean, Red Sea (type location: Karek Island, Iran; Silva et al. 1996), Indo-Pacific, and Australia.

Taylor 1960: p. 210; Earle 1969: p. 144; Dawes 1974: p. 99; Schneider and Searles 1991: p. 152; Littler and Littler 2000: p. 252.

Sphacelaria tribuloides Meneghini 1840a: p. (2), no. 1; 1840b: p. 512 (Gr. *tribulus*: three-pointed + *oides*: resembling, similar to). Plate XVI, fig. 2.

Filaments are multiseriate, have 1–4 longitudinal walls, occur solitarily or in dense tufts, and are 4–10 mm tall; prostrate axes are attached by short interwoven rhizoids. Erect axes are cylindrical with branching irregular or radial; segments are 25–60 μm in diam. by 1.0–1.5 diam. long. Deciduous hairs are 10–15 μm in diam. Propagules are stalked, with 2–3 arms, broadly triangular in side view, and 200 μm long; arms are short, thick, and have a span of 140–165 μm; each arm has an obvious apical cell.

Common, on coralline algae and forming a turf in wave-exposed sites in the Keys; also known from North Carolina, Bermuda, the Caribbean, Gulf of Mexico, and Brazil, as well as the eastern North and tropical Atlantic, Mediterranean (type location: La Spezia, Italy; Lipkin and Silva, 2002), Indo-Pacific, and Australia.

Taylor 1960: p. 211; Earle 1969: p. 146; Dawes 1974: p. 99; Schneider and Searles 1991: p. 153; Littler and Littler 2000: p. 252.

Stypocaulaceae

Plants in this family form dense tufts, to 30 cm tall, and are attached by rhizoidal discs or by a loose rhizoidal mass. Erect axes are much branched and end in an apical cell. The medulla has large cells, while the cortex has small pigmented cells that may be rhizoidal. Cells have many discoid plastids. Sporophytes have unilocular sporangia while gametophytes have plurilocular sporangia; gametes are anisogamous or oogamous.

Halopteris Kützing 1843: p. 292 (Gr. *halos*: salt, the sea + *pteris*: a fern, wing).

Plants are freely branched, in dense tufts, and usually attached by a matted holdfast of descending rhizoids from axial cells; a few species may have a basal discoid holdfast (Womersley, 1987). Axes are parenchymatous, multiseriate, and derived from a single large apical cell. Axes are densely corticated by descending rhizoids. Unilocular and plurilocular sporangia occur in clusters at the base of branches.

Halopteris filicina **(Grateloup) Kützing 1843: p. 293 (L. *filicina*: fernlike, resembling a fern). Plate XVI, fig. 3.**

Basionym. *Ceramium filicinum* Grateloup 1806: (1), fig. 1.

Plants are 5–10 cm tall, densely branched, naked below, and attached by rhizoids. Branchlets are multipinnate; primary axes have 4 large cells surrounded by small cortical ones. Unilocular and plurilocular sporangia are ovoid, sessile or short stalked, and occur in the axils of small branchlets.

Rare, reported from 46–98 m off the east coast (Hanisak and Blair, 1988); also known from the Bahamas (Ballantine and Aponte, 2003), eastern North and tropical Atlantic (type location: Montpellier, France; Abbott and Huisman, 2004), the Mediterranean, Indo-Pacific, and New Zealand.

Sedis incertae: Onslowiaceae

Draisma and Prud'homme van Reine (2001) established the family Onslowiaceae, designating it *sedis incertae* and placing *Onslowia* and *Verosphacela* in this family and not in the Choristocarpaceae (Prud'homme van Reine, 1993) or Sphacelariaceae (Searles and Leister, 1980). Draisma et al. (2003) suggested that a new order may be required for the Onslowiaceae as it does not belong in the Sphacelariales *sensu strictu* (Wynne, 2005a).

Plants are filamentous epiphytes or endophytes, irregularly branched, and have isomorphic haplo-diplontic life histories. Growth is by an apical cell that lacks transverse divisions of subapical cells. Filamentous cells may divide longitudinally once or twice to become multiseriate.

Onslowia Searles in Searles and Leister 1980: p. 38 (named for the type locality, Onslow Bay, North Carolina).

Plants are endophytic and reproduce asexually by propagules that usually have four apical cells.

Onslowia endophytica **Searles in Searles and Leister 1980: p. 37–38, fig. 6–15 (L. *endophytica*: endophytic, growing within plants). Plate XVI, fig. 6, 7.**

Filaments are endophytic, uniseriate to biseriate, 8–11 μm in diam., and with irregular to alternate branching. Emerging filaments form multicellular, colorless hairs with basal trichothallic growth or stalked globose cells that divide longitudinally to form 4 cells; each cell has a lens-shaped cell at its outer face. Branches produce rectangular cushion-shaped propagules that lack large apical cells. Unilocular sporangia are ovoid, sessile or stalked, and 30–45 μm in diam. by 45–60 μm long. The life history is isomorphic, with some specimens having unilocular sporangia and others plurilocular-like structures (Henry 1987a).

Rare, endophytic in *Halymenia* spp. and *Soliera filiformis* on the east coast; also collected subtidally (17–32 m) in Onslow Bay, North Carolina (type location: Searles and Leister, 1980) and Georgia.

Schneider and Searles 1991: p. 151.

Verosphacela E. C. Henry 1987b: p. 183 (named for the type locality, Vero Beach, Florida + Gr. *sphakelos*: gangrene; tips are dark when alive).

Filaments are endophytic, branched, and have one large apical cell; subapical cells undergo 1–2 longitudinal divisions to form multiseriate filaments. Propagules are globose, lack apical cells and arms, and often germinate from their basal cell.

Verosphacela ebrachia **E. C. Henry 1987b: p. 183–189, fig. 18–27 (L. *e*: a negative condition, lacking + *brachi*: arm; thus, without branches). Plate XVI, fig. 8.**

Filaments are endophytic, 20–30 µm in diam., irregularly branched, compressed beneath the outer layer of the host's wall, and have apical cells, 30–60 µm in diam. Emergent filaments often end in colorless endogenous hairs or unilocular and plurilocular sporangia. Propagules are globose, lack branches, and have 3–7 cells.

Rare, endophytic in *Spatoglossum schroederi* from Vero Beach, Florida (type location) during June–September; also recorded from Mexico (Mateo-Cid et al., 2002).

SYRINGODERMATALES

Henry (1984) established the order based upon the plant's unique heteromorphic life histories and extreme reduction of gametophytes.

Life histories are heteromorphic; sporophytes are small, and erect fronds have distinct apical cells and laterally fused filaments. Unilocular sporangia produce zoospores with two flagella or nonmotile spores that develop into gametophytes. Gametophytes are microscopic prostrate filaments with apical growth and plurilocular gametangia or spherical gametophytes that germinate from spores, lack vegetative growth, and form plurilocular gametangia directly.

Syringodermataceae

The family has the same characteristics as the order.

Syringoderma Levring 1940a: p. 6 (Gr. *Syringos*: a pipe, tube + *derma*: skin, hide).

The genus has the same characteristics as the order and family.

Syringoderma floridana **E. C. Henry 1984: p. 420–423, fig. 4–19 (L. *floridana*: of or from the state of Florida). Plate XVI, fig. 4, 5.**

Sporophytes are flat, fan-shaped blades, 10–25 mm by 7–25 mm wide, and composed of one layer (monostromatic) of laterally fused filaments that often split axially. The stipe is formed from downgrowing rhizoids that produce an irregular holdfast. Blade filament cells are rectangular, 13–22 µm in diam., 20–90 µm long, and have many discoid plastids that lack pyrenoids. Unilocular sporangia are 15–22 µm in diam., elliptical, sessile, and usually on the dorsal surface of fronds. Biflagellate zoospores are weakly motile and form spherical 2-celled gametophytes; each cell releases one biflagellate isogamete having a single plastid and eyespot.

Rare, type locality is in 56–90 m off the east coast at Jupiter Inlet, Florida; also known from the Canary Islands.

ECTOCARPALES

Synonyms. Chordariales, Dictyosiphonales, Mesogloiales, Scytosiphonales.

See Wynne (2005a) regarding ordinal synonymy and various molecular studies of brown algal orders (Draisma et al., 2001, 2003; Peters and Ramírez, 2001; Rousseau et al., 2001; Cho et al., 2004). The Ectocarpales were more narrowly defined earlier (Fritsch, 1945), but now include the Acinetosporaceae, Adenocystaceae, Chordariaceae, Ectocarpaceae, and Scytosiphonaceae (Peters and Ramírez, 2001). The traditional interpretation of an ancestral Ectocarpales has not been supported by several molecular studies outlined above (Wynne, 2005a).

Plants range from uniseriate, branched filaments to fleshy pseudoparenchymatous thalli. Cell divisions are diffuse, intercalary, or trichothallic, while plastids are discoid, starlike, or laminate. Life histories are usually an alternation of heteromorphic or isomorphic generations. Diploid sporophytes often have plurilocular and unilocular sporangia, with multiples of four zoospores arising from unilocular ones after meiosis. Haploid gametophytes have only plurilocular sporangia, which produce zoospores or gametes.

Acinetosporaceae

Peters and Ramírez (2001) established this monotypic family.

Plants are filamentous, uniseriate, and branched; growth is diffuse with defined meristematic areas at the bases of lateral branchlets.

Acinetospora Bornet 1891: p. 370 (Gr. *akinetos*: fixed + *spora*: seed).

Plants are filamentous, in tufts, epiphytic or saxicolous, and attached by rhizoids. Branching is irregular and sparse; most branches are short and occur at right angles to the axes; cells have many discoid plastids. Unilocular, plurilocular, and monosporangia may occur. They are stalked or sessile; plurilocular sporangia are conical and monosporangia spherical.

***Acinetospora crinita* (Carmichael *ex* Harvey in W. J. Hooker) Kornmann 1953: p. 205, fig. 1–14 (L. *crinita*: long haired or with a tuft of long hairs). Plate XVI, fig. 9–11.**

Basionym. *Ectocarpus crinitus* Carmichael *ex* Harvey in W. J. Hooker 1833: p. 326, pl. 330.

Synonym. *Acinetospora pusilla* (Griffiths and Harvey) Bornet 1891: p. 370.

Kornmann (1953) and Knoepffler-Péguy (1970, 1974) considered *Acinetospora crinita* to be a life history phase of other brown algae. M. Kim and Lee (1994) and Pedersen and Kristiansen (2001) cited several characters that justified the independent status of the genus and species.

Filaments are uniseriate, in tufts 4–8 cm tall, and are attached by lateral rhizoids or free-floating. Axes are entangled below, flexuous above, and with free tips. Branching is irregular to opposite. Branches are at right angles to axes and mostly short. Growth is intercalary and not trichothallic; cells are 15–56 µm in diam., 1–6 diam. long, and have many discoid plastids. Reproductive structures are variable. Unilocular sporangia are spherical, often stalked, and 25–50 µm in diam. Monosporangia are also spherical, 30–55 µm in diam., and produce one nonmotile spore. Plurilocular sporangia (acinetosporangia) are conical, 75–130 µm long, 24–42 µm in diam., and of two types: megasporangia have large locules and meiosporangia small ones. Both sporangia produce motile cells.

Common, in late winter and summer on rocks, jetties, and entangled with algae in Tampa Bay; also known from North Carolina, Bermuda, the Antilles, and Gulf of Mexico, as well as the eastern North and tropical Atlantic (type location: Appin, Scotland; Womersley, 1987), the Mediterranean, southern Africa, Japan, and Australia.

Taylor 1960: p. 214; Earle 1969: p. 128; Dawes 1974: p. 96; Schneider and Searles 1991: p. 111.

Feldmannia Hamel 1939b: xi (named for J. Feldmann, a French phycologist who conducted taxonomic, floristic, and culture studies of Mediterranean seaweeds).

Filaments are uniseriate, in small tufts, mostly basally branched, and attached by rhizoids. Branchlets have intercalary growth and end in vegetative hairs. Plastids are discoid, many per cell, and have single pyrenoids. Unilocular sporangia are rare and ovoid or spherical. Plurilocular sporangia are multiseriate, symmetrical, sometimes stalked, and produced below distinct growth zones on erect filaments.

Key to species (2) of *Feldmannia*

1. Plurilocular sporangia blunt-tipped, cylindrical, and not widest at the base . . . . *F. indica*
1. Plurilocular sporangia conical to fusiform. . . . . *F. irregularis*

Feldmannia indica **(Sonder in Zollinger) Womersley and A. Bailey 1970: p. 288–289 (L. *indica*: Indian, of or from India). Plate XVI, fig. 12.**

Basionym. *Ectocarpus indicus* Sonder in Zollinger 1854: p. 2–3, footnote.
Synonyms. *Giffordia indica* (Sonder in Zollinger) Papenfuss and Chihara in Papenfuss 1968b: p. 30.
Giffordia duchassaingiana (Grunow) W. R. Taylor 1960: p. 207–208; pl. 29, fig. 10.
Hincksia indica (Sonder in Zollinger) Tanaka 1990: p. 24, fig. 6, 7.

Filaments occur in soft masses or tufts, are irregularly branched, to 2–5 cm tall, originate from prostrate axes, and are attached by fine rhizoids. Filaments are uniseriate, 20–34 µm in diam.; cells vary from 0.5 diam. long below to 2.0 above; each cell has many discoid plastids. Unilocular sporangia are sessile, ovoid, to 70 µm in diam., and 110 µm long. Plurilocular sporangia are cylindrical, have blunt tips, are sessile or short stalked and 20–50 µm wide by 100–250 µm long.

Common, inconspicuous, on all coasts (to 20 m); in Tampa Bay during summer, growing on limestone, shells, sponges, or seagrasses; also recorded from Maryland, Bermuda, the Caribbean, Gulf of Mexico, eastern tropical Atlantic, Red Sea, Indo-Pacific (type location: Bima Bay, Indonesia; Silva et al., 1996), and New Zealand.

Taylor 1960: p. 207; Earle 1969: p. 136; Dawes 1974: p. 98; Littler and Littler 2000: p. 240.

Feldmannia irregularis **(Kützing) Hamel 1939b: XVII, 69, fig. 61F (L. *irregularis*: having parts of dissimilar size). Plate XVI, fig. 13.**

Basionym. *Ectocarpus irregularis* Kützing 1845: p. 234, pl. 62, fig. 1; 1849: p 454.
Synonyms. *Ectocarpus guadalupensis* P. Crouan and H. Crouan in Schramm and Mazé 1865: p. 1; P. Crouan and H. Crouan in Mazé and Schramm 1878 (1870–1877): p. 110.
Ectocarpus conifera Børgesen 1914b: p. 8–13, fig. 5, 6 (as '*coniferus*').
Giffordia conifera (Børgesen) W. R. Taylor 1960: p. 207.
Giffordia irregularis (Kützing) Joly 1965: p. 72–73, pl. VIII, fig. 111–119.
Hincksia conifera (Børgesen) Abbott 1989: p. 226.
Hincksia irregularis (Kützing) Amsler in Schneider and Searles 1991: p. 120–122, fig. 129–131.

M. Kim and Lee (1994) gave a detailed description of the species. Greer and Amsler (2004) found several cryptic forms of this species (as *Hincksia irregularis*) based on culture studies of North Carolina and Florida isolates.

Filaments are uniseriate, in tufts 5–7 cm long, and attached by prostrate filaments with rhizoids. Branching is alternate, secund to irregular, and at right angles to axes; branches are hair tipped. Main axes are 25–40 µm in diam. and have cells 0.5–4 diam. long and having many discoid plastids. Unilocular sporangia are rare, ovoid, sessile, and 24–45 µm wide by 50–100 µm long. Plurilocular sporangia are common, conical to fusiform, 20–40 µm wide by 40–200 µm long, and solitary or in a series near the base of main axes. The polymorphic species is distinguished by its conical to fusiform plurilocular sporangia that are limited to basal areas of the main axes.

Common, on hard substrata, seagrasses, and algae in Tampa Bay and the Keys (to 25 m); also known from Northumberland Strait (Canada) through the eastern North and tropical Atlantic, the Mediterranean (type location: Adriatic Sea; Coppejans et al., 2001b), and Indo-Pacific.

Taylor 1960: p. 207; Earle 1969: p. 135; Dawes 1974: p. 98; Schneider and Searles 1991: p. 120; Littler and Littler 2000: p. 242.

Hincksia J. E. Gray 1864: p. 12 (named for Miss Hincks, an Irish seaweed collector).

Synonym. *Giffordia* E.A.L. Batters 1898: p. 86.

Plants are filamentous, in tufts, and usually with dense irregular branching; epiphytic, epizoic, or on hard substrata, with plants attached by rhizoids and prostrate filaments. Plastids are discoid, numerous, and have 0–2 pyrenoids; cell divisions are diffuse or in an intercalary meristem. Unilocular sporangia are rare, while plurilocular sporangia are common; both types of sporangia are sessile.

Key to the species (2) of *Hincksia* (modified from Schneider and Searles, 1991)

1. Branches inserted at right angles to axes in middle of axial cells; plurilocular sporangia conical to ovoid . . . . *H. onslowensis*
1. Branches not as above; plurilocular sporangia cylindrical . . . . *H. mitchelliae*

Hincksia mitchelliae **(Harvey) P. C. Silva in Silva et al. 1987: p. 73, 130 (named for Miss A. Mitchell, an American collector from Cape Cod who sent specimens of this plant to Harvey). Plate XVI, fig. 14.**

Basionym. *Ectocarpus mitchelliae* Harvey 1852: p. 142–143, pl. XII.G, fig. 1–3.

Synonym. *Giffordia mitchelliae* (Harvey) Hamel 1939a: p. 66; 1939b: p. XIV, fig. 61c,d.

Filaments are uniseriate, in tufts or mats, 2–3(20) cm long, and attached by basal rhizoids or adrift. Branching is abundant, irregular, alternate, or spiraled; upper branches may be hair tipped. Main axes 17–50 µm in diam., 1–4 diam. long, and with tips tapering to 15 µm in diam. Cells have many discoid plastids. Plurilocular sporangia are often sessile, cylindrical, slightly tapering at tips, 40–220 µm long, 15–40 µm wide, and usually of two types: microsporangia (6–7 µm in diam.) and megasporangia (10–17 µm in diam.). Unilocular sporangia are rare, sessile or stalked, ovoid, and 25–50 by 50–100 µm.

Common, in summer–fall in Tampa Bay and year-round in the Keys; on jetties and limestone or entangled in drift algae on all coasts; also known from the Canadian Maritimes to the Caribbean (type location: Nantucket, Massachusetts), Bermuda, the Gulf of Mexico, Brazil, and Uruguay, as well as the eastern North and tropical Atlantic, Mediterranean, Indo-Pacific, and Australia.

Taylor 1960: p. 206; Earle 1969: p. 138; Dawes 1974: p. 98; Schneider and Searles 1991: p. 122; Littler and Littler 2000: p. 242.

Hincksia onslowensis (Amsler and Kapraun) P. C. Silva in Silva et al. 1987: p. 130 (named for Onslow Bay, North Carolina, the type locality). Plate XVI, fig. 15, 16.

Basionym. *Giffordia onslowensis* Amsler and Kapraun 1985: p. 94, fig. 1–3.

Filaments are uniseriate, entangled in tufts to 5 cm long, and attached by fine rhizoids of prostrate axes. Growth is diffuse; basal axes are irregularly branched. Cells are 20–50 µm in diam. and to 4 diam. long. Lower branches are sparse, short, and arise midway on cells; upper branches are profuse and arise on the upper ends of cells, 4–22 µm in diam. Plurilocular sporangia are ovoid, 20–50 µm long, 10–20 µm in diam., and sessile or stalked.

Uncommon, inconspicuous, forming a brown fuzz on rocks, seaweeds, mangrove roots, and to 50 m in the Keys; also known from North Carolina (type location: Onslow Bay), Georgia, the Lesser Antilles, Brazil, and Canary Islands.

Schneider and Searles 1991: p. 123; Littler and Littler 2000: p. 244.

Kuetzingiella Kornmann in Kuckuck 1956: p. 293, 314 (named for F.T. Kützing, a German phycologist who published books on algae in the middle of the nineteenth century + *ella*: a diminutive; thus a small ectocarpoid alga).

Based on ontogenetic and morphological studies M. Kim and Lee (1994) placed *Kuetzingiella* with *Acinetospora*, *Feldmannia*, and *Hincksia*.

Plants are uniseriate prostrate filaments with short cells; erect axes are simple or sparsely branched. Basal axes form pseudoparenchymatous mats. Growth is diffuse or by small scattered meristems. Cells have numerous discoid plastids, each with a single pyrenoid. Plurilocular and unilocular sporangia are sessile and on basal filaments or short (1–3-celled) lateral stalks on erect axes.

Kuetzingiella elachistaeformis (Heydrich) M. Balakrishnan and Kinkar 1981: p. 25–26, fig. 20, 99–100 (Gr. _elachistos_: very small + _formis_: shape; also the brown algal genus _Elachista_; thus, formed like _Elachista_). Plate XVI, fig. 17, 18.

Synonym. *Ectocarpus elachistaeformis* Heydrick 1892: p. 470–471, pl. 25, fig. 14.

A number of taxa have been confused under this name (Silva et al. 1996); plants from the Indian Ocean lack hairs and have discoid plastids (Balakrishnan and Kinkar, 1981), while those from North Carolina have band-shaped ones (Amsler, 1991).

Plants are filamentous, uniseriate, in small fine tufts to 1.5 mm tall, and with prostrate axes attached by penetrating haptera. Branching is sparse and mostly from axial bases. Main axial cells are 10–18 µm in diam. by 1–3 diam. long and have parietal and ribbonlike plastids. Plurilocular sporangia are elongate to fusiform, 15–25 µm in diam, 60–200 µm long, and taper to the tips.

Common, on all coasts, in fall and winter in Tampa Bay; entangled with other drift algae, or attached to buoys, fishing lines, and mangrove roots, or epiphytic on other seaweeds or seagrass blades via penetrating rhizoids. Also known from Virginia southward, Bermuda, the Caribbean, Gulf of Mexico, Brazil, the Mediterranean, and Indo-Pacific (type location: Mandang, Papua New Guinea; Lipkin and Silva, 2002).

Taylor 1960: p. 202; Earle 1969: p. 133; Dawes 1974: p. 97; Schneider and Searles 1991: p. 115; Littler and Littler 2000: p. 240.

Chordariaceae

Synonyms. Dictyosiphonaceae, Elachistaceae, Leathesiaceae, Myrionemataceae, Myriotrichiaceae, Punctariaceae, Spermatochnaceae, Stilophoraceae, and Striariaceae.

Based on molecular studies, Peters and Ramírez (2001) transferred the Chordariales and Dictyosiphonales to the Ectocarpales and reduced them to families.

Sporophytes are macroscopic and erect, cylindrical, branched, and have uniaxial or multiaxial construction. Growth is mostly trichothallic, with intercalary meristems near branch tips and terminal hairs. The medulla consists of colorless elongate filaments, while the cortex has pigmented filaments that produce colorless hairs with basal meristems. Plastids are many and discoid. Unilocular sporangia are formed at bases of cortical filaments. Gametophytes, where known, are small, microscopic, and similar to *Ectocarpus*.

Asperococcus J.V. Lamouroux 1813: p. 277 (L. *asper*: rough + *coccus*: berry-shaped).

Sporophytes are tubular and attached by discoid holdfasts. Outer cortical cells are small and pigmented and inner ones large and colorless. Tufts of colorless multicellular hairs occur on the surface in sori mixed with unilocular and plurilocular sporangia. Gametophytes are microscopic, filamentous or discoid, and have erect filaments with plurilocular sporangia.

***Asperococcus fistulosus* (Hudson) W. J. Hooker 1833: p. 277 (L. *fistulosus*: hollow, tubular, pipelike). Plate XVI, fig. 19–21.**

Basionym. *Ulva fistulosa* Hudson 1778: p. 569.

See Taylor (1960), Earle (1969), Fletcher (1987), Womersley (1987), and Wynne (2005a) for synonymy.

Sporophytes are irregularly cylindrical, hollow to 50 cm tall, 2–5(20) mm in diam., basally branched, tapering to narrow stalks, and attached by rhizoidal discs. Surface cells are elongate, 8–26 µm wide by 10–36 µm long, and 9–22 µm in cross section. Sori are scattered and have elongate, multicellular, sterile hairs; unilocular sporangia are oval to spherical, 38–70 µm long, and either sessile or on 1-celled stalks.

Rare and questionable for Florida (Taylor, 1960; Earle, 1969); known from Baffin Island, northern Labrador to Virginia, North Carolina, Iceland, Norway to Spain (type location: England; Womersley, 1987), the Baltic Sea, Mediterranean, Madeira Archipelago, Canary Islands, and Australia.

Taylor 1960: p. 259; Earle 1969: p. 199 (both as uncertain records).

Cladosiphon Kützing 1843: p. 329 (Gr. *clado*: a branch + *siphon*: tubular, a pipe).

See Sansón et al. (2006) for a comparison of the species in this genus.

Sporophytes are wormlike, gelatinous, and irregularly branched; axes are cylindrical, with multiaxial construction, and sometimes hollow. Medullary filaments are 4–8 times longer than wide and united to form pseudoparenchyma. The subcortex is 1–3 layers thick and composed of small cells, plus simple or basally branched photosynthetic filaments. Colorless hairs are present. Unilocular sporangia are at bases of filaments, while plurilocular sporangia are terminal.

***Cladosiphon occidentalis* Kylin 1940: p. 27–28, fig. 15f, pl. 3, fig. 8 (L. *occidentalis*: west, western, or occidental). Plate XVII, fig. 1, 2.**

Synonyms. *Eudesme zosterae sensu auct.*, non (J. Agardh) Kylin 1907: p. 85, fig. 20; also *sensu auct.*, non 1933: p. 57; 1947: p. 57, fig. 49, 50.

Cladosiphon zosterae sensu auct., non (J. Agardh) Kylin 1940: p. 28, pl. 4, fig. 9.

See Wynne (1998a) regarding confusion of *Cladosiphon zosterae* (and its various synonyms) with

Eudesme virescens (Carmichael *ex* Berkeley) J. Agardh. Guiry and Nic Dhonncha (2006) included *C. occidentalis* as a synonym of *C. zosterae*, while Sansón et al. recognized both species and outlined diagnostic features.

Plants are soft, gelatinous, wormlike, often densely branched, 10–30 cm long, and attached by a basal disc. Branches are cylindrical, 0.7–3 mm in diam., hollow or solid at bases, and covered by fine colorless hairs, 8–15 µm in diam. and to 150 µm long. Inner medullary cells are compact, longitudinally elongate, cylindrical, 50–130 µm in diam., and 200–400 µm long. Surface filaments are pigmented, radially oriented, 100–300 µm long, and basally branched. Unilocular sporangia are at bases of surface filaments, ovoid, 30–40 µm wide, and 50–80 µm long. Plurilocular sporangia are near branch tips and to 50 µm long.

Common, on seagrass blades and seaweeds to 12 m on all coasts (type location: Dry Tortugas; Kylin 1940); also known from Labrador to the Caribbean, Bermuda, the Bahamas, Gulf of Mexico, Mediterranean, and Indian Ocean.

Taylor 1960: p. 247, 248; Earle 1969: p. 180, 182; Dawes 1974: p. 105, 106; Schneider and Searles 1991: p. 136; Littler and Littler 2000: p. 244.

Elachista Duby 1830: p. 972, as '*Elachistea*,' *nom.* and *orth. cons.* (Gr. *elachistos*: very small).

Plants are small brush- to cushionlike epiphytes with bases of densely packed colorless filaments that may penetrate the host and cause a tumorlike appearance. Erect filaments (i.e., assimilators) arise from basal branches and are long, free, uniseriate, and photosynthetic. Sterile hairs are uncommon; each cell has multiple discoid plastids. Unilocular sporangia are ovoid to pyriform, while plurilocular sporangia are cylindrical; both types of sporangia occur mixed with sterile hairs (paraphyses).

***Elachista minutissima* W. R. Taylor 1928: p. 113, pl. 14, fig. 17; 1960: p. 245, pl. 29, fig. 11 (L. *minutissima*: extremely small). Plate XVII, fig. 3.**

Plants form minute tufted epiphytes with basal filaments in cryptostomata of *Sargassum* and produce erect, basally branched, photosynthetic filaments, plus sterile hairs and sporangia. Erect filaments are to 425 µm tall and 9–13 µm in diam.; cell lengths are equal or twice their diam.; sterile hairs are 40–56 µm long and 4.5–5.0 µm in diam. Plurilocular sporangia are 4–6 µm in diam., mostly uniseriate, and 38–81 µm long.

Uncommon, from the Dry Tortugas (type location) and Mississippi Sound.

Taylor 1960: p. 245; Earle 1969: p. 179.

Hecatonema Sauvageau 1897: p. 248; 1898: p. 88 reprint (Gr. *hecato*: many, a hundred + *nema*: thread).

See Fletcher (1987) regarding life history affiliations with *Punctaria* (Clayton, 1974), *Myriotrichia* (Loiseaux, 1969), and *Asperococcus* (Pedersen, 1984).

Crusts are disclike with distromatic basal filaments and sometimes with rhizoids. Erect uniseriate filaments and colorless hairs arise from discs, as well as pigmented cysts (ascocysts) and sporangia. Unilocular sporangia are uncommon, while plurilocular sporangia are more abundant, biseriate, and sessile or stalked.

Hecatonema floridanum (W. R. Taylor) W. R. Taylor 1960: p. 241, as '_floridana_' (L. _floridanum_: of or from the state of Florida). Plate XVII, fig. 4, 5.

Basionym. *Phycocelis floridana* W. R. Taylor 1928: p. 109, pl. 14, fig. 19.

Plants form minute clusters of loose prostrate filaments or compact discs that are attached by fine rhizoids. Disks are usually 2–3 cells thick and produce erect uniseriate photosynthetic filaments to 12 cells in length; cells are 9–12 µm in diam. and 18–25 µm long. Hairs are rare, with basal cells 7–11 µm in diam. and 11–20 µm long. Plurilocular sporangia are 13–30 µm in diam., 50–80 µm long, spindle-shaped, and have blunt or pointed tips.

Rare, an epiphyte on "*Zonaria*" (probably *Stypopodium zonale* or *Taonia abbottiana*) in shallow waters of the Dry Tortugas (type location); also known from North Carolina and Brazil.

Taylor 1960: p. 241; Schneider and Searles 1991: p. 130.

Hummia Fiore 1975: p. 497–498 (named for Harold J. Humm, an American phycologist who studied seaweeds from Newfoundland to Florida and the Caribbean)

Synonym. *Farlowiella* Kornmann in Kuckuck 1956: p. 293.

Sporophytes are macroscopic, cylindrical, usually unbranched, parenchymatous, and have trichothallic growth with pigmented uniseriate hairs that arise from discoid crusts or cushions of cells. The medulla has four rows of large, colorless, isodiametric cells, while the cortex is composed of 1–3 layers of small pigmented cells, plus deciduous, colorless hairs. Gametophytes are uniseriate filaments with alternate to opposite branches, colorless terminal hairs, and stalked plurilocular sporangia.

Hummia onusta (Kützing) Fiore 1975: p. 497–498 (L. _onusta_: overloaded, full). Plate XVII, fig. 6–9.

Basionym. *Ectocarpus onustus* Kützing 1849: p. 457; 1855: p. 22, pl. 74, fig. 1.
Synonyms. *Stictyosiphon subsimplex* Holden 1899: p. 198, pl. 9, fig. a–f.
Ectocarpus subcorymbosus Holden *ex* Collins 1905: p. 227, *nomen nudum*.
Myriotrichia scutata Blomquist 1954: p. 37, pl. 6.
Myriotrichia subcormybosa (Holden *ex* Collins) Blomquist 1958b: p. 24.
Farlowiella onusta (Kützing) Kornmann in Kuckuck 1956: p. 321, fig. 14.

The species combines two taxa: the sporophyte *Stictyosiphon subsimplex* and gametophyte *Myriotrichia subcorymbosa* (Fiore, 1975, 1977).

Sporophytes have soft erect axes to 5(12) cm tall and 0.4–1.0 mm in diam. Axes are simple or with occasional short branches and attached by rhizoids. Unilocular sporangia are globular, 30–60 µm in diam., and 50–80 µm long, while plurilocular sporangia are stalked, cone-shaped, 15–25 µm, and 25–45 µm.

Gametophytes are small cushion or discoid crusts, with erect filaments, and attached by rhizoids. Erect filaments are 1–5 mm tall, 10–15 µm in diam., with cells 2–4 diam. long. Filaments have plurilocular sporangia, 20–30 µm in diam., 100–30 µm long, and have hairlike microgametangia, 3–5 cells wide, or obtuse macrogametangia, 2–3 cells wide.

Common, both phases occur on seagrasses on the west coast during winter; also known from the Canadian Provinces to the Caribbean, and Gulf of Mexico (type location: Galveston, Texas).

Taylor 1960: p. 201; Earle 1969: p. 193, 197; Dawes 1974: p. 110, 111; Schneider and Searles 1991: p. 141.

Myrionema Greville 1827a: pl. 300 (Gr. *myrio*: numerous, myriad + *nema*: thread).

Plants form small, epiphytic discs; the prostrate radiating filaments form unistratose discs, and each cell bears a short, erect, photosynthetic filament, a hair, cyst (ascocysts), or sporangia. Growth is marginal; plurilocular sporangia are uniseriate or biseriate.

Key to the species (2) of *Myrionema*

1. Ascocysts (pigmented cysts) and endogenous hairs present . . . . *M. magnusii*
1. Ascocysts and endogenous hairs absent. . . . . *M. strangulans*

Myrionema magnusii **(Sauvageau) Loiseaux 1967a: p. 338 (named for Paul W. Magnus, a German botanist who studied plants of the North Sea). Plate XVII, fig. 10.**

Basionym. *Ascocylus magnusii* Sauvageau 1927: p. 13.
Synonyms. *Myrionema orbiculare sensu auct.*, non J. Agardh 1848 (1848–1901): p. 48.
Ascocyclus orbicularis sensu auct., non (J. Agardh) Kjellman 1890: p. 80.

Loiseaux (1967b) reported two forms of non-discoid plants in culture: one had filaments 8–10 µm in diam. and plurilocular sporangia similar to those on the discs, the other had filaments 12–15 µm in diam. and lacked hairs or cysts.

Plants are small epiphytic discs, 0.5–1.5 mm in diam. Each disc cell has 1–3 plastids; edge cells are rectangular, 3–7 µm by 6–13 (20) µm, while central cells are rounded and 4–9 µm in diam. Endogenous hairs, if present, are 6–10 µm in diam., sheathed, and have basal meristems; pigmented cysts (ascocysts) are 8–11 µm in diam. by 45–150 µm long. Plurilocular sporangia are uniseriate (rarely biseriate), 7–10 µm in diam., and 30–70 µm long. Unilocular sporangia are in clusters of 2–3, and are 14–16 µm in diam. by 35 µm long.

Uncommon, easily overlooked, on seagrasses in winter and spring within Tampa Bay; also known from Newfoundland to North Carolina, Bermuda, and the Gulf of Mexico, as well as the eastern North and tropical Atlantic, Mediterranean, and Indo-Pacific.

Taylor 1960: p. 242; Earle 1969: p. 178; Dawes 1974: p. 106; Schneider and Searles 1991: p. 132.

Myrionema strangulans **Greville 1827a: pl. 300 (L. *strangulo*: to choke, strangle). Plate XVII, fig. 11, 12.**

Synonym. *Myrionema vulgare* Farlow 1881 ('1882'): p. 79.

Plants are small discoid epiphytes, 1–2 mm in diam.; cells of basal filaments are 5–9 µm in diam. and 1–3 diam. long. Erect photosynthetic filaments, 50–100 µm tall and to 9 cells long; lower cells are cylindrical and upper ones subspherical. Hairs, if present, are 8–13 µm in diam.; cells have 1–3 discoid plastids, and each has a single pyrenoid. Unilocular sporangia, which are at the bases of filaments, ovoid, 20–35 µm in diam., and 35–65 µm long. Plurilocular sporangia are sessile or stalked, cylindrical, 7–11 µm in diam., 15–50 µm long, and with blunt tips.

Uncommon, inconspicuous, on seagrasses on the west coast and at the Dry Tortugas; also known from the Canadian Arctic to the Caribbean and Gulf of Mexico, as well as the eastern North and tropical Atlantic (syntype locations: Appin and Isles of Iona and Staffia, Scotland; Womersley, 1987), Mediterranean, Indo-Pacific, Australia, and Antarctica.

Taylor 1960: p. 240; Earle 1969: p. 179; Dawes 1974: p. 106; Schneider and Searles 1991: p. 133.

Myriotrichia Harvey 1834: p. 299 (Gr. *myrio*: numerous + *trichon*: a small hair).

Plants are filamentous with bases of prostrate uniseriate filaments attached by rhizoids. Erect axes are uniseriate below, parenchymatous above, and covered with short radial branchlets; growth is by intercalary divisions. Unilocular sporangia are solitary or in opposite to whorled clusters; plurilocular sporangia are uniseriate or multiseriate, laterally attached, and solitarily or in clusters; both types of sporangia may occur on the same plant.

Key to the species (2) of *Myriotrichia*

1. Erect filaments, often uniseriate below, multiseriate and parenchymatous above; plurilocular sporangia to 8 µm in diameter . . . . *M. clavaeformis*
1. Some erect filaments pluriseriate; sporangia 18–20 µm in diameter. . . . .*M. occidentalis*

Myriotrichia clavaeformis Harvey 1834: p. 300, pl. 188 (L. *clava*: club + *formis*: shaped). Plate XVII, fig. 13.

Synonyms. *Streblonema sphaericum* (Derbès and Solier in Castagne) Thuret in Le Jolis 1863: p. 73?

Myriotrichia repens Hauck 1879: p. 242.

According to Pedersen (1978), *Myriotrichia repens*, *M. filiformis* Harvey, and *M. clavaeformis* represent progressive stages in the development of the same species for which the name *M. clavaeformis* has priority. It is a microscopic epiphyte of various hosts (Earle, 1969). Under suboptimal light and salinity conditions one growth form, *M. repens*, is prevalent (Fletcher, 1987). Pedersen (1978) noted that *M. repens* resembled *Streblonema sphaericum*, which is usually placed in the Ectocarpaceae (Womersley, 1987).

Filaments form small tufts, 0.25–1 mm tall, with prostrate filaments attached by rhizoids. Erect axes are simple and uniseriate, with cells 12–25 µm in diam. and 2–4 diam. long. Hairs may be on basal filaments or at the tips of erect axes. Unilocular sporangia are spherical or ovoid, 25–50 µm in diam., sessile or stalked, often terminate in erect filaments, and are solitary or in whorls around axes. Uniseriate plurilocular sporangia are clustered at tips of erect axes and to 8 µm in diam.

Rare, on *Sargassum* near the mouth of the Homosassa River on the west coast of Florida (as *M. repens*); also known from Bermuda and the Gulf of Mexico, as well as the North Atlantic (type location: Torquay England; Womersley, 1987), Mediterranean, Black Sea, Canary Islands, and Australia.

Taylor 1960: p. 257; Earle 1969: p. 196.

Myriotrichia occidentalis Børgesen 1920; p. 437–440, fig. 413–414 (L. *occidentalis*: west, western, occidental). Plate XVII, fig. 14.

The species may be part of the life history of a taxon provisionally placed in the genus *Hecatonema* (Fletcher, 1987; Womersley, 1987).

Plants are filamentous, to 1 mm tall, with basal filaments attached by irregularly branched rhizoids. Erect filaments are multiseriate above with cells 18 µm in diam. and 18 µm long, often hair-tipped, with cells about 12 µm in diam. Sporangia are sessile or stalked; unilocular sporangia are spherical and 18–20 µm in diam.

Common locally, in winter, on seagrasses and seaweeds near Bayport, and to 10 m on the west coast; also known from the Antilles (type location: Virgin Islands) and Gulf of Mexico.

Taylor 1960: p. 257; Earle 1969: p. 195; Dawes 1974: p. 110.

Nemacystus Derbès and Solier 1850: p. 269 (Gr. *nema*: a thread + *cystus*: a cyst).

Plants are branched irregularly to dichotomously and have wide forks. The central filament is visible near the tips of branches and axes are solid and cylindrical above. The central tissue of older axes is pseudoparenchymatous as a result of the outgrowth of corticating filaments, while the outer tissues consist of unbranched photosynthetic filaments with sporangia at their bases.

Nemacystus howei (W. R. Taylor) Kylin 1940: p. 49 (named for M. Howe, a botanist and director of the New York Botanical Garden, who studied seaweeds from Bermuda and the Bahamas and North American bryophytes). Plate XVII, fig. 15, 16.

Basionym. *Castagnea howei* W. R. Taylor 1928: p. 111, pl. 15, fig. 1–8.

Plants are soft, mucilaginous, and to 40 cm long; main axes are to 1 mm in diam. and attached by a rhizoidal holdfast. Branching is abundant, to 3 orders, and alternate to irregular; branches are slender, cylindrical, and often entangled. Axes are cylindrical and their uniaxial structure is visible below apical cells. Central medullary cells are 60–170 µm in diam. and 130–1400 µm long. Surface filaments are unbranched, 8–13 cells long, with rounded cells (10–15 µm in diam.), that have several discoid plastids per cell; these cells produce colorless hairs, 8–10 µm in diam. Unilocular sporangia are ovoid, 20–40 µm in diam., and 20–45 µm long; multiple sporangia (to 6) occur at the bases of filaments or on central cells; plurilocular sporangia are 6–8 µm by 40–80 µm, and on medullary cells.

Uncommon, epiphytic, entangled with other algae, or on limestone on the west coast (13–55 m) during winter (type location: Dry Tortugas); also known from North Carolina, Bermuda, and the Canary Islands.

Taylor 1960: p. 249; Earle 1969: p. 184; Dawes 1974: p. 107; Schneider and Searles 1991: p. 138.

Stilophora J. Agardh 1841: p. 6, *nom. cons.* (Gr. *stilo*: a point + *phora*: produce, carry).

Sporophytes are macroscopic, erect, with dichotomous to irregular branching, and attached by discoid holdfasts. Axes have a cluster of apical cells, a pseudoparenchymatous medulla, and unbranched cortical branchlets. Unilocular and plurilocular sporangia arise from short tufts of filaments. Gametophytes are microscopic filaments.

Stilophora tenella (Esper) P. C. Silva in Silva et al. 1996: p. 624, 928–929 (L. *tenella*: delicate, very tender, soft). Plate XVII, fig. 17, 18.

Basionym. *Fucus tenellus* Esper 1800 (1797–1800): p. 197, pl. CIX.

Synonym. *Chordaria rhizodes* (C. Agardh) C. Agardh 1817: p. 15.

Plants are erect and 10–30 cm tall; axes are slender, terete, to 3 mm wide, with dichotomous to irregular branching, and attached by rhizoids. Axes are hollow below, solid above with 4–5 closely packed filaments; the cortex is pseudoparenchymatous and firm. Surface filaments are 75–85 µm long by 3.5 µm in diam. at bases and 9–13 µm in diam. and photosynthetic near tips. Unilocular sporangia are club-shaped, 22–32 µm in diam., and 36–60 µm long. Plurilocular sporangia are uniseriate, with 4–10 cells, and 10–12 µm in diam. by 30–50 µm long.

Common, during winter and early spring, adrift and to 3 m on limestone, algae, and seagrasses on the west coast; also known from Prince Edward Island to North Carolina, Bermuda, and the Gulf of Mexico, as well as eastern North and tropical Atlantic, Mediterranean (type location: Trieste, Italy), Indo-Pacific, and Australia.

Taylor 1960: p. 250; Earle 1969: p. 185; Dawes 1974: p. 107; Schneider and Searles 1991: p. 139.

Ectocarpaceae

See previous comments regarding the phylogeny of the Ectocarpales (Peters and Ramírez, 2001) and the inclusion of other families and orders.

Plants are filamentous, endophytic, epiphytic, epizoic, or epiplithic. Filaments are erect or prostrate and lack basal crusts; plastids are rod-, band-, or disc-shaped. Life histories are isomorphic or slightly heteromorphic.

Bachelotia (Bornet) Kuckuck ex Hamel 1939a: p. 66; 1939b: p. X (named for A. J. Bachelot de la Pylaie, an early French explorer and naturalist).

Synonym. *Pylaiella* subg. *Bachelotia* Bornet 1889: p. 8.

Pedersen (1984) placed *Bachelotia* and *Pylaiella* in the Pylaiellaceae (Tilopteridales) based on the pseudoparenchymatous construction derived from longitudinal divisions and sporangia formed by transformation of vegetative cells. Silva et al. (1996) and Wynne (2005a) retained both genera in the Ectocarpaceae and Draisma et al. (2003) listed *Bachelotia* as having doubtful taxonomic affinities.

Filaments are uniseriate, sparsely branched, and attached by prostrate axes (when epiphytic) or free-floating. Plastids are rod-shaped and in 1–2 stellate clusters per cell; sporangia are intercalary and in branched uniseriate filaments.

Bachelotia antillarum **(Grunow) Gerloff 1959, p. 38 (L. *antillarum*: of or from the Antilles, West Indies). Plate XVII, fig. 19, 20.**

Basionym. *Ectocarpus antillarum* Grunow 1867, p. 46–47 (footnote), pl. IV, fig. 2.

Synonyms. *Ectocarpus hooperi* Harvey 1852: p. 143 status uncertain.

Ectocarpus hooperi sensu P. Crouan and H. Crouan in Mazé 1868: p. 25

Pilayella antillarum (Grunow) De Toni 1895: p. 535–536.

Bachelotia fulvescens (Bornet) Kuckuck 1929: p. 21.

Blomquist (1958a), Taylor (1960), and Earle (1969) considered *Bachelotia antillarum* and *B. fulvescens* to be synonymous, while Gerloff (1959) noted that the former has priority over the latter. By contrast, Silva et al. (1996) listed them as distinct taxa.

Filaments are epiphytic or in entangled mats attached by prostrate axes. Main axes are 10–50 µm in diam. and cells 0.5–2 diam. long; branching is sparse and branchlets spinelike. Plastids are stellate with radiating rodlike arms and 1–2 per cell. Unilocular sporangia are intercalary and in a series in uniseriate filaments.

Common, on all coasts during spring and early summer, growing on rocks, shells, and entangled with other algae; also known from North Carolina, Bermuda, the Bahamas, Caribbean (type location: Guadeloupe, French West Indies), Gulf of Mexico, and Brazil, as well as the eastern North and tropical Atlantic, Indo- Pacific, Australia, and Antarctica.

Taylor 1960: p. 197–198; Earle 1969: p. 129; Dawes 1974: p. 96; Schneider and Searles 1991: p. 112.

Ectocarpus Lyngbye 1819: p. 130, *nom. cons.* (Gr. *ecto*: external + *carpus*: fruit).

Plants are uniseriate filaments, moderately to densely branched, with diffuse growth, epiphytic or epizoic, and attached to substrata by rhizoids. Plastids are parietal, band- or ribbonlike, solitary to several per cell, and usually have pyrenoids. Unilocular sporangia are often sessile and less common than sessile or stalked plurilocular ones.

Key to species (3) of *Ectocarpus* (modified from Womersley, 1987)

1. Plants 1–5 mm tall; plurilocular sporangia fusiform, not hair-tipped. *E. rallsiae*
1. Plants more than 5 mm tall; hair-tipped; plurilocular sporangia sometimes present 2
   2. Axes have branches throughout; main axes with subdichotomous to irregular branching and branches not in fascicles; plurilocular sporangia elongate, scattered, usually not in a series, and often ending in a hair . . . . *E. siliculosus*
   2. Plants with axes arising from a clumped base, with branchlets in fascicles above; plurilocular sporangia elongate-ovoid to tapering, often in series of 2–4 on axes of branchlets and not ending in a hair . . . . *E. fasciculatus*

Ectocarpus fasciculatus Harvey 1841: p. 40 (L. *fasciculatus*: clustered). Plate XVIII, fig. 1–3.

Synonym. *Ectocarpus fasciculatus* var. *draparnaldioides* P. L. Crouan and H. M. Crouan 1867: p. 162.

According to Russell (1967), *Ectocarpus fasciculatus* is a later synonym for *E. distortus* Carmichael *ex* Harvey in W. J. Hooker (1833).

Filaments are erect, in tufts 2–20 cm tall, epiphytic or saxicolous, olive to brown, and from a disk-like or penetrating holdfast. Main axes are obvious, usually bare below, and with fasciculate, lateral branches and branchlets above. Some basal branches may be twisted and compacted by rhizoids; terminal axes are free. Main axial cells are 30–70 µm in diam. and to 1.5 diam. long. Plurilocular sporangia are elongate-conical, 60–150 µm long, 18–40 µm wide, and on the inner sides of small branchlets.

Rare, the plant's occurrence in Florida is based upon Smithsonian Institution Herbarium specimens from St. Augustine Inlet and the Gulf Coast; also reported from the North Atlantic (type location: Britain; Womersley 1987), the Mediterranean, eastern tropical Atlantic, and Antarctic.

Taylor 1957: p. 107; Schneider and Searles 1991: p. 115.

Ectocarpus rallsiae Vickers 1905: p. 59 (L. *rallus*: fine textured, thin + *siae*: free). Plate XVIII, fig. 4.

Synonyms. *Giffordia rallsiae* (Vickers) W. R. Taylor 1960; p. 208.
Hincksia rallsiae (Vickers) P. C. Silva in Silva et al. 1987: p. 73, 130.

Kristiansen et al. (1993) questioned placing this species in the Ectocarpaceae, as drawings by Vickers (1908: pl. XXXII) lacked meristematic zones.

Filaments are uniseriate, minute, erect to 5 mm tall, and attached by creeping basal filaments with rhizoids. Branching is sparse, irregular, and branches may be hair-tipped. Cells are 25–40 µm in diam. by 2–5 diam. long. Unilocular sporangia are ovoid and 40–45 µm by 70 µm. Plurilocular sporangia are sessile or stalked and at the bases of branches; they are fusiform and tapered to an acute apex, 25–40 µm in diam., and 80–120 µm long.

Common, inconspicuous, epiphytic on red algae in Tampa Bay; also known from Bermuda, the Caribbean (type location: Barbados; Silva et al. 1996), and Gulf of Mexico, as well as the eastern tropical Atlantic and Indo-Pacific.

Taylor 1960: p. 208; Earle 1969: p. 140; Dawes 1974: p. 99.

Ectocarpus siliculosus (Dillwyn) Lyngbye 1819: p. 131–132, pl. 43 B,C (L. *siliculosus*: producing silicles, an elongate pod or silicle). Plate XVIII, fig. 5.

Basionym. *Conferva siliculosa* Dillwyn 1809 (1802–1809): p. 69, suppl. pl. E, fig. 121, 122.
Synonyms. *Ectocarpus arctus* Kützing 1843: p. 289.

Ectocarpus confervoides (Roth) Le Jolis 1863: p. 75.
Ectocarpus dasycarpus Kuckuck 1891: p. 97, fig. 4.

Filaments are uniseriate, in tufts to 30 cm long or forming entangled mats; plants are epiphytic, epizoic, or lithophytic and attached via prostrate filaments. Branching is alternate to secund; main axes are 40–60 μm in diam. and cells 1–5 diam. long, each with a band-shaped parietal plastid. Plurilocular sporangia are stalked, elongate to conical, 10–50 μm in diam. by 50–400(600) μm long, and often end in a hair. Unilocular sporangia are ovoid, sessile or stalked, 20–50 μm in diam., and 30–140 μm long.

Common, on all coasts, during winter to early summer, on limestone, coarse algae, and seagrass blades; also known from the Canadian Arctic through the Caribbean, Bermuda, Gulf of Mexico, and Brazil, as well as eastern North and tropical Atlantic (syntype locations: Cromer and Hastings, England; Silva et al. 1996), Indo-Pacific, Australia, and Antarctica.

Taylor 1960: p. 199, 200; Earle 1969: p. 130, 132, 133; Dawes 1974: p. 97, 98; Schneider and Searles 1991: p. 115.

Herponema J. Agardh 1882: p. 55 (Gr. *herpo*: to creep + *nema*: thread).

Filaments are minute endophytes with irregular branching. Cell divisions are diffuse to intercalary; filaments are hair-tipped, and their plastids band-shaped. Unilocular and plurilocular sporangia are terminal on short filaments near the bases of erect filaments.

***Herponema tortugense* (W. R. Taylor) W. R. Taylor 1960: p. 204 (named for the type locality, the Dry Tortugas, Florida). Plate XVIII, fig. 6.**

Basionym. *Ectocarpus tortugensis* W. R. Taylor 1928: p.108, pl. 14, fig. 1.

The plant may be a life history stage of another taxon, such as *Spongonema tomentosum* (Kuckuck, 1960).

Plants are small epiphytes (to 0.2 mm tall), with prostrate filaments in mats and slightly penetrating algal hosts. Erect filaments are short and to 10 cells long; cells are 7.5–13 μm in diam. and 1.5 to 2 diam. long. Plurilocular sporangia are 1–3 per filament, sessile, ovoid, 7.5–15 μm in diam., and 25–50 μm long; unilocular sporangia are unknown.

Rare, reported on "*Zonaria*" (probably *Stypopodium zonale* or *Taonia abbottiana*) at the Dry Tortugas (type location); also known from Puerto Rico and Mexico (Mateo-Cid et al., 2002).

Taylor 1960: p. 204; Earle 1969: p. 14.

Phaeostroma Kuckuck in Reinbold 1893: p. 43 (Gr. *phaeo*: brown + *stroma*: a layer).

Filaments are epiphytic; prostrate plants form small monostromatic discoid mats with 1–3 layers in the center, with these producing erect colorless hairs that originate from basal meristems. Each cell has several platelike plastids. Unilocular and plurilocular sporangia arise from vegetative cells after multiple subdivisions.

***Phaeostroma pusillum* M. Howe and Hoyt 1916: p. 109, pl. II, fig. 1–9 (L. *pusillum*: dwarf, small, weak, or slender). Plate XVIII, fig. 7.**

Filamentous epiphytes form irregular discoid colonies, to 1 mm in diam., and with colorless hairs arising from basal meristems; prostrate filaments are attached by rhizoids. Cells are irregular to rectangular, 5–10 μm in diam., and 10–16 μm long. Unilocular sporangia are spherical to globular and 8–20(25) μm in diam. Plurilocular sporangia are ovoid to conical, sessile, and 20–40 μm long by 15–20 μm wide.

Uncommon, and easily overlooked, forming small brown spots on *Dictyota*, other algae, and hydroids in the Keys; also known from North Carolina (type location), Bermuda, the Bahamas, Gulf of Mexico, and Canary Islands.

Taylor 1960: p. 205; Earle 1969: p. 142; Schneider and Searles 1991: p. 125.

Spongonema Kützing 1849: p. 461 (L. *spongia*: sponge + *nema*: thread).

Plants form small tufts or entangled ropelike strands; filaments are uniseriate, densely branched, and attached by prostrate branches with rhizoids. Main axes are irregularly, frequently branched, and with short, hooklike (recurved) entangled branchlets. Chloroplasts are band-shaped, 1–few per cell, and with 1–2 pyrenoids. Unilocular sporangia are rare and stalked; plurilocular sporangia are sessile and lateral.

Spongonema tomentosum **(Hudson) Kützing 1849: p. 461 (L.** ***tomentosum*****: densely covered with matted wool or short hairs). Plate XVIII, fig. 8.**

Basionym. *Conferva tomentosa* Hudson 1762: p. 480; 1778, p. 594.

Synonym. *Herponema luteolum* (Sauvageau) Hamel 1939b: p. XIX.

Kuckuck (1960) suggested that *Herponema luteolum* was a life history stage of *Spongonema tomentosum*.

Plants have two possible growth phases: (1) minute prostrate filamentous epiphytes that form a pseudoparenchymatous cushion, and (2) erect filaments that are simple to densely branched from the base, in entangled strands, 2.5–20.0 cm long, and with many recurved or strongly arched branchlets. Branches of erect plants are hair-tipped and 100–400 µm long; cells 5–8 µm in diam. and 2–4 diam. long. Cells of both forms have ribbonlike plastids with 1–2 pyrenoids. Erect branches produce plurilocular sporangia that are basal, sessile or stalked, 7–13 µm in diam. and 30–80 µm long. Unilocular sporangia are stalked and basally attached.

Rare, as entangled strands in the Gulf of Mexico (Earle, 1969); also known from the Prince Edward Island to Virginia and Bermuda, as well as the eastern North and tropical Atlantic, northeastern and southeastern Pacific, and Arctic and Antarctic Oceans.

Taylor 1960: p. 204; Earle 1969: p. 142.

Sedis incertae: *Xanthosiphonia* J. Agardh 1892: p. 112–113 (Gr. *xantho*: yellow + *siphon*: a pipe).

Xanthosiphonia was described based on *X. wattsii* (Australia) and *X. halliae* (Florida). Womersley (1967: p. 424) considered *X. wattsii* to be a sterile species of *Ectocarpus*, and *X. halliae* an "ectocarpoid alga with very elongate plurilocular sporangia". He lectotyped the genus based on the latter species.

Filaments are alternately branched with young uniseriate branches and older multiseriate ("polysiphonous") branches. Plurilocular sporangia are limited to special multiseriate branches and surround a central row of sterile cells.

Xanthosiphonia halliae **J. Agardh 1892: p. 113 (named for Mrs. G. A. Hall, an American who collected seaweeds in Florida including the type material from St. Augustine).**

The Florida specimen was described by J. Agardh (1894) as a 'brown *Polysiphonia*' with elongate plurilocular sporangia. Womersley (1967) believed it had uncertain status.

Filaments are slender with abundant, widespread branching. Primary axes are irregular, with upper parts clustered to fasciculate and spreading to form narrow plumes. Filaments are "polysiphonous" at forks and have elongate plurilocular sporangia.

A questionable record from Florida that is based on the original collection from St. Augustine.

Scytosiphonaceae

Synonym. Chnoosporaceae

See Wynne (2005a) for synonymy and previous comments on the designation of the Scytosiphonaceae as a family in an expanded Ectocarpales (Peters and Ramírez, 2001).

Gametophytes are macroscopic, dioecious, and parenchymatous; fronds are erect and unbranched, bladelike, globular, or hollow, and with many plurilocular sporangia. Growth is apical or intercalary; cells have one parietal plastid and a single pyrenoid. Sporophytes are crustose or erect, parenchymatous or pseudoparenchymatous, and have unilocular or plurilocular sporangia. Life histories appear to be heteromorphic (Cho et al., 2003).

Chnoospora J. Agardh 1847: p. 7 (Gr. *chnoos*: wool, fine down + *spora*: a seed).

Plants are bushy, firm, and branches taper to pointed or blunt tips; axes are mostly compressed and attached by discoid bases. The medulla is pseudoparenchymatous with colorless, elongate, and large cells; cortical cells decrease in size toward the surface and those of the outer layers have a single plastid (Fotos, 1981). Unilocular sporangia occur in a series on moniliform filaments that arise from the cortex.

Chnoospora minima **(K. Hering) Papenfuss 1956: p. 69–70 (L. *minima*: small). Plate XVIII, fig. 9.**

Basionym. *Fucus minimus* K. Hering 1841: p. 92.

Plants are bushy, erect, and to 15 cm tall; axes are compressed, widest at forks, repeatedly dichotomously branched, and attached by discoid holdfasts. Axes are strap-shaped, to 2 mm wide, 7–40 mm long, and to 1 mm thick; tips are tapered or blunt. Medullary cells are rounded, colorless, and to 40 µm in diam.; cortical cells decrease in size to the surface; surface cells are to 10 µm in diam. and have a single plastid. Surface hairs are in distinct tufts, to 1 mm long, with cells to 10 µm in diam., and 80 µm long.

Rare, on wave-exposed intertidal limestone in the Keys; also known from the Caribbean and Brazil, as well as West and South Africa (type location: Port Natal, Durban; Silva et al., 1996), the Indo-Pacific, and Australia.

Taylor 1960: p. 263; Littler and Littler 2000: p. 246.

Colpomenia (Endichler) Derbès and Solier in Castagne 1851: p. 95 (Gr. *kolpos*: a packet + *meno*: to remain, to be lasting).

Gametophytes are macroscopic, saccate to globular, irregularly lobed to hollow when mature, and attached by rhizoids. The medulla has large colorless cells that grade into small, pigmented surface cells. Plurilocular sporangia are in sori intermixed with colorless hairs. Sporophytes are microscopic filaments (*Ectocarpus*-like) and have plurilocular and unilocular sporangia.

Colpomenia sinuosa **(Mertens *ex* Roth) Derbès and Solier in Castagne 1851: p. 95; Derbès and Solier 1856: p. 95 (L. *sinuosa*: curved, wavy). Plate XVIII, fig. 10, 11.**

Basionym. *Ulva sinuosa* Mertens *ex* Roth 1806: p. 327, pl. 12.

Synonym. *Soranthera leathesiaeformis* P. Crouan and H. Crouan in Schramm and Mazé 1865: p. 4.

The heteromorphic life history includes a macroscopic erect gametophyte and a microscopic, prostrate *Ectocarpus*-like sporophyte (Kogame, 1997; Toste et al., 2003a).

Gametophytes are hollow, crisp, sessile, saccate or irregularly lobed, 6–10 (20) cm in diam., and attached by rhizoids. Walls of saccate thalli are 0.3–0.5 mm thick and 4–6 cell layers; inner cells are colorless, to 240 µm in diam., and grade into small pigmented surface cells 4–8(20) µm in diam. Surface sori are dark, in raised patches, and 1.0–1.5 mm in diam. Plurilocular sporangia are biseriate, cylindrical to club-shaped, 4–8 µm in diam., and 18–30 µm long. Colorless sterile hairs are 10 µm in diam., to 50 µm long, and occur on the surface.

Common, on limestone and epiphytic on various algae at wave-exposed intertidal sites in the Keys and Dry Tortugas and from 15 m on the west coast; also known from North Carolina through the Caribbean, Bermuda, Gulf of Mexico, and Brazil, as well as the eastern North and South Atlantic, Mediterranean (type location: Cádiz, Spain; Silva et al., 1996), Red Sea, Indo-Pacific, and Australia.

Taylor 1960: p. 260; Earle 1969: p. 200; Dawes 1974: p. 110; Schneider and Searles 1991: p. 146; Littler and Littler 2000: p. 248.

Hydroclathrus Bory 1825a (1822–1831): p. 419–420 (Gr. *hydro*: water + *clathrus*: a lattice).

Gametophytes are macroscopic, sessile, and attached by rhizoids. Young plants are spherical, porous, and hollow, while older ones are irregular, with subcylindrical strands and enrolled margins. Outer cortical cells are small, pigmented, and cover the large inner colorless cells. Plurilocular sporangia are scattered over the surface and in sori mixed with hairs having basal meristems. Sporophytes are microscopic *Ectocarpus*-like filaments with plurilocular and unilocular sporangia.

***Hydroclathrus clathratus* (C. Agardh) M. Howe 1920a: p. 590 (Gr. *clathratus*: lattice, network, lacy). Plate XVIII, fig. 12, 13.**

Basionym. *Encoelium clathratum* C. Agardh 1823 (1820–1828): p. 412

The life history is heteromorphic. The erect parenchymatous gametophyte has plurilocular sporangia, while the prostrate filamentous sporophyte has both plurilocular and unilocular sporangia. Toste et al. (2003b) described a direct and a heteromorphic, monophasic type of life history.

Young gametophytes are pale golden brown, spherical, highly perforated, sessile, and attached by rhizoids. Older plants (5–10 cm in diam.) are irregularly expanded, often enrolled into compressed strands, and have many pores, 0.1–3 cm in diam. The walls of saccate thalli are parenchymatous, 100–200 µm thick, and have 3–6 cell layers; the inner cells are colorless and 50–150 µm in diam., while the surface cells are pigmented and 5–10 µm in diam. Plurilocular sporangia are ovoid, uniseriate or biseriate, 4–20 µm in diam., 13–24 µm long, and scattered over young plants. Sori are depressed and contain dense tufts of hairs with basal meristems.

Common, on limestone or adrift at the Dry Tortugas and Keys (to 10 m); also known from Bermuda, the Caribbean, Gulf of Mexico, and Brazil, as well as the eastern North and tropical Atlantic (type location: Belle Îsle, France; Abbott and Huisman, 2004), the Mediterranean, Indo-Pacific, and Australia.

Taylor 1960: p. 261; Earle 1969: p. 202; Littler and Littler 2000: p. 248.

Petalonia Derbès and Solier 1850: p. 265, *nom. cons.* (Gr. *petalon*: a leaf, petal).

Synonyms. *Ilea* Fries 1835 (1835–1837): p. 321.
Phyllitis Kützing 1843: p. 342.

Guiry and Nic Dhonncha (2006) cited species of *Ilea* as having uncertain taxonomic status.

Gametophytes are macroscopic, foliose, with one or more erect, stalked blades, and attached by a discoid base. The medulla is parenchymatous with large, colorless cells grading into smaller pigmented outer cortical cells. Plurilocular sporangia are in dense patches on both blade surfaces. Sporophytes are crustose with a prostrate hypothallus and an erect epithallus; unilocular sporangia are rare and occur laterally at the bases of paraphyses.

Petalonia fascia (O. F. Müller) Kuntze 1898: p. 419, pl. 28 (L. _fascia_: a bundle, band, or surgical bandage) XVIII, fig. 14, 15.

Basionym. *Fucus fascia* O. F. Müller 1778: p. 7, pl. 768.
Synonyms. *Ilea fascia* (O. F. Müller) Kützing f. *caespitosa* (J. Agardh) Farlow 1881.
Petalonia fascia var. *caespitosa* (J. Agardh) W. R. Taylor 1937: p. 173.

Blades in culture produce zooids that grow into prostrate forms without sexual fusion and produce new blades (Reinke, 1878; Sauvageau, 1929; Kylin, 1933; Dangeard, 1962; Fletcher, 1978). However, *Ralfsia*- or *Microspongium*-like crusts also produce erect *Petalonia fascia* and *Scytosiphon lomentarian*–like plants (Wynne, 1969; Edelstein et al., 1970; McLachlan et al., 1970; Rhodes and Connell, 1973; Roeleveld et al., 1974; Fletcher, 1978).

Gametophytes are erect, 5–30 (45) cm tall, dark to olive brown, with one or more flat strap-like blades (140–200 µm thick), and attached by a discoid base. Blades have deciduous hairs and uniseriate plurilocular sporangia of 6–8 cells, 4–10 µm in diam., and 30–75 µm long.

Uncommon, blades occur in winter on limestone in shallow waters in northwestern Florida and on coquina at Marineland; also known from the Arctic to Georgia, Texas, and Brazil, as well as the eastern North and South Atlantic (type location: Christiansund, Norway; Womersley, 1987), Mediterranean, Indo-Pacific, Australia, and Antarctica.

Taylor 1960: p. 258; Earle 1969: p. 205; Schneider and Searles 1991: p. 147.

Rosenvingea Børgesen 1914b: p. 22; continuous pagination = p. 178 (named for L. Kolderup Rosenvinge, a Danish phycologist who studied North Atlantic seaweeds).

Axes are tubular, cylindrical, or compressed; branching is sparse, densely alternate, or dichotomous. Axes are usually hollow, with 3–4 cubical cell layers, and attached by a small basal disc. Inner cells are large, colorless, and grade into small, pigmented surface cells with hairs. Plurilocular sporangia arise from surface cells in conspicuous sori.

Key to the species (3) of *Rosenvingea*

1. Axes erect, cylindrical, and not entangled 2
1. Axes often flat to compressed and entangled . . . . *R. intricata.*
    2. Axes sparingly branched, mostly from base; tips acute. . . . . *R. antillarum*
    2. Axes often branched; branching is cervicorn to subdichotomous; branches with blunt to hairlike tips. . . . . *R. orientalis.*

Rosenvingea antillarum (P. Crouan and H. Crouan in Schramm and Mazé) M. J. Wynne 1997: p. 331–336 (L. _antillarum_: of the Antilles, West Indies). Plate XVIII, fig. 16, 17.

Basionym. *Stilophora antillarum* P. Crouan and H. Crouan in Schramm and Mazé 1865: 2.
Synonym. *Cladosiphon floridanus* W. R. Taylor 1928: p. 113, pl. 15, fig. 10–14 (as '*floridana*').
Rosenvingea floridana (W. R. Taylor) W. R. Taylor 1955: p. 72–74, pl. 4; 1960: p. 262, pl. 29, fig. 7, 8.

In contrast to the above synonymy Guiry and Nic Dhonncha (2006) listed *Rosenvengia antillarum* and *R. floidana* as distinct taxa.

Plants are bushy, golden brown, and to 30 cm tall; axes are terete to slightly compressed, hollow, and with solid discoid holdfasts composed of rhizoidal filaments. Axes are 0.3–1.5 mm in diam.; branches are slender, irregular, tapering to acute tips, and usually on lower parts. Surface cells are to 12 µm in diam. and have hairs; plurilocular sporangia are stalked, 3–4 cells long, to 14 µm wide, and in sori or scattered on the surface of thalli.

Uncommon, on shallow (2–3 m) limestone during spring to early summer in the Keys and at the Dry Tortugas; also known from the Gulf of Mexico, Lesser Antilles (type location: Guadeloupe), Canary Islands, and Angola.

Taylor 1960: p. 262; Earle 1969: p. 206.

Rosenvingea intricata (J. Agardh) Børgesen 1914b: p. 26; continuous pagination = p. 182 (L. *intricata*: involved, entangled). Plate XVIII, fig. 18, 19.

Basionym. *Asperococcus intricatus* J. Agardh 1847: p. 7.

Synonyms. *Asperococcus schrammii* P. Crouan and H. Crouan in Mazé and Schramm 1878 (1870–1877): p. 115.

Striaria intricata (J. Agardh) Vickers 1905: p. 59.

Although most specimens of *Rosenvengia intricata* are distinct from *R. orientalis*, the species do intergrade (Abbott and Huisman, 2004).

Plants are hollow, to 40 cm tall, and highly branched; axes are matted, entangled, and attached by small discs. Axes are soft, irregular, and taper at bases; branches are contorted, often adhering to one another, taper to tips, are 1–10(20) mm in diam., 70–150 µm thick, and have 2–4 cell layers. Inner cells are colorless, 30–100 µm in diam., and 55–130 µm long; outer cells are angular and 10–20 µm in diam. Plurilocular sporangia are ovoid, to 20 µm in diam., 50 µm long, and produced from surface cells in obvious sori.

Common, in spring as drift or attached to coquina, limestone, or seagrasses on the west coast and the Keys (2–35 m); also known from Bermuda, North Carolina, the Caribbean, Gulf of Mexico (type location: Veracruz, Mexico; Abbott and Huisman, 2004), and Brazil, as well as the Mediterranean, eastern tropical Atlantic, Red Sea, Indo-Pacific, and Australia.

Taylor 1960: p. 262; Earle 1969: p. 207; Dawes 1974: p. 111; Littler and Littler 2000: p. 250.

Rosenvingea orientalis (J. Agardh) Børgesen 1914b: p. 26; continuous pagination = p. 182 (L. *orientalis*: of the east, eastern, or oriental). Plate XIX, fig. 1–3.

Basionym. *Asperococcus orientalis* J. Agardh 1848 (1848–1901): p. 78.

Synonym. *Rosenvingea sanctae-crucis* Børgesen 1914b: p. 22–28, fig. 14–17; continuous pagination = p. 178–181, fig. 141–144.

Based on comments by Earle (1969), Littler and Littler (2000), and Wynne (2005a), plus our observations of morphological overlap, *Rosenvengia sanctae-crucis* is merged with *R. orientalis*.

Plants are golden to dark brown and 10–40 cm tall; axes hollow, cylindrical to compressed, to 2 mm wide, and attached by inconspicuous discoid bases. Branching is dichotomous to irregular; branches are to 2 mm in diam. and taper slightly to blunt or hairlike tips; their cortex consists of 3–4 cells and is 120–280 µm thick. Inner cells are rounded, 80–200 µm in diam., and grade into polygonal surface cells, 10–20 µm in diam. Hairs are scattered on the surface and 8–9 µm in diam. Plurilocular sporangia are in ovoid sori on the surface; sporangia are ovoid, cylindrical to clavate, and 5–12 by 20–40 µm.

Uncommon, on seagrasses off Cape Romano and on shallow limestone at the Content Keys; also known from North Carolina, the Caribbean, Gulf of Mexico, and Brazil, as well as the Canary Islands, Indo-Pacific (type location: Manila, Philippines; Abbott and Huisman, 2004), and Australia.

Taylor 1960: p. 263; Earle 1969: p. 210; Schneider and Searles 1996: p. 148; Littler and Littler 2000: p. 250.

Scytosiphon C. Agardh 1820 (1820–1828): p. 160, *nom. cons.* (Gr. *skytos*: hide, skin; also *scyto*: leather + *siphon*: tubular, a tube).

See Tatewaki (1966), Clayton (1980), Kogame (1996, 1998), and Camus et al. (2005) for a synopsis of life history data.

Gametophytes are dioecious (Kogame et al., 2005), erect, unbranched tubes with or without regular constrictions, and attached by discoid bases. Cells are parenchymatous and grade from large colorless inner ones to small, pigmented surface cells. Plurilocular sporangia occur in sori mixed with sterile hairs and fucosan-filled single-celled paraphyses; sori are limited to small areas or spread over the surface. Sporophytes are prostrate crusts that produce unilocular sporangia. The life history is usually biphasic, with erect tubular gametophytes that produce plurilocular sporangia (gametangia) and prostrate sporophytes that produce unilocular sporangia. However, the sporophytes can recycle themselves or form erect plants; the gametophyte can also recycle itself presumably by parthenogenetic gametes.

Scytosiphon lomentaria **(Lyngbye) Link 1833: p. 233, pl. 28, *nom. cons.* (L. *lomentaria*: legumelike; also *loment*: a legumelike fruit or a pod with constrictions). Plate XIX, fig. 4, 5.**

Basionym. *Chorda lomentaria* Lyngbye 1819: p. 74, pl. 18E.
Synonyms. *Ulva simplicissima* Clemente y Rubio 1807: p. 320.
Scytosiphon simplicissumus (Clemente y Rubio) Cremades in Cremades and Pérez-Cirera 1990: p. 492.

Cremades (in Cremades and Pérez-Cirera, 1990) noted that *Ulva simplicissima* Clemente y Rubio (1807) predated *Chorda lomentaria* Lyngbye (1819), the basionym of *Scytosiphon lomentaria*, and made the combination *S. simplicissimus*. Because of the widespread use of *S. lomentaria*, Pedersen and Kristiansen (1994) argued that the combination be rejected according to Article 56.1. Their proposal was recommended (Compère, 1997).

Gametophytes are erect, tubular, usually in clusters, and attached by discoid bases. Axes are often constricted at irregular intervals, 15–70 cm tall, 5–10(30) mm wide, and taper to both the base and apex. Plurilocular sporangia (gametangia) are elongate, uniseriate, 3–10 µm in diam., to 65 µm long, usually in dense sori, and have abundant hairs that are vegetative, 8–13 µm in diam., and 25–30 µm long. Sporophytes are unknown in Florida. Although typically isogamous and dioecious, some strains are asexual and genetically distinct (Kogame et al., 2005).

Rare, on coquina at Marineland in winter and spring; Florida plants are smaller than ones from North Carolina (Schneider and Searles, 1991); also known from the Canadian Arctic to Brazil, as well as the eastern North and South Atlantic, Mediterranean, Red Sea, Pacific Ocean, Australia, and Antarctica. Syntype localities include the Faeroe Islands and Denmark (Womersley, 1987; Silva et al., 1996).

Taylor 1960: p. 259; Schneider and Searles 1991: p. 149.

SPOROCHNALES

Sporophytes are macroscopic, bushy, and with tufts of trichothallic hairs at the tips of young branches. Life histories are heteromorphic; gametophytes are filamentous, microscopic, and have oogamous reproduction. The order has six genera in one family.

Sporochnaceae

The family has the same features as the order.

Nereia Zanardini 1846: p. 899 (Gr. *Nereis*: the daughter of the ancient sea god *Nereus*).

Plants are bushy, have indistinct main axes, and are attached by discoid holdfasts. Branchlets are short, spurlike, and end in tufts of brown filaments arising from cortical cells. Unilocular sporangia are on papillae below tufts of sessile filaments. *Nereia* differs from *Sporochnus* by its indistinct axis and lateral tufts of sessile filaments.

Nereia filiformis **(J. Agardh) Zanardini 1846: p. 899–900 (L. *filiform*: threadlike). Plate XIX, fig. 6, 7.**

Basionym. *Desmarestia filiformis* J. Agardh 1841: p. 6.

Synonyms. *Stilophora tropica* W. R. Taylor 1928: p. 115, pl. 3, fig. 9; pl. 14, fig. 5, 6; pl. 19, fig. 2. *Nereia tropica* (W. R. Taylor) W. R. Taylor 1955: p. 74, pl. 5; 1960: p. 252, pl. 29, fig. 3, 4; pl. 35, fig. 1.

Prud'homme van Reine et al. (1994) compared herbarium specimens of *Nereia tropica* (Caribbean) and *N. filiformis* (Mediterranean) and could not justify their separation.

Plants are erect, bushy, and to 35 cm tall. Axes are alternately to irregularly branched, with main axes indistinct and attached by discoid holdfasts. Branches are 0.3–2.0 mm in diam. and end in tufts of pigmented filaments; lateral tufts occur on scattered spurlike branchlets; filaments are 15–25 µm in diam. and 3–6 mm long. Internal medullary cells are colorless and 30–50 µm in diam., while outer cortical cells are pigmented and 10–20 µm in diam. Unilocular sporangia are cylindrical to ovoid, 10–20 µm in diam., 25–45 µm long, and at the bases of sterile hairs that are club-shaped, unicellular, and 10–30 µm in diam.

Common, during winter, adrift and on limestone to 30 m on the west coast (type location: Dry Tortugas); also known from Bermuda, the Bahamas, Antilles, Gulf of Mexico, Brazil, and Canary Islands.

Taylor 1960: p. 252; Earle 1969: p. 188; Littler and Littler 2000: p. 246.

Sporochnus C. Agardh 1817: p. XII (Gr. *spora*: a seed, spore + *chnoos*: wool, down).

Plants are erect, wiry, with abundant alternate to irregular branching, and have a distinct primary axis. Ultimate branchlets are determinate, stalked, and terminate in filamentous tufts similar to apices. Unilocular sporangia and sterile branched hairs are restricted to swollen terminal parts of stalked branchlets (unlike *Nereia*).

Key to the species (2) of *Sporochnus*

1. Stalks of branchlets cylindrical and to 10 mm long . . . . *S. bolleanus*
1. Stalks of branchlets swollen, oval, or fusiform and to 2 mm long . . . . *S. pedunculatus*

Sporochnus bolleanus **Montagne 1856: p. 393 (L. *bolli*: a fruit, cotton capsule + *anus*: pertaining to). Plate XIX, fig. 8.**

Plants are dark brown, wiry, and coarse; axes are filiform, to 1 m long, and 0.5 mm in diam.; main axes are distinct, with alternate to irregular branching, and attached by discoid holdfasts. Tips of branches and branchlets have tufts of filaments that are light olive to brown, 10–15 mm long, and deciduous; branchlets are stalked, 1.5–10 mm long, cylindrical, and not swollen.

Uncommon, in drift during spring and early summer, on limestone in 20 m offshore Homosassa Springs (west coast) and at the Dry Tortugas; also known from Bermuda, the Greater Antilles, Brazil, eastern tropical Atlantic, and Galapagos Islands.

Taylor 1960: p. 253; Earle 1969: p. 190; Dawes 1974: p. 108.

Sporochnus pedunculatus **(Hudson) C. Agardh 1820 (1820–1828): p. 149 (L. *pedunculatus*: stalked). Plate XIX, fig. 9, 10.**

Basionym. *Fucus pedunculatus* Hudson 1778: p. 587.

Plants are wiry, with filiform axes, regularly branched, to 30 cm tall, and attached by discoid holdfasts. Branches are to 20 cm long and have many stalked branchlets that end in tufts of olive brown filaments. Main axes are distinct, cylindrical, and to 1 mm in diam.; branchlets are 0.7–3.0 mm long, 0.5 mm in diam., and with a terminal swelling 700–850 μm in diam. Young branchlets end in a tuft of filaments to 5 mm long and 15–22 μm in diam. Unilocular sporangia are cylindrical to club-shaped, 10–20 μm in diam., and 25–45 μm long; they occur at the bases of filaments mixed with sterile hairs that are club-shaped to ovoid, 10–28 μm in diam., and 25–50 μm long.

Uncommon, in winter on seagrass blades or attached to limestone (5 to 55 m) in the northeastern Gulf of Mexico; also recorded from North Carolina (14–45 m), Bermuda, the Antilles, Gulf of Mexico, and Brazil, as well as the eastern North and South Atlantic, Mediterranean, and Pacific Mexico. Type location: Isle of Portland, Dorsetshire, southern England (Hudson, 1778).

Taylor 1960: p. 253; Earle 1969: p. 190; Dawes 1974: p. 109; Schneider and Searles 1991: p. 165.

FUCALES

Plants often have a holdfast, stipe, blade, and vesicles (also called air bladders, floats, or pneumatocysts). Axes are parenchymatous with an inner and outer cortex and a filamentous medulla with cells that translocate. Growth is by division of one or more apical cells. Life histories are diplontic; antheridia and oogonia occur in sunken conceptacles on fertile tips (receptacles) or specialized branchlets. Meiosis occurs before gamete production to form haploid motile uniflagellated sperm and nonmotile eggs that are released from the conceptacles prior to fertilization.

Sargassaceae

Synonym. Cystoseiraceae.

Molecular studies by Rousseau and de Reviers (1999b) showed that the Cystoseiraceae was paraphyletic and should be merged with the Sargassaceae (Wynne, 2005a).

Plants have one or more erect axes, a perennial base, and upper deciduous reproductive branches; axes can produce spine or leaflike projections, vesicles, and receptacles. Growth is by a single three-sided apical cell. Gametangia occur in conceptacles; oogonia have one viable egg nucleus with seven others being expelled; each antheridium produces 64 sperms.

Cystoseira C. Agardh 1820 (1820–1828): p. 50, *nom. cons.* (Gr. *kusteos*: a bladder, pouch + *seira*: a chain, rope, series).

Plants are bushy, light brown, and have one or more primary (often perennial) axes attached by a basal disc. Axes are terete to flat, smooth to spiny, and with alternate to irregular branching; upper axes have stalked vesicles and flat, leaf to spinelike foliar projections. Growth is by a large three-sided apical cell. Cryptostomata (hair pits) are present or absent on branches and foliar parts. Receptacles are usually bisexual and located near the tips of branchlets.

Cystoseira myrica **(S. G. Gmelin) C. Agardh 1820 (1820–1828): p. 53–54 (Gr. *myrike*: ancient name for tamarisk, a fine-leaved deciduous shrub tolerant of salt spray, heat, and drought; *myro*: to flow). Plate XIX, fig. 11–13.**

Basionym. *Fucus myrica* S. G. Gmelin 1768: p. 88, pl. III, fig. 1.

Plants are clustered, erect, and to 30 cm tall; axes are tough and wiry, terete, with a rough surface, and attached by small discoid holdfasts. Foliar organs are spinelike and 0.5–1 mm long; vesicles are 2 mm in diam., 3–4 mm long, tipped by a short spine, solitary or in clusters, and on spinelike stalks, 1–4 mm long. Receptacles are spiny, terete, 1 mm in diam., 2–9 mm long, and end in branchlets. Cryptostomata are on branches and spines.

Common, forming dense patches in tide pools at the Keys (for example, Content Keys) and Dry Tortugas; also known from the Bahamas, Caribbean, Red Sea, and Indian Ocean. Type location is unknown (Lipkin and Silva, 2002).

Taylor 1960: p. 267; Earle 1969: p. 214; Littler and Littler 2000: p. 280.

Sargassum C. Agardh 1820 (1820–1828): p. 1, *nom. cons.* (Spanish: *sargaza*: seaweed).

Plants are erect, with axes having blades and vesicles and attached by a solid or fibrous holdfasts. Basal axes may be perennial and produce fertile, terete, annual axes. Blades are flat, broad to thin, simple or forked, entire or serrate, and usually have a midrib; vesicles are stalked and often abundant. Receptacles occur in the axils of blades, are much branched, and usually terete or less often compressed to flat. Conceptacles are monoecious or dioecious. Oogonia each have one viable egg nucleus after the expulsion of seven nuclei. Each antheridium produces 64 sperms.

Key to the species (12) of *Sargassum* (modified from Earle, 1969; Taylor, 1960)

1. Plants free-floating, without a holdfast, and not initially attached; cryptostomata usually absent or sparse on blades; receptacles absent 2
1. Plants usually attached by a basal holdfast; cryptostomata and receptacles usually present, particularly on mature plants 3
 2. Blades linear, margins with slender teeth; air bladders often have a spine. . . . . *S. natans*
 2. Blades lanceolate and with broad teeth; vesicles usually lack a spine . . . . . *S. fluitans*
3. Midrib of blades not prominent; ridged or winged near base 4
3. Midrib prominent; toothed, ridged, or winged near base . . . . . *S. pteropleuron*
 4. Blade margins serrate or if smooth axes with spines. 5
 4. Blade margins smooth or with fine teeth; axes with or without spines . . . . . *S. cymosum*
5. Blades near tips usually forked; axes spiny 6
5. Blades near tips rarely forked; axes smooth or with spines 7
 6. Upper blades often forked 1–4 times; midribs not prominent or raised; with 1–2 vesicles per blade; axes spiny . . . . . *S. wolfii*
 6. Upper blades pinnate or with tips forked; midribs prominent and raised; with few vesicles per blade; axes smooth. . . . . *S. ramifolium*

7. Blades lanceolate to linear, length 7–50 times width 8
7. Blades oblong to lanceolate, length usually less than 7 times width. 10
   8. Axes smooth; blade margins serrate; cryptostomata scattered 9
   8. Axes spiny and rough; blade margins entire to serrate; cryptostomata usually aligned along each side of midrib . . . . *S. acinarium*
9. Blades 20–43 mm long; vesicles nearly sessile and often with a spine . . . . *S. mathiesonii*
9. Blades 18–41(150) mm long; vesicles stalked and lacking a spine . . . . *S. filipendula*
   10. Blades oval to lanceolate, receptacle branches not lobed; axes smooth to spiny 11
   10. Blades oblong to elliptical; receptacles lobed; axes smooth . . . . *S. hystrix*
11. Blade length 2–4 times width, often crisp; axes usually spiny . . . . *S. polyceratium*
11. Blade length 4–8 times width, not crisped; axes smooth to weakly spiny . . . . *S. vulgare*

Sargassum acinarium (Linneaus) Setchell 1933: p. 208, pl. 39 (L. *acinarium*: a stem covered with vesicles). Plate XIX, fig. 14–16.

Basionym. *Fucus acinarius* Linnaeus 1753: p. 1160.

Plants are erect and to 30 cm tall; basal axes are perennial and attached by tough discoid holdfasts; reproductive axes are annual. Blades of perennial axes are thin, lanceolate to linear, and 3–8 mm wide; blades of upper annual axes narrow, 1–3 mm wide, 3–7 cm long, serrate, and produced on short stalks (to 1 mm long). Cryptostomata are 100–150 µm in diam. and either scattered or in irregular rows on both sides of midrib. Vesicles are ovoid, abundant, and on flattened (winged) stalks, 0.5–1.0 mm wide by 1–6 mm long. Receptacles are 0.5–0.7 mm in diam., 0.5–1.5 cm long, often forked, and have a wartlike surface.

Uncommon, on limestone in lagoons in the Keys (to 2 m); also known from Bermuda, the Caribbean, and Gulf of Mexico, as well as the Mediterranean (type location; Silva et al. 1996), eastern tropical Atlantic, and India.

Taylor 1960: p. 271; Earle 1969: p. 216; Littler and Littler 2000: p. 280.

Sargassum cymosum C. Agardh 1820 (1820–1828): p. 20 (L. *cymosum*: producing cymes, relating to cymes). Plate XX, fig. 1, 2.

Synonym. *Sargassum rigidulum* Kützing 1849: p. 615.

We follow Schnetter (1976), who merged *Sargassum rigidulum* with *S. cymosum* because of overlapping morphologies. Others have described pronounced morphological variation within *S. cymosum* (De-Paula, 1988; De Széchy and De-Paula, 1998) but retained both taxa. Moreira and Suárez (2002c) reported possible hybridization of *S. cymosum* and *S. filipendula*.

Plants are erect, 10–30(100) cm tall, and attached by discoid holdfasts. Basal axes are perennial, short, and sometimes covered by lateral branches. Blades are sparingly pinnately divided and have a conspicuous midrib and scattered cryptostomata. Blades of annual, upper axes are elliptical, oblong, or linear, 2–4 cm long by 2.0–3.5 cm wide, and have smooth or crenulate margins. Vesicles are sparse, rounded, usually spiny, and on a stalk equal in length to the vesicle's diam. Receptacles are dichotomously branched, filiform or tumor- shaped, up to half the length of subtending blades, and usually have a rough surface.

Uncommon, on limestone in shallow waters at the Dry Tortugas; also known from the Caribbean, Brazil (type location; Silva et al., 1996), and the eastern tropical Atlantic.

Taylor 1960: p. 272, 278; Earle 1969: p. 216.

Sargassum filipendula C. Agardh 1824: p. 300 (L. *filipendula*: thread-suspended, a hanging thread). Plate XX, fig. 3.

Synonyms. *Sargassum filipendula* Harvey 1852: p. 61.
Sargassum vulgare Farlow 1881 ('1882'): p. 103.

The species is highly variable (Earle, 1969; Schneider and Searles, 1991; De Széchy and De-Paula, 1998). Moreira and Suárez (2002c) reported evidence for hybridization between *Sargassum filipendula* and *S. cymosum*, as well as between *S. filipendula* and *S. f.* var. *montagnei*. Kilar (1992) questioned records from the west coast.

Plants are erect and to 1 m tall; basal perennial axes are 5–20 cm tall, tough, smooth, rarely forked, and attached by tough, spreading holdfasts. Perennial axes have large blades that are usually divided, serrate, and with prominent midribs. Annual axes are slender, smooth, and terete, with alternating branches and reproductive axes. Blades of annual axes are on short slender stalks, are linear, thin, serrate, 0.5–1.9 mm wide, 18–41 mm long, and have a distinct midrib. Cryptostomata are abundant, oval, 50–100 μm in diam., and scattered on blades, vesicles, and stalks. Vesicles are on stalks, abundant, 2–8 mm long, spherical, often apiculate, and 2–5 mm in diam. Receptacles are 0.5–1.0 mm in diam. and occur on simple or forked axillary branches with wartlike surfaces. Distinct male and female conceptacles occur on the same receptacle.

var. *laxum* J. Agardh 1848 (1848–1901): p. 315 (as '*laxa*') (L. *laxum*: loose, weak).

Plants are slender, sparingly branched, and have vesicles 1–4 mm in diam. The perennial axial blades are 0.2 mm wide by 4 cm long, while annual axial blades are 0.5–1.0 mm wide and 0.5–3.0 cm long.

var. *montagnei* (J. W. Bailey in Harvey) Grunow 1916: p. 171 (named for the French phycologist J. F. C. Montagne, a friend of Harvey and J. W. Bailey, who studied Caribbean algae in the mid-1800s).

Basionym. *Sargassum montagnei* J. W. Bailey in Harvey 1852: p. 58, pl. IA.
Synonyms. *Sargassum vulgare* C. Agardh var. *montagnei* (J. W. Bailey) Farlow 1881 ('1882'): p. 103

See Moreira and Suárez (2002c) regarding possible hybridization between *Sargassum filipendula* and *S. f.* var. *montagnei*.

Annual reproductive axes are long, slender, delicate, smooth, and laxly branched. Blades of annual axes are linear, 3–6 mm wide, to 15 cm long, with a distinct midrib and a few small cryptostomata. Vesicles are spherical, 2–3 mm in diam., and end in a minute blade or spine.

var. *pinnatum* Grunow 1916: p. 170 (L. *pinnatum*: plumed with feathers).

Blades of upper annual axes are forked, sparingly pinnate, to 1 mm wide, 3.3 cm long, and have vesicles, 2.0–4.5 mm in diam.

Common, on all coasts and to 33 m; the species is also known from southern Massachusetts to the Caribbean (type location: West Indies; Silva et al. 1996), Bermuda, Gulf of Mexico, and Brazil, as well as the eastern tropical Atlantic and Sri Lanka. The var. *laxum* is known from the Content and Sawyer Keys (Grunow, 1916; Dawes et al. 1967), as well as North Carolina, Brazil, and the Caribbean. The var. *montagnei* occurs on all coasts (6 to 33 m) and is also known from Massachusetts to Georgia, Bermuda, the Caribbean, and Brazil. The var. *pinnatum* is reported from Key West (Grunow, 1916), North Carolina, the Caribbean, and Brazil.

Taylor 1960: p. 270, 271; Earle 1969: p. 217; Dawes 1974: p. 112; Schneider and Searles 1991: p. 170; Littler and Littler 2000: p. 282.

Sargassum fluitans (Børgesen) Børgesen 1914b: p. 66; continuous pagination = p. 222, plus footnote (L. *fluitans*: floating, swimming). Plate XX, fig. 4.

Basionym. *Sargassum hystrix* J. Agardh var. *fluitans* Børgesen 1914a: p. 11, fig. 8.

Plants are free-floating and pelagic; axes are to 1 m long, widely forked, smooth to slightly spiny, and lack holdfasts. Blades are numerous, short-stalked, firm, flat, 3–8 mm wide, 2–6 cm long, serrate, and with pointed tips. Cryptostomata are uncommon or absent. Vesicles are numerous, 1–2 at blade bases, round to ovoid, lack a spine, 3–6 mm in diam., may be winged, and on stalks 2–3 mm long. Although usually vegetative, monoecious reproductive plants were reported in Cuba (Moreira and Suárez, 2002d).

Common, plants occur in drift on all coasts; also known in drift from Bermuda, the Sargasso Sea (type location; Silva et al. 1996), Newfoundland through the Caribbean, Gulf of Mexico, Brazil, and Indo-Pacific.

Taylor 1960: p. 281; Earle 1969: p. 221; Dawes 1974: p. 112; Schneider and Searles 1991: p. 172; Littler and Littler 2000: p. 282.

Sargassum hystrix J. Agardh 1847: p. 7; 1889: p. 91, pl. 7 (Gr. *hystrix*: bristly, porcupine-like, a porcupine). Plate XX, fig. 5, 6.

Synonym. *Carpacanthus spinulosum* Kützing 1861: p. 15, pl. VII, fig. 1–5.

Plants are bushy, erect, and to 40 cm tall; perennial axes are smooth, cylindrical, and attached by distinct fibrous holdfasts. Blades of annual axes are closely set, radially arranged, oblong to elliptical, 1–4 cm long, and to 1 cm wide, with strongly serrate margins and raised (costate) midribs; blade bases are asymmetrical. Cryptostomata are scattered on blades, small, to 160 µm in diam., and 120–160 µm deep. Vesicles are spherical, numerous, on short stalks, and have sharp spines. Receptacles are terete, palmately branched, to one-third as long as subtending blades, and have a rough surface due to many conceptacles.

var. *buxifolium* Chauvin *ex* J. Agardh 1848 (1848–1901): p. 322 (L. *Buxus*: the generic name for boxwood + *folium*: blade; thus with leaves similar to boxwood).

Plants are to 50 cm long and attached by tough distinct holdfasts. Perennial axes are smooth, cylindrical, and covered by short lateral branchlets and blades. Blades of annual axes are radially arranged, oval to oblong, 1.5 cm wide, 6 cm long, and with smooth to slightly serrate margins. Midribs are obscure on mature blades. Vesicles are spherical, stalked, with scattered spines, and with one per blade. Cryptostomata are scattered, sparse, and have pores less than 100 µm in diam. Receptacles are irregular, short, less than a third of a blade length, clustered, and rough due to many conceptacles.

Uncommon, dredged from 57 m at the Dry Tortugas; reports of plants offshore Tampa Bay (Phillips and Springer, 1960) are questionable. The species is also known from Bermuda, the Caribbean (Martinique), Gulf of Mexico (type location: Campeche Banks, Yucatan, Mexico; J. Agardh 1847), Brazil, North Senegal, and Sierra Leone. The var. *buxifolium* is common locally in drift and on limestone in the Keys and also known from Massachusetts, Brazil, the Caribbean, and the Gulf of Mexico.

Taylor 1960: p. 279; Earle 1969: p. 223, 225; Dawes 1974: p. 114; Littler and Littler 2000: p. 282, 284.

Sargassum mathiesonii Kilar 1992: p. 115–121, fig. 1–18, table 1, 2 (named for the American phycologist A. C. Mathieson, who studies New England seaweeds). Plate XX, fig. 7, 8.

Plants are bushy and form vertical columns to 2 m high; perennial axes 5–7(35) cm tall, with conical or lobed massive holdfasts, and 1–10 annual branches that are slender, smooth, and branched. Blades of annual axes are simple, rarely forked, fusiform to lanceolate or polymorphic, stalked, flat, 3.8–6.2 mm wide, 20–43 mm long, and with symmetrical bases, pointed tips, distinct midribs, and serrated margins. Cryptostomata are small and scattered. Vesicles subtend blades and are nearly sessile, elliptical to spherical, 2.7–3.6 mm in diam., and lack spines. Plants are dioecious, with separate male and female plants forming receptacles basipetally on annual axes; the receptacles are slightly roughened by many conceptacles. The species is separated from *S. filipendula* by its short-stalked vesicles, basipetally formed receptacles, and dioecious conceptacles.

Known from 1–3 m at the mouth of the Homosassa River on the west coast of Florida (type locality).

Sargassum natans (Linnaeus) Gaillon 1828: p. 355 (L. _natans_: swimming, floating on the surface). Plate XX, fig. 9.

Basionym. *Fucus natans* Linnaeus 1753: p. 1160.

Plants are pelagic, lack a holdfast, and often occur in entangled clumps. Branches are wiry, thin, 10–50 cm long, and have wide-angled forks and a smooth surface. Blades are acutely serrate, 1–4 mm wide, 2–10 cm long, have long stalks, and lack cryptostomata. Vesicles are 3–5 mm in diam. and tipped by a spine, hook, or minute blades; they are common on axes, the bases of blades, and on stalks, with their lengths equal to those of vesicles. Although they are usually vegetative, monoecious reproductive plants were reported in Cuban plants, which are also described as having pronounced morphological variation (Moreira and Suárez, 2002d).

Common, plants occur in drift on all coasts; also known in drift from Newfoundland to Brazil, Bermuda, the Sargasso Sea, Jamaica (type location?; Børgesen 1914a), Caribbean, Gulf of Mexico (Mexico), as well as the eastern tropical Atlantic, Indo-Pacific, and Australia.

Taylor 1960: p. 281; Earle 1969: p. 225; Dawes 1974: p. 114; Schneider and Searles 1991: p. 172: Littler and Littler 2000: p. 284.

Sargassum polyceratium J. Agardh 1847: p. 356 (Gr. _polus, poly_: many + _cerato_: horned, horn-shaped). Plate XX, fig. 10.

Plants are erect and 45–90 cm tall; axes tough, muriculate to spiny except near bases, and with many fronds arising from irregular disclike holdfasts. Blades of annual axes are closely set, almost sessile, ovate, 3–8 mm wide, and 0.5–2.0 cm long; they are also asymmetrical, with deeply seriated margins, and have obtuse or pointed tips. Cryptostomata are small and scattered on vesicles and blades. Vesicles are spherical, 3–6 mm in diam., sessile or attached to stalks 2–6 mm long, and arise near the bases of blades. Receptacles occur on short, forked auxiliary branches within axils of blades.

var. _ovatum_ (Collins) W. R. Taylor 1928: p. 129, pl. 18, fig. 7, 10; pl. 19, fig. 16 (L. _ovatum_: egg-shaped).

Basionym. *Sargassum vulgare* C. Agardh f. *ovatum* Collins 1901: p. 248.

Plants are to 130 cm long with main axes densely covered by short spur branches and crisp small blades that are 1.5–2.0 cm long, 3–5 mm wide, and usually forked. Vesicles are spherical, common, inconspicuous, 3–5 mm in diam., and on short stalks that are to 2 mm long. Receptacles are densely branched and produced within the axils of blades. Kilar (1988) describes a pronounced seasonal variation in morphology.

Common, the species occurs on limestone and rubble in the Keys (to 1.5 m) and as drift on all coasts; also known from Bermuda, the Bahamas, Caribbean (type location: Havana, Cuba; J. Agardh 1847), and Brazil. The variety occurs in protected shallow reefs at the Dry Tortugas and in Bermuda, the Bahamas, Caribbean (type location: Havana Harbor, Cuba), and Brazil.

Taylor 1960: p. 276; Earle 1969: p. 227; Dawes 1974: p. 114.

Sargassum pteropleuron Grunow 1867: p. 55, pl. 6, fig. 1 (Gr. *ptero*: winged, fernlike + *pleuro*: ribbed-edged). Plate XX, fig. 11, 12.

Plants are erect and to 4 m tall; axes branch alternately and are attached by large lobed to cushion-like holdfasts. Perennial bases are 10–20 cm long, terete to angular, spiny, and with blades longer on estuarine than open coastal plants. Annual axes have rigid receptacles and blades that are narrow, linear, and 2–7 mm wide by 25–95 mm long. Midribs are raised (costate) and prominent; cryptostomata are small, few, scattered, and sometimes aligned along the midribs. Vesicles are common, spherical, 3–8 mm in diam., usually in groups of 4 or more, and sessile or on short stalks, 2–3 mm long. Receptacles are in loose clusters, at blade bases, and dichotomously branched.

Common, on limestone in estuaries, on all open coasts to 20 m, and sometimes form dense canopies, 0.5–2.0 m tall; also known from Bermuda, the Bahamas (type location: Nassau; Guiry and Nic Dhonncha, 2006), Caribbean, Gulf of Mexico, and the Philippines.

Taylor 1960: p. 274; Earle 1969: p. 228; Dawes 1974: p. 115; Littler and Littler 2000: p. 286.

Sargassum ramifolium Kützing 1861: p. 10, pl. 32, fig. 1a,b. (L. *rami*: branch + *folium*: a leaf). Plate XX, fig. 13.

Plants are erect, to 30 cm tall, and with 1–3 main axes arising from a short (1–2 cm) stalk attached by tough padlike holdfasts. Stalks are smooth, cylindrical, and sparsely branched. Blades are elongate, linear, forked or sparingly pinnate, 1–3 mm wide by 2–5 cm long, and radially arranged. Margins of blades are usually smooth and have a few teeth when young; midribs are distinct and raised (costate). Cryptostomata are small, 80–100 µm in diam., and occur in one series along blade margins. Vesicles are 1–2 per blade, spherical, 3–8 mm in diam., smooth, lack spines, and have stalks 2.5–3.5 mm long. Receptacles are in clusters within blade axils, spindle-shaped to cylindrical, and forked or alternately branched.

Rare, reported from the Keys on limestone near mangals; also known from the Caribbean, Brazil, Canary Islands, Mauritania, and Guinea-Bissau.

Taylor 1960: p. 278; Littler and Littler 2000: p. 286.

Sargassum vulgare C. Agardh 1820 (1820–1828): p. 3–4, *nom. illeg.* (L. *vulgare*: common, ordinary). Plate XXI, fig. 1, 2.

Silva et al. (1996) noted that the name *Sargassum vulgare* was superfluous and illegitimate.

Plants are erect, to 40 cm tall, have one to several perennial axes with long side branches, and are attached by tough discoid holdfasts. Annual axes are smooth to slightly spiny with spines 0.3–0.7 mm long. Blades are numerous, firm, oblong, 3–8 mm wide, and 1–3 cm long; they have scattered teeth on their margins, asymmetrical bases, rounded to notched tips, and distinct midribs.

Cryptostomata are numerous, scattered, and 100–160 µm in diam. Vesicles are spherical, 2.5–8.0 mm in diam., with cylindrical stalks 0.8–1.0 mm in diam. by 0.5–2 mm long; vesicles are located near blade bases and lack a terminal spine. Receptacles occur within blade axils, are 2–3 cm long, branchlike, and have vesicles and small blades.

var. *foliosissimum* (J. V. Lamouroux) J. Agardh 1889: p. 108 (L. *folio*: a leaf + *sissimus*: very; much).

Basionym. *Fucus foliosissimus* J. V. Lamouroux 1813: p. 36, pl. 7, fig. 1.

The variety has smaller, broader, and more obtuse blades than the species; its axes are also undulate, twisted, and have closely arranged blades with shorter receptacles.

Uncommon, the species occurs on limestone at Content and Sawyer Keys (to 6 m) and in drift off Tampa Bay, while the variety occurs in drift off the west coast. The species is known from Bermuda, the Caribbean, Gulf of Mexico, Brazil, the Mediterranean (syntype localities include Rodosto, Turkey; Silva et al., 1996), eastern tropical Atlantic, and Indo-Pacific. The var. is also known from tropical West Africa.

Taylor 1960: p. 272; Earle 1969: p. 229; Littler and Littler 2000: p. 288.

Sargassum wolfii S. Earle 1969: p. 229, fig. 123, 126 (named for the American botanist Frederick A. Wolfe, who taught at Duke University). Plate XXI, fig. 3, 4.

Plants are erect, 15–25 cm tall, with 2 or more main axes, and attached by small holdfasts. Branches are spiny to rough and to 1 mm in diam. Blades are numerous, linear, 1–2 mm wide, to 70 mm long, simple or usually forked 1–4 times and have margins that may be dentate. Cryptostomata occur along both sides of midribs; vesicles are sparse and 2–3 mm in diam., with stalks 1.5–2.0 times longer than vesicle length. Receptacles are in the axils of blades of many branches, 2–3 mm long, and on slender stalks. The species differs from *S. acinarium* by its wider, less frequently forked blades, and by receptacles that are longer than the blades.

Rare, known only from its original site on the north central west coast of Florida (type location: Bayport) in 1–2 m and on pilings and limestone 15 km off St. Theresa.

Earle 1969: p. 229.

Turbinaria J.V. Lamouroux 1825: p. 71 (L. *turbinatus*: cone-shaped, shaped like a top).

Plants have many sparsely to highly branched erect axes that arise from creeping holdfasts. Axes are smooth, except for blade or branchlet scars. Fronds are pyramidal-peltate, vesicle-like, and may have 1–2 rings of teeth near their tips and/or longitudinally serrated ridges. Cryptostomata are common and scattered; receptacles are densely forked.

Key to the species (2) of *Turbinaria*

1. Stalks of turbinate blades dentate, tops flat to convex . . . . *T. tricostata*
1. Stalks of turbinate blades mostly smooth, tops rounded to concave . . . . *T. turbinata*

Turbinaria tricostata E. S. Barton 1891: p. 218, pl. 54, fig. 3, 4 (L. *tri*: three + *costa*: ribbed; or three-sided). Plate XXI, fig. 5–7.

Plants are dark to yellow brown and to 17 cm tall; axes are tough, sparsely branched, and attached by tough haptera-like, branching holdfasts. Blades are clustered, pyramidal in shape, 0.8–1.2 cm wide, and taper sharply to stalks twice as long as their diam. Tops of blades are triangular, flat to convex, have small, sharply dentate teeth, and a vesicle embedded in the center. Cryptostomata are common, scattered over blades and stalks, cone-shaped, and have pores that roughen the surface. Receptacles are densely branched and clustered in the axils of blades.

Rare, on rubble in exposed sites at the Dry Tortugas (to 1 m); also known from Bermuda and the Caribbean (type location: Guadeloupe; Wynne, 2002).

Taylor 1960: p. 284; Littler and Littler 2000: p. 288.

Turbinaria turbinata (Linneaus) Kuntze 1898: p. 434 (L. *turbinata*: shaped like a top, triangular). Plate XXI, fig. 8–10.

Basionym. *Fucus turbinatus* Linnaeus 1753: p. 1160.

Synonyms. *Sargassum turbinatum* (Linnaeus) C. Agardh 1820 (1820–1828): p. 41.

Plants are erect, to 40 cm tall, and attached by tough haptera-like holdfasts. Axes are dark brown, tough, and may be naked below. Upper parts of axes are covered with short spur branchlets that produce blades, which are shaped like inverted pyramids, to 1.2 cm wide, and taper lengthwise to smooth triangular stalks. Blade tops are triangular, flat or concave, and have an embedded vesicle and nearly smooth ridges. Cryptostomata are common, scattered, and with pores mostly flush with surface. Receptacles are abundant, forked, and occur within the axils of stalks.

Uncommon, in tide pools on wave-exposed reefs at the Dry Tortugas and Keys, as well as adrift in northwestern Florida; also recorded from the Caribbean (type location: Jamaica; Wynne, 2002), Gulf of Mexico, Brazil, and the Indian Ocean.

Taylor 1960: p. 285; Earle 1969: p. 231; Littler and Littler 2000: p. 290.

Sedis incertae: Asteronema Delépine and Asensi 1975: p. 296, *nom. cons. prop.* (Gr. *astero*: starry, stellate + *nema*: thread).

Draisma et al. (2003) considered *Asteronema* to be a taxon of uncertain taxonomic status, while others retained it in the Ectocarpaceae (Abbott and Huisman, 2003). See Silva (1996) regarding the conservation of *Asteronema* Delépine and Asensi, a later homonym of the Hyphomycete fungal genus *Asteronema* Trevisan (1845: p. 46).

Filaments are uniseriate, tufted, much branched, have long unbranched rows of cells, and are attached by multicellular rhizoids. Growth is by both apical and intercalary divisions. Cells have several elongate, stellate plastids, each with a terminal pyrenoid.

Asteronema breviarticulatum (J. Agardh) Ouriques and Bouzon 2000: p. 271, fig. 2–12, tab. 1 (L. *brevi*: short + *articulatum*: segments, joints). Plate XXI, fig. 11–13.

Basionym. *Ectocarpus breviarticulatus* J. Agardh 1847: p. 7.

Synonyms. *Ectocarpus hamatus* P. Crouan and H. Crouan in Schramm and Mazé 1865: p. 1; P. Crouan and H. Crouan in Mazé and Schramm 1878 (1870–1877): p. 111.

Ectocarpus spongodioides P. Crouan and H. Crouan in Mazé and Schramm 1878 (1870–1877): p. 111.

Filaments are produced in twisted to ropelike tufts 2–4 cm tall; they are irregularly branched and have prostrate and erect axes that produce short-hooked branchlets, which may form basal rhizoids. Cells are 25–55 µm in diam., 1–2 diam. long, and have many discoid plastids. Plurilocular sporangia are ovoid to spherical, stalked, 55 µm wide by 62 µm long, and at nearly right angles to the axes.

Rare, on wave-exposed intertidal limestone and coral rubble in the Keys; also known from the

Caribbean, Brazil, West and South Africa, and the Pacific Ocean (type location: St. Augustine [San Agustín], Oaxaca, Mexico; N'Yeurt and Payri, 2006).

Taylor 1960: p. 201.

RALFSIALES

The affinities of the order and family are unclear because of variable chloroplast shapes and life history traits (Fletcher, 1987). When Nakamura (1972) introduced the ordinal name it was not validated with a Latin diagnosis (Wynne, 2005a). The family has been placed in the Chordariales (Taylor, 1960; Womersley, 1987; Wynne, 1998a), Ectocarpales (Schneider and Searles, 1991), and Scytosiphonales (Fletcher, 1987). Based on molecular studies of Tan and Druehl (1994), the type species of *Ralfsia*, *R. fungiformis* Setchell and Gardner (1924), was placed outside the present delineation of the Ectocarpales. Draisma et al. (2003) believed that more *rbcL* sequences of *Ralfsia* were needed to clarify its phylogenetic relationships.

Plants form fleshy brown crusts with pseudoparenchymatous construction. Basal layers are prostrate and several cells thick; upper filaments are erect, closely adherent, and densely packed. Growth is marginal (apical) or by intercalary divisions. Plastids are numerous and discoid (*Petroderma*, *Pseudolithoderma*) or single and platelike (*Ralfsia, Sorapion, Symphyocarpus*). Nakamura (1972) described the Ralfsiales as having an apparent isomorphic haplo-diplophasic life history, a discoid-type of germination, and cells with a single parietal, platelike chloroplast that lacks a pyrenoid. He did not include members of the Lithodermataceae that had many discoid chloroplasts (see below).

Ralfsiaceae

Synonym. Lithodermataceae.

The family has the same features as the order.

Pseudolithoderma Svedelius in Kjellman and Svedelius 1910 ('1911'): p. 175 (Gr. *pseudo*: false + *litho*: stone + *derma*: skin, covering).

Plants form fleshy adherent crusts on rocks or shells. The basal layer is indistinct and erect filaments are densely packed. Each cell has many plastids. Ascocysts (cystlike cells) may be present. Plurilocular and unilocular sporangia arise from surface cells.

***Pseudolithoderma extensum* (P. Crouan and H. Crouan) S. Lund 1959: p. 84 (L. *extensum*: drawn out, widely spread). Plate XXI, fig. 14–17.**

Basionym. *Ralfsia extensa* P. Crouan and H. Crouan 1860: p. 368; 1867, p. 166.

Synonym. *Lithoderma fatiscens* Kuckuck 1894: p. 238, fig. 11A-B, *nom. illeg.*; non Areschoug 1875: p. 22.

Crusts are thin, dark brown, to 8 cm in diam., and with poorly defined basal layers. Erect filaments are 8–12 µm in diam. and to 150 µm long. Unilocular and plurilocular sporangia are terminal on outer cortical cells. Unilocular sporangia are globose; while plurilocular sporangia are uniseriate or biseriate and occur on the same or different crusts.

Rare, inconspicuous, on rocks and shells to 10 m in the Keys; also known from the Canadian Arctic to North Carolina, the Bahamas, Caribbean, Gulf of Mexico, and Brazil, eastern North Atlantic (type location: Brest, France; Athanasiadis, 1996b), the Mediterranean, and Black Sea.

Taylor 1960: p. 244 (uncertain records); Schneider and Searles 1991: p. 128; Littler and Littler 2000: p. 244.

Ralfsia M. J. Berkeley in J. E. Smith and J. Sowerby 1843: pl. 2866, 1 Nov. 1841 (named for John Ralfs, a British botanist and phycologist).

Plants are fleshy brown crusts, to 5 cm in diam., and usually firmly attached, except at margins. The basal layer has prostrate radiating filaments attached by rhizoids and produces erect photosynthetic filaments that are laterally fused and have inconspicuous tufts of hairs. Unilocular and plurilocular sporangia are in clusters (sori) at the surface; the former are lateral at the base of hairs and the latter are terminal on erect filaments.

***Ralfsia expansa* (J. Agardh) J. Agardh 1848 (1848–1901): p. 63 (L. *expansa*: expanded, spread out). Plate XXI, fig. 18, 19.**

Basionym. *Myrionema expansum* J. Agardh 1847: p. 7.

Crusts are dark brown with prostrate filaments attached by rhizoids below and erect filaments above. Erect filaments are densely packed, adherent, 100–170 µm tall, and composed of 5–14 cells; cells to 12 µm in diam. and 4–7 diam. long at their bases. Unilocular sporangia are ovoid, 30 µm in diam., 75–120 µm long, and lateral at the bases of hairs. Plurilocular sporangia are terminal on erect filaments and may be biseriate in upper parts; cells are 15–20 µm long by 5–6 µm in diam.

Uncommon, on limestone and coral rubble at the Dry Tortugas; also known from the Caribbean, Brazil, the Mediterranean, eastern tropical Atlantic, and Indo-Pacific (type location: Veracruz, Mexico; Abbott and Huisman, 2004).

Taylor 1960: p. 243; Earle 1969: p. 143.

4

Subkingdom Rhodoplantae

Cavalier-Smith (1998) placed the red algae in the subkingdom Billiphyta within the kingdom Glaucophyta. Doweld (2001) argued that the red algae should not be included with the Glacucophyta and Saunders and Hommersand (2004) placed the red algae in the subkingdom Rhodoplantae of the kingdom Plantae.

The subkingdom contains eukaryotic organisms that lack flagella and store photosynthetic reserves in the cytoplasm. Plastids have nonaggregated thylakoids with accessory pigments (phycoerythrin and phycocyanin) in phycobilisomes on their surfaces.

Phylum RHODOPHYTA

Synonym. Rhodymeniophyta Doweld 2001.

The red algal phylum of Wettstein (1901) contains 4,000 to 6,000 species (Norton et al., 1996) of which 3 percent occur in freshwater habitats and the rest in saline environments (Sheath, 1984). Cells are eukaryotic, with one to many nuclei, and lack flagella. The reserve food is floridean starch that is insoluble in boiling water, appears refractive under a light microscope, and is a branched amylopectin consisting of α-1–4 linked glucans with ß-1–6 linked side glucan chains. Photosynthetic pigments include water-soluble phycobilins (blue-reflecting phycocyanin and red-reflecting phycoerythrin), chlorophyll *a*, and various carotenes and xanthophylls. In addition to cellulose, the walls often contain sulfated galactan polymers, some of which are economically important phycocolloid extracts (for example, agar, carrageenan, furcellaran). More than 300 species of red algae are also used as a direct source of food known as sea vegetables. Pit connections and pit plugs are characteristic of red algal cell walls (mostly in the class Florideophyceae) and usually visible under the light microscope. Primary pit connections result from the division of two sister cells, while secondary ones are formed through contact of neighboring cells. Several species are calcified with a calcite form of calcium carbonate that often plays a major role in the formation of tropical reefs (Littler and Littler, 2000).

Asexual reproduction is common throughout the phylum, including nonmotile spores (bispores, carpospores, conchospores, endospores, monospores, zygotospores; Guiry, 1990) and vegetative fragmentation. Spores may be the only mode of reproduction for many members of the class Rhodellophyceae; sexual reproduction is known for several members of the class Bangiophyceae and common for most members of the advanced class Florideophyceae. In addition to tetraspores and carpospores, some members of the Florideophyceae may produce other types of sexual nonmotile spores (bispores, polyspores, paraspores; Guiry, 1990). Life histories in the Bangiophyceae are often biphasic but may vary; those in the Florideophyceae may also be variable but are often triphasic, with two diploid phases (carposporophyte and tetrasporophyte) and one haploid phase (gametophyte). Meiosis occurs in free-living diploid tetrasporophytes and results in the production of four haploid tetraspores. The pattern of division within tetrasporangia varies among orders (cruciate, zonate, tetrahedral, zonate; Guiry and Irvine, 1989; Guiry, 1990). Upon release, haploid tetraspores germinate to form free-living male and female gametophytes. Male gametophytes produce nonmotile spermatia, while the females produce carpogonia (with an egg nucleus) that usually have an elongate hairlike extension or trichogyne. Fertilization occurs after a spermatium attaches to the trichogyne and its nucleus fuses with the egg nucleus in the carpogonium. In contrast to the Bangiophyceae that lack a diploid parasitic carposporophyte generation, the zygote in most red algae undergoes a series of mitotic divisions that result in a diploid carposporophyte that will produce diploid carpospores while attached to the female gametophyte. The carposporophyte may be enclosed in a protective encasement of branches or a pericarp that is part of the female gametophyte. Cystocarps may be visible to the naked eye. Carpospores grow into free-living tetrasporophytes.

Red algal lineages

Previously, the red algae were placed in one class (Rhodophyceae), two subclasses (Bangiophycidae and Florideophycidae), and 10–18 orders (Woelkerling, 1990; Wynne, 1998a). Based on molecular and morphological studies, Saunders and Hommersand (2004) designated three subphyla within the Rhodophyta: the Rhodellophytina Cavalier-Smith (1998), Metarhodophytina Saunders and Hommersand (2004), and Eurhodophytina Saunders and Hommersand (2004). Further studies by Yonn et al. (2006) recognized seven red algal lineages of which four include species found in Florida: Bangiophyceae (*Bangia*, *Porphyra*), Compsopogonophyceae (*Erytrotrichia*, *Erythrocladia*, *Sahlingia*), Florideophyceae (containing most of the red algae), and Stylonematophyceae (*Stylonema*, *Chroodactylon*).

RHODELLOPHYTINA

The subphylum has unicellular and pseudofilamentous algae with varying types of plastids and organelle mixes (Cavalier-Smith, 1998). Sexual reproduction is unknown.

RHODELLOPHYCEAE

The features of this class are the same as the subphylum (Cavalier-Smith, 1998).

STYLONEMATALES

The order contains unicellular forms, palmelloid colonies, or pseudofilaments that are uni- to multiseriate and with or without irregular branches (Drew, 1956). Growth is intercalary and pit connections are absent. Cells are in a matrix, uninucleate, and have one stellate or many discoid plastids. Asexual reproduction is by monospores, akinetes, or vegetative cells; sexual reproduction is unknown.

Stylonemataceae

Synonym. Phragmonemataceae Skuja 1939.

Members are multicellular, filamentous, prostrate or erect, with basal cells little modified, and all cells are in a mucilaginous matrix (Brodie and Irvine, 2003). Plastids are single, stellate, and central or lobed and parietal; pyrenoids are single and central. Asexual reproduction is by monospores or akinetes, while sexual reproduction is unknown.

Chroodactylon Hansgirg 1885: p. 15 (Gr. *chroos*: color, skin + *daktylos*: digit, finger).

Synonyms. *Allogonium* sect. *Asterocytis* Hansgirg 1892: p. 132.
Asterocytis (Hansgirg) Gobi *ex* F. Schmitz in Engler and Prantl 1896 (1890–1897): p. 314.

Uniseriate filaments are epiphytic, with simple, irregular, or pseudodichotomous branching, and attached by a basal cell. Cells are irregularly spaced in a matrix; each cell has one blue-green stellate plastid and a single pyrenoid. Asexual reproduction is by monospores and akinetes formed from vegetative cells.

Chroodactylon ornatum (C. Agardh) Basson 1979: p. 67, pl. IX, fig. 52 (L. *ornatum*: decorated, equipped, adorned). Plate XXII, fig. 1, 2.

Basionym. *Conferva ornata* C. Agardh 1824: p. 104 (as '*ornatum*').
Synonyms. *Asterocytis ornata* (C. Agardh) Hamel 1924: p. 451, fig. 7B–E.
Asterocytis ramosa (Thwaites in Harvey) Gobi 1879: p. 85.

Filaments are uniseriate, simple or subdichotomously branched, in soft pale green tufts, 1–10 mm tall, and 12–20(25) µm in diam. Cells are 5–9 by 8–20 µm, ovoid to elongate, and in irregular rows in a matrix. Akinetes are produced from vegetative cells, have dense cytoplasm, 16 µm long, and have walls 2 µm thick.

Common, inconspicuous, on seagrasses near Anclote Key (Ballantine, 1972) and Content Keys; also known from the Canadian Maritime Provinces through the Caribbean, Bermuda, as well as the eastern North and tropical Atlantic (type location: Lake Mälaren, Sweden; Abbott, 1999), Mediterranean, Red Sea, Indo-Pacific, and Australia.

Taylor 1960: p. 287; Dawes 1974: p. 117; Schneider and Searles 1991: p. 177.

Stylonema Reinsch 1875 ('1874/1875'): p. 40 (Gr. *stylos*: a pillar + *nema*: thread).

Synonym. *Goniotrichum* Kützing 1843: p. 244 (in part), *excl. typ.* (Wynne 1985c).

Wynne (1985c) and Silva et al. (1996) discussed the replacement of *Goniotrichum* Kützing by *Stylonema* Reinsch, and a proposal by Demoulin and Hoffmann (1992) to conserve *Goniotrichum* (not recommended; Compère, 1997).

Plants are epiphytic, usually simple to irregularly branched uniseriate filaments, and often attached by a basal cell. Cells are narrow and widely spaced in a matrix; each cell has one pink, red, to purple stellate plastid with a pyrenoid. The plant reproduces asexually by monospores from vegetative cells.

Stylonema alsidii (Zanardini) K. Drew 1956: p. 72 (L. *alsidena*: like an onion; also a genus of ferns *Alsidium*). Plate XXII, fig. 3, 4.

Basionym. *Bangia alsidii* Zanardini 1839: p. 136; also 1840a: p. 136.
Synonyms. *Goniotrichum alsidii* (Zanardini) M. A. Howe 1914: p. 75.
Stylonema elegans (Chauvin) V. May 1965: p. 352, 354

Cells are produced in chainlike filaments, <1 mm to 6 mm long, in a matrix, and are attached by a basal cell. Filaments are uniseriate to multiseriate, abundantly subdichotomously to irregularly branched, rarely simple, 12–35 µm in diam. Cells are globose to irregular, 6–13 µm in diam., and 4–13 µm long.

Common, inconspicuous, on macroalgae and seagrasses on all coasts; also known from Labrador through the Caribbean, Bermuda, Gulf of Mexico, Brazil, and Uruguay, as well as the eastern North and tropical Atlantic, Mediterranean (type location: Trieste, Italy; Silva et al., 1996), Indo-Pacific, and Australia.

Taylor 1960: p. 288; Dawes 1974: p. 118; Schneider and Searles 1991: p. 178.

METARHODOPHYTINA

The subphylum (Saunders and Hommersand, 2004) has the same features as the subclass Metarhodophycidae (Magne, 1989).

Monosporangia and spermatia are usually formed by curved walls from ordinary vegetative cells. A golgi-endoplasmic reticulum association encircles the thylakoids in the plastid. Life histories, if known, are biphasic.

COMPSOPOGONOPHYCEAE

The class has the same features as the subphylum.

ERYTHROPELTIDALES

See Garbary et al. (1980), Silva et al. (1996), and Wynne (2005a) for descriptions of the order. Zuccarello et al. (2000) noted that the Erythropeltidales and Rhodochaetales, which are members of the Compsopogonophyceae, are as divergent at the molecular level as some members of the class Florideophyceae (Saunders and Hommersand, 2004).

Plants are filamentous, crustose or membranous, and grow on or penetrate animals, plants, and hard substrata. Filaments are erect, uniseriate or multiseriate, and irregularly branched; if saccate or bladelike, they are entire or split lengthwise at maturity. Growth is by intercalary cell divisions. Cells have stellate, axial, band-shaped parietal plastids. Pit connections are absent. Asexual reproduction is by monospore production from vegetative cells formed from an obliquely curving, unequal division. Sexual reproduction is poorly known.

Erythrotrichiaceae

Synonym. Erythropeltidaceae Skuja 1939: p. 33.

Plants are erect or prostrate, filamentous or membranous, and are attached by a single cell, by a prostrate uni- to multilayered disc, or by rhizoidal down-growths from the lower cells. Sexual reproduction is poorly known.

Erythrocladia Rosenvinge 1909: p. 71 (Gr. *erythros*: red + *klados*: a branch, shoot).

Plants are filamentous, have apical growth, and grow on or in various plants and animals. The monostromatic discs may be multilayered in the center and marginally free. Branching is irregular to subdichotomous. Cells are polymorphic, lack pit connections, and have one single bowl-shaped plastids that lack pyrenoids. Monosporangia are cut off from vegetative cells by obliquely curving walls.

Erythrocladia endophloea M. Howe 1914: p. 81, pl. 30, fig. 1–7 (Gr. _endon_: within, below + _phlois_: skin, bark). Plate XXII, fig. 5.

Synonyms. *Erythrocladia recondita* M. Howe and Hoyt 1916: p. 112, pl. 12, fig. 1–5; pl. 13, fig. 1. *Erythocladia vagabunda* M. Howe and Hoyt 1916: p. 115, pl. 12, fig. 6–11; pl. 13, fig. 2.

Discs are minute, on or in plants and animals, and usually in the outer walls of algae. Branching is irregular, spreading, and pseudoparenchymatous, with bases (pads) having radiating filaments, 0.75–2.26 mm in diam. Cells are polymorphic, 3–15 µm in diam., and 8–40 µm long; monospores are globose, 4–5 µm in diam., and arise from intercalary cells.

Common, inconspicuous, endophytic, endozoic or partially epiphytic on or in various plants and animals on all coasts; also known from eastern Connecticut, the Carolinas, Puerto Rico, and Peru (type location: Bay of Sechura).

Taylor 1960: p. 290, 291; Schneider and Searles 1991: p. 180.

Erythrotrichia Areschoug 1850: p. 435, *nom. cons*, see Wynne, 1985c (Gr. *erythro*: red + *trichia*: hair).

Plants are epiphytic or lithophytic and simple erect filaments or monostromatic blades that arise from multilayered basal discs or expanded basal cells. Erect filaments are uniseriate below and bi- to multiseriate above; lower cells have rhizoids. Growth is intercalary. Cells have a single stellate plastid with one pyrenoid. Monosporangia and spermatia are separated from vegetative cells by oblique walls. Spermatia are smaller than monospores. Carpogonia arise from modified vegetative cells. The heteromorphic life history includes a sporophytic "conchocelis" stage that penetrates shells.

***Erythrotrichia carnea* (Dillwyn) J. Agardh 1883: p. 15, pl. 1, fig. 8–10 (L. *carnea*: pink, flesh colored). Plate XXII, fig. 6, 7.**

Basionym. *Conferva carnea* Dillwyn 1807 (1802–1809): p. 54, pl. 84.

Plants are epiphytic filaments which are 0.5 to 8 cm tall, uniseriate below, biseriate or multiseriate above, 15–60 µm in diam., and attached by basal cells with lobed projections or by minute prostrate discs. Filamentous cells are swollen, polymorphic, 15–20 µm in diam. by 12–30 µm long above and 9–13 µm in diam. by 12–40 µm long below. Monospores are globose and 13–18 µm in diam.

Common, on algae and seagrasses on all coasts; also known from the Canadian Arctic through the Caribbean, Bermuda, Bahamas, Gulf of Mexico, and Brazil, as well as the eastern North and tropical Atlantic (type location: near Loughor, Wales; Dawson, 1953), Mediterranean, Red Sea, and Indo-Pacific.

Taylor 1960: p. 292; Dawes 1974: p. 117; Schneider and Searles 1991: p. 182.

Sahlingia Kornmann 1989: p. 227, fig. 1, 6–13 (named for P. H. Sahling, a German phycologist and collaborator with P. Kornmann).

Womersley (1994) did not separate this genus from *Erythrocladia*.

Plants are epiphytic or epizoic; discs are monostromatic, prostrate, and with filaments irregularly to subdichotomously branched. Growth is marginal, with edge cells bifurcate and inner ones polymorphic. Each cell has a single band- to bowl-shaped plastid lacking pyrenoids; monosporangia are formed by oblique divisions of vegetative cells.

***Sahlingia subintegra* (Rosenvinge) Kornmann 1989: p. 227, fig. 1, 6–13 (L. *sub*: beneath + *integra*: whole, entire). Plate XXII, fig. 8.**

Basionym. *Erythrocladia subintegra* Rosenvinge 1909: p. 73–75, fig. 13, 14.

Synonym. *Erythropeltis subintegra* (Rosenvinge) Kornmann and Sahling 1985: p. 224, fig. 8.

Plants are epiphytic or epizoic discs, which are minute, monostromatic, 300 µm in diam., and compact. The central cells of discs are coalesced, while marginal ones are free and forked. Cells are rectangular, oblong to irregular, and 2–5 μm in diam. by 7–13 μm long. Monospores are globose, ca. 6 μm in diam., and arise from inner disc cells.

Common, on algae and seagrasses (to 16 m) on all coasts; also known from the Canadian Arctic through the Caribbean, Bermuda, Gulf of Mexico, and Brazil, as well as the eastern North and tropical Atlantic (type location: Møllegrund, Denmark; Womersley, 1994), Mediterranean, Pacific Ocean, and Australia.

Taylor 1960: p. 290; Dawes 1974: p. 117; Schneider and Searles 1991: p. 184.

EURHODOPHYTINA

The features of this subphylum (Saunders and Hommersand, 2004) are similar to the subclass Eurhodophycidae of Magne (1989). Transformed cells produce spores or gametes. Plants have a golgi-endoplasmic reticulum-mitchondrial association, biphasic or triphasic life histories (where known), and pit plugs in at least one life history stage.

BANGIOPHYCEAE

Life histories in the class are biphasic and heteromorphic; gametophytes are macroscopic, initially uniseriate, and become pluriseriate or foliose by diffuse growth (Wettstein, 1901). Cells often are in a gelatinous matrix. Plants are endophytic, endozoic, epiphytic, or attached to hard substrata. Cells are cylindrical, spherical, or ellipsoid and uninucleate and with many small discoid or a few stellate plastids that are axial or parietal. Pyrenoids are present or absent. Pit connections are usually absent. Asexual reproduction is by monospores, akinetes, or the release of entire vegetative cells. Sexual reproduction, if known, is oogamous and spermatia fertilize nuclei of carpogonia that may or may not have a trichogyne. Fertilized carpogonia form zygotospores or carpospores; no carposporophyte generation occurs (Guiry, 1990). Sporophytes are filamentous; their pit plugs have one cap layer without a membrane, and conchospores are usually formed in fertile cell rows.

BANGIALES

> Müller et al. (2001) and Nelson et al. (2006) confirmed the monophyletic nature of the order Bangiales, with the latter study reporting that it would contain at least ten genera instead of the present two (*Bangia*, *Porphyra*).

Plants are erect filaments or blades and epiphytic, endophytic, or grow on hard substrata. Upright erect axes are unbranched or split and attached by rhizoidal discs. Cells have one stellate axial plastid and a central pyrenoid. Asexual reproduction is by monospores from vegetative cells. Sexual reproduction, where known, is by small spermatia and carpogonia with trichogynes arising from transformed vegetative cells. Fertilized carpogonia produce pigmented zygotospores (Guiry, 1990) directly without a carposporophyte generation. Zygotospores form shell-boring, uniseriate, branched filaments (conchocelis stage) that produce monospores that grow into erect axes or repeat the conchocelis stage; pit connections are known only in the shell-boring stage.

Bangiaceae

The family has the same characteristics as the order.

Bangia Lyngbye 1819: p. 82 (named for Niels Hoffman Bang of Copenhagen, a Danish botanist and squire who was a patron of H. G. Lyngbye and a friend of Hans Christian Andersen).

Several authors (Oliveira et al., 1995; Brodie et al., 1998; Müller et al., 1998) have stated that *Bangia* belongs in the genus *Porphyra*. Niwa et al. (2003) noted that they are paraphyletic. However, Brodie and Irvine (2003) argued that multiple genera may be involved and so retained both taxa until more molecular and morphological data are available.

Filaments are erect, uniseriate to multiseriate, cylindrical, unbranched or expanded into narrow fronds above, and attached by rhizoidal outgrowths of basal cells. Growth is intercalary; cells are in irregular or regular rows, in a firm gelatinous sheath; they lack pit connections and are cylindrical, quadrate, or polyhedral. Each cell has one axial, stellate plastid with a pyrenoid. Reproductive cells develop from vegetative ones. Neutral spores are produced singly or 2–4 per cell. Spermatia are in packets and produced by multiple divisions in 3 planes. Trichogynes are present or absent. Zygotospores (carpospores) are formed directly from a fertilized carpogonium. Sporophytes are filamentous, shell-boring (i.e., conchocelis stage), and have pit connections.

***Bangia fuscopurpurea* (Dillwyn) Lyngbye 1819: p. 82–83, pl. 24C (L. *fuscopurpurea*: dark purple or brownish purple). Plate XXII, fig. 9.**

Basionym. *Conferva fuscopurpurea* Dillwyn 1807 (1802–1809): p. 54, pl. 92 (as '*fusco-purpurea*').
Synonym. *Bangia lutea* J. Agardh 1842: p. 14.

Den Hartog (1972) and Geesink (1973) suggested that the freshwater *Bangia atropurpurea* (Dillwyn) Lyngbye and saltwater *B. fuscopurpurea* were conspecific. However, Müller et al. (2001, 2003) and Niwa et al. (2003) provided molecular evidence to maintain the two species, with the marine taxon designated *B. fuscopurpurea*. Culture (Drew, 1952) and cytological studies (Cole et al., 1983) suggested the possibility of multiple species of *Bangia*.

Filaments are erect, rose red to brown or black, 5–10 cm tall, multiseriate, unbranched, and attached by a basal cell with descending rhizoids. Filaments are 20–60 µm wide below, to 220 µm above, they taper to their bases, and are often irregularly constricted or contorted. Each cell has a stellate plastid and a central pyrenoid. Plants alternate with a shell-boring "conchocelis" stage.

Common, on rubble, wood, and other substrata in the upper intertidal zone during winter on all coasts; also known from Greenland through the Caribbean, Bermuda, Texas, Uruguay, and Brazil, as well as the eastern North and tropical Atlantic (type location: Wales; Silva et al., 1996), Mediterranean, Pacific Ocean and Australia.

Taylor 1960: p. 293; Schneider and Searles 1991: p. 186.

Porphyra (C. Agardh) C. Agardh 1824: p. xxxii, 190, *nom. cons.* (Gr. *porphyra*: purple, purple dye).

Basionym. *Ulva* Tribe *Porphyra* C. Agardh 1822 (1820–1828): p. 2.
Synonyms. non *Porphyra* Loureiro 1790: p. 69 = *Callicarpa* Linnaeus 1753, Verbenaceae.
Conchocelis Batters 1892: p. 27.

Florida *Porphyra* should be reexamined based on studies of Coll and Cox (1977), Kapraun and Freshwater (1987), Broom et al. (2000, 2002), Klein et al. (2003), and West et al. (2005). Taylor's (1960) records of *Porphyra umbilicalis* (Linnaeus) J. Agardh are questionable. Specimens identified as *P. leucosticta* Thuret in Le Jolis may be *P. rosengurttii* Coll and Cox (Schneider and Searles, 1991) or other taxa (Neefus et al., 2000; West et al., 2005).

Gametophytic blades are mono- or distromatic, narrow to broad, solid to perforate, and attached by small rhizoidal discs. In surface view vegetative cells are randomly arranged or in packets; they are in a firm gelatinous matrix, lack pit connections, square to polymorphic, and have 1–2 stellate plastids with one central pyrenoid. Monosporangia, spermatia, and zygotosporangia arise from vegetative cells often in distinct patches. Spermatial packets arise from divisions in 3 planes with up to 128 cells. Carpogonia may have trichogynes and form 4–64 zygotospores after fertilization. Sporophytes (conchocelis stage) are poorly known, filamentous, and have pit connections.

Key to the species (3) of *Porphyra*

1. Plants large, to 15 cm tall . . . . *P. leucosticta*
1. Plants smaller and to 4 cm tall 2
   2. Frond margins have minute teeth and a rhizoidal base. . . . . *P. suborbiculata*
   2. Frond margins lack minute teeth and a discoid base . . . .*P. rosengurttii*

Porphyra leucosticta Thuret in Le Jolis 1863: p. 100 (Gr. *leuco*: white + *sticta*: spotted or dotted; also *leukostiktos*: grizzled, slippery). Plate XXII, fig. 10–12.

The identification and distributional records of this species need confirmation with molecular studies as it may represent a complex of cryptic taxa (Bird and McLachlan, 1992; Neefus et al., 2000; West et al., 2005).

Blades are rose to purple, to 15 cm tall, 13 cm wide, and with 1–3 fronds arising from short, stalked, rhizoidal holdfasts. Blades are oblong to cordate, monostromatic, 25–50 µm thick, usually entire, and with undulate to folded margins. Cells are subquadrate in surface view, 12–15(20) µm in diam., and 1.5–2 times as high as wide; each cell has one lobed to stellate plastid. Gametophytes are monoecious. Spermatia are in 8 tiers, with 64–128 per cell, and 3–6 µm in diam. Spermatangial sori form elongate colorless patches along margins and are 1–1.5 by 5–10 mm. Zygotospores, which are from fertilized carpogonia, are 4–32 in number, 6–9 µm in diam., pink in hue, and in marginal bands.

Uncommon, on mangrove roots in the Keys and coquina at Marineland; also known from the Arctic to North Carolina, Bermuda, Texas, Brazil, the western South Atlantic, as well as the eastern North and Central Atlantic (type location: Cherbourg France; Brodie and Irvine, 2003), Mediterranean, Black Sea, and Angola.

Taylor 1960: p. 295; Schneider and Searles 1991: p. 188.

Porphyra rosengurttii Coll and Cox 1977: p. 157–159, fig. 9–16 (named for the Uruguayan botanist Bernardo Rosengurtt). Plate XXII, fig. 13–17.

The species was previously misidentified as *Porphyra linearis* Greville and *P. umbilicalis* (L.) Kützing (Wynne, 2005a), as well as *P. leucosticta* (Coll and Cox, 1977).

Blades are membranous, usually ovoid, lanceolate, or irregular, in clusters, brown red to pale violet, 4.0 cm long by 2.5 cm wide, and arise from basal holdfasts. Blades are monostromatic, entire, ruffled, lobed or torn, with cordate or overlapping lobed bases, and 30 µm thick in vegetative, spermatangial, and zygotosporangial areas. In surface view, vegetative cells are 9.0 by 10.0–17.5 µm and have one stellate chloroplast with a central pyrenoid. In cross section, cells are 17.5–22.5 µm tall. Plants are monoecious Spermatangial sori occur in whitish patches near edges, each packet 7.5 x 12.5 µm with 16 (or 32) spermatia in 4 tiers of 4 (or 8) cells. Spermatia are 3.5 to 5.0 µm in diam. Zygotosporangia are in vegetative areas near spermatangia. Carpogonia have small distinct trichogynes. Zygotospores are produced in packets, 10.0 by 17.0 µm with 4–8 in 2 tiers of 2–4. Zygotospores are 8.0–12.0 µm by 11.0–15.0 µm and 8.0–12.0 µm tall.

Although not recorded for Florida, it probably was misidentified as *P. leucosticta* from Marineland (for example, Humm, 1952a) and Georgia (Chapman, 1971); known from Long Island Sound, the Carolinas (type location: Pivers Island, North Carolina), and Georgia.

***Porphyra suborbiculata* Kjellman 1897: p. 10–13, pl. 1, fig. 1–3; pl. 2, fig. 5–9; pl. V, fig. 4–7 (L. *suborbiculata*: nearly orbicular). Plate XXII, fig. 18.**

Synonyms. *Porphyra areolata* Kjellman 1897: p. 8, pl. 2, fig. 1–4; pl. 5, fig. 1–3.
Phyllona areolata (Kjellman) Kuntze 1898: p. 420.
Phyllona suborbiculata (Kjellman) Kuntze 1898: p. 420.
Porphyra carolinensis Coll and J. Cox 1977: p. 155–156, fig. 1–8.
Porphyra lilliputiana Nelson, Knight, and Hawkes 1998: p. 57–58, fig. 2–15.

See Broom et al. (2000, 2002) and Klein et al. (2003) for discussions and synonymy. Humm (1979) suggested that plants designated as *Porphyra carolinensis* may have been introduced to North Carolina after 1960, presumably from Asia (Kjellman, 1897) and the western Pacific (Broom et al., 2002). Freshwater and Kapraun (1989) cultured this species using plants from Sebastian Inlet, Florida. Monotilla and Notoya (2004) noted that this widely distributed species acclimated and differentiated under varying environmental conditions.

Blades are small, 4 cm tall, 2.5 cm wide, and to 30 μm thick; they are usually brown to red, single or clustered, and attached by rhizoidal holdfasts. Blades are monostromatic, oval, elongate, or irregular, and typically folded with small marginal teeth. Cells in surface view are oval to irregular, in rosette clusters, 5.0–12 μm in diam. by 7.5–20.0 μm tall, and have one stellate plastid with a single central pyrenoid. Plants are monoecious. Spermatia and zygotospores arise from marginal vegetative cells. Monospores are common and in marginal patches. Spermatangia have 32 spermatia in 4 tiers of 8; 16 zygotospores are formed per carpogonium in 2 tiers of 8.

Uncommon, during winter on hard substrata, epiphytic or epizoic in the upper intertidal at Sebastian Inlet and Marineland; also known from Long Island Sound to Florida, as well as the Canary Islands, Sri Lanka, Indo-Pacific (type location: Goto-retto, Japan; Silva et al., 1996), and Australia.

Schneider and Searles 1991: p. 188.

FLORIDEOPHYCEAE

Synonym. Rhodymeniophyceae Doweld 2001.

Based on molecular studies, four primary phylogenetic lineages were recognized in Cronquist's (1960) class Florideophyceae (Saunders and Bailey, 1997, 1999; Saunders and Kraft, 1997; Harper and Saunders, 2001; Saunders and Hommersand, 2004). Saunders and Hommersand (2004) designated three subclasses (Ahnfeltiophycidae, Hildenbrandiophycidae, and Rhodymeniophycidae) along with the Nemaliophycidae (Christensen, 1978).

Plants are filamentous, crustose, or foliose and sometimes calcified. Growth is apical or marginal with a few species having intercalary cell divisions. Cells have pit connections of varied shapes, one or more nuclei per cell, and a few to many plastids. Asexual reproduction is usually by spores (monospores, bispores, polyspores, and paraspores), with a few species reproducing by fragments or specialized propagules. Life histories are usually triphasic, with gametophytic, carposporophytic, and tetrasporophytic phases. Reproductive cells (monosporangia, spermatangia, carposporangia, and tetrasporangia) are generally terminal or lateral on filaments. Carpogonia are also terminal or lateral and have an apical extension (trichogyne) on which spermatangia attach. Carposporophytes develop directly from the carpogonium or from one of its derivatives.

HILDENBRANDIOPHYCEAE

The subclass (Saunders and Hommersand, 2004) has the same features as the Hildenbrandiales of Pueschel and Cole (1982). Plants are crustose and smooth to tubular or with erect branches and have a basal layer of laterally adherent branched filaments, plus laterally appressed simple or branched erect filaments. Tetrasporangia are zonately or irregularly divided, apomeiotic, and in ostiolate conceptacles. Pit plugs have a single cap layer and a membrane. Sexual reproduction is unknown.

HILDENBRANDIALES

Plants are crustose, smooth to tubercular, or with erect branches. Crusts are tightly adherent, lack rhizoids, and compact. Perithallial filaments are erect, arise from a basal hypothallus of radiating, branched filaments, and may be horizontally stratified. Tetrasporangia are zonately or cruciately divided and in pitlike conceptacles with or without sterile hairs. The order is recognized by conceptacles that enlarge via erosion and by the presence of pit plugs (Pueschel and Cole, 1982).

Hildenbrandiaceae

The family has the same features as the monotypic order.

Hildenbrandia Nardo 1834: p. 676 ('675'), *nom. cons.*, as '*Hildenbrandtia*' (named for Professor F. E. Hildenbrandt of Vienna, an Austrian physician and naturalist).

Plants are thin, wide-spreading, noncalcified crusts with the lower side firmly adherent. The hypothallus consists of horizontal filaments and produces a perithallus of strongly united vertical cell rows. Tetrasporangia have irregular zonate divisions and occur in sunken conceptacles with one large pore.

Hildenbrandia rubra (Sommerfeld) Meneghini 1841: p. 426, as '_Hildenbrandtia rubra_' (L. _rubra:_ red). Plate XXII, fig. 19.

Basionym. *Verrucaria rubra* Sommerfelt 1826: p. 140.

Synonym. *Hildenbrandia prototypus* Nardo 1834: p. 675–676 (as '*Hildenbrandtia prototypus*').

Based on molecular and morphological studies, Sherwood and Sheath (2003) noted that the species was paraphyletic, with morphologically similar but evolutionarily distinct lineages.

Crusts are thin, firmly attached, noncalcified, 0.2–0.5 mm thick, and orange to brownish red in sun to deep red in shade. Hypothallial cells are in horizontal rows. Perithallial cells are 4–7 µm in diam. and in compact vertical rows that infrequently branch. Conceptacles are scattered, to 100 µm in diam., and uniporate. Tetrasporangia are 16–30 µm in diam., 9–14 µm long, obliquely to irregularly divided into 4 cells, and in conceptacles that lack sterile hairs.

Uncommon, on hard substrata at exposed coastal sites and to 20 m on all coasts; also known from the Canadian Arctic through the Caribbean, Brazil, and Uruguay, as well as the eastern North and South Atlantic (type location: Nordland, Norway; Silva et al., 1996), Mediterranean, and Indo-Pacific.

Taylor 1960: p. 369; Dawes 1974: p. 122.

NEMALIOPHYCIDAE

See Harper and Saunders (2002) for a review of this subclass that contains the Acrochaetiales, Balbianiales, Balliales, Batrachospermales, Corallinales, Nemaliales, Palmariales, Rhodogorgonales, and Thoreales (Sheath et al., 2000).

Plants of the subclass Nemaliophycidae Christensen (1978) have two caps on the cytoplasmic faces of their pit plugs.

ACROCHAETIALES

See Harper and Saunders (2002) for a revision of this order and the Acrochaetiaceae.

Plants are simple or branched uniseriate filaments and germinate from persistent, large spores that produce irregular, prostrate filamentous systems. Erect axes are simple to abundantly branched, and with or without hairs; branching is irregular, radial, secund, or opposite. Cells of marine species are uninucleate and with multiple parietal discoid to band-shaped plastids that lack pyrenoids or have one stellate plastid with a pyrenoid. Plants mostly reproduce by monosporangia. Life histories, if known, are triphasic.

Acrochaetiaceae Feldmann

The family has the same characteristics as the order.

Acrochaetium Nägeli in Nägeli and Cramer 1858: p. 52; Nägeli 1862: p. 402 (Gr. *acro*: apex, tip + *chaetium*: a bristle, hair).

Synonym. *Liagorophila* Yamada 1944b: p. 16.

Lee and Lee (1988) and Harper and Saunders (2002) emphasized the importance of chloroplast morphology and designated *Liagorophila* as congeneric with *Acrochaetium*, while Afonso-Carrillo et al. (2003) noted variable plastid morphology and recognized *Liagorophila*. Wynne (1998a) described two possible type species: *A. daviesii* (Dillwyn) Nägeli (Woelkerling, 1971) and *A. secundatum* (Lyngbye) Nägeli (Stegenga, 1985; Lee and Yoshida, 1997). Based on nuclear ribosomal DNA data, Harper and Saunders (2002) removed 8 of 14 species of *Audouinella* and *Acrochaetium* and transferred them to *Colaconema*.

Filaments are clustered and arise from basal cells, discs, or prostrate axes; erect filaments are tapered, hairlike, or end in true hairs. Each cell has one axial stellate plastid with a central pyrenoid. Plants reproduce primarily by monospores that are terminal, lateral, scattered, clustered, or in rows. Where known, life histories are triphasic. Cells of *Acrochaetium* have one stellate, platelike, axial or parietal plastid; those in *Audouinella* have many plastids per cell and lack pyrenoids (Feldmann, 1962; Silva et al., 1996; Wynne, 1998a).

Key to the species (7) of *Acrochaetium*

1. Plastid platelike to lobed and parietal 2
1. Plastid axial, lobed, stellate, or radiating.......... 3
    2. Plants endophytic; erect axes more than 250 μm tall .... *A. liagorae*
    2. Plants mostly external to host; erect axes less than 220 μm tall .... *A. microscopicum*
3. Base creeping, forming loosely woven filaments or a fused disc.......... 4
3. Base a single cell modified from the original spore .... *A. hallandicum*
    4. Lower cells of erect filaments more than 6 μm in diam.......... 5
    4. Lower cells of erect filaments 2–3(6) μm in diam. .... *A. leptonema*

5. Base conspicuously filamentous and often with a congested center 6
5. Base a parenchymatous disc, not obviously filamentous .... *A. savianum*
  6. Monosporangia on short lateral branchlets .... *A. avrainvilleae*
  6. Monosporangia often stalked in a long series on branchlets...................... *A. hypneae*

Acrochaetium avrainvilleae Børgesen 1915: p. 48–50, fig. 47–49 (named for its green algal host _Avrainvillea_, which commemorates M. Avrainville, a French naturalist who collected seaweeds in Guadeloupe). Plate XXIII, fig. 1, 2.

Synonym. *Audouinella avrainvilleae* (Børgesen) Garbary 1987: p. 24.

Filaments are uniseriate, to 1 mm tall, with basal portions having short axes that penetrate its host. Its main axes are erect, simple below, and alternately branched above with 1–2 branchlets, 5 µm in diam.; axial cells are to 9 µm in diam. by 33 µm long. Plastids are parietal with one large pyrenoid. Monospores are sessile, in short series on dorsal sides of lower branchlets, ovoid, and 11 by 22 µm.

Uncommon, on *Avrainvillea* in seagrass beds (2–3 m) on the west coast; also known from the Virgin Islands (type location: St. John).

Taylor 1960: p. 311; Dawes 1974: p. 118.

Acrochaetium hallandicum (Kylin) Hamel 1927: p. 20, fig. 19–21 (L. _hallandicum_: from the northern and southern provinces of the Netherlands). Plate XXIII, fig. 3, 4.

Basionym. *Chantransia hallandica* Kylin in Kjellman 1906: p. 123, fig. 8.
Synonyms. *Acrochaetium dufourii* Collins in Collins, Holden, and Setchell (1895–1919): ex. no. 1594 (Børgesen 1915: p. 19–20).
Audouinella hallandica (Kylin) Woelkerling 1973a: p. 82, fig. 1–4.
Audouinella sargassi (Børgesen) Garbary 1979: p. 490.
Audouinella dufourii (Collins in Collins, Holden, and Setchell) Garbary 1987: p. 62.

Filaments are uniseriate, to 1 mm tall, with 1–3 erect axes arising from an ovoid, subglobose, or rectangular spore that has a persistent wall. Its base is a pad of cells to 20 µm in diam.; prostrate cells are 5–8 µm in diam. and similar to cells of erect axes; erect axes have irregular, lateral or secund branching. Branches are well spaced and may taper to tips; hairs are absent. Lower cells are 3–7 µm in diam. and 10–30 µm long. All cells have a parietal, lobed plastid with one pyrenoid. Monosporangia are ovoid, 5–10 µm in diam., 8–16 µm long, solitary or in pairs, lateral, secund, or rarely terminal, and sessile or on 1-celled stalks. Gametophytes are rare and dioecious. Spermatangia are in opposing pairs on short lateral branches. Carpogonia are sessile.

Common, inconspicuous, on macroalgae and seagrasses (to 35 m) on all coasts; also known from North Carolina through the Caribbean, Sargasso Sea, Brazil, and Uruguay, as well as the eastern North and tropical Atlantic and tropical Pacific. Type location: "halläsndische kuste" (Kylin in Kjellman, 1906) or the province of Halland in Sweden.

Taylor 1960: p. 305, 306; Dawes 1974: p. 118; Schneider and Searles 1991: p. 202.

Acrochaetium hypneae (Børgesen) Børgesen 1915: p. 51–52, fig. 53, 54 (named for the red algal host genus _Hypnea_; Gr. _hypnon_, a moss; also, a genus of moss, _Hypnum_). Plate XXIII, fig. 5–7.

Basionym. *Chantransia hypneae* Børgesen 1909b: p. 2, fig. 2.
Synonyms. *Acrochaetium seriatum* Børgesen 1915: p. 32–34, fig. 25–28.
Audouinella seriata (Børgesen) Garbary 1979: p. 490.

Audouinella hypneae (Børgesen) Lawson and John 1982: p. 172: 1987: p. 159.

Colaconema seriata (Børgesen) Stegenga 1985: p. 323, fig. 24.

Filaments are uniseriate, to 1 mm tall, and arise from compact or spreading prostrate filaments or from a pseudoparenchymatous disc. Erect axes are radially branched below, alternate to secund and abundant above, and may end in unicellular colorless hairs to 60 µm in length. Cells are cylindrical, 6–10 µm in diam., and 16–22 µm long. Branchlet tips are 16–30 µm long and taper to 4–7 µm in diam. Each cell has a parietal plastid and a large pyrenoid. Monosporangia are single, lateral, in secund series or scattered, sessile or on a 1-celled stalk, ovoid, and 6–10 µm in diam. by 9–13 µm long.

Common, on macroalgae on the west coast and the Keys; also known from North Carolina through the Caribbean (type location St. Thomas), Bermuda, Gulf of Mexico, and Brazil, as well as Mauritania, Namibia, the Hawaiian Islands, and East Indies.

Taylor 1960: p. 310, 313; Dawes 1974: p. 118; Schneider and Searles 1991: p. 204.

Acrochaetium leptonema (Rosenvinge) Børgesen 1915: p. 31 (Gr. *leptonema*: delicate threads or fine-threaded). Plate XXIII, fig. 8, 9.

Basionym. *Chantransia leptonema* Rosenvinge 1909: p. 118, fig. 47–48.

Synonyms. *Audouinella leptonema* (Rosenvinge) Garbary 1979: p. 490.

Plants are erect, minute, to 0.3 mm tall, uniseriate, and from radiating prostrate filaments or a pseudoparenchymatous disc. Prostrate filaments are irregularly branched and mask the original spore; cells are swollen, 3–4 µm in diam., and 4–12 µm long. Erect axes are simple or branch sparsely; cells are cylindrical to swollen, 2–3 µm in diam., 6–15 µm long, and have one parietal, platelike plastid. Monosporangia and tetrasporangia are unilateral or terminal, on 1st- and 2nd-order branches, solitary or in pairs, and sessile or on 1-celled stalks. Monosporangia are ovoid, 5.5–6.5 µm in diam., and 10–12.5 µm long.

Rare, on macroalgae at the Dry Tortugas; also known from Bermuda, the eastern North Atlantic (type location: Hanstholm, Denmark), Mediterranean, and Alaska.

Taylor 1960: p. 308.

Acrochaetium liagorae Børgesen 1915: p. 57–59, fig. 60–62; non *Acrochaetum liagorae* (Weber-van Bosse) Hamel 1927: p. 92; 1928: p. 185 (named for the red algal host *Liagora*, which was named after a sea nymph of the God Nerius). Plate XXIII, fig. 10.

Synonyms. *Acrochaetium collinsianum* Børgesen 1920: p. 454; 1937: p. 43; 1952: p. 15.

Kylinia liagorae (Børgesen) Papenfuss 1947: p. 438.

Audouinella liagorae (Børgesen) Woelkerling 1971: p. 31–33, fig. 9.

Plants are endophytic, prostrate, and less than 1 mm tall. Axes are sparingly and irregularly branched with wide forks and may end in hyaline hairs. Cells are cylindrical to barrel-shaped, 6–14 µm in diam., and to 35 µm long. Plastids are axial, radiate, and have one pyrenoid. Sporangia are sessile, terminal, on extensions of supporting cells or 1-celled stalks, ovoid to spherical, 14 µm in diam., and 15–20 µm long.

Rare, inconspicuous, endophytic in *Liagora pinnata* in the Keys and Dry Tortugas (to 10 m); also known from Bermuda, Puerto Rico, and the Virgin Islands, as well as the eastern tropical Atlantic, Indo-Pacific (type location: Mahabalipuram, India; Silva et al., 1996), and Australia.

Taylor 1960: p. 301.

Acrochaetium microscopicum (Nägeli _ex_ Kützing) Nägeli 1858: p. 532, footnote; 1862: p. 406, fig. 24–25 (L. _microscopicum_: microscopic, tiny). Plate XXIII, fig. 11, 12.

Basionym. *Acrochaetium trifilum* (Buffham) Batters 1902: p. 58.
Synonyms. *Chantransia crassipes* Børgesen 1909b: p. 1, fig. 1.
Acrochaetium crassipes (Børgesen) Børgesen 1915: p. 20, fig. 12.
Acrochaetium parvulum (Kylin) Hoyt 1920: p. 470, fig. 25.
Acrochaetium compactum Jao 1936: p. 241, pl. 10, fig. 6–14.
Kylinia crassipes (Børgesen) Kylin 1944: p. 13.
Audouinella microscopica (Nägeli *ex* Kützing) Woelkerling 1971: p. 33–35, fig. 10, 23A; 1972: p. 85, fig. 1–14; 1973a: p. 86, fig. 46–51.
Audouinella crassipes (Børgesen) Garbary 1979: p. 490.

Plants are minute; erect axes are 20–220 µm tall, uniseriate, may end in hyaline hairs to 50 µm long, and arise from subglobose to cylindrical thick-walled spores, 5–15 µm in diam. Erect axes are 1–6 cells long; cells are subspherical, 3–10 µm in diam., 3–11 µm long, and have a lobed to stellate plastid and one pyrenoid. Monosporangia are terminal, lateral, or secund, ovoid, 4–15 µm in diam. by 6–22 µm, and sessile or on 1-celled stalks. Tetrasporangia are rare, sessile, tetrahedrally divided, 20 µm in diam., and 30 µm long. Gametophytes are dioecious. Spermatangia are ovoid, 2–4 µm long, on erect axes, and sessile or on 1-celled stalks. Carpogonia are sessile and terminal or lateral; carpospores are terminal or lateral, 14–16 in a cluster, each 6–8 µm in diam. and 7–10 µm long.

Common, on macroalgae, mangrove roots, and seagrass blades in the Keys and west coast; also known from Newfoundland through the Caribbean, Bermuda, Gulf of Mexico, and Brazil, as well as the eastern North and South Atlantic, Mediterranean (type location: Gulf of Naples, Italy; Silva et al., 1996), Indo-Pacific, and Australia.

Taylor 1960: p. 300; Dawes 1974: p. 119; Schneider and Searles 1991: p. 206.

Acrochaetium savianum (Meneghini) Nägeli 1862 ('1861'): p. 405, 414 (L. _savianum_: of or from Savon, a seaport on the Gulf of Genoa). Plate XXIII, fig. 13, 14.

Basionym. *Callithamnion savianum* Meneghini 1840a: (2); 1840b: p. 511.
Synonyms. *Acrochaetium sagraeanum sensu auct.*, non (Montagne) Bornet 1904: p. XXI.
Acrochaetium thuretii (Bornet) Collins and Hervey 1917: p. 98.
Audouinella thuretii (Bornet) Woelkerling 1971: p. 9, 36, 37, fig. 12, 24; 1973a: p. 88.
Audouinella saviana (Meneghini) Woelkerling 1973b: p. 560, fig. 56–60.

Filaments are erect, to 4 mm tall, uniseriate, and with pseudoparenchymatous discs 60–120 µm in diam. The original spore is not evident. Erect axes have irregular, alternate, or secund branching and rarely unicellular hairs. Cells are cylindrical, 7–14 by 20–60 µm, and have a parietal, lobed plastid. Monosporangia are single or paired, ovoid, 10–15 by 18–27 µm, in irregular or secund patterns, and sessile or on 1–3-cell stalks. Tetrasporangia are cruciately divided, 17–24 by 16–34 µm, arranged like monosporangia, and often in small groups. Gametophytes are monoecious. Spermatangia and carpogonia are uncommon and on short branches near their bases. Carpospores are 9–13 by 18–21 µm.

Common, on *Codium*, other algae, and seagrasses on all coasts; also known from Newfoundland through the Caribbean, Bermuda, the Gulf of Mexico, and Brazil, as well as the eastern North and tropical Atlantic, Arctic Sea, Indo-Pacific, and Australia. Type locations cited as Cherbourg, France (Woelkerling and Womersley, 1994) and Genoa, Italy (Silva et al., 1996).

Taylor 1960: p. 310, 312; Dawes 1974: p. 118; Schneider and Searles 1991: p. 207.

COLACONEMATALES

Harper and Saunders (2002) separated the order and its single family Colaconemataceae from the Acrochaetiales based on molecular, morphological, and reproductive features.

Plants are uniseriate, simple or branched filaments. Cells have one to several parietal plastids of varying shapes (lobed, spiral, or irregular, but not stellate) with or without pyrenoids. Asexual reproduction is primarily by monosporangia; life histories where known are triphasic.

Colaconemataceae

The family has the same characteristic as the order

Colaconema Batters 1896: p. 8; see H.T. Harper and G.W. Saunders 2002: p. 473.

The genus has the same features as the order and family.

Key to the species (3) of *Colaconema*

1. Original spore persistent and visible at base of erect filaments 2
1. Original spore not easily visible, penetrating surface cells of host . . . . *C. dasyae* (in part)
  2. Base consisting of widely spreading and branched filaments . . . . *C. ophioglossum*
  2. Base usually consisting of a mass of congested filaments . . . . *C. dasyae* (in part)

Colaconema dasyae (Collins) Stegenga, I. Mol, Prud'homme van Reine and Lokhorst 1997: p. 26 (named for the red algal host _Dasya_; Gr. _dasys_: shaggy, hairy). Plate XXIII, fig. 15–17.

Basionym. *Acrochaetium dasyae* Collins 1906b: p. 191–192.
Synonym. *Acrochaetium zosterae* Papenfuss 1945: p. 307.
Audouinella dasyae (Collins) Woelkerling 1973b: p. 545, fig. 10–31.
Colaconema dasyae (Collins) J. T. Harper and G. W. Saunders 2002: p. 473.

Guiry and Nic Dhonncha (2006) noted that *Colaconema dasyae* may be the gametophytic stage of other taxa such as *Acrochaetium seriatum* or *C. savianum* (Meneghini) R. Nielsen (1994).

Filaments are uniseriate, to 3 mm tall, and attached by a spore that produces spreading, prostrate axes. Spores have persistent walls and several erect axes with many irregular branches that taper toward tips. Cells are cylindrical and 6–18 µm in diam. by 25–90 µm long in main axes and 6–10 µm in diam. by 15–60 µm long in branchlets. Each cell has a parietal plastid with one pyrenoid. Monosporangia are single or clustered, in secund or scattered groups, sessile or on 1-celled stalks, ovoid, 7–12(16) µm in diam., and 16–27 µm long. Gametophytes are monoecious or dioecious. Spermatangia are terminal on short lateral branches, globose to ovoid, and 2–4 by 3–5 µm. Carpogonia are scattered and sessile or on 1-celled stalks. Carposporophytes have many terminal or lateral ovoid carposporangia that are 9–12 µm in diam. by 16–24 µm long.

Common, inconspicuous, on macroalgae such as *Dasya* and *Sargassum*, on all coasts; also known from Prince Edward Island to the Virgin Islands (type location, Woods Hole, Massachusetts; Cordero, 1981), as well as the Netherlands, eastern tropical Atlantic, and Indo-Pacific.

Taylor 1960: p. 314–316; Schneider and Searles 1991: p. 198.

***Colaconema ophioglossum* (Schneider) Afonso-Carrillo, Sansón, and Sangil 2003: p. 109, fig. 1–11 (Gr. *ophio*: snake, snakelike + *glossum*: tongue-shaped; also the generic name for the Adder's tongue fern, *Ophioglossum*). Plate XXIII, fig. 18, 19.**

Basionym. *Audouinella ophioglossa* Schneider 1983: p. 13, fig. 3a–p.

Synonym. *Acrochaetium ophioglossum* (Schneider) D. L. Ballantine and Aponte 1997a: p. 151–152.

Filaments are uniseriate, 1–3 mm tall, with 1–2 erect axes arising from a persistent, spherical spore (8–13 μm in diam.) that penetrates algal hosts. Rhizoids, if present, are contorted and 2–4 μm in diam. Erect axes are irregularly branched above with terminal or lateral unicellular hairs (to 2 mm long). Cells of lower axes are cylindrical, 4–5 μm in diam., and 45–55 μm long. Each cell has a lobed, spiral, or platelike parietal plastid with one pyrenoid. Monosporangia are terminal or lateral, on short branchlets, sessile or on 1-celled stalks, ovoid to elliptical, 6–13 μm in diam., 13–20 μm long, and often associated with gametangia. Gametophytes are monoecious. Spermatangial branches are in whorls of 4, in opposite pairs, and solitary or in series above. Spermatia are on 1-celled stalks and 2.5 μm in diam. Carpogonia are lateral, opposite, or near monosporangia, and one per node. Carpospores are terminal, spherical, and 10–15 μm in diam.

Rare, in 9 m at the Dry Tortugas (Ballantine, 1996); also known from North Carolina (type location: Onslow Bay and on *Dudresnaya crassa* in 22–35 m), Georgia, Puerto Rico, and the Canary Islands.

Schneider and Searles 1991: p. 207.

CORALLINALES

See Silva and Johansen (1986) regarding the separation of this order from the Cryptonemiales. Florida corallines are poorly understood and identifications of many crustose species need further study (Woelkerling, pers. comm.).

Plants are calcified with calcium carbonate (calcite) and form crusts or erect axes that are rigid (nonarticulate) or jointed (articulate). The calcified segments of jointed branches and noncalcified joints (genicula) are flat or terete, and have dichotomous, oppositely pinnate, or irregular branching. Crustose species usually have three zones: a lower hypothallus of one to several layers of compact filaments; a middle perithallus with vertical filaments arising from the hypothallus that may contain trichocytes (large clear heterocysts) and large thick-walled hair cells; and an upper epithallus with one or more layers of compressed cover cells. Jointed species have two layers, a medulla of parallel cells that often occurs in arching tiers and an outer cortex of round to rectangular pigmented cells. The noncalcified joints of articulated species have one or more tiers of long, thick-walled medullary cells and lack a cortex. Life histories are isomorphic, with tetrasporangia and gametangia in elevated conceptacles having one (uniporate) to many pores (multiporate). Tetrasporangia are usually zonately divided and rarely 2-celled. Gametophytes are dioecious. Male conceptacles have short compact filaments that produce spermatia, while female ones are procarpial with 2-celled carpogonial filaments and one or two branches on the support cells that function as auxiliary cells after fertilization. Gonimoblastic filaments arise from large fusion cells and have large terminal carposporangia. Crustose species are difficult to identify and decalcification (for example, 10% hydrochloric acid) is needed for sectioning (Johansen, 1981).

Hapalidiaceae

The family (Gray, 1864), as emended by Harvey et al. (2003a,b), includes members whose tetraspores are zonately divided and have tetrasporangia and/or bisporangia in conceptacles with apical plugs. Their sporangia also develop under multiporate plates and do not occur individually in calcified sporangial compartments.

Melobesioideae

The subfamily contains crustose forms and plants with erect branches that lack joints. Some vegetative filament cells are fused and secondary pit connections are absent or rare. Tetrasporangial and bisporangial conceptacles have apical pit plugs and are multiporate.

Lithothamnion Heydrich 1897b: p. 413, *nom. cons.* (Gr. *lithos*: a stone + *thamnion*: a shrub).

Synonym. *Lithothamnion* Philippi 1837: p. 387; rejected (Woelkerling, 1985a).
See Woelkerling (1985a) for a taxonomic discussion.

Crusts are thick, heavily calcified, and usually firmly attached or form detached fragments. The hypothallus is multilayered and develops from an 8-celled initial. The perithallus has several tiers, trichocytes and secondary pit connections are absent, and cell fusions may occur between adjacent filaments. The epithallus usually has one layer with cells angular in cross section. Conceptacles are superficial to slightly immersed, conical or rounded, and multiporate. Tetrasporangia are zonately divided.

Key to the species (2) of *Lithothamnion*

1. Crusts have knobs, wartlike structures, or short thick branches . . . . *L. heteromorphum*
1. Crusts have obvious nonjointed branches that often fuse . . . . *L. occidentale*

Lithothamnion heteromorphum **(Foslie) Foslie 1907 ('1908') p. 10; 1929: p. 42, pl. XII, fig. 20 (Gr. *heteromorphum*: heteromorphic, variable in form). Plate XXIII, fig. 20.**

Basionym. *Lithothamnion brasilense* Foslie f. *heteromorphum* Foslie 1900: p. 4.

Crusts initially firmly attached and becoming detached when older. Branches are short, 3–4 mm in diam., irregularly dichotomously branched, and with rounded and thickened tips. Hypothallial cells are 9–14 µm in diam. by 14–32 µm long; perithallial cells are square, 9–11 µm in diam. or vertically elongate, and 7–11 µm wide by 9–22 µm long. Tetrasporangial conceptacles are 350–600 µm in diam.

Rare, on limestone or as free fragments in the Keys; also recorded from Brazil (type location: coast of Sâo Sebastião; Foslie, 1900).

Taylor 1960: p. 383.

Lithothamnion occidentale **(Foslie) Foslie 1908: p. 3–4 (L. *occidentale*: western, or occidental). Plate XXIV, fig. 1, 2.**

Basionym. *Lithothamnion fruticulosum* (Kützing) Foslie var. *occidentale* Foslie 1904a: p. 12.

Young crusts are often fragmented, detached, and irregularly branched. Branches are smooth, glossy, heavily calcified, terete to compressed, 1–2(3) mm in diam., and often fused. The hypothallus is overgrown by a multilayered perithallus. Perithallial filament cells are swollen, rectangular or elongate, 6–10 µm in diam., 10–32 µm long, without secondary pit connections, and may fuse to form globose darkly pigmented cells. The epithallus lacks trichocytes and is one layer thick; cells are compressed, angular, 8–12 µm in diam., 4–7 µm long, and polygonal in surface view extensions. Tetrasporangia are zonately divided, elliptical, 75 µm in diam., 180 µm long, and have indistinct conceptacles.

Common, often as detached fragments in the Keys and on the west coast; also known from the Carolinas (15–45 m), Gulf of Mexico, Caribbean, and Brazil.

Taylor 1960: p. 385; Dawes 1974: p. 126; Schneider and Searles 1991: p. 239.

Melobesia J.V. Lamouroux 1812: p. 186; 1816: p. 313 (named for Melobosis, the daughter of Oceanus, a Titan in Greek mythology who ruled over the sea).

Crusts are delicate, firmly attached, mostly monostromatic in vegetative parts, and polystromatic around conceptacles. Each cell divides obliquely to produce a small associated cell. Tetrasporangia are in multiporate conceptacles with sterile filaments. Spermatangia are lateral on 2-celled filaments that spread over the conceptacle base. Carpogonial branches are 3-celled; they arise from basal layers in the center of conceptacles and are surrounded by vegetative filaments.

Melobesia membranacea **(Esper) J. V. Lamouroux 1812: p. 186; 1816: p. 315 (L. *membranacea*: thin, like a skin or membrane). Plate XXIV, fig. 3–5.**

Basionym. *Corallina membranacea* Esper 1796 (1794–1797): pl. Corallina XII; 1806 (1805–1830), pl. XII.

Crusts are epiphytic, thin, to 90 µm, lightly calcified, red to light rose, to 5cm in diam., with entire margins and are monostromatic, except at conceptacles where they have 4–5 layers. Cells in the monostromatic layer are 9–20 µm in diam., 5–16 µm long in surface view, and 5–18 µm high in cross section. Conceptacles are hemispherical and multiporate. Tetrasporangial conceptacles are scattered, wartlike, 110–140(350) µm in diam., and have 8–27 pores. Gametophytic conceptacles are conical, to 250 µm in diam., and have filiform cells projecting from the pores.

Common, on seagrasses and seaweeds on all coasts; also known from the Canadian Maritimes through the Caribbean, Gulf of Mexico, and Brazil, as well as the eastern North and South Atlantic, Mediterranean (neotype location: France; Silva et al., 1996), Indo-Pacific, and Australia.

Taylor 1960: p. 379.

Mesophyllum Lemoine 1928: p. 251 (Gr. *meso*: middle, intermediate + *phyllum*: leaflike).

Crusts are thick, heavily calcified, and firmly or loosely attached; loose crusts appear leafy. The hypothallus has thick, multilayered filaments in parallel or arching tiers. The perithallus has a few layers of similar-sized cells, with the layers thickest near the conceptacles. Trichocytes and secondary pit connections are absent. If present, the epithallus is monostromatic and has rectangular or rounded cells. Conceptacles are elevated, circular, and uniporate. Tetrasporangia are zonately divided and in conceptacles with apical plugs that block the pore before sporangial release.

Key to the species (4) of *Mesophyllum*

1. Crusts attached; knoblike protuberances absent 2
1. Crusts attached or becoming free; knoblike bumps or branches present 3
    2. Crusts firmly attached, foliar, and with overlapping layers; tetrasporangial conceptacles 100–140 µm in diam. . . . . *M. mesomorphum*

2. Crusts loosely attached, often with superimposed layers; tetrasporangial conceptacles 400–600 µm in diam. . . . . *M. syntrophicum*
3. Crust with knobs or wartlike bumps; branches short . . . . *M. floridanum*
3. Crust with obvious branches; branches often fused . . . . *M. incertum*

Mesophyllum floridanum (Foslie) W. H. Adey *ex* M. J. Wynne 1986: p. 2242, 2257 (L. *floridanum*: of or from Florida). Plate XXIV, fig. 6.

Basionym. *Lithothamnion floridanum* Foslie 1906: p. 11.

Crusts are purple red, heavily calcified, firmly attached, and conform to limestone or shell surfaces. Its surface is nodulose with mammillate projections, 0.2–4.6 mm thick, 4–7 mm tall, and 3–6 mm in diam. The hypothallus has filaments in arching layers, 8–15 filaments thick; cells are 5–10 µm in diam. by 2–14 µm long. The perithallus is tiered and folded; cells are elongate to rectangular, 7–14 µm in diam., 9–22 µm long, and often fused between filaments. The epithallus is monostromatic, with cells rounded, 5–8 µm in diam., 3–5 µm long, and lacking trichocytes. Tetrasporangia are zonately divided, ovoid, 45–85 µm in diam., and 105–150 µm long. Conceptacles are on projections 400–600 µm in diam., 290–360 µm wide, and 140–180 µm deep; they are elliptical in cross section and have one pore and plug.

Uncommon, as thick crusts on shallow exposed limestone in the Keys (lectotype location; Foslie, 1906); also known from North Carolina (18–27 m), the Caribbean, and Sao Tomé.

Taylor 1960: p. 383; Schneider and Searles 1991: p. 240.

Mesophyllum incertum (Foslie) Lemoine 1928: p. 252 (L. *incertum*: uncertain, doubtful). Plate XXIV, fig. 7.

Synonym. *Lithothamnion incertum* Foslie 1929: p. 42, pl. XIV, fig. 20–22.

Crusts are firmly attached, heavily calcified, and in hemispherical cushions, 2.5–5.0 cm high by 7–15 cm wide. Projections are erect, spreading, branched, and fused; erect parts are subterete (1.0–1.5 mm in diam.) or flat (2 mm wide). Conceptacles occur near tips of projections, are depressed to hemispherical, and 300–400 µm in diam.

Rare, on shallow limestone (0.3–1.2 m) at wave-exposed sites in the Keys; also known from Bermuda (type location) and the Bahamas.

Taylor 1960: p. 384.

Mesophyllum mesomorphum (Foslie) Adey 1970: p. 25 (Gr. *meso*: middle, intermediate + *morphum*: form, shape). Plate XXIV, fig. 8, 9.

Basionym. *Lithothamnion mesomorphum* Foslie 1901 ('1900'): p. 5–6.

Crusts are heavily calcified, loosely attached, lobed, 300–500 µm thick, and consist of thin overlapping brittle sheets that can be separated. The hypothallus represents one-half to two-thirds of the crust's thickness with rectangular cells 4–10 µm wide and 10–25 µm high. Subsurface perithallial cells are round to square and 4–9 µm in diam. Surface cells are nearly square and 5–8 µm wide. Tetrasporangial conceptacles are hemispherical, 100–140 µm in diam., to 60 µm high, partially sunken, and multiporate. Cystocarpic conceptacles are slightly sunken and cone-shaped.

Common, on limestone and algae in shady areas in the Keys; also known from Bermuda (type location; Silva et al., 1996), the Bahamas, and Caribbean (to 35 m), as well as the Comoro and Seychelles islands.

Taylor 1960: p. 382; Littler and Littler 2000: p. 34.

Mesophyllum syntrophicum **(Foslie) Adey 1970: p. 26 (Gr. *syn*: joined, grown together + *trophos*: nourishment). Plate XXIV, fig. 10.**

Basionym. *Lithothamnion syntrophicum* Foslie 1901 ('1900'): p. 6; 1929: p. 46, pl. V, fig. 18–19.
Crusts are heavily calcified, with multiple layers and nodules, 2–3 cm in diam., and conforming to substrata. New layers are irregular and grow loosely over one another; each layer is 200–800 µm thick. The hypothallus is thin, with cells 6–9 µm in diam. and 12–25 µm long. The perithallus is thicker, its cells are quadrate to rounded, 4–7 µm in diam., and taller than wide. Tetrasporangial conceptacles are partially immersed to convex and 400–600 µm in diam.; tetrasporangia are ovoid and 60 by 120 µm.

Uncommon, in 0.1–0.3 m on limestone at wave-exposed sites in the Keys and in areas with tidal currents off Tampa Bay (to 5 m) and from Bermuda (type location).

Taylor 1960: p. 381; Dawes 1974: p. 126.

Corallinaceae

Note previous comments on separation of the Haplididaceae by Harvey et al. (2003a,b). The subfamilies Mastophoroideae, Corallinoideae, and Lithophylloideae were described based on morphological and molecular data (Wynne, 2005a). Previously, the Amphiroideae was recognized (Woelkerling, 1996b; Wynne, 1998a), but the type genus *Amphiroa* J. V. Lamouroux has a close phylogenetic relationship to *Titanoderma* (Bailey, 1999) in the Lithophylloideae.

The family and ordinal features (i.e., corallinales) are the same.

Mastophoroideae.

Bailey et al. (2004) suggested that the subfamily was polyphyletic and some genera may eventually be placed elsewhere.

Plants are crustose with erect branches that lack joints. Some vegetative cells are fused and secondary pit connections are rare or absent. Tetrasporangia and bisporangia lack apical plugs and occur in uniporate conceptacles.

Hydrolithon (Foslie) Foslie 1909: p. 55, *sensu lato*, emended Penrose and Woelkerling 1992: p. 87 (Gr. *hydro*: water, liquid + *lithon*: rock, stone).

Synonym. *Fosliella* M. Howe 1920a: p. 587.

Guiry and Nic Dhonncha (2006) listed *Fosiella* as a distinct genus, while other authors (Penrose and Woelkerling, 1988, 1992; Penrose and Chamberlain, 1993; Irvine and Chamberlain, 1994) merged *Porolithon* Foslie 1909 with *Hydrolithon*. However, Littler and Littler (2000, 2003) have stressed several unique ecological features and microscopic traits, including horizontal surface "megacell" plates that differ from *Hydrolithon* (Wynne, 2005a).

Crusts are calcified; the hypothallus is monostromatic and has isodiametric cells. The perithallus is thick with cells that are irregular in pattern and size, and trichocytes. Secondary pit connections are absent and intercellular fusions present. Conceptacles are uniporate.

Key to the species (3) of *Hydrolithon*

1. Crusts lithophytic, many cells thick, massive, and with knobs. . . . . *H. boergesenii*
1. Crusts epiphytic, 1–2 cells thick, delicate, and lack knobs. 2
    2. Crusts are fan- to disc-shaped; germination disc is 4-celled and surrounded by 12 cells. . . . . *H. farinosum.*
    2. Crusts mostly have creeping unconsolidated filaments; germination disc is 4-celled and surrounded by 4 cells . . . . *H. boreale.*

Hydrolithon boergesenii (Foslie) Foslie 1909: p. 56 (named for the Danish phycologist F. Børgesen, who studied the seaweeds of the Virgin and Canary Islands). Plate XXIV, fig. 11–14.

Basionym. *Goniolithon boergesenii* Foslie 1901 ('1900'): p. 19.

Synonym. *Porolithon boergesenii* (Foslie) Lemoine in Børgesen 1917: p. 178–180, fig. 168, 169.

Crusts are firmly attached and heavily calcified. Older crusts are thick, massive, and have nodules; their surfaces are smooth to rough, chalky, granular, and with a checkerboard pattern. Knoblike nodules are present, 5–8 mm in diam., and 3–5 mm high. The hypothallus is monostromatic; with cells 8–20 µm in diam., 7–25 µm high, and sometimes vertically elongate. The perithallus is multilayered, with its cells 8–12(20) µm in diam., 7–18(25) µm high, and in poorly organized rows. Surface (cap) cells are 6–18 µm in diam. and 4–10 µm high. Trichocytes are rare, 9–25 µm in diam., and 12–30 µm high. Conceptacles are uniporate, hemispherical, 300–400 µm in diam., and 50–160 µm tall. Tetrasporangia are zonately divided, ovoid, 60–80 µm in diam., and 120–160 µm long.

Uncommon, on wave-exposed limestone ridges below low tide in the Keys; also known from the Caribbean (to 67 m; type location: St. Croix) and Sao Tomé.

Taylor 1960: p. 397; Littler and Littler 2000: p. 28.

Hydrolithon boreale (Foslie) Y. M. Chamberlain in Irvine and Chamberlain 1994: p. 116–120, fig. 3B, 50, 51, 56 (L. _boreale_: boreal, northern). Plate XXIV 15, 16.

Basionym. *Melobesia farinosa* J. V. Lamouroux f. *borealis* Foslie 1905: p. 96.

Synonym. *Fosliella farinosa* f. *callithamnioides* (Foslie) Y. Chamberlain 1983: p. 351, fig. 20b.

Wynne (2005a; pers. comm.) noted that there are two heterotypic "*callithamnioides*," *Hapalidium callithamnioides* and *Fosliella farinosa* f. *callithamnioides* P. L. Crouan and H. M. Crouan and that the latter is a synonym for *H. boreale*.

Crusts are calcified, gray green to pink, epiphytic, to 3 mm in diam., and with individual lobes to 20 µm thick. Prostrate creeping filaments are irregularly branched and often not consolidated into crusts. Crusts are 2-layered; cells in the lower layer are 14–29 µm in diam., while surface cells are rounded and conspicuous. Trichocytes are terminal on basal filaments. Germination discs are 4-celled and surrounded by 4 cells. Tetrasporangial conceptacles are domelike and 166–200 µm in diam. Tetrasporangia are 35–65 µm long by 23–40 µm in diam. Bisporangia are common, 55–75 µm long, and 25–40 µm in diam.

Common, on seagrasses and algae on all coasts; also known from Bermuda, North Carolina, the Caribbean, and Gulf of Mexico, as well as the British Isles (lectotype location: Roundstone Ireland; Irvine and Chamberlain, 1994), Canary Islands, Italy, the Arabian Gulf, and Australia.

Taylor, 1960: p. 389; Schneider and Searles, 1991: p. 234; Littler and Littler 2000: p. 30.

Hydrolithon farinosum (J. V. Lamouroux) Penrose and Y. M. Chamberlain 1993: p. 295–296, fig. 1–19 (L. _farinosum_: mealy, covered with powder). Plate XXIV, fig. 17–19.

Basionym. *Melobesia farinosa* J. V. Lamouroux 1816: p. 315, pl. XII, fig. 3.

Synonyms. *Fosliella farinosa* (J. V. Lamouroux) M. Howe 1920a: p. 587.

Fosliella farinosa var. *solmsiana* (Falkenberg) W. R. Taylor 1939: p. 11, pl. 1, fig. 11.

Crusts are epiphytic, pink, lightly calcified, 2–5 mm in diam. by 20 µm thick, and have irregular margins. Prostrate filaments form radiating fan- or disc-shaped thalli, and arise from a 4-celled germination disc that is surrounded by 12 cells. Crusts are 1(2) cells thick. Hypothallial filaments are compact to loosely associated, radially organized, and dichotomously branched; cells are variable, 5–18 µm wide, and 10–30 µm long. Trichocytes are ovoid, 8–30 µm wide by 22–40 µm long, and at the tips of filaments. The perithallus is limited to cells around conceptacles and occasional cap cells on filament surfaces; cells are oval, 3–10 µm wide, and 7–14 µm long. Conceptacles are uniporate. Cystocarpic and tetrasporangial conceptacles are hemispherical and 140–250 µm in diam. Male conceptacles are smaller. Tetrasporangia are zonately divided, elongate, 20–50 µm wide, and 35–90 µm long; bisporangia are ovoid, 25–40 µm wide, and 54–80 µm long.

Common, on seaweeds and seagrasses in sheltered sites on all coasts; also known from Quebec through the Caribbean, Bermuda, the Gulf of Mexico, and Brazil, as well as the Mediterranean, eastern tropical Atlantic, Indo-Pacific, and Australia.

Taylor 1960: p. 388; Dawes 1974: p. 123; Schneider and Searles 1991: p. 234; Littler and Littler 2000: p. 28, 30.

Lithoporella (Foslie) Foslie 1909: p. 58 (Gr. *litho*: rock, stone + *porella*: a small pore).

Crusts are thin, lightly calcified, often foliose, usually epiphytic, lack protuberances, and are attached by cell adhesion or rhizoids. The basal layer consists of radially, fused filaments with columnar cells. Crusts are 3–5 cells thick, except near margins where they are monostromatic. Trichocytes are single or in clusters. Apical cells are not flared. Tetrasporangial conceptacles are uniporate, lack a columella, and their roofs are formed from filaments interspersed among developing tetrasporangia.

Lithoporella atlantica **(Foslie) Foslie 1909: p. 59 (L. *atlantica*: of or from the Atlantic Ocean). Plate XXIV, fig. 20.**

Basionym. *Mastophora atlantica* Foslie 1906: p. 27.
Synonyms. *Melobesia atlantica* (Foslie) Lemoine in Børgesen 1917: p. 175, fig. 166.
Fosliella atlantica (Foslie) W. R. Taylor 1960: 387.

Crusts are irregular, mostly monostromatic to rarely distromatic, and attached by cell adhesion and occasional rhizoids. Cells are 10–25 µm in diam. and 18–32 µm long (Lemoine, 1917) or 18–40 µm in diam. and 32–60 µm long (Foslie, 1906). Conceptacles are conical, elevated, and 500–800 µm in diam.

Common, on seagrasses and seaweeds on the west coast; also known from Hispaniola (type location: St. John, Virgin Islands; Foslie, 1906).

Taylor 1960: p. 387; Dawes 1974: p. 123.

Neogoniolithon Setchell and L. R. Mason 1943: p. 89 (*neo*: new + *gonio*: a joint, knee + *lithon*: rock, stone; thus a new form of the red algal genus *Goniolithon*).

Synonym. *Paragoniolithon* Adey, Townsend and Boykins 1982: p. 12 (cf. Woelkerling, 1987)

Many species assigned to this genus, based on the type species *Neogoniolithon fosliei* (Woelkerling, 1988; Penrose, 1991; 1992), may be transferred to *Spongites* (Woelkerling, 1985b, 1988; Woelkerling and Penrose, 1988).

Crusts are thick, heavily calcified, and firmly attached. The hypothallus has multilayered parallel or bent filaments. The perithallus is multilayered and thickest near conceptacles; cells are similar in size, lack secondary pit connections, and may have fused filaments. Trichocytes are scattered or in vertical rows. The epithallus is composed of one layer of cells, which are round in cross section. Conceptacles are slightly raised, round to oval, and uniporate. Tetrasporangia are zonately divided. Because of its similar structure to species of *Neogoniolithon*, *Porolithon pachydermum* is included in this key.

Key to the species (4) of *Neogoniolithon* and *Porolithon pachydermum*

1. Plants crustose, lacking erect branches 2
1. Plants crustose and with erect nonjointed branches. . . . .*N. strictum*
   2. Surface of crusts smooth, conforming to the contour of substrata 3
   2. Surface of crusts rough due to wart- to pimplelike bumps . . . . *N. mamillare*
3. Conceptacles more than 250 µm in diameter. 4
3. Conceptacles 150–250 µm in diameter . . . . *Porolithon pachydermum*
   4. Conceptacles 300–400 µm in diam. . . . . *N. accretum*
   4. Conceptacles 600–1200 µm in diam. . . . . *N. fosliei*

Neogoniolithon accretum (Foslie and M. Howe) Setchell and L. R. Mason 1943: p. 90 (L. *accretum*: piled up, grown together). Plate XXIV, fig. 21.

Basionym. *Goniolithon accretum* Foslie and M. Howe 1906: p. 131–132, pl. 85, fig. 2, pl. 91.
Synonym. *Lithophyllum accretum* (Foslie and M. Howe) Lemoine in Børgesen 1917: p. 159.

Masaki (1968) suggested that this widely distributed and variable species could be separated into Atlantic and Pacific taxa.

Crusts are rose to brownish red, heavily calcified, 80–340 µm thick, firmly attached, conforming to substrata, and with some overlapping layers. Its hypothallus is multilayered, 6–15 layers thick, and has compact and ascending filaments; cells are 6–10 µm in diam., 15–20 µm tall, and form a few cell fusions. The perithallus is tiered; its cells are elongate to rectangular, 4–8 µm in diam., 6–10 µm tall, and with many cell fusions. Trichocytes are frequent, solitary or in vertical rows of 2–3 cells, 10–13 µm in diam., and 10–14 µm tall. Tetrasporangial conceptacles are slightly raised, scattered, and 300–400 µm in diam. Tetrasporangia are ovoid to elliptical, zonately divided, and 20–25 by 55–70 µm. Gametophytes are monoecious. Female conceptacles are raised and 350–400 µm in diam. Carposporangia arise from large fusion cells at the bases of conceptacles and are in chains. Male conceptacles are only slightly raised and 250–270 µm in diam.; spermatangial mother cells are branched.

Common, on limestone in wave-exposed sites and rubble in protected areas of the Keys (type location Sand Key; Silva et al., 1996); also known from North Carolina (18–21 m), the Caribbean, Canary Islands, West Africa, Kenya, and Japan.

Taylor 1960: p. 396; Schneider and Searles 1991: p. 242.

Neogoniolithon fosliei (Heydrich) Setchell and L. R. Mason 1943: p. 90 (named after M. Foslie, a Norwegian phycologist who studied coralline red algae). Plate XXIV, fig. 22.

Basionym. *Lithothamnion fosliei* Heydrich 1897a: p. 58–60, fig. 1, pl. III: fig. 9–11.
Synonyms. *Goniolithon solubile* Foslie and M. Howe in Foslie 1907 ('1908'): p. 21.
Goniolithon propinquum f. *solubile* Foslie 1908: p. 4 (as '*solubilis*').
Lithophyllum propinquum (Foslie) Lemoine in Børgesen 1917: p. 166–167, fig. 161–162.
Paragoniolithon solubile (Foslie and M. Howe in Foslie) Adey, Townsend and Boykins 1982: p. 12.

Babbini and Bressan (1997) considered *Neogoniolithon foslie* to be a synonym of *N. brassica-florida* (Harvey) Setchell and Mason, while Wynne (2005a) cites *N. propinquum* (Foslie) Lemoine as a distinct taxon with *Goniolithon propinquum* a synonym.

Crusts are closely adherent, almost purple, 0.2–0.8 mm thick, with irregular to nodulose surfaces, and layers overgrowing themselves. The hypothallus is multilayered with filaments in parallel series;

cells are rectangular, 7–12 µm wide, and 15–35 µm long. Perithallial cells are in vertical compact rows, rectangular, 5–12 µm wide, and 7–25 µm tall. Tetrasporangial conceptacles are on the surface, 0.6–1.2 mm in diam., conical, and have raised pores. Tetrasporangia are ovoid to elongate, 30–40 µm in diam., and 70–80 µm long.

Uncommon, on limestone at wave-exposed and protected sites near Key West; also known from the Bahamas, Antilles, Brazil, and Indonesia.

Taylor 1960: p. 396.

Neogoniolithon mamillare (Harvey) Setchell and L. R. Mason 1943: p. 91. (L. *mamillare*: with breasts, nipples, protuberances). Plate XXIV, fig. 23, 24.

Basionym. *Melobesia mamillaris* Harvey 1849 (1847–1849): p. 109, pl. XLI (in part).
Synonyms. *Goniolithon mamillare* (Harvey) Foslie 1898b: p. 9.
Porolithon mamillare (Harvey) Lemoine in Børgesen 1917: p. 177–178.

Crusts form semicircular thin lobes with many fine, mammillate projections. Older crusts have terete, nodule-like branches 0.75–1.5 mm tall. Hypothallus is uniform in structure. Perithallial cells are 15 by 10–20 µm; surface cells 5–15 µm in diam. and 5–7 µm long. Conceptacles are large, conical, to 1.4 mm in diam., and in projections.

Rare, on limestone in the Keys; also known from the western Antilles, Brazil, Argentina, and eastern tropical Atlantic. Syntype localities: Bahia, Brazil; Tierra del Fuego; Cape Province, South Africa (Silva et al., 1996).

Taylor 1960: p. 397.

Neogoniolithon strictum (Foslie) Setchell and L. R. Mason 1943: p. 92 (L. *strictum*: strict, erect, straight). Plate XXV, fig. 1–3.

Basionym. *Goniolithon strictum* Foslie 1901 ('1900'): p. 14; Foslie in Foslie and M. Howe 1906: p. 131, pl. 82, fig. 1.
Synonym. *Lithophyllum strictum* (Foslie) Lemoine var. *nana* Foslie in Foslie and M. Howe 1906: p. 131, pl. 82, fig. 1.

Plants are heavily calcified and usually form unattached fragile clumps of branches, 14–20 cm in diam. Branches are cylindrical to compressed, rigid, to 1.5 cm in diam., and taper to tips; branching is irregular to subdichotomous, spreading or recurved, and with internodes 4–10 in diam. Hypothallial filaments are arching and compact; cells are rectangular, 15–20 µm wide, and 20–40 µm tall. Perithallus is multilayered with curved ascending filaments; cells are 12–20 µm wide by 12–25 µm tall. Trichocytes are large, numerous, and scattered at the surface.

Common, as unattached clumps in seagrass beds and tide pools in the Keys (to 3 m); also known from the Bahamas, Antilles, Belize, and Venezuela where it forms massive ridges on turbulent coral reefs. Type locations: Florida and the Bahamas (Foslie, 1901).

Taylor 1960: p. 399; Littler and Littler 2000: p. 36.

Pneophyllum Kützing 1843: p. 385 (Gr. *pneo*: lung, air, to breathe + *phyllum*: leaf, blade).

Synonym. *Heteroderma* Foslie 1909: p. 56, *pro parte*.

Penrose and Woelkerling (1991) separated *Pneophyllum* and *Spongites* on the basis of tetrasporangial conceptacle formation.

Crusts are thin, lightly calcified, and closely adherent to hosts. The hypothallus has one layer of tightly united filaments that may have intercalary trichocytes. Pit connections are absent and cells between some filaments are fused. The perithallus is uni- to multilayered and thickest near conceptacles. The epithallus has one layer of rounded cells. Conceptacles are uniporate and slightly elevated.

Pneophyllum fragile Kützing 1843: p. 385 (L. *fragile*: fragile). Plate XXV, fig. 4, 5.

Synonyms. *Heteroderma lejolisii* (Rosanoff) Foslie 1909: p. 56.
Fosliella lejolisii (Rosanoff) M. Howe 1920a: p. 588.
Pneophyllum lejolisii (Rosanoff) Chamberlain 1983: p. 359, fig. 28–32.

Crusts are epiphytic, delicate, lightly calcified, pink to white, 0.5–2.0 mm in diam., 15–30 µm thick, and often coalesced. Older crusts are to 4 cells thick. The hypothallus is compact; its filaments are radially organized, dichotomously branched, and radiating from an original 8-celled structure. Cells are square to rectangular, 5–10 µm in diam., and 5–20 µm long. Trichocytes are rare, 8–13 µm in diam., and at tips of surface cells. Crust margins have superficial cells derived from the hypothallus. The perithallus is formed mostly near the center and has 1–3 layers. Surface cells are 6–12 µm wide, 3–12 µm long and have cap cells 3–8 µm wide by 1.5–4 µm long. Tetrasporangial and cystocarpic conceptacles are flat or slightly raised, circular, and 60–300 µm in diam. Male conceptacles are 75–100 µm in diam. Tetrasporangia are zonately divided, elongate, 30–50 µm in diam., and 50–80 µm long. Bisporangia are 17–51 µm in diam. by 30–65 µm long.

Common, on seagrasses and seaweeds on all coasts; also known from western Greenland and Newfoundland through the Caribbean, Gulf of Mexico, and Brazil, as well as the eastern North and tropical Atlantic, Mediterranean (type location; Penrose, 1996b), Red Sea, Pacific Ocean, and Australia.

Taylor 1960: p. 387; Dawes 1974: p. 124; Schneider and Searles 1991: p. 244; Littler and Littler 2000: p. 38.

Porolithon Foslie 1909: p. 57 (L. *poro*: porous + *lithos*: rock).

Contrary to Guiry and Nic Dhonncha (2006), Wynne (2005a) cites it as *Porolithon* (Foslie) Foslie 1908.

Plants are calcareous and form thin to massive horizontal crusts with or without erect branches. The hypothallus is polystromatic; cells are elongate and parallel to the substratum and noncoaxial. Perithallial cells are polystromatic, often variable in morphology, with many intercellular fusions, and lack secondary pit connections. Trichocytes occur in transverse disks of distinctly large cells. Tetrasporangial conceptacles are uniporate and have paraphyses in their center; tetrasporangia are lateral.

Porolithon pachydermum (Foslie) Foslie 1909: p. 57 (Gr. *pachyderm*: thick-skinned). Plate XXV, fig. 6–8.

Basionym. *Lithophyllum onkodes* (Heydrich) Heydrich f. *pachydermum* Foslie 1904b: p. 4.

Penrose and Woelkerling (1992) moved the species to *Hydrolithon* (Foslie) Foslie, while Bailey et al. (2004) designated it as *H. pachydermum* (Foslie) J. C. Bailey, J. E. Gabel, and D. W. Freshwater.

Plants are thin or thick calcified crusts 0.1 to 1 mm thick. Crusts are pink gray and have a chalky surface. The perithallus is compact; cells are small, 5–12 µm in diam., 4–11 µm tall, and occur in rows with scattered groups of 8 large (17–20 by 8–15 µm) cells in the midregion . The hypothallus is poorly formed; cells are 4–10 µm in diam. and 4–17 µm long. Basal layer cells are horizontal and parallel to the substrata. Trichocytes are 8–16 µm in diam., 17–29 µm tall, common, and in horizontal clusters. Cystocarpic conceptacles are convex and 200–300 µm in diam. Tetrasporangial conceptacles are convex, not obvious, partially embedded, and 150–250 µm in diam. Tetrasporangia are 30–40 µm in diam., 60–70 µm long, zonately divided, and formed at the periphery of conceptacles.

Uncommon, a major component of algal reef crests; known in Florida only from a Smithsonian Institution herbarium specimen from Looe Key; also recorded from the Bahamas and Caribbean.

Taylor 1960: p. 400; Littler and Littler 2000: p. 38.

Corallinoideae

Members of the subfamily are erect, with noncalcified joints (genicula), and attached by crusts. Joints usually lack a cortex and have a medulla with one tier of elongate cells. Lateral branches and secondary pit connections are absent and fusions between cells of filaments are common. Tetrasporangia and bisporangia are in uniporate conceptacles; pore plugs are absent.

Corallina Linnaeus 1758: p. 805 (L. *corallina*: coral red; also Gr. *corallion* a coral).

Plants are calcified, erect, repeatedly branched, and attached by crustose discs. Axes are articulate, joints cylindrical to flat and with one tier of elongate medullary cells that are thick walled. A cortex is absent. Segments are terete to compressed, oppositely and pinnately branched and usually in one plane. The segment medulla has arching transverse tiers of elongate cells that lack secondary pit connections and is covered by a cortex of a few to several layers of small, rounded cells that radiate from the medulla. Conceptacles are uniporate and on unbranched segments. Carpogonial branches are 2-celled and on supporting cells at the base of conceptacles that serve as auxiliary cells. Most gonimoblastic filaments become carposporangia.

Corallina officinalis Linnaeus 1758: p. 805; 1761: p. 539 (L. _officinalis_: medicinal, herbal). Plate XXV, fig. 9.

Synonym. *Corallina officinalis* f. *profunda* Farlow 1881: p. 179.

Axes are heavily calcified, erect, to 10 cm tall, and often in dense tufts with one to several axes arising from a basal crust. Main axes are terete below and flat above. Segments are 0.2–1.8 mm wide, 0.2 to 1 mm thick, and 0.3–2 mm long. Axes branch pinnately, with up to 5 branchlets at top of each segment and mostly in one plane; branchlets are terete and blunt tipped. Segment medulla has tiers of parallel, long (36–60 µm) cells and a cortex 50–75 µm thick with rounded, pigmented cells, 7–12 µm in diam., that radiate outward obliquely. Joints are not calcified; medullary cells are in one tier and 80–210 µm long. A cortex is absent. Tetrasporangia are 30–50 µm in diam. by 70–125 µm long; conceptacles are uniporate, oval, and terminal or lateral on unbranched segments.

Uncertain and with no confirmed records for Florida; known from Greenland to the Carolinas, Western Antilles, Brazil, Uruguay, and Argentina, as well as the eastern North and South Atlantic (type location "Oceanus Europae"; Womersley and Johansen, 1996), Mediterranean, Indo-Pacific, and Antarctica.

Taylor 1960: p. 410; Schneider and Searles 1991: p. 228.

Haliptilon (Decaisne) Lindley 1846: p. 25 (Gr. *hali*: the sea + *ptilon*: a feather).

Basionym. *Jania* sect. *Haliptylon* Decaisne 1842b: p. 123 (reprint p. 111; as '*Haliptylon*').

Plants are erect, heavily calcified, articulate, repeatedly branched, and with several axes arising from a basal crust. Branching is dichotomous to oppositely pinnate; segments are terete to compressed. Joints have one tier of elongate, thick-walled medullary cells and no cortex. Segment medulla has arching, transverse tiers of elongate cells that lack pit connections and a cortex of small round cells that radiate from well-defined medullary cells. Conceptacles are uniporate and on branched segments. Tetrasporangia occur in apical conceptacles. Gametophytes are dioecious. Spermatangia are in terminal conceptacles with deep narrow cavities. Carpogonial branches are 2-celled and on supporting cells that form the floor of the conceptacles and function as auxiliary cells. Carposporophytes are in conceptacles at branch tips; most gonimoblast cells become carposporangia.

Key to the species (2) of *Haliptilon*

1. Branching is irregular and loosely pinnate to whorled; main axes terete . . . . *H. cubense*
1. Branching is regular and usually closely pinnate; main axes flat . . . . *H. subulatum*

Haliptilon cubense (Montagne *ex* Kützing) Garbary and H. W. Johansen 1982: p. 218, fig. 10, as '*cubensis*' (L. *cubense*: of or from Cuba). Plate XXV, fig. 10–12.

Basionym. *Jania cubensis* Montagne *ex* Kützing 1849: p. 709–710.

Synonym. *Corallina cubensis* (Montagne *ex* Kützing) Kützing 1858: p. 37, pl. 77, fig. II.

Plants are heavily calcified, delicate, erect, 1–3 cm tall, often in dense tufts, and attached by a crust and calcareous pads of branches. Several axes arise from a crust, are terete, and taper to blunt tips; branching is irregular, dichotomous, or pinnate, in multiple planes, and also adventitious. Lower axial segments are 100–400 µm wide by 150–660 µm long; middle segments are 80–180 µm wide by 600 µm long. The medulla of segments have tiers of parallel, long cells (35–100 µm by 12 µm wide); cortical cells are round, pigmented, and 5–8 µm in diam. Joints are not calcified, have one tier of elongate cells (80–210 µm), and lack a cortex. Tetrasporangia are zonately divided, ovoid, and 20–70 µm in diam. by 80–120 µm long; conceptacles are uniporate, urn-shaped and at ends of segments between branches (horns).

Common, in delicate clumps on limestone and in seagrass beds (to 25 m) on all coasts; also known from North Carolina, Bermuda, the Bahamas, Caribbean (type location: Cuba; Silva et al., 1996), Gulf of Mexico, and Brazil, as well as the eastern tropical Atlantic, Pakistan, and the Philippines.

Taylor 1960: p. 409; Dawes 1974: p. 122; Schneider and Searles 1991: p. 229; Littler and Littler 2000: p. 26.

Haliptilon subulatum (J. Ellis and Solander) H. W. Johansen 1970: p. 79, 83–84, fig. 10, 12, as '*subulata*' (L. subulatum: awl-shaped). Plate XXV, fig. 13.

Basionym. *Corallina subulata* J. Ellis and Solander 1786: p. 119–120, pl. 21, fig. b.

Axes are erect, bushy, to 4 cm tall, heavily calcified, with dichotomous branching and opposite lateral branches in one plane, and attached by small padlike crusts. Main axes are terete to flat and 100–240 µm in diam. Middle segments are compressed, 0.7–1.0 mm in diam., and 0.3–0.7 mm long. Branchlets are flat to cylindrical, to 60 µm in diam., and taper to tips. Segment medulla have tiers of elongate cells to 10 µm in diam. and a cortex of one tier of rounded to rectangular pigmented cells, 5–8 µm in diam. Joints are not calcified, have cells parallel in one tier, and lack a cortex. Conceptacles are oval, on smooth swollen tip segments, and usually between hornlike projections.

Uncommon, on seaweeds in tide pools in the Keys; also known from the West Indies (type location; Abbott, 1999), southern Caribbean, Gulf of Mexico, and Brazil, as well as the eastern tropical Atlantic and Indo-Pacific.

Taylor 1960: p. 410; Littler and Littler 2000: p. 26.

Jania J.V. Lamouroux 1812: p. 186, as '*Iania*' (L. *ianus*: a covered passage; also *Janus*: a two-faced Roman God of mythology).

Plants are erect, bushy, articulate, heavily calcified, and attached by basal crusts. Branching is mostly dichotomous, in one plane, and with segments mostly terete. Segment medulla has arching transverse tiers of elongate cells that lack secondary pit connections and a cortex with few to several layers of small rounded cells that radiate from distinct medullary cells. Joints are not calcified and lack a cortex; medullary cells are in one tier, narrow, and thick walled. All conceptacles are uniporate and produced near segment tips between forks (horns). Gametophytes are dioecious. Spermatangial conceptacles have deep narrow pits. Carpogonial branches are 2-celled and occur on supporting cells that function as auxiliary cells and fuse after fertilization to form a fertile floor. Carposporophytes are produced in conceptacles, with most gonimoblastic filaments becoming carposporangia.

Key to the species (4) of *Jania*

1. Plants taller than 3 mm 2
1. Plants less than 3 mm tall . . . . *J. pumila*
  2. Forks dichotomous, wide spreading, and with angles greater than 30° 3
  2. Forks dichotomous, narrow, and with angles of 10–20° . . . . *J. rubens*
3. Plants to 4 cm tall; segments greater than 100 µm in diameter . . . . *J. adhaerens*
3. Plant to 1 cm tall; segments 100 µm or less in diameter . . . . *J. capillacea*

Jania adhaerens J. V. Lamouroux 1816: p. 270 (L. *adhaerens*: adhering). Plate XXV, fig. 14.

Synonym. ?*Jania comosa* P. Crouan and H. Crouan in Schramm and Mazé 1865: p. 18; P. Crouan and H. Crouan in Mazé and Schramm 1878 (1870–1877): p. 209.

Wynne (2005a) discussed several distinctions between *Jania adhaerens* and *J. capillacea.*

Plants are calcified, articulate, delicate, erect, to 4 cm tall, and with one or more axes arising from prostrate axes. Plants are initially attached by basal crusts and secondarily by calcareous pads. Segments are cylindrical, 90–200 µm in diam., and 0.4–1.0 mm long. Axes are dichotomously branched, mostly in one plane, and with conical and blunt tips. Segment medulla is tiered with elongate cells (60–85 µm long) in parallel rows; cortical cells are rounded, pigmented, and 6–8 µm in diam. Joints are not calcified, at regular intervals between forks, and near segment bases. The medulla has one tier of elongate cells 7–10 µm in diam. by 40–120 µm long; a cortex is absent. Tetrasporangia are zonately divided, ovoid, 60–70 µm in diam., and 80–100 µm long; conceptacles are uniporate, urn-shaped, 240–300 µm in diam., 300–340 µm high, and between two terminal branches (horns). Surface cells have a unique pattern of turnover involving localized wall decalcification (Pueschel et al., 2005).

Common, in pink clumps or mats on limestone, seagrasses, and coarse algae (to 18 m) and adrift on all coasts; also known from North Carolina (17–35 m), Bermuda, the Bahamas, Caribbean, Gulf of Mexico, and Brazil, as well as the Mediterranean (type location; Abbott, 1999), eastern tropical Atlantic, Red Sea, Indo-Pacific, and Australia.

Taylor 1960: p. 413; Dawes 1974: p. 124; Schneider and Searles 1991: p. 231; Littler and Littler 2000: p. 30.

Jania capillacea Harvey 1853: p. 84–85 (L. *capillacea*: hairlike). Plate XXV, fig. 15.

Cribb (1983) placed *Jania capillacea* in synonymy with *J. adhaerens*, while Schneider and Searles (1991) and Littler and Littler (2000) recognized both taxa based on robustness and size.

Plants are delicate, articulate, erect (4–10 mm tall) or creeping, and lightly calcified; branching is widely dichotomous and mostly in one plane. One to several axes arise from basal crusts and branches also become attached by calcareous pads. Segments are terete; lower ones are 45–100 µm in diam. by 300–600 µm long; upper ones are smaller and have recurved branches with conical and blunt to acute tips. Segment medullary cells are in parallel tiers, elongate, and 25–40 µm long; cortical cells are rounded, pigmented, and 5–8 µm in diam. Joints are not calcified, are at bases of segments, and lack a cortex; medullary cells are elongate, 40–75 µm long, and in one tier. Tetrasporangia are zonately divided, ovoid, 30–40 µm in diam., and 80–105 µm long; conceptacles are uniporate, urn-shaped, on swollen tips of branches (240–300 µm wide), and occur terminally between 2(4) branches (horns).

Common, on seaweeds and seagrasses and adrift (to 15 m) on all coasts (type location: Bahia Honda Key; Silva et al., 1996); also known from Virginia (in drift), North Carolina, Bermuda, the Caribbean, Gulf of Mexico, Brazil, eastern tropical Atlantic, Red Sea, and Indo-Pacific.

Taylor 1960: p. 412; Dawes 1974: p. 125; Schneider and Searles 1991: p. 232; Littler and Littler 2000: p. 32.

Jania pumila J. V. Lamouroux 1816: p. 269, pl. IX, fig. 2 (L. *pumila*: little, dwarf, low). Plate XXV, fig. 16.

South and Skelten (2003) placed *Jania pumila* into synonymy with *J. tenella* (Kützing) Grunow.

Plants are minute, calcified, delicate, articulate, 2–3 mm tall or prostrate and 0.5–1 cm in diam.; branching is widely dichotomous (2–3 times) with one or more axes arising from basal discs. Segments are terete, 60–150 µm in diam., and to 200 µm long; medullary cells elongate, 20–30 µm long, and in parallel tiers; cortical cells are round, pigmented, and 5–8 µm in diam. Joints are not calcified, at the bases of forks, and lack a cortex; medullary cells are elongate, 4–8 µm in diam., and in one tier. Tetrasporangial and female conceptacles are uniporate, swollen, 250 µm in diam., 800 µm long, and on triangular segments at forks between 2 branchlets, each with 2–4 segments (horned). Tetrasporangia are zonately divided. Spermatangial conceptacles are thick, spindle-shaped, and 140 by 250 µm.

Common, a minute epiphyte of seaweeds and seagrasses on all coasts; also known from Bermuda, the Caribbean, Gulf of Mexico, and Brazil, as well as the eastern tropical Atlantic, Red Sea (type location: "Mer rouge"; Abbott 1999), and Indo-Pacific.

Taylor 1960: p. 414; Dawes 1974: p. 125; Littler and Littler 2000: p. 32.

Jania rubens (Linnaeus) J. V. Lamouroux 1812: p. 186; 1816: p. 272, pl. IX, fig. 6, 7 (L. *rubens*: reddening, blush red). Plate XXV, fig. 17.

Basionym. *Corallina rubens* Linnaeus 1758: p. 806; 1766 (1766–1768): p. 1304, 1305.

Plants are calcified and in tufts or cushions to 8 cm tall; branching is dichotomous with narrow forks (10–20°) and branches comblike at the top. One or more axes arise from discoid crusts and some branches are attached by calcareous pads. Older segments are mostly terete, 125–280 µm in diam., and 0.6–2.5 mm long; branchlets are strongly club-shaped, 65–100 µm in diam., and have acute tips. Segment medullary cells are elongate, 80–150 µm long, and in tiers; cortical cells are rounded, pigmented, and 7–13 µm in diam. Joints are not calcified, at bases of branches, and have regular intervals between branches; medullary cells are elongate, 80–170 μm long, and in one tier; no cortex is present. Tetrasporangia are zonately divided, ovoid, 40–60 μm in diam., and 100–150 μm long; conceptacles are uniporate, urn-shaped, 250–300 μm in diam., and terminal on swollen segments between 2–3 branches (horns). Gametophytes are monoecious; female conceptacles are horned, while male conceptacles are terminal, fusiform, and lack horns.

Common, on rubble in calm waters (often overgrown by sponges) and seaweeds at exposed sites on all coasts; also known from North Carolina, Bermuda, the Caribbean, Gulf of Mexico, and Brazil to Argentina, as well as the eastern North and tropical Atlantic (type location: Europe; Silva et al., 1996), Mediterranean, Red Sea, and Indo-Pacific.

Taylor 1960: p. 413; Schneider and Searles 1991: p. 233; Littler and Littler 2000: p. 32.

Lithophylloideae.

With the exception of the articulated *Amphiroa*, which has a close phylogenetic relationship to *Titanoderma* (Bailey, 1999), the subfamily primarily contains crustose forms plus a few taxa with erect branches that lack joints. Some filaments are connected by secondary pit connections and cell fusions are rare or absent. Tetrasporangia and bisporangia lack apical plugs and are in uniporate conceptacles.

Amphiroa J.V. Lamouroux 1812: p. 185; 1816: p. 294 (Gr. *amphi*: on both sides, around).

Plants are erect, tufted or matted, and attached by basal crusts. Branching is repeatedly dichotomous to irregular; segments are heavily calcified and terete to flat. Segment medulla have small ovoid cells that alternate with longer ones and a cortex with a few layers of small rounded cells that radiate from distinct medullary cells. The joint medulla has 1–3 tiers of arching elongate cells with secondary pit connections and no cortex. Conceptacles are small and lateral on segments. Tetrasporangial conceptacles are hemispherical and have a central pore formed by the breakdown of roof cells. Gametophytes are dioecious. Spermatangia are in convex conceptacles. Carpogonial branches are 2-celled; supporting cells form the floor of the female conceptacles and act as auxiliary cells; female conceptacles are uniporate. The cell arrangement of intergenicula medullary tiers is useful for separating some species of *Amphiroa* (Dolan, 2001).

Key to the species (5) of *Amphiroa*

1. Axes cylindrical or slightly flattened and less than 2 mm wide 2
1. Axes mostly flat and greater than 2 mm wide; branching irregular . . . . *A. tribulus*
   2. Branches with some compression or flattening. 3
   2. Branches cylindrical throughout. 4
3. Thallus in loose clumps; branching strictly dichotomous, rarely fusing with adjacent branches . . . . . *A. beauvoisii*
3. Thallus in tightly crowded clumps; branching irregular to dichotomous, usually fusing with adjacent branches. . . . .*A. brasiliana*
   4. Branches 150–600 μm in diam; joints often swollen at forks . . . . *A. fragilissima*
   4. Branches 1.0–2.0 mm in diam; joints rarely swollen at forks . . . .*A. rigida*

***Amphiroa beauvoisii* J. V. Lamouroux 1816: p. 299–300 (named for Ambroise Marie Francois Joseph Palisot de Beauvois, a French botanist who named *Psilotum nudum*). Plate XXV, fig. 18–21.**

Plants are in sparse to dense clumps, 2–5 cm tall, calcified, white pink, and attached by inconspicuous crustlike holdfasts. Axes are abundantly dichotomously branched, mostly at joints, usually in one plane, and with branches often recurved. Segments vary in size and shape (1–5 mm long); lower ones are terete, 400–650 μm in diam., and 4–5 times as long as broad; upper ones are 400–650 μm in diam., compressed, and often cuneate-linear; apical ones are bifurcate, often rounded, and enlarged. Cortical segments have several layers of short cells that increase in number and thickness with age; the medulla usually has 6–7 transverse bands of thin-walled cells, 70–80 μm long, that alternate with a single row of short cells (15–25 μm long). Joints have thick-walled medullary cells arranged in 4 bands. Uniporate conceptacles are lateral, hemispherical, and 17–25 μm in diam. Tetrasporangia are zonately divided, 17–25 μm in diam., and 38–63 μm long.

Rare, to 2m on hard substrata and mangrove roots; known only in Florida from a Smithsonian Institution herbarium specimen from Brevard County (as "Oceanus"); also known from the Carolinas, Venezuela, Columbia, and Mexico, as well as the eastern North and South Atlantic (type location: Portugal; Silva et al., 1996), Mediterranean, and Indo-Pacific.

Schneider and Searles 1991: p. 225; Littler and Littler 2000: p. 20.

Amphiroa brasiliana Decaisne 1842b: p. 125 (L. *brasiliana*: of or from Brazil). Plate XXV, fig. 22, 23.

Plants are erect, jointed, to 5 cm tall, heavily calcified, and attached by small crusts. Branching is irregular to dichotomous, wide-angled in compacted forms and narrow-angled in elongated ones. Segments are terete to flat, 0.3–1.5 mm wide, 1–4 mm long, may be branched, and often fused laterally to other branches. Segment medullary cells are small and in tiers that curve into a cortex. The cortex has 2–5 rows of cells 60–80 μm long that alternate with 1–2 rows of short (15–20 μm) ones that are 8–15 μm in diam. Joints are mostly covered on the lateral margins by expansion of noncalcified segment cells in an anticlinal series. Conceptacles are obvious, abundant on the surface, convex, 270–420 μm in diam., and have a terminal pore. Tetrasporangia are zonately divided, ovoid, 25–35 μm in diam., and 35–65 μm long.

Common, on limestone and rubble in protected sites and seagrass beds in the Keys; also known from the Lesser Antilles, southern and western Caribbean, Brazil (type location: Rio de Janeiro; Silva et al., 1996), Uruguay, and India.

Taylor 1960: p. 405; Littler and Littler 2000: p. 20.

Amphiroa fragilissima (Linnaeus) J. V. Lamouroux 1816: p. 298 (L. *fragilissima*: very fragile). Plate XXV, fig. 24–26.

Basionym. *Corallina fragilissima* Linnaeus 1758: p. 806; 1759: p. 1305.

Plants are erect (to 4 cm tall) or in entangled mats and attached by small crusts. Branches are widely forked; branching is usually regularly di-, or trichotomous, at joints, and may be adventitious. Segments are thin, brittle, cylindrical, 150–600 μm in diam., and 8–20 diam. long. Segment medullary cells are in tiers that curve in irregular horizontal rows, with 4–8 transverse rows of long (55–90 μm) cells alternating with 1–2 rows of shorter ones (15–30 μm). Cortical cells are short and in rows. Joints are not calcified and flexible; surface cells 8–10 μm in diam. and thick walled. Conceptacles are uniporate, hemispherical, 300–400 μm in diam., and lateral on segments. Tetrasporangia are zonately divided, ovoid, 20–25 μm in diam., and 50 μm long.

Common, weakly attached to seaweeds and seagrasses and in clumps in moderately exposed to sheltered areas in the Keys (to 7 m); also known from Bermuda, the Caribbean (type location: Jamaica; Silva et al., 1996), Gulf of Mexico, and Brazil, as well as the eastern tropical Atlantic, Mauritius, Pacific Ocean, and Australia.

Taylor 1960: p. 403; Dawes 1974: p. 122; Littler and Littler 2000: p. 22.

Amphiroa rigida J. V. Lamouroux 1816: p. 297, pl. XI, fig. 1 (L. *rigida*: stiff). Plate XXV, fig. 27–29.

Synonym. *Amphiroa rigida* var. *antillana* Børgesen 1917: p. 182–185, fig. 171–173.

Plants are strongly calcified, in dense clumps 10–15 cm in diam., and attached by small crusts. Branching is regularly dichotomous and with a few wide adventitious branches that are blunt or tapered. Segments are cylindrical, 1–2 mm in diam., with forks usually below joints, and branches that are unequal in length near apices. Segment medullary cells are in tiers, are gradually reduced in size, curve into the cortex, and in 2 transverse rows of long (80–100 µm) cells that alternate with a layer of short medullary cells (20 µm long). The cortex has several layers of short cells. Joints are not calcified and have two transverse rows of long, thick-walled cells. Conceptacles are uniporate, hemispherical, 250–400 µm in diam., and lateral on segments. Tetrasporangia are ovoid and zonately divided.

Common, on rubble or limestone and often as detached clumps in seagrass beds in the Keys (to 11 m) and on all coasts; also known from the Caribbean, Gulf of Mexico, Mediterranean (type location; Abbott, 1999), eastern tropical Atlantic, and Indo-Pacific.

Taylor 1960: p. 404; Dawes 1974: p. 122; Littler and Littler 2000: p. 22.

Amphiroa tribulus (J. Ellis and Solander) J. V. Lamouroux 1816: p. 302 (Gr. *tribulus*: 3-pointed; also L. *Tribulus*: a plant with thorns and spiny flower heads; a caltrop). Plate XXVI, fig. 1, 2.

Basionym. *Corallina tribulus* J. Ellis and Solander 1786: p. 124, pl. 21, fig. e.

Plants are in fragile clumps, to 10 cm in diam., and attached by small crusts. Axes are di-, trichotomously, or irregularly branched at joints. Segments are brittle, thin, 2–4 mm wide, terete, flat or triangular, and with distinct midribs and dentate margins. Segment medullary cells are in straight tiers with 4–8 transverse rows of long cells (60–70 µm) that alternate with a row of short cells (30 µm). Cortical cells, which are cut off abruptly from medullary ones, are short, and in several layers. Joints are not calcified, flexible, and constructed like segments. Conceptacles are dome-shaped and common on segment surfaces; tetrasporangia are ovoid and zonately divided.

Uncommon, on limestone and in spur areas of coral reefs in the Keys (to 18 m); also known from the Caribbean (type location: West Indies; Silva et al., 1996), Gulf of Mexico, and Indo-Pacific.

Taylor 1960: p. 406; Littler and Littler 2000: p. 24.

Lithophyllum Philippi 1837: p. 387 (Gr. *litho*: stone + *phyllum*: a leaf).

Some authors (Campbell and Woelkerling, 1990; Woelkerling and Campbell, 1992; Woelkerling, 1996b; John et al., 2004) have considered *Lithophyllum* and *Titanoderma* to be congeneric; however, molecular data (Bailey, 1999) indicate they are distinct. Morphological studies (Chamberlain et al., 1991; Irvine and Chamberlain, 1994) also separated *Titanoderma* from *Lithophyllum* based on a dimerous thallus margin and basal filamentous cells in the former and their absence in the latter.

Plants are calcified crusts that adhere closely to substrata. The hypothallus has one layer of cells and may be covered by a perithallus that can produce a second hypothallus of horizontal overarching filaments. The perithallus has several layers of quadrate cells with secondary pit connections between adjacent filaments; cell fusions are absent. The epithallus has one to a few layers of rounded cells that lack trichocytes. Conceptacles are elevated, partly immersed, and uniporate. Tetrasporangia are zonately divided.

Lithophyllum intermedium Foslie 1906: p. 24 (L. *intermedium*: between, in between). Plate XXVI, fig. 3, 4.

Littler and Littler (2000) and Guiry and Nic Dhonncha (2006) cited *Goniolithon intermedium* Foslie as a synonym, while Wynne (2005a) considered the former a heterotypic synonym for *Neogoniolithon spectabile* (Foslie) Setchell and L. R. Mason.

Crusts are 0.3–3.0 mm thick, firmly attached, and conform to the substrata with smooth surfaces and lobed to crenulate margins. The hypothallus is initially one-layered, with vertically elongate cells 7–10 µm in diam. and 8–20 µm tall, and then becomes multilayered. The perithallus is multilayered; its cells are elongate or rectangular, 5–9 by 10–20 µm, and with many secondary pit connections. The epithallus has 1–3 layers of rounded cells 2.5–5.0 µm in diam. Tetrasporangia are 15–40 µm in diam., 40–75 µm high, in uniporate conceptacles that are slightly raised, numerous, scattered, and 200–250 µm in diam. Female conceptacles are unknown. Spermatangial conceptacles are slightly raised, 100–160 µm in diam., and have elongate spermatia.

Common, on limestone at wave-exposed shores in the Keys; also known from North Carolina (18 m), Bermuda (type location), the Antilles, western Caribbean, and Galapagos Islands.

Taylor 1960: p. 391; Schneider and Searles 1991: p. 237.

Titanoderma Nägeli in Nägeli and Cramer 1858: p. 532.

Synonym. *Dermatolithon* Foslie 1898: p. 11.

Crusts are attached or unattached, flat, and have regenerating overlapping layers or nonjointed branches. Plants are dorsiventral; thallus is pseudoparenchymatous, dimerous, or dimerous and monomerous. Branches, if present, are mostly radial and monomerous. Dimerous thalli have a single layer of basal filaments with or without erect filaments that may become very long; epithallial cells are on basal cells or terminal on erect nonflared filaments. Margins of dimerous thalli have palisade basal cells or only epithallial cells. Basal cells are palisade-like and often sinuate and oblique; cells of erect filaments are joined by secondary pit connections and usually in the upper third of cells. Cell fusions are rare and trichocytes uncommon. Spore germination patterns show 2 rows of 4 cells. Conceptacles are uniporate. Gametophytes are dioecious; spermatangial branches are simple and restricted to the conceptacle floor. Carposporangial conceptacles have central fusion cells that produce gonimoblastic filaments. Tetrasporangial conceptacles have peripheral zonate tetrasporangia and a central columella; bisporangial conceptacles are similar but contain bisporangia with 2 uninucleate bispores.

Key to the species (3) of *Titanoderma*

1. Old crusts more than 2 cells thick; perithallus has vertical tiers of cells. 2
1. Old crusts mostly 2 cells thick. . . . . *T. prototypum*
    2. Surface of crust smooth; margins lobed, entire, not foliose . . . .*T. pustulatum*
    2. Surface of crust irregular; margins flakelike, foliose, overlapping. . . . . *T. bermudense*

Titanoderma bermudense (Foslie and M. Howe) Woelkerling, Chamberlain, and Silva 1985: p. 333 (L. *bermudense*: of or from Bermuda). Plate XXVI, fig. 5, 6.

Basionym. *Lithophyllum bermudense* Foslie and M. Howe 1906: p. 132, pl. 81, fig. 3, pl. 85, fig. 3, pl. 982.

Synonyms. *Dermatolithon bermudense* (Foslie and M. Howe) Foslie and M. Howe in Foslie 1909: p. 58.

Tenarea bermudensis (Foslie and M. Howe) Adey 1970: p. 6.

Crusts are smooth, conform to substrata, and are 1–9 mm thick with foliose overlapping layers. Construction is usually dimerous. The hypothallus has one layer of cells that are somewhat oblique, 12–20 µm wide, and 15–50 µm tall. The perithallus usually has vertical tiers of cells, 10–20 µm in diam. by 15–45 µm long with terminal and lateral secondary pit connections. The epithallus (surface) has 1–3 layers; cells are 10–16 µm in diam. by 4–8 µm long. Tetrasporangial conceptacles are flush to slightly hemispherical, uniporate, and 330–600 µm in diam.; tetrasporangia are 40–60 µm in diam. and 85–120 µm long.

Uncommon, on rubble in shallow waters at the Keys; also known from Bermuda (type location: Spanish Point; Woelkerling, 1996b), the Mediterranean, Madeira Archipelago, and Australia.

Taylor 1960: p. 393.

Titanoderma prototypum (Foslie) Woelkerling, Chamberlain, and Silva 1985: p. 333 (L. *prototypum*: original, first type). Plate XXVI, fig. 7, 8.

Basionym. *Lithothamnion prototypum* Foslie 1897: p. 18.
Synonyms. *Lithophyllum prototypum* (Foslie) Foslie 1905: p. 129; 1909: p. 49.
Tenarea prototypa (Foslie) Adey 1970: p. 7.

Crusts are calcified, stonelike, and glossy. Young crusts are circular and have thin margins. Older crusts are irregular, to 300 µm thick, and appear like layered flakes. Crusts are mostly 2 cells thick and overgrow themselves. Surface (cap) cells are square to round and 10–15 µm wide by 2–9 µm tall; lower cells are elongate, 10–15 µm wide by 10–25 µm tall, and with pit connections between vertical walls. Tetrasporangial and bisporangial conceptacles are uniporate, domelike, to 400 µm in diam., and scattered. Cystocarpic conceptacles are 450–550 µm in diam. and partially sunken.

Uncommon, on razor clam shells (*Pinna*) and pebbles in the Keys; also known from the Bahamas, Virgin Islands (type location: St. Croix; Woelkerling, 1996b), Puerto Rico, and Jamaica, as well as Indo-Pacific, and Australia.

Taylor 1960: p. 392; Littler and Littler 2000: p. 40.

Titanoderma pustulatum (J. V. Lamouroux) Nägeli 1858: p. 352 (L. *pustulatum*: blistered, with blisters). Plate XXVI, fig. 9, 10.

Basionym. *Melobesia pustulata* J. V. Lamouroux 1816: p. 315, pl. 12, fig. 2.
Synonyms. *Dermatolithon pustulatum* (J. V. Lamouroux) Foslie 1898b: p. 11.
Lithophyllum pustulatum (J. V. Lamouroux) Foslie 1904b: p. 8.

Crusts are moderately calcified. Young crusts are flat, rounded or lobed, and completely adherent. Older crusts are thickened and convex; their margins are entire, 2–10 mm in diam., and 2–8 cells and 350 µm thick. Cap cells are 8–15 µm in diam. with convex tops. Perithallial cells are elongate, in vertical rows, and 8–15 µm wide by 17–55 µm tall. Hypothallial cells are rectangular to square and 8–15 by 12–55 µm. Tetrasporangial and bisporangial conceptacles are uniporate, prominent, hemispherical, 300–500 µm in diam., and scattered. Tetrasporangia are ovoid, 20–70 µm in diam., and zonately divided.

Common, on seaweeds and seagrasses in the Keys (to 5 m); also known from the Canadian Maritimes to North Carolina, Bermuda, the Caribbean, Gulf of Mexico, and Brazil, as well as the eastern North and tropical Atlantic (type location: France; Silva et al., 1996), Mediterranean, Red Sea, Thailand, and Fiji.

Taylor 1960: p. 392; Littler and Littler 2000: p. 40.

Sedis incertae: Goniolithon Foslie 1898a: p. 5 (Gr. *gonio*: a joint, knee + *lithon*: rock, stone).

The genus needs further taxonomic evaluation as Florida plants may belong to *Neogoniolithon* (Woelkerling, 1985b, 1988; Wynne, 1998a; pers. comm).

Crusts are calcified and usually firmly attached. Their hypothallus and perithallus are composed of several cell layers, with the latter having scattered trichocytes. Tetrasporangia are in flask-shaped uniporate conceptacles with a prolonged apex.

Goniolithon decutescens **(Heydrich) Foslie in M. Howe 1918: p. 238 (L. *de*: separate, peel + *cutis*: skin). Plate XXVI, fig. 11.**

See note above regarding uncertain taxonomic status of the genus and species.

Crusts are calcified, cushion-shaped, and 0.4–1 mm thick; branches are simple or erect, not jointed, terete, sometimes fused, 5–10 mm long, and to 1 mm in diam. The hypothallus has subquadrate to elongate cells, 8–15 by 12–30 µm. The perithallus is uniform; its cells are 8–15 µm in diam., to 24 µm tall, and have trichocytes that are scattered, large, clear, and with thickened walls. Tetrasporangial conceptacles are on the surface and 0.8–1 mm in diam.

Common, on mangrove roots and limestone in the Keys (5–10m) and on the west coast; also known from Bermuda, the Bahamas, and Caribbean (syntype locations: Puerto Rico, the Bahamas, and West Calicos; Foslie, 1929).

Taylor 1960: p. 398; Dawes 1974: p. 124.

Fossil records of coralline algae

Based on studies of Eocene deposits in Duval County, Johnson and Ferris (1948) found seven fossilized coralline algae, including three new taxa: *Lithophyllum compactum* Johnson and Ferris, *L. zonatum* Johnson and Ferris, and *Amphiroa americana* Johnson and Ferris. Fragments of other species of *Lithophyllum* and *Lithothamnion* were also observed. Numerous fragments of *Lithophyllum* were found with their tissues having well-defined strongly arched layers or zones, each composed of numerous rows of cells.

NEMALIALES Schmitz 1892

Based on molecular evaluations, Huisman et al. (2004a) identified three lines in this order: Liagoraceae (including the Dermonemataceae and probably Nemaliaceae) and two others. The Galaxauraceae includes *Actinotrichia*, *Dichotomaria*, *Galaxaura*, and *Tricleocarpa*; a new family, the Scinaiaceae, includes *Scinaia* and *Nothogenia* and has noncalcified members of the old Galaxauraceae (Huisman, 1985, 1986; Huisman and Womersley, 1992).

The order is characterized by a simple postfertilization development in which the gonimoblast phase develops directly from the zygote or postfertilized carpogonium. By contrast, the zygote nucleus in several other orders is transferred to an auxiliary cell from which gonimoblast filaments arise.

Liagoraceae

Synonyms. Dermonemataceae, Helminthocladiaceae, and Nemaliaceae.

Kraft (1989) added genera from the Nemaliaceae (*Nemalion*) and Dermonemataceae (*Cumagloia*, *Dotyphycus*, *Patenocarpus*, and *Yamadaella*).

Plants are erect, soft, gelatinous, slimy, or lightly calcified. Axes are terete or slightly compressed, multiaxial, and with a cortex of loosely arranged filaments. Central medullary filaments produce lateral pigmented branches whose cells have one large parietal plastid. Life histories may be triphasic and heteromorphic; carpospores may form a filamentous "*Acrochaetium*-like" stage that can produce monospores or tetraspores. Tetraspores may form protonema-like filaments with erect "buds" that grow into large gametophytes with carpogonial and spermatangial branches. Gametophytes are monoecious or dioecious. Spermatangia are clustered and usually terminal. Carpogonial branches have 3 to many cells; the hypogynous cell is naked or rarely has short filaments. Cystocarps are immersed and involucres may be present. Carposporangia are usually terminal on gonimoblastic filaments.

Ganonema K. C. Fan and Y. C. Wang 1974: p. 492, pl. 1. (Gr: *ganos*: stomach, bag + *nema*: thread).

See Fan and Wang (1974), Huisman and Kraft (1994), Huisman and Wynne (1999), Huisman and Schils (2002), Ballantine and Aponte (2002), and Huisman et al. (2004a) regarding separation of *Ganonema* from *Liagora*.

Plants are erect, dichotomously to trichotomously branched, encrusted with calcium carbonate, and attached by discoid holdfasts. The medulla has longitudinal filaments and thin descending rhizoids; cortical filaments branch di- or trichotomously and have cylindrical, pigmented cells. Spermatangia are in dense heads associated with photosynthetic apical filaments. Carpogonial branches are slender and restricted to special short fertile filaments. Fertile axes are from basal cells of vegetative filaments and are different from those of *Liagora* (see below). Carposporophytes initially divide transversely. Cystocarps are subspherical and surrounded by sterile filaments.

Key to the species (3) of *Ganonema*

1. Branching mainly dichotomous, often with proliferations . . . . *G. farinosum*
1. Branching mainly alternate to irregular 2
    2. Plants gooey to gelatinous, branching mostly in one plane . . . . *G. pinnatum*
    2. Plants soft but not gooey; branching irregular, not in one plane . . . . *G. dendroideum*

Ganonema dendroideum **(P. Crouan and H. Crouan in Mazé and Schramm) D. L. Ballantine and Aponte 2002: p. 220 (L. *dendroideum*: treelike). Plate XXVI, fig. 12.**

Basionym. *Helminthora dendroidea* P. Crouan and H. Crouan in Mazé and Schramm 1878 (1870–1877): p. 178.

Synonyms. *Helminthora antillarum* P. Crouan and H. Crouan in Mazé and Schramm 1878 (1870–1877): p. 178.

Liagora mucosa M. Howe 1920a: p. 556–557.

Liagora dendroidea (P. Crouan and H. Crouan in Mazé and Schramm) Abbott 1990b: p. 317.

Plants are soft, gelatinous or gooey, to 20 cm tall, pink to white, with light uniform calcification, and attached by small discoid holdfasts. Main axes are to 1.3 mm in diam. with irregular, spreading branches; central filaments are longitudinal, terete, and 20–80 µm in diam., with cells to 6 diam. long. Cortical filaments are radial, clustered, pigmented, beadlike, and branched 3–4 times; cells are 13–19 µm in diam. by 13–24 µm long; tip cells are 13–20 µm in diam. and can produce fine colorless hairs. Gametophytes are dioecious. Spermatangia form compact, rounded tufts 25–50 µm in diam. on outer cortical cells. Carpogonial branches are 4-celled and 15–30 µm in diam. Cystocarps are compact, 100–200 µm wide, terminal, and with few to many involucral filaments.

Common, on rubble and shells (to 15 m) in the Keys; also known from Bermuda, the Bahamas, Caribbean (type location: Point-a-Pitre, Guadeloupe), and Gulf of Mexico.

Taylor 1960: p. 328; Littler and Littler 2000: p. 52.

Ganonema farinosum (J. V. Lamouroux) K. C. Fan and Y. C. Wang 1974: p. 492, pl. 1, as '_farinosa_' (L. _farinosum_: powdered, covered with farina). Plate XXVI, fig. 13, 14.

Basionym. *Liagora farinosa* J. V. Lamouroux 1816: p. 240, fig. 23–29.

Synonyms. *Liagora bipinnata* P. Crouan and H. Crouan in Mazé and Schramm 1878 (1870–1877): p. 183.

Liagora paniculata J. Agardh 1896: p. 106, fig. 29.

Brodie and Norris (1996) described its life history from culture studies of plants collected in the Florida Keys.

Plants are entangled, to 12 cm in diam., 10–20 cm tall, soft to fleshy but not gooey, moderately calcified, and attached by small discoid rhizoidal holdfasts. Branching is loose to irregular below, widely dichotomous above, with axes to 2 mm in diam. and tapering to 0.5 mm at the tips. Central filaments are compact, vertical, cylindrical, to 45 µm in diam., and have cells 5–10 diam. long. Cortical filaments are radial and grow from the calcified medulla; cells are terete, swollen or moniliform, and to 40 µm in diam. Gametophytes are dioecious. Spermatangia are 40–60 µm in diam. and arise in terminal tufts from cortical filaments. Carpogonial branches are erect and on special short fertile filaments that arise from the basal cells of radial branchlets.

Common, on rubble in seagrass beds and to 10 m in the Keys; also known from Bermuda, the Caribbean, Gulf of Mexico, and Brazil, as well as the Canary Islands, Sao Tomé, the Red Sea (type location near Suez, Egypt; Abbott, 1999), Pacific Ocean, and Australia.

Taylor 1960: p. 326; Littler and Littler 2000: p. 50.

Ganonema pinnatum (Harvey) Huisman 2002: p. 828 (L. _pinnatum_: featherlike, winged). Plate XXVI, fig. 15, 16.

Basionym. *Liagora pinnata* Harvey 1853: p. 138–139, pl. XXXI.B.

Plants are soft to fleshy but not gooey, moderately calcified, 20–30 cm in diam., and attached by small rhizoidal holdfasts. Branching is moderate to dense, widespread, and mostly in one plane. Central filaments are longitudinal, to 60 µm in diam. below and 15 µm near tips; cortical filaments are radially oriented, dichotomously branched 3–4 times, with cells slightly swollen and 17–28 µm in diam. Gametophytes are monoecious; spermatangia are clustered and 25–40 µm in diam. Carpogonial branches are 4(5)-celled and slightly curved; cystocarps have ascending involucral filaments.

Uncommon, on rubble in shallow waters near coral reefs in the Keys (type location: Sand Key); also known from the Bahamas, Caribbean, and the Indo-Pacific.

Taylor 1960: p. 329.

Helminthocladia J. Agardh 1851 (1848–1901): p. 412, *nom. cons.* (Gr. *helmins*: a worm + *klados*: a branch, sprout).

Plants are soft, irregularly, alternately or radially branched, and attached by small discoid holdfasts. Axes are terete, gelatinous, and become firm with age. The medulla has loose longitudinal filaments; cortical filaments are dichotomously branched and pigmented. Branches are loose to compact and may be hair-tipped. Tetraspores are known for some species based on culture studies. Gametophytes are monoecious or dioecious. Spermatangia are in terminal clusters. Carpogonial filaments are 3(5)-celled and lateral on cortical filaments; carpogonia are conical. Gonimoblasts are branched and from the upper 1–2 cells of carpogonial filaments after fertilization; carpospores are terminal. Cystocarps have loosely arranged involucral filaments.

Helminthocladia calvadosii (J. V. Lamouroux _ex_ Duby) Setchell in Collins, Holden, and Setchell 1915 (1895–1919): ex. no. 2035 (L. _calvus_: hairless + _dosis_: endowed with). Plate XXVI, fig. 17, 18.

Basionym. *Dumontia calvadosi* Lamouroux *ex* Duby 1830: p. 941.

Plants are erect, bushy, gelatinous, 20–60 cm tall, 5–15 cm in diam., and attached by small conical holdfasts that expand into thick primary axes. Branches are terete, more slender than main axis, thick in the middle, tapering to tips, and branched alternately or radially 2–3 times. The medulla is multiaxial and has longitudinal filaments. The cortex has radially arranged pigmented filaments with pyriform tips; cells are 20–25 µm in diam. and each has a large radially lobed plastid with a pyrenoid.

Uncommon, on rubble in quiet areas (0.1 to 20.0 m) in the Keys; also known from Bermuda, Puerto Rico, the eastern North and tropical Atlantic, and India.

Taylor 1960: p. 324.

Liagora J.V. Lamouroux 1812: p. 185 (L. *liagora*: a sea nymph of the Sea God Nerius).

Abbott (1990a,b) reviewed the genus, while Brodie and Norris (1996) have described life histories and morphology of *Liagora ceranoides* Lamouroux, *L. pinnata* Harvey, and *L. valida* Harvey.

Plants are bushy, soft to firm, sometimes gelatinous and lightly calcified, and attached by discoid holdfasts. Axes are multiaxial; its medulla has compact longitudinal filaments that branch radially to form pigmented cortical filaments. Spermatangia are clustered and lateral or terminal on cortical filaments. Carpogonial branches are from middle cells of cortical filaments and usually 4-celled (or 3–6-celled). Cystocarps may be entangled with cortical filaments; gonimoblasts are slender and the outer cells produce carpospores.

Key to the species (3) of *Liagora*

1. Branch tips soft; >500 µm in diam.; older axes lightly to heavily calcified 2
1. Branches stiff, wiry, <500 µm in diam.; axes heavily calcified . . . . *L. albicans*
   2. Calcification heavy in older axes and light near tips; axes closely branched; medullary filaments 20–40 µm in diam. and cells 20–40 diam. long. . . . . *L. valida*
   2. Calcification moderate; branching widely dichotomous; medullary filaments 20–40 µm in diam. and cells to 5 diam. long. . . . . *L. ceranoides*

Liagora albicans J. V. Lamouroux 1816: p. 240, pl. 7, fig. 7 (L. _albicans_: whitish, off-white). Plate XXVI, fig. 19–21.

Synonym. *Liagora decussata* Montagne 1849: p. 64.

Plants are erect, stiff or wiry, to 10 cm tall, heavily calcified and whitish, except for pink tips, and attached by small padlike holdfasts. Branching is dichotomous or opposite; branches are tapered, to 500 µm in diam., and 0.7–1.5 cm long. Medullary filaments are longitudinal, cylindrical or swollen, 15–35 µm in diam., and with cells 3–15 diam. long. Cortical filaments are radial, cylindrical, 6–12 µm in diam., and dichotomously branched 3–5 times, with spherical to pyriform apical cells 7–15 µm in diam. and to 20 µm long. Spermatangia are 2–4 µm in diam, clustered, and at the tips of cortical filaments. Carpogonial branches are incurved, 2–5-celled, and 10–20 µm in diam. Carposporophytes appear as red dots, to 300 µm in diam., and are terminal on cortical filaments. Carpospores are pyriform and 5–6 µm in diam.

Uncommon, on limestone to 2 m at wave-exposed sites in the Keys; also known from the Bahamas and Antilles (type location: "Indes Orientales"; Silva et al., 1996), as well as the Canary and Cape Verde Islands and Indo-Pacific.

Taylor 1960: p. 330; Littler and Littler 2000: p. 50.

Liagora ceranoides J. V. Lamouroux 1816: p. 239 (Gr. *ceras*: horn + *oides*: similar to, resembling). Plate XXVII, fig. 1, 2.

Synonyms. *Liagora patens* P. Crouan and H. Crouan in Schramm and Mazé 1865: p. 14; P. Crouan and H. Crouan in Mazé and Schramm 1878 (1870–1877): p. 184.
Liagora prolifera P. Crouan and H. Crouan in Mazé and Schramm 1878 (1870–1877): p. 185.
Liagora opposita J. Agardh 1896: p. 101.
Liagora pilgeriana Zeh 1912: p. 272.

Abbott (1990a, 1999) recognized f. *ceranoides* J. V. Lamouroux and f. *leprosa* (J. Agardh) Yamada based on the presence or absence of short hairs, respectively (Yamada, 1938b). The forms differ morphologically, but are anatomically similar.

Plants are in compact tufts, soft, pink white, to 8 cm in diam., moderately calcified, and attached by discoid holdfasts. Branching is sparse below and abundantly dichotomous above. Branches are 1–2 mm in diam. and taper to tips; medullary filaments are longitudinal, 20–40 μm in diam., and with cells slightly swollen to 5 diam. long. Cortical filaments extend radially into the calcified mucus and branch at wide angles; cells are beadlike; tip cells are 5–12 μm in diam. Spermatangia are 1–3 μm in diam., on outer cortical cells, and clustered in platelike discs. Carpogonial branches are 3(5)-celled and 12–24 μm in diam. Cystocarps are spherical, 100–500 μm in diam., and have involucres.

Common, on rubble in shallow water and to 20 m in the Keys; also known from Bermuda, the Caribbean (type location: St. Thomas; Abbott, 1999), Gulf of Mexico, and Brazil, as well as the Mediterranean, eastern tropical Atlantic, Indo-Pacific, and Australia.

Taylor 1960: p. 326; Littler and Littler 2000: p. 50.

Liagora valida Harvey 1853: p. 138, pl. XXXI.A, fig. 1–5 (L. *valida*: well developed, strong, robust). Plate XXVII, fig. 3.

Synonyms. *Liagora annulata* J. Agardh 1876 (1848–1901): p. 518.
Liagora rosacea Zeh 1912: p. 271.

Plants are heavily calcified below, lightly calcified and soft above, to 10 cm tall, with reddish brown tips, dichotomously and closely branched, and attached by discoid holdfasts. Axes are to 1.3 mm in diam. and tips taper to 0.8 mm. The medulla has longitudinal filaments 20–40 μm in diam., slightly swollen cells 10–20 diam. long, and rhizoidal filaments 5–10 μm in diam. Cortical filaments are radial, branched 4–5 times, and to 16 μm in diam. Outer cells are beadlike to ovoid with one stellate plastid and a pyrenoid. Spermatangia are terminal and clustered on cortical filaments. Carpogonial branches are 3(5)-celled, straight to curved, and 10–20 μm in diam. Cystocarps appear as terminal red spots, are to 400 μm in diam., and have involucral filaments of ovoid cells.

Common, on shallow limestone in the Keys (type location: Sand Key); also known from Bermuda, the Caribbean, Gulf of Mexico, and Brazil, as well as the eastern tropical Atlantic, Indo-Pacific, and Australia.

Taylor 1960: p. 327; Littler and Littler 2000: p. 54.

Trichogloea Kützing 1847a: p. 53 (L. *tricho*: hairlike + *gloeo*: glue, sticky).

Plants are bushy, with radial and alternate branching, gelatinous to slimy, with the axial core calcified, and attached by discoid holdfasts. The medulla has loosely interwoven filaments. Cortical filaments are radial, dichotomously branched, and have large beadlike surface cells with a band-shaped plastid and pyrenoid. Gametophytes are dioecious or monoecious. Spermatangia are in whorls on tip cells. Carpogonial branches are 3(5)-celled, branches fuse after fertilization, and most gonimoblast cells form carposporangia.

Trichogloea requienii (Montagne) Kützing 1847a: p. 52–55 (L. *requies*: rest, repose; also *requium*: song for the dead, deathlike). Plate XXVII, fig. 4.

Basionym. *Batrachospermum requienii* Montagne 1843: p. 355–356.

Plants are bushy, to 15 cm tall, gelatinous to gooey, pink to white with calcification limited to the medulla, and attached by discoid holdfasts. Branching is irregular, alternate, or dichotomous to pinnate (to 3 times). Main axes are cylindrical, to 4 mm in diam., and have branches to 2 mm in diam. Medullary filaments are cylindrical, longitudinal, and 18–35 µm in diam. Cortical filaments are radial, dichotomously branched, and 12–25 µm in diam. Plants are usually dioecious. Spermatangia are 2–5 µm in diam. and whorled on subapical cortical filaments. Carpogonial branches are terminal and have 8 or more cells. Carposporophytes are 45–154 µm in diam.; carposporangia are ovoid and 5–9 by 10–12 µm.

Rare, on limestone (to 15 m) in the Keys; also known from the Bahamas, Antilles, and Gulf of Mexico, as well as Cape Verde Islands, Red Sea (type location; Abbott, 1999), Indo-Pacific, and Australia.

Taylor 1960: p. 323; Littler and Littler 2000: p. 56.

Trichogloeopsis Abbott and Doty 1960: p. 636 (L. *tricho*: hairlike + *gloeo*: sticky + *opsis*: similar to; thus similar to the genus *Trichogloea*).

Plants are erect, with spreading monopodial branching, lightly calcified, gelatinous, with multiaxial construction, and attached by discoid holdfasts. Medullary filaments are longitudinal and cylindrical, while cortical filaments are radial and pigmented. Gametophytes are usually dioecious. Spermatangia form crowded heads. Carpogonial branches are 3(6)-celled and either replace vegetative lateral branches or are on terminal cortical filaments. Cystocarps are compact, usually short, and not covered by involucral filaments, except for descending sterile filaments from gonimoblastic axes. Carposporangia are terminal on gonimoblastic filaments.

Trichogloeopsis pedicellata (M. Howe) Abbott and Doty 1960: p. 638, fig. 18–20 (L. *pedicellata*: slender-stalked, provided with a stalk). Plate XXVII, fig. 5, 6.

Basionym. *Liagora pedicellata* M. Howe 1920a: p. 556.

Plants are pyramidal, soft, flaccid, to 16 cm tall, pink to white, with calcified granules in the outer mucus, irregularly alternately branched, and attached by discoid holdfasts. Axes show monopodial growth; lower axes are 1.5–2.5 mm in diam. and usually denuded. Medullary filaments are longitudinal, cylindrical, 30–100 µm in diam., and mixed with rhizoids 8–13 µm in diam. Cortical branches are radial, dichotomously branched 4–7 times, and moniliform near tips. Apical cells are ovoid, 12–20 µm in diam., 15–26 µm long, and often have fine colorless hairs. Gametophytes are monoecious. Spermatangia are 13–20 µm in diam. and form small tufts on outer cortical cells. Carpogonial branches are erect, 3(6)-celled, and 20–25 µm in diam. Carposporophytes are on outer cortical cells and 90–215 µm in diam. Cystocarps lack an involucre, but have a few rhizoids.

Common, on limestone in shallow water of moderately exposed sites in the Keys; also known from the Bahamas (type location), Caribbean, Gulf of Mexico, and Canary Islands.

Taylor 1960: p. 329; Littler and Littler 2000: p. 54.

Galaxauraceae

See Wang and Chiang (2001), Wang et al. (2005), and Huisman et al. (2004a) regarding the inclusion of the genera *Actinotrichia, Dichotomaria, Galaxaura,* and *Tricleocarpa* in this family. The first genus has not been reported in Florida.

Plants are erect and bushy; branches flat or terete, usually dichotomous, soft to firm, lightly to moderately calcified, and attached by discoid holdfasts. The inner cortex is filamentous or pseudoparenchymatous and has multicellular, pigmented hairs that extend to the surface. Outer cortical cells are clear or pigmented, swollen, square or elongate in cross section, and form a parenchymatous-like surface. Pigmented surface cells have one stellate plastid. Asexual reproduction is by monospores or tetraspores. Life histories are isomorphic; gametophytes and tetrasporophytes either differ in cortical structure or have a filamentous tetrasporophyte (*Tricleocarpa*). Spermatangia are in cavities. Carpogonial branches are 3-celled, at forks of the inner cortex, and associated with nutritive cells. Cystocarps are embedded, ostiolate, and have pericarps of slender compact filaments or gonimoblastic filaments. Carpospores are in a series or at tips of gonimoblastic filaments.

Dichotomaria Lamarck 1816: p. 145, *gen. emend.* Wang et al. (2005): p. 688.

Heterotypic synonym: *Brachycladia* Sonder 1855: p. 514.

See Huisman et al. (2004a,b) for synonymy and reinstatement of the genus.

Plants are calcified, subdichotomously branched, with terete or flat axes, and attached by discoid holdfasts. Erect axes are multiaxial, pseudodichotomously branched, and may have distinct segments. The cortex of gametophytes is mostly 3-layered, with an inner 1–2 layers of large hyaline elongate cells that are closely packed and an outer layer of small, pigmented cells. The cortex of tetrasporophytes is pseudoparenchymatous and has 3–6 layers; an inner 1–3 layers of large hyaline cells and outer layer of pigmented cells on stalks. The medulla is filamentous. Spermatangia are in cavities. Carpogonial branches are 3-celled with a trichogyne and a hypogynous cell, they have 3–4 sterile branches and 4–5 involucral filaments. Cystocarps are dome-shaped, ostiolate, and sunken; carposporangia are ovoid and project into the cavity. Tetrasporangia are cruciately divided.

Key to the species (2) of *Dichotomaria*

1. Branches flat to subterete, usually with bands of pigmented multicellular hairs; tetrasporangia terminal on short filaments on outer cortical cells . . . .*D. marginata.*
1. Branches cylindrical, not covered with hairlike filaments; tetrasporangia on stalk cells cut off from superficial cortical cells . . . . *D. obtusata*

Dichotomaria marginata **(J. Ellis and Solander) Lamarck 1816: p. 145 (L. *marginata*: with a border, margin). Plate XXVII, fig. 7–9.**

Basionym. *Corallina marginata* J. Ellis and Solander 1786: p. 115, pl. 22, fig. 6.

Synonyms. *Galaxaura marginata* (J. Ellis and Solander) J. V. Lamouroux 1816: p. 264.

Galaxaura angustifrons Kjellman 1900: p. 82–83, pl. 15, fig. 11–25; pl. 20, fig. 27.

Galaxaura stupocaula Kjellman 1900: p. 75, pl. 14, fig. 1–9; pl. 20, fig. 28 (as *G. stupicaulon*).

Galaxaura veprecula Kjellman 1900: p. 80–81, pl. 16, fig. 17–33; pl. 20, fig. 20.

Galaxaura occidentalis Børgesen 1916: p. 109–113, fig. 118–123.

Wynne (2005a) listed *Galaxaura apiculata* Kjellman as a possible synonym (Papenfuss et al., 1982), while Guiry and Nic Dhonncha (2006) cited it as a distinct taxon. The gametophytic (*G. occidentalis*) and sporophytic (*G. marginata*) stages had been previously cited as distinct species.

Plants are rose white, to 14(20) cm tall; axes are smooth, subdichotomously branched, and attached by terete holdfasts. Branches are flat or terete, 1–5 mm wide, 300–500 μm thick, often lack hairs, and usually have transverse bands. Medullary filaments are 8–15 µm in diam. and occur in dense mucilage. The cortex is 2–3 cells thick, with its inner cells 25–65 µm in diam. and lightly pigmented. Tetrasporophyte surface cells are often apiculate and stalked, 22–40 µ m in diam., 30–60 µm long, and laterally free. Surface cells of gametophytes are oval to columnar, 10–16 µm in diam., and 25–45 µm long; subsurface cells are triangular, compressed, and laterally adherent. Cystocarps are spherical to flat and 500–600 µm in diam. Spermatangial branches occur in cavities 200–300 µm in diam.; spermatangia are 6–8 by 4–6 µm. Tetrasporangia are spherical to ovoid, cruciately divided, 35–55 by 30–45 µm, and in sori on flat surfaces of branches.

var. *dilatata* (Kützing) M. J. Wynne 2005a: p. 83 (L. *dilatata*: spread out).

Basionym. *Galaxaura dilatata* Kützing 1858: p. 17, pl. 37, fig. 1.
Synonym. *Galaxaura marginata* (J. Ellis and Solander) J. V. Lamouroux var. *dilatata* (Kützing) J. Agardh 1876 (1848–1901): p. 534.

Axes are broad, to 5 mm wide or more at forks, 7–15 mm long, and not constricted. Surface cells are polygonal, 30–46 µm in diam., 18–20 µm tall, and with thick lateral walls.

var. *linearis* (Kützing) M. J. Wynne 2005a: p. 84 (L. *linearis*: narrow).

Basionym. *Galaxaura linearis* Kützing 1858: p. 37.

Segments are slender and constricted, with tufts of filaments at constrictions.

Common, the species and varieties occur in tide pools and shallow waters in the Keys and to 55 m on both coasts; also known from Bermuda, the Bahamas (type location of species; Abbott, 1999), Caribbean, Gulf of Mexico, and Brazil, as well as West Africa, the Red Sea, Indo-Pacific, and Australia.

Taylor 1960: p. 343; Littler and Littler 2000: p. 58.

Dichotomaria obtusata (J. Ellis and Solander) Lamarck 1816: p. 145 (L. *obtusata*: obtuse, blunt, rounded). Plate XXVII, fig. 10–12.

Basionym. *Corallina obtusata* J. Ellis and Solander 1786: p. 113, pl. 22, fig. 2.
Synonyms. *Galaxaura obtusata* (J. Ellis and Solander) J. V. Lamouroux 1816: p. 262–263
Galaxaura umbellata (Esper) J. V. Lamouroux 1816: p. 262.
Galaxaura decaisnei J. Agardh 1876 (1848–1901): p. 526.
Galaxaura moniliformis Kjellman 1900: p. 83, pl. 17, fig. 15–30; pl. 20, fig. 54.

See Wynne (2005a) regarding confusion of generations.

Plants are erect, 7–10 cm tall, stiff, bushy, light pink to rose red, lightly calcified, and attached by small discoid holdfasts. Branching is mostly dichotomous, branches are not calcified, 1.5–4.0 mm in diam., cylindrical, smooth or with pigmented multicellular hairs, and with constricted and flexible forks. Medullary filaments are dichotomously branched, few, 10–20 µm in diam., and in dense mucilage. The cortex has 3 layers; inner cells are ovoid, 60–120 µm in diam., and colorless; outer cells are polyhedral in surface view, 20–45 µm in diam., and densely pigmented. Mid cortical cells of tetrasporophytes are stalklike, 8–15 µm in diam., and 15–25 µm long. Midcortical cells of gametophytes are swollen and 20–40 µm in diam. Sporophytes and gametophytes are isomorphic. Tetrasporangia are scattered. Spermatangial branches are in spherical cavities 270–400 µm in diam. Spermatia are 8–9 µm in diam. and have thick walls. Carpogonial branches are 3-celled and replace vegetative filaments. Cystocarps are immersed and 160–600 µm in diam.; carposporangia are ovoid and 15–28 µm in diam.

var. *major* (W. R. Taylor 1960) M. J. Wynne 2005a; p. 84 (L. *major*: greater).

Basionym. *Galaxaura obtusata* var. *major* W. R. Taylor 1960: p. 633, pl. 44, fig. 5.

Plants are loosely branched, to 17 cm tall, with branches 3–4 mm in diam., and sometimes constricted at forks. Older segments are 3–9 diam. long, lightly calcified below, and mostly noncalcified above.

Common, the species and variety grow on limestone in shallow water and to 53 m in the Keys; also known from North Carolina, Bermuda, the Caribbean (type location: Jamaica; Papenfuss et al., 1982), Gulf of Mexico, and Brazil, as well as the eastern tropical Atlantic, Red Sea, Indo-Pacific, and Australia.

Taylor 1960: p. 342; Dawes 1974: p. 119; Schneider and Searles 1991: p. 213; Littler and Littler 2000: p. 60.

Galaxaura J.V. Lamouroux 1812: p. 185; 1816: p. 259 (Gr. *gala*: milky white + *auros*: gold).

Howe (1917, 1918) noted that the cortex of sexual and tetrasporic plants differed, which has caused confusion (Papenfuss and Chiang, 1969) and resulted in the descriptions of many extra species (Kjellman, 1900; Kurihara et al., 2005).

Plants are wiry, bushy, soft to firm, usually with dichotomous branching, and attached by basal discs. Branches are smooth or with pigmented hairs and may appear segmented due to a lack of carbonate at forks. Medullary filaments are slender and interwoven. Cortical filaments are radial and radiate outward. Gametophytes have a 3-layered cortex, with the inner cells large and colorless and the outer two layers pigmented, progressively smaller, and often fused. Tetrasporophytes have a cortex of photosynthetic branches or 1–3 inner layers of large colorless cells and one layer of outer cells on small pigmented supporting cells with carbonate in the gaps. Tetrasporangia are cruciately divided and stalked or sessile. Spermatangia are in conceptacles. Carpogonial branches are usually 3-celled, they replace vegetative filaments, and are associated with nutritive cells. Gonimoblastic filaments are lateral and form the cystocarp wall. Pericarps are ostiolate and immersed. Carposporangia grow into the cavity from all sides of the cystocarp wall.

Galaxaura rugosa (J. Ellis and Solander) J. V. Lamouroux 1816: p. 263 (L. *rugosa*: wrinkled, rugose). Plate XXVII, fig. 13–16.

Basionym. *Corallina rugosa* J. Ellis and Solander 1786: p. 115, pl. 22, fig. 3.
Synonyms. *Galaxaura lapidescens* (J. Ellis and Solander) J. V. Lamouroux 1816: p. 264.
Galaxaura lichenoides (J. Ellis and Solander) J. V. Lamouroux 1816: p. 265.
?*Galaxaura plicata* Kützing 1858: p. 15, pl. 33, fig. II.
Galaxaura tomentosa Kützing 1858: p. 18, pl. 38, fig. II.
Galaxaura delabida Kjellman 1900: p. 49.
Galaxaura flagelliformis Kjellman 1900: p. 47, pl. 3, fig. 2–11; pl. 20, fig. 16.
Galaxaura glabriuscula Kjellman 1900: p. 56–57, pl. 7, fig. 1, 2; pl. 20, fig. 26.
Galaxaura intricata Kjellman 1900: p. 57.
Galaxaura liebmannii (Areschoug) Kjellman 1900: p. 46, pl. 1; pl. 20, fig. 41.
Galaxaura ramulosa Kjellman 1900: p 5, pl. 3, fig. 24, 25; pl. 4, fig. 1–3; pl. 20, fig. 18.
Galaxaura rudis Kjellman 1900: p. 43–44, pl. 2, fig. 1–9; pl. 20, fig. 11.
Galaxaura squalida Kjellman 1900: p. 55–56, pl. 6, fig. 1–12; pl. 20, fig. 9.
Galaxaura subverticillata Kjellman 1900: p. 48–49, pl. 3, fig. 12–14; pl. 20, fig. 17.
Galaxaura subfruticulosa R. Chou in W. R. Taylor 1955: p. 140–141.

See Huisman et al. (2004b) and Wynne (2005a) regarding morphological differences between reproductive stages.

Gametophytes are bushy, stiff, compact, to 12 cm tall, loosely dome-shaped, cream red to rose red, and attached by discoid holdfasts. Branches are short, crowded, cylindrical, 1–3 mm in diam., dichotomously branched, and calcified, except for flexible joints that are not calcified. Medullary filaments are few, 8–20 µm in diam., and in a dense mucilage. The cortex is pseudoparenchymatous and has 3–4 layers; its surface is smooth or with a few hairs; inner cortical cells are 30–60 µm in diam., colorless, and often fused laterally; surface cells are 18–30 µm in diam., fused, and polyhedral in surface view. Surface hairs are 10–18 µm in diam., to 1 mm long, and on lower axes. Carpogonial branches are 3-celled. Cystocarps are round or flat, 400–900 µm in diam., with ovoid carposporangia, and 55–65 µm in diam. Spermatangia occur in cavities 200–320 µm in diam. Spermatia are ovoid and 6–9 µm in diam.

Sporophytes are bushy, stiff, compact, dome-shaped, creamy red, 4–12 cm tall, and attached by discoid holdfasts. Branches are short, crowded, cylindrical, 1–3 mm in diam., and slightly calcified. The cortex is filamentous and has a dense covering of stiff, pigmented, hairlike filaments in whorls or in alternating short and long bands. The short bands are 100–200 µm long and 3-celled, while the long bands are 300–900 µm long, to 10 cells, 15 µm in diam. below, and to 20 µm in diam. at tips. Both types of filaments occur on a bulbous basal cell, which is 30–55 µm in diam. and lightly pigmented. Medullary filaments are colorless and 7–28 µm in diam. Tetrasporangia are cruciately divided, ovoid to spherical, 27–33 µm in diam., and terminal on filaments.

Common, on limestone in shallow water to 18 m in the Keys; also known from Bermuda, the Bahamas, Antilles (type location: Jamaica; Abbott, 1999), Gulf of Mexico, and Brazil, as well as the eastern tropical Atlantic, Red Sea, Japan, and Indo-Pacific.

Taylor 1960: p. 337–340; Littler and Littler 2000: p. 60–62.

Tricleocarpa Huisman and Borowitzka 1990: p. 164 (L. *tricleo*: a bower + *carpa*: fruit; refers to sterile hairs that mix with gonimoblastic filaments).

Gametophytes are calcified, dichotomously branched, and attached by discoid holdfasts. The medulla has longitudinal filaments, while the cortex has 3–4 cell layers, with the inner ones inflated and grading into outer small pigmented ones that lack secondary fusions. Plants are monoecious or dioecious. Spermatangia occur in spherical cavities. Carpogonial branches are 3-celled. Pericarps arise from basal cells of carpogonial branches before the formation of gonimoblasts; vegetative filaments project into cystocarp cavities and mix with gonimoblastic ones. Tetrasporophytes are known only in culture. *Tricleocarpa* is distinguished from *Galaxaura* by: (1) a pericarp of sterile filaments that develop from the basal cell of the carpogonial branch and not from gonimoblastic filaments; (2) sterile hairs or paraphyses of the pericarp project into the cavity and mix with gonimoblastic ones; (3) heteromorphic and filamentous tetrasporophytes (in culture) have terminal or lateral tetrasporangia; and (4) no lateral fusions occur between cortical cells on the gametophyte.

Key to the species (2) of *Tricleocarpa*

1. Branches 0.4–1.0 mm in diam., constricted at forks; tips not tapered . . . . *T. cylindrica*
1. Branches 1–2 mm in diam., little constricted at forks; tips tapered . . . . *T. fragilis*

Tricleocarpa cylindrica (J. Ellis and Solander) Huisman and Borowitzka 1990: p. 164–168, fig. 40–45, 50–52 (L. *cylindrica*: cylindrical). Plate XXVII, fig. 17, 18.

Basionym. *Corallina cylindrica* J. Ellis and Solander 1786: p. 114, pl. 22, fig. 4.
Synonyms. *Galaxaura cylindrica* (J. Ellis and Solander) J. V. Lamouroux 1821: p. 22, pl. 22, fig. 4.
Galaxaura fastigiata Decaisne 1842b: p. 116.

Gametophytes are bushy, to 10 cm tall, heavily calcified except at noncalcified joints, and attached by small discoid holdfasts. Branching is dichotomous, with internodes 0.3–2.5 cm long; axes are cylindrical, strongly constricted at forks, and 0.5–1.5 mm in diam. Medullary filaments are few, longitudinal, 3–15 μm in diam, and with dichotomous branches that radiate outward to form the cortex. The cortex has 2–4 layers of inflated cells that lack secondary fusions. Inner cortical cells are 18–75 μm in diam. and longer than broad; subsurface cells are ovoid and to 30 μm in diam.; surface cells are 10–20 μm in diam., 5–6-sided in surface view, and usually less pigmented than supporting cells. Cultured filamentous tetrasporophytes have cells about 25 μm in diam. Gametophytes are mostly monoecious. Pericarp walls have gonimoblastic filaments and inward growing sterile hairs. Cystocarps are 200–300 μm in diam. Spermatangia are ovoid, 6–8 μm long, 4–6 μm in diam., and on special filaments in cavities, 250–300 μm in diam.

Common, on limestone in protected sites and to 15 m in the Keys; also known from Bermuda, the Caribbean (type location: West Indies; Womersley, 1994), and Gulf of Mexico, as well as the eastern tropical Atlantic, Indo-Pacific, and Australia.

Taylor 1960: p. 341; Littler and Littler 2000: p. 64.

Tricleocarpa fragilis (Linnaeus) Huisman and Townsend 1993: p. 100, tab. 2 (L. *fragilis*: fragile, brittle). Plate XXVII, fig. 19–21.

Basionym. *Eschara fragilis* Linnaeus 1758: p. 805.
Synonyms. *Galaxaura oblongata* (J. Ellis and Solander) J. V. Lamouroux 1816: p. 262.
Galaxaura fragilis (Lamarck) J. V. Lamouroux *ex* Decaisne 1842b: p. 116.
Tricleocarpa oblongata (J. Ellis and Solander) Huisman and Borowitzka 1990: p. 168–169, fig. 46–49, 53–56.

Gametophytes are to 15 cm tall, lightly calcified, dichotomously branched, and attached by small discoid holdfasts. Axes have internodes 0.6–1.1 cm, with segments tapering to tips; axes are cylindrical, weakly constricted at forks, and 1.5–2 mm in diam. Medullary filaments are longitudinal, 10–20 μm in diam., in a dense mucilage, and dichotomously branched. The cortex has radiating dichotomously branched filaments that lack secondary fusions and have 3–4 layers; inner cells are inflated, colorless, and 35–60 μm in diam.; subsurface ones are ovoid and 25–40 μm in diam.; surface cells are pigmented, compressed, and 8–17 μm in diam. Gametophytes are monoecious or dioecious. Spermatangia are in cavities, 300–350 μm in diam., and have spermatia 6–8 by 4–6 μm. Cystocarpic cavities are 400–500 μm in diam. with gonimoblast filaments at their bases mixed with hairs 3–4 cells long. Tetrasporophytes are unknown.

Common, on limestone in shallow protected sites and to 30 m in the Keys; also known from Bermuda, the Bahamas, Caribbean (type location: Jamaica; Huisman and Townsend, 1993), Gulf of Mexico, and Brazil, as well as the Adriatic Sea, eastern tropical Atlantic, Indo-Pacific, and Australia.

Taylor 1960: p. 341; Littler and Littler 2000: p. 66.

Scinaiaceae

Huisman et al. (2004a) separated the genera from the Galaxauraceae. The family, which includes *Scinaia*, *Nothogenia*, and probably *Whidbeyella* and *Gloiophloea* (Huisman, 1985, 1987), may be an early divergence in the Nemaliales, and is related to the Liagoraceae.

Plants are noncalcified and erect, axes terete to slightly compressed, branching dichotomous to subdichotomous or rarely irregular, and attached by discoid holdfasts. Axes are multiaxial with longitudinal medullary filaments; the cortex is filamentous with anticlinal, subdichotomously branched assimilators or it is pseudoparenchymatous. Spermatangia occur in surface nemathecia or pits. Carpogonial branches are on inner cortical filaments, 3-celled, and their hypogynous cell bears nutritive cells. Gonimoblast filaments develop directly from carpogonia. Cystocarps occur in the outer medulla, do not project, and contain branched gonimoblast filaments and terminal carposporangia. Pericarps are ostiolate, distinct, and arise from filaments on basal cells of the carpogonial branch. Tetrasporophytes are heteromorphic (where known), reduced, filamentous (*Scinaia*, *Gloiophloea)*, or crustose (*Nothogenia*); tetrasporangia are cruciately divided.

Scinaia Bivona-Bernardi 1822: p. 232 (named for Domenico Scina of Palermo, an Italian naturalist).

Axes are bushy, soft, gelatinous or firm, terete to flat, repeatedly dichotomously branched, lack constrictions at the forks, and are attached by small discoid holdfasts. Medullary filaments are few, colorless, and longitudinally oriented in a clear gel. The cortex has 3 regions: an inner area with a few dichotomously branched filaments, a middle region with 2–3 layers of pigmented subcortical cells, and surface cells. The surface cells are bulbous, polyhedral in surface view, form a smooth pseudoparenchymatous cover and may have small regularly spaced pigmented cells. Monosporangia are spherical and scattered between surface cells. Gametophytes are monoecious or dioecious. Spermatangia are spherical and in small sori or scattered between epidermal cells. Carpogonial branches are 3-celled, in the inner cortex, and with scattered nutritive cells. Cystocarps have a thin-walled pericarp and an obscure pore. Carpospores are in a series at the tips of gonimoblasts. Tetrasporophytes, if known, are small branched uniseriate filaments similar to those of *Acrochaetium*.

Key to the species (4) of *Scinaia*

1. Branches cylindrical, compressed, or slightly flattened; edges not enrolled 2
1. Branches flattened; edges enrolled to form raised margins. . . . . *S. incrassata.*
   2. Axes 1–6 mm in diam.; branches not constricted at forks 3
   2. Axes 0.5–2.0 mm in diam., branches constricted at forks . . . . *S. caribaea*
3. Axes weakly constricted at forks, 1–6 mm in diam. . . . . *S. complanata*
3. Axes not constricted at forks, 3–4 mm in diam. . . . . *S. halliae*

***Scinaia caribaea* (W. R. Taylor) Huisman 1985: p. 417 (L. *caribaea*: of or from the Caribbean). Plate XXVII, fig. 22.**

Basionym. *Gloiophloea caribaea* W. R. Taylor 1943 ('1942'): p. 148, pl. 1, fig. 2.

Synonym. *Pseudogloiophloea caribaea* (W. R. Taylor) Joly et al. 1965: p. 81.

Plants are bushy, soft, gelatinous and slippery, 3–4 cm tall, pale reddish brown to clear, and attached by conical holdfasts. Branching is dichotomous, 7–13 times; upper branches are constricted below forks, cylindrical to slightly flat, smooth, and with segments 0.5–1.8 mm in diam. by 3–4 diam. long. The medulla has a loose bundle of 30–50 filaments in clear mucilage; these are thick walled, 6–14 µm in diam., and mixed with slender dichotomously branched filaments, 2–6 µm in diam. The cortex has 2 layers; inner cells are ovoid, to 10 µm in diam., and pigmented, while the surface cells are clear, polyhedral in surface view, to 30 µm in diam., and with slender pigmented cells between the larger surface cells. Gametophytes are monoecious. Spermatangia are spherical and 2.5–7.5 µm in diam. Cystocarps are spherical, 145–175 µm in diam., and in cortical cavities. The pericarp has 5–8 layers of filaments.

Rare, on rubble in shallow water and to 20 m in the Dry Tortugas; also known from the Caribbean (type location: Haiti; Taylor, 1942, 1943) and Canary Islands.

Taylor 1960: p. 332; Littler and littler 2000: p. 62.

Scinaia complanata (Collins) Cotton 1907: p. 260 (L. *complanata*: leveled, flat in one plane). Plate XXVIII, fig. 1, 2.

Basionym. *Scinaia furcellata* (Turner) J. Agardh f. *complanata* Collins in Collins, Holden and Setchell 1901 (1895–1919): fasc. 17, ex. no. 836; Collins 1906a, p. 110.

Plants are bushy, soft and gooey, pale to rose red, to 8 cm tall, with dichotomous branching 7–9 times, and attached by small discoid holdfasts. Branches are cylindrical to compressed, 1–6 mm in diam., and 4–7 diam. long; forks rarely constricted, tips obtuse, and branchlets fusiform. The medulla has 6–8(30) filaments in a clear gel that are to 4 µm in diam. The cortex has 2 layers; subsurface cells are subspherical to pyriform and densely pigmented; surface cells are colorless, polygonal in surface view, 12–44 µm in diam., rectangular to elongate in section, and mixed with small pigmented cells. Gametophytes are monoecious. Spermatangia are spherical, 2.5–7.5 µm in diam., and solitary or in patches between surface cells. Cystocarps are spherical to oval, 145–350 µm in diam., and in the cortex. Pericarps have 2–3 filamentous layers. Carposporangia are ovoid to ellipsoid, 4–5 by 7.5–12.5 µm, and in a series of 2–3. Tetrasporophytes are known only from culture.

Common, on limestone in 2–10 m on both coasts (type location: Indian River Inlet) and 55 m in the Keys; also known from North Carolina (17–21 m) southward, Bermuda, the Caribbean, Gulf of Mexico, and Brazil, as well as the Mediterranean, eastern tropical Atlantic, and Indo-Pacific.

Taylor 1960: p. 334; Dawes 1974: p. 119; Schneider and Searles 1991: p. 215; Littler and Littler 2000: p. 64.

Scinaia halliae (Setchell) Huisman 1985: p. 417 (named for Mrs. G. A. Hall, an American collector who sent Florida seaweeds to W. A. Setchell). Plate XXVIII, fig. 3, 4.

Basionym. *Gloiophloea halliae* Setchell 1914: p. 116, fig. 13.

Synonym. *Pseudogloiophloea halliae* (Setchell) Joly et al. 1965: p. 81.

Plants are 11–13 cm tall, branched dichotomously 5–7 times, and attached by discoid bases. Branches are cylindrical, 3–4 mm in diam., and not constricted at forks. The medulla has loosely arranged longitudinal filaments in a gel and radial branches that form the cortex. The outer cortical layer is 22–40 µm thick and has anticlinal pigmented branches alternating with globose, clear, utricle-like cells. Gametophytes are monoecious. Cystocarps are pyriform, 65–80 µm in diam., and 100–120 µm long. Pericarps are ostiolate with 5–6 layers of filaments.

Rare, on limestone in shallow water in the Keys; also known from Venezuela.

Taylor 1960: p. 332 (as *Gloiophloea halliae*).

Scinaia incrassata Eiseman 1979: p. 356–359; fig. 2–6 (L. _incrassata_: thickened). Plate XXVIII, fig. 5.

Plants are rose pink and to 30 cm tall; axes are dichotomously branched, flat, with inflated margins and apices, 0.1–0.2 mm thick in center and 0.4–0.6 mm at margins, and attached by discoid holdfasts. Growth is multiaxial; apical cells are sunken in branch tips and coarse filaments that extend toward the inflated margins. Medullary filaments are loosely arranged, in a gel, and 2.5 µm in diam. The inner cortex has one layer of pigmented spherical cells 7–15 µm in diam. Surface cells are in a continuous layer, colorless, angular in surface view, 17–26 µm long, and mixed with small pigmented cells. Gametophytes are monoecious. Cystocarps are pyriform and project from inflated margins. Spermatangia occur in sori 50–75 µm in diam. Spermatia are 2–4 by 4–8 µm.

Rare, from 20–46 m off the east coast (holotype location: 27°11.95'N 80°0.12'W in 37.5 m) between April and June.

RHODOGORGONALES

Plants lack defined meristems and a distinct cortex and their medulla has interconnected photosynthetic filaments that form rhizoidal filaments with thin-walled cells coated with calcite (calciferous cells). Carpogonial branches are 2-celled and auxiliary cells and ooblast filaments are absent. Gonimoblastic filaments arise from fertilized carpogonia and become diffusely branched. Terminal cells of rhizoidal filaments fuse with intercalary gonimoblastic filaments.

Rhodogorgonaceae

The family contains two genera, *Renouxia* and *Rhodogorgon* (Fredericq and Norris, 1995). Plants are slippery, loose or compact, and have distinct photosynthetic and rhizoidal filaments. Photosynthetic branches are indeterminate and pseudodichotomous or determinate and pseudotrichotomous. Apical cells are pigmented or clear and dome-shaped or swollen. Calciferous cells are sessile or stalked and have calcite husks. Spermatangial mother cells occur in small subterminal clusters.

Rhodogorgon J. N. Norris and Bucher 1989: p. 1053, 1055 (Gr. *Gorgonia*: a genus of soft branching anthozoan corals that the plant resembles).

Plants are multiaxial, calcified, densely branched, and attached by irregular discoid holdfasts. The medulla has a central filament; each axial cell can produce one or more cortical filaments that form fascicles. Calciferous cells are scattered in an organic matrix, with calcium carbonate accretions often fused together. Tip cells of cortical fascicles are bulbous, embedded in a firm gel-like substance, and form a smooth outer surface.

Rhodogorgon ramosissima J. N. Norris and Bucher 1989: p. 1059, fig. 6–7 (L. _ramosissima_: much branched or densely branched). Plate XXVIII, fig. 6–8.

Synonym. *Rhodogorgon carriebowensis* J. N. Norris and Bucher 1989: p. 1057, fig. 1–5, 8–31.
Based on studies by Ogden (1992), only one species is recognized.

Plants are erect, bushy, to 50 cm tall, fleshy to leathery, whitish pink, and attached by small disclike holdfasts. Primary axes are irregularly to alternately branched, radial, distichous or secund above and denuded below. Medullary filaments are mixed with cells to 4 µm in diam. and 90 µm long. Cortical filaments are distinct, 70–105 µm in diam., radially branched, cylindrical, and pigmented;

cells are to 6 μm in diam. and 18 μm long. Calciferous cells are in surface patches and 55–80 μm long. Tetrasporangia are unknown and gametophytes dioecious. Spermatangia are 2–3 by 3–6 μm and in refractive clusters on terminal cortical filaments. Procarps are sessile; carpogonial branches replace vegetative cortical filaments. Carposporangia are 12 by 35 μm. Monosporangia, which occur on both male and female plants, are terminal and to 40 μm in diam.

Rare, resembling octocorals and in 9 m at the Dry Tortugas (Ballantine, 1996); also known from the Bahamas and Caribbean (to 12 m; type location: Carlisle Bay, Antigua), as well as Papua New Guinea and Fiji.

Littler and Littler 2000: p. 70.

AHNFELTIOPHYCIDAE

The subclass has plants with terminal carpogonia and carposporophytes that grow outward (Saunders and Hommersand, 2004). Pit plugs lack caps or membranes.

PIHIELLALES

Based on morphological and molecular studies, Huisman et al. (2003) created the order, family, and species that is endophytic in Liagoraceae species (*Ganonema farinosum, Liagora ceranoides, L. valida,* and *Trichogloeopsis* sp.). The plants had been described as autonomous (Børgesen, 1920), propagules, and "monosporangial discs" of the host plant (Kützing, 1858; Howe, 1920b; Yamada, 1938a; Abbott, 1945; Huisman, 2002).

Plants reproduce by 1-celled (possibly) monosporangia or zygotosporangia, or both that are undivided but have a basal stalk cell. Carposporophytes and tetrasporangia are absent along with secondary pit connections and cell fusions.

Pihiellaceae

The family is monotypic with the same characteristics as the order.

Pihiella Huisman, Sherwood, and Abbott 2003: p. 981. (L. *pihi*: derived from the Hawaiian word meaning button; also Gr. *phil*: loving, fond).

Plants are endophytic or epiphytic, minute, discoid to spherical, and attached by filaments or rhizoids. The thallus is pseudoparenchymatous and cells lack secondary pit connections. Surface cells have elongate basal hairs that are connected by pit connections. Sporangia are undivided and basally stalked. Spermatangia occur on the surface of mother cells. Carpogonia are sessile, produced from subsurface cells, and have an elongate trichogyne.

Pihiella liagoraciphila Huisman, Sherwood, and Abbott 2003: p. 981–987, fig. 1–4 (L. _liagorac_: refers to host species in the Liagoraceae; also _Liagora_: a Nereid, a sea nymph of the Sea God Nerius). Plate XXVIII, fig. 9–11.

Plants are minute, endophytic to epiphytic, to 400 μm in diam., 150 μm thick, and attached by filaments and rhizoids. Sporangia are elliptical, with walls to 5 μm thick and spores 20–25 by 15–20 μm in diam. Spermatangia are elliptical and 5 by 3 μm in diam.

Common(?), but not reported for Florida; probably epiphytic or endophytic in or on *Ganonema farinosum, Liagora ceranoides* and *L. valida*; also known from Puerto Rico, the Lesser Antilles, Indo-Pacific (type location: Mokulëa, Hawaii), and Australia.

RHODYMENIOPHYCIDAE

The subclass contains most red algal taxa and they have triphasic life histories. Carposporophytes develop directly from the carpogonium or carpogonial fusion cell, or indirectly from an auxiliary cell that receives a post-fertilization diploid nucleus. Pit plugs have membranes and lack caps.

BONNEMAISONIALES

Species have heteromorphic life histories with uninucleate cells. Gametophytes are macroscopic, uniaxial, and pseudoparenchymatous. Sporophytes are smaller, erect or creeping, filamentous, and with tetrahedrally divided tetrasporangia. Spermatangia occur in clusters that spiral around the axial row. Carpogonial branches are 2(4)-celled, from basal lateral axes, and have hypogynous cells with nutritive cells. Gonimoblasts grow directly from fertilized carpogonia or hypogynous cells. Carposporangia are solitary, terminal, and yield one carpospore.

Bonnemaisoniaceae

Gametophytes are densely and bilaterally branched, soft, gelatinous, and attached by discoid holdfasts or prostrate axes with clusters of rhizoids. Growth is by one apical cell; the axial cell row is covered by a pseudoparenchymatous cortex. Tetrasporophytes, if known, are uniseriate or polysiphonous-like branched filaments, or have creeping-branched crusts. After fertilization, the carpogonia fuse with branch cells and gonimoblasts to form a superficial cystocarp. The pericarp is formed from adjacent tissue.

Asparagopsis Montagne in Barker-Webb and Berthelot 1841 (1839–1841): p. XIV (Gr. *asparagos*: a sprout + *opsis*: similar to).

Synonyms. *Falkenbergia* F. Schmitz in Engler and Prantl 1897 (1890–1897): p. 479.

Gametophytes (*Asparagopsis*) are soft, bushy, alternately to radially branched, and with prostrate axes attached by rhizoidal clusters. Erect axes are terete and have recurved or hooked branches and vesicle cells. Plants are dioecious. Spermatangia are in dense sori on the surface of short, thickened branches. Carpogonial branches are 2(4)-celled and on short branches. Cystocarps occur on short stalks, are ovoid, and produce carpospores.

Sporophytes (*Falkenbergia*) are filamentous with 3 pericentral cells; erect axes are ecorticate and irregularly and frequently branched; prostrate axes are attached by multicellular rhizoids. Tetrasporangia are tetrahedrally divided, sessile, and arise from pericentral cells.

Asparagopsis taxiformis (Delile) Trevisan de Saint-Léon 1845: p. 45 (L. _taxi_: the genus _Taxus_, a yew + _formis_: form; thus, branching in one plane like a yew). Plate XXVIII, fig. 12–15.

Gametophytic stage:
Basionym. *Fucus taxiformis* Delile 1813 (1813–1826): p. 295–296, pl. 57, fig. 2.
Synonym. *Falkenbergia hillebrandii* (Bornet) Falkenberg 1901: p. 689–690.

Dixon and Irvine (1977) and Lawson and John (1987) questioned separating *Asparagopsis taxiformis* and *A. armata* Harvey. However, molecular and morphological studies confirm that they are distinct (Andreakis et al., 2004), with the gametophyte of the former being paler and having more spiny branches, while the latter is more densely branched. According to Chualáin et al. (2004) there are two clades of *A. taxiformis* that may be recognized as species or subspecies (Tittley and Neto,

2005). In culture, *A. taxiformis* alternates with the sporophytic *Falkenbergia* stage (Chihara, 1961; Magne, 1964; Rojas et al., 1982).

Gametophytes are delicate, soft, gray to deep red, erect, 5–20 cm tall, and attached by prostrate axes with clusters of rhizoids. Branches are 2–5 cm long, lax, terete, to 2 mm in diam., and with dense alternate branching to several orders. Axes are pyramidal above with lower parts naked. Axial filaments are to 45 µm in diam. and show 3 pericentral cells near tips. The cortex is 3–6 cells thick; subsurface cells are rounded and to 60 µm in diam.; surface cells are to 15 µm in diam. Branchlets are long, three cells in diam., and ecorticate. Cystocarps are 400–550 µm in diam. and on short, swollen branches.

Sporophytes are prostrate or erect filamentous tufts, pink to red, to 2 cm tall, irregularly branched, and attached by discoid holdfasts in entangled masses. Axes are ecorticate 50–80 µm in diam.; axial cells are covered by three pericentral cells arranged in a 60° spiral; segments are 0.5–1.2 diam. long. Tetrasporangia are solitary on branches.

Common, gametophytes occur in shallow water on limestone and adrift at protected sites in the Keys; sporophytes occur on algae to 9 m on all coasts; also known from North Carolina (17–35 m), through the Caribbean, Bermuda, Gulf of Mexico, and Brazil, as well as the Mediterranean (type location: Alexandria, Egypt; Abbott, 1999), eastern tropical Atlantic, Red Sea, Indo-Pacific, and Australia.

Taylor 1960: p. 348, 571; Dawes 1974: p. 157; Schneider and Searles 1991: p. 220; Littler and Littler 2000: p. 66, 68.

Naccariaceae

Abbott (1999) placed the family in the Gigartinales rather than the Nemaliales. Schils et al. (2003) suggested that the family was not monophyletic (cf. Huisman et al., 2004a).

Plants are gelatinous, with apical growth, irregular to radial branching, and attached by discoid holdfasts. Axes are uniseriate and corticated by descending filaments of short lateral deciduous branches. Carpogonial branches are 2(3)-celled; carpogonia fuse with the hypogynous cell after fertilization and gonimoblasts arise from carpogonia in swollen cystocarps that lack a pericarp. Carpospores are terminal. Tetrasporophytes (if known) are small, branched, uniseriate filaments and *Acrochaetium*-like.

Naccaria Endlicher 1836 (1836–1840): p. 6 (named for Aloysio Fortunato Naccari, an Italian botanist).

Plants are bushy and erect; axes are terete, gelatinous, tapered, and attached by discoid holdfasts. Apical cells divide by oblique divisions to form a central axial filament that produces short, lateral, determinate, and deciduous photosynthetic filaments. Basal cells of lateral filaments produce rhizoidal down-growths to form a pseudoparenchymatous cortex that may separate from the central axes. Gametophytes are monoecious or dioecious. Spermatangia are clustered at bases of lateral axes and spiral around the axes. Carpogonial branches are 2(4)-celled. Cystocarps are often subterminal and form distinct swellings.

***Naccaria corymbosa* J. Agardh 1899: p. 109 (L. *corymbosa*: flower-clustered, corymbose). Plate XXVIII, fig. 16.**

Plants are bushy, to 7 cm tall, abundantly and alternately branched, soft to firm below, appearing nodulose due to large axial cells, and attached by discoid holdfasts. Axial cells are 4–6 µm wide, 95–200 µm long, and covered by large pericentral cells, 80–100 µm in diam. Main axes and large branches are covered by lateral branchlets that often are deciduous; branchlets are 27–40(150) µm long and have moniliform, swollen, ovoid, or cask-shaped cells to 13 µm in diam. Cystocarps occur in the central branch region and covered by ecorticate branches constricted at the base and 120–140 µm in diam. Carposporangia are terminal, ovoid, 10–13 µm in diam., and 17–20 µm long.

Rare, adrift at Key West; also known from North Carolina (24–30 m), Bermuda, and the Salvage Islands.

CERAMIALES

Plants in this order are filamentous, foliose, fleshy, and erect or prostrate. Growth is by a single apical cell and is monopodial or sympodial; morphology is radial, bilateral, or dorsiventral; axes are uniseriate, corticated, or polysiphonous. Cells have 1 to many nuclei and plastids lack pyrenoids. Asexual reproduction includes monospores, tetraspores, polyspores, paraspores, vegetative fragmentation, and distinct propagules. Life histories are isomorphic. Tetrasporangia are covered by cortical cells and in specialized branches (stichidia) that are solitary or clustered in sori. Gametophytes are monoecious or dioecious. Spermatangia occur in branched clusters. Carpogonial branches are 2(4)-celled and auxiliary cells are produced from support cells after fertilization. Carposporophytes arise from auxiliary cells or fusion cell complexes. Cystocarps have or lack pericarps or involucral filaments and each carposporangia produces one carpospore.

Ceramiaceae

A subfamily and multiple tribes have been suggested for the family (Choi et al., 2005).

Plants are mostly erect, bushy, and filamentous; a few species have netlike blades. Axes are uniaxial and ecorticate or covered by rhizoids or cortical cells that are smaller than axial filaments. Growth is by an apical cell that may be large. Tetrasporangia are tetrahedrally divided, sessile or stalked, and solitary or in clusters on uniseriate axes; if present, polysporangia contain 8–64 spores. Spermatangia are in hyaline clusters on short fertile branchlets or in cortical patches. Carpogonial branches are 2–4-celled, on support cells, and with or without sterile cells. Auxiliary cells (1–2) are formed after fertilization. Ooblasts often join fertilized carpogonia and auxiliary cells to form large fusion cells that produce gonimoblast cells and carposporangia. Cystocarps are naked or partially enclosed by involucral filaments.

Aglaothamnion Feldman-Mazoyer 1941 ('1940'): p. 451 (Gr. *aglaia*: splendor, beauty; also: *Algaia*: one of three mythological Greek graces + *thamnion*: a small bush).

Plants are erect with uniseriate axes and uninucleate cells. Branching is usually in a spiral or radial sequence, and rarely alternately distichous. Prostrate axes are attached by rhizoids. Tetrasporangia are tetrahedrally divided, adaxial on branches, and one per cell. Spermatangial clusters are formed on 1(3)-celled fertile axes and occur adaxially on lateral branches. Procarps are lateral on intercalary cells. Gonimoblastic lobes are triangular and project on both sides of axes.

Key to the species (4) of *Aglaothamnion* (modified from Taylor, 1960, as *Callithamnion*).

1. Axes uncorticated or with a few rhizoidal filaments that do not cover the axial cell 2
1. Axes mostly heavily corticated throughout . . . . *A. herveyi.*

2. Axes completely uncorticated. 3
2. Older axes have a few rhizoidal filaments that descend from branch bases and form secondary holdfasts but do not cover axial cells . . . . *A. cordatum*
3. Distal branches of axes always in one plane; branchlets 33–37 µm in diam. near bases and taper sharply to acute tip cells . . . . *A. uruguayense*
3. Distal branches of axes mostly in one plane; branchlets 10–20 µm in diam. and tapering gradually to obtuse apical cells . . . . *A. halliae*

Aglaothamnion cordatum (Børgesen) Feldmann-Mazoyer 1941 ('1940'): p. 459 (L. *cordatum*: heart-shaped). Plate XXVIII, fig. 17, 18.

Basionym. *Callithamnion cordatum* Børgesen 1909b: p. 10–11, fig. 5, 6; 1917: p. 216–218, fig. 201–204.

Synonyms. *Aglaothamnion neglectum* Feldmann-Mazoyer 1941 ('1940'): p. 459–464, fig. 181–183.

Callithamnion neglectum (Feldmann-Mazoyer) M. J. Wynne 1986a: p. 2246, 2260.

Filaments are uniseriate, uncorticated, 4–6 cm tall, and attached by fine rhizoidal outgrowths from basal cells. Main axes are alternately branched below and pseudodichotomous above; branches are flexuous and abundant; branchlets may form unicellular hairs, 50–450 µm long. Cells are uninucleate; main axial cells are to 200 µm in diam., 300 µm long, and thick walled; younger (upper) axial cells are 50–160 µm in diam., 1–1.5 mm long, and enlarged at nodes. Branchlet cells are 7–10 µm in diam. by 12–30 µm long; apical cells are to 50 µm in diam. and 400 µm long. Tetrasporangia are tetrahedrally divided, sessile, ovoid, 25–50 µm in diam., 40–60 µm long, and with 1–3 sporangia on adaxial surfaces of branchlets. Spermatangial sori are on short branchlets, with 2–4 spermatangia per branchlet. Carposporophytes are heart-shaped to trilobed, near tips, at forks, and with globular to irregular carposporangia, 30–50 µm in diam.

Common, during summer on coarse subtidal algae on the west (to 10 m) and east coasts (25–74 m); also known from North Carolina (19–32 m), Bermuda, the Caribbean (type location: Virgin Islands; Dawson, 1962), and Brazil, as well as the Mediterranean, Canary Islands, South Africa, Indo-Pacific, and Australia.

Taylor 1960: p. 507; Dawes 1974: p. 143; Schneider and Searles 1991: p. 371; Littler and Littler 2000: p. 140.

Aglaothamnion halliae (Collins) Aponte, D. L. Ballantine, and J. N. Norris 1997: p. 81–87, fig. 1–3 (named for Mrs. G. A. Hall, an American collector of seaweeds who sent Florida specimens to Harvey). Plate XXVIII, fig. 19, 20.

Basionym. *Callithamnion halliae* Collins in Collins, Holden and Setchell 1900 (1895–1919): ex. no. 698; 1906a: p. 111.

Synonyms. *Callithamnion byssoides sensu auct.* (*pro parte*), non Arnott *ex* Harvey in W. J. Hooker 1833: p. 342.

Callithamnion pseudobyssoides sensu auct., non P. Crouan and H. Crouan 1852: no. 132; 1867: p. 136.

Callithamnion ellipticum Montagne var. *major* P. Crouan and H. Crouan in Mazé and Schramm 1878 (1870–1877): p. 144.

Callithamnion byssoides Arnott ('as *byssoideum*') var. *jamaicensis* Collins 1901: p. 258.

Aglaothamnion pseudobyssoides sensu auct., non (P. Crouan and H. Crouan) L'Hardy-Halos 1965: p. 117.

Aglaothamnion byssoides sensu auct., non (Arnott *ex* Harvey in W. J. Hooker) L'Hardy-Halos and Rueness 1990: p. 352.

Aglaothamnion westbrookiae Rueness and L'Hardy-Halos 1991: p. 650, fig. 1–4.

Womersley and Wollaston (1998) cited *Callithamnion pseudobyssoides* and *C. byssoides* as distinct species. Furnari et al. (1998) interbred Mediterranean *Aglaothamnion tenuissimum* (Bonnemaison) Feldmann-Mazoyer with Atlantic *A. byssoides* (Arnott *ex* Harvey in W. J. Hooker) Boudouresque and Perret-Boudouresque and *A. pseudobyssoides* and suggested that *A. tenuissimum* was conspecific with *A. byssoides*, while *A. pseudobyssoides* was a distinct species.

Filaments are uniseriate, to 5 cm tall, ecorticate or with a few descending rhizoids, and attached by rhizoidal outgrowths. Main axes are straight below, flexuous above, alternately distichously branched, and with incurved branchlets. Cells are uninucleate; older cells are 40–200 µm in diam. by 110–440 µm long; branchlet cells are 10–20 µm in diam. by 25–50 µm long, while apical cells are longer and obtuse. Tetrasporangia are tetrahedrally divided, sessile, ovoid to elliptical, and 25–50 µm in diam. by 40–65 µm long. Spermatangia occur in tufts and spermatangia and tetrasporangia are adaxial on branchlets. Carposporophytes are distal and divided, with each half globular and 170 µm in diam. Carposporangia are also globular and 25–40 µm in diam.

Common, epiphytic or epizoic in shallow waters in the Keys (type location: Key West) and from 34–65 m off the east coast; also known from Nova Scotia to Georgia, the Caribbean, Gulf of Mexico, and eastern North and tropical Atlantic.

Taylor 1960: p. 505, 506; Dawes 1974: p. 143; Schneider and Searles 1991: p. 372.

Aglaothamnion herveyi (M. Howe) Aponte, D. L. Ballantine, and J. N. Norris 1994: p. 232, fig. 1, 4–15 (named for A. B. Hervey, an American amateur phycologist and president of St. Lawrence University who studied Bermuda algae with F. S. Collins). Plate XXVIII, fig. 21, 22.

Basionym. *Callithamnion herveyi* M. Howe 1918: p. 52.

Axes are to 5 cm tall, uniseriate, pale purple, slightly gelatinous, radially alternately branched (rarely dichotomous above), and attached by fine rhizoids. Branches are densely corticated below by rhizoids and ecorticate above. Main axial cells are uninucleate, 40–80 µm in diam., to 160 µm long, and with cell walls 12–25 µm thick; branchlet cells tapered, 8–12 µm in diam., and 1–3 diam. long. Tetrasporangia are tetrahedrally divided, ovoid to spherical, 38–40 µm in diam., and lateral at forks. Monosporangia are ovoid, 35–65 µm in diam., and scattered or in terminal clusters. Spermatangial clusters are oval to spherical, 30–50 µm in diam., and occur laterally on 1-celled stalks. Carposporophytes are spherical and 100–220 µm in diam.

Rare, on seagrasses and seaweeds to 1 m in the Keys; also known from Bermuda (type location: Smith Bay), the Greater Antilles, and southern Caribbean.

Taylor 1960: p. 508; Littler and Littler 2000: p. 140.

Aglaothamnion uruguayense (W. R. Taylor) Aponte, D. L. Ballantine, and J. N. Norris 1994: p. 237, fig. 2 (L. *uruguayense*: of or from Uruguay). Plate XXIX, fig. 1–4.

Basionym. *Callithamnion uruguayense* W. R. Taylor 1939 ('1938'): p. 150, pl. III, fig. 1; pl. VI, fig. 1–4; pl. VII, fig. 1.

Filaments are erect, 2.5–3.5 cm tall, bushy, uniseriate, ecorticate, and pale pink to purple; main axes are distinct, stiff, and attached by long basal rhizoidal cells. Lower branches are entangled, divergent, and recurved, while upper branches are free. Branching is often irregular to alternate and in one plane; tips are tapered or acute. Cells are uninucleate; older axial cells are quadrate and 115 µm in diam. by 100 µm long; upper axial cells are elongate, 50–60 µm in diam., and 200–220 µm long; branchlet cells are 30–37 µm in diam., 55–60 µm long, and taper sharply to acute tips. Tetrasporangia are tetrahedrally divided, solitary, in clusters, or in series, adaxial on distal branches, ovoid, and 40–45 μm in diam. by 50–55 μm long. Carposporophytes are heart-shaped, lateral, and with each half 95–100 μm in diam. and 175–210 μm long.

Rare, reported from 39–53 m off the east coast (Hanisak and Blair, 1988); also known from the Greater Antilles, Brazil, and Uruguay (type location: Punta del Este).

Taylor 1960: p. 506.

Anotrichium Nägeli 1862: p. 159–163, 393–397 (Gr. *ano*: upward + *trichinos*: hairy).

Filaments are uniseriate, bushy, ecorticate or corticated by rhizoids in older parts, appearing segmented, and attached by rhizoids. Branching is sparse to subdichotomous or lateral. Cells are cylindrical, multinucleate, and with a large central vacuole and many discoid plastids. Branchlet tips have whorls or clusters of transparent trichoblasts with di-, tri-, or polychotomous branching. Tetrasporangia are tetrahedrally divided. Gametophytes are dioecious; procarps are near branchlet tips. Carpogonial branches are 4-celled, with 1(2)-celled involucral filaments forming from a large hypogynous cell.

Key to the species (2) of *Anotrichium*

1. Older erect axes 200–500 μm in diam. and 1.2–4.0 mm long; tetrasporangia solitary near branchlet tips on basal cells of whorled trichoblasts. . . . . *A. barbatum*
1. Older erect axes 200–300 μm in diam. and 0.5–1.1 mm long; tetrasporangia clustered around nodes on short pedicels . . . . *A. tenue*

Anotrichium barbatum (C. Agardh) Nägeli 1862: p. 398 (L. *barbatum*: bearded, with tufts of hair). Plate XXIX, fig. 5, 6.

Basionym. *Griffithsia barbata* C. Agardh 1828 (1820–1828): p. 132 (as '*Griffitsia*').

Filaments are delicate, soft, bushy, 1–6(12) cm tall, and attached by unicellular rhizoids of prostrate axes. Erect axes are uniseriate, dichotomously branched, and ecorticate; lower cells are 200–500 μm in diam. by 1.2–4.0 mm long; upper cells are tapered, narrow, and with whorls of trichoblasts at distal ends. Tetrasporangia are tetrahedrally divided, solitary, near branchlet tips, and on basal cells of trichoblasts. Gametophytes are dioecious. Spermatangial branches occur in whorls on upper axes; cystocarps are terminal on short 1-celled stalks and surrounded by 1–5 large incurved involucral cells.

Rare, reported from 59 m off the east coast (Hanisak and Blair, 1988); also known from the Lesser Antilles, as well as the British Isles (type location: Sussex; Maggs and Hommersand, 1993) to Atlantic Spain, the Mediterranean, and eastern tropical Atlantic.

Taylor 1960: p. 515.

Anotrichium tenue (C. Agardh) Nägeli 1862 ('1861'): p. 399 (L. *tenue*: thin, delicate, or slender). Plate XXIX, fig. 7.

Basionym. *Griffithsia tenuis* C. Agardh 1828 (1820–1828): p. 131 (as '*Griffitsia*').

Synonym. *Callithamnion tenue* (C. Agardh) Harvey 1853: p. 130.

Anotrichium tenue and *A. secundum* (Harvey *ex* J. Agardh) G. Furnari are separated by tetrasporangial features (Cormaci et al., 1994; Wynne, 1998a; Schneider, 2003).

Filaments are erect, uniseriate, delicate and soft, bushy, to 7 cm tall, and attached by unicellular rhizoids of prostrate spreading axes. Branching on erect axes is sparse, lateral, and at right angles or slightly secund; branches also arise from lower ends of axial cells. Older axial cells are cylindrical, 200–300 μm in diam., and 0.5–1.1 mm long; upper cells are oblong and 100–200 μm in diam.

Cells near the tips of branchlets are shorter and more spherical than basal ones that have whorls of short, dichotomously branched trichoblasts. Tetrasporangia are tetrahedrally divided, globular, 50–100 μm in diam., and in whorls of 8–50 on 1-celled stalks. Spermatangia are to 3 μm in diam. and in terminal clusters (to 7 per node) on 1(2)-celled stalks. Carposporophytes arise from large fusion cells and have 1–3 gonimoblastic lobes covered by a whorl of 6–14 one-celled involucral filaments. Most gonimoblastic cells become carposporangia.

Jompa and McCook (2003) reported that *A. tenue* can overgrow and kill corals. Common, on algae, limestone, and shells in tidal marshes on the west coast and in 36–44 m off the east coast (Hanisak and Blair, 1988); also known from Massachusetts through the Caribbean, Bahamas, and Brazil, as well as the Mediterranean (type location: Venice, Italy; Abbott, 1999), eastern tropical Atlantic, Red Sea, Indo-Pacific, and Australia.

Taylor 1960: p. 516; Dawes 1974: p. 146; Schneider and Searles 1991: p. 358; Littler and Littler 2000: p. 142.

Antithamnion Nägeli 1847: p. 202 (Gr. *anti*: opposite + *thamnion*: a small bush, shrub).

Synonym. *Callithamnion* subg. *Antithamnion* (Nägeli) Farlow 1881 ('1882'): p. 122.

Filaments are erect, tufted, uniseriate, ecorticate, and attached by multicellular rhizoids. Branching is alternate to opposite below and opposite, distichous, or whorled above. Cells are uninucleate with disc-like to band-shaped plastids; gland cells, if present, occur adaxially on special short cells. Basal cells of branchlet are quadrate and distinct from other branch cells. Tetrasporangia are cruciately or tetrahedrally divided and sessile or stalked on lower cells of branchlets. Gametophytes are dioecious. Spermatangia are in clusters on special branches. Procarps are formed on basal cells of branchlets; carpogonial branches are 4-celled; carposporophytes are terminal and most gonimoblast cells produce carposporangia.

Key to the species (3) of *Antithamnion*

1. Main axial cells to 50 μm in diam.; axes repeatedly branched. 2
1. Main axial cells 40–90 μm in diam.; axes little branched . . . . *A. cruciatum*
   2. Whorled branchlets of prostrate axes distinct from vertical (ventral) branchlets that are usually simple and unbranched . . . . *A. antillanum*
   2. Whorled branchlets on prostrate axes similar in form and size and have erect determinate laterals . . . . *A. decipiens*

***Antithamnion antillanum* Børgesen 1917: p. 226–229, fig. 213–216 (L. *antillanum*: of or from the Antilles). Plate XXIX, fig. 8, 9.**

Synonyms. *Callithamnion lherminieri* P. Crouan and H. Crouan in Mazé and Schramm 1878 (1870–1877): p. 144.

Antithamnion cruciatum (C. Agardh) Nägeli var. *profundum sensu auct.*, non Feldmann-Mazoyer 1941 ('1940'): p. 258, fig. 91.

Antithamnion iherminieri (p. Crouan and H. Crouan) Bornet in Nasr 1941: p. 66, fig. 9, 10.

Abbott (1979) recognized *Antithamnion antillanum* and *A. lherminieri* as distinct taxa based on tetrasporangial and morphological features in Caribbean and Hawaiian material. Abbott (1999) and others (Athanasiadis, 1985, 1996a; Athanasiadis et al., 1990) placed *A. lherminieri* in synonymy with *A. antillanum*. Guiry and Nic Dhonncha (2006) listed *A. antillanum* as a synonym for *A. iherminieri*, because *Callithamnion iherminieri* is the oldest basionym (Silva et al., 1996; Schneider and Searles, 1997). Athanasiadis (1997) proposed *Antithamnion antillanum* for conservation against *A. iherminieri*, and this was subsequently approved (Grueter et al., 2000).

Filaments are delicate fine tufts, to 8(70) mm tall, and pink or red; they arise from prostrate spreading axes with unequal lateral branches, and are attached by thick-walled multicellular rhizoids with discoid pads. Erect filaments are uniseriate, 20–50 µm in diam., and with cells 50–200 µm long. Basal cells of branchlets are quadrate, while all others are terete. Prostrate axes are 40–60 µm in diam.; cells are 70–200 µm long and thick walled. Erect axes are sparingly, oppositely to alternately branched, and usually occur in one plane. Branchlets of first order alternate on every cell. Second order branchlets are short and 1–3 cells long. Gland cells are 13–18 µm in diam., 20–25 µm long, adaxial, and cover two branchlet cells. Tetrasporangia are cruciately divided, elliptical, and 40 by 90 µm.

Rare, reported on algae in 25–54 m off the east coast (Hanisak and Blair, 1988); also known from Bermuda, the Bahamas, Caribbean (type location: St. Thomas, Virgin Islands; Price and Scott, 1992), Gulf of Mexico, and Brazil, as well as the Mediterranean, eastern tropical Atlantic, Pacific Ocean and Australia.

Littler and Littler 2000: p. 142.

Antithamnion cruciatum (C. Agardh) Nägeli 1847: p. 200 (L. *cruciatum*: cross-shaped). Plate XXIX, fig. 10–12.

Basionym. *Callithamnion cruciatum* C. Agardh 1827: p. 637.

Synonyms. *Callithamnion cruciatum* C. Agardh var. *radicans* J. Agardh 1841: p. 44.

Antithamnion cruciatum var. *radicans* (J. Agardh) Collins 1900: p. 48; Collins and Hervey 1917: p. 141–142.

Filaments are in tufts to 5 cm tall and attached by prostrate axes with multicellular rhizoids. Branching of erect axis is sparse, alternate below, opposite and in one or two planes above, and with congested tips. Branchlet basal cells are short, rectangular, and produce rhizoids. Lower axial cells are 40–90 µm in diam. by 90–300 µm long; cells near tips 9–28 µm in diam. Gland cells, if present, occur adaxially on special 1–3-celled branchlets. Tetrasporangia are cruciately divided, elliptical, 38–70 µm in diam., 60–100 µm long, are sessile or stalked, and replace vegetative branchlets. Spermatangia occur in clusters of 4–8 per branchlet. Carposporophytes have two gonimoblastic lobes to 400 µm in diam.

Rare, on algae in shallow waters at the Keys; also known from Newfoundland to the Caribbean, Brazil, and Argentina, as well as the eastern North and tropical Atlantic, Mediterranean (type location: Trieste, Italy; Maggs and Hommersand, 1993), Black Sea, India, and Australia.

Taylor 1960: p. 498; Schneider and Searles 1991: p. 360.

Antithamnion decipiens (J. Agardh) Athanasiadis 1996a: p. 151–153, fig. 70A–G (L. *decipiens*: deceptive; species that resemble one another). Plate XXIX, fig. 13.

Basionym. *Callithamnion decipiens* J. Agardh 1842: p. 70–71.

Synonym. *Antithamnion ogdeniae* I. A. Abbott. 1979: p. 218, fig. 11–14.

Filaments are short, 4–12 mm tall, uniseriate, and with freely branched erect axes; prostrate axes are not entangled, have lateral branches of equal length, and are attached by simple rhizoids. Older axial cells are to 50 µm in diam. by 9–10 diam long; upper cells are smaller and 25–30 µm in diam. by 2 diam long. Branching is opposite or whorled; branchlets are subdichotomously divided, to 200 µm long, and with tip cells 8 µm in diam. Gland cells are often abundant, on 2(3)-celled branches, and cover 2 branch cells. Tetrasporangia are cruciately divided, ovoid, 40 µm in diam. by 80 µm long, and occur adaxially on the 2nd or 3rd cells of branches. Spermatangial branches appear as corn cobs, 25 by 42 µm.

Rare, on rubble among shallow turf algae (5–6 m) at Molasses Reef (Bucher et al., 1990); also known from Bermuda, Belize, Antilles (type location: St. Croix; Abbott, 1999), as well as the Mediterranean, Canary Islands, Hawaiian Islands, and New Zealand.

Antithamnionella Lyle 1922: p. 348 (Gr. *anti*: opposite + *thamnion*: a small bush + *ella*: lesser; thus, smaller than the red algal genus *Antithamnion*).

Filaments are erect or prostrate, uniseriate, ecorticate, and attached by rhizoids. Branchlets are whorled with 1–4 arising from each axial cell; basal cells of branches are not quadrate but similar in length to other cells. Cells are uninucleate with small, discoid to band-shaped plastids. Gland cells, if present, are adaxial. Tetrasporangia are cruciately or tetrahedrally divided and sessile or stalked. Gametophytes are dioecious. Spermatangia are in adaxial clusters on special branches. Procarps are solitary on basal cells of branchlets near tips; carpogonial branches are 4-celled. Carposporophytes are terminal; gonimoblastic lobes are not covered by involucral branches; carposporangia arise from most gonimoblastic cells.

Key to the species (3) of *Antithamnionella*

1. Branchlets (determinate branches) of older axes to 10 cells long 2
1. Branchlets of older axes more than 10 cells long 3
 2. Branchlets opposite and in one plane . . . . *A. atlantica*
 2. Branchlets usually in whorls of 3; older axes may have opposing pairs of branchlets . . . . *A. boergesenii* (in part)
3. Branchlets whiplike, 2 per axial cell; basal cells constricted in center; gland cells single or paired, in series, or scattered on branchlets . . . . *A. graeffei*
3. Branchlets not whiplike, 2–3 per axial cell; basal cells elongate; gland cells single on second to sixth proximate cell of branchlets . . . . *A. boergesenii* (in part)

***Antithamnionella atlantica* (E. C. Oliveira) Schneider 1984: p. 455, fig. 1 (L. *atlantica*: of or from the Atlantic Ocean). Plate XXIX, fig. 14.**

Basionym. *Antithamnion atlanticum* E. C. Oliveira 1969: p. 37, fig. 9, 10.

Filaments are uniseriate, minute (0.2–1.2 mm tall), prostrate, and dorsiventrally organized; prostrate axes can produce erect branches dorsally and multicellular rhizoids ventrally. Erect axes arise from most cells of prostrate axes; indeterminate axes are oppositely branched, in one plane, and 15–20 µm in diam. Branchlets of erect axes are 2–7 cells long by 8–10 µm in diam. Cells of prostrate axes are 18–30 µm in diam., 45–115 µm long, and have upward curved tips covered with branchlets. Gland cells are common and overlap the 2nd or 3rd basal cell of branchlets. Tetrasporangia are tetrahedrally divided, globose, to 45 µm in diam., and sessile on basal cells of branchlets. Carposporophytes are terminal with two elongated gonimoblastic lobes to 38 µm in diam.

Rare, reported from 50–53 m off the east coast (Hanisak and Blair, 1988); also known from North Carolina (17–50 m) to Georgia, and Brazil (type location: Island of Frances; Athanasiadis, 1996a).

Schneider and Searles 1991: p. 362.

***Antithamnionella boergesenii* (Cormaci and Furnari) Athanasiadis 1996a: p. 104–105, fig. 45A–E (named for the Danish phycologist F. Børgesen, who studied the seaweeds of the Danish West Indies and Canary Islands). Plate XXIX, fig. 15, 16.**

Basionym. *Antithamnionella elegans* (Berthold) J. Price and D. John in J. Price et al. var. *boergesenii* Cormaci and Furnari 1988: p. 345.

Synonyms. *Antithamnion elegans* Berthold 1882: p. 516.

Antithamnion elegans sensu auct., non Berthold 1882: p. 516; also *sensu* Børgesen 1930a: p. 56–60, fig. 21–23.

Antithamnionella elegans var. elegans *sensu auct.*, non (Berthold) J. Price and D. John in J. Price et al. 1986: p. 16; also *sensu* Bucher et al. 1990: p. 303

Filaments are uniseriate, delicate, ecorticate, erect, 0.5–5.0 mm tall, and pale rose red; prostrate axes are attached by many rhizoids that replace one or a pair of dorsal branches. Erect indeterminate axes are 0.5–5 mm tall and may replace branchlets; most axial cells produce 1–3 branchlets in a whorl that have up to 4 orders of branching, 8–13 cells, and tapering tips. Axial cells are 20–36 µm in diam. and 45–75 µm long; prostrate cells are 30–50 µm in diam. and 60–150 µm long. Gland cells cover most or all of branchlet basal cells. Tetrasporangia are tetrahedrally or cruciately divided, ovoid, 17–28 µm in diam., and single on basal cells of branchlets. Spermatangia are terminal on the 2nd to 5th cell of branchlets; carposporophytes are terminal with 2 or more ovoid gonimoblastic lobes that are to 63 µm in diam.

Uncommon, on algae in 5–6 m at Molasses Reef (Bucher et al., 1990) and in 26–87 m off the east coast (Hanisak and Blair, 1988); also known from North Carolina (17–50 m) to Georgia, Bermuda, the Greater Antilles, and Brazil, as well as Atlantic Spain, the Mediterranean, eastern tropical Atlantic (type location: Bahia del Confital, Canary Islands; Cormaci and Furnari, 1988), Indo-Pacific, and Gulf of California.

Schneider and Searles 1991: p. 363; Littler and Littler 2000: p. 142.

Antithamnionella graeffei (Grunow) Athanasiadis 1996a: p. 107, fig. 47 (named for E. Graeffee, a German botanist and collector of seaweeds in Tonga and the Samoa Islands). Plate XXIX, fig. 17.

Basionym. *Callithamnion graeffei* Grunow 1874: p. 27.

Synonyms. ?*Antithamnion tristicum* Joly and Yamaguishi in Joly et al. 1963: p. 57, fig. 1, 2.

?*Antithamnionella ternatum* Joly and Cordeiro in Joly et al. 1966 ('1965'): p. 166, pl. III, fig. 5–10.

Antithamnionella flagellata (Børgesen) Abbott 1979: p 222, fig. 18–20.

Filaments are uniseriate, ecorticate, minute (to 1 mm tall), and epiphytic; prostrate axes are attached by rhizoids arising opposite erect axes. Rhizoids are multicellular, branched, elongate, 10–30 µm in diam, and penetrate hosts. Prostrate axes are indeterminate and curve upward at tips. Most cells of erect indeterminate axes produce opposite pairs or whorls of 3 usually determinate branchlets. Branchlets arise at distal ends of axial cells in decussate or spiral patterns. Prostrate and erect cells are 20–60 µm in diam. by 10–130 µm long. Branchlets are 10–20 cells long, simple, alternate, or subdichotomously branched; cells are 12–20 µm in diam. and taper to 3–4 µm at tips; basal cells are short and to 25 µm long. Gland cells are rare to common, adaxial or abaxial on branchlets, single, paired, seriate, or scattered, and cover most or all of a support cell. Tetrasporangia are sessile, cruciately or tetrahedrally divided, ovoid, 27–40 µm in diam., 37–60 µm long, adaxial, and in secund series of 1–3. Spermatangia occur in elongate loose sori on branchlets. Carposporophytes are terminal with 3–5 gonimoblastic lobes that are to 150 µm in diam. Carposporangia are spherical to elliptical and to 40 µm in diam.

Rare, on algae in 50–71 m off the east coast (Hanisak and Blair, 1988); also known from North Carolina (29–45 m), Bermuda, the Caribbean, and Brazil, as well as Mauritius, the Red Sea, Pacific Ocean, and Australia (type location: Tonga; Huisman, 2000).

Schneider and Searles 1991: p. 364.

Balliella Itono and Tanaka 1973: p. 249 (L. *Ballia*: a genus of red algae named for Anne E. Ball, an Irish phycologist, by Harvey + *ella*: lesser; thus smaller than *Ballia*).

Synonym. *Bakothamnion* Hoek 1978: p. 48.

Filaments are uniseriate, mostly erect, and irregularly branched, with lower cells partially covered by rhizoids, and prostrate axes attached by rhizoids. Basal cells of lateral branches have many branchlets. Gland cells are spherical and abaxial on basal cells of lateral branches and branchlets. Tetrasporangia are cruciately divided, oblong to ovoid, sessile or stalked, and adaxial on lateral branches. Procarps are on basal cells of upper and opposite lateral branches. Carposporophytes are partially covered with sterile filaments and have compact rounded carposporangia. Spermatangia are produced on special short projections of opposite lateral branches.

Balliella pseudocorticata (E. Y. Dawson) D. N. Young 1981: p. 94 (L. *pseudo*: false + *corticata*: covered with bark, cortex). Plate XXIX, fig. 18, 19.

Basionym. *Antithamnion pseudocorticatum* E. Y. Dawson 1962: p. 20–21, pl. 7. fig. 1–5.

Synonym. *Bakothamnion curassavicum* Hoek 1978: p. 48, fig. 1, 2.

Filaments are uniseriate, erect, 10–25 mm tall, in tufts, pale red, delicate, and arise from prostrate axes attached by long, multicellular rhizoids. Erect axes have alternate branching; secondary axes are oppositely branched with branch pairs uneven. Branchlets are slightly incurved, mostly in one plane, and arise from distal parts of cells. Main axial cells are to 60 μm in diam., 350 μm long, and loosely corticated below by rhizoids. Branchlets are to 20 μm in diam. and taper to tips; basal cells are small and rectangular. Gland cells, if present, are 1–2 per basal cell, spherical, to 15μm in diam., and occur abaxially on basal cells of lateral branches. Tetrasporangia are cruciately divided, spherical, sessile, and occur abaxially on branchlet basal cells.

Rare, reported in 44 m off the east coast (Hanisak and Blair, 1988); also known from Bermuda, the Caribbean, Maldives and Seychelles Islands, Baja California (type location: San Lorenzo Channel, La Paz; Adams, 1994), Pacific Mexico, the Galapagos Islands, and New Zealand.

Littler and Littler 2000: p. 144.

Callithamniella Feldmann-Mazoyer 1938a: p. 1119; 1941 ('1940'): p. 428 (L. *calli*: beautiful + *thamnion*: a shrub + *ella*: lesser; thus smaller than the genus *Callithamnion*).

Filaments are uniseriate, ecorticate, and have uninucleate cells; erect (determinate) branchlets and indeterminate axes arise irregularly from prostrate axes that are attached by simple ventral rhizoids. Basal cells of erect branchlets are 7–30 cells long and may produce a rhizoid and an indeterminate axis. Some prostrate axial cells produce a rhizoid and an indeterminate erect axis opposite each other; prostrate axial tips curve upward. Tetrasporangia are cruciately divided, stalked, and in an alternate or irregular pattern on the distal ends of erect indeterminate axes. Spermatangia occur in elongate clusters on branchlets. Carposporophytes are unknown.

Callithamniella tingitana (Schousboe *ex* Bornet) Feldmann-Mazoyer 1938a: p. 1119; 1941 ('1940'): p. 428 (L. *tingitana*: of or from Tangier, Africa). Plate XXIX, fig. 20.

Basionym. *Callithamnion tingitanum* Schousboe *ex* Bornet 1892: p. 329–330.

Synonym. *Grallatoria tingitana* (Schousboe *ex* Bornet) Abbott 1976: p. 148.

Erect filaments are uniseriate, to 4 mm tall, delicate, and pale red; prostrate axes are indeterminate, have erect filaments, and produce rhizoids from the distal ends of ventral cells. Rhizoids are simple, 20 µm in diam., and may have discoid haptera. Prostrate axial cells are cylindrical, 24–35 µm in diam., and 100–310 µm long. Erect axial cells are 15–25 µm in diam., 75–140 µm long, and with tips tapering to 5 µm in diam. Erect indeterminate axes are alternately spirally branched. Branchlets are simple to dichotomously branched 1–2 times and have 7–20 cells; basal cells are 10–15 µm in diam. by 270–470 µm long; cells near tips taper to 5–10 µm in diam. Tetrasporangia are cruciately divided, elliptical, 25–38 µm in diam. by 38–55 µm long, and solitary on 1-celled stalks. Spermatangia occur in whorled clusters, in elongate heads, and from prostrate and erect axes.

Rare, reported from 29–42 m off the east coast (Hanisak and Blair, 1988); also known from North Carolina (18–60 m), Bermuda, and Brazil, as well as Atlantic Spain and Portugal, the Mediterranean, and eastern tropical Atlantic (type location: Tangier, Morocco; Feldmann-Mazoyer, 1938b).

Schneider and Searles 1991: p. 369.

Centroceras Kützing 1841 ('1842'): p. 731; 1843: p. 381 (Gr. *centro*: middle + *ceras*: horned, hornlike projection)

Based upon a molecular systematics study of Brazilian *Centroceras* and *Ceramium* taxa, Barros-Barreto et al. (2006) concluded that *Ceramium* was not monophyletic with respect to *Centroceras* and *Corallophila* Weber-van Bosse; further, all three *Centroceras* sequences they evaluated were found to be a single species, *C. clavulatum*.

Filaments form in mats or bushy tufts, are usually dichotomously branched with lateral proliferations, and attached by basal and lateral multicellular rhizoids that arise from nodes. Two acropetal and one basipetal cortical filament are produced by each periaxial cell; a second basipetal filament forms from one of the two acropetal cortical filaments and, together with the primary basipetal filament, forms the continuous cortication of longitudinal rows of rectangular cells. The periaxial cells arise from the distal ends of central large axial cells and nodes usually have hyaline spines. Cells are uninucleate with many discoid plastids. Tetrasporangia are cruciately or tetrahedrally divided, in whorls of 8–20 at nodes, and project above cortical bands. Gametophytes are dioecious. Spermatangia occur in terminal clusters on branches at nodes. Procarps are lateral at the nodes and have 4-celled carpogonial branches. Cystocarps have two gonimoblastic lobes surrounded by short involucral branches.

Key to the species (2) of *Centroceras* (modified from Won et al., 2006)

1. Nodes with obvious 2-celled spines; gland cells at nodes inconspicuous . . . . *C. clavulatum*
1. Nodes lack spines; gland cells obvious and in whorls . . . . *C. internitens*

***Centroceras clavulatum* (C. Agardh in Kunth) Montagne in Durieu de Maisonneuve 1846: p. 140 (L. *clavulatum*: club-shaped). Plate LI, fig. 15–17.**

Basionym. *Ceramium clavulatum* C. Agardh in Kunth 1822: p. 2.
Synonym. *Centroceras clavulatum* var. *brachyacanthum* (Kützing) P. Crouan and H. Crouan in Mazé and Schramm 1878 (1870–1877): p. 151.

Filaments are delicate, rose pink to deep red brown, in erect tufts 10–20 cm tall, and attached by uniseriate rhizoids of basal and nodal cells from erect and interwoven prostrate axes. Axes are subdichotomously (pseudo- to trichotomously) branched every 9–10 axial nodes, with older parts often having lateral proliferations. Axial cells are corticated by a layer of 12–16 periaxial cells that produce the corticating filaments. Segments are 50–200 µm in diam. and 300–750 µm long. Apices are pincer-shaped, incurved, and with exposed apical cells. Tetrasporangia occur in whorls at nodes, divided tetrahedrally, projecting, ovoid or spherical, 45–70 µm in diam., 50–100 µm long, thick walled, and without involucral filaments. Spermatangia are produced in terminal clusters. Carposporophytes are lateral at joints.

Common, intertidal and to 5 m on mangrove roots, shells and limestone on all coasts; also known from North Carolina, Bermuda, the Bahamas, Caribbean, Gulf of Mexico, and Brazil (see Barros-Barreto et al., 2006), as well as Atlantic Spain and Portugal, the Mediterranean, eastern tropical Atlantic, Red Sea, Indo-Pacific (type location: Callao de Lima, Peru; Womersley and Bailey, 1970), and Australia.

Taylor 1960: p. 537; Dawes 1974: p. 143; Schneider and Searles 1991: p. 374; Littler and Littler 2000: p. 144.

Centroceras internitens Gallagher and Humm 1983: p. 261, fig. 1–16 (L. *inter*: among + *nitens*: shining; named for its host, *Ceramium nitens*). Plate LI, fig. 27.

Filaments are epiphytic and epizoic, to 5 cm tall, in sparse tufts, alternately to irregularly branched, and attached by rhizoids that penetrate the cortex of their hosts. Main axes are to 350 µm in diam., darkly pigmented at nodes, and with segments 100–500 µm long. Periaxial cells (16–20) are in one layer, rectangular, and cover nodes and internodes. Apices are strongly tapered and sharply hooked at tip. Nodal spines are absent and gland cells occur at nodes. Tetrasporangia are tetrahedrally divided, ovoid to spherical, 40–50 µm in diam., 50–60 µm long, thick walled, and in whorls at nodes. Spermatangia are in nodal whorls. Carposporophytes form laterally at joints and have several lobes and involucral filaments.

Rare or overlooked, on *Ceramium nitens* and other algae, seagrasses, soft corals, and sponges in the Keys to 18 m (type location: West Summerland Key); also known from Bermuda, the Bahamas, Greater Antilles, and Belize.

Ceramium Roth 1797: p. xxvi, 146, *nom. cons.* (Gr. *keras*: a horn + L. *ium*: a collection).

Filaments are erect, bushy, or matted, and attached by simple multicellular rhizoids that arise from bases and prostrate axes. Axes are uniseriate and with pseudodichotomous or dichotomous branching and corticated at cross walls (nodes). Periaxial cells arise from the upper ends of axial cells and give rise to a strict pattern of acropetal (ascending) and basipetal (descending) cortical filaments that form the 1 to many bands of pigmented cells at the nodes. In a few species, the internodes are completely corticated. Cells are uninucleate and have many discoid to elongate plastids; gland cells may be present. Tetrasporangia are tetrahedrally or cruciately divided, sessile, and at nodes or scattered on fully corticated species. Spermatangia form a hyaline layer at nodes. Carpogonial branches are 4-celled; 1–2 branches arise from a support cell. Carposporophytes are globular and have 1–3 gonimoblastic lobes, plus a few sterile involucral filaments.

Key to the taxa (14) of *Ceramium*

1. Axial row in older parts completely corticated by small cells 2
1. Nodes of axial cells corticated by small cells and naked at least in older parts 3
   2. Branch tips pincer-like; nodal banding faintly visible under small surface cells; axes appear banded and lumpy . . . . *C. virgatum*
   2. Branch tips not pincer-like; nodal banding not evident; axes smooth . . . . *C. nitens*

3. Nodal bands in older axes have mostly 1–2 transverse rows of small cells 4
3. Nodal bands in older axes always have more than 2 transverse cell rows that often are irregular 6
   4. Plants minute, epiphytic; erect axes less than 1.5 cm tall; nodal cortication usually 2–3 cell rows, older axes may have up to 4 rows 5
   4. Plants larger, not epiphytic; erect axes to 3 cm tall; nodal cortication usually with 1 full cell row plus a few scattered cells . . . . *C. leptozonum*
5. Axial nodes have 5 periaxial cells . . . . *C. reptans*
5. Axial nodes have 4 periaxial cells . . . . *C. codii*
   6. Plants primarily erect and bushy 7
   6. Plants primarily prostrate but with some erect branches . . . . *C. cruciatum*
7. Branching mostly dichotomous, sometimes unequal and with lateral branches 9
7. Branching alternate 8
   8. Branching alternate and branches in one plane (distichous) . . . .*C. floridanum*
   8. Branching alternate but branches not in one plane. . . . .*C. flaccidum*
9. Nodal cortication not continuous with adjacent band on axes 10
9. Nodal cortication continuous near tips and limited to nodes below . . . . *C. corniculatum*
   10. Tips not strongly pincer-like; nodes not projecting; gland cells absent 11
   10. Tips pincer-like; nodes projecting; gland cells present. . . . . *C. diaphanum* (in part)
11. Cortical cells at nodes irregularly disposed in older axial bands 12
11. Cortical cells of nodes in a few fairly distinct transverse rows. 14
   12. Tetrasporangia immersed or slightly emerged from nodal bands 13
   12. Tetrasporangia obviously projecting from nodal band . . . . *C. cimbricum*
13. Old nodal cells 13 µm in diam.; proliferations common . . . . *C. diaphanum* (in part)
13. Old nodal cells 10–22 µm in diam.; proliferations uncommon . . . . *C. deslongchampsii*
   14. Branches spurlike; nodal cells in 2–3 irregular bands; top row with small, rounded cells; mid row with large, rounded cells . . . .*C. brevizonatum* var. *caraibicum.*
   14. Branches not spurlike; nodal cells in 3 bands; top band with 2–4 irregular rows of small angular cells, mid band with one row of large cells, and lower band with 2–4 rows of small irregular cells . . . . *C. subtile*

Ceramium brevizonatum H. E. Petersen var. *caraibicum* H. E. Petersen and Børgesen in Børgesen 1924: p. 29, fig. 11 (L. *brevi*: short + *zonatum*: banded; *caribicum*: of or from the Caribbean). Plate XXX, fig. 1, 2.

Molecular studies by Barros-Barreto et al. (2006) showed that there may be cryptic variability within plants designated as this taxon, as phylogenetic analyses of Brazilian material differed from those from Yucatan, Mexico that they cite as the type locality (cf. comments below by Silva et al., 1996).

Filaments are uniseriate, in soft tufts to 15 cm tall, rose to pinkish-red, and attached by rhizoids from basal cells that form tight mats. Branching is dichotomous; tips pincer-shaped and lateral branches spurlike. Segments are 80–120 µm in diam. and to 1.3 mm long; nodes are (40) 80–120 µm in diam. by 90–136 µm long. Nodal cortication has 2–3 irregular bands; top row of cells are the smallest and middle row the largest and irregularly rounded; gland cells are absent. Tetrasporangia are tetrahedrally divided, projecting, 30–60 µm in diam., one per node, and often in a lateral vertical series. Carposporophytes occur at upper forks and have several involucral filaments.

Rare, on rubble and algae at the Dry Tortugas; also known from Bermuda, the Caribbean (type location: Dominican Republic; Silva et al., 1996), Gulf of Mexico, Brazil, Tanzania, the Seychelles, and Fiji.

Taylor 1960: p. 527; Littler and Littler 2000: p. 146.

Ceramium cimbricum H. E. Petersen in Rosenvinge 1924: p. 378, fig. 318, 319 (L. *cimbricum*: Cimbrian, a German tribe that invaded the Roman Empire and was defeated by Marius). Plate XXX, fig. 3–5.

Synonym. *Ceramium fastigiatum* Harvey 1834: p. 303–304 *nom. illeg.*; non *Ceramium fastigiatum* Roth 1800a: p. 463.

See comments below regarding *Ceramothamnion adriaticum* Schiller and *Ceramium codii* (H. Richards) Feldmann-Mazoyer.

Filaments are uniseriate, in delicate tufts to 10 cm tall, rose to purple red, and with prostrate axes attached by simple rhizoids that may form haptera. Axes are irregularly dichotomously branched and wide spreading; branchlets are upturned, spurlike, and clustered above; tips are erect and rarely pincer-shaped. Cortication is nodal, with 2 traverse bands in young branches and up to 6 bands in older axes; segments are 60–155 µm in diam. and to 600 µm long; nodes to 170 µm in diam. and 65 µm long. Cells of mid nodal bands are large and rectangular; cells of upper bands are small and globose. Gland cells sometimes present. Tetrasporangia are tetrahedrally divided, spherical, 1–6 per node, in series or whorls, projecting, to 65 µm in diam., 70 µm long, and usually partially covered by a few sterile filaments. Carposporophytes have 1–2 ovoid-shaped gonimoblastic lobes near tips of short branchlets and are subtended by 2–4 short corticated laterals that are longer than the cystocarp.

f. *flaccidum* (H. E. Petersen) Furnari and Serio in Cecere et al. 1996: p. 212 (L. *flaccidum*: weak, feeble, flabby, soft).

Synonym. *Ceramium fastigiatum* Harvey f. *flaccidum* H. E. Petersen in Børgesen 1918: p. 241–243, fig. 231.

Filaments of form are uniseriate, in delicate tufts to 10 cm tall, pink to rose, irregularly dichotomously branched with straight tips, and attached by a sparse rhizoidal mat. Cortical cells occur in 1–2 transverse bands at nodes with the largest cells in the lower row. Segments are to 60 µm in diam. by 400 µm long and nodes 50–70 µm in diam. by 40–40 µm long. Spermatangia cover the nodes. Carposporophytes have 3 gonimoblastic lobes that are overtopped by involucral filaments.

Common, species and form occur on mangrove roots, seagrasses, and macroalgae (to 40 m) on all coasts; also known from Newfoundland through the Caribbean (type location of form: Virgin Islands; Silva et al., 1996), and the Gulf of Mexico, as well as the eastern North Atlantic (type location of species: Ejerslev Røn, Denmark; Maggs and Hommersand, 1993), Mediterranean, Red Sea, and Indo-Pacific.

Taylor 1960: p. 526, 527; Dawes 1974: p. 145; Schneider and Searles 1991: p. 378; Littler and Littler 2000: p. 148.

Ceramium codii (H. Richards) Feldmann-Mazoyer 1938b: p. 324–325 (Gr. *codium*: fleece or skin; named for its host, the green algal genus *Codium*). Plate XXX, fig. 6–8.

Basionym. *Ceramothamnion codii* H. Richards 1901: p. 264–265, pl. 21, 22.

Maggs and Hommersand (1993) suggested that *Ceramothamnion codii* and *C. adriaticum* Schiller may be older names for *Ceramium cimbricum* rather than *C. codii*. Abbott (1999) noted that *C. serpens* Setchell and Gardner is similar to *C. codii*, except that its short basal filaments are more common, while *C. mazatlanense* E. Y. Dawson resembles both *C. serpens* and *C. codii*. Further molecular evaluations are needed to resolve these species.

Plants are minute, delicate, mostly prostrate, pink to deep red tufts to 1 cm tall, with nodal cortication, and attached by rhizoids from prostrate and lower nodes. Nodes are 20–35 µm in diam., corticated by 2–3 irregular bands, and have 4 periaxial cells that cut off 2 acropetal cells and 1 basipetal cortical cell that expands horizontally. Axial cells are spherical to cylindrical and to 16 µm in diam. and 49 µm long. Tetrasporangia are tetrahedrally divided, 20–30 by 30–45 µm, secund, 1 per node, and partially covered by involucral filaments. Spermatangia are clustered and usually cover two nodes below the axial tips. Cystocarps are near branch tips and partially covered by 4 involucral branches.

Uncommon, epiphytic on *Codium* spp. in shallow waters on the west coast as well as on *Hypnea spinella* in the Florida Keys; also known from Bermuda (type location: Cooper's Island; Cho and Fredericq, 2006), the Caribbean, Bahamas, and Gulf of Mexico, as well as Spain, Portugal, the Mediterranean, eastern tropical Atlantic, Red Sea, Indo-Pacific, and Australia.

Taylor 1960: p. 526; Dawes 1974: 145.

Ceramium corniculatum Montagne 1861 ('1860'): p. 172, pl. 11, fig. B1–B3 (L. *corniculatum*: horned, curved like a horn). Plate XXX, fig. 9.

Synonym. *Ceramium strictoides* P. Crouan and H. Crouan in Mazé and Schramm 1878 (1870–1877): p. 148.

Filaments are uniseriate, in rose red tufted mats to 3 cm long, and attached by rhizoids. Branching is mostly alternate above and irregularly dichotomous below; tips are straight to slightly incurved. Axial cells are 90–200 µm in diam. and shorter above and longer below; nodes are 100–210 µm in diam. and 160 µm long. Cortical cells occur in 3–4 rows; all cells are similar in size and often longer than broad. Cortication is confluent at tips and limited to nodes below. Tetrasporangia are tetrahedrally divided, spherical, to 50 µm in diam., and in whorls on short lateral branches from nodes or near tips.

Common, on seagrasses, mangrove roots, seaweeds, and rubble (to 20 m) on all coasts; also known from Bermuda, the Bahamas, Caribbean (type location: Guadeloupe), Gulf of Mexico, Brazil, Uruguay, and Senegal.

Taylor 1960: p. 530; Dawes 1974: p. 145; Littler and Littler 2000: p. 148.

Ceramium cruciatum Collins and Hervey 1917: p. 144–145, pl. IV, fig. 27, 28 (L. *cruciatum*: a cross, cross-shaped). Plate XXX, fig. 10.

Filaments are uniseriate, pink to rose red, with small creeping tufts to 1.5 cm tall, unequally dichotomously branched (hornlike, cervicorn), and attached by unicellular rhizoids of prostrate axes. Prostrate axes are to 200 µm in diam. and slightly constricted at nodes; erect axes are 80–120 µm in diam. and have pincer-like incurved tips. Nodes are 120–140 µm in diam., corticated with several bands, and have the largest cells in the mid band. Tetrasporangia are cruciately divided, ovoid, 35–50 µm in diam., with 6 µm thick walls, 1(4) per node, and do not project above cortication. Spermatangia are clustered, 5–8 µm in diam., and cover the nodes.

Uncommon, on algae in the Keys; also known from Bermuda (type location: Gibbet Island; Silva et al., 1996), the Caribbean, Gulf of Mexico, and Indian Ocean.

Taylor 1960: p. 530; Littler and Littler 2000: p. 150.

Ceramium deslongchampsii Chauvin *ex* Duby 1830: p. 967 (named for L. A. Deschamps, a French naturalist and botanist). Plate XXX, fig. 11–13.

Synonym. *Ceramium strictum* (Kützing) Harvey 1849: p. 164, non. *Ceramium strictum* Roth 1806: p. 130.

Guiry and Nic Dhonncha (2006) noted the incorrect spelling, *C. deslongchampii.*

Uniseriate filaments form delicate tufts 2–8 cm tall and are attached by rhizoids of basal cells. Branching is regularly dichotomous and with a few proliferations; branches are spreading below, clustered above, and have pincer-shaped tips. Segments are 0.6–1.5 mm long; older nodes are 210–335 µm in diam., 85–210 µm long, and with internodes 185–270 µm long. Cortication is sharply defined at margins with cells in irregular series; upper marginal cells are small and lower marginal cells are 10–22 µm in diam. Gland cells are absent. Tetrasporangia are cruciately divided, immersed or partially exposed in the mid region of nodal bands, and 50–60 µm in diam. by 60–70 µm long. Cystocarps are lateral on upper branches and overtopped by 4–6 involucral branches.

Rare, on seagrasses and oyster shells in shallow water in the Keys; also known from Newfoundland to Virginia, the Antilles, Gulf of Mexico, and Brazil (see Barros-Barreto, 2006), as well as the eastern North and tropical Atlantic (type location: Normandy, France; Silva et al., 1996), Mediterranean, Red Sea, and Indo-Pacific.

Taylor 1960: p. 530, 536 (as an uncertain record).

Ceramium diaphanum (Lightfoot) Roth 1806: p. 154–155 (L. *diaphanum*: colorless or transparent). Plate XXX, fig. 14–15.

Basionym. *Conferva diaphana* Lightfoot 1777: p. 996.

Synonym. *Ceramium tenuissimum* (Roth) J. Agardh 1851 (1848–1901): p. 120; non *Ceramium tenuissimum* Bonnemaison 1828: p. 132.

Taylor (1960) questioned if *Ceramium diaphanum* and f. *strictoides* H. E. Petersen were in Florida, but recorded *C. tenuissimum* from the Keys. Schneider and Searles (1991) found morphological overlap between *C. diaphanum*, *C. deslongchampsii*, and *C. fastigiatum* (Wulfen *ex* Roth) Harvey in Hooker.

Filaments are uniseriate, in large erect tufts to 20 cm, and attached by rhizoids from ventral cells of prostrate axes or basal nodes of erect axes. Axes are irregularly dichotomously branched, spreading below and clustered above, with few to many lateral proliferations, and pincer-shaped to straight tips. Nodal cortication is projecting near tips and strongly serrate; lower axial nodes are not serrate, 210–450 µm in diam., and broader than long. Internodes are 185–460 µm in diam. by 0.5–1.25 mm long in older segments. Gland cells may be present. Cortication is obvious, not confluent at tips, and in four or more transverse bands; mid band cells are large, round, unequal in size, and partially covered by small cells. Tetrasporangia are tetrahedrally divided, elliptical, 50–75 µm in diam., and in whorls at nodes. Spermatangia arise as tufts from cortical cells. Carposporophytes have 1–2 gonimoblastic lobes; cystocarps are near tips of branches and have 1–6 involucral branches.

Uncommon, on seagrasses and seaweeds in shallow water at the Keys; also known from Newfoundland to the Carolinas (intertidal to 29 m), Georgia, the Caribbean, Gulf of Mexico, Brazil, and Argentina, as well as the eastern North and tropical Atlantic (type location: Scotland; Silva et al., 1996), Mediterranean, Red Sea, Indo-Pacific, Australia, and Antarctica.

Taylor 1960: p. 531, 532; Schneider and Searles 1991: p. 377.

Ceramium flaccidum (Kützing) Ardissone 1871: p. 40 (L. *flaccidus*: flaccid, soft) Plate XXX, fig. 28–30.

Basionym. *Hormoceras flaccidum* Harvey *ex* Kützing 1862: p. 21, pl. 69, Fig. a–d

Synonyms: *Ceramium byssoideum* Harvey 1853: p. 218

Ceramium transversale Collins and Hervey 1917: p. 143, pl. V, Fig. 29, 31

Ceramium taylorii E. Y. Dawson 1950: p. 127–128, pl. 2, fig. 13, pl. 4, fig. 31–33

Plants are 0.3–12 cm tall and corticated only at nodes by 3–6 rows of cells; erect axes arise from interwoven prostrate ones whose ventral periaxial cells produce unicellular rhizoids with digitate pads. Erect axes are alternately branched and may be determinate; branching is every 10th to 14th axial cell; tips are pincer-like and incurved. Cortication is nodal with 3–6 transverse rows of small, round to angular cells and an occasional gland cell near nodal edges. Axes have 4–7 periaxial cells and the basal cortical filament is 1–4 cells long. Gland cells are rare, ovoid, and 12 by 8 µm.

Probably common, on seagrasses and seaweeds in shallow waters on all coasts and in 33 m off the east coast; probably in the Carolinas (18–21 m), Bermuda, the Bahamas, Caribbean, Gulf of Mexico, Brazil, the British Isles (type location: Kilkee, Ireland; Silva et al., 1996).

Ceramium floridanum J. Agardh 1894: p. 46 (L. *floridanum*: of or from the state of Florida). Plate XXX, fig. 16–18.

Filaments are uniseriate, delicate, to 8 cm tall, pink to pale red, with prostrate and erect axes, and attached by rhizoids from basal cells and nodes of prostrate axes. Branching is irregular to alternate below; above spreading, more regular, alternate to almost pinnate, and frondlike. Main axes are 2–3 mm wide and 10 mm long; older segments are 200–300 µm in diam. and 330–400 µm long; older nodes are 55–430 µm in diam., 95–470 µm long, and not swollen; younger nodes are 30–60 µm in diam. and 150 µm long. Nodal cortication is compact with 1–3 bands of large rounded cells covered by smaller angular ones. Tetrasporangia are tetrahedrally divided, spherical, to 75 µm in diam., in whorls on branchlets, projecting, and swollen by cortication. Cystocarps are terminal on short lateral branches and subtended by a few involucral branches.

Uncommon, in shallow waters on seagrasses, seaweeds, and limestone at Point of Rocks, Marineland, and the Keys (type location: Florida), as well as Venezuela.

Taylor 1960: p. 534; Dawes 1974: p. 145; Schneider and Searles 1991: p. 380.

Ceramium leptozonum M. Howe 1918: p. 531 (Gr. *lepto*: slender, thin + *zonum*: girdle).

Filaments are delicate, to 3 cm tall, pale red, and attached by rhizoids from lower axial cells and nodes of prostrate axes. Branching is regularly pseudodichotomous and abundant above; forks are acute and tips pincer-shaped or straight. Segments are 40–70 µm in diam., 2.5–4 diam. long, terete below, and swollen above. Cortication is nodal, usually with a single band of 4 vertically elongated cells, 20–40 µm long around the cross walls of young axial cells. A second, incomplete band of small cortical cells may also be present. Tetrasporangia are tetrahedrally divided, projecting, usually 50–65 µm in diam., 1(3) per node, usually secund on outer edges of nodes, and subtended by 2–4 small cells.

Rare, reported from 26–92 m off the east coast (Hanisak and Blair, 1988); also known from Georgia (19–21 m), Bermuda (type location in a salt pond; Schneider, 2003), and Mauritania.

Taylor 1960: p. 525; Schneider and Searles 1991: p. 380.

Ceramium nitens (C. Agardh) J. Agardh 1851 (1848–1901): p. 130 (L. *nitens*: gleaming, shining). Plate XXX, fig. 19, 20.

Basionym. *Ceramium rubrum* (Hudson) C. Agardh var. *nitens* C. Agardh 1824: p. 136.

Filaments are uniseriate, rust to rose red, in erect tufts or interwoven strands 5–10 cm tall, and attached by rhizoids with discoid haptera that arise from basal cells of erect axes and nodes of prostrate ones. Branching is regular, dichotomous to alternate below, and spreading or secund above; tips are not pincer-like. Main axes are 130–430 µm in diam. and axial cells 2–3 times as long as broad. Nodal banding is not evident and cortication by small round cells is present throughout all axes. Hairs, if present, are straight, to 1 mm long, and at nodes. Tetrasporangia are tetrahedrally divided, randomly dispersed, and within the cortication. Cystocarps are partially covered by 2–4 involucral branches.

Rare or infrequent, on algae and rubble (to 10 m) at Big Pine Key and Key West; may form blooms on soft corals during spring (Lapointe et al., 2004); also known from Bermuda, the Bahamas, Antilles (type location: "ad Antillas"; C. Agardh, 1824), Cape Verde Island, and the Philippines.

Taylor 1960: p. 535; Littler and Littler 2000: p. 150.

Ceramium reptans Cho and Fredericq 2006: p. 496, fig. 1–23 (L. *reptans*: creeping). Plate XXX, fig. 21–23.

Plants are minute, epiphytic, prostrate, and produce short (to 15 segments long) erect unbranched filaments 0.8–1.2 mm tall. Prostrate axes are attached by short rhizoids with 3–4-celled digitate pads and branch pseudodichotomously. Nodes are incompletely corticated with 2 narrow bands of cells, including 5 periaxial cells derived from oblique divisions of the parent axial cell. Each periaxial cell produces three cortical cells, 2 above and 1 below; the first acropetal cortical cell becomes a pseudoperiaxial cell that extends longitudinally. Tetrasporangia are in whorls at nodes, cruciately divided, spherical to ellipsoidal, and 41 by 40 µm. Spermatangia occur in whorls around several nodes, are colorless, spherical to elliptical, and 1 by 2 µm. Female plants are unknown.

Common(?), described from the Missouri-Ohio Key Channel in the Florida Keys growing on *Hypnea spinella* with *Ceramium codii*.

Ceramium subtile J. Agardh 1851 (1848–1901): p. 120 (L. *subtile*: thin, delicate). Plate XXX, fig. 24.

Synonyms. *Ceramium chilense* (Kützing) P. Crouan and H. Crouan in Mazé and Schramm 1878 (1870–1877): p. 149.

Ceramium cornigerum P. Crouan and H. Crouan in Mazé and Schramm 1878 (1870–1877): p. 149.

? *Ceramium nodiferum* (Kützing) P. Crouan and H. Crouan in Mazé and Schramm 1878 (1870–1877): p. 149.

Filaments are uniseriate, in soft entangled tufts to 15 cm long, pink to rose red, and attached by rhizoids. Branching is mostly dichotomous with strongly incurved tips. Lower segments are to 120 µm in diam. and 1.25 mm long; nodes are to 130 µm in diam. and mostly shorter than wide. Cortical cells are in three distinct bands; an upper one has 2–4 irregular rows of small angular cells; the mid band has a row of large rounded cells; the lower band has 2–4 rows of small irregular cells. Tetrasporangia are tetrahedrally divided, spherical, 60 µm in diam., mostly solitary, and project from nodes. Cystocarps occur on forks of upper branches and usually have a few involucral filaments.

Rare, in shallow water in the Keys; also known from Bermuda, the Bahamas, Caribbean, and Texas.

Taylor 1960: p. 527.

Ceramium virgatum Roth 1797: p. 148, pl. VIII, fig. 1 (L. *virgatum*: twiggy, long and slender, streaked, or rodlike). Plate XXX, fig. 25–27.

Synonyms. *Ceramium nodulosum* (Lightfoot) Ducluzeau 1805: p. 61.

Ceramium rubrum (Hudson) C. Agardh 1811 (1810–1812): p. 17 (*nom. cons. prop.* Silva et al. 1996: p. 403).

Ceramium rubriforme Kylin 1907: p. 183, pl. 7, fig. 7.

Some have referred this cosmopolitan species to *Ceramium rubrum* (Garbary et al., 1978; Womersley, 1978; South and Tittley, 1986), while Maggs and Hommersand (1993) designated it as *C. nodulosum* (Lightfoot) Ducluzeau. Silva et al. (1996) considered the latter name illegitimate because of the existence of *C. nodulosum* De Candole (in Lamarck and De Candole 1805: p. 45) and suggested that *C. rubrum* be retained because of its wide use (*nom. cons. prop.*). Maggs et al. (2002) showed that there was a complex of fully corticated *Ceramium* species in the British Isles having the same *rbcL* sequences as *C. virgatum*. Skage et al. (2005) confirmed the same pattern after evaluating the phylogenetic basis of three complexes ('*C. rubrum*' and '*C. strictum sensu* Harvey groups,' plus one with cortical spines). We follow the designation by Guiry and Nic Dhonncha (2006) of *C. virgatum*, while Wynne (2005a) listed it as a questionable record from the tropical and subtropical western Atlantic because of its uncertain taxonomic status.

Filaments are uniseriate, stiff, coarse, black to rose red, to 40 cm tall, alternately to dichotomously branched, wide spreading, and attached by rhizoidal mats. Axes are completely corticated; main axes have weak to strong macroscopic bands and slightly swollen segments, to 1 mm in diam.; tips are straight, incurved, and pincer-shaped. Axial cells are 1–3 diam. long. Cortical cells are angular to rounded; subcortical cells are elongate and arise from 8–16 large internal nodal cortical cells. Tetrasporangia are tetrahedrally divided, ovoid or spherical, 30–80 μm in diam. by 40–80 μm long, immersed, scattered or confined to nodes and making axes nodulose. Spermatangia occur in tufts. Carposporophytes are spherical, near the tips of short lateral branchlets, and have 2–6 involucral filaments.

Uncommon, reported from Marineland, Florida (Humm, 1952a); also known from Hudson Bay, Canada, to the Carolinas (adrift), Antilles, and northern Gulf of Mexico, as well as the eastern North and tropical Atlantic (type location: Oldenburg, Germany), Mediterranean, Black and Red Seas, Indo-Pacific, and Australia.

Taylor 1960: p. 535 (with doubt); Schneider and Searles 1991: p. 381; Littler and Littler 2000: p. 152.

Compsothamnion (Nägeli) F. Schmitz 1889: p. 450 (Gr. *kompsos*: elegant + *thamnion*: a small shrub).

Filaments are uniseriate, ecorticate, erect, bushy, and attached by rhizoids. Branching is abundant, to 3 times, alternate, in one plane, and each axial cell produces a branch. Tetrasporangia are tetrahedrally divided and terminate lateral branches. Gametophytes are dioecious. Spermatangia and procarps occur on lateral branches. Carpogonial branches are 2-celled. Carposporophytes have 1–2 gonimoblastic lobes and lack sterile involucral filaments.

Compsothamnion thuyoides **(J. E. Smith in J. E. Smith and J. Sowerby) F. Schmitz 1889: p. 450 (L. *Thuja*: the generic name for arborvitae, a conifer with dense, flattened branches + *oides*: like; thus, similar to *Thuja*). Plate XXXI, fig. 1.**

Basionym. *Conferva thuioides* J. E. Smith in J. E. Smith and J. Sowerby 1810 (1790–1814), pl. 2205 (as '*thuyoides*').

Filaments are uniseriate, in delicate soft tufts, 2(8) cm tall, pink to brown, distinctively alternately branched in one plane on opposite sides of axes, and attached by delicate rhizoids growing from lower cells. Axes are flexuous and lower cells 100–230 μm in diam., to 400 μm long, and with walls to 20 μm thick. Basal cells of lateral branches are 50–100 μm in diam. and with multicellular rhizoids growing down the axes. Branchlet cells are to 50 μm in diam., they taper to tips, and have walls to 10 µm thick. Tetrasporangia are tetrahedrally to irregularly divided, slightly ovoid, 30–50 µm in diam., 50–60 µm long, and have short branchlets at their tips. Procarps are terminal or lateral. Carposporophytes are solitary, exposed, lobed, and on outer branches.

Rare, reported as an epiphyte in 37–93 m off the east coast (Hanisak and Blair, 1988); also known from the Carolinas (40–60 m), Bermuda, and Bahamas, as well as the eastern North and tropical Atlantic (type location: Yarmouth Beach, England) and Mediterranean.

Taylor 1960: p. 511; Schneider and Searles 1991: p. 382; Littler and Littler 2000: p. 152.

Crouania J. Agardh 1842: p. 83; 1851 (1848–1901): p. 104 (named for the two French brothers H. M. Crouan and P. L. Crouan, who studied Caribbean seaweeds).

Plants are uniseriate, erect, bushy, and attached by rhizoids from basal cells. Axes are in a jelly-like matrix that may contain calcium carbonate. Branching is abundant and irregular to alternate; branchlets are cylindrical and taper to tips. Axial cells are large and produce whorls of 3–4 branches. Tetrasporangia are irregularly tetrahedrally divided and occur on distal ends of lower branch cells. Cystocarps are lateral or terminal, solitary, immersed, and lack sterile involucral filaments.

Key to the species (3) of *Crouania*

1. Main axes ecorticate 2
1. Main axes corticated by down-growing rhizoids . . . . *C. pleonospora*
    2. Rhizoids small and not obvious; axial cells to 170 µm in diam.; branchlets bend upward on central axes; tetrasporangia are 1 per branchlet fascicle. . . . . *C. attenuata*
    2. Axes attached by rhizoidal haptera; axial cells 70–180 µm in diam.; branchlets not curled upward; tetrasporangia are 1–3 per branchlet fascicle . . . . *C. elisiae*

Crouania attenuata (C. Agardh) J. Agardh 1842: p. 83 (L. *attenuata*: drawn out, reduced, narrowed). Plate XXXI, fig. 2, 3.

Basionym. *Mesogloia attenuata* C. Agardh 1824: p. 51.

Synonym. *Crouania bispora* P. Crouan and H. Crouan 1848: p. 374, pl. 12, fig. 21–23.

Plants are uniseriate, pale pink to white, delicate, gelatinous, to 5 cm tall, and attached by inconspicuous rhizoids or entangled with other algae. Gel of older axes is whitened by calcium carbonate. Axes are irregularly alternately branched and 0.2–0.5 mm in diam.; main axes are ecorticate, to 170 µm in diam., and 300µm long. Branchlets are in whorls and curve upward. Basal cells of branchlets are 25–30 µm in diam. by 30 µm long; upper cells are 5–8 µm in diam. by 12–48 µm long and have blunt to hair-tipped tips. Tetrasporangia are irregularly tetrahedrally divided, ovoid, 40–80 µm in diam., 60–85 µm long, and single on basal cells of branchlets. Spermatangia occur in small clusters (2–3) on tips of branchlets. Cystocarps are solitary and on short branches.

Common, entangled with seaweeds and on seagrass blades or limestone (to 20 m) on all coasts; also known from Bermuda, the Bahamas, Caribbean, and Gulf of Mexico, as well as the British Isles to the Mediterranean (type location: Brittany, France; Maggs and Hommersand, 1993), eastern tropical Atlantic, Red Sea, Indo-Pacific, and Australia.

Taylor 1960: p. 495; Dawes 1974: p. 146; Littler and Littler 2000: p. 154.

Crouania elisiae Schneider 2004: p. 566–569, fig. 10–30 (named for Dr. Elise M. Wollaston, an Australian phycologist who studied *Crouania*). Plate XXXI, fig. 4.

Plants are uniseriate, erect, to 3 cm tall, covered with mucilage, lightly calcified, and attached by rhizoidal haptera modified from whorled branchlets and by uniseriate branched rhizoids from basal cells of branchlets. Branching is irregularly sympodial (ramisympodial). Main axes are slender and ecorticate; axial cells are 70–180 µm in diam., 130–275 µm long below, with fibril-like plastids, and have three branchlets in a whorl. Branchlets are widely di- or trichotomously branched, taper to tips, do not curl upward, and have basal cells 10–30 µm in diam. by 23–45 µm long. Tetrasporangia are formed sequentially (1–3) from basal cells of branchlets, are tetrahedrally divided, and 35–67 µm in diam. Cystocarps (1–2) occur at tips, 120–200 µm in diam., and have elongated branchlets that form loose involucres. Carposporangia are ovoid and 25–55 µm long.

Possibly common in Florida (Schneider, 2004), from the Keys near Bahia Honda Bridge and Key Largo canal cut, as well as Bermuda (type location: John Smith's Bay), Haiti, and Antigua.

Crouania pleonospora W. R. Taylor 1928: p. 193, pl. 27, fig. 10; pl. 30, fig. 5; pl. 32, fig. 8 (Gr. *pleon*: many, more + *spora*: seed, spore). Plate XXXI, fig. 5.

Filaments are uniseriate, in soft rose red to white gelatinous tufts, to 7 cm tall, lightly calcified in older parts, and attached by rhizoids from basal axial cells. Branching is irregular to alternate with main axes 0.2–2.0 mm in diam.; axial cells are 250–400 µm in diam. and 0.5 to 2.0 times as long as wide. Older (lower) main axes are loosely to densely corticated by descending rhizoids arising from basal cells of branchlets. Lateral branchlets occur in whorls and arise at right angles above the midpoint of axial cells. Branchlets branch irregularly or alternately and two to many times; basal cells are to 30 µm in diam. by 50 µm long; tip cells are to 8 µm in diam. by 24 µm long. Tetrasporangia are irregularly tetrahedrally divided, spherical, 60–90 µm in diam., and have one to several per fork of lateral branches. Cystocarps are in groups of 1–2 on branches of small plants.

Uncommon, from 9 m on limestone, seagrasses, and algae in the Keys (type location: Dry Tortugas); also known from Bermuda and the Caribbean.

Taylor 1960: p. 496; Littler and Littler 2000: p. 154.

Diplothamnion A. B. Joly and Yamaguishi in Joly et al. 1966 ('1965'): p. 169, pl. IV, fig. 1–5 (Gr. *diplo*: paired + *thamnion*: a shrub).

Filaments are uniseriate, ecorticate, delicate, and arise from stolonlike creeping axes that are attached by rhizoids. Erect axes are decussate; branchlets occur in opposing pairs that alternate at right angles above and below one another. The apical cell cuts off segments by horizontal walls and the third distal cell divides to form two opposite, slightly unequal initials that produce 2(3)-celled whorled branchlets. Basal cells of some branchlets produce pairs of 2-celled filaments or mixtures with one an indeterminate axis. Tetrasporangia are tetrahedrally divided and on erect axes or basal cells of branchlets.

Diplothamnion jolyi Hoek 1978: p. 51 (47–61) (named for A. B. Joly, a Brazilian scientist who studied Brazilian seaweeds, especially red algae). Plate XXXI, fig. 6, 7.

Filaments are uniseriate, 0.3–0.5(1.0) cm tall, ecorticate, and have erect axes arising from prostrate ones that are attached by unicellular rhizoids with lobed discs. Each erect axial cell has two short, opposing pairs of branchlets at right angles to the pairs above and below to form four vertical rows (decussate); tip cells are rounded. Basal cells of some branchlets have pairs of 2-celled filaments or indeterminate axes. Prostrate axes are 30–50 µm in diam. Branchlets are 25–40µm in diam., with basal cells 18–30 µm in diam. Tetrasporangia are tetrahedrally divided, 25–35 µm in diam., sessile, solitary, and on the distal, adaxial side of basal branchlet cells. Gametophytes are unknown. The species resembles *Monosporus indicus* Børgesen, but it has opposite instead of subdichotomous branching and tetrasporangia rather than propagules.

var. *ecellulare* Bucher and J. N. Norris in Bucher et al. 1990: p. 303–305, fig. 16 (L. *e-*: a negative condition, destitute of + *cellulare*: celled).

The variety is taller (to 1 cm) than the species and has rounded to abruptly pointed tip cells. Axial cells are 75–110 µm in diam. and branchlet basal cells 30–54 µm in diam.

Common, the species and variety grow in 5–6 m at Molasses Reef (type location of variety), while the species grows from 26–42 m on the east coast (Hanisak and Blair, 1988); also known from Bermuda, Puerto Rico, Curaçao (type location), Columbia, and the Canary and Hawaiian Islands; the variety is known from the Bahamas (Ballantine and Aponte, 2005).

Grallatoria M. Howe 1920a: p. 560 (L. *grallator*: to walk on stilts; also *gralliae*: stilts).

Filaments are uniseriate, dorsiventrally organized, ecorticate, and attached by rhizoidal haptera. Prostrate axes have 2–3 short branchlets at nodes, 2 lateral and one dorsal. Lateral branchlets divide at bases with one fork forming a short haptera and the other an erect branch; thus, the plants appear stiltlike. Erect (dorsal) branches are stiff and repeatedly subdichotomously to trichotomously branched. Plastids are irregular to spiral. Tetrasporangia are tetrahedrally divided and dorsal on branchlets.

Grallatoria reptans M. Howe 1920a: p. 560 (L. *reptans*: creeping, prostrate, and rooting). Plate XXXI, fig. 8–10.

See Abbott (1976) and Wynne and Ballantine (1984) for descriptions of the plant.

Filaments are uniseriate, uninucleate, ecorticate, and in red velvety mats to 5 mm high. Erect lateral axes are simple or once divided and arise from simple or sparingly dichotomously branched prostrate axes attached by rhizoids. Rhizoids are unicellular or multicellular, 20–26 µm in diam., 150–450 µm (4–6 cells) long, and simple or once forked. Erect axes are sub- to trichotomously branched and 1–4 mm tall; cells are 34–46 µm in diam. and 1.5–2 diam. long. Upper axes have clusters of secund branchlets inside the final forks. Axial cells are 30–100 µm in diam. by 1.5–2.0 diam. long; most prostrate cells produce 3 opposing branches: 2 dorsal branchlets and a ventral hapteral branch. Dorsal branchlets are 200–500 µm long and have 4–8 cells; basal cells are 20 µm in diam.; tip cells are 8 µm in diam. and filiform. Tetrasporangia are tetrahedrally divided, on 1-celled stalks, ovoid, 25–50 µm in diam., 40–64 µm long, and near tips of repent branchlets. Spermatangial sori are ovoid to elongate, stalked, and on indeterminate branches. Cystocarps occur at tips of branches and are to 400 µm in diam. Carposporangia are tear-shaped and 20–23 by 50–55 µm.

Rare, reported from 53 to 55 m off the east coast (Hanisak and Blair, 1988); also known from Bermuda, the Bahamas (type location), Puerto Rico, and Canary Islands.

Taylor 1960: p. 496.

Griffithsia C. Agardh 1817: p. xxviii (as '*Griffitsia*,' orth. mut.), W. J. Hooker 1821: p. 84, *orth. cons.* (named for Amelia Warren Griffiths of Torquay, an ardent British collector of seaweeds who sent Harvey, J. Agardh, and D. Turner her collections).

Filaments are uniseriate, ecorticate, erect or prostrate, bushy, with sparse to dense and subdichotomous or lateral branching, and attached by rhizoids. Axes appear beaded. Cells near tips have whorls of hyaline trichotomous to polychotomous hairs. Axial cells are large, elongate to ovoid, multinucleate, and have many discoid plastids and a large vacuole. Tetrasporangia are tetrahedrally or cruciately divided, in whorls at nodes, and may have involucral branches. Gametophytes are mostly dioecious. Spermatangia are in whorls at nodes and covered by involucral branches or in caplike clusters on terminal cells. Procarps are subapical and single or in pairs on 3-celled axes. Carpogonial branches are 4-celled with 2-celled involucral branches that arise from hypogynous cells. Carpospores occur in a gelatinous matrix.

Key to the species (2) of *Griffithsia*

1. Erect axes attached by rhizoids; hairs trichotomously branched . . . . *G. globulifera*
1. Erect axes arise from prostrate ones; hairs dichotomously branched . . . . *G. heteromorpha*

Griffithsia globulifera Harvey *ex* Kützing 1862: p. 10, pl. 30, fig. a–d (L. *globulifera*: with clusters of small globes or spheres). Plate XXXI, fig. 11, 12.

Synonyms. *Griffithsia corallina* C. Agardh var. *globifera* Harvey 1853: p. 228, pl. XXXV, fig. A.
Griffithsia globifera (Harvey) J. Agardh 1876 (1848–1901): p. 67.
Griffithsia bornetiana Farlow 1881 ('1882'): p. 131, pl. X, fig. 4, pl. XI, fig. 3, 5.

Axes are uniseriate, ecorticate, to 6.5 cm tall, in bushy or matlike pink to rose red tufts, and attached by fibrous rhizoids of basal cells. Branching is mostly dichotomous; branches are clustered and elongate (fastigiate) or proliferous; axes are beadlike to segmented. Upper cells produce whorls of hyaline filaments that branch trichotomously. Axial cells are large, elongate to ovoid, multinucleate, and have many discoid plastids and a large vacuole. Older cells are 0.4–0.9 mm in diam. by 0.6–3.2 mm long; cells near tips are globular. Branches of tetrasporic and female plants taper near apices, while those of male plants end in large cells. Tetrasporangia are tetrahedrally divided, spherical, to 110 µm in diam., in whorls of 8–20 at nodes, up to 3 on 1-celled stalks, and covered by incurved sausage-shaped involucral cells. Gametophytes are usually dioecious. Spermatangia are elliptical, 2.5–3.2 µm in diam., 3.8 µm long, and in caplike sori that cover 30–50 percent of terminal cells. Procarps are near branchlet tips. Carposporophytes have a central fusion cell and gonimoblastic filaments. Carposporangia are elliptical to spherical, 15–40 µm in diam., and surrounded by sausage-shaped involucral cells.

Common, on algae, salt marsh culms, mangrove roots, and limestone, and to 34 m on all coasts; also known from warm-water sites in the southern Gulf of St. Lawrence and southward (type location: New York; Silva et al., 1996) to North Carolina (20–40 m), Georgia, Bermuda, the Caribbean, Gulf of Mexico, Brazil, and Indian Ocean.

Taylor 1960: p. 517; Dawes 1974: p. 146; Schneider and Searles 1991: p. 384; Littler and Littler 2000: p. 156.

Griffithsia heteromorpha Kützing 1863a: p. 2, pl. 3, fig. a–b (L. *heteromorpha*: variable or different in form). Plate XXXI, fig. 13, 14.

Filaments are uniseriate, ecorticate, delicate, gelatinous, in tufts of 8–10 axes to 2 cm tall, and from prostrate axes attached by multicellular rhizoids that arise from distal ends of cells. Branching is dichotomous to irregular; branches often are fused by short lateral cells. Prostrate axial cells are 500–1700 µm in diam.; lower axial cells are 300–900 µm in diam., to 2 mm long, and ovoid to terete. Upper axial cells are bulbous, beadlike, 450–900 µm in diam., 500–1000 µm long, and have spherical tip cells. Hairs are hyaline, in whorls on distal ends of terminal cells, dichotomously branched, and deciduous. Reproductive structures lack involucral filaments. Tetrasporangia are tetrahedrally divided, spherical to tear-shaped, 110 µm in diam., and occur on short stalks at nodes. Spermatangia form dense collars on nodes between distal cells and cystocarps near tips.

Rare, reported from 33–41 m off the east coast (Hanisak and Blair, 1988); also known from Bermuda, Puerto Rico, and Belize, as well as the Indo-Pacific (type location: New Caledonia; Abbott, 1999), and Australia.

Gymnothamnion J. Agardh 1892: p. 27 (Gr. *gymno*: naked + *thamnion*: a small bush).

Filaments are uniseriate and ecorticate; fronds are plumose, arise from prostrate axes, and attached by rhizoids. Main axes are oppositely pinnately to bipinnately branched. Tetrasporangia are tetrahedrally divided, stalked, and distal on tip cells. Spermatangia are produced from lower branch cells and in secund to comblike clusters. Cystocarps are elongate, clustered on axial tips, and lack involucral filaments.

Gymnothamnion elegans **(Schousboe *ex* C. Agardh) J. Agardh 1892: p. 178 (L. *elegans*: graceful or elegant). Plate XXXI, fig. 15.**

Basionym. *Callithamnion elegans* Schousboe *ex* C. Agardh 1828 (1820–1828): p. 162–163.

Synonym. *Gymnothamnion bipinnatum* Collins and Hervey 1917: p. 139–141, pl. IV, fig. 26.

Filaments are uniseriate, ecorticate, plumose and erect (0.5–4.0 cm tall) or in thin mats with pinnate branching. Prostrate axes produce erect ones dorsally every 2nd to 5th cell and opposite a rhizoid. Prostrate axial cells are 20–25(55) µm in diam.; erect cells are 2–5 diam. long and cylindrical to club-shaped. Erect axes are fernlike with upward curving branchlets (pinnae), 10–15 µm in diam., 60–210 µm long, and often regularly pinnate or secund. Branching is variable, reduced, absent, or alternating at 90° in succeeding nodes; branchlets often arise from distal ends of erect axial cells. Tetrasporangia are tetrahedrally divided, ovoid to spherical, 30–35 µm in diam., and at or near tips. Gametophytes are dioecious. Spermatangia are to 1 µm in diam, distal, and in pinnate or bipinnate series. Procarps are elongate, subterminal, and on short laterals. Cystocarps have 2–3 ovoid lobes.

Rare, inconspicuous, on intertidal turf algae at wave-exposed sites and from 58–92 m off the east coast; also known from Bermuda, the Bahamas, Caribbean, and Brazil, as well as Atlantic Spain and Portugal, the Mediterranean (type location: Tangier, Morocco; Womersley, 1998), eastern tropical Atlantic, Indo-Pacific, and Australia.

Taylor 1960: p. 522.

Monosporus Solier in Castagne 1845: p. 242; 1851: p. 119, as '*Monospora*' (Gr. *mono-*: single or one + *sporus*: a seed or spore.

Synonyms. non *Monospora* Hochstetter 1844: p. 660 (a flowering plant genus in the family Euphorbiaceae).

Neomonospora Setchell and Gardner 1937: p. 86.

Corynospora J. Agardh 1851 (1848–1901).

Filaments are uniseriate, ecorticate, erect or prostrate, and with uninucleate cells; main axes are dichotomously branched and attached by unicellular or multicellular rhizoids of prostrate axes that often arise opposite an erect axis. Filaments are mostly dorsiventral; erect axes are determinate or indeterminate and mostly oppositely branched. Growth is monopodial or sympodial. Monosporangia, tetrasporangia, or polysporangia occur at forks on distal ends of vegetative cells. Spermatangia are produced in ovoid heads on the distal ends of vegetative cells. Procarps have 4-celled carpogonial branches, with subapical cells of the fertile branches acting as support cells.

***Monosporus indicus* Børgesen 1931: p. 12–13, fig. 8, as '*Monospora indica*' (L. *indicus*: of or from India or the East Indies). Plate XXXI, fig. 16–18.**

Filaments are uniseriate, ecorticate, prostrate or erect, 1–2 cm tall, and attached by elongate unicellular or multicellular rhizoids that arise from distal ends of prostrate axial cells and terminate in clawlike to lobed pads. Cells are uninucleate. Prostrate axes have sympodial growth; the subapical cells cut off new apical cells obliquely and alternately from the upper corner of every segment and produce an opposing ventral rhizoid and dorsal lateral branch. Prostrate axial tips curve upward, cells are to 130 μm in diam. and to 500 μm long. Branching of erect axes is mostly dichotomous; cells are 50–150 μm in diam. and to 350 μm long; tip cells are blunt and to 28 μm in diam. Lateral branches are mostly determinate, dichotomous, and fork 2–3 times to form branchlets or a few indeterminate axes at irregular intervals. Monosporangia are stalked, ovoid, 90–180 μm in diam., 90–200 μm long, and at the distal ends of branchlet cells; stalks are 30–40 μm in diam. by 25–35 μm long. Sexual reproduction or tetrasporangia have not been observed.

Rare, reported from 9 m on algae at the Dry Tortugas (Ballantine, 1996); also known from Puerto Rico, the eastern North Atlantic, and Indo-Pacific (type location: Bombay, India; Silva et al., 1996).

Perikladosporon Athanasiadis 1996a: 137–138 (Gr. *peri*: encircling + *klados*: a branch + *sporo*: a spore).

Filaments are uniseriate, ecorticate, and with erect and prostrate axes attached by rhizoids. Axes are straight and grow by transverse divisions of an apical cell; lower axial cells are abruptly larger than upper ones and form unequal whorled branches. Erect axes have distichous, opposite branching; branch pairs of prostrate axes may be at right angles to one another (decussate). Gland cells are uncommon and adaxial on intercalary cells of whorled branchlets. Gametophytes are unknown. Tetrasporangia are spherical to ovoid, sessile, and usually on the basal cells of branchlets. The monotypic genus is distinguished by the adaxial position of its gland cells (Athanasiadis, 1996a).

***Perikladosporon percurrens* (E. Y. Dawson) Athanasiadis 1996a: p. 137–138 (L. *percurrens*: extending through or continuous). Plate XXXI, fig. 19–21.**

Basionym. *Antithamnion percurrens* E. Y. Dawson 1957: p. 116, fig. 24A,B.

Filaments are uniseriate, ecorticate, minute, and with erect (to 6 mm tall) and prostrate axes attached by short lateral rhizoids. Prostrate cells are 15–18 μm in diam. and mostly 2–3 diam. long. Erect axes are branched in one plane (distichous) with lateral branchlets formed oppositely from every cell; main axes are visible and percurrent. Indeterminate branches are absent; determinate laterals may bear 1–3 adaxial short blunt-tipped branchlets of 1(2) cell(s). Gland cells may be present. Tetrasporangia are tetrahedrally divided, sessile, to 20 μm in diam., and in large, obvious gelatinous envelopes on the adaxial sides of basal branchlet cells.

Rare, reported from 8 m at Molasses Reef (Bucher et al., 1990); also known from the Pacific Ocean (type location: Eniwetok Atoll; Abbott, 1999) and Australia.

Pleonosporium Nägeli 1862: p. 326, 339, *nom. cons.* (Gr. *pleon*: many + *sporos*: spores).

Filaments are uniseriate, ecorticate or slightly corticated, erect, and bushy; branching is alternate and distichous to multistichous with axes attached by multicellular rhizoids. Branching of laterals is similar to main axis, pinnate or unilateral to several orders. Tetrasporangia are polyhedrally or tetrahedrally divided, sessile or stalked, adaxial and secund, or alternately adaxial and abaxial on laterals. Gametophytes are dioecious. Spermatangia are sessile or stalked and in elongate heads at tips of branches. Procarps are subapical on indeterminate axes with 4-celled carpogonial branches. Carposporophytes have 4–8 gonimoblastic lobes surrounded by involucral filaments.

Pleonosporium flexuosum (C. Agardh) Bornet *ex* De Toni 1903: p. 1305 (L. *flexuosum*: tortuous, wavy, zig-zag). Plate XXXI, fig. 22, 23.

Basionym. *Ceramium flexuosum* C. Agardh 1824: p. 141.

Filaments are uniseriate, ecorticate or with a few rhizoids below, bushy and erect, to 6 cm tall, brownish red, and attached by uniseriate branched rhizoids (45–105 µm in diam.) from lower axial cells. Branching is pseudodichotomous below and alternate and distichous above; lateral branches are smaller than primary axes. Axial cells are club-shaped to terete, 300–625 µm in diam., 930–1560 µm long, and taper to 40–60 µm in diam. at tips. Branchlets are incurved and with obtuse tips. Polysporangia are ovoid, 65–110 µm in diam., 85–160 µm long, with 8–32 spores, and on branch cells. Spermatangia form ovoid heads on branchlet tips. Carposporophytes are spherical, to 510 µm in diam., on branchlet bases, and covered by incurved involucral filaments. Carposporangia are spherical, elongate, and 30–55 µm long.

Rare, a questionable record for Florida; known from North Carolina (18–27 m), France to Portugal, the Mediterranean, Morocco (type location: North Africa; Silva et al., 1996), and South Africa.

Schneider and Searles 1991: p. 390; Littler and Littler 2000: p. 160.

Pterothamnion Nägeli in Nägeli and Cramer 1855: p. 66 (Gr. *ptero*: feather, wing + *thamnion*: a small bush).

Filaments are uniseriate, ecorticate or sparsely covered by rhizoids from bases of branches, erect, bushy, and attached by multicellular rhizoids from prostrate axes and bases of erect branches. Each axial cell produces unequal branchlets in opposite, distichous, or whorled patterns. Gland cells are common to rare, spherical to angular and often on the upper surfaces of intercalary branch cells. Tetrasporangia are tetrahedrally divided, spherical, sessile or stalked, and single or clustered on branch cells. Gametophytes are dioecious. Spermatangia are clustered and produced by 1–3 successive orders of whorled mother cells that replace branchlet cells. Procarps (5–20) are on lateral axes; carpogonial branches are solitary or in a series on branchlet basal cells. Carposporophytes are single.

Pterothamnion plumula (J. Ellis) Nägeli in Nägeli and Cramer 1855: p. 66 (L. *plumula*: a small feather). Plate XXXII, fig. 1, 2.

Basionym. *Conferva plumula* J. Ellis 1768: p. 425, table XVIII, fig. g,G,G1.

Synonym. *Antithamnion plumula* (J. Ellis) Thuret in Le Jolis 1863: p. 112.

Filaments are uniseriate, ecorticate or sparsely covered by rhizoids from bases of branches, erect and bushy, delicate or stiff, to 12 cm tall, and attached by basal rhizoids from erect axes. Main axes are alternately to irregularly branched; branches are 100–200 µm in diam. with cells 3–5 diam. long. Branches produce opposite pairs of short, upward-curved final branches, 0.5–0.7 mm long; cells are 50 µm in diam. by 2–3 diam. long. Final branches have spinelike branchlets on their upper sides that taper from 20 µm to acute points. Gland cells are adaxial on final branches and cover one cell.

Tetrasporangia are tetrahedrally divided, ovoid, 25–30 by 35–40 µm, and sessile or on short stalks on lower branchlets near tips. Spermatangia are in ovoid patches on the upper sides of branchlets. Cystocarps are large, in clusters of 2–4, and surrounded by supporting branchlets.

Uncommon, on seagrasses, algae, limestone, and pilings in shallow waters in the Keys and in 42–93 m off the east coast (Hanisak and Blair, 1988); also known from Nova Scotia to New Jersey, Bermuda, the eastern North and tropical Atlantic (type location: Sussex, England; Silva et al., 1996), Mediterranean, Pakistan, and Australia.

Taylor 1960: p. 500.

Ptilothamnion Thuret in Le Jolis 1863: p. 118 (Gr. *ptilotos*: a wing or feather of a bird + *thamnion*: a small bush).

Filaments are uniseriate, ecorticate, and with prostrate axes attached ventrally by unicellular rhizoids with lobed discs. Branching of erect axes is simple, pinnate, unilateral, or irregular. Cells are multinucleate and have multiple discoid plastids. Tetrasporangia are tetrahedrally divided, sessile or on short stalks, and usually in terminal clusters on lateral branches. Some species have polysporangia. Gametophytes are dioecious. Spermatangia are in ovoid to spherical clusters and stalked or terminate lateral branches. Procarps are subapical; carpogonial branches are 4-celled with one auxiliary cell. Carposporophytes are globular and have one or more lobes surrounded by a loose cluster of filaments.

Ptilothamnion speluncarum **(Collins and Hervey) D. L. Ballantine and M. J. Wynne 1998b: p. 227, fig. 1–6 (L. *speluncarum*: of caves or in caves). Plate XXXII, fig. 3.**

Basionym. *Rhodochorton speluncarum* Collins and Hervey 1917: p. 147–148.

Synonym. *Spermothamnion speluncarum* (Collins and Hervey) M. Howe 1920a: p. 578.

Filaments are uniseriate, ecorticate, in dark red brown velvety mats, and attached by short rhizoids arising from the midregion of prostrate axial cells and opposite erect axes. Rhizoids are 30–40 µm in diam., 2–4 diam. long, and terminate in irregular pads. Erect axes are cylindrical to tapering and have none to a few alternate or secund branchlets. Most cells are 22–40 µm in diam; lower axial cells are 2–2.5 diam. long; upper ones are larger, 3–4 diam. long, and have rounded tips. Tetrasporangia are tetrahedrally divided, 50–65 µm in diam., solitarily or opposite, and sessile or on 1(2)-celled stalks at lower forks. Spermatangial clusters are terminal, ovoid, 20–30 µm in diam, and to 40 µm long. Procarps are on subapical cells.

Rare, reported on algae from 34–93 m off the east coast (Hanisak and Blair, 1988); also known from Bermuda (type location: from under ledges; Silva et al., 1996), the Bahamas, Caribbean, Brazil, Canary Islands, Ghana, and India.

Taylor 1960: p. 521; Littler and Littler 2000: p. 160.

Seirospora Harvey 1846 (1846–1851): p. 21 (Gr. *seira*: a chain + *spora*: seed).

Filaments are uniseriate, delicate, ecorticate or with rhizoidal growth, alternately branched below and subdichotomous above; axes terminate in slender branchlets and are attached by unicellular rhizoids from basal cells. Cells are cylindrical, uninucleate, and have many discoid plastids. Tetrasporangia are tetrahedrally or cruciately divided and lateral on upper branches. Seirospores (paraspores) are in radiating clusters that replace branch tips; terminal spores are largest. Carpogonial branches are 3(4)-celled and have two auxiliary cells. Gonimoblastic filaments are spreading and develop completely into carposporangia.

Key to the species (2) of *Seirospora*

1. Plants pink to red, with axes ecorticate . . . . *S. occidentalis*
1. Plants purple in hue; axes corticated below by rhizoidal filaments . . . . *S. purpurea*

Seirospora occidentalis Børgesen 1909b: p. 14; 1917: p. 222–225, fig. 209–212 (L. *occidentalis*: west, western, or occidental). Plate XXXII, fig. 4.

Filaments are uniseriate, ecorticate, bushy, 1–2 cm tall, pink to rose red, mostly alternately branched below and subdichotomous above, and attached by short rhizoids arising from thick-walled quadrate basal cells to 200 µm in diam. Axial cells are to 85 µm in diam. by 3–5 diam. long; tip cells are 8–11 µm in diam. by 2–10 or more diam. long. Tetrasporangia are mostly tetrahedrally divided, sessile, and occur adaxially on branchlet tips. Seirospores (paraspores) are on branchlet tips and in simple to branched rows of 3–5 short dark subspherical cells 18–20 µm in diam. Spermatangia are clustered and located like sporangia. Cystocarps have two groups of elongate gonimoblastic filaments on the lower cells of branches. Carpospores are 40–42 µm in diam.

Uncommon, on algae in shallow waters at the Keys and in 27 m on the west coast; also known from the Bahamas, Puerto Rico, Virgin Islands (type location: Christianfort), French Polynesia, and Australia.

Taylor 1960: p. 510; Dawes 1974: p. 147.

Seirospora purpurea M. Howe 1918: p. 529 (L. *purpurea*: purple, reddish purple). Plate XXXII, fig. 5, 6.

Synonym. *Callithamnion byssoideum* Arnott var. *jamaicense* Collins 1901: p. 258; Collins, Holden and Setchell 1900 (1895–1919): ex. no. 2045.

Filaments are uniseriate, corticated below by rhizoids, bushy, to 2.5 cm tall, usually pale purple, with alternate branching below and subdichotomous above, and attached by rhizoids. Axes are 100–250 µm in diam., tapered near tips, swollen at joints, and have walls 8–20 µm thick. Branchlets taper, are 6–13 µm in diam. near obtuse tips, and have cells 2–4 diam. long. Tetrasporangia are tetrahedrally divided, ovoid, 50–60 µm in diam., sessile, and scattered at forks. Spermatangial clusters are ovoid to irregular, 25–40 µm in diam., 45–65 µm long, and at forks. Cystocarps are domelike and 300–400 µm in diam. Gonimoblastic filaments are erect, spreading, and beadlike; carpospores are ovoid and 35–40 µm long.

Rare, on seagrasses and algae in the Keys and Bermuda (type location).

Taylor 1960: p. 511; Littler and Littler 2000: p. 162.

Spermothamnion Areschoug 1847: p. 334 (Gr. *spermo*: seed or sperm + *thamnion*: a shrub).

Synonym. *Herpothamnion* Nägeli 1862: p. 345.

Cystocarpic plants are needed to separate *Spermothamnion* and *Tiffaniella* as the involucral filaments of the former are a defining character (see Abbott, 1999).

Filaments are uniseriate, ecorticate, with erect axes oppositely or unilaterally branched, and attached by unicellular, elongate rhizoids with lobed pads. Cells are uninucleate and have numerous discoid plastids. Tetrasporangia are tetrahedrally divided; polyspores, if present, are in ovoid to cylindrical clusters and adaxial and terminal on branchlets. Carpogonial branches are 4-celled, from middle cells of lateral branchlet tips, and have two auxiliary cells that divide to form gonimoblastic filaments; the outermost cells produce carposporangia. Cystocarps lack involucral filaments.

Key to the species (4) of *Spermothamnion*

1. Filaments >1 cm tall and forming tufts or cushions 2
1. Filaments 2–8 mm tall and in feltlike low cushions or entangled with other algae 3
   2. Tetrasporangia 45–65 μm in diam.; axes oppositely branched . . . . *S. repens*
   2. Tetrasporangia 60–80 μm in diam.; axes irregularly branched . . . . *S. gymnocarpum*
3. Basal cells of erect filaments 10–20 μm in diameter . . . . *S. investiens*
3. Basal cells of erect filaments mostly 40–65 μm in diameter . . . . *S. macromeres*

Spermothamnion gymnocarpum M. Howe: 1920a: p. 579 (Gr. *gymnos*: naked, lightly clad + *carpum*: fruit, fruited). Plate XXXII, fig. 7–9.

Filaments are uniseriate, ecorticate, 1–3 cm tall, in purple tufts, and with prostrate axes attached by simple rhizoids with pads. Erect axes are irregularly to dichotomously branched below, opposite above, and secund to unilateral near tips. Erect axes are 50–115 μm in diam., 12–20 μm in diam. near tips, with thick walls 5–40 μm, and obtuse terminal cells. Prostrate axes are obvious; cells 50–130 μm in diam., 130–140 μm long, and with walls 13–50 μm thick. Tetrasporangia are tetrahedrally divided, single or in clusters (to 5) near forks, ovoid, 60–80 μm in diam., 40–80 μm long, and on 1-celled stalks. Gametophytes are dioecious. Spermatangial clusters are ovoid, 25–40 μm in diam., and 40–80 μm long. Cystocarps are hemispherical, 80–150 μm in diam., and solitary or in clusters; involucral filaments absent.

Uncommon, inconspicuous, and on algae, mangrove roots, and seagrasses in the Keys (to 10 m); also known from Bermuda, the Bahamas (type location), and Caribbean.

Taylor 1960: p. 518; Littler and Littler 2000: p. 162.

Spermothamnion investiens (P. Crouan and H. Crouan in Schramm and Mazé) Vickers 1905: p. 64 (L. *investiens*: clothed, covered). Plate XXXII, fig. 10, 11.

Basionym. *Callithamnion investiens* P. Crouan and H. Crouan in Schramm and Mazé 1865: p. 7; P. Crouan and H. Crouan in Mazé and Schramm 1878 (1870–1877): p. 141.

Filaments are uniseriate, ecorticate, minute, 1–3 mm tall, pale pink to red; prostrate axes are attached to hosts by unicellular rhizoids with lobed discs. Erect axes are numerous, 10–25 μm in diam., mostly simple or sparingly alternately branched, and 30–100 μm long. Tetrasporangia are tetrahedrally divided, ovoid, opposite or secund, and usually occur on 1-celled stalks, 30–40 μm in diam. by 37–45 μm long. Gametophytes are dioecious. Spermatangia are ovoid to elongate and solitary at branch tips. Cystocarps lack involucral filaments.

Rare, reported from 39–46 m off the east coast (Hanisak and Blair, 1988); also known from Bermuda, the Bahamas, Caribbean (type location: Guadeloupe; Silva et al., 1996), as well as Mauritania, Ghana, Gabon, and the Seychelles.

Taylor 1960: p. 520.

Spermothamnion macromeres Collins and Hervey 1917: p. 132–134; Collins, Holden, and Setchell 1900 (1895–1919): ex. no. 2044. (Gr. *macro*: great, large + *mere*: a part, a segment).

Filaments are uniseriate, ecorticate, and in minute mats to 8 mm tall; prostrate axes are to 65 μm in diam., with cells 3–5 diam. long, and attached by short rhizoids. Lower cells of erect filaments are 40–65 μm in diam.; upper cells are terete to club-shaped, to 30 μm in diam., and 5–20 diam. long. Branching is sparse, irregular to subdichotomous. Tetrasporangia are tetrahedrally divided, ovoid, 50–55 µm in diam., and terminate in short di- or trichotomous branchlets. Gametophytes are dioecious. Spermatangia are clustered on upper sides of branches, they are ovoid to cylindrical, 50–60 µm in diam., and 125–150 µm long. Cystocarps are terminal on short clavate stalk cells at forks and lack involucral filaments.

Rare, reported in 41 m off the east coast (Hanisak and Blair, 1988); also known from Bermuda (type location: Smith Bay), the Bahamas, Caribbean, and Mauritania.

Taylor 1960: p. 521.

Spermothamnion repens (Dillwyn) Rosenvinge 1924: p. 298 (L. *repens*: creeping, trailing). Plate XXXII, fig. 12.

Basionym. *Conferva repens* Dillwyn 1802 (1802–1809): pl. 18.

Synonym. *Spermothamnion turneri* (Mertens in Roth) Areschoug in Fries 1845: p. 127.

Taylor (1960) questioned Florida records of this common cold-water species (as *Spermothamnion turneri*). Schneider and Searles (1991) did not record it from the Carolinas.

Filaments are uniseriate, ecorticate, in erect tufts, 2–5 cm tall; erect axes arise from prostrate filaments growing on various hosts that are attached by unicellular rhizoids with discoid bases. Prostrate axial cells are 30–45 µm in diam. and 3–6 diam. long. Erect axes are oppositely pinnately branched, with spreading, slender branches; cells are 20–80 µm in diam. by 3–8 diam. long; branchlet cells are 20–45 µm in diam. Tetrasporangia are tetrahedrally divided, spherical, 45–65 µm in diam., in a series on adaxial sides of branchlets, and sessile or on 1-celled stalks. Procarps occur below tips of axes. Cystocarps are surrounded by small branchlets and lack involucral filaments.

var. *variabile* (C. Agardh) M. J. Wynne 1986: p. 2247, 2260 (L. *variabile*: variable).

Basionym. *Callithamnion variabile* C. Agardh 1828 (1820–1828): p. 163.

Taylor (1957) questioned the validity of this variety, as variable branching patterns can occur on the same axis.

Branching is alternate or unilateral and not opposite as in the species.

Rare, the species is recorded from 94 m off the east coast (Hanisak and Blair, 1988) and the variety on *Codium* (9 m) off Anclote Key (Dawes, 1974); the species is also known from Nova Scotia to Virginia, the eastern North Atlantic (type location: England; Maggs and Hommersand, 1993), Mediterranean, and Canary Islands; the variety is also known from Jamaica (with doubt; Taylor, 1960), the Mediterranean, and Mauritania.

Taylor 1960: p. 519; Dawes 1974: p. 148.

Spyridia Harvey in W. J. Hooker 1833: p. 259, 336 (Gr. *spyris*: a basket, reel).

Filaments are uniseriate and heavily corticated; young axes have many short, slender, and deciduous (determinate) branchlets; older ones are denuded below and attached by basal rhizoidal discs. Main axes are evident (percurrent) and abundantly branched; axial cells are large, corticated by a layer of small cells, and have weak nodal banding. Young axes and deciduous branchlets are corticated only at nodes (cross walls) by elongated pigmented cortical cells in transverse bands; the internodes (axial cells) are naked. Tips are blunt or acute and may have recurved spines arising below tip cells. Tetrasporangia are tetrahedrally divided and at nodes of branchlets. Gametophytes are dioecious. Spermatangia occur in hyaline patches on nodes of branchlets. Procarps are lateral or terminal on short axes. Carpogonial branches are 4-celled. Carposporophytes have 1–3 lobes and slender involucral filaments; carposporangia arise from most lobes.

Key to the species (3) of *Spyridia*

1. Axes of smaller branches not club-shaped or swollen 2
1. Smaller branches club-shaped or swollen near tips . . . . *S. clavata*
   2. Apical branch not hooked; tip cells of branchlet only have an erect spine 3
   2. Apical branch often have a thickened hook (crozier); tip cells of branchlets have an erect spine and one or more hooked spines . . . . *S. hypnoides*
3. Branchlets mostly distichous and incurved . . . . *S. ceramioides*
3. Branchlets usually radial around axes and not incurved . . . . *S. filamentosa*

Spyridia ceramioides J. Agardh 1894: p 144 (L. (Gr. *keras*: a horn + L. *oides*: similar to; thus similar to the genus *Ceramium*).

Filaments are uniseriate, corticated, and have radial, irregular, and widespread branching. Branchlets are similar, abundant, distichous, and incurved, have two rows of nodal cortication, with internodes twice as long as broad, and tips tapering to a point.

Uncertain record; although collected by Floretta A. Curtiss in Florida (type location) and sent to Agardh, no other collections have been recorded (Taylor, 1960; p. 542).

Taylor 1928: p. 196.

Spyridia clavata Kützing 1841 ('1842'): p. 744–745: 1849: p. 667 (L. *clavata*: club-shaped). Plate XXXII, fig. 13.

Filaments are uniseriate, bushy, erect, and 15–20 cm tall; axes are terete, to 1 mm in diam. below, 1–2 mm wide and flat above, and attached by rhizoidal mats. Branching of indeterminate and determinate (branchlets) axes is usually in one plane (distichous) and irregular, alternate to secund or opposite, and with a mix of large and small branches. Short branches are club-shaped, 1–4 mm long, and usually with acute tips. Axial cells are to 500 µm long and corticated to their tips by short oval cells and also layers of long cells over internodes. Branchlets are corticated only at nodes (cross walls) by bands of cells with internodes naked; branchlet cells are 30–55 µm in diam., 0.6–1.75 mm long, taper to 17–25 µm in diam., and have straight terminal spines. Tetrasporangia are tetrahedrally divided, spherical or ovoid, to 50 µm in diam., and on lower nodes of branchlets.

Common, on limestone (to 12 m) in the Keys; also known from North Carolina (in salt marshes and to 60 m), South Carolina, the Caribbean (type location: St. Thomas; Silva et al., 1996), Gulf of Mexico, Brazil, West Africa, the Red Sea, and Sumatra.

Taylor 1960: p. 541; Schneider and Searles 1991: p. 395.

Spyridia filamentosa (Wulfen) Harvey in W. J. Hooker 1833: p. 337; Harvey 1841: p. 101; 1849 (1846–1851): pl. 46; 1853: p. 204 (L. *filamentosa*: threadlike, filamentous). Plate XXXII, fig. 14, 15.

Basionym. *Fucus filamentosus* Wulfen 1803: p. 64

Molecular studies by Won et al. (2005) suggest that cryptic forms identified as *Spyridia filamentosa* may require the use of older names, creation of new taxa, and a reevaluation of the distribution of this species.

Filaments are uniseriate, bushy, to 20 cm tall, straw-colored to pale pink, with radial and alternate branching, and attached by discoid rhizoidal holdfasts. Main axes are corticated and 1–2 mm in diam.; lower axial cells are to 340 µm in diam. and upper ones 80–120 µm in diam. Cortication in younger axes is in alternating transverse bands of long narrow cells 6–10 µm in diam. by 50–70 µm long, and short broad ones 15–20 µm in diam. by 30–40 µm long; cortical bands are obscured with age. Branchlets (determinate axes) are radial, variable, delicate, deciduous, and 20–45 µm in diam. by 0.5–1.5 mm long; nodes (cross walls) are corticated by bands of cells, internodes are naked, and tips have a hyaline spine. Tetrasporangia are tetrahedrally divided, spherical, 40–70 µm in diam., and sessile at forks of branchlets. Gametophytes are dioecious. Spermatangia are to 5 µm in diam. and occur in sori on lower parts of branchlets. Cystocarps are terminal and on short branches with incurved involucral filaments, 180–225 µm in diam.

Common, on seagrasses in estuaries and limestone and in drift (to 8 m) on all coasts; also as scattered populations from James Bay, Quebec, Nova Scotia, and Maine and then continuous from Cape Cod through the Caribbean, Bermuda, Gulf of Mexico, and Brazil, as well as the eastern North and tropical Atlantic, Mediterranean (type location: Adriatic Sea; Abbott, 1999), Red Sea, Indo-Pacific, and Australia.

Taylor 1960: p. 539; Dawes 1974: p. 148; Littler and Littler 2000: p. 164.

Spyridia hypnoides (Bory in Bélanger) Papenfuss 1968b: p. 281–282 (L. *hypno*: moss + *oides*: like; thus, resembling the moss genus *Hypnum*). Plate XXXII, fig. 16, 17.

Basionym. *Thamnophora hypnoides* Bory in Bélanger 1834: p. 175.

Synonym. *Spyridia aculeata* (Schimper) Kützing 1843: p. 377; 1849: p. 668.

Schneider and Searles (1991) could not distinguish *Spyridia aculeata* from *S. hypnoides* in North Carolina.

Filaments are uniseriate, erect or entangled, 5–25 cm tall, straw-colored to pink, and attached by rhizoidal discs. Branching is radially alternate and sometimes in one plane (distichous). Main axes are cylindrical to slightly compressed and 1–2 mm broad; lower central axial cells are to 340 µm in diam. and 500 µm long; upper ones are 80–120 µm in diam. Indeterminate and determinate (branchlet) axes are mixed, taper from base to apex, and have acute tips. Some axes may end in thickened, hooklike (crozier) tips. Cortication is continuous to branch tips of main axes with transverse bands of long narrow cells, 6–10 µm in diam. by 50–70 µm long, that alternate with short broad cells, 15–20 µm in diam. by 30–40 µm long; bandings are lost with age. Branchlets are deciduous below, corticated only in bands at nodes, and 0.3–2 mm long by 30 µm in diam.; apical cells are acute, spinelike, and usually with 1–2(3) recurved spines. Tetrasporangia are tetrahedrally divided, spherical, 50–85 µm in diam., and solitary or in clusters of 2–3 on the lower nodes of branchlets. Carposporophytes have 2–3 gonimoblastic lobes, they develop at ends of short branchlets, and have incurved involucral filaments.

var. *disticha* (Børgesen) G. W. Lawson and D. M. John 1982: p. 300, pl. 47, fig. 1–3 (L. *disticha*: two-ranked, in two opposed ranks).

Basionym. *Spyridia aculeata* (C. Agardh *ex* Decaisne) Kützing var. *disticha* Børgesen 1917: p. 238–240, fig. 229.

Branches are featherlike and branchlets occur in one plane (distichous).

var. *inerme* (Børgesen) P. C. Silva in Silva et al. 1996: p. 429 (L. *inermis*: unarmed, without spines or prickles).

Basionym. *Spyridia aculeata* (C. Agardh *ex* Decaisne) Kützing var. *disticha* f. *inermis* Børgesen 1917: p. 239–240, fig. 230.

Synonym. *Spyridia hypnoides* f. *inermis* (Børgesen) Lawson and John 1982: p. 300.

Branchlets are opposite and oppose the pairs above and below them at a 90° angle (decussate) and lack terminal hyaline spines.

Common, the species and both varieties occur on seagrasses, mangrove roots, and limestone, as well as adrift on all coasts; also known from North Carolina southward, Bermuda, the Caribbean, Gulf of Mexico, and Brazil, as well as the Mediterranean, eastern tropical Atlantic, Red Sea, and Indo-Pacific (type location: Kanyakumari, India; Silva et al., 1996).

Taylor 1960: p. 541, 542; Schneider and Searles 1991: p. 396; Littler and Littler 2000: p. 164.

Tiffaniella Doty and Meñez 1960: p. 135 (named for L. H. Tiffany, an American phycologist who studied freshwater algae, especially the green algal genus *Oedogonium*).

Filaments are uniseriate and ecorticate; erect axes are little branched and similar to prostrate ones that have inflated unicellular or multicellular rhizoids and penetrate various hosts, particularly species of *Codium*. Polysporangia are naked and contain 16–64 polyspores. Carpogonial branches are near tips of erect axes. Carposporophytes lack fusion cells and gonimoblasts produce terminal carpospores and are covered with involucral filaments. The genus is distinguished by its pigmented swollen rhizoids, naked polysporangia, and terminal carpospores arising on gonimoblastic filaments.

Key to the species (2) of *Tiffaniella*

1. Plants purple red; erect filaments 2–5 mm tall, 20–50 μm in diam.; tips to 14 μm in diam.; rhizoids multicellular and often swollen . . . . *T. gorgonea*
1. Mats pink red, erect filaments 2–3 mm tall, 25–30 μm in diam.; tips to 10 μm in diam.; rhizoids mostly unicellular, swollen in middle, and irregularly lobed . . . . *T. saccorhiza*

Tiffaniella gorgonea (Montagne) Doty and Meñez 1960: p. 138 (Gr. *gorgon*: a creature in Greek mythology who had snakes as hair; also *Gorgonia*, a genus of soft corals). Plate XXXII, fig. 18.

Basionym. *Callithamnion gorgoneum* Montagne 1857: p. 289–290.
Synonyms. *Spermothamnion gorgoneum* (Montagne) Vickers 1897 ('1896'): p. 305.
Spermothamnion gorgoneum (Montagne) Bornet in Vickers 1905: p. 64.

Filaments are uniseriate, ecorticate and in soft purple feltlike masses with interwoven, rhizomelike, prostrate axes 3–5 mm long. Rhizoids are pigmented, multicellular, often swollen, they arise on ventral sides of prostrate axes, and penetrate hosts by lobed attachments that may have secondary pit connections. Erect axes are free and alternately branched; cells are 20–50 μm in diam., 2–7 diam. long, and have walls 6–10 μm thick. Branches taper to 14 μm in diam. at tips. Polysporangia are solitary, with 6–32 spores, and occur on 1–2-celled stalks. Spermatangial clusters are 30–40 μm in diam. and to 100 μm long. Cystocarps are spherical, to 200 μm in diam., and lateral on 1-celled stalks; carposporophytes have involucral filaments.

Rare, reported from 42–90 m off the east coast on *Codium* (Hanisak and Blair, 1988); also known from Bermuda, the Bahamas, Caribbean, and eastern tropical Atlantic.

Taylor 1960: p. 521; Littler and Littler 2000: p. 166.

Tiffaniella saccorhiza (Setchell and Gardner) Doty and Meñez 1960: p. 138 (Gr. *sakkos*: a sac + *rhiza*: a root). Plate XXXII, fig. 19.

Basionym. *Pleonosporium saccorhiza* Setchell and Gardner 1930: p. 168–169, pl. 10, fig. 39.

Filaments are uniseriate, sparingly branched, ecorticate and arise from branched creeping axes attached to algal hosts by ventral rhizoids. Rhizoids are pigmented, elongate, unicellular, 300–500 µm long, swollen in the middle, and with irregular lobed to bulbous tips. Erect axes are 2–3 mm tall, subdichotomously branched, 25–30 µm in diam. at the base, and taper to blunt tips; axial cells are 8–12 diam. long. Polysporangia are spherical to ovoid, solitary or paired on unicellular stalks, 75–85 µm in diam., and with 16–32 thick-walled spores. Spermatangia are solitary, on stalks, in ovoid heads, and arise from the lower parts of erect branches. Carpogonial branches are near tips and lack an obvious central fusion cell. Cystocarps are to 125 µm in diam.

Rare, reported on *Codium* spp. from 43–90 m off the east coast (Hanisak and Blair, 1988); also known from Puerto Rico, Venezuela, South Africa, and the Indo-Pacific (type location: Guadalupe Island, Mexico; Abbott, 1999).

Wrangelia C. Agardh 1828 (1820–1828): p. 136 (named for Fredrik A. Wrangel, a Swedish phycologist and Royal Secretary).

Synonym. *Haliacantha* J. Agardh 1899: p. 109

Filaments are uniseriate with large central axial cells, and attached by rhizoidal masses. Indeterminate branches arise from the axial cells mostly in two ranks; axes are ecorticate or corticated by filaments from bases of lateral (determinate) branchlets; the cortication may be multilayered. Tetrasporangia are tetrahedrally divided and stalked. Spermatangia are terminal and clustered on specialized short branchlets from nodes. Cystocarps are terminal on branchlets. Reproductive structures are on branchlets near tips of primary branches and usually covered by slender involucral filaments.

Key to the species (3) of *Wrangelia*

1. Plants usually bushy or tufted, not forming a turf, and more than 1.5 cm tall. 2
1. Plants usually plumose, often forming a turf, and less than 1.5 cm tall . . . . *W. argus*
   2. Branching irregular to sparse, not in 2 ranks; axes to 400 µm in diam.; older axes not corticated, except for a few irregularly tangled filaments. . . . . *W. bicuspidata*
   2. Branching regular, in 2 ranks; axes to 200 µm in diam.; older axes completely corticated . . . . . . *W. penicillata*

Wrangelia argus **(Montagne in Barker-Webb and Berthelot) Montagne 1856: p. 444 (Gr. *argus*: a hundred-eyed monster in Greek legend; a guardian). Plate XXXII, fig. 20, 21.**

Basionym. *Griffithsia argus* Montagne in Barker-Webb and Berthelot 1841 (1839–1841): p. 176, pl. 8, fig. 4.

Synonym. *Callithamnion beauii* P. Crouan and H. Crouan in Schramm and Mazé 1865: p. 7; P. Crouan and H. Crouan in Mazé and Schramm 1878 (1870–1877): p. 142.

Plants are soft, turflike, to 1.5 cm tall, plumose, with iridescent purple-red tips and white to gray axes; prostrate axes are attached by rhizoids that form multicellular haptera. Erect filaments are cylindrical, to 400 µm in diam., ecorticate or with a few filaments around nodes; cells are to 800 µm long. Branchlets are subdichotomously branched and in whorls at nodes; tip cells are attenuated to spinelike and may be deciduous. Tetrasporangia are tetrahedrally divided, terminal, spherical, 60–70 µm in diam., at nodes, and surrounded by slender, incurved, short-celled involucral filaments.

Common, on limestone in shallow water in the Keys and from 26–48 m off the east coast (Hanisak and Blair, 1988); also known from Bermuda, the Caribbean, Gulf of Mexico, and Brazil, as well as the eastern tropical Atlantic (type location: Roque del Gando, Canary Islands; Price and Scott, 1992), Indo-Pacific, and Australia.

Taylor 1960: p. 502; Littler and Littler 2000: p. 166.

Wrangelia bicuspidata Børgesen 1916: p. 118–120: fig. 127–130 (L. *bicuspidata*: having two sharp points). Plate XXXIII, fig. 1, 2.

Plants forming soft, lax, rose-white tufts to 8 cm tall, with primary branching irregularly alternate. Plants attached to algal hosts by long, slender rhizoids from basal cells. Main axes are terete with axial cells 120–400 µm in diam. by 2–10 diam. long. Branchlets are thin, subdichotomously branched, in whorls, tapered at tips, and with 1–3 pointed deciduous apical spines. Branchlets near base are slender, they extend both up and down, and loosely encircle main axes. Tetrasporangia are tetrahedrally divided, 60–90 µm in diam., and have a few incurved and tapered involucral filaments, 25 µm in diam. Spermatangial clusters are spherical, 25–80 µm in diam., and surrounded by a few involucral filaments.

Common, on seagrasses and seaweeds in shallow waters in the Keys and in 40 m off the east coast (Hanisak and Blair, 1988); also known from Bermuda, the Caribbean (type location between St. John and St. Thomas, Virgin Islands; Silva et al., 1996), and Gulf of Mexico, as well as the Red Sea and Indian Ocean.

Taylor 1960: p. 503; Littler and Littler 2000: p. 168.

Wrangelia penicillata (C. Agardh) C. Agardh 1828 (1820–1828): p. 138 (L. *penicillata*: brushlike or having tufts of hairs). Plate XXXIII, fig. 3–5.

Basionym. *Griffithsia penicillata* C. Agardh 1824: p. 143 (as '*Griffitsia*').

Plants are bushy, firm, 5–20 cm tall, light pink, and attached by long slender branched rhizoids. Main axes are widely and regularly alternately branched; secondary branching is alternate and in two ranks from every segment. Main axes are terete and pseudodichotomously branched; cells are 180–200 µm in diam. and to 800 µm long. Older axes are corticated by filaments from branchlet basal cells and axial nodes to form a pseudoparenchymatous cortex. Secondary branches are slender and taper to tip cells that are hornlike and curve upward with blunt ends. Tetrasporangia are tetrahedrally divided, spherical, to 80 µm in diam., and covered by short involucral filaments. Spermatangia are terminal, in spherical clusters (55–60 µm in diam.), on short lateral branchlets, and with a few involucral filaments. Cystocarps are terminal on short branchlets and surrounded by slender incurved involucral filaments to 400 µm in diam.

Common, in 6–30 m on seaweeds and seagrasses in the Keys; also known from Bermuda, the Bahamas, Caribbean, and Gulf of Mexico, as well as the Mediterranean (type location: Italy; Adams, 1994), eastern tropical Atlantic, and Indo-Pacific.

Taylor 1960: p. 503; Littler and Littler 2000: p. 168.

Delesseriaceae

Based on molecular studies Lin et al. (2001) recognized three subfamilies, the Delesserioideae, Nitophylloideae, and Phycodryoideae. Because three Florida genera (*Calloseris, Cottoniella*, and *Taenioma*) were not evaluated, the subfamilies are not presented. Using DNA sequence data, Choi et al. (2002) reported that the Delesseriaceae was polyphyletic.

Plants are usually foliose; branches may be slender, filamentlike, with simple, alternate, or dichotomous branching, and may arise from midribs. Axes are initially attached by discoid holdfasts and secondarily by rhizoids. Blades are linear to oval and may have midribs and veins. Growth is by an apical cell that forms an obvious axial cell row. Tetrasporangia are tetrahedrally divided and in sori. Spermatangia are on blades or in marginal sori. Procarps are formed on axial cells of the midrib or blade cells. Support cells form 3(4)-celled carpogonial branches and 1–2 groups of attached sterile branches 1–5 cells long. Auxiliary cells arise after fertilization and combine with the carpogonia and surrounding cells to form large fusion cells. Gonimoblast filaments have monopodial branching and terminal carposporangia. Cystocarps are ostiolate. Pericarps have one to several layers.

Apoglossum J. Agardh 1898 (1848–1901): p. 194 (Gr. *apo*: separate + *glossum*: tongue).

Blades are monostromatic, highly branched, with obvious midribs, and attached by discoid holdfasts. Secondary blades arise as buds from midrib (axial) cells, are smaller than the main blades, and usually branch 2–3 times. Midribs are covered by small rhizoidal cells and connect to lateral microscopic veinlets. Apical cells divide transversely and do not extend to margins. Tetrasporangia are tetrahedrally divided, elongate, and in elevated sori along the midrib. Gametophytes are dioecious. Spermatangia occur in elongated sori that are elevated on both surfaces between the midrib and margin. Procarps are at the tips of blades and usually solitary.

***Apoglossum ruscifolium* (Turner) J. Agardh 1898 (1848–1901): p. 190 (L. *russ*: reddish + *folium*: a leaf; also *Rusci*: a genus of rustlike fungus and *Ruscus*: butcher's broom in the family Liliaceae that produces purple foliage). Plate XXXIII, fig. 6.**

Basionym. *Fucus ruscifolius* Turner 1802: p. 127–130, pl. 8, fig. 1, 2.

Plants are erect or creeping, 2–10 cm tall, pinkish red, and irregularly branched (2–3 times) with branches arising from or near midribs. Older blades are 4–6 mm broad, have lateral veinlets, and are attached by discoid holdfasts. Secondary blades are small, linear, oblong, or oval, 1–6 mm wide, monostromatic, with pointed tips, and entire, usually undulate margins. Tetrasporangia are tetrahedrally divided, ovoid, 30–55 µm long, and in paired sori to 1 mm long on both sides of midribs. Spermatangial sori occur between the midrib and margin in parallel, oblique rows. Cystocarps are hemispherical, 500–800 µm in diam., on midribs, one per blade, and near bases of young blades. Carpospores are spherical and 22–30 µm in diam. Although similar, the species has more obvious veins and smaller blade cells (<20 µm long) than *Hypoglossum hypoglossoides*, where microscopic veins are absent and blade cells are >60µm long.

Rare, reported from 38–98 m off the east coast (Hanisak and Blair, 1988); also known from the Carolinas, Bermuda, Lesser Antilles, and Gulf of Mexico, the eastern North and tropical Atlantic (type location: Yarmouth, England; Maggs and Hommersand, 1993), Mediterranean, and Australia.

Schneider and Searles 1991: p. 401; Littler and Littler 2000: p. 182.

Branchioglossum Kylin 1924: p. 8 (Gr. *branchion*: a fin, gill + *glossum*: a tongue).

Blades are erect and simple or marginally divided, with irregular, opposite, alternate or secund branching; branches arise from prostrate axes that are attached by discoid or rhizoidal holdfasts. Blades are monostromatic with a polystromatic midrib and may have lateral veins. Apical cells divide transversely to form lateral rows that reach the blade margins; cells become compressed, crowded, and lack intercalary divisions in axial and lateral cell rows. Tetrasporangia are tetrahedrally divided and in elongated elevated sori on blades. Gametophytes are dioecious. Spermatangia occur in irregular transverse elongate sori midway between the margins and midribs. Procarps form sequentially on midribs below blade tips and cystocarps project on both surfaces.

Key to the species (2) of *Branchioglossum* (modified from Schneider and Searles, 1991)

1. Plants erect; blades oval-elongate with wavy margins . . . . *B. minutum*
1. Plants prostrate; blades linear to lanceolate with smooth margins. . . . .*B. prostratum*

Branchioglossum minutum **Schneider in Schneider and Searles 1975: p. 92, fig. 13, 15, 16 (L. *minutum*: very small or tiny). Plate XXXIII, fig. 7–9.**

Blades are small, clustered, monostromatic, erect, to 2 cm tall, pinkish red, oval to oblong, 1–3 mm wide by 1–6 mm long, and attached by discoid and rhizoidal holdfasts. Blades are mostly marginal, branching irregular or opposite 1–3 times, and also arise from damaged midribs; margins are undulate. Blades are corticated at bases, have conspicuous midribs; tips are rounded to indented and have a prominent apical cell. Axial cells are 8–40 µm in diam., 80–120 µm long, and covered by short, broad pericentral cells. Tetrasporangia are tetrahedrally divided, spherical, 20–40 µm in diam., and in semicircular sori on opposite sides of midribs near blade tips. Spermatangia occur in irregular sori midway between the midrib and margin on the upper half of blades. Cystocarps are hemispherical and projecting, 250–650 µm in diam., and 1–2 per blade on midribs near blade tips. Carpospores are spherical and 25–32 µm in diam.

Rare, inconspicuous, on seaweeds from 32–43 m off the east coast (Hanisak and Blair, 1988); also known from North Carolina (type location), Georgia, the Bahamas, and western Caribbean.

Schneider and Searles 1991: p. 402; Littler and Littler 2000: p. 182.

Branchioglossum prostratum **Schneider 1974: p. 1094, fig. 1A, 2 (L. *prostratum*: prostrate, procumbent). Plate XXXIII, fig. 10, 11.**

Blades are mostly prostrate, profusely branched, spreading, pink to red, with linear pointed tips, and usually attached by rhizoidal holdfasts and multicellular uniseriate rhizoids produced from blade margins. Branching is mostly alternate and marginal; blades are 0.8–1.6 mm in diam., branches to 0.7 mm and have smooth margins. Midribs are absent or difficult to see; axial cells are to 40 µm in diam., 185 µm long, and covered by rows of short cells; cortication is limited to frond bases. Tetrasporangia are tetrahedrally divided, spherical, 30–40 µm in diam., and in elongated sori on both sides of midribs. Spermatangia occur in elongated sori that are scattered between the midribs and margins. Procarps are produced in a series and extend from the apex along the midrib.

Rare, reported from 32–92 m off the east coast (Eiseman, 1979; Hanisak and Blair, 1988); also known from North Carolina (type location in 40–60 m), the Antilles, and the Canary and Hawaiian Islands.

Schneider and Searles 1991: p. 403.

Calloseris J. Agardh 1898 (1848–1901): p. 210 (Gr. *callos*: beautiful + *seris*: endive).

Blades are membranous, entangled or erect, marginally branched, and attached by rhizoidal holdfasts. Branching is dichotomous, trichotomous, or irregular; blades are often monostromatic and sometimes polystromatic to tubular near bases; midribs and veins are absent. Growth is by apical and intercalary cell divisions; the apical cells divide transversely, with the cell rows obscured by periclinal divisions. Apical margins have fine teeth near tips; older margins are dentate and irregularly proliferous. Tetrasporangia are tetrahedrally divided and occur in elevated corticated sori on both surfaces and at bases of proliferations. Cystocarps are produced in similar positions, but project only on one side of a blade.

Calloseris halliae J. Agardh 1898 (1848–1901): p. 213 (named for Mrs. G. A. Hall, an American collector who sent Florida seaweeds to J. Agardh). Plate XXXIII, fig. 12, 13.

Fronds are delicate, to 7 cm tall, pink to red, and from irregular prostrate rhizoidal holdfasts. Blades are monostromatic, narrow, strap-shaped, to 1 cm wide, 40–60 μm thick, irregularly branched, with marginal proliferations, lack midribs and veins, and have scattered patches of densely pigmented cells and acute tips. Blade cells are mostly rounded and 50–120 μm in diam.; apical cells divide transversely and the axial cell row is hidden by periclinal cell divisions. Tetrasporangia are tetrahedrally divided, obvoid, 45–110 μm in diam., and in oval sori that are irregularly corticated, 200 μm thick, 1.3 mm in diam., and parallel to the blade margins. Cystocarps are hemispherical, to 1.3 mm in diam., project above the surface, and occur at bases of marginal proliferations. Carpospores are pear-shaped, 26–32 μm in diam., and 56–70 μm long.

Rare, entangled with seaweeds and on limestone on the east coast of Florida (type location: Jupiter Inlet); also known from North Carolina (30 m) and the Lesser Antilles.

Schneider and Searles 1991: p. 404; Littler and Littler 2000: p. 184.

Caloglossa (Harvey) G. Martens 1869: p. 234, 237, *nom. cons.* (Gr. *calos*: beautiful + *glossa*: a tongue, tongue-shaped).

Synonyms. *Caloglossa* W. J. Hooker and Harvey 1845a: p. 270; 1845b: p. 542 (not defined).
Aparium Durant 1850 (cf. Silva, 2004; Compère, 2005; Wynne, 2005a).
Delesseria subg. *Caloglossa* Harvey 1853: p. 98.
Caloglossa J. Agardh 1876 (1848–1901): p. 498.

See Silva et al. (1996) and Wynne (2005a) for generic authorship plus Schneider (2003) and King and Puttock (1994) for detailed morphological characterizations.

Blades are regularly dichotomously branched, usually prostrate and spreading, nodes are constricted, secondary branches arise from midribs, and fronds are attached by rhizoids from the ventral surface of midribs. Blades are monostromatic, have prominent midribs, and lack lateral veins. Growth is by intercalary and apical cell divisions; apical cells are wider than long, divided transversely, and all cell rows extend from midribs to margins in the blade. Tetrasporangia are tetrahedrally divided, in oblique mostly distal series, and from mother cells derived from lateral cells of blades. Procarps and cystocarps are sessile and on midribs; pericarps are thin walled. Spermatangial sori are near midribs, on terminal or subterminal blades, and elevated on both surfaces.

Key to the species (2) of *Caloglossa*

1. Blades little constricted at nodes; with one rhizoid per blade cell; axial cell at lateral branch produces 1 cell row . . . . *C. monosticta*
1. Blades constricted at nodes; having more than one rhizoid per blade cell at nodes; axial cell at a lateral branch produces 2–6 cell rows . . . . *C. leprieurii*

Caloglossa leprieurii (Montagne in Barker-Webb and Berthelot) G. Martens 1869: p. 234, 237 (named for M. Leprieur, a French naturalist who collected seaweeds in French Guiana). Plate XXXIII, fig. 14–16.

Basionym. *Delesseria leprieurii* Montagne in Barker-Webb and Berthelot 1840 (1839–1841): p. 196–197, pl. 5, fig. 1.

Kamiya (2004) presented evidence for three reproductively incompatible "morphotypes" in the *Caloglossa leprieurii* complex that included many mating groups (Wynne, 2005a).

Plants are spreading, mostly prostrate, to 2 cm tall, 4–5 cm wide, purplish red to almost black, and attached by rhizoids from the ventral surface at constrictions. Blades are linear to broadly oval, to 2 cm wide, and constricted at forks or elsewhere; tips are acute and pincer-like. Branches are formed at nodes or midribs. Tetrasporangia are tetrahedrally divided, spherical, to 45 µm in diam., and near blade tips. Cystocarps are 460–800 µm in diam. and arise at midribs on underside of blades.

Common, on mangrove roots on all coasts and salt marsh plants, limestone, wood pilings, and jetties in salinities of 5 to 35 ppt.; also known from Connecticut through the Caribbean, Bermuda, the Gulf of Mexico, French Guiana (type location: Cinerary; Silva et al., 1996), and Brazil, as well as West and South Africa, and the Indo-Pacific.

Taylor 1960: p. 544; Dawes 1974: p. 149, 150; Schneider and Searles 1991: p. 406; Littler and Littler 2000: p. 184.

Caloglossa monosticha Kamiya in Kamiya et al. 1997: p. 104, fig. 5 (L: *mono-*: single + *sticha*: in a row; i.e., having a single row). Plate XXXIII, fig. 17, 18.

See Wynne and De Clerk (1999b) for a discussion of the complex that includes *Caloglossa continua* (Okamura) R. J. King and Puttock, *C. monosticha*, and *C. saigonensis* T. Tanaka and Pham-Hoàng Hô.

Blades are spreading, mostly prostrate, from axial cells, and attached by divergent rhizoids on the ventral side of axial cells with one rhizoid per cell. The first axial cell of a lateral branch axis produces adaxial cells. Asexual sporophytes may be produced from out-crossing of certain sexual isolates of this species (Kamiya and West, 2005).

Common(?), on mangroves at Plantation Key (Kamiya et al., 2003); also known from estuarine and sheltered coastal areas in Brazil, Peru, Singapore, Indonesia, Malaysia, and Australia (type location: Derby, Western Australia).

Calonitophyllum Aregood 1975: p. 348 (L. *calo*: beauty + *nitor*: bright + *phyllon*: leaf).

Blades are erect, monostromatic above and polystromatic near bases, strap-shaped, subdichotomously branched, and attached by large, irregular cartilaginous holdfasts. Blades are entire and lack midribs; margins are undulate, proliferous, and dentate; veins are multicellular, abundant, and fusing. Apical cells divide transversely and intercalary divisions are common in the axial rows. Tetrasporangia are tetrahedrally divided, scattered in corticated areas, and swell both blade surfaces. Gametophytes are dioecious, with procarps scattered over nonaxial parts of blade. Cystocarps are ostiolate and project from both blade surfaces. Spermatangia occur in raised sori on both surfaces.

Calonitophyllum medium (Hoyt) Aregood 1975: p. 348, fig. 1–23, 24b (L. *medium*: middle, intermediate). Plate XXXIII, fig. 19, 20.

Basionym. *Nitophyllum medium* Hoyt 1920: p. 494, fig. 35, pl. CV, CXIV, fig. 4, 5.

Synonym. *Hymenena media* (Hoyt) W. R. Taylor 1960: p. 553–554.

Fronds are 22–25 cm tall, rose red, and on short stipes; holdfasts are perennial, cartilaginous, irregular, creamy white, and lightly calcified. Fronds are alternately to subdichotomously branched, formed along margins, and have rounded to acute tips and margins that may be thick and undulate. Blades are monostromatic; veins are 3–6 cells thick, microscopic, anastomosing, and fused at stipes.

Apical cells are broad to dome-shaped and divide transversely; axial cells produce anticlinal walls to form two lateral pericentral cells that produce lateral filaments. Tetrasporangia are tetrahedrally divided, subspherical, 45–60 µm in diam., and in scattered oval to elliptical sori. Spermatangia are subspherical and in irregular sori. Male plants are smaller than females, more regularly dichotomous, and lack macroscopic veins. Carposporangia are irregular and in chains of 1–3 cells, 15–20 µm in diam.; pericarps have 8–10 layers of irregular cells.

Rare, reported from 38–58 m off the east coast (Hanisak and Blair, 1988); also known from the Carolinas (type location: North Carolina) and Caribbean.

Taylor 1960: p. 553; Schneider and Searles 1991: p. 407.

Cryptopleura Kützing 1843: p. 444, *nom. cons.* (Gr. *crypto-*: covered + *pleura*: a rib).

Plants are bushy; lower axes are often stalklike and attached by discoid holdfasts; upper axes are dichotomously, palmately, or laterally branched. Axes are membranous, flat, and often have proliferous margins; apical cells are inconspicuous. Blades are monostromatic with many multiseriate veins and small fused veinlets. Tetrasporangia divide tetrahedrally. Spermatangia occur in elongated marginal sori or on proliferations. Cystocarps are projecting.

Cryptopleura ramosa **(Hudson) Kylin *ex* L. Newton 1931: p. 332, as '*ramosum*' (L. *ramosa*: branching, branched). Plate XXXIII, fig. 21.**

Basionym. *Ulva ramosa* Hudson 1762: p. 476.
Synonyms. *Cryptopleura lacerata* (S. G. Gmelin) Kützing 1843: p. 444.
Cryptopleura fimbriata (Greville) Kützing 1849: p. 872.

According to Wynne (1998a), *Nitophyllum reptans* P. Crouan and H. Crouan is a growth form of *Cryptopleura ramosa*.

Fronds are bushy, delicate, 7–15 cm tall, pale to deep red, variable in width (to 2.5 cm wide), repeatedly irregularly branched or palmately to subdichotomously branched, and attached by small discoid holdfasts. Ultimate blades are oblong, and monostromatic except at veins, with rounded tips and undulate proliferous margins. Veins are large below and delicate to inconspicuous above. Tetrasporangia are tetrahedrally divided and in marginal sori or leaflets of pinnate fronds. Cystocarps are scattered and near margins.

Rare, reported from 38–65 m off the east coast (Hanisak and Blair, 1988); also known from Brazil and Uruguay, as well as the eastern North and tropical Atlantic (type location: Yorkshire, England; Maggs and Hommersand, 1993) and the Mediterranean.

Taylor 1960: p. 554.

Grinnellia Harvey 1853: p. 91 (named for Henry Grinnell, an American botanist from New York, who promoted the search for the missing Arctic expedition of Sir John Franklin).

Plants are large, membranous, monostromatic, and irregularly branched blades; they are often proliferous near their bases, have short stipes, and are attached by small discoid holdfasts. Blade midribs are obvious and multilayered below, while inconspicuous or absent above. Blades lack veins and have entire, undulate margins. Growth is by an apical cell that divides transversely in young blades and also by intercalary cell divisions in older blades. Tetrasporangia are tetrahedrally divided, in elongate sori that are corticated, scattered, and may swell either surface. Spermatangia occur in scattered sori. Procarps are produced from axial cells that are in special clusters (islets), usually one per cluster. Cystocarps are stalked and project from either surface. Gonimoblastic filaments are produced from large fusion cells. Cystocarps have thin-walled, hemispherical pericarps with one apical pore.

Grinnellia americana (C. Agardh) Harvey 1853: p. 92, pl. XXI, B, fig. 1–8 (L. *americana*: of or from America). Plate XXXIII, fig. 22, 23.

Basionym. *Delesseria americana* C. Agardh 1822 (1820–1828): p. 173.

Synonym. *Grinnellia americana* var. *caribaea* W. R. Taylor 1942: p. 127.

The variety *caribaea*, with ovate to cordate blades, is not recognized as a distinct taxa by Schneider and Searles (1991).

Blades are usually clustered, mostly simple, 1–50(100) cm tall by 1–17 mm wide, and on short stalks attached to small discoid holdfasts. Fronds are narrow to broad, oval, oblong, or elongate, and mostly monostromatic, except at basal midribs. Tetrasporangia are tetrahedrally divided, 50–65 μm in diam., and in elongate sori, 0.3–1 mm in diam. by 0.5–2 mm long. Spermatangia are ovoid to spherical and in sori that are to 0.5 mm long on blades 1–3 cm long. Cystocarps are 0.3–1.5 mm in diam. and have thin-walled pericarps. Carposporangia are ovoid, 5–15 μm in diam., and in terminal chains.

Uncommon, on intertidal jetties and coquina on both coasts, and in 40–74 m off the east coast; also known from Massachusetts to Georgia, the Virgin Islands, and Venezuela.

Taylor 1960: p. 547; Dawes 1974: p. 150; Schneider and Searles 1991: p. 409.

Hypoglossum Kützing 1843: p. 444 (Gr. *hypo*: less than + *glossum*: tonguelike).

Plants are usually bushy, foliose, prostrate or erect, attached by small discoid holdfasts, and secondarily by marginal rhizoids. Blades are slender, flat, and branched. They have an obvious midrib, lack veins, may have stalks, and are usually monostromatic except for the midribs that are 3 cells thick and corticated below by rhizoids. Lateral blades arise from midribs by endogenous budding of central axial cells. Apical cells are visible in young blades and divide transversely; apical initials of all lateral cell rows reach the margins. Tetrasporangia are tetrahedrally divided and in elevated sori on both sides of the midrib and blades. Gametophytes are dioecious. Spermatangia occur in interrupted transverse sori. Procarps are sequential on midribs and below the tips of blades. Cystocarps are ostiolate, a few per blade, sessile, and project on either side.

Key to the species (8) of *Hypoglossum*

1. Prostrate axes produce erect blades 2
1. Prostrate axes produce few to no erect blades 5
    2. Erect blades lack enrolled tips 3
    2. Blades have enrolled tips . . . . *H. involvens*
3. Plants large and to 20 cm tall 4
3. Plants minute and to 6 mm tall . . . . *H. subsimplex*
    4. Blades linear to oval and 1–5 cm long; tetrasporangia 40–80 μm in diam., random, and in elongated sori . . . . *H. hypoglossoides*
    4. Blades linear and to 5 mm long; tetrasporangia 70–120 μm in diam., and in a series with oldest ones nearest the bases of blades . . . . *H. tenuifolium*
5. New blades emerge directly from midrib, not midway from margin 6
5. New blades emerge midway between midrib and margin . . . . *H. anomalum*
    6. Plants minute; blades 1–3 mm long and arising from a rhizome . . . . *H. rhizophorum*
    6. Plants larger, blades arising from prostrate blades 7
7. Blades usually arising in pairs and without fringing rhizoids . . . . *H. caloglossoides*
7. Blades arising singly, alternately, and with fringing rhizoids . . . . *H. barbatum*

Hypoglossum anomalum M. J. Wynne and D. L. Ballantine 1986: p. 189, fig. 1–15 (L. _anomalum_: unusual, abnormal). Plate XXXIV, fig. 1.

Blades are small, repent, to 30 mm long, 1–1.8 mm wide, delicate, undulate, and monostromatic except for midribs. Blades occur on short, cylindrical corticated stalks that are attached by discoid rhizoidal pads and multicellular rhizoids on the abaxial side of midribs or blade margins. Lateral blades arise endogenously from axial cells of the midrib in opposing pairs and emerge between midribs and margins. Tetrasporangia are tetrahedrally divided, to 70 µm in diam., and in oval sori (to 500 µm wide) along midribs.

Rare, dredged at the Dry Tortugas and from 27–70 m off the east coast; also known from the Bahamas, Puerto Rico (type location: in 61 m off La Paguera), Indo-Pacific, and Australia.

Hypoglossum barbatum Okamura 1901 (1900–1902): p. 19, pl. VII (L. _barbatus_: bearded, with tufts of long weak hairs). Plate XXXIV, fig. 10, 11.

Plants are delicate, decumbent, and to 2 cm long; blades are monostromatic, linear-lanceolate, to 1 cm wide, and initially attached by rhizoids from tips and secondarily by a fringe or 'beard' of marginal rhizoids. Branches are endogenous, single and alternate, and arise from the axial row of cells; midribs are 3 cells thick. Tips have an obvious apical cell and all cells of 2nd-order rows produce 3rd-order ones. Tetrasporangia arise from 2nd- and 3rd-order cells and are 74–80 µm in diam.

Questionable presence; Schneider (2000) suggested that it may be in Florida; also recorded from Bermuda and the Indo-Pacific (type location: Hyuga, Japan; Silva et al., 1996).

Hypoglossum caloglossoides M. J. Wynne and Kraft 1985: p. 9–19, fig. 1–19 (Gr. _calo_: beautiful + _glossum_: tongue-shaped + _oides_: similar to; thus, resembling the red algal genus _Caloglossa_). Plate XXXIV, fig. 2, 3.

Blades are delicate, repent, 6–12 mm long, 540–630 µm wide, have regular nodal constrictions and internodes 0.5–1.5(2) mm long, and are attached by multicellular discoid holdfasts arising at nodes. Lateral blades are erect, usually in pairs, they arise endogenously at nodes and have one or both laterals that form indeterminate prostrate axes. Blades are monostromatic, have midribs that are 3 cells thick, and lack lateral veins. Blade tips have one apical cell; all 2nd-order rows produce 3rd-order ones. Tetrasporangia are tetrahedrally divided, 30–35 µm in diam., lack cover cells, and are in median sori. Spermatangia occur in paired sori along midribs. Carpogonial branches are 4-celled. Cystocarps are spherical, 460 µm in diam., and one per blade; carposporangia are 25–30 µm in diam.

Rare, on crustose corallines in 4.6 m at Molasses Reef (Bucher et al., 1990); also known from Bermuda, the Bahamas, Pacific Ocean, and Australia (type location: Lord How Island; Abbott, 1999).

Hypoglossum hypoglossoides (Stackhouse) Collins and Hervey 1917: p. 116 (Gr. _hypo_: less than + _glossum_: tonguelike + _oides_: similar to). Plate XXXIV, fig. 4.

Basionym. *Fucus hypoglossoides* Stackhouse 1801 (1795–1801): p. 76, pl. 13.
Synonyms. *Delesseria hypoglossum* (Woodward) Harvey 1846 (1846–1851): p. 51, pl. II.
Hypoglossum tenuifolium (Harvey) J. Agardh var. *carolinianum* Williams 1951: p. 156.

The var. *carolinianum*, which is described as rarely 1 cm long, is not recognized because of a continuum in size within the species (Wynne and Ballantine, 1986).

Blades are delicate, pale red, in erect tufts or prostrate mats, to 20 cm tall, and attached by small padlike holdfasts and secondarily by marginal multicellular rhizoids. Blades are undivided, monostromatic, linear to oval, 1–5 cm long, and 1–4 mm wide; lateral blades arise endogenously from midribs and divide up to 3 orders. Margins are entire and undulate; tips are acute with one obvious apical cell; midribs are 3 cells thick with 1–2 axial rows of elongate pericentral cells. Older parts are corticated. Tetrasporangia are tetrahedrally divided, spherical to ovoid, 40–80 µm in diam., and in ovoid to elongated corticated sori that occur along midribs near blade tips. Gametophytes are dioecious. Spermatangia occur in irregular to confluent sori on the upper sides of blades. Cystocarps are sessile, spherical, to 100 µm in diam., 1–2 per blade, and central on midribs.

Uncommon, on coquina in shallow water at St. Augustine and from 42 m off the east coast; also known from the Carolinas (to 60 m), Georgia, Bermuda, the Caribbean, Gulf of Mexico, and Brazil, as well as the British Isles (type location: Cornwall; Wynne, 1984) to Portugal, the Mediterranean, and eastern tropical Atlantic.

Schneider and Searles 1991: p. 412; Littler and Littler 2000: p. 186.

Hypoglossum involvens (Harvey) J. Agardh 1898 (1848–1901): p. 185 (L. *involvens*: enveloping). Plate XXXIV, fig. 5.

Basionym. *Delesseria involvens* Harvey 1853: p. 97–98, pl. XXII A, fig. 1–4.

Blades are monostromatic, pale red, in small, delicate tufts, and attached to a small discoid holdfast and secondarily by multicellular rhizoids. Blades are 1.0–2.5 cm long, to 1.5 mm wide, narrow, linear, with obtuse bases, undulate margins, and attenuated tips that curl inward. Midribs consist of 3 rows of large cells; central cells are narrow, elongate, and cylindrical. Pericentral cells are mostly hexagonal and produce 3rd-order oblique rows. Lateral blades arise endogenously from the midrib every 3–10 mm. Blade margins have rows of narrow terete cells in oblique ascending lines (bullations). Tetrasporangia are tetrahedrally divided, and 50–70 µm in diam.; sori occur along midribs on upper portions of blades and in oblique ascending rows. The species is distinguished by its strongly incurved apices and patches of cells that form marginal swellings.

Rare, at Knight Key (Wynne and Ballantine, 1986) and Key West (type location); probably a deep-water species.

Taylor 1960: p. 545.

Hypoglossum rhizophorum D. L. Ballantine and M. J. Wynne 1988a: p. 8–9, fig. 1–14 (Gr. *rhizo*: rootlike + *phorum*: bearing or bearer). Plate XXXIV, fig. 6.

Blades are small, erect, single or paired, and arise from midribs of prostrate rhizomes that are polysiphonous. Rhizomes produce rhizoids ventrally and blades dorsally. Rhizoids branch only at the point of contact with substrata. Erect blades usually arise from every 12th to 16th pericentral cell of rhizomes, and are 1.3–3.2 mm long and 400–1100 µm wide. Its apical cell is obvious and 2nd-order cell rows produce 3rd-order ones. Rhizomes are mostly terete, often branched, spreading, to 200 µm in diam., and 15 cells in width; nodes of erect blades are constricted or cylindrical.

Rare, tetrasporic plants are from 9 m at the Dry Tortugas (Ballantine, 1996); also from Bermuda, the Bahamas, and Puerto Rico (type location), as well as the Hawaiian and Marshall Islands.

Hypoglossum subsimplex M. J. Wynne 1994: p. 257, fig. 1–10 (L. *subsimplex*: simple, plain, partially divided). Plate XXXIV, fig. 7.

Plants are small, delicate, clustered, to 6 mm tall, and 1 mm wide; they are simple to subsimple, initially attached by small discoid bases, and secondarily by 1(3)-celled rhizoids from the smooth blade margins. Midribs are uncorticated; pericentral cells are 8–18 μm in diam., 40–66 μm long, and with all 2nd-order row cells producing 3rd-order ones. Tetrasporangia are tetrahedrally divided, 40–48 μm in diam., and arise from second- and third-order rows near midribs. Spermatangial sori are on both surfaces and in diagonal or irregular groups. Cystocarps are small, ostiolate, on midribs, and 1–2 per blade.

Rare, on *Halimeda tuna* at the Content Keys (type location) and also Bermuda.

Hypoglossum tenuifolium (Harvey) J. Agardh 1898 (1848–1901): p. 186 (L. *tenuifolium*: slender-leaved or with slender leaves). Plate XXXIV, fig. 8, 9.

Basionym. *Delesseria tenuifolia* Harvey 1853: p. 97, pl. XXII B, fig. 1–4.

Plants are delicate, erect, to 20 cm tall, prostrate and spreading, monostromatic except at midribs, and attached by rhizoidal holdfasts. Main blades are linear, 1–3 mm wide, 1–5 mm long, and have midribs with one row of terete cells flanked by elongate and often hexagonal cells. Blade tips are blunt or indented and with an obvious apical cell. Older blades are corticated by divisions of pericentral cells that are in oblique rows and mostly elongate near margins. Lateral blades arise endogenously from midribs and branch 3–4 times, are not divided or lobed, and have entire margins. Tetrasporangia are tetrahedrally divided, spherical, 70–120 μm in diam., and are produced in 2–3 rows along the midribs. Gametophytes are dioecious. Spermatangial sori occur in oblique rows, are irregular to confluent, on upper surfaces, and midway between the margins and midribs. Cystocarps are urceolate, to 100 μm in diam., and sessile on midribs.

Common, in shallow waters, on mangrove roots, coarse algae, and limestone on all coasts, plus in 24–94 m off the east coast (type location: Key West); also known from North Carolina (40–45 m), Bermuda, the Caribbean, Gulf of Mexico, and Brazil.

Taylor 1960: p. 545; Dawes 1974: p. 151; Schneider and Searles 1991: p. 414; Littler and Littler 2000: p. 186.

Martensia Hering 1841: p. 90, *nom. cons.* (named for G. Martens from Germany, who studied the seaweeds of Brazil, eastern Asia, and India).

Synonyms. *Hemitrema* R. Brown in Endlicher 1843: p. 50.
Mesotrema J. Agardh 1854: p. 110.
Capraella De Toni (Giuseppe, filius) 1936: p. 3.

Molecular studies by Lin et al. (2005) have suggested that *Martensia* is monophyletic and has two natural assemblages that, at the generic level, would correspond to *Martensia sensu* Hering (thallus has one network band) and to *Mesotrema* J. Agardh (thallus has many network bands).

Blades are monostromatic and arise from a thick stalk attached by a discoid holdfast. Blades are lobed, dichotomously to irregularly branched, broadly strap-shaped, undulate, and crisped. Apical growth is by a marginal row of cells with secondary intercalary divisions that produce grid- or net-like zones of thin ribs aligned at right angles to the blade's axes. Veins are absent. Tetrasporangia are tetrahedrally divided and occur in sori on upper blade surfaces, usually on ribs. Cystocarps are produced in gridlike bands on margins and project above the surface.

Martensia pavonia (J. Agardh) J. Agardh 1863 (1848–1901): p. 831–832 (named for Dr. Joseph Pavon of Peru). Plate XXXIV, fig. 12–14.

Basionym. *Mesotrema pavonia* J. Agardh 1854: p. 110

Millar (1990) combined Pacific *Martensia pavonia* with *M. fragilis* Harvey, while Littler and Littler (2000) retained the two as distinct taxa based on observations of Caribbean and Pacific plants. We follow Littler and Littler (2000) and Wynne (2005a) and designate Florida material as *M. pavonica* rather than *M. fragilis* (Millar, 1990).

Plants are delicate, 3–4 cm tall, pale blue to pink, sessile, and attached by rhizoids from the margins and elsewhere. Blades are widely lobed or branched; segments are 2–10 mm wide with marginal proliferations; solid parts are 1–2 cells thick and alternate with netlike grid areas. Grid areas have parallel or fused bands of 1-cell thick ribs and rectangular to square holes. Tetrasporangia are tetrahedrally divided, occur alone or in small sori, and mostly in older gridlike parts. Spermatangia are clustered on the sides of grids.

Rare, on subtidal algae from 30 m in the Keys; also known from the Caribbean (type location: Guadeloupe; Price and Scott, 1992) and Gulf of Mexico, as well as the tropical western Atlantic, Indo-Pacific, and Australia.

Taylor 1960: p. 555; Littler and Littler 2000: p. 186.

Nitophyllum Greville 1830: p. xlvii, 77, *nom. cons.* (L. *nitor*: bright + Gr. *phyllum*: leaf).

Plants are delicate, with straplike segments that are dichotomous to palmately branched, and attached by small discoid holdfasts. Blades are monostromatic, except at bases or in reproductive areas, and lack obvious apical cells, veins, or veinlets. Tetrasporangia are tetrahedrally divided and in scattered circular sori. Spermatangia occur in elongated sori. Gonimoblastic filaments have 1 or rarely 2 terminal carposporangia. Cystocarps are sessile, ostiolate, and scattered over blade surfaces.

Key to the species (2) of *Nitophyllum*

1. Plants irregularly lobed; margins smooth and without teeth . . . .*N. punctatum*
1. Plants dichotomously to multi branched; margins with teeth . . . . *N. wilkinsoniae*

Nitophyllum punctatum **(Stackhouse in Withering) Greville 1830: pp. xlvii, 79, pl. 12 (L. *punctatum*: dotted, with minute glands). Plate XXXIV, fig. 15, 16.**

Basionym. *Fucus punctatus* Stackhouse in Withering 1796: p. 405.

Synonym. *Nitophyllum ocellatum* (J. V. Lamouroux) J. Agardh 1842: p. 156.

Taylor (1928) recognized the var. *ocellatum* that Harvey (1853) identified from Key West, while Wynne (2005a) did not.

Blades are thin, delicate, 2–20 cm tall, bright pink, palmately branched below versus irregularly dichotomous (2–3 times) or lobed above, and attached by discoid holdfasts. Blades are monostromatic, oval to strap-shaped, 3–20 mm wide, and lack veins and stalks. Margins are smooth to slightly undulate and have blunt tips. Tetrasporangia are tetrahedrally divided, 45–60 μm in diam., and occur in oblong to round sori on margins near tips of lobed blades; sori are 6.0–6.5 mm long and to 2 mm wide. Cystocarps are ostiolate, sessile, to 1 mm in diam., and scattered on blades.

Rare, reported from 27–92 m off the east coast (Hanisak and Blair, 1988); known from the Caribbean, England (type location: Weymouth; Maggs and Hommersand, 1993), Mediterranean, Madeira Archipelago, Canary Islands, Pakistan, and India.

Taylor 1960: p. 551; Littler and Littler 2000: p. 188.

Nitophyllum wilkinsoniae **Collins and Hervey 1917: p. 115, pl. 1, fig. 8; pl. 2, fig. 9; pl. 5, fig. 32–33 (named for Miss Wilkinson of Ripleigh, Bermuda, who collected the type specimens and shared her collections with Collins). Plate XXXIV, fig. 17, 18.**

Blades are in delicate tufts to 10 cm tall, pale pink to red, monostromatic, lack veins, have di- to polychotomous branching, are segmented, and attach by basal filamentlike rhizoids. Segments are strap-shaped, oval or lobed, 2–10 mm wide, and 40–70 μm thick; margins are undulate and toothed and tips obtuse and 15–20 μm thick. Cells are angular and 20–80 μm in diam. in surface view. Tetrasporangia are tetrahedrally divided, spherical or irregular, 45–60 μm in diam., and in small, scattered, elevated sori.

Rare, reported from 27–58 m off the east coast (Hanisak and Blair, 1988); also known from Bermuda (type location) and the Lesser Antilles.

Taylor 1960: p. 552; Littler and Littler 2000: p. 190.

Polyneura Kylin 1924: p. 33, *nom. cons.* (Gr. *polus*: many + *neura*: nerved, veined).

Synonym. *Searlesia* Schneider and Eiseman 1979: p. 323, fig. 1–16.

Searlesia was transferred to *Polyneura* by Yoshida and Mikami (1991).

Blades are membranous, monostromatic, erect, marginally branched, initially attached by small discoid holdfasts, secondarily by multicellular rhizoids that arise from the margins and apices of prostrate portions. Midribs are branched, 3 cells thick, and veins are absent or not obvious. Tetrasporangia are tetrahedrally divided, in cortical sori elevated on surfaces, and scattered or midway between midribs and margins. Gametophytes are dioecious. Spermatangia occur in similar sori. Procarps are in monostromatic parts of blades; cystocarps are ostiolate and have beaked pericarps projecting from either surface.

Polyneura subtropica **(Schneider) T. Yoshida and Mikami 1991 p. 128 (L. *subtropica*: subtropical). Plate XXXIV, fig. 19–21.**

Basionym. *Membranoptera subtropica* Schneider 1974: p. 1097, fig. 1B, 3

Synonym. *Searlesia subtropica* (Schneider) Schneider and Eiseman 1979: p. 324, fig. 1–16

Blades are delicate, to 4 cm tall, pink to red, entire or lobed, dichotomously branched below and alternately opposite to irregular above, and attached by discoid holdfasts and multicellular rhizoids of prostrate axes. Blades are monostromatic, 0.3–2 cm wide, and to 75 μm thick; tips are obtuse and midribs 3 cells thick. Tetrasporangia are tetrahedrally divided, ovoid to spherical, 30–75 μm in diam., in oblong sori, 210–220 μm in diam., and 625 μm long. Spermatangia are ovoid to elliptical, 4–5 μm in diam., 8–9 μm long, and in irregular sori, 0.8–1.2 mm long, on upper side of frond. Carposporangia are ovoid and to 50 μm in diam. Cystocarps are spherical and to 1 mm in diam.

Rare, reported from 37–98 m off the east coast (Hanisak and Blair, 1988); also known from North (type location) and South Carolina.

Schneider and Searles 1991: p. 417.

Taenioma J. Agardh 1863 (1848–1901): p. 1256 (Gr. *taeni*: band, strap, ribbon).

Plants are filamentous, polysiphonous, and with erect and creeping axes; prostrate axes are abundantly alternately branched and attached by rhizoids. Erect axes are compressed and end in narrow determinate blades with a midrib. Midribs have an axial cell and 4 pericentral cells of equal length; lateral pericentral cells have two marginal cells half their length forming winglike edges on both sides. Apical cells are large, they divide transversely, and are replaced by obvious hairs with basal growth. Tetrasporangia are tetrahedrally divided and in rows along the midrib. Spermatangia occur on both surfaces; cystocarps are urceolate.

***Taenioma nanum* (Kützing) Papenfuss 1952: p. 179 (L. *nanum*: dwarf, small). Plate XXXIV, fig. 22, 23.**

Basionym. *Polysiphonia nana* Kützing 1849: p. 897; 1863a: p. 10, 29.

Synonym. *Taenioma macrourum* Thuret in Bornet and Thuret 1876: p. 69, pl. XXV, fig. 1–7.

Sartoni and Salghetti (1987) separated *Taeonima nanum* from *T. perpusillum* J. Agardh. Papenfuss (1952) merged *T. nanum* with *T. macrourum* as both have 2 hairs per tip, while *T. perpusillum* has 3; Abbott (1999) noted that there were 2–5 hairs per tip on *T. perpusillum*.

Plants are small (to 2 mm tall), filamentous, and brownish red; prostrate axis are terete, 60–75(150) µm in diam., have upturned tips, and are attached by uniseriate rhizoids that are swollen (50–130 µm in diam.) or terete (20–40 µm in diam.) and end in lobed discs. Short branchlets and indeterminate axes are formed alternately along the upper sides of every 4th to 6th segment of prostrate axes. Determinate branchlets are flat, linear, 60–75 µm wide, and with 8–20 segments 240–400 µm long that end in two long, thick, hyaline, multicellular hairs (20–40 µm in diam.) or two large apical cells. Tetrasporangia are tetrahedrally divided and on both sides of midribs of fertile branchlets (80–90 µm wide) on 1(2)-celled stalks.

Uncommon, in shallow waters under coquina ledges at Point of Rocks on the west coast and on limestone in the Keys; also known from Bermuda, the Bahamas, Caribbean, and Brazil, as well as the Mediterranean (type location: Naples, Italy; Womersley and Bailey, 1970), eastern tropical Atlantic, Indo-Pacific, and Australia.

Taylor 1960: p. 548; Dawes 1974: p. 151.

Sarcomeniaceae

Plants have erect, much-branched thalli; branches are terete or compressed, gelatinous, and decay rapidly (Womersley, 2003). Apical cells are prominent and protrude; axial cells have 4 pericentral cells. Lateral pericentral cells are slightly larger and each cuts off 2 half-length flanking cells. Branching is endogenous. Trichoblasts are absent with some genera having simple, pigmented monosiphonous filaments that arise from pericentral or flanking cells below apices or endogenously from axial cells. Gametophytes are dioecious. Procarps are adaxial on pericentral cells, with a 4-celled carpogonial branch and 2 sterile cells. Carposporophytes have a basal fusion cell. Cystocarps are external to branches and ovoid to urceolate; pericarps arise after fertilization and are ostiolate, with many erect filaments. Spermatangia occur in sori on the surface of shorter, compressed, lateral blades. Tetrasporangia are in stichidia derived from lateral pericentral cells and in 2 longitudinal distichous rows with postsporangial cover cells occurring on both sides.

Cottoniella Børgesen 1919: p. 333; 1930a: p. 144 (named for A. D. Cotton, a British phycologist who studied Irish seaweeds, including those from Clair Island).

Plants are mostly filamentous; prostrate axes curve upward, become erect, have alternate to secund branching, and are attached by rhizoids and multicellular haptera. Growth is by a prominent apical cell that divides transversely. Erect filaments are pigmented and arise endogenously from axial cells. Axes have 4–5 pericentral cells, and the lower axes are corticated, compressed, and denuded. Branchlets are determinate, uniseriate, in secund patterns above, one per segment, and arise from the convex sides of curved branch tips. Tetrasporangia are tetrahedrally divided.

Cottoniella filamentosa (M. Howe) Børgesen 1920: p. 477–479 (L. *filamentosa*: threadlike, filamentous). Plate XXXIV, fig. 24–26.

Basionym. *Sarcomenia filamentosa* M. Howe 1905b: p. 571, pl. 27; pl. 29, fig. 1–11.

Plants are filamentous, delicate, bushy, 4–16 cm tall, pink to rose red, and with prostrate axes attached by rhizoids that form haptera. Erect axes are corticated for 0.5–0.8 of their length and have curved tips with apical cells. Axes are compressed to flat, dorsiventral, and 250–900 μm wide. Secondary branches arise from the midline of upper axes and are narrow, linear, and with two parallel rows of outer cells along two large pericentral cells. Branchlets are determinate, uniseriate, and in one secund row on the upper side near tips that are 0.2–1.0 mm long and with 7–20 cells, each 2–6 diam. long.

Rare, on seaweeds in shallow water in the Keys (type location: Biscayne Key; Silva et al., 1996) and the Dry Tortugas; also known from Bermuda, the Greater Antilles (to 92 m), and Brazil, as well as the Mediterranean, eastern tropical Atlantic, Indian Ocean, and Australia.

Taylor 1960: p. 550.

Dasyaceae

Choi et al. (2002) determined that the Dasyaceae was polyphyletic and named the subfamilies Dasyoideae and Heterosiphonioideae based on molecular (Lin et al., 2001), anatomical, and morphological studies (Falkenberg, 1901; Hommersand, 1963; Parsons, 1975; Maggs and Hommersand, 1993; Choi, 1996; de Jong et al., 1998; Parsons and Womersley, 1998).

Plants are erect or prostrate and dorsiventral, foliaceous, and exhibit mostly sympodial growth and radial or distichous branching. Axes are polysiphonous, with 4–5 pericentral cells, and may be corticated by rhizoidal growth. Axes usually produce pigmented branchlets that are uniseriate near their tips. Branching is simple to dichotomous and spirally arranged; branches arise from each axial cell of the main axes. Tetrasporangia are tetrahedrally divided and in groups of 4–6 in whorls on specialized stichidia or ramelli. Stichidia are oval, oblong, or linear, and have 2–4 cover cells covering the tetrasporangia. Gametophytes are dioecious. Spermatangia occur in conical to linear stichidia. Procarps are produced on pericentral cells, one per node, and near branchlet tips. Carpogonial branches are 4-celled. Carposporophytes have one monopodial gonimoblast. Fusion cells are large and in pericarps with ostioles. Carposporangia are terminal and in short chains.

Dasyoideae

Members of this subfamily have 5 (rarely 4) pericentral cells. Tetrasporangia are covered by cells from the mother cell after spore production. Procarps are from the 2nd to 3rd intercalary pericentral cells cut off from the primary axial row. Carposporangia occur in chains and are associated with vegetative (sterile) tissue formed after fertilization.

Dasya C. Agardh 1824: p. xxxiv, 211, *nom. cons.* (Gr. *dasys*: shaggy, hairy).

Synonym. *Eupogonium* Kützing 1843: p. 415.

Adanson (1763) erected the genus *Baillouviana*, a pre-Linnaean title of an alga named *La Baillouviana* by Griselini (1750). Thus, the generic name *Dasya* C. Agardh is conserved.

Plants are erect or partly prostrate, bushy to sparsely branched, and attached by fleshy discoid holdfasts. Main axes have sympodial growth with tips displaced by lateral branches. Axes are terete, polysiphonous, and corticated by rhizoids or divisions of pericentral cells. Main axes are usually covered by false lateral branchlets (ramelli) that arise from most pericentral cells. Ramelli are pigmented, radial, distichous, mostly uniseriate, simple, and branched or netlike. Tetrasporangia are tetrahedrally divided and in compressed or cylindrical stichidia. Spermatangial clusters occur on special lateral branchlets. Carpogonial branches are 4-celled and flanked by two groups of sterile cells. Support cells produce auxiliary cells after fertilization that are connected to the fertilized carpogonia by 1–2 ooblast filaments forming large fusion cells. Carposporangia are produced in short chains on monopodial gonimoblastic filaments of fusion cells. Pericarps are distinct, often pointed, and ostiolate. Life histories are typically triphasic (López-Piñero and Ballantine, 2001).

Key to the species (12) of *Dasya* (after Taylor, 1960 and Littler and Littler, 2000)

1. Major axes corticated 2
1. Cortication absent or limited to older axes . . . . *D. rigidula*
    2. Branchlets not tapering abruptly but gradually decrease in diam. to tips 3
    2. Branchlets taper abruptly to form fine hairlike tips . . . . *D. corymbifera*
3. Mature plants less than 12 cm tall. 4
3. Mature plants more than 12 cm tall 9
    4. Branchlets (ramelli) less than 2 mm long 5
    4. Ramelli more than 2 mm long 7
5. Upper axes abundantly covered by branchlets. 6
5. Upper axes sparsely covered by branchlets . . . . *D. crouaniana*
    6. Ramelli in close whorls usually on all axes; main axes not obvious . . . . *D. mollis*
    6. Ramelli in close spirals on young axes; main axes obvious. . . . . *D. collinsiana*
7. Plants pale to dark rose or purple red; branching alternate or dichotomous below 8
7. Plants pale brown purple; branching irregular below . . . . *D. antillarum*
    8. Main axes alternately branched below; branches to 0.3 mm in diam.; ramelli in spiral to irregular patterns plus dense ocellate clusters near tips. . . . . *D. ocellata*
    8. Main axes dichotomously branched below; branches to 1 mm in diam.; ramelli whorled and tufted on older axes, not forming ocellate clusters . . . . *D. punicea*
9. Main axes alternately branched; ramelli 2–4 mm long 10
9. Main axes dichotomously branched; ramelli less than 2 mm long. . . . . *D. ramosissima*
    10. Ramelli not clustered near branch tips, usually randomly arranged 11
    10. Ramelli forming ocellate, spiraled clusters near branch tips. . . . . *D. caraibica*
11. Branching on main axis sparse; plants soft and delicate . . . . *D. baillouviana*
11. Branching on main axis abundant; plants firm and stiff . . . . *D. harveyi*

Dasya antillarum (M. Howe) A. Millar 1996: p. 155 (L. *antillarum*: of or from the Antilles or West Indies). Plate XXXV, fig. 1–4.

Basionym. *Dasyopsis antillarum* M. Howe 1920a: p. 577.
Synonym. *Eupogodon antillarum* (M. Howe) P. C. Silva in Silva et al. 1987: p. 129.

Plants are tough, bushy, 4–12 cm tall, pale to dark red, with branching irregular below and alternate to subdichotomous above, and attached by small fleshy holdfasts. Main axes are terete to compressed, 1.0–2.5 mm in diam., corticated, and densely covered by spinelike ramelli, 1–3 mm long. Ramelli are alternately to dichotomously branched with cells variable in size; basal cells are 10–15 µm in diam., middle ones 40–50 µm in diam. and to 150 µm long, and apical cells are 8–10 µm in diam., 100–120 µm long, and taper to points. Tetrasporangia are tetrahedrally divided; stichidia are fusiform or swollen, sometimes hair-tipped, 75–150 µm in diam., 300–450 µm long, and sessile or stalked. Spermatangial stichidia are 50–60 µm in diam., to 1 mm long, and terminal; cystocarps are to 1 mm in diam.

Uncommon, on limestone in shallow water in the Keys and to 30 m on the west coast; also known from the Bahamas (type location) and Caribbean.

Taylor 1960: p. 564; Dawes 1974: p. 165; Littler and Littler 2000: p. 170.

Dasya baillouviana (S. G. Gmelin) Montagne in Barker-Webb and Berthelot 1841 (1839–1841): p. 165 (named for Chevalier Johan de Baillou by Griselini in 1750. Baillou, the director-general of all mines and factories in Parma and Piacenze, lectured on experimental physics and was in the court of Duke Antonio Farnese. His collections are in the Museum of Natural History in Vienna). Plate XXXV, fig. 5.

Basionym. *Fucus baillouviana* S. G. Gmelin 1768: p. 165.

Synonyms. *Dasya pedicellata* (C. Agardh) C. Agardh 1824: p. 211.

Eupogodon mazei P. Crouan and H. Crouan in Schramm and Mazé 1865: p. 28; P. Crouan and H. Crouan in Mazé and Schramm 1878 (1870–1877): p. 267.

The species name is from the pre-Linnaean nonbinomial generic *Baillouviana* that was erected by Adanson in 1763 for the same alga. The post-Linnaean name was *Fucus baillouviana* Gmelin 1768, which was transferred to *Dasya* by Montagne in 1841.

Plants are delicate, soft, to 1 m tall, and bright rose to red; axes are alternately sparsely branched and have small discoid holdfasts. Main axes are corticated, slippery, 2–6 mm in diam., and covered by ramelli that are hairlike, pigmented, deciduous, dichotomously branched (2–3 times), and taper to both the base and tips. Ramelli cells are 10–40 μm in diam. and 20–50 μm long at bases; tip cells are obtuse, 7–12 μm in diam., and to 200 μm long. Tetrasporangia are tetrahedrally divided, spherical, 40–80 μm in diam., and in stichidia near branchlet bases. Stichidia are pale red, little branched, lance-shaped, 80–160 μm in diam., 1.2 mm long, and on 1(2)-celled stalks. Spermatangial stichidia are red, 60–75 μm in diam., and 0.2–0.6 mm long. Cystocarps are urn-shaped, solitary, near tips of ramelli, to 1.1 mm in diam., and have necks 100–200 μm in diam.

Common, in drift and on limestone, shells, or macroalgae (to 110 m) on all coasts; also known as scattered (often estuarine)populations from the Canadian Maritime Provinces to Cape Cod and continuous from Long Island Sound through the Caribbean, Bermuda, Gulf of Mexico, and Brazil, as well as from Sweden and Holland (introduced), the Mediterranean (type location; Dixon and Irvine, 1970), eastern tropical Atlantic, and Indo-Pacific.

Taylor 1960: p. 562; Dawes 1974: p. 165; Schneider and Searles 1991: p. 420; Littler and Littler 2000: p. 170.

Dasya caraibica Børgesen 1919: p. 319–322, fig. 322, 323 (L. _caraibica_: of or from the Caribbean). Plate XXXV, fig. 6, 7.

Plants are bushy to pyramidal, to 20 cm tall, dark rose purple, and attached by small discoid holdfasts. Main axes are filiform, often naked below, and with a few primary forks; secondary axes are abundant, alternately or radially branched, and covered terminally with ramelli. Axial segments are 400–500 μm long and 1 mm in diam.; each axial cell is surrounded by 5 pericentral cells that are corticated by rhizoids in older parts. Ramelli are 2–4 mm long, spirally arranged, and dichotomously branched near tips; basal cells are 50 μm in diam. and 60–70 μm long; branch cells are to 24 μm in diam. and 200 μm long on lower parts and to 8 μm in diam. near tips. Reproductive plants are unknown in Florida. A *Polysiphonia*-type life history is known for Puerto Rican plants (López-Piñero and Ballantine, 2001).

Uncommon, on limestone in 1–2 m at Content Keys and in 30–50 m at the Dry Tortugas; also known from Puerto Rico, the Virgin Islands (type location: between St. John and St. Thomas: Silva et al., 1996), eastern tropical Atlantic, Indo-Pacific, and Australia.

Taylor 1960: p. 560.

Dasya collinsiana M. Howe 1918: p. 524 (named for Frank Shipley Collins, an American phycologist who studied the green algae of North America and seaweeds of New England and Bermuda). Plate XXXV, fig. 8, 9.

Plants are soft, bushy, to 4 cm tall, pale red, and irregularly, alternately, or dichotomously branched 5–10 times, and attached by small discoid holdfasts. Main axes are 0.5–1.5 mm in diam. near bases with most axes corticated up to the youngest 2–3 forks. Axes have 5 pericentral cells 2–3 diam. long with overlapping ends. Ramelli are clustered and cover the upper stiff branches; upper ramelli are comma-shaped, 0.5–0.9 mm long, spirally arranged, alternately to dichotomously branched 4–5 times, and taper to a sharp point or short hairs. Basal cells of ramelli are 100–130 µm in diam.; tip cells are smaller, 45–55 µm in diam., and 2 diam. long. Tetrasporangia are tetrahedrally divided; stichidia are spinelike, 90–130 µm in diam., 300–500 µm long, and sessile or on 1-celled stalks. Cystocarps are solitary, spherical to ovoid, and 670–920 µm in diam. by 600–750 µm long.

Uncommon, on algae and limestone in shallow water in the Keys and from 25–33 m off the east coast; also known from Bermuda (type location).

Taylor 1960: p. 558; Littler and Littler 2000: p. 170.

Dasya corymbifera J. Agardh 1841: p. 31; 1863 (1848–1901): p. 1219 (L. *corymbifera*: producing a cluster). Plate XXXV, fig. 10, 11.

Plants are bushy to pyramidal, 5–12 cm tall, pink to purple, densely alternately to irregularly branched, and attached by small discoid holdfasts. Older axes are to 0.8 mm in diam., sparsely covered by ramelli, and corticated by rhizoids. Young (upper) axes are densely covered with ramelli and sparsely corticated. Ramelli are in whorls or spirals, 1–2 mm long, and dichotomously branched; lower cells are 50–70 µm in diam., 70–90 µm long, and taper to delicate upper segments; tip cells are 10 µm in diam., 100–120 µm long, and strongly incurved. Tetrasporangia are tetrahedrally divided, spherical, 30–40 µm in diam., and in stichidia that are lanceolate to cylindrical, 350–600 µm in diam., and on 1(3)-celled stalks. Cystocarps are ostiolate, oval, sometimes beaked, 550–650 µm in diam., 600–800 µm long, and sessile on short branches.

Uncommon, on shells, limestone, and algae in shallow waters on the west coast and at the Keys, plus 24–58 m off the east coast; also known from Bermuda, the Caribbean (type location; Abbott, 1999), as well as the British Isles to Spain, the Mediterranean (lectotype location: Tangier, Moracco; Maggs and Hommersand, 1993), Madeira Archipelago, Canary Islands, and Hawaiian Islands.

Taylor 1960: p. 559; Dawes 1974: p. 164; Littler and Littler 2000: p. 172.

Dasya crouaniana J. Agardh 1890: p. 95 (named for the two French brothers H. M. and P. L. Crouan, who studied Caribbean seaweeds). Plate XXXV, fig. 12–14.

Plants are bushy, firm, to 6.5 cm tall, and greenish to rose red; axes are terete, widely dichotomously to irregularly branched, and attached by discoid holdfasts. Main axes are corticated throughout by rhizoids (0.5–0.7 mm in diam.) and sparsely covered by ramelli that are 1–2 mm long and 2–3 times dichotomously branched, arise from cortical cells, and taper from basal to acute terminal cells. Basal cells of ramelli are 8–20 µm in diam. and 20–40 µm long; apical cells are obtuse, 5–12 µm in diam. and 4 or more diam. long. Tetrasporangia are tetrahedrally divided, spherical, 18–30 µm in diam., and in oval to lanceolate stichidia (80–120 µm in diam., to 1 mm long), and on lower forks of ramelli. Spermatangial stichidia are similar, 60–70 µm in diam. and 150–225 µm long.

Uncommon, on limestone at wave-exposed intertidal and shallow subtidal areas in the Keys and from 29–72 m off the east coast; also known from the Bahamas, Caribbean, and Canary Islands.

Taylor 1960: p. 561; Littler and Littler 2000: p. 172.

Dasya harveyi Ashmead in Harvey 1858: p. 127, pl. 50, A (named for W. H. Harvey, a British phycologist from Trinity College Dublin who published on seaweeds from Europe, North America, and Australia). Plate XXXV, fig. 15–17.

Plants are rose red, delicate, bushy, to 25 cm tall, with abundant alternate to secund (above) branching, and attached by discoid holdfasts. Axes are firm to stiff and densely corticated; the youngest branches are covered by slender, lax ramelli. Branches are to 2 mm in diam., 1.0–2.5 cm long, and densely corticated by elongate surface cells with slender tips. Ramelli are 15–20 µm in diam., 2–5 mm long, they taper to 8 µm in diam. near tips, and are pseudodichotomously branched with cells several times longer than broad. Tetrasporangia are tetrahedrally divided, spherical, and in two rows near tips of stichidia that are elongate, tapered, and at forks on branches. Cystocarps are urn-shaped, to 600 µm in diam., and near branch tips.

Rare, on limestone (to 10 m) in the Keys (type location: Key West).

Taylor 1960: p. 561; Littler and Littler 2000: p. 172.

Dasya mollis Harvey 1853: p. 62–63. (L. *mollis*: soft). Plate XXXV, fig. 18, 19.

Plants are delicate, bushy, 3–4 (15) cm tall, pale red, with irregular branching, an inconspicuous main axes, and attached by discoid holdfasts. Axes are corticated, to 1 mm in diam., and taper to tips; upper axes are densely covered by whorls of ramelli. Ramelli are corticated at bases, dichotomously branched 3–4 times, and with cells to 80 µm in diam. and 1–2 diam. long; tip cells are to 6 µm in diam. and 10 diam. long. Tetrasporangia are tetrahedrally divided, spherical, and to 50 µm in diam. Stichidia are oval, they occur near bases of branchlets, and are to 135 µm in diam. by 300 µm long.

Rare, on limestone in sheltered areas at Key West (type location); also known from Bermuda, the Caribbean (8–33 m), Gulf of Mexico, and Indo-Pacific.

Taylor 1960 p. 562; Littler and Littler 2000: p. 174.

Dasya ocellata (Grateloup) Harvey in W. J. Hooker 1833: p. 335 (L. *ocellata*: eyelike, with small eyes). Plate XXXV, fig. 20.

Basionym. *Ceramium ocellatum* Grateloup 1806: pl. I, fig. 2; 1807: p. 34, fig. 11.

Plants are delicate, soft, 2–8 cm tall, reddish purple, sparsely alternately branched, and attached by small discoid holdfasts. Main axes are corticated and to 0.3 mm in diam.; upper parts are densely covered by ramelli that form ocellate (eye-shaped) tips. Ramelli arise from short pericentral cells, are hairlike, fine, and to 5 mm long; they also occur in spiraled or irregular patterns, are basally corticated, and dichotomously branched 4–5 times. Basal cells of ramelli are 30–80 µm in diam. by 25–50 µm long; mid cells are 20–30 µm in diam. by 60–140 µm long; tip cells taper to 10 µm in diam. Tetrasporangia are tetrahedrally divided, spherical, 60–70 µm in diam. Stichidia are 100–200 µm in diam., to 1 mm long, and on lower branchlets. Spermatangial stichidia are 60–70 µm in diam. and to 200 µm long.; cystocarps are spherical.

Uncommon, on mangrove roots, seaweeds, and limestone in the Keys (to 40 m); also known from North Carolina, Bermuda, the Bahamas, and Caribbean, as well as the British Isles to Morocco (type location: Sète, France; Silva et al., 1996), Mediterranean, and eastern tropical Atlantic.

Taylor 1960: p. 559; Littler and Littler 2000: p. 174.

Dasya punicea (Zanardini) Meneghini in Zanardini 1841: p. 168 (L. _punicea_: blood red, crimson, reddish purple). Plate XXXV, fig. 21, 22.

Basionym. *Baillouviana punicea* Meneghini in Zanardini 1840b: p. 204.

Plants are delicate, conical, bushy, 5–10 cm tall, pale rose red, dichotomously branched, with intermixed long and short branches, and attached by discoid holdfasts. Main axes are slender and 0.4–1.0 mm in diam.; lower axes are naked with upper axes brushlike and covered with soft ramelli or minor branchlets. Ramelli are hairlike, filamentous, slender, 2–4 mm long, in whorls at close intervals, dichotomously forked 2–3 times near bases, and from axial and cortical cells. Most ramelli cells are 20–40 µm in diam., 50–80 µm long, and narrow to 12 µm in diam. near tips. Pericentral cells are visible only in young axes with cortication almost to tips. Tetrasporangia are tetrahedrally divided and in stichidia on 1–2-celled stalks; stichidia are oval to lanceolate, 70–90 µm in diam., and 300–400 µm long. Cystocarps are ostiolate, ovoid; attenuate, to 1 mm in diam., and occur at the tips of short stalks.

Rare, reported from 33–71 m off the east coast (Hanisak and Blair, 1988); also known from Bermuda, the Bahamas, and Caribbean, as well as England, France, the Mediterranean (type location: Venice, Italy; Maggs and Hommersand, 1993), Japan, and the Philippines.

Taylor 1960: p. 563; Littler and Littler 2000: p. 174.

Dasya ramosissima Harvey 1853: p. 61 (L. _ramosissima_: much or highly branched). Plate XXXV, fig. 23.

Plants are delicate, bushy, 15–30 cm tall, and brown to red; lower axes are widely and repeatedly alternately branched and attached by discoid holdfasts. Branches are to 1 mm in diam., corticated below, and naked near tips. Ramelli are dense, hairlike, and cover branch tips; they (i.e. ramelli) are to 2.5 cm long, abundantly dichotomously branched, from cortical cells, to 25 µm in diam., 4–6 diam. long, and taper to apices; tip cells are to 60 µm long and 10 µm in diam. Tetrasporangia are tetrahedrally divided and in linear to oblong stichidia that are to 50 µm in diam., 500 µm long, and on lower cells of ramelli. Cystocarps are oval, sessile, and subapically on short branchlets.

Uncommon, on limestone in shallow waters in the Keys (type location), in 35–55 m at the Dry Tortugas, and from 33–71 m off the east coast; also known from Bermuda.

Taylor 1960: p. 561; Littler and Littler 2000: p. 176.

Dasya rigidula (Kützing) Ardissone 1878: p. 140 (L. _rigidula_: rigid, stiff). Plate XXXV, fig. 24, 25.

Basionym. *Eupogonium rigidulum* Kützing 1843: p. 415.

Plants are delicate, soft, to 7 cm tall, and pink to purple; axes are stiff, irregularly to freely alternately branched, initially attached by small discoid holdfasts, and secondarily by adventitious rhizoids. Main axes are 300–500 µm in diam., lightly corticated below, ecorticate above, and with tips covered by slender ramelli; cells of central axes are 70–180 µm in diam. by 350–400 µm long. Ramelli are in loose spirals on each segment, incurved, dichotomously branched, ecorticate, 0.4–1.0 mm long, and form tufts at branch tips. Basal cells of ramelli are 50–140 µm in diam.; tip cells are 30–100 µm in diam., 70–190 µm long, and taper to hairlike tips that are long, blunt, 10–20 µm in diam., and to 10 diam. long. Tetrasporangia are tetrahedrally divided, ovoid, 20–50 µm in diam., and in stichidia (30–170 µm in diam. by 80–400 µm long) that replace the lower forks of ramelli. Spermatangial stichidia are to 100 µm in diam., and 300 µm long. Cystocarps are urn-shaped, 450–520 µm in diam., and 700–900 µm long.

The plant is often overlooked as it is small and similar to *Heterosiphonia crispella*. Common, on algae and limestone in shallow waters in the Keys and 34–55 m off the east coast; also known from North Carolina, Bermuda, the Caribbean, Gulf of Mexico, Brazil, Mediterranean (type location: Split, Croatia), and eastern tropical Atlantic.

Taylor 1960: p. 558, Schneider and Searles 1991: p. 422; Littler and Littler 2000: p. 176.

Heterosiphonioideae

Members of this subfamily have 4 or 7–13 pericentral cells. Tetrasporangia are covered by cells derived from mother cells produced before spore production. Procarps are on modified determinate branches from the 4th or last pericentral cell to be cut off in the primary axial row. Carposporangia are in chains or terminal and associated with sterile tissue produced before fertilization.

Dictyurus Bory in Bélanger 1834: p. 170 (Gr. *dictyo*: net, netted + *urus*: taillike).

Plants are erect, radially branched, and attached by rhizoidal clusters arising from bases and prostrate axes. Primary axes have 4 pericentral cells, dense cortication, and a cover of polysiphonous branchlets that arise alternately at every second node. Branchlets are uniseriate and connected to form a superficial and prismatic network. Tetrasporangia are tetrahedrally divided and in stichidia arising from ramelli with polysiphonous bases.

Dictyurus occidentalis **J. Agardh 1847: p. 17 (L. *occidentalis*: west, western, or occidental). Plate XXXVI, fig. 1, 2.**

Plants are soft, often in large tufts or colonies (4–12 cm tall), pink to rose red, and attached by rhizoidal clusters from their bases and prostrate branches. Primary axes have 4 pericentral cells; older axes are densely corticated and surrounded by large (to 60 μm in diam.) irregularly rounded cells. Branches are flexuous and have sympodial growth. Young axes appear 4-sided due to a pattern of short spiraled to whorled branchlets that form weblike nets with round or angular holes. Branchlets are uniseriate, 60–80 μm in diam., and 2–3 times as long as wide.

Common, on shallow limestone (1–2 m) in the Keys and from 24–39 m off the east coast; also known from the Bahamas, Caribbean, Gulf of Mexico (type location: Vera Cruz, Mexico; Silva et al., 1996), Brazil, and the Seychelles.

Taylor 1960: p. 567; Littler and Littler 2000: p. 178.

Heterosiphonia Montagne 1842c: p. 4, *nom. cons.* (Gr. *heteros*: different + *siphon*: tube).

Plants have sympodial growth; prostrate axes produce erect ones with plumose to bushy branching dorsally and rhizoids with lobed pads ventrally. Indeterminate axes are terete or compressed, polysiphonous, and have 4–12 pericentral cells; rhizoidal cortication is present or absent at branch bases. Branching is alternate to dichotomous and may be distichous. Prostrate axes produce determinate erect axes every 2–9 segments; erect axes are polysiphonous at bases, with uniseriate tips, and alternate to irregularly dichotomous branching. Tetrasporangia are tetrahedrally divided and in groups of 4–6; stichidia are oblong, pointed, whorled, and replace normal branchlets. Procarps occur on branchlets below forks; carpogonial branches are 4-celled and have monopodial gonimoblastic filaments that arise from fusion cells. Carposporangia are terminal or lateral and with heavily corticated ostiolate cystocarps.

Key to the species (3) of *Heterosiphonia*

1. Plants more than 10 cm tall; axes densely corticated 2
1. Plants to 6 cm tall; axes ecorticate . . . . *H. crispella*
   2. Branching sparse; axes not featherlike . . . . *H. gibbesii*
   2. Branching abundant; axes featherlike or distichous . . . . *H. hummii*

Heterosiphonia crispella (C. Agardh) M. J. Wynne 1985b: p. 87 (L. *crispella*: very crisped or curled). Plate XXXVI, fig. 3–5.

Basionym. *Callithamnion crispellum* C. Agardh 1828 (1820–1828): p. 183.

Synonym. *Dasya wurdemannii* J. W. Bailey *ex* Harvey 1853: p. 64, pl. XV C, fig. 1–3 (as '*wurdemani*').

Plants are soft, entangled, to 6 cm tall, and red to orange red; prostrate axes are spreading and attached by rhizoidal holdfasts. Branches are ecorticate, with 4–6 pericentral cells, and to 175 µm in diam.; cells near branch bases are 35–75 µm in diam. by 70–250 µm long. Branchlets are polysiphonous near bases, uniseriate and filiform above, and arise from every other segment. Branchlets are dichotomously branched 4–7 times and taper to 25 µm in diam. near tips; cells are thick walled and shorter than broad or less than equal. Tetrasporangia are tetrahedrally divided, spherical, to 75 µm in diam., and in terminal stichidia that are ovoid, tapered to tips, to 200 µm in diam., and 200–750 µm long.

var. *laxa* (Børgesen) M. J. Wynne 1985b: p. 87 (L. *laxa*: loose, lax, weak).

Basionym. *Heterosiphonia wurdemannii* (J. W. Bailey in Harvey) Falkenberg var. *laxa* Børgesen 1919: p. 327, fig. 327, 328.

Falkenberg (1901), Dawson (1963b), and Oliveira (1969) treated the variety as a distinct species (*Heterosiphonia laxa* Kylin).

Plants are light pink, to 6 cm tall; main axes are ecorticate, with 4 pericentral cells, and to 170 µm in diam. Axes are loosely, sparingly, and dichotomously or alternately branched 2–6 times. Branchlets are 0.9–1.3 mm long, deciduous, incurved, and on every other segment. Basal branchlet cells are multiseriate, 30–60 µm in diam., and 40–70 µm long; tip cells are tapered, blunt, 12–26 µm in diam., and 20–50 µm long. Tetrasporangia are tetrahedrally divided, spherical, and 25–40 µm in diam.; stichidia are terminal, conical to lanceolate, 110–120 µm in diam., 70–400 µm long, and covered by cortical cells.

Common, the species and variety occur in drift and on limestone, algae, seagrasses, and mangrove roots in shallow sites at the Keys and from 25–58 m off the east coast; also known from North Carolina, Bermuda, the Bahamas, Caribbean, Gulf of Mexico, and Brazil, as well as Spain (type location: Cádiz; Wynne, 1985b), the Mediterranean, eastern tropical Atlantic, Indo-Pacific, and Australia.

Taylor 1960: p. 565; Schneider and Searles 1991: p. 428; Littler and Littler 2000: p. 180.

Heterosiphonia gibbesii (Harvey) Falkenberg 1901: p. 653 (named for Professor Lewis R. Gibbes of Charleston, South Carolina, an American who sent W. H. Harvey specimens from Key West). Plate XXXVI, fig. 6–9.

Basionym. *Dasya gibbesii* Harvey 1853: p. 59, pl. XV A, fig. 1–8.

Plants are often in large soft tufts, to 20 cm tall, pink to orange red, sparsely dichotomously branched, and with erect axes arising from prostrate ones; fronds are initially attached by rhizoidal holdfasts and secondarily by rhizoids from repent branches. Erect axes are to 1.5 mm in diam. and with 6–9 pericentral cells; older axes are corticated and cross walls of axial cells are visible. Branches of main axes are long, flexuous, usually naked below, and with alternate, pinnate, or bipinnate branching above. Branchlets (ramelli) are congested near axial tips, alternately branched near bases, polysiphonous with 6 pericentral cells, and uniseriate and dichotomously branched near tips with cells that are to 30 µm in diam. Tetrasporangial and spermatangial stichidia are lanceolate and carpogonia occur near bases of branchlets.

Common, on shallow limestone (to 5 m) in protected to moderately exposed sites in the Keys (type location); also known from Bermuda, the Bahamas, Caribbean, and Brazil.

Taylor 1960: p. 566; Littler and Littler 2000: p. 180.

Heterosiphonia hummii Aregood 1977: p. 393–395, fig. 1–12 (named for Harold J. Humm, an American phycologist who studied seaweeds from Newfoundland to Florida, and the Caribbean). Plate XXXVI, fig. 10, 11.

Plants are soft, featherlike, pyramidal or oval, to 15 cm tall, pale red to pink, and attached by rhizoidal discs. Main axes have regular sympodial growth, 7–9 pericentral cells, and rhizoidal cortication. Older segments are connected with disclike structures. Branching is alternate, with branches usually distichous or featherlike and having pseudolateral branches that arise by sympodial growth. Lower axes have many uniseriate branchlets (ramelli). Tetrasporangia are tetrahedrally divided and occur in stichidia that are short (with 12–16 segments), lanceolate, pointed, and have 5 pericentral cells. The stichidia are on 2-celled uniseriate stalks that replace 3rd- or 4th-order branches, to 5 sporangia per segment, and with 2–4 cover cells.

Rare, in Florida known only from 30 m off Tampa Bay (type location); also recorded from the Gulf of Mexico and Puerto Rico.

Sedis incertae: Halydictyon Zanardini 1843: p. 52 (Gr. *hals*: the sea + *dictyon*: net, reticulated).

Synonym. *Halodictyon* Montagne 1847: p. 53 *orth. var.*

Choi et al. (2002) placed the genus in the Dasyaceae with uncertainty, while Womersley (2003) included it within the Rhodomelaceae. Guiry and Nic Dhonncha (2006) spell the generic name as *Halodictyon*.

Plants form irregular networks, lack distinct primary axes, and are attached by rhizoidal clusters from prostrate axes. Filaments are dichotomously branched, uniseriate, and fused. Reproductive structures are on short polysiphonous branchlets with four pericentral cells. Tetrasporangia are tetrahedrally divided and in 2 rows in short compressed stichidia. Spermatangia are in flat to ovoid clusters with a marginal band of sterile cells. Procarps are from the 2nd pericentral cell of a branchlet. Cystocarps are urn-shaped and have a thin-walled pericarp.

Halydictyon mirabile Zanardini 1843: p. 52 (L. *mirabile*: wonderful or remarkable). Plate XXXVI, fig. 12, 13.

Plants are soft, delicate, pink to rust red, in small net- to meshlike clumps, to 8 cm tall, dichotomously branched, and attached by rhizoidal mats from lower axial cells and prostrate axes. Filaments are uniseriate, to 110 µm in diam., and 3–6 diam. long below; above, the filaments taper and fuse laterally to form a 3-dimensional irregular polygonal net. Tetrasporangia are tetrahedrally divided; stichidia are short, compressed, with 4 pericentral cells, usually sessile, and solitary or in pairs near branch tips. Cystocarps are on older axes and project from nets.

Uncommon, adrift and on algae in shallow water in the Keys and 92 m at the Dry Tortugas; also known from Bermuda, the Bahamas, Caribbean, Mediterranean, and eastern tropical Atlantic.

Taylor 1960: p. 567; Littler and Littler 2000: p. 178.

Rhodomelaceae

Plants are usually erect, bushy to sparingly branched or partly to entirely prostrate, and attached by basal holdfasts and rhizoids. Indeterminate axes are terete or flattened to frondlike, with polysiphonous construction and monopodial growth. Axes are ecorticate or corticated by divisions of pericentral cells or descending rhizoids and may have uniseriate branched or unbranched pigmented or colorless filaments (trichoblasts). Tetrasporangia are tetrahedrally divided, often from pericentral cells, and are in long series on fertile branches. Spermatangia are usually on fertile trichoblasts with polysiphonous bases. Carpogonial branches are 4(5)-celled and often on the second basal cell of polysiphonous trichoblasts that have 2 vegetative (sterile) cells. Auxiliary cells form after fertilization and have 1–2 connecting cells (ooblasts) that fuse with the fertilized carpogonia and auxiliary cells forming fusion cells. Carposporangia are in short chains on sympodial gonimoblastic filaments that grow from fusion cells. Cystocarps are ostiolate with distinct pericarps.

Acanthophora J.V. Lamouroux 1813: p. 44 or 132 (L. *acanthophora*: spine bearing).

Synonym. *Acanthochondria* Weber-van Bosse 1910: p. 29; 1923: p. 348.

Plants are erect and bushy; axes are coarse to stiff, alternately and radially branched, and attached by irregular, lobed, or discoid holdfasts. Indeterminate axes are terete with 5 pericentral cells visible near the tips; older axes are heavily corticated to parenchymatous in cross section. Axes are covered with branchlets that are spirally arranged, spurlike, polysiphonous, and have deciduous terminal trichoblasts. Tetrasporangia are tetrahedrally divided and in swollen spiny branchlets or determinate branches. Gametophytes are dioecious. Spermatangia are stalked, on trichoblasts of spinelike branchlets, and in hyaline sori with sterile margins. Pericarps are sessile, oval to vase-shaped, and attached at forks or bases of spinelike branchlets.

Key to the species (2) of *Acanthophora*

1. Main axes and branches have spurlike branchlets . . . . *A. muscoides*
1. Spurlike branchlets limited to side branches and not on main axes . . . . *A. spicifera*

***Acanthophora muscoides* (Linnaeus) Bory 1828 (1826–1829): p. 156 (L. *muscoides*: mosslike, resembling a moss). Plate XXXVI, fig. 14.**

Basionym. *Fucus muscoides* Linnaeus 1753: p. 1161; 1763: p. 1630.

Synonym. *Acanthophora delilei* Harvey 1853: p. 16.

Plants are bushy, tough or brittle, to 25 cm tall, and vary in color; main axes are irregularly and radially branched, to 2.5 mm in diam., and attached by irregularly lobed holdfasts. Axes are polysiphonous, with 5 pericentral cells visible near tips; older axes are heavily corticated and parenchymatous in cross section. Spurlike branches are on indeterminate and determinate branches, to 1.0 mm long, spirally disposed, and with clusters of deciduous trichoblasts at tips. Tetrasporangia are tetrahedrally divided and in rows within small spurlike branchlets. Cystocarps are ostiolate, vase-shaped, 0.5–1.0 mm in diam., solitary, and lateral on spurlike branchlets.

Uncommon, on shallow limestone on all coasts and from 25–57 m off the east coast; also known from Bermuda, the Caribbean, Gulf of Mexico, and Brazil, as well as the Mediterranean, eastern tropical Atlantic (type location: Ascension Island; Silva et al., 1996), Red Sea, Indo-Pacific, and Australia.

Taylor 1960: p. 619; Dawes 1974: p. 151; Littler and Littler 2000: p. 190.

Acanthophora spicifera (Vahl) Børgesen 1910: p. 201, fig. 18, 19; 1915: p. 529, fig. 253, 254 (L. *spicifera*: spike-bearing, to bear spears). Plate XXXVI, fig. 15–18.

Basionym. *Fucus spiciferus* Vahl 1802: p. 44 (as '*spicifer*' and '*spicifera*')
Synonym. *Acanthophora intermedia* P. Crouan and H. Crouan in Schramm and Mazé 1865: p. 24; P. Crouan and H. Crouan in Mazé and Schramm 1878 (1870–1877): p. 247.

Plants are red to brown, tough, bushy, erect, to 25 cm tall, and attached by irregular, lobed holdfasts. Main axes are spreading, cylindrical, polysiphonous with 5 pericentral cells, heavily corticated, and lack spurlike branches; older axes are 0.6–3.0 mm in diam. Only short branches produce spurlike branchlets that are to 0.5 mm long and have deciduous trichoblasts. Tetrasporangia are tetrahedrally divided, ovoid, 40–50 µm in diam., 60–80 µm long, and in rows within swollen spurlike branchlets. Spermatangial clusters are disclike. Cystocarps are urn-shaped, 0.5–1 mm in diam., and on short stout stalks within axils of spurlike branchlets.

Common, adrift, as mats in seagrass beds, and attached to limestone and mangrove roots in sheltered bays (to 8 m) on all coasts; also known from Bermuda, North Carolina (drift), the Caribbean (type location: St. Croix; Silva et al., 1996), Gulf of Mexico, and Brazil, as well as the eastern tropical Atlantic, Indo-Pacific (introduced into the Hawaiian Islands), and Australia.

Taylor 1960: p. 620; Dawes 1974: p. 151; Schneider and Searles 1991: p. 431; Littler and Littler 2000: p. 192.

Amansia J.V. Lamouroux 1809c: p. 133; 1813: p. 54, cf. Kylin, 1956 (named for M. de Saint-Amans, a student of natural history and friend of J. Lamouroux).

Plants are erect and pinnately branched; axes are flat, narrow, membranous, with transverse bands, enrolled tips, and attach by small discoid holdfasts. Midribs have 5 pericentral cells and often are corticated. Tetrasporangia are tetrahedrally divided and within two rows in incurved stichidia on marginal teeth or midribs. Spermatangia are ovoid and in sori near enrolled apices. Cystocarps are oval, in two rows, and with pear-shaped carpospores.

Amansia multifida J. V. Lamouroux 1809d: p. 332, pl. 6, fig. 2 C–E (L. *multifida*: cleft, cut by many divisions). Plate XXXVI, fig. 19.

Plants are bushy, tough, to 20 cm tall, and red or purple; fronds are flat, pinnately branched, and with cylindrical stipes attached by discoid holdfasts. Blades are oppositely pinnately to tripinnately branched, strap-shaped, 2.0–5.0 mm wide, 2 cells thick, and with serrate margins and enrolled tips. Cells are hexagonal and in transverse lines; midribs have 5 pericentral cells. Tetrasporangia are tetrahedrally divided and in 2 rows on incurved stichidia in long marginal teeth or from midribs. Cystocarps are oval, to 75 µm in diam. by 250 µm long, and along ventral sides of midribs on marginal branchlets.

Rare, dredged at the Dry Tortugas (Taylor, 1928) and adrift at Key West (Harvey, 1853); also known from the Caribbean (type location: Santo Domingo, Dominican Republic; Silva et al., 1996), Gulf of Mexico, Brazil, and Madagascar.

Taylor 1960: p. 608; Littler and Littler 2000: p. 192.

Bostrychia Montagne 1842c: p. 39, *nom. cons.* (Gr. *bostrychos*: a small curl, twisted).

Based on molecular studies, Zuccarello and West (2006) confirmed that *Stictosiphonia* W. J. Hooker and Harvey (1845b: p. 177) should be retained in *Bostrychia*.

Indeterminate axes are usually creeping, have erect determinate branches dorsally and rhizoids ventrally, and are attached by random bundles of rhizoids or transformed branches. Axes are polysiphonous, ecorticate or corticated, cylindrical or flat, and with 4–10 pericentral cells. Pericentral cells divide one or more times to form tiers of cells, with the inner tier connected by pits to the axial cells. Branching is alternately distichous, subdichotomous, or radial (rarely); erect determinate axes are polysiphonous, branched 2 or more times, and have incurved tips. Apical cell divisions are oblique and alternate from side to side. Tetrasporangia are tetrahedrally divided, from pericentral cells, and in whorls of 2–6 per segment in stichidia. Gametophytes are dioecious. Spermatangia are in sori and from surface cells of branchlets. Procarps arise from pericentral cells in successive series on segments of branchlets. Carpogonial branches are 4-celled; carposporophytes are sessile. Pericarps are subspherical and 1–2 per branch.

Key to the species (5) of *Bostrychia*

1. Axes ecorticate 2
1. Axes corticated 3
 2. Branchlets entirely uniseriate . . . . *B. moritziana*
 2. Branchlets polysiphonous near base and uniseriate near tips . . . . *B. radicans*
3. Branchlets uniseriate for at least the last 5 segments to the tip cell 4
3. Branchlets entirely polysiphonous, except for the tip cell . . . . *B. calliptera*
 4. Tip cell 18–25 μm in diam.; branchlets uniseriate for last 6–60 cells . . . . *B. tenella*
 4. Tip cell 25–30 μm in diam.; branchlets uniseriate for last 2–5 cells . . . . *B. montagnei*

Bostrychia calliptera **(Montagne) Montagne 1842d: p. 661 (L. *calli*: beautiful + *ptera*: winged). Plate XXXVI, fig. 20–24.**

Basionym. *Rhodomela calliptera* Montagne 1840: p. 161, pl. 5, fig. 2.

Synonym. *Bostrychia pinnata* J. Tanaka and M. Chihara 1984: p. 122–125, fig. 4, 5.

See Silva et al. (1996), Zuccarello and West (2002), and Wynne (2005a) regarding the merger of *Bostrychia pinnata* with *B. calliptera*. Zuccarello and West (2006) suggested that *B. calliptera* was polyphyletic based on molecular data.

Plants are delicate, feathery, 2–11 cm tall, and dark red to purple; prostrate axes are indeterminate and produce erect determinate or indeterminate axes dorsally and disclike haptera ventrally. Haptera are formed randomly from dense clusters of rhizoids and usually occur on the lower side of forks. Erect axes are featherlike, with 2 to many branches per fork, pinnately branched, and in one plane; older branches are often denuded and irregularly branched. Main axes have 6 short pericentral cells that are covered by rhizoidal filaments. Branchlets are simple, ecorticate, and entirely polysiphonous with 4 pericentral cells. Tetrasporangia are tetrahedrally divided and in 2 rows within stalked stichidia that are lanceolate and attenuate.

Common, forming dense turfs on mangrove roots on all coasts; also known from the Gulf of Mexico, Caribbean, Guiana, French Guiana (type location: Cayenne; Silva et al., 1996), Brazil, Indo-Pacific, and Australia.

Taylor 1960: p. 597; Dawes 1974: p. 152.

Bostrychia montagnei Harvey 1853: p. 55, pl. XIV B, fig. 1–4 (named for the French phycologist J. F. C. Montagne, a friend of Harvey and of J. W. Bailey who studied Caribbean algae during the 1800s). Plate XXXVI, fig. 25, 26.

Plants are delicate to coarse, in fernlike tufts to 7.5 cm tall, and dark to dull purple; erect branches arise randomly from prostrate fibrous axes that produce random haptera ventrally at forks. Haptera form disclike rhizoidal clusters that grow from peripheral cells. Branching of erect axes is sparse and alternate below and distichous and spreading above. Older erect axes are determinate, stiff, denuded, to 650 µm in diam. below, and with flexible, pinnate, featherlike, incurved branchlets above. Main axes have 6–7 pericentral cells and 5–7 layers of cortical cells. Branchlets are ecorticate, 50–80 µm in diam., taper to 25–30 µm, and have 5 pericentral cells; the last 2–5 segments are uniseriate and 0.8–1.2 diam. long. Tetrasporangia are tetrahedrally divided, 80–100 µm in diam., and within 2 rows in swollen stichidia that are short-stalked and partially corticated. Cystocarps are ovoid, to 800 µm in diam., 700 µm long, terminal, and solitary or in pairs.

Common, on mangrove roots on all coasts (type location: Key West); also known from Bermuda, the Caribbean, Gulf of Mexico, Brazil, and Sierra Leone.

Taylor 1960: p. 598; Dawes 1974: p. 153; Littler and Littler 2000: p. 194.

Bostrychia moritziana (Sonder _ex_ Kützing) J. Agardh 1863 (1848–1901): p. 862–863 (named for Johannes Moritz, a German botanical explorer who collected in the West Indies). Plate XXXVII, fig. 1–3.

Basionym. *Polysiphonia moritziana* Sonder *ex* Kützing 1849: p. 838

Synonyms. *Polysiphonia bostrychioides* P. Crouan and H. Crouan in Schramm and Mazé 1865: p. 26 (as "*bostrychioidies*" in index); P. Crouan and H. Crouan in Mazé and Schramm 1878 (1870–1877): p. 262.

Bostrychia gaudelupensis P. Crouan and H. Crouan in Schramm and Mazé 1865: p. 26; P. Crouan and H. Crouan in Mazé and Schramm 1878 (1870–1877): p. 253.

Bostrychia polysiphonioides P. Crouan and H. Crouan in Schramm and Mazé 1865: p. 25; P. Crouan and H. Crouan in Mazé and Schramm 1878 (1870–1877): p. 252.

Bostrychia radicans Montagne f. *moniliformis* Post 1936 p. 15.

Zuccarello and West (2003, 2006) plus Zuccarello and colleagues (Zuccarello et al., 2006a, b) have suggested that the *Bostrychia moritziana/B. radicans* complex may represent several cryptic taxa that are morphologically similar. Zuccarello et al. (2006a, b) also noted that distribution of the haplotypes agreed with the existence of a geological strait that separated southern Georgia and northern Florida.

Plants form a turf or mosslike tufts, are to 6.5 cm tall, and dull brown to black purple; prostrate axes are indeterminate and produce haptera ventrally and erect branches dorsally. Haptera, which replace branches, are 4–9 cells long, with padlike bases, and arise from peripheral and axial cells of prostrate axes. Erect branches are denuded below, 75–100 µm in diam., and with 7–8 ecorticate pericentral cells. Branchlets are often upcurved, regularly alternate, 20–50 µm in diam., and to 2.2 mm long. Bases of branchlets are polysiphonous, and with segments broader than long, while tips are uniseriate with segments longer than broad. Tetrasporangia are tetrahedrally divided, 25–40 µm in diam., and in two rows within stichidia with 2–4 per segment. Stichidia are long, to 200 µm in diam. by 900 µm long, with 5–12 axial cells, and 5 pericentral cells per axial cell. Cystocarps are spherical or urn-shaped, to 500 µm in diam., and 640 µm long.

Common, on mangrove roots and as a turf on limestone in sheltered areas on all coasts; also known from the Gulf of Mexico, Caribbean, Guiana (syntype localities: West Indies and French

Guiana; Kützing, 1849), and Brazil, as well as West and South Africa, the Indo-Pacific, and Australia.

Taylor 1960: p. 596; Dawes 1974: p. 153; Littler and Littler 2000: p. 194.

Bostrychia radicans (Montagne) Montagne in Orbigny 1842d: p. 661 (L. *radicans*: rooting stems or aerial roots). Plate XXXVII, fig. 4, 5.

Basionym. *Rhodomela radicans* Montagne 1840: p. 198, pl. 5, fig. 3.

Synonym. *Bostrychia rivularis* Harvey 1853: p. 57–58, pl. XIV.D, fig. 1–5.

Zuccarello and West (2003) suggested that there may be 7 cryptic species in the *Bostrychia radicans/B. moritziana* complex.

Plants form mosslike or feathery turfs to 4 cm tall that are dark purple to black; prostrate axes are indeterminate, ecorticate, and with erect determinate branches dorsally and haptera ventrally. Haptera are transformed branches with padlike endings and are 4–7 axial cells long. Prostrate and erect axes are cylindrical to flat. Main axes are ecorticate, have 2 tiers of 6–8 pericentral cells per axial segment, are regularly alternately branched below, and subdichotomously branched with axes to 200 µm in diam. above. Branchlets are 50–100 µm in diam., 1–2 mm long, incurved, and uniseriate for the final 2–5 tip cells that are 30–40 µm in diam. Tetrasporangia are tetrahedrally divided, spherical, 50–80 µm in diam., and in whorled subapical stichidia (800–1200 µm in diam.) with 4–5 pericentral cells per axial cell. Cystocarps are ovoid and terminal on lower branchlets.

Common, on mangrove roots, stems of marsh plants, and limestone in protected sites on all coasts; also known from Massachusetts through the Caribbean, Bermuda, Gulf of Mexico, French Guiana (type location: Cayenne; Silva et al., 1996), and Brazil, as well as West Africa, the Indo-Pacific, and Australia.

Taylor 1960: p. 595; Dawes 1974: p. 153; Schneider and Searles 1991: p. 432; Littler and Littler 2000: p. 194.

Bostrychia tenella (J. V. Lamouroux) J. Agardh 1863 (1848–1901): p. 869–871 (L. *tenella*: delicate, tender, soft). Plate XXXVII, fig. 6–8.

Basionym. *Plocamium tenellum* J. V. Lamouroux 1813: p. 138

Synonyms. *Bostrychia calamistrata* (Montagne) Montagne 1842d: p. 661

Bostrychia binderi Harvey 1849 (1847–1849): p. 68, pl. XXVIII (upper figs.)

Bostrychia elegans P. Crouan and H. Crouan in Schramm and Mazé 1865: p. 25; P. Crouan and H. Crouan in Mazé and Schramm 1878 (1870–1877): p. 252

Bostrychia mazei P. Crouan and H. Crouan in Schramm and Mazé 1865: p. 26; P. Crouan and H. Crouan in Mazé and Schramm 1878 (1870–1877): p. 253

Bostrychia muscoides P. Crouan and H. Crouan in Schramm and Mazé 1865: p. 25; P. Crouan and H. Crouan in Mazé and Schramm 1878 (1870–1877): p. 252

Plants form blackish prostrate mats to 5 cm tall; prostrate axes are attached by rhizoidal haptera and produce erect determinate branches dorsally every 1–2 axial cells. Haptera have dense clusters of rhizoids that arise randomly from ventral cells at forks and do not replace branches. Erect axes are denuded below and densely and alternately branched 2–4 times above. Branchlets are pinnate and featherlike with incurved tips. Prostrate and main erect axes have 6–8 pericentral cells that are corticated by 1–4 layers of cells. Branchlets are variable, 25–300 µm in diam., and taper to uniseriate filaments of 6–60 cells in length and 18–25 µm in diam. Tetrasporangia are tetrahedrally divided, spherical, 30–70 µm in diam., and in linear stichidia 200–1500 µm in diam. by 2–20 axial cells long. Stichidia are subapical or intercalary on axial cells of branchlets with 5 pericentral cells. Cystocarps are ostiolate, oval to spherical, to 700 µm in diam., and at tips of branchlets.

Common, on mangrove roots, jetties, and limestone outcrops in protected intertidal areas on all coasts; also known from Bermuda, the Caribbean (type location: St. Croix, Virgin Island; Silva et al., 1996), Gulf of Mexico, and Brazil, as well as West Africa, the Indo-Pacific, and Australia.

Taylor 1960: p. 598, 599; Dawes 1974: p. 152; Littler and Littler 2000: p. 196.

Bryocladia F. Schmitz in Schmitz and Falkenberg 1897: p. 442 (Gr. *bryo*: relating to mosses + *cladia:* a small branch, shoot).

Plants are erect, bushy, radially organized, arise from a creeping base, and are attached by rhizoidal haptera. Erect axes are cylindrical, ecorticate, polysiphonous, and with 6–16 pericentral cells; they are also alternately and laterally branched, and covered with spirally arranged simple branchlets or with pinnately branched indeterminate branches. Branchlets are polysiphonous, short, simple, stiff, erect or recurved, and often with uniseriate, branched, and deciduous trichoblasts. Tetrasporangia are tetrahedrally divided, in rows with one per pericentral cell, and formed acropetally on branchlets. Carpogonial branches are 4-celled; procarps are near branchlet tips. Cystocarps are stalked and ovoid or urceolate.

Key to the species (2) of *Bryocladia*

1. Axes densely covered with short, straight, or recurved polysiphonous branchlets; main axes have 6–8 pericentral cells . . . . *B. cuspidata*
1. Axes have a sparse to loose cover of lax, pinnately branched polysiphonous branchlets; main axes have 9–12 pericentral cells . . . . *B. thyrsigera*

***Bryocladia cuspidata* (J. Agardh) De Toni 1903: p. 968 (L. *cuspidata*: with a cusp or a sharp stiff point). Plate XXXVII, fig. 9, 10.**

Basionym. *Polysiphonia cuspidata* J. Agardh 1852 (1848–1901): p. 953.

Plants are dark purple to black and in erect bushy tufts to 8 cm tall; erect axes are covered by short branchlets and arise from prostrate ones attached by discoid holdfasts. Erect axes are alternately branched, cylindrical, 195–350 µm in diam., and polysiphonous with 6–8 pericentral cells; segments are broader than long, ecorticate, and covered with short, spurlike branchlets. Branchlets are polysiphonous, stiff, radially arranged, 0.7–1.2 mm long, straight or recurved, strongly tapering from base to apex, and simple or once branched at bases. Trichoblasts, if present, are uniseriate, deciduous, simple to once branched, and taper from base to apex. Tetrasporangia are tetrahedrally divided, spherical, to 48 µm in diam., and in linear series in upper parts of branchlets. Cystocarps are ovoid to urceolate and to 320 µm in diam.

Common, in shallow water (to 3 m) at wave-exposed sites on all coasts; also known from Georgia, the Caribbean, Gulf of Mexico, Brazil, and the eastern tropical Atlantic.

Taylor 1960: p. 586; Dawes 1974: p. 154; Schneider and Searles 1991: p. 434.

***Bryocladia thyrsigera* (J. Agardh) F. Schmitz in Falkenberg 1901: p. 169 (L. *thyrs*: a stalk + *ger*: bearing; a thyrse, a mixed inflorescence). Plate XXXVII, fig. 11, 12.**

Basionym. *Polysiphonia thyrsigera* J. Agardh 1847: p. 17.

Plants are erect, solitary or in tufts, to 10 cm tall, red to black, and arise from creeping bases attached by discoid holdfasts. Erect axes are cylindrical, spreading, alternately and laxly branched, and 165–350 µm in diam.; their axes are covered with branchlets that are polysiphonous with 9–12 pericentral cells one-half as long as broad and 70–165 μm in diam, and ecorticate. Branchlets are polysiphonous, determinate, radially arranged, 0.6–3.0 mm long, alternately pinnately branched, and tapering to tips. Trichoblasts, if present, are uniseriate, deciduous, simple or once branched, and taper to tips. Tetrasporangia are tetrahedrally divided, spherical, 35–45 μm in diam., and on short, stiff, usually recurved branchlets. Cystocarps are stalked and oval or urceolate.

Uncommon, on shallow limestone in wave-exposed sites at Content Keys and Marineland; also known from the Caribbean, Gulf of Mexico (type location: Mexico), Brazil, and West Africa.

Taylor 1960: p. 585; Schneider and Searles 1991: p. 435.

Bryothamnion Kützing 1843: p. 433 (Gr. *bryo*: mosslike + *thamnion*: a small bush).

Plants are erect and bushy; axes are tough to cartilaginous, radially or bilaterally organized, and attached by discoid holdfasts. Main axes are cylindrical, triangular or compressed, polysiphonous with 6–9 pericentral cells, and parenchymatous in cross section. Main axes have a few indeterminate branches that produce many branchlets, which are short, corticated, simple or branched, alternate to spiral, from every 2nd axial cell, and in the axils of hyaline, uniseriate, deciduous trichoblasts. Tetrasporangia are tetrahedrally divided and formed acropetally on nodulose branchlets that spiral around upper pericentral cells. Gametophytes are dioecious. Spermatangia are on trichoblasts in oval sori. Carpogonial branches are 4-celled, on the 2nd cell of trichoblasts, and procarpial; cystocarps are oval to spherical and on short stalks.

Key to the species (2) of *Bryothamnion*

1. Axes and branches cylindrical to compressed in cross section; branchlets 2–3 mm long and in 2 to 4 vertical rows .... *B. seaforthii*
1. Axes and branches triangular in cross section; branchlets to 2 mm long and usually in 3 vertical rows. .... *B. triquetrum*

Bryothamnion seaforthii **(Turner) Kützing 1843: p. 433 (named for Lord Francis Seaforth, the governor of Barbados and a patron of natural history who sent seaweeds to D. Turner). Plate XXXVII, fig. 13–15.**

Basionym. *Fucus seaforthii* Turner 1808 (1808–1809): p. 129–130, pl. 120; 1811 (1809–1811): pl. 190

Plants are erect, bushy, and 8–20 cm tall; axes are tough to cartilaginous, pale red to deep purple, covered with branchlets, and attached by discoid holdfasts. Main axes are indeterminate, compressed or cylindrical, to 2 mm in diam., polysiphonous with 8–9 pericentral cells and densely corticated. Indeterminate branches are similar to primary axes. Branchlets are in axils of trichoblasts and have pinnate, alternate, distichous, or, less often, radial branching. Branchlets are 2–3 mm long, in 2 to 4 vertical rows, and with short, acute tips that end in one to several polysiphonous spines. Trichoblasts are deciduous, once branched, 4 cells long above forks, and taper to apex. Tetrasporangia are tetrahedrally divided, spherical, 60–100 μm in diam., in spiral series on branchlets, and clustered at tips. Cystocarps are oval and subapical on branchlets.

f. *imbricatum* J. Agardh 1863 (1848–1901): p. 849 (L. *imbricatum*: overlapping in a regular order like tiles on a roof).

See J. Agardh 1863 (1848–1901), who named three forms under *Bryothamnion seaforthii*: (a) "*disticha*" (p. 848), (b) "*fastigiata*" (p. 848), and (c) "*imbricata*" (p. 849).

Forma *imbricatum* has cylindrical axes covered by radially arranged branchlets.

Common, the species is often in drift, in 2–4 m on limestone in the Keys, and in 28–29 m off the east coast; the form occurs on limestone in wave-exposed sites in the Keys. The species also is known from North Carolina, the Caribbean (type location: Barbados; Silva et al., 1996), Gulf of Mexico, Guiana, Brazil, the eastern tropical Atlantic, and Sri Lanka. The form is recorded from other Caribbean sites.

Taylor 1960: p. 587; Schneider and Searles 1991: p. 435.

Bryothamnion triquetrum (S. G. Gmelin) M. Howe 1915: p. 222 (L. *triquetrum*: three-sided or three-angled, the angles usually sharp). Plate XXXVII, fig. 16, 17.

Basionym. *Fucus triqueter* S. G. Gmelin 1768: p. 122, pl. VIII, fig. 4.

Plants are stiff, tough, bushy, to 25 cm tall, and dark brown to deep red; axes are covered by short stiff branchlets and attached by small discoid holdfasts. Primary axes are cylindrical below, triangular or quadrate above, irregular to alternately branched, polysiphonous with 7–9 pericentral cells, and heavily corticated. Branchlets are often in 3 rows, spirally arranged, heavily corticated, to 2 mm long, and with simple branching below and trichotomous above. Branchlet tips are stiff, incurved, forked, and clustered near tips; they are more widely spaced below and in axils of short, once-forked trichoblasts. Tetrasporangia are tetrahedrally divided and stalked. Spermatangia are ovoid and on special branchlets. Cystocarps are stalked and oval.

Uncommon, on limestone in 1–5 m at exposed sites and to 20 m in the Keys; also known from the Caribbean (type location: probably Jamaica; Silva et al., 1996), Gulf of Mexico, Brazil, and São Tomé.

Taylor 1960: p. 587; Littler and Littler 2000: p. 196.

Chondria C. Agardh 1817: xviii, *nom. cons.* (Gr. *chondros*: cartilaginous, grainy).

Plants are erect, often bushy, and with one or more erect axes arising from a common discoid holdfast; plants are prostrate, spreading, and attached by multicellular rhizoids. Primary axes are cylindrical to compressed; branching is radial, alternate, or irregular and branches arise from basal cells of trichoblasts near apices. Branchlets are often constricted or tapered at bases and club- to spindle-shaped. Growth is by an exposed apical cell on an acute tip or by one hidden in a pit on obtuse tips and with or without a terminal tuft of branching trichoblasts. Axes are corticated and polysiphonous with the 5 pericentral cells visible only in sections of young branches. Tetrasporangia are tetrahedrally divided and from pericentral cells of branchlets. Gametophytes are dioecious. Spermatangia are on lowest branches of trichoblasts in hyaline oval sori with sterile margins. Carpogonial branches are 4-celled; procarps are on the 2nd cell of a 3-celled hair. Carposporophytes arise from large fusion cells. Cystocarps are urceolate to spherical, with pseudoparenchymatous pericarps that are lateral on branchlets.

Key to the species (12) of *Chondria*

Note: Young branches of *Chondria* have 5 pericentral cells visible in cross section, while they are not visible in similar branches of *Laurencia* spp.

1. Plants prostrate or cushionlike; rhizoids in tight random packets on axes 2
1. Plants erect or sprawling; rhizoids not in tight packets along axes 3
   2. Axes show a faint banding pattern from thickened end walls of pericentral cells; tips blunt and apical cell usually in a terminal pit . . . . *C. curvilineata*
   2. Axes lack bands; tips acute, apical cell visible and not in a pit . . . . *C. polyrhiza*
3. Branch tips pointed and with apical cell exposed and not in a pit 4
3. Branch tips blunt and apical cell in a pit 7

4. Plants usually not epiphytic; 10–40 cm tall; branchlets 250–500 µm in diameter 5
4. Plants often epiphytic; 2–5 cm tall; branchlets 80–200 µm in diameter . . . . *C. baileyana*
5. All axes have a regular decrease in diam.; branchlets little tapered at tips 6
5. Branchlets much thinner than main axes and tapering at both ends . . . . *C. littoralis*
6. Plants pale yellow to dull purple; apical cells not flush at truncated tips of branchlets; hair tips sparse and usually not obscuring apical cells . . . . *C. capillaris*
6. Plants dark purple to red; apical cells often flush with blunt tips of branchlets; hair tips abundant and usually covering apical cells in young branchlets . . . . *C. atropurpurea*
7. Branchlets usually constricted at bases; tips often have a tuft of fine hairs 8
7. Branchlets not constricted at bases; tips often lack a tuft of hairs . . . . *C. floridana*
8. Axes lacking a banding pattern 9
8. Axes with faint bands caused by thick end walls of pericentral cells . . . . *C. collinsiana*
9. Branches more than 0.5 mm in diam.; branchlets 200–600 µm in diameter 10
9. Branches 0.2–0.4 mm in diam.; branchlets to 0.25 mm in diameter . . . . *C. leptacremon*
10. Older branchlets produce few branchlets and these not clustered at nodes. 11
10. Older branchlets produce clusters of 2–6 new branchlets present at nodes. . . . . *C. cnicophylla*
11. Plants soft, 5–30 cm tall; staining paper on drying; main axes 1.0–1.25 mm in diam.; branchlets 2–10 mm long and with a dense tuft of hairs at branch tips . . . . *C. dasyphylla*
11. Plants tough, 10–15 cm tall; dried plants not staining paper; main axes 1.0–2.5 mm in diam.; branchlets 1–5 mm long and lacking dense tufts of terminal hairs . . . . *C. sedifolia*

Chondria atropurpurea Harvey 1853: p. 22, pl. XVIII E, fig. 1–3 (L. *atropurpurea*: dark purple, blackish purple). Plate XXXVII, fig. 18–20.

Plants are sparsely branched to bushy, 20–40 cm tall, and dark purplish red; axes are cylindrical, firm, 0.8–2 mm in diam., often denuded below, and attached by discoid holdfasts. Branching is sparse and either alternate or irregular above (1–2 times). Branchlets are unbranched, 0.25–0.40 mm in diam., 0.2–1.0 cm long, linear to fusiform, constricted at bases, and tapering to acute tips. Apical cells project from tufts of trichoblasts that are short, to 200 µm long, and dichotomously branched. Tetrasporangia are tetrahedrally divided, spherical, 40–110 µm in diam., and in small branchlets. Cystocarps are crowded on branchlets; pericarps are obvious, subspherical to urceolate, and to 1.5 mm in diam.

Common, on seagrasses and limestone (intertidal to 5 m) in the Keys; also known from Bermuda, the Carolinas (type location: Charleston, South Carolina), Georgia, the Caribbean, Gulf of Mexico, and Brazil.

Taylor 1960: p. 613; Schneider and Searles 1991: p. 438; Littler and Littler 2000: p. 198.

Chondria baileyana (Montagne in Bailey) Harvey 1853: p. 20–21, pl. XVIII A, fig. 1–6 (L.: named for J. W. Bailey, an American scientist who studied diatoms and desmids plus seaweeds of the southeastern United States). Plate XXXVII, fig. 21, 22.

Basionym. *Laurencia baileyana* Montagne in Bailey 1848: p. 63.

Synonyms. *Chondriopsis tenuissima* J. Agardh var. *baileyana* (Montagne) Farlow 1881 ('1882'): p. 166.

Chondria tenuissima (Goodenough and Woodward) C. Agardh var. *baileyana* (Montagne) Farlow in Farlow, Anderson and Eaton 1889 (1877–1889), ex. no. 18.

Plants are bushy, soft, solitary or clustered, to 25 cm tall, pale yellow to deep reddish purple, cylindrical, 0.2–1.0 mm in diam. below, and attached by discoid holdfasts. Main axes are densely corticated, with 5 pericentral cells and a few long branches. Branches have many branchlets that are short, slender, club-shaped, to 5 mm long by 80–200 µm in diam., constricted at bases, tapered or not, and with obtuse tips. Apical cells are visible or partially obscured by short to long tufts of trichoblasts. Tetrasporangia are tetrahedrally divided, spherical, 55–160 µm in diam., and in the mid to upper regions of branchlets or in subapical bands. Spermatangial sori are ovoid, 250–400 µm in diam., and 240–250 µm long. Cystocarps are spherical or urceolate, to 1 mm in diam., and solitary on short branchlets.

Uncommon, on seagrass, jetties, and limestone in the Keys (to 5 m); also known as scattered disjunct populations in Northumberland Strait (Canada) and the Gulf of Maine, but more common from Long Island Sound to North Carolina (type location: Newport, Rhode Island, to Fort Hamilton, New York; Lipkin and Silva, 2002), the Caribbean, Mediterranean, and Red Sea.

Taylor 1960: p. 614; Schneider and Searles 1991: p. 439; Littler and Littler 2000: p. 198.

Chondria capillaris (Hudson) M. J. Wynne 1991: p. 317 (L. *capillaris*: hairlike, very slender). Plate XXXVII, fig. 23.

Basionym. *Ulva capillaris* Hudson 1778: p. 571; non *Fucus capillaris* Hudson 1778: p. 591.
Synonyms. *Alsidium tenuissima* f. *crassa* P. Crouan and H. Crouan in Schramm and Mazé 1865. *Chondriopsis tenuissima* f. *crassa* P. Crouan and H. Crouan in Mazé and Schramm 1878 (1870–1877): p. 245.

Plants are soft, bushy, solitary or clustered, erect, to 25 cm tall, and pale yellow to dull purple; axes are terete, coarse, 0.5–2.5 mm in diam., and attached by discoid holdfasts. Primary axes are heavily corticated and have 5 large irregular pericentral cells. Branching is often simple and limited; main branches are 5–8 cm long while secondary ones are sparse and widely divergent. Branchlets are fusiform to linear, 250–500 µm in diam., 12–15 mm long, constricted at bases, and taper to acute or blunt tips. Apical cells are partially hidden by trichoblasts that are short, to 200 µm long, and dichotomously branched. Tetrasporangia are tetrahedrally divided, spherical, 40–150 µm in diam., and scattered near branchlet tips. Cystocarps are oval to urn-shaped, to 1 mm in diam., sessile or stalked, and on short laterals near the tips of main axes.

Common, on intertidal coquina at Marineland and on shells, limestone, and seagrass (to 3m) on all coasts; also known from southern Massachusetts through the Caribbean and Gulf of Mexico, as well as the eastern North and tropical Atlantic (syntype locations: near Christchurch and Margate, England; Silva et al., 1996), Mediterranean, and Indo-Pacific.

Taylor 1960: p. 613; Dawes 1974: p. 156; Schneider and Searles 1991: p. 445; Littler and Littler 2000: p. 198.

Chondria cnicophylla (Melvill in G. Murray) De Toni 1903: p. 849 (L. *cnicus*: thistle + Gr. *phyllon*: leaf, leaves). Plate XXXVII, fig. 24, 25.

Basionym. *Chondriopsis cnicophylla* Melvill in G. Murray 1888: p. 333, pl. 284, fig. 1a,b.

Plants are soft, entangled to bushy, to 20 cm tall, and straw yellow with pale red tips; main axes are indistinct and attached by discoid holdfasts. Branching is irregular to alternate, spreading, and almost at right angles. Branches are cylindrical, with 6–8 pericentral cells, heavily corticated, and 0.5–1.3 mm in diam. Branchlets are common, radially arranged, cylindrical, to 500 µm in diam., 2–5 mm long, and often in clusters of 1–2(–6) at nodes; bases are constricted and tips tapering or obtuse. The apical cell is in a terminal pit and covered by dichotomously branched trichoblasts. Tetrasporangia are tetrahedrally divided, spherical, and in clusters near the tips of fertile branchlets. Cystocarps are oval, subsessile, and alone or in small groups on upper branchlets.

Common, on limestone, shells, and seagrass beds in shallow waters of all coasts (type location:

Cedar Key, west coast of Florida) and 29–36 m off the east coast; also recorded from the Gulf of Mexico and the Caribbean.

Taylor 1960: p. 614; Dawes 1974: p. 154; Littler and Littler 2000: p. 200.

Chondria collinsiana M. Howe 1920a: p. 568, fig. 9 (named for Frank Shipley Collins, an American phycologist, who studied the green algae of North America and seaweeds of New England and Bermuda). Plate XXXVII, fig. 26.

Plants are soft, erect, to 8 cm tall, yellow to red, with a few obvious primary axes that are little branched, and attached by small discoid holdfasts. Primary axes are 0.4–0.8 mm in diam., cylindrical, heavily corticated, and have 5–6 large pericentral cells with thickened cross walls visible as faint bands near branch tips. Branchlets are fusiform, cylindrical to cone-shaped, 0.2 to 0.5 mm in diam., 2–5 diam. long, constricted at bases, and with obtuse to truncate tips. Branchlet apical cells are hidden in terminal pits and covered by tufts of trichoblasts; surface cells are 26–40 μm in diam. and 65–160 μm long at apices. Tetrasporangia are tetrahedrally divided, spherical, to 140 μm in diam., and scattered near branchlet tips. Spermatangial sori are at trichoblast bases and to 500 μm in diam. Cystocarps are spherical or ovoid, to 700 μm in diam., and stalked.

Uncommon, on seagrasses near Anclote Key and algae in shallow waters of all coasts and in 24 m off the east coast; also known from the Bahamas (type location: Mariguana; Silva et al., 1996), Caribbean, Greece (introduced), Turkey, the Red Sea, and Indian Ocean.

Taylor 1960: p. 617; Dawes 1974: p. 154; Littler and Littler 2000: p. 200.

Chondria curvilineata Collins and Hervey 1917: p. 120–121, pl. II, fig. 10, 11 (L. *curvi*: curved + *lineata*: lined, with many lines). Plate XXXVII, fig. 27–29.

Plants are prostrate, spreading, 1–3 cm tall, and light to deep red; main axes are cylindrical, lax, 200–500 μm in diam., and attached by tufts of simple multicellular rhizoids. Branching is irregular and rarely more than 2 orders; axes are heavily corticated and have 6 large pericentral cells with thickened, curved cross walls that appear as faint bands near tips. Branchlets are short, cylindrical to cone-shaped, 90–200 μm in diam., constricted at bases, and with obtuse tips. The apical cell is in a terminal pit and may project through a tuft of short trichoblasts and coarse filaments. Trichoblasts are dichotomously branched; surface cells at tips are 10–30 μm in diam. by 40–100 μm long. Tetrasporangia are tetrahedrally divided, globose, to 70 μm in diam., and near branchlet tips. Cystocarps are oval and sessile.

Rare, reported from 34 m off the east coast (Hanisak and Blair, 1988); also known from Bermuda (type location: Heron Bay), North Carolina (29–40 m), the Bahamas, Antilles, and Gulf of Mexico, as well as the Salvage Islands, Mediterranean (introduced into Egypt), Maldives, and Philippines.

Schneider and Searles 1991: p. 440; Littler and Littler 2000: p. 200.

Chondria dasyphylla (Woodward) C. Agardh 1817: p. xviii; 1822 (1820–1828): p. 350–351; 1824: p. 205 (Gr. *dasy*: hairy, shaggy + *phyllus*: leaf). Plate XXXVII, fig. 30–32.

Basionym. *Fucus dasyphyllus* Woodward 1794a: p. 239–241, pl. 23, fig. 1–3.
Synonyms. *Chondria dasyphylla* (Woodward) Harvey 1853: p. 20.
Alsidium dasyphyllum P. Crouan and H. Crouan in Schramm and Mazé 1865: p. 25; in Mazé and Schramm 1878 (1870–1877): p. 245.

Chondria dasyphila (C. Agardh) Farlow 1881 ('1882'): p. 166.

According to Wynne (pers. comm.), *Alsidium dasyphyllum* from Florida and Bermuda is a heterotypic synonym for *Chondria dasyphylla*. The type specimen of *A. dasyphyllum* needs confirmation (Norris, 1994) as it may be the same as the Mediterranean *A. helminthochorton* (La Tourette) Kützing.

Plants are soft or firm, bushy, 5–30 cm tall, and pale yellow to red; axes are cylindrical, 1.0–2.5 mm in diam., and attached by discoid holdfasts. Main axes are few and covered with many alternate or opposite short to long secondary branches. Branchlets are short, strongly constricted at bases, club- to spindle-shaped, 200–600 µm in diam., 2–10 mm long, and in alternate, opposite, or spiraled patterns. Older branchlets are irregularly swollen. Tips are obtuse; the apical cell occurs in a pit and may project though dichotomously branched trichoblasts. Surface cells near tips are 20–30 µm in diam. and may elongate with age. Tetrasporangia are tetrahedrally divided, spherical, 40–165 µm in diam., and mostly near branchlet tips. Spermatangial sori are ovoid, 400–600 µm in diam., and occur near basal cells of trichoblasts. Cystocarps are ovoid, sessile, to 1 mm in diam., and near branchlet tips.

Common, on limestone, seagrasses, and adrift on all coasts (to 5 m); also known from Cape Cod southward to North Carolina, Bermuda, the Bahamas, Caribbean, Gulf of Mexico, Brazil, and Uruguay, plus the eastern North and tropical Atlantic (type location: Yarmouth, England; Gordon Mills, 1987), Mediterranean, Red Sea, and Indo-Pacific.

Taylor 1960: p. 616; Dawes 1974: p. 155; Schneider and Searles 1991: p. 441; Littler and Littler 2000: p. 202.

Chondria floridana (Collins) M. Howe in W. R. Taylor 1928: p. 170, pl. 34, fig. 3 (L. *floridana*: of or from the state of Florida). Plate XXXVIII, fig. 1, 2.

Basionym. *Chondria dasyphylla* (Woodward) C. Agardh f. *floridana* Collins in Collins, Holden and Setchell 1902 (1895–1919): ex. no. 996; Collins 1906a: p. 111.

Plants are bushy to pyramidal, erect, to 20 cm tall, and pale red; axes are lax to firm, terete, 0.5–3 mm in diam., denuded below, and attached by discoid holdfasts. Main axes have 5–6 pericentral cells, heavy cortication, and regular and progressively alternate branching to 4 or 5 times. Branchlets are abundant, straight to curved, cylindrical to club-shaped, not constricted at bases, 0.3–0.6 mm in diam., and 5–10 mm long. Branchlet tips are blunt; trichoblasts are inconspicuous and dehisce early; the apical cell occurs in a terminal pit. Tetrasporangia are tetrahedrally divided, spherical, and in bands 1–2 mm below branchlet tips or scattered below.

Uncommon, on limestone in 15 m near Anclote Key and in 2–36 m at the Dry Tortugas (type location: Jupiter Inlet); also known from North Carolina, the Bahamas, Caribbean, Gulf of Mexico, Brazil, and South Africa.

Taylor 1960: p. 616; Dawes 1974: p. 155; Schneider and Searles 1991: p. 442; Littler and Littler 2000: p. 202.

Chondria leptacremon (Melvill) De Toni 1903: p. 848 (Gr. *leptos*: thin, small, weak + *cremor*: cream-colored). Plate XXXVIII, fig. 3–5.

Basionym. *Chondriopsis leptacremon* Melvill in G. Murray 1888: p. 333, pl. 284, fig. 2a,b.

Plants are stiff, erect, to 15 cm tall, pale yellow to red with a number of axes that are sparsely branched and attached by hemispherical discoid holdfasts. Main axes are slender, cylindrical, 0.2–0.4 mm in diam., heavily corticated, and with 4–6 rounded pericentral cells to 60 µm in diam.

Branching is radial and alternate; branchlets are to 1 mm long by 250 µm in diam., abundant, club-shaped, constricted at bases, and 1–5 per node. Branchlet tips are blunt with an apical cell in a pit and hidden by a tuft of dichotomously branched trichoblasts. Surface cells near tips are elongate and 20–30 µm in diam.; older surface cells are to 60 µm in diam. Tetrasporangia are tetrahedrally divided, spherical, and 40–80 µm in diam.

Common, on mangrove roots, seagrasses, algae, and limestone (to 5 m) on all coasts; also known from the Bahamas, Caribbean, and Gulf of Mexico. Type location: "low water on a coral slab below Fort Taylor at southwest extremity of the islands of Key West, South Florida"; Melvill in G. Murray (1888).

Taylor 1960: p. 615; Dawes 1974: p. 155; Littler and Littler 2000: p. 202.

Chondria littoralis Harvey 1853: p. 22 (L. *littoralis*: the seashore, on or near the shore). Plate XXXVIII, fig. 6–8.

Plants are soft or firm, erect, to 10–35 cm tall, sparsely branched or bushy, yellow to dark red, and attached by discoid holdfasts. Main axes are 1–2 mm in diam., denuded, sparingly branched, heavily corticated below, and with 5–6 pericentral cells. Upper axes have irregular clusters of branchlets that are long, spindle-shaped, to 500 µm in diam., 1–15(25) mm long, with tapered tips, constricted bases, and may rebranch. Branch tips are acute or obtuse and the apical cell is visible or covered by dichotomously branched trichoblasts. Tetrasporangia are tetrahedrally divided, spherical, 100–120 µm in diam., and near tips. Spermatangial sori are ovoid, 450 µm in diam., and lateral on branchlets. Cystocarps are ostiolate, urceolate, to 850 µm in diam., solitary, and lateral on branchlets. The species is similar to *Chondria capillaris*, except for its larger size.

Common, on shells, limestone, and rubble (to 9 m) in oceanic and estuarine sites on all coasts (type location: Keys); also known from Bermuda, North Carolina, Georgia, the Caribbean, Gulf of Mexico, and Brazil.

Taylor 1960: p. 612; Dawes 1974: p. 155; Schneider and Searles 1991: p. 443; Littler and Littler 2000: p. 204.

Chondria polyrhiza Collins and Hervey 1917: p. 121–122, pl. II, fig. 12 (Gr. *polus*, *poly*: many + *rhiza*: a root). Plate XXXVIII, fig. 9–11.

Plants are firm, prostrate and spreading, to 8 mm tall, and pale to rose red; prostrate axes are attached ventrally by dense clusters of short unicellular to multicellular rhizoids. Erect axes are cylindrical, to 525 µm in diam., indistinct, sparsely alternate to irregularly branched, covered with 1–2 orders of short, slender branchlets, and with 5–6 pericentral cells that may be visible in young branches. Branchlets are constricted at bases; they are also fusiform, cylindrical, or club-shaped, and have acute tips. The apical cell is visible or covered by short trichoblasts. Surface cells near branch tips are angular, 5–15 µm in diam., and 25–60 µm long; older surface cells are more rectangular. Tetrasporangia are tetrahedrally divided, spherical, to 140 µm in diam., and swell branchlet tips.

Uncommon, on shallow limestone in the Keys, in 18 m at the Dry Tortugas, and 37 m off the east coast; also known from Bermuda (type location: Shelly Bay; Silva et al., 1996), North Carolina, Georgia, the Caribbean, Gulf of Mexico, and Brazil, as well as Greece and Italy (both introduced), and the Indo-Pacific.

Taylor 1960: p. 617; Schneider and Searles 1991: p. 444; Littler and Littler 2000: p. 204.

Chondria sedifolia Harvey 1853: p. 19–20, pl. XVIII G, fig. 1, 2 (L. *sedes* (*sedis*): a seat, position + *folia*: a leaf). Plate XXXVIII, fig. 12, 13.

Synonyms. *Laurencia thujoides* Kützing 1865: p. 26, pl. 74 *sensu* Taylor 1960.
Laurencia chondryopsides P. Crouan and H. Crouan in Mazé and Schramm 1878 (1870–1877): p. 242.
Chondriopsis dasyphylla (Woodward) C. Agardh var. *sedifolia* Farlow 1881 ('1882'): p. 166.

Records of *Chondria sedifolia* from North Carolina and southward are misidentifications of *C. dasyphylla* and *C. littoralis* and need confirmation (Schneider and Searles, 1991). Sears (2002) noted that *C. dasyphylla* and *C. sedifolia* have obtuse branch tips and overlapping features.

Plants are firm, cartilaginous, densely bushy, 10–15 cm tall, and pale yellow to light red; main axes are coarse, 1.0–2.5 mm in diam., and arise from discoid holdfasts. Branching is broadly pyramidal and spreading 1–2 times. Branchlets are abundant, single, opposite, or clustered at nodes; they are also strongly constricted at bases, elongate to club-shaped, and 0.3–0.6 mm in diam. by 1–5 mm long. Tips are obtuse; the apical cell occurs in a terminal pit and is covered by dichotomously branched trichoblasts. Tetrasporangia are tetrahedrally divided, spherical, and near branchlet tips. Cystocarps are oval, subsessile, and with 1 to a few produced on a branch. In contrast to *Chondria dasyphylla*, *C. sedifolia* stains paper while drying and lacks dense tufts of apical trichoblasts.

Rare, from 10 m off Anclote Key and near Key West (type location); also known from southern Massachusetts to the Caribbean, Brazil, the Red Sea, and Indo-Pacific.

Taylor 1960: p. 615; Dawes 1974: p. 155.

Chondrophycus (Tokida and Saito in Saito) Garbary and J.T. Harper 1998 emend. Nam 2006: p 695(Gr. *chondros*: cartilage, granular + *phycus*: alga, seaweed).

Basionym. *Laurencia* subgenus *Chondrophycus* Tokida and Saito in Saito 1967: p. 72.

Garbary and Harper (1998) segregated *Chondrophycus* from *Laurencia* Lamouroux following Nam and colleagues's (1994) earlier resurrection of *Osmundea* Stackhouse as a distinct genus. The three genera are morphologically similar with apical cells sunken in terminal pits of branches, a recognizable axial cell row only near the apex, and an extensive cortex (Kylin, 1956; Nam, 2006). *Chondrophycus* and *Laurencia* are distinguished from *Osmundea* with trichoblast-type rather than filament-type spermatangial development and tetrasporangial production by pericentral cells rather than random epidermal cells (Nam, 1999; Nam et al., 1994). *Chondrophycus* is also separated from *Laurencia* by having 2 instead of 4 pericentral cells. Nam (2006) has placed four Caribbean taxa (*C. gemiferous*, *C. iridescens*, *C. papillosus*, and *C. poiteaui*) in a new genus, *Palisada* (Yamada) Nam based on a cladistic analysis of 47 species of *Laurencia*, *Chondrophycus*, and *Osmundaria*.

Plants of *Chondrophycus* are erect, cartilaginous, with alternate opposite, or irregular branching, and attach by discoid holdfasts. Apical cells are in terminal pits and two pericentral cells are usually visible at apices. Axes are heavily corticated with secondary pit connections between epidermal cells and a *corps en cerise* (cherry red spons visible in living or freshly preserved cells) may be present in some species Nam, 1999). The epidermal layer may be columnar or not. Spermatangial pits are cup-shaped and at the base of trichoblasts. Tetrasporangia are tetrahedrally divided. Use the key to *Palisada* for a morphological separation of *Chondrophycus corallopsis* from species of *Palisada*.

Chondrophycus corallopsis (Montagne) Nam 1999: p. 463 (L. *corallum*: coral + *opsis*: like, thus, resembling a coral). Plate XXXVIII, fig. 14.

Basionym. *Sphaerococcus corallopsis* Montagne 1842c: p. 49–51, pl. III, fig. 1.
Synonym. *Laurencia corallopsis* (Montagne) M. Howe 1918: p. 519.

Littler and Littler (2000) and Wynne (2005a) disagreed with Howe's (1918) merger of *Laurencia cervicornis* Harvey 1853 with *L. corallopsis* and included it under *Laurencia.*

Plants are bushy, tough, brown to rose red, and 4–16 cm tall; axes are terete below to compressed above and attached by branching holdfasts. Axes are subdichotomously to dichotomously branched below, irregularly alternate above, and 0.75–2 mm in diam. Young plants are 2–6 cm tall, compactly branched, and lack club-shaped branchlets. Older plants are 10–14 cm tall and have more widely spaced branches and club-shaped branchlets that are short, upcurved, 1–4 mm long, usually swollen, often forked, and with slightly constricted bases. The tips are obtuse with an apical cell in a terminal pit and two pericentral cells visible in section near tips. Surface cells are compact, angular to rectangular, and to 50 μm in diam. Tetrasporangia are tetrahedrally divided, ovoid, 60–100 μm long, and near branchlet tips. Cystocarps are also near branchlet tips, sessile, oval, subglobose to urn-shaped, and with obvious necks.

Uncommon, on shells and rubble (to 14 m) in coral reefs at the Keys; often covered with calcareous red algal crusts or bryozoans; also known from Bermuda, North Carolina (20–60 m), the Caribbean (type location: Havana, Cuba; Silva et al., 1996), as well as the eastern tropical Atlantic and Indo-Pacific.

Taylor 1960: p. 623; Schneider and Searles 1991: p. 452.

Digenea C. Agardh 1822 (1820–1828): p. 388 (L. *dignus*: worthy, fit, deserving).

Plants are bushy, erect, and cartilaginous; axes are widely dichotomous to irregularly branched and attached by discoid holdfasts. Axes are polysiphonous, with 6–8 (10–12) pericentral cells, and a cortex of large inner and small cortical cells. Its main axes is usually covered by short, stiff, mostly simple branchlets that have a thin cortex, visible pericentral and apical cells, and tips with deciduous trichoblasts. Tetrasporangia are tetrahedrally divided, covered by 3 pericentral cells, and often occur near the tips of ecorticate and irregularly swollen branches. Spermatangial sori are in small oval-shaped clusters at branchlet tips. Cystocarps are ovoid and terminal or lateral on small branchlets.

***Digenea simplex* (Wulfen) C. Agardh 1822 (1820–1828): p. 389 (L. *simplex*: simple, undivided). Plate XXXVIII, fig. 23, 24.**

Basionym. *Conferva simplex* Wulfen 1803: p. 17.

Plants are stiff, cartilaginous, 3–25 cm tall, pale red to dark red brown, widely dichotomously to irregularly branched, and attached by coarse, short rhizoids that form discoid holdfasts. Axes are naked below and covered above by short, stiff branchlets that are 3–5(15) mm long, spindle-shaped, and densely corticated. Axes have 6–8(10–12) pericentral cells visible at tips and deciduous trichoblasts. Axial cells of branchlets are distinct and 40–55 μm in diam.; surface cells are 25–35 μm in diam. Tetrasporangia are tetrahedrally divided, to 80 μm in diam., and in swollen branchlets. Spermatangia are in oval clusters at branchlet tips. Cystocarps are oval and lateral or at tips. The species has kaenic acid and has been used as a vermifuge to treat intestinal disorders (Dawes, 1981).

Common, on shells, limestone, and adrift in seagrass beds (to 20 m) on all coasts; often heavily epiphytized. Also known from Bermuda, the Caribbean, Gulf of Mexico, and Brazil, as well as the Mediterranean (type location: Trieste, Italy; Silva et al., 1996), eastern tropical Atlantic, Red Sea, Indo-Pacific, and Australia.

Taylor 1960: p. 588; Dawes 1974: p. 156; Littler and Littler 2000: p. 204.

Enantiocladia Falkenberg in F. Schmitz 1889: p. 447; F. Schmitz in Engler and Prantl 1890–1897: p. 466 (Gr. *enantio*: opposite + *cladia*: a small branch or shoot).

Plants are erect, fleshy, and with opposite branching that is mostly in one plane; main axes are elongate and attached by discoid holdfasts. Primary axes have bladelike wings and corticated midribs with 5 pericentral cells. Blades have large medullary and smaller cortical cells. The cortex is thicker in older parts of plants due to a subcortex. Blade margins have opposing spines. Young branches are slender and with marginal branchlets or small teeth. Tetrasporangia are tetrahedrally divided and in crowded stichidia on upper parts of proliferations; cystocarps are oval and have a thick-walled pericarp.

Enantiocladia duperreyi **(C. Agardh) Falkenberg in F. Schmitz 1889: p. 447 (named for M. Duperrey, a French collector who sent J. Agardh seaweed specimens from Martinique for identification). Plate XXXVIII, fig. 25.**

Basionym. *Rytiphlaea duperreyi* C. Agardh 1828 (1820–1828): p. 52; 1828 (1828–1835): pl. 20

Plants are firm, fleshy, 10–18 cm tall, and dark red to red brown; main axes have irregular opposite branching, are flat, and attached by discoid bases. Basal stalks are short, terete, and produce 1–3 primary fronds that are 2.0–4.5 mm wide and pinnately branched 1–3 times. Branching is irregularly opposite and pinnate; fronds are narrow, flat or grooved above, and with oval bases. Branch margins have opposing serrate teeth that taper to fine points, curve inward, and are often forked. Fertile branch margins are often fringelike. Tetrasporangia are tetrahedrally divided and in 2 rows on the outside of pinnately arranged stichidia that are strongly incurved and in rows.

Rare, on shells of living *Strombus* dredged at the Dry Tortugas; also known from the Antilles and Brazil.

Taylor 1960: p. 610.

Herposiphonia Nägeli 1846: p. 238 (Gr. *herpo*: to creep + *siphon*: tubular, pipe).

Plants are polysiphonous, prostrate and spreading, often detached or attached by unicellular rhizoids and fingerlike lobes from distal ends of ventral pericentral cells. Prostrate axes are indeterminate, have 6–18 ecorticate pericentral cells, and produce erect indeterminate axes dorsally every fourth segment that do not develop or branch much. Younger branches have enrolled tips and lack trichoblasts. Erect branchlets are determinate, simple or branched, and usually alternate on every segment with an opposing indeterminate branch. Tetrasporangia are tetrahedrally divided and one per segment in a linear series in the middle of branchlets. Gametophytes are dioecious. Spermatangia are in conical or linear sori on trichoblasts or replace them. Procarps arise from the second segment of reduced trichoblasts on determinate branches. Carpogonial branches are 4- celled; cystocarps are oval to urn-shaped and sessile or on short stalks.

Key to the species (4) of *Herposiphonia* (modified from Littler and Littler, 2000)

1. Plants usually prostrate; erect branchlets not alternately offset along axes 2
1. Plants mostly erect; branchlets alternately offset along erect axes . . . . *H. pecten-veneris*
   2. Erect branchlets usually arise from every segment 3
   2. Erect branchlets mostly arise from every 5th or 6th segment . . . . *H. secunda*
3. Tips of erect branchlets blunt and with 2–3 trichoblasts . . . . *H. parca*
3. Tips of erect branchlets not blunt and usually with 3 or more trichoblasts . . . . *H. tenella*

Herposiphonia parca Setchell 1926: p. 103, pl. 20, fig. 2 (L. *parca*: sparse, limited). Plate XXXVIII, fig. 26–29.

Filaments are polysiphonous, mostly prostrate, brown to purplish red, ecorticate, and often penetrate algal hosts by swollen rhizoids (to 100 µm in diam.) that end in fingerlike pads. Prostrate axes are irregularly to alternately branched, with upcurved tips, 100–150 µm in diam., and have branchlets in almost every segment. Branchlets are erect, to 6 mm tall, 40–90 µm in diam., and with 8–10 pericentral cells 0.5–2.5 diam. long; branchlets are strongly arched when young and have blunt, truncated tips. Trichoblasts branch dichotomously 3–5 times and are in clusters of 2–3 at branchlet tips. Tetrasporangia are tetrahedrally divided, spherical, 80–120 µm in diam., and in linear series on branchlets. Spermatangial stichidia are oblong, to 80 µm in diam., and end in 1–5 uniseriate cells. Cystocarps are terminal and to 400 µm in diam.

Rare, in the Keys(?); known from the western Caribbean (to 15 m at Belize), Mediterranean, and Indo-Pacific (type location: Arue Reef, Tahiti; Silva et al., 1996).

Littler and Littler 2000: p. 208.

Herposiphonia pecten-veneris (Harvey) Falkenberg 1901: p. 315 (L. *pecten*: comb + *veneris*: Venus, the Roman goddess of love). Plate XXXIX, fig. 1, 2.

Basionym. *Polysiphonia pecten-veneris* Harvey 1853: p. 46, pl. XVIC, fig. 1–3, pl. XVID, fig. 1–5.

Filaments are polysiphonous, mostly erect to 10 cm tall, with recurved tips, and yellow to light red brown; prostrate axes are wide spreading and attached by unicellular rhizoids with fingerlike lobes that arise distally from ventral pericentral cells. Erect axes are recurved, comblike, 70–165 µm in diam., with 9–12 pericentral cells (2–3 diam. long), and produce branchlets in secund series. Branchlets are short, 60–80 µm in diam., 8–10 segments (1 mm) long, unbranched, perpendicular to axes, and alternately offset in secund series. Trichoblasts are deciduous, in spirals at tips, dichotomously branched 2–5 times, and leave a scar cell. Tetrasporangia are tetrahedrally divided, spherical, 35–75 µm in diam., and in straight series on branchlets. Spermatangial stichidia are cylindrical. Cystocarps are spherical, 0.3–0.6 mm in diam., and terminal.

var. *laxa* W. R. Taylor 1960: p. 603, 635 (L. *laxa*: loose, lax, or weak).

Filaments are polysiphonous; prostrate axes are straight, yellow, and mostly free, except for older attached parts that are alternately to strongly bilaterally branched and usually produce an indeterminate (main) erect axis or rudiment on every fourth segment with the other three segments having branchlets. Erect main axes are long and have 9–10 pericentral cells that are 120–160(250) µm in diam.; branchlets are short, with 6–7 pericentral cells, 8–12 segments, and 35–60 µm in diam.

Common, on algae, seagrasses, and under ledges (to 2 m) in the Keys (type locations: Key West for species, Dry Tortugas for variety); the species is known from Bermuda, the Caribbean, and Gulf of Mexico, while the variety is reported from Jamaica, the Calicos Islands, and Bahamas.

Taylor 1960: p. 603; Littler and Littler 2000: p. 208.

Herposiphonia secunda (C. Agardh) Ambronn 1880: p. 197–198, pl. 4, fig. 8, 11–12 (L. *secunda*: behind, second). Plate XXXIX, fig. 3.

Basionym. *Hutchinsia secunda* C. Agardh 1824: p. 149; 1828 (1820–1828): p. 106.

Synonym. *Herposiphonia tenella* (C. Agardh) Ambronn f. *secunda* (C. Agardh) Hollenberg 1968b: p. 556–557.

Filaments are polysiphonous, mostly prostrate and in entangled clusters 1–2 cm in diam.; they are brown to bright red and attached by lobed pads of multicellular or unicellular rhizoids from ventral pericentral cells. Prostrate axes are 70–150 µm in diam., with 8(7–10) pericentral cells, repetitive branching, and an erect axis every 5th or 6th(8th) segment. Erect indeterminate axes are usually rudimentary, with branching similar to prostrate axes. Erect (determinate) branchlets are short, simple, tapering toward bases, 0.5–1.4 mm tall, 60–125 µm in diam., with 12–14 segments (each 125 µm long), and produce a tuft or spiral of trichoblasts near branchlet tips. Trichoblasts are deciduous and dichotomously branched 2–4 times. Tetrasporangia are tetrahedrally divided, spherical, to 110 µm in diam., and one per segment near branchlet tips. Cystocarps are ovoid, 0.3–0.5 mm in diam., and solitary near tips.

Common, on all coasts, in mats on limestone and mangrove roots, adrift, and in 26–34 m off the east coast; also known from Bermuda, the Carolinas, Bahamas, Caribbean, Gulf of Mexico, and Brazil, as well as Atlantic Spain, the Mediterranean (type location: Sicily; Abbott, 1999), eastern tropical Atlantic, Red Sea, Indo-Pacific, and Australia.

Taylor 1960: p. 604; Dawes 1974: p. 157; Schneider and Searles 1991; Littler and Littler 2000: p. 208.

Herposiphonia tenella (C. Agardh) Ambronn 1880: p. 197, pl. 4, fig. 9–11, 13–17 (L. *tenella*: delicate, soft). Plate XXXIX, fig. 4.

Basionym. *Hutchinsia tenella* C. Agardh 1828 (1820–1828): p. 105.

Synonym. *Herposiphonia secunda* (C. Agardh) Ambronn f. *tenella* (C. Agardh) M. J. Wynne 1985a: p. 173, 175, fig. 11, 12.

Some authors (Wynne, 1985a, 1998a; Silva et al., 1996) recognized the plant as *Herposiphonia secunda* f. *tenella*, while others treated it as *H. tenella* (Morrill, 1976; Cordeiro-Marino, 1978; Wynne, 2005a).

Filaments are polysiphonous, mostly prostrate and spreading, with slightly upcurved tips, 7–10 pericentral cells 50–150 µm in diam., and attached by unbranched rhizoids from every segment. Most prostrate axial segments produce erect indeterminate, determinate, or rudimentary branches with few bare segments. Indeterminate erect branches are 25 µm in diam., to 0.2 mm long, and usually have 1–3 branchlets that alternate irregularly with a determinate branchlet. Erect (determinate) branchlets are often strongly incurved dorsally, 11–45 segments long, 1–2(8) mm tall, 30–90 µm in diam., and 2–5 times as long as broad.

Common, the species occurs on all coasts, as mats on limestone and mangrove roots, adrift, and from 26 to 34 m off the east coast. The species is also known from Bermuda, the Carolinas, Bahamas, Caribbean, Gulf of Mexico, Brazil, and the Indo-Pacific.

Taylor 1960: p. 604; Dawes 1974: p. 157, 158; Schneider and Searles 1991: p. 449; Littler and Littler 2000: p. 208, 210.

Heterodasya Joly and Oliveira 1966: p. 118 (Gr. *heteros*: different + Gr. *dasys*: shaggy, hairy).

Heterodasya is similar to *Micropeuce* and considered to be a distinct genus (Parsons, 1975; Millar, 2000; Wynne, 2005a) or congeneric with *Micropeuce* J. Agardh 1899 (Oliveira, 1977). *Heterodasya* was placed in the Dasyaceae (Joly and Oliveira, 1966), Lophothalieae (Parson, 1975), or Brongniartelleae with *Micropeuce* (Millar, 2000) based on its monopodial rather than sympodial growth.

Plants are bushy and polysiphonous, with 5 pericentral cells; main axes are terete, densely corticated by rhizoidal growth, alternately radially branched, and attached by discoid holdfasts. Branches arise from basal cells of trichoblasts, which are pigmented, persistent, uniseriate, alternately to subdichotomously forked, in one plane, and with an inverse U-shaped junction of adjacent end wall segments. Growth is monopodial. Stichidia have branched laterals from their surfaces. Tetrasporangia are tetrahedrally divided, in spirally arranged nodulose segments at the tips of primary axes, and have 1–3 tetrasporangia per segment. Gametophytes are dioecious. Spermatangia are in terete sori, replace trichoblasts or branches, and are on short uniseriate stalks. Procarps are distal and from basal cells of reduced trichoblasts; carpogonial branches are 4-celled. Cystocarps are ovoid to urceolate.

Heterodasya mucronata (Harvey) M. J. Wynne 2005a: p. 89 (L. *mucronata*: pointed, ending in a sharp point). Plate XXXIX, fig. 5–7.

Basionym. *Dasya mucronata* Harvey 1853: p. 63.
Homotypic synonym. *Brongniartella mucronata* (Harvey) F. Schmitz 1893: p. 218,
Heterotypic synonyms. *Dasya sertularioides* M. Howe and W. R. Taylor 1931: p. 21–22, fig. 12, 13.
Heterodasya sertularioides (M. Howe and W. R. Taylor) Joly and Oliveira 1966: p. 118–120, pl. III, fig. 1–6; pl. IV, fig. 1–4.

Plants are bushy, plumose, erect, 15–20 cm tall, rose red to pale pink, and with one or more main axes arising from discoid holdfasts. Main axes are terete, to 1.5 cm in diam., and have 5 pericentral cells. Branching is irregular and radial; upper axes are covered by uniseriate, pigmented trichoblasts, giving tips a bush- or eyelike (ocellate) appearance. Trichoblasts are persistent, 0.6–1.0 mm long, one per axial cell, in a spiral pattern, alternately or subdichotomously branched (4–6 orders), and taper to pointed apices that may end in rhizoidlike pigmented filaments. Basal cells of trichoblasts are club-shaped, 30–70 µm in diam., and 50–130 µm long; middle cells are club- shaped to cylindrical, 35–50 µm in diam., and 90–120 µm long. Tetrasporangia are tetrahedrally divided, spherical, 90–150 µm in diam., and spiraled at tips of irregularly swollen axes. Spermatangial stichidia are conical to cylindrical, 60–90 µm in diam., 30–150 µm long, and on trichoblasts near primary axial tips. Cystocarps are ostiolate, ovoid to urn-shaped, 0.3–0.7 mm in diam., and 0.3–1 mm long. The species can be separated from *Dasya* spp. by the persistent short trichoblasts with pointed tips and monopodial apical growth.

Common, on shallow limestone in 2–5 m in the Keys (type location: Key West) and in 32 m and 25 m off the east and west coasts, respectively; also known from Bermuda, North Carolina (14–35 m), the Caribbean, Gulf of Mexico, and Brazil, as well as North Senegal, Ghana, and the Indo-Pacific.

Taylor 1960: p. 590; Dawes 1974: p. 154; Schneider and Searles 1991: p. 456.

Laurencia J.V. Lamouroux 1813: p. 42 (130), *nom. cons.* (probably named for Jean Baptiste Auguste Francois Marie la Laurencie, a French naturalist in the late 1800s).

Synonyms. *Pinnatifida* Stackhouse 1816: p. ix.
Corynecladia J. Agardh 1876 (1848–1901): p. 642.

See Nam et al. (1994), Garbary and Harper (1998), and Furnari et al. (2001) and Nam (2006) regarding the *Laurencia* complex. Plants of *Laurencia* are similar in morphology to species of *Chondrophycus*, *Osmundea*, and *Palisada* and difficult to separate using vegetative features.

Plants are erect, usually bushy, fleshy to cartilaginous, and with one or more main axes arising from a common discoid holdfast. Axes are polysiphonous, the four pericentral cells are usually covered by dense cortication. The medulla contains large hyaline cells; cortex has small cells with pigmented surface cells that may have secondary pit connections and a *corps en cerise*. Branching is radial or pinnate and branches arise from basal cells of apical trichoblasts. Branchlets are usually short and club-shaped or basally constricted and expanded to blunt tips that have a pit with an apical cell and tuft of branched trichoblasts. Tetrasporangia are tetrahedrally divided, from specific pericentral cells, and scattered near tips of branchlets. Gametophytes are dioecious. Spermatangia are hyaline and in sori on trichoblasts in apical pits. Procarps occur on the 2nd cell of a 3-celled trichoblast in apical pits. Carpogonial branches are 4-celled. Carposporophytes arise from basal fusion cells. Pericarps are multilayered; cystocarps are sessile, oval to spherical, and lateral.

Nam et al. (1994) noted that *Laurencia* has spermatangia formed on trichoblasts in apical pits, while in *Osmundea* spermatangia occur on branches of epidermal cells also in apical pits. Further, in *Laurencia* tetrasporangia arise from specific pericentral cells and not from random ones as in *Osmundea*. *Laurencia* has four pericentral cells per segment in vegetative axes, while there are 2 in *Chondrophycus*, *Osmundea*, and *Palisada*.

Key to the species (8) of *Laurencia*

1. Plants erect and more than 2 cm tall 2
1. Plants crustose and with erect axes to 1.2 cm . . . . *L. coelenterata*
   2. Axes compressed to flat 3
   2. Axes mostly cylindrical or slightly compressed. 4
3. Branchlets often fused with one another and also attached to substrata . . . . *L. caraibica*
3. Branches not fused with other branchlets nor attached to substrata . . . . *L. brongniarti*
   4. Surface cells at tips of branchlets mostly larger than 40 μm in diameter. 5
   4. Surface cells near tips mostly less than 40 μm in diameter 6
5. Plants not pale to purple green; branching mostly alternate and radial; cross walls of main branches not thickened; branches lack lenslike bandings. . . . . *L. intricata*
5. Plants pale to purple green; branching alternate, radial to irregular; cross walls of main branches thickened; branches with lenslike bandings . . . . *L. microcladia* (in part)
   6. Plants not dark to bright green; branching variable; a bright red cellular inclusion (*corps en cerise*) absent from each living or freshly preserved surface cell 7
   6. Plants dark to bright green; branching sparse below and dense above; a *corps en cerise* visible in each living or freshly preserved surface cell . . . . *L. obtusa*
7. Axial cell walls in main branches lack lenslike thickenings; tetrasporangia are unknown or to 80 μm in diameter. 8
7. Axial cell walls in main branches with refractive lenslike thickenings visible through the cortex; tetrasporangia 65–100 μm in diameter . . . . *L. microcladia* (in part)
   8. Plants in clumps to 4 cm tall; axes 1–2 mm in diam.; branchlets to 1 mm in diam., club-shaped; surface cells compact, to 40μm wide, and angular . . . . *L. cervicornis*
   8. Plants pyramidal, to 10 cm tall, and not clumped; axes 0.2–1.5 mm in diam., branchlets 0.2–0.5 mm in diam., surface cells 30–60 μm in diam., oval . . . . *L. chondrioides*

Laurencia brongniartii J. Agardh 1841: p. 20–21; 1863 (1848–1901): p. 768 (named for A. Brongniart, a French naturalist who collected the type specimen from Martinique). Plate XXXIX, fig. 8.

Synonym. *Laurencia concinna* Montagne 1842a: p. 6; non *Laurencia concinna sensu* Okamura 1913 (1912–1915): p. 38, pl. 40, fig. 1–6.

Plants are bushy, firm, 5–18 cm tall, and yellow red to deep red; fronds are pyramidal and main axes clearly visible (percurrent), compressed to flat, and attached by discoid holdfasts. Secondary branches are flat; older axes with raised midribs; branching irregularly alternate below and pinnate above. Branchlets are short and club-shaped; tips are blunt and have one apical cell in the terminal pit. Surface cells are pigmented, quadrate, with surface projections, and smaller than outer cortical cells. Medullary cells have thick walls and may appear lenslike. Tetrasporangia are tetrahedrally divided, ovoid, 55 μm in diam. by 65 μm long, and in rows on swollen branchlets. The species contains antibiotic compounds and can be mass cultured in tanks (Nishihara et al., 2004, 2005).

Rare, in drift during March and April near Hillsboro Beach on the east coast (Hackett, 1974); also known from the Caribbean (type location: Martinique; Silva et al., 1996), Canary Islands, Ghana, Indo-Pacific, and Australia.

Laurencia caraibica P. C. Silva 1972: p. 205 (L. *caraibica*: of or from the Caribbean). Plate XXXIX, fig. 9–11.

Synonym. *Laurencia nana* M. A. Howe in Britton and Millespaugh 1920a: p. 556, *nom. illeg.*

Plants are cartilaginous but fragile, densely tufted, entangled, 4–7 cm tall, vary from dark red to green brown to pink purple, and are initially attached by a padlike holdfast and secondarily by branches with numerous lateral rhizoids. Axes are dichotomously branched below to irregular above. Main axes are compressed, slightly flat, 0.15–1.0mm in diam., more or less coherent, and commonly regenerate from centers of broken branches. Branchlets are 0.1–0.5 mm in diam., to 2 mm long, sparse, and terete; tips are broad, blunt, and with tufts of short, inconspicuous filaments. Branchlets often fuse with adjacent branchlets and/or attach to substrata. Surface cells are 30–40 μm in diam. and angular; secondary pit connections are rare; medullary cells are 75–100 μm in diam., spherical, and have thick end walls (lens-shaped thickenings). Apical cells are in a deep pit. Tetrasporangial branchlets are slightly enlarged; tetrasporangia are tetrahedrally divided and in a whorl near tips.

Rare, on dead coral, coralline algae, or other hard surfaces; known in Florida only from Smithsonian Institution herbarium specimens from Looe Key and Cay Sal Bank. Also recorded from the Bahamas, Jamaica, Caribbean, Brazil, and Indo-Pacific.

Littler and Littler 2000: p. 212.

Laurencia cervicornis Harvey 1853: p. 73, pl. 18.C (L. *cervicornis*: with branches on one side, like the horns of a deer). Plate XXXIX, fig. 12–14.

Littler and Littler disagreed with Howe's (1918) merger of *Laurencia cervicornis* with *L.* (*Chondrophycus*) *corallopsis* (Montagne) Nam 1999 because it was much smaller and had small surface cells.

Plants are tough, cartilaginous, in clumps to 4 cm tall, red brown to olive green, with branching basal, spreading, and irregular to dichotomous or whorled, and attached by padlike bases. Branches are 1–2 mm in diam. and curve upward; branchlets are numerous, club-shaped, often forked, to 1 cm long, and tips are slightly swollen with an apical pit and with a tuft of dichotomously branched trichoblasts. Surface cells of tips are compact, angular, to 40 μm wide, and 35 μm long; medullary cells are spherical, thin-walled, and to 150 μm in diam. Conceptacles are near tips, partially embedded, and with an opening.

Rare, known in Florida only from Harvey's original description of plants from Key West (type location); also known from the Bahamas and Lesser Antilles.

Littler and Littler 2000: p. 202.

Laurencia chondrioides Børgesen 1918: p. 252–253, fig. 243–246 (Gr. _chondros_: cartilaginous, grainy + _oides_: like; thus resembling the red algal genus _Chondria_). Plate XXXIX, fig. 15–17.

Plants are cartilaginous, erect (to 6 cm) or sprawling, pale pink and with red tips, alternately to irregularly and radially branched, and with discoid holdfasts. Axes are terete, slender, irregular in thickness, and to 500 µm in diam. Branchlets are numerous, club- to spindle-shaped, and taper to both ends. Tips are blunt; an apical cell occurs in a terminal pit with a tuft of deciduous, dichotomously branched trichoblasts. Surface cells near tips are compact, swollen, to 40 µm in diam., and 140 µm long; the 4 pericentral cells and central axial cell are visible only in young branchlets. Tetrasporangia are tetrahedrally divided, ovoid to spherical, to 80 µm in diam., and infrequent at branch tips.

Rare, on rubble in seagrass beds in the Keys and from 26–40 m off the east coast; also known from the Bahamas, Caribbean (type location: Virgin Islands), Mediterranean, Canary Islands, and Philippines.

Taylor 1960: p. 627; Littler and Littler 2000: p. 212.

Laurencia coelenterata D. L. Ballantine and Aponte 1995: p. 417–418, fig. 1–8 (Gr. _koilos_: hollow + _enteron_: intestine; also Coelenterata, a phylum of invertebrates with radially symmetrical bodies and internal cavities). Plate XXXIX, fig. 18–20.

Plants are crusts to 2 mm thick, 1 cm in diam., and attached by multicellular rhizoids of ventral cortical cells. Erect axes are to 12 mm tall, terete, 0.6–1.0 mm in diam., sparingly branched with 3–4 unequal branches, have 4 pericentral cells, and enlarged tips containing an apical cell and trichoblasts in the terminal pit. Cortical cells are 25–42 µm wide, 23–31 µm long, and with secondary pit connections. Tetrasporangia are tetrahedrally divided, ovoid, 115–140 µm in diam., and on the abaxial side of branchlets in irregular bands around apical pits. Spermatangial branches arise from basal cells of trichoblasts in apical pits and produce ovoid to globular cells that are 15 by 20 µm in diam. Cystocarps are at tips of axes, embedded, and with beaked pericarps.

Rare, on subtidal limestone (9 m) at the Dry Tortugas (type location) and from 26–40 m off the east coast (Hanisak and Blair, 1988); also known from the Bahamas.

Laurencia intricata J. V. Lamouroux 1813: p. 131, pl. 9, fig. 8, 9 (L. _intricata_: entangled, complicated). Plate XXXIX, fig. 21, 22.

Synonym. *Laurencia implicata* J. Agardh 1852 (1848–1901): p. 745.

Some authors (J. Agardh, 1842; Yamada, 1931) have cited the plant as *Laurencia obtusata* var. *intricata* (J. V. Lamouroux) J. Agardh.

Plants are fleshy to crisp, single or in loose mats, 5–25 cm tall, and pale yellow to rose red. Main axes are indistinct below, conspicuous in the last 2–3 forkings, and attached by entangled prostrate branches with holdfasts. Axes are polysiphonous with 4 pericentral cells visible in section near tips. Branching is sparse, irregularly alternate or opposite, and often at right angles to axes or in loose clusters; axes are 0.4–0.8 mm in diam. Branchlets are terete, rarely club-shaped, and have blunt tips; surface cells are 40–70 µm in diam. The apical cell and dichotomously branched trichoblasts are in a terminal pit. Tetrasporangia are tetrahedrally divided, to 120 µm in diam., and near branchlet tips. Spermatangial clusters are ovoid and in apical pits. Cystocarps are partially embedded. Rigid plants of *L. intricata* resemble *Chondrophycus poiteaui*, while lax forms are similar to drift *L. obtusa*.

Common, in drift or mats on shells, limestone, and seagrasses (to 36 m) on all coasts; also known from Bermuda, the Caribbean (type location: Antilles; Silva et al., 1996), and Gulf of Mexico, as well as the Mediterranean, eastern tropical Atlantic, Indo-Pacific, and Australia.

Taylor 1960: p. 626; Dawes 1974: p. 158; Littler and Littler 2000: p. 214.

Laurencia microcladia Kützing 1865: p. 22, pl. 60, fig. b, c (Gr. *micro*: small + *klados*: a branch, shoot). Plate XXXIX, fig. 23.

Plants are bushy, wiry, to 10 cm tall, narrowly pyramidal, usually gray green to purple green with red tips, and attached by small padlike or discoid holdfasts. Branching is alternate to irregular and usually dense above; axes are terete, polysiphonous with 4 pericentral cells visible in section at tips, and 0.2–1.5 mm in diam.; primary axes are distinct, long, and slender. Branchlets are terete, elongate or blunt to wartlike, 0.2–0.5 mm in diam., and 0.2–3.0 mm long; tips are often swollen and clustered. Apical pits contain an apical cell and deciduous, dichotomously branched trichoblasts. Surface cells are ovoid, to 30–60 µm in diam., and with secondary pit connections; medullary cells are 100–120 µm in diam. Cross walls of axial and pericentral cells are lens-shaped and thick. Tetrasporangia are tetrahedrally divided, spherical to ovoid, and 65–100 µm in diam. Cystocarps are spherical to urn-shaped, 500–700 µm in diam., and near branchlet tips.

Common, in drift and on limestone (intertidal to 4 m) on all coasts; also known from Bermuda, the Caribbean (type location: West Indies; Silva et al., 1996), Gulf of Mexico, Brazil, the Mediterranean, eastern tropical Atlantic, and tropical Pacific.

Taylor 1960: p. 627; Dawes 1974: p. 158; Littler and Littler 2000: p. 216.

Laurencia obtusa (Hudson) J. V. Lamouroux 1813: p. 130 (L. *obtusa:* blunt, rounded at the end). Plate XXXIX, fig. 24, 25.

Basionym. *Fucus obtusus* Hudson 1778: p. 586.

Yamada (1931) cited *Laurencia intricata* as a variety of *L. obtusa*.

Plants are bushy, firm, to 25 cm tall, olive green to pale red, and attached by discoid holdfasts. Axes are polysiphonous and with 4 pericentral cells; main axes are obvious, sparingly alternately branched below, often densely branched above, terete, and 0.7–1.5 mm in diam. Branchlets are spiraled or opposite, 0.5–0.8 mm in diam., 1–2 mm long, in irregular clusters, and with obtuse tips; the apical cell occurs in a terminal pit with deciduous trichoblasts. Surface cells are 24–30 µm in diam., spherical, pigmented, and with cherry red spherical bodies (*corps en cerise*) visible in fresh tissue. Tetrasporangia are tetrahedrally divided, spherical, and in bands near tips of branchlets. Spermatangia are ovoid and clustered in apical pits. Cystocarps are oval and 450–650 µm in diam. Drift *Laurencia obtusa* may be lax and similar to *L. intricata*, but they have obvious main axes, *corps en cerise* in living surface cells, and lack prostrate systems.

Common, in drift, on jetties and limestone (to 8 m) on all coasts; also known from Bermuda, the Caribbean, Gulf of Mexico, and Brazil, as well as the eastern North and tropical Atlantic (type location: Hastings, England; Lipkin and Silva, 2002), the Mediterranean, Red Sea, Indo-Pacific, and Australia.

Taylor 1960: p. 626; Dawes 1974: p. 158; Littler and Littler 2000: p. 216.

Lophocladia F. Schmitz 1893: p. 222 (Gr. *lopho*: crested + *cladia*: a small branch).

Plants are filamentous, erect, repeatedly dichotomously to alternately branched above and denuded below. Prostrate axes are attached by multicellular rhizoids that form haptera. Erect axes have 4 pericentral cells and numerous radially arranged uniseriate filaments that are dichotomously to alternately branched and deeply pigmented. Tetrasporangial stichidia are flexuous, irregular in shape, and from transformed branchlets. Tetrasporangia are tetrahedrally divided, solitary, and spirally disposed.

Lophocladia trichoclados (C. Agardh) F. Schmitz 1893: p. 223, as footnote (Gr. *tricho*: hairy, hairlike + *clados*: a branch, shoot). Plate XXXIX, fig. 26–28.

Basionym. *Griffithsia trichoclados* C. Agardh 1828 (1820–1828): p. 132 (as '*Griffitsia*').

Plants are bushy, filamentous, soft and lax, to 10 cm tall, rose pink, and with wide dichotomous branching above; prostrate axes are attached by multicellular rhizoids of most segments and some polysiphonous rhizoids form haptera. Erect axes are 0.5–2.0 mm in diam., taper to tips, and have 4 pericentral cells, are corticated by rhizoids below, and have hairlike uniseriate filaments on their upper axes. Filaments are deciduous, pigmented, to 60 μm in diam., 2 mm long, and dichotomously branched; cells are 2–3 diam. long near the base and 6–8 diam. near tips. Tetrasporangia are spherical, to 60 μm in diam., and in groups of 6–8 in a spiral pattern in stichidia that are to 120 μm in diam., 500 μm long, and from basal cell of trichoblasts. Spermatangial branches are cylindrical to spindle-shaped and on 2–3-celled uniseriate stalks of the basal cells of trichoblasts. Cystocarps are ostiolate and oval to vase-shaped.

Rare, from 20 m at the Dry Tortugas and 33–38 m off the east coast; also known from Bermuda, the Caribbean (type location: West Indies; Silva et al., 1996), Gulf of Mexico, and Guyana, as well as the Mediterranean, eastern tropical Atlantic, Indo-Pacific, and Australia.

Taylor 1960: p. 590; Littler and Littler 2000: p. 218.

Lophosiphonia Falkenberg in Engler and Prantl 1897 (1890–1897): p. 459; Falkenberg 1901: p. 495 (Gr. *lopho*: crest, crested + *siphon*: pipe, tubular).

Filaments are polysiphonous with 6–10 pericentral cells, ecorticate, and tufted; main axes are prostrate, spreading, and attached by rhizoidal holdfasts. Prostrate axes are cylindrical and produce erect axes dorsally; prostrate and erect apices may curve upward and have secund clusters of deciduous hyaline trichoblasts that branch dichotomously on outer surfaces. Erect axes are endogenous and sparingly branched. Tetrasporangia are tetrahedrally divided and in upper branches.

Key to species (2) of *Lophosiphonia* (after Littler and Littler, 2000)

1. Trichoblasts not noticeably offset, on every 3rd, 4th, to 8th or more segments . . . . *L. cristata*
1. Trichoblasts slightly alternate (offset) on every upper erect segment . . . . *L. obscura*

Lophosiphonia cristata Falkenberg 1901: p. 499, pl. 9, fig. 7–10 (L. *cristata*: crested, with tassel-like tips). Plate XL, fig. 1–4.

Filaments are polysiphonous with 6–13 pericentral cells; they occur in brown to deep red tufts, and are to 3 cm tall. Prostrate axes are mostly attached ventrally by 1–(3)-celled rhizoids with lobed pads. Prostrate apices are to 120 μm in diam. and curve upward. Erect axes are 80–100 μm in diam., they arise from every 8th to 10th dorsal prostrate segment, and taper to down-turned tips. Apical trichoblasts are hyaline, deciduous, in an obvious crest on every 3rd to 8th or more segments, are pseudodichotomously branched, and often have a spherical basal cell. Tetrasporangia are tetrahedrally divided, spherical, to 50 μm in diam., and occur in straight to spiral series near tips of swollen branchlets.

Uncommon, inconspicuous, on limestone and algae in the Keys (to 5 m) and at Anclote Key on the west coast (in 9 m); also known from Bermuda, the Bahamas, Antilles, and Gulf of Mexico as well as the Mediterranean (type location: Punta del Posilipo, Italy; Abbott, 1999), eastern tropical Atlantic, and tropical Pacific Ocean.

Taylor 1960: p. 606; Dawes 1974: p. 160; Littler and Littler 2000: p. 220.

Lophosiphonia obscura **(C. Agardh) Falkenberg in F. Schmitz and Falkenberg 1897: p. 460 (L. *obscura*: indistinct, hidden, dark). Plate XL, fig. 5–8.**

Basionym. *Hutchinsia obscura* C. Agardh 1828 (1820–1828): p. 108.
Synonym. *Lophosiphonia subadunca* (Kützing) Falkenberg 1901: p. 496.

Filaments are erect, polysiphonous, 1.0–3.0 cm tall, and deep red to purple; they arise from prostrate axes 140 µm in diam., and are attached by many short haptera. Prostrate axes have irregular branching and upcurved tips. Rhizoids are 1-celled, 30–50 µm in diam., end in lobed discs, and are extensions of pericentral cells. Erect filaments branch infrequently and irregularly, arise from every 4th, 8th, or 10th segment, and have upcurved tips. Branchlet tips are 100 µm in diam., are tapered, and bend downward; older branchlet segments are longer than broad and have 6–13 pericentral cells. Trichoblasts are crested on outer margins of branchlets, they occur on every segment, are slightly alternate (offset), deciduous, pseudodichotomously branched, and have spherical basal cells. Tetrasporangia divide tetrahedrally and are spherical to ovoid, 50 µm in diam., one per segment, and in swollen branchlets.

Uncommon, inconspicuous, and grow intertidally in protected sites on hard surfaces; known in Florida only from Smithsonian Institution herbarium specimens from Anastasia Island and St. Augustine. Also known from the Bahamas, Antilles, and Gulf of Mexico, as well as the eastern North and tropical Atlantic (type location: Cádiz, Spain; Silva et al., 1996), the Mediterranean, Red Sea, Iran, Oman, and Micronesia.

Taylor 1960: p. 606; Littler and Littler 2000: p. 220.

Murrayella F. Schmitz 1893: p. 227 (named for G. Murray, a British phycologist who catalogued the seaweeds of the West Indies and Ceylon).

Filaments are polysiphonous and with 4 ecorticated pericentral cells; they are prostrate and have erect indeterminate axes that branch subdichotomously, radially, or alternately. Prostrate axes are attached by multicellular, lobed rhizoids at irregular intervals. Erect axes are denuded below and abundantly branched above. Branchlets become uniseriate after 1–2 forks. Tetrasporangia are tetrahedrally divided and occur in pairs or oblique whorls in stichidia on upper parts of branchlets. Cystocarps are ovoid to urn-shaped and on uniseriate stalks.

Murrayella periclados **(C. Agardh) F. Schmitz 1893: p. 227, footnote (Gr. *peri*: around + *klados*: shoot, branch). Plate XL, fig. 9.**

Basionym. *Hutchinsia periclados* C. Agardh 1828 (1820–1828): p. 101–102.
Synonym. *Bostrychia tuomeyi* Harvey 1853: p. 58, pl. XIV E, fig. 1–6.

Filaments are polysiphonous and with 4 pericentral cells; they are ecorticate, often in dense tufts to 5 cm tall, and dark purple to reddish brown. Prostrate axes are spreading and attached by multicellular branched rhizoids that arise from the ends of pericentral cells and have lobed tips. Erect axes are subdichotomously branched below, alternately near tips, and covered by branchlets. Branchlets are determinate, deciduous, radially and alternately branched, 25–60 µm in diam., uniseriate above and often polysiphonous near bases. Tetrasporangia are 55 µm in diam. and in whorls of four in stichidia, which are nearly terete, tapered to tips, 100 µm in diam., to 900 µm long, and have polysiphonous bases. Spermatangial stichidia are 40–60 µm in diam. Cystocarps are ovoid, terminal, and to 400 µm in diam. *Lophocladia* and *Brongniartella* differ from *Murrayella* because their tetrasporangial stichidia have whorls of four tetrasporangia.

Common, on mangrove roots, on stalks of salt marsh plants, and as a turf under intertidal ledges on all coasts; also known from Bermuda, the Caribbean (type location: St. Croix, Virgin Islands; Silva et al., 1996), Gulf of Mexico, Guiana, and Brazil, as well as West Africa, Indo-Pacific, and Australia.

Taylor 1960: p. 593; Dawes 1974: p. 160; Littler and Littler 2000: p. 222.

Neosiphonia M. S. Kim and I. K. Lee 1999: p. 271–281 (Gr. *neo*: new + *siphon*: tubular, pipe).

Molecular studies support separation of *Neosiphonia* and *Polysiphonia* (Kim and Lee, 1999; Kim et al., 2005) based upon the following features: (1) lateral branch initials are formed on successive segments and not separated by one or more naked segments as in *Polysiphonia*; (2) erect indeterminate branches are from main axes and not a creeping base as in *Polysiphonia*; (3) trichoblasts are abundant, while scarce to absent in *Polysiphonia*; (4) procarps have a three-celled carpogonial branch versus a four-celled one in *Polysiphonia*; (5) spermatangial branches arise from a trichoblast branch and not from lateral branch initials as in *Polysiphonia*; (6) tetrasporangia occur in a spiral series and not in a straight series as in *Polysiphonia*; (7) rhizoids are separated by a cross wall from pericentral cells versus being "open" in *Polysiphonia*.

Filaments are polysiphonous with 4–9 pericentral cells; they are ecorticate or corticated, solitary, epiphytic or saxicolous, and with holdfasts of tightly clumped rhizoids cut off from basal pericentral or cortical cells. Branches are mostly exogenous and some arise from scar cells of trichoblast remnants. Trichoblasts are present at branch tips, one per segment, and in a spiral pattern; they are slender, elongate, simple or forked, deciduous, and often leave obvious scar cells. Procarps have 3-celled carpogonial branches. Cystocarps are scattered on branchlets, narrow to globose or ovate, and have short stalks. Spermatangial branches are on forks of trichoblasts, cylindrical, and may have a sterile tip. Tetrasporangia are tetrahedral and in spiral series in upper branchlets.

Key to species (4) of *Neosiphonia*

1. Axes with 4 pericentral cells 2
1. Axes with 7–8 pericentral cells . . . . *N. tepida*
   2. Plants prostrate and spreading or initially erect and secondarily attached 3
   2. Plants erect, from discoid holdfasts, never prostrate and spreading . . . . *N. gorgoniae*
3. Plants to 15 cm tall; branching distinctly alternate . . . . *N. ferulacea*
3. Plants to 2 cm tall; branching mostly pseudodichotomous . . . . *N. sphaerocarpa*

Neosiphonia ferulacea (Suhr *ex* J. Agardh) S. M. Guimarães and M. T. Fujii in Guimarães et al. 2004: p.165 (L. *ferulacea*: a hollow reed, cane; also, the giant fennel genus, *Ferula*). Plate XL, fig. 10–14.

Basionym. *Polysiphonia ferulacea* Suhr *ex* J. Agardh 1863 (1848–1901): p. 980–981.

Filaments are erect and prostrate, polysiphonous (4 pericentral cells), ecorticate, in bushy tufts, to 15 cm tall, brown to purple, and attached by unicellular rhizoids at distal ends of prostrate segments. Prostrate axes are spreading, lack apical trichoblasts, and produce erect branches every 4th to 6th segment. Erect axes are indeterminate, 80–300 µm in diam., and clearly alternately branched. Branches are narrow angled, often constricted at bases, tapered to points, and replace trichoblasts that are obvious, forked, and leave residual scar cells. Tetrasporangia are tetrahedrally divided, spherical, 50–60 µm in diam., in spiral series, and cause branches to swell. Spermatangial sori are short, cylindrical, to 60 µm in diam., 50 µm long, and have 1–2 uniseriate tip cells; they are also lateral on lower segments and occur in acropetal series at tips of trichoblasts. Cystocarps are ostiolate, ovoid, to 200 µm in diam., sessile or stalked, occur at branch tips, and lack necks.

Common, on shallow limestone, seaweeds, and seagrasses on all coasts and from 39–42 m off the east coast; also known from Bermuda, North Carolina, the Caribbean, Gulf of Mexico, Guadalupe Island, Mexico, and Brazil, as well as from the eastern North and tropical Atlantic, Mediterranean, Japan, and Australia. Syntype localities are Atlantic Mexico, Guadeloupe, the Hawaiian Islands, and Australia (Silva et al., 1996).

Taylor 1960 p. 578; Dawes 1974: p. 161; Schneider and Searles 1991: p. 465; Littler and Littler 2000: p. 226.

Neosiphonia gorgoniae (Harvey) S. M. Guimarães and M. T. Fujii in Guimarães et al. 2004: p. 168 (Gr. *gorgon*: a mythical Greek figure; also the host *Gorgonia*, a genus of soft corals). Plate XL, fig. 15–18.

Basionym. *Polysiphonia gorgoniae* Harvey 1853: p. 39.

Filaments are erect and only secondarily prostrate, polysiphonous, with 4 pericentral cells, and ecorticate. They form bushy tufts to 4 cm tall, are pale yellow to brown red, and attached by small discoid holdfasts with many unicellular rhizoids. Erect axes are to 120 µm in diam., with branches every 5th or 6th segment; branching below is wide, pseudodichotomous, and more irregular, and above to several orders. Branches are tapered from base to tips, to 105 µm in diam. below and to 30 µm above; segments are 0.7–1 diam. long below, 1.5–2 diam. long above, and their tip segments are short. Trichoblasts are near branch tips, sparse, deciduous, and leave obvious scar cells. Tetrasporangia are tetrahedrally divided, spherical, to 75 µm in diam., and in spiral series near branch tips. Cystocarps are ostiolate, ovoid, 250–500 µm in diam., occur on upper branches, and lack necks; they are sessile or on short stalks. Spermatangial branches are oblong, 25–30 by 149–160 µm, and subtended by a branched trichoblast.

Uncommon, on shallow (1–3 m) rubble and soft corals in the Keys (type location: Key West) and on seagrass and coarse algae on the west coast; also known from Georgia, the Bahamas, Antilles, Gulf of Mexico, Brazil, Mauritius, and the Philippines.

Taylor 1960: p. 576; Dawes 1974: p. 161; Schneider and Searles 1991: p. 467; Littler and Littler 2000: p. 228.

Neosiphonia sphaerocarpa (Børgesen) M. S. Kim and I. K. Lee 1999: p. 280 (Gr. *sphairo*: globose, spherical + *karpos*: a fruit). Plate XL, fig. 19–21.

Basionym. *Polysiphonia sphaerocarpa* Børgesen 1918: p. 271–274, fig. 267–271.

Filaments are erect and prostrate, polysiphonous with 4 pericentral cells, ecorticate, bushy, to 2 cm tall, and brown to rose red; erect axes are attached by discoid holdfasts and secondarily by many unicellular rhizoids of prostrate axes. Rhizoids arise from distal ends of ventral pericentral cells of prostrate axes and have pit connections. Prostrate axes are spreading and 100–210 µm in diam.; erect axes are alternately to pseudodichotomously branched, 60–200 µm in diam., and with segments 100–125 µm long. Branches are 60 µm in diam., with segments 175 µm long, often turgid, constricted at bases, and with meshlike rhizoidal haptera that are connected to other branches. Haptera are narrowly forked; they replace trichoblasts, and taper to acute tips. Trichoblasts are deciduous, rarely 800 µm long, arise at tips, have several orders of branching, and leave obvious scar cells. Tetrasporangia are tetrahedrally divided, spherical, 50–60 µm in diam., often in short spirals, and swell upper segments of erect axes. Spermatangial sori are at the bases of trichoblasts, are terete, sometimes forked, and are 50–60 µm in diam. by 50–180 µm long. Cystocarps are ostiolate, spherical, 200–300 µm in diam., lack necks, and are on short stalks.

Rare, reported from 38–45 m off the east coast (Hanisak and Blair, 1988); also known from North Carolina, the Bahamas, and Caribbean (type location: Magens Bay, Virgin Islands; Price and Scott, 1992), as well as the Mediterranean, eastern tropical Atlantic, Indo-Pacific, and Australia.

Taylor 1960: p. 576; Schneider and Searles 1991: p. 474; Littler and Littler 2000: p. 232.

Neosiphonia tepida (Hollenberg) S. M. Guimarães and M. T. Fujii in Guimarães et al. 2004: p. 171 (L. *tepida*: warm, lukewarm). Plate XL, fig. 22–23.

Basionym. *Polysiphonia tepida* Hollenberg 1958: p. 65–67, fig. 1.

Synonym. *Polysiphonia taylorii* Williams 1948: p. 694 , *nom. nudum*.

Filaments are erect and prostrate, polysiphonous with 7–8 pericentral cells, ecorticate, and form reddish brown mats 1–8 cm tall; prostrate axes are attached by rhizoids that arise from the proximal ends of ventral pericentral cells and have pit connections. Prostrate axes are spreading, have erect indeterminate axes every 6th to 8th segment or less often, are 80–140 µm in diam., and have segments 1–2 diam. long. Erect axes are 80–130 µm in diam., often alternately branched in one plane, and with obvious scar cells. Lateral branches are common and replace trichoblasts, which are hyaline, slender, deciduous, and forked 2–3 times. Scar cells from trichoblasts are mostly 3–4 segments apart and form a right-handed spiral. Tetrasporangia are tetrahedrally divided, spherical, 30–35 µm in diam., scattered or in straight to spiraled series, and cause the middle of branches to swell. Spermatangial sori are on lower portions of trichoblasts, cylindrical, 60–80 µm in diam., and 250 µm long. Cystocarps are rare and to 160 µm in diam.

Common, on limestone in the low intertidal of estuarine and protected sites on the west coast and at Key West; also known from the Carolinas (type location: Beaufort; Abbott, 1999), Greater Antilles, Gulf of Mexico, Brazil, as well as the Madeira Archipelago, Canary, Cape Verde, Maldives, and Hawaiian Islands, plus the Philippines.

Taylor 1960: p. 581; Dawes 1974: p. 162; Schneider and Searles 1991: p. 476.

Osmundaria J.V. Lamouroux 1813: p. 22 or 110 (L. *Osmunda*: Saxon name for Thor, the god of thunder; also, the genus of cinnamon ferns, *Osmunda*, + *aria*: derived from).

Synonym. *Vidalia sensu* J. Agardh 1863 (1848–1901): p. 1117; non Lamouroux 1824: p. 387.

Plants are erect; with fronds fleshy, flat, elongate, and have faint to conspicuous midribs with 5 pericentral cells; fronds are also twisted, spiraled, or straight and attached by discoid holdfasts. The medulla has large cells and the cortex smaller ones; secondary pit connections may occur between subapical surface cells. A *corps en cerise* is absent. Branches arise at irregular intervals from the margins of fronds and cells in oblique rows. Branch tips are curled and margins undulate, serrate, or pinnate; minute veinlets connect the teeth to midribs. Tetrasporangia are tetrahedrally divided, in 2 rows, and arise from surface cells of stichidia-like projections on the tips of marginal teeth. Spermatangia are clustered at bases of trichoblasts. Cystocarps are formed near the tips of teeth.

Osmundaria obtusiloba (C. Agardh) R. E. Norris 1991a: p. 14, fig. 9, 21 (L. _obtusiloba_: obtusely lobed). Plate XL, fig. 24, 25.

Basionym. *Rytiphlaea obtusiloba* C. Agardh 1824: p. 161.

Synonym. *Vidalia obtusiloba* (C. Agardh) J. Agardh 1863 (1848–1901): p. 1123.

Guiry and Nic Dhonncha (2006) cited the plant as *Osmundaria obtusiloba* (Mertens *ex* C. Agardh) R. E. Norris and its basionym as *Rytiphlaea obtusiloba* Mertens *ex* C. Agardh.

Plants are bushy, tough, to 15 cm tall, and arise from short stalks attached by discoid holdfasts. Fresh fronds are dark brown to deep red, while dried ones are dark red. Axes are denuded near bases, while upper parts have strap-shaped blades that are twisted, 7–12 mm wide, pinnately to irregularly branched (2–3 times), and narrowed at forks. Midribs are indistinct and have 5–8 pericentral cells. Branchlets are flat and have rounded inwardly curled tips; margins are undulate to alternately serrate with deltoid to linear teeth. Branchlets are incurved with faint veins connecting marginal spines to their midribs. Surface cells are elongate and 20–50 µm in diam. The medulla is 2 cells thick near tips and 4 cells thick near bases. Tetrasporangia are tetrahedrally divided, 60–80 µm in diam., and in two rows within stichidia that are incurved, fingerlike, and on the inner sides of marginal branchlets. Cystocarps are subsessile, solitary, and on ventral sides of midribs on marginal teeth.

Rare, in drift at the Dry Tortugas; possibly a deep water plant. Also known from the Caribbean, Gulf of Mexico, Brazil (type location; Abbott, 1999), South Africa, Japan, and the Hawaiian Islands.

Taylor 1960: p. 609; Littler and Littler 2000: p. 222.

Osmundea Stackhouse 1809: p. 56, 79 (L. *Osmunda*: Saxon name for Thor, the god of thunder; also, the genus of cinnamon fern, *Osmunda*).

Nam et al. (1994) resurrected *Osmundea* and gave a new description of *Laurencia*. Also see Nam (2006) for an emended description of *Chondrophycus*, plus the separation of *Palisada*, *Laurencia*, and *Osmundea*, with these being difficult to separate based upon vegetative features.

Plants are erect, usually bushy, fleshy to cartilaginous, and have one or more axes arising from a common discoid holdfast. Axes are flat, compressed or terete, and polysiphonous, with the two pericentral cells usually obscured by dense cortication. Branching is radial or pinnate; branchlets are usually short, club-shaped, or basally constricted and expanded near their blunt tips with a pit containing an apical cell and a tuft of branched trichoblasts. The medulla has large hyaline cells, while the cortex has smaller cells; surface cells are pigmented and may have secondary pit connections. Tetrasporangia are tetrahedrally divided and arise from random epidermal cells near branchlet tips. Gametophytes are dioecious. Spermatangia are hyaline and produced from apical and epidermal cells at branchlet tips. Procarps occur are on the second cell of a 3-celled trichoblast in terminal pits. Carpogonial branches are 4-celled; carposporophytes arise from basal fusion cells. Pericarps are multilayered; cystocarps are sessile, oval to spherical, and lateral.

Osmundea pinnatifida (Hudson) Stackhouse 1809: p. 79 (L. _pinnatifida_: pinnately cut, divided). Plate XL, fig. 26, 27.

Basionym. *Fucus pinnatifidus* Hudson 1762: p. 473; 1778: p. 473.

Synonym. *Laurencia pinnatifida* (Hudson) J. V. Lamouroux 1813, p. 130.

Furnari and Serio (1993) questioned whether this taxon occured in the tropical western Atlantic.

Plants are bushy, firm to cartilaginous, to 15 cm tall, and purple to rose red; axes are strongly compressed and attached by discoid holdfasts. Branching is pinnate, opposite or alternate, and to 4 orders. Branchlets are short, either tapering to both ends or club-shaped, 2–3 mm long, and 1–3 mm in diam. Tips are blunt and have an apical cell and trichoblasts in a terminal pit. Surface cells are round to elongate and to 45 µm in diam. Tetrasporangia are tetrahedrally divided, ovoid, 25–35 µm in diam., clustered in long subapical bands on branchlets, and arise from epidermal cells. Spermatangia are in 1–4 cryptlike pits in the apical pit and arise from apical and epidermal cells. Cystocarps are sessile, ovoid or urceolate, to 1.2 mm in diam., and near branchlet tips.

Rare, reported from 24–37 m off the east coast (Hanisak and Blair, 1988); also known from North Carolina (24–45 m) and Brazil, as well as the eastern North and tropical Atlantic (type location: Essex, England; Maggs and Hommersand, 1993), Mediterranean, Red Sea, and Indian Ocean.

Schneider and Searles 1991: p. 453.

Palisada (Yamada) Nam 2006: p. 693; 2007: p. 52.

Plants have apical cells always sunken in a terminal pit of a branchlet; the central axial cells are visible only near branch tips; cortication is extensive. Two pericentral cells are visible near branchlet tips. Spermatangial branches are produced from one of two laterals on a suprabasal cell of the trichoblast. Segments that bear procarps have 4–5 pericentral cells; auxiliary cells are formed after fertilization. Tetrasporangia form from pericentral cells with one pericentral cell being fertile while the other is sterile. Because of the morphological similarities, *Chondrophycus corallopsis* is included in this key.

Key to the species (4) of *Palisada* and *Chondrophycus corallopsis.*

1. Axes erect and not iridescent; upper axes covered by short, knobby branchlets . . . . 2.
1. Axes prostrate and with a brilliant blue iridescence; upper axes sparsely covered by branchlets . . . . *P. iridescens.*
2. Surface cells at tips larger than 40 µm in diam., palisade-like, and with a surface spine (mammillate) . . . . *P. gemmiferus.*
2. Surface cells at tips less than 50 µm in greatest diam. and lack a spine . . . . 3.
3. Branching alternate to highly irregular above . . . . 4.
3. Branching subdichotomous to dichotomous throughout . . . . *P. papillosus.*
4. Surface cells of branch tips to 50 µm in diam. . . . . *C. corallopsis.*
4. Surface cells of branch tips to 20 µm in diam. . . . . *P. poiteaui.*

Palisada gemmifera (harvey) Nam 2006: p. 694 (L. *gemmifera* having jewels; botanical: having buds) Plate XXXVIII, fig. 15–17.

Basionym. *Laurencia gemmifera* Harvey 1853: p. 73–74, pl. XVIII B, fig. 1, 2

Synonyms. *Laurencia poiteaui* (J. V. Lamouroux) M. Howe var. *gemmifera* (Harvey) Yamada 1931: p. 220

Chondrophycus gemmiferus (Harvey) Garbary and J. T. Harper 1998: p. 194

Plants are bushy, stiff, cartilaginous, 5–15 cm tall, and straw-yellow to rose-red; main axes are indistinct and attached by discoid holdfasts. Branching is usually dense, alternate or irregular, and may be in one plane. Axes are rigid, cylindrical or compressed with age, 1–2 mm in diam., and taper to 0.5 mm near tips. Branchlets are numerous, cylindrical, 0.2–0.9 mm in diam., to 2 mm long, and irregular, wart-, bud-, or tumor-like. Branchlet tips are obtuse with the apical cell in a pit and two pericentral cells visible at tips. Surface cells of main axes are 40–130 µm in diam., square to columnar in side view, with secondary pit connections, and a single spine. Tetrasporangia are tetrahedrally divided, ovoid, 10–12 µm in diam., 14–12 µm long, and in small branchlets.

Common, on limestone and adrift (to 30 m) on all coasts (type location: Key West); also known from Bermuda, the Caribbean, and Gulf of Mexico.

Taylor 1960: p. 624; Dawes 1974: p.158; Littler and Littler 2000: p. 214.

Palisada iridescens (M. J. Wynne and D. Ballantine) Nam 2006: p. 694 (L. *iridescens*: iridescent, a metallic sheen) Plate XXXVIII, fig. 18.

Basionym. *Laurencia iridescens* M. J. Wynne and D. L. Ballantine 1991: p. 395, fig. 1–11
Synonym: *Chondrophycus iridescens* (M. J. Wynne and D. L. Ballantine) Garbary and J. T. Harper 1998: p. 195

Plants are tough, rigid, cartilaginous, to 2 cm tall, and an iridescent red blue on the dorsal surface; prostrate axes are tightly aggregated, irregularly branched, and firmly attached by pad-like rhizoids. Young branches are terete and 1.0–1.3 mm in diam.; older branches are compressed. Branchlets are few, irregularly dispersed, peg-like, 1–4 mm long, and with blunt tips; apical cells are in pits with the 2 pericentral cells visible in sections at tips. Surface cells near tips are columnar, 28–50 µm in diam., and with a spine and secondary pit connections. Medullary cells are elongate, 175–275 µm in diam., and 330–475 µm long. Tetrasporangia are tetrahedrally divided, ovoid or spherical, 40–90 µm in diam., and at tips of outer branchlets.

Rare and apparently in the Keys; also known from the Bahamas, Caribbean (type location: Guadeloupe; Wynne and Ballantine, 1991), and Gulf of Mexico.

Littler and Littler 2000: p. 216.

Palisada papillosa (C. Agardh) Nam 2006: p. 694 (L. *papillosus*: covered with buds or protuberances) Plate XXXVIII, fig. 19, 20.

Basionym. *Chondria papillosa* C. Agardh 1822 (1820–1828): p. 344; 1824: p. 203
Synonyms. *Laurencia papillosa* (C. Agardh) Greville 1830: p. lii
Chondrophycus papillosus (C. Agardh) Garbary and J. T. Harper 1998: p. 195

Plants are solitary or in entangled masses, 5–8 (-16) cm tall, straw yellow to purple green, and attached by overlapping discoid holdfasts. Main axes are cylindrical to slightly compressed, 1–2 mm in diam., denuded below, alternately to irregularly branched, and with 2 pericentral cells visible near tips. Secondary axes are short, radially spreading, mostly distal, and usually covered with branchlets. Branchlets are short, club-shaped, truncate, wart-like to knobby, 0.5–0.7 mm in diam., and 0.5–2.5 mm long; branchlets may be longer and less frequent in plants of protected or shaded sites. Branchlet tips are blunt and an apical cell occurs in a pit with a few to many trichoblasts and two pericentral cells are visible in sections at tips. Young surface cells are 12–27 µm in diam. and 24–30 µm long on main axes and 8–15 µm in diam. by 20–30 µm wide on branchlets. Surface cells are thick-walled and lack secondary pit connections. Tetrasporangia are tetrahedrally divided, spherical, to 100 µm in diam., and in irregular swollen branchlets. Cystocarps are subspherical and usually clustered on branchlets.

Common, on limestone on all coasts (intertidal to 7 m); also known from Bermuda, the Caribbean, Gulf of Mexico, and Brazil, as well as the Mediterranean, eastern tropical Atlantic, Red Sea (type location: Mokha, Yemen; Silva *et al.*, 1996), Indo-Pacific, and Australia.

Taylor 1960: p. 623; Dawes 1974: p. 159; Littler and Littler 2000: p. 218.

Palisada poiteaui **(J. V. Lamouroux) Nam 2006: p. 694 (Named for Pierre Antoine Poiteau, a French soldier stationed on Hispaniola, who sent specimens to J. V. Lamouroux and also became a famous artist) Plate XXXVIII, fig. 21, 22.**

Basionym. *Fucus poiteaui* J. V. Lamouroux 1805: p. 63–64, pl. XXXI, fig. 2, 3 (as '*poitei*').
Synonyms. *Laurencia poiteaui* (J. V. Lamouroux) M. Howe 1905b: p. 583; 1918: p. 518, (as '*poitei*').
Chondrophycus poiteaui (J. V. Lamouroux) K. W. Nam 1999: p. 463.

Plants are bushy, stiff to cartilaginous, to 30 cm tall, and gray green to deep red; main axes are obvious, have alternate to irregular branching, and are attached by small discoid holdfasts. Main axes are cylindrical, 0.5–2.0 mm in diam. and decrease in diameter from one branching order to another; upper branches are slightly flattened at nodes and with a few to many branchlets. Branchlets are short, terete, wart-like to swollen, 0.3–1.0 mm in diam., and 0.5–2 mm long. Tips are obtuse; the apical cell is hidden within a pit and covered by dichotomously branched, deciduous trichoblasts. Propagules or propagule-like branches occur in this species (Cruz Adames and Ballantine, 1996) and other related species (Cassano et al., 2006). Branchlet surface cells are compact, round to elongate, and to 20 µm in diam.; secondary pit connections are absent. Two pericentral cells are visible near tips; the inner medullary cells are 80–135 µm in diam. Tetrasporangia are tetrahedrally divided, ovoid, 90–130 µm in diam., and near tips. Spermatangia are clustered in apical crypts; cystocarps are spherical, sessile, and near branchlet tips.

Common, probably the most common species of *Palisada* in Florida; forming large drifting masses (1–2 m) that contain invertebrates and fish (Dawes, 1974, 1998). On shell and limestone (to 20–40 m) off the west coast and often heavily epiphytized; also known from Bermuda, North Carolina, the Caribbean (type location: Santo Domingo, Dominican Republic; Silva et al., 1996), Gulf of Mexico, and Brazil, as well as West Africa, Reunion Island, Kenya, Tanzania, India, Sri Lanka, and Australia.

Taylor 1960: p. 625; Dawes 1974: p. 159; Schneider and Searles 1991: p. 454; Littler and Littler 2000: p. 218.

Polysiphonia Greville 1823 (1823–1824); pl. 90; 1824: p. 308, *nom. cons.* (Gr. *polus, poly*: many + *siphon*: tube, pipe).

Synonyms. *Carradoria* Kylin 1956: p. 494, 503; non *Carradoria* Martius 1833: p. 15.
Carradoriella P. C. Silva in Silva et al. 1996: p. 920

Although vegetative features are useful (Hollenberg, 1968a) species identifications can be difficult (Taylor, 1960; Kapraun, 1977, 1979; Kapraun and Norris, 1982; Kapraun et al., 1983; Schneider and Searles, 1991). As noted above, species of *Polysiphonia* can be separated from *Neosiphonia* using morphological (Kim and Lee, 1999) and molecular data (Kim et al., 2005).

Filaments are polysiphonous with 4–14 pericentral cells; they are often abundantly branched, with prostrate axes producing erect axes and attached by rhizoidal holdfasts from basal cells of erect axes or ventral cells of prostrate axes. Rhizoids of some species have an open cytoplasmic connection with pericentral cells while others have a cross wall (see also *Neosiphonia*). Erect axes are terete and cortication, if present, is usually derived by divisions of pericentral cells. Secondary pit connections are present or absent. Growth is by a single apical cell; branches produce secondary ones or hyaline determinate trichoblasts that are branched or unbranched, often deciduous, and produce scar cells. Tetrasporangia are tetrahedrally divided with one per segment near tips of main axes. Gametophytes are dioecious. Spermatangial sori are conical or cylindrical and on short, uniseriate stalks of trichoblasts. Procarps are near branch tips on the second cell of reduced trichoblasts; carpogonial branches are 4-celled; cystocarps are ostiolate and ovoid or vase-shaped.

Key to the species (16) of *Polysiphonia*

1. Plants minute, less than 2 mm tall and forming minute cushions. 2
1. Plants larger, 1.0 cm or taller and not forming minute cushions 3
 2. Prostrate axes 50–70 µm in diam.; rhizoids swollen or saclike, to 160 µm in diam., and usually pigmented. . . . . *P. saccorhiza*
 2. Prostrate axes 30–40 µm in diam.; rhizoids not swollen, less than 30 µm in diam., and not pigmented . . . . *P. anomala*
3. Axes have 4 pericentral cells 4
3. Axes have more than 4 pericentral cells. 12
 4. Main axes covered with slender or spinelike branchlets to 5 mm long 5
 4. Main axes lacking slender or spinelike branchlets 8
5. Main axes ecorticate or with limited cortication on older basal axes. 6
5. Main axes lightly corticated above, heavily below. . . . . *P. ramentacea*
 6. Plants less than 5 cm tall; spinelike branchlets 2–4 mm long 7
 6. Plants 6–15 cm tall; stiff spinelike branchlets to 1 mm long. . . . . *P. echinata*
7. Plants matted, spreading below, and to 1.5 cm tall; spinelike branchlets to 2 mm long; axes subdichotomously branched . . . . *P. fracta*
7. Plants not matted or spreading and to 5 cm tall; thornlike branchlets 2–4 mm long; axes dichotomously branched below and irregular above . . . . *P. haplacantha*
 8. Axes usually attached by rhizoids that arise from a distinct prostrate axis. 9
 8. Axes usually attached by a basal, disclike holdfast and secondarily by rhizoids; plant with or without a prostrate axis . . . . *P. binneyi*
9. Branching irregular to alternate. 10
9. Branching dichotomous. . . . . *P. atlantica*
 10. Branches replacing apical hairs or trichoblasts. 11
 10. Branches formed within the axils of trichoblasts . . . . *P. havanensis*
11. Segments 1–2 diam. long; trichoblasts often obscuring apical cells; erect axes with more than 3 tetrasporangia in straight rows . . . . *P. scopulorum*
11. Segments 1–8 diam. long; trichoblasts usually sparse and not obscuring apical cells; erect axes with 1–3 tetrasporangia in spiral series. . . . . *P. subtilissima*
 12. Axes with 8 or more pericentral cells. 13
 12. Axes mostly with 5–6 pericentral cells, rarely 5–8 . . . . *P. denudata*
13. Plants in small, creeping mats to 5 cm tall; axes with 8–16 pericentral cells 14
13. Plants erect, to 10 cm tall; axes with 12–24 pericentral cells. . . . . *P. opaca*
 14. Mats to 2.5 cm tall; axes with 8–10 pericentral cells . . . .*P. exilis*
 14. Filamentous mats to 5 cm tall; axes with 8–16 pericentral cells . . . . *P. howei*

Polysiphonia anomala Hollenberg 1968a: p. 59, fig. 1A–C (L. *anomala*: irregular, uneven, unusual). Plate XL, fig. 28, 29.

Filaments are polysiphonous with 4 pericentral cells, ecorticate, in minute turflike patches to 1.5 mm tall, and with prostrate filaments attached by numerous rhizoids. Rhizoids are unicellular, near centers of ventral pericentral cells, and with cross walls. Prostrate axes are indeterminate, 30–40 µm in diam., and 1.0–1.5 diam. long. Erect axes are determinate, have segments mostly shorter than wide, and arise at frequent intervals from prostrate axes. Trichoblasts are to 1.1 mm long, one per segment, spirally arranged on prostrate and erect axes, and branch 3–4 times; tips are long, tapering, deciduous, and with scar cells, 9–11 µm in diam. Tetrasporangia are tetrahedrally divided and in spiral series of 10–12 near branch tips. Cystocarps are oval, 120–140 µm in diam., and terminal on short erect branches. The species is distinguished by irregularly arranged trichoblasts, a prostrate habit, and tetrasporangia that usually occur in a series of 10–12 below branch tips.

Rare, collected by E. Y. Dawson from Lake Surprise at Key Largo; also known from the central Pacific Ocean (type location: Bikini Atoll; Abbott, 1999).

Polysiphonia atlantica Kapraun and J. N. Norris 1982: p. 226, fig. 107 a–c (L. *atlantica*: of or from the Atlantic Ocean). Plate XL, fig. 30–34.

Synonym. *Polysiphonia macrocarpa* Harvey in Mackay 1836: p. 206–207 *nom. illeg.*; non *P. macrocarpa* (C. Agardh) Sprengel 1827: p. 350.

Filaments are polysiphonous with 4 pericentral cells, ecorticate, in mats to 2 cm tall, brown to deep purplish red, and attached by unicellular rhizoids with lobed pads. Rhizoids arise from the center of ventral pericentral cells and lack cross walls. Prostrate axes are 60–100 µm in diam., with segments 2–3 diam. long, tips that curve upward, and obvious apical cells. Erect axes arise from every 4th to 6th segment, are sparsely branched above and subdichotomously branched below, and have segments 1–3 diam. long and 40–90 µm in diam. Branches usually replace trichoblasts that are at tips, simple or branched to 2 times, deciduous, and leave large scar cells. Tetrasporangia divide tetrahedrally, are spherical, 40–60 µm in diam., in straight series, and cause swelling of upper segments. Spermatangial sori are on short stalks, terete, 40–100 µm in diam., 180–500 µm long, incurved, and terminal or lateral. Cystocarps are ovoid, 150–330 µm in diam., and sessile or on short stalks; cystocarps are ostiolate, urn-shaped, and conspicuous.

Common, as a turf on mangrove roots and limestone in shallow waters on all coasts; also known from Bermuda, North Carolina through the Caribbean, Gulf of Mexico, and Brazil, as well as the eastern North and tropical Atlantic (syntype localities: Port Stewart and Miltown, Ireland; Silva et al., 1996), Mediterranean, India, and Australia.

Taylor 1960: p. 578; Dawes 1974: p. 162; Schneider and Searles 1991: p. 460; Littler and Littler 2000: p. 224.

Polysiphonia binneyi Harvey 1853: p. 37 (named for M. Binney, an American collector who sent Harvey seaweeds from Key West). Plate XLI, fig. 1, 2.

Filaments are polysiphonous, ecorticate, have 4 pericentral cells, 2–15 cm tall, reddish brown, and with or without prostrate axes. Lower segments of erect axes are attached by discoid holdfasts and descending rhizoids of pericentral cells; rhizoids lack cross walls. Erect axes are indeterminate, stiff, tapered to tips, to 380 µm in diam., and with segments 1.0–2.5 diam. long. Branching is subdichotomous below, alternate and bushy above, and with branches to 150 µm in diam. Trichoblasts are deciduous, from each segment at branch tips, and in alternate and radial patterns. Tetrasporangia are tetrahedrally divided, spherical, 55–80 µm in diam., in short spirals, and cause tips to swell. Spermatangial branches are cylindrical, on lower segments of trichoblasts, 25–50 µm in diam. by 100–200 µm long, and end in 2–3 uniseriate cells. Cystocarps are ostiolate, oval to spherical, lack necks, are 90–420 µm in diam., and on 1(2)-celled stalks near branch tips.

Common, on seagrasses in the Indian River, on limestone or rubble on all coasts, and to 9 m at the Dry Tortugas (type location: Key West); also known from Bermuda, the Bahamas, Caribbean, and Gulf of Mexico.

Taylor 1960: p. 577; Schneider and Searles 1991: p. 461; Littler and Littler 2000: p. 226.

Polysiphonia denudata (Dillwyn) Greville ex Harvey in W. J. Hooker 1833: p. 332 (L. denudata: naked, bare, stripped). Plate XLI, fig. 3–5.

Basionym. *Conferva denudata* Dillwyn 1809 (1802–1809): p. 85, pl. G.

Filaments are polysiphonous, have 5–6 pericentral cells, are ecorticate or slightly corticated near bases, in dark brown to purple red tufts 15–25 cm tall, and initially attached by discoid bases and secondarily by rhizoids that arise from basal ends of pericentral cells with cross walls. Erect axes are indeterminate, sparsely branched, 100–750 µm in diam., and with segments 1–3 diam. long. Branching is alternate to subdichotomous and usually divergent (90–120°); branches are narrower than main axes, 100–170 µm in diam., and in axils of trichoblasts; young branches are 35–34 µm in diam. Trichoblasts, if present, are deciduous and leave scar cells; apical cells are conspicuous and divide transversely. Tetrasporangia are tetrahedrally divided, ovoid, 50–100 µm in diam., in straight rows and cause tips to swell. Spermatangial sori occur laterally on lower segments of trichoblasts, are conical to cylindrical, 30–80 µm in diam. and 100–300 µm long. Cystocarps are ostiolate and without necks. They are also ovoid, 330–540 µm in diam., and sessile or on short stalks.

Common, on mangrove roots, seagrasses, and algae; sometimes with penetrating rhizoids. Growing in estuaries and to 35 m on limestone, on all coasts; also known from Canadian Maritime Provinces southward through the Caribbean, Gulf of Mexico, and Brazil, as well as the eastern North and tropical Atlantic (type location: Hampshire, England; Maggs and Hommersand, 1993), Mediterranean, Black and Red Seas, Indian Ocean, and Australia.

Taylor 1960: p. 580; Dawes 1974: p. 160; Schneider and Searles 1991: p. 463; Littler and Littler 2000: p. 226.

Polysiphonia echinata Harvey 1853: p. 38 (L. echinata: prickly, beset with bristles, covered with prickles like a hedgehog). Plate XLI, fig. 6–8.

Filaments are polysiphonous, have 4 pericentral cells, are ecorticate, stiff, robust, 6–15 cm tall, dark brown, and covered with spinelike branchlets. They are initially attached by rhizoidal discs and secondarily by unicellular rhizoids with cross walls. Axial segments are shorter than broad and have thick walls. Branching is widely divergent, subdichotomous or alternate. Branchlets are stiff, simple to forked, spinelike, at right angles to axes, about 1 mm long, and closely spaced to about 1 mm apart.

Rare, on seagrasses in the Anclote River Basin (Ballantine, 1972) and at Key West (type location); also known from Texas.

Taylor 1960: p. 579; Dawes 1974: p. 161.

Polysiphonia exilis Harvey 1853: p. 47 (L. exilis: small). Plate XLI, fig. 9.

Filaments are polysiphonous with 8–11 pericentral cells, ecorticate, densely entangled or matted, 0.3–2.5 cm tall, dark brown, and have prostrate axes attached by unicellular rhizoids. Rhizoids are often lobed or bulbous, they arise from the mid or proximal ends of pericentral cells, and have cross walls. Apical cells are obvious and to 170 µm in diam.; subapical segments are mostly 0.5–0.8 diam. long and often have stratified walls to 20 µm thick. Trichoblasts are hyaline, radial near tips, and branched 3–5 times; branches may replace trichoblasts that are 28–40 µm wide, to 2 mm long near the base, deciduous, and leave small scar cells. Tetrasporangia are tetrahedrally divided, ovoid, and swell branches. Spermatangial branches are oblong, 50 µm in diam., and 150 µm long.

Rare, in the subtidal fringe on limestone at the Dry Tortugas and Key West (type location; Abbott, 1999); also known from Bermuda, Belize, the Indo-Pacific, and Australia.

Taylor 1960: p. 581.

Polysiphonia fracta Harvey 1853: p. 38 (L. *fracta*: broken, feeble).

Filaments are polysiphonous with 4 pericentral cells, ecorticate, rigid, in mats, spreading below and tufted to bundlelike above, to 1.5 cm high, dull brownish purple, and imperfectly adhering to paper. Axes have numerous, thornlike branchlets and prostrate axes attached by branching rhizoids. Erect axes are straight or irregularly curved and subdichotomously branched; forks are distant and spreading, with segments shorter than broad. Upper branches may be secund and widely forked; branchlets are spinelike and to 2 mm long. Trichoblasts are deciduous. Reproduction is unknown.

Rare, known only from the type specimen (type location: Key West) and the Lesser Antilles.

Taylor 1960: p. 576.

Polysiphonia hapalacantha Harvey 1853: p. 39 (Gr. *hapalos*: soft, tender + *acantha*: a spine, thorn). Plate XLI, fig. 10–12.

Filaments are polysiphonous with 4 pericentral cells that are ecorticate above and lightly corticated below; they form dense tufts, are to 5 cm tall, and black to dark red. Erect axes are attached by rhizoidal discs and the prostrate parts by unicellular rhizoids. Branching is usually wide, dichotomous below, and more clustered or irregular above; branches are covered by spinelike branchlets. Pericentral cells are 1.5 diam. long below and shorter than broad above. Branchlets are numerous, slender, spinelike, 2–4 mm long, 4 per node, and at right angles to axes. Trichoblasts are common at apices. Reproduction is unknown.

Common, adrift and on rubble, coarse algae, mangrove roots, and stalks of salt marsh plants on all coasts (type location: Key West) plus the Bahamas.

Taylor 1960: p. 579; Dawes 1974: p. 161.

Polysiphonia havanensis Montagne 1837: p. 352 (L. *havanensis*: of or from Havana, Cuba). Plate XLI, fig. 13, 14.

Schneider and Searles (1991) believed that the species might be a synonym for *Polysiphonia sertularioides* (Grateloup) J. Agardh from the Mediterranean.

Filaments are polysiphonous with 4 pericentral cells, ecorticate, mat-forming and spreading, to 9 cm tall, yellow brown to red, and attached by unicellular rhizoids from prostrate axes that arise from the ventral pericentral cells and with an open connection. Erect axes are alternately to irregularly branched, 50–100 μm in diam., and with segments to 3 diam. long; prostrate axes are 80–200 μm in diam. Branches are mostly short, narrow at bases, and arise in the axils of trichoblasts; adventitious branches are abundant and arise from scar cells along the entire axes. Trichoblasts are deciduous, much branched, near tips, and leave obvious scar cells. Tetrasporangia are tetrahedrally divided, round, 50–70 μm in diam., and solitary or in rows of 2–3 in outer branches. Spermatangial branches are on lower cells of trichoblasts, cylindrical, conical, 30–60 μm in diam. by 80–300 μm long, and lack uniseriate tip cells. Cystocarps are small, to 300 μm in diam., oval, have narrow ostioles, and sessile or on short stalks.

Common, on limestone, shells, and coarse algae in shallow waters on the west coast and at the Keys; also known from Bermuda, North Carolina, Georgia, the Bahamas, (type location: Havana, Cuba), Gulf of Mexico, Brazil, and the eastern tropical Atlantic.

Taylor 1960: p. 577; Dawes 1974: p. 161; Schneider and Searles 1991: p. 469; Littler and Littler 2000: p. 228.

Polysiphonia howei Hollenberg in Taylor 1945: p. 302–303, fig. 3 (named for M. Howe, director of the New York Botanical Garden, who studied seaweeds from Bermuda and the Bahamas as well as North American bryophytes). Plate XLI, fig. 15, 16.

Filaments are polysiphonous with 8–16 pericentral cells, ecorticate, erect and mat forming, 0.5–1(5) cm tall, pale brown to rose, and attached by rhizoids. Rhizoids arise from the distal ends of pericentral cells of prostrate axes and have cross walls with pit connections. Erect indeterminate axes are 70–150 µm in diam., may arch toward prostrate tips when young, and have simple to sparse, irregular branching. Prostrate axes are 100–190 µm in diam., with segments 0.7–1 diam. long, tips slightly upturned, and have obvious trichoblast scar cells. Branches replace trichoblasts at axial tips; adventitious branches arise from scar cells in older axes. Trichoblasts are abundant at tips, short, deciduous, branched to 3 times, and with large, rounded basal cells that become scar cells. Tetrasporangia are tetrahedrally divided, spherical, 40–55 µm in diam, in long spirals, and cause tips to swell. Spermatangial sori are 22–25 µm in diam., 100–170 µm long, and on lower segments of trichoblasts. Cystocarps are oval and 140–190 µm in diam.

Common, on limestone, shells, seagrasses, and algae in shallow waters, and 34 m off the east coast; also known from Bermuda, the Bahamas (type location: Whale Cay; Hollenberg in Taylor, 1945), North Carolina, Caribbean, Gulf of Mexico, and Brazil, as well as West Africa, the Indo-Pacific, and Australia.

Taylor 1960: p. 582; Dawes 1974: p. 161; Schneider and Searles 1991: p. 470; Littler and Littler 2000: p. 230.

Polysiphonia opaca (C. Agardh) Moris and De Notaris 1839: p. 208, 264, pl. 5, II (L. _opaca_: shaded, darkened, dull). Plate XLI, fig. 17, 18.

Basionym. *Hutchinsia opaca* C. Agardh 1824: p. 148; 1828 (1820–1828): p. 97

Filaments are polysiphonous with 12–24 pericentral cells, ecorticate, in loose mats or cushions to 10 cm tall, dull purple, and become brownish black after drying. Prostrate axes produce erect, indeterminate axes at irregular intervals and are attached by unicellular rhizoids with lobed ends. Erect axes are 150–200 µm in diam., alternately branched above, and dichotomously below; branches are in axils of trichoblasts, abundant at tips, with segments 1–15 diam. long, and covered by short, widely forked branchlets, 1–4 mm long. Tetrasporangia are tetrahedrally divided, spherical, to 90 µm in diam., in spiral series, and swell branches. Cystocarps are oval, 380–420 µm in diam., and sessile.

Uncommon, forming dark cushions in tide pools on wave-exposed shores at Point of Rocks, Content Keys, and the Dry Tortugas; also known from Bermuda, the Bahamas, Lesser Antilles, and Gulf of Mexico, as well as the eastern North and tropical Atlantic, Mediterranean (type location: Adriatic Sea; Maggs and Hommersand, 1993), and Australia.

Taylor 1960: p. 583; Dawes 1974: p. 162; Littler and Littler 2000: p. 230.

Polysiphonia ramentacea Harvey 1853: p. 42, pl. XVI A, fig. 1–5 (L. _ramentacea_: producing hairs, scales). Plate XLI, fig. 19–21.

Filaments are polysiphonous with 4 pericentral cells, ecorticate above, lightly to densely corticated below, bushy, 3–10 cm tall, and dark brown to reddish black. Erect axes are indeterminate and attached by discoid holdfasts and secondarily by prostrate axes with unicellular rhizoids. Erect axes and branches may be coarse to stiff and are covered by twiglike determinate branchlets that are simple or forked, 2–5 mm long, and mostly uniform in length, except for shorter ones near the tips. Trichoblasts occur at tips of axes and are deciduous. Tetrasporangia are tetrahedrally divided, ovoid, 50–90 µm in diam., sparse, and swell upper branches. Spermatangial sori are oval to cylindrical and clustered at branch tips. Cystocarps are sparse and spherical to ovoid.

Common, on seagrasses, algae, and limestone in estuaries and protected areas on the west coast and at the Keys (type location: Key West), as well as the Bahamas.

Taylor 1960: p. 580; Dawes 1974: p. 162.

Polysiphonia saccorhiza (Collins and Hervey) Hollenberg 1968a: p. 76, fig. 4 C,D (Gr. *sakkos*; a sac, strainer + *rhiza*: a root). Plate XLI, fig. 22, 23.

Basionym. *Lophosiphonia saccorhiza* Collins and Hervey 1917: p. 127.

Womersley (1979) suggested that the species might be an immature form of *Polysiphonia sertularioides* (Grateloup) J. Agardh with its rhizoids being modified by the host's surface.

Filaments are polysiphonous with 4 pericentral cells, ecorticate, rose red, and in feltlike mats to 2 mm tall. Prostrate axes are attached by rhizoids, which are unicellular, swollen (saccate), pigmented, to 160 µm in diam. by 800 µm long, arise from the center of ventral pericentral cells, and have cross walls. Prostrate axes are 50–70 µm in diam. and 1–2 diam. long. Erect axes are numerous, to 40 µm in diam., 2 mm tall, 6–12 segments long, and constricted at bases; segments are 1.5–3.0 diam. long and with distal trichoblasts that are hyaline, deciduous, to 300 µm long, with 1–2 forks, in spiral series, and leaving residual scar cells. Tetrasporangia are tetrahedrally divided, ovoid, 40–50 µm in diam., in rows, and swell upper parts of branches. Spermatangial sori are ovoid and from primary branches of trichoblasts. Cystocarps are ovoid, to 100 µm in diam., and have one or more short, subtending branches.

Rare, on *Codium taylorii* at Point of Rocks; also known from Bermuda (type location; Abbott, 1999), Midway, and the Hawaiian Islands.

Taylor 1960: p. 606; Dawes 1974: p. 160.

Polysiphonia scopulorum Harvey 1855a: p. 540 (L. *scopulorum*: growing on cliffs). Plate XLI, fig. 24, 25.

Filaments are polysiphonous with 4 pericentral cells, ecorticate, in a thick brown to reddish turf to 3 cm tall, have prostrate axes attached by rhizoids growing from distal ends of ventral pericentral cells, and lack cross walls. Prostrate axes are spreading, 60–180 µm in diam., with segments 1.0–1.5 diam. long; erect indeterminate axes arise from every 2nd to 6th segment. Erect axes are 30–100 µm in diam., with segments to 2 diam. long, and scar cells absent. Branches are simple to sparingly divided and replace trichoblasts; older axes have adventitious branches. Trichoblasts are deciduous, sparse, dichotomously branched to 4 times, and leave inconspicuous scar cells. Tetrasporangia are tetrahedrally divided, spherical, 50–60 µm in diam., in straight rows, and swell branches below their tips. Spermatangial sori are cylindrical and replace trichoblasts near branch tips. Cystocarps are oval and 150–190 µm in diam.

var. *villum* (J. Agardh) Hollenberg 1968a: p. 81, 83, fig. 7a (L. *villum*: a drop of wine).

Basionym. *Polysiphonia villum* J. Agardh 1863 (1848–1901): p. 941.

Abbott (1999) suggested that var. *villum* may be an ecovariant as intermediate types exist. Others (Abbott and Hollenberg, 1976; Scagel et al., 1989; Schneider and Searles, 1991; Silva et al., 1996; Abbott, 1999; Littler and Littler, 2000; Womersley, 2003) have listed both taxa.

The erect filaments of the variety are positioned more irregularly and at greater intervals from the prostrate axes than in the species. Scar cells resulting from deciduous trichoblasts are also more obvious and tetrasporangia larger (65–75 µm in diam.) and in spiral rows. Rhizoids arise from the middle of ventral pericentral cells.

Rare, the species is known from southern Florida and the variety from Biscayne Bay. The species is known from the Adriatic Sea, eastern tropical Atlantic, Indo-Pacific, and Australia (type location: Rottnest Island; Silva et al., 1996). The variety is known from Bermuda, North Carolina, the Caribbean, Brazil, Red Sea, and Indo-Pacific.

Schneider and Searles 1991: p. 473; Littler and Littler 2000: p. 232.

Polysiphonia subtilissima Montagne 1840: p. 199 (L. *subtilissima*: most slender, very delicate). Plate XLI, fig. 26, 27.

Synonym. *Polysiphonia subtilissima* var. *wespointensis* Harvey 1853: p. 35.

Filaments are polysiphonous with 4 pericentral cells, ecorticate, in brown green to deep red mats, turfs, or tufts to 15 cm tall, and attached by numerous rhizoids. Rhizoids are simple, unicellular, arise from the proximal ends of pericentral cells, and lack cross walls. Prostrate axes are indeterminate, spreading, with upturned tips, 70–120 µm in diam., with segments 1–4 diam. long, and produce erect axes arising from every 7th to 10th segment or more often. Erect axes are indeterminate, irregularly to alternately branched, 30–60 µm in diam., with segments 1–8 diam. long, and rarely have adventitious branches. Branches are 30–45 µm in diam., with segments to 130 µm long, and replace trichoblasts that are in a regular sequence at axial tips. Trichoblasts are deciduous, often forked, and not covering the large apical cell. Tetrasporangia are spherical, 65–75 µm in diam., in rows near midregions of erect branches, and cause segments to swell. Spermatangial branches are cylindrical and may have uniseriate tip cells. Cystocarps are unknown.

Common, probably the most common species of *Polysiphonia* in Florida; in drift, on limestone and jetties in estuaries, and in protected bays on all coasts; also known as isolated (often estaurine) populations from the Canadian Maritimes to Cape Cod, then southward through the Caribbean, Bermuda, the Gulf of Mexico, French Guiana (type location; Abbott, 1999), Brazil, and Uruguay, as well as the Mediterranean, West Africa, Indo-Pacific, and Australia.

Taylor 1960: p. 575; Dawes 1974: p. 162; Schneider and Searles 1991: p. 475; Littler and Littler 2000: p. 234.

Waldoia W. R. Taylor 1962a: p. 58–59 (named for Professor Waldo Schmitt, an American zoologist, a colleague of Taylor who also collected seaweed for him at the University of Michigan).

Plants are flat, fleshy with the polysiphonous construction inconspicuous; axes are constricted to stalk-like bases attached by discoid holdfasts. Main fronds are alternately and marginally branched and lack midribs or axial cells. Tetrasporangia are tetrahedrally divided and in undivided cylindrical stichidia scattered over the surface. Cystocarps are ostiolate, sessile or stalked, and on ordinary branches.

Waldoia antillana W. R. Taylor 1962a: p. 59, pl. 3, fig. 1, 2; pl. 4, fig. 2 (L. *antillana*: of or from the Antilles or West Indies). Plate XLI, fig. 28, 29.

Fronds are soft, flat, to 13 cm tall, and rose to dull purple; one or more fronds occur on short, cylindrical stalks (2–15 mm long) that are attached by coarse fibrous holdfasts. Axes expand gradually from stalks into linear or oblong blades 4–7 mm wide; blades are without veins, with undulate to crenate margins, and have obtuse to indented tips. The apical cell is sunken in a notch or projecting; pericentral cells are inconspicuous. Fronds are branched 2–3 times and have marginal or alternate proliferations that are stalked or contracted at base. Cortication is limited to irregular cortical cells or rhizoids. Tetrasporangia are tetrahedrally divided, in cylindrical stichidia 4–5 mm long, 250–400 µm in diam., and on the surface. Cystocarps are ovoid, to 700 µm in diam., sessile or on short stalks, and scattered over blades.

Rare, reported from 25–56 m off the east coast (Hanisak and Blair, 1988); also known from the Lesser Antilles (type location: Tobago) and Ghana.

Wrightiella F. Schmitz 1893: p. 221 (named for Professor E. Percival Wright of Trinity College in Dublin).

Plants are erect, firm to cartilaginous, radially branched, and attached by haptera or rhizoidal discs. Axes are cylindrical, indeterminate, and have 4–5 pericentral cells and cortication by descending rhizoids. All axes are covered by spinelike branchlets that arise endogenously from axial cell rows. Branchlets are short, simple or forked; they occur in a spiral pattern, are basally polysiphonous, and taper abruptly to an apical cell. Branchlets are covered by pigmented uniseriate filaments that arise from apical cells before formation of pericentral and axial cells. Filaments are deciduous, determinate, in spiral sequences, and with basal cells smaller than those above and below. Tetrasporangia are tetrahedrally divided and in a spiral series in corticated swollen stichidia on determinate uniseriate branchlets. Procarps occur near tips of truncated uniseriate branchlets and have 4-celled carpogonial branches. Cystocarps are oval to urceolate and on short stalks.

Key to the species (2) of *Wrightiella*

1. Plants stiff, spiny; to 20 cm tall; with many spur branchlets 1–5 mm long; filaments are deciduous, uniseriate, and to 30 µm in diameter . . . . *W. blodgettii*
1. Plants usually soft, lax; to 75 cm tall; with few spur branchlets; filaments are deciduous, uniseriate, and to 60 µm in diameter . . . . *W. tumanowiczii*

***Wrightiella blodgettii* (Harvey) F. Schmitz 1893: p. 221 (named for Dr. Blodgett, who sent specimens from Key West to Harvey). Plate XLI, fig. 30–32.**

Basionym. *Alsidium blodgettii* Harvey 1853: p. 16, pl. XV, B, fig. 1–5.

Plants are stiff, spiny, erect, to 20 cm high, and pale yellow to bright red; main axes are visible (percurrent), repeatedly alternately to pinnately branched, and attached by discoid holdfasts. Erect axes are 0.3–1.3 mm in diam. and smooth below; upper parts produce 4 spiral rows of spur- to spinelike branchlets that are to 350 µm in diam., 1–5 mm long, simple to once forked, and with soft acute tips. Young branchlet tips have pigmented filaments that are uniseriate and to 30 µm in diam. Tetrasporangia are tetrahedrally divided, spherical, 30–50 µm in diam., and in spiral series on filaments. Cystocarps are oval to urn-shaped, to 900 µm in diam., on short stalks, and have beaked ostioles.

Rare, adrift at St. Marks and Key West (type location), in 7–36 m at the Dry Tortugas, and in 27 m off the east coast (Hanisak and Blair, 1988); also known from Bermuda, the Bahamas, and the Caribbean.

Taylor 1960: p. 591; Littler and Littler 2000: p. 234.

Wrightiella tumanowiczii **(Gatty *ex* Harvey) F. Schmitz 1893: p. 222 (named for Mr. Tumanowicz, a Polish marine biologist who lived in England, collected in the West Indies, and sent specimens to Mrs. Gatty). Plate XLII, fig. 1, 2.**

Basionym. *Dasya tumanowiczii* Gatty *ex* Harvey 1853: p. 64, plus supplement p. 128.

Schneider (1975) noted that specimens of *Wrightiella blodgettii* and *W. tumanowiczii* from North Carolina were difficult to differentiate and might be conspecific.

Plants are erect and bushy, to 75 cm tall, straw yellow to bright red, with sparse, alternate to irregular branching, and with one or more axes attached by discoid bases that are reinforced by rhizoidal branchlets. Indeterminate (main) axes are 0.5–2.0 µm in diam., with 4–5 pericentral cells, and corticated below by descending short-celled rhizoids from basal cells of lateral branches. Main axes have a few filiform to rigid branches and many branchlets that branch repeatedly above, are to 4 cm long, tapering, and produce short spines with uniseriate filaments. Spines are in irregular spirals, to 150 µm in diam., and 300 µm long at bases. Uniseriate filaments are pigmented, inconspicuous, deciduous, to 2 mm long, and 60 µm in diam.; they are produced from small basal cells, are dichotomously branched below, slender and whiplike above, and tapering to 10 µm. Tetrasporangia are tetrahedrally divided, spherical, 30–60 µm in diam., and in spiral series on the upper parts of filaments. Cystocarps are ovoid to urceolate and on short polysiphonous stalks.

Common, on limestone (1–3 m) in the Keys, from 9 m at Anclote Key, 18–27 m at the Dry Tortugas, and 32–39 m off the east coast; also known from Bermuda, North Carolina (34–40 m), the Bahamas (type location), Antilles, Gulf of Mexico, and Brazil.

Taylor 1960: p. 592; Dawes 1974: p. 163; Schneider and Searles 1991: p. 480; Littler and Littler 2000: p. 236.

GELIDIALES

The order was established by Kylin (1923) and variously treated thereafter (Saunders and Hommersand, 2004). Dixon (1973) placed it in an emended Nemaliales by deemphasizing life history traits. Pueschel and Cole (1982) supported the order based on the structure of pit plugs. Hommersand and Fredericq (1988) reinstated it based on anatomical studies, while Freshwater et al. (1994) confirmed the order's uniqueness using plastid *rbcL* patterns.

Plants are erect and prostrate; axes are slender, wiry to cartilaginous, flat or terete, and attached by rhizoidal holdfasts. Tips have an apical cell; axes are uniaxial but hidden by long axial filaments and thick-walled rhizoidlike cells (rhizines). The cortex has one or more layers of small pigmented cells. Life histories are isomorphic; gametophytes are often unknown or dioecious. Spermatangia are in hyaline sori near tips of branches. Carpogonia are sessile or on short branches. Gonimoblasts arise directly from fertilized carpogonia, in association with fused hypogenous cells or long nutritive filaments (fusion cells). Auxiliary cells are absent. Several species are harvested for the phycocolloid agar and others are farmed as "sea vegetables."

Gelidiaceae

Plants are tough, wiry, branched, and usually have an isomorphic life history. Tetrasporangia are embedded in the cortex and scattered on axes or in compressed axial tips. Sporangial divisions are cruciate, tetrahedral, or irregular. Cystocarps are scattered or on axial tips, often below a branch.

Gelidium J.V. Lamouroux 1813: p. 128, *nom. cons.* (L. *gelidus*: ice-cold, gelatinous).

Plants are erect and pinnately to irregularly branched; axes are tough, terete or flattened, and attached by rhizoidal holdfasts. Tips have one apical cell with the uniaxial axes obscured by secondary filaments. The medulla has a compact mix of thick-walled rhizoids (rhizines) and filaments. The cortex has round to angular or elongated pigmented cells. Tetrasporangia are tetrahedrally divided and occur in the cortex of branches with restricted growth. Spermatangia form extensive patches on the branches of male plants. Carpogonial branches are 3-celled; cystocarps have two locules and protrude on both surfaces of fertile branches.

Key to the species (5) of *Gelidium*

1. Rhizines or refractive thick-walled filaments are limited to the medulla 2
1. Rhizines are present in the medulla and inner (sub) cortex 3
 2. Axes often longer than 10 cm; branching pinnate to irregular . . . . *G. pusillum*
 2. Axes to 6 cm long; branching irregular, simple, sparse or palmate . . . . *G. americanum*
3. Older axes compressed to flat, 1–2 mm wide 4
3. Older axes compressed, to 0.5 mm wide . . . . *G. crinale*
 4. Older axes 1 mm wide; branching pinnate . . . . *G. floridana*
 4. Older axes strap-shaped, 0.5–2 mm wide; branching sparse and irregular . . . . *G. corneum*

Gelidium americanum **(W. R. Taylor) Santelices 1976: p. 173, fig. 28–33 (L. *americanum*: of or from America). Plate XLII, fig. 3–6.**

Basionym. *Pterocladia americana* W. R. Taylor 1943: p. 154, pl. IV, fig. 1.

Plants are tough; erect axes are 0.5–6.0 cm tall, terete, compressed to strap-shaped, and to 1 mm wide; prostrate axes are attached by rhizoids. Branching is simple, sparse, or irregular, with regenerating axes that are alternate, pinnate, or palmate. Tips have one apical cell. The medulla has long, thin-walled filaments 10–20 µm in diam. and slender, thick-walled rhizines, 3–6 µm in diam. The cortex has one to several layers, with surface cells rounded to angular and 10–20 µm in diam. Tetrasporangia are ovoid, 30–45 µm in diam., and in scattered patches on flattened blades or in V-shaped rows on branchlets. Cystocarps have two locules, each with an ostiole.

Common, forming a turf in the intertidal and shallow subtidal on wave-exposed limestone on the west (Point of Rocks) and east (Vero Beach) coasts and the Keys; also known from Bermuda, North Carolina to Georgia, the Caribbean (type location: Port Antonio, Jamaica), Gulf of Mexico, and Brazil.

Taylor 1960: p. 360; Dawes 1974: p. 120; Schneider and Searles 1991: p. 217; Littler and Littler 2000: p. 42.

Gelidium corneum **(Hudson) J. V. Lamouroux 1813: p. 129 (L. *corneum*: horny, hard, and close-textured but not brittle). Plate XLII, fig. 7, 8.**

Basionym. *Fucus corneus* Hudson 1762: p. 474; 1778: p. 585.

Synonym. *Gelidium sesquipedale* (Clemente) Thuret in Bornet and Thuret 1876: p. 61.

Lee and Kim (1995) suggested that the name *Gelidium corneum* was controversial and should be changed to *G. sesquipedale*. Silva et al. (1996) and Wynne (2005a) believed *G. corneum* was a distinct species that was misidentified as *G. sesquipedale* and should be reinstated. Taylor (1960) considered Florida records of *G. corneum* as untrustworthy.

Plants are bushy and tough; main erect axes obvious, 3–50 cm long, flat, and pinnately branched 1–3 times; prostrate axes are attached by rhizoids. Axes are 0.5–2 mm wide; branchlets are slightly constricted at bases and broad and rounded at apices. Tetrasporic axes are less branched than gametophytic ones. Thick-walled rhizines occur in the inner cortex and outer medulla; the cortex has one or more layers of rounded to elongate cells.

Rare, and a questionable record for Florida (Taylor, 1960); also known from South Carolina, Bermuda, the Caribbean, Gulf of Mexico (Texas), and Brazil, as well as the eastern North and tropical Atlantic, Mediterranean (syntype locations, "Mare mediterraneum et anglicum"; Silva et al., 1996), Indian Ocean, and Australia.

Taylor 1960: p. 356.

Gelidium crinale (Hare *ex* Turner) Gaillon 1828: p. 362 (L. *crinale*: hairy). Plate XLII, fig. 9, 10.

Basionym. *Fucus crinalis* Hare *ex* Turner 1815 (1811–1819): p. 4–5, pl. 198, fig. a–c, e–g.

Gelidium crinale and *G. pusillum* are either designated as distinct taxa (Børgesen, 1927; Guiry and Nic Dhonncha, 2006) or combined as *G. pusillum* (Hoyt, 1920; Schneider and Searles, 1991).

Plants form a tough turf and are attached by prostrate axes with rhizoids. Axes are subterete, 1.5–7.5 cm tall, with irregular to alternate branching below and irregular to pinnate above, and with one apical cell. Branches of tetrasporic plants are flat to strap-shaped and have acute or obtuse tips. The medulla has axial filaments and thick-walled rhizines that also occur in the inner cortex. The cortex is composed of one or more layers; cells are round to elongate, 15–16 µm in diam., and 6–16 µm long. Tetrasporangia are 10–22 µm in diam. and 13–25 µm long. Gametophytes are dioecious; spermatangia occur on branchlet surfaces. Cystocarps are ostiolate, single, on spindle-shaped branchlets, and have 2 locules.

var. *platycladum* W. R. Taylor 1943: p. 152, pl. II, fig. 1 (G. *platycladum*: flat-branched).

Plants are 7–10 cm tall; main axes are twisted, flat, to 550 µm wide above, and often widest at forks. Branching is irregularly dichotomous below and alternate above; branchlets are flat, comb-like, crowded, and have acute tips. The inner cortex has a few obscure rhizines. Tetrasporangia are scattered and in pointed expanded branchlets.

Common, on all coasts; the species grows on jetties near Tampa Bay and the variety at exposed sites on both coasts (Point of Rocks and Marineland); the species is also known in the Caribbean, Gulf of Mexico, Brazil, and Uruguay, as well as the eastern North and tropical Atlantic (lectotype location: Devonshire, England; Dixon and Irvine, 1977), Adriatic Sea, Indo-Pacific, and Australia. The variety is known from Texas.

Taylor 1960: p. 355; Dawes 1974: p. 120.

Gelidium floridanum W. R. Taylor 1943: p. 153, pl. III, fig. 1, 2 (L. *floridanum*: of or from Florida). Plate XLII, fig. 11.

Plants are bushy, to 13 cm tall, spreading, entangled, repent, strap-shaped, 1 mm wide, and with terete bases attached by rhizoidal holdfasts. Branching is sparingly pinnate below and more regularly distichous above. The medulla has axial filaments with thick-walled rhizines that are also in the inner cortex. The cortex has one or more layers of round to elongate cells. Tetrasporangia occur in sori on the upper parts of fertile pinnate branchlets. Cystocarps are ostiolate, single, and have 2 locules that swell both surfaces.

Rare, from the Indian River (type location) and the Dry Tortugas; also known from Puerto Rico, Trinidad, and Venezuela.

Taylor 1960: p. 357.

Gelidium pusillum (Stackhouse) Le Jolis 1863: p. 139 (L. *pusillum*: minute, weak). Plate XLII, fig. 12–14.

Basionym. *Fucus pusillus* Stackhouse 1795 (1795–1801): p. 16, 17, pl. VI; 1816: p. 9, p. 6.

See Silva et al. (1996), Abbott (1999), and Lipkin and Silva (2002) for comments on the separation of *Gelidium crinale* and *G. pusillum*.

Plants form tough, dense turfs to 15 cm tall; axes are to 1.5 mm wide and attached by padlike holdfasts from a creeping base. Erect axes are terete to strap-shaped above; branching is sparsely pinnate; tips are trifurcate, to 0.5 mm wide, acute, and have an apical cell. The medulla is compact, with axial filaments (to 25 µm in diam.) and thick-walled rhizines (4–10 µm in diam.). The cortex has one or more layers; cells are round and surface ones 4–10 µm in diam. Tetrasporangia are mostly cruciately divided, spherical to ovoid, 25–40 µm in diam., terminal or lateral, and swell the branchlets. Cystocarps are solitary, with 2 locules, and occur below tips or on short spindle-shaped branchlets.

Common, on intertidal limestone and coquina at exposed sites on all coasts; known also from disjunct populations in the Gulf of St. Lawrence to Long Island Sound and more continuous southward through the Caribbean, Bermuda, Gulf of Mexico, Brazil, Uruguay, and Falkland Islands; as well as the eastern North and tropical Atlantic (type location: Devon, England; Dixon and Irvine, 1977), Mediterranean, Black Sea, and Indo-Pacific.

Taylor 1960: p. 354; Dawes 1974: p. 120; Schneider and Searles 1991: p. 218; Littler and Littler 2000: p. 44.

Gelidiellaceae

Gelidiella has been placed in its own family based on the lack of rhizines and sexual plants (Fan, 1961; Hommersand and Fredericq, 1988; Norris, 1992b) or in the Gelidiaceae (Maggs and Guiry, 1988; Womersley and Guiry, 1994). Reports of cystocarpic *G. diuens* Sreenivasa Rao and Trivedi (1980) and the occurrence of a few medullary rhizines at attachment areas in the repent *G. calcicola* Maggs and Guiry (1988) have confused the family designation.

Plants are bushy and wiry, uniaxial in construction, and have one apical cell that divides transversely to form an axial row of cells that enlarge and elongate. Surface cells are short, small, and have secondary pit connections. The tetrasporangial stichidia of *Gelidiella acerosa* are swollen or conical and have compact sporangia; those of *G. pannosa* J. Feldman and Hamel are compressed and have sporangia in transverse series.

Gelidiella J. Feldmann and Hamel 1934: p. 529 (L. *gelidus*: cold, congealed + *ella*: lesser; thus, a small *Gelidium*).

Plants are bushy and wiry, with slender terete branches, and prostrate axes attached by rhizoids. Axes are uniaxial and have an obvious apical cell and a medulla of many longitudinal filaments. The cortex has 1(2–3) layers of ovoid cells, with each cell containing several chloroplasts. Axial filaments and lateral cell rows may be visible; rhizines are absent or rare (Womersley and Guiry,

1994). Tetrasporangia divide tetrahedrally or irregularly, are in special branchlets or enlarged tips of the main axes, and in parallel rows.

Gelidiella acerosa (Forsskål) J. Feldmann and Hamel 1934: p. 533 (L. *acerosa*: pointed, needlelike). Plate XLII, fig. 15–18.

Basionym. *Fucus acerosus* Forsskål 1775: p. 190

Plants are usually clustered, tough or wiry, to 15 cm tall, and have prostrate axes attached by rhizoidal clusters or pads. Erect axes are terete to compressed, elongate, or recurved; branches are secund, sparingly alternate or bilaterally arranged, 2–6 mm long, and have acute to obtuse tips and one apical cell. The medulla has thick-walled filaments (30–40 µm) and lacks rhizines. The cortex has one or more layers of round to elongate cells with surface cells 8–10 µm in diam. Tetrasporic branches are club-shaped and irregularly disposed; tetrasporangia are formed progressively from tips downward. Cystocarps form obvious unilateral swellings on fertile branches.

Common, on all coasts on limestone in wave-exposed sites, in tide pools, and to 7.5 m in the Keys; also known from Bermuda, the Caribbean, Brazil, as well as the eastern tropical Atlantic, Red Sea (type location: Al Mukha, Yemen; Abbott, 1999), Indo-Pacific, and Australia.

Taylor 1960: p. 351; Dawes 1974: p. 119; Littler and Littler 2000: p. 46.

GIGARTINALES

The order contains 29 families, including taxa from Kylin's (1956) Cryptonemiales, with the auxiliary cells being in specialized accessory filaments, versus the Gigartinales, with intercalary vegetative cells acting as auxiliary cells (Saunders et al., 2004). The ordinal separation was not effective (Kraft and Robins, 1985). Later, small homogenous orders were erected including the Ahnfeltiales (Maggs and Pueschel, 1989), Corallinales (Silva and Johansen, 1986), Gracilariales (Fredericq and Hommersand, 1989a), Halymeniales and Nemastomatales (Saunders and Kraft, 1996; Kraft and Saunders, 2000), and Plocamiales (Saunders and Kraft, 1994).

Plants are erect, prostrate, or crustose, and gooey, fleshy, wiry, or cartilaginous. Thalli are uniaxial or multiaxial and pseudoparenchymatous. Life histories are triphasic and include isomorphic and heteromorphic tetrasporophytes and gametophytes. Tetrasporangia are cruciately or zonately divided. Gametophytes may be monoecious or dioecious. Spermatangia are scattered or in discrete superficial or cryptlike sori. Carpogonial branches range from 2 to 22 cells. Auxiliary cells may be supporting cells, derivatives from supporting cells, or intercalary vegetative cells; if distant, a connecting filament (ooblast) may extend from a fertilized carpogonium or fusion cell. Gonimoblastic filaments form inwardly or outwardly from auxiliary cells or from ooblast junctions or fusion cells. Cystocarps usually have ostiolate pericarps that are embedded or more often projecting. The order is distinguished by auxiliary cells that exist before fertilization as vegetative or specialized filaments (nonprocarpic) or as closely related cells (procarpic) of a carpogonial branch. Several species are harvested and farmed for the phycocolloid carrageenan (Tseng, 1946).

Caulacanthaceae

Plants are terete or flat and radially or bilaterally branched. The medulla is filamentous, while the cortex has close, radial rows of cells with inner colorless larger cells and smaller pigmented surface ones. Tetrasporangia are zonately divided and scattered over the surface. Carpogonial branches are 2(5)-celled long and cystocarps are immersed. Carposporangia radiate from lobed fusion cells and the overlying cortex functions as a specialized pericarp with a discharge pore.

Catenella Greville 1830: p. lxiii, 166, *nom. cons.* (L. *catena*: a chain, to connect in a series + *ella*: a diminutive; thus, a small chain or series).

Plants are prostrate to erect and irregularly branched; segments are jointed to spindle-shaped, terete, narrow, or broadly oval, and attached by haptera at joints. The medulla is hollow, uniaxial, or has a central cluster of 5–6 slender longitudinal filaments. The cortex has anticlinal rows of cells and small outer pigmented ones. Tetrasporangia are zonately divided, in specialized terminal branchlets, and scattered in the cortex. Cystocarps have a distinct pore and are solitarily on short, thick terminal segments.

***Catenella caespitosa* (Withering) L. M. Irvine in Parke and Dixon 1976: p. 590 (L. *caespitosa*: turfy, tufted, matted). Plate XLII, fig. 19, 20.**

Basionym. *Ulva caespitosa* Withering 1776: p. 735.
Synonyms. *Catenella opuntia* (Goodenough and Woodward) Greville 1830: p. lxiii, 166. nom illeg.
Catenella pinnata Harvey 1853: p. 201, pl. 29B.
Catenella repens (Lightfoot) Batters 1902: p. 69.

Plants are mostly prostrate, to 3 cm tall, di- to trichotomously or oppositely branched, and attached by short, stalked cells from constricted branches that form uncorticated rhizoidal haptera. Segments are usually elongate and dorsiventrally compressed, while estuarine plants axes may be nearly terete. Segments are oval to spindle-shaped, 3–5 times longer than broad, 0.2–0.5 cm in diam., and constricted. Growth is by an apical cell, with the central filament often obscured by 5–6 thick-walled longitudinally oriented filaments in the medullary cavity. The cortex has 2–3 layers of cells (6–10 µm in diam.) in radially arranged filaments that branch from the medulla. Tetrasporangia are ovoid, zonately divided, and scattered. Cystocarps are ovoid, 200–360 µm in diam., and on terminal segments.

Common, in estuaries on limestone, stems of salt marsh plants, and roots of mangroves on all coasts; also known from Bermuda, the Caribbean, Gulf of Mexico, and Brazil, as well as the eastern North and tropical Atlantic, Mediterranean (type location: Anglesey, England; Silva et al., 1996), and Indo-Pacific.

Taylor 1960: p. 462; Dawes 1974: p. 133; Littler and Littler 2000: p. 72.

Cystocloniaceae

Synonym. Hypneaceae

The Hypneaceae was merged with the Cystocloniaceae based on molecular studies (Saunders et al., 2004). Min-Thein and Womersley (1976) also proposed the move using vegetative and reproductive features.

Plants are bushy, erect or prostrate, cartilaginous, and with subsimple alternate to irregular and radial branching; axes are attached by small discoid or crustose holdfasts and secondarily by spreading axes. Parasitic plants are cushionlike or lobed. Axes are terete to flattened and uniaxial with a visible axial filament. The medulla and cortex are compact and pseudoparenchymatous; outer cortical cells are pigmented and small, while inner ones are larger and isodiametric. Life histories are isomorphic. Tetrasporangia are zonately divided and from cortical cells in swollen sori occurring on short branchlets. Gametophytes are dioecious. Spermatangia are in superficial sori. Female plants are procarpic; carpogonial branches are 3-celled and occur laterally on cortical cells; auxiliary cells arise from support cells before fertilization. Fertilized carpogonia do not form fusion cells. Gonimoblastic filaments produce seriate carposporangia. Cystocarps develop from enlarged auxiliary cells that project from branchlets and have multilayered pericarps.

Hypnea J.V. Lamouroux 1813: p. 131 (Gr. *hypnon*: a moss; also cf. the moss genus *Hypnum*).

The genus probably includes a number of cryptic species and a worldwide study is needed (Price et al., 1992; Rull Lluch, 2002).

Plants are bushy, entangled, and erect or spreading; axes are covered by short, linear, spinelike to stellate branchlets and attached by discoid holdfasts. Branch tips have an obvious apical cell. Cystocarps are hemispherical and lack a pore.

Key to the species (4) of *Hypnea* (modified from Schneider and Searles, 1991)

1. Axes cylindrical. 2
1. Axes flat or compressed. . . . . *H. volubilis*
   2. Swollen hooklike (crozier) tips absent 3
   2. Swollen hooklike tips present. . . . . *H. musciformis*
3. Axes covered with simple to branched spines and stellate branchlets; branching alternate below and not cervicorn above; cystocarps usually clustered. . . . . *H. cornuta*
3. Axes have only simple spinelike branchlets; branching pseudodichotomous below and cervicorn above; cystocarps solitary . . . . *H. spinella*

Hypnea cornuta (Kützing) J. Agardh 1851 (1848–1901): p. 449–450 (L: *cornuta*: horned or horn-shaped). Plate XLII, fig. 21, 22.

Basionym. *Chondroclonium cornutum* Kützing 1849: p. 741

Some authors consider *Hypnea cornuta* to be a distinct taxon (Mshigeni and Chapman, 1994; Yamagishi et al., 2003), while others (Børgesen, 1920, 1943) have listed it as a synonym for *H. valentiae* Turner 1809 (1808–1809). Yamagishi et al. (2000) separated *H. valentiae* from *H. cornuta* based on the former having horizontal, short, subulate, and simple branchlets.

Plants are bushy and wiry, entangled, 4–12(40) cm tall, and attached by discoid holdfasts. Branches are terete, 0.4–1.5 mm in diam, repeatedly alternately branched, and have tips that taper to a visible apical cell. Axes are covered by simple, sessile, spinelike branchlets, 2.2 mm long; some branchlets are stalked and starlike with 3–6 tips. The axial cell row is visible in cross section near the tips of branches; axial cells are thick-walled, surrounded by thick-walled medullary cells, and 50–300 μm in diam. The cortex has 1–2 layers of rounded pigmented cells, 7.5–12.5 μm in diam. Tetrasporangia are elliptical, zonately divided, 15–25 μm in diam. by 40–50 μm long, and in sori around spines. Gametophytes are dioecious. Cystocarps are ostiolate, projecting, 325–500 μm in diam., and have multilayered pericarps. Carpospores are 5–10 μm in diam. The species resembles *Hypnea spinella* (C. Agardh) Kuntze but has stellate to crownlike spines.

Common, on mangrove roots, shells, and rubble in shallow waters on all coasts; also known from North Carolina, Bermuda, the Caribbean, Gulf of Mexico, and Brazil, as well as the Mediterranean (introduced into Egypt and Israel), Red Sea (type location; Abbott, 1999), eastern tropical Atlantic, and Pacific Ocean.

Taylor 1960: p. 467; Dawes 1974 p. 138; Schneider and Searles 1991: p. 308; Littler and Littler 2000: p. 78.

Hypnea musciformis (Wulfen in Jacquin) J. V. Lamouroux 1813 p. 43 (p. 131 in reprint) (L. *musci*: moss + *formis*: resembling, formed). Plate XLII, fig. 23–25.

Basionym. *Fucus musciformis* Wulfen in Jacquin 1791 ('1789'): p. 154, pl. 14, fig. 3.

Synonym. *Hypnea arborescens* P. Crouan and H. Crouan in Schramm and Mazé 1865: p. 15; P. Crouan and H. Crouan in Mazé and Schramm 1878 (1870–1877): p. 195.

Plants are bushy, entangled, wiry, 10–50 cm tall, and attached by discoid holdfasts. Erect axes are alternately to irregularly branched, 0.5–2.0 mm in diam.; smaller axes have secund branching. Branchlets are numerous, simple or branched, curved upward, and spurlike, with a few being swollen and strongly hooked (crozierlike). Tips are acute and have an obvious apical cell. The axial cell row is visible in sections near tips of branchlets and obscured below. Medullary cells are irregular, thick-walled, and 40–300 µm in diam.; the outer two cortical cell layers are pigmented and 7–18 µm in diam. Tetrasporangia are ovoid, zonately divided, 22–30 µm in diam. by 40–85 µm long, and in raised sori at bases of spines. Gametophytes are dioecious. Cystocarps are clustered on branchlets or spines, 0.3–1.0 mm in diam., ostiolate, and protruding; pericarps have several layers of cortical cells. Carpospores are 19–30 µm in diam. by 20–45 µm long. Crozier tips are common on epiphytic or attached plants at exposed sites and usually absent in drift material.

Common, adrift in estuaries, as entangled mats in seagrass beds, and on limestone at exposed sites (to 10 m) on all coasts; also known from southern Massachusetts through the Caribbean, Bermuda, the Gulf of Mexico, Brazil, and Uruguay, as well as the eastern North and tropical Atlantic, Mediterranean (type location: Trieste, Italy; Silva et al., 1996), and Indo-Pacific.

Taylor 1960: p. 467; Dawes 1974: p. 138; Schneider and Searles 1991: p. 306; Littler and Littler 2000: p. 76.

Hypnea spinella (C. Agardh) Kützing 1847b: p. 23; 1849: p. 759; 1868: pl. 26 (L. *spinella*: a prickle, small spine, thorn). Plate XLII, fig. 26, 27.

Basionym. *Sphaerococcus spinellus* C. Agardh 1822 (1820–1828): p. 323–324.
Synonym. *Hypnea cervicornis* J. Agardh 1851 (1848–1901): p. 451–452.
Xia and Zhang (1999) combined *Hypnea spinella* with *H. cervicornis* J. Agardh.

Plants are wiry, to 15 cm tall or in dense mats; they are radially branched with lower axes decumbent, and attached by discoid holdfasts and secondarily by rhizoidal pads of prostrate axes. Branches are 0.4–1.0 mm in diam., cylindrical, alternately branched to hornlike above and subdichotomous below. Branchlets are spinelike and have pointed tips with an obvious apical cell. The axial cell row is visible in cross section near tips; cells are thick-walled and 70–80 µm in diam. Medullary cells are thick-walled and 100–350 µm in diam. The cortex has 1–2 layers of rounded, pigmented cells, 7.5–25 µm in diam. Tetrasporangia are ovoid, zonate, 10–25 µm in diam., 40–50 µm long, and in raised sori. Gametophytes are dioecious. Cystocarps are ostiolate, hemispherical, 0.1–0.9 mm in diam., and protruding. Pericarps have radiating layers of cortical cells; carposporangia are 5–12 µm in diam.

Common, on limestone, rubble, and algae on all coasts (to 7 m); plants form loose clumps in protected areas and compact tufts in wave-exposed areas; also known from North Carolina southward, Bermuda, the Bahamas, Caribbean (type location: West Indies; Dawson, 1961), Gulf of Mexico, and Brazil, as well as the Mediterranean, eastern tropical Atlantic, Indo-Pacific, and Australia.

Taylor 1960: p. 465, 466; Dawes 1974: p. 138; Schneider and Searles 1991: p. 306; Littler and Littler 2000: p. 78.

Hypnea volubilis Searles in Schneider and Searles 1976: p. 53, fig. 3–7, 9–10 (L. *volubilis*: bending, turning, twisting). Plate XLII, fig. 28.

Plants are prostrate, spreading, to 6 cm tall, and entangled with other algae or attached by rhizoidal pads that arise from margins, blades, and tips of branches. Axes are flattened and 0.4–0.9(2.0) mm in diam.; main axes are sparingly subdichotomously branched and with alternate to irregular branching from margins. Marginal branchlets are sparse and secund above; apices are acute and have a visible apical cell. The central axial cell row is visible in cross section near tips. Medullary cells are 42–180 µm in diam. The cortex has a layer of round pigmented cells 6–48 µm in diam. Tetrasporangia are elliptical, zonately divided, 17–24 µm in diam., 25–42 µm long, and in sori around branchlets. Spermatangia are in sori or scattered. Cystocarps are lateral, nearly spherical, and to 350 µm in diam.

Rare, entangled with algae in 25–45 m off the east coast (Hanisak and Blair, 1988); also known from the Carolinas (type location: 23–24 m off North Carolina), Puerto Rico, and Gulf of California.

Schneider and Searles 1991: p. 309.

Dumontiaceae

Plants are soft and gelatinous or firm and fleshy, erect or prostrate, and attached by small discoid holdfasts. Upright plants are terete to compressed and have a crustose stage. Growth is often by an apical cell. Each axial cell produces four branching filaments at right angles to the axis. The medulla is filamentous, with descending rhizoids arising from basal cells of the lateral filaments. Surface cells are pigmented and form a loose to compact layer. Life histories are isomorphic or heteromorphic. Tetrasporangia (if known) are cruciately or zonately divided. Gametophytes are monoecious or dioecious. Spermatangia occur in cortical sori. Gametophytes lack procarps; carpogonial branches are 4–24-celled, distinct from obvious auxiliary cells, and on long special intercalary or terminal branches. Gonimoblastic filaments arise directly from carpogonia or from connecting (ooblast) filaments. Cystocarps are embedded, scattered, and lack ostioles.

Dudresnaya P. Crouan and H. Crouan 1835: p. 98, pl. 2, fig. 2, 3, *nom. cons.* (named for Dudresnay de St. Pol-de-Leon, a French soldier and collector of seaweeds).

See Tai et al. (2001) regarding the possible resurrection of J. Agardh's 1876 (1848–1901) family Dudresnayaceae for *Dudresnaya*, *Acrosymphyton*, and other genera.

Plants are erect, gelatinous, bushy or wormlike, and alternately and radially branched; axes are terete to compressed and attached by discoid holdfasts. Determinate axes are loosely arranged and uniaxial growth is obvious near tips, but obscured by descending rhizoids below. Each axial cell produces 3–4 determinate or indeterminate radial branches. The medulla is filled with gel. Tetrasporangia are zonately divided and on determinate filaments. Gametophytes are monoecious or dioecious. Spermatangia are subspherical and on outer cortical cells in superficial sori. Carpogonial branches have 5(9) cells and arise laterally on determinate branches; intercalary auxiliary cells are conspicuous and on branches with 9(14) cells. Gonimoblastic filaments arise from auxiliary cells or connecting filaments. Pericarps are minimal to lacking, immersed, and lack a cortical pore.

Key to the species (2) of *Dudresnaya*

1. Axes cylindrical, to 10 mm in diam.; branching radial and dense . . . . *D. crassa*
2. Axes flattened, to 2 cm wide by 3 mm thick; branches lobed . . . . *D. patula*

***Dudresnaya crassa* M. Howe 1905b: p. 572, pl. XXVIII, XXIX, fig. 12–26 (L. *crassa*: stout, fleshy, thick). Plate XLII, fig. 29, 30.**

Plants are soft, gelatinous, 5–25 cm tall, and densely and radially branched; axes are cylindrical, to 10 mm in diam., and attached by small, fleshy discoid holdfasts. Tips taper to 1 mm and the apical cell is obscured; the uniaxial filament is visible, terete, 50–165 µm in diam., 200–430 µm long, and surrounded by rhizoidal filaments from cortical cells. Each axial cell produces 4 radiating cortical filaments that branch alternately to dichotomously. Outer cortical cells are pigmented, elongate, terete, and 3–5 µm in diam. Tetrasporangia are zonately divided, cylindrical, 10–13 µm in diam., 35–50 µm long, and occur at the ends of cortical filaments. Gametophytes are dioecious; male plants are smaller and paler than females. Spermatangia are cylindrical to ovoid, 2–3 µm in diam., and terminal or in lateral clusters. Carpogonial branches have 6(10) cells. Auxiliary cells are globose to rectangular and 8–13 µm in diam. Carposporangia are ovoid and 7–15 µm in diam. The plant differs from *Dudresnaya bermudensis* Setchell, a Caribbean species, by its terete cortical filaments versus moniliform subspherical cells of the latter (Bula-Meyer and Norris, 2001).

Uncommon, growing on subtidal coral fragments (to 20 m) in the Keys; also known from North Carolina, Georgia, Bermuda (type location), the Bahamas, Caribbean, Canary Islands, and North Senegal.

Taylor 1960: p. 364; Schneider and Searles 1991: p. 261; Littler and Littler 2000: p. 76.

***Dudresnaya patula* Eiseman and J. N. Norris 1981: p. 188–189; fig. 1–9 (L. *patula*: spreading, opened out). Plate XLII, fig. 31.**

Plants are light pink, soft and gelatinous, and to 10 cm tall; axes are flat, to 2 cm wide, 3 mm thick, 3–10 mm in diam., and on short stalks attached by discoid holdfasts. Plants are uniaxial; axial filaments are to 75 µm in diam., obscured in older parts, and with each cell producing dichotomously branched laterals that radiate to form a cortex of 2–3 layers. Cortical cells are 2–3 µm in diam., rounded, and pigmented. Gametophytes are dioecious. Spermatangia are 1–3 µm in diam., terminal on cortical filaments, and often in a triangular pattern. Carpogonial branches are 13(20)-celled, unbranched, and the distal 4–6 cells stain densely. Carpogonia are slightly recurved and with a long trichogyne. Auxiliary cell filaments are 27(43)-celled, unbranched, and in forks of lateral filaments.

Rare, on rubble in 32–58 m on the east coast (type location off Singer Island).

Gigartinaceae

Plants are erect and wiry to cartilaginous with one or more axes arising from discoid or crustose holdfasts; a few taxa have secondary attachments. Plants are multiaxial; the cortex is obscurely filamentous with files of small cells radiating at right angles to axes; outer cells are pigmented. Life histories are isomorphic. Tetrasporangia are in swollen sori or nemathecia, cruciately divided, in chains of 4 or more, and from inner cortical cells or special medullary filaments. Gametophytes are dioecious. Spermatangia are spherical and in surface sori. Female plants are procarpic with 3-celled carpogonial branches on large supporting cells that also function as auxiliary cells. Gonimoblastic filaments are slender, formed after fertilization, and branch into the medulla. If present, pericarps are diffuse, lack ostioles, and scattered or on papillae.

Chondracanthus Kützing 1843: p. 396–399 (Gr. *chondros*: cartilage + *acanthus*: a spine).

Plants are simple, foliose, irregularly to subdichotomously branched, terete to compressed, with a few to many papillae, and attached by holdfasts. Its anatomy is similar to that described for the family. Tetrasporangia are tetrahedrally divided, often arise from inner cortical cells, and are in sori on raised papillae. Carposporophytes are surrounded by dense rings of sterile involucral filaments. Cystocarps are on papillae.

Chondracanthus acicularis (Roth) Fredericq in Hommersand et al. 1993: p. 117 (L. _acicularis_: shaped like a needle). Plate XLII, fig. 32, 33.

Basionym. *Ceramium aciculare* Roth 1806: p. 114.

Synonym. *Gigartina acicularis* (Roth) J. V. Lamouroux 1813: p. 136 (reprint p. 44).

According to Schneider and Lane (2005) Florida plants may be *Chondracanthus saundersii* Schneider and Lane.

Plants are wiry, cartilaginous, erect, 3–10 cm tall, spreading, entangled, and initially attached by discoid holdfasts, and secondarily by rhizoidal pads. Axes are cylindrical and irregularly, alternately, or subdichotomously branched below; branches are 0.5–1.0 mm in diam., short, alternate to secund, recurved above, and tapered to acute tips. Medullary filaments are 2–15 µm in diam. and entangled. Inner cortical cells are stellate, fused, and 5–18 µm in diam.; outer cortical cells are round and 3–7 µm in diam. Tetrasporangia are elliptical, cruciately divided, 15–18 µm in diam., and 22–40 µm long. Cystocarps are subspherical, on short papillae, project on one surface, lack an ostiole, and are single or in short series near the middle of recurved branchlets.

Common, on limestone at moderately exposed shallow reefs off Vero Beach, as well as in the intertidal at Marineland; also known from North Carolina, Bermuda, the Antilles, Gulf of Mexico, Brazil, and Uruguay, as well as the eastern North and tropical Atlantic, Mediterranean (type location: the Adriatic Sea; Dixon and Irvine, 1977), Black Sea, Indo-Pacific, and Australia.

Taylor 1960: p. 473; Schneider and Searles 1991: p. 315.

Kallymeniaceae

Plants are erect, gelatinous or firm, and with simple, irregular, dichotomous, or pinnate branching; axes are flat to foliaceous and attached by discoid holdfasts or by irregular lobed cushions (if parasitic). The medulla is thick and filamentous. The inner cortex has refractive stellate cells with filamentous arms or pseudoparenchyma mixed with rhizoidal filaments. Life histories are isomorphic or heteromorphic. Tetrasporangia are cruciately divided and on upper axes or prostrate ones. Gametophytes are dioecious. Spermatangia arise from outer cortical cells and are scattered or in patches. Carpogonial branches are 3-celled and one or more arise from a support cell. If procarpial, the auxiliary cell is the supporting cell; if nonprocarpial it is connected by ooblast filaments after fertilization. Gonimoblastic filaments grow outward. Cystocarps are scattered over the blade surfaces, protruding, and have obvious pericarps.

Cirrulicarpus Tokida and Masaki 1956: p. 70 (L. *cirrus*: a curl, lock + *carpus*: fruit).

Perestenko (1994) assigned *Cirrulicarpus* to the Crossocarpaceae, while Womersley (1994) and Wynne (2005a) retained it in the Kallymeniaceae.

Plants are erect and tough; blades are flat, narrow to broadly strap-shaped, often expanded from base to apex, and attached by small discoid holdfasts. Medullary filaments are densely entangled and often have stellate ganglia. The cortex has several layers and small surface cells. Life histories are isomorphic. Tetrasporangia are cruciately divided and scattered or in a subcortical layer. Gametophytes are dioecious. Spermatangia occur in irregular sori. Carpogonial branches are 3-celled, nonprocarpial, and have 1–5 branches arising from a support cell that also produces sterile subsidiary cells. The outer gonimoblastic cells produce carposporangia and are mixed with vegetative filaments. Cystocarps are immersed and visible on one or both sides of blades.

Cirrulicarpus carolinensis Hansen 1977: p. 1–22, fig. 1–23 (L. _carolinensis_: of or from North Carolina). Plate XLIII, fig. 1, 2.

Plants identified as *Cryptonemia crenulata* J. Agardh *sensu* Hoyt (1920) may belong to this species (Schneider and Searles, 1991).

Fronds are erect and 2–16 cm tall; young blades are fleshy, older ones firm, strap-shaped, to 2 cm wide, 115–400 µm thick, and on one or more short stalks that are attached by discoid holdfasts. Blades are subdichotomously branched, with undulate, entire, or eroded margins and obtuse tips. The medulla has densely entangled small filaments (2–3 µm in diam.) and larger stellate ganglia (10–40 µm in diam.). The inner cortex has 1–2 layers of ovoid cells 2–40 µm in diam., while the outer cortex has 1–2 layers of round cells 4–8 µm in diam. and irregular in surface view. Gametophytes are dioecious. Spermatangia are 2–3 µm in diam. and in patches. Each support cell has 1–2 carpogonial branches. Carposporophytes are diffuse, near branch tips, elliptical, and to 3 mm in diam. Gonimoblastic filaments are mixed with vegetative tissue. Carposporangia are spherical to irregular and 8–12 µm in diam. Tetrasporophytes are unknown.

Rare, reported from 5 to 37 m off the east coast (Hanisak and Blair, 1988); also known from North Carolina in 4–94 m (type location).

Schneider and Searles 1991: p. 283.

Kallymenia J. Agardh 1842: p. 98–99 (Gr. *kallos*: beauty, beautiful + *humen*: a membrane).

Blades are erect, usually stalked, broad, membranous to firm, and attached by small discoid holdfasts. Medullary filaments and stellate ganglia are in a gel. The cortex has 2–4 layers and small outer pigmented cells. Life histories are isomorphic. Tetrasporangia are cruciately divided and among cortical cells. Gametophytes are dioecious. Spermatangia occur in large irregular sori. Carpogonial branches are 3-celled, solitary or clustered, and from support cells with subsidiary cells. Carposporophytes are nonprocarpial, immersed, and scattered. Gonimoblasts produce carposporangia that are separated by vegetative filaments. Cystocarps are swollen on one or both surfaces of blades because of thickening of the cortex; ostioles are absent.

Key to species (2) of *Kallymenia*

1. Blades crisp, 0.5–1.5 mm in diam., simple or lobed, not perforated . . . . *K. limminghii*
1. Blades soft, to 20 cm or more in diam., lobed, with regular perforations . . . . *K. westii*

Kallymenia limminghii Montagne 1861 ('1860'): p. 173, pl. 11A (L., named for Comte Limminghe, who collected the type material from Guadeloupe, French West Indies). Plate XLIII, fig. 3.

Plants are 1–10 cm tall, foliaceous, and with one to many subspherical membranous blades on terete axes, which are 0.5–1.5 mm in diam., branched or not, 1–3 mm long, and attached by discoid holdfasts. Blades are crisp, lobed to proliferous, strap-shaped, 3–12 mm wide, to 5 cm long, and 150 µm thick; margins are usually strongly dentate. The compact medulla has terete, intertwined filaments (7.5–10.0 µm in diam.) and stellate ganglia in a dense gel. The inner cortex has 2–3 layers of hyaline cells, 25–30 µm in diam.; the outer cortex has 1–2 layers of pigmented cells 5–7 µm in diam.

Common, on limestone in 2–10 m in the Keys and 20–30 m off the west coast; often deep red; also known from Bermuda, the Bahamas, and Caribbean.

Taylor 1960: p. 432; Dawes 1974: p. 128; Littler and Littler 2000: p. 80.

Kallymenia westii Ganesan 1976: p. 169, fig. 1–12, 14, 16–21 (L., named for John A. West, an American phycologist who has studied the life histories and physiology of seaweeds, especially mangrove algae). Plate XLIII, fig. 4, 5.

Synonym. *Kallymenia perforata sensu auct.*, non J. Agardh 1872: p. 9; 1876 (1848–1901): p. 219.
Fronds are delicate, soft to gelatinous, perforated, lobed, 10–50(100+) cm in diam., 200–500 μm thick, and attached by discoid holdfasts. Medullary filaments are in a dense gel, loose, spreading, 4–6(20) μm in diam., and mixed with stellate ganglia 60–90 μm in diam. The inner cortex has 1–2 layers of inflated elongate cells, 15–90 μm in diam., while the outer cortex consists of 1–2 layers of round surface cells, 3–12 μm in diam. Tetrasporangia are ovoid, irregularly cruciate, 10–15 μm in diam., 12–25 μm long, and scattered in the outer cortex. Gametophytes are dioecious. Spermatangia are spherical, 1–3 μm in diam., and in sori on the outer cortex. Female plants are nonprocarpial; carpogonial branches are 2–4(–8)-celled; auxiliary cells are distinct, surrounded by 8–15 subsidiary cells, and form large fusion cells after fertilization. Cystocarps are clustered, immersed, spherical, 0.5–1 mm in diam., and lack involucral filaments and ostioles.

Common, epiphytic, on limestone in the Keys, and in 10–90 m off the west coast, where it may overgrow benthic communities; also known from North Carolina, Bermuda, the Caribbean (type location: Isla de Margarita, Venezuela), Gulf of Mexico, and Canary Islands.

Taylor 1960: p. 432; Dawes 1974: p. 128; Schneider and Searles 1991: p. 285; Littler and Littler 2000: p. 80.

Peyssonneliaceae

Saunders et al. (2004) retained the family in the Gigartinales but suggested that molecular and cell wall chemistry data may raise it to ordinal status. Saunders and Hommersand (2004) also treated the family as having "equivocal taxonomic affinity."

Plants are crustose, fleshy or calcified (throughout or basally), and completely or partially attached by unicellular to multicellular rhizoids or not. Growth is by marginal rows of apical cells of the hypothallus that divide transversely to form vertical, erect, compact cortical filaments. Life histories are isomorphic. Tetrasporangia are cruciately divided and in raised or sunken nemathecia. Gametophytes are monoecious or dioecious. Spermatangia are formed in terminal sori or in raised to immersed nemathecia. Plants are procarpial or nonprocarpial. Carpogonial and auxiliary cell branches are 2(6)-celled and lateral on erect filaments in nemathecia. Connecting filaments (ooblasts) often connect with many auxiliary cells to form broad fusion cells; gonimoblasts are short, chainlike, and from one or more auxiliary cells.

Metapeyssonnelia Boudouresque, Coppejans, and Marcot 1976: p. 288 (L. *meta*: after, associated with + *peysonnelia*: a crustose genus of red algae; thus, like *Peyssonnelia*).

Crusts are orbicular, partially calcified, adhering and conforming to substrata, and have unicellular rhizoids. The hypothallus is monostromatic. Perithallial filaments are erect, firmly united, and with ascending rows of cells arising from each hypothallial cell (as in *Peyssonnelia*). Unlike *Peyssonnelia*, marginal perithallial filaments also arise from the ventral side of hypothallial cells to form an ascending (superior) and descending (inferior) fountainlike branching pattern in nonparallel rows (a generic feature). Tetrasporangia are cruciately divided and in inconspicuous nemathecia. Sexual plants are unknown.

Metapeyssonnelia corallepida Verlaque, Ballesteros, and Antonius 2000: p. 191–192, fig. 2–27 (Gr. _corallion_: a coral + _lepidos_: scale, peel, husk). Plate XLIII, fig. 6–8.

The species is keyed out with *Peyssonnelia* because of strong morphological similarities. Crusts are dark purple to violet, orbicular, and completely adhering by unicellular rhizoids, 18–36 µm in diam. The hypothallus is calcified, branched, and in a spreading pattern visible from below; cells are 11–32 µm in diam. from below and 20–63 µm tall in cross section. Perithallial filaments are laterally united, in ascending branched filaments, and arise from the entire upper surface of a hypothallial cell. Near crust margins, perithallial filaments arise from both ventral and dorsal surfaces of perithallial cells to form fountainlike "mesothallial" regions above and below the hypothallial layer (a generic feature). Cystoliths occur in lower to midparts of crusts and form vertical pillars 2–10 cells thick. Hairs are 18–20 µm in diam and terminal on perithallial filaments. Tetrasporangia are 90–150 µm in diam. and mixed with paraphyses 6–7 cells long and 2.5–6.5 µm in diam. Sexual plants are unknown.

Rare, but may be common and confused with *Peyssonnelia* spp.; known from Grecian Rocks off Key Largo (type location; Velaque et al., 2000) on corals in 0.6–2.2 m, Puerto Rico (Ballantine et al., 2004), and Belize (1.4 m).

Peyssonnelia Decaisne 1841: p. 168, 196, 197 (named for J. A. Peyssonnel, a French naturalist).

Synonym. *Cruoriella* P. L. Crouan and H. M. Crouan 1859: p. 289.

Calcification is variable, ranging from the entire crust, limited to the hypothallus, or absent. Crusts are attached wholly or partly by unicellular or multicellular rhizoids and conform to the surface or are free. Concentric growth rings or radial lines may be present on crust surfaces. The hypothallus has one cell layer. Perithallial filaments are erect and multi-layered. Tetrasporangia, spermatangia, and carpogonial branches are in superficial nemathecia and usually associated with sterile multicellular paraphyses. Both tangential and radial vertical sections are necessary for identification of species.

Key to the species (4) of *Peyssonnelia* and (1) *Metapeyssonnelia*

1. Marginal perithallial filaments arise from dorsal and ventral sides of hypothallial cells to form ascending and descending fountainlike patterns . . . . *M. corallepida*
1. Marginal perithallial filaments arise only from the dorsal side of hypothallial cells 2
    2. Crusts loosely attached throughout, margins often free 3
    2. Crusts firmly attached, although margins may be free 4
3. Crusts lacking radial lines or concentric rings in perithallus . . . . *P. simulans*
3. Crusts with faint radial lines and concentric rings in perithallus . . . . *P. inamoena*
    4. Hypothallial filaments are in small fan-shaped clusters . . . . *P. rosenvingei*
    4. Hypothallial filaments are in straight dichotomously branched rows . . . . *P. rubra*

Peyssonnelia inamoena Pilger 1911: p. 311, fig. 23–29 (L. _in_: not + _amoena_: beautiful, pleasing). Plate XLIII, fig. 9–11.

Crusts are membranous, 2–5 cm in diam, with weak concentric lines; thallus is almost foliose, round, irregular, or lobed, and loosely attached by rhizoids that are unicellular, 3–14 µm in diam. by 120 µm long, and from the center of hypothallial cells. Hypothallial and lower perithallial cells are calcified. Vegetative crusts are to 215 µm thick versus 385 µm in reproductive thalli. The hypothallus has one layer of cells in horizontal, staggered, and parallel rows; cells are elongate to irregularly rectangular, 7–15 µm in diam., 18–50 µm long, and to 26 µm high. Upper walls of hypothallial cells are oblique and produce perithallial cells at angles of 60° to 80° in lower parts. The first perithallial cells are "boot-shaped" and support two filaments, each with 3–6 cells. Surface cells are 6–15 µm in diam. and 3–12 µm long. Tetrasporangia are ovoid, cruciately divided, 27–57 µm in diam., 70–100 µm long, and in raised nemathecia. Spermatangia are in chains and round to angular. Carposporophytes are in raised pigmented nemathecia. Carpogonial and auxiliary branches are 4(6)-celled, lateral on erect filaments, and densely stained. Carposporangia are in terminal chains of 2–4 or at the tips of gonimoblasts, 105 µm in diam.

Rare, in 20–60 m off the east coast; also known from the Carolinas, Bahamas, Caribbean, and Brazil, as well as the Mediterranean, eastern tropical Atlantic, Cameroon (type location: Goss-Batanga; Womersley, 1994), tropical Pacific Ocean, and Tasmania.

Schneider and Searles 1991: p. 255; Littler and Littler 2000: p. 88.

Peyssonnelia rosenvingei F. Schmitz in Rosenvinge 1893: p. 782, fig. 8; 1894: p. 61–62, fig. 8 (named for L. K. Rosenvinge, a Danish phycologist who studied the seaweeds of Denmark). Plate XLIII, fig. 12–14.

Synonym. *Peyssonnelia dubyi sensu auct.*, non P. Crouan and H. Crouan 1844: p. 368, pl. 11, fig. 6–10.

Crusts are strongly adherent, radially lined, slightly colored, to 20 cm in diam., 0.5 mm thick, with undulate margins, calcified below, and attached by rhizoids. The hypothallus is monostromatic, fan-shaped, and with radiating filaments of large cells, 3–4 diam. long. Perithallial filaments are vertical, compact, simple or sparingly dichotomously branched, and with 6–30 cells per filament; lower cells are 20–40 µm in diam. and upper ones smaller and isodiametric. Cystocarps are in slightly elevated pigmented nemathecia; monosporangia are large and solitary or in chains.

A questionable record; in shallow waters off the Dry Tortugas (Taylor, 1928); also known from Greenland to Massachusetts, Bermuda, the Caribbean, and Brazil, as well as the eastern North Atlantic and Madeira Archipelago.

Taylor 1960: p. 373.

Peyssonnelia rubra (Greville) J. Agardh 1851 (1848–1901): p. 502 (L. _rubra_: red). Plate XLIII, fig. 15–17.

Basionym. *Zonaria rubra* Greville 1827b: p. 340–341, pl. 3, fig. 3.

Most Caribbean and American plants correspond to *Peyssonnelia inamoena* (Schneider and Reading, 1987; Schneider, 2003). The presence of calcified bodies (crystoliths) in surface cells distinguishes this species (Abbott, 1999).

Crusts are membranous, 1–3 cm in diam., and grow irregularly with rounded, overlapping lobes; the lower part is lightly calcified and attached by short rhizoids throughout. The hypothallus is mostly monostromatic, with long, dichotomously branched filaments not in parallel rows; cells are oblong, 10–20 µm in diam., and 12–16 µm high; marginal tip cells are tall and radially elongate. Perithallial filaments are erect and 6–10 cells long; lower cells are taller than wide and upper ones shorter than wide (10–14 µm in diam.); some surface cells are enlarged and have calcified bodies (crystoliths). Tetrasporangia are cruciately divided, 15–30 µm in diam., 40–60 µm long, and in terminal or lateral sori. Carpogonial and auxiliary cells occur on lateral branches of 4–5 cells each; carposporangia are 30–40 µm in diam. and in branched rows. Spermatangia are terminal.

Rare, with questionable records from the west coast (10m) and Keys (33m); also known from North Carolina, Bermuda, and Puerto Rico, as well as the British Isles to Spain, the Mediterranean (type location: Ionian Islands, Greece; Abbott, 1999), eastern tropical Atlantic, Red Sea, Indo-Pacific, and Australia.

Taylor 1960: p. 371; Dawes 1974: p. 122.

Peyssonnelia simulans **Weber-van Bosse in Børgesen 1916: p. 142–144, fig. 138, 140, 148 (L. *simulans*: similar to). Plate XLIII, fig. 18–20.**

Crusts are membranous, basally calcified, round to irregular, without concentric lines, 1–3 cm broad, loosely adherent, and attached by unicellular rhizoids that are 10–13 μm in diam. by 65 μm long. Vegetative thalli are to 200 μm thick and reproductive ones to 510 μm. The hypothallus is monostromatic with filaments in horizontal, staggered, and parallel rows; cells are elongate, irregularly rectangular, 5–11 μm in diam., 20–50 μm long, and 20–45 μm thick. Upper hypothallial cell walls are oblique and produce perithallial cells at angles of 60–70° in lower parts. The first perithallial cell is "boot-shaped" and produces two erect filaments of 3–6 cells; surface cells are irregularly arranged, angular, 6–20 μm in diam., and 6–22 μm long. Tetrasporangia are cruciately divided, 12–30 μm in diam., 40–80 μm long, occur laterally on filaments in raised nemathecia, and have sterile, slender, forked hairs 80–120 μm long. Carpospores occur in raised pigmented nemathecia, 200 μm thick. Carposporangia are in chains (to 100 μm long) of 2–3 spherical cells that arise from transverse divisions of gonimoblastic filaments. Spermatangia are unknown.

Uncommon, on subtidal limestone (18–46 m) on the west coast; also known from Bermuda, North Carolina, the Lesser Antilles (syntype locations, Walter Island, St. Cruz Bay, and Ram's Head, St. Thomas; Silva et al., 1996), Brazil, and Indian Ocean.

Taylor 1960 p. 372; Schneider and Searles 1991: p. 257; Littler and Littler 2000: p. 90.

Phyllophoraceae

Synonym. Petrocelidaceae

Plants are erect, crustose, and free-living or small tuberous to lobed parasites of red algae. Erect plants are bushy, wiry to cartilaginous, simple, and branched or lobed; axes are cylindrical to flat, often proliferous, and attached by discoid holdfasts or crusts. Plants are multiaxial with cortical filaments radiating at right angles to medullary ones. Life histories are heteromorphic, isomorphic, or modified. Tetrasporangia are cruciately divided (if known), in chains of 3 or more, and from cortical cells or carposporangia (carpotetrasporangia) that are in swollen sori or pits. Gametophytes are monoecious or dioecious. Spermatangia are terete and in embedded or surficial sori. Carpogonial branches are 3-celled and from support-auxiliary cells; fusion cells are absent. Auxiliary cells form gonimoblastic filaments that branch toward the medulla. Carposporophytes lack special involucral filaments; carpospores are in chains. Cystocarps are embedded, swollen on one or both surfaces, and marginal or on proliferations.

Gymnogongrus Martius 1833: p. 27 (Gr. *gymno*: naked + *gongrus*: a tubercular swelling or excrescence).

Lewis and Womersley (1994) noted that the type species *Gymnogongrus griffithsia* had biphasic erect gametophytes and tetrasporophytes as pustules ("tetrasporoblasts") on females. Bird and McLachlan (1992) described upright gametophytes and crustose sporophytes in *Gymnogongrus* sp. from the Canadian Maritime Provinces. However, the Canadian plants may belong to *Ahnfeltiopsis*, a cystocarpic taxon (Silva and DeCew, 1992; Masuda, 1993).

Plants are bushy with dichotomous to irregular branching; branches are terete or flattened and have anatomical features typical of the family. Gametophytes are dioecious. Spermatangia are in superficial sori. Carposporophytes are immersed or swell one or both sides because of a thickened cortex; ostioles are absent. Carpotetrasporangia arise from meiotic cruciate divisions of terminal carposporangia on gonimoblastic filaments and form pustules ("tetrasporoblasts") on the surface of females.

Gymnogongrus griffithsiae (Turner) Martius 1828 (1828–1834): p. 27; 1833: p. 27 (named for Amelia W. Griffiths, a British phycologist). Plate XLIII, fig. 21–23.

Basionym. *Fucus griffithsiae* Turner 1808 (1807–1808): p. 80, pl. 37 (as '*griffithsii*').

Plants are wiry, turflike, 1–6 cm tall, and attached by discoid pads. Axes are terete to slightly compressed, 0.3–0.8 mm in diam., dichotomously, irregularly, or pinnately branched and with acute tips. Medullary cells are thick-walled, elongate, and to 30 µm in diam. Cortical cells are in radiating series, elongate, and fused; inner cells are to 8 µm in diam. and surface cells rounded and to 2 µm in diam. Only female plants are known and have raised pustules 1–3 mm long that produce carpotetrasporangia that are elliptical, often partially divided, 7–20 µm in diam. by 10–42 µm long, and in chains on outward gonimoblast filaments.

Uncommon, on intertidal coquina at Marineland and worm reefs at Vero Beach; also known from southern Massachusetts to the Carolinas, Caribbean, Brazil, and Uruguay, as well as the eastern North and tropical Atlantic (type location: Devon, England; Lewis and Womersley, 1994), Mediterranean, Vietnam, Australia, and Fuegia Island, Antarctica.

Taylor 1960: p. 470; Schneider and Searles 1991: p. 311.

Solieriaceae

Synonym. Wurdemanniaceae Taylor 1960

Based on molecular studies (Saunders et al., 2004) the monotypic Wurdemanniaceae was placed in the family Solieriaceae.

Plants are firm to cartilaginous, flexible, and attached by discoid or branching holdfasts or rhizoidal clusters. Parasitic plants are padlike. Erect axes are terete, compressed, or foliose with limited, irregular, alternate, or pinnate branching. Plants are uniaxial or multiaxial; medullary filaments are loose to compact and the cortex has radiating files of smaller cells. Life histories are isomorphic; tetrasporangia are zonately divided and arise from outer cortical cells near branch tips. Gametophytes are dioecious. Spermatangia are in slightly raised sori. Plants are nonprocarpial, with carpogonial branches 3(rarely 4)-celled and auxiliary cells on supporting cells in the inner cortex. Cells around the auxiliary cells divide before gonimoblast formation and one initial grows inward to the medulla. Carposporophytes arise from fusion cells or small-celled vegetative tissue. Cystocarps are embedded or swollen, they occur on one or both surfaces, and may be marginal, scattered, or in papillae. Pericarps and ostioles are present.

Agardhiella F. Schmitz in Schmitz and Hauptfleish 1896 ('1897'): p. 371; F. Schmitz 1889: p. 441 (named for J. G. Agardh, a Swedish phycologist and taxonomist who published extensively during the 19th century).

Synonym. *Neoagardhiella* M. J. Wynne and W. R. Taylor 1973: p. 102–103.

See Silva et al. (1996) on the nomenclature of *Agardhiella* and *Solieria*. Guiry and Nic Dhonncha (2006) listed *Neoagardhiella* as a distinct genus, with *N. baileyi* (Kützing) Wynne and Taylor as the type species and placed *Agardhiella* in the Areschougiaceae, not the Solieriaceae.

Plants are firm, with terete to flattened axes, radially alternate to pinnately branched, and attached by discoid holdfasts. Medullary filaments are loose to compact and interwoven; inner cortical cells are hyaline, small, and the outer ones pigmented. Tetrasporangial and spermatangial sori are in raised sori. Carpogonial branches are 3(4)-celled; trichogynes are bent backward and gonimoblasts extend inward from the inner cortex. Carposporophytes have a mixture of small and large vegetative cells surrounded by gonimoblastic filaments that produce distal chains of 3–4 carposporangia that are mixed with sterile involucral strands. Cystocarps are embedded, swollen on one surface, or on axial margins; pericarps are thick, ostiolate, and have involucral filaments.

Key to the species (3) of *Agardhiella*

1. Axes cylindrical to compressed, but not flat 2
1. Axes foliaceous with flat blades.......... *A. floridana*
   2. Branching mostly in one plane; larger axes compressed .... *A. ramosissima*
   2. Branching radial; all axes cylindrical .... *A. subulata*

Agardhiella floridana (Kylin) P. W. Gabrielson *ex* Guimarães and Oliveira 1996: p. 656, fig. 1–18 (L. *floridana*: of or from the state of Florida). Plate XLIV, fig. 1, 2.

Basionym. *Meristotheca floridana* Kylin 1932: p. 29, fig. 6.

Synonym. *Meristotheca gigartinoides* Joly and Ugadim *ex* Joly et al. 1965: p. 82.

Blades are pink to rose red, fleshy, oblong-elliptical, 12–100 cm long, 0.5–30.0 cm wide, 0.5–1.4 mm thick above to 2.8 mm below, and with stalks 1.5–4.0 mm in diam. attached by discoid holdfasts. Medullary filaments are interconnected, loose, 2–10 µm in diam., and have many rhizoids. The inner cortex is 2–4 layered, pseudoparenchymatous, and with rhizoids 400 µm below the apex; cells are thick-walled, with secondary pits and starch grains, and to 300 µm in diam. The outer cortex has 2–4 layers of rounded pigmented cells; surface cells are 10–25 µm in diam. by 4–13 µm long. Tetrasporangia are zonately divided, 44–52 µm long, 20–35 µm in diam., and occur laterally on outer cortical cells. Spermatangia are scattered, 3–4 µm in diam., and from outer cortical cells. Carpogonial branches are 3-celled, grow inward, are in papillae, and produce an outwardly bent trichogyne. Auxiliary cell complexes arise from inner cortical cells that receive ooblast filaments of fertilized carpogonia. Up to four ostiolate cystocarps (500–1500 µm in diam.) occur on papillae that are 1–5 by 1–15 mm. Carposporangia are 14–27 µm in diam.

Uncommon, with one collection in 10 m off Tampa Bay and adrift at the Keys; also known from North Carolina, Bermuda, the Bahamas, Caribbean, and Brazil.

Taylor 1960: p. 460; Dawes 1974: p. 138.

Agardhiella ramosissima (Harvey) Kylin 1932: p. 17 (L. *ramos*: branch + *issima*: much, highly). Plate XLIV, fig. 3, 4.

Basionym. *Chrysymenia ramosissima* Harvey 1853: p. 190, pl. XXX, fig. b.

Synonyms. *Agardhiella ramosissima* var. *dilatata* J. Agardh 1855: p. 85.

Agardhiella ramosissima var. *harveyana* J. Agardh 1855: p. 85.

Neoagardhiella ramosissima (Harvey) M. J. Wynne and W. R. Taylor 1973: p. 103.

Guimarães and Oliveria (1996) did not recognize var. *dilatata* and var. *harveyana* (see Taylor, 1960). Guiry and Nic Dhoncha (2005) spelled all of the above as "*ramossissima*."

Plants are fleshy to firm, 5–30 cm tall, rose red to brown, pinnately alternate to oppositely branched to 3–4 orders, and attached by small discoid holdfasts. Main axes are flat to compressed, strap-shaped, and 2–6(15) mm wide. Branchlets are subterete, spindle-shaped, and taper toward the bases as well as the tips. Medullary filaments are loose to densely entangled and 2–10 µm in diam. The inner cortex has 3–4 layers of large, hyaline, thick-walled cells 50–250 µm in diam.; the outer cortex has 1–2 layers of elongate, pigmented cells 7–12 µm in diam. Surface cells are spherical and 7–13 µm in diam. Tetrasporangia are ovoid, zonately divided, 30–50 µm in diam. by 45–80 µm long, and in patches in the outer cortex. Gametophytes are dioecious. Carposporangia are ovoid, 17–38 µm in diam., 20–45 µm long, and in terminal chains of 3–4 cells. Gonimoblastic filaments are short and in a central mass of vegetative tissue. Cystocarps are spherical, 1–1.7 mm in diam., and have an ostiolate, thick-walled pericarp.

Common, in winter and spring, on jetties, as drift in estuaries, and to 30 m on the west coast and in the Keys; also known from North Carolina, Bermuda, the Caribbean, Gulf of Mexico, and Brazil.

Taylor 1960: p. 457; Dawes 1974: p. 133; Schneider and Searles 1991: p. 294; Littler and Littler 2000: p. 94.

Agardhiella subulata (C. Agardh) Kraft and M. J. Wynne 1979: p. 329 (L. *subulata*: awl-shaped). Plate XLIV, fig. 5, 6.

Basionym. *Sphaerococcus subulatus* C. Agardh 1822 (1820–1828): p. 328; 1824: p. 239.

Synonyms. *Agardhiella tenera* (J. Agardh) F. Schmitz in F. Schmitz and Hauptfleish 1896 ('1897'): p. 371.

Agardhiella baileyi (Harvey in Kützing) W. R. Taylor in W. R. Taylor and Rhyne 1970: p. 13.

Neoagardhiella baileyi (Harvey in Kützing) M. J. Wynne and W. R. Taylor 1973: p. 101, fig. 7–11.

Guiry and Nic Dhonncha (2006) accepted *Neoagardhiella* as a distinct taxon and cited its type species as *N. baileyi*.

Plants are rose red, fleshy to firm, and to 45 cm tall; axes are terete, 0.5–5 mm in diam., and attached by discoid holdfasts. Branching is radial, alternate, and to several orders; branchlets are subpinnate or secund, often tapering to bases and to rounded tips. Medullary filaments are loosely to densely entangled; longitudinal filaments are 10–15 µm in diam. and lateral ones 8–12 µm. The inner cortex has 2–4 layers of hyaline thick-walled cells, 60–250 µm in diam., while the outer cortex has 1–2 layers of elongate pigmented cells 5–20 µm in diam. Tetrasporangia are ovoid, zonately divided, 20–45 µm in diam., 44–50 µm long, and in the outer cortex. Gametophytes are dioecious. Gonimoblasts are short and in a diffuse central vegetative tissue. Carposporangia are terminal, in chains of 3–4, ovoid, 15–30 µm in diam., and 35–45 µm long. Cystocarps are spherical, 1–2 mm in diam., and have ostiolate, thick-walled pericarps.

Common, on limestone, mangrove roots, adrift, and to 30 m on all coasts; also known from the Gulf of St. Lawrence (Canada) to the Carolinas, Caribbean, Gulf of Mexico, and Brazil, as well as France, Italy (introduced), Mauritius, India, and Sri Lanka.

Taylor 1960: p. 456; Dawes 1974: p. 135; Schneider and Searles 1991: p. 295.

Eucheuma J. Agardh 1847: p. 16 (Gr. *eu*: true, well formed + *cheuma*: molten, poured).

Plants are erect or prostrate, terete or flattened, strongly cartilaginous, radially to bilaterally branched, often with papillae, and attached by discoid holdfasts and rhizoidal clusters from branches (Pacheco-Ruíz et al., 2005). The medulla is compactly filamentous or a mix of large cells and thin, densely intertwined rhizoidal filaments. The inner cortex has large, hyaline cells and an outer cortex of small, pigmented ones. Tetrasporangial and spermatangial sori occur in the cortex. Carpogonial branches are 3-celled with trichogynes bent outward. Gonimoblastic filaments are from inner cortical cells and bend inward. Carposporophytes are from large fusion cells and surrounded by short gonimoblasts. Carpospores are terminal and mixed with sterile involucral filaments. Cystocarps are embedded and swell one surface. Pericarps are ostiolate with thick walls.

Eucheuma isiforme (C. Agardh) J. Agardh 1847: p. 16, as '_isiformis_'(L. _isis_: Egyptian goddess of motherhood and fertility, wife of Osiris + _formis_: shape, form). Plate XLIV, fig. 7, 8.

Basionym. *Sphaerococcus isiformis* C. Agardh 1822 (1820–1828): p. 271.
Synonym. *Hypnea wurdemannii* Harvey 1853: p. 119, *nom. nudum*

Plants are erect, tough, cartilaginous, 30–70 cm tall, and gold to pale yellow; axes are terete, 2–8 mm in diam., irregularly and radially branched 1–5 times, and attached by discoid holdfasts and secondarily by rhizoidal clusters from branches (Pacheco-Ruíz et al., 2005). Surfaces are smooth or densely covered with spines or nodules. Medullary filaments are compact, longitudinal, and 3–6 µm in diam. The cortex has 2–3 layers; its inner cells are hyaline, spherical to stellate, thin-walled, ovoid, and 12–80 µm in diam. and grade into smaller outer cortical cells. Surface cells are pigmented, ovoid, 5–7 µm in diam., and have thick outer walls (to 20 µm). Tetrasporangia are scattered in the cortex of spines and branches, are ovoid, zonately divided, 12–20 µm in diam., and 35–55 µm long. Gametophytes are dioecious. Carpogonial branches occur in nemathecia on the upper portions of papillae or among surface cells. Carposporangia are ovoid, 9–14 µm in diam., and in biseriate series on gonimoblasts around central fusion cells. Cystocarps are in the tips of papillae, stalked, and 0.5–2.0 mm in diam. A "Bahia Honda form" (Dawes et al., 1974) has densely branched prostrate axes with many secondary attachments.

var. _denudatum_ D. P. Cheney 1988: p. 214–217, fig. 7 (L. _denudatum_: naked). Plate XLIV, fig. 9, 10.

Synonym. *Eucheuma isiforme* (C. Agardh) J. Agardh var. *denudatum* D. P. Cheney in Moon and Dawes 1976: p. 165, *nom. nudum*; non *Eucheuma nudum* J. Agardh 1852 (1848–1901): p. 625.

Axes are smooth, lax, slender, deep red under low light, often procumbent, initially attached by small discoid holdfasts, and secondarily by rhizoidal mats from prostrate axes.

Common, all forms become deep red under low light; the species grows in the Keys on limestone to 5 m, the variety in 10–50 m on the east and west coasts, and the "Bahia Honda form" in 0.5–2.0 m on the north side of Bahia Honda Key; the variety and species are known from North Carolina, Bermuda, the Caribbean, Gulf of Mexico, and Philippines. Type location of the species is "In mari Indiae occidentalis" (Silva et al., 1996) and the variety is Riomar Reef, Vero Beach (Cheney, 1988).

Taylor 1960: p. 459; Dawes 1974: p. 135, 137; Schneider and Searles 1991: p. 297; Littler and Littler 2000: p. 94.

Meristotheca J. Agardh 1872: p. 36 (Gr. *meristos*: divided + *-theca*: cover, case, or container).

Synonym. *Meristiella* D. P. Cheney in Gabrielson and Cheney 1987: p. 482.

Based on molecular, phylogenetic, and morphological analyses, Fredericq et al. (1999) suggested merging *Meristiella* into *Meristotheca*. Molecular and anatomical studies by Faye et al. (2004b) supported these findings, with the former genus reduced to synonymy with the latter. According to Guimarães and Oliveira (1996), *Meristotheca* is distinct from *Eucheuma* by having nemathecia and forming an auxiliary cell complex after fertilization.

Plants are flat, linear to bladelike, erect to 25 cm tall or prostrate, and attached by a main discoid holdfast plus numerous secondary discs. Axes are multiaxial, with a broad filamentous medulla of mostly unbranched rhizoids. Inner cortical cells are large and often stellate. Carpogonial branches are 3-celled and the basal cell may have a sterile cell. Fertilization results in the production of an unbranched connecting filament; gonimoblast initials are formed laterally from the auxiliary cell. Carposporophytes have a central placenta or a mixture of fused gametophytic and gonimoblastic cells surrounded by carpospores, gonimoblastic filaments, and narrow sterile gonimoblasts that fuse with cells of encircling filaments. Carposporangia are single, terminal, and in a thick, ostiolate, and projecting pericarp. Gametophytes are monoecious. Spermatangial mother cells occur at the surface of carpogonial nemathecia. Tetrasporangia are cut off laterally.

Meristotheca gelidium (J. Agardh) Faye and Masuda in Faye et al. 2004b: p. 254 (L. *gelidus*: gelatinous, congealed; also the red alga *Gelidium* L.). Plate XLIV, fig. 11–13.

Basionym. *Sphaerococcus gelidium* J. Agardh 1841: p. 17.

Synonyms. *Eucheuma gelidium* (J. Agardh) J. Agardh 1847: p. 16; 1852 (1848–1901): p. 627: p. 602.

Chrysmenia? *acanthocladia* Harvey 1853: p. 191.

Eucheuma echinocarpum Areschoug 1854: p. 349.

Mychodea guadelupensis P. Crouan and H. Crouan in Schramm and Mazé 1865: p. 10; P. Crouan and H. Crouan in Mazé and Schramm 1878 (1870–1877): p. 164.

Mychodea schrammii P. Crouan and H. Crouan in Schramm and Mazé 1865: p. 10; P. Crouan and H. Crouan in Mazé and Schramm 1878 (1870–1877): p. 163.

Eucheuma acanthocladum (Harvey) J. Agardh 1876 (1848–1901): p. 600.

Mychodea pennata P. Crouan and H. Crouan in Mazé and Schramm 1878 (1870–1877): p. 165.

Mychodea polyacantha P. Crouan and H. Crouan in Mazé and Schramm 1878 (1870–1877): p. 164.

Nemastoma jardinii J. Agardh var. *antillarum* P. Crouan and H. Crouan in Mazé and Schramm 1878 (1870–1877): p. 152.

Eucheuma schrammii (P. Crouan and H. Crouan) J. Agardh 1892: p. 124.

Meristiella echinocarpa (Areschoug) D. P. Cheney and P. W. Gabrielson in P. W. Gabrielson and D. P. Cheney 1987: p. 483, fig. 6.

Meristiella gelidium (J. Agardh) D. P. Cheney and P. W. Gabrielson in Gabrielson and Cheney 1987: p. 483, fig. 1–4, 7–26.

Meristiella schrammii (P. Crouan and H. Crouan) D. P. Cheney and P. W. Gabrielson in Gabrielson and Cheney 1987: p. 483.

Meristotheca echinocarpa (Areschoug) Faye and Masuda in Faye et al. 2004b: p. 254.

Meristotheca schrammii (P. Crouan and H. Crouan) Faye and Masuda in Faye et al. 2004b: p. 254.

Gabrielson and Cheney (1987) and Littler and Littler (2000) separated three species (*Meristiella gelidium*, *M. echinocarpa*, and *M. schrammii*) using frond width and main branch dominance (Wynne 1998a), while Schneider (2003) recognized *M. gelidium* and *M. schrammii*. Guimarães and Oliveria (1996) suggested that *Meristiella* (= *Meristotheca*) *gelidium* is variable and includes "*echinocarpum*" and "*schrammii*" types. We follow Guimarães and Oliveria (1996) based on overlapping morphologies of deep and shallow water Florida material and interpret *M. gelidium* as a polymorphic species.

Plants are gelatinous when young, tough to leathery when mature, pale yellow to deep red, with prostrate (mound-forming) and erect fronds to 40 cm tall, and attached by fleshy fibrous holdfasts. Main axes are flattened, 2–10 mm wide, and strap-shaped in wave-exposed sites or cylindrical in quiet or deep water sites. Branching is irregular, lobed to palmate, and mostly in one plane; branchlets arise from margins and are short, spurlike, cylindrical or flat, and 4–36 mm long. The medulla has three components: spherical cells 80–180 µm in diam., loosely intertwined filaments 4–15 µm in diam. by 300–400 µm long, and thick-walled rhizoids <2 µm in diam. The cortex has 3–9 layers; inner cells are 150–300 µm in diam. and mixed with stellate cells 80–100 µm in diam.; surface cells are ovoid, pigmented, and 5–20 µm in diam. Tetrasporophytes and female gametophytes are isomorphic; male plants are unknown. Tetrasporangia are ovoid, zonately divided, 10–30 µm in diam. by 35–50 µm long, and lateral on outer cortical cells. Carpogonial branches and auxiliary cells are on separate branches in the cortex (nonprocarpial). Fertilized carpogonia produce ooblasts that connect with auxiliary cells. Cystocarps are ostiolate, spherical, and 0.8–1.0 mm in diam. Carposporangia are terminal, ovoid, and 20–25 µm in diam. by 30–40 µm long.

Common, on limestone in 1–3 m in the Keys and 20–40 m on the east and west coasts; also known from North Carolina (30–32 m), Bermuda, the Caribbean (type location: West Indies), and Gulf of Mexico.

Taylor 1960: p. 458, 459; Dawes 1974: p. 135; Schneider and Searles 1991: p. 299; Littler and Littler 2000: p. 96, 98.

Sarcodiotheca Kylin 1932: p. 15 (L. *sarcodes*: fleshlike + *theca*: a cover, container).

Plants are firm and attached by discoid holdfasts; axes are narrow to wide, flat, and dichotomously to irregularly branched or terete with radial or alternate branching. Medullary filaments are loose or compact; inner cortical cells are large, thin-walled, and grade into small pigmented surface ones. Tetrasporangial and spermatangial sori are scattered. Carpogonial branches are 3–4-celled; trichogynes bend inward. Gonimoblastic filaments grow inward from the inner cortex. Carposporophytes are compact and have a central area of large cells surrounded by gonimoblastic filaments (to 8 cells long) that produce terminal carposporangia mixed with involucral filaments. Cystocarps protrude on one surface and the ostiolate pericarps have thick walls.

Sarcodiotheca divaricata W. R. Taylor 1945: p. 223–224, pl. 74, fig. 1 (L. *divaricata*: wide spreading, diverging). Plate XLIV, fig. 14, 15.

Synonyms. *Chrysymenia dichotomo-flabellata sensu* Collins, Holden and Setchell 1897 (1895–1919), fasc. VIII, #385; non *Chrysymenia dichotoma flabellata* P. *et.* H. Crouan in Mazé and Schramm 1878 (1870–1877): p. 162.

Sarcodiotheca tenuis W. R. Taylor 1945: p. 223, pl. 74, fig. 2.

Main axes are 7–15 cm tall, flat, 300–650 µm thick, widely dichotomously branched (to several orders), and with rounded forks and acute tips; stipes are short, 2–7 mm in diam., and attached by discoid holdfasts. Medullary filaments are loose and 2–20 µm in diam. The inner cortex has 2–3 layers; cells are elongate, thin-walled, 12–60 µm in diam., and fused; outer cells are pigmented, thick-walled, and 5–20 µm in diam. Tetrasporangia are ovoid, zonately divided, 24–30 µm, 30–40 µm long, and in the outer cortex. Spermatangia are unknown. Cystocarps are pyriform, ostiolate, and marginal or on stalks. Only tetrasporic plants are known from the Atlantic (Schneider and Searles 1991).

Rare, from 24–53 m off the east coast (Hanisak and Blair, 1988); also known from North and South Carolina (35–45 m), Colombia, the Galapagos (type location: Santa Cruz), and Canary Islands.

Schneider and Searles 1991: p. 302.

Solieria J. Agardh 1842: p. 156 (named for A. J. J. Solier, a French phycologist who studied the reproductive biology of several seaweeds, including the green alga *Derbesia*).

See Silva et al. (1996) and above on the taxonomy of *Solieria* and *Agardhiella*.

Plants are erect and firm; axes are cylindrical to compressed, radially, alternately or pinnately branched, and attached by discoid holdfasts. Medulla may appear hollow because of loosely arranged filaments. The inner cortex consists of thin-walled cells that grade into smaller pigmented surface cells. Tetrasporangial and spermatangial sori are scattered. Carpogonial branches are 3(4)-celled and trichogynes bend inward. Carposporophytes grow from a large central fusion cell. Gonimoblastic filaments are short and grow inward from inner cortical cells. Carposporangia are terminal and interspersed with involucral filaments. Cystocarps are embedded and project on one surface or are on the margins of blades; pericarps are thick walled and ostiolate.

Solieria filiformis (Kützing) P. W. Gabrielson 1985: p. 278, fig. 1b, 1d, 2 (L. *filiformis*: threadlike, like a thread). Plate XLIV, fig. 16, 17.

Basionym. *Euhymenia filiformis* Kützing 1863b: p. 15.

Synonyms. *Agardhiella tenera sensu auct. pro parte*, non (J. Agardh) F. Schmitz in F. Schmitz and Hauptfleish 1896 ('1897'): p. 371.

Solieria tenera sensu auct., non (J. Agardh) M. J. Wynne and W. R. Taylor 1973: p. 100, fig. 1–6.

Plants are bushy, fleshy to firm, 5–30 cm tall, and attached by haptera-like holdfasts. Axes are slender, 0.5–1 mm in diam., often constricted at forks, alternately radial to irregularly branched above, and with dichotomous branching below. The medulla may appear hollow or loosely organized with a few longitudinal filaments 10–20 µm in diam. and descending rhizoids 2–12 µm in diam. The inner cortex has 3–4 layers; inner cells are ovoid and 60–160 µm in diam.; middle cells are small, pigmented, 4–10 µm in diam., and 9–15 µm long; surface cells are irregular to round and 5–15 µm in diam. Tetrasporangia are ovoid, zonately divided, 15–35 µm in diam., 30–55 µm long, and in the outer cortex. Gametophytes are dioecious. Spermatangia are ovoid, 2–3 µm in diam., 3–4 µm long, and in scattered patches in the outer cortex. Carposporangia are spherical to ovoid, 10–20 µm in diam., 50 µm long, and terminal on short gonimoblastic filaments that surround a large fusion cell. Cystocarps are globose, 0.5–1 mm in diam., and have ostiolate thick-walled pericarps.

Common, adrift in estuaries and on limestone at wave-exposed sites (to 30 m) on all coasts; also known from North Carolina southward, the Caribbean (type location: West Indies; Womersley, 1994), Gulf of Mexico, and Brazil, as well as the eastern North Atlantic, Mediterranean, Canary Islands, and West Africa.

Taylor 1960: p. 456; Dawes 1974: p. 135; Schneider and Searles 1991: p. 304; Littler and Littler 2000: p. 98.

Wurdemannia Harvey 1853: p. 245 (named for Dr. Wurdemann, an American naturalist who collected in Florida and sent specimens to Harvey).

Synonym. *Pseudogelidium* Schiffner 1933: p. 292.

See Saunders et al. (2004) and comments above regarding generic placement within the Solieriaceae.

Plants are clustered and often form a dense wiry turf; erect axes are slender and terete, with sparse irregular, alternate, or pinnate branching; prostrate axes are attached by rhizoidal clusters. The medulla is compact, with large, thick-walled elongate cells in rows. Cortical cells are in irregular patterns, radially elongate, polyhedral in surface view, and thick-walled. Tetrasporangia are zonately divided and in swollen tips; cystocarpic plants are uncommon.

Wurdemannia miniata **(Sprengel) Feldmann and Hamel 1934: p. 544, fig. 9–11; 1936: reprint p. 260 (L. *miniata*: vermillion, cinnabar red). Plate XLV, fig. 1–4.**

Basionym. *Sphaerococcus miniatus* Sprengel 1827: p. 340, *nom. novum.*

Synonym. *Wurdemannia setacea* Harvey 1853: p. 245, 246.

Plants are dull red to pink, bushy, tufted to turflike, and to 3 cm tall; axes are entangled and prostrate with rhizomelike branches attached by rhizoidal holdfasts. Erect axes are 150–400 µm in diam.; branching is limited, irregular, at right angles to axes, and branchlets taper to the tips. Medullary filaments are longitudinal, thick walled, and to 250 µm in diam.; cortical cells are ovoid and surface cells 6–12 µm in diam., thick walled, and pigmented. Tetrasporangia are ovoid, zonately divided, 23–26 µm in diam., 48–55 µm long, and in slightly swollen tips. Spermatangial sori also occur in swollen tips.

Uncommon, on intertidal limestone at Point of Rocks, mangrove roots on the west coast, and in shallow waters in the Keys; also known from Bermuda, North Carolina, the Caribbean, Gulf of Mexico, and Brazil, as well as the eastern North and tropical Atlantic (type location: Montpellier, France; Silva et al., 1996), Mediterranean, Indo-Pacific, and Australia.

Taylor 1960: p. 361; Dawes 1974: p. 120.

Sedis incertae: Acrosymphytaceae

Lindstrom (1987) removed *Acrosymphyton* from the Dumontiaceae because of its unique female reproductive structures and postfertilization development and recognized the family. Wilce and Davis (1984) expressed similar opinions using morphological data. Tai et al. (2001) gave molecular support for the separation and suggested resurrecting J. Agardh's 1876 (1848–1901) family Dudresnayaceae for *Dudresnaya*, *Acrosymphyton*, and other taxa. Withall and Saunders (2006) have raised this family to ordinal level because it failed to join other families in the original Gigartinales.

Plants are erect, gelatinous, alternately and radially branched, and attached by discs. Axes are terete, uniaxial, and have a medulla of longitudinal filaments. The cortex is pigmented and has radial whorls of di- or trichotomous filaments. Tetrasporangia are unknown. Spermatangia occur whorled on the last 2–3 cells of cortical filaments. Auxiliary cell and carpogonial branches are moniliform. Carpogonial branches are curved, pinnately branched, and occur laterally on lower forks of cortical filaments. Carpogonial trichogynes are basally coiled and extend beyond the cortex. Auxiliary cells are enlarged and terminal on cortical branch filaments. Cystocarps are embedded in the cortex and lack a pericarp.

Acrosymphyton L. G. Sjöstedt 1926: p. 8 (Gr. *acro*: apex + *sym*: joined + *phyton*: a plant)

The genus has the same features as the family.

Acrosymphyton caribaeum **(J. Agardh) L. G. Sjöstedt 1926: p. 9 (L. *caribaeum*: of or from the Caribbean). Plate XLV, fig. 5.**

Basionym. *Calosiphonia caribaea* J. Agardh 1899: p. 84.

Plants are bushy, to 16 cm tall, alternately branched 5–6 times, and with attenuate branchlets less than 1 mm in diam.; main axes are terete, 5–8 mm in diam., and attached by discoid holdfasts. Axial filaments are clustered and 60–130 µm in diam.; outer filaments are pigmented, whorled, branched 5–7 times, and arise from the midregion of axial cells. Surface cells are moniliform, ovoid, and 6.5–10 μm in diam. Gametophytes are monoecious or dioecious. Spermatangia are clustered in the outer cortex. Carpogonial branches are 7–9 cells long, curved, and on lateral branchlets, 1–4 cells long. Auxiliary cells are 12–14 μm in diam. and on the tips of filaments, 5–12 cells long. Cystocarps are 45–90 μm in diam.

Rare, growing on limestone (12 m) in the Dry Tortugas; also known from Bermuda, the Bahamas, and Lesser Antilles.

Taylor 1960: p. 366; Littler and Littler 2000: p. 74.

GRACILARIALES

See Fredericq and Hommersand (1989a) regarding this order and Gurgel and Fredericq (2004) on generic designations.

Plants are erect, wiry to firm, branched, and with one or more axes from discoid or crustose holdfasts. A few species are lobed or hemispherical colorless parasites on other related genera (adelphoparasites). Axes are multiaxial and pseudoparenchymatous. Medullary cells are large, isodiametric, hyaline, and grade into small cortical cells. The cortex has a few to many layers and is compact; surface cells are small and pigmented. Hyaline hairs may be formed under low nutrient conditions. Life histories are isomorphic. Tetrasporangia are cruciately divided and on intercalary cortical cells. Spermatangia occur in surface sori or pits. Carpogonial branches are 2–3-celled and lateral on intercalary cortical cells. Sterile cell branches arise from supporting cells and fuse with carpogonia. After fertilization the branch cells form enlarged fusion cells, gonimoblasts radiate outward, and outer cells become carposporangia. Auxiliary cells are absent; pericarps are multilayered and ostiolate.

Gracilariaceae

Gurgel and Fredericq (2004) included seven genera in the family: *Congracilaria* Yamamoto 1986; *Curdiea* Harvey 1855b; *Gracilaria* Greville 1830; *Gracilariophila* Setchell and Wilson in Wilson, 1910; *Gracilariopsis* Dawson 1949a; *Hydropuntia* Montagne 1842a; *Melanthalia* Montagne, 1843. All genera have fewer than 20 species, except for *Gracilaria*, with more than 150 and the third largest genus of red algae (Byrne et al., 2002). Two Florida genera have nutritive cells in their cystocarps (*Gracilaria, Hydropuntia*), while *Gracilariopsis* lacks them. The morphology of nutritive cells and the presence of spermatangial conceptacles were used by Fredericq and Hommersand (1990) to separate genera. Based on molecular data, Gurgel and Fredericq (2004) identified three clades, including *Curdiea-Melanthalia*, *Gracilariopsis*, and *Gracilaria* (Iyer et al., 2005), with the last genus (*sensu lato*) having nine distinct lineages.

The family has the same characteristics as the order.

Gracilaria Greville 1830: p. liv, 121, *nom cons.* (L. *gracilis*: thin, slender, graceful).

Some authors recognized *Gracilaria*, *Gracilariopsis*, and *Hydropuntia* as distinct genera (Wynne 1989, 2005a; Gurgel and Federicq, 2004), while others listed *Hydropuntia* as a synonym of *Gracilaria* (Abbott et al., 1991; Bellorin et al., 2002) or all three in synonymy with *Gracilaria* (Gargiulo et al., 1992; Abbott 1995, 1999).

Plants are often bushy, branched, and attached by small discoid holdfasts. Axes are pseudoparenchymatous, with large, hyaline medullary cells and a cortex of 1–6 layers of smaller cells. Tetrasporangia are cruciately divided and from intercalary cortical cells that divide obliquely. Gametophytes are dioecious. Spermatangia are often in pits on outer cortical cells. Carpogonial branches are 2-celled and have support cells. Fusion cells are formed from two sterile cell branches and fertilized carpogonia. Gonimoblasts grow directly from fusion cells; auxiliary cells are absent. Nutritive cells are thin, tubular, and occur between the carposporophyte and pericarp. Carposporangia are in terminal clusters or chains on the gonimoblasts. Cystocarps are ostiolate, dome-shaped, protruding, and have a large central cellular placenta. Some species are polymorphic, have similar anatomy, and are difficult to identify using vegetative features only (Destombe and Douglas, 1991; Bird et al., 1992, 1994; Goff et al., 1994; Byrne et al., 2002).

Key to the species (19) of *Gracilaria*

1. Main axes flat (except possibly the stipe or proliferations) and wider than 4 mm 2
1. Axes cylindrical, compressed or flat, and less than 4 mm wide 8
  2. Mature fronds usually wider than 2 cm 3
  2. Mature fronds usually less than 2 cm wide 5
3. Axes straplike, more than 150 µm thick; branchlets arise from margins 4
3. Axes spreading, less than 150 µm thick; branchlets not from margins. . . . . *G. hayii*
  4. Mature fronds 1.2–3.5 cm wide; cortex of 1–2 layers, with a gradual transition between medullary and cortical cells . . . . *G. curtissiae*
  4. Mature fronds 1.0–4.0 cm wide; cortex of 2–3 layers, with an abrupt transition between medullary and cortical cells . . . . *G. intermedia*
5. Branchlets not concentrated at tips; main axes less than 500 µm thick 6
5. Branchlets mostly at tips; main axes to 1400 µm thick . . . . *G. isabellana*
  6. Fronds not crisp nor with ruffled margins 7
  6. Fronds usually crisp and with ruffled margins . . . . *G. cuneata*
7. Axes straplike and flat; branchlets in one plane and from margins . . . . *G. domingensis*
7. Axes compressed to flat; branchlets not in one plane or marginal . . . . *G. tikvahiae* (in part)
  8. Axes cylindrical to slightly compressed, less than 0.2 cm wide. 13
  8. Axes compressed to flat, more than 0.2 cm wide 9
9. Cortex of 1–3 layers and with the transition to medullary cells gradual 10
9. Cortex mostly 2 layered and with their cells abruptly smaller than medullary cells. 11
  10. Main axes indistinct above, compressed to flat, 1.7–2.0 mm thick. . . . . *G. flabelliformis*
  10. Main axes distinct, cylindrical to compressed, ca. 0.25 mm thick . . . . *G. galetensis*
11. Fronds compressed to flat; cortex 1–2 layers, cell walls thick 12
11. Fronds terete to compressed; cortex 1–3 layers, cell walls thin . . . . *G. tikvahiae* (in part)
  12. Axes 0.5–1.0 cm wide, initial branching widespread, regularly dichotomous, and not narrowing at forks; surface cells elongate . . . . *G. mammillaris*
  12. Axes compressed to flat, to 0.4 cm wide; branching repeatedly irregular and narrowing at forks; surface cells rounded to angular. . . . . *G. occidentalis*
13. Axes completely cylindrical. 18
13. Axes compressed at least at forks or bases, but may be terete elsewhere 14
  14. Branches widely spread to mostly on sides of axes (subdistichous); branches irregularly dichotomous to "zigzag" or secund, and all in one plane . . . . *G. divaricata*
  14. Branching otherwise and not divaricate; branches not in a "zigzag" pattern 15
15. Cortex of mature axes 1–3 layers 16
15. Cortex of mature axes 3–7 layers . . . . *G. cervicornis*
  16. Transition between cortical and medullary cells abrupt 17
  16. Large medullary cells grading into smaller cortical ones. 19
17. Main axes cylindrical to slightly compressed at forks; branching alternate above; cortex usually with 3 cell layers . . . . *G. tikvahiae* (in part)
17. Main axes oval to compressed below, above flat; branching regularly dichotomous above; cortex with 1–2 cell layers . . . . *G. venezuelensis*

18. Branching of main axes irregularly dichotomous below, above alternate to secund; medullary cells to 300 µm in diameter . . . . *G. bursa-pastoris*
18. Main axial branching dichotomous below; above irregular to subsecund; medullary cells to 200 µm in diameter . . . . *G. damaecornis*
19. Main axes 1–4 mm in diam.; branchlets constricted at bases. 20
19. Main axes 3–5 mm in diam.; branchlets not constricted at bases . . . . *G. armata*
20. Main axes 1–2 mm in diam.; branching alternate to irregular . . . . *G. blodgettii*
20. Main axes 2–4 mm in diam.; branching sparse to irregular . . . . *G. cylindrica*

Gracilaria armata (C. Agardh) Greville 1830 p. liv (L. *armata*: thorny, spiny, armed). Plate XLV, fig. 6–8.

Basionym. *Sphaerococcus armatus* C. Agardh 1827: p. 645

Plants are fleshy or cartilaginous and to 30 cm tall; axes terete, 3–5 mm in diam., and attached by discoid holdfasts. Branches are spreading, 4–8 cm long, and with radial and alternate to dichotomous branching; branchlets are simple to forked, pointed, often secund on outside of incurved branches, and not constricted at bases. Medullary cells are thick-walled, 210–260 µm in diam. and have an abrupt transition to the cortex. The cortex has 3 layers and surface cells are 7–11 µm in diam. Tetrasporangia are elongate, cruciately divided, and in outer cortex. Cystocarps are large, spherical, and mostly on the surface; carposporangia are oval to elongate.

Uncommon, on limestone in shallow waters at wave-exposed sites in the Keys and adrift in Tampa Bay; also known from Bermuda, the Caribbean, Gulf of Mexico, and Brazil, as well as the Mediterranean, eastern tropical Atlantic, and Indian Ocean.

Taylor 1960: p. 441; Dawes 1974: p. 128.

Gracilaria blodgettii Harvey 1853: p. 111 (named for Dr. Blodgett, who sent specimens to Harvey, including the type material from Key West). Plate XLV, fig. 9, 10.

Several authors (Norris and Fredericq, 1990; Fredericq and Norris, 1992; Withell et al., 1994; Silva et al., 1996; Wynne, 1998a; Guiry and Nic Dhonncha, 2006) considered *Gracilaria blodgettii* and *G. cylindrica* Børgesen (1920) to be conspecific. Others (Taylor, 1960; Dawes, 1974; Schneider and Searles, 1991; Littler and Littler, 2000; Wynne, 2005a) retained both taxa and stressed differences in nemathecial and branching features.

Plants are cartilaginous, pale red, erect, to 20 cm tall, and attached by discoid holdfasts. Axes are terete, 1–2 mm in diam. and with radial, alternate, or irregular branching to 5 times; branchlets are 0.2–1.0 mm in diam., tapering to acute tips, and constricted at bases. Medullary cells are large, thin-walled, and to 600 µm in diam.; the transition to the cortex is abrupt. The cortex has 1–3 layers; inner cells are thick-walled and to 35 µm in diam.; surface cells are small, round, 7–25 µm in diam., and pigmented. Tetrasporangia are spherical, cruciately divided, 20–40 µm in diam., 30–50 µm long, and cause swelling of the cortex above constrictions. Spermatangia occur in concave sori in the outer cortex. Carposporangia are ovoid to spherical and to 40 µm long. Cystocarps are scattered, prominent, wartlike, and to 1 mm in diam.

Common, on limestone in 2–3 m in estuaries and in 15–30 m on all coasts (type location: Key West; Silva et al., 1996); also known from the Carolinas (16–35 m), Georgia, the Caribbean, Gulf of Mexico, Brazil, Indo-Pacific, and Australia.

Taylor 1960 p. 449; Dawes 1974: p. 128; Schneider and Searles 1991: p. 320; Littler and Littler 2000: p. 110.

Gracilaria bursa-pastoris (S. G. Gmelin) P. C. Silva 1952: p. 265 (L. *bursa-pastoris*: shepherd's purse). Plate XLV, fig. 11–13.

Basionym. *Fucus bursa-pastoris* S. G. Gmelin 1768: p. 121, pl. VIII, fig. 3 (as '*bursa pastoris*').
Synonym. *Gracilaria compressa* (C. Agardh) Greville 1830: liv, 54, 125.

Plants are fleshy, erect, 15–25 cm tall, reddish brown to reddish purple, and attached by discoid holdfasts. Basal branching is irregularly dichotomous and alternate to unilateral and secund above. Main axes and forks are slightly compressed and 1–4 mm wide; branchlets are terete, tapered to tips, and not constricted at bases. Medullary cells are thin-walled, to 300 µm in diam., and grade into a cortex of 1–3 layers. Inner cortical cells are 9–25 µm in diam., while surface ones are round, pigmented, and 8–9 µm in diam. Cystocarps are scattered, oval to hemispherical, protruding, and have ostioles.

Common, in drift and attached to jetties or rubble (to 5 m) on all coasts; also known from the Caribbean, Gulf of Mexico, Brazil, British Isles, Mediterranean (type location; Silva et al., 1996), eastern tropical Atlantic, and Indo-Pacific.

Taylor 1960: p. 444; Dawes 1974: p. 129; Littler and Littler 2000: p. 110.

Gracilaria cervicornis (Turner) J. Agardh 1852 (1848–1901): p. 604 (L. *cervicornis*: with branches on one side, like the horns of a deer). Plate XLV, fig. 14, 15.

Basionym. *Fucus cervicornis* Turner 1808 (1808–1809): p. 131–132, pl. 121.
Synonyms. *Gracilaria ferox* J. Agardh 1852 (1848–1901): p. 592; 1876 (1848–1901): p. 414.
Gracilaria patens var. *gracilis* P. Crouan and H. Crouan in Schramm and Mazé 1865: p. 19; P. Crouan and H. Crouan in Mazé and Schramm 1878 (1870–1877): p. 216.
Gracilaria acanthophora (Kützing) P. Crouan and H. Crouan in Mazé and Schramm 1878 (1870–1877): p. 218.

Gurgel et al. (2004c) suggested that '*Gracilaria ferox*' from Brazil may be a distinct taxon. *Gracilaria apiculata* P. Crouan and H. Crouan in Schramm and Mazé is related but distinct and often identified as *G. cervicornis* (López-Bautista and Kapraun, 1995).

Plants are erect, 8–40 cm tall, fleshy, bushy, reddish brown to green brown, and with one or a few axes arising from small discoid holdfasts. Branching is dichotomous below and pinnate to hornlike, dense, and irregular above. Main axes are terete to compressed and 1–5 mm wide; branchlets are short, cylindrical, and acute. Medullary cells are 175–350 µm in diam. and irregularly spherical; the transition to the cortex is gradual. The cortex has 3–7 layers; inner cells are 20–30 µm in diam., while surface cells are rounded, 6–10 µm in diam., and pigmented. Tetrasporangia are ovoid, cruciately divided, 20–25 µm in diam., 30–40 µm long, and clustered among surface cells. Spermatia occur in small cavities. Cystocarps are prominent, hemispherical, and 0.5–1.25 mm in diam.

Common, in drift and on hard substrata (to 10 m) on all coasts; also known from Bermuda, the Caribbean (type location: West Indies; Silva et al., 1996), Gulf of Mexico, Brazil and Argentina, as well as the Mediterranean, Canary Islands, West Africa, Kenya, and the Seychelles.

Taylor 1960: p. 444, 445; Dawes 1974: p. 129; Littler and Littler 2000: p. 112.

Gracilaria cuneata Areschoug 1854: p. 351 (L. *cuneata*: wedge-shaped). Plate XLV, fig. 16, 17.

Plants are tough, fleshy, foliaceous, erect, 12–18 cm tall, reddish brown to reddish purple, repeatedly dichotomously branched and on short stipes attached by discoid holdfasts. Fronds are strap-shaped, membranous, 4–12 mm wide, 200–300 µm thick, and with ruffled, crisp, often proliferous margins. Medullary cells are ovoid, and 100–180 µm in diam.; the transition to the cortex is gradual. The cortex has 1–3 layers; subsurface cells are 15–35 µm in diam. and surface cells are pigmented and 6–12 µm in diam. Cystocarps are scattered, dome-shaped, and swell one side of fronds.

Rare, on subtidal limestone (5–20 m) off Cape Canaveral (Hanisak and Blair, 1988) and misidentified as *Petroglossum undulatum* Schneider by Eiseman (1979); also known from the Caribbean and Brazil.

Taylor 1960: p. 448; Littler and Littler 2000: p. 112.

Gracilaria curtissiae J. Agardh 1885: p. 61 (named for Floretta A. Curtiss, an American collector who sent specimens to J. Agardh). Plate XLV, fig. 18, 19.

Plants are foliaceous, pale pink to yellow, erect, and to 40 cm tall; axes are strap-shaped, membranous, 0.5–1 mm thick, and attached by discoid holdfasts. Branching is palmate to irregular below and more dichotomous above. Fronds are lanceolate, 3–22 cm long, 1.0–3.5 cm wide, and constricted slightly at bases; branchlets are on margins and irregularly placed. The medulla has 3–4 layers of irregular, hyaline, thin-walled cells (120–600 µm in diam.) with an abrupt transition to the cortex. The cortex is thin, with 1–2 layers, and has obvious but not common gland cells. The inner cortical cells are often compressed, while surface cells are isodiametric, dome-shaped, and 7–13 µm in diam. Tetrasporangia are cruciately divided and scattered below the outer cortical cells. Spermatangia occur in vase-shaped cavities and form light patches on the surface. Carposporophytes are elliptical to globular; carposporangia are ovoid or terete and to 38 µm long. Cystocarps are abundant, hemispherical, slightly pointed, to 1.2 mm in diam., and on margins or surfaces of fronds.

Rare, in 6–9 m at the Dry Tortugas (Taylor, 1928) and from 39 m off the east coast (Hanisak and Blair, 1988); also known from North Carolina (27–40 m), Bermuda, the Bahamas, and Caribbean. Type location: Florida (J. Agardh, 1885).

Taylor 1960; p. 449; Schneider and Searles 1991: p. 321; Littler and Littler 2000: p. 114.

Gracilaria cylindrica Børgesen 1920: p. 375–377, fig. 364, 365 (L. *cylindrica*: cylindrical). Plate XLV, fig. 20, 21.

See earlier comments regarding possible synonymy of *Gracilaria blodgettii* and *G. cylindrica*.

Plants are fleshy, deep red to light pink, and to 38 cm tall; branching is irregular and sparse; axes are terete, 2–4 mm in diam. and tapering to small discoid holdfasts. Branchlets are similar in size to main axes, with constricted or stalked bases. Axes are radially alternately branched, have blunt tips, and arch inward. Medullary cells are hyaline, thin-walled, and 200–525 µm in diam.; the transition to the cortex is abrupt. The cortex has 1–2 layers of pigmented, elongate cells, 5.0–7.5 µm in diam. Tetrasporangia are elongate to spherical, cruciately divided, 20–40 µm in diam., 40–60 µm long, and scattered or in small clusters in the outer cortex. Carposporangia are ovoid to spherical and to 40 µm long; cystocarps are scattered, wartlike, and to 2 mm in diam.

Common, on limestone (15–110 m) on all coasts and the Dry Tortugas; also known from North Carolina (27–35 m), the Bahamas, Caribbean (syntype locations: St. John and St. Thomas, Virgin Islands), Gulf of Mexico, Brazil, and the Philippines.

Taylor 1960: p. 450, Dawes 1974: p. 129; Schneider and Searles 1991: p. 322; Littler and Littler 2000: p. 114.

Gracilaria damaecornis J. Agardh 1852 (1848–1901): p. 597; 1876 (1848–1901): p. 415 (L. *damae*: a fallow-deer + *cornis*: having horns). Plate XLV, fig. 22–24.

Synonyms. *Gracilaria circinnata* P. Crouan and H. Crouan in Schramm and Mazé 1865: p. 19; P. Crouan and H. Crouan in Mazé and Schramm 1878 (1870–1877): p. 217.
Gracilaria flabellata P. Crouan and H. Crouan in Schramm and Mazé 1865: p. 20; P. Crouan and H. Crouan in Mazé and Schramm 1878 (1870–1877): p. 217–218.
?*Plocaria complanata* P. Crouan and H. Crouan in Mazé and Schramm 1878 (1870–1877): p. 226, *pro parte*.
Plocaria damaecornis (J. Agardh) P. Crouan and H. Crouan in Mazé and Schramm 1878 (1870–1877): p. 225.
Hydropuntia albornozii (Rodriguez) M. J. Wynne 1989: p. 476.
Gracilaria albornozii (Rodriguez) Abbott, Zhang, and Xia 1991: p. 24 (in part).
Plastino and Oliveira (1997) compared this taxon to several other species.

Plants are bushy, fleshy to cartilaginous, 7–15 cm tall, pale red to brown green, and attached by fleshy discoid holdfasts. Axes are dichotomously branched, terete to slightly compressed, and 1.5–4 mm in diam. Branching is dichotomous below and irregular above; branchlets are slightly secund, short, and have blunt to tapered tips. Medullary cells are thin- walled, to 200 µm in diam., and lightly pigmented; the transition to the cortex is gradual. The cortex has 2–3 layers; surface cells are 8–20 µm in diam. and have thick outer walls. Tetrasporangia are cruciately divided and below the surface. Spermatangia occur in vaselike cavities near the surface; cystocarps are hemispherical.

Uncommon, on limestone in 1–3 m at Pigeon Key and adrift in Tampa Bay; also recorded from Bermuda, the Caribbean, Gulf of Mexico, North Senegal, and the Philippines. Type location: "in the Atlantic on the shores of North America" [Agardh, 1852 (1848–190)].

Taylor 1960: p. 443; Dawes 1974: p. 129; Littler and Littler 2000: p. 114.

Gracilaria divaricata Harvey 1853: p. 109 (L. *divaricata*: spreading at a wide angle).

Fronds are cartilaginous to succulent, purple to red, to 10 cm long, attenuated above, 2 mm wide or more near bases, and attached by a rhizoidal holdfast. Branching is abundant, irregular, and dichotomous or secund. Main axes are subterete; branching is irregularly dichotomous, zigzag or secund, and wide spreading; axes are covered with lateral branchlets (ramelli) that are irregularly inserted, nearly at right angles to axes, 0.4–0.6 cm long, simple or branched, acute, and with forked tips.

Rare, in Florida known only from Key West (type location); also known from Bermuda (Collins and Hervey, 1917), but doubtful (Schneider, 2003), and Jamaica.

Gracilaria domingensis (Kützing) Sonder *ex* Dickie 1874a: p. 149 (L. *domingensis*: of or from the Dominican Republic of Hispaniola). Plate XLV, fig. 25, 26.

Basionym. *Sphaerococcus domingensis* Kützing 1869: p. 8, pl. 22a,b.
Synonyms. *Plocaria bipennata* P. Crouan and H. Crouan in Mazé and Schramm 1878 (1870–1877): p. 228.
Plocaria polymorpha P. Crouan and H. Crouan in Mazé and Schramm 1878 (1870–1877): p. 228.
Plocaria squarrosa P. Crouan and H. Crouan in Mazé and Schramm 1878 (1870–1877): p. 227, *pro parte*.
Polycavernosa domingensis (Kützing) Price and John in Price et al. 1988: p. 236, 243–244.
Guiry and Nic Dhonncha (2006) cited the plant as '*Gracilaria domingiensis*.'

Plants are bushy, in clumps 10–35 cm tall, and greenish brown to rose red; axes are flat, strap-shaped, marginally branched, and attached by discoid holdfasts. Main axes are 2–12 mm wide, to

350 µm thick, sparingly divided, and irregularly branched. Branchlets are short, slender, mostly opposite to pinnately divided, and marginal. Medullary cells are compressed, thin-walled, and to 600 µm in diam.; the transition to the cortex is gradual. The cortex has 2–4 layers; inner cortical cells are 20–50 µm in diam. and outer ones rounded and 6–10 µm in diam. Hyaline hairs are rare. Tetrasporangia are ovoid to spherical, cruciately divided, to 45 µm in diam., and occur at the surface. Spermatangia are in vase-shaped cavities and to 40 µm in diam. Cystocarps are dark red, hemispherical, and conspicuous on lesser branches.

Uncommon, on limestone at Key West and the Dry Tortugas (to 15 m); also known from the Caribbean, Gulf of Mexico, Brazil, and West Africa.

Taylor 1960: p. 446; Littler and Littler 2000: p. 116.

Gracilaria flabelliformis (P. Crouan and H. Crouan in Mazé and Schramm) Fredericq and Gurgel in Gurgel and Fredericq 2004: p. 155, fig. 2d, as *G. flabelliforme* (L. *flabellum*: a small fan + *formis*: form of). Plate XLVI, fig. 1–4.

Basionym. *Plocaria flabelliformis* P. Crouan and H. Crouan in Schramm and Mazé 1865: p. 21, #131 (as '*flabelliforme*'); Schramm and Mazé 1866: p. 48; Mazé and Schramm 1878 (1870–1877): p. 224.

Synonyms. *Sphaerococcus lacinulatus* Kützing 1868: p. 30, pl. 87, *nom. illeg.*

Plocaria lacinulatus P. Crouan and H. Crouan in Mazé and Schramm 1878 (1870–1877): p. 226.

Gracilaria lacinulata (P. Crouan and H. Crouan in Mazé and Schramm) Piccone 1886: p. 71, *nom. illeg.* [non *G. lacinulata* (Vahl) Howe 1920a: p. 562; non *G. lacinulata* (P. Crouan and H. Crouan in Mazé and Schramm) Piccone *sensu* Ortega, Godínez and Garduño-Solórzano, 2001: p. 124].

Gracilaria flabelliformis (P. Crouan and H. Crouan in Schramm and Mazé) Fredericq and Gurgel in Gurgel and Fredericq 2004: p. 155, fig. 2d; Gurgel et al. 2004a: p. 162.

Gurgel et al. (2004c) found that *Gracilaria lacinulata* (P. Crouan and H. Crouan) Piccone was conspecific with *G. flabelliformis* and noted that it was common but rarely identified. Its morphology is highly variable and includes three subspecies (*aionana*, *flabelliformis*, *simplex*) with only the nominate subspecies being in Florida.

Plants are 5–7 cm tall; main axes are tough, 2–8 mm wide, 1.7–2.0 mm thick, flat to compressed, straplike above, and tapering to a stipe below; stipes are short, terete, 0.2–2.0 mm long, and with one or more arising from a discoid holdfast. Main axes are indistinct after 1st or 2nd order branching, which is highly variable (di-, tri-, poly-, or subdichotomous), mostly near tips, marginal, and always in the same plane. The medulla has 10–15 layers; cells are small, elliptical, 150 by 220 µm in diam., and thick walled; the transition to the cortex is gradual. The cortex has 1–3 cell layers (to 21 layers at bases); cells are radially elongate, 5–7 µm in diam., and have hyaline hairs. Tetrasporangia are cruciately divided and in 3–5 layers of radially elongate cortical cells on both surfaces. Cystocarps are obvious and constricted at bases. Pericarps are 350 µm thick and with inner star-shaped (stellate) cells.

Possibly common, on limestone and in seagrass beds, and within the intertidal and shallow subtidal on the east coast (Walton Rocks, Riomar Reefs); also known from the Caribbean (type location: Vieux-Bourg, Gauadeloupe) and Brazil.

Gracilaria galetensis Gurgel, Fredericq, and J. N. Norris 2004a: p. 181–184, fig. 8A–H (named for the reefs and mangrove forests of Isla Galeta, Panama). Plate XLVI, fig. 5, 6.

Plants described as "*Rhodymenia divaricata*" *sensu* Schneider and Searles 1991, non E. Y. Dawson from Florida (Hanisak and Blair, 1988) may be this species (Gurgel et al., 2004a).

Fronds are erect, delicate, rose to red colored, solitary or clustered, 4–5(20) cm tall, flat, strap-shaped, on short (0.8–1.0 cm) cylindrical to compressed stalks, and attached by discoid or lobed holdfasts. Blades are 5–8 mm wide in midregions and branch 1–3 times at wide angles (80–100°); branching is dichotomous to subdichotomous below and trichotomous (to 5 times) above. Fronds are 5–30 mm wide and 250–275(600) μm thick; margins are smooth and have few proliferations and obtuse tips, with proliferations being most common at damaged sites. The medulla has 4–5(–7) irregular layers; inner cells are rounded, 81–133(200) μm in diam., hyaline, and polygonal; outer cells are strongly compressed; the transition to the cortex is gradual. The cortex has 1–3 layers of pigmented anticlinal elongate cells; surface cells are 5–8 μm in diam. by 8–17 μm long. Tetrasporangia are scattered in midregion of blades, cruciately divided, ovoid, 12–20 μm in diam., and 20–25 μm long. Cystocarps are ostiolate, scattered, dome-shaped, and to 1 mm in diam. Carpospores are in hemispherical masses, darkly pigmented, round to elliptical, and 7–18 μm in diam.

Rare, in 27–50 m off the east coast (as *Rhodymenia divaricata*, Hanisak and Blair, 1988); also known from the Carolinas (40–54 m), Caribbean (type location: Galeta Point, Panama), Gulf of California, Pacific Panama, and Galapagos Islands.

Schneider and Searles 1991: p. 347; Littler and Littler 2000: p. 138.

Gracilaria hayi Gurgel, Fredericq, and J. N. Norris 2004a: p. 178–181, fig. 7A–F (named for Mark E. Hay, an American marine phycologist who studies seaweed herbivory). Plate XLVI, fig. 7.

Synonyms. *Gracilaria mammilaris sensu* Schneider and Searles 1991; *sensu* Littler and Littler 2000; non *G. mammillaris* (Montagne) M. Howe 1918.

Gracilaria cuneata sensu pro parte Taylor 1960; non *G. cuneata* Areschoug 1854.

See Gurgel et al. (2004a) regarding the confusion between *Gracilaria cuneata* and *G. mammillaris*. The identification of older or damaged plants can be difficult.

Fronds are 5–15 cm tall, flat, dichotomously or subdichotomously branched, and on terete stipes (1–2 cm long) attached by round holdfasts. Blades are cuneate at bases, strap-shaped, 5–25 mm wide, thin (to 100 μm thick), and with indented to obtuse tips. The medulla has 3–5 layers; cells are round to ovoid, 50–85 μm in diam.; the transition to the cortex is abrupt. The cortex has 1–2 periclinal layers of compressed cells and a few small gland cells; surface cells are 2.5–6.5 by 6.3–10 μm. Tetrasporangia and spermatangial cavities are scattered on the surface. Cystocarps are also scattered on surfaces, hemispherical, 1–2 mm in diam., ostiolate, and with a pericarp of 12–14 layers

Rare(?), in drift within the intertidal and to 25 m off the east coast; also known from Panama in 6–15 m (type location: Galeta Point).

Taylor 1960: p. 448; Schneider and Searles 1991: p. 323; Littler and Littler 2000: p. 116.

Gracilaria intermedia J. Agardh 1901: p. 95–96 (L. *intermedia*: intermediate, in between). Plate XLVI, fig. 8, 9.

Synonym. *Gracilaria foliifera sensu* Taylor 1960, pro parte; non Børgesen 1932b.

See Gurgel et al. (2004a) for a summary of misidentified associated taxa.

Fronds are flat and to 20 cm tall; main fronds are dichotomously branched in one plane, 3–4 cm wide, 685–884 μm thick, and from a discoid holdfast. Branches are strongly compressed; branching is irregular, opposite, or alternate, and from margins of main fronds. Young branches are often basally constricted and have expanded and rounded or obtuse tips. The medulla has two parts; inner cells are 116 by 250 μm and outer ones small, thick-walled, refractive, and 54 μm in diam.; the transition to the cortex is gradual. The cortex has 2–3 layers in younger parts and 3–7 layers below with cells in rows; surface cells are 6 μm in diam. and isodiametric or to 9 μm long and in anticlinal rows. Hyaline hair cells are obvious and 9 by 14 µm; gland cells are uncommon and 10 µm in diam. Tetrasporangia are unknown; spermatangia occur in cavities. Cystocarps are ostiolate, hemispherical, protruding, 900–1700 µm in diam., and on the dorsal side and margins of fronds.

Rare(?), in 2.5–4.0 m on the east coast at Riomar Reef and Walton Rocks (lectotype location: St. Augustine, "Indiae occidentalis"; probably collected by Mrs. G. A. Hall and sent to J. G. Agardh); also known from Venezuela.

Taylor 1960: p. 447, *pro parte.*

Gracilaria isabellana Gurgel, Fredericq and J. N. Norris 2004c: p 384, fig. 34–42 (named for Isabella Abbott, an American specialist on red algal taxonomy and the seaweed floras of California and Hawaii). Plate XLVI, fig. 10–12.

Synonyms. *Fucus lacinulatus* Vahl 1802: p. 39.

Gracilaria multipartita sensu auct., non (Clemente) Harvey 1846 (1846–1851): pl. XV.

Gracilaria lacinulata (Vahl) Howe in Britton and Millspaugh 1920a (26 June): p. 562, *nom. illeg.* [non *G. lacinulata* (Kützing) Piccone 1886: p. 71].

Gracilaria lacinulata (Vahl) Børgesen 1920 (23 October): p. 379, fig. 366 [non *G. lacinulata* (P. Crouan and H. Crouan) Piccone 1886: p. 71].

Gracilaria foliifera sensu Taylor 1960: p. 446 (*pro parte*); Ortega, Godínez & Garduño-Solórzano 2001: p. 124 [non *G. foliifera* (Forsskål) Børgesen 1932b: p. 7].

Gracilaria lacinulata sensu Ortega, Godínez & Garduño-Solórzano 2001: p. 124 [as '*G. lacinulata* (P. Crouan and H. Crouan) Piccone; non *G. lacinulata* (Kützing) Piccone 1886: p. 71].

See Gurgel et al. (2004c) regarding the replacement of *Gracilaria isabellana* for the illegitimate *G. lacinulata* (Vahl) M. Howe, a later homonym of *G. lacinulata* (P. Crouan and H. Crouan) Piccone (1886).

Plants are erect, to 12 cm long, and light green or dark brown; main axes are compressed to flat, linear to lanceolate, to 1.8 cm wide, and 1.4 mm thick; stipes are short, terete, and attached by cushionlike holdfasts. Axes are branched in one plane; branches vary in width and shape; branching is simple, irregular, di-, or trichotomous. Branchlets are simple, from margins, irregularly distributed, mostly near tips of main axes, and may be secund, compressed at bases, and with bent rounded tips. The medulla has 3–5 layers of large cells, 175 by 320 µm; the transition to the cortex is abrupt. The cortex has 1–2 layers of cells; hyaline hairs and gland cells are rare; cortical cells vary in size and are mostly anticlinally elongate and 13 by 20 µm. Tetrasporangia are ovoid, cruciately divided, 20–35 µm in diam., 30–45 µm long, and on the surface. Cystocarps are obvious, project on both sides of fronds or on margins, are constricted at bases and to 1 mm in diam. Pericarps are multilayered, to 175 µm thick, and with 9–13 cell layers.

Possibly common in Florida and often misidentified as *G. foliifera*; also known from the Virgin Islands (type location: St. Croix) and Venezuela.

Taylor 1960: p. 446; Dawes 1974: p. 130.

Gracilaria mammillaris (Montagne) M. Howe in Britton 1918: p. 515 (L. *mammillaris*: a breast, nipple, protuberance). Plate XLVI, fig. 13, 14.

Basionym. *Rhodymenia mammillaris* Montagne 1842b: p. 242.

Synonym. *Gracilaria dichotomoflabellata* P. Crouan and H. Crouan in Schramm and Mazé 1865: p. 20; Mazé and Schramm 1878 (1870–1877): p. 218 (as '*dichotomo-flabellata*').

Collins and Hervey (1917) and Silva et al. (1996) thought that *Gracilaria dichotomoflabellata* was a distinct species. Records of *G. mammillaris* in Florida probably include other species recently designated by Gurgel et al. (2004a).

Fronds are clustered or solitary, tough, membranous to leathery, pink to bright red, 7–10(15) cm tall, and from short stipes; one or more stipes arise from tough irregular holdfasts. Fronds are strap-shaped, flat, 5–8 (-10) mm wide, 250–330 µm thick, and have shield-shaped bases and obtuse or notched tips. Branching is initially dichotomous, in one plane, to 5 times, and with rounded forks; older fronds are irregularly branched and proliferous in damaged areas. The medulla has 4–5 layers; cells are spherical to slightly compressed, thick-walled, and 50–170 µm in diam.; the transition to the cortex is abrupt. The cortex has 1–2 layers; inner cells are pigmented and compressed and surface cells are ovoid and 4–7 by 6– 18 µm in diam. Tetrasporangia are in the cortex, ovoid or spherical, cruciately divided, and 25–35 µm in diam. Spermatangia are in small, scattered concave sori at the surface. Carposporangia are spherical to irregular and to 30 µm in diam.; cystocarps are 1.3 mm in diam., projecting, and hemispherical.

Common, on subtidal limestone (to 70 m) on all coasts; older plants are often covered by bryozoans and crustose algae; also known from North Carolina, Georgia, Bermuda, the Caribbean (type location: Martinique, Lesser Antilles), Louisiana, and Brazil, as well as Senegal and the Pacific Ocean.

Taylor 1960: p. 447; Dawes 1974: p. 131; Schneider and Searles 1991: p. 323; Littler and Littler 2000: p. 116.

Gracilaria occidentalis (Børgesen) Bodard 1965: p. 874–878; fig. 1a–e (L. *occidentalis*: west, western, or occidental). Plate XLVI, fig. 15, 16.

Basionym. *Rhodymenia occidentalis* Børgesen 1920: p. 387–389, fig. 371, 372.

Plants are erect and to 25 cm tall; axes are flat, membranous, to 4 mm wide, narrow at forks and base, and on stalks attached by discoid bases. Fronds are smooth, linear, 350–800 µm thick near margins, repeatedly forked, and with irregular, di-, or trichotomous branching; tips are rounded, obtuse, or forked. The medulla has 3–4 layers; inner cells are thick-walled and 100–270 µm in diam., while outer cells are pigmented; the transition to the cortex is abrupt. The cortex has 1–2(3) layers of thick-walled cells; surface cells are 5–10 by 9–16 µm and round or polygonal. Gland cells are few and small. Tetrasporangia and spermatangia are unknown; cystocarps are to 500 µm in diam.

Rare, from 27–93 m off the east coast (Hanisak and Blair, 1988); also known from the Virgin Islands (type location) in 18–27 m, Venezuela, Mauritania, and Senegal.

Taylor 1960: p. 485.

Gracilaria tikvahiae McLachlan 1979: p. 19–20, fig. 1 (L., named for Tikvah Edelstein, a Canadian phycologist from Israel who conducted floristic and taxonomic studies of seaweeds). Plate XLVI, fig. 17–19.

Synonym. *Gracilaria foliifera* (Forsskål) Børgesen var. *angustissima* (Harvey) W. R. Taylor 1937b: p. 230.

McLachlan (1979) designated plants known as *Gracilaria foliifera* var. *angustissima* (Taylor, 1960; Dawes, 1974) as *G. tikvahiae*. Based on *rbcL* gene sequences, Gurgel et al. (2001, 2004b) identified four distinct haplotypes of *G. tikvahiae*: (1) Canadian and northeastern USA; (2) southeastern Florida; (3) eastern Gulf of Mexico; and (4) western Gulf of Mexico. They found no evidence for the species occurring in the Caribbean.

Plants are erect, to 40 cm tall, solitary or clustered, and green to deep red; axes are highly polymorphic, delicate to coarse, and attached by discoid holdfasts. Fronds are cylindrical at bases, compressed to markedly flat (1–15 mm wide) above, and 0.2–1 mm thick at forks. Drift plants are often terete and slightly compressed at forks. Branches are spreading, irregular, and with 2 to many per fork. Branching is dichotomous below, alternate above, usually in one plane, and proliferous; tips are acute. Medullary cells are irregular, compressed, and 70–270 µm in diam.; the transition to the cortex is abrupt. The cortex has 1–4 layers; outer cortical cells are pigmented, rounded to angular, and 5–13 µm in diam. Tetrasporangia are ovoid, cruciately divided, 10–35 µm in diam. by 20–45 µm long, and in the outer cortical layer. Spermatangia occur in concave sori. Gonimoblast filaments branch and have elongated cells that arise from a rounded basal cell. Carposporangia are ovoid to spherical and 15–40 µm in diam. Cystocarps are obvious, numerous, dome-shaped and to 1 mm in diam. The species appears similar to *G. foliifera,* except that the gonimoblast filaments have long branched cells arising from a rounded basal cell (McLachlan, 1979).

Common to ubiquitous, in drift and intertidally to 50 m on limestone in protected to wave-exposed sites on all coasts; known from warm water embayments in the Canadian Maritime Provinces (type location: Nova Scotia) to Cape Cod and southward to Florida, Bermuda, the Gulf of Mexico, Brazil, and Hawaii (introduced).

Taylor 1960: p. 447; Dawes 1974: p. 131; Schneider and Searles 1991: p. 325; Littler and Littler 2000: p. 116.

***Gracilaria venezuelensis* W. R. Taylor 1942 p. 110, pl. 3, fig. 10; pl. 15, fig. 1, 2 (L., of or from Venezuela). Plate XLVI, fig. 20, 21.**

Synonym. *Gracilaria tikvahiae sensu* Schneider and Searles (1991) and *sensu* Littler and Littler (2000).

Attached plants are bushy, erect, 10–15 cm tall, rose red to dark purple, and from discoid holdfasts; drift plants are more fragile. Branching is to 6 orders, dichotomous to alternate below, more regularly dichotomous above, mostly in one plane, and with proliferations. Axes are oval to compressed and 2–3 mm wide in lower parts, versus thin, flat, and 0.5–1.0 mm wide near tips. Forks may be to 3.5 mm wide and with an angle of >45°. Tips are acute and often forked. The medulla has 5–6 layers of large and long thin-walled cells; the transition to the cortex is abrupt. The cortex has 1–2 layers of small, pigmented cells. Cystocarps are obvious, 0.75 to 1.5 mm in diam., and marginal or on the surface.

Rare, but probably overlooked; epiphytic on seagrasses in Tampa Bay and the Indian River near Fort Pierce; also known from Mexico, Belize, Venezuela (type location: Cubagua Island), and the Philippines.

Taylor 1960: p. 448.

Gracilariopsis E.Y. Dawson 1949a: p. 40 (L. *gracilis*: thin, graceful + *opsis*: similar to; thus, resembling the red algal genus *Gracilaria*).

Dawson (1949a) separated *Gracilariopsis* from *Gracilaria* using cystocarp features, while others have not (Papenfuss, 1967; Abbott, 1985). However, morphological (Fredericq and Hommersand, 1989b) and molecular studies (Bird et al., 1994; Goff et al., 1994; Gurgel et al., 2003; Gurgel and Fredericq, 2004) have confirmed the separation of both genera.

Plants are erect with branched cylindrical axes that are irregular, alternate, or proliferous, and attached by discoid holdfasts. Axes are pseudoparenchymatous with a multiaxial medulla and a cortex of 2–3 layers of radiating cells. Tetrasporangia are formed by oblique divisions of inner cortical cells. Gametophytes are dioecious with spermatangia in surface sori. Carpogonial branches are 2-celled and from supporting cells formed obliquely from inner cortical cells. Fertilized carpogonia fuse with 2(3)-celled sterile branches and form fusion cells. Cystocarps are ostiolate, hemispherical, have a large central cellular placenta, and lack tubular nutritive cells between the carposporophytes and pericarps. Inner cells of the pericarp form dark-staining nutritive tissue; gonimoblast filaments fuse to the cystocarp floor by connecting cells; carposporangia occur in straight terminal chains.

Gracilariopsis lemaneiformis (Bory) E. Y. Dawson, Acleto and Foldvik 1964: p. 59, pl. 56, fig. A, as '*lemanaeformis*' (L. *Lemanea*: a genus of freshwater red algae named for Dominique Sébastian Léman, a French cryptogamic botanist + *formis*: shaped like; thus, similar to *Lemanea*). Plate XLVI, fig. 22–26.

Basionym. *Gigartina lemaneiformis* Bory 1828b (1826–1829): p. 151 (as '*lemanaeformis*').

Molecular studies by Gurgel et al. (2003) have shown that *Gracilariopsis lemaneiformis* from the Carolinas is similar to *G. carolinensis* Liao and Hommersand (2003). We are uncertain if Florida plants are *G. lemaneiformis* or *G. carolinensis*, although the morphology is similar to that described by Schneider and Searles (1991) for the former taxon.

Plants are entangled or erect, 50–100(200) cm tall, greenish brown to deep red, initially attached by discoid holdfasts, and secondarily by prostrate rhizomelike branches. Branching is sparse, radial, alternate or irregular, and with distinct main axes. Branches are terete, 0.5–2 mm in diam., with acute tips, constricted at bases, and have a few short laterals. Medullary cells are thin-walled and to 400 μm in diam.; the transition to the cortex is gradual. The cortex has 3–6 layers in anticlinal rows and pigmented surface cells that are 2.5–7.5 μm. Tetrasporangia are ovoid, 25–30 μm in diam. by 30–48 μm long, and in the outer cortex. Spermatangia occur in open, scattered sori on the surface of blades. Carposporophytes lack filaments that radiate between gonimoblasts and pericarps; carposporangia are ovoid to spherical, 12–25 μm in diam., and 20–40 μm long. Cystocarps are prominent, hemispherical, and 0.8–1.8 mm in diam. The species resembles *Hydropuntia caudata* Gurgel and Fredericq, but has a dominant axis, cortical cells in anticlinal rows, and spermatangia in open sori, not concave depressions.

Uncommon, on shells and limestone on all coasts between 10 and 20 m; also known from the Carolinas, Caribbean, Gulf of Mexico, and Brazil, as well as West Africa, Indo-Pacific (type location: Paitá, Peru; Abbott, 1999), and Australia.

Taylor 1960: p. 449; Dawes 1974: 132; Littler and Littler 2000: p. 118.

Hydropuntia Montagne 1842a: p. 7 (Gr. *hydro-*: water or of water + L. *Opuntia*: a genus of cactus; thus a cactuslike aquatic plant.

Synonym. *Polycavernosa* Chang and Xia 1963: p. 120.

Guiry and Nic Dhonncha (2006) retained *Polycavernosa* as a distinct genus.

Plants are erect or recumbent and attached by a rhizomelike holdfast derived from a discoid holdfast. Axes are polymorphic, terete to flat, and parenchymatous. The cortex has small cells and the medulla has large cells with abundant fusions. Plants are usually dioecious. Carpogonial branches are 2-celled and may have multiple sterile branches. After fertilization, the cells of sterile branches fuse with the carpogonium and then with vegetative cells to form highly dissected fusion cells. Cystocarpic cavities have many fusions of multinucleate tubular nutritive cells with floor cells; carposporangia are aligned in straight chains. Spermatangial mother cells are in filaments, from intercalary cortical cells, and often in confluent pits. Spermatangia arise from convex divisions of spermatangial mother cells. Tetrasporangia are cruciately divided and superficial.

Key to the species (4) of *Hydropuntia*

1. Plants erect, with branches spreading; axes sparingly to densely branched 2
1. Plants prostrate to creeping; axes irregularly branched and thick . . . . *H. crassissima*
   2. Plants slender; branching abundant, irregular and radial 3
   2. Plants coarse and with few main axes; upper branching often secund . . . . *H. cornea*
3. Branching is abundant, 5–7 times . . . . *H. caudata*
3. Branching is sparse, 2–3 times. . . . . *H. secunda*

Hydropuntia caudata (J. Agardh) Gurgel and Fredericq 2004: p. 155 (L. *caudata*: having a tail, tapering to a tail). Plate XLVI, fig. 27, 28.

Basionym. *Gracilaria caudata* J. Agardh 1852 (1848–1901): p. 598.
Gracilaria andersonii sensu auct., non (Grunow) Kylin 1941: p. 21.
Gracilariopsis sjoestedtii sensu auct., non (Grunow) Dawson 1949a: p. 43.
Gracilaria verrucosa sensu auct., non (Hudson) Papenfuss 1950a: p. 195.
See Taylor (1960) and Wynne (1989, 2005a) for historical data.

Plants are lax to rigid, bushy, 20–60 cm tall, with one or more main branches, and attached by discoid holdfasts. Its morphology is highly variable, ranging from laxly branched soft forms to sparsely branched firm ones. Branching is abundant, 5–7 times, radial, and alternate or irregular; main axes are terete, 1–3 mm in diam. and tapering basally. Branchlets are 0.3–1.0 mm in diam., loosely branched, congested above, and not pinched at bases. Medullary cells are irregular, thin walled, and to 650 µm in diam. The cortex has 2–6 layers; the outer 3 layers may be in distinct anticlinal rows; surface cells are pigmented, 5–15 µm in diam., and often with hairs. Tetrasporangia are ovoid, cruciately divided, 20–40 µm in diam. by 35–40 µm long, and arise from outer cortical cells. Spermatangia occur in open, concave sori. Cystocarps are oval, warty, and 0.8–1.8 mm in diam. Carposporophyte filaments radiate between the gonimoblastic filaments and pericarp.

Common, sometimes abundant as drift in estuaries, on limestone (to 30 m), and on all coasts; also known from Bermuda, the Caribbean, Gulf of Mexico, and Brazil.

Taylor 1960: p. 441; Dawes 1974: p. 132; Littler and Littler 2000: p. 112.

Hydropuntia cornea (J. Agardh) M. J. Wynne 1989: p. 476 (L. *cornea*: horny, tough). Plate XLVII, fig. 1–3.

Basionym. *Gracilaria cornea* J. Agardh 1852 (1848–1901): p. 598.
Synonyms. *Fucus debilis sensu auct.*, non Forsskål 1775: p. 191.
Fucus wrightii sensu auct., non Turner 1811 (1809–1811): p. 31, pl. CXLVIII.
Gracilaria wrightii sensu auct., non (Turner) J. Agardh 1852 (1848–1901): p. 599.
Gracilaria debilis sensu auct., non (Forsskål) Børgesen 1932b: p. 7; *sensu auct.*, non *Fucus debilis* Forsskål 1775: p. 191.
Polycavernosa debilis sensu auct., non (Forsskål) Fredericq and J. N. Norris 1985: p. 152.

Plants are tough, coarse and cartilaginous, bushy, and often appear gnarled; axes are 3–6 mm in diam., to 30 cm tall, pale yellow to deep red, and attached by fleshy holdfasts. Branching is abundant and irregular or alternate; branchlets are short, acute to recurved, and secund. Medullary cells are thick-walled, 100–320 µm in diam., and grade into small cortical cells. The cortex has 3–7 layers; inner cells are 35–70 µm in diam., while surface cells are ovoid to tear-shaped and 5–15 µm in diam. Tetrasporangia are cruciately divided. Cystocarps are obvious, wartlike and on branchlets. Dried plants appear dull gray to brown, are "horny," and do not adhere well to herbarium paper.

Common, on limestone at semi-exposed to protected habitats (to 30 m) on all coasts; often in clumps on the south side of Bahia Honda Key; also known from Bermuda, the Caribbean, Gulf of Mexico, Brazil, West Africa, and Indian Ocean.

Taylor 1960: p. 442; Dawes 1974: p. 129; Littler and Littler 2000: p. 118.

Hydropuntia crassissima (P. Crouan and H. Crouan in Schramm and Mazé) M. J. Wynne 1989: p. 477 (Gr. *crassi*: fleshy, thick + *issima*: much). Plate XLVII, fig. 4, 5.

Basionym. *Plocaria crassissima* P. Crouan and H. Crouan in Schramm and Mazé 1865: p. 20.
Synonyms. *Gracilaria crassissima* (P. Crouan and H. Crouan) P. Crouan and H. Crouan in Schramm and Mazé 1866: p. 46; also P. Crouan and H. Crouan in Mazé and Schramm 1878 (1870–1877): p. 218.
Gracilaria horizontalis Collins and Hervey 1917: p. 111.
Polycavernosa crassissima (P. Crouan and H. Crouan in Schramm and Mazé) Fredericq and J. N. Norris 1985: p. 152.

Plants are prostrate to creeping, rubbery, tough, mottled green red to golden copper, in masses to 30 cm in diam., attached by discoid holdfasts and also by rhizoids of prostrate lobes. Axes are flat, entangled, usually recurved, to 5 mm thick, and 2 cm wide; branching is irregularly palmate to alternate. Marginal branches are terete to flat and curve downward along edges of lobes. Medullary cells are thick-walled and 80–460 µm in diam. The cortex has 5–7 layers; outer cells are radially compressed; surface cells are in short chains, pigmented, ovoid to tear-shaped, and 8–15 µm long by 3–5 µm in diam. Tetrasporangia are cruciately divided and to 20 µm in diam.

Uncommon, firmly attached to intertidal limestone in areas of moderate to high wave action (to 11 m) in the Keys; also known from Bermuda, the Caribbean (type location: Guadeloupe), and on exposed limestone in southwest Yucatan.

Taylor 1960: p. 443; Littler and Littler 2000: p. 120.

Hydropuntia secunda Gurgel and Fredericq 2004: p. 156 (L. *secundus*: following, second). Plate XLVII, fig. 6, 7.

Synonym. *Gracilaria secunda* P. and H. Crouan in Schramm and Mazé 1865: p. 19, *nom. illeg.*; non. *G. secunda* (C. Agardh) Zanardini 1840a: p. 214 (= *Sphaerococcus secundus* Agardh).

Attached plants are bushy, erect, 7–15 cm tall, fleshy to brittle, not adhering well to paper when dry, pale yellow to dark brown, and attached by discoid holdfasts. Main axes are cylindrical to slightly compressed, 1.6–2.0 mm in diam., filiform, and mostly obtuse; branches taper to tips. Branching is sparse, subdichotomous to irregular, wide angled, and may be secund. The medulla has 8–9 layers; inner cells are polygonal to round, 130 by 180 µm in diam., and thick-walled; outer cells are smaller, round to radially elongate, and 50 by 80 µm in diam.; cells are often rich in floridean starch. The cortex has multiple layers; inner cells are 4–12 by 15–20 µm and form a distinct subcortex (especially in tetrasporic plants); cells of outer two layers are dome-shaped, wider than tall (10 by 12 µm), and rarely compressed.

Drift plants are bushier than attached ones and have profuse subdichotomous to irregular branching in all directions. Main axes are inconspicuous, 750–925 µm in diam., and terete throughout. The cortex usually has two layers of cells.

Common? Identified from Tampa Bay using molecular tools (Gurgel and Fredericq, 2004); previously confused with *Hydropunta caudata*. Also collected in drift on the east and west coasts, from Texas, and the French West Indies (type location: Guadeloupe).

HALYMENIALES

The order was established based on molecular studies (Saunders and Kraft, 1996; Kraft and Saunders, 2000; Saunders et al., 2004) and included the Halymeniaceae, Sebdeniaceae, and Tsengiaceae (Saunders and Kraft, 2002) with several genera from Kylin's (1956) Cryptonemiales (De Smedt et al., 2001). There is debate on which name to use, Halymeniales or Cryptonemiales (Silva, 2002). The phylogenetic position of the Sebdeniaceae was thought to molecularly cluster with both the Halymeniaceae and Rhodymeniales (Gavio et al., 2005). However the family was raised to ordinal status and thought to be intermediate between the Halymeniales and Rhodymeniales (Withall and Saunders, 2006) based on small and large subunit ribosomal DNA analyses.

Plants are foliaceous or terete and attached by discoid holdfasts. The cortex is pseudoparenchymatous and the medulla is filamentous; secondary pit connections are common between inner medullary and inner cortical cells. Life histories are triphasic, with isomorphic gametophytes and tetrasporophytes. Carpogonial branches have 2–4 cells and grow outward. Plants are nonprocarpial, with auxiliary and supporting cells on different cortical branches. Fusion cells are small and have branched connecting filaments that arise directly from fertilized carpogonia. The gonimoblast initial is 1-celled; carposporophytes are lobed and most cells produce carposporangia. Tetrasporangia are cruciately divided.

Halymeniaceae

Synonym. Corynomorphaceae

Molecular studies (Saunders et al., 2004) suggested that *Corynomorpha* should be in the Halymeniaceae and not the Corynomorphaceae (Balakrishnan, 1962). The Halymeniaceae contains about 20 genera (Womersley and Lewis, 1994).

Plants are soft to cartilaginous and erect or crustose; axes are terete, compressed, or foliaceous with simple, alternate, dichotomous, or pinnate branching, and attached by discoid holdfasts. Axes are multiaxial; medullary filaments are fused to and adjacent to refractive stellate cells with connecting arms. Cortical cells are in branched anticlinal rows and stellate or round. Outer cortical cells are pigmented and form a compact surface. Tetrasporangia are scattered or on papillae. Gametophytes are monoecious or dioecious. Spermatangia arise from outer cortical cells. Carpogonial branches are 2-celled; auxiliary cells occur in ampullae or separate branch systems (Chiang, 1970). Connecting filaments (ooblasts) arise from fertilized carpogonia and hypogenous cells often fuse to one or more auxiliary cells in lower parts of ampullae. Cystocarps are elevated and on the surface.

Corynomorpha J. Agardh 1872: p. 3, *nom. cons.* (Gr. *coryno*: clublike + *morpha*: shaped).

Plants are solitary or clustered and on cylindrical stalks; axes are club-shaped, terete or prismatic, and attached by discoid holdfasts. The medulla has loose filaments embedded in mucilage that fuse with one another. The cortex has rows of small cells derived from radially spreading, dichotomously branched filaments. Cystocarps are immersed in swollen tips of club-shaped branches.

Corynomorpha clavata **(Harvey) J. Agardh 1872: p. 4 (L. *clavata*: club-shaped). Plate XLVII, fig. 8, 9.**

Basionym. *Acrotylus clavatus* Harvey 1853: p. 196.

Plants are solitary or clustered, firm to fleshy, erect, to 6.5 cm tall, and on slender stalks 2–5 mm long; the stalks produce one to several cylindrical club-shaped axes and are attached by discoid holdfasts. Axes are 3–7 mm in diam., 2.0–4.5 cm long, and have obtuse tips. Medullary filaments are slender, few, and in a gel that may contain Cyanobacteria. The cortex is 50–100 µm thick with cells tightly packed in radial rows and surface cells 3–5 µm in diam. Cystocarps are embedded in swollen tips.

Uncommon, but locally abundant on shallow limestone (2–3 m) in the Keys (type location: Key West); also known from Bermuda and the Caribbean.

Taylor 1960: p. 429; Littler and Littler 2000: p. 72.

Cryptonemia J. Agardh 1842: p. 100 (Gr. *crypto*: obscure, covered + *nema*: thread).

Plants are erect, stalked, and attached by holdfasts. Blades are firm, strap-shaped, palmately to dichotomously lobed or branched, and may have a basal midrib. The medulla has a mixture of thick, slender filaments and stellate cells. The compact cortex has large inner and small outer cells. Tetrasporangia are tetrahedrally divided and from cortical cells. Gametophytes are dioecious; spermatangia are in large sori. Carpogonial branches are 2-celled and in ampullae. Carposporophytes are immersed and slightly elevated on one surface of the blade; cystocarps have ostioles and involucral filaments.

Key to the species (2) of *Cryptonemia*

1. Blades usually lack a midrib; margins with star-shaped teeth . . . . *C. crenulata*
1. Lower parts of blades with a midrib; margins smooth. . . . . *C. seminervis*

Cryptonemia crenulata (J. Agardh) J. Agardh 1851 (1848–1901): p. 225–226 (L. *crenulata*: notched, scalloped). Plate XLVII, fig. 10–13.

Basionym. *Phyllophora crenulata* J. Agardh 1841: p. 18.

Plants are often bright red and 5–15 cm tall; blades are strap-shaped and on short stalks attached by small lobed holdfasts. Blades are 0.3–1.5 cm wide, 60–150 µm thick, irregular, dichotomous, or palmately divided, lack midribs, and have indented tips; margins are undulate, toothed, forked, star-shaped, or proliferous. The medulla has a network of filaments and stellate ganglia; filaments are 2–10 µm in diam. and have ganglia extensions to 10 µm in diam. The cortex has 1–3 layers; inner cells are spherical and 10–30 µm in diam.; outer cells are angular, deeply pigmented, and to 5–7 µm in diam. Tetrasporangia are cruciately divided, 8–10 µm in diam., 10–13 µm long, and occur near blade tips.

Uncommon, on limestone (14–30 m) on the west coast and in shallow waters (2–3 m) in the Keys; also known from Bermuda, North Carolina, the Bahamas, Caribbean, Gulf of Mexico, and Brazil (type location: Bahia; Silva et al., 1996), as well as the eastern tropical Atlantic and Indo-Pacific.

Taylor 1960: p. 427; Littler and Littler 2000: p. 100.

Cryptonemia seminervis (C. Agardh) J. Agardh 1846: pl. XVII; 1876 (1848–1901): p. 165 (L. *semi*: partially + *nervis*: nerved, with veins). Plate XLVII, fig. 14.

Basionym. *Sphaerococcus seminervis* C. Agardh 1821: p. 2, pl. XVII.

Synonym. *Cryptonemia luxurians* (C. Agardh) J. Agardh 1851 (1848–1901): p. 228.

Plants are foliaceous, deep red, 5–25 cm tall, and on cylindrical stipes attached by discoid holdfasts. Stipes are 1.0–1.5 mm long and extend halfway up the blades, forming a midrib. Blades are irregularly to palmately branched 3–7 times, 5–15 mm wide, and 100–200 µm thick; margins are undulate, entire, eroded, or proliferous, lack teeth, and have blunt tips. Medullary filaments are compact and 2–8 μm in diam. The cortex has 2–3 layers; inner cells are spherical and 10–25 μm in diam.; surface cells are pigmented, rectangular, and 5–7 μm in diam. Tetrasporangia are cruciately divided, ovoid, 7–10 μm in diam., 10–13 μm long, and often in marginal leaflets at tips. Gametophytes are dioecious. Spermatangia are 2–3 μm in diam. and on marginal blades. Carposporophytes are spherical, 100–150 μm in diam., have involucral filaments, and are in marginal blades or at blade tips. Carposporangia are angular and 7–10 μm in diam.

Rare, inconspicuous, in cracks or under ledges subtidally (10–25 m) in the Dry Tortugas and usually encrusted by bryozoans and coralline algae; also known from North Carolina (17–40 m), Bermuda, the Caribbean, and Brazil, as well as the eastern North and tropical Atlantic (type location: Cádiz, Spain; Silva et al., 1996), and Indo-Pacific.

Taylor 1960: p. 428; Schneider and Searles 1991: p. 272; Littler and Littler 2000: p. 102.

Grateloupia C. Agardh 1822 (1820–1828): p. 221, *nom. cons. prop.*, non *Grateloupia* Bonnemaison 1822: p. 189 (named for J. P. F. Grateloup, a French naturalist).

Synonyms. *Prionitis* J. Agardh 1851 (1848–1901): p. 185.
Dermocorynus P. L. and H. M. Crouan 1858: p. 69.
Sinotubimorpha Li and Ding 1998: 1–5.

See Dixon (1959), Silva (1980), Parkinson (1980a, 1981), and Scagel et al. (1989) for a discussion of the genus. Wilkes et al. (2005a) considered *Grateloupia* to be the largest genus of the family, with more than 50 species, and one of the most difficult complexes to separate.

Plants are erect and gelatinous to firm; axes are foliose to strap-shaped, compressed, elongate, and on short stipes attached by discoid holdfasts. Bladelets are flat, pinnate, and arise from margins. Medullary filaments are longitudinal, interconnected, and in a dense gel. The cortex has 4–8 layers of cells in anticlinal rows and is subtended by 1–2 layers of stellate cells; rhizoids grow from inner cells. Tetrasporangia are tetrahedrally divided and from outer cortical cells. Gametophytes are monoecious or dioecious. Spermatangia are in superficial sori. Carpogonial branches are 2-celled and in ampullae. Carposporophytes are scattered and immersed, with most gonimoblasts producing carposporangia. Cystocarps are ostiolate, protrude from one surface, and have involucral filaments.

Key to the species (3) of *Grateloupia*

1. Axes linear, with many marginal pinnate branches; bases symmetrical 2
1. Axes narrow to wide, with few pinnate branches; bases asymmetrical . . . . *G. cuneifolia*
   2. Mains axes more than 200 μm thick, flat and strap-shaped . . . . *G. gibbesii*
   2. Main axes to 200 μm thick, cylindrical to compressed. . . . . "*G. filicina*"

Grateloupia cuneifolia J. Agardh 1849: p. 85 (L. *cuneifolia*: leaves wedge-shaped or tapering toward the base). Plate XLVII, fig. 15–17.

Synonym. *Grateloupia doryphora sensu auct.*, non (Montagne) M. A. Howe 1914: p. 169.

Ardré and Gayral (1961) assigned most of the foliose Atlantic and Pacific forms to *Grateloupia lanceola* (J. Agardh) J. Agardh, while others (Taylor, 1960; Schneider and Searles, 1991; Wynne, 2005a) have recognized multiple species.

Plants are polymorphic, foliose, and 4–40 cm tall; stipes are short, irregularly divided, expanding asymmetrically into blades, and attached by discoid holdfasts. Blades are linear, oval, lanceolate, or crescent-shaped, sometimes highly divided to palmately lobed, 4–5 cm in width, and 160–250 μm thick. Blade bases are shield-shaped; margins are smooth, dentate, or eroded; tips are acute. Medullary filaments are thin, 1–3 μm diam., entangled in a dense gel, and surrounded by stellate cells, 5–10 µm in diam. The cortex has 3–7 layers and surface cells are pigmented and 2–4 µm in diam. Tetrasporangia are ovoid, cruciately divided, 8–19 µm in diam., 26–35 µm long, and scattered in the inner cortex. Cystocarps are spherical, to 130 µm in diam., and with inconspicuous ostioles; carposporangia are 8–13 µm in diam.

Uncommon, but occurring year-round, on wave-exposed intertidal coquina at Marineland; also known from the western Caribbean, Brazil, Uruguay, and Italy.

Taylor 1960: p. 425; Schneider and Searles 1991: p. 273.

Grateloupia gibbesii Harvey 1853: p. 199, pl. XXVI (named for Lewis R. Gibbes, an American collector who sent seaweed specimens to Harvey for identification). Plate XLVII, fig. 18, 19.

Ardré and Gayral (1961) placed *Grateloupia gibbesii* in *G. lanceola* (J. Agardh) J. Agardh, while Dawson et al. (1964) considered *G. gibbesii* to be a synonym of *G. doryphora (*Montagne) Howe (= *G. turuturu* Yamada). Wiseman (1966) and Schneider and Searles (1991) recognized *G. gibbesii* and *G. cuneifolia* as distinct taxa based on morphological features.

Plants are polymorphic, 5–60 cm tall, and pale to dark purple or red; blades are erect, with one to several fronds arising from short stipes attached by discoid holdfasts. Blades are simple, linear to lanceolate, irregularly branched, marginally pinnate, 1–5 cm wide, and 240–950 µm thick; pinnae are elongate and 2–3 mm wide by 9 cm long. Medullary filaments are loosely entangled in a gel, 1–4 µm in diam., and from inner cortical cells. The cortex has 3–8 layers; inner cells are stellate and 5–10 µm in diam.; surface cells are pigmented, rounded, and 2–6 µm in diam. Tetrasporangia are cruciately divided, ovoid, 10–18 µm in diam., 28–45 µm long, and in marginal pinnae. Gametophytes are monoecious. Spermatangia are 2–3 µm in diam., in scattered sori, and are from outer cortical cells. Carposporophytes are immersed; carposporangia are ovoid and 9–16 µm in diam. Cystocarps are ostiolate, scattered, spherical, and to 150 µm in diam.

Uncommon, on jetties and limestone on all coasts; also known from South Carolina (type location), the Lesser Antilles, and Gulf of Mexico.

Taylor 1960: p. 425; Schneider and Searles 1991: p. 275; Littler and Littler 2000: p. 104.

Sedis incertae: "*Grateloupia filicina* (J. V. Lamouroux) C. Agardh" 1822 (1820–1828): p. 223 (L. *filicina*: fernlike). Plate XLVII, fig. 20, 21.

Basionym. *Delesseria filicina* J. V. Lamouroux 1813: p. 125 (reprint p. 37).
Synonyms. *Grateloupia concatenata* Kützing 1849: p. 731.
Grateloupia filicina var. *elongata* Kützing 1849: p. 730.
Grateloupia filicina var. *bipinnata* P. Crouan and H. Crouan in Mazé and Schramm 1878 (1870–1877): p. 154.
Grateloupia filicina var. *congesta* P. Crouan and H. Crouan in Mazé and Schramm 1878 (1870–1877): p. 155.

Using molecular tools, De Clerck et al. (2005a) found that a number of cryptic taxa were originally identified originally as *Grateloupia filicina*, while the true species was limited to the Mediterranean. Plants from Florida and tropical areas formed a single clade, but the present data were too limited to separate them. Wang et al. (2000, 2001), Kawaguchi et al. (2001), Faye et al. (2004a), and Wilkes et al. (2005) have also evaluated this complex.

Plants are slick, purple to red, and 5–10(75) cm tall; stipes are short, with one or more blades, and attached by small discoid holdfasts. Axes are compressed to flat, 0.5–5 mm wide, and 150–200 µm thick. Branching is irregular to marginally pinnate with lower branches often highly divided to palmately lobed; branchlets are abundant, elongate, and 0.3–1 mm wide. Medullary filaments are terete, 3–4(7) µm in diam., and loosely entangled in a gel. The cortex has 3–7 layers; inner cells are irregular and 10–20 µm in diam., while surface cells are pigmented, ovoid, and 3–5 µm in diam. Tetrasporangia are cruciately divided, ovoid, 11–27 µm in diam., 25–47 µm long, and in marginal bladelets. Gametophytes are monoecious. Spermatangia are in scattered patches in the outer cortex. Carposporangia are ovoid and 7–20 µm in diam. Cystocarps are mostly spherical, 150–300 µm in diam., and ostiolate.

Common, in sheltered sites on jetties of the east and west coasts and to 20 m in the Keys; also known from the Carolinas, Georgia, the Bahamas, Antilles, Brazil, and Uruguay, as well as the eastern North and tropical Atlantic, Mediterranean (type location of true species: Trieste, Italy; Abbott, 1999), Red Sea, Pacific Ocean, and Australia.

Taylor 1960: p. 424: Dawes 1974: p. 126; Schneider and Searles 1991: p. 274; Littler and Littler 2000: p. 104.

Halymenia C. Agardh 1817: p. XIX, 35, *nom. cons.*, cf. Parkinson 1980b (Gr. *hals*: the sea; *hali*: salt + *humen*: a membrane).

Plants are erect or sprawling, mucilaginous to firm, and foliose or bushy; fronds are entire, lobed, or branched, with flat to compressed forks, and on short stipes attached by discoid holdfasts. Medullary filaments are loosely organized in a gel; some filaments traverse the medulla and others radiate from inner cortical cells or stellate ganglia. Inner cortical cells are large and hyaline; surface cells are small and pigmented. Tetrasporangia are cruciately divided and from cortical cells. Gametophytes are monoecious, nonprocarpial, with spermatangia in large irregular sori. Carpogonial branches are 2-celled, distinct from auxiliary cells, and in ampullae. Carposporophytes are immersed, scattered, with most gonimoblasts becoming carposporangia; cystocarps cause swelling of one or both sides of blades. Frond size and thickness, plus the presence or absence of a stipe, basal rib, and ganglia are used to separate species (Guimarães and Fujii, 1998).

Key to the species (8) of *Halymenia*

1. Plants have one or more foliaceous to flat blades 2
1. Plants have cylindrical axes; forks may be somewhat compressed .... *H. elongata*
 2. Blades branched or lobed 3
 2. Blades irregularly lobed to proliferous 4
3. Blades divided several times; final branches narrow or filiform .... *H. floresii*
3. Blades not divided several times; final branches strongly serrate .... *H. pseudofloresii*
 4. Blades entire, with few to no lobes; margins entire 5
 4. Blades with lobes; margins dentate to serrate .... *H. floridana*
5. Subcortex and medulla lack or rarely have stellate ganglia .... 6.
5. Subcortex and medulla have stellate ganglia .... *H. rosea.*
 6. Blades firm to membranous, thin (to 35–50 µm thick) .... *H. hancockii.*
 6. Blades gelatinous to membranous, thick (125–300 µm thick) .... *H. integra*

Halymenia elongata C. Agardh 1822 (1820–1828): p. 209 (L. *elongata*: long, elongate). Plate XLVII, fig. 22.

Synonyms. *Halymenia trigona sensu auct.*, non (Clement y Rubio) C. Agardh 1822: p. 211. *Halymenia agardhii sensu auct.*, non De Toni 1905: p. 1542.

Silva (1991) noted that *Halymenia elongata* is the proper name for Florida plants called *H. agardhii* De Toni and *H. trigona* (Clement y Rubio) C. Agardh.

Plants are gelatinous, fleshy, pink to deep red, 5–20 cm tall, and attached by discoid holdfasts. Axes are cylindrical and slightly compressed at forks; branching is dichotomous, 4–12 times, and in different planes. Sinuses are rounded and broad below; segments are 2–7 mm in diam. and taper to rounded tips. Medullary filaments are loosely arranged, longitudinal, 2–6 µm in diam., and mixed with a few stellate ganglia. The cortex has 1–4 layers; inner cells are 20–30 µm in diam. and loosely arranged; surface cells are pigmented, 3–9 µm in diam., elongate to club-shaped, 3–13 µm long, and ovoid in surface view. Tetrasporangia are ovoid, cruciately divided, 10–15 µm in diam., 15–20 µm long, and in the outer cortex. Gametophytes are dioecious. Carposporophytes are embedded and carposporangia 5–22 µm in diam. Cystocarps are spherical, 70–180 µm in diam., and in the medulla. Cystocarps do not project and have inconspicuous ostioles.

Common, on limestone (5–20 m) on all coasts; also known from the Antilles, as well as Spain (type location: Cádiz; Silva et al., 1996), the Mediterranean, eastern tropical Atlantic, Pakistan, and Australia.

Taylor 1960: p. 417; Dawes 1974: p. 126; Schneider and Searles 1991: p. 281; Littler and Littler 2000: p. 106.

Halymenia floresii (Clemente y Rubio) C. Agardh 1817: XIX; 1822 (1820–1828): p. 209; 1824: p. 243; with an orthographic change from '*floresia*' by Schneider, 2003 (probably named for D. Francisco Flores, a Mexican landowner). Plate XLVIII, fig. 1, 2.

Basionym. *Fucus floresius* Clemente y Rubio 1807: p. 312

Guiry and Nic Dhonncha (2006) cited the basionym as *Fucus floresii*. See Kawaguchi (2004) for a morphological description of type material.

Blades are gelatinous, to 50 cm tall, pink to rose red and on short stipes attached to small discoid holdfasts. Blades are to 30+ cm wide and 120–400 µm thick. Branching is basal, pinnate to subdichotomous, and 1–2 times; branchlets are slender, marginal, linear, and with acute tips. Medullary filaments are longitudinal, loose, in a gel, 3–10 µm in diam., and mixed with a few stellate ganglia. The cortex has 2–3 layers; inner cells are 10–25 µm in diam.; surface cells are pigmented and 5–8 µm in diam. Tetrasporangia are ovoid, cruciately divided, 10–18 µm in diam., 13–30 µm long, and among surface cells. Gametophytes are dioecious. Carposporangia are ovoid and 7–12 µm in diam. Cystocarps are 150–230 µm in diam., spherical, lack ostioles, and do not project.

Common, on limestone (10–30 m) and adrift on all coasts; also known from North Carolina (30–50 m), Bermuda, the Caribbean, Gulf of Mexico, and Brazil, as well as the eastern North and tropical Atlantic (type location: Cádiz, Spain; Kawaguchi, 2004), Mediterranean, Red Sea, Indo-Pacific, and Australia.

Taylor 1960: p. 418; Dawes 1974: p. 126; Schneider and Searles 1991: p. 277; Littler and Littler 2000: p. 108.

Halymenia floridana J. Agardh 1892: p. 59 (L. *floridana*: of or from Florida). Plate XLVIII, fig. 3–6.

Synonyms. *Halymenia bermudensis* Collins and M. Howe 1916: p. 169.

Halymenia gelinaria Collins and M. Howe 1916: p. 173–175.

According to Schneider and Searles (1991), *Halymenia gelinaria* and *H. floridana* overlap with respect to blade thickness, orientation of medullary filaments, structure of the outer cortex, and number of stellate ganglia. Guimarães and Fujii (1998) noted that *H. bermudense* is conspecific with *H. floridana* and that some North Carolina specimens identified by Schneider and Searles (1991) are *H. brasiliana* Guimarães and Fujii 1998. Thus, *H. brasiliana* may be in Florida. Further, *H. vinacea* M. Howe and Taylor reported from deep water off the east coast (Hanisak and Blair, 1988) may be conspecific with *H. floridana* (Schneider and Searles, 1991).

Blades are fleshy to membranous, 5–20(60) cm tall, and with one or more on short stipes attached by small discoid holdfasts. Blades are 4–20 cm wide, oval, cordate, or elliptical, often lobed, and 60–550 µm thick, with entire, dentate, or lacerated margins and obtuse tips. Medullary filaments are loose to densely packed, longitudinal, and 3–14 µm in diam. The inner cortex has 1–3 layers, with transverse anastomosing filaments and a few to many ganglia, 15–65 µm in diam. and with 3–10 arms. The outer cortex has up to 3 layers; surface cells are 5–10 µm in diam., angular, and pigmented; inner ones are larger and 13–25 µm in diam. Tetrasporangia are ovoid, cruciately divided, 8–13 µm in diam., 12–26 µm long, and in the outer cortex. Gametophytes are dioecious; spermatangia are 2–3 µm in diam. and in surface sori. Carposporophytes are immersed and scattered; carposporangia are round and 4–8 µm in diam. Cystocarps are globose, 100–250 µm in diam., swollen on one or both sides of blades, and lack ostioles.

Uncommon, from 10–72 m or adrift off the east (type location: Gilberts Bar; Silva et al., 1996) and west coasts; also known from the Carolinas, Georgia, Bermuda, the Caribbean, Gulf of Mexico, Brazil, the Salvage and Canary Islands, and Yemen.

Taylor 1960: p. 419, 420; Dawes 1974: p. 127, 128; Schneider and Searles 1991: p. 277, 279.

Halymenia hancockii W. R. Taylor 1942: p. 98–99, pl. 3, fig. 6; pl. 14, fig. 1–5 (named for Captain Allan Hancock, sponsor of the Caribbean Hancock Expedition of 1939 in which W. R. Taylor participated and published the seaweed flora). Plate XLVIII, fig. 7–9.

Blades are firm, purple to red, 3–10 cm tall, to 2 cm wide, 35–50 µm thick, and on stipes 1–3 mm long that are attached by small discoid holdfasts. Blades are lanceolate to oval, tapering to bases and tips, and with bilobed tips and marginal bladelets. Medullary filaments are few, longitudinal, loose in a gel, 3–8 µm in diam., and connected to subcortical cells with 3–4 arms; stellate ganglia are absent. The cortex has 1–2 layers; surface cells are round and 8–10 µm in diam. Tetrasporangia are ovoid, cruciately divided, 7–13 µm in diam., 12–16 µm long, and among surface cells. Spermatangia are unknown. Carposporangia are round to irregular and 5–7 µm in diam. Cystocarps are compressed, spherical, 50–280 µm in diam., swell both surfaces, and lack ostioles.

Rare, reported from 42–93 m off the east coast (Hanisak and Blair, 1988); also known from North (17–50 m) and South Carolina, Georgia, Puerto Rico, Colombia (type location), and the eastern tropical Atlantic.

Taylor 1960: p. 421; Schneider and Searles 1960: p. 280.

Halymenia integra M. Howe and W. R. Taylor 1931: p. 25–26, fig. 14, pl. 1, fig. 5, 6; pl. 2, fig. 8 (L. *integra*: whole, entire). Plate XLVIII, fig. 10–12.

Blades are gelatinous, 3–20 cm tall, and on short stalks (1–6 mm long) that are attached by discoid holdfasts. Blades are oval or elongate, 1–6 cm wide, 125–300 µm thick, with pointed or rounded tips and shield-shaped bases. Medullary filaments are loose in a gel and 2–6 µm in diam. The inner cortex has 2–3 layers; inner cells are ovoid to spherical and 100–180 µm in diam.; surface cells are 20–50 µm in diam., ovoid to angular in surface view, 6–18 µm in diam., widely spaced, and in a surface gel 8–10 µm thick.

Rare, reported from 44 m off the east coast (Hanisak and Blair, 1988); also dredged offshore Brazil (type locality).

Taylor 1960: p. 422.

Halymenia pseudofloresii Collins and M. Howe 1916: p. 177; as an orthographic change from '_pseudofloresia_' by Schneider 2003 (_pseudo_: false + _floresii_; thus, not the species _H. floresii_, which was named for D. Francisco Flores). Plate XLVIII, fig. 13, 14.

Blades are membranous, gelatinous, pink to deep red, to 40 cm tall, 25 cm wide, and on stipes (1–10 mm long) that are attached by small discoid holdfasts. Blades are oval or elliptical, 7–36 cm long, and often deeply divided with irregular to subpalmate lobes or 1–8 cm wide divisions. Blade margins often have lanceolate, triangular, or pinnate proliferations with acute tips. Medullary filaments are cylindrical, 10–16 µm in diam., loose, and oriented transversely and longitudinally. The cortex has 2–6 layers with a few inner stellate cells, 20–55 µm in diam. The outer cortical cells are elongate, pigmented, 4–10 µm in diam., and 2–5 times taller than broad; three surface cells cover one subcortical cell. Tetrasporangia are spherical to ovoid, cruciately divided, 10–15 µm in diam., 14–26 µm long, and among surface cells.

Uncommon, on limestone (to 30 m) and adrift on all coasts; also known from Bermuda (type location) and the Caribbean.

Taylor 1960: p. 418; Dawes 1974: p. 127; Littler and Littler 2000: p. 108.

Halymenia rosea M. Howe and W. R. Taylor 1931: p. 28–29, fig. 15; pl. 1, fig. 3, 4; pl. 2, fig. 7 (L. _rosea_: rose, rose red). Plate XLVIII, fig. 15–17.

John et al. (2004) noted that *Halymenia rosea* may belong to *Sebdenia*.

Blades are moderately gelatinous, rose to dull red, oval to round, entire, 4–6 cm tall, 3–4 cm wide, and on short stipes (2–4 mm long) attached by small discoid holdfasts. Medullary filaments are loose and in transverse and longitudinal patterns; stellate cells are common in the medulla and inner cortex, 20–55 µm in diam. and have 7–16 arms. The cortex has 2–3 layers; inner cells are ovoid to elliptical, 40–65 µm in diam., and have short filaments radiating into the medulla; surface cells are small, round, pigmented, and 4–10 µm in diam. Reproduction is unknown.

Rare, reported from 27–88 m off the east coast (Hanisak and Blair 1988); also known from deep water off Brazil (type location) and the eastern tropical Atlantic.

Taylor 1960: p. 422.

Sebdeniaceae

See Saunders and Kraft (1996) for a discussion of the family's taxonomy.

Plants are soft to firm, erect, and attached by discoid holdfasts. Axes are terete to flat or blade-like. The medulla has loose, intertwined, longitudinal filaments in a clear gel mixed with stellate cells (ganglia); the latter also occur in the inner cortex. Cortical cells grade into small surface cells. Life histories are isomorphic. Tetrasporangia are cruciately divided and laterally on outer cortical cells. Gametophytes are monoecious or dioecious; spermatangia are subglobose, from outer cortical cells, and in surface sori. Plants are nonprocarpial; carpogonial branches are 3(4)-celled and lateral on inner cortical cells; auxiliary cells are intercalary. Gonimoblasts grow toward the surface. Carposporophytes lack special involucral filaments and most gonimoblastic cells become carposporangia. Cystocarps are ostiolate, embedded, and from periclinal divisions of the cortex.

Sebdenia (J. Agardh) Berthold 1882: p. 530; 1884: p. 21 [source of name unknown; *Sebdenia* is a section of *Halymenia* described by J. Agardh 1876 (1848–1901): p. 134 that Harvey considered using for a new genus; Berthold (1882) cited *Sarcodia (Sebdenia) platycarpa* from Harvey's "Algal Exsic. Ceylon, Friendly Island Algae" no 52].

The genus has the same characteristics as the family.

Sebdenia flabellata (J. Agardh) P. G. Parkinson 1980b: p. 12 (L. _flabellata_: fanlike, fan-shaped). Plate XLVIII, fig. 18, 19.

Basionym. *Isymenia flabellata* J. Agardh 1899: p. 66.
Synonyms. *Halymenia agardhii* De Toni 1905: p. 1542.
Halymenia polydactyla Børgesen 1932a: p. 122–124, fig. 10, pl. IV.
Sebdenia polydactyla (Børgesen) M. Balakrishnan 1961 ('1960'): p. 89–90.

Schneider and Searles (1991) reviewed the terete *Halymenia* taxa of the western Atlantic.

Plants are gelatinous to firm, pink to deep red, erect, 4–30 cm tall, and attached by small discoid holdfasts. Axes are cylindrical, 3–10 mm in diam, and compressed at forks that are equidistant (0.7–3.5 cm) along branches. Branching is dichotomous, usually in one plane, and with obtuse or forked tips. The inner medulla has thin-walled axial filaments (3–10 µm in diam.) and long-armed stellate ganglia, 18–40 µm in diam. The outer medulla has loose ganglia, 12–25 µm in diam., and dark-staining glandlike cells. The cortex has 3–7 layers; inner cells are 20–35 µm in diam. and irregularly lobed; surface cells are pigmented, polygonal in surface view, thick-walled, ovoid, 3–9 µm in diam., and 3–13 µm long. Tetrasporangia are ovoid, cruciately divided, 13–27 µm in diam., and 19–30 µm long. Spermatangia are pyriform, 2–3 µm in diam., and terminal on outer cortical cells, near branch tips. Carpogonial branches are 3–4-celled and lateral on central cortical cells; auxiliary cells and nearby vegetative cells stain deeply with aniline blue. Nutritive cells are formed after fertilization. Gonimoblasts are derived from auxiliary cells. Cystocarps are ostiolate, 150–300 µm in diam., and lack involucral filaments. Pericarp walls have 4–7 layers of flat cells. Carposporangia are subspherical to irregular and 12–20 µm in diam.

Uncommon, on limestone and rubble on all coasts; also known from North Carolina (17–55 m), Georgia, Bermuda, the Caribbean (type location: Guadeloupe; Silva et al., 1996), Gulf of Mexico, and Brazil, as well as Spain, the Mediterranean, eastern tropical Atlantic, Indo-Pacific, and Australia.

Taylor 1960: p. 417; Dawes 1974: p. 126; Schneider and Searles 1991: p. 269; Littler and Littler 2000: p. 108.

NEMASTOMATALES

The order was established by Kylin (1925), merged with the Gigartinales (Kylin, 1932), and was resurrected by Saunders and Kraft (2002) for the Nemastomataceae and Schizymeniaceae.

Plants are gelatinous to slippery, multiaxial, nonprocarpical, and mostly lack pit connections. Tetrasporophytes, where known, are filamentous clusters or adhesive crusts that produce zonate tetrasporangia. Carpogonial branches are 3-celled and may represent a cortical filament of the inner supporting cells. Auxiliary cells are intercalary and arise in vegetative filaments or unbranched rhizoids. Gonimoblasts are lateral or directed outward and arise from auxiliary cells. Carposporophytes are mostly carposporangia.

Nemastomataceae

Synonym. Gymnophloeaceae (Wynne, 1986a).

The four genera in the family (*Adelophycus*, *Itonoa*, *Nemastoma*, and *Predaea*) have heteromorphic life histories (Masuda and Guiry, 1995; Gavio et al., 2005).

Plants are erect, gelatinous to firm and fleshy, with a few taxa lightly calcified. Fronds are simple, irregularly lobed, or with peltate branching, and attached by discoid holdfasts. Blades are compressed or cylindrical, with a multiaxial medulla of parallel hyaline filaments. Cortical filaments branch at right angles to axes and may have gland cells. Tetrasporangia (if known) are cruciately or zonately divided and from inner cortical cells of erect or prostrate (i.e., crustose) plants. Gametophytes are monoecious or dioecious. Spermatangia are in sori and from outer cortical cells. Plants are nonprocarpical, with carpogonial branches 2(5)-celled, and lateral on inner cortical cells. Carposporophytes lack fusion cells; gonimoblastic filaments extend inward after fusing with obvious auxiliary cells. Cystocarps are embedded, scattered, and may have ostioles or involucral filaments.

Nemastoma J. Agardh 1842: p. 89, *nom. cons.* (Gr. *nema*: a thread + *stoma*: a mouth)

Plants are erect, gelatinous or fleshy, and attached by discoid holdfasts. Axes are foliaceous and lobed or with terete branches. Plants are multiaxial with a loose medulla of slender filaments. The cortex has compact, pigmented, and branched filaments mixed with rhizoidlike threads.

***Nemastoma gelatinosum* M. Howe 1918: p. 536 (L. *gelatinosum*: gelatinous). Plate XLVIII, fig. 20, 21.**

Plants are soft, gooey, 6–16 cm tall, and in bushy tufts; the main axes are flat, 4–7 mm thick, 9–15 mm wide, lobed with blunt tips, and attached by discoid holdfasts. Branching is mostly in one plane and dichotomous to palmate. Medullary filaments are loose in a gel and 7–12 μm in diam. Cortical filaments are at right angles to the medulla, unequally to dichotomously branched (4–7 times), and have swollen cells. Apical cells are spherical and 3–9 μm in diam.; gland cells are also spherical and 20–30 μm in diam.

Uncommon, on rubble (to 15 m) in sheltered areas of the Keys; also known from Bermuda (type location: Howe, 1918).

Taylor 1960: p. 435; Littler and Littler 2000: p. 82.

Predaea De Toni (fil.) 1936: p. 5, unnumbered (named for A. Preda, an Italian phycologist).

Plants are gelatinous and erect or peltate; axes are compressed, simple or lobed, and on stalks attached to discoid holdfasts. Medullary filaments are loose and mostly longitudinal. Cortical filaments are pigmented, they extend at right angles to the medulla, and may have gland cells. Life histories are heteromorphic (Vergés et al., 2004). Tetrasporophytes often form compact, monostromatic crusts of branched filaments, with tetrasporangia cruciately, zonately, or irregularly divided. Spermatangia are subglobose, from outer cortical cells, and in superficial sori. Plants are nonprocarpial; carpogonial branches are 2–3-celled; auxiliary cells are distant, obvious, and mixed with nutritive filaments. Gonimoblasts have 1–3 terminal lobes and arise from an auxiliary cell or connecting filament after fusion. Carposporophytes are embedded and lack involucral filaments and an ostiole.

Key to the species (2) of *Predaea*

1. Plants are multilobed, to 30 cm tall; outer cortical cells cylindrical . . . . *P. feldmannii*
1. Plants are flattened blades, to 8 cm tall; outer cortical cells beadlike . . . . *P. masonii*

Predaea feldmannii Børgesen 1950b: p. 7, fig. 1–3 (named for J. Feldmann, a French phycologist who conducted taxonomic, floristic, and culture studies of Mediterranean seaweeds). Plate XLVIII, fig. 22, 23.

Plants are gelatinous, slippery, 2–30 cm tall, with irregular branching, and attached by small fleshy discoid holdfasts. Axes are compressed to flat, irregularly lobed in various planes, 0.5–6 cm wide, to 1 cm thick, and have a few to many surface papillae and blunt tips. The medulla is multiaxial with thin-walled longitudinal filaments (2–3 µm in diam.) embedded in a gel. Cortical filaments are pigmented, they radiate at right angles to the medulla, branch dichotomously, and are 4–5 µm in diam. Rhizoidal filaments arise from cortical and axial filaments. Gametophytes are dioecious and dimorphic, with males being more highly lobed than females. Spermatangia are spherical, 2–3 µm in diam., and have 1–2 terminal surface cells. Carpogonial branches are 3-celled and on inner cortical cells; auxiliary cells are ovoid, 10–30 µm in diam. by 12–35 µm long, densely stained, and have 4–6 branching chains of spherical nutritive cells, 5–6 µm in diam. Gonimoblast initials occur at junctions of connecting filaments and auxiliary cells and have 2–4 ovoid lobes, are 40–80 µm in diam., and 55–190 µm long. Carposporangia are ovoid or irregular and 7.5–15 µm in diam.

Rare, reported from 26–56 m off the east coast (Hanisak and Blair, 1988); also known from North Carolina, Georgia (18–55 m), the Antilles (15–55 m), Venezuela, and the eastern tropical Atlantic.

Schneider and Searles 1991: p. 266; Littler and Littler 2000: p. 82.

Predaea masonii (Setchell and Gardner) De Toni (fil.) 1936: p. 5, with pages unnumbered (named for Herbert L. Mason, an American botanist who made the original collection). Plate XLVIII, fig. 24, 25.

Basionym. *Clarionea masonii* Setchell and Gardner 1930: p. 174–175, pl. 9, fig. 34–38.

Plants are gelatinous, 2–12 cm tall, and have compressed axes attached by fleshy discoid holdfasts. Blades are subspherical, simple or lobed, to 9 cm wide, 98 mm thick, and may have papillae and faint veins. Medullary filaments are thin-walled and 2–4 µm in diam.; cortical filaments radiate and branch subdichotomously. Inner cortical cells are cylindrical and forked; surface cells are globular, pigmented, and 2–5 µm in diam. Gland cells, if present, are 8–28 µm in diam. Gametophytes are dioecious. Carpogonial branches are 2-celled and on forks of cortical cells. Auxiliary cells are ovoid, 8–30 µm in diam., 15 µm long, and mixed with 5–9 branched chains of nutritive cells; each chain is up to 7 cells long with cells, 2–4 µm in diam. Auxiliary and nutritive cells stain densely. Gonimoblast initials are at the junctions of connecting filaments and auxiliary cells and form one lobe (50–200 µm in diam. by 50–100 µm long). Carposporangia are angular, ovoid, and 6–15 µm in diam.

Rare, reported from 34–59 m off the east coast (Hanisak and Blair, 1988); also recorded from North Carolina, the Caribbean, and Brazil, as well as the Canary Islands, Ghana, California, and Pacific Mexico (type location: Isla Clarion, Revillagigedo).

Schneider and Searles 1991: p. 267.

Schizymeniaceae

Plants are erect, bladelike, slippery to soft, and attached by crusts or holdfasts. Medullary filaments are entangled. The compact cortex has radially branched filaments. Tetrasporophytes are unknown or crustose with zonate tetrasporangia. Spermatangia are in superficial patches and from outer cortical cells. Carpogonial branches are 3-celled and lateral on inner cortical cells. Auxiliary cells are scattered and intercalary on inner cortical filaments. After fertilization, the carpogonium fuses with a nutritive cell of a support cell and produces connecting filaments that fuse with auxiliary cells. Gonimoblasts become converted into carposporangia. No involucral filaments are produced, but cortical cells around cystocarps become modified.

Platoma Schousboe ex F. Schmitz 1894: p. 627 (Gr. *platy*: flat, broad, wide).

Plants are soft and gelatinous, bladelike, lobed, or pinnately divided, and attached by thick holdfasts. The medulla is large, usually with loose, slender filaments interwoven with rhizoids, and in a soft gel. The cortex has dichotomously branched pigmented filaments; inner filaments are loose; outer ones are beadlike and in anticlinal rows. Gland cells are large and often present. Tetrasporangia are unknown. Carpogonial branches are 3-celled; cystocarps are small, scattered, immersed, with obscure pores, and composed of carposporangia.

Key to the species (2) of *Platoma*

1. Plants irregularly lobed or branched 2–3 times . . . . *P. cyclocolpum*
1. Plants bladelike with marginal lobes . . . . *P. tenue*

***Platoma cyclocolpum* (Montagne in Barker-Webb and Berthelot) F. Schmitz 1889: p. 453; 1894: p. 627; as an orthographic change from '*cyclocolpa*' (Schneider, 2003). (Gr. *kyklos*: a circle, wheel + *kolpos*: bosom or a shallow depression). Plate XLIX, fig. 1.**

Basionym. *Halymenia cyclocolpa* Montagne in Barker-Webb and Berthelot 1841 (1839–1841): p. 163; 1846 p. 116, pl. XI, fig. B.

Guiry and Nic Dhonncha (2006) suggested conservation of *Platoma cyclocolpa* in place of the neuter *P. cyclocolpum* listed by Athanasiadis (1988). See comments by Furnari et al. (1999).

Plants are 4–8 cm tall, soft, gelatinous, lobed, and red with orange-yellow tips; main axes are flat, to 8 mm thick, 9 cm wide, and attached by small discoid holdfasts. Plants are initially flat oval blades that form irregular to palmately branched lobes. Medullary filaments are loose, 8–14 µm in diam., and in a gel. Cortical filaments are radial, forked 2–4 times, 2–7 µm in diam. by 10–15 µm long, and pigmented. Surface cells are ovoid, 5–7 µm in diam., and 7–9 µm long. Cystocarps are spherical and 60–110 µm in diam. Auxiliary cells are ovoid, 10–15 µm in diam. by 25–30 µm long. Nutritive cells are spherical and 8–12 µm in diam. Carposporangia are ovoid and 6–10 µm in diam.

Rare, reported from 9–48 m off the east coast (Hanisak and Blair, 1988); also known from Bermuda, the Lesser Antilles, Mediterranean, eastern tropical Atlantic, Fiji, and Australia.

Taylor 1960: p. 437; Littler and Littler 2000: p. 92.

***Platoma tenue* M. Howe and W. R. Taylor 1931: p. 32, text-fig. 15b, pl. 2, fig. 1 (L. *tenue*: thin, slender, delicate). Plate XLIX, fig. 2, 3.**

Synonym. *Predaea masonii sensu auct.*, non (Setchell and Gardner) De Toni (fil.) 1936: p. 5.

Plants are subsessile, flat, soft and gelatinous, and attached by thick holdfasts. Blades are entire or sparingly lobed, oval, 1.5–6.5 cm long or wide, 60–90 µm thick, and with undulate or crenulate margins. Medullary filaments are compact and homogeneous. Cortical filaments radiate, are loosely clustered and usually dichotomously branched (3–4 times). Cortical cells are ovoid; inner cells are 6–10 µm in diam. and surface cells 3–5 µm. Cystocarps are embedded, to 50 µm in diam., and naked or with a few filaments.

Rare, reported from 35–59 m off the east coast (Hanisak and Blair, 1988; as *Predaea masonii*); also known from the Bahamas, Lesser Antilles, and Brazil (type location: Cabo Frio).

Taylor 1960: p. 436.

Titanophora (J. Agardh) Feldmann 1942: p. 111 (L. *titanus*: very large + *phora*: to carry).

Plants are slippery, soft, white to dull red, and have needlelike crystals of aragonite in their medulla. Axes are compressed to bladelike, irregularly to dichotomously branched, with cleft to entire margins, and attached by basal discs. The uniaxial construction is visible near branch tips but obscured below by thick-walled filaments that fuse irregularly and branch. Cortical filaments are subdichotomously branched, radial, and short. Gland cells are ovoid to elongate and at tips of cortical filaments. Gametophytes are monoecious. Spermatangia are at tips of cortical filaments. Carpogonial branches are 2(3)-celled and from inner cortical cells. Cystocarps are ostiolate, from cortical cells, and surrounded by elevated involucral filaments.

***Titanophora incrustans* (J. Agardh) Børgesen 1949b: p. 4, pl. 1 (L. *incrustans*: encrusting, covering the surface). Plate XLIX, fig. 4.**

Basionym. *Halymenia incrustans* J. Agardh 1885: p. 15.

Norris (1992a) suggested that there was one variable species of *Titanophora*, *T. pikeana* (Dickie) Feldmann, while Masuda and Guiry (1995), Silva et al. (1996), and Wynne (1998a) have listed up to nine species. Further study is needed (cf. Bucher and Norris, 1992).

Plants are bladelike, soft, slippery, lightly calcified, to 14 cm tall, and attached by discoid holdfasts. Blade surfaces are rough, initially pink, and then grayish green when dried. Main axes are subsimple, to 4–20 mm wide, 0.5–1.3 mm thick, and deeply lobed with age; older plants are irregularly or pinnately branched at margins. Branchlets are terete to flat and polymorphic, with forked tapering tips. One or more medullary filaments are formed in a gel; they are loose, longitudinal, 15–30 µm in diam., and 200–400 µm long. The cortex is 30–80 µm thick, has trichotomously branched (2–4 times) filaments with spherical cells, 10–12 µm in diam. by 14 µm long. Gland cells are spherical or irregular, 30–40 µm in diam., and 35–70 µm long.

Rare, reported by J. Agardh (1885) from Lake Worth, Florida (type location); also known from the Caribbean and Philippines.

Taylor (1960): p. 437; Littler and Littler 2000: p. 92.

RHODYMENIALES

Plants are erect, prostrate, or crustose; axes are soft and gelatinous, firm, or membranous, also cylindrical, foliose, or strap-shaped, and attached by discoid or rhizoidal holdfasts. Axes are multiaxial with loose to compact filaments, with hollow cavities traversed by filaments, or solid and pseudoparenchymatous. Life histories are isomorphic. Tetrasporangia are cruciately or tetrahedrally divided and variously placed. Spermatangia are scattered or in nemathecia and from outer cortical cells. Plants are procarpial; carpogonial branches are also scattered or in nemathecia, 3–4-celled, and with 1–3-celled auxiliary cell branches on the same support cells. After fertilization, carpogonia connect to auxiliary cells by ooblast filaments; gonimoblastic filaments extend outward from resulting fusion cells. A reticulum of attenuated filaments (*tela arachnoidea*) may form in the pericarp around the carposporophyte. Cystocarps are ostiolate and projecting; each carposporangia produces one carpospore.

Rhodymeniaceae

Plants are erect, gelatinous, firm or membranous, with irregular, pinnate, palmate, or dichotomous branching, and attached by discoid holdfasts. Parasitic species are whitish and tumorlike. Axes are multiaxial, cylindrical to foliose or with bulbous vesicles. The medulla is solid and pseudoparenchymatous or hollow and gelatinous; if hollow, its axes always lack longitudinal filaments, in contrast to the Champiaceae. The cortex has large inner cells and small pigmented surface ones or compact to loosely organized dichotomously branched filaments. Life histories are isomorphic. Tetrasporangia are mostly intercalary, cruciately divided, and scattered or in sori. Gametophytes are dioecious. Spermatangia arise from outer cortical cells. Plants are procarpial; carpogonial branches are mostly 4-celled and arise from a support cell with many 2–3-celled auxiliary branches. Auxiliary cells produce gonimoblasts with compact lobes after receipt of a diploid nucleus from a fertilized carpogonia; fusion cells are elongate and slender. Carposporophytes have cortical pericarps. Cystocarps are ostiolate, scattered, and protruding; gonimoblastic filaments become carposporangia.

Asteromenia Huisman and A. Millar 1996: p. 138–139, tab. 1, fig. 1–17 (Gr. *astero*: star-shaped, stellate + *humen*: a membrane).

Blades are peltate and circular when young, versus stellate to lobed at maturity; they occur on stalks with discoid holdfasts. Blades may fuse with one another and to the substrata. The medulla has large and small cells that grade into 3–4 layers of pigmented cortical cells. Tetrasporangia are cruciately divided and from intercalary cortical cells. Plants are procarpial, with 4-celled carpogonial branches on the same inner cortical cells as auxiliary cells. Cystocarps are ostiolate and project; pericarps have extended necks and lack a network of filaments (the *tela arachnoidea*).

Asteromenia peltata **(W. R. Taylor) Huisman and A. Millar 1996: p. 139–144, fig. 2, 3, 7–9, 11, 14 (L. *peltata*: armed with a shield, shield-shaped). Plate XLIX, fig. 5, 6.**

> Basionym. *Fauchea peltata* W. R. Taylor 1942: p. 113–114, pl. 3, fig. 9, pl. 16; fig. 1–5.
> Synonyms. *Weberella peltata* (W. R. Taylor) Schneider 1975: p. 133–134, fig. 2, 5, 6.
> *Halichrysis peltata* (W. R. Taylor) P. Huvé and H. Huvé 1977: p. 105.
>
> The plant was first named *Fauchea peltata* based on vegetative features (Taylor, 1942), then transferred to *Weberella* using reproductive data (Schneider, 1975), then to *Halichrysis* on the basis of its habit and anatomical and cystocarpic features (Huvé and H. Huvé, 1977), and finally to *Asteromenia*, based on tetrasporangial and carpogonial branch data (Huisman and Millar, 1996).

Plants are deep red, fleshy to gelatinous, prostrate or peltate, stalked, and 1–3 cm tall; stalks are terete, 1–2 mm in diam., to 15 mm long, initially attached by small lobed holdfasts, and secondarily by blade margins. Blades are circular, irregularly lobed, or stellate, to 10 cm in diam., and 200–480 µm thick. The medulla has 5–6 layers of ovoid, hyaline cells that are variable in size (60–400 µm in diam. by 80–160 µm long). The cortex has 3–4 layers with pigmented surface cells 3–6 µm in diam. and irregularly spaced in translucent cell packets. Tetrasporangia are cruciately divided, scattered, ovoid, 20–30 µm in diam., and 15–25 µm long. Gametophytes are dioecious. Spermatangia arise from outer cortical cells and are scattered, ovoid, and 2–3 µm in diam. Carpogonial branches are 4-celled and from the same inner cortical cells as auxiliary cell branches. Cystocarps are mammillate and ostiolate with extended necks to 0.5 mm in diam. Carposporangia are irregular to globular and 12–18 μm in diam.

Uncommon, on limestone off the east (32–55 m) and west (25–60 m) coasts; also known from North Carolina (22–45 m), the Bahamas, Caribbean (type location: Tortuga Island, Venezuela), Gulf of Mexico, and Brazil, as well as the Canary Islands, West Africa, Indo-Pacific, and Australia.

Taylor 1960: p. 477; Dawes 1974: p. 142; Schneider and Searles 1991: p. 345; Littler and Littler 2000: p. 128.

Botryocladia (J. Agardh) Kylin 1931: p. 17, *nom. cons.* (Gr. *botrys*: bladder + *klados*: branch).

Wilkes et al. (2006) reviewed the morphological features of various species of *Botryocladia* and the closely related genus *Irvinea*.

Plants are erect, with axes simple or branched, cylindrical, tough, and attached by lobed or discoid holdfasts. Branching is irregular to subdichotomous and radial or bilateral. Branches have one or more soft, ovoid vesicles that are stalked and mucilage filled; inner cells are large, colorless, and may have dark-staining gland cells that project into the cavity; cortical cells are small and pigmented. Tetrasporangia are cruciately divided and from subsurface cortical cells of vesicles. Spermatangia are on vesicles and in small superficial sori. Plants are procarpial; cystocarps are ostiolate, scattered on vesicles, and elevated as a result of cortical thickening. Gonimoblast cells are attached to the bases of cystocarps with most becoming carposporangia; a network of filaments (the *tela arachnoidea*) is absent.

Key to the species (5) of *Botryocladia*

1. Gland cells, if present, 1–2 in the subcortex 2
1. Gland cell present, 4–8(16) in the subcortex 3
   2. Gland cells rare; surface cells of vesicles have a spine. . . . . *B. spinulifera*
   2. Gland cells 1–2 per cortical cell; vesicle surface cells lack a spine. . . . . *B. occidentalis*
3. Surface cells form an incomplete or nearly complete layer over the subcortex, which has 4–8(–17) gland cells per cell 4
3. Surface cells cover the subcortex, which has 2–5 gland cells per cell . . . . *B. caraibica*
   4. Surface cells form an incomplete layer over larger subcortical cells with 4–8 gland cells that are not modified. . . . . *B. pyriformis*
   4. Surface cells limited to junctions of subcortical cells and do not cover cortical cell faces; subcortical cells have 3–6(1–17) gland cells that may be stellate . . . . *B. wynnei*

***Botryocladia caraibica* Gavio and Fredericq 2003: 95–97, fig. 1–5 (L. *caraibica*: of or from the Caribbean). Plate XLIX, fig. 7, 8.**

See Gavio and Fredericq (2003) regarding misidentifications as *Botryocladia pyriformis*.

Plants are to 3–4 cm tall, with 2–30 deep red vesicles on stalks that are branched, 1–3 cm long, 1.2 mm thick, and attached by discoid holdfasts. Axes and stalks are solid, with medullary cells 23–50 μm in diam. in the middle and smaller cells near the surface. Vesicles are leathery, ovoid to pyriform, 2–12 mm long, and contain mucilage; cortical walls are 68–93 μm thick. The cortex of vesicles has 1–2 layers; surface cells are round to elliptical, 2–5 by 4–8 μm in diam., pigmented, and covering subcortical cells. The vesicle's medulla has 2–3 layers of elongate (in cross section) colorless cells; inner cells are 35–45 by 75–85 μm, while outer ones are 8–18 by 16–32 μm. Gland cells are 32–42 um in diam., in clusters of 2–5, and on medullary cells or modified supporting cells. Gland cells are ovoid and 16–32 by 32–40 μm. Reproductive plants are unknown.

Rare, on rubble in 5–8 m at the Dry Tortugas; also known from 12–15 m off Panama (type location).

Taylor 1960: p. 483.

Botryocladia occidentalis (Børgesen) Kylin 1931: p. 18 (L. *occidentalis*: west, western, or occidental). Plate XLIX, fig. 9–11.

Basionym. *Chrysymenia ulvaria* var. *occidentalis* Børgesen 1920: p. 403, fig. 388

Axes are wiry, to 25 cm tall, with many pale pink to deep red vesicles, and attached by small to large holdfasts. Branching is alternate to subdichotomous; axes and stalks are solid, 0.5–1.5 mm in diam., and pseudoparenchymatous. Vesicles are ovoid to globular, 2–8 by 3–11 mm, and with mucilage; inner cortical cells are polygonal, 50–150 µm in diam., and with 1–2 globular gland cells that are 7–20 µm in diam.; surface cells are irregular to globular, 8–15 µm in diam., and cover the inner cells. Tetrasporangia are unknown. Cystocarps are ostiolate and to 1.5 mm in diam.; carposporangia are rounded and 12–25 µm in diam.

Common, on limestone in 5–72 m on all coasts; often epiphytized by other algae and bryozoans; also known from the Carolinas, Georgia, Bermuda, Caribbean (type location: between St. John and St. Thomas, Virgin Islands), Gulf of Mexico, Brazil, and the Canary Islands.

Taylor 1960: p. 483; Dawes 1974: p. 141; Schneider and Searles 1991: p. 335; Littler and Littler 2000: p. 130.

Botryocladia pyriformis (Børgesen) Kylin 1931: p. 18 (L. *pyriformis*: pear-shaped). Plate XLIX, fig. 12–15.

Basionym. *Chrysymenia pyriformis* Børgesen 1910: p. 187, fig. 8, 9; 1920: p. 400–402, fig. 384–387.

Previous identifications of Caribbean plants may include *Botryocladia caraibica*, which is a more common species in shallow water (Gavio and Fredericq, 2003).

Plants are to 4 cm tall; axes are little to much branched, with a few to many light pink vesicles, and attached by lobed or discoid holdfasts. Axes and stalks are solid; the outer 1–2 layers of stalks consist of small cells that abruptly change to large inner ones. Vesicles are soft and slimy, ovoid, 4–12 mm long, and contain mucilage; the cortex is 85–110 µm thick and has 2 layers of cells; surface cells are pigmented, 3–8 µm in diam., and nearly cover the inner layer; inner cortical cells are larger, thick walled, and rounded-polygonal (30–50 µm in diam.). Gland cells are in clusters of 4–8, 5–25 µm in diam., and project into the vesicle cavity. Tetrasporangia are ovoid, 34 µm in diam., and 44 µm long. Spermatangia arise from outer cortical cells and are 1–3 µm in diam. Cystocarps are ostiolate, spherical, and to 1 mm in diam. Carposporangia are irregular to spherical and 7–23 µm in diam. The species resembles *Bostrychia caraibica*, but has soft, slimy vesicles, a light pink color, an abrupt transition in cell size between the cortex and medulla of stalks, and vesicle walls with only two cell layers.

Rare, reported from the Dry Tortugas; also known from North Carolina (18–55 m), South Carolina, Bermuda, the Virgin Islands (type location: St. John; Afonso-Carrillo and Sobrino, 2004), Gulf of Mexico, and Brazil, as well as the Canary, Reunion, and Seychelles islands, China, and the Philippines.

Taylor 1960: p. 483 (as *B. pyriformis*, but probably *pro parte B. caraibica*); Schneider and Searles 1991: p. 336; Littler and Littler 2000: p. 130.

Botryocladia spinulifera W. R. Taylor and Abbott 1973: p. 410, fig. 1–4 (L. _spinulifera_: having small spines). Plate XLIX, fig. 16, 17.

Plants are 1–3 cm tall and have dark red vesicles; axes are solid, cylindrical, sparsely and irregularly branched, and attached by discoid holdfasts. Vesicles are compressed to ovoid, 1–4 mm in diam., 1–5 mm long, filled with mucilage, and 1–3 per stalk that are hollow. Vesicle walls have 2–3 layers; inner cells are round to polygonal, 40–150 µm in diam., and thin walled; subsurface cells are 20–30 µm in diam.; surface cells are pigmented, 3–17 µm in diam., ovoid, and each has a spinelike projection, 11–17 µm long. Gland cells are rare, 8–10 µm in diam., number 1–2 on inner medullary cells, and project into the vesicle cavity. Tetrasporangia are cruciately divided, ovoid, 15–18 µm in diam., 28 µm long, and in sori near the tips of vesicles. Gametophytes are unknown.

Rare, on rubble in 4–8 m at Molasses Reef (Bucher et al., 1990); also known from the Bahamas (40–49 m), Antilles (type location: Grass Point, St. Croix), western Caribbean, Papua New Guinea, and Fiji.

Littler and Littler 2000: p. 132.

Botryocladia wynnei D. L. Ballantine 1985: p. 199, fig. 1–5 (named for M. J. Wynne, an American phycologist who studies seaweeds from the Caribbean and worldwide and specializes in the family Delesseriaceae). Plate XLIX, fig. 18, 19.

Plants are 0.6–3.5 cm tall; axes (stalks) are short, unbranched, have 1–5 pale pink to red vesicles, and are attached by small discoid holdfasts. Vesicles are ovoid, 3–14 mm in diam., 6–35 mm long, and with mucilage. The vesicle cortex has 2–3 layers; inner cells are round to polygonal, 30–90 µm in diam., and some have 3–6(1–17) gland cells that are ovoid, 15–33 µm in diam., 17–26 µm long, project into the cavity, and may be stellate. Surface cells are pigmented, irregular to subspherical, 7–14 µm in diam., do not cover the inner cells but align anticlinally to form rosette patterns. Tetrasporangia are cruciately divided, spherical, and 18–30 µm in diam. Gametophytes are monoecious. Spermatangia are 1–3 µm in diam. and with outer cortical cells. Cystocarps are ostiolate and 440–670 µm in diam. Carposporangia are irregular to subspherical and 7–23 µm in diam.

Rare, from 9 m at the Dry Tortugas (Ballantine, 1996); also known from North Carolina (17–21m), Georgia, the Bahamas, Puerto Rico (type location: near La Parguera), and the Canary Islands.

Schneider and Searles 1991: p. 337.

Chrysymenia J. Agardh 1842: p. 105 (Gr. _chrysos_: gold, yellow + _humen_: a membrane).

Wynne (2005b) gave a census of the 21 recognized species of _Chrysymenia_.

Plants are erect and soft; axes are terete, compressed, or foliose, alternately branched, and attached by discoid holdfasts. Axes are hollow, filled with mucilage, and may have longitudinal filaments. The inner cortex has large compact cells with some having sessile gland cells. The outer cortex has 1–3 layers; cells are small, in anticlinal rows, and with surface ones pigmented. Tetrasporangia are cruciately divided, scattered, and from cortical cells. Gametophytes are dioecious. Spermatangia are in irregular sori. Females are procarpial; cystocarps are scattered and partly immersed or raised on a thickened cortex. Gonimoblasts are attached to a flat base by filaments and most become globular masses of carposporangia. A _tela arachnoidea_ is absent in the pericarp.

Key to the species (5) of *Chrysymenia*

1. Plants locally hollow; axes foliaceous to bladed 2
1. Plants hollow throughout; axes cylindrical to compressed 3
  2. Blades shield-shaped; medullary filaments few to numerous .... *C. agardhii*
  2. Blades broad, irregular, and lobed; medullary filaments few .... *C. planifrons*
3. Axes cylindrical or swollen and not segmented nor constricted 4
3. Axes cylindrical and with constricted segments .... *C. enteromorpha*
  4. Branching mostly dichotomous or with adventitious branches .... *C. halymenioides*
  4. Branching alternate and radial, often irregular .... *C. ventricosa*

Chrysymenia agardhii Harvey 1853: p. 189, pl. XXXA, fig. 1–4 (named for J. G. Agardh, a Swedish phycologist and taxonomist who published extensively between the 1830s and 1901). Plate XLIX, fig. 20, 21.

Synonym. *Cryptarachne agardhii* (Harvey) Kylin 1931: p. 11

Plants are foliaceous, gelatinous to membranous, 10–20 cm tall, and attached by discoid holdfasts. Axes are strap-shaped, 1–5 cm wide, 0.7–1.1 mm thick, and simple or divided; lobes are narrowly shield-shaped and taper to bases. Tips are rounded and obtuse; margins are smooth to unevenly dentate, undulate, and often proliferous. The central cavity has mucilage and a few longitudinal filaments, 5–20 µm in diam. The cortex has 1–5 layers; cells of inner 1–2 layers are compact, elongate, 170–320 µm in diam., and a few have a single gland cell that projects into the cavity. Cells of outer 1–3 cortical layers are small and radiate outward; surface cells are 4–5 µm in diam., ovoid, and scattered in an outer matrix. Cystocarps are scattered, hemispherical, and have small ostioles. Young, simple blades are similar to *Halymenia floridana*, but lack stellate ganglia in the inner cortex and have few gland cells (Schneider and Searles, 1991).

Rare, in drift at Key West (type location) and from 33–58 m off the east coast; also known from North Carolina (17–50 m), Bermuda, and the Antilles.

Taylor 1960 p. 480; Schneider and Searles 1991: p. 338.

Chrysymenia enteromorpha Harvey 1853: p. 187–188 (Gr. *enteron*: intestine + *morpha*: shape; thus, shaped like intestines). Plate XLIX, fig. 22–24.

Plants are gelatinous, erect and bushy, 7–30 cm tall, yellow to pale red, and with short stalks attached by discoid holdfasts. Axes are cylindrical, 4–10 mm in diam. below, and have frequent, alternate to irregular and radial to distichous branching; branches are often constricted at bases, taper to obtuse tips, and produce branchlets to 2 mm long. The axial cavity is filled with mucilage. The inner cortex has 1–2 layers of large, hyaline, elongate cells (100–250 µm in diam.) that produce 3–18 ovoid gland cells, which are 10–18 µm in diam. and project into the cavity. The outer cortex has 1–2 layers; surface cells are pigmented, ovoid to spherical, and 5–10 µm in diam. Tetrasporangia are ovoid, cruciately divided, 27–40 µm in diam., 35–45 µm long, and scattered. Spermatangia are to 1 µm in diam. and in sori with surface cells. Ostiolate cystocarps are scattered, obvious, rounded or conical, and to 800 µm in diam. Carposporophytes are lobed. Carposporangia are irregular and 15–20 µm in diam.

Uncommon, on limestone in 25–90 m off the east (Hanisak and Blair, 1988) and west coasts (Cheney and Dyer, 1974) plus the Keys (type location: Key West); also known from North Carolina (19–45 m), South Carolina, Georgia, Bermuda, the Caribbean, Gulf of Mexico, Brazil, eastern tropical Atlantic, and Bangladesh.

Taylor 1960: p. 479; Dawes 1974: p. 141; Schneider and Searles 1991: p. 339; Littler and Littler 2000: p. 134.

Chrysymenia halymenioides Harvey 1853: p. 188, pl. XXA, fig. 1–8 (Gr. *hals*: the sea + *humen*: a membrane + *oides*: like; thus, similar to the red algal genus *Halymenia*). Plate XLIX, fig. 25.

Plants are 7–10 cm tall, soft and gelatinous, pale yellow to pink red, and on short cylindrical stalks attached by discoid holdfasts. Axes are repeatedly dichotomously divided and little to strongly compressed; lobes are 0.5–1 cm wide and little constricted. The axial cavity has mucilage and usually lacks longitudinal filaments. The inner cortex has 2–3 layers; inner cells are large, hyaline, polygonal, and have 1–2 dense gland cells projecting into the cavity. The outer cortex has 1–2 layers; surface cells are small, pigmented, and form an incomplete network over the thallus surface. Tetrasporangia are cruciately divided and spherical to elliptical. Cystocarps are rounded-conical and obvious.

Rare, recorded from Key West (type location) and in 41 m off the east coast (Hanisak and Blair, 1988); also known from Bermuda, the Antilles, and Gulf of Mexico.

Taylor 1960: p. 479.

Chrysymenia planifrons (Melvill) J. Agardh 1876 (1848–1901): p. 319 (L. *plani*: flat + *frons*: blade). Plate L, fig. 1, 2.

Basionym. *Chrysymenia agardhii* var. ß *planifrons* Melvill 1875: p. 263.

Synonym. *Cryptarachne planifrons* (Melvill) Kylin 1931: p. 12.

Kylin (1931; p. 12) suggested that *Chrysymenia curtissiana* J. Agardh might be a large form of *C. planifrons* (also Taylor 1960: p. 481). See Wynne (2005b, 2006) for a description of this poorly known species that is similar to *C. tigillum* Wynne from the Sultanate of Oman.

Plants are soft, gelatinous, and to 40 cm tall; blades are expanded, thick, and on short stalks attached to small discoid holdfasts. Blades are large, simple to somewhat irregular with short broad lobes, undulate margins, and smooth surfaces. The cavity is narrow, filled with mucilage, and has a few rhizoidal filaments. The cortex has 2–4 layers; the outer 1–2 layers have small pigmented cells, while the inner 1–2 layers have large hyaline cells with one to a pair of gland cells that project into the central cavity. Cystocarps are abundant and scattered on the surface of blades.

Rare, from 25–72 m off the east coast (Hanisak and Blair, 1988), Key West (type location; Wynne, 2005b), and in 30 m at the Dry Tortugas; also the Antilles and Brazil.

Taylor 1960: p. 481.

Chrysymenia ventricosa (J. V. Lamouroux) J. Agardh 1842: p. 106 (L. *ventricosa*: distended, expanded, swollen). Plate L, fig. 3.

Basionym. *Dumontia ventricosa* J. V. Lamouroux 1813: p. 45, pl. 4, fig. 6.

Plants are bushy, soft, gelatinous, and pale to deep purple red; axes have irregular, opposite, alternate, or radial branching, and are on stalks (3–15 mm long) attached by lobed or discoid holdfasts. Branches are cylindrical to slightly flattened below, constricted at bases, and with blunt tips. The axial cavity is filled with mucilage. The cortex has 3–5 layers; cells of two inner layers are large (to 600 µm in diam.), ovoid, and with 1–5 gland cells in clusters. Gland cells are tear-shaped, 12–20 µm in diam., 20–30 µm long, and project into the cavity. Cells of outer 2–3 cortical layers are small, compact, ovoid, 4–8 µm in diam., and to 12 µm long. Tetrasporangia are cruciately divided, ovoid, 15–20 µm in diam., and scattered. Cystocarps are also scattered and dome-shaped.

Rare, in 90 m at the Dry Tortugas; also known from Bermuda, the Antilles (20–130 m), and southern Caribbean, as well as Atlantic Spain, the Mediterranean (type location: French coast; Wynne, 2005b), eastern tropical Atlantic, and French Polynesia.

Taylor 1960: p. 479; Littler and Littler 2000: p. 136.

Coelarthrum Børgesen 1910: p. 192 (Gr. *coelos*: hollow + *arthron*: a joint, jointed).

Plants are usually gelatinous, erect, and dichotomously branched; axes are segmented (swollen and constricted by cellular partitions), hollow, filled with mucilage, and attached by small discoid holdfasts. The cortex usually has two layers; inner cells are large, closely packed, and have up to 5 gland cells in clusters that project into the cavity; outer cells are small, pigmented, and not completely covering the inner cells. The cellular plates (partitions) in the axial cavity consist of large cells. Tetrasporangia are tetrahedrally divided; they occur between surface cells and project outward. Cystocarps are ostiolate, hemispherical, and scattered.

Coelarthrum cliftonii **(Harvey) Kylin 1931: p. 15, fig. 4c, as '*cliftoni*' (named for G. Clifton, who collected algae for Harvey, including the type specimen from Western Australia). Plate L, fig. 4–7.**

Basionym. *Chylocladia cliftonii* Harvey 1855a: p. 556 (as '*cliftoni*').

Synonym. *Coelarthrum albertisii* (Piccone) Børgesen 1910: p. 189, 192, fig. 11, 12; 1920: p. 404. Silva et al. (1996) recognized *Coelarthrum albertisii* as a distinct species.

Plants are soft, delicate, gelatinous, to 5 cm tall, and pink to transparent red; axes are dichotomously branched and attached by discoid holdfasts. Segments are swollen to bulbous, hollow with mucilage, to 6 mm in diam., and to 10 mm long; branches are basally constricted. The cortex has 1–2 layers; inner cells are large, compressed, angular, to 100 µm in diam., and have 1–5 spherical gland cells (to 40 µm in diam.) that project into the axial cavity. The outer cortical cells are pigmented, ovoid, and smaller (to 25 µm in diam.) than inner ones; they also occur between the inner cells and do not cover their surfaces. The cellular plates (partitions) in the axial cavity consist of large cells. Tetrasporangia are cruciately to irregularly divided, spherical, 25–45 µm in diam., and cause swelling. Cystocarps are scattered and dome-shaped.

A questionable record for Florida; known from Bermuda, the Bahamas, Caribbean (to 35 m), and Gulf of Mexico, as well as the Canary Islands, Indo-Pacific, and Australia (type location: Fremantle, Australia; Silva et al., 1996).

Taylor 1960: p. 482; Littler and Littler 2000: p. 136.

Coelothrix Børgesen 1920: p. 389 (Gr. *coleos*: hollow + *thrix*: hair).

Plants form wiry turfs that are attached by dense rhizoidal clusters and secondarily by procumbent branches with rhizoids. Axes are slender, terete, firm, and often irregularly branched. The axial cavity is small and often filled with mucilage. Inner cortical cells are thick walled, vertically elongate, and have single, sessile gland cells that project into the cavity. The outer cortex has 1–2 layers of small, pigmented cells. Tetrasporangia are tetrahedrally divided and clustered on swollen branch tips. Spermatangia are in sori. Cystocarps are scattered and spherical.

Coelothrix irregularis **(Harvey) Børgesen 1920: p. 389–391, fig. 373, 374 (L. *irregularis*: irregular, asymmetrical). Plate L, fig. 8–10.**

Basionym. *Cordylecladia irregularis* Harvey 1853: p. 156.

Plants are 2–3 cm tall, in bushy to turflike cushions, and attached by rhizoidal clusters and proliferations. Axes are entangled, bright iridescent blue, 0.5–0.8 mm in diam., irregularly branched, hollow, and with a small axial cavity filled with mucilage. Branches are cylindrical, tapered, and fused together. Tetrasporangial plants have thin branches, 300–400 µm in diam. The cortex has 4–6 layers; cells of the inner 2–3 layers are large, 45–80 µm in diam., 120–200 µm long, hyaline, thick walled, and with ovoid gland cells that project into the axial cavity. The outer cortex has 1–2 layers; cells are irregular to elongate, 18–27 µm in diam., and 45–54 µm long. Tetrasporangia are tetrahedrally divided, subspherical, 40–60 µm in diam., and in swollen branches 1–4 mm long and to 650 µm in diam. Spermatangia are in sori on swollen branch tips and to 4 µm in diam. Cystocarps are spherical and to 600 µm in diam.

Uncommon, forming a turf on wave-exposed limestone to 9 m in the Keys (type location: Key West); also known from Bermuda, the Caribbean, Gulf of Mexico, and Brazil, as well as Portugal, the eastern tropical Atlantic, Indo-Pacific, and Australia.

Taylor 1960: p. 488; Littler and Littler 2000: p. 136.

Maripelta E.Y. Dawson 1963a: p. 446 (L. *mari*: of the sea + *pelta*: a small shield).

Based on gene-sequence data, Wilkes et al. (2005b) concluded that the type species, *Maripelta rotata* E. Y. Dawson, was correctly placed in the Rhodymeniaceae.

Plants have peltate deciduous blades that are symmetrical, simple, and radially expanded from cylindrical, simple or branched stalks that are attached by discoid holdfasts. The blade's medulla has one layer of large, polygonal cells and 2–3 cortical layers of small, pigmented ones. Tetrasporangia are cruciately divided and in small surface sori. Spermatangia are in pale, superficial patches. Cystocarps are ostiolate, spherical, projecting, and lack a *tela arachnoidea* in the pericarp.

Maripelta atlantica Eiseman and R. L. Moe 1981: p. 299–300, fig. 1–22 (L. *atlantica*: of or from the Atlantic Ocean). Plate L, fig. 11.

Plants are erect, to 6 cm tall, and have peltate blades on irregular, cylindrical stalks (2–5 mm in diam) that are attached by lobed holdfasts and have annular scars. Blades are circular, to 70 mm in diam., and have a medulla of one layer of thick-walled hyaline cells (200–400 µm in diam.) that is visible through the cortex. The cortex has 2 layers; inner cells are 25–35 µm in diam. and surface ones 3–5 µm in diam. Tetrasporangia are cruciately divided, from cortical cells, and in marginal nemathecia on the underside of blades. Gametophytes are dioecious. Spermatangia are 2 µm in diam. and in marginal sori. Female plants are procarpial and have 3-celled carpogonial branches. Cystocarps are ostiolate, spherical, projecting, and to 2 mm in diam.

Rare, on limestone in 60 to 100 m off the east coast (Hanisak and Blair, 1988; type location: 3.6 km east of Singer Island) and 20–60 m off the west coasts (Dawes and van Breedveld, 1969, as *Fauchea peltata*); also known from the Greater Antilles.

Dawes 1974: p. 142.

Rhodymenia Greville 1830: p. 84, pl. XLVIII, *nom. cons.*, as '*Rhodomenia*' (*rhodon*: red or rose colored + *humen*: a membrane).

Blades are erect and membranous, initially attached by discoid holdfasts and secondarily by cylindrical, prostrate axes. Blades are firm, flat, with irregular to repeatedly dichotomous branching, and with holes or marginal proliferations. Axes are multiaxial; the medulla has a few to several layers of axially elongate, hyaline cells. The cortex has 1–5 layers; cells are progressively smaller and pigmented toward the surface. Tetrasporangia are cruciately divided and scattered or in sori in the outer cortex. Gametophytes are dioecious. Female plants are procarpial; carpogonial branches 3(4)-celled; gonimoblastic cells form masses of carposporangia. Cystocarps are ostiolate, elevated, mammillate or hemispherical, and lack a *tela arachnoidea*.

Key to the species (2) of *Rhodymenia*

1. Plants 4–18 cm tall; blades 3–5 mm wide, 100–250 µm thick; branching palmate at wound sites. . . . . *R. pseudopalmata*
1. Plants 4–5 cm tall; blades 5–10 mm wide, 375–600 µm thick; branching not palmate at wound sites. . . . . *R. divaricata*

Rhodymenia divaricata E. Y. Dawson 1941: p. 141, pl. 23, fig. 31 (L. *divaricata*: wide spreading, forked, divergent). Plate L, fig. 12, 13.

Records of *Rhodymenia divaricata* (Schneider and Searles, 1991; Hanisak and Blair, 1988) may be *Gracilaria galetensis*; thus, Florida records are uncertain (Gurgel et al., 2004).

Plants are erect, 4–5 cm tall, solitary or in clusters, tough, rose to red, on short stalks or not, initially attached by lobed discoid holdfasts and sometimes by rhizoids. Blades are flat, strap-shaped, widely (80–100°) and regularly dichotomously branched, 5–30 mm wide, and 100–600 µm thick; margins are smooth or have a few proliferations; tips are obtuse. Medullary cells are in 3–7 irregular layers, 90–200 µm in diam., to 250 µm thick, axially elongate, hyaline, and polygonal. The cortex has 2–3 layers of irregular to ovoid cells; surface cells are angular, pigmented, 5–8 µm in diam., and 8–17 µm long. Tetrasporangia are cruciately divided, ovoid, 12–20 µm in diam., 20–25 µm long, scattered, and from inner cortical cells. Cystocarps are ostiolate, scattered, hemispherical, and to 0.8 mm in diam. Carpospores occur in hemispherical masses and are darkly pigmented, round to elliptical, and 7–18 µm in diam.

Rare and a questionable record from 27–50 m off the east coast (Hanisak and Blair, 1988); also known from the Carolinas (40–54 m), Puerto Rico, Gulf of California (type location: Guaymas Bay, Mexico), Galapagos Islands, and Fiji.

Schneider and Searles 1991: p. 347; Littler and Littler 2000: p. 138.

Rhodymenia pseudopalmata (J. V. Lamouroux) P. C. Silva 1952: p. 265 (Gr. *pseudo*: false + L. *palmata*: lobed, like a hand). Plate L, fig. 14–17.

Basionym. *Fucus pseudopalmata* J. V. Lamouroux 1805: p. 29, fig. 2.

Synonym. *Rhodymenia pseudopalmata* var. *caroliniana* W. R. Taylor 1960: p. 485, 486, 634.

Plants are erect, foliaceous, and 4–18 cm tall; blades are tough, membranous, pale to dark red, and on stalks that are cylindrical, 0.3–0.7 mm in diam., and attached by fibrous or discoid holdfasts. Blades are 3–5 mm wide, 100–250 µm thick, and dichotomously branched 1–4 times at angles of 30–70°; margins are smooth to proliferous; tips are obtuse or pointed. Blades are multiaxial; medulla has 3–7 layers of spherical, axially elongate cells, 45–90 µm in diam., and to 120 µm long. The cortex has 2–3 layers; surface cells are pigmented, ovoid, axially elongate, 5–12 µm in diam. by 9–14 µm long, and not completely covering inner cells. Tetrasporangia arise from cortical cells, are cruciately divided, spherical to ovoid, 15–25 µm in diam., 25–40 µm long, and in pigmented bands at tips. Gametophytes are dioecious. Spermatangia are 2–3 µm in diam., in pairs, and in ovoid sori near tips. Carpogonial branches are 3-celled. Cystocarps are ostiolate, hemispherical to mammillate, to 1 mm in diam., and on blade margins. Carposporangia are pigmented and 5–20 µm in diam.

Rare, reported from 34–89 m off the east coast (Hanisak and Blair, 1988); also known from North Carolina (40–54 m) southward, the Caribbean, Gulf of Mexico, and Brazil, as well as the eastern North and tropical Atlantic (lectotype location: La Corunna, Spain; Guiry, 1977), and Mediterranean.

Taylor 1960: p. 485; Schneider and Searles 1991: p. 348; Littler and Littler 2000: p. 138.

Faucheaceae

Saunders et al. (1999) established the family using molecular and morphological data. Gavio and Fredericq (2005) confirmed its segregation from the Rhodymeniaceae.

Plants vary in construction and are attached by discoid holdfasts. The medulla is solid and pseudoparenchymatous or hollow. The cortex has small cells in anticlinal rows and cruciately divided tetrasporangia. Carpogonial branches are 3-celled; gonimoblasts are globular with distinct lobes that become carposporangia. Fusion cells are narrow.

Fauchea Montagne and Bory in Durieu de Maisonneuve 1846: p. 64 (named for Jean-Baptise Fauchè, a French naturalist who wrote the section on grasses in *Expedition Scientifique de Morée to the Peloponnesus Peninsula, Greece*).

Plants are erect, foliaceous, soft or firm, alternately branched, and attached by discoid discs. Axes have a medulla of one or more layers of large cells. The inner cortical cells are large and irregularly arranged, while surface ones are pigmented and in anticlinal rows. Tetrasporangia are tetrahedrally divided and in sori. Cystocarps are prominent; pericarps lack papillae or horns and have a *tela arachnoidea* in their cavity.

Fauchea hassleri **M. Howe and W. R. Taylor 1931: p. 13–14, fig. 6, pl. 2, fig. 2, 3 (named for the *R.V. Hassler* Expedition of 1872 that collected the type specimen by dredging). Plate L, fig. 18, 19.**

Plants are foliaceous, erect, 5–10 cm tall, and attached by discoid holdfasts. Blades are pseudoparenchymatous, dichotomously branched 1–4 times, and spreading; segments are 1–12 mm wide at forks and 150–280 µm thick; tips are obtuse and margins entire. The medulla is composed of 1–3 layers; cells are irregular to ovoid, hyaline, 60–130 µm in diam., and 100–260 µm long. The inner cortex has one layer of cells 15–22 µm in diam. The outer cortex has 2–4 layers; cells are in a gel, in anticlinal rows, pigmented, and 3–6 µm long. Tetrasporangia are tetrahedrally divided and in scattered irregular surface sori; cystocarps are marginal, domelike, and 0.7–1 mm in diam.

Rare, from 20–60 m off Tampa Bay; also known from Brazil (type location: Cabo Frio) and North Senegal.

Taylor 1960: p. 476; Dawes 1974: p. 142.

Gloiocladia J. Agardh 1842: p. 87 (Gr. *gloios*: gelatinous + *clados*: a small branch).

Synonym. *Gloioderma* J. Agardh 1851 (1848–1901): p. 243.

The genus includes taxa previously placed in *Gloioderma* (Gavio and Frederica, 2005). The two genera were separated by the type of tetrasporangial divisions (i.e., tetrahedral in *Gloiocladia* and cruciate in *Gloioderma*) and shape of cystocarps (coronate in *Gloioderma* and spherical in *Gloiocladia*). However, Irvine and Guiry (1983) questioned the separation after finding cruciate tetrasporangia and coronate cystocarps in the type species (*Gloiocladia furcata*). The separation of *Gloiocladia* and *Fauchea* is also unclear (Gavio and Fredericq, 2005; Sánchez and Rodríguez-Prieto, 2005), with *Gloiocladia* usually having scattered tetrasporangia (sometimes in raised sori) and *Fauchea* distinct nemathecia; thus, some authors (Irving and Guiry, 1983; R. E. Norris, 1991b) do not separate the two genera.

Plants are gelatinous, prostrate or erect, tubular or foliose, with irregular, alternate, or dichotomous branching that may fuse. Plants are attached by small basal holdfasts. The medulla has large, rounded, hyaline cells, with inner cortical cells smaller and forming an anastomosing network. The outer cortical cells are pigmented and in dichotomously branched anticlinal rows in a gel. Tetrasporangia are terminal, scattered, cruciately divided, and occur on inner lateral cortical branches. Gametophytes are dioecious. Spermatangia are on outer cortical branches. Carpogonial branches are 3-celled; carposporophytes are scattered; most gonimoblastic cells form masses of carposporangia that are attached to the bases of cystocarps. A *tela arachnoidea* is present and surrounds the carposporophyte. Cystocarps are obvious, ostiolate, and have a thickened raised cortex.

Key to the species (3) of *Gloiocladia* (modified from Schneider and Searles, 1991)

1. Axes flattened to strap-shaped; branching lateral 2
1. Axes cylindrical; branching radial . . . . *G. rubrispora*
   2. Axes 1–2 mm in diam.; branching mostly pinnate . . . . *G. atlantica*
   2. Axes 2–5 mm in diam.; branching mostly dichotomous . . . . *G. blomquistii*

Gloiocladia atlantica (Searles) R. E. Norris 1991b: p. 592 (L. *atlantica*: of or from the Atlantic Ocean). Plate L, fig. 20, 21.

Basionym. *Gloioderma atlanticum* Searles 1972: p. 23, fig. 1b, 4.

Plants are prostrate and erect (1–5 cm tall), gelatinous, and rosy red; axes are flat, highly branched, initially attached by discoid holdfasts and secondarily by rhizoids from the margins or tips of prostrate axes. Axes are 1–2 mm in diam.; they taper to bases of forks, and branch pinnately 3–4 times; branchlets are in alternate or opposite positions, 0.5–2.5 mm long, and have rounded tips. The medulla has 4–6 layers of compact, longitudinal, polygonal cells 75–240 µm long. The cortex has 2 layers; inner cells are axially elongate and in a network; outer cells are ovoid, 3–4 µm in diam., 5–10 µm long, and in dichotomously branched, anticlinal filaments. Tetrasporangia are cruciately divided, 27–38 µm in diam., 35–50 µm long, and scattered in the outer cortex. Spermatangia are 1–2 µm in diam. and in sori. Cystocarps are ostiolate, scattered, spherical or mammillate, to 1.1 mm in diam., and may have 1–3 horns. Carposporangia are irregular to spherical and 2–20 µm in diam.

Rare, reported by Hanisak and Blair (1988) from 30–55 m off the east coast; plants reported by Eiseman (1979) from the same area were identified as *G. rubrispora* (Schneider and Searles, 1991); also known from Bermuda, North Carolina (type location from 24–54 m), the Bahamas, Greater Antilles, and Canary Islands.

Schneider and Searles 1991: p. 342.

Gloiocladia blomquistii (Searles) R. E. Norris 1991b: p. 592 (named for the American phycologist H. L. Blomquist, who studied the seaweeds of North Carolina). Plate L, fig. 22, 23.

Basionym. *Gloioderma blomquistii* Searles in Schneider and Searles 1975: p. 87, fig. 6–9.

Plants are erect, foliaceous, gelatinous, 2–4 cm tall, and attached by discoid holdfasts. Main axes are 2–5 mm wide and 240–320 µm thick; branching is dichotomous to irregular (3–4 times), with branches overlapping, fusing, and having obtuse tips. Medullary cells are in one layer, quadrate, and 140–290 µm in diam. The inner cortical cells are axially elongate, interconnected, and form a network; outer cells are ovoid, 4–6 µm in diam., 8–10 µm long, and in anticlinally oriented dichotomous filaments. Surface cells are pigmented and irregular. Tetrasporangia are scattered, cruciately divided, 19–20 µm in diam., and 30–34 µm long. Cystocarps are ostiolate, spherical, 0.7–1.5 mm in diam., and may have 1–3 horns. Carposporangia are irregular to spherical, 17–35 µm in diam., and 17–40 µm long.

Rare, from 38–59 m off the east coast (Eiseman, 1979; Hanisak and Blair, 1988); also known from North Carolina (39–45 m; type location) and the Canary Islands.

Schneider and Searles 1991: p. 343.

Gloiocladia rubrispora **(Searles) R. E. Norris 1991b: p. 592 (L.** ***rubri*****: red + Gr.** ***spora*****: seed or spore). Plate L, fig. 24.**

Basionym. *Gloioderma rubrisporum* Searles 1984: p. 217, fig. 1–5.

Plants are gelatinous, to 2.5 cm tall, and rosy red; axes are terete, highly branched, initially attached by discoid holdfasts, and secondarily by rhizoids from blade margins or branch tips. Axes are 0.4–1.0 mm in diam. and radially alternately branched (3–4 times); branchlets are to 1.8 mm long and have rounded tips. The medulla has 1–3 layers of polygonal cells 60–100 µm in diam. and 120–200 µm long. The inner cortical cells are axially elongate, interconnected, and form a network; outer cells are ovoid, small, 10 µm in diam., 14 µm long, and occur in anticlinal dichotomous filaments. Surface cells are pigmented, 2 µm in diam., and 3–8 µm long. Tetrasporangia are scattered, elliptical, cruciately divided, and 20–24 µm in diam. by 30–34 µm long. Cystocarps are ostiolate, spherical, 450–900 µm in diam., and with 2–5 horns. Carposporangia are oval, 10–14 µm in diam., and 40 µm long.

Rare, reported from 26–45 m off the east coast (Eiseman, 1979, as *G. atlantica*; Schneider and Searles, 1991); also known from North Carolina (type location: in 24–54 m), the Bahamas, Puerto Rico, and Australia.

Schneider and Searles 1991: p. 343.

Leptofauchea Kylin 1931: p. 9 (Gr. *lepto*: slender, thin + *Fauchea*, a genus of red algae named for Jean-Baptise Fauchè, a French naturalist who wrote the section on grasses in *Expedition Scientifique de Morée*; thus, a slender form of *Fauchea*).

Plants are erect, foliaceous, soft, irregular to dichotomously branched, and attached by discoid holdfasts. The medulla has a few layers of large polygonal cells, while the cortex is monostromatic and has small pigmented cells. Tetrasporangia are cruciately divided and terminal or in chains in raised nemathecia. Plants are procarpial; cystocarps are ostiolate and on blade margins. Most gonimoblastic cells form masses of carposporangia; a filamentous reticulum (*tela arachnoidea*) occurs in the pericarp and surrounds the carposporophyte.

Leptofauchea brasiliensis **A. B. Joly 1957b: p. 130–133, text fig. 2, pl. 12, fig. 7a,b, pl. 19, fig. 4 (L.** ***brasiliensis*****: of or from Brazil). Plate L, fig. 25, 26.**

Plants are foliaceous, soft, gelatinous, 3–4 cm tall, rosy red, and on short stalks attached by discoid holdfasts. Branching is irregular to dichotomous with constricted nodes; blades are narrow, strap-shaped, 2–4(5) mm wide, 160–290 µm thick, and with smooth margins. Blades near the tips are narrow and do not overlap or fuse in one plane. Medullary cells are in one layer and 60–140 µm in diam.; cortical cells are also in one layer with small, pigmented, round to irregular cells 5–10 µm in diam. Tetrasporangia are cruciately divided, ovoid to angular, 10–20 µm in diam., 10–30 µm long, in raised sori, and terminal or in chains of 3–4. Gametophytes are unknown.

Rare, from 27–93 m off the east coast (Hanisak and Blair, 1988); also known from North Carolina (50 m), Brazil (type location: São Paulo), and the Canary Islands.

Taylor 1960: p. 477; Schneider and Searles 1991: p. 346.

Champiaceae

Plants are soft, erect or prostrate, and attached by small discoid holdfasts. Axes are radially, irregularly, or alternately branched, also cylindrical or compressed, and hollow, except at bases where longitudinal filaments border the axial cavity. The medullary cavity is divided by monostromatic plates of cells. The cortex has a few to several layers of cells. Tetrasporangia are tetrahedrally divided and intercalary within cortical filaments; some species produce polysporangia. Gametophytes are dioecious. Spermatangia arise from outer cortical cells and are in sori that girdle branch axes. Plants are procarpial; carpogonial branches are 4-celled and on support cells with 1–2 auxiliary cell branches. Gonimoblasts are globular and mostly have hemispherical carposporangia that arise on fusion cells or may radiate and only produce carposporangia from terminal cells. Fusion cells are small and compact. Pericarps are obvious, ostiolate, and corticated.

Champia Desvaux 1809: p. 245 (named for L. A. Deschamps, a French naturalist and botanist).

Synonym. *Mertensia* Roth 1806: p. 318, *nom. illeg.*; non *Mertensia* Roth 1797: p. 34; non *Mertensia* Willdenow 1804: p. 165.

Plants are bushy, erect or prostrate, and attached by rhizoidal bases. Axes are repeatedly and alternately branched, multiaxial, hollow with mucilage, and constricted at regular intervals by monostromatic plates of small cells. The cortex has 2–3 layers, an inner layer of large cells, and 1–2 outer incomplete layers of small, pigmented cells. The medulla is hollow and has a few longitudinal filaments that produce small gland cells. Tetrasporangia are tetrahedrally divided, scattered, and intercalary within cortical filaments. Gametophytes are dioecious and procarpic. Spermatangia are in sori that girdle the axes with 2–3 spermatangia from each mother cell. Carposporophytes are scattered; outer gonimoblastic cells form ovoid to spherical masses of carposporangia. Ostiolate cystocarps are raised on the surface by a thickened cortex.

Key to the species (3) of *Champia*

1. Axes cylindrical and with small discoid holdfasts 2
1. Axes flattened and with prostrate bases. . . . . *C. vieillardii*
  2. Main axes 2–4 mm in diam.; segments strongly constricted. . . . . *C. salicornioides*
  2. Main axes often less than 1.5 mm in diam.; segments little constricted. . . . . *C. parvula*

Champia parvula **(C. Agardh) Harvey 1853: p. 76 (L. *parvula*: very small, minute). Plate L, fig. 27, 28.**

Basionym. *Chondria parvula* C. Agardh 1824: p. 207.

Synonym. *Champia intricata* Cremades in Cremades and Pérez-Cirera 1990: p. 489.

Silva (1991) noted that *Champia intricata*, which was proposed to replace *C. parvula*, is a later homonym. Womersley (1996) recognized two varieties of *C. parvula* based partially on their habitat; the epilithic var. *parvula* and epiphytic var. *amphibolis* (Reedman and Womersley 1976).

Plants are soft or gelatinous, pale yellow to red, in clumps to 10 cm tall and attached by small discoid holdfasts. Axes are constricted at regular intervals, densely branched, terete, 0.5–2 mm in diam., and with obtuse tips. Axial segments are hollow, cask-shaped, and 1–5 diam. long. Branching is variable, mostly alternate, and with branches sometimes fused. The medullary cavity has a gel matrix and is divided by monostromatic plates of cells. Longitudinally oriented filaments are 8–10 µm in diam., visible through cortical layers, evenly spaced, and may produce ovoid gland cells. The inner cortical cells are 22–40 µm in diam. and 50–130 µm long. Surface cells are irregular, 7–13 μm in diam., and cover the anticlinal walls of inner cells. Tetrasporangia are tetrahedrally divided, globose, 50–100 μm in diam., and scattered. Spermatangia are elliptical, 1–2 μm in diam., and in sori. Cystocarps are scattered, urn-shaped, ostiolate, and to 0.9 mm in diam. Carposporangia are ovoid to irregular and 40–50 μm in diam.

Common, on seagrasses, coarse algae, and limestone on all coasts and to 30 m on the west coast; also known from the Canadian Maritimes to Georgia, Bermuda, the Bahamas, Caribbean, Gulf of Mexico, and Brazil, as well as the eastern North and tropical Atlantic, Mediterranean (type location: Cádiz, Spain; Irvine and Guiry, 1983), Red Sea, Indo-Pacific, and Australia.

Taylor 1960: p. 490; Dawes 1974: p. 139; Schneider and Searles 1991: p. 350; Littler and Littler 2000: p. 120.

Champia salicornioides Harvey 1853: p. 76–77, pl. XIX B (L. *sali*: salt + *corneus*: horny + *oides*: like; *Salicornia*, a genus of succulent, segmented flowering plants of salt marshes; thus, similar to *Salicornia*). Plate LI, fig. 1.

Plants are gelatinous, pale yellow to red, and 4–12 cm tall; axes are strongly constricted at nodes, irregularly oppositely branched, and attached by short stalks with discoid holdfasts. Segments are swollen, cylindrical to slightly compressed, 1–4 mm in diam., and as long as broad. The medullary cavity is filled with a gel and divided by monostromatic plates of cells. The inner cortex is lined with longitudinal filaments 10–12 μm in diam. and 10–20 diam. long, and may have ovoid gland cells. The outer cortex has 1–2 layers; inner cells are large, rounded, 30–60 μm in diam., and 50–140 μm long; surface cells are small, 10–20 μm in diam., and align along anticlinal walls of the inner cortical cells. Tetrasporangia are tetrahedrally divided, scattered, and to 90 μm in diam. Cystocarps are urn-shaped, scattered, and to 0.9 mm in diam.; carposporangia are 40–50 μm in diam.

Uncommon, in shallow water on limestone and seagrasses in the Keys (type location: Key West); also known from Bermuda, the Caribbean, and Brazil, as well as the eastern tropical Atlantic, and Indo-Pacific.

Taylor 1960: p. 491; Littler and Littler 2000: p. 122.

Champia vieillardii Kützing 1866: p. 14, pl. 37, fig. e,f (Named for Eugené Vieillard, a French navel surgeon and botanist who collected in New Caledonia and Tahiti and was director of the Jardin Botanique de Caen). Plate LI, fig. 2.

Synonym. *Champia compressa sensu auct.*, non Harvey 1838: p. 402.

Grunow (1874) combined *Champia vieillardii* with *C. compressa*, while Dawson (1954) and Millar (1990) recognized both taxa. Millar (1990) noted that type specimens of both species were dissimilar. According to Abbott (1999), *C. compressa* can be separated from *C. vieillardii* by the number of cells in the partitions, its compressed versus flat axes, and segment constriction.

Plants are gelatinous, pale to reddish purple, and form tufts 2–3(12) cm tall or soft robust patches to 30 cm in diam.; prostrate bases are attached by discoid holdfasts. Branching is irregular to alternate, in one plane, and mostly basal. Axes are flat and regularly constricted to spindle-shaped, tapering to blunt tips, and have some fusion between lower branches. Segments are swollen to barrel-shaped, 0.7–1.5 mm in diam., and 0.5–2.0 mm long. The medullary cavity is filled with a gel and divided by monostromatic plates of cells. The inner walls of the cortex are lined with longitudinal filaments 8–10 μm in diam. by 200–250 μm long that may form ovoid gland cells. The outer cortex has 1–2 layers; inner cells are large, rounded, and 40–50 μm in diam. by 50–90 μm long; surface cells are small, sparse, spherical to ovoid, to 10 μm in diam., and have hair cells. Tetrasporangia are tetrahedrally divided, spherical, scattered, and 80–100 μm in diam. Spermatangia are 1–2 μm long and in sori that girdle segments. Cystocarps are urn-shaped, scattered, and 0.6–0.6 μm in diam.

Rare, only recorded once from 9 m at the Dry Tortugas (Ballantine, 1996); also known from the Bahamas, Caribbean, and Brazil, as well as the eastern tropical Atlantic, Indo-Pacific, (type location: New Caledonia, Silva et al., 1996; Abbott, 1999), and Australia.

Littler and Littler (2000): p. 122.

Lomentariaceae

See Lee (1978) regarding designation of this family

Plants are soft to gelatinous, prostrate or erect, and attached by discoid holdfasts. Branching is radial to distichous, irregular, and alternate or dichotomous. Axes are hollow (except at bases) and often have multilayered cell plates at nodes. The medulla has loosely arranged longitudinal filaments and the cortex is pseudoparenchymatous. Life histories are isomorphic. Tetrasporangia are tetrahedrally divided, at tips of outer cortical filaments, with sterile filaments, and in depressed sori. Gametophytes are dioecious. Spermatangia arise from outer cortical cells in sori on branchlets. Plants are procarpial; carpogonial branches are 3-celled and from supporting cells with 1(2)-celled auxiliary cell branches. Fusion cells are large; gonimoblastic filaments are globular, grow outward, and most cells become carposporangia. Ostiolate cystocarps are prominent and scattered.

Gelidiopsis F. Schmitz 1895: p. 148 (L. *gelidus*: ice cold or congealed + *opsis*: similar to; thus, resembling the red algal genus *Gelidium*).

Price and Kraft (1991) and Silva et al. (1996) placed *Gelidiopsis* in the Rhodymeniaceae, while Wynne (2005a) and Schneider (2003) placed it in the Lomentariaceae based on molecular studies by Saunders et al. (1999).

Axes are tufted, turflike, or in clusters, with irregular branching, also slender, entangled, and attached by rhizoidal pads. Axes are terete, often fused, and have blunt tips; apical meristems are multicellular and the medulla is pseudoparenchymatous with small, elongate, and thick-walled cells. The cortex is 2 layered with large thick-walled inner cells and small rounded and pigmented outer cells. Tetrasporangia are tetrahedrally divided and swell branch tips. Plants are procarpial; carpogonial branches are 3-celled and from supporting cells with 2-celled auxiliary cell branches. Cystocarps are obvious and scattered on upper branches. Carposporophytes arise from fusion cells.

Key to the species (3) of *Gelidiopsis* (modified from Littler and Littler, 2000).

1. Axes cylindrical, to 0.3 mm in diameter 2
1. Upper branches compressed to flattened, to 1.1 mm in diam. . . . . *G. planicaulis*
   2. Plants to 8 cm tall; branches 210–310 μm in diam., and often fused . . . . *G. intricata*
   2. Plants to 4 cm tall; branches 150–300 μm in diam. and not fused . . . . *G. variabilis*

***Gelidiopsis intricata* (C. Agardh) Vickers 1905: p. 61 (L. *intricata*: entangled, involved). Plate LI, fig. 3–5.**

Basionym. *Sphaerococcus intricata* C. Agardh 1822 (1820–1828): p. 333–334.

Synonym. *Ceratodictyon intricatum* (C. Agardh) R. E. Norris 1987: p. 245.

Price et al. (1988) and Abbott (1999) suggested that *Gelidiopsis intricata* and *G. variabilis* (Greville *ex* J. Agardh) F. Schmitz were conspecific, while Silva et al. (1996) and Wynne (2005a) considered them to be distinct.

Plants are red to greenish brown; axes are wiry, entangled, to 8 cm tall, and attached by rhizoids with pads and prostrate axes. Branching is sparse and irregular or dichotomous. Axes are multiaxial, cylindrical, and often fused; tips are obtuse and have multicellular meristems. Erect axes are 230–310 μm in diam., while prostrate ones are 200–300(425) μm in diam. Medullary cells are elongate, thick walled, and 20–30 μm in diam. The cortex has 5–7 layers with small, spherical surface cells 8–12 μm in diam., 8–14 μm long, and with thickened (8–10 μm) outer walls.

Uncommon, as a turf on wave-exposed intertidal limestone, mats on mangrove roots, and to 10 m in the Keys; also known from Bermuda, the Caribbean, Gulf of Mexico, and Brazil, as well as the eastern tropical Atlantic, Indo-Pacific, and Australia. Syntype locations are Mauritius, Hawaii, and Indonesia (Silva et al., 1996).

Taylor 1960: p. 353; Littler and Littler 2000: p. 124.

Gelidiopsis planicaulis (W. R. Taylor) W. R. Taylor 1960: p. 353 (L. *plani*: flat + *caulis*: a stem or stalk). Plate LI, fig. 6, 7.

Basionym. *Wurdemannia miniata* (Sprengel) J. Feldmann and Hamel var. *planicaulis* W. R. Taylor 1943 ('1942'): p. 158.

Norris (1987) combined *Gelidiopsis planicaulis* and *G. gracilis* (Kützing) J. Feldmann with *G. variabilis*. Littler and Littler (2000) agreed with *G. gracilis* but kept *G. planicaulis* distinct because of its variable branching and tetrasporangial features (Price et al., 1988). Guiry and Nic Dhonncha (2006) listed *G. planicaulis* as a synonym of *Ceratodictyon variabile*.

Plants are pale to deep red, tough, wiry, to 8 cm tall, sparsely irregularly branched, and attached by prostrate axes and rhizoids that terminate in pads. Branches are terete near tips, compressed to strap-shaped below, 0.5–1.1 mm wide, and 125–185 μm thick; tips are rounded and have multicellular meristems. Prostrate axes are 200–400 μm in diam. Medullary cells are 10–20 μm in diam., 80 μm long, and grade to smaller, shorter, outer cells. The cortex has 2–3 layers; surface cells are ovoid, 8–10 μm in diam., and 10–15 μm long. Tetrasporangia are spherical, tetrahedrally divided, 20–40 μm in diam., and swell branch tips.

Rare, reported from 29 m off the east coast (Hanisak and Blair, 1988); also known from Bermuda, the Caribbean (type location: Jeremie, Haiti; Schneider and Lane, 2005), Brazil, Canary Islands, and West Africa.

Taylor 1960: p. 353; Littler and Littler 2000: p. 124.

Gelidiopsis variabilis (Greville *ex* J. Agardh) F. Schmitz 1895: p. 148 (L. *variabilis*: variable or changeable). Plate LI, fig. 8, 9.

Basionym. *Gelidium variabile* Greville *ex* J. Agardh 1851 (1848–1901): p. 468.

Synonyms. *Gelidiopsis gracilis* (Kützing) J. Feldmann 1931: p. 157.

Ceratodictyon variabile (J. Agardh) R. E. Norris 1987: p. 243–245, fig. 1–9 (as '*variabilis*').

Gelidiopsis tenuis Setchell and Gardner 1924: p. 749, pl. 22, fig. 2.

Guiry and Nic Dhonncha (2006) regarded *Gelidiopsis intricata* as a synonym for *G. variabilis* and suggested the inclusion of *G. planicaulis* and *G. gracilis* in *G. variabilis* (Norris, 1987; Littler and Littler, 2000), and cited *G. tenuis* as a distinct taxon.

Plants are red to reddish green, stiff and wiry, entangled, to 4 cm tall, and with prostrate axes attached by rhizoids with pads. Branching is sparse and irregular; erect axes are cylindrical, 150–300 μm in diam., and have blunt tips and multicellular meristems. Prostrate axes are 220–350 μm in diam. The medulla has 6–8 layers of cells to 16 μm in diam. Surface cells are spherical to ovoid, 8–12 μm in diam., and with thick outer walls 10–20 μm in diam.

Uncommon, on limestone in 3–5 m at the Biscayne National Park (Lirman and Biber, 2000); also known from the Caribbean and Brazil, as well as the Indo-Pacific (type location: Madras, India; Abbott, 1999) and Australia.

Littler and Littler 2000: p. 126.

Lomentaria Lyngbye 1819: p. 101 (L. *lomentum*: a legumelike fruit or pod with constrictions).

Plants are tufted, erect, and irregularly, alternately, or oppositely branched; axes are terete to compressed, hollow, filled with mucilage, initially attached by discoid holdfasts, and secondarily by rhizoids. The medulla has a few loose, longitudinal filaments and is divided by multilayered septa located at constrictions of branch bases. The inner cortical cells are larger and surface cells incompletely cover inner ones. Tetrasporangia are tetrahedrally divided and in the outer cortex within sunken sori with pores. Spermatangia are produced in sori on branchlets; each mother cell yields 2–3 spermatangia. Carposporophytes are scattered; cystocarps are ostiolate and swollen due to cortical thickening; gonimoblasts form spherical masses of carposporangia.

***Lomentaria baileyana* (Harvey) Farlow 1876: p. 698 (named for J. W. Bailey, an American phycologist who studied diatoms and desmids as well as seaweeds of the east coast USA). Plate LI, fig. 10–12.**

Basionym. *Chylocladia baileyana* Harvey 1853: p. 185, pl. XXc, fig. 1–5.

Some authors placed *Lomentaria baileyana* in *L. uncinata* Meneghini in Zanardini (Farlow, 1881; Bodard and Mollion, 1974). We have followed Taylor (1960), Cribb (1983), and Wynne (2005a) and consider them distinct species.

Plants are soft to firm, turflike, pale to deep red, 3–20 cm tall, densely branched, initially attached by padlike rhizoidal holdfasts, and secondarily by recurved branch tips. Axes are terete, 0.5–1.5 mm in diam., taper from base to apex, and have obtuse tips. Branching is irregular, radial, or secund. Axes are hollow and filled with a gel; they are divided by multilayered cell plates, sometimes constricted at nodes, and with only a few longitudinal medullary filaments (2–3 μm in diam.). The cortex has 2–3 layers; inner cells are ovoid, elongate or irregular, 25–40 μm in diam., and 50–115 μm long; spherical gland cells may be present. The surface cell layer is incomplete, with cells aligned along anticlinal walls of inner ones; cells are rounded and 7–13 μm in diam. Tetrasporangia are tetrahedrally divided and 30–70 μm in diam. Spermatangia are in raised sori on branchlets; spermatia are ovoid and 1–2 μm in diam. Cystocarps are ostiolate, oval, and 0.2–0.6 mm in diam.; carposporangia are spherical and 17–23 μm in diam.

Common, on seagrasses, seaweeds, limestone (to 33 m), and adrift on all coasts (type location: Key West); also known from scattered warm water sites in the Canadian Maritime Provinces to Cape Cod, and southward to Georgia, Bermuda, the Bahamas, Caribbean, Gulf of Mexico, and Brazil, as well as the eastern tropical Atlantic and Indo-Pacific.

Taylor 1960: p. 487; Dawes 1974: p. 140; Schneider and Searles 1991: p. 353; Littler and Littler 2000: p. 126.

Sedis incertae: *Agardhinula* De Toni 1897: p. 64 (named for J. G. Agardh, a Swedish phycologist and taxonomist who published from the mid-1830s to 1901 + *ula*: a diminutive; thus, a small form of the red algal genus *Agardhiella*).

According to Saunders et al. (1999), the systematic position of *Agardhinula* within the Rhodymeniales is uncertain.

Plants are erect; blades are fleshy to soft, flat, irregularly, di- to trichotomously or palmately branched, and attached by discoid holdfasts. Axes are multiaxial with a medulla of a few layers of large cells interspersed with smaller ones. The cortex has 2–3 layers of small cells in anticlinal rows, with the outermost cells not covering inner ones. Tetrasporangia are from surface cells, cruciately divided, and in reticulated gelatinous sori scattered over the surface, and cause a spotted appearance. Carposporophytes are scattered; gonimoblastic cells form masses of carposporangia attached by a few filaments. Cystocarps swell one surface because of cortical thickening; pericarps lack both ostioles and a *tela arachnoidea*.

Agardhinula browneae (J. Agardh) De Toni 1897: p. 64; 1900: p. 523 (named for "_D:na_" = "Mrs."? Browne). Plate LI, fig. 13, 14.

Basionym. *Callophyllis browneae* J. Agardh 1885: p. 36

Fronds are fleshy to gelatinous, 10–30 cm tall, pink to red, flat, palmately branched below, dichotomous to irregularly branched at margins, and attached by discoid holdfasts. Blades are straplike, 1–5 cm wide below nodes, wider in older areas, 350–625 µm thick, and have rounded, obtuse, or truncate tips. The medulla has a mix of large and small cells, 35–370 µm in diam. The cortex has 1–3 layers with surface cells irregular to spherical, pigmented, and 4–10 µm in diam. Tetrasporangia are cruciately divided, ovoid or spherical, 12–22 µm in diam., and 7–38 µm long. Spermatangia are ovoid or spherical and 1–2 µm in diam. Cystocarps are scattered, hemispherical, to 1 mm in diam., and lack obvious ostioles. Carposporangia are oval to irregular and 10–15 µm in diam.

Uncommon, in 34–90 m on hard substrata off both the east and west coasts (type location: "ad oras Floridae"; J. Agardh, 1885); also known from the Carolinas (17–55 m), Lesser Antilles, and northern Gulf of Mexico.

Taylor 1960: p. 482; Schneider and Searles 1991: p. 333; Littler and Littler 2000: p. 128.

5

Keys to the Algal Genera

Key to the Genera of the Phylum Chlorophyta (green algae), plus *Chrysophaeum*, *Chrysonephos*, and *Vaucheria*

1. Plants cellular, divided into uninucleate or multinucleate cells by cross walls 2
1. Plants coenocytic (not composed of cells, vegetative parts lacking cross walls) 42
2. Plants filamentous, uniseriate, branched or unbranched, not forming blades. 3
   2. Plants are tubes, discs, or blades; if filamentous, axes multiseriate or forming blades. 13
3. Filaments unbranched, occasional cells may produce rhizoids 4
3. Filaments moderately to abundantly branched 7
   4. Diameter of filaments less than 30 µm 5
   4. Diameter of filaments more than 30 µm. . . . . *Chaetomorpha*
5. Chloroplasts platelike, lateral, extending halfway around cell. . . . .*Stichococcus*
5. Chloroplasts collarlike or covering inner wall and netlike (reticulate) 6
   6. Cells uninucleate, chloroplasts collarlike . . . . *Ulothrix*
   6. Cells multinucleate, chloroplasts mesh or netlike (reticulate) . . . .*Rhizoclonium*
7. Filaments bearing hairs that lack basal cross walls (aseptate) . . . .*Phaeophila*
7. Filaments lacking aseptate hairs 8
   8. Filaments irregular, penetrating shells, limestone, tests, or algae . . . . *Gomontia*
   8. Filaments regular, not penetrating shells, limestone, tests, or algae. 9
9. Filaments creeping, epi- or endophytic, less than 10 µm in diameter . . . . *Entocladia*
9. Filaments erect or matted, not creeping, diameter mostly more than 10 µm 10
   10. Plants with parenchymatous filamentous bases; erect filaments cylindrical, little branched, irregular in shape, with plugs in cross walls . . . . *Smithsoniella*
   10. Plants not as above, lacking pseudoparenchymatous bases 11
11. Cross walls absent from bases of branches . . . . *Cladophoropsis*
11. Cross walls present at bases of branches 12
   12. Filaments often irregular in shape, branching unilateral, secund; cells 0.3–1.0 mm in diameter . . . . *Valoniopsis*
   12. Filaments usually regular in shape, branching varied, sometimes unilateral near tips; cells rarely more than 0.5 mm in diameter . . . . *Cladophora*
13. Plants forming colonies; cells uninucleate, embedded in a gelatinous matrix. 14
13. Plants otherwise, thallus more substantial and organized 15
   14. Colonies yellow green; cells usually in groups of four . . . . *Pseudotetraspora*
   14. Colonies sulfur yellow; cells stalked, anterior indented . . . .*Chrysophaeum*
15. Plants disclike, cells uninucleate 16
15. Plants not disclike, cells uninucleate or multinucleate 18

16. Pyrenoids absent in plastids; marginal cells of discs indented or forked . . . . *Ulvella*
16. Pyrenoids present in plastids; marginal cells of discs not indented 17
17. Cell with hairs lacking basal cross wall (aseptate) . . . . *Pringsheimiella*
17. Cells with hairs having basal cross wall. . . . . *Pseudendoclonium*
18. Plants tubular (at least in part) . . . . *Ulva* (in part)
18. Plants not tubular 19
19. Plants bladelike 20
19. Plants not bladelike 26
20. Blades mesh- or netlike (reticulate), consisting of filaments 21
20. Blades not mesh- or netlike 23
21. Blades stalked, stalks consisting of a single filament 22
21. Blades not stalked, forming a net, prostrate or erect . . . . *Microdictyon*
22. Central stalk with 14–20 paired branchlets; laterals formed by segregative cell divisions, intercalary cell divisions absent . . . . *Struvea*
22. Central stalk with fewer than 14 paired branchlets; laterals formed by centripetal cross walls, intercalary cell division often present . . . . *Phyllodictyon*
23. Blades with two layers of cells. . . . . *Ulva* (in part)
23. Blades monostromatic, with one cell layer. 24
24. Blades with tubular bases. . . . . *Ulva* (in part)
24. Blades lacking tubular bases. 25
25. Blades of compact, similar sized cells, lacking veins . . . . *Gayralia*
25. Blades with cells of different sizes, larger cells forming veins. . . . . *Anadyomene*
26. Plants partially to completely composed of biseriate to multiseriate filaments 27
26. Plants not composed of biseriate or multiseriate filaments 28
27. Plants green, staining deep blue with I_2KI; filaments mostly biseriate, sometimes to 4 cells wide. . . . . *Percursaria*
27. Plants yellow green, not staining deep blue with I_2KI; filaments irregularly multiseriate *Chrysonephos*
28. Plants calcified 29
28. Plants not calcified 34
29. Plants calcified only in annular constrictions of stalks, capped by large tuft of coarse filaments, filaments not calcified nor in whorls . . . . *Chamaedoris*
29. Plants calcified at least in part; branchlets in whorls that may not be calcified 30
30. Plants large, 10–30 cm tall, distinctly segmented, older segments cylindrical; branch tips terminating in green, noncalcified, filamentous tufts . . . . *Cymopolia*
30. Plants smaller, less than 4 cm tall, not segmented. 31
31. Plants cylindrical to club-shaped, often with fine green filaments projecting beyond calcified axes . . . . *Neomeris*
31. Plants umbrellalike, the stalk bearing a cap of raylike filaments 32
32. Spores in rays of mature caps not calcified. . . . . *Acetabularia*
32. Spores in rays of mature caps calcified 33
33. Spores embedded in calcium carbonate, sporangial membrane not calcified; corona inferior present . . . . *Acicularia* (see *Acetabularia*)
33. Spores calcified, free and not embedded, sporangial membrane also calcified; corona inferior absent . . . . *Chalmasia*
34. Individual cells macroscopic, easily visible, bulbous or club-shaped. 35
34. Individual cells not macroscopic, visible under 10X magnification. 36
35. Branching whorled near apices of parent branches . . . . *Ernodesmis*
35. Branching radiating laterally from stalks, not whorled . . . . *Siphonocladus*
36. Branchlets arranged in whorls around uniseriate stalk 37
36. Branchlets not arranged in whorls, plants not on uniseriate stalks 38

37. Main axes with loose whorls of branchlets every 0.2–2.0 mm; each mature branchlet with many gametangia . . . . *Batophora*
37. Main axes with tight whorls of branchlets every 100–150 μm; each mature branchlet with a single gametangia. . . . .*Dasycladus*
38. Plants small, crustose, with evident adherent radial filaments . . . . *Petrosiphon*
38. Plants not crustose, cells larger, lacking adherent radial filaments 39
39. Plants are single large spherical cells with rhizoids 40
39. Plants are clusters of large macroscopic cells with rhizoids 41
40. Cells a refractive green, growing on any hard surface, with distinctive rhizoidal cells having cross walls . . . . *Ventricaria*
40. Cells a glossy green, growing on crustose coralline algae, with rhizoids lacking cross walls. . . . . *Halicystis* stage of *Derbesia osterhoutii*
41. Cells round to oval, more than 3 mm in diam.; if less, then elongated and 5–20 mm long . . . *Valonia*
41. Cells angular to polygonal, small, less than 3 mm in diameter . . . .*Dictyosphaeria*
42. Plants are coenocytes, calcified. 43
42. Plants are coenocytes, not calcified. 47
43. Thallus distinctly segmented, segments flat, disclike, or spherical . . . .*Halimeda*
43. Thallus not segmented. 44
44. Plants with stalk bearing coarse, fibrous siphons, appearing brushlike. 45
44. Plants with stalks bearing fan- or bladelike tops, not brushlike 46
45. Cap siphons in one plane, tapering distally to half diameter, dichotomies usually not constricted at first forking; lower cap margin straight. . . . . *Rhipocephalus*
45. Cap siphons not in one plane, forming a brush, minimally tapered from base to apex, deeply constricted at forkings; lower cap margin rounded . . . . *Penicillus*
46. Plants less than 1 cm tall; stipe with one siphon, lacking cortication . . . . *Rhipidosiphon*
46. Plants more than 1 cm tall; stipe corticated with lateral branchlets . . . .*Udotea*
47. Plants usually differentiated into fronds, rhizomes, and rhizoids; with microscopic internal transverse wall projections (trabeculae) . . . . *Caulerpa*
47. Plants not so differentiated, lacking trabeculae 48
48. Siphons clearly constricted at bases of branchlets; plants usually forming loosely woven mats, not organized into macroscopic thalli. 49
48. Siphons not deeply constricted at branch bases; plants in loose tufts, mats, or organized into macroscopic thalli 51
49. Branching usually dichotomous or pseudodichotomous . . . . *Boodleopsis*
49. Branching otherwise 50
50. Branching abundant, lateral, often opposite, feather- or plumelike; gametangia distinct from vegetative branches . . . .*Bryopsis*
50. Branching limited and irregular; gametangia are vegetative branches. . . . . *Trichosolen*
51. Siphons in loose tufts or mats, branching dichotomous to irregular 52
51. Siphons not in tufts or mats but forming macroscopic thalli or filaments 53
52. Plants usually oceanic; siphons staining dark blue with I_2KI . . . .*Derbesia*
52. Plants usually estuarine; siphons not staining dark blue with I_2KI . . . .*Vaucheria*
53. Plants penetrating limestone, shells, and echinoderm tests, endophytic or endozoic; filaments twisted, irregular in diameter . . . . *Ostreobium*
53. Plants not penetrating; filaments more regular 54
54. Plants lacking a distinct stipe; erect or prostrate; cortex compact and composed of firmly attached surface utricles. 55
54. Plants with a distinct stipe that bears a paddle- or umbrella-shaped blade; or plants prostrate and matlike; cortex lacking firmly attached utricles 56

55. Plants erect, axes with 1–6 medullary siphons; chloroplasts restricted to medullary siphons........ .... *Pseudocodium*
55. Plants erect or prostrate, axes with many medullary siphons; chloroplasts restricted to surface utricles .... *Codium*
56. Surface siphons colorless, thin, contorted, tightly interwoven, forming a cortex, distinctly different from medullary siphons .... *Cladocephalus*
56. Surface siphons absent or if present, loosely interwoven, and surface not distinct from medullary siphons.. 57
57. Cortex of lateral branchlets, terminating in fingerlike (hapteral) apices.................. *Rhipilia*
57. Cortex of lateral branchlets lacking specialized hapteral apices..................... *Avrainvillea*

Key to the Genera of the Class Phaeophyta (brown algae; Phylum Heterokontophyta)

1. Plants primarily creeping, crustose, or endophytic... 2
1. Plants erect except for attachment or prostrate bases 11
2. Plants growing on hard substrata including shells 3
2. Plants growing on algae or vascular plants.. 5
3. Plants parenchymatous, blades creeping, loosely attached....................*Lobophora* (in part)
3. Plants pseudoparenchymatous to filamentous, closely adherent to substrata................... 4
4. Margins of crusts loosely attached, easily lifted; thallus with arched erect filaments arising from distinct basal layer; apical cells of filaments not obvious .... *Ralfsia*
4. Margins of crusts firmly attached; thallus with arched erect filaments arising from an indistinct basal layer; apical cells clearly present....................................... *Pseudolithoderma*
5. Basal filaments endophytic 6
5. Basal filaments epiphytic... 8
6. Endophytic filaments uniseriate throughout, attenuated, apical cells not large, lacking multicellular propagules .... *Herponema*
6. Endophytic filaments multiseriate, not attenuated, apical cells large; with multicellular propagules 7
7. Endophytic filaments infrequently biseriate, emergent propagules globose, with four bulbous apical cells .... *Onslowia*
7. Endophytic filaments usually multiseriate, emergent propagules rectangular, lacking evident apical cells .... *Verosphacela*
8. Epiphytes with basal crust and erect filaments 9
8. Epiphytes discoid, flattened, erect filaments absent 10
9. Thalli brush- to cushionlike; basal portion uniseriate with densely packed colorless filaments, often penetrating host; erect filaments long, uniseriate .... *Elachista*
9. Thalli discoid crusts; basal filaments biseriate; not penetrating host; erect filaments and colorless hairs arising from crusts .... *Hecatonema*
10. Filamentous thalli with radial growth; plastids platelike .... *Phaeostroma*
10. Filamentous thalli with irregular growth; plastids discoid .... *Myrionema*
11. Erect portions with uniseriate filaments; not forming blades or complex bodies 12
11. Erect portions at least multiseriate at forks, parenchymatous or filamentous in construction; usually forming tubes, blades, or complex plant bodies 18
12. Unilocular and plurilocular sporangia intercalary .... *Bachelotia*
12. Unilocular sporangia terminal; plurilocular sporangia hair-tipped 13
13. Growth by meristems near tips that form hairs (trichothallic); rhizoids produced at right angles to filaments .... *Acinetospora*
13. Growth not trichothallic but diffuse or apical; rhizoids not arising at right angles 14
14. Branches often hooked or bearing hooklike, recurved branchlets 15
14. Branches not often hooked or bearing recurved branchlets 16

15. Growth apical; filaments with many short hooklike branchlets . . . . *Spongonema*
15. Growth intercalary; recurved branchlets absent; filaments often with long hooked branches (not branchlets) . . . . *Asteronema*
16. Plastids elongate, ribbon- or band-shaped . . . . *Ectocarpus*
16. Plastids discoid. 17
17. Meristems distinct, intercalary at base of long unbranched filaments; sporangia usually stalked. . . . . *Feldmannia*
17. Meristems diffuse, rarely basal to branches; sporangia usually sessile . . . . *Hincksia*
18. Plants compressed, flattened, or bladelike throughout 19
18. Plants not bladelike; axes cylindrical or inflated; hollow or solid 27
19. Fronds lobed, irregularly split, may be fan-shaped; all growing from a row of apical cells 20
19. Blades regularly forked; neither fan-shaped nor frequently split; branching often dichotomous; all branches growing from one or a few apical cells 26
20. Sporophytes small, to 25 mm long; blade monostromatic and with a single layer of filaments; stipe composed of rhizoids . . . . *Syringoderma*
20. Plants more than 25 mm long, blades multiseriate, parenchymatous; stipe not composed of rhizoids 21
21. Plants growing in the upper intertidal, bases of blades often cylindrical . . . . *Chnoospora*
21. Plants growing subtidally or in tide pools; bases of blades flattened or fibrous 22
22. Blades broad, initially fan-shaped, older ones may be highly cleft 23
22. Blades subpalmately to pinnately lobed and proliferous. . . . . *Spatoglossum*
23. Apical margin of fan-shaped blades enrolled (curled) . . . . *Padina*
23. Apical margin of fan-shaped blades not enrolled. 24
24. Blades prostrate, creeping or erect; sori without sterile hairs . . . . *Lobophora* (in part)
24. Blades erect; sori with sterile hairs 25
25. Blades split into segments, medulla lack covering surface cells near apical margin; sporangia sessile. . . . . *Stypopodium*
25. Blades often torn, segments irregular, medulla covered by surface cells close to apical cells; sporangia on 2-cell stalk. . . . . *Taonia*
26. Midribs present; branch tips have a group of apical cells . . . . *Dictyopteris*
26. Midribs absent; branch tips with one apical cell. . . . . *Dictyota*
27. Plants bladderlike, hollow or bearing hollow air bladders (vesicles) 28
27. Plants solid throughout, lacking hollow structures 35
28. Plants spherical or subspherical 29
28. Plants cylindrical or more complex, bearing air bladders (vesicles) 30
29. Plants sacklike or bulbous, often lobed, hollow when mature; entire and lacking holes, margins not enrolled . . . . *Colpomenia*
29. Plants hollow when young, soon flattening and covering substrata; with many holes, margins enrolled. . . . . *Hydroclathrus* (in part)
30. Branches bearing distinctive individual air bladders (vesicles) 31
30. Branches lacking distinctive air bladders, all or most of axes hollow 33
31. Blades minute, spinelike; conceptacles on ordinary branches. . . . . *Cystoseira*
31. Blades larger, flattened or triangular; conceptacles on specialized branches. 32
32. Axes bearing leaflike blades; vesicles distinct, bulbous . . . . *Sargassum*
32. Axes bearing pyramid-shaped blades; vesicles embedded in blades . . . . *Turbinaria*
33. Branching irregular to dichotomous; axes inflated or collapsed. *Rosenvingea*
33. Plants unbranched, axes tubular to club-shaped; axes inflated 34
34. Thallus linear, hollow, usually with constrictions; plurilocular sporangia sessile, in irregular confluent patches; unilocular sporangia absent . . . . *Scytosiphon*
34. Thallus linear to club-shaped, lacking constrictions; plurilocular sporangia and unilocular sporangia stalked, in small regular sori . . . . *Asperococcus*

35. Plants soft, construction filamentous; filaments radially, not laterally united 36
35. Axes firm, at least multiaxial at forks; construction pseudoparenchymatous or parenchymatous . . 37
36. Axes to 0.3 mm in diam.; growth uniaxial and with one apical cell . . . . *Nemacystus*
36. Axes diam. more than 0.7 mm; growth multiaxial and with many apical cells. . . . . . . *Cladosiphon*
37. Axes with single apical cells at tips 38
37. Axes with multiple apical cells at tips 40
38. Apical cells not large; axes multiseriate only near forks . . . . *Xanthosiphonia*
38. Apical cells large; axes multiseriate throughout, except at tips 39
39. Plants to 3 cm tall, in tufts; branching limited; asexual reproduction by vegetative propagules . . . . *Sphacelaria*
39. Plants to 10 cm tall, freely branched, multiaxial; propagules absent . . . . *Halopteris*
40. Branches terminating in tufts of 50 or more filaments 41
40. Branches not ending in tufts with numerous filaments 42
41. Terminal tufts of filaments on spurlike branchlets only . . . . *Nereia*
41. Terminal tufts of filaments on branchlets with bulbous apices . . . . *Sporochnus*
42. Medulla with four large cells in cross section . . . . *Hummii*
42. Medulla filamentous or parenchymatous, not with four large cells 43
43. Erect axes cylindrical; attached by uniseriate filaments with rhizoids 44
43. Erect axes bearing one or more blades; attached by discoid holdfasts . . . . *Petalonia*
44. Erect axes with intercalary growth; erect axes uniseriate below to parenchymatous above, filamentous base attached by rhizoids . . . . *Myriotrichia*
44. Erect axes with apical growth; medulla pseudoparenchymatous, attached by discoid holdfasts . . . . *Stilophora*

Key to the Genera of the Phylum Rhodophyta (red algae)

1. Plants calcified, partially or completely impregnated with calcium carbonate (dissolves in 1% hydrochloric acid [HCl] and releases H_2 as bubbles) 2
1. Plants not calcified, lacking calcium carbonate, not reactive with HCl 25
2. Thallus firm to rock hard 3
2. Thallus soft, gelatinous, containing calcified material 10
3. Plants forming crusts, branches inflexible, sometimes unattached and forming loose calcified spheres (rhodoliths) 16
3. Plants erect, may develop from crusts; branches flexible, with joints (genicula) 4
4. Resembling coral growths (octocorals); calciferous cells with calcium carbonate accretions scattered throughout the plant . . . . *Rhodogorgon*
4. Not looking like octocorals, calciferous cells absent 5
5. Erect axes appearing segmented; composed of calcified segments (intergenicula) that alternate with uncalcified joints (genicula) 6
5. Erect axes of similar morphology throughout; not partitioned into calcified and uncalcified segments 9
6. Conceptacles formed laterally on the surfaces of segments; joints usually more than one tier of cells; cells of adjacent segment filaments linked by secondary pit connections . . . . *Amphiroa*
6. Conceptacles formed axially at the distal ends of segments; joints with one tier of cells; cells of adjacent segment filaments linked by cell fusions 7
7. Branching mostly dichotomous . . . . *Jania*
7. Branching mostly pinnate, but sometimes with adventitious branchlets 8
8. Tetrasporangia usually fewer than 15 per conceptacle; most segments with fewer than 10 tiers of cells; primary axes (in Florida plants) usually less than 500 µm in diameter . . . . *Haliptilon*
8. Tetrasporangia usually more than 30 per conceptacle; most segments with more than 10 tiers of cells; primary axes (in Florida plants) usually more than 500 µm in diameter. . . . . *Corallina*
9. Surface hairlike filaments common at least at base of plants . . . . *Galaxaura*
9. Surface hairlike filaments absent . . . . *Tricleocarpa*

10. Plants foliose, with bladelike parts 11
10. Plants cylindrical to slightly compressed, lacking bladelike parts 12
11. Plants little calcified, calcification limited to perennial, prominent holdfast; veins present, fusing at bases . . . . *Calonitophyllum* (in part)
11. Plants calcified throughout, veins absent . . . . *Titanophora*
12. Axes uniseriate, each axial cell bearing a whorl of branchlets . . . . *Crouania*
12. Axes not uniseriate 13
13. Plants pink, usually with a white halo, calcification primarily around axial core; carpogonial branches terminal on cortical branchlets . . . . *Trichogloea*
13. Plants pink, lacking a white halo, tips often red, calcification not limited to axial core; carpogonial branches lateral on cortical branchlets 14
14. Plants moderately to strongly calcified . . . . *Ganonema*
14. Plants little to lightly calcified 15
15. Branching dichotomous to pinnate; medullary filaments compact . . . . *Liagora*
15. Branching irregularly alternate; medullary filaments loose . . . . *Trichogloeopsis*
16. Crusts calcified only in the hypothallus . . . . *Peyssonnelia*, *Metapeyssonnelia*
16. Crusts calcified throughout 17
17. Tetrasporangial conceptacles each with several to many pores 18
17. Tetrasporangial conceptacles with one pore 20
18. Construction dimerous (one basal layer of cells, each bearing an upright filament of one or more cells) . . . . *Melobesia*
18. Construction monomerous (one set of filaments that form a central ventral core and peripheral region, where they curve outward toward the thallus surface) 19
19. Filaments end in cells with distal walls flat and flared . . . . *Lithothamnion*
19. Filaments end in cells with distal walls rounded or flat, not flared . . . . *Mesophyllum*
20. Cells of adjacent filaments have secondary pits; cell fusions absent . . . . *Lithophyllum*
20. Cells of adjacent filaments have cell fusions; secondary pits absent 21
21. Crusts have a ventral layer of vertically elongate, palisadelike cells . . . . *Lithoporella*
21. Plants lack a ventral layer of vertically elongate, palisadelike cells 22
22. Pore canals of tetrasporangial conceptacles lined by more or less vertical cells that do not protrude laterally into the pore canal . . . . *Hydrolithon*
22. Pore canals of tetrasporangia conceptacles lined by cells arranged more or less at an angle and protrude laterally into the canal to varying extents 23
23. Roofs of tetrasporangial conceptacles formed both by filaments interspersed among the sporangia and by filaments at the margin of the chamber; normally epiphytic, unbranched, and less than 50 µm thick . . . . *Pneophyllum*
23. Roofs of tetrasporangial conceptacles formed only by filaments at the margin of the chamber; normally on rocks or unattached, may have upright projections or branches usually more than 60 µm to 1 mm thick 24
24. Hypothallus thin, perithallus thick, trichocytes scattered . . . . *Goniolithon*
24. Hypothallus and perithallus thick, with numerous cells in rows; trichocytes solitary, arranged in platelike groups, or in vertical rows . . . . *Porolithon*, *Neogoniolithon*
25. Plants discoid to crustose; one to several layers, compact or spreading 26
25. Plants erect or prostrate, not discoid or crustose 28
26. Crusts of radiating filaments, ends free at margins . . . . *Erythrocladia* (in part)
26. Crusts compact, marginal cells united to form a continuous edge 27
27. Crusts multiseriate; thallus with horizontal hypothallial and vertical perithallial filaments . . . . *Hildenbrandia*
27. Crusts only one cell layer thick, growth marginal . . . . *Sahlingia*

28. Plants filamentous, corticated in bands, with continuous cortication, polysiphonous (with a central cell row and pericentral cells), or fleshy 29
28. Plants pseudofilamentous or filamentous; uniseriate throughout, lacking regular cortication except for possible rhizoidal growth near bases 32
29. At least parts of plants are composed of uniseriate branchlets that fuse to form meshlike networks 30
29. Plants not composed in part of meshlike networks of uniseriate branchlets. 59
30. Solid parts of blade interrupted with netlike zones or bands . . . . *Martensia* (in part)
30. Plants not blades with solid and netlike zones 31
31. Plants are delicate clumps of interlocking filaments. . . . . *Halydictyon*
31. Plants with central supporting axes encircled by branchlets that form a delicate network of filaments . . . . *Dictyurus*
32. Cells loosely embedded in a matrix, lacking pit-connections between cells. 33
32. Cells in regular series, with pit-connections between cells 38
33. Cells endophytic, endozoic, or prostrate and spreading. . . . . *Erythrocladia* (in part)
33. Plants erect, branched or unbranched 34
34. Plants unbranched 35
34. Plants sparingly to frequently branched 37
35. Axes uniseriate throughout; cells spaced irregularly and widely, axes bluish green to gray green in color . . . . *Chroodactyon* (in part)
35. Axes initially uniseriate, becoming multiseriate above; cells closely and regularly spaced, axes rose red to purple red 36
36. Plants epiphytic . . . . *Erythrotrichia*
36. Plants growing on limestone, unbranched . . . . *Bangia*
37. Plants with limited branching; cells blue green to gray green. . . . . *Chroodactyon* (in part)
37. Plants repeatedly branched; cells pinkish red, rose, to purple red . . . . *Stylonema*
38. Axial cells of filaments more than 250 µm in diameter 39
38. Axial cells of filaments less than 250 µm in diameter. 40
39. Cells mostly cylindrical; apical and subapical cells spherical to caplike; tetrasporangia on conspicuous stalks . . . . *Anotrichium*
39. Cells mostly swollen, filaments beadlike; subapical cells globose; tetrasporangia sessile or inconspicuously stalked . . . . *Griffithsia*
40. Plants mostly creeping and dorsiventrally organized; attached by stiltlike lateral branchlets . . . . *Grallatoria*
40. Plants mostly erect, any creeping axes lacking stiltlike lateral branchlets 41
41. Plants small, erect axes to 1 cm tall. 42
41. Erect axes more than 1 cm tall 47
42. Branching in opposite pairs (decussate); branchlets distichous, opposite, and often at right angles to one another 43
42. Branching opposite, whorled, alternate, or irregular, not decussate 44
43. Erect axes to 1 cm tall, prostrate axes with rhizoids that terminate in lobed discs; axial cells each bearing two short oppositely placed branchlets . . . . *Diplothamnion*
43. Erect axes to 0.6 cm tall, prostrate axes with short, lateral rhizoids; axial cells dividing to produce unequal whorled branchlets. . . . . *Perikladosporon*
44. Erect axes consistently bearing opposite or whorled branches, gland cells usually present. . . . . *Antithamnionella* (in part)
44. Erect axes bearing few to many unilateral to irregular or alternate branches, sometimes with opposite branches; gland cells absent 45
45. Filaments bear monosporangia; cells have one stellate plastid (*Acrochaetium*) or one or more nonstellate plastids (*Colaconema*) . . . . *Acrochaetium*, *Colaconema*
45. Plants lacking monosporangia; each cell with many discoid plastids 46

46. Erect indeterminate axes alternately branched, spiraled; branchlets simple to dichotomous; tetrasporangia cruciate . . . . *Callithamniella*
46. Erect indeterminate axes sparingly branched, with few alternate or secund branchlets; tetrasporangia tetrahedral . . . . *Ptilothamnion* (in part)
47. Plants mostly prostrate, creeping; rhizoids terminating in claw- or lobelike pads; reproduction by monospores . . . . *Monosporus*
47. Plants mostly erect; rhizoids not terminating in clawlike pads; reproduction rarely by monospores (but see *Acrochaetium*) 48
48. Branching irregular to unilateral 49
48. Branching pinnate, opposite (not trichotomous), or alternate 50
49. Rhizoids covering the basal cell; lower cells of erect axes less than 15 µm in diam.; cells bearing many discoid plastids. . . . . *Balliella* (in part)
49. Basal cell not corticated, lower cells of erect axes more than 20 µm in diameter; cells bearing a single large stellate plastid . . . . *Acrochaetium* (in part)
50. Gland cells present. 51
50. Gland cells absent or seldom present 53
51. Branching plumose, distichous; each axial cell bearing opposite to whorled branches; branches bearing spinelike branchlets on upper side . . . . *Pterothamnion*
51. Branching opposite, alternate, or whorled; spinelike branchlets absent. 52
52. Basal cells of branches distinctly shorter than others; gland cells covering more than one vegetative cell, borne on specialized branchlets. . . . . *Antithamnion*
52. Basal cells of branches similar to others; gland cells covering part or all of a vegetative cell, not borne on specialized branchlets . . . . *Antithamnionella* (in part)
53. Branching mostly opposite or unilateral, rarely alternate 54
53. Branching alternate, main axes appearing dichotomous, rarely unilateral. 56
54. Erect axes plumose; branching pinnate to bipinnate. . . . . *Gymnothamnion*
54. Erect axes not plumose; branching opposite or unilateral 55
55. Prostrate axes with erect filaments arising from one end and rhizoids descending from opposite end of the cells . . . . *Spermothamnion*
55. Prostrate axes with erect filaments and descending rhizoids arising in the middle of the cells and opposite one another . . . . *Ptilothamnion* (in part)
56. Rhizoids coarse; plants bearing polysporangia, rarely tetrasporangia. 57
56. Rhizoids delicate, fibrous; plants bearing only tetrasporangia 58
57. Main axial cells 150–625 µm in diam.; rhizoids elongate . . . . *Pleonosporium*
57. Main axial cells 20–50 µm in diam.; rhizoids inflated . . . . *Tiffaniella*
58. Erect axes with spiral or radial branching, rarely in one plane; tetrasporangia sessile on the adaxial (upper) side of branchlets. . . . . *Aglaothamnion*
58. Erect axes abundantly alternately branched, in one plane; tetrasporangia terminal on short branchlets . . . . *Compsothamnion* (in part)
59. Plants filamentous; uniseriate or visibly polysiphonous; not fleshy; any cortication not completely obscuring central filament, at least in branchlets or young axes 60
59. Plants fleshy or parenchymatous; axes usually thickened; if polysiphonous, pericentral cells not visible except near branch tips 81
60. Axes uniseriate; cortication only at nodes, covering all axes with small cells, or only on older axes and then rhizoidal; axes not polysiphonous 61
60. Axes polysiphonous, at least visible at tips, often corticated in older regions 68
61. Axes corticated throughout, cortical cells thinly cover all axes 62
61. Cortication by small cells in bands at nodes of all axes or limited to branchlets, absent or by rhizoidal cells on older axes 63

62. Cortication continuous on all axes, by small rectangular cells arranged in vertical rows; spines present at nodes . . . . *Centroceras*
62. Cortication continuous on all axes, by polygonal to rounded cells not arranged in vertical rows, spines absent . . . . *Ceramium* (in part)
63. Cortication by small cells in bands on all axes . . . . *Ceramium* (in part)
63. Cortication not in bands on all axes but on branchlets, absent, or by rhizoidal cells. 64
64. Cortication complete on primary axes, in bands at nodes of branchlets, with rectangular short and wide cells alternating with long and thin ones . . . . *Spyridia*
64. Cortication absent or, if present, not as above 65
65. Older axes surrounded by free, unattached rhizoids near bases; plants bearing spores in chains (seirospores) . . . . *Seirospora*
65. Rhizoids not free-floating or unattached; plants bearing tetrasporangia 66
66. Main axes to 60 µm in diameter . . . . *Balliella* (in part)
66. Main axes more than 60 µm in diameter 67
67. Cortication partial, by descending rhizoids along axes; tetrasporangia less than 60 µm in diameter, terminal, involucral filaments absent . . . .*Compsothamnion* (in part)
67. Cortication partial, by descending filaments; tetrasporangia larger than 60 µm, near nodes, surrounded by involucral filaments. . . . .*Wrangelia*
68. Axes having only three pericentral cells . . . . *Asparagopsis*
68. Axes with four or more pericentral cells, sometimes corticated below 69
69. Axes bearing pigmented, monosiphonous determinate branches 70
69. Axes naked or bearing colorless monosiphonous determinate branches that may be deciduous (trichoblasts) 72
70. Branching mostly in two rows (distichous). . . . .*Heterosiphonia* (in part)
70. Branching radial. 71
71. Determinate branchlets short, persistent; with sharply pointed branchlet tips; growth is monopodial and apical . . . .*Heterodasya*
71. Determinate branchlets long, often deciduous at base, sharp-pointed branchlet tips absent; growth is sympodial . . . .*Dasya* (in part)
72. Plants with dorsiventral organization and obvious creeping axes. 73
72. Plants primarily erect; organization not dorsiventral although prostrate axes may be present . . 76
73. Plants often featherlike; intertidal; usually in estuarine waters (salinities of 10–25 ppt); often epiphytic on salt marsh stalks and mangrove roots . . . . *Bostrychia* (in part)
73. Plants not featherlike, intertidal or not; usually in oceanic waters (salinities of 30 ppt or more) . . . 74
74. Erect axes ending in narrow monostromatic axes with a midrib; branch tips with one to two large apical cells or paired long colorless hairs . . . . *Taenioma*
74. Erect axes cylindrical, lacking a midrib, branch tips with one apical cell; trichoblasts not limited to two long hairs 75
75. Erect branch arising from prostrate axes every 8th to 10th segment, recurved; trichoblasts on erect branches forming a secund (comblike) crest. . . . .*Lophosiphonia*
75. Erect branch arising from prostrate axes every 4th segment, enrolled; trichoblasts on erect branches forming a terminal tuft or spiraled. . . . . *Herposiphonia* (in part)
76. Axes polysiphonous, compressed to flattened, with two parallel rows of cells aligned along two large pericentral cells; monosiphonous filaments forming a single secund row on upper sides of branch tips. . . . .*Cottoniella*
76. Axes not compressed or flattened or having parallel rows of outer cells; monosiphonous filaments, if present, not in a single secund row 77
77. Axes with 4 pericentral cells; branchlets uniseriate at least after 1–2 forks and pigmented; colorless trichoblasts may also be present. 78
77. Axes with 4 or more pericentral cells; uniseriate pigmented branchlets absent; colorless trichoblasts may be present. 79

78. Branchlets delicate; axes covered by completely uniseriate pigmented filaments; tetrasporangial stichidia have monosiphonous axes . . . . *Lophocladia*
78. Branchlets stiff; branchlets polysiphonous at bases and uniseriate after 1–2 forks; tetrasporangial stichidia have polysiphonous axes . . . . *Murrayella*
79. Main erect axes arise from prostrate ones; rhizoids form discoid holdfasts or remain distinct and have open or closed connections with the pericentral cells 80
79. Main erect axes arise from other erect ones; discoid holdfasts absent, rhizoids distinct and always separated from the pericentral cells by a cell wall. . . . . *Neosiphonia*
80. Axes arise from prostrate ones that attach by rhizoidal holdfasts. . . . . *Bryocladia*
80. Axes arise from prostrate ones that attach by simple rhizoids with open or closed connections with pericentral cells . . . . *Polysiphonia*
81. Plants bladelike, with inwardly curled margins 82
81. Plants bladelike or not, lacking enrolled margins 84
82. Main axes irregularly branched; blades 1–3 times irregularly opposite to pinnately branched; midribs or veins absent . . . . *Enantiocladia*
82. Main axes alternately branched; blades opposite to pinnately branched; midribs or veins present. 83
83. Blades lacking cortication, two cells thick; midribs distinct, raised; margins having small teethlike branchlets (serrated). . . . . *Amansia*
83. Blades corticated, more than two cells thick; midribs indistinct, connected to marginal spines by faint veins; margins with pointed inwardly curled teeth . . . . *Osmundaria*
84. Plants monostromatic or distromatic blades (1 or 2 cells thick) 85
84. Plants various morphologies and more than two cells thick 96
85. Blades with a midrib and/or veins at least near base of main axes. 86
85. Blades lacking a midrib or veins anywhere 93
86. Blades not arising from central midrib or veins 87
86. Blades usually arising from central midrib or veins 91
87. Blades usually unbranched or a few arising from base, not proliferating from margins, having a midrib near base; usually on a short stalk . . . . *Grinnellia*
87. Blades branching, proliferating from margins, one or more growing from bases; with veins and/or a midrib; lacking a stalk 88
88. Blades each with single midrib, extending into branches, and often arising from prostrate axes . . 89
88. Blades each lacking a midrib; with multiple veins arising near the base, usually not arising from prostrate axes. 90
89. Plants to 2 cm tall; fronds mostly with irregular to alternate branching below; tips indented or pointed. . . . . *Branchioglossum*
89. Plants to 4 cm tall; fronds mostly entire, often lobed, blades mostly with dichotomous branching below; tips rounded to obtuse . . . . *Polyneura*
90. Blades subdichotomously branched from margins; apical cells domelike; holdfast perennial, prominently branched, lightly calcified . . . . *Calonitophyllum* (in part)
90. Fronds irregularly branched from margins, proliferating; apical cells not evident; attached by small, discoid, noncalcified holdfasts . . . . *Cryptopleura*
91. Lateral veins extending from midrib to margin in older blades . . . . *Apoglossum*
91. Lateral veins absent. 92
92. Main branching pseudodichotomous and arising from blade tips; secondary blades arising from midrib, tips usually forked; blades dark purple red . . . . *Caloglossa*
92. Main branching from midrib; blade tips pointed or rounded, not forked; plants transparent, pink, red, or yellow. . . . . *Hypoglossum*
93. Blade surface interrupted by netlike rows of cells . . . . *Martensia* (in part)
93. Blades solid, continuous, lacking netlike rows of cells. 94

94. Blades monostromatic; growth intercalary (lacking apical cells). *Porphyra*
94. Blades monostromatic, growth apical (with apical cells) 95
95. Growth by a single transversely dividing apical cell; margins undulate and with macroscopic teeth . . . . *Calloseris*
95. Growth by a zone of obliquely dividing apical cells; margins not undulate, smooth or with microscopic teeth. . . . . *Nitophyllum*
96. Plants usually fleshy, gelatinous, or gooey; composed of slender filaments in mucilage; medulla and cortex filamentous (if calcified see first leg of key) 97
96. Plants firm to cartilaginous; not composed of slender filaments embedded in mucilage; cortex not filamentous 103
97. Branching radial, alternate, or irregular; axes cylindrical to compressed. 98
97. Axes strongly compressed, branching lobed or dichotomous. 101
98. Axes with many loosely arranged longitudinal filaments; cortical filaments dichotomously branched and pigmented . . . . *Helminthocladia*
98. Axes uniaxial, a single central filamentous strand surrounded by longitudinal filaments; cortex of whorls of di- or trichotomous filaments 99
99. Main axes 1 mm or less in diam., corticated by down-growing rhizoids arising from basal cells of lateral, indeterminate branches . . . . *Naccaria*
99. Main axes more than 1 mm in diam., cortication by lateral filaments arising from axial cell of central filaments 100
100. Cortical filaments radiating laterally, in distinct whorls; axial cells large and resulting in branches appearing nodulose (beadlike) . . . . *Acrosymphyton*
100. Cortical filaments not always radiating laterally, in distinct whorls; axial cells not unusually large nor branches appearing nodulose. . . . . *Dudresnaya*
101. Gland cells spherical, 20–30 μm in diam.; cortical filaments dichotomously and unequally branched, with spherical tip cells . . . . *Nemastoma*
101. Gland cells usually absent; cortical filaments equally dichotomously branched, tip cells cylindrical or elongate 101
102. Branch tips usually orange-yellow iridescent, bluntly rounded; medullary filaments 2–3 μm in diameter. . . . . *Predaea*
102. Branch tips not orange to yellow iridescent, tips bluntly pointed; medullary filaments 8–14 μm in diameter. . . . . *Platoma*
103. Plants with branching or peltate blades; medulla one layer of cells 104
103. Plants with blades or not; medulla not a single layer of cells 105
104. Blades irregularly dichotomous; medullary cells 60–140 μm in diameter, cortex one layer of cells, 5–10 μm in diameter . . . . *Leptofauchea*
104. Blades peltate; medullary cells 200–400 μm in diameter; cortex of 2 layers, inner cells 25–35 μm and outer cells 3–5 μm in diameter . . . . *Maripelta*
105. Plants are flat, strap-shaped fronds, dichotomously branched; medulla with up to 3 layers of pseudoparenchymalike cells near base. . . . . *Fauchea*
105. Plants frondlike or not, dichotomously branched or not; medulla not consisting of 1–3 layers of pseudoparenchyma-like cells. 106
106. Cortex distinct from the medulla, which has loosely arranged filaments or large cells 107
106. Cortex and medulla showing a gradation of cell size; medullary filaments or central cells are not loosely arranged. 124
107. Plants firm; lacking mucilage or gel in medulla 108
107. Plants soft to gelatinous, with mucilage or gel in medulla 109
108. Plants subtidal; fronds not segmented, tough; branching pseudodichotomous; the medulla has interwoven filaments traversing cavity and stellate cells . . . . *Cirrulicarpus*
108. Plants intertidal, axes segmented, not tough, compressed to flattened; medullary cavity has a central cluster of 5–6 longitudinal filaments . . . . *Catenella*

109. Plants or plant parts hollow, usually filled with mucilage; medullary filaments are absent or limited to axes; older parts at base or stalk 111
109. Plants not hollow, medullary filaments common, embedded in gel 115
110. Plants with wiry, tough axes, which bear one or more bulbous to flattened blades filled with mucilage . . . . . *Botryocladia*
110. Plants lacking wiry axes and bulbous blades 111
111. Plants with internal septa that partition the main axes 112
111. Plants lacking internal partitions, axes hollow or with a few filaments 113
112. Septa evenly spaced in all branches; axial segments cylindrical to barrel-shaped; branching opposite, whorled, or irregular . . . . . *Champia*
112. Septa only at bases of branches; axes lacking regular segmentation; branching mostly dichotomous . . . . . *Coelarthrum*
113. Plants to 30 cm tall, soft, delicate; branches cylindrical to flattened, 2–12 mm in diameter. . . . . *Chrysymenia*
113. Plants to 7 cm tall, often turflike or tangled; branches cylindrical, to 1.5 mm in diameter. 114
114. Plants 3–7 cm tall; cortex 2–3 cells thick . . . . . *Lomentaria*
114. Plants to 3 cm tall; cortex 4–6 cells thick . . . . . *Coelothrix*
115. Branches repeatedly dichotomously branched 116
115. Branches not repeatedly dichotomously branched 118
116. Cortex two cells thick, delicate; subcortical cells small, tear-shaped; medullary filaments longitudinal, loosely bundled, and in 6–50 strands . . . . . *Scinaia*
116. Cortex 1–7 cells thick, decreasing in size toward surface; medullary filaments scattered, entangled 117
117. Gland cells present at intersections of medullary filaments; surface cells spherical, branching repeatedly dichotomous . . . . . *Sebdenia*
117. Gland cells absent; surface cells radially elongated; branching mostly unilaterally dichotomous . . . . . *Halymenia* (in part)
118. Plants with cylindrical stalks bearing club-shaped, cylindrical to prismatic axes; medulla with a few loose filaments in a gel . . . . . *Corynomorpha* (in part)
118. Plants not with cylindrical stalks and club-shaped like axes; medulla with many loosely to compactly arranged filaments 119
119. Branches usually constricted at bases, tetrasporangia zonate. 120
119. Branches not constricted at bases, tetrasporangia cruciate. 122
120. Plants foliaceous; blades flat, thin, 300–650 µm thick; axes repeatedly dichotomously branched. . . . . *Sarcodiotheca* (in part)
120. Axes cylindrical or flat, more than 1 mm thick; axis radially alternate to pinnately branched 121
121. Plants clumped; medulla with scattered filaments; cystocarps 0.5–1.0 mm in diameter, swelling branches, and with a central fusion cell . . . . . *Solieria* (in part)
121. Plants solitary; medulla with intertwined filaments; cystocarps 1–2 mm in diameter, sunken, and the sterile central tissue is multicellular . . . . . *Agardhiella* (in part)
122. Plant blades stalked and peltate to bladelike or as delicate perforated sheets; carpogonia and auxiliary cells on specialized filaments . . . . . *Kallymenia*
122. Plants neither peltate nor delicate blades; carpogonial and auxiliary cells not borne on specialized filaments 123
123. Fronds fleshy to cartilaginous; cortex 4–6 cells thick and arranged in obvious rows; tetrasporangia in clusters. . . . . *Grateloupia*
123. Fronds delicate to gelatinous; cortex 2–4 cells thick, not arranged in obvious rows; tetrasporangia scattered. . . . . *Halymenia* (in part)

124. Plants have a polysiphonous core; pericentral cells at least visible at tips in cross section and/or branch tips with an apical pit or projecting apical cell 125
124. Polysiphonous core and apical pit absent; no pericentral cells visible 136
125. Plants common in tidal marshes; with dorsiventral organization; apical pit and tufts of pigmented or colorless filaments absent . . . . *Bostrychia* (in part)
125. Plants not common to tidal marshes; erect and lacking dorsiventral organization; apical pit and tufts of pigmented or colorless filaments sometimes present 126
126. Plants flat; main blades linear, 4–7 mm wide, bearing marginal bladelets with constricted to stalked bases . . . . *Waldoia*
126. Plants not flat, linear blades although axes may be compressed; marginal bladelets and stalks absent 127
127. Tips bear deciduous colorless filaments (trichoblasts), at least in young plants 128
127. Tips bear persistent, pigmented filaments (branchlets) 134
128. Branchlets peg-, club-, spindle-shaped, or compressed 129
128. Branchlets spine- or spike-shaped 132
129. Branchlet tips have 4–5 pericentral cells obscured with age 130
129. Branchlet tips usually have 2 pericentral cells (species of *Chondrophycus*, *Osmundea*, *Palisada*, and *Laurencia* are difficult to identify using vegetative features; see also *L. poiteaui*) 131
130. Branchlet tips have 4 pericentral cells . . . . *Laurencia*
130. Branchlet tips have 5 pericentral cells . . . . *Chondria*
131. The *Corps en cerise* may be present in cortical cells of fresh material; secondary pit connections absent between epidermal cells; tetrasporangia arise from particular pericentral cells . . . . *Chondrophycus corallopsis* and *Palisada*
131. The *Corps en cerise* absent in cortical cells of fresh material; secondary pit connections common between epidermal cells; tetrasporangia arise from random epidermal cells . . . . *Osmundea*
132. Main axes dichotomously branched; axes cylindrical, brushlike; covered with stiff, tough, unbranched, cylindrical branchlets throughout; tips blunt . . . . *Digenea*
132. Main axes not dichotomously branched; axes not brushlike; branchlets spinelike, branching or not, wider at bases and tapering to acute tips 133
133. Axes cylindrical or flattened; branchlets arranged in 2–4 ranks; branches cylindrical, triangular or compressed, and on all axes . . . . *Bryothamnion*
133. All axes cylindrical; branchlets radially disposed; spinelike branchlets mostly clustered on short axes . . . . *Acanthophora*
134. Lateral filaments alternately branched, basal cells of branchlets short, cuboidal; tetrasporangia in spiral series on main axes . . . . *Wrightiella*
134. Lateral filaments dichotomously branched; basal cells of branchlets elongated; tetrasporangia whorled within specialized branchlets (stichidia) 135
135. Branching mostly alternate; main growth dorsiventral; tetrasporangia almost completely covered by protective cells . . . . *Heterosiphonia* (in part)
135. Branches radially arranged; main growth erect; tetrasporangia exposed at surface of stichidia and not covered completely by protective cells . . . . *Dasya* (in part)
136. Plants tough and wiry; erect axes arising from runners; medulla and cortex with similar-sized cells except for surface ones 137
136. Plants not tough or wiry, erect axes not arising from runners; cell size differing between medulla and cortex 141
137. Branch tips with one apical cell 138
137. Branch tips with multiple apical cells 139
138. Medulla lacking thick-walled, slender, rhizoidlike filaments (rhizines) . . . . *Gelidiella*
138. Medulla with thick-walled, slender rhizines . . . . *Gelidium*

139. Plants attached by discoid pads; branching basically dichotomous; medulla with thick-walled filamentous cells (not rhizines) to 250 μm in diameter. . . . . *Gymnogongrus*

139. Plants attached by rhizoids; branching not dichotomous; medulla with thick- or thin-walled filaments (not rhizines), to 30 μm in diameter. 140

140. Cells of central medulla thick-walled, tightly packed, cell walls angular in cross section; tetrasporangia zonate . . . . *Wurdemannia*

140. Cells of central medulla thin-walled, not tightly packed; cell walls circular in cross section; tetrasporangia cruciate . . . . *Gelidiopsis*

141. Plants firm, cartilaginous; the narrow stipe bearing one to a few cylindrical or club-shaped axes, little branched . . . . *Corynomorpha* (in part)

141. Plants not having club-shaped axes on narrow stipes 142

142. Medulla filamentous, with filaments either tightly packed or loosely arranged, sometimes mixing with inner cortical cells 143

142. Medulla with parenchymalike cell structure, not filamentous. 150

143. Plants are fronds or have distinct blades; flattened or strap-shaped. 144

143. Axes cylindrical to compressed 146

144. Main axes with dichotomous to palmate branching. *Agardhiella* (in part)

144. Branching opposite or irregular 145

145. Plants deep red, blade margins rippled or undulate; midribs often near base; medullary filaments tightly packed and distinct from cortical cells . . . . *Cryptonemia*

145. Plants rose to purple red; blade margins not undulate; midribs absent; medulla with loosely scattered filaments, mixing with cortical cells . . . . *Sarcodiotheca* (in part)

146. Plants lacking spine- or spurlike proliferations. . . . . *Solieria* (in part)

146. Plants often with spine- or spurlike proliferations on margins of compressed axes or as short branchlets on cylindrical axes 147

147. Axes compressed or flattened . . . . *Meristotheca*

147. Axes cylindrical. 148

148. Axes with spines or spurlike branches; medulla with compact, large cells surrounding a core of densely packed longitudinal slender filaments . . . . *Eucheuma*

148. Spines and spurlike branchlets absent; medulla with entangled transverse and longitudinal filaments. 149

149. Medulla in part hollow; filaments mostly longitudinal, bases of branches often slightly compressed; inner cortical cells large, ellipsoidal . . . . *Agardhiella* (in part)

149. Medulla not hollow; with loosely to densely entangled filaments; producing radially organized cortex . . . . *Chondracanthus*

150. Fronds usually peltate, medulla usually with 2–3 cell layers of pseudoparenchyma; cortex 2–3 layers of small cells, these are scattered in matrix. . . . . *Asteromenia*

150. Plants not peltate, medulla more than 2–3 layers; cortex lacking matrix 151

151. Plants flattened; branching abundantly pinnate; medulla composed of 4–6 layers of closely packed, axially elongate pseudoparenchyma. . . . . *Gloiocladia*

151. Plants cylindrical or flattened; branching rarely pinnate; medulla lacking axially elongate parenchyma cells 152

152. Axes cylindrical; branchlets short, spurlike, and sometimes hooked. 153

152. Axes cylindrical or flattened; lacking short, spurlike branchlets 154

153. Branches with a single apical cell; central axial cell row visible in cross section of young branches. . . . . *Hypnea*

153. Branches with multiple apical cells; axial row absent . . . . *Gracilaria* (in part)

154. Axes flattened; medullary cells large and mixed with thin-walled small cells; blades fused to adjacent ones . . . . *Agardhinula*

154. Axes flattened or cylindrical; medullary cells uniformly large, surface cells small; blades not fused to adjacent ones 155

155. Axes flattened; medullary cells often rectangular; cortical cells often compressed to slightly flat; auxiliary cells present. . . . .*Rhodymenia*
155. Axes flattened or cylindrical; medullary cells spherical to irregular; cortical cells mostly spherical to elongated outward; auxiliary cells absent. 156
156. Axes cylindrical, dominant, single; cortical cells in rows . . . .*Gracilariopsis*
156. Axes cylindrical, compressed or flattened, lacking a single dominant axes; cortical cells not in rows. . . . .*Gracilaria* (in part)

PLATE I

Stichococcus marinus. 1. Two uniseriate unbranched filaments with each cell having a single parietal plastid; scale 2 µm (after Humm and Taylor, 1961).

Pseudotetraspora marina. 2. Portion of a gelatinous mass of cells each having a single lobed plastid; scale 5 µm.

Verdigellas peltata. 3. The bright green gelatinous thallus has two lobes and a short stalk (*arrow*); scale 5 mm. 4. Cross section shows scattered cells scattered throughout the gelatinous thallus; scale 50 µm (both after Littler and Littler, 2000).

Ulothrix subflaccida. 5. Three uniseriate unbranched filaments, each cell with a collar-shaped (parietal) plastid; one with an elongated basal cell (*left*); and two upper parts, the middle one vegetative and the right one with zooids; scale 25 µm (after Womersley, 1984).

Gomontia polyrhiza. 6. Two stalked unicellular sporophytes; scale 20 µm (after Burrows, 1991). 7. Irregularly branched filaments with sporangia (*arrow*) in a decalcified shell; scale 20 µm (after Humm and Taylor, 1961).

Gayralia oxysperma. 8. A delicate, ruffled blade; scale 10 cm (after Schneider and Searles, 1991). 9. Surface view shows polygonal cells; scale 25 µm (after Humm, 1979). 10. Cross section of monostromatic blade; scale 50 µm (after Womersley, 1984).

Percursaria percursa. 11. Portion of an elongate biseriate filament; scale 20 µm. 12. Each cell of the biseriate filament has one parietal plastid and one or more pyrenoids; scale 50 µm (both after Stegenga and Mol, 1983).

Ulva chaetomorphoides. 13. Portions of two filaments, each with 3 rows of cells; scale 50 µm (after Littler and Littler, 2000).

Ulva compressa. 14. A cluster of axes with compressed mid-upper regions and basal branching; scale 1 mm (after Schneider and Searles, 1991). 15. Surface view shows each cell with a parietal plastid and one pyrenoid; scale 10 µm (after Villalard-Bohnsack, 1995).

Ulva fasciata. 16. Habit of several narrow distromatic blades; scale 2 cm (L. Baumhart).

Ulva flexuosa. 17. The axis is compressed and branchlets tubular; scale 1 mm (after Schneider and Searles, 1991). 18. Surface view of rectangular cells with a dissected (stellate) parietal plastid and one or more pyrenoids; scale 10 µm (after Villalard-Bohnsack, 1995).

Ulva intestinalis. 19. Five unbranched, tubular, contorted axes; scale 1 mm (L. Baumhart).

Ulva lactuca. 20. Blade is elongate, irregular, and distromatic; scale 2 cm (L. Baumhart).

Ulva linza. 21. Three strap-shaped blades have ruffled margins; scale 10 mm (after Scagel, 1966). 22. Surface view shows unordered polygonal cells; scale 50 µm (after Stegenga and Mol, 1983). 23. Cross section with hollow margin; scale 20 µm (after Scagel, 1966).

Ulva clathrata. 24. Plant silhouette with abundantly branched axes; scale 1 cm (after Humm and Taylor, 1961). 25. An older axis has unordered cells and spinelike branchlets; scale 100 µm (after Stegenga and Mol, 1983). 26. The quadrate cells have one stellate parietal plastid and multiple pyrenoids; scale 10 µm (after Villalard-Bohnsack, 1995).

Ulva flexuosa subsp. *paradoxa*. 27. Cells are mostly polygonal and contain one stellate parietal plastid and multiple pyrenoids; scale 10 µm (after Villalard-Bohnsack, 1995). 28. Plant has tubular axes covered with many uniseriate branchlets; scale 2 mm (after Schneider and Searles, 1991).

Ulva profunda. 29. Marginal portion of blade has pores of various sizes; scale 1 cm (after Taylor, 1960). 30. In cross section the distromatic blades show angular cells; scale 20 µm.

Ulva prolifera. 31. A cluster of much branched tubular axes; scale 2 cm (after Schneider and Searles, 1991). 32. Surface view shows rectangular to irregularly shaped cells with one parietal plastid and a central pyrenoid; scale 10 µm (after Villalard-Bohnsack, 1995).

Ulva rigida. 33. Five blades are from a common rhizoidal holdfast; scale 2 cm. 34. Two teethlike marginal projections of blades; scale 50 µm. 35. Cross section of a distromatic blade showing rounded cells; scale 10 µm (after Schneider and Searles, 1991).

Ulva flexuosa subsp. *flexuosa*. 36. Cross section shows the hollow tubular axis with a single layer of rectangular cells; scale 50 µm.

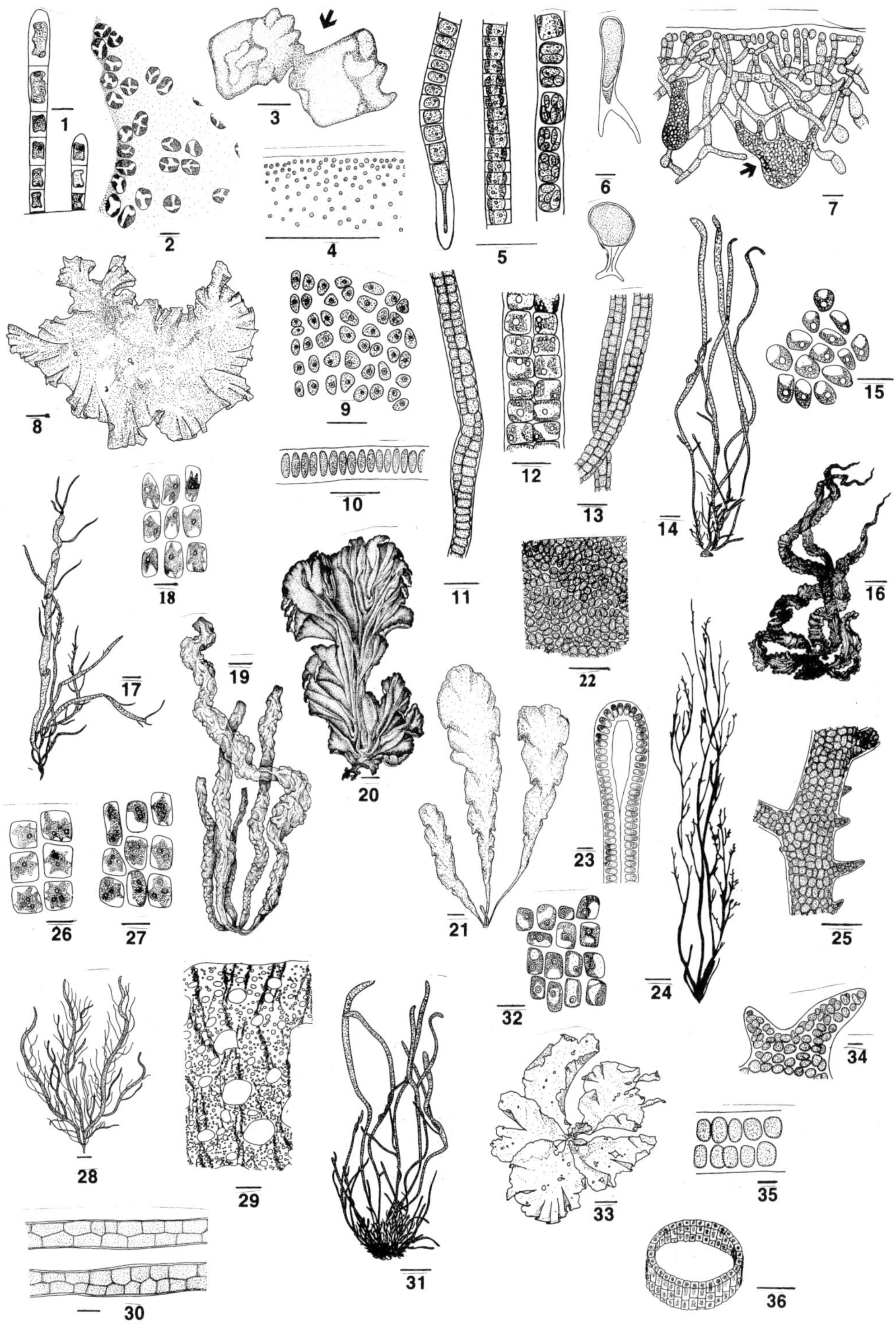
1
2
3
4
5
6
7
8
9
10
11
12
13
14
15
16
17
18
19
20
21
22
23
24
25
26
27
28
29
30
31
32
33
34
35
36

PLATE II

Epicladia flustrae. 1. Entangled uniseriate branched filaments growing on a sponge; scale 50 μm (after Humm and Taylor, 1961).
Entocladia wittrockii. 2, 3. Unbranched (2) and branched (3) uniseriate filaments endophytic in the wall of *Ectocarpus* sp.; scales 20 μm (both after Wille, 1890).
Entocladia viridis. 4. Creeping uniseriate and irregularly branched filaments in the mucilaginous layer of a host seaweed; scale 20 μm (after Humm, 1979).
Pringsheimiella scutata. 5. The discoid epiphyte has radiating cells and central columnar hairs; scale 100 μm (after Littler and Littler, 2000).
Pseudendoclonium submarinum. 6. A crust with irregularly arranged and variably shaped cells; scale 10 μm.
Ulvella lens. 7. A pseudoparenchymatous crust with radiating cells; scale 50 μm (after Littler and Littler, 2000).
Valoniopsis pachynema. 8. A cluster of irregularly branched uniseriate filaments with basal rhizoids (*arrow*); scale 2 mm (after Littler and Littler, 2000).
Phaeophila dendroides. 9. Prostrate, irregularly branched uniseriate filaments have erect twisted or spiraled setae; scale 40 μm (after Humm, 1979).
Anadyomene menziesii. 10. Plant has a crisp, ruffled blade and rhizoidal holdfast; scale 1 cm. 11. The frond's veins consist of 2–6 parallel filaments; scale 1 mm (both after Littler and Littler, 1991).
Anadyomene pavonina. 12. Plant has a porous blade attached by a rhizoidal holdfast; scale 2 cm. 13. The blade's veins consist of large-celled filaments that radiate from different centers; scale 1 mm (L. Baumhart).
Anadyomene saldanhae. 14. Upper and lower parts of a blade with radiating veins of swollen cells; scale 200 μm. 15. Base of blade with rhizoids arising from veins; scale 2 mm (both after Littler and Littler, 1991).
Anadyomene stellata. 16. Plant has a crisp, ruffled blade and a fibrous holdfast; scale 1 cm. 17. Frond's veins consist of large branched filaments; scale 200 μm (both after L. Baumhart in Dawes, 1974).
Microdictyon boergesenii. 18. Part of a network of uniseriate filaments; scale 250 μm (after Schneider and Searles, 1991). 19. The network consists of uniseriate filaments joined together without attachment cells; scale 200 μm (A. L. King).
Phyllodictyon anastomosans. 20. A small netlike blade on a uniseriate stalk; scale 2 mm (after Taylor, 1960).
Phyllodictyon pulcherrimum. 21. A young netlike blade with branches and uniseriate stalk; scale 2 mm (after Schneider and Searles, 1991). 22. A mature blade on a uniseriate stalk with parallel veins and a rhizoidal base; scale 2 cm (L. Baumhart). 23. Part of the filamentous network; scale 1 mm. 24. The uniseriate filaments fuse via hapteroid cells; scale 100 μm (23, 24 after Littler and Littler, 1997). 25. The uniseriate branchlets attach by small clear hapteroid cells; scale 200 μm (after Littler and Littler, 2000).
Struvea elegans. 26, 27. Young (26) and older (27) parts of netlike blade on uniseriate stalks; scales 2 mm (after Taylor, 1960).
Chaetomorpha aerea. 28. Three unbranched uniseriate filaments attached by single elongated rhizoidal cells; scale 500 μm (L. Baumhart).
Chaetomorpha brachygona. 29. Two unbranched uniseriate filaments showing slightly swollen cells; approximately 100 μm in diam. and with scale 50 μm (after Schneider and Searles, 1991).
Chaetomorpha crassa. 30. Two apical portions of unbranched uniseriate filaments; approximately 500 μm in diam. and with scale 200 μm (J. Futch).
Chaetomorpha gracilis. 31. Basal parts of unbranched uniseriate filaments with a holdfast cell on the left and an upper part of another filament on the right; scale 100 μm (J. Futch).
Chaetomorpha linum. 32. Two filaments with cells 1–2 diam. long and approximately 250 μm in diam.; scale 200 μm (J. Futch).

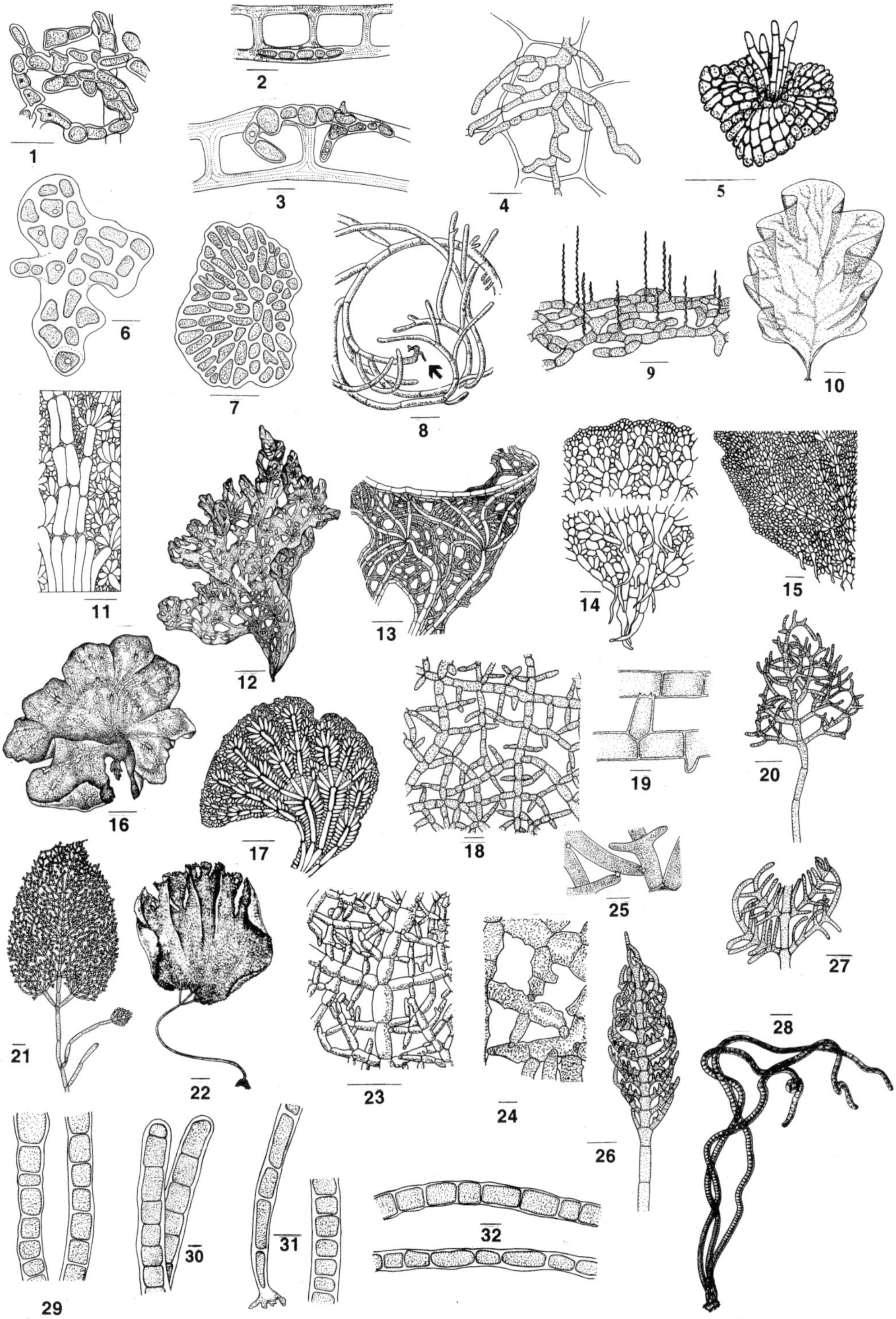
1
2
3
4
5
6
7
8
9
10
11
12
13
14
15
16
17
18
19
20
21
22
23
24
25
26
27
28
29
30
31
32

PLATE III

Chaetomorpha minima. 1. Unbranched uniseriate filaments showing a basal rhizoidal cell (*left*) and an upper axis (*right*); scale 50 μm (after Schneider and Searles, 1991).

Chaetomorpha picquotiana and *C. linum*. 2. Cell sizes of *C. picquotiana* (*left*) and *C. linum* (*right*), with the former being larger (200–400 μm in diam and 3–5 diam. long) than the latter; scale 100 μm (after Blair, 1983).

Chaetomorpha nodosa. 3. Lower portion of a uniseriate unbranched filament with a long basal cell; scale 100 μm (after Lawson and John, 1987).

Cladophora albida. 4, 5. Portion of an old axis (4) and a branch tip (5) with secund, curved branchlets; scales 5 mm and 100 μm (after Schneider and Searles, 1991).

Cladophora blomquistii. 6. Terminal portion of filaments with irregular, abundant branching and multiple branches per node; scale 1 mm (after Hoek, 1982).

Cladophora catenata. 7. Filaments are coarse and curved with most axial cells bearing a branch; scale 1 mm (after Hoek, 1982).

Cladophora coelothrix. 8. The wave-exposed frond has congested acropetal branching; scale 200 μm (after Hoek, 1982).

Cladophora dalmatica. 9. Upper part shows pseudodichotomous or trichotomous branching with the branches clustered and curved at tips; scale 200 μm (after Schneider and Searles, 1991).

Cladophora jongiorum. 10. Upper axes have thornlike branches and short tapered tip cells; scale 200 μm (Hoek, 1982).

Cladophora liniformis. 11. Upper axes are laxly branched, have angled cross walls, and elongate and narrow cells; scale 200 μm (Hoek, 1982).

Cladophora longicellulata. 12. Terminal portion of a plant with conspicuous main axis, alternate branching, and wide forks. 13. Elongated thick-walled basal cell with rhizoids; scales 200 μm (both after Hoek, 1982).

Cladophora montagneana. 14. A small plant with short, club-shaped cells attached by basal rhizoids; scale 200 μm (after Hoek, 1982).

Cladophora pellucidoidea. 15. Upper axis with abundant, narrow angled (30{dg}) and elongated cells (to 48 diam. long); scale 1 mm (after Hoek, 1982).

Cladophora prolifera. 16. A young plant with a densely branched axis and basal rhizoids; scale 1 mm. 17. Thick-walled basal cells with annular thickenings and rhizoids; scale 1 mm (J. Futch).

Cladophora ruchingeri. 18. The upper axis is coarse, distinct, and each cell often bears a short straight branch; scale 1 mm.

Cladophora sericea. 19. Tips of filament with a dense cluster of branches; scale 200 μm (after Schneider and Searles, 1991). 20. Silhouette of main axis with pseudodichotomous branching; scale 2 mm (after Humm and Taylor, 1961).

Cladophora socialis. 21. Upper axes have irregular, entangled, and contorted branches; scale 200 μm (after Hoek, 1982).

Cladophora vadorum. 22. Upper axes with irregular, wide angled (>45{dg}) branching; apical cells are also long and tapered; scale 200 μm (after Schneider and Searles, 1991).

Cladophora vagabunda. 23. An upper axis with recurved branches arising from most cells; scale 200 μm (after Schneider and Searles, 1991). 24. A mature plant with numerous clusters of branches; scale 1 mm (L. Baumhart). 25. Upper axis with a fascicle of branches; scale 200 μm (after Hoek, 1982).

Rhizoclonium africanum. 26. Uniseriate unbranched filament with a hyaline lateral rhizoid; scale 100 μm. 27. Filaments have variable diameters and two long rhizoids arising from cell ends; scale 100 μm (both after Taylor, 1960).

Rhizoclonium riparium. 28. Uniseriate unbranched filament with short cells and long branched rhizoids; scale 200 μm (after Stegenga and Mol, 1983). 29. A single rhizoid arises from the end of a cell; scale 200 μm.

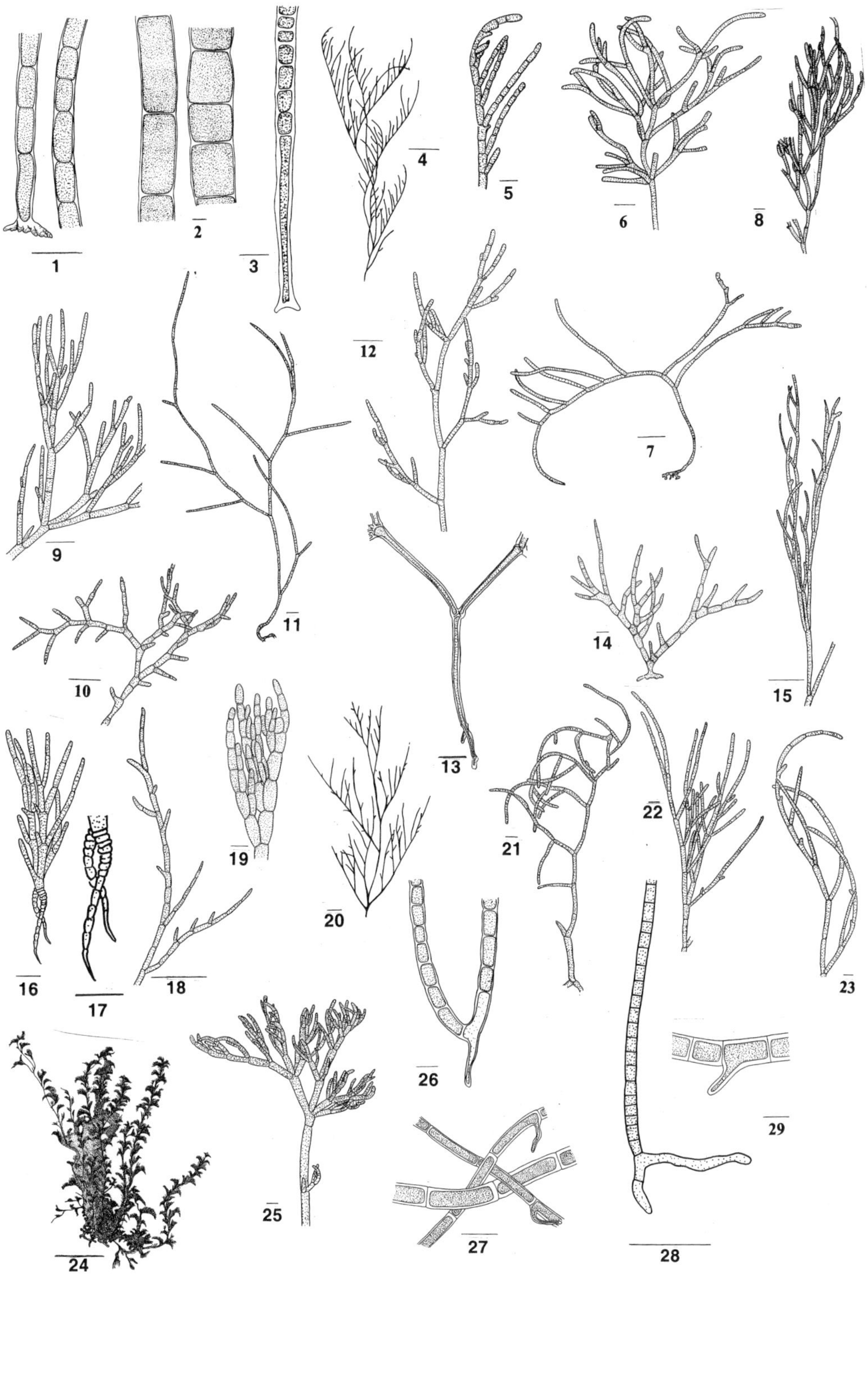
1
2
3
4
5
6
8
7
9
10
11
12
13
14
15
16
17
18
19
20
21
22
23
24
25
26
27
28
29

PLATE IV

Chamaedoris peniculum. 1. Plant has a stalk with annular thickenings, basal rhizoids, and filamentous cap; scale 2 mm (R. Holbrook). 2. Stalk base with annular constrictions and branched rhizoids; scale 3 mm (after Littler and Littler, 2000).

Cladophoropsis macromeres. 3. Filament tip with secund branches that lack basal cross walls; scale 2 mm (after Taylor, 1960). 4. A colorless basal cell with rhizoids; scale 200 μm (after Littler and Littler, 2000).

Cladophoropsis membranacea. 5. Tip of two filaments with secund branching; branches lack basal cross walls; scale 1 mm (L. Baumhart).

Dictyosphaeria cavernosa. 6. Plant with an irregular saclike body and large cells; scale 5 mm (R. Holbrook). 7. Polygonal surface cells connected by microscopic tenacular cells; scale 1 mm. 8. A row of tenacular cells; scale 100 μm (7, 8 after Littler and Littler, 2000).

Dictyosphaeria ocellata. 9. Plant is an irregular spherical mass of swollen cells with a cluster of basal rhizoids; scale 1 mm (L. Baumhart).

Siphonocladus rigidus. 10. Coarse, irregularly branched filaments have internal cells; scale 2 mm (J. Futch). 11. The new branch lacks a cross wall; scale 200 μm.

Siphonocladus tropicus. 12. Axis with irregular branches arising from internal cells (*left*) and a branch tip with internal cells (*right*); scale 1 mm (after Børgesen, 1913). 13. Basal part of the axis with irregularly lobed rhizoids; scale 1 mm (J. Futch).

Ventricaria ventricosa. 14. Plant is a macroscopic saclike cell with basal rhizoids; scale 1 cm (R. Holbrook).

Ernodesmis verticillata. 15. Plant with bulbous elongate cells and irregular whorls of branches; scale 2 cm.

Petrosiphon adhaerens. 16. Margin of crust with parallel, forked filaments; scale 1 mm. 17. Tip cells of filaments with marginal forks; scale 200 μm (both after Taylor, 1928).

Valonia aegagropila. 18. Base has irregular to polygonal cells and minute rhizoids; scale 200 μm. 19, 20. Axes and various shaped branches are interwoven; scales 2 mm (all after Kuckuck, 1907).

Valonia macrophysa. 21. Portion of an axis with irregular swollen branches; scale 2 mm (after Littler and Littler, 2000).

Valonia utricularis. 22. Base of a swollen cell with multicellular and irregularly branched rhizoids; scale 200 μm (after Littler and Littler, 2000). 23. Two plants with curved, swollen branches; scale 2 mm (after Kuckuck, 1907).

Bryopsis halliae. 24. Axis is bushy and covered with short branches with fine pigmented branchlets; scale 1 cm. 25. A branch with radial branchlets; scale 2 mm (both after Montagne, 1861).

Bryopsis hypnoides. 26. Part of terminal axis with radially arranged branchlets having the same diameter; scale 50 μm (after Stegenga and Mol, 1983).

Bryopsis pennata. 27, 28. Creeping and erect axes of the species (27) and the var. *secunda* (28); scales 2 cm (J. Futch). 29. Axial tip of var. *secunda* with secund branching; scale 1 mm (after Taylor, 1960). 30. Axial tip of the species with opposite branchlets; scale 100 μm (after Schneider and Searles, 1991).

Bryopsis plumosa. 31. Axial tip with branchlets in a triangular pattern; scale 200 μm (after Taylor, 1960). 32. Horizontal rhizoids of an erect axis; scale 1 mm (after Littler and Littler, 2000).

Bryopsis ramulosa. 33, 34. Segments of two plants showing prostrate axes and erect branches with a few branchlets; scale 2 mm (33 after J. Futch; 34 after Littler and Littler, 2000).

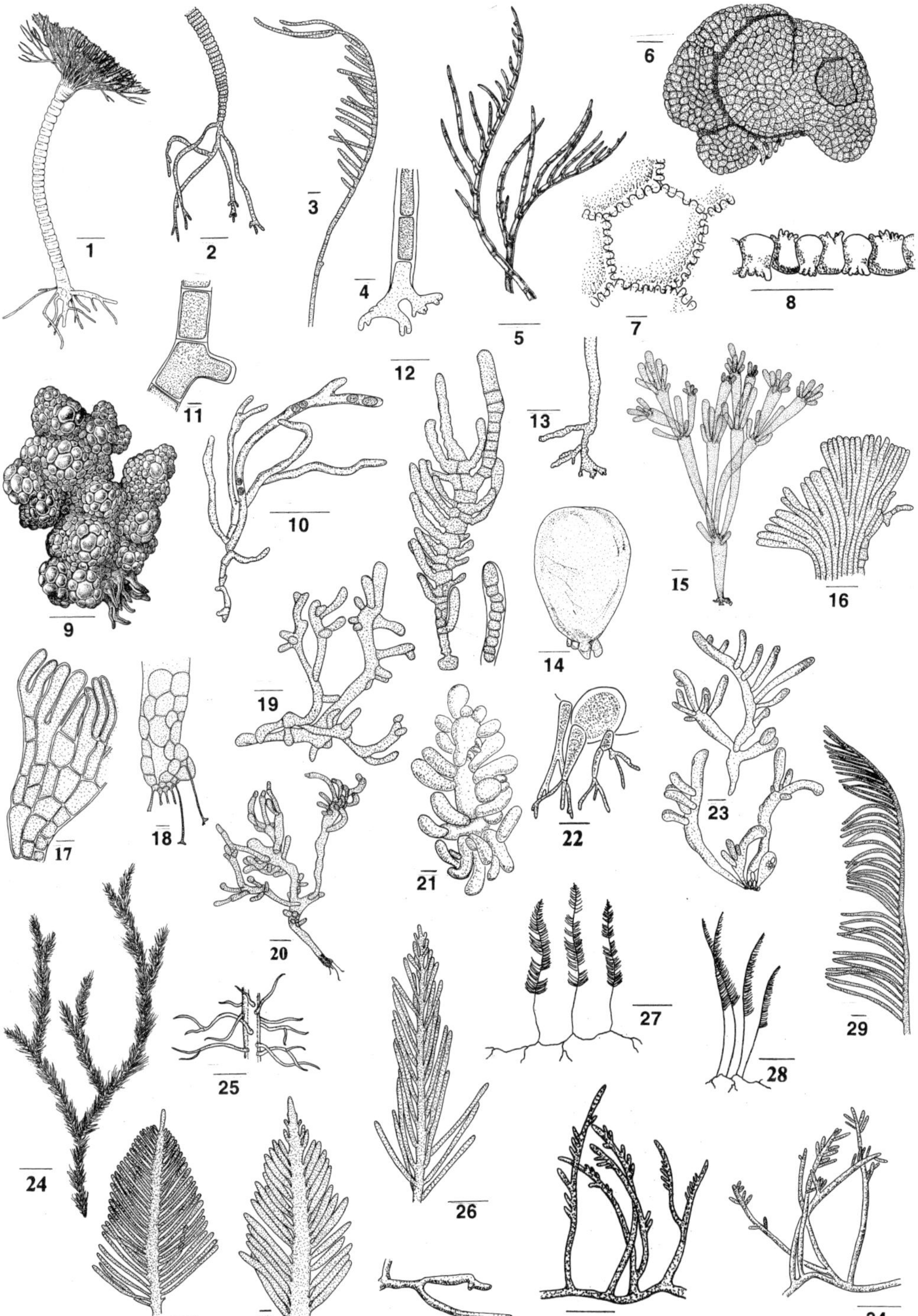
1
2
3
4
5
6
7
8
9
10
11
12
13
14
15
16
17
18
19
20
21
22
23
24
25
26
27
28
29
30
31
32
33
34

PLATE V

Derbesia fastigiata. 1. Erect axes are dichotomously branched and arise from prostrate siphons; scale 200 µm (after Littler and Littler, 2000).

'*Halicystis*' stage of *Derbesia osterhoutii*. 2. A balloonlike gametophytic cell; scale 1 mm. 3. Cell base with blunt-tipped rhizoids; scale 100 µm (both after J. Futch).

Derbesia vaucheriaeformis. 4. Upper axes with dichotomous branching; scale 100 µm (after Harvey, 1858). 5. A lateral sporangium with two cross walls; scale 50 µm (after Farlow, 1881).

Trichosolen duchassaingii. 6. Upper part of a dichotomously branched axis is densely covered with fine pigmented branchlets; scale 1 mm (J. Futch).

Trichosolen molassensis. 7. Part of an axial segment with swollen, basally constricted, and radial branchlets; scale 100 µm (J. Futch).

Codium carolinianum. 8. Plant is prostrate and with rhizoidal clusters (*arrows*) formed at irregular intervals; scale 2 cm. 9, 10. Two pairs of elongate utricles with thin apical walls and lateral hair scars; scales 200 µm (all after Schneider and Searles, 1991).

Codium decorticatum. 11. Plant with dichotomous to irregular branching, compressed forks, and a rhizoidal base; scale 2 cm (after Schneider and Searles, 1991). 12. Three swollen utricles with lateral hair scars and thin-walled tips; scale 200 µm (after Littler and Littler, 2000).

Codium intertextum. 13. Plant is lobed and epiphytic on a host seaweed; scale 1 cm (after Taylor, 1928). 14. Three utricles with indented tips and lateral hair scars; scale 100 µm (Littler and Littler, 2000).

Codium isthmocladum. 15. Plant is dichotomously branched with a crustose holdfast; scale 2 cm (L. Baumhart). 16. Four utricles with thickened apical walls; scale 200 µm.

Codium repens. 17. The creeping compressed axis is dichotomously to irregularly branched; scale 1 cm (J. Futch). 18. Two utricles with an immature gametangium (*left*) and a hair cell (*right*); scale 100 µm (after Littler and Littler, 2000).

Codium taylorii. 19. Plant is erect, dichotomously branched, and with a fibrous holdfast; scale 2 cm (after Schneider and Searles, 1991). 20. Plant is lax, dichotomously branched, and with a crustose holdfast (*arrow*); scale 2 cm (after L. Baumhart in Dawes, 1974). 21. Two utricles with hair scars (*arrow*) and thin-walled tips; scale 100 µm (after Schneider and Searles, 1991).

Caulerpa ashmeadii. 22. Tip of a pinnately branched frond; branchlets have swollen tips; scale 1 cm. 23. Part of a rhizome with branching rhizoids and an upright pinnate frond; scale 1 cm (both after Taylor, 1960).

Caulerpa brachypus f. *parvifolia*. 24. The plant has multibranched rhizomes and oval to elliptic blades; scale 2.5 cm. 25. Blade tip with a slightly toothed margin; scale 5 mm (both after Coppejans and Beeckman, 1990).

Caulerpa cupressoides. 26. Part of the species rhizome bearing dense rhizoidal clusters and two upright fronds; scale 1 cm (L. Baumhart). 27. Frond tip of var. *cupressoides*; scale 2 mm. 28. Flattened frond tip of var. *flabellata*; scale 2 mm. 29. Frond tip of var. *mamillosa*; scale 2 mm. 30. Frond tip of var. *lycopodium*; scale 2 mm (27, 29 after Taylor, 1960; 28, J. Futch; 30 after Littler and Littler, 2000).

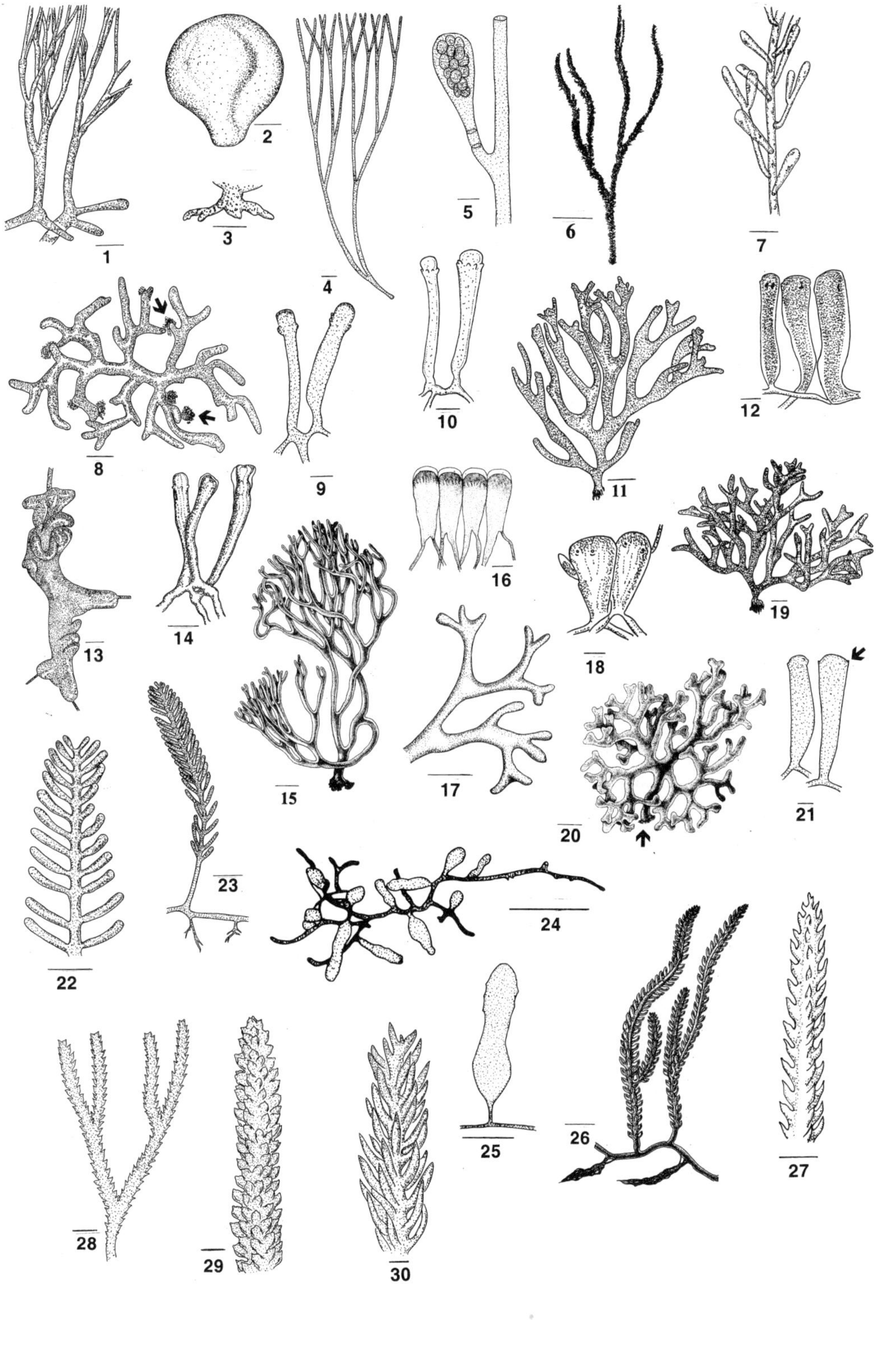
1
2
3
4
5
6
7
8
9
10
11
12
13
14
15
16
17
18
19
20
21
22
23
24
25
26
27
28
29
30

PLATE VI

Caulerpa fastigiata. 1. Plant is filiform with a rhizome bearing rhizoids and erect axes; scale 1 mm. 2. Longitudinal section of a rhizome showing internal trabeculae; scale 100 µm (both, R. Holbrook).
Caulerpa floridana. 3. A rhizome with rhizoids and three erect unbranched axes with pinnate branchlets; scale 1 cm. 4. Branchlet tips are basally constricted and end in a minute spine; scale 2 mm (both after Taylor, 1960).
Caulerpa lanuginosa. 5. Rhizome and erect axes are densely covered with spine-tipped branchlets and the rhizoids encrusted with sediment; scale 1 cm (L. Baumhart).
Caulerpa macrophysa. 6. The erect axis has spherical branchlets and is from a rhizome with a single cluster of rhizoids; scale 1 cm.
Caulerpa mexicana. 7. The two fronds are on a rhizome bearing multiple clusters of rhizoidal; scale 1 cm (L. Baumhart). 8. Branchlet tips are pointed and arise on a flat, basally constricted frond; scale 1 cm (after Littler and Littler, 2000).
Caulerpa microphysa. 9. The rhizome bears branched rhizoids and eleven erect axes in grapelike clusters; scale 1 cm. 10. Four beadlike branchlets; scale 1 mm (both after Taylor, 1960).
Caulerpa ollivieri. 11. Plants from the Gulf of Mexico with a more patent growth habit and lower density of rhizoids than those in the Mediterranean; scale 1 cm. 12. The type specimen from Villefranche-sur-Mer (Mediterranean) with small thin stolons and erect branches; scale 1 cm (both after Hine and Humm, 1971).
Caulerpa paspaloides. 13. Part of var. *compressa* showing a rhizome, multiple rhizoids, and two palmlike erect axes; scale 1 cm. 14. Frond tip of var. *compressa* with pinnate branchlets; scale 2 mm. 15. Frond tip of var. *wurdemannii* with secund pinnae on one side of axis; scale 1 cm. 16. Frond tip of the species with pinnate branchlets; scale 1 mm. 17. Frond tip of var. *laxa* with stubby branchlets in 3 ranks; scale 200 µm (13–15, 17, J. Futch; 16 after Littler and Littler, 2000).
Caulerpa prolifera. 18, 19. Two plants with blades, rhizomes and rhizoidal clusters that may bear delicate (18) narrow blades (previously f. *zosterifolia*) or robust (19) broad, proliferating blades (previously f. *obovata*); scales 1 cm (L. Baumhart).
Caulerpa racemosa. 20. A robust form of the species with erect axis, rhizome, and rhizoidal clusters; scale 1 cm (L. Baumhart). 21. Tip of var. *peltata* with flat topped branchlets. 22. Tip of var. *racemosa* with densely packed club-shaped branchlets. 23. Tip of var. racemosa. 24. Frond tip of var. *occidentalis*. 25. Frond tip of typical species with large elliptical branchlets (21–25, scales 2 mm; after Taylor, 1960).
Caulerpa sertularioides. 26. Frond tip of f. *farlowii* with a dense covering of branchlets; scale 2 mm. 27. Portion of f. *longiseta* with branched erect axes, opposite branchlets, and a rhizome with rhizoidal clusters; scale 1 cm. 28. The rhizome, rhizoids, and erect axes of the species; scale 1 cm. 29. Terete branchlets and acute tips of the species; scale 1 cm (26, 28, L. Baumhart; 27, 29, J. Futch).

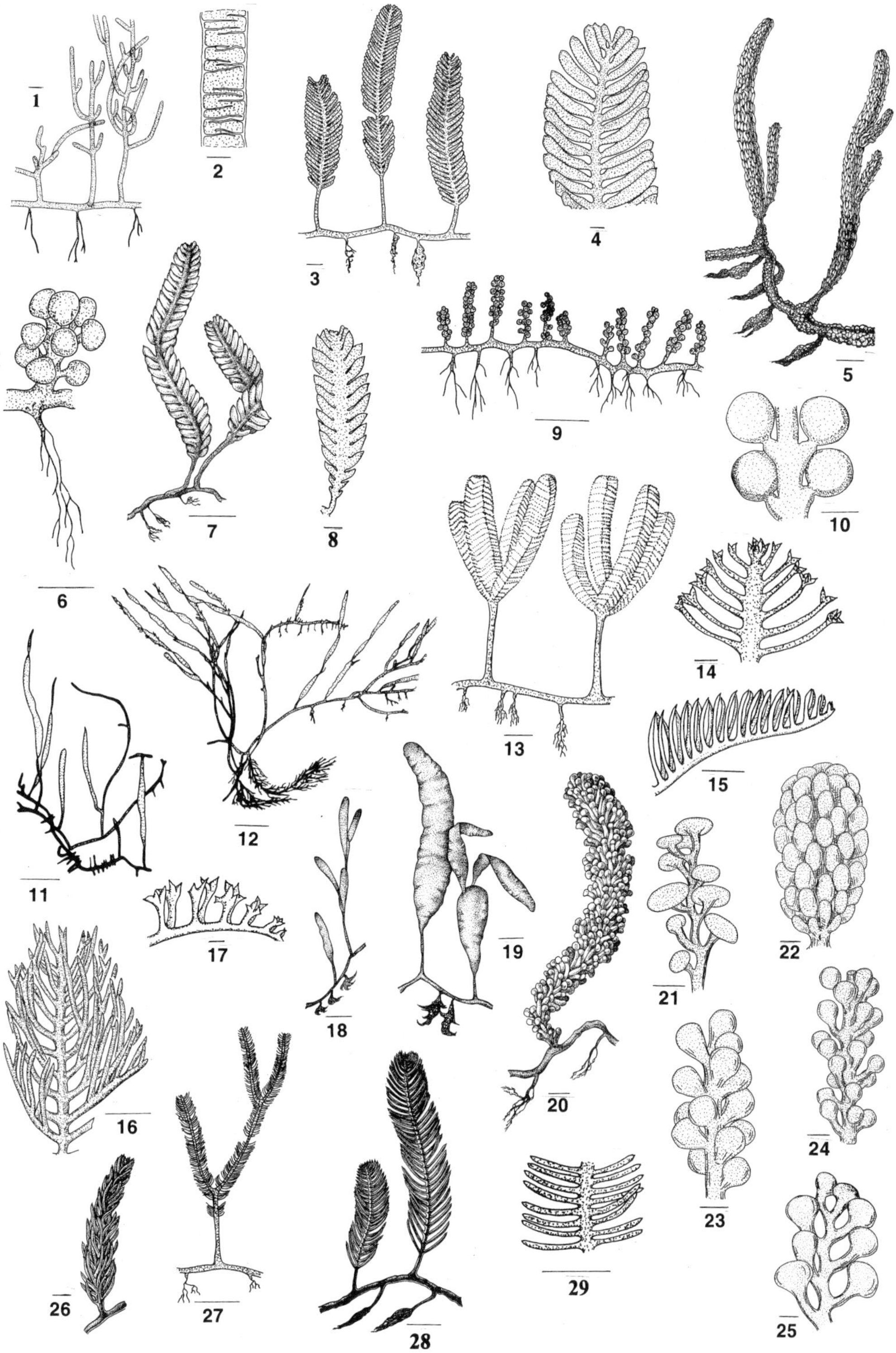
1
2
3
4
5
6
7
8
9
10
11
12
13
14
15
16
17
18
19
20
21
22
23
24
25
26
27
28
29

PLATE VII

Caulerpa verticillata. 1. Two erect fronds with delicate whorled branchlets arising from a horizontal rhizome with small rhizoidal clusters; scale 1 mm. 2. A longitudinal section of a rhizome showing internal trabeculae; scale 400 µm (R. Holbrook).

Caulerpa vickersiae. 3. Silhouette shows the rhizome has two minute fronds with a few, mostly opposite, branchlets; scale 2 mm (after Taylor, 1960). 4. Frond tip of var. *furcifolia* with pinnate, whorled, or irregular branchlets; scale 1 mm (after Littler and Littler, 2000). 5. Frond tip of var. *luxurians* with short, pinnate branchlets; scale 1 mm (J. Futch).

Caulerpa webbiana. 6. Frond tip with whorled, spiny branchlets; scale 200 µm (J. Futch).

Ostreobium quekettii. 7. The filamentous coenocyte was extracted from a decalcified shell and has branches and sporangia (*arrow*); scale 100 µm (after Taylor, 1957).

Avrainvillea asarifolia. 8. Plant has a rhizoidal holdfast, stalk, and blade with zonal banding; scale 2 cm (L. Baumhart). 9, 10. Moniliform-shaped siphons from blade margin (9) and cortex (10); scale 100 µm (J. Futch).

Avrainvillea elliottii. 11. Plant with a fan-shaped blade, short stalk, and rhizoidal holdfast; scale 1 cm (J. Futch). 12. Surface siphon of blade is slightly moniliform to irregular; scale 100 µm (after Littler and Littler, 2000).

Avrainvillea fulva. 13. Blade is coarse, fibrous, with ragged margins, and on a short stalk with a rhizoidal holdfast: scale 2 cm (after Littler and Littler, 1992). 14. Siphons of surface stalk are branched and moniliform to irregular; scale 100 µm (after Littler and Littler, 2000).

Avrainvillea geppiorum. 15. Three plants with thin ragged fronds on stalks arising from one holdfast; scale 1 cm. 16. A weakly moniliform medullary siphon of the blade; scale 100 µm. 17. Stalk surface siphon is strongly moniliform; scale 100 µm (all, J. Futch).

Avrainvillea levis. 18. Plant has a smooth blade less than 1 mm thick and a short stalk with a holdfast; scale 1 cm. 19. A slightly moniliform blade siphon; scale 50 µm (J. Futch).

Avrainvillea longicaulis. 20. The blade of f. *laxa* is small, thin, and on a long stalk; scale 2 cm. 21. Blade of the species is thick, rounded, and on a thick stalk; scale 2 cm (after Littler and Littler, 1992). 22, 23. Blade siphons partly moniliform; scale 100 µm (20, 22 after Littler and Littler, 1997; 23 after J. Futch).

Avrainvillea mazei. 24. Blade is thick, fibrous, loosely woven, and on a compressed stalk with a rhizoidal holdfast; scale 1 cm. 25. A smooth blade siphon; scale 100 µm (both after Littler and Littler, 1992).

Avrainvillea nigricans. 26. Plant has a thick, fibrous blade on a stout, cylindrical stalk with a rhizoidal holdfast; scales 2 cm (L. Baumhart). 27. The same, but with two fronds from a single holdfast; scales 2 cm (after Taylor, 1960). 28. Moniliform surface siphons; scale 100 µm (after Littler and Littler, 1992).

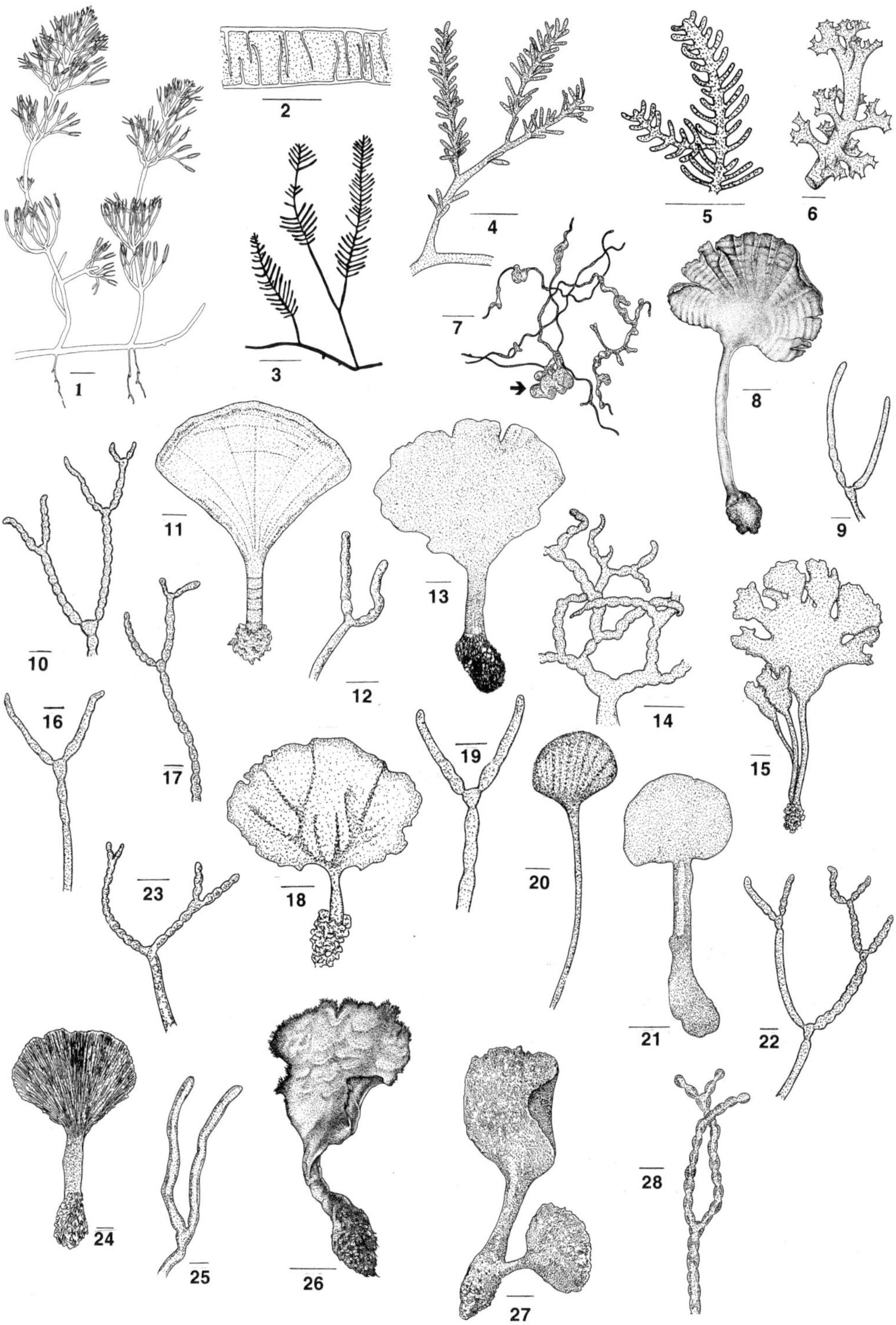
1
2
3
4
5
6
7
8
9
10
11
12
13
14
15
16
17
18
19
20
21
22
23
24
25
26
27
28

PLATE VIII

Avrainvillea rawsonii. 1. Erect axes are lobed, spongy, and lack blades; scale 1 cm (L. Baumhart). 2. Surface siphons are slightly moniliform; scale 100 μm (after Littler and Littler, 1992).
Avrainvillea silvana. 3. Blade is oval, slightly ragged to eroded, and has a fibrous holdfast; scale 5 cm. 4. Internal siphons from margin of blade; scale 200 μm. 5. Internal siphons from surface of blade; scale 100 μm (all after Littler and Littler, 2000).
Avrainvillea sylvearleae. 6. Blade is oval, thin, fibrous, and on a cylindrical stalk with a rhizoidal holdfast; scale 1 cm (after Littler and Littler, 1992). 7. Two blade siphons that are mostly smooth; scale 100 μm (J. Futch).
Boodleopsis pusilla. 8. Filaments are dichotomously branched, smooth, and constricted at forks. 9. Siphons moniliform and beadlike. 10. Siphons are coralloid and irregular in branching and shape; all scales 100 μm (8, 10 after Taylor, 1960; 9, R. Holbrook).
Cladocephalus luteofuscus. 11. Blade is semi-circular, spongy, and on an irregular stalk; scale 1 cm. 12. Blade surface siphons are irregular and repeatedly branched; scale 100 μm (both after Børgesen, 1913).
Halimeda discoidea. 13. Segments of upper axes are lightly calcified, smooth, and oval; scale 1 cm (after Humm and Taylor, 1961). 14. Two large subsurface urticles with many flat smaller surface ones; scale 100 μm (J. Futch).
Halimeda goreauii. 15. Part of a pendent deep water plant with thin, brittle, and deltoid segments; scale 1 cm (J. Futch). 16. Medullary filaments are fused, branched, and have small surface utricles (*arrow*); scale 100 μm (after Taylor, 1962b).
Halimeda gracilis. 17. The branch tip has flat, discoid brittle segments; scale 2 mm. 18. Small surface utricles arise from elongate club-shaped subsurface ones; scale 100 μm (both after Littler and Littler, 2000).
Halimeda hummii. 19. Subsurface utricles with 3 short surface ones; scale 100 μm. 20. Tips of two axes with irregular, flat to cylindrical segments; scale 2 mm (both after Ballantine, 1982).
Halimeda incrassata. 21. Plant with multiple branched axes, a rhizoidal basal holdfast, and narrow, trilobed segments; scale 1 cm (L. Baumhart). 22. Utricles are in three tiers, with the lowest ones large and club-shaped; scale 100 μm (after Littler and Littler, 2000).
Halimeda lacrimosa. 23. Axes with subglobose segments (A. L. King). 24. Portion of an axis with irregular segments; scale 2 mm (after Taylor, 1960). 25. Portion of an axis with tear-shaped segments; scale 2 mm (after Taylor, 1960). 26. Surface utricles arise from elongate subsurface ones; scale 100 μm (after Littler and Littler, 2000).
Halimeda monile. 27. Plant is highly branched, with a rhizoidal base, cylindrical segments above and fused ones basally to form a flat stalk; scale 1 cm (L. Baumhart). 28. Surface utricles in three tiers; the lowest ones oval; scale 100μm (after Littler and Littler, 2000).
Halimeda opuntia. 29. Axial tip has flat, ribbed, or rounded segments; scale 1 cm. 30. The tip of f. *triloba* has deeply trilobed segments; scale 2 mm. 31. Surface utricles are in four tiers; scale 100 μm (all by J. Futch).

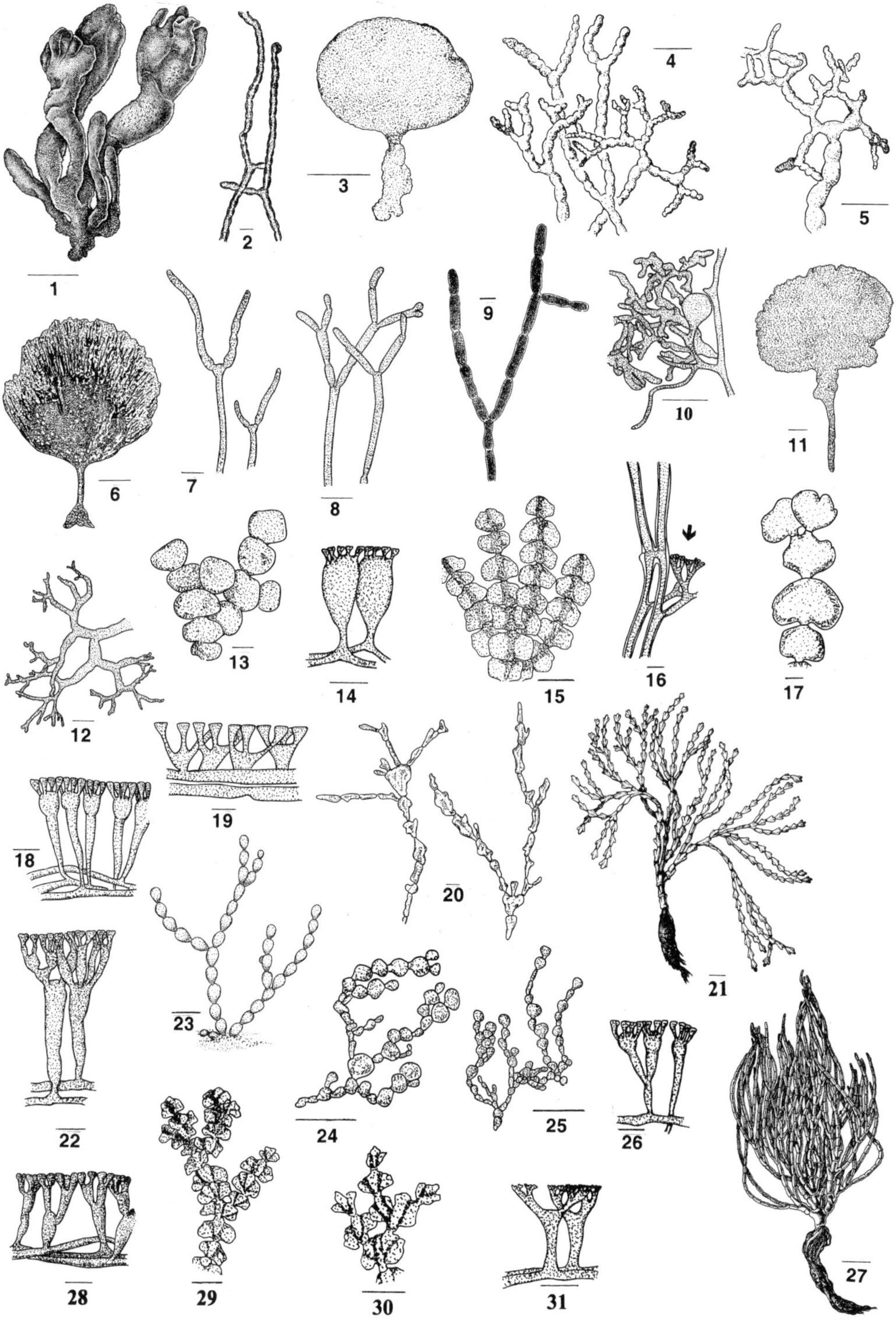
1
2
3
4
5
6
7
8
9
10
11
12
13
14
15
16
17
18
19
20
21
22
23
24
25
26
27
28
29
30
31

PLATE IX

Halimeda scabra. 1. A small attached plant with 2–6 rounded segments per axes; scale 1 cm (J. Futch). 2. Five surface utricles each with a single apical spine; scale 10 μm (A. L. King).

Halimeda simulans. 3. An upper axial part with ribbed to lobed segments; scale 1 cm. 4. Three swollen subsurface utricles with each bearing 4 short surface ones; scale 100 μm (J. Futch).

Halimeda tuna. 5. Upper part of a young plant with kidney-shaped segments above and fused ones below; scale 2 cm. 6. Utricles in three tiers with subsurface ones club-shaped; scale 100 μm (both, J. Futch). 7. Upper part of f. *platydisca* with disc-shaped, flat segments; scale 1 cm (after Taylor, 1928).

Penicillus capitatus. 8. Plant with a long stalk, rhizoidal base, and a filamentous cap; scale 1 cm (L. Baumhart). 9. Surface siphons of the stalk end in thick walled tips; scale 100 μm. 10. The cap filament is dichotomously branched and constricted at forks; scale 1 mm (9, 10 after J. Futch).

Penicillus dumetosus. 11. Plant with a thick stalk, rhizoidal base, and filamentous cap; scale 1 cm (L. Baumhart). 12. Surface appendages of stalk siphon with many acute tips; scale 200 μm (A. L. King). 13. Cap filament is constricted at forks; scale 2 mm (J. Futch).

Penicillus lamourouxii. 14. Plant with a thick stalk, rhizoidal base, and filamentous cap; scale 1 cm (A. L. King). 15. Cap filament is constricted at forks; scale 1 mm (A. L. King). 16. Surface appendages of stalk siphon with flat tips; scale 100 μm (after Taylor, 1960).

Penicillus pyriformis. 17. Plant with a flat indented cap, short stalk, and a rhizoidal base; scale 2 cm (J. Futch). 18. Surface appendages of stalk siphon taper to blunt tips; scale 200 μm (A. L. King).

Pseudocodium floridanum. 19. Plant is dichotomously branched with a rhizoidal base; scale 1 cm. 20. The main axis has clear surface utricles and central pigmented siphons with reproductive bodies (both after L. Baumhart in Dawes, 1974). 21. Branch tip grows by its central siphons; scales 200 μm (after Dawes and Mathieson, 1972).

Rhipilia tomentosa. 22. Blade is spongy, fan-shaped, and on a flat stalk with a rhizoidal base; scale 1 cm. 23. Blade siphons with haptera-like tips; scale 100 μm (L. Baumhart).

Rhipocephalus oblongus. 24. Cap is long, cylindrical, composed of whorls of filaments, and on a cylindrical stalk with a rhizoidal base; scale 1 cm (after Taylor, 1960). 25. Stalk siphon with flat surface appendages; scale 200 μm (after Taylor, 1960). 26. Cap filament with constrictions at forks; scale 100 μm (after Littler and Littler, 2000).

Rhipocephalus phoenix. 27. Cap consists of small flat bladelets and is on a cylindrical stalk with a rhizoidal base; scale 1 cm (L. Baumhart). 28. Two blades with fused cap siphons; scale 2 cm (L. Baumhart). 29. Surface appendages of stalk siphon; scale 100 μm (J. Futch). 30. Fused blade filaments; scale 100 μm (after Littler and Littler, 2000).

Rhipidosiphon floridensis. 31. Blade is monostromatic, composed of fused siphons, and on a short stalk with multiple rhizoids; scale 1 mm (after Littler and Littler, 1990b).

Udotea abbottiorum. 32. Blade is fan-shaped, with loosely woven siphons, and on a short stalk with a rhizoidal base; scale 1 cm. 33. Surface appendages of truncated stalk siphons; scale 200 μm (after Littler and Littler, 1990a).

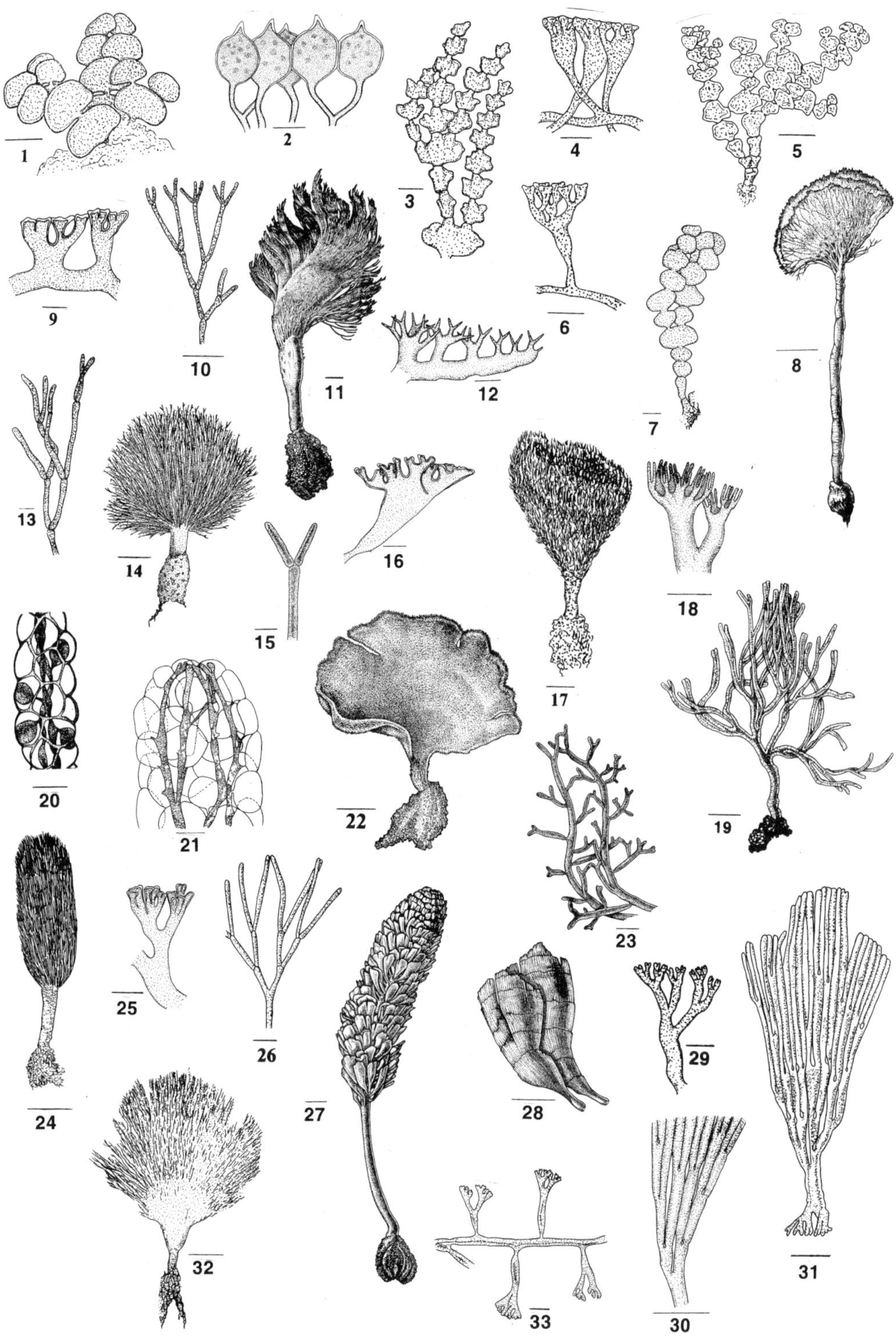
1
2
3
4
5
6
7
8
9
10
11
12
13
14
15
16
17
18
19
20
21
22
23
24
25
26
27
28
29
30
31
32
33

PLATE X

Udotea caribaea. 1. Plant has a fan-shaped blade with a lateral frond on its stalk, plus a rhizoidal base; scale 1 cm (after Littler and Littler, 1990a). 2. Surface appendages of stalk siphon are fingerlike; scale 200 µm (J. Futch).

Udotea conglutinata. 3. Blade is zonate, with frayed margins, and on a stalk with a rhizoidal base; scale 1 cm (L. Baumhart). 4. Surface appendages of the stalk siphon are irregular and fingerlike; scale 100 µm (Littler and Littler, 2000).

Udotea cyathiformis. 5, 6. Two cup-shaped blades with short cylindrical stalks and rhizoidal bases; scale 1 cm (L. Baumhart). 7. Surface appendages of stalk siphons are swollen and irregularly branched; scale 100 µm (after Littler and Littler, 1990a).

Udotea dixonii. 8. Three older blades arise from a common rhizoidal base; one on the left broken and residual; the other two intact, with concentric bands, and on individual stalks; scale 2 cm. 9. Surface appendages of a blade siphon are swollen and knoblike; scale 200 µm. 10. Surface appendages of a stalk siphon are stubby and dichotomously branched; scale 100 µm (all after Littler and Littler, 1990a).

Udotea flabellum. 11. Three highly divided zonate blades with cylindrical stalks and a common rhizoidal base; scale 2 cm (L. Baumhart). 12. A stalk siphon with multiple rounded surface appendages; scale 100 µm. 13. A blade siphon with multiple rounded surface appendages; scale 100 µm (12, 13 after Littler and Littler, 1990a).

Udotea looensis. 14. Blade is fan-shaped, faintly zonate, fibrous, and on a short stalk; scale 1 cm. 15. Surface appendages of stalk siphon are long and tapered; scale 100 µm (both after Littler and Littler, 1990a).

Udotea luna. 16. Blade is thin, fibrous, and on a short stalk; scale 1 cm. 17. A stalk siphon with acute surface appendages; scale 100 µm (after Littler and Littler, 1990a).

Udotea occidentalis. 18. Plant with a multilobed blade, short stalk, and rhizoidal base; scale 1 cm (after Littler and Littler, 1990a). 19. Lateral appendages of stalk siphon with stubby, rounded tips; scale 100 µm (J. Futch). 20. Surface appendages of blade siphon end in conspicuous knobs; scale 50 µm (after Littler and Littler, 1990a).

Udotea spinulosa. 21. Blade is smooth, on a stalk, and with a rhizoidal base; scale 1 cm. 22. Surface appendages of stalk siphon are long and pointed; scale 100 µm. 23. Surface appendages of blade siphon are spinelike; scale 50 µm. 24. Two surface appendages of blade siphon with many spines; scale 100 µm (all after Littler and Littler, 1990a).

Udotea unistratea. 25. Habit of a plant with a monostromatic blade produced on a cylindrical stalk and with basal rhizoids; scale 1 cm. 26. Surface appendages of stalk siphons that are fingerlike and blunt; scale 200 µm (both after Littler and Littler, 1990a).

Udotea wilsonii. 27. Plant with multiple fan-shaped blades radiating from a central axis, a short stalk, and rhizoidal base; scale 1 cm. 28. Surface appendages of stalk are stubby with blunt tips; scale 100 µm. 29. Knoblike surface appendages of blade siphons; scale 50 µm (all after Littler and Littler, 1990a).

Smithsoniella earleae. 30. Small portion of erect and prostrate filaments with lateral sporangia; scale 20 µm. 31. Erect axis with sporangia; scale 30 µm (R. Holbrook). 32. Two lateral sporangia with spores; scale 20 µm (30, 32 after Sears and Brawley, 1982).

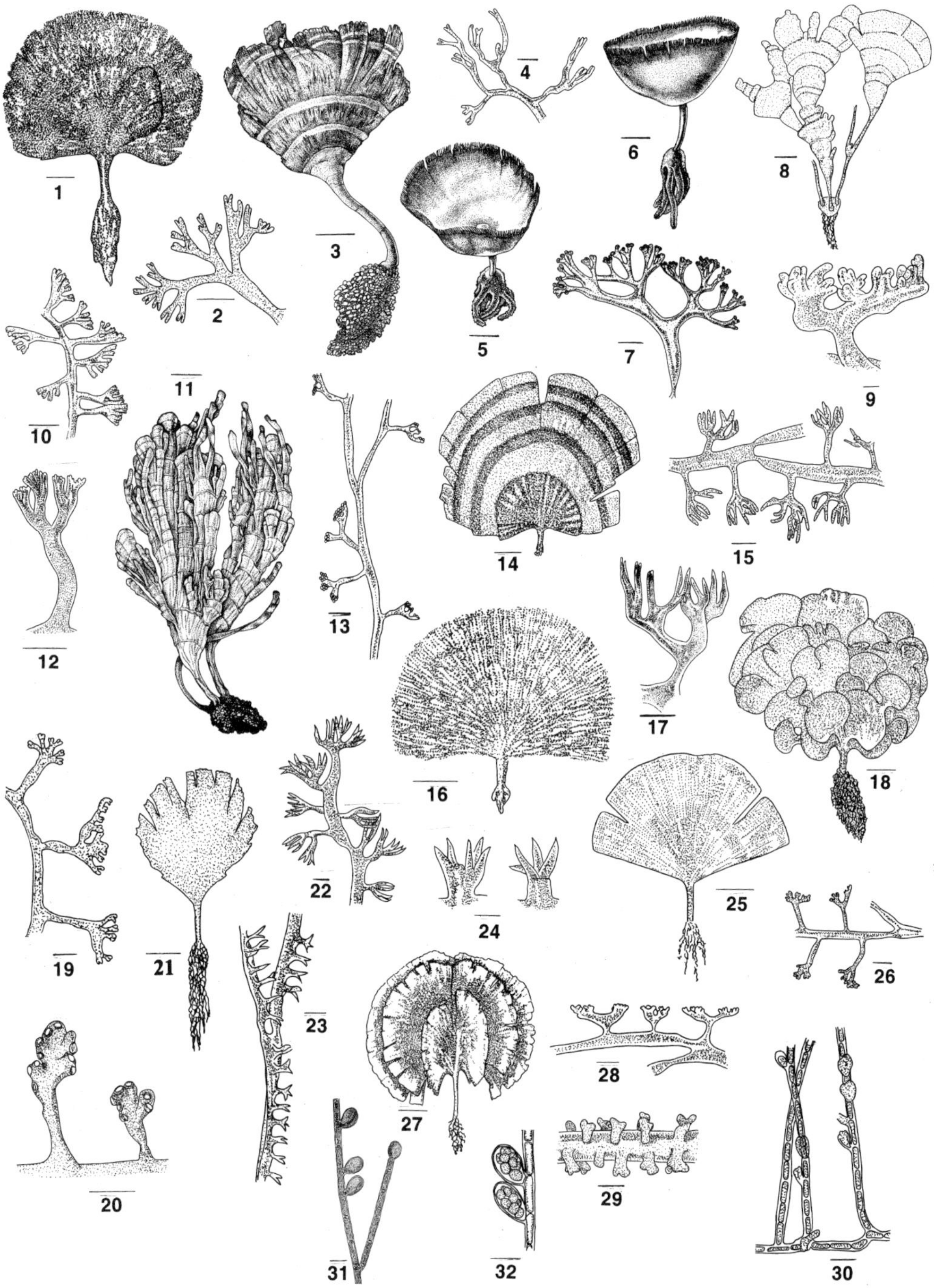
1
2
3
4
5
6
7
8
9
10
11
12
13
14
15
16
17
18
19
20
21
22
23
24
25
26
27
28
29
30
31
32

PLATE XI

Batophora occidentalis. 1. Two plants with central axial cells and whorls of branches; scale 2 mm. 2. A branch of var. *largoensis* with two gametangia; 200 µm (J. Futch).

Batophora oerstedii. 3. A vegetative plant with whorled branches and a central axial cell; scale 1 cm. 4. A reproductive plant with gametangia; scale 5 mm (3, 4 after L. Baumhart in Dawes, 1974). 5. A branch with three gametangia; scale 1 mm. 6. Cross section of a branch showing the central axial cell, a whorl of branchlets, and several gametangia; scale 2 mm (5, 6 after Littler and Littler, 2000).

Cymopolia barbata. 7. Five axes with cylindrical segments that end in whorls of filaments and arise from a common rhizoidal base; scale 2 cm. 8. Branch tips have whorls of filaments with calcified lower segments; scale 2 mm (7, 8 after L. Baumhart in Dawes 1981). 9. Partial cross section of an axial cell has club-shaped branchlets with gametangia; scale 100 µm (after Littler and Littler, 2000).

Dasycladus vermicularis. 10. Seven club-shaped, densely branched axes with rhizoidal bases; scale 1 cm (L. Baumhart). 11. Frond cross section shows the axial cell, whorled branchlets with 3 acute tips, and multiple gametangia (densely stippled); scale 1 mm (after Littler and Littler, 2000).

Neomeris annulata. 12. Two axes with whorls of calcified branchlets and tufts of apical noncalcified filaments; scale 1 mm (L. Baumhart). 13. A branchlet with two polyhedral cells and a gametangium (deeply stippled); scale 100 µm (after Littler and Littler, 2000).

Acetabularia calyculus. 14. The two plants have a disc of rays, long stalks, and basal rhizoids; scale 2 mm. 15. Corona superior with hair scars. 16. Corona inferior; scales 100 µm. 17. Four rays with notched tips; one with gametangia; scale 1 mm (all after Børgesen 1913).

Acetabularia crenulata. 18. Five plants with concave caps and single stalks on a shell fragment; scale 1 cm. 19. Cap rays bear one apical spine each; scale 2 mm (18, 19 after L. Baumhart in Dawes, 1981). 20. Corona inferior; scale 100 µm. 21. Corona superior with faint hair scars; scale 100 µm (20, 21 after Taylor, 1960).

Acetabularia farlowii. 22. Cap of a young plant with adherent rays; scale 2 mm. 23. Cap of an older plant with free rays; scale 2 mm (both after Humm and Taylor, 1961).

Parvocaulis pusilla. 24, 25. Two plants have caps with separate rays and lobed basal rhizoids; scales 2 mm (after Taylor, 1960).

Acetabularia schenckii. 26. Plant with a convex cap of rays, a long stalk, and rhizoids. 27. A cap with 40 rays; scales 1 cm (26, 27, R. Holbrook). 28. Ray tips with one central spine each and gametangia; scale 100 µm. 29. Corona superior with hair scars; scale 100 µm. 30. Corona inferior with forked tips; scale 100 µm (28–30 after Taylor, 1960).

Chalmasia antillana. 31. Three plants with long stalks, basal rhizoids, and concave caps of 32 rays; scale 1 mm. 32. Three rays with wedge-shaped tips; scales 1 mm (31, 32, A. L. King). 33. Corona superior with hair scars (*arrow*) and rays at the base; scale 100 µm (after Valet, 1969).

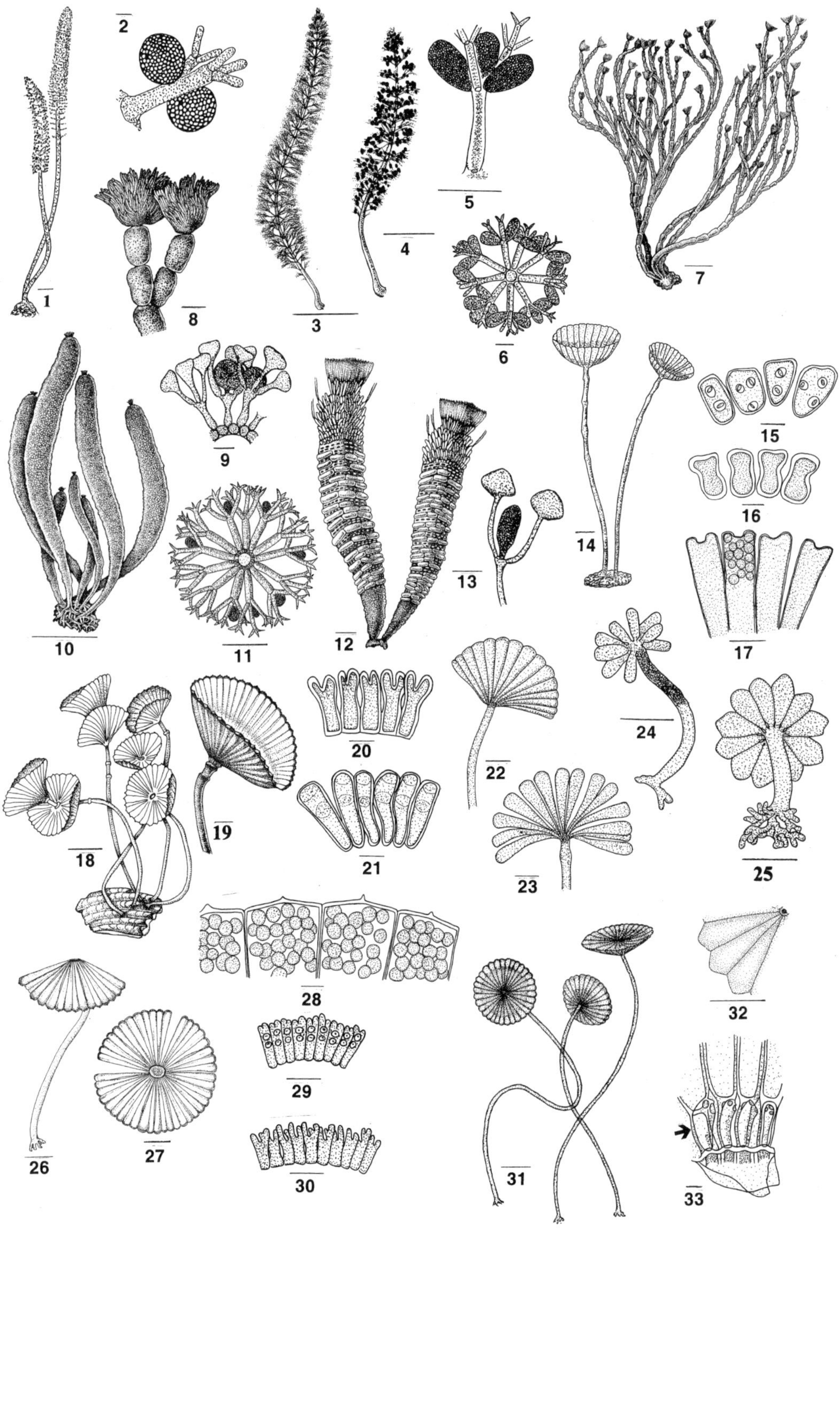
1
2
3
4
5
6
7
8
9
10
11
12
13
14
15
16
17
18
19
20
21
22
23
24
25
26
27
28
29
30
31
32
33

PLATE XII

Chrysonephos lewisii. 1. Part of an erect, dichotomously branched filament; scale 100 μm. 2. Axes are multiseriate at forks; scale 20 μm. 3. Cells at a uniseriate tip contain one yellow, parietal plastid that is lobed 2–4 times or bandlike; scale 10 μm (all after Taylor, 1951).

Vaucheria litorea. 4. An antheridium with two cross walls at tip of siphon and two lateral discharge pores; scale 100 μm. 5. A coenocytic siphon with two oogonia separated by cross walls; each oogonium has one oospore; scales 100 μm (both after Ott and Hommersand, 1974).

Vaucheria longicaulis. 6. An oogonium at the end of a side branch with two cross walls and one oospore; scale 100 μm. 7. One terminal antheridium with two cross walls; scale 100 μm (both after Ott and Hommersand, 1974).

Vaucheria velutina. 8. A siphon with four sessile, empty antheridia and one oogonium with an oospore; both have cross walls; scale 100 μm (after Ott and Hommersand, 1974).

Chrysophaeum taylorii. 9. Tip of a branch streamer with elliptical stalked cells; scale 100 μm. 10. A stalked cell with a gulletlike cavity, nucleus, and many irregularly oval parietal plastids; scale 10 μm (both after Lewis and Brian, 1941).

Dictyopteris delicatula. 11. Two plants with fibrous holdfasts, flat fronds, midribs, and scattered sori; scale 2 cm (after L. Baumhart in Dawes, 1981). 12. Cross section of a frond margin with distromatic construction; scale 100 μm (after Littler and Littler, 2000).

Dictyopteris hoytii. 13. Two strap-shaped dichotomously branched fronds with midribs and weak lateral veins; scale 2 cm (after Taylor, 1960).

Dictyopteris justii. 14. An older plant that is marginally eroded with narrow blades and obvious midribs; scale 2 cm (after Earle, 1969). 15. Cross section of a frond with a distromatic medulla surrounded by one layer of small surface cells, plus a cluster of hairs; scale 100 μm (after Littler and Littler, 2000).

Dictyopteris plagiogramma. 16. Plant is dichotomously branched, and with a midrib and weakly pinnate veinlets; scale 2 cm (after Earle, 1969). 17. Cross section shows a midrib with multiple layers of variably sized cells; scale 100 μm (after Littler and Littler, 2000).

Dictyopteris polypoioides. 18. Frond is dichotomously branched, with a conspicuous basal midrib and a few marginal proliferations; scale 2 cm (after Misra, 1966). 19. Frond cross section showing 4 cell layers at the margin and distromatic construction inward; scale 100 μm (after Littler and Littler, 2000).

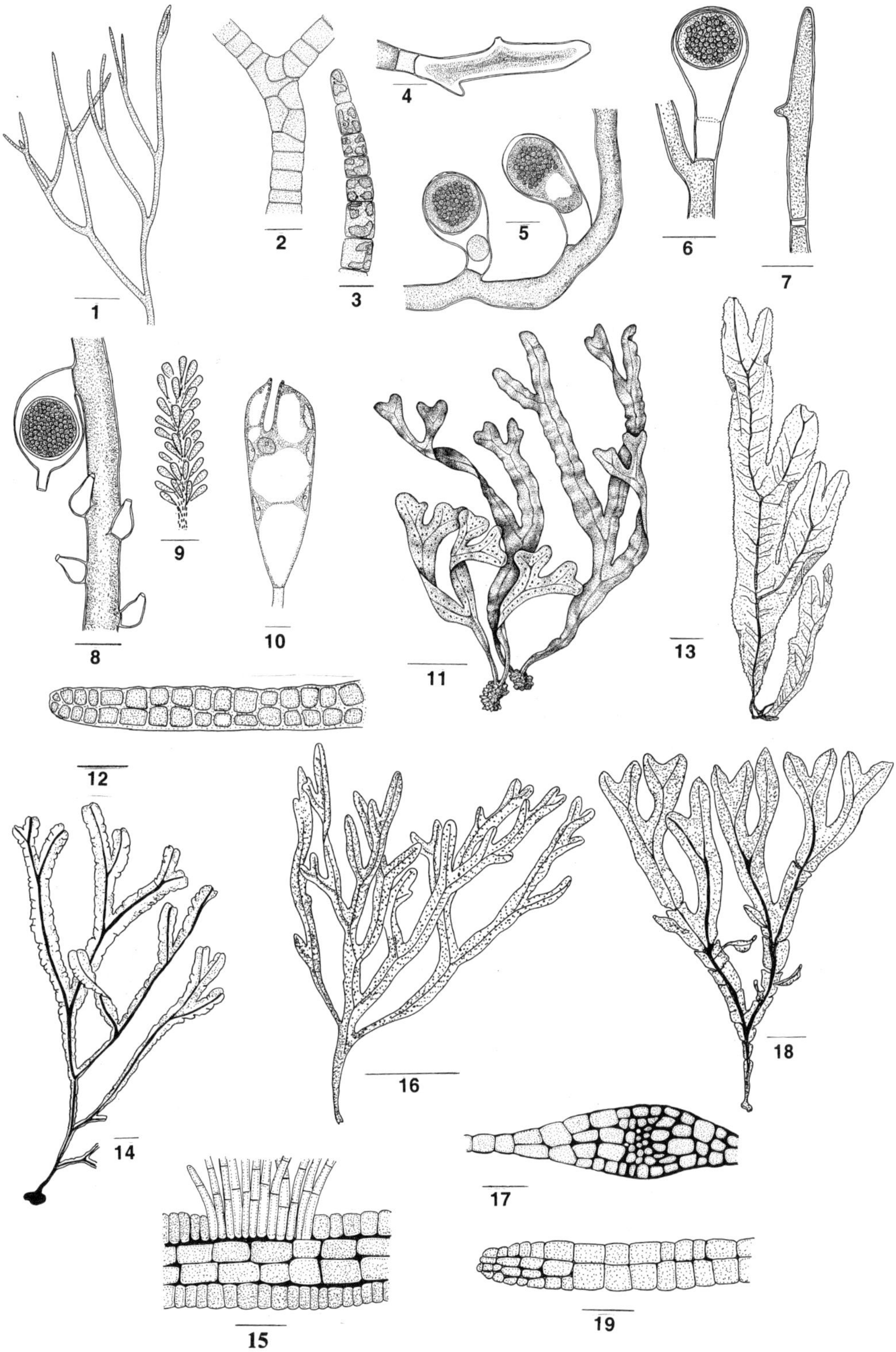
1
2
3
4
5
6
7
8
9
10
11
12
13
14
15
16
17
18
19

PLATE XIII

Dictyota bartayresiana. 1. A plant with a broad, flat, dichotomously branched blade; scale 2 cm (after Misra, 1966).
Dictyota caribaea. 2. Tip of an entangled, narrow axis with cervicorn to dichotomous branching; scale 1 mm (after Littler and Littler, 2000).
Dictyota cervicornis. 3. Upper part of a frond with a narrow axis and dichotomous to cervicorn branching; scale 2 cm (after Earle, 1969).
Dictyota ciliolata. 4. Part of a broad, dichotomously branched frond with scattered marginal teethlike projections; scale 1 cm (after Misra, 1966).
Dictyota crenulata. 5. Upper part of frond is strap-shaped, often twisted, and with small marginal teeth; scale 5 mm. 6. Cross section of a tetrasporic frond showing an immature tetrasporangium; scale 100 μm. 7. Cross section of a male frond showing an antheridial sorus; scale 50 μm. 8. Cross section of a female frond showing an oogonial sorus (top) and marginal (lateral) rhizoids; scale 100 μm (all after Littler and Littler, 2000).
Dictyota crispata. 9. Upper part of a frond showing alternate branching; scale 5 mm. 10. Frond tip with one large apical cell; scale 100 μm (both after Littler and Littler, 2000).
Dictyota guineensis. 11. Upper part of an alternately branched frond with dichotomously branched lateral axes; scale 5 mm (after Taylor, 1960). 12. Cross section showing a medulla of 2–3 cell layers and a surface layer of small cells; scale 25 μm (after Earle, 1969).
Dictyota menstrualis. 13. Fronds are flat, dichotomously branched, and with a rhizoidal holdfast; scale 2 cm (after L. Baumhart in Dawes, 1981). 14. Cross section of a tetrasporic frond showing a single layer of large medullary cells and a surface layer of small cells, plus an undivided sporangium; scale 25 μm (after Schneider and Searles, 1991).
Dictyota mertensii. 15. Part of a twisted upper axis with alternate branching and spurlike branchlets; scale 2 cm (after Earle, 1969).
Dictyota pinnatifida. 16. Upper part of a densely and irregularly branched frond; scale 1 cm (after Taylor, 1960). 17. Cross section shows the medulla with one layer of large cells in the center, 2–3 layers at the margin, and one layer of small surface cells; scale 50 μm (after Earle, 1969).

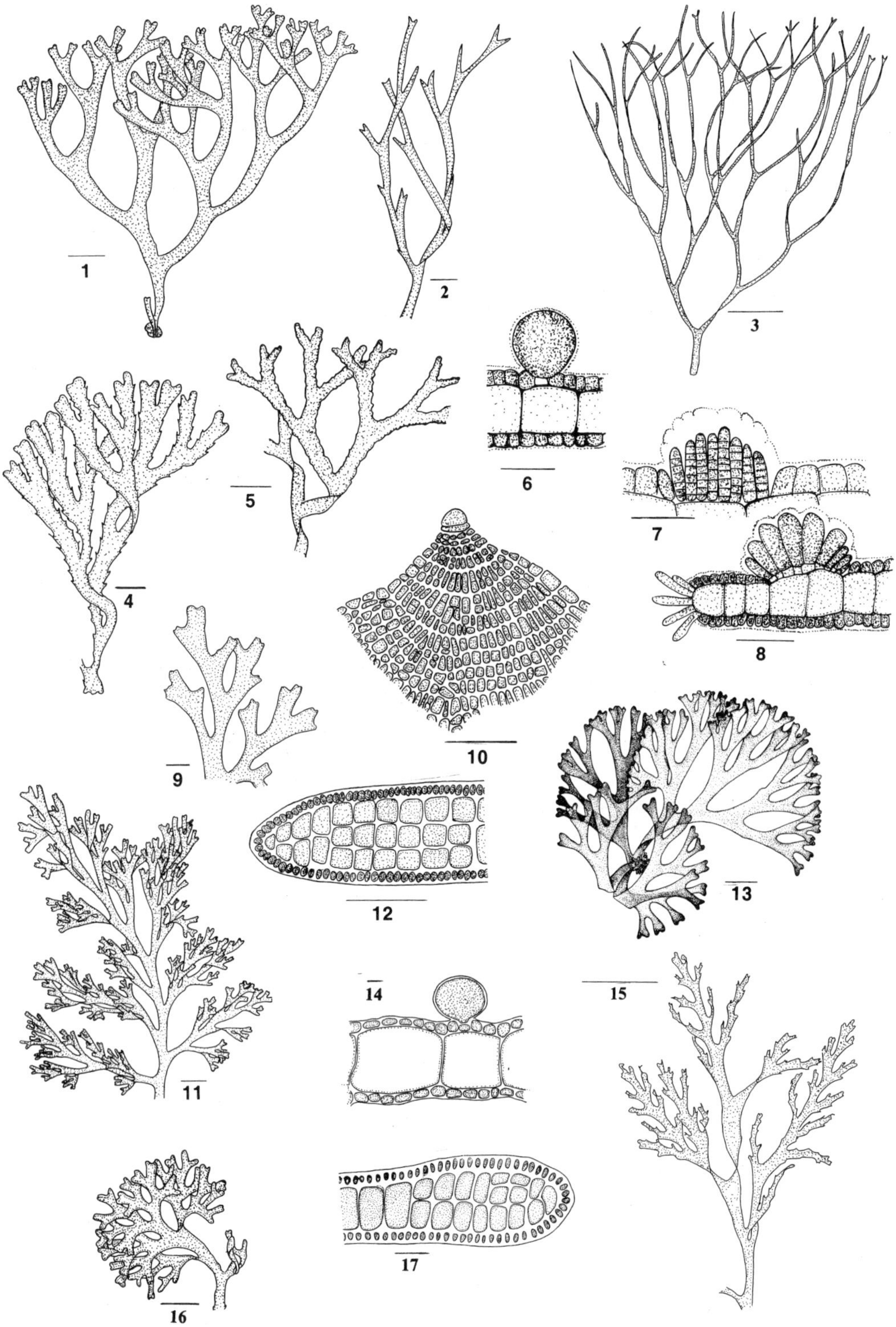
1
2
3
4
5
6
7
8
9
10
11
12
13
14
15
16
17

PLATE XIV

Dictyota pulchella. 1. A plant with twisted, narrow, and dichotomous axes and a fibrous holdfast; scale 1 cm (L. Baumhart). 2. Cross section at frond margin with one row of large medullary cells covered by small surface cells and a young hair on the upper right surface; scale 100 µm (after Earle, 1969). 3. Part of a lateral branch with a narrow axis, irregular to dichotomous branching, and tufts of hairs; scale 2 mm (after Littler and Littler, 1997). 4. Part of two axes with slightly twisted branching and progressive narrowing at each forking; scale 2 mm (after Littler and Littler, 2000).

Lobophora variegata. 5. Plant with many fan-shaped and zonately banded fronds; scale 2 cm (after Earle, 1969). 6. Longitudinal section of a crustose frond showing the dorsal surface with an oogonial sorus that lacks an indusium and a ventral surface with branched rhizoids; scale 50 µm (after Littler and Littler, 2000).

Dictyerpa jamaicensis (a stage of *Padina*). 7. Axes are entwined, terete to compressed, and end in hooks or a small fan-shaped blade; scale 2 mm (after Littler and Littler, 2000).

Padina antillarum. 8. Fronds are highly dissected and with enrolled margins and concentric bands of hairs on the surface; scale 2 cm (after Wynne and De Clerk, 1999).

Padina boergesenii. 9. Longitudinal section of blade showing three cell layers (a colorless medulla and pigmented cortical cells), a hair band, and a sporangial sorus that lacks an indusium; scale 500 µm (after Allender and Kraft, 1983).

Padina glabra. 10. Cross section of blade with mostly one layer of large medullary cells covered by small pigmented surface ones. 11. Longitudinal section of an enrolled blade apex with an apical cell that forms 3 cell layers subapically. 12. Blade base is multilayered and has multicellular rhizoids; all scales 50 µm (all after Gaillard, 1966).

Padina gymnospora. 13. Blades are mostly intact, with concentric hair bands, enrolled apical margins, and a rhizoidal base; scale 2 cm (L. Baumhart). 14. Longitudinal section of blade with a distromatic medulla, 1-celled cortex, and an oogonial sorus covered by an indusium; scale 50 µm (after Littler and Littler, 2000).

Padina pavonica. 15. Cross section of a blade showing three layers of similar sized cells; scale 100 µm (after Littler and Littler, 2000).

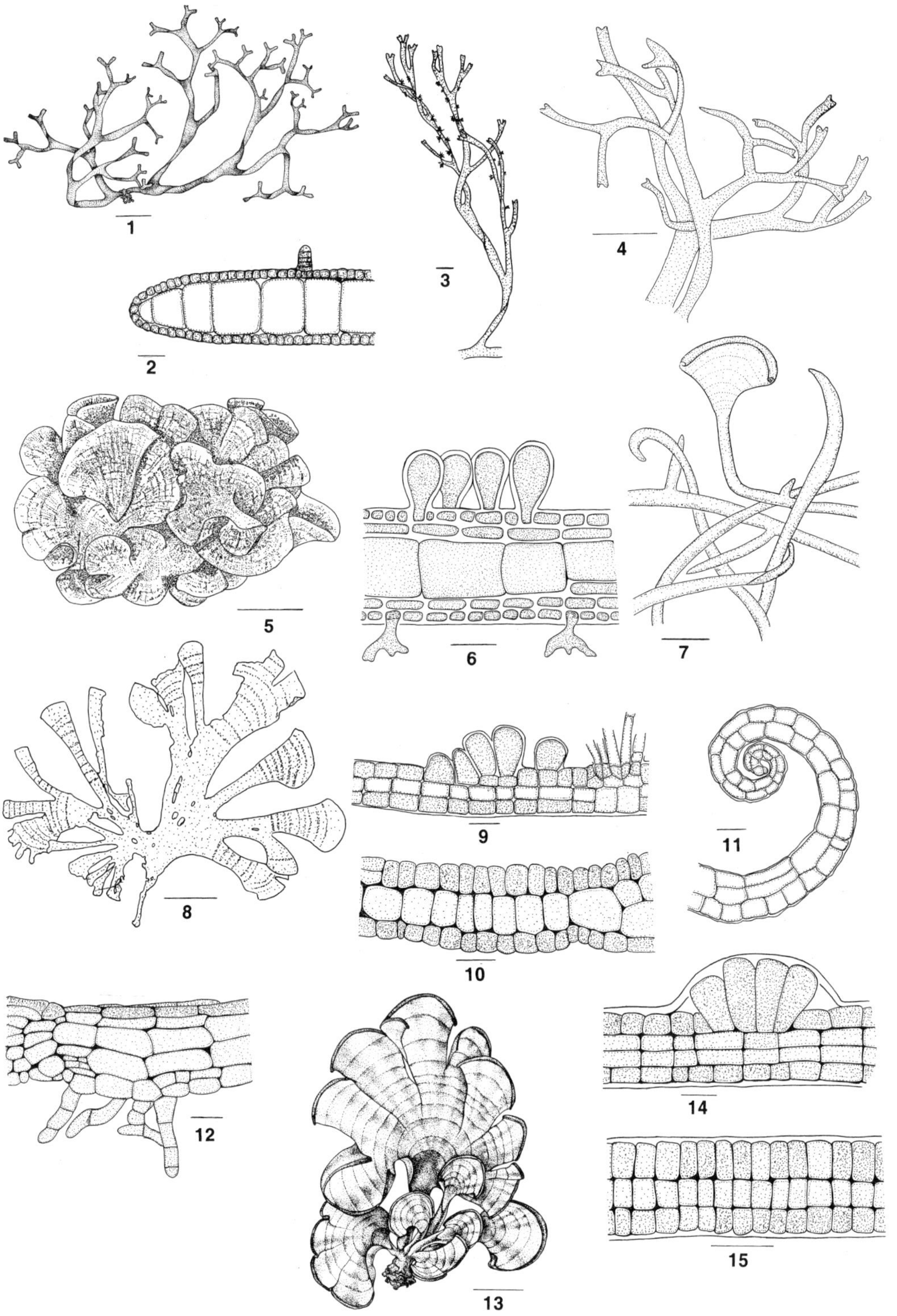
1
2
3
4
5
6
7
8
9
10
11
12
13
14
15

PLATE XV

Padina perindusiata. 1. Longitudinal section of a distromatic blade with an oogonial sorus covered by an indusium; scale 100 µm (after Littler and Littler, 2000).

Padina profunda. 2. Part of a blade with longitudinal splits and an enrolled apical margin; scale 5 mm (after Schneider and Searles, 1991). 3. Cross section near base of blade showing 2–4 layers of medullary cells, small surface cells, and colorless hairs; scale 1 mm (after Earle, 1969). 4. Longitudinal section of a young frond with immature hairs, an indusial cover, and a row of large medullary cells covered by small surface (cortical) cells; scale 50 µm (after Littler and Littler, 2000).

Padina sanctae-crucis. 5. Longitudinal section of a distromatic blade with an enrolled margin, an apical cell, and a hair band covered by a membrane; scale 100 µm (after Børgesen, 1914b). 6. Blade with many narrow segments, concentric hair bands, and a rhizoidal base; scale 1 cm (after Taylor, 1960).

Spatoglossum schroederi. 7. Frond is irregularly divided, with many marginal projections and a rhizoidal holdfast; scale 2 cm (L. Baumhart). 8. Cross section showing a distromatic medulla covered with a multilayered cortex of small cells; scale 50 µm (after Earle, 1969).

Stypopodium zonale. 9. Part of a frond with multiple splits and zonate bands of hairs at irregular intervals; scale 2 cm (after Misra, 1966). 10. Longitudinal section of a frond with a distromatic medulla covered by small surface cells and a concentric hair band; scale 100 µm. 11. Longitudinal section of a blade apex with one apical cell and a distromatic medulla covered by small surface cells subapically; scale 100 µm (10, 11 after Littler and Littler, 2000).

Taonia abbottiana. 12. Drawing of the holotype showing a highly dissected blade and small fibrous holdfast; scale 2 cm. 13. Longitudinal section of the blade tip with a distromatic medulla, no small surface cells near the apical cell, and the presence of one cortical layer subapically; scale 100 µm. 14. Cross section of an older blade with 4–5 layers of medullary cells covered by one layer of pigmented surface cells; scale 100 µm. 15. A sporangium on a two-celled stalk; scale 100 µm (all after Littler and Littler, 2000).

Sphacelaria fusca. 16. A stalked triradiate propagule with long arms; two cross sections of multiseriate axes are also visible; scale 100 µm (after Prud'homme van Reine, 1982). 17. Mature (*left*) and immature (*right*) propagules; scale 100 µm (after Earle, 1969).

Sphacelaria novae-hollandiae. 18. The triangular-shaped propagule has a three-celled stalk, three short arms (two are visible), and an obvious central apical cell; scale 50 µm (after Littler and Littler, 2000).

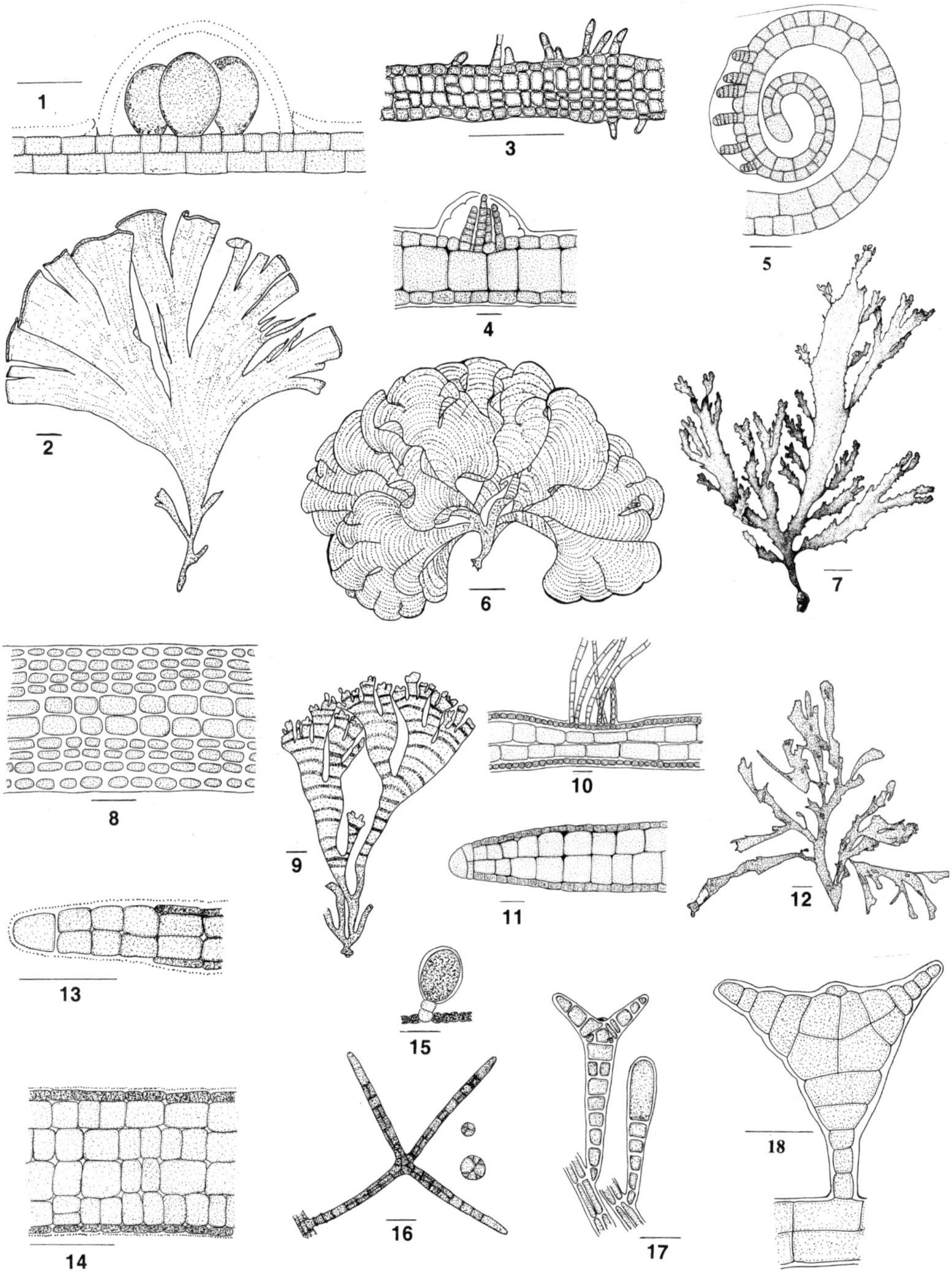
1
2
3
4
5
6
7
8
9
10
11
12
13
14
15
16
17
18

PLATE XVI

Sphacelaria rigidula. 1. Part of a biseriate filament with a biradiate propagule having long arms; scale 100 µm (R. Holbrook).
Sphacelaria tribuloides. 2. Part of a multiseriate axis with lateral branches and a squat, triangular-shaped propagule (*arrow*); scale 50 µm (R. Holbrook).
Halopteris filicina. 3. Tip of a biseriate axis with alternate branches and stalked unilocular sporangia in the axils of branchlets; scale 100 µm (after Newton, 1931).
Syringoderma floridana. 4. Blade is eroded, fan-shaped, and monostromatic through lateral fusion of filaments; scale 5 mm. 5. Many two-celled gametophytes growing on the sporangial wall with a central gametophyte releasing two gametes (*arrows*) and others being empty; scale 10 µm (after Henry, 1984).
Onslowia endophytica. 6. Filaments are endophytic, a few have biseriate segments, and one a young swollen propagule; scale 25 µm. 7. A stalked propagule with four lens-shaped cells on a host plant; scale 10 µm (both after Searles and Leister, 1980).
Verosphacela ebrachia. 8. A prostrate biseriate filament with spherical propagules (*arrow*) grown in culture; scale 50 µm (after Henry, 1987a).
Acinetospora crinita. 9. Filaments are uniseriate and branch at right angles; one has immature plurilocular sporangia (*arrow*); scale 200 µm (after Stegenga and Mol, 1983). 10. Some prostrate branches; scale 20 µm. 11. A sessile unilocular sporangium on a branched uniseriate filament; scale 20 µm (10, 11 after Schneider and Searles, 1991).
Feldmannia indica. 12. The erect uniseriate filament has branches with plurilocular sporangia that are sessile and cylindrical blunt-tipped; scale 50 µm (after Earle, 1969).
Feldmannia irregularis. 13. Part of an erect uniseriate filament with sessile fusiform plurilocular sporangia on the base of branches; scale 100 µm (after Misra, 1966).
Hincksia mitchelliae. 14. Filament is alternately branched, uniseriate, and with sessile cylindrical plurilocular sporangia; scale 100 µm (after Misra, 1966).
Hincksia onslowensis. 15. Part of a prostrate uniseriate filament with branches at right angles and a few oval-shaped plurilocular sporangia (*arrow*); scale 500 µm. 16. A stalked oval plurilocular sporangium; scale 50 µm (both after Amsler and Kapraun, 1985).
Kuetzingiella elachistaeformis. 17. Several erect axes arising from prostrate filaments and bearing three plurilocular sporangia; scale 50 µm (after Earle, 1969). 18. The plurilocular sporangia are fusiform to elongate; scale 25 µm (after Schneider and Searles, 1991).
Asperococcus fistulosus. 19. A plant with multiple tubular upright axes arising from a basal holdfast; scale 1 cm (after Schneider and Searles, 1991). 20. A small reproductive plant with sori containing paraphyses, a few hairs, and oval unilocular sporangia; scale 50 µm. 21. Part of a cross section of a small plant showing a sorus with paraphyses, a single young hair (*center*), and three unilocular sporangia; scale 50 µm (20, 21 after Taylor, 1957).

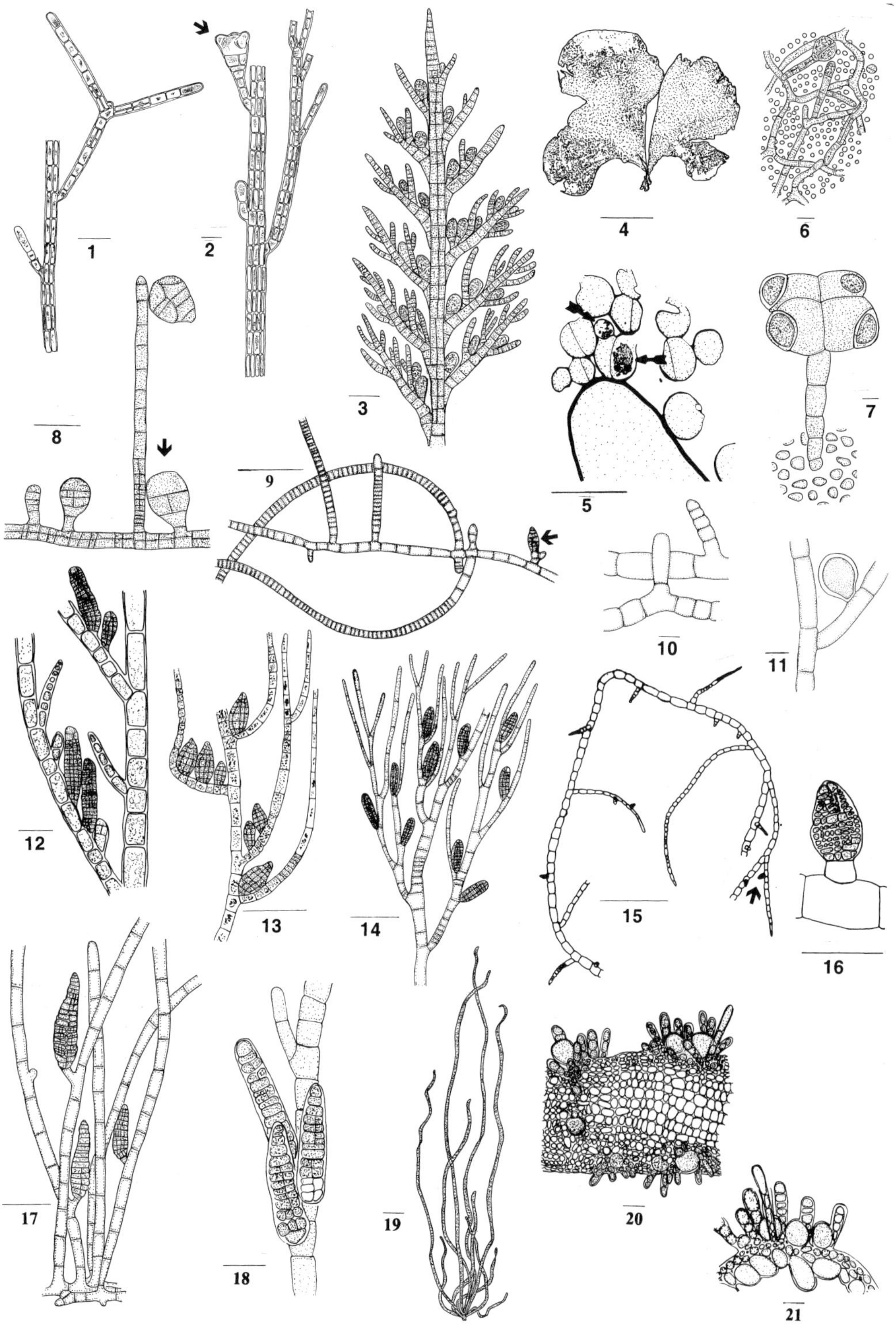
1
2
3
4
5
6
7
8
9
10
11
12
13
14
15
16
17
18
19
20
21

PLATE XVII

Cladosiphon occidentalis. 1. A plant with wormlike irregular branching growing on a seagrass leaf; scale 2 cm (after L. Baumhart in Dawes, 1981). 2. Axial cross section with a hollow center, lateral filaments, and colorless hairs; scale 100 μm (after Littler and Littler, 2000).
Elachista minutissima. 3. Part of an erect filament with a cluster of plurilocular sporangia, a uniseriate assimilatory filament (i.e., larger filament), and one sterile hair; scale 50 μm (after Taylor, 1928).
Hecatonema floridanum. 4. Cross section showing a prostrate base with erect filaments and plurilocular sporangia; scales 20 μm. 5. The same, showing two unilocular sporangia (*center*), erect filaments and two hairs (*left*) with intercalary growth; scales 20 μm (both after Schneider and Searles, 1991).
Hummia onusta. 6. Sporophytes are erect, wormlike, and on a seagrass blade; scale 1 cm (after L. Baumhart in Dawes, 1981). 7. Sporophyte cross section showing four large medullary cells, pigmented outer cortical cells, developing hairs (deciduous), and young stalked plurilocular sporangia; scale 2 cm. 8. Gametophyte is filamentous, branched, erect, and with terminal gametangia; scale 100 μm. 9. Part of a gametophytic filament with two "stalked" terminal plurilocular sporangia; scale 50 μm (7–9, R. Holbrook in Dawes, 1981).
Myrionema magnusii. 10. Side view of a filamentous disc with several clear, erect sterile hairs, a longer pigmented filament, and uniseriate plurilocular sporangia (*arrow*); scale 50 μm (after Earle, 1969).
Myrionema strangulans. 11. Surface view of a young epiphytic disc with forked marginal cells; scale 20 μm (after Stegenga and Mol, 1983). 12. Cross section of disc showing basal filaments, three pigmented assimilatory filaments, and two short plurilocular sporangia (*arrow*); scale 10 μm (after Misra, 1966).
Myriotrichia clavaeformis. 13. Erect unbranched filaments arise from prostrate axes and bear plurilocular sporangia (*left*) and unilocular sporangia (*right*); scale 100 μm (after Earle, 1969).
Myriotrichia occidentalis. 14. Part of two erect axes with irregular multiseriate sections (*left*) and two sessile unilocular sporangia (*right*); scale 50 μm (after Earle, 1969).
Nemacystus howei. 15. Plant is soft, mucilaginous, with irregular, slender branches, and a rhizoidal holdfast; scale 5 mm (after Schneider and Searles 1991). 16. Longitudinal section of outer cortical tissue showing pigmented surface filaments and a few hairs; scale 50 μm (after Earle, 1969).
Stilophora tenella. 17. Plant with many raised sori and a small discoid holdfast; scale 1.0 mm (L. Baumhart). 18. Surface view of a sorus with pigmented filaments and long colorless hairs; scale 100 μm (after Earle, 1969).
Bachelotia antillarum. 19. Part of a uniseriate branched filament with intercalary (*left*) and terminal (*right*) plurilocular sporangia; scale 100 μm (after Earle, 1969). 20. Part of a plant with prostrate and erect axes; the erect ones show intercalary cell divisions; scale 20 μm (after Schneider and Searles, 1991).

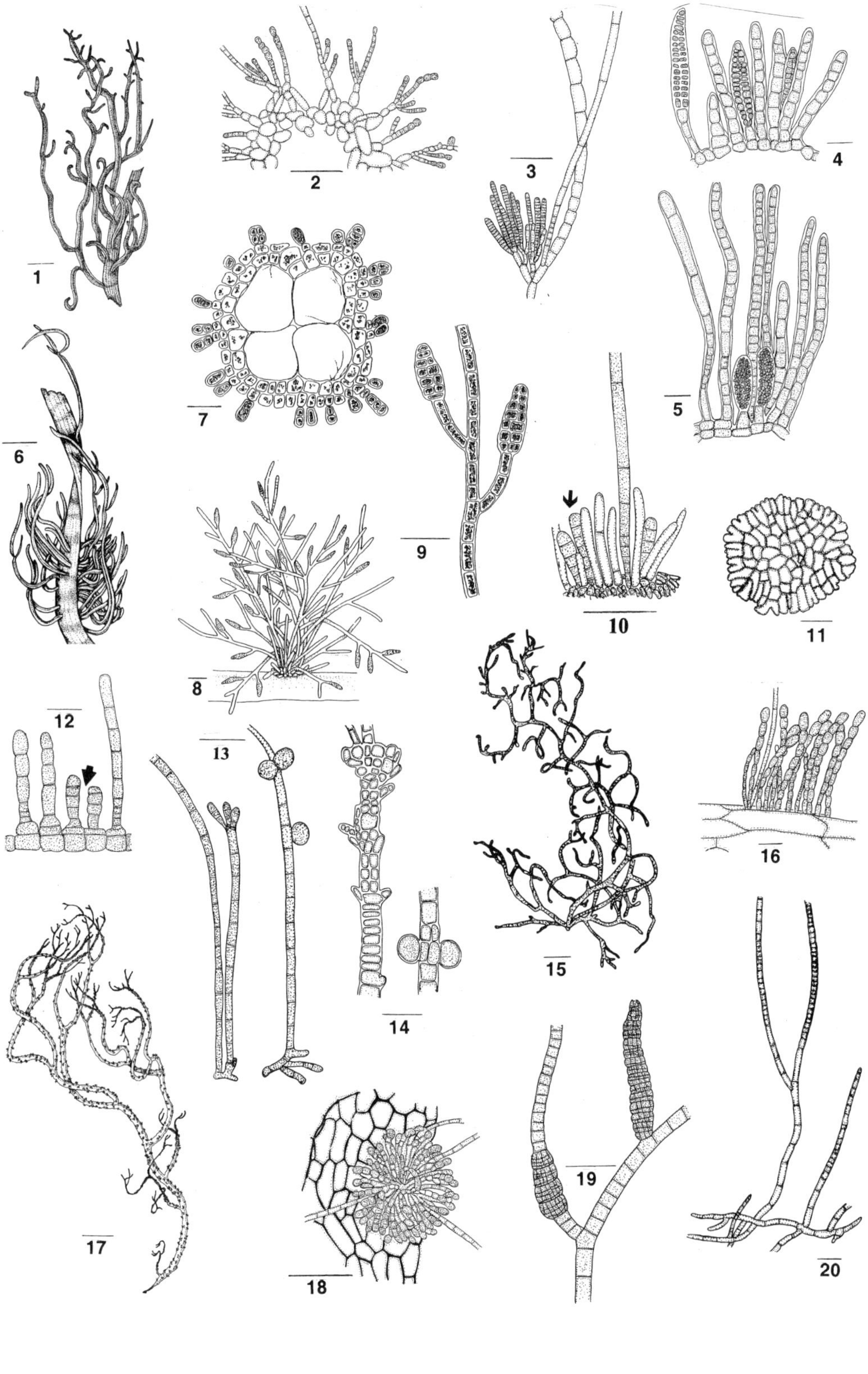
1
2
3
4
5
6
7
8
9
10
11
12
13
14
15
16
17
18
19
20

PLATE XVIII

Ectocarpus fasciculatus. 1. Three fronds have many plurilocular sporangia and fasciculate branching; scale 1 mm. 2. Upper part of a branched axis with lateral plurilocular sporangia; scale 200 μm. 3. A branch with six lateral plurilocular sporangia; scale 50 μm (all after Womersley, 1987).

Ectocarpus rallsiae. 4. Part of a filament with sessile and stalked fusiform plurilocular sporangia and one unilocular sporangium (*arrow*); scale 60 μm (after Børgesen, 1914b).

Ectocarpus siliculosus. 5. An upper erect filament bears one hair-tipped plurilocular and one oval unilocular sporangia; scale 25 μm (after Stegenga and Mol, 1983).

Herponema tortugense. 6. Plant has prostrate and erect filaments, plus sessile cylindrical plurilocular sporangia; scale 20 μm (after Taylor, 1928).

Phaeostroma pusillum. 7. The irregularly branched prostrate filaments have seven sessile unilocular sporangia and a colorless hair (setae), growing on a seagrass blade; scale 20 μm (after Hoyt, 1920).

Spongonema tomentosum. 8. The prostrate filament has sessile and lateral plurilocular sporangia; scale 50 μm (after Stegenga and Mol, 1983).

Chnoospora minima. 9. The plant has loose dichotomous branching, compressed axes, and flattened forks; scale 2 cm (after Taylor, 1960).

Colpomenia sinuosa. 10. The gametophyte is irregularly lobed and hollow; scale 2 cm (L. Baumhart). 11. The cross section of the cortex shows a surface sorus with three long colorless hairs and many plurilocular sporangia (gametangia); scale 50 μm (after Earle, 1969).

Hydroclathrus clathratus. 12. A fragment of a porous, irregularly expanded plant; scale 2 cm (after Earle, 1969). 13. Cross section of the cortex shows a surface sorus with many plurilocular sporangia (gametangia); scale 50 μm (after Misra, 1966).

Petalonia fascia. 14. A cluster of blades arising from one discoid holdfast; scale 1 cm (after Schneider and Searles, 1991). 15. Cross section of outer blade with densely packed uniseriate plurilocular sporangia (gametangia), two uniseriate hairs, plus parenchymatous cortical and medullary cells; scale 100 μm (after Taylor, 1957).

Rosenvingea antillarum. 16. The upper part of the plant has irregularly branched axes and tapered tips; scale 1 cm (L. Baumhart). 17. Three stalked plurilocular sporangia (gametangia) on the surface; scale 10 μm (after Taylor, 1928).

Rosenvingea intricata. 18. The axial part has irregular branching and contorted laterals; scale 2 cm (L. Baumhart). 19. Cross section shows colorless hairs, outer pigmented and inner colorless cortical cells, and a hollow center; scale 200 μm (after Earle, 1969).

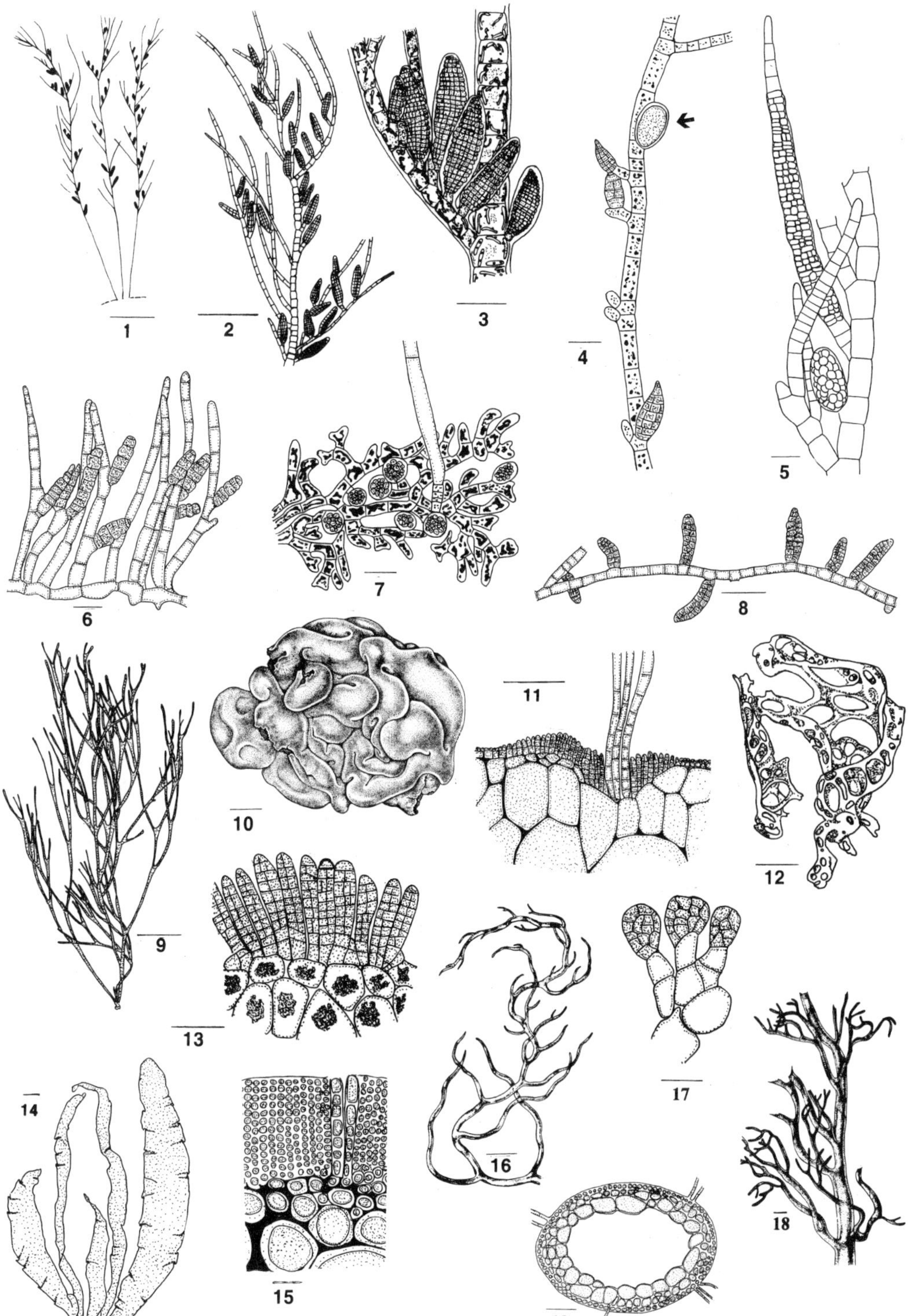
1
2
3
4
5
6
7
8
9
10
11
12
13
14
15
16
17
18
19

PLATE XIX

Rosenvingea orientalis. 1. Plant has irregular to cervicorn branching and a discoid holdfast; scale 1 cm (after Børgesen, 1914b). 2. Upper axes have compressed cylindrical branches and irregular to dichotomous branching; scale 1 cm (after Schneider and Searles, 1991). 3. Cross section shows a hollow center, a surface sorus with many plurilocular sporangia, and four colorless multicellular hairs; scale 50 μm (after Børgesen, 1914b).

Scytosiphon lomentaria. 4. A cluster of five unbranched tubular axes with constrictions with all arising from one discoid base; scale 5 mm. 5. Cross section shows a hollow center, a parenchymatous cortex and medulla, uniseriate plurilocular sporangia (gametangia), single-celled paraphyses, and three multicellular colorless hairs; scale 20 μm (both after Schneider and Searles, 1991).

Nereia filiformis. 6. Plant has irregular branching, a discoid base, and many filamentous tufts; scale 2 cm. 7. Part of an axes with two short filamentous tufts; scale 50 μm (both after L. Baumhart).

Sporochnus bolleanus. 8. Two stalked, cylindrical branchlets end in tufts of filaments, one is unbranched (*left*) and the other branched (*right*); scale 5 mm (after Earle, 1969).

Sporochnus pedunculatus. 9. Plant has regular branching, a basal disc, and many swollen branchlets with tufts; scale 2 cm (after L. Baumhart in Dawes, 1981). 10. The stalked swollen branchlet ends in a filamentous tuft; scale 2 cm (after Taylor, 1960).

Cystoseira myrica. 11. Plant has spiny or serrated axes, a rhizoidal holdfast, multiple vesicles, and spinelike foliar organs; scale 2 cm (after L. Baumhart in Dawes, 1981). 12. One spiny branch tip with two vesicles; scale 2 mm (after Littler and Littler, 2000). 13. Silhouette of an axial tip showing alternate branching; scale 2 cm (after Earle, 1969).

Sargassum acinarium. 14. Two vesicles with wartlike or bumpy surfaces. 15. Part of a main axis with blades, fertile branches, and vesicles. 16. A single blade with a serrated margin and cryptostomata in irregular rows; scales 5 mm (all after Littler and Littler, 2000).

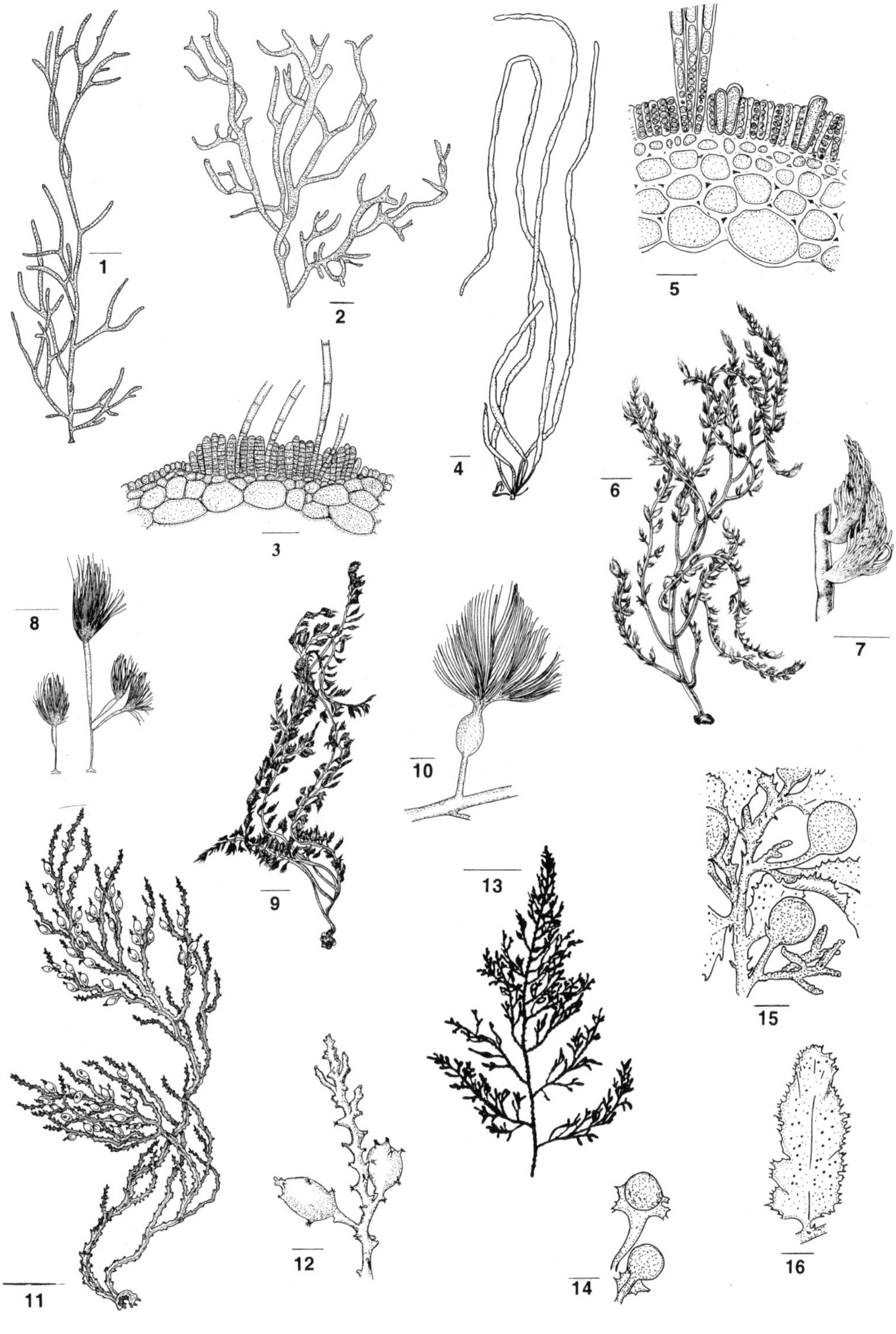
1
2
3
4
5
6
7
8
9
10
11
12
13
14
15
16

PLATE XX

Sargassum cymosum. 1. Plant has a discoid holdfast and linear blades. 2. An axial tip with densely branched fertile branches in the blade axils; scales 1 cm (both, R. Holbrook).

Sargassum filipendula. 3. Upper branch with many vesicles and elongate fertile branchlets in the axils of linear blades; scale 1 cm (after L. Baumhart in Dawes, 1981).

Sargassum fluitans. 4. Upper part of a free-floating plant with mostly smooth axes, rounded vesicles, and broad serrated blades that lack cryptostomata; scale 2 cm (L. Baumhart).

Sargassum hystrix. 5. Tip of var. *buxifolium* with densely set blades, a few marginal teeth and cryptostomata, and many spherical vesicles; scale 1 cm (after Taylor, 1960). 6. Basal part with a discoid holdfast and three branches with spherical vesicles and serrated elliptical blades with scattered cryptostomata; scale 2 cm (R. Holbrook).

Sargassum mathiesonii. 7. Tip of axes with scattered cryptostomata. 8. Silhouette of a perennial massive holdfast (base); scales 1 cm (both after Kilar, 1992).

Sargassum natans. 9. Upper part of an unattached axis with narrow serrated blades lacking cryptostomata and vesicles tipped with a spine or a blade; scale 2 cm (L. Baumhart).

Sargassum polyceratium. 10. Axial tip with stalked vesicles and sessile, finely serrated ovate blades with scattered cryptostomata; scale 1 cm (after Taylor, 1928).

Sargassum pteropleuron. 11. The plant tip has spherical vesicles and elongate blades with raised, toothed midribs; scale 1 cm. 12. The blade has a raised, toothed midrib; scale 2 cm (both after L. Baumhart in Dawes, 1981).

Sargassum ramifolium. 13. The central part of the plant has slender blades, a smooth main axis, and two stalked spherical vesicles; scale 5 mm (after Littler and Littler, 2000).

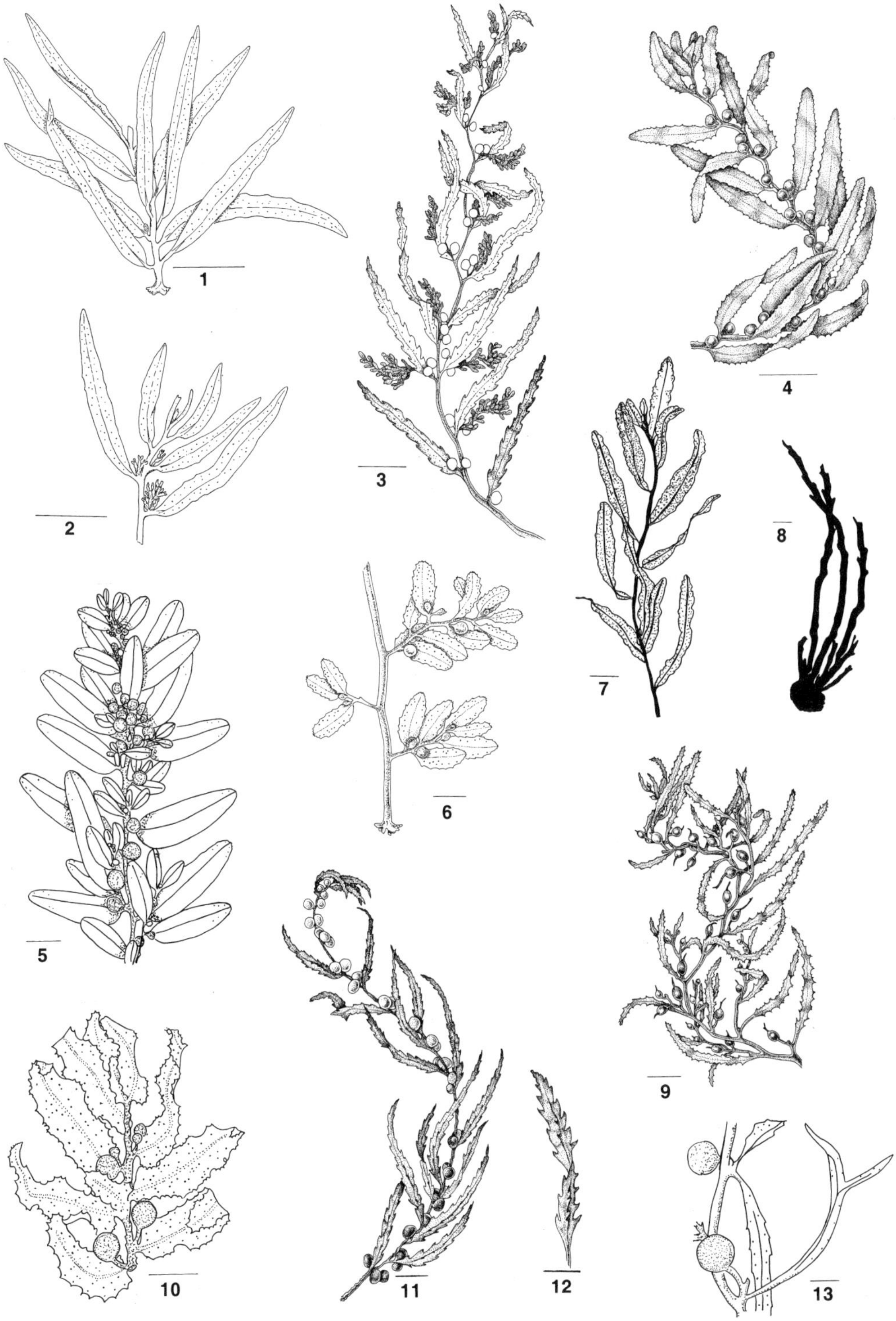
1
2
3
4
5
6
7
8
9
10
11
12
13

PLATE XXI

Sargassum vulgare. 1. The upper part of an axis has a smooth axis, many spherical vesicles, and oblong, dentate blades; scale 1 cm (after Taylor, 1960). 2. Cross section of a blade showing one cryptostomata with many sterile hairs; scale 50 µm (after Littler and Littler, 2000).

Sargassum wolfii. 3. Holotype silhouette showing abundant narrow blades; scale 2 cm. 4. Branch tip with spiny axes and narrow forked blades; scale 1 cm (both after Earle, 1969).

Turbinaria tricosta. 5. The small plant has axes densely covered with pyramidal blades and a discoid holdfast with irregular stolenlike haptera; scale 1 cm (after Schnetter, 1976). 6. Blade with a convex top and marginal teeth; scale 2 mm. 7. Blade cross section showing a cryptostomata with abundant sterile hairs; scale 50 µm (6, 7 after Littler and Littler, 2000).

Turbinaria turbinata. 8. A branch tip with clusters of pyramidal blades and fertile branches; scale 1 cm. 9. The two triangular-shaped blades have concave tops, scattered cryptostomata, and basal clusters of fertile branches; scale 5 mm (8, 9 after L. Baumhart in Dawes, 1981). 10. The blade is stalked, pyramidal, has smooth edges, and a concave top; scale 1 cm (after Taylor, 1960).

Asteronema breviarticulatum. 11. The uniseriate filaments have a branchlet ending in a hook and two immature unilocular sporangia; scale 10 µm. 12. A hooked branchlet. (11, 12 after Børgesen, 1914b). 13. A stalked plurilocular sporangium and contiguous vegetative cells with many discoid plastids.

Pseudolithoderma extensum. 14–16. Cross section of vegetative crusts; scales 50 µm (14–16 after Fletcher, 1987). 14. Erect filaments closely adjoined. 15. Terminal plurilocular sporangia. 16. Terminal (spherical) unilocular sporangia. 17. Cross section showing compact vertical filaments with surface hairs; scale 100 µm (after Littler and Littler, 2000).

Ralfsia expansa. 18. A lateral unilocular sporangium at the base of a paraphysis; scale 50 µm. 19. Cross section of a crust showing compact basal and erect filaments that have lateral unilocular sporangia; scale 100 µm (both after Børgesen, 1914b).

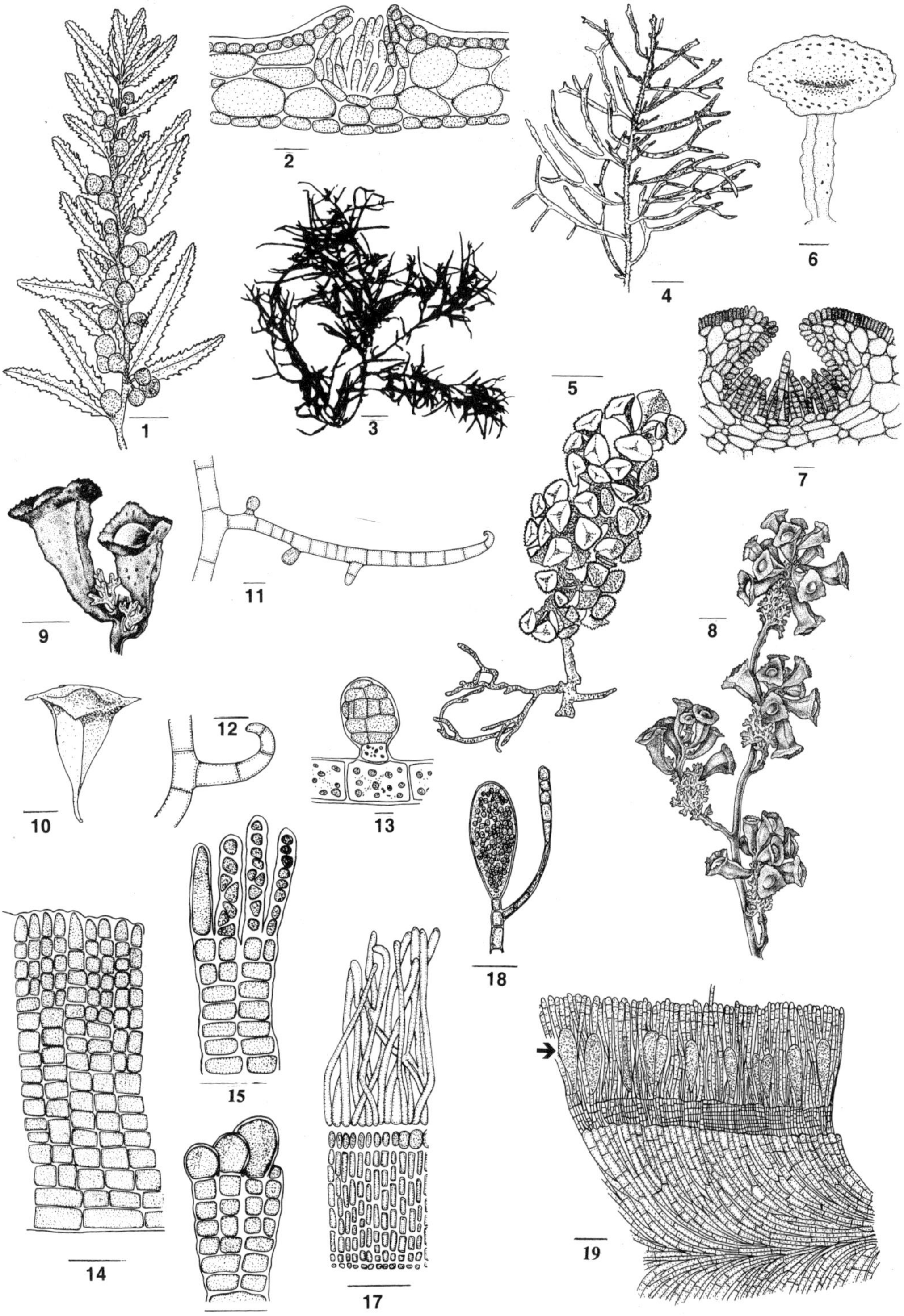
1
2
3
4
5
6
7
8
9
10
11
12
13
14
15
16
17
18
19

PLATE XXII

Chroodactylon ornatum. 1. Part of a branched filament with cells in a gelatinous matrix; scale 20 μm. 2. A short branching axis; scale 100 μm (2 after Abbott, 1999).

Stylonema alsidii. 3. A branched filament with a basal holdfast, uni-{#}to multiseriate axes, and cells in a gelatinous matrix; scale 20 μm. 4. The four vegetative cells each have one stellate plastid and a central pyrenoid; scale 5 μm.

Erythrocladia endophloea. 5. Filament is irregularly branched and growing on and in *Gracilaria tikvahiae*; scale 25 μm (after Schneider and Searles, 1991).

Erythrotrichia carnea. 6. The three filaments are erect, unbranched, uniseriate, and with holdfast cells on a seagrass blade; scale 25 μm. 7. An old filament with 4 small monospores on the right side; scale 25 μm (7 after Schneider and Searles, 1991).

Sahlingia subintegra. 8. Disc is epiphytic, compact, monostromatic, and with forked marginal cells; scale 10 μm (after Taylor, 1960).

Bangia fuscopurpurea. 9. Axis is unbranched, multiseriate, and with cells in a gelatinous matrix; scale 25 μm (after Schneider and Searles, 1991).

Porphyra leucosticta. 10. Four blades produced from the same holdfast; scale 3 cm. 11. Cross section of a monostromatic blade; scale 25 μm. 12. Surface view of a blade margin with cells in an irregular pattern; scale 25 μm (all after Stegenga and Mol, 1983).

Porphyra rosengurttii. 13. Blades are ruffled and cover a basal holdfast; scale 2 cm. 14. Cross section of a monostromatic blade with elliptical cells; scale 20 μm. 15–17. Surface views; scales 50 μm. 15. Spermatangia with 8 or 16 visible cells. 16. Usually 4 cells are visible in mature zygotosporangia. 17. Monospores (all after Coll and Cox, 1977).

Porphyra suborbiculata. 18. Two small blades with short basal holdfasts; scale 2 cm (after Schneider and Searles, 1991).

Hildenbrandia rubra. 19. Two cross sections of fleshy crusts; the tetrasporangial crust (*left*) has a conceptacle with irregularly divided tetrasporangia; the vegetative crust (*right*) has vertical rows of perithallial cells and horizontal rows of basal hypothallial cells; scale 20 μm (after Taylor, 1957).

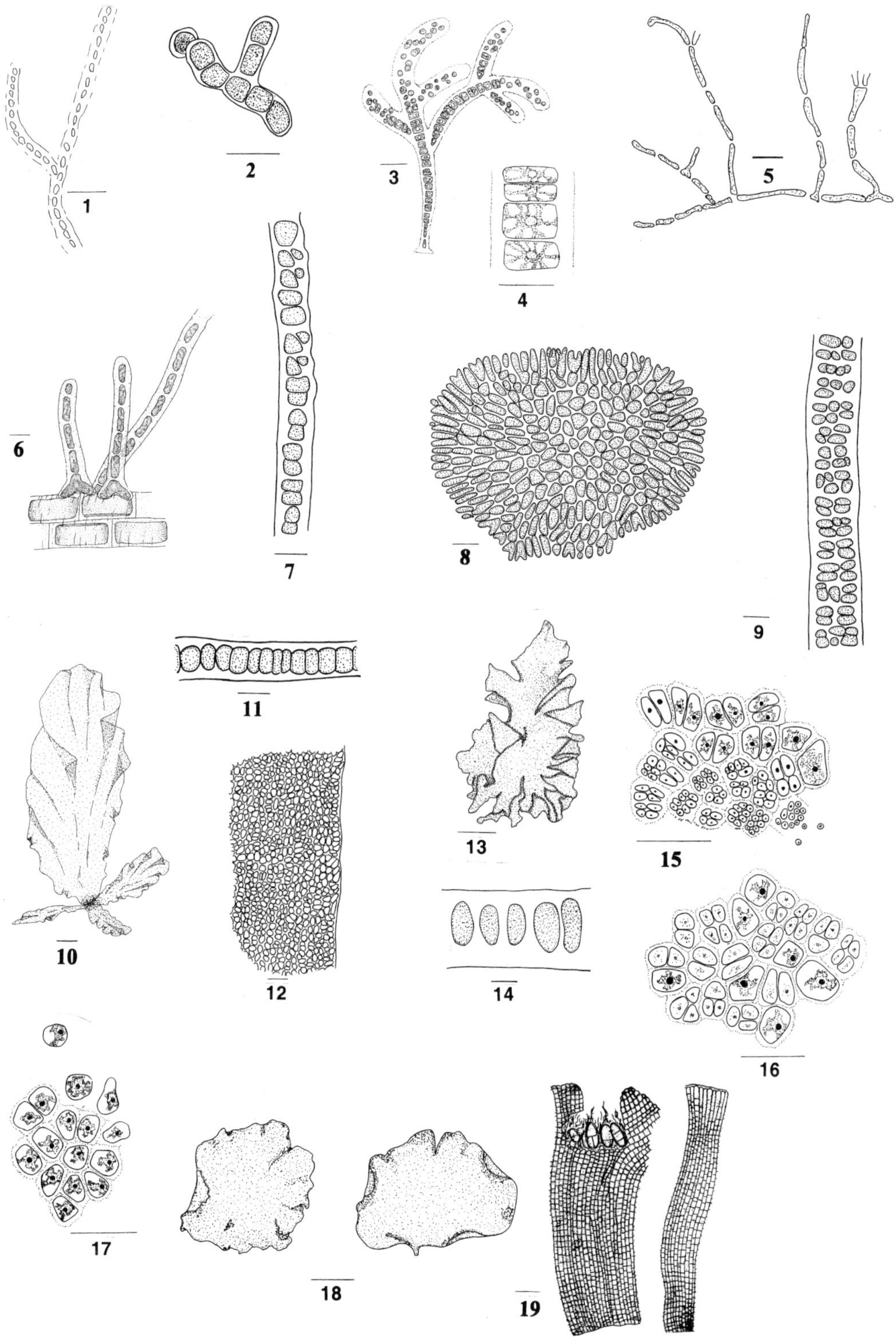
1
2
3
4
5
6
7
8
9
10
11
12
13
14
15
16
17
18
19

PLATE XXIII

Acrochaetium avrainvilleae. 1. Filamentous base and parts of two erect axes; scale 10 μm. 2. Branching axis with two sessile monosporangia; scale 10 μm (both after Børgesen, 1915).

Acrochaetium hallandicum. 3, 4. Filaments with sessile and stalked monosporangia on individual branches; scales 10 μm (3 after Børgesen, 1915; 4, R. Holbrook).

Acrochaetium hypneae. 5. Filaments are epiphytic, uniseriate, branched, and with sessile monosporangia; scale 10 μm (A. L. Paul). 6. A filamentous basal attachment; scale 10 μm. 7. Upper branched axis with sessile seriate monosporangia; scale 20 μm (6, 7 after Børgesen, 1915).

Acrochaetium leptonema. 8. The prostrate filaments have erect axes, monosporangia (*arrow*) and a terminal hair; scale 10 μm. 9. Prostrate filaments growing on an algal surface and bearing erect filaments; scale 10 μm (both after Rosenvinge, 1909).

Acrochaetium liagorae. 10. A prostrate, branched filament with sessile or terminal sporangia and one lateral hair; scale 10 μm (after Børgesen, 1915).

Acrochaetium microscopicum. 11. A prostrate multiseriate epiphyte with setae; scale 10 μm (after Schneider and Searles, 1991). 12. The erect axis has one sessile sporangium (*arrow*), three setae, and a single basal cell; scale 10 μm (after Børgesen, 1915).

Acrochaetium savianum. 13. The upper part of the axis has sessile monosporangia and cruciate tetrasporangia, plus some sporangia with released tetraspores; scale 20 μm. 14. The pseudoparenchymatous base has four erect axes with two bearing monosporangia; scale 50 μm (both after Schneider and Searles, 1991).

Colaconema dasyae. 15. Part of an erect branched axis with sessile monosporangia. 16. Base of three axes with an ovoid cell (*arrow*). 17. Base of four erect axes with an enlarged ovoid cell (*arrow*) and growing within an algal host; all scales 20 μm (all after Schneider and Searles, 1991).

Colaconema ophioglossum. 18. Part of an endophytic filament with a basal spore and a rhizoid. 19. Part of an axis that extends above the host and bears sessile monosporangia, some of which are empty; scales 10 μm (both after Schneider and Searles, 1991).

Lithothamnion heteromorphum. 20. Fragment of an irregularly lobed crust; scale 5 mm (after Foslie, 1929).

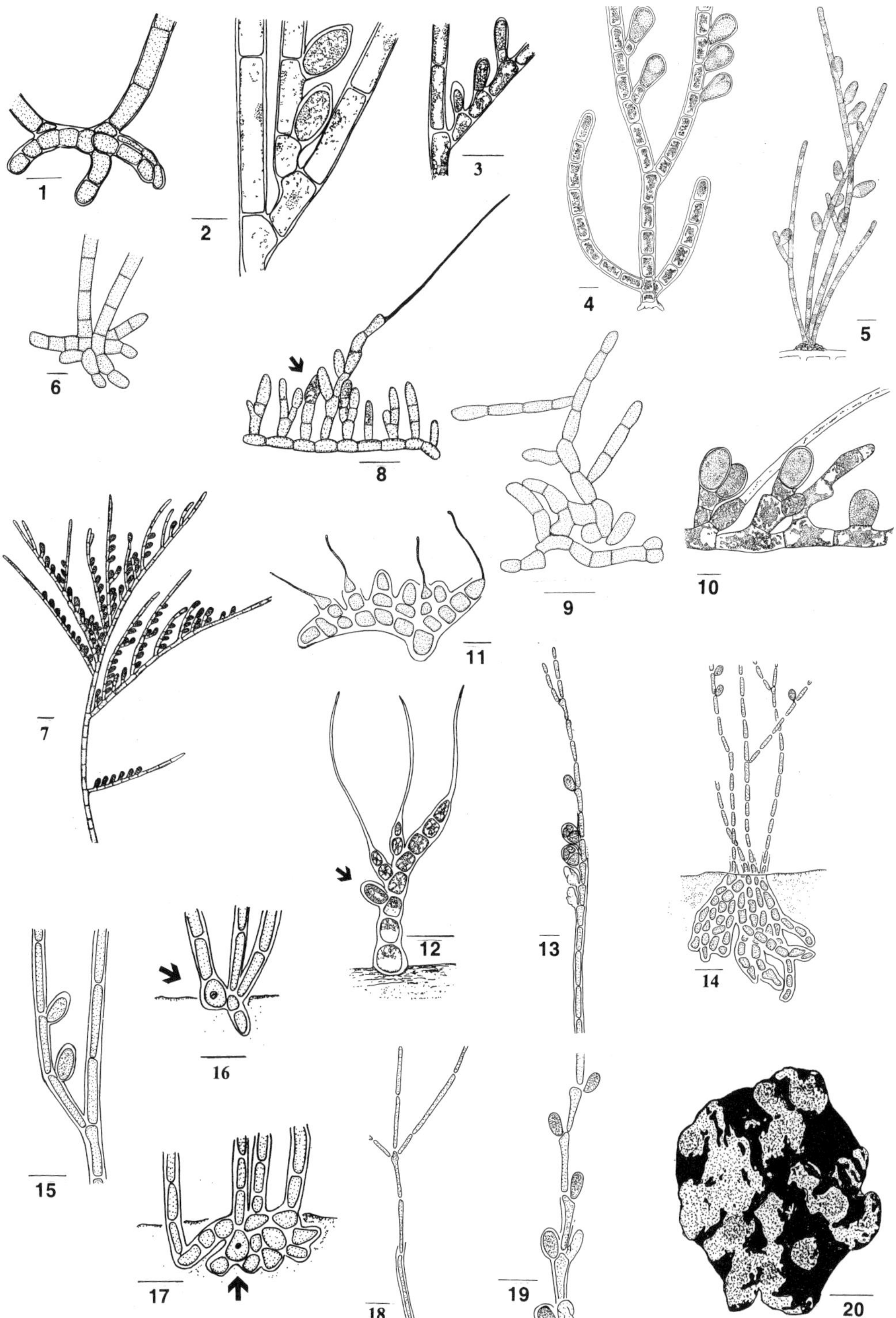
1
2
3
4
5
6
7
8
9
10
11
12
13
14
15
16
17
18
19
20

PLATE XXIV

Lithothamnion occidentale. 1. An unattached crust with irregular branching; scale 1 cm (L. Baumhart). 2. In cross section, the perithallus has vertical rows of rectangular branching cells; scale 20 µm (after Børgesen, 1917).

Melobesia membranacea. 3. Surface view of crust showing cone-shaped conceptacles; scale 500 µm. 4. A vertical section showing perithallial filaments and terminal (small) cap cells; scale 100 µm. 5. A cross section showing a conceptacle with tetrasporangia zonately divided, vertical rows of perithallial filaments, and cap cells; scale 50 µm (all after Irvine and Chamberlain, 1994).

Mesophyllum floridanum. 6. A calcified crust with irregular knobs; scale 1 cm (after Foslie, 1929).

Mesophyllum incertum. 7. Surface view of a rugose to knobby calcified crust overgrowing the coral *Porities porities* (*arrow*); scale 1 cm (after Taylor, 1960).

Mesophyllum mesomorphum. 8. A calcified crust with overlapping scales; scale 1 cm. 9. A cross section showing hypothallial filaments (*arrow*) plus cell fusions and cap cells in the perithallus; scale 50 µm (both after Littler and Littler, 2000).

Mesophyllum syntrophicum. 10. Surface view of thick crusts overgrowing one another and with conspicuous growth rings; scale 1 cm (L. Baumhart).

Hydrolithon boergesenii. 11. Crust is calcified, irregular, and with knobs; scale 5 mm. 12. A cross section near the surface of a crust showing a layer of cap cells and one flask-shaped trichocyte; scales 50 µm. 13. The same, showing the perithallial base with cubical cells below and laterally fused medullary cells above; scales 50 µm. 14. The same, showing basal columnar hypothallial cells (*arrow*) and cubical perithallial cells above; scale 20 µm (11, 14 after Børgesen, 1917; 12, 13 after Littler and Littler, 2000).

Hydrolithon boreale. 15. The germinating spore has four initials (*dark*) and four others around them (*stippled*); scale 25 µm (after Irvine and Chamberlain, 1994). 16. Filaments are free and with spherical trichocytes; scale 50 µm (after Littler and Littler, 2000).

Hydrolithon farinosum. 17. Each calcified crust has a uniporate conceptacle; scale 2 mm (L. Baumhart). 18. Germinating spore with four initials (*shaded*) and 12 others around them; scale 25 µm (after Irvine and Chamberlain, 1994). 19. Cross section of a thick crust with a trichocyte (*center*); scale 10 µm (after Børgesen, 1917).

Lithoporella atlantica. 20. Cross section of a monostromatic crust; scale 100 µm (after Børgesen, 1917).

Neogoniolithon accretum. 21. Surface view of a firmly adherent crust that conforms to the substrata; scale 1 cm (after Taylor, 1928).

Neogoniolithon fosliei. 22. A cross fracture of a crust showing trichocytes (*arrows*) in vertical rows and small basal hypothallial cells; scale 50 µm (after Woelkerling, 1987).

Neogoniolithon mamillare. 23. A crust surface with mammillate-like projections; scale 5 mm (after Foslie, 1929). 24. A cross section showing an indistinct (monomerous) hypothallus and perithallus; scale 50 µm (after Børgesen, 1917).

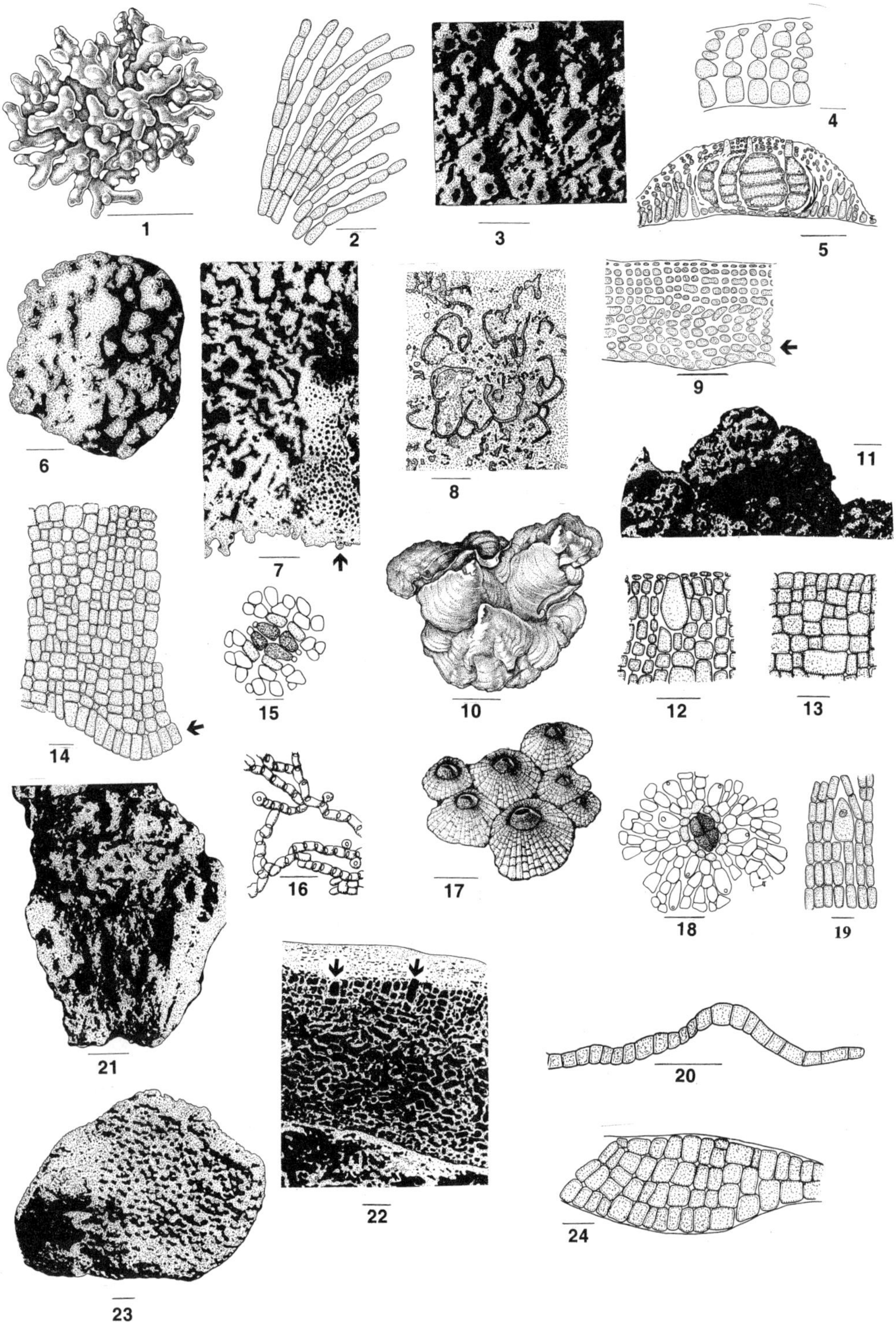
1
2
3
4
5
6
7
8
9
10
11
12
13
14
15
16
17
18
19
20
21
22
23
24

PLATE XXV

Neogoniolithon strictum. 1. The axis is cylindrical, nonarticulate, and with irregular branching; scale 1 cm (after Taylor, 1960). 2. A cross section near surface showing an apical trichocyte; scales 100 μm. 3. A longitudinal section showing lateral fusions between perithallial cells; scale 100 μm (2, 3 after Littler and Littler, 2000).
Pneophyllum fragile. 4. A crust surface showing the original 8-celled cluster (*shaded*) that was formed from a spore; scale 50 μm (after Littler and Littler, 2000). 5. A cross section of a conceptacle with zonately divided tetrasporangia; scale 20 μm (after Schneider and Searles, 1991).
Porolithon pachydermum. 6. A cross section of a stony crust with many embedded conceptacles; scale 200 μm. 7. A cross section at surface with five large terminal trichocytes; scale 50 μm. 8. A cross section of a tetrasporangial conceptacle showing zonately divided tetrasporangia; scale 50 μm (all after Littler and Littler, 2000).
Corallina officinalis. 9. Upper part of an erect flat axis with pinnate branching and cylindrical branchlets; scale 1 mm (after Schneider and Searles, 1991).
Haliptilon cubense. 10. Tip of a sparsely branched axis; scale 5 mm (after Børgesen, 1917). 11. Tips of a compact plant with three conceptacles at forks; scale 5 mm (after Taylor, 1960). 12. A tetrasporangial conceptacle between two branchlets (horns); scale 25 μm (after Schneider and Searles, 1991).
Haliptilon subulatum. 13. Upper part of a compressed axis with opposite lateral branchlets, some have conceptacles between hornlike projections (*arrows*); scale 1 cm (after Taylor, 1960).
Jania adhaerens. 14. Upper part of a dichotomously branched axis with blunt tips; scale 5 mm (after Taylor, 1960).
Jania capillacea. 15. Dichotomously branched axes with conceptacles at forks (*arrow*); scale 1 mm (after Taylor, 1960).
Jania pumila. 16. A small (3 mm tall) epiphytic plant with dichotomous branching, conceptacles at forks, and a discoid holdfast; scale 500 μm (after Børgesen, 1917).
Jania rubens. 17. Part of a vegetative axis with cylindrical, elongate segments and dichotomous branching; scale 2 mm (after Taylor, 1960).
Amphiroa beauvoisii. 18. An upper axes bearing lateral conceptacles on subterete segments; scale 0.5 cm (after Schneider and Searles, 1991). 19. A typical branch has terminal striations and dichotomous forks; scale 2 mm. 20. A geniculum with small spherical cells; scale 100 μm. 21. A longitudinal section showing tiered medullary cells (*right*) and small spherical cortical cells (*left*); scale 100 μm (all after Littler and Littler, 2000).
Amphiroa brasiliana. 22. Upper axes are mostly dichotomously branched, cylindrical, and with conceptacles; scale 1 cm (after Taylor, 1960). 23. Segments of two flat fronds with conceptacles; scale 5 mm (R. Holbrook).
Amphiroa fragilissima. 24. Upper axes are cylindrical and di-{#}or trichotomously branched; scale 2 mm. 25. A segment with hemispherical conceptacles; scale 1 mm. 26. A noncalcified geniculum (joint) and a fork; scale 1 mm (all after Taylor, 1960).
Amphiroa rigida. 27. Upper axes are cylindrical and dichotomously branched; scale 5 mm (after Taylor, 1960). 28. Dome-shaped conceptacles on cylindrical segments; scale 1 mm. 29. A longitudinal section of a segment showing small surface cells and alternating rows of long and short medullary ones; scale 50 μm (28, 29 after Littler and Littler, 2000).

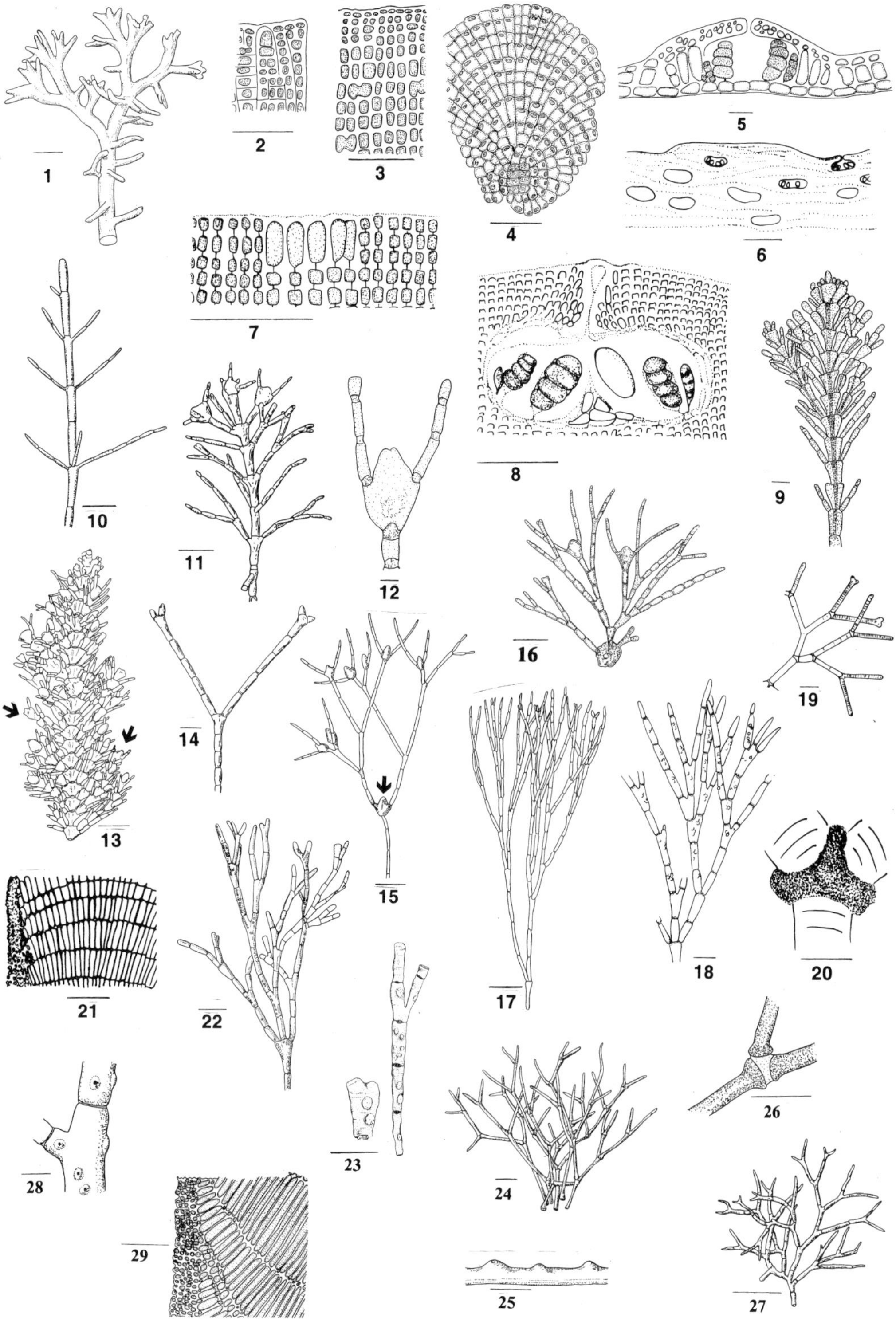
1
2
3
4
5
6
7
8
9
10
11
12
13
14
15
16
17
18
19
20
21
22
23
24
25
26
27
28
29

PLATE XXVI

Amphiroa tribulus. 1. Axes are irregularly branched, subterete, and older segments flat; scale 1 cm. 2. A flat segment with conceptacles on wings; scale 5 cm (both after Taylor, 1960).
Lithophyllum intermedium. 3. The crust has lobed margins; scale 5 mm (after Foslie, 1929). 4. A cross section of a crust showing horizontal hypothallial filaments (*arrow*) and vertical rows of compact perithallial cells; scale 20 µm (after Børgesen, 1917).
Titanoderma bermudense. 5. The crust has overlapping scalelike layers; scale 5 mm. 6. A cross section showing the dimerous construction with horizontal hypothallial filaments (*arrow*) and vertical rows of perithallial cells; scale 20 µm (both after Furnari et al., 1996).
Titoderma prototypum. 7. A crust surface with two flakes (*arrows*); scale 5 mm. 8. A cross section showing four overgrowths, each with a layer of elongate and smaller cap cells; scale 50 µm (both after Littler and Littler, 2000).
Titanoderma pustulatum. 9. A crust surface with dome-shaped conceptacles; scale 1mm. 10. A cross section of a developing conceptacle with compact elongated vertical cells and a surface cover of convex cap cells; scale 25 µm (both after Irvine and Chamberlain, 1994).
Goniolithon decutescens. 11. Part of a crust with erect knobs (5–8 mm long); scale 1 cm (after Foslie and Howe, 1906).
Ganonema dendroideum. 12. The medullary filaments and cortical filaments end in beadlike cells; scale 100 µm (after Littler and Littler, 2000).
Ganonema farinosum. 13. Axes soft, fleshy, and with dichotomous branching; scale 1 cm (after Taylor, 1960). 14. The cortical filaments are hair-tipped and have three rhizoids growing down on axial filament; scale 100 µm (after Littler and Littler, 2000).
Ganonema pinnatum. 15. Branch tip with pinnate branching; scale 2 cm (after Børgesen, 1915). 16. A soft, densely branched plant; scale 2 cm (L. Baumhart).
Helminthocladia calvadosii. 17. Plant is gelatinous and irregularly branched; scale 2 cm (L. Baumhart). 18. Axis has radial, alternate branching; scale 1 cm (after Taylor, 1960).
Liagora albicans. 19. Tip of a densely branched axis; scale 1 cm (after Abbott, 1999). 20. Cortical filaments are from a medullary axis and have a carpogonial branch (*arrow*) with a conspicuous trichogyne; scale 50 µm. 21. Spermatangia are terminal on cortical cells (20, 21 after Littler and Littler, 2000).

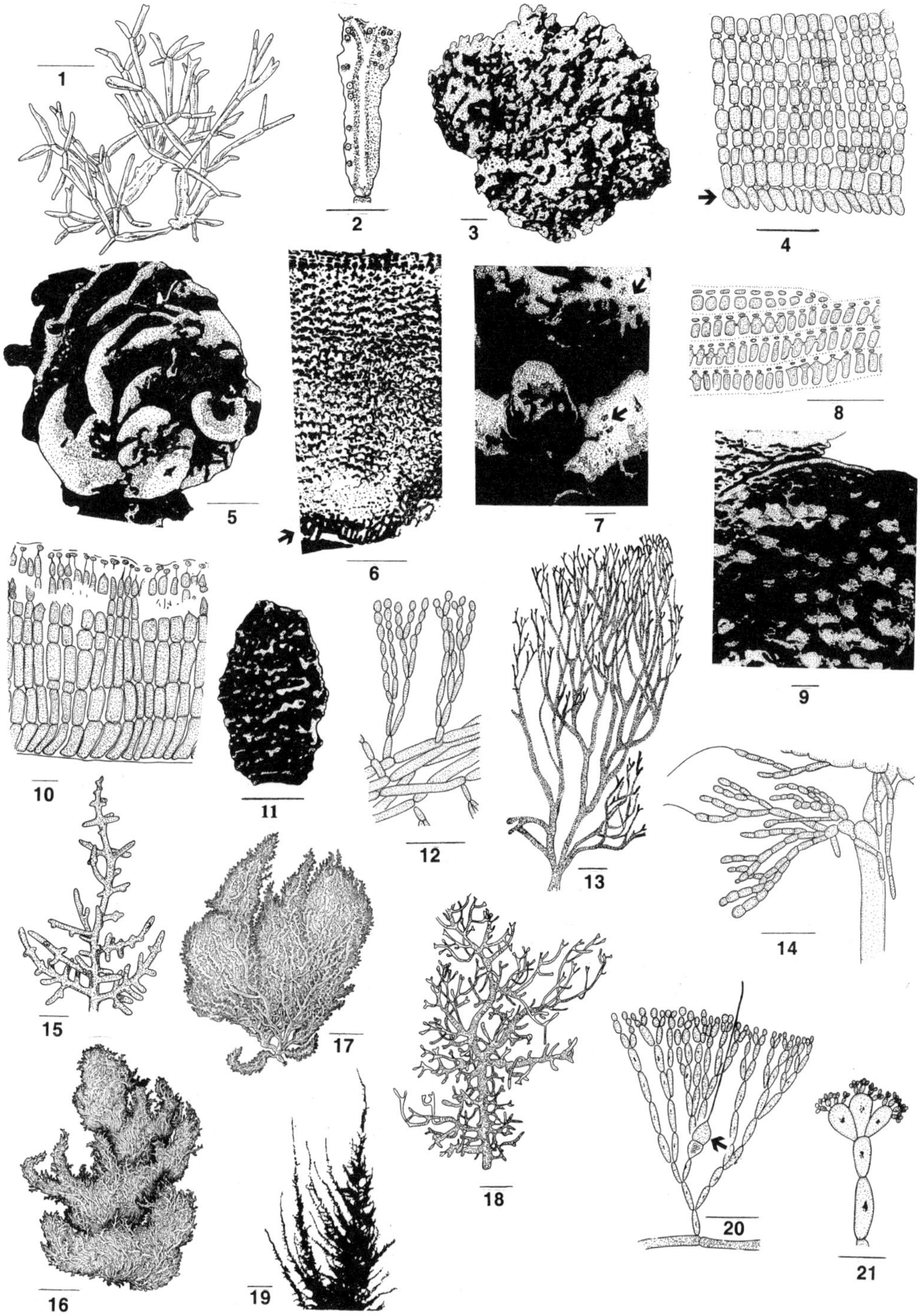
1
2
3
4
5
6
7
8
9
10
11
12
13
14
15
16
17
18
19
20
21

PLATE XXVII

Liagora ceranoides. 1. Upper part of a stiff, dichotomously branched plant; scale 1 cm. 2. The cortical branchlet bears terminal spermatangia; scale 20 μm (both after Taylor, 1960).

Liagora valida. 3. Plant is dichotomously branched, calcified, and with a discoid holdfast; scale 5 cm (A. L. Paul).

Trichogloea requienii. 4. Silhouette of a small gelatinous plant with irregular to pinnate branching and a discoid holdfast; scale 2 cm.

Trichogloeopsis pedicellata. 5. Some branches with soft, pyramidal axes; scale 1 cm. 6. The medullary axis has cortical filaments ending in spherical cells; scale 100 μm (both after Littler and Littler, 2000).

Dichotomaria marginata. 7. A plant with compressed dichotomously branched axes, cross banding and a discoid holdfast; scale 2 cm (L. Baumhart). 8, 9. Cross sections showing the sporophytic cortex (8) with stalked elliptical cells and gametophytic cortex (9) with columnar cells; scales 20 μm (both after Taylor, 1960).

Dichotomaria obtusata. 10. The plant has dichotomous branching, constrictions, transverse banding, and a discoid holdfast; scale 2 cm (after L. Baumhart in Dawes, 1981). 11, 12. Cross sections showing the sporophytic cortex with stalked subsurface cells (11) and gametophytic cortex (12) with sessile surface cells and swollen subsurface ones; scales 50 μm (both after Huisman and Borowitzka, 1990).

Galaxaura rugosa. 13. Plant is young, stiff, dichotomously branched, and covered with whorls of hairlike filaments; scale 1 cm (after Taylor, 1960). 14. An older dichotomously branched plant with smooth, hairless axes and a discoid holdfast; scale 1 cm. 15, 16. Cross sections showing a gametophytic cortex (15) with fused polyhedral surface cells and sporophytic cortex with a filamentous surface (16); scales 25 μm (15) and 50 μm (16); 14–16 after Huisman and Borowitzka, 1990.

Tricleocarpa cylindrica. 17. Upper axis is dichotomously branched and segmented; scale 10 mm. 18. A cross section of the cortex showing inflated subsurface cells and fused polygonal surface cells; scale 25 μm (both after Huisman and Borowitzka, 1990).

Tricleocarpa fragilis. 19. Plant is dichotomously branched and with a rhizoidal holdfast; scale 2 cm (A. L. Paul). 20. Tips are cylindrical and dichotomously branched; scale 2 mm (after Littler and Littler, 1997. 21. A cross section showing the branched cortical filaments with flat surface cells; scale 20 μm (after Taylor, 1928).

Scinaia caribaea. 22. Upper portion of a cylindrical to compressed axis with dichotomous branching and constrictions at forks; scale 3 mm (after Littler and Littler, 2000).

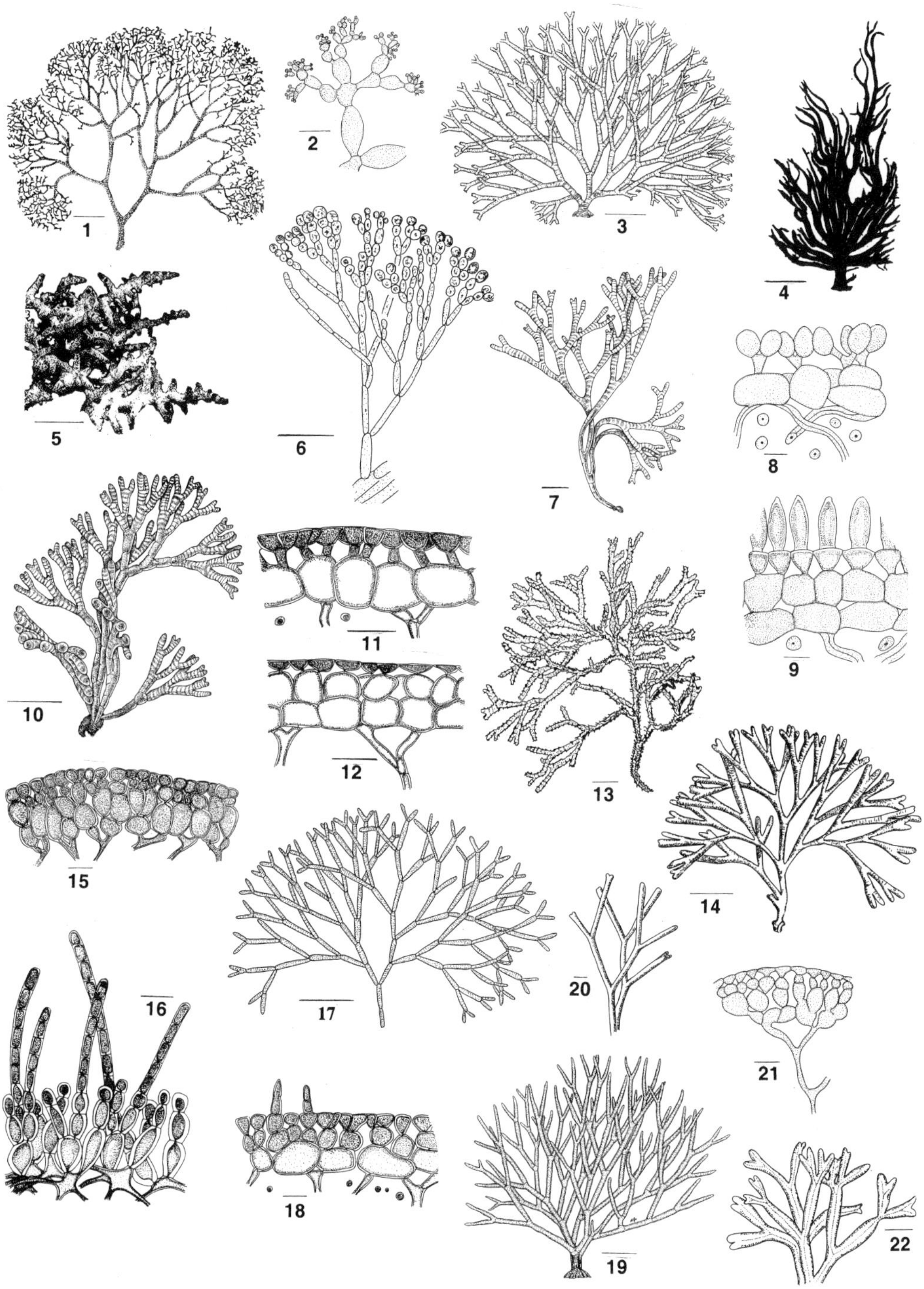
1
2
3
4
5
6
7
8
9
10
11
12
13
14
15
16
17
18
19
20
21
22

PLATE XXVIII

Scinaia complanata. 1. Plant is dichotomously branched, slightly constricted at forks, and with a discoid holdfast; scale 1 cm (L. Baumhart). 2. A cross section showing the fused polygonal surface cells; scale 25 μm (after Setchell, 1914).

Scinaia halliae. 3. The plant has dichotomous branching, cylindrical axes, and a discoid holdfast; scale 1 cm (L. Baumhart). 4. The globular surface cells alternate with pigmented cortical filaments from the outer medulla; scale 10 μm (after Setchell, 1914).

Scinaia incrassata. 5. Upper part of a flattened branch with inflated margins; scale 5 mm (after Eiseman, 1979).

Rhodogorgon ramosissima. 6. Upper part of a fleshy erect plant with irregular branching; scale 2 cm. 7. Cortical filaments with two calciferous cells (*arrows*); scale 50 μm (6, 7 after Norris and Bucher, 1989). 8. Branched cortical filaments; scale 50 μm (after Littler and Littler, 2000).

Pihiella liagoraciphila. 9. Possible carpogonia with trichogynes; scale 20 μm (after Børgesen, 1920). 10. A "sporeling" with a young discoid thallus; scale 20 μm (after Huisman et al., 2003). 11. Surface view of a discoid plant with possible carpogonia and trichogynes; scale 20 μm (after Howe, 1920b).

Asparagopsis taxiformis. 12. Part of two filamentous sporophytes with "polysiphonous-like" construction; scale 100 μm (after Littler and Littler, 2000). 13. Tip of a "polysiphonous-like" sporophytic filament with a branch; scale 20 μm. 14. A cross section of a sporophyte axis showing 3 pericentral cells; scale 20 μm (13, 14, R. Holbrook). 15. The gametophyte has a prostrate base, erect axis, and dense alternate branching to several orders; scale 1 cm (after Taylor, 1960).

Naccaria corymbosa. 16. The plant is bushy, alternately branched, and with a small discoid holdfast; scale 5 mm (after Schneider and Searles, 1991).

Aglaothamnion cordatum. 17. Plant is densely branched and with a discoid holdfast; scale 1 cm. 18. The upper axis is uniseriate and pseudodichotomously branched; scale 300 μm. (17, 18, L. Baumhart).

Aglaothamnion halliae. 19. Axial tip is uniseriate and alternately distichously branched; scale 200 μm. 20. Part of an upper ecorticate axis with alternate branching; scale 300 μm (both, A. L. Paul).

Aglaothamnion herveyi. 21. Branch on the main axis with alternate, radial branching; scale 100 μm. 22. An older axis with a few cortical rhizoids; scale 40 μm (both, A. L. Paul).

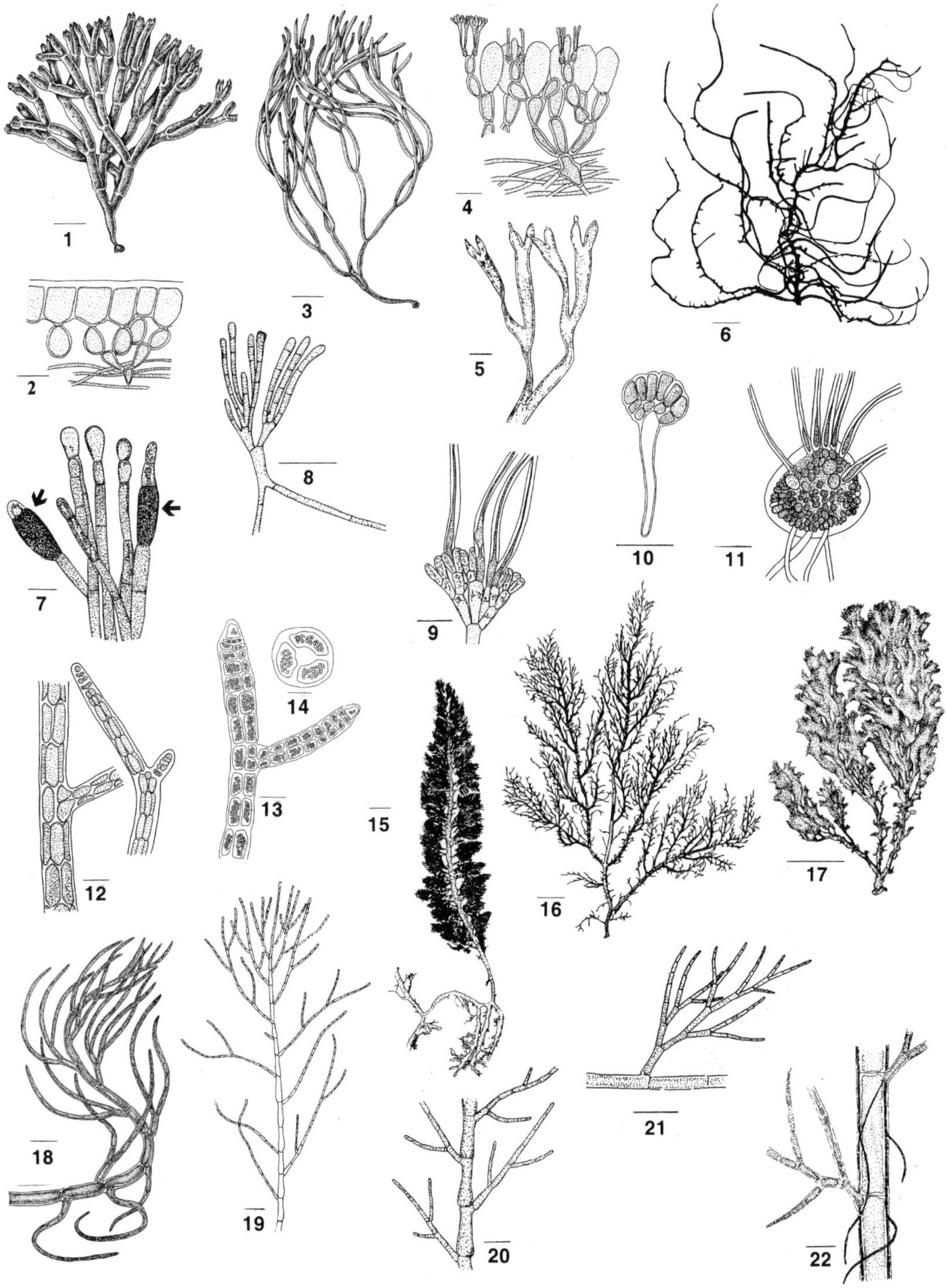
1
2
3
4
5
6
7
8
9
10
11
12
13
14
15
16
17
18
19
20
21
22

PLATE XXIX

Aglaothamnion uruguayense. 1. Part of an old axis with repent branchlets; scale 100. 2. Tip with alternate branching; scale 100 μm. 3. Uniseriate tip with branchlets in one plane and adaxial tetrasporangia; scale 100 μm. 4. Tetrahedrally divided tetrasporangia in an adaxial series on branchlets; scale 50 μm (1–3 after Taylor, 1960).

Anotrichium barbatum. 5. Part of a prostrate axis with one rhizoid and an erect branch; scale 50 μm (A. L. Paul). 6. The branch tip has bulbous cells and whorls of trichoblasts; scale 50 μm (after Feldmann-Mazoyer, 1941).

Anotrichium tenue. 7. The erect axial tip has bulbous cells and whorls of dichotomously branched slender trichoblasts; scale 50 μm (after Schneider and Searles, 1991).

Antithamnion antillanum. 8. Branchlet is uniseriate and has one cruciate tetrasporangium; scale 100 μm. 9. The prostrate axis is uniseriate, has thick-walled rhizoids with pads, and two erect axes that vary in size and have alternate branching; scale 200 μm (both after Børgesen, 1917).

Antithamnion cruciatum. 10. The tip of a uniseriate branch has opposite branching and two gland cells (*arrow*); scale 200 μm. 11. Axis is uniseriate and with whorled branching; scale 200 μm (A. L. Paul). 12. Branchlet with stalked cruciate tetrasporangia and two small adaxial gland cells (*left side*); scale 50 μm (after Schneider and Searles, 1991).

Antithamnion decipiens. 13. The tip of an axis has opposite to whorled branching and adaxial gland cells (*arrow*); scale 100 μm (after Athanasiadis, 1996a).

Antithamnionella atlantica. 14. The prostrate axis has multicellular rhizoids and erect axes with opposite branching and gland cells; scale 20 μm (after Schneider and Searles, 1991).

Antithamnionella boergesenii. 15. Uniseriate prostrate axis bears an erect filament with whorled branchlets and multicellular rhizoids with basal pads. 16. Branchlet tip with a gland cell; scales 20 μm (both after Børgesen, 1930a).

Antithamnionella graeffei. 17. The prostrate uniseriate axis has branched multicellular rhizoids and an erect axis with opposite branching and gland cells (*arrow*); scale 25 μm (after Schneider and Searles, 1991).

Balliella pseudocorticata. 18. Tip of an erect uniseriate axis with opposite branching; scale 100 μm (after Harvey, 1840). 19. An older axis with a branchlet and clasping cortical cells; scale 50 μm (after Dawson, 1962).

Callithamniella tingitana. 20. The prostrate axis has simple rhizoids and erect uniseriate axes; the largest erect one (*left*) has cruciate tetrasporangia (*arrow*) and conspicuous alternate branching; scale 50 μm (after Schneider and Searles, 1991).

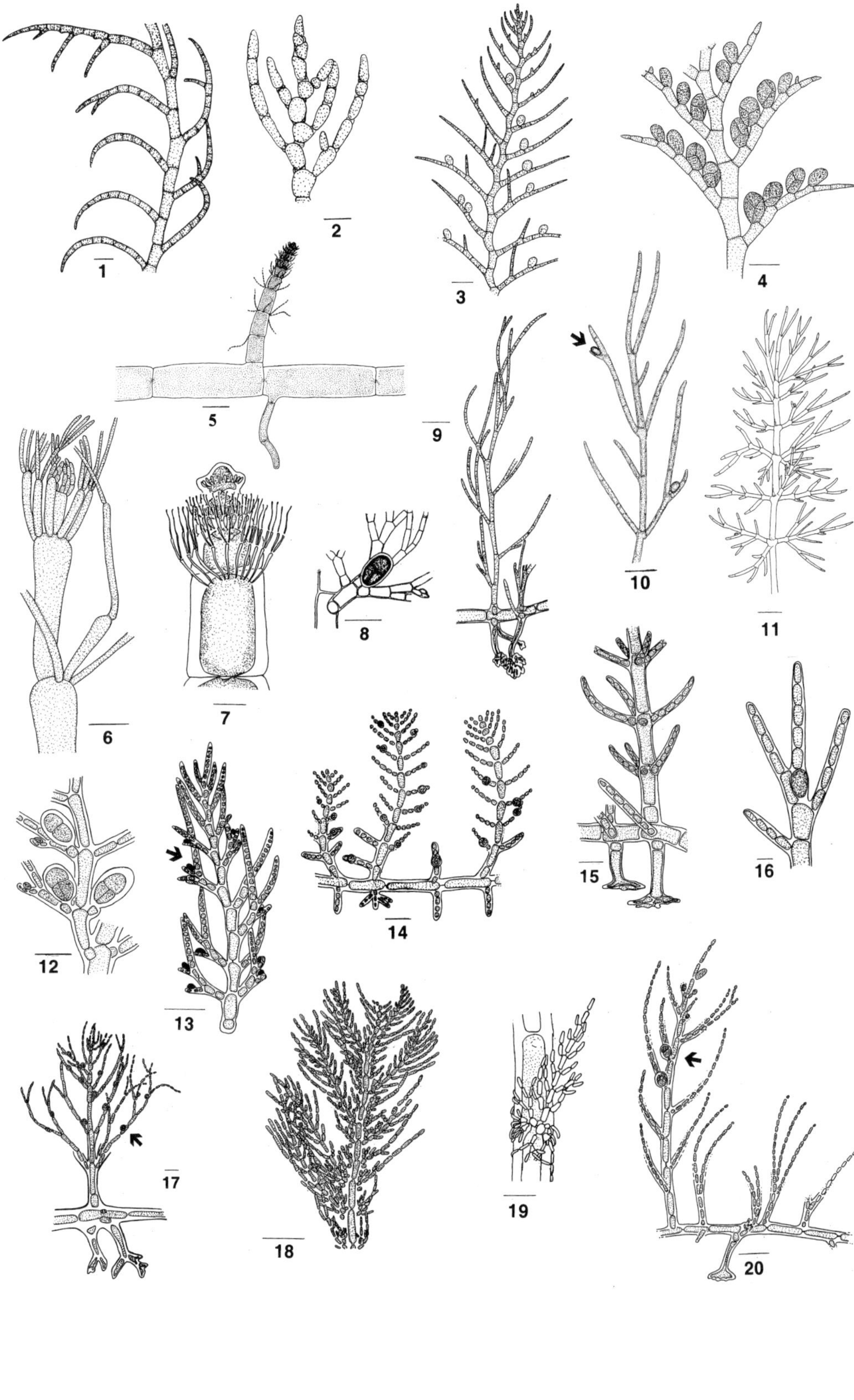
1
2
3
4
5
6
7
8
9
10
11
12
13
14
15
16
17
18
19
20

PLATE XXX

Ceramium brevizonatum var. *caraibicum*. 1. The axis has an incurved branchlet and is corticated by three rows of nodal cells with larger central (periaxial) cells; scale 50 µm (after Littler and Littler, 2000). 2. A younger axis with smaller cortical bands; scale 100 µm (Gary Landtrip).

Ceramium cimbricum. 3. The older axis has 6 rows of cortical cells; scale 25 µm (after Schneider and Searles, 1991). 4, 5. Younger axes with 2 rows of cortical cells; 4: scale 25 µm (after Taylor, 1960) 5: scale 50 µm (Gary Landtrip).

Ceramium codii. 6. Older axial nodes with two irregular rows of cortical cells; scale 20 µm. 7. The periaxial cell (*P*) has produced 2 acropetal cells and 1 basipetal cortical cell; scale 5 µm. 8. A cross section at the axial node reveals the axial cell (*AX*), 4 periaxial cells, and 8 cortical cells; scale 20 µm (all after Cho and Fredereicq, 2006).

Ceramium corniculatum. 9. Nodes at the axial tip are corticated by four rows of small cells with nodes almost confluent; scale 50 µm (after Littler and Littler, 2000).

Ceramium cruciatum. 10. Prostrate axis has rhizoids with pads, an erect filament base, and uneven, elongate nodal cells; scale 100 µm (after Collins and Hervey, 1917).

Ceramium deslongchampsii. 11. Tip is pincer-like and a protruding tetrasporangium (*arrow*) is on a lower node; scale 100 µm. 12. The branch is dichotomously branched and has pincer-like tips and bands of nodal cortication; scale 1 cm (11, 12, A. L. Paul). 13. An older axis has nodal cells in irregular rows and larger cells in the center and below; scale 50 µm (after Børgesen, 1918).

Ceramium diaphanum. 14. An older node has projecting hairs and more than four irregular cell rows with the middle (periaxial) cells the largest; scale 20 µm (after Taylor, 1957). 15. Tip is pincer-like and has nodes corticated by three irregular cell rows with central (periaxial) ones the largest; scale 25 µm (after Schneider and Searles, 1991).

Ceramium floridanum. 16. The axis is mostly pinnately branched and in one plane; scale 500 µm. 17. The tip is forking and has four exposed cruciate tetrasporangia; scale 200 µm. 18. Nodes have four irregular rows of angular cells; scale 100 µm (all, A. L. Paul).

Ceramium nitens. 19. The axis is completely corticated, the nodal banding is obscured, and has projecting hairs; scale 50 µm (after Taylor, 1960). 20. Cross section of an internode showing a large axial cell (*AX*), an inner layer of periaxial cells and two outer cortical layers; scale 100 m (after Littler and Littler, 1997).

Ceramium reptans. 21. The mature erect axis has two rows of cortication; scale 20 µm. 22. A periaxial cell (*P*) produces two acropetal cells and one basipetal cell; one acropetal cell has a two-celled filament; scale 10 µm. 23. A nodal cross section showing the axial cell (*AX*), five periaxial cells (*P*), and cortical cells; scale 20 µm (all after Cho and Fredericq, 2006).

Ceramium subtile. 24. An older node with irregular small cells above and below a central row of larger periaxial cells; scale 50 µm (after Taylor, 1928).

Ceramium virgatum. 25. Silhouette of a densely branched plant with dichotomous to alternate branching and pincer-like tips; scale 1 cm (after Hillson, 1977). 26. The tip is completely corticated; scale 20 µm (after Taylor, 1957). 27. The cortication completely covers all axes and almost obscures the nodal bands; scale 200 µm (after Schneider and Searles, 1991).

Ceramium flaccidum. 28. Axis is dichotomously branched and has distinct nodal cortication; scale 1 cm (F. Hoff). 29. The mature node has two periaxial cells (*P*) and five irregular rows of cells; scale 25 µm (after Taylor, 1928). 30. Two mature nodes each have a periaxial cell (*P*) and four irregular cortical bands (3 acropetal, 1 basipetal); scale 20 µm.

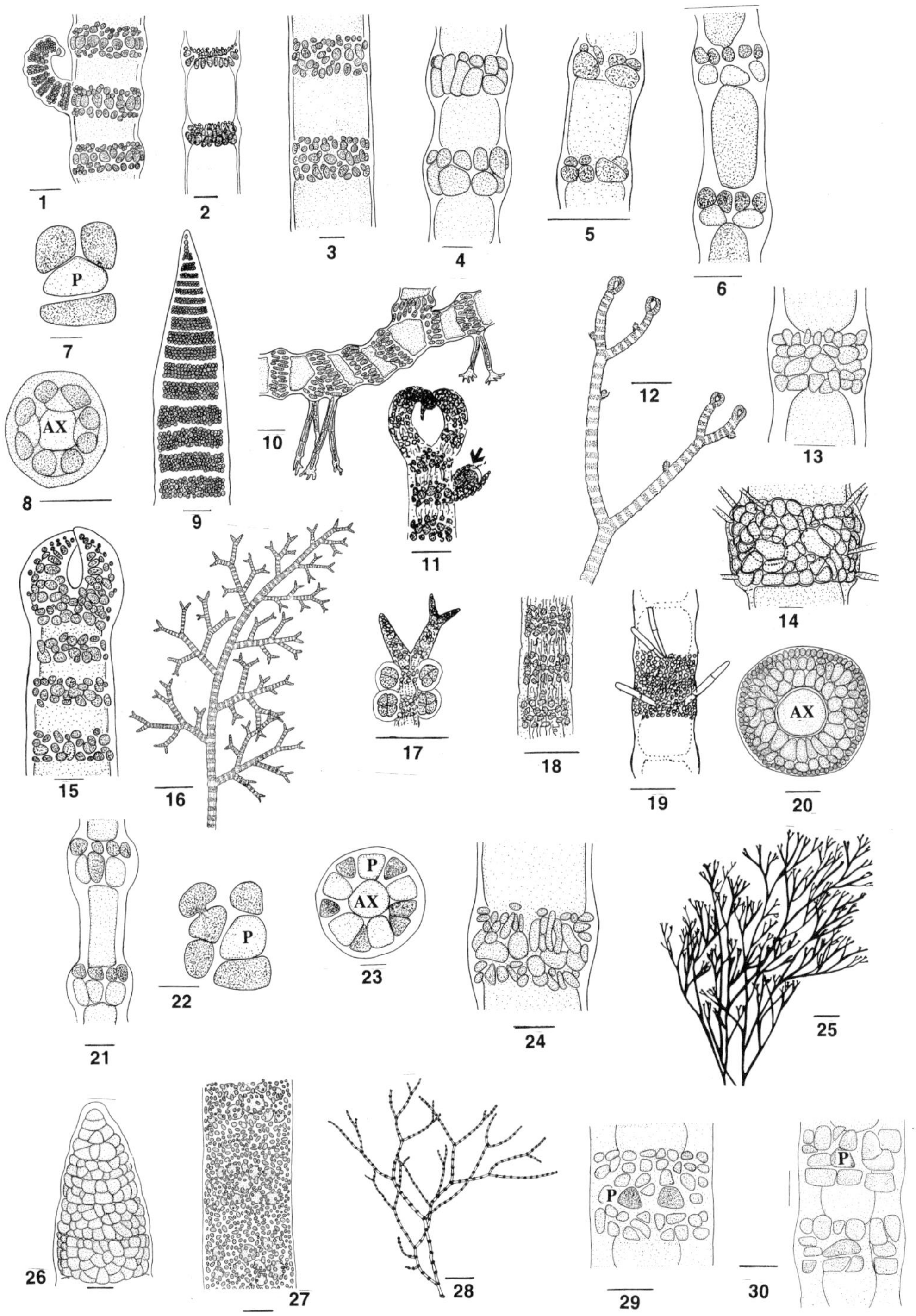
1
2
3
4
5
6
P
7
AX
8
9
10
11
12
13
14
15
16
17
18
19
AX
20
21
P
22
P
AX
23
24
25
26
27
28
P
29
P
30

PLATE XXXI

Compsothamnion thuyoides. 1. Part of a uniseriate main axis with branchlets alternating in one plane, tetrasporangia, and descending rhizoids; scale 200 µm (after Schneider and Searles, 1991).
Crouania attenuata. 2. Upper part of a main axis is alternately branched and with whorled branchlets; scale 5 mm. 3. The unseriate main axis with whorls of upward curving branchlets; scale 100 µm (both, R. Holbrook).
Crouania elisiae. 4. A cross section of main axis with a whorl of branchlets perpendicular to the axis; scale 50 µm (after Schneider, 2004).
Crouania pleonospora. 5. Part of a uniseriate main axis with whorls of lateral branchlets and two tetrasporangia (*arrow*); scale 50 µm (after Littler and Littler, 2000).
Diplothamnion jolyi. 6. Part of a uniseriate prostrate axis with lobed unicellular rhizoids; scale 150 µm. 7. Upper part of a uniseriate erect axis with 2–4 branchlets per axial cell and several tetrasporangia; scale 100 µm (both after van den Hoek, 1978).
Grallatoria reptans. 8. Silhouette showing part of a creeping plant with multicellular rhizoids and erect axes; scale 500 µm. 9. The prostrate axis has multicellular digitate rhizoids plus branched and unbranched erect axes; scale 100 µm. 10. A tip of an erect axis with stalked tetrasporangia (*arrow*); scale 50 µm (after Abbott, 1976).
Griffithsia globulifera. 11. The axial tips have bulbous cells; scale 5 mm (F. Hoff). 12. The tip has whorls of trichotomously branched hairs and bulbous cells; scale 100 µm (after Schneider and Searles, 1991).
Griffithsia heteromorpha. 13. The forked filament has bulbous cells; scale 500 µm (after Abbott, 1999). 14. A bulbous apical cell has a whorl of tetrasporangia (*arrow*) clustered around the node; scale 200 µm (after Littler and Littler, 1997).
Gymnothamnion elegans. 15. Part of a prostrate axis with digitate, multicellular rhizoids and two erect, pinnately branched, uniseriate axes; scale 100µm (after Abbott, 1999).
Monosporus indicus. 16. A silhouette showing part of a prostrate plant with multiple erect determinate branches and one indeterminate branch (*middle*); scale 3 mm (after Abbott, 1999). 17. The uniseriate prostrate axial tip has sympodial growth and erect branches arising opposite rhizoids; scale 200 µm. 18. A rhizoid that has a digitate, multicellular pad and is opposite an erect axis; scale 200 µm (17, 18 after Ballantine, 1996).
Perikladosporon percurrens. 19. The erect uniseriate axis has opposite branching and one tetrasporangium (*arrow*); scale 50 µm. 20. Tip of a laxly branched axis; scale 100 µm (19, 20 after Dawson, 1957). 21. Part of a uniseriate prostrate axis with rhizoidal clusters opposite erect oppositely branched axes; scale 500 µm (after Abbott, 1999).
Pleonosporium flexuosum. 22. A uniseriate axial tip with alternate branching and a few mature polysporangia; scale 200 µm. 23. A mature polysporangium (*arrow*) on the adaxial side of a branchlet; scale 100 µm (after Schneider and Searles, 1991).

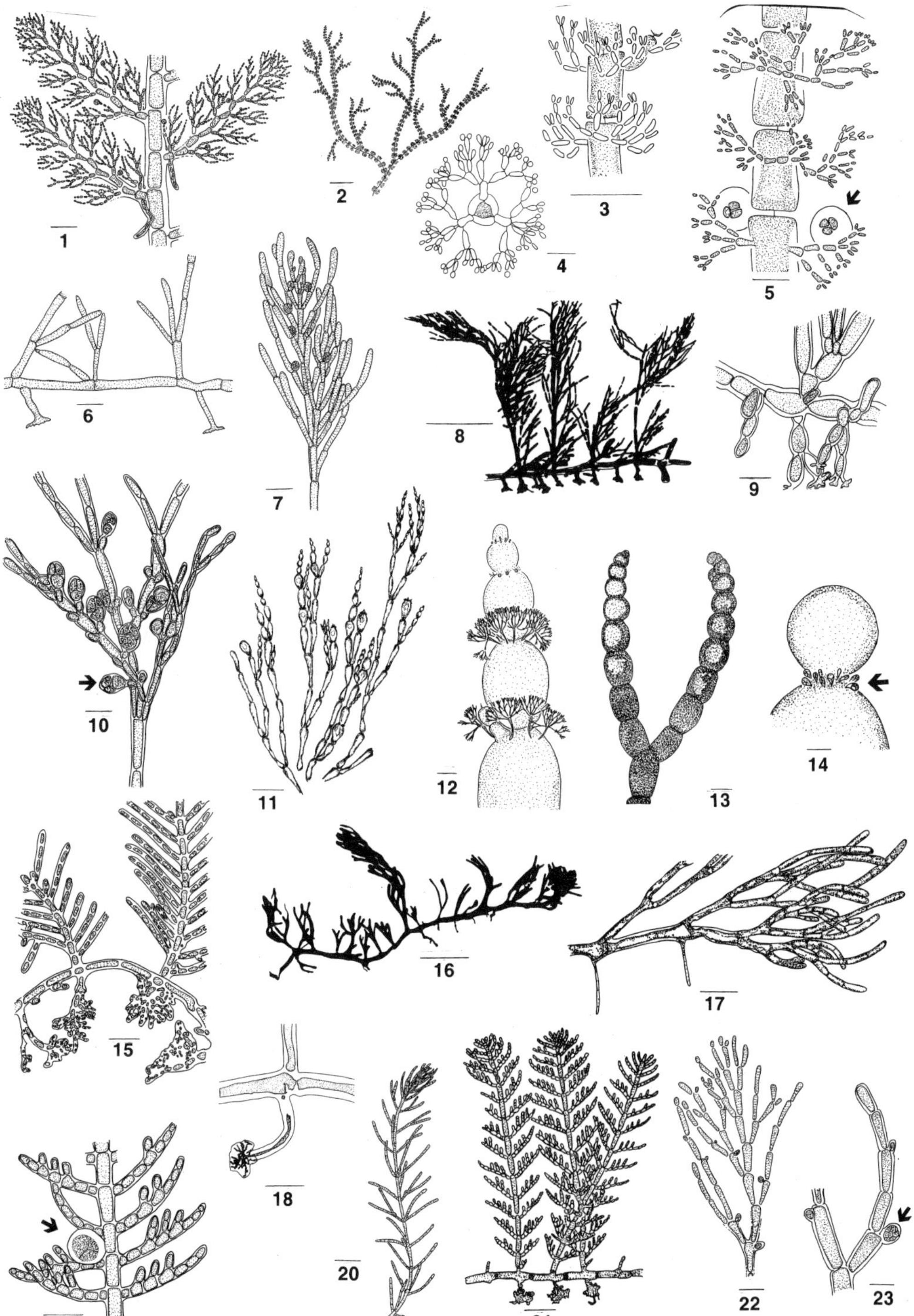
1
2
3
4
5
6
7
8
9
10
11
12
13
14
15
16
17
18
19
20
21
22
23

PLATE XXXII

Pterothamnion plumula. 1. Upper part of a uniseriate axis with opposite branchlets in one plane, many tetrasporangia, and some smaller gland cells on small stalked branchlets; scale 100 µm. 2. A branchlet with one terminal tetrasporangium and three lateral gland cells; scale 50 µm (both after Feldmann-Mazoyer, 1941).

Ptilothamnion speluncarum. 3. Basal part of a uniseriate prostrate axis with two erect filaments opposite rhizoids and a tip of an erect filament (*middle*); scale 100 µm (after Børgesen, 1930a).

Seirospora occidentalis. 4. An erect uniseriate tip with alternate branching and rows of 3–4 seirospores (*arrow*); scale 100µm (after Børgesen, 1917).

Seirospora purpurea. 5. Axial tip has alternate to subdichotomous branching and is uniseriate; scale 50 µm. 6. Main axis is uniseriate and has descending corticating rhizoids from bases of branchlets; scale 100 µm (both after Littler and Littler, 2000).

Spermothamnion gymnocarpum. 7. Part of a uniseriate prostrate axis with rhizoidal pads and erect filaments that arise at one end of parent cell; scales 100 µm. 8. A branch tip with unilateral branchlets; scale 100 µm. 9. The uniseriate erect axis has one tetrasporangium on a unicellular stalk; scale 50 µm (all after Littler and Littler, 2000).

Spermothamnion investiens. 10. The uniseriate prostrate axis has unicellular rhizoids and erect sparingly branched upright filaments; scale 100 µm. 11. The erect uniseriate axis has two lateral tetrasporangia; scale 50 µm (both after Børgesen, 1920).

Spermothamnion repens. 12. The uniseriate prostrate axis has unicellular rhizoids, erect pinnately branched filaments, and tetrasporangia (*arrow*); scale 200 µm (after Feldmann-Mazoyer, 1941).

Spyridia clavata. 13. The axial tip has club-shaped final branches, uniseriate branchlets, and nodal bands visible through the cortication; scale 20 µm (after Joly, 1965).

Spyridia filamentosa. 14. Branchlet tip bears a spine and nodal cortication by a single row of cells; scale 25 µm (after Taylor, 1960). 15. Tips of axes have hairlike branchlets and nodal bands visible through the cortication; scale 1 cm (F. Hoff).

Spyridia hypnoides. 16. The upper axes are covered by uniseriate branchlets and end in hooked tips; scale 500 µm (after Schneider and Searles, 1991). 17. Branchlet tip with an erect spine, a lateral hook, and single rows of cortical cells at the nodes; scale 25 µm (after Taylor, 1960).

Tiffaniella gorgonea. 18. The uniseriate prostrate plant is epiphytic on algae and has erect filaments with polysporangia (*arrow*) and swollen multicellular rhizoids (*left*); scale 1 mm (after Børgesen, 1930a).

Tiffaniella saccorhiza. 19. The uniseriate prostrate axes is epiphytic on algae and has erect unbranched filaments (*top*) and swollen unicellular rhizoids (*bottom*); scale 1 mm (after Doty and Meñez, 1960).

Wrangelia argus. 20. The uniseriate axis has corticating rhizoidal branchlets; scale 100 µm (after Taylor, 1960). 21. The branchlet on a uniseriate axis is spinelike and produces corticating rhizoids; scale 50 µm (after Børgesen, 1916).

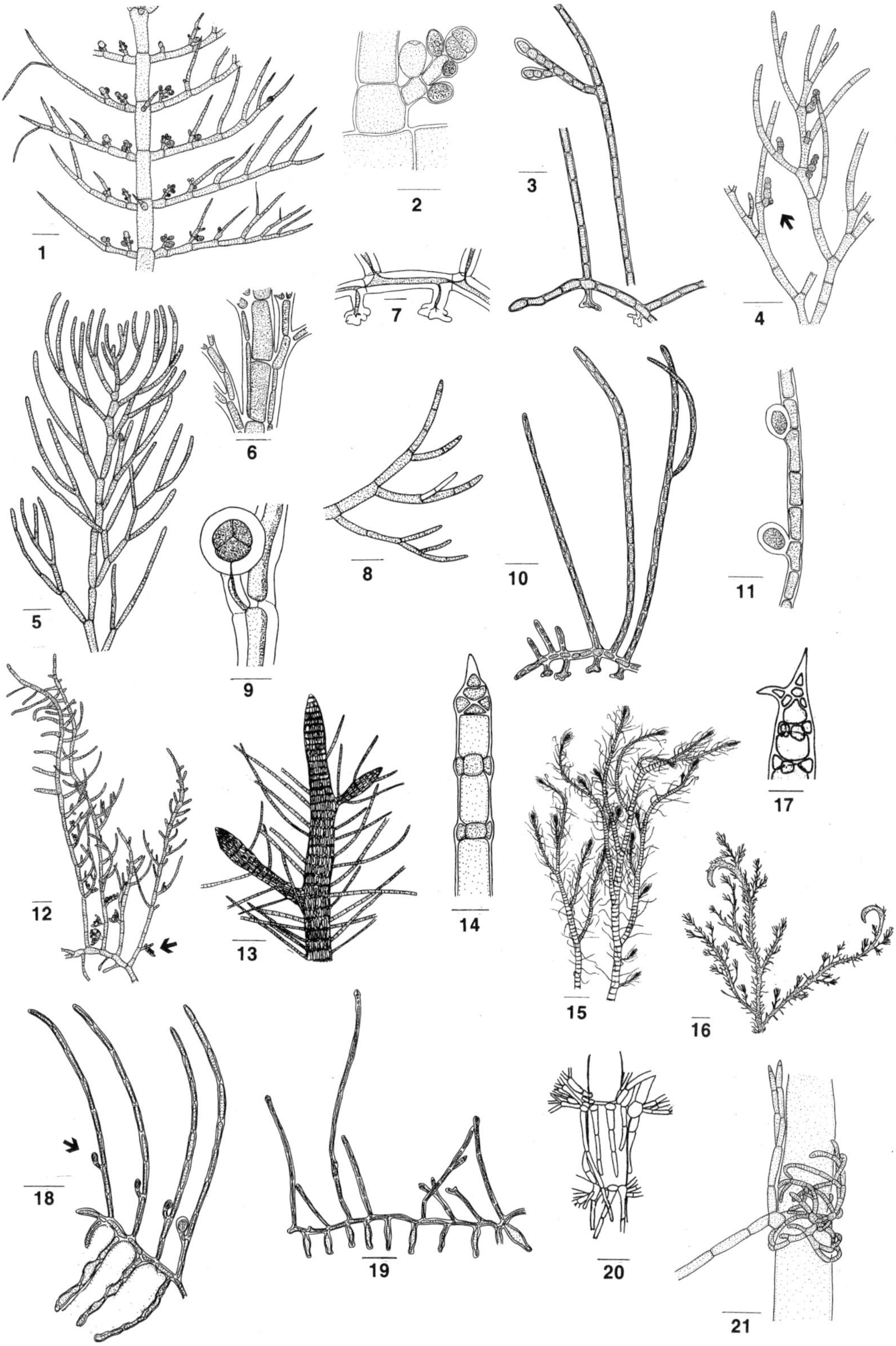
1
2
3
4
5
6
7
8
9
10
11
12
13
14
15
16
17
18
19
20
21

PLATE XXXIII

Wrangelia bicuspidata. 1. Branchlet tip is uniseriate and each tip cell has two spines; scale 100 μm (after Børgesen, 1916). 2. Nodes of the uniseriate axis bear whorled branchlets that produce corticating rhizoids; scale 200 μm (R. Holbrook).

Wrangelia penicillata. 3. Uniseriate main axis has whorled branchlets, initial rhizoidal cortication, and a tetrasporangium (*arrow*); scale 100 μm (R. Holbrook). 4. The old axis is completely corticated by down-growing rhizoids; scale 100 μm. 5. Upper part of plant is plumose; scale 2 mm (4, 5 after Taylor, 1960).

Apoglossum ruscifolium. 6. The frond tip has many secondary blades arising from midribs; scale 5 μm (after Schneider and Searles, 1991).

Branchioglossum minutum. 7. Five ovate blades arising from a rhizoidal holdfast; scale 1 mm (after Schneider and Searles, 1991). 8. The blade margin produces rhizoids; scale 100 μm (after Littler and Littler, 2000). 9. Blade apex has one apical cell with the lateral cell rows extending to margins; scale 10 μm (after Schneider and Searles, 1991).

Branchioglossum prostratum. 10, 11. Two blade morphologies. 10. The broad frond lacks marginal branching; scale 1 mm (after Abbott, 1999). 11. The narrow fronds has marginal branching; scale 2 mm (after Schneider and Searles, 1991).

Calloseris halliae. 12. The monostromatic blade margin has one apical cell at each acute tip; scale 10 μm. 13. Upper part of a tetrasporic plant with marginal proliferations; scale 2 mm (both after Schneider and Searles, 1991).

Caloglossa leprieurii. 14. Each frond tip has one apical cell; scale 20 μm (after Schneider and Searles, 1991). 15. Frond has dichotomous branching with constrictions at forks; scale 2 mm (after Taylor, 1960). 16. The axial cell at the junction of a lateral branch (*right*) produces four rows of cells running to the edge of the blade (*bracket, arrow*); scale 100 μm (after Kamiya et al., 1997).

Caloglossa monosticha. 17. Fronds are constricted at forks and have an obvious axial row; scale 2 mm. 18. The axial cell at the fork of a lateral branch (*right*) produces one row of cells running to the blade edge (*arrow*); scale 100 μm (both after Kamiya et al., 1997).

Calonitophyllum medium. 19. Frond is narrow and has veins at the base and a discoid holdfast; scale 1 cm (after Schneider and Searles, 1991). 20. Silhouette of a lobed frond; scale 1 cm (after photo of W. R. Taylor's specimen from St. Kitts).

Cryptopleura ramosa. 21. Frond tips have wide lobes and multiple veins; scale 1 cm (after Taylor, 1960).

Grinnellia americana. 22. Three blades arising from a discoid holdfast; blades have a midrib and ruffled edges; scale 2 cm (L. Baumhart). 23. Blade tip is monostromatic and has one apical cell; scale 10 μm (after Schneider and Searles, 1991).

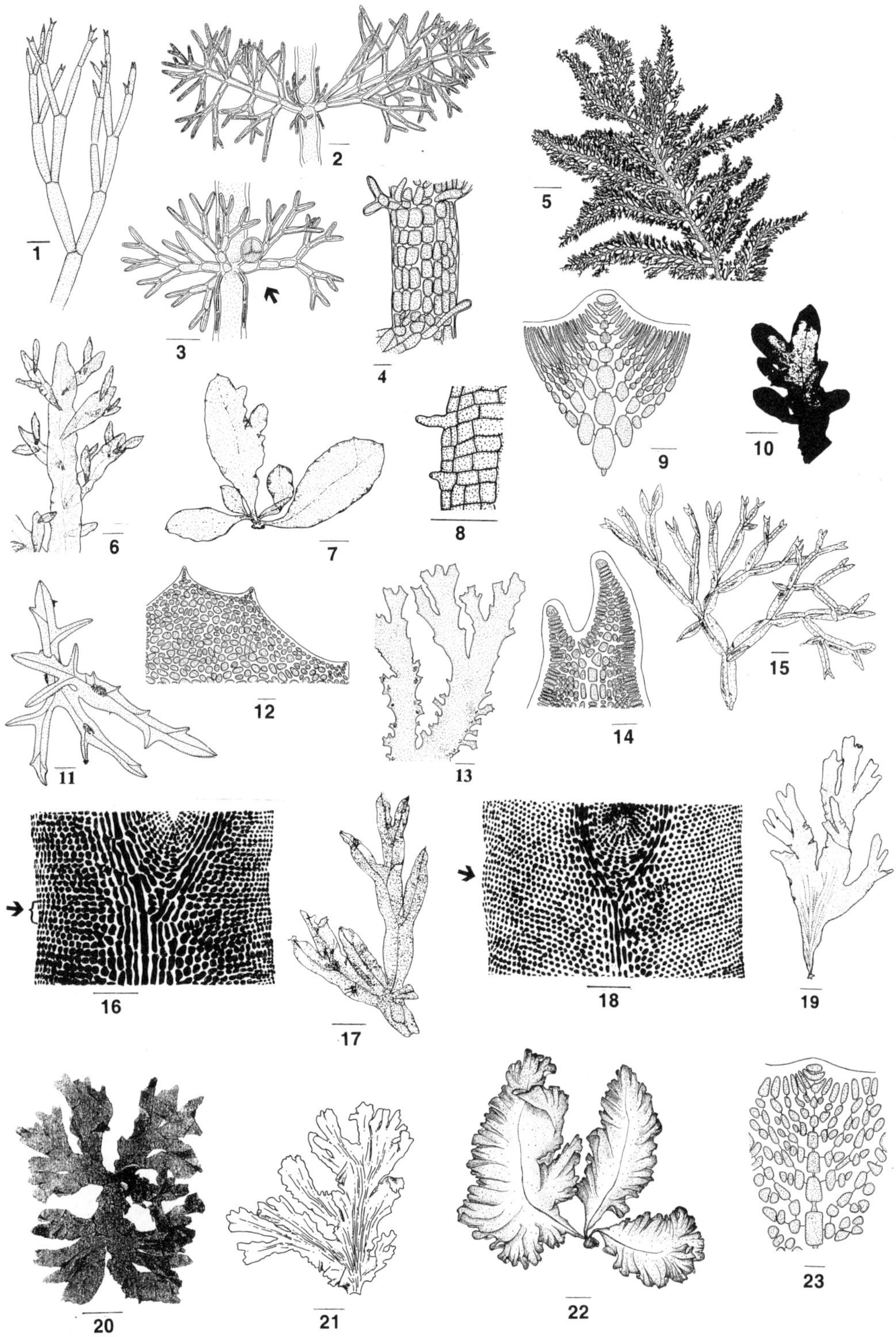
1
2
3
4
5
6
7
8
9
10
11
12
13
14
15
16
17
18
19
20
21
22
23

PLATE XXXIV

Hypoglossum anomalum. 1. Blade produces lateral fronds endogenously from the axial cells; scale 1 mm (after Wynne and Ballantine, 1986).
Hypoglossum caloglossoides. 2. Frond is narrow, prostrate, regularly constricted, and has many rhizoidal clusters (*arrow*); scale 1 mm. 3. Blade tip has one apical cell and all second order rows form third order rows; scale 50 µm (both after Wynne and Kraft, 1985).
Hypoglossum hypoglossoides. 4. Frond tip has lateral blades arising endogenously from the axial row (midrib) on alternate sides; scale 1 mm (after Schneider and Searles, 1991).
Hypoglossum involvens. 5. Upper part of a frond has a midrib and curled tips; scale 2 mm (after Taylor, 1960).
Hypoglossum rhizophorum. 6. A minute blade with a midrib, polysiphonous rhizome, and rhizoids on the opposing side; scale 100 µm (after Ballantine and Wynne, 1988a).
Hypoglossum subsimplex. 7. The four minute blades are wide and arise from a discoid holdfast; cystocarps are on midribs of two blades; scale 500 µm (after Wynne, 1994).
Hypoglossum tenuifolium. 8. Plant has a prostrate axis with rhizoids, strap-shaped blades with midribs, and bladelets arising from midribs; scale 5 mm (L. Baumhart). 9. Blade is monostromatic, has one apical cell, and all second order cells form third order rows; scale 10 µm (after Schneider and Searles, 1991).
Hypoglossum barbatum. 10. Blade tip with one apical cell; all second order cells produce third order rows; scale 25 µm. 11. Blade has elongate midrib cells (*above*), fused cells at margins, and rhizoidal tufts (*below*); scale 50 µm (both after Schneider, 2000).
Martensia pavonia. 12. Frond is spreading, lobed, thin, and with gridlike pattern below; scale 1 mm. 13. The gridlike multicellular network has tetrasporangia (*arrow*) (12, 13 after Børgesen, 1919). 14. Basal area of frond forms marginal rhizoids; scales 100 µm.
Nitophyllum punctatum. 15. Fronds are monostromatic, dichotomously branched, blunt-tipped, and with a discoid holdfast; scale 2 cm (R. Holbrook). 16. Upper part of a cystocarpic frond with rounded bifurcate tips; scale 1 cm (after Taylor, 1960).
Nitophyllum wilkinsoniae. 17. A cluster of monostromatic, lobed blades attached by a discoid holdfast; scale 1 cm (R. Holbrook). 18. Blade margin has proliferations formed by single apical cells; scale 100 µm (after Collins and Hervey, 1917).
Polyneura subtropica. 19. The tetrasporic plant has divided fronds, conspicuous midribs, and a discoid holdfast; scale 5 mm (after Schneider and Searles, 1991). 20. Tetrasporic frond has broad lobes and obvious venation; scale 5 mm. 21. Blade tip of blade shows transverse divisions of individual apical cells and intercalary divisions in the axial row; scale 20 µm (20, 21 after Schneider and Eiseman, 1979).
Taenioma nanum. 22. The polysiphonous prostrate axis has swollen uniseriate rhizoids and erect fronds with apical trichoblasts; scale 200 µm (R. Holbrook). 23. Branch tip is polysiphonous and has two thick-walled trichoblasts; scale 20 µm (after Ganesan and Lemus, 1969).
Cottoniella filamentosa. 24. Two cross sections of frond tips include a young ecorticate axis (*left*) and an older corticated (compressed) polysiphonous axis (*right*); scale 100 µm. 25. Tip of ecorticate axis has secund filaments; scale 100 µm. 26. Three segments of a single axis include a polysiphonous axial tip with an obvious apical cell (*top*), uncorticated lower axis (*right*), and heavily corticated basal axis (*below*); scale 20 µm (24–26 after Howe, 1905b).

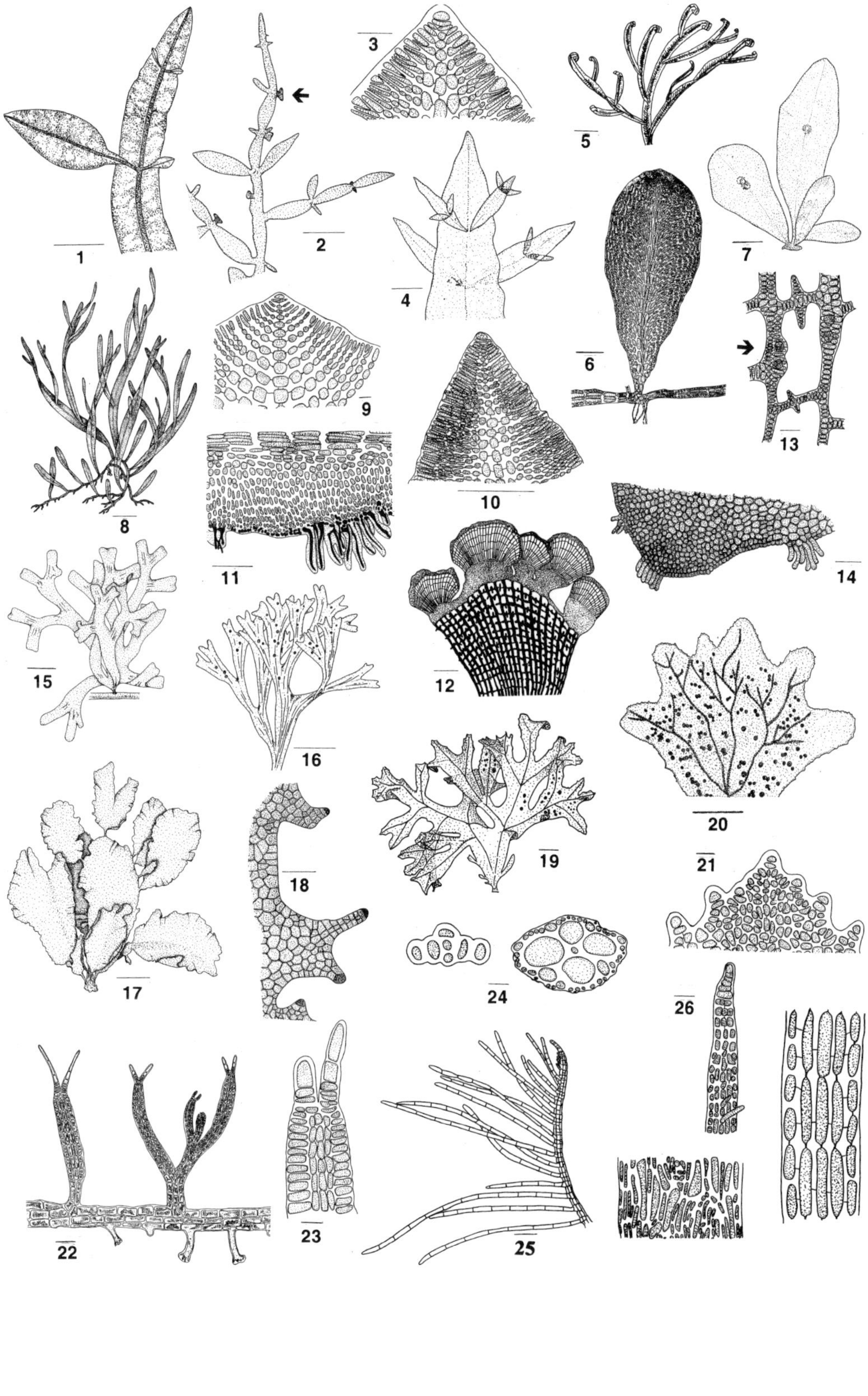
1
2
3
4
5
6
7
8
9
10
11
12
13
14
15
16
17
18
19
20
21
22
23
24
25
26

PLATE XXXV

Dasya antillarum. 1. Silhouette of an older plant showing alternate branching and a discoid holdfast; scale 5 cm (after Ballantine and Aponte, 1997b). 2. Part of a young axis with a few narrow descending rhizoids between pericentral cells; scale 100 µm. 3. A cross section of a young axis with five pericentral cells, an axial cell, and three corticating rhizoids (*arrow*); scale 100 µm. 4. The uniseriate ramelli end in fine pigmented filaments; scale 50 µm (2–4 after Littler and Littler, 2000).

Dasya baillouviana. 5. The young plant is covered by uniseriate pigmented ramelli and attached by a discoid holdfast; scale 2 cm (L. Baumhart).

Dasya caraibica. 6. Upper axes with dense (ocellate) clusters of branches; scale 1 cm 7. Main axis with pigmented uniseriate branched ramelli; scale 2 mm (both by L. Baumhart).

Dasya collinsiana. 8. A young plant with uniseriate, branched ramelli and a discoid holdfast; scale 5 mm. 9. Main axis with alternate uniseriate branches and initial rhizoidal cortication; scale 1 cm (both by R. Holbrook).

Dasya corymbifera. 10. Ramelli are uniseriate and dichotomously branched. 11. Cross section of a main axis showing the central axial cell, five pericentral cells, and outer cortical cells; scales 100 µm (both after Littler and Littler, 2000).

Dasya crouaniana. 12. Upper axes are dichotomously branched and bear pigmented ramelli; scale 1 cm (after Taylor, 1960). 13. Older axis is corticated and has several older tetrasporangial stichidia; scale 100 µm. 14. Tip of an old stichidium that is empty, except for some terminal tetrasporangia; scale 100 µm (13, 14 after Littler and Littler, 2000).

Dasya harveyi. 15. Silhouette of an upper branch showing brushlike ramelli; scale 2 cm (after Harvey, 1858). 16. Part of a younger axis that has a ramellus and initial rhizoidal cortication along pericentral cells; scale 100 µm. 17. Axial tip with two uniseriate ramelli; scale 100 µm (16, 17 after Littler and Littler, 2000).

Dasya mollis. 18. Part of a main axis showing whorled uniseriate ramelli with basal cells that are wider than long; scale 100 µm (after Littler and Littler, 1997). 19. Axis bearing a swollen basal cell of a uniseriate ramellus; scale 50 µm (after Børgesen, 1919).

Dasya ocellata. 20. Branch tip with a cluster of immature uniseriate ramelli; scale 50 µm (after Littler and Littler, 2000).

Dasya punicea. 21. The upper axes are covered with uniseriate ramelli; scale 1 mm. 22. Older axis is corticated and has bases of uniseriate ramelli; scale 100 µm (21, 22 after Littler and Littler, 2000).

Dasya ramosissima. 23. Part of a main axis with initial rhizoidal cortication and bases of two uniseriate ramelli; scale 100 µm (after Taylor, 1928).

Dasya rigidula. 24. Silhouette of upper axis showing clusters of ramelli near tips; scale 2 mm (after Schneider and Searles, 1991). 25. Axis is ecorticate and has uniseriate and alternately radially arranged ramelli; scale 400 µm (R. Holbrook).

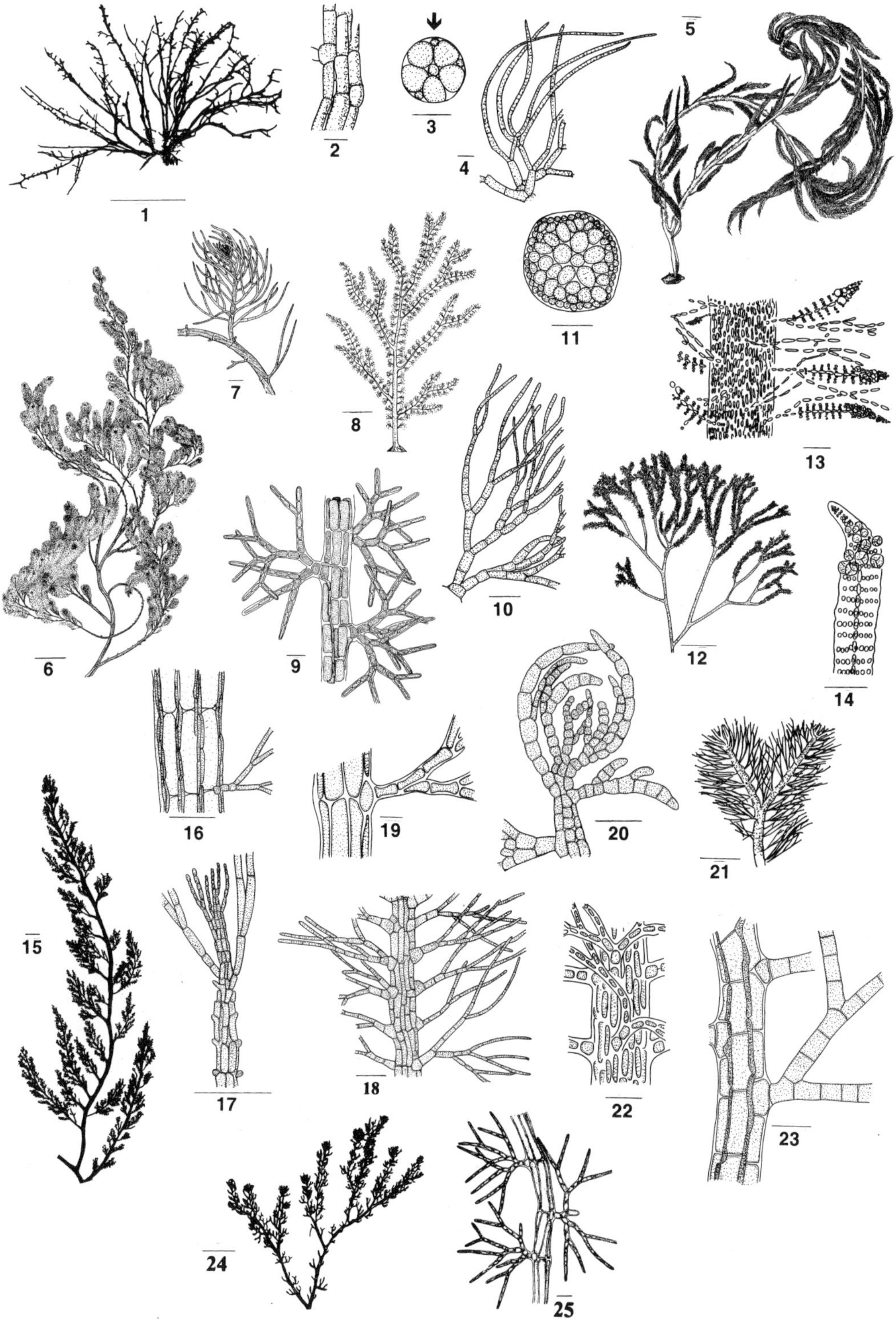
1
2
3
4
5
6
7
8
9
10
11
12
13
14
15
16
17
18
19
20
21
22
23
24
25

PLATE XXXVI

Dictyurus occidentalis. 1. A cluster of plants arising from a rhizoidal mat; scale 1 cm. 2. The frond tip has weblike whorls of branchlets; scale 2 mm (both by L. Baumhart).

Heterosiphonia crispella. 3. Axial tip of var. *laxa* is ecorticate, polysiphonous and has uniseriate branches that are dichotomously branched and with young stichidia; scale 200 µm. 4. Cross section showing 5 pericentral cells and an axial cell; scale 20 µm. 5. Base of the main axis with rhizoidal pads (*arrow*) and alternately to irregularly arranged uniseriate branchlets; scale 20 µm (all after Børgesen, 1919).

Heterosiphonia gibbesii. 6. The three erect plants are closely attached and have obvious tufts of apical uniseriate filaments; scale 2 cm (L. Baumhart). 7. Uniseriate filaments are clustered at tips; scale 2 mm (L. Baumhart). 8. Cross section of a main axis with nine pericentral cells, an axial cell, and outer cortication; scale 100 µm (after Littler and Littler, 2000). 9. Branchlet has basal polysiphonous segments and uniseriate tips; scale 200 µm (after Taylor, 1960).

Heterosiphonia hummii. 10. Plant silhouette is plumose; scale 1 cm (after Ballantine and Aponte, 1997b). 11. Tip of main axis with sympodial growth; scale 20 µm (after Aregood, 1975).

Halydictyon mirabile. 12. Filaments are uniseriate and form nets; scale 200 µm (after Taylor, 1960). 13. A mature half empty tetrasporangial stichidium; scale 50 µm (after Børgesen, 1930a).

Acanthophora muscoides. 14. Branch tip has spurlike spines on the axis and branchlets; scale 1 mm (after Littler and Littler, 2000).

Acanthophora spicifera. 15. The robust plant has spurlike spines only on its branchlets and a rhizoidal holdfast; scale 2 cm. 16. Branch tip with spurlike spines restricted to branchlets; scale 2 mm. 17. Close up of a branchlet with spurlike spines; scales 2 mm (15–17, L. Baumhart). 18. Cross section showing an axial cell surrounded by five pericentral cells and dense outer cortication; scale 100 µm (after Børgesen, 1918).

Amansia multifida. 19. Upper axis is flat, pinnately branched, and with a midrib and enrolled tips; scale 5 mm (after Taylor, 1960).

Bostrychia calliptera. 20. Erect axes are pinnately branched and arise from a rhizome; scale 1 mm (after Lawson and John, 1987). 21. Branchlet tip is short, ecorticate, mostly polysiphonous, and with a large apical cell; scale 20 µm (after Joly, 1954). 22. The erect axis has pinnate branchlets, a rhizomatous base, and an opposing rhizoidal haptera; scale 2 mm. 23. Branchlet tips are polysiphonous and end in large apical cells; scale 75 µm. 24. Two parts of a branchlet: a tip with one large apical cell (*above*) and a subapical part that is entirely polysiphonous (*below*); scale 50 µm (22–24 after Tanaka and Chihara, 1984).

Bostrychia montagnei. 25. An erect plant with alternate branching, a rhizome, and a large rhizoidal haptera; scale 1 mm. 26. Axial tips are polysiphonous and have uniseriate branchlets and tetrasporangial stichidia (*arrow*); scale 200 µm (both by L. Baumhart).

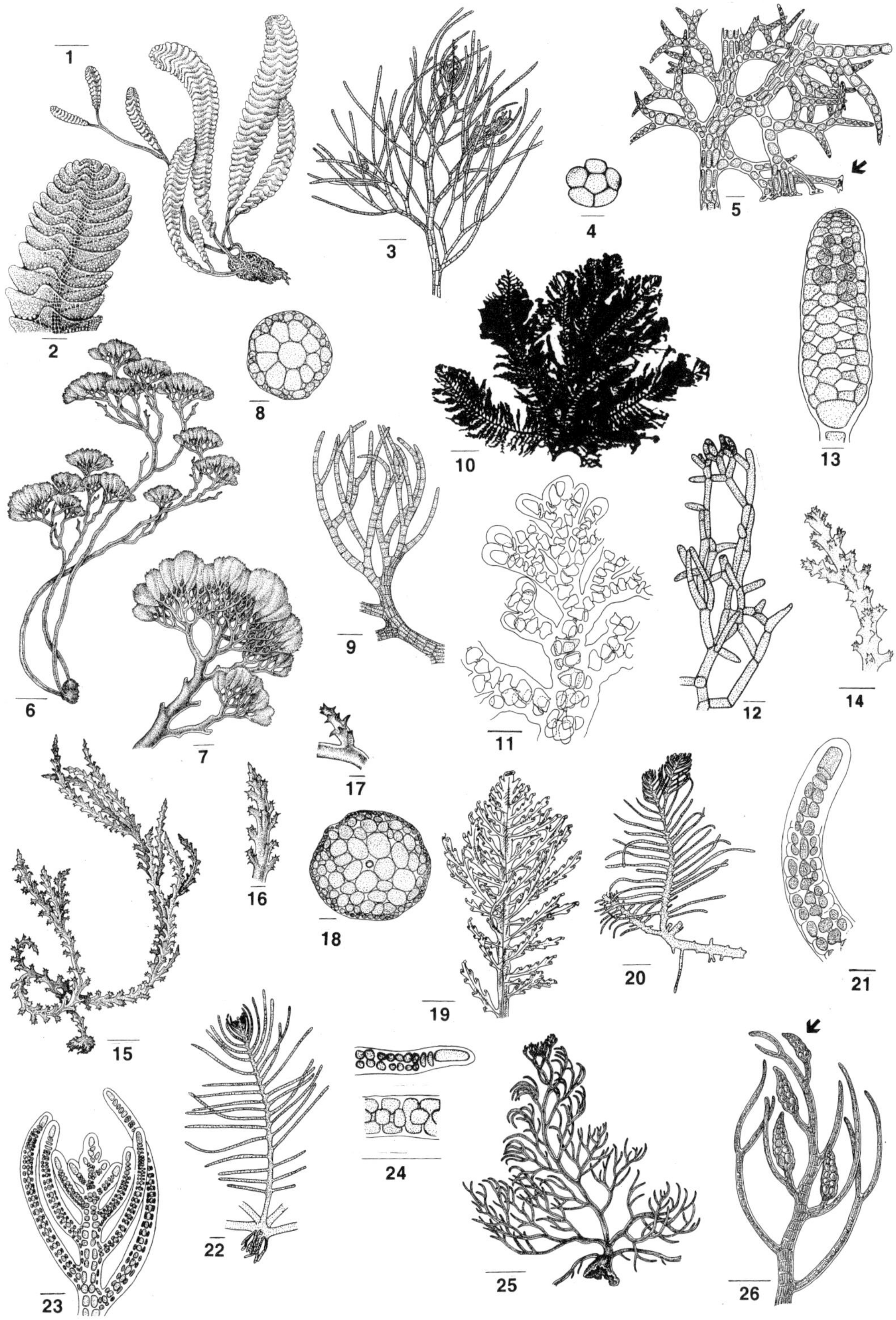
1
2
3
4
5
6
7
8
9
10
11
12
13
14
15
16
17
18
19
20
21
22
23
24
25
26

PLATE XXXVII

Bostrychia moritziana. 1. A young polysiphonous branch with alternating uniseriate branchlets; scale 100 µm (after Littler and Littler, 2000). 2. The base of an erect axis has a rhizoidal haptera and branchlets with uniseriate tips and polysiphonous bases; scale 200 µm (after Lawson and John, 1987). 3. A tetrasporangial stichidium on a branchlet with a polysiphonous base and a uniseriate tip; scale 100 µm (after Littler and Littler, 2000).

Bostrychia radicans. 4. The prostrate axis has a rhizoidal haptera alternating and a mostly polysiphonous erect axis; scale 200 µm. 5. A cross section of an older ecorticate axis showing 8 pericentral cells and an axial cell; scale 50 µm (both after Joly, 1954).

Bostrychia tenella. 6. Plant has plumose axes, a rhizome, and discoid haptera; scale 1 cm. 7. An axial tip with pinnate uniseriate branchlets; scale 1 mm (6, 7, L. Baumhart). 8. The axial tip has a conspicuous apical cell and uniseriate branchlets; scale 50 µm (after Tanaka and Chihara, 1984).

Bryocladia cuspidata. 9. The many erect axes are densely covered with branchlets and arise from massive basal rhizoidal clusters; scale 1 cm. 10. Branch tip is covered by short polysiphonous branchlets; scale 5 mm (both, L. Baumhart).

Bryocladia thyrsigera. 11. Silhouette of a branch tip showing lax and alternately branched axes; scale 25 mm (after Schneider and Searles, 1991). 12. Silhouette showing the upper axes covered with pinnately divided branchlets; scale 2 mm (after Lawson and John, 1987).

Bryothamnion seaforthii. 13. Branch tip has compressed to flattened branchlets in one plane; scale 2 mm. 14. A polysiphonous branchlet tip; scale 1 mm (13, 14, A. L. Paul). 15. A cross section showing 8 pericentral cells, an axial cell, and extensive cortication; scale 200 µm (after Børgesen, 1918).

Bryothamnion triquetrum. 16. The main axes are alternately branched, densely covered by branchlets in three rows, and with a rhizoidal holdfast; scale 1 cm. 17. Branch tip with three rows of polysiphonous corticated branchlets; scale 2 mm (16, 17, L. Baumhart).

Chondria atropurpurea. 18. Silhouette showing irregular branching that occurs mostly near tips; scale 5 mm. 19. A cross section showing 5 pericentral cells, an axial cell, and dense cortication; scale 100 µm (18, 19 after Schneider and Searles, 1991). 20. A branchlet with embedded tetraspores (*arrow*), a basal constriction, blunt tip, and apical trichoblasts; scale 200 µm (after Littler and Littler, 2000).

Chondria baileyana. 21. Silhouette of an upper frond showing pyramidal branching with branchlets clustered near axial tips; scale 1 cm (after Taylor, 1957). 22. Branchlet tip with tetrasporangia and trichoblasts; scale 100 µm (after Schneider and Searles, 1991).

Chondria capillaris. 23. Upper part of a main axis with abundant fusiform branchlets; scale 1 cm (L. Baumhart).

Chondria cnicophylla. 24. Silhouette of an older tip showing irregular branching and clusters of branchlets; scale 2 mm (after Taylor, 1960). 25. The branchlet is club-shaped and has apical trichoblasts; scale 200 µm (after Littler and Littler, 2000).

Chondria collinsiana. 26. Tip is obtuse and has a lateral club-shaped branchlet and apical trichoblasts; scale 200 µm (after Littler and Littler, 2000).

Chondria curvilineata. 27. The prostrate main axis has blunt-tipped branchlets; scale 1 mm (after Schneider and Searles, 1991). 28. Cross section showing 6 pericentral cells, an axial cell, and extensive cortication; scale 50 µm (after Littler and Littler, 2000). 29. In the older axis the thick end walls of pericentral cells appear as dark bands; scale 100 µm (after Collins and Hervey, 1917).

Chondria dasyphylla. 30. The young plant is erect, cystocarpic, and has alternate branching, short constricted branchlets, and a discoid holdfast; scale 2 mm (after Børgesen, 1918). 31. The older plant has a rhizoidal base, branchlets limited to upper parts, and many apical tufts of trichoblasts; scale 1 cm (L. Baumhart). 32. The club-shaped tetrasporangial branchlet has a dense tuft of apical trichoblasts; scale 500 µm (after Børgesen, 1918).

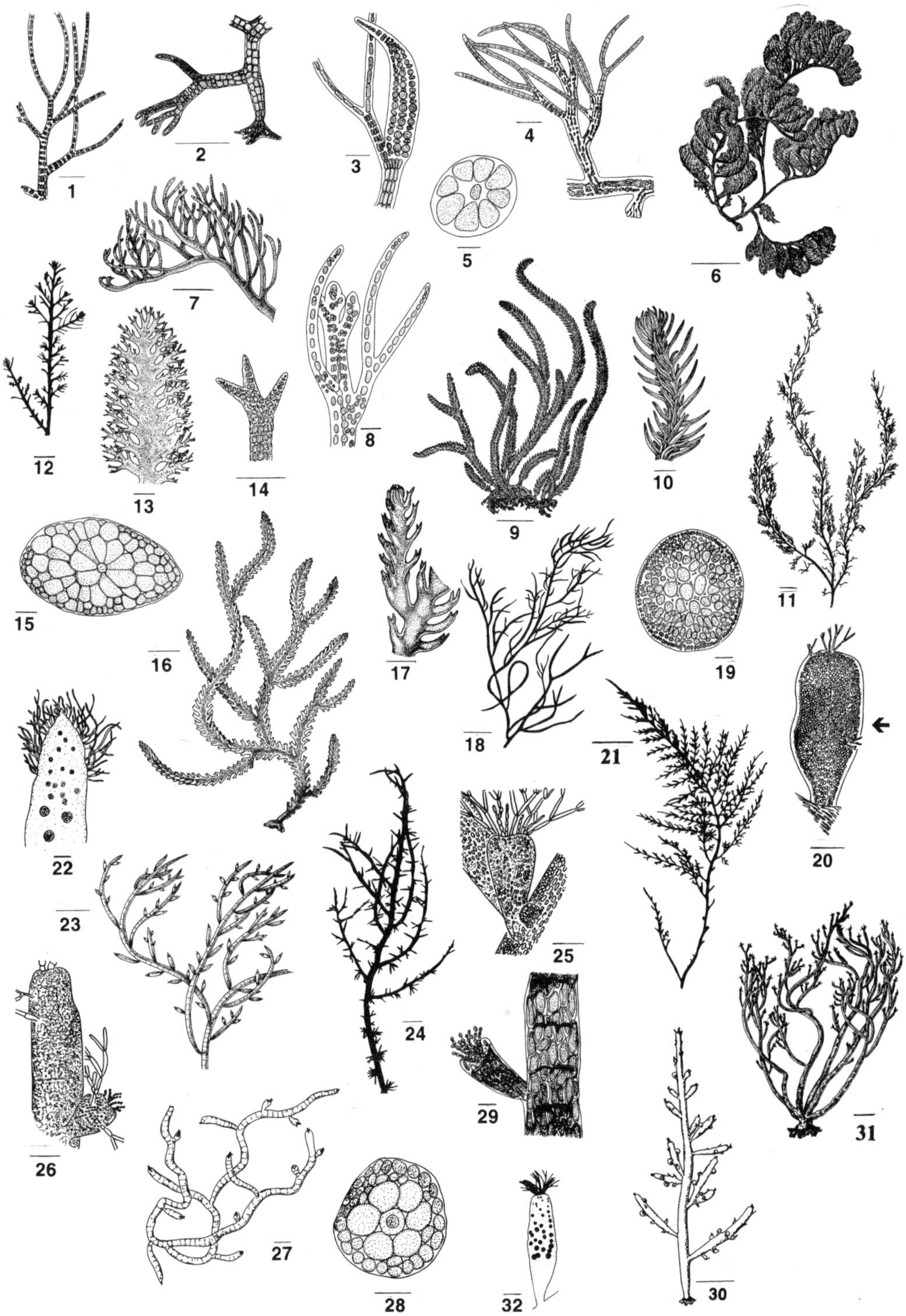
1
2
3
4
5
6
7
8
9
10
11
12
13
14
15
16
17
18
19
20
21
22
23
24
25
26
27
28
29
30
31
32

PLATE XXXVIII

Chondria floridana. 1. Main axial tip with irregular branching; scale 2 mm. 2. The branch tip has cylindrical branchlets; scale 1 cm (both after Schneider and Searles, 1991).

Chondria leptacremon. 3. Silhouette of six erect axes with many branchlets; scale 2 cm. 4. The tip is irregularly branched and has club-shaped branchlets; scale 1 mm (3, 4 after Melville in Murray, 1888). 5. The branch tip has a cluster of branchlets bearing tetrasporangia and trichoblasts; scale 200 μm (after Littler and Littler, 2000).

Chondria littoralis. 6. The upper axis has fusiform branchlets with apical tufts of trichoblasts; scale 1 cm (L. Baumhart). 7. The fusiform branchlet has tetrasporangia and trichoblasts; scale 200 μm (after Børgesen, 1918). 8. The plant has alternate branching and a rhizoidal holdfast; scale 2 cm (L. Baumhart).

Chondria polyrhiza. 9. Axis is prostrate with multiple rhizoidal clusters (*arrow*); scale 1 mm (after Schneider and Searles, 1991). 10. A rhizoidal cluster on the axis; scale 50 μm (after Børgesen, 1918). 11. An axial tip with terminal trichoblasts and a fusiform lateral branchlet with apical trichoblasts; scale 200 μm (after Schneider and Searles, 1991).

Chondria sedifolia. 12. The upper part of the main axis is pyramidal in shape and densely branched; scale 2 mm. 13. Branch tip with club-shaped branchlets and wartlike cystocarps (*arrow*); scale 2 mm (both after Taylor, 1957).

Chondrophycus corallopsis. 14. Portion of a frond with dichotomous branching mostly limited to upper parts; scale 1 cm (after Schneider and Searles, 1991).

Palisada gemmifera. 15. Upper part of axis with stubby branchlets; scale 1 cm (after Harvey, 1853). 16. The branch tip has cylindrical, blunt branchlets; scale 2 mm (A. L. Paul). 17. Cross section shows spinelike surface cells, a dense cortex and medulla, 5 pericentral cells, and the axial cell; scale 200 μm (after Littler and Littler, 2000).

Palisada iridescens. 18. Plant has irregular lumpy axes, irregular branching, and a cartilaginous texture; scale 200 μm (after Wynne and Ballantine, 1991).

Palisada papillosa. 19. Plant has irregular branching, a dense covering of wartlike branchlets, and a discoid holdfast; scale 1 cm. 20. The branch tip has tuberculate branchlets, each with an apical pit; scale 1 cm (both, L. Baumhart).

Palisada poiteaui. 21. The plant has many club-shaped branchlets and a discoid holdfast; scale 1 cm. 22. Two club-shaped branchlets, each with one apical pit and a basal constriction; scale 1 mm (both, L. Baumhart).

Digenea simplex. 23. The plant is irregularly branched, covered by polysiphonous branchlets, and attached to a shell fragment by a rhizoidal holdfast; scale 1 cm (L. Baumhart). 24. A cross section showing dense cortication that obscures pericentral cells; scale 50 μm (after Littler and Littler, 2000).

Enantiocladia duperreyi. 25. The tip of a flattened frond has marginal teeth and an obvious banding pattern; scale 1 cm (after a herbarium specimen of W. R. Taylor from St. Kitts).

Herposiphonia parca. 26. Prostrate axis with an erect branchlet arising from every segment; scale 200 μm. 27. A cross section showing 9 pericentral cells, an axial cell, and no cortication; scale 50 μm. 28. A swollen rhizoid with a digitate attachment; scale 200 μm. 29. A branchlet tip with many secund trichoblasts; scale 50 μm (all after Littler and Littler, 2000).

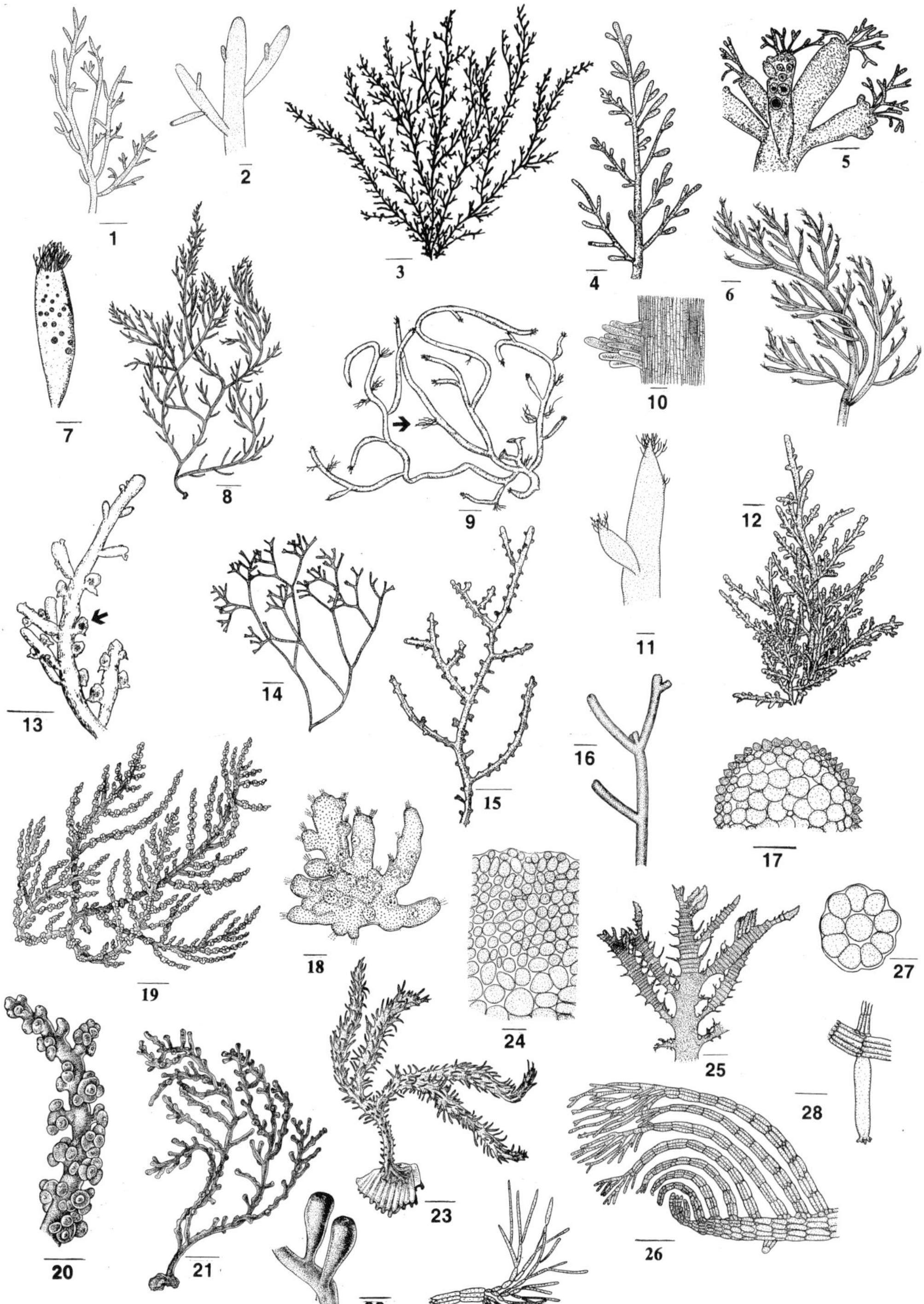
1
2
3
4
5
6
7
8
9
10
11
12
13
14
15
16
17
18
19
20
21
22
23
24
25
26
27
28
29

PLATE XXXIX

Herposiphonia pecten-veneris. 1. The tip of a prostrate axis is ecorticate and has erect axes from each segment with spiral trichoblasts; scale 2 mm (after Littler and Littler, 2000). 2. The erect branch has terminal, uniseriate trichoblasts; scale 100μm (after Taylor, 1928).

Herposiphonia secunda. 3. The prostrate axis is polysiphonous and has indeterminate branchlets arising from every 5th or 6th segment; these are preceded by a simple determinate branch; scale 1 mm (after Lawson and John, 1987).

Heterosiphonia tenella. 4. The polysiphonous prostrate axis has indeterminate branchlets arising from most segments; scale 300 μm (after Lawson and John, 1987).

Heterodasya mucronata. 5. Upper part of axis is alternately to irregularly branched and covered with spirally arranged ramuli; scale 1 cm (after Lawson and John, 1987). 6. The older corticated axis has a dichotomously branched uniseriate ramulus; scale 50 μm (after Howe and Taylor, 1931). 7. The branchlet is alternately divided, uniseriate, and with mucronate tips; scale 50 μm (after Taylor, 1928).

Laurencia brongniartii. 8. Silhouette of an upper flat axis showing pinnate branching; scale 1 cm (after Hackett, 1974).

Laurencia caraibica. 9. An upper axes with slightly swollen tips; scale 3 mm. 10. A cross section with a slightly convoluted outer cortical layer; scale 200 μm. 11. Longitudinal section of a branch tip with short, apical inconspicuous filaments and a terminal depression; scale 200 μm (all after Littler and Littler, 2000).

Laurencia cervicornis. 12. The tough, cartilaginous plant forms clumps to 4 cm tall and is attached by a padlike base; scale 5 mm. 13. The upper part of an axis has irregular to dichotomous branching and slightly swollen tips with apical pits; scale 1 cm (after Harvey, 1853). 14. Surface cells are thick walled and angular; scale 100 μm. (12, 14 after Littler and Littler, 2000).

Laurencia chondrioides. 15. Upper part of main axis has slightly fusiform branchlets with blunt tips; scale 1 mm. 16. Axis is cylindrical and has fusiform tetrasporic branchlets and terminal trichoblasts; scale 1 mm. 17. Cross section showing pigmented outer cortical cells and internal cortication that obscures the six pericentral cells and axial cell; scale 50 μm (all after Børgesen, 1918).

Laurencia coelenterata. 18. A crust with cylindrical branchlets; scale 200 μm. 19. Cross section of crust showing branched rhizoids and internal cortical cells; scale 100 μm. 20. Longitudinal section of an erect branch showing an apical pit (*left*), a developing cystocarp (*right*), and internal cortication; scale 50 μm (all after Ballantine and Aponte, 1995).

Laurencia intricata. 21. Upper axis with cylindrical branchlets and irregular to alternate branching; scale 2 mm (after Børgesen, 1918). 22. The branchlet is blunt, cylindrical, and has internal tetrasporangia (*arrow*) and apical trichoblasts; scale 200 μm (after Littler and Littler, 1997).

Laurencia microcladia. 23. Upper axial branch with many cylindrical, blunt-tipped branchlets; scale 2 mm (after Littler and Littler, 1997).

Laurencia obtusa. 24. The young plant has many radial branchlets and a holdfast; scale 5 mm. 25. The two branchlets are blunt-tipped, cylindrical, basally constricted, and with apical pits; scale 1 mm (both by L. Baumhart).

Lophocladia trichoclados. 26. The prostrate axis has a mixture of simple and complex rhizoidal haptera, plus uniseriate trichoblasts; scale 100 μm. 27. Tip of an erect axis with young trichoblasts; scale 50 μm. 28. Part of an erect polysiphonous axis with uniseriate trichoblasts and young descending rhizoids; scale 200 μm (all after Børgesen, 1919).

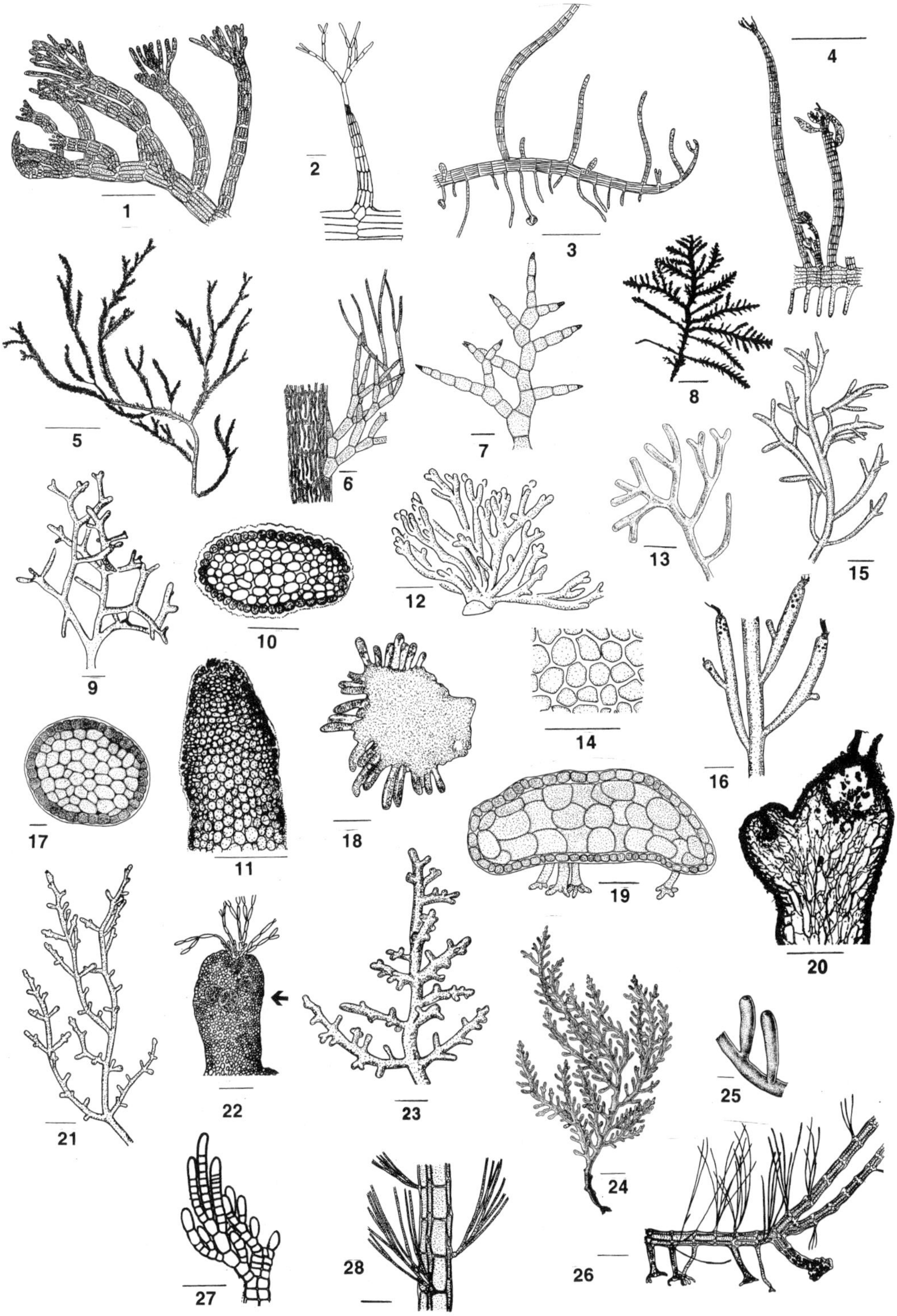
1
2
3
4
5
6
7
8
9
10
11
12
13
14
15
16
17
18
19
20
21
22
23
24
25
26
27
28

PLATE XL

Lophosiphonia cristata. 1. Apex is curved downward and has developing tetrasporangia and uniseriate apical filaments that form a crest; scale 100 µm. 2. Cross section of an erect filament showing 13 pericentral cells and an axial cell; scale 100 µm. 3. The prostrate axis has one-celled rhizoids and indeterminate erect axes with apical trichoblasts every 6–8 segments; scale 1 cm (after Børgesen, 1918). 4. Part of a prostrate filament with single-celled rhizoids (1, 2, 4 after Littler and Littler, 2000).

Lophosiphonia obscura. 5. Apex is curved downward with uniseriate apical filaments offset alternately at every segment to form a crest; scale 50 µm. 6. Upper axial branches usually have recurved tips; scale 1 cm (after Børgesen, 1918). 7. Tip of a prostrate polysiphonous filament with a single apical cell; scale 100 µm. 8. Cross section of an erect filament showing 14 pericentral cells; scale 50 µm (5, 7, 8 after Littler and Littler, 2000).

Murrayella periclados. 9. The prostrate axis has multicellular, uniseriate rhizoids and an erect branch with 4 pericentral cells; scale 200 µm (after Børgesen, 1919).

Neosiphonia ferulacea. 10. Prostrate axis has an erect branching axis and rhizoidal cluster (*right*); scale 200 µm. 11. The erect polysiphonous axis and branches have tetrasporangia with two apical trichoblasts at the top; scale 50 µm. 12. The apex has an obvious apical cell and no trichoblasts; scale 20 µm. 13. Cross section showing 4 pericentral cells, an axial cell, and no cortication; scale 100 µm (10–13 after Børgesen, 1918). 14. Prostrate axis has 4 rhizoids with cross walls; scale 200 µm (after Kapraun et al., 1983).

Neosiphonia gorgoniae. 15. Branch tip is polysiphonous and has spermatangial sori in stichidia (*arrow*) subtended by uniseriate trichoblasts; scale 50 µm. 16. Cross section showing 4 pericentral cells, an axial cell, and no cortication; scale 100 µm (after Littler and Littler, 2000). 17. Prostrate axis has digitate rhizoids without cross walls; scale 50 µm. 18. Base of erect axis has rhizoids without cross walls; scale 50 µm (15, 17, 18 after Kapraun, 1979).

Neosiphonia sphaerocarpa. 19. The lateral branch on an erect tetrasporic plant has young trichoblasts and a distinct apical cell; scale 20 µm. 20. Cross section showing 4 pericentral cells, an axial cell, and no cortication; scale 50 µm. 21. Prostrate axis has long rhizoids with cross walls; scale 100 µm (all after Børgesen, 1918).

Neosiphonia tepida. 22. Upper part of an erect axis has a lateral branch initial growing in the axil of a uniseriate trichoblast. 23. Prostrate axis has a septate digitate rhizoid (*arrow*); scales 50 µm (both after Kapraun, 1979).

Osmundaria obtusiloba. 24. Frond tip has incurved branchlets and is faintly banded by the thick-walled pericentral cells; scale 5 mm. 25. Plant is flat, pinnately branched, and has a discoid holdfast; scale 2 cm (both, L. Baumhart).

Osmundea pinnatifida. 26. The flat upper axis is pinnately branched; scale 1 cm (after Stegenga and Mol, 1983). 27. Upper part of a compressed, alternately branched axis; scale 1 cm (after Schneider and Searles, 1991).

Polysiphonia anomala. 28. The cystocarp is oval and on a short polysiphonous axis in the axil of a branch; scale 50 µm. 29. Prostrate axis bears a branch initial at the base of a branched uniseriate hair and a rhizoid at the center of the ventral pericentral cell; scale 50 µm (both after Hollenberg, 1968a).

Polysiphonia atlantica. 30. The prostrate plant has padlike unicellular rhizoids and erect axes; scale 200 µm. 31. Tip of an erect axis with 6 tetrasporangia; scale 100 µm. 32. Cross section showing 4 pericentral cells, an axial cell, and no cortication; scale 50 µm. 33. The branch tip has trichoblasts and a young cystocarp (*arrow*); scale 50 µm. 34. A branch bearing a lateral spermatangial sorus; scale 20 µm (all after Børgesen, 1918).

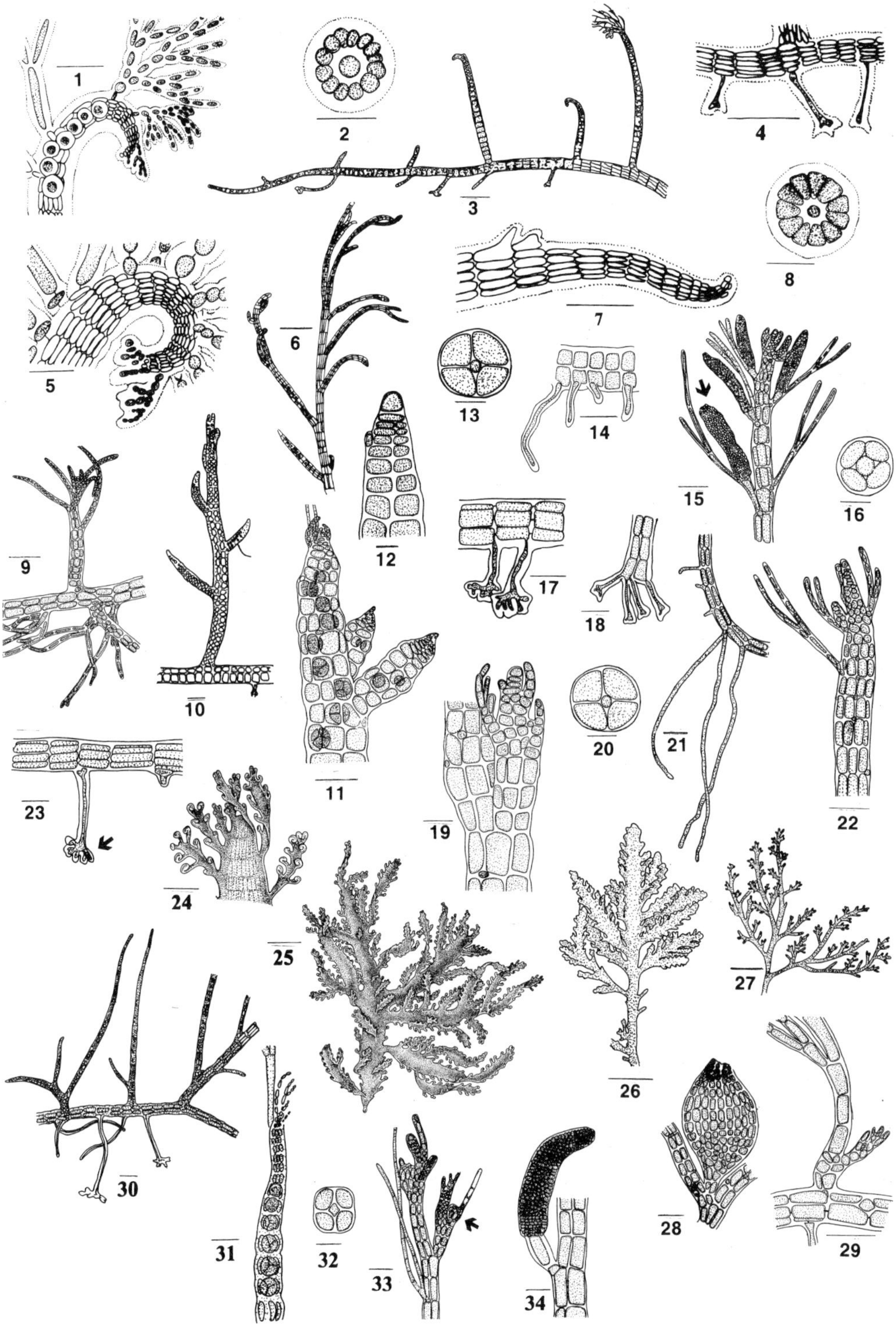
1
2
3
4
5
6
7
8
9
10
11
12
13
14
15
16
17
18
19
20
21
22
23
24
25
26
27
28
29
30
31
32
33
34

PLATE XLI

Polysiphonia binneyi. 1. The polysiphonous branch tips each have a large apical cell and dichotomously branched trichoblasts; scale 20 µm. 2. The axial base has rhizoids that lack cross walls with pericentral cells; scale 100 µm (both after Kapraun et al., 1983).
Polysiphonia denudata. 3. The polysiphonous branch tip has a large apical cell and young lateral branch with a branched uniseriate trichoblast; scale 20 µm (after Børgesen, 1918). 4. Cross section showing 6 pericentral cells, an axial cell, and no cortication; scale 50 µm (after Taylor, 1957). 5. The prostrate axis has long rhizoids that lack cross walls with pericentral cells and digitate attachment pads; scale 20 µm (after Børgesen, 1918).
Polysiphonia echinata. 6. The prostrate polysiphonous axis has an adventitious branchlet with uniseriate branching trichoblasts. 7. The two unicellular rhizoids on a prostrate axis have cross walls with pericentral cells and with digitate attachment pads; scales 50 µm (6, 7 after Kapraun, 1979). 8. The erect tip has a few trichoblasts and 6 branches with visible apical cells; scale 100 µm (A. L. Paul).
Polysiphonia exilis. 9. The prostrate polysiphonous axis has a young erect branch with trichoblasts and two unicellular rhizoids having cross walls with the pericentral cells; scale 50 µm (after Kapraun and Norris, 1982).
Polysiphonia haplacantha. 10. Upper axis has spinelike branchlets, some swollen by tetrasporangia; scale 2 mm. 11. The polysiphonous lateral branchlet has tetrasporangia (*arrow*) and trichoblasts; scales 100 µm. 12. Cross section showing 4 pericentral cells, an axial cell, and no cortication; scale 200 µm (all, A. L. Paul).
Polysiphonia havanensis. 13. Tip of a polysiphonous axis has obvious apical cells and lateral branches arising in the axils of trichoblasts; scale 25 µm (after Børgesen, 1918). 14. The polysiphonous prostrate axis has rhizoids that lack cross walls with the pericentral cells; scale 50 µm (after Kapraun et al., 1983).
Polysiphonia howei. 15. The tip of an erect polysiphonous axis has dichotomously branched trichoblasts and a young branch; scale 100 µm. 16. Tip of the polysiphonous prostrate axis has long rhizoids with cross walls; scale 100 µm (both after Kapraun et al., 1983).
Polysiphonia opaca. 17. The tip of an erect polysiphonous axis has short lateral branches covered by trichoblasts; scale 20 µm. 18. Cross section showing 19 pericentral cells, an axial cell, and no cortication; scale 100 µm (both after Børgesen, 1930a).
Polysiphonia ramentacea. 19. Silhouette showing alternate branching, a covering of many short bristlelike branchlets, and a rhizoidal holdfast; scale 2 cm (after Harvey, 1853). 20. The branchlet on an erect axis is delicate, polysiphonous, and distinctly smaller; scale 100 µm (after Taylor, 1928). 21. Cross section showing 4 pericentral cells, an axial cell, and 2 cortical layers; scale 100 µm (after Harvey, 1853).
Polysiphonia saccorhiza. 22. Tip of an erect polysiphonous axis with young trichoblasts; scale 20 µm (after Collins and Hervey, 1917). 23. The polysiphonous prostrate axis has swollen rhizoids with cross walls; scale 50 µm (after Hollenberg, 1968a).
Polysiphonia scopulorum. 24. The prostrate polysiphonous axis has 3 erect branches and 2 rhizoids without cross walls; one is mature and digitate (*left*) and the other immature (*right*). 25. The tip of an erect polysiphonous axis has 2 dichotomously branched trichoblasts and internal tetrasporangia; scales 100 µm (both after Kapraun and Norris, 1982).
Polysiphonia subtilissima. 26. The prostrate polysiphonous axis has 3 erect branches and one rhizoid without a cross wall; scale 100 µm (after Kapraun, 1979). 27. The tip of an erect polysiphonous axis has large apical cells, a lateral branch, and one uniseriate trichoblast; scale 50 µm (after Kapraun et al., 1983).
Waldoia antillana. 28. Plant is an alternate to irregularly branched frond with a discoid holdfast; scale 2 cm. 29. Frond tip with a large apical cell and transverse cell divisions; scale 25 µm (both after Taylor, 1962a).
Wrightiella blodgettii. 30. Upper axes with spinelike branchlets and transverse bandings; scale 1 mm (after Littler and Littler, 2000). 31. Axial tip with rhizoidal cortication; scales 100 µm. 32. Cross section of a young axis showing 4 pericentral cells, an axial cell, and limited rhizoidal growth (*arrow*); scale 100 µm (31, 32 after R. Holbrook).

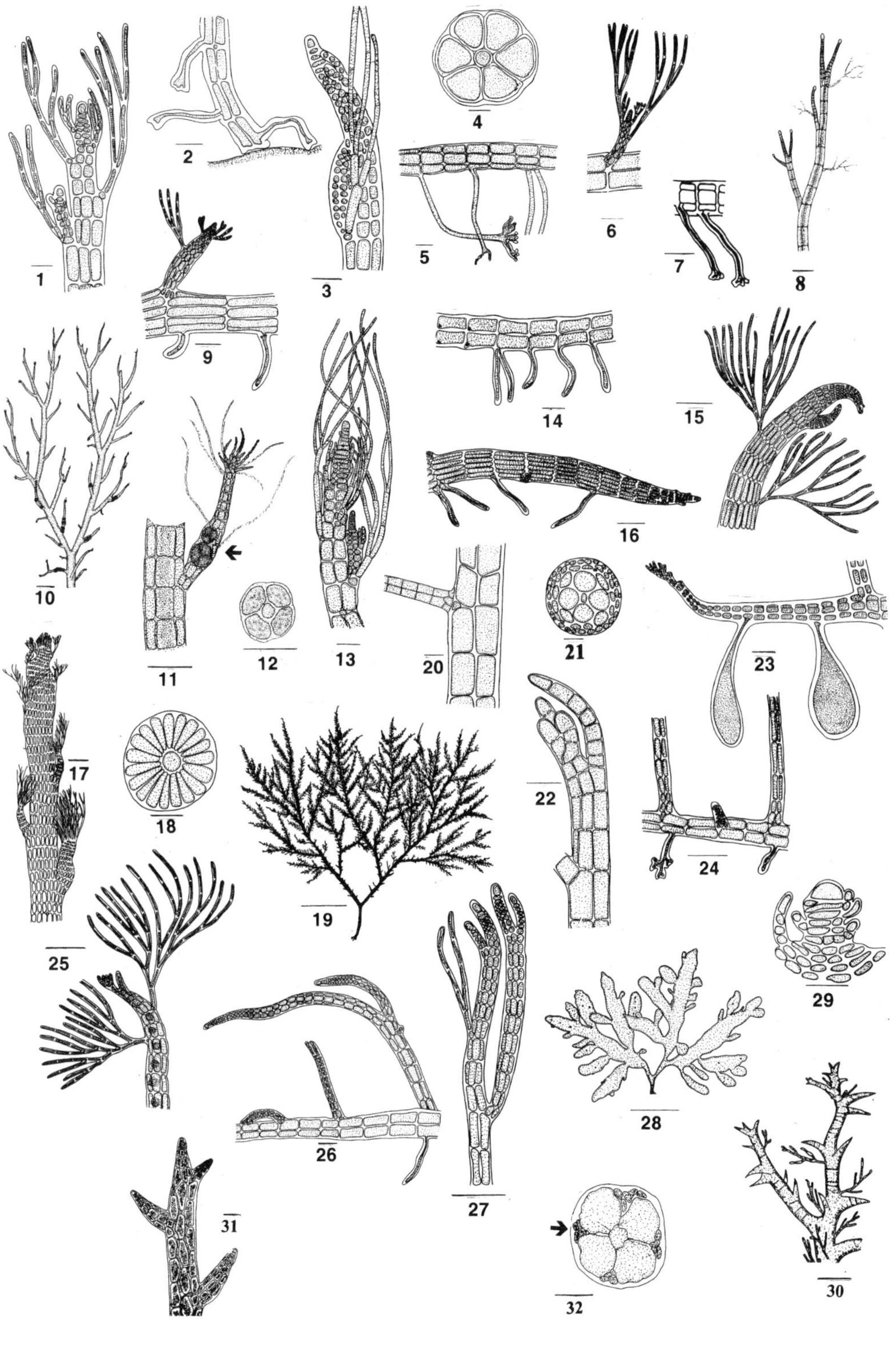
1
2
3
4
5
6
7
8
9
10
11
12
13
14
15
16
17
18
19
20
21
22
23
24
25
26
27
28
29
30
31
32

PLATE XLII

Wrightiella tumanowiczii. 1. Plant is bushy and feathery and has a discoid holdfast; scale 2 cm (L. Baumhart). 2. The young axis has numerous trichoblasts and spinelike branchlets plus new rhizoidal cortication; scale 100 µm (after Børgesen, 1919).

Gelidium americanum. 3. Upper axes are cylindrical to slightly compressed; scale 1 cm (F. Hoff). 4. The upper fronds are compressed to straplike; scale 2 cm (after Schneider and Searles, 1991). 5. Cross section showing a medulla with large filaments interspersed with small thick-walled cells (rhizines) and a row of small surface cells; scale 50 µm. 6. Frond tip with one apical cell; scale 50 µm (5, 6 after Littler and Littler, 2000).

Gelidium corneum. 7. Silhouette showing a bushy plant with an obvious main axis, pinnate branching (1–3 times), and prostrate rhizoidal axis; scale 5 cm. 8. The tip of a compressed axis is pinnately branched and has basally constricted branchlets; scale 1.0 cm (both after Guiry and Nic Dhonncha, 2005).

Gelidium crinale. 9. Silhouette showing a prostrate base and erect axes with irregular to alternate branching; scale 5 mm. 10. Cross section showing a medulla with many small thick-walled cells (rhizines) mixed with large medullary filaments; scale 20 µm (9, 10 after Taylor, 1957).

Gelidium floridanum. 11. Silhouette showing an upper tetrasporic axis with pinnate branching and dense clusters of branches; scale 5 mm (after Taylor, 1943).

Gelidium pusillum. 12. Plant has a compressed axes with alternate to sparsely pinnate branching; scale 5 mm (after Schneider and Searles, 1991). 13. The two small plants have strap-shaped axes and rhizoidal holdfasts; scale 2mm (after Taylor, 1960). 14. Branch tip with one apical cell; scale 50 µm (after Littler and Littler, 2000).

Gelidiella acerosa. 15. Plant forms a turf and has prostrate and erect axes with radial to secund, alternate branching; scale 2 cm (L. Baumhart). 16. A young axis with irregular pinnate branching; scale 2 mm. 17. A partial cross section showing thick-walled inner medullary cells grading into smaller outer ones and the oval surface cells; scale 50 µm. 18. A branch tip with one apical cell; scale 50 µm (16–18 after Littler and Littler, 2000).

Catenella caespitosa. 19. The tetrasporic plant has constricted axes and elongated segments; scale 2 cm (L. Baumhart). 20. Cross section showing a central core of medullary filaments, contiguous hollow area, and filaments radiating outward to the cortex; scale 50 µm (after Littler and Littler, 2000).

Hypnea cornuta. 21. Tip of a young branch with sessile starlike and simple spines; scale 4 mm (L. Baumhart). 22. Cross section showing a central axial cell, parenchymatous medulla, and outer pigmented cortical cells; scale 100 µm (after Børgesen, 1920).

Hypnea musciformis. 23. Upper axis with crozierlike swollen tips and spinelike branchlets; scale 1 cm. 24. Base of a thickened axis of a plant from high energy habitat with coarse spines and one crozier tip on upper axis; scale 1 cm (23, 24 after L. Baumhart in Dawes, 1981). 25. Branch tip cross section showing a central axial cell, parenchymatous medulla, and outer small cortical cells; scale 25 µm (after Schneider and Searles, 1991).

Hypnea spinella. 26. The upper axis of a drift plant is moderately spiny; scale 0.5 cm. 27. Upper axes of a turf-forming plant with irregular branching and many radial spinelike branchlets; scale 0.5 cm (both after Schneider and Searles, 1991).

Hypnea volubilis. 28. Plant is decumbent with flat axes and irregular to subdichotomous branching; scale 15 mm (after Schneider and Searles, 1991).

Dudresnaya crassa. 29. The young gelatinous plant has cylindrical axes, radial branching, and discoid holdfast; scale 5 mm (after Schneider and Searles, 1991). 30. Axil filament has uniseriate corticating ones ending in hairs; scale 50 µm (after Littler and Littler, 2000).

Dudresnaya patula. 31. The holotype is flat, gelatinous, and irregularly lobed; scale 1 cm (after Eiseman and Norris, 1981).

Chondracanthus acicularis. 32. Silhouette showing the wiry, cylindrical axes, irregular to alternate branching, and discoid holdfast; scale 2 cm (after L. Baumhart in Dawes, 1981). 33. Cross section showing a dense filamentous cortex and medulla; scale 20 µm (after Schneider and Searles, 1991).

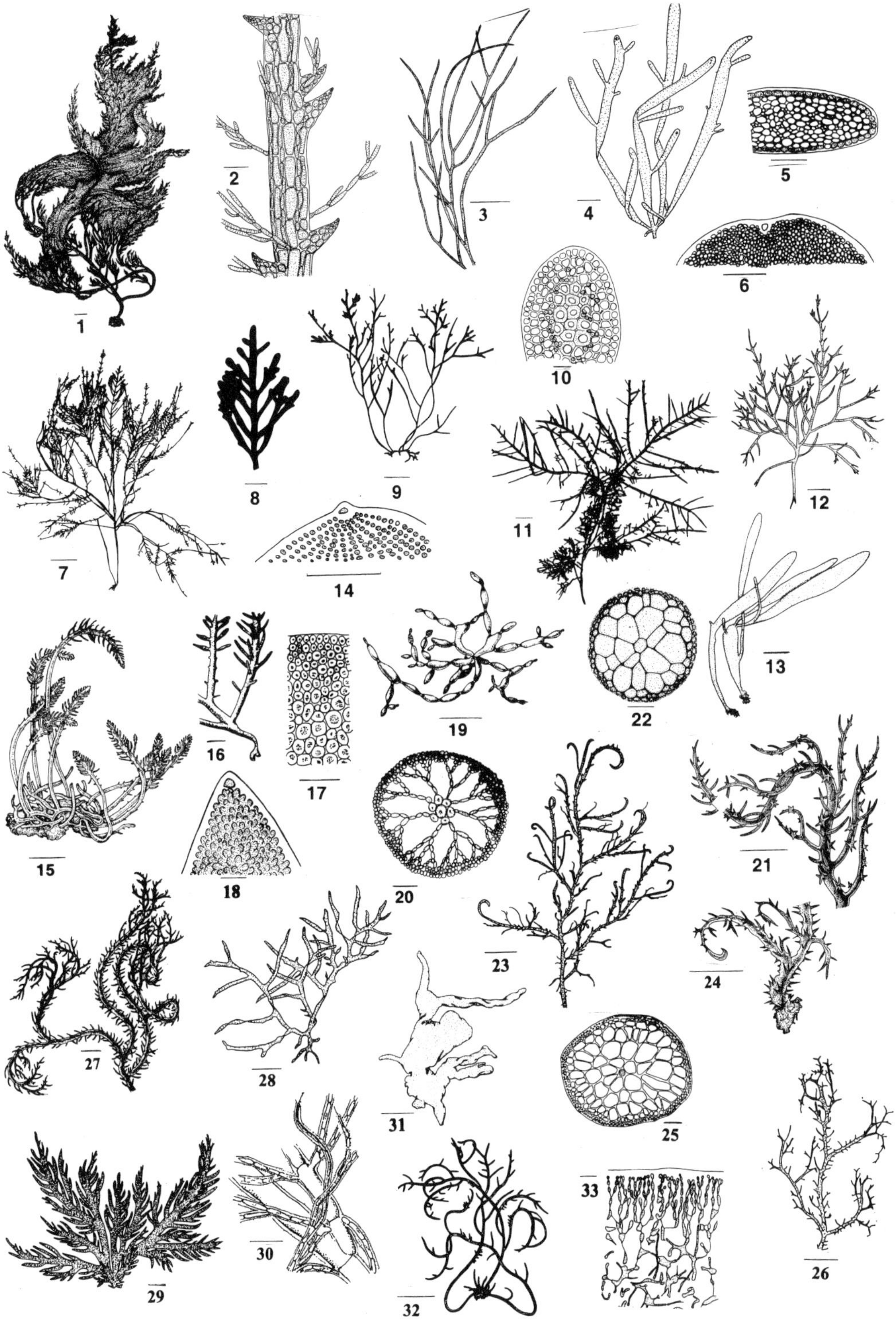
1
2
3
4
5
6
7
8
9
10
11
12
13
14
15
16
17
18
19
20
21
22
23
24
25
26
27
28
29
30
31
32
33

PLATE XLIII

Cirrulicarpus carolinensis. 1. Plant with crisp, tough blades and a cylindrical stalk; scale 2 cm. 2. Cross section showing a medulla of small filaments and larger stellate ganglia, plus a multilayered cortex of oval cells; scale 20 µm (both after Schneider and Searles, 1991).

Kallymenia limminghii. 3. Plants are membranous blades and epiphytic on a stipe of *Sargassum filipendula*; scale 5 mm (L. Baumhart).

Kallymenia westii. 4. Plant is a gelatinous, porous blade with a discoid holdfast; scale 2 cm (L. Baumhart). 5. Cross section showing small surface cells, large subcortical cells, and filaments that traverse the medulla; scale 40 µm (after Børgesen, 1919).

Metapeyssonnelia corallepida. 6, 7. Radial-vertical sections of crusts. 6. Marginal region has fountainlike growth, a perithallial zone (*arrow*), and a hypothallial row (*h, stippled*). 7. Central tissue region has a *Peyssonnelia*like structure with a basal hypothallial row (*h, stippled*) and developing hairs (*arrows*). 8. Crust surface with hairs that arise from wide perithallial filaments (*stippled*); all scales 50 µm (all after Veerlaque et al., 2000).

Peyssonnelia inamoena. 9. The fleshy crust has raised margins; scale 5 mm. 10, 11. Radial (10) and tangential (11) sections showing the hypothallus with a layer of rectangular cells and rhizoids, the perithallus with vertical rows of irregular to rectangular cells, and small surface cells; scales 20 µm (all after Schneider and Reading, 1987).

Peyssonnelia rosenvingi. 12. The two crusts are fleshy with radial lines and white edges; scale 1 cm (after Villalard-Bohnsack, 1995). 13. Radial section of a thick crust showing the hypothallus and vertical rows of perithallial cells; scale 50 µm (after Bird and McLachlan, 1992). 14. Cross section of a thin crust showing nemathecial filaments with a developing tetrasporangium plus a mature one; scales 10 µm (after Børgesen, 1929).

Peyssonnelia rubra. 15. Crust is fleshy, lobed, and with concentric lines and raised edges; scale 5 mm (R. Holbrook). 16. Crust surface with a few clear, large, crystal-containing cells; scale 50 µm (after Abbott, 1999). 17. Longitudinal section of margin showing a perithallus with an apical cell and hypothallus forming a rhizoid; scale 20 µm (after Børgesen, 1929).

Peyssonnelia simulans. 18. Crust is fleshy, membranous, and loosely adherent; scale 5 mm. 19. Cross section of crust showing large basal hypothallial cells with rhizoids; each cell bears multiple perithallial rows; scale 20 µm (18, 19 after Schneider and Searles, 1991). 20. Cross section of a fertile crust showing vertical rows of perithallial cells, nemathecial filaments, and carpospores; scale 100 µm (after Børgesen, 1917).

Gymnogongrus griffithsiae. 21. Plant has tough coarse axes, irregular branching, and blunt tips; scale 2 cm (after L. Baumhart in Dawes, 1981). 22. A young, irregularly branched plant with cystocarps; scale 1 cm (R. Holbrook). 23. Cross section showing the nemathecia filaments of the cortex radiating to the surface and internal medullary filaments; scale 10 µm (after Schneider and Searles, 1991).

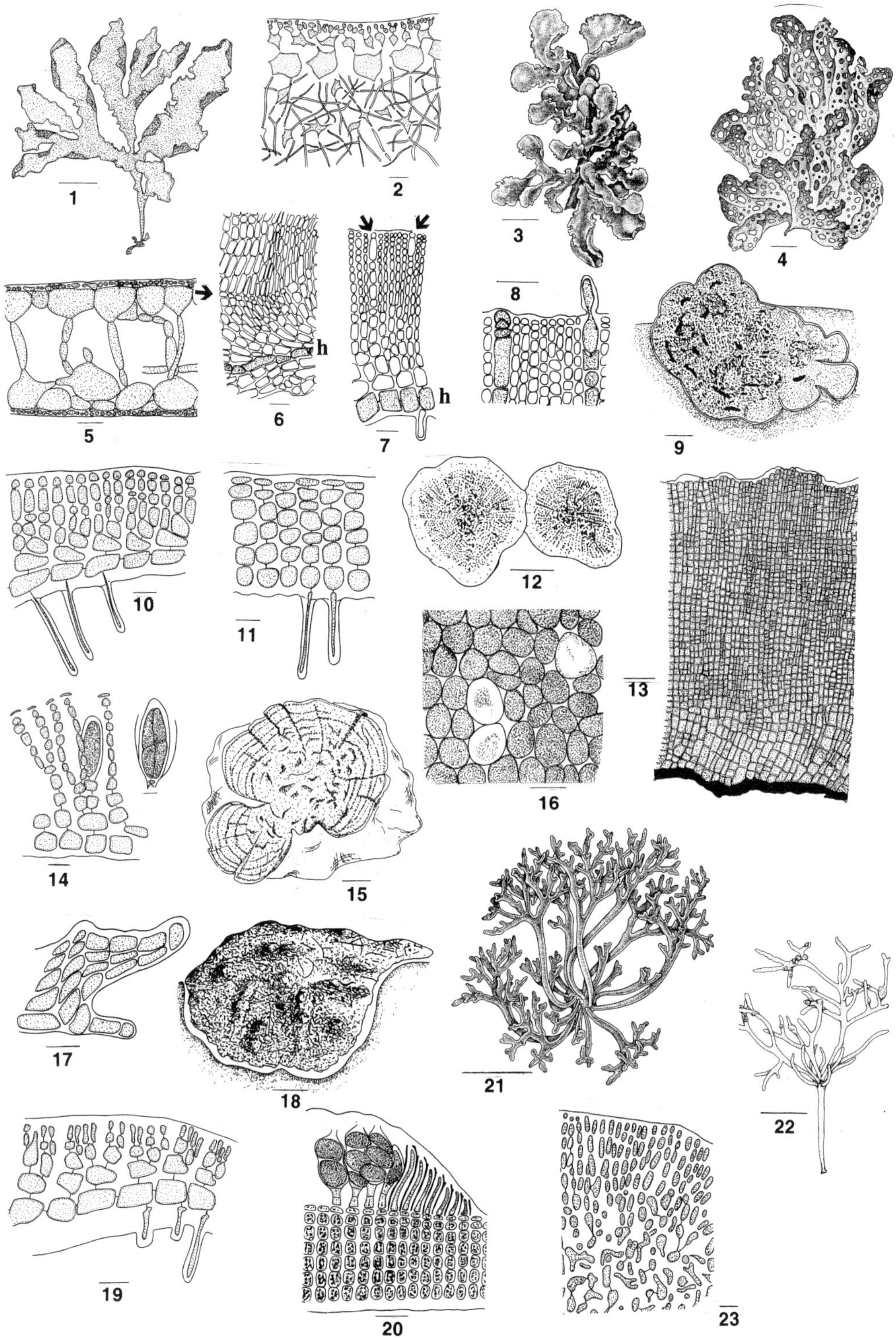
1
2
3
4
5
6
h
7
h
8
9
10
11
12
13
14
15
16
17
18
21
22
19
20
23

PLATE XLIV

Agardhiella floridana. 1. The plant has three irregular fleshy frond segments, a flat stipe, and discoid holdfast; scale 10 cm (after Guimarães and Oliveira, 1996). 2. Cross section showing loosely arranged medullary filaments, radially organized cortical cells of variable shapes, and small pigmented surface cells; scale 100 µm (after Schneider and Searles, 1991).

Agardhiella ramosissima. 3. The young plant has a compressed main axis and alternate to opposite branching; scale 1 cm. 4. Cross section showing a medulla of loosely arranged filaments, layers of cortical cells of variable size, and five zonately divided tetrasporangia in the outer cortex of tiny cells; scale 100 µm (both after Schneider and Searles, 1991).

Agardhiella subulata. 5. The plant has a fleshy, cylindrical axis, radial and alternate branching, and cystocarps that swell the branchlets; scale 2 cm (after Schneider and Searles, 1991). 6. Cross section showing loosely arranged medullary filaments, large inner cortical cells, and small pigmented outer ones; scale 20 µm (after Børgesen, 1919).

Eucheuma isiforme. 7. A plant from Bahia Honda Key that is spiny, decumbent, and has many haptera-like holdfasts (*arrow*); scale 2 cm. 8. A plant from Molasses Key that is erect, bushy, with one holdfast, and has many thick, short, knobby spines; scale 2 cm (both by L. Baumhart).

Eucheuma isiforme var. *denudatum*. 9. A plant from deep water (i.e., 20 m) that lacks spines, has many lateral cystocarps and a discoid holdfast; scale 2 cm (L. Baumhart). 10. Cross section showing a dense filamentous medulla and a cortex of larger inner and smaller outer ones that is typical of the species; scale 50 µm (C. Torres).

Meristotheca gelidium. 11. The plant has compressed main axes and branchlets covered by spines; scale 2 cm. 12. A plant with cylindrical main axes covered with spines; scale 2 cm (11, 12, L. Baumhart). 13. Cross section showing a loose filamentous medulla and large inner and smaller outer cortical cells; scale 200 µm (after Littler and Littler, 1997).

Sarcodiotheca divaricata. 14. Plant is flat, dichotomously branched, and with discoid holdfast; scale 1 cm. 15. Cross section showing a loose filamentous medulla with larger inner and smaller outer cortical cells; scale 50 µm (both after Schneider and Searles, 1991).

Solieria filiformis. 16. A young plant with cylindrical axes, radial alternate branching, and a discoid holdfast; scale 2 cm (L. Baumhart). 17. Cross section showing a loose filamentous medulla, plus larger inner and smaller outer cortical cells; scale 200 µm (after Schneider and Searles, 1991).

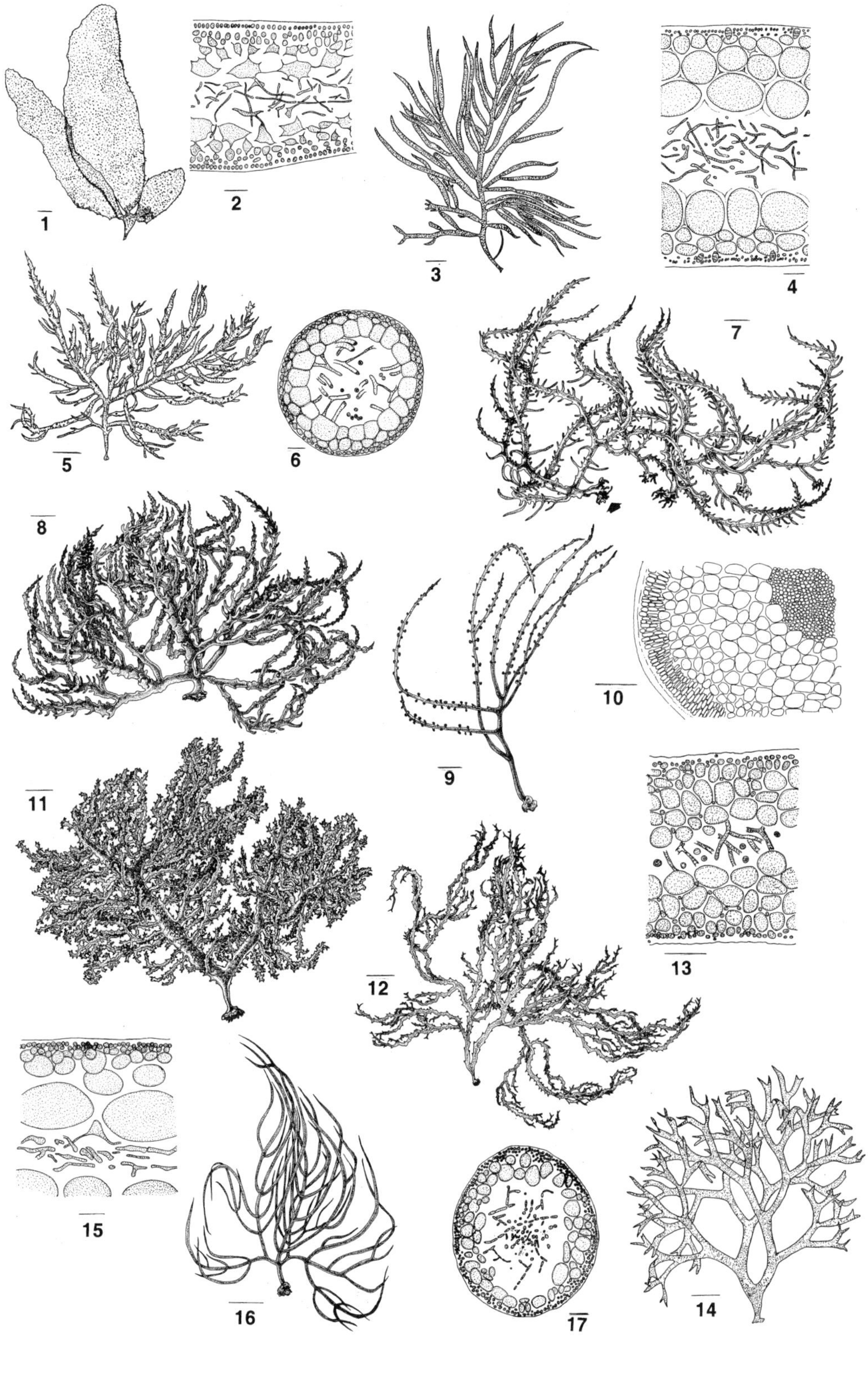
1
2
3
4
5
6
7
8
9
10
11
12
13
14
15
16
17

PLATE XLV

Wurdemannia miniata. 1. The dense turf consists of many wiry plants; scale 6 cm. 2. Three axes have irregular branching and short branchlets; scale 2 mm (1, 2, L. Baumhart). 3. The branch tip has multiple apical cells; scale 20 µm. 4. Cross section showing the densely packed medullary filaments, oval-shaped cortical cells, and thick-walled surface cells; scale 100 µm (3, 4 after Børgesen, 1919).

Acrosymphyton caribaeum. 5. Upper axes of a large plant are compressed and alternately branched; scale 2 cm (after Setchell, 1912).

Gracilaria armata. 6. Tip of a compressed axis with spherical cystocarps; scale 1 cm. 7. A cross section showing cruciate tetrasporangia in the outer cortex (*arrow*); scale 50 µm. 8. A longitudinal section showing rectangular medullary cells that grade into smaller cortical cells that end in tiny pigmented surface cells; scale 100 µm (all after Kützing, 1843).

Gracilaria blodgettii. 9. The plant has cylindrical axes, basally constricted branchlets, and a discoid holdfast; scale 2 cm (L. Baumhart). 10. A cross section showing medullary cells of variable sizes that grade into small cortical cells and tiny pigmented surface cells; scale 100 µm (after Schneider and Searles, 1991).

Gracilaria bursa-pastoris. 11. The upper part of an older plant has compressed axes and irregular branching; scale 2 cm (after Taylor, 1928). 12. The tip of an older plant has secund branching; scale 2 mm (after Littler and Littler, 2000). 13. A cross section showing a parenchymatous medulla that grades into small cortical cells, and two tiny pigmented surface cells; scale 200 µm (after Littler and Littler, 2000).

Gracilaria cervicornis. 14. Upper axes are compressed, tapered toward tips, and with dense irregular branching; scale 1 cm (after Taylor, 1960). 15. A cross section showing a gradual transition between the parenchymatous medulla, inner and outer cortex, and tiny pigmented surface cells; scale 100 µm (after Littler and Littler, 2000).

Gracilaria cuneata. 16. The upper axes are tough, flat, dichotomously branched, and with ruffled margins; scale 1 cm. 17. A cross section showing a gradual transition between the parenchymatous medulla, inner and outer cortex, and tiny pigmented surface cells; scale 50 µm (both after Littler and Littler, 2000).

Gracilaria curtissiae. 18. A cross section showing an abrupt transition between the large-celled parenchymatous medulla and cortex with pigmented surface cells; scale 40 µm (after Gurgel et al., 2004a). 19. Plant is cystocarpic, flat, strap-shaped, has few proliferations, and a discoid holdfast; scale 1 cm (after Schneider and Searles, 1991).

Gracilaria cylindrica. 20. The plant has thick cylindrical axes, basally constricted branches, and a discoid holdfast; scale 2 cm (L. Baumhart). 21. A cross section showing large medullary cells and an abrupt transition to the small pigmented cortical cells; scale 500 µm (after Børgesen 1920).

Gracilaria damaecornis. 22. The upper frond has regular dichotomous branching; scale 1 cm (after Littler and Littler, 2000). 23. An older irregularly branched axis with blunt tips and many cystocarps; scale 1 cm (after Taylor, 1960). 24. A cross section showing a gradual transition from large medullary cells to smaller cortical cells to tiny pigmented surface ones; scale 25 µm (after Littler and Littler, 2000).

Gracilaria domingensis. 25. The upper axes are strap-shaped, compressed, and with marginal, mostly pinnate branching; scale 2 cm (after Taylor, 1960). 26. A tip of an older flat axis has marginal branching; scale 2 cm (Gary Landtrip).

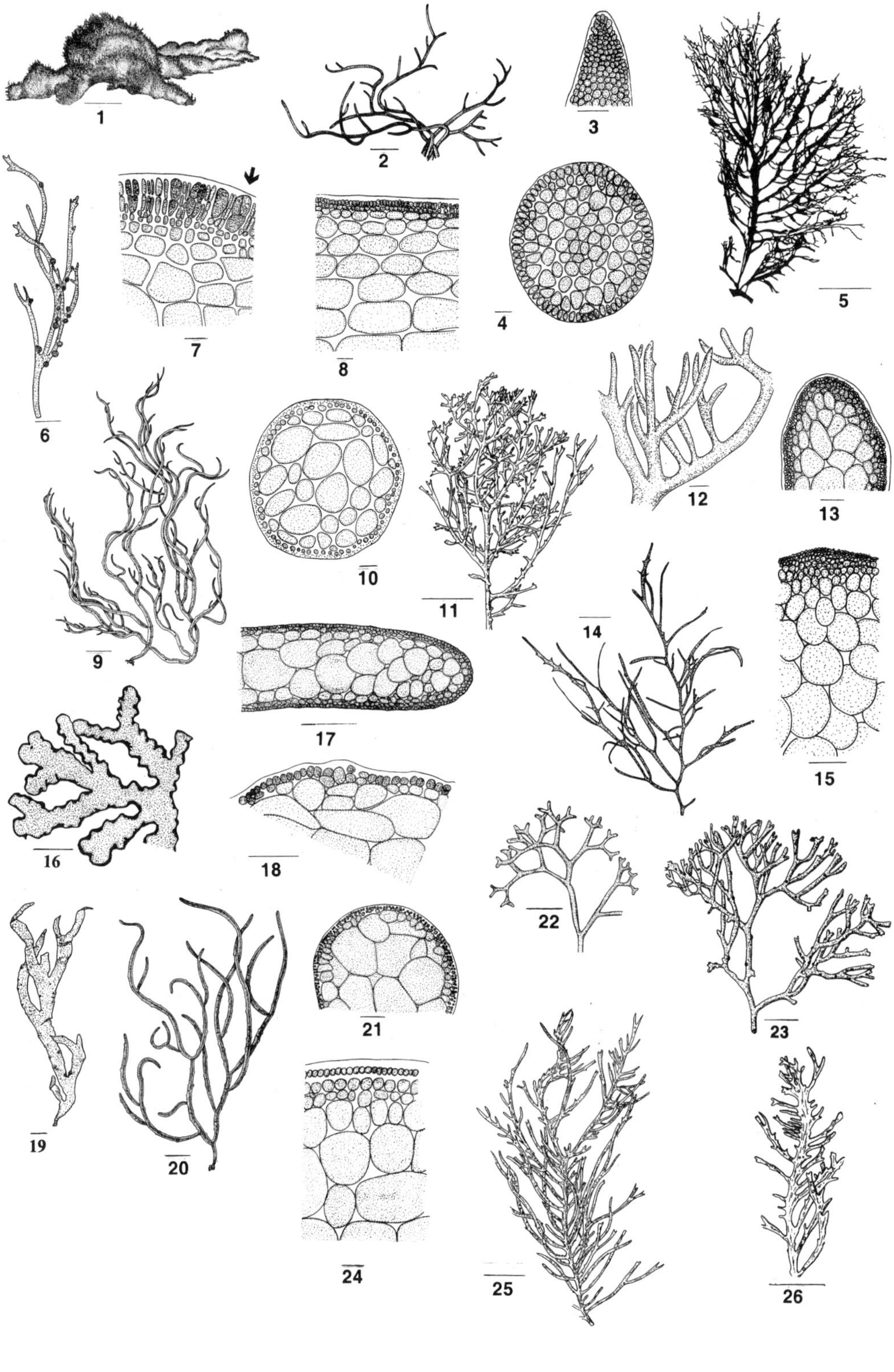
1
2
3
4
5
6
7
8
9
10
11
12
13
14
15
16
17
18
19
20
21
22
23
24
25
26

PLATE XLVI

Gracilaria flabelliformis. 1. Silhouette of a young plant showing irregular branching, strap-shaped axes, and a discoid holdfast; scale 5 mm (after Gurgel et al., 2004c). 2. Silhouette of an older plant showing irregular branching, strap-shaped axes, and a discoid holdfast; scale 2 cm (Frank Hoff). 3. Silhouette of a small plant showing broad axes and rounded tips; scale 5 mm. 4. A cross section showing a gradual transition between the medulla, cortex, and tiny pigmented surface cells; scale 200 μm (3, 4 after Gurgel et al., 2004c).
Gracilaria galetensis. 5. Silhouette of a tip showing flat, subdichotomous branching; scale 4 cm. 6. A cross section showing a gradual transition between the medulla and the two-layered cortex with pigmented surface cells; scale 100 μm (both after Gurgel et al., 2004a).
Gracilaria hayi. 7. A silhouette showing two flat axes, dichotomous to subdichotomous branching, and rounded tips; scale 2 cm (after Gurgel et al., 2004a).
Gracilaria intermedia. 8. A silhouette showing a compressed main axis, irregular to alternate to opposite branching, and many marginal branches; scale 2 cm. 9. A cross section of an old axis showing gradual transition between the large parenchymatous medullary cells, irregular cortex of 5–7 layers, and surface cells; scale 100 μm (both after Gurgel et al., 2004a).
Gracilaria isabellana. 10. Upper axis is wide, strap-shaped, and cystocarpic; scale 1 mm. 11. Upper axis is narrow, strap-shaped, and cystocarpic; scale 1 mm. 12. A cross section showing an abrupt transition from the medulla to the 2-layered pigmented cortex; scale 200 μm (both after Gurgel et al., 2004c).
Gracilaria mammillaris. 13. The plant has many flat, dichotomously branched axes, rounded tips, and a fibrous holdfast; scale 2 cm (L. Baumhart). 14. A cross section showing an abrupt change from the medulla to a 2-layered pigmented cortex; scale 50 μm (after Taylor, 1928).
Gracilaria occidentalis. 15. Upper axis is flat, with irregular branches narrow at forks; scale 1 cm. 16. A cross section showing a medulla of thick-walled cells, an abrupt transition to the cortex, and pigmented surface cells; scale 50 μm (both after Børgesen, 1920).
Gracilaria tikvahiae. 17. Upper axes are compressed and flat at forks; scale 1 cm (after Schneider and Searles, 1991). 18. The drift plant is cystocarpic with axes compressed only at forks (holdfast still present); scale 2 cm. 19. A cross section showing compressed medullary cells and an abrupt transition to a 2–3-layered cortex with pigmented surface cells; scale 50 μm (18, 19 after Littler and Littler, 2000).
Gracilaria venezuelensis. 20. A silhouette showing upper axes that are compressed to flattened at forks and alternately to dichotomously branched; scale 2 cm. 21. A cross section showing an abrupt change between the medulla and the tiny pigmented surface cells; scale 100 μm (both after Taylor, 1942).
Gracilariopsis lemaneiformis. 22. The tips of two older cystocarpic plants have distinct main axes and little branching; scale 1 cm (after Dawson 1949a). 23. A tip with cystocarps; scale 1 cm (after Schneider and Searles, 1991). 24. The upper part of a young plant has an indistinct main axis and many alternating branches; scale 5 mm (R. Holbrook). 25. A cross section of a young axis showing large parenchymatous medullary cells that grade into 2–3 layers of small cortical cells; scale 200 μm. 26. A cross section of an old axis showing parenchymatous medullary cells that grade into many cortical layers with pigmented surface cells; scales 200 μm (25, 26 after Dawson, 1949a)
Hydropuntia caudata. 27. The plant has irregular to subdichotomous branching and a discoid holdfast (*arrow*); scale 2 cm (L. Baumhart). 28. A cross section showing the large medullary parenchyma cells grading into multiple cortical layers, including pigmented surface cells; scale 100 μm (after Børgesen, 1920).

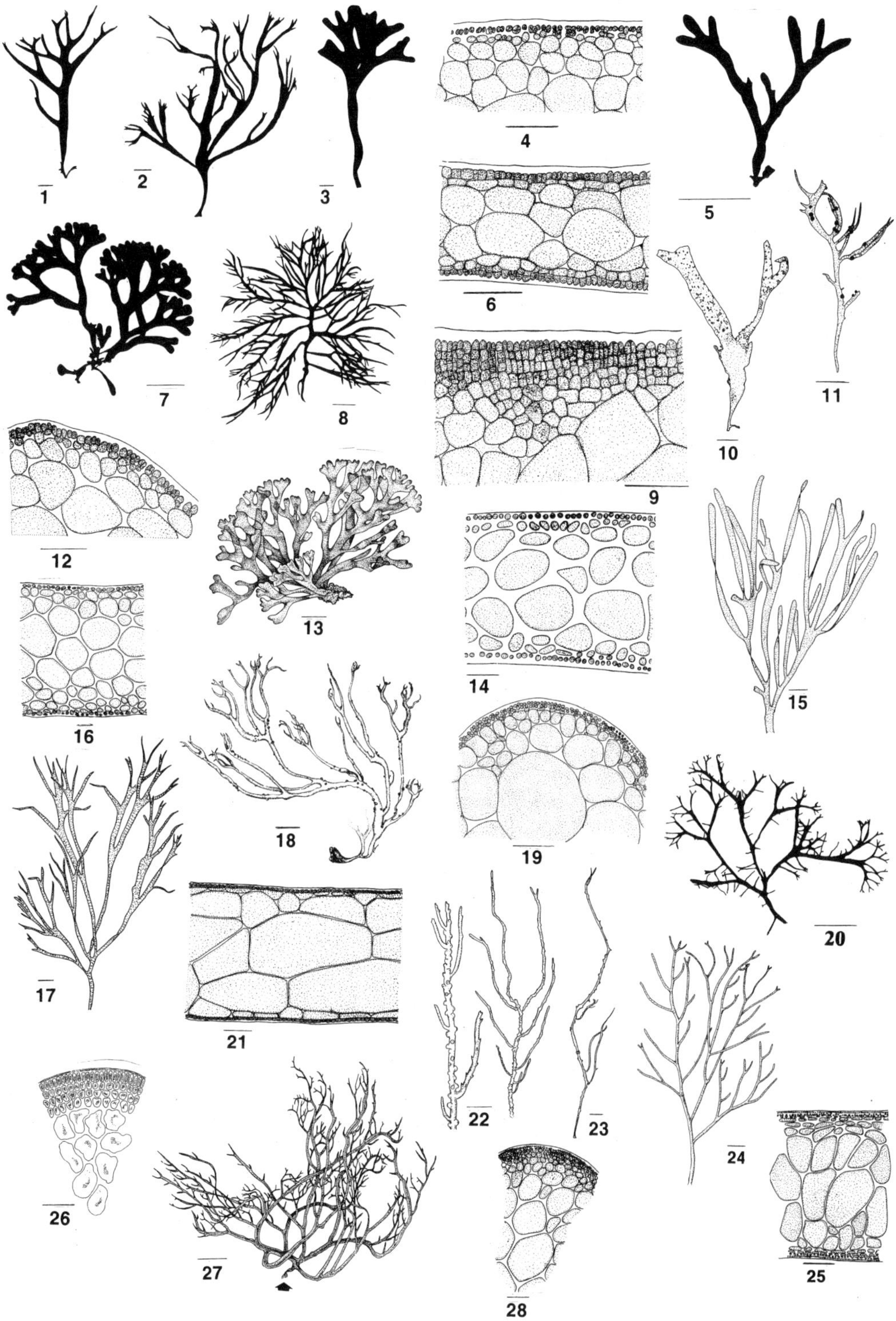
1
2
3
4
5
6
7
8
9
10
11
12
13
14
15
16
17
18
19
20
21
22
23
24
25
26
27
28

PLATE XLVII

Hydropuntia cornea. 1. The plant has coarse axes, secund branching, and a discoid holdfast; scale 2 cm (after L. Baumhart in Dawes, 1981). 2. A cross section of a young axis showing thin-walled medullary cells grading into smaller cortical and pigmented surface cells; scale 300 μm. 3. The same, with thick-walled medullary cells; scale 100 μm (both after Taylor, 1960).

Hydropuntia crassissima. 4. The upper part of a flat, rugose frond has narrow irregular branches; scale 1 cm. 5. A cross section showing thick-walled medullary, smaller cortical, and oval pigmented surface cells; scale 100 μm (both after Taylor, 1928).

Hydropuntia secunda. 6. The upper cylindrical axis has irregular to subdichotomous branching; scale 6 mm. 7. A cross section showing 2 pit connections between round surface cells and medullary cells; scale 5.0 μm (both after Gurgel and Fredericq, 2004).

Corynomorpha clavata. 8. Two unbranched club-shaped plants with discoid holdfasts; scale 1 cm (after Taylor, 1960). 9. A cross section showing inner medullary filaments and outer cortical and surface cells; scale 50 μm (after Littler and Littler, 2000).

Cryptonemia crenulata. 10. Frond is strap-shaped, mostly basally branched, and with a discoid holdfast; scale 1 cm. 11. The frond tip is highly branched; scale 1 cm (10, 11 after Taylor, 1960). 12. Marginal proliferations of frond are stellate; scale 50 μm. 13. A cross section showing a dense medulla of thick-walled filaments, thin-walled cortical cells, and elliptical, pigmented surface cells; scales 50 μm (12, 13 after Littler and Littler, 2000).

Cryptonemia seminervis. 14. Plant is foliose, and has a midrib, cylindrical stipe, discoid holdfast (*left*); also a ragged terminal segment (*right*); scale 1 cm (R. Holbrook).

Grateloupia cuneifolia. 15. A cross section showing one tetrasporangium (*arrow*) in the cortex plus medullary and cortical filaments; scale 20 μm (after Børgesen 1916). 16. The upper part of the frond is flat, branched, and has a cylindrical stipe; scale 2 cm (after Taylor, 1960). 17. Stellate medullary cells that form "ganglia"; scale 20 μm (after Børgesen 1916).

Grateloupia gibbesii. 18. Fronds are basally branched, with marginal proliferations, and on a short stipe with a discoid holdfast; scale 2 cm (after Harvey, 1853). 19. A cross section showing a few transverse medullary filaments connected to stellate cortical cells (*arrow*) and small pigmented surface cells; scale 50 μm (after Littler and Littler, 2000).

"*Grateloupia filicina.*" 20. The plant has cylindrical axes, numerous secondary branches, and a discoid holdfast; scale 2 cm (L. Baumhart). 21. The base of an old (injured) plant has a compressed main axis and discoid holdfast; scale 2 cm (after Taylor, 1960).

Halymenia elongata. 22. The plant has gelatinous cylindrical axes, dichotomous branching, rounded tips, and a discoid holdfast; scale 2 cm (L. Baumhart).

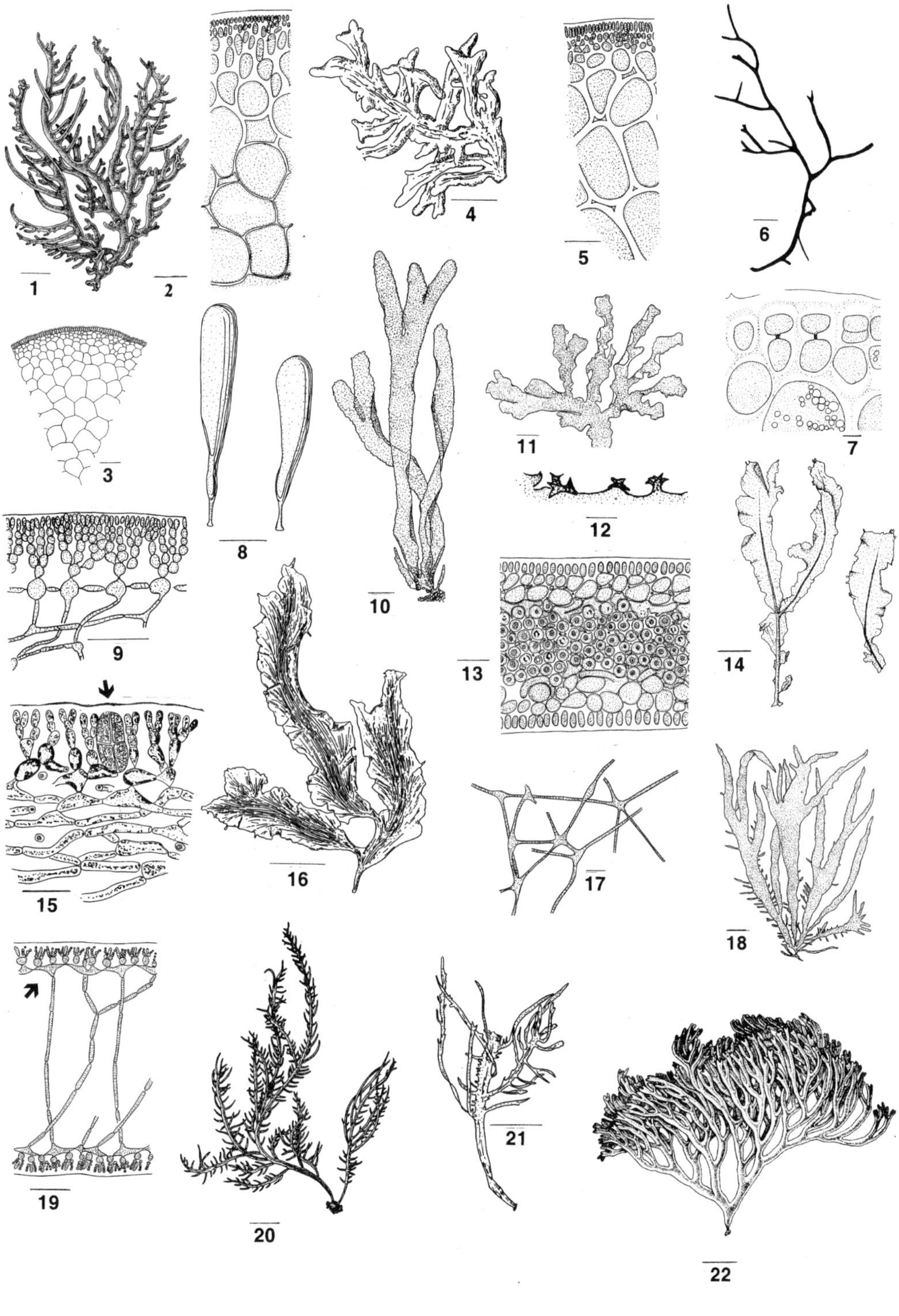
1
2
4
5
6
3
8
10
11
7
12
9
13
14
15
16
17
18
19
20
21
22

PLATE XLVIII

Halymenia floresii. 1. Fronds are gelatinous, strap-shaped, and with many marginal proliferations, a short stipe, plus a discoid holdfast; scale 2 cm (L. Baumhart). 2. A cross section showing a loose medulla with transverse filaments, a few stellate ganglia, and cruciately divided tetrasporangia (*arrow*) in the cortex; scale 100 μm.

Halymenia floridana. 3. A plant with numerous membranous blades on a cylindrical stipe; scale 2 cm (after Taylor, 1960). 4. Frond is fleshy, apically divided, and with a short stipe and discoid holdfast; scale 2 cm (L. Baumhart). 5. A cross section showing transverse medullary filaments connected to stellate subcortical cells (*arrow*); scale 50 μm (after Schneider and Searles, 1991). 6. Subcortical stellate cells or "ganglia"; scale 50 μm (after Taylor, 1928).

Halymenia hancockii. 7. Two small blades growing on a coral; scale 1 cm. 8. Frond is old and damaged; scale 1 cm (7, 8 after Taylor, 1942). 9. A cross section showing a few medullary filaments and many cruciate tetrasporangia (*arrow*) in the cortex; scale 20 μm (after Schneider and Searles, 1991).

Halymenia integra. 10. Blade is shield-shaped and on a short stipe; scale 1 cm. 11. A cross section showing a filamentous medulla plus larger inner and smaller outer cortical cells. 12. Fused medullary filaments; scales 20 μm (all after Howe and Taylor, 1931).

Halymenia pseudofloresii. 13. Plant is divided into many blades, sessile, membranous, and with abundant marginal proliferations and a discoid holdfast; scale 2 cm (L. Baumhart). 14. A cross section showing small pigmented outer cortical cells, larger and stellate inner ones, and filaments traversing the medulla; scale 50 μm (after Littler and Littler, 2000).

Halymenia rosea. 15. The frond is oval and on a short stipe with a discoid holdfast; scale 1 cm. 16. A stellate cell from the inner cortex; scale 50 μm. 17. A cross section showing a filamentous medulla plus large inner and smaller outer cortical cells; scale 50 μm (all after Howe and Taylor, 1931).

Sebdenia flabellata. 18. The cylindrical to slightly compressed axes are dichotomously branched; scale 2 cm (after Taylor, 1960). 19. Stellate ganglia and gland cells are in the medulla; scale 50 μm (after Schneider and Searles, 1991).

Nemastoma gelatinosum. 20. Branch is gelatinous, compressed, and with lobed tips; scale 1 cm (after Taylor, 1960). 21. A cross section showing a dichotomously branched cortical filament with a basal gland cell (*arrow*); scale 50 μm (after Littler and Littler, 2000).

Predaea feldmannii. 22. Plant is gelatinous, flat, and with many small lobes; scale 1 cm. 23. Cortical filaments are dichotomously branched and bear a young carposporophyte near nutritive cells (*arrow*); scale 20 μm (both after Schneider and Searles, 1991).

Predaea masonii. 24. Blade is simple, subspherical, covered with papillae, and with a discoid holdfast; scale 5 mm. 25. Cortical filaments bear an auxiliary cell surrounded by dense nutritive cells (*arrow*); scale 20 μm (after Schneider and Searles, 1991).

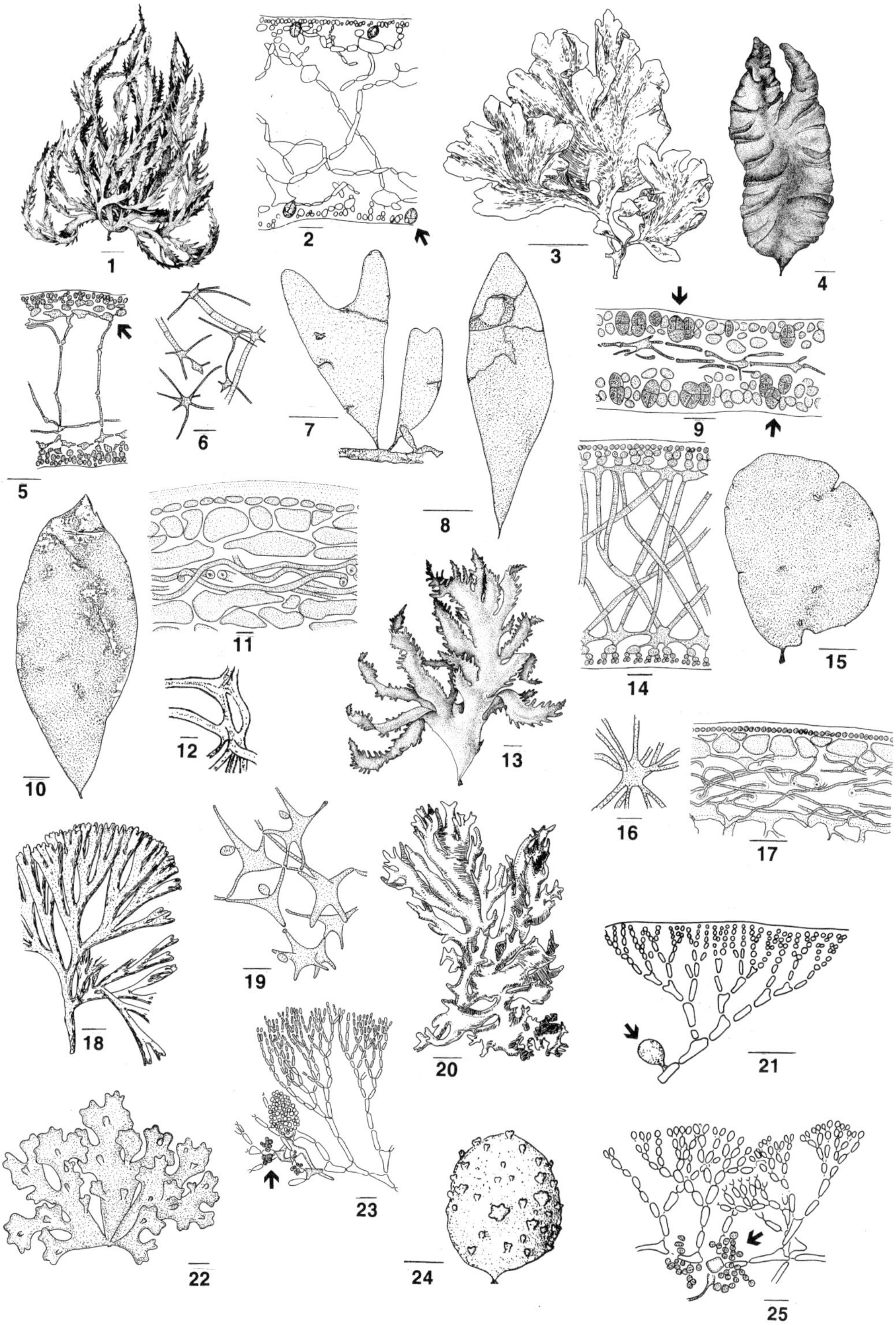
1
2
3
4
5
6
7
8
9
10
11
12
13
14
15
16
17
18
19
20
21
22
23
24
25

PLATE XLIX

Platoma cyclocolpum. 1. The upper part of a gelatinous flat frond has irregular to palmate lobes; scale 1 cm (after Taylor, 1960).

Platoma tenue. 2. Frond is flat, lobed, and gelatinous; scale 1 cm. 3. Cortical filaments extracted from a macerated blade; scale 20 µm (both after Howe and Taylor, 1931).

Titanophora incrustans. 4. A silhouette showing a highly divided flat blade; scale 1 cm (after Børgesen, 1949b).

Asteromenia peltata. 5. Two fleshy peltate blades with one showing a short stipe (*left*) and the other a top view (*right*); scale 1 cm (both after Taylor, 1960). 6. A cross section showing large medullary cells, small cortical cells, and very small pigmented surface cells with cruciate tetrasporangia (*arrow*); scale 50 µm (after Schneider and Searles, 1991).

Botryocladia caraibica. 7. The stalk has two terminal vesicles; scale 1 cm. 8. Cross section of a vesicle membrane showing a 3–4 celled cortical layer and pigmented surface cells; scale 20 µm (after Gavio and Fredericq, 2003).

Botryocladia occidentalis. 9. The branch tip has many vesicles; scale 2 cm (after Littler and Littler, 2000). 10. The erect plant has many vesicles, cylindrical axes, and a discoid holdfast; scale 2 cm (L. Baumhart). 11. Cross section of vesicle membrane showing 3–4 cortical layers, pigmented surface cells, and one gland cell (*arrow*) on a large inner cell; scale 100 µm (after Littler and Littler, 2000).

Botryocladia pyriformis. 12. Three small plants growing on limestone, each with 1–2 vesicles; scale 2 cm (after Taylor, 1960). 13. A larger plant with 20 mostly lateral vesicles; scale 2 cm. 14. Cross section of a vesicle stalk showing an abrupt transition between outer cortical and larger medullary cells; scale 500 µm (13, 14 after Børgesen, 1920). 15. A cross section of the vesicle membrane showing 1–2 layers of large cortical cells, pigmented surface cells, and a cluster of gland cells (*arrow*) on an inner cortical cell; scale 100 µm (after Littler and Littler, 2000).

Botryocladia spinulifera. 16. The plant is decumbent, irregularly and sparsely branched, and with several vesicles; scale 2 mm (after Taylor and Abbott, 1973). 17. Cross section of a vesicle membrane with spinelike surface cells and one gland cell (*arrow*) on an inner cortical cell; scale 50 µm (after Littler and Littler, 1997).

Botryocladia wynnei. 18. Each of the two plants have one vesicle; scale 1 cm. 19. The vesicle surface has small outer cells that vary in size and line the edges of larger inner cells; scale 20 µm (18, 19 after Schneider and Searles, 1991).

Chrysymenia agardhii. 20. The blade is thick, gelatinous, dissected, and with a discoid holdfast; scale 1 cm. 21. A cross section showing small cortical and pigmented surface cells, plus a single gland cell on a large inner cortical cell (*arrow*); scale 50 µm (after Schneider and Searles, 1991).

Chrysymenia enteromorpha. 22. The plant has swollen gelatinous axes, irregular radial branching, and a discoid holdfast; scale 2 cm (L. Baumhart). 23. A cross section showing a hollow medulla, multiple gland cells (*arrow*) on an inner cortical cell, and small pigmented surface cells; scale 50 µm. 24. An internal view of inner medullary cells shows many gland cells; scale 50 µm (23, 24 after Littler and Littler, 2000).

Chrysymenia halymenioides. 25. The plant has gelatinous swollen axes, dichotomous branching, a short cylindrical stipe, and a discoid holdfast; scale 2 cm (L. Baumhart).

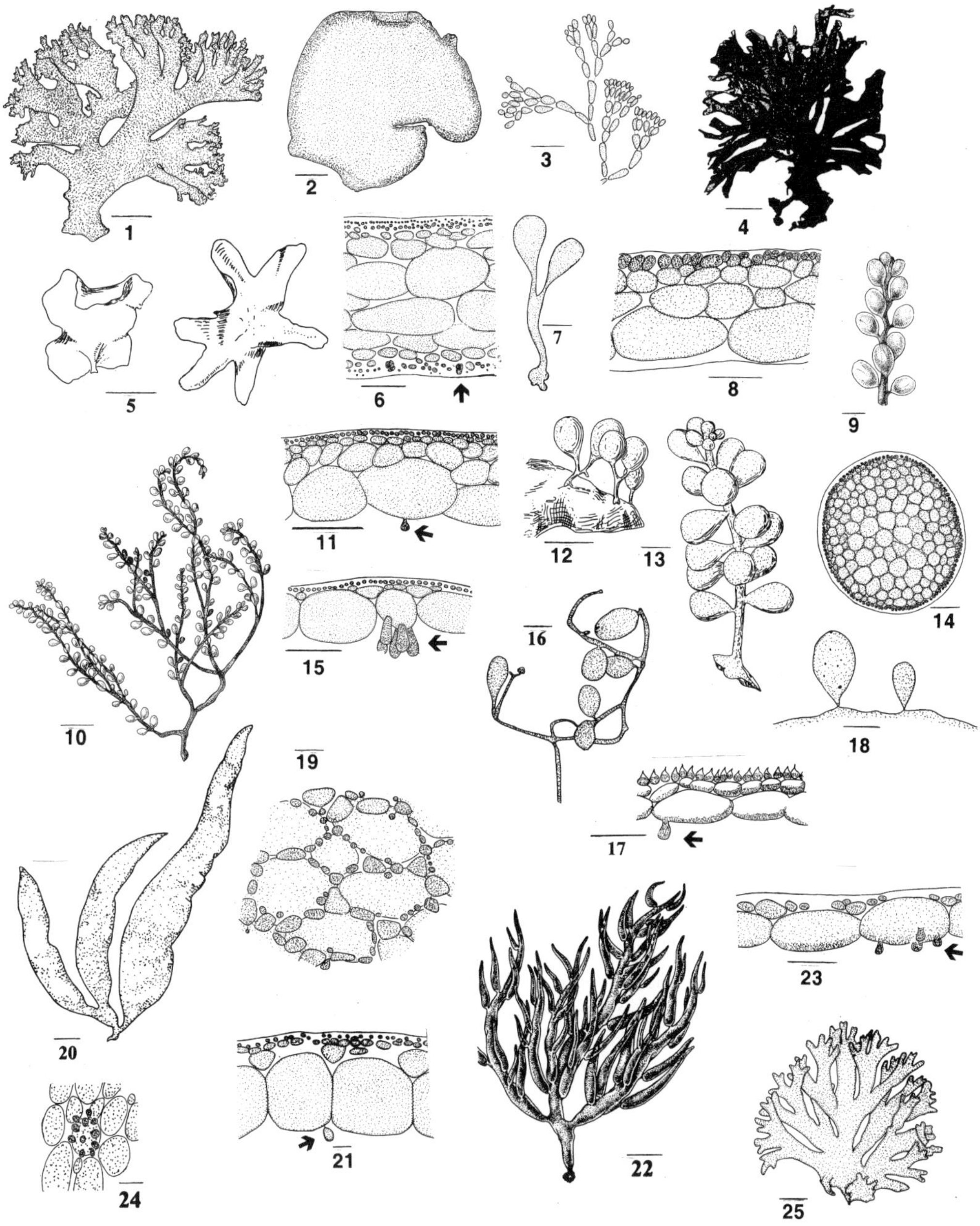
1
2
3
4
5
6
7
8
9
10
11
12
13
14
15
16
17
18
19
20
21
22
23
24
25

PLATE L

Chrysymenia planifrons. 1. The blade is gelatinous, lobed, and on a short stalk; scale 2 cm (after Taylor, 1960). 2. A cross section showing a loose filamentous medulla, two gland cells attached on an inner cortical cell (*inner right*), and small outer pigmented surface cells with tetrasporangia (*arrow*); scale 50 μm (after Børgesen, 1920).
Chrysymenia ventricosa. 3. The upper axes are irregularly branched, fleshy, and cylindrical to compressed; scale 2 cm (after Taylor, 1960).
Coelarthrum cliftonii. 4. The upper axes are dichotomously branched, regularly constricted, and gelatinous; scale 2 mm. 5. An internal view of inner cortical cells showing a stellate cell and two gland cells (*center*); scale 50 μm (4, 5 after Børgesen, 1920). 6. The cross section of the thallus membrane shows a spherical gland cell (*arrow*) below and pigmented surface cells aligned on the edges of large cortical cells; scale 100 μm. 7. The surface view of the thallus membrane shows an interior gland cell (*arrow*) and small surface cells aligned along edges of large inner cells; scale 100 μm (6, 7 after Littler and Littler, 2000).
Coelothrix irregularis. 8. The prostrate axis has one basal tuft of rhizoids and three irregularly branched, cylindrical, erect axes; scale 2 mm. 9. A cross section showing the hollow axis with spherical cortical and small pigmented surface cells; scale 400 μm (8, 9, A. L. Paul). 10. A longitudinal section of an axis showing elongate inner cortical cells, two gland cells (*arrow*), and outer cortical plus small surface cells; scale 100 μm (after Børgesen, 1920).
Maripeltia atlantica. 11. The two peltate plants have fleshy stalks and discoid holdfasts; the lower plant has abundant marginal cystocarps; scale 2 cm (L. Baumhart).
Rhodymenia divaricata. 12. The plant has four broad dichotomously branched fronds and a discoid holdfast; scale 1 cm (after Schneider and Searles, 1991). 13. A cross section of the cortex shows small pigmented surface cells and larger inner cortical cells; scale 100 μm (after Littler and Littler, 2000).
Rhodymenia pseudopalmata. 14. The upper part of a strap-shaped, dichotomously branched frond has tips with bands of tetrasporangial sori; scale 2 cm (after Schneider and Searles, 1991). 15. The upper axes are irregularly branched; scale 2 cm. 16. A cross section showing large medullary cells that grade into small cortical and pigmented surface cells; scale 50 μm. 17. A longitudinal section showing a gradual change from elongate medullary cells to oval outer cortical and pigmented surface cells; scale 50 μm (15–17 after Littler and Littler, 2000).
Fauchea hassleri. 18. The plant is flat, dichotomously branched, and with entire margins and a discoid holdfast; scale 2 cm (L. Baumhart). 19. A cross section showing large irregular or oval medullary cells that grade into anticlinal rows of cortical and surface cells; scale 20 μm (after Howe and Taylor, 1931).
Gloiocladia atlantica. 20. The upper axis is gelatinous and pinnately branched; scale 2 mm (after Schneider and Searles, 1991). 21. A cross section at frond margin showing a cortex of small radiating cortical cells (3–4 μm in diam.) and large polygonal medullary cells; scale 200 μm (after Searles, 1972).
Gloiocladia blomquistii. 22. Upper axes are compressed, gelatinous, and dichotomously to irregularly branched; scale 1 cm. 23. A cross section showing one layer of large medullary cells and a cortex of smaller cells (4–6 μm in diam.); scale 50 μm (both after Schneider and Searles, 1991).
Gloiocladia rubrispora. 24. Upper branches are fleshy, cylindrical, and covered by knobby branchlets and horned cystocarps (*arrows*); scale 1 mm (after Searles, 1984).
Leptofauchea brasiliensis. 25. The plant has irregularly lobed to strap-shaped axes; scale 1 cm. 26. A cross section showing large inner medullary cells of variable size surrounded by smaller cortical and surface cells; scale 100 μm (both after Joly, 1965).
Champia parvula. 27. The upper axis has a cystocarp (*arrow*) and regular constrictions; scale 4 mm (F. Hoff). 28. The same, with the plant being vegetative; scale 4 mm (after Taylor, 1960).

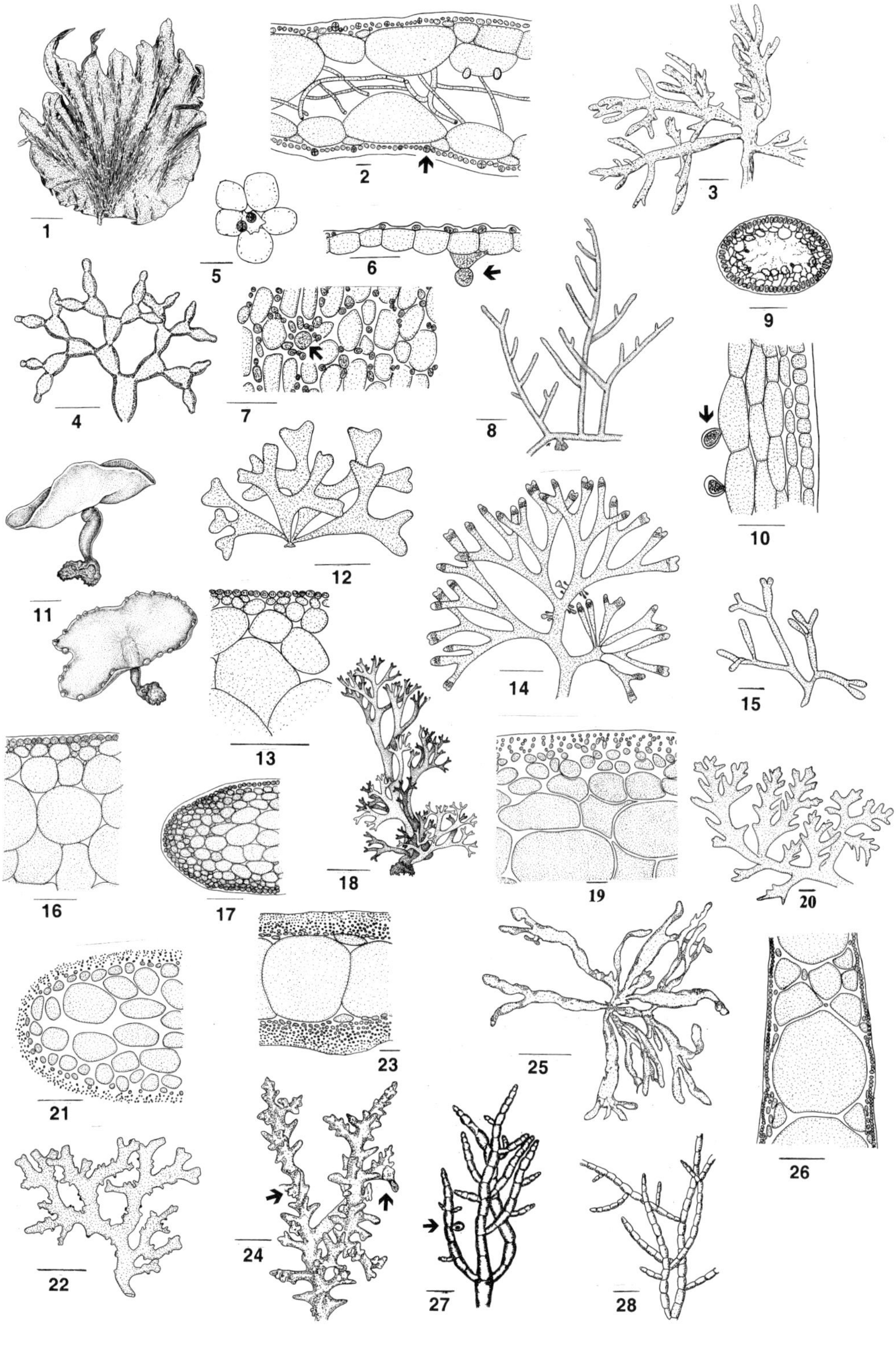
1
2
3
4
5
6
7
8
9
10
11
12
13
14
15
16
17
18
19
20
21
22
23
24
25
26
27
28

PLATE LI

Champia salicornioides. 1. The plant has large bulbous segments, irregular branching, and a discoid holdfast; scale 1 cm (L. Baumhart in Dawes, 1981).

Champia vieillardii. 2. Axes are segmented, compressed to flattened, and irregularly branched; scale 5 mm (after Lawson and John, 1987).

Gelidiopsis intricata. 3. Upper axes are compressed and sparsely subdichotomously branched; scale 2 cm (after Lawson and John, 1987). 4. A longitudinal section of a branch tip showing a densely packed medulla and many apical cells; scale 50 µm. 5. A cross section showing thick-walled medullary cells that grade into thin-walled cortical and pigmented surface cells; scale 50 µm (4, 5 after Littler and Littler, 1997).

Gelidiopsis planicaulis. 6. Upper axes are compressed and mostly irregularly branched near the tips; scale 1 cm (after Lawson and John, 1987). 7. A cross section of a mature branch showing a densely packed medulla surrounded by small cortical and pigmented surface cells; scale 100 µm (after Littler and Littler, 1997).

Gelidiopsis variabilis. 8. The upper axes are tough, cylindrical, and subdichotomously branched; scale 2 mm. 9. The branch tip has many apical cells and a densely packed medulla; scale 100 µm (both after Littler and Littler, 1997).

Lomentaria baileyana. 10. The large cluster of plants with irregular branching is epiphytic on a seagrass blade (L. Baumhart). 11. The five young plants on coral rubble have secund branching; scales 2 mm (after Taylor, 1928). 12. A cross section showing a hollow axis and gland cells (*arrow*) on the inner cortical cells; scale 50 µm (after Littler and Littler, 1997).

Agardhinula browneae. 13. The plant is fleshy and palmately branched; scale 1 cm. 14. A cross section showing variation in medullary cell size, plus small cortical and pigmented surface cells; scale 100 µm (both after Schneider and Searles, 1991).

Centroceras clavulatum. 15. Part of an upper axis showing 2 nodes, each with a single 2-celled hyaline spine; scale 50 µm. 16, 17. Cross sections of an internode showing 13 periaxial cells and 27 cortical cells (16), plus a node with 30 cortical cells (17); scales 100 µm (after Littler and Littler, 2000).

Centroceras internitens. 18. The branch tip has a straight apex, whorls of gland cells at nodes, and no spines; scale 100 µm (after Littler and Littler, 2000).

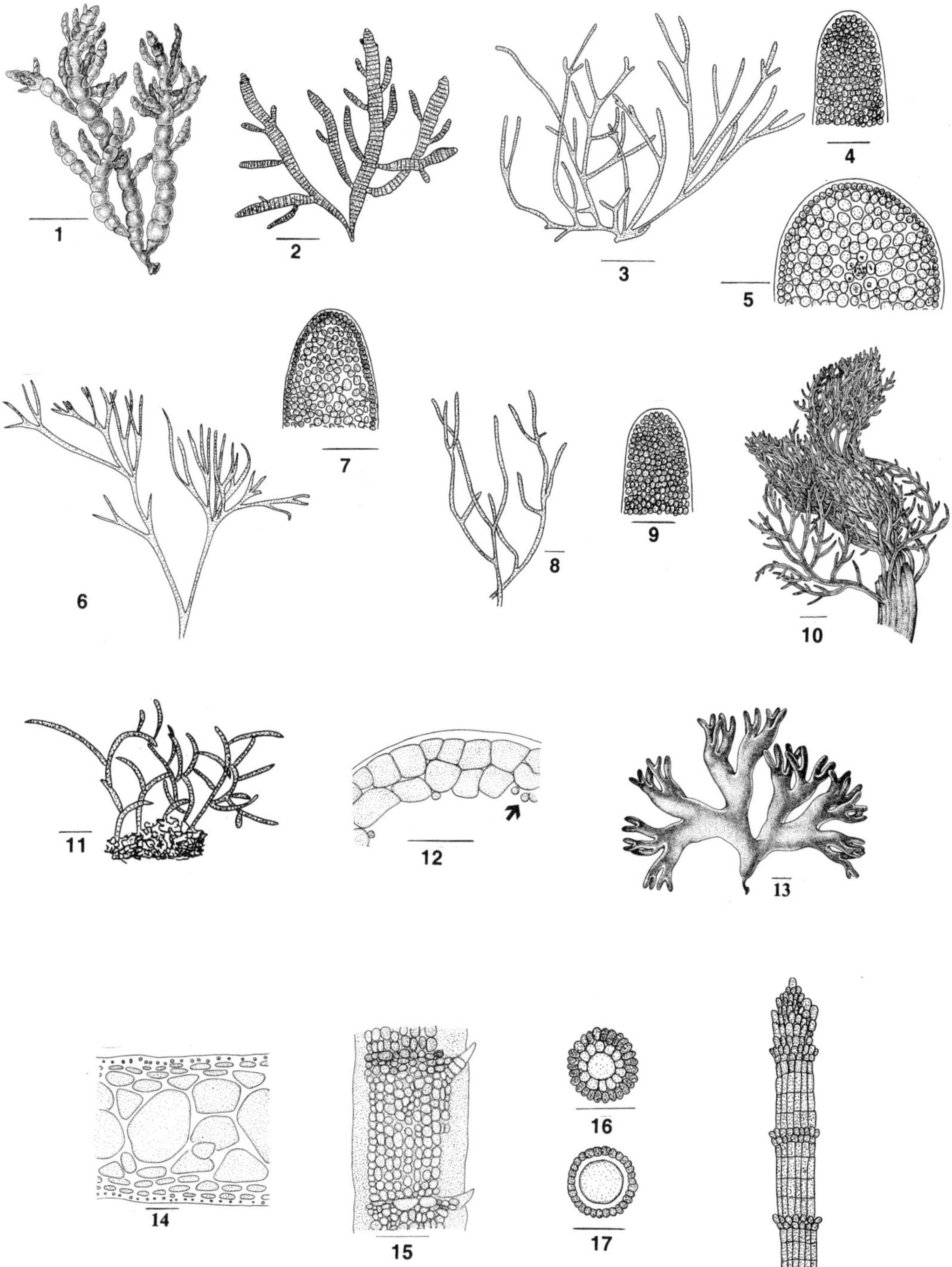
1
2
3
4
5
6
7
8
9
10
11
12
13
14
15
16
17
18

Appendix

Sources for Seaweed Distributions Worldwide

Records throughout the world are summarized based on regional texts and checklists using the following sources:
Abbott, 1999; Abbott and Hollenberg, 1976; Abbott and Huisman, 2003, 2004; Abbott and Littler, 1969; Adams, 1994; Aguilar-Rosas et al., 1998; Aguilar-Rosas and Aguilar-Rosas, 2003; Allender and Kraft, 1983; Ardré and Gayral, 1961; Athanasiadis, 1985, 1990; Aziz, 1965; Ballantine and Aponte, 1997a,b, 2003, 2005; Ballesteros and Romero, 1982; Ben Maiz et al., 1987; Benhissoune et al., 2001, 2002a,b, 2003; Betancourt and Herrera-Moreno, 2001; Bird and McLachlan, 1992; Børgesen, 1935, 1937, 1939, 1941, 1942, 1945, 1952; Bressan and Babbini, 2003; Brodie and Irvine, 2003; Bula-Meyer, 1995; Bula-Meyer and Norris, 2001; Burrows, 1991; Cardinal, 1964; Cassano and Yoeshigue-Valentin, 2001; Chamberlain, 1983, 1994; Chapman, 1952, 1956, 1961, 1963, 1977b; Coll and Oliveira, 1976; Conde et al., 1996; Coppejans, 1992; Coppejans et al., 2001a,b; Coppejans and Beeckman, 1990; Cordero, 1981; Cribb, 1966, 1996; Dawes, 1967, 1974; Dawson, 1944, 1949b, 1953, 1954, 1956, 1957, 1961, 1962, 1963a,b; Dawson et al., 1964; De Clerk, 2003; De-Paula, 1988; De Smedt et al., 2001; Díaz-Martin and Espinoza-Avalos, 2000; Díaz-Piferrer, 1970; Díaz-Pulido and Bula-Meyer, 1997; Dixon and Irvine, 1977; Edwards, 1970; Egerod, 1975; Eiseman and Blair, 1982; Etcheverry, 1986; Feldmann, 1937, 1941; Feldmann-Mazoyer, 1938b; Fletcher, 1987; Furnari et al., 1999, 2001; Garbielson et al., 2004; Gallardo et al., 1985, 1993; Ganesan, 1990; García and Gómez, 2001; Garreta, 2001; Garreta et al., 2001; Gavio and Fredericq, 2005; Giaccone, 1968, 1969; Gilbert, 1962, 1965; Gómez Garreta et al., 2001; Gordon, 1972; Guiler, 1952; Guimarães, 1990; Guiry and Nic Dhonncha (2005); Guiry and Nic Dhonncha, 2005; Güner et al., 1983–1984; Gunnarsson and Jónsson, 2002; Güven and Oztig, 1971; Hamel, 1931a,b, 1935, 1937, 1938; Hardy and Guiry, 2003; Haroun et al., 2002; Hatta and Prud'Homme van Reine, 1991; Heerebout, 1968; Hillson, 1977; Hodgson et al., 2004; Hoffman and Santelices, 1999; Hopkins et al., 1977; Hornig et al., 1992a,b, 1993; Huang, 2000; Huisman, 1996; Huisman and Walker, 1990; Humm, 1979; D. Irving, 1956; L. Irving, 1983; L. Irving and Farnham, 1983; L. Irving and Guiry, 1983; Isaac and Chamberlain, 1958; Islam, 1976; Jassund, 1976, 1977; John et al., 1979, 2003, 2004; Joly, 1957a,b, 1965; Juett et al., 1976; Kamiya et al., 2003; Kang, 1965; Kawaguchi, 2004; King et al., 1988; King and Puttock, 1989; Kornmann and Sahling, 1977; Koster, 1955; Krishnamurthy and Joshi, 1970; Kristiansen et al., 1993; Kylin, 1947, 1949; Lawson and John, 1987; Leliaert and Coppejans, 2003; Lewis and Norris, 1987; Levring, 1940b, 1945, 1953; Lewmanomont et al., 1995; Lipkin and Silva, 2002; Littler, et al., 1989; Littler and Littler, 1997, 2000; Lucas and Perrin, 1947; Maggs and Hommersand, 1993; Marcot-Coqueugniot, 1991; Mario et al., 2004; Martínez-Daranas et al., 2001; Mateo-Cid et al.,

2002; Mendoza- González et al., 2000, 2001; Millar and Kraft 1993; Min-Their and Womersley, 1976; Misra, 1966; Moreira and Suárez, 2002a; Nam, 2004; Neto et al., 2001; Nielsen et al., 2001; Nizamuddin, 1964; Nizamuddin and Gerloff, 1980; Noda, 1969, 1970, 1987; J. Norris and Bucher, 1982; R. Norris, 1986; N'Yeurt et al., 1996; N'Yeurt and Payri, 2006; O'Clair and Lindstrom, 2000; Oliveira et al., 1983; Otero-Schmitt and Pérez-Cirera, 2002; Otero-Schmitt and Sanjuan, 1992; Pacheco-Ruíz and Zertuche-González, 2002; Papenfuss, 1964, 1968; Park, 1980; Payr et al., 2000; Penrose, 1996a; Pham-Hoang, 1969; Phillips, 1997; I. Price and Scott, 1992; J. Price et al., 1986, 1988, 1992; Prud'homme van Reine, 1982, 1993; Ramírez and Santelices, 1991; Rezak et al., 1985; Ribera et al., 1992; Rindi et al., 2002; Riosmena-Rodríguez and Woelkerling, 2000; Rodríguez-Prieto et al., 1999; Rosenvinge and Lund, 1924; Rueness, 1977; Rull Lluch, 2002; Rull Lluch and Gómez Garreta, 2001, 2002; Sahoo et al., 2001; Saito and Womersley, 1974; Sartoni, 1978; Scagel, 1957, 1966; Scagel et al., 1989; Schneider, 2000; Schneider and Searles, 1991; Schnetter, 1976; Searles, 1988; Sears, 2002; Setchell and Gardner, 1903; Silva et al., 1987, 1996; Silva and Lipkin, 2003; Skelton and South, 1999, 2004; Skottsberg, 1923; Smith, 1944; South et al., 2001; South and Hooper, 1980; South and Keats, 1996; South and N'Yeurt, 1993; South and Skelton, 2003; South and Tittley, 1986; Steentoft, 1967; Stegenga, 1985; Stegenga et al., 1997, 2001; Stegenga and Mol, 1983, 1988; Stegenga and Vroman, 1987, 1988; Stenth, 2001; Stewart and Norris, 1981; Takamatsu, 1943; Taylor, 1933, 1954a,b, 1957, 1960, 1967, 1969; Tittley and Neto, 2005; Tseng, 1936, 1937, 1941, 1945, 1983; Tseng and Lu, 2000; Tsutsui et al., 2005; Villalard-Bohnsack, 1995; Weber-van Bosse, 1928; Welten et al., 2002; West and Zuccarello, 1995; 1975, , 1996a; Woelkerling and Nelson, 2004; Womersley, 1948, 1950, 1956, 1979, 1984, 1994, 1996, 1998, 2003; Womersley and Bailey, 1994; Womersley and Cartledge, 1975; Womersley and Guiry, 1994; Womersley and Kraft, 1997; Womersley and Lewis, 1994; Womersley and Wollaston, 1998; Wynne, 1986, 1998a, 2001, 2005a,b,c; Wysor, 2004, 2005; Wysor and De Clerck, 2003; Wysor and Kooistra, 2003; Xia and Zhang, 1999; Yamada, 1934, 1944a, 1950; Yoneshigue-Valentin and Loivos, 2005; Yoshida, 1998; Yoshida et al., 1990, 1995; Zinova, 1967.

Glossary

Abaxial Located on the side, facing away from an axis or branch, or on lower side of a branch.
Accessory. In addition to; also, appended to a main or central structure (e.g., accessory branches in red algae in addition to vegetative ones).
Acropetal. Proceeding from the base or below it toward the apex.
Acute. Sharp-pointed or tapering sharply to a point.
Adaxial. Located on the side or face next to the main axis or branch; toward the axis on the upper side of a branch.
Adnate. Attached or fused to another; pertaining to unlike parts that grow together.
Agar. A phycocolloid; a sulfated polysaccharide (i.e., a galactan) found in the cell walls and intercellular spaces of different Rhodophyta (e.g., *Ahnfeltia*, *Gelidium*, *Gracilaria*).
Akinete. A single-celled, nonmotile resting spore resulting from the metamorphosis of a vegetative cell with its thickened wall not separated from the parent cell.
Alginate. A phycocolloid; a sulfated carbohydrate extracted from brown algae (e.g., *Sargassum*); also a general term for salts of alginic acid.
Alternate. Formed singly at regular intervals along an axis; arranged in two rows but not opposite or secund on the axis.
Ampulla (ampullae). A special cluster of filaments that produces carpogonial branches or auxiliary cells as in some members of the Cryptonemiaceae.
Anastomosing. The joining together of adjacent parts so that the connections form a reticulum.
Androgynous: The presence of separate male and female conceptacles in the same receptacle and plant versus one where male and female reproductive organs occur together.
Annular. Ringlike, in the form of a ring, or with ringlike thickenings.
Antheridium (antheridia). The male reproductive organ (gametangium) that produces flagellated sperm in oogamous sexual reproduction; the mother-cell for antherizoids.
Anticlinal. Organized perpendicular to the surface of a structure (cf. periclinal).
Apiculate. To terminate in a sharp point or spine; also, with a point or apicula.
Aplanospore. A nonmotile spore or an arrested zoospore with its wall separated from the parent plant (cf. akinete).
Articulate (articulated). Segmented, jointed, or appearing as segmented, flexible branches (i.e., in the Corallinales).
Attenuate (attenuated). Narrow and tapering to a long slender point; also, slender or thin.
Auxiliary cell. A specialized cell in the advanced red algal class Florideophyceae that receives the diploid nucleus or its division product from a fertilized carpogonium.
Axial. The main or morphological axis; also, being along an axis or occupying a central position within a cell, as with chloroplasts.

Axial filament. The central filament of a monosiphonous axis; also, a cellular filament or strand that runs through the morphological axis of a thallus (cf. multiaxial and uniaxial).
Axil. The angle between two structures, such as between a branch and a main axis.
Axillary. To be positioned between a main axis and lateral branch (e.g., in the axil).
Basionym. The original or oldest and most valid name of a taxon now used in a new combination because of a change in taxonomic status or rank.
Bifurcate. Forked, dichotomously branched, or divided into two equal branches or forks.
Biseriate. Arranged in two rows, series, or columns; also, composed of two cells in width.
Bisporangium (bisporangia). A sporangium whose contents divide once, usually via mitosis, to produce two spores (i.e., bispores).
Bispore. One of two spores produced by division of a bisporangium.
Bistromatic. A blade composed of two layers of cells (also distromatic).
Branchlet. A small secondary branch that is usually the ultimate part of a branch system.
Bullate (bullations). A surface appearing swollen and blistered (e.g., species of *Hypoglossum*); also, having bullae or bulging of a thallus surface.
Calcareous. Covered, impregnated, or containing lime (i.e., calcium carbonate); also, chalky, limy, or stony.
Calcified. Involving the deposition of calcium carbonate in or on cell walls; also, to be encrusted or impregnated with calcium carbonate (lime).
Carotene. A family of pigments with oxygen-free, unsaturated hydrocarbons (e.g., isoprenoid polyene pigments, $C_{40}H_{56}$) that are fat soluble and reflect yellow, orange, or red light.
Carotenoid. A general term for carotenes and xanthophyll pigments that are long hydrocarbon molecules with terminal benzene rings.
Carpogonium (carpogonia). A female sexual cell in the Rhodophyta and certain fungi that includes the egg and an apical process, the trichogyne, that receives the male cell.
Carposporangium (carposporangia). A single-celled reproductive structure (i.e., sporangium) produced on the minute, parasitic carposporophyte generation in the red algae; it produces nonmotile (diploid) carpospores that give rise to the macroscopic tetrasporophyte generation.
Carpospore. A spore released from the carposporangium and resulting from fertilization; typically it is diploid.
Carposporophyte. The diploid, partially parasitic phase attached to the female gametophyte in the triphasic red algae life history; it develops by mitotic divisions of the zygotic nucleus.
Carpotetrasporangium. A sporangium in which four carpotetraspores are produced via meiosis in place of a carpospore in the carposporophyte or gonimoblast in red algae (see *Gymnogongrus*).
Carrageenan. A phycocolloid or a sulfated polysaccharide found in the cell walls of some red algae (see *Eucheuma* and *Hypnea*).
Cartilaginous. Firm, tough, and elastic, or with a gristly or tough texture (like cartilage).
Cervicorn. With branching more or less one-sided or with unequal forks where one fork is smaller and not divided any further (e.g., suggesting the horns of deer, antlerlike).
Coenocyte (coenocytic). An organism (e.g., *Caulerpa*) with multiple nuclei that lacks transverse (cross) walls and has continuous cytoplasm throughout.
Columella. In the Corallinales (red algae), a filamentous, sometimes calcified structure, in the central floor tissue of tetrasporangial (bisporangial) conceptacles. In brown algae (Fucales), it refers to the central conical cellular mass in conceptacles that bears a terminal tuft of hairs.
Compressed. Slightly flattened laterally; in cross section appearing elliptical and not completely flattened like a blade.

Conceptacle. A cavity in the surface tissues of a thallus that contains reproductive structures; found in the brown (Fucales) and red (Corallinaceae) algae.
Conchocelis stage. A diploid, filamentous, microscopic sporophyte phase in the life histories of some red algae (e.g., *Porphyra*); it inhabits calcareous material (e.g., dead mollusc shells).
Conchospore. A uninucleate spore produced within a conchosporangia; it ultimately produces a foliose *Porphyra* thallus, with meiosis being recorded in some species.
Connecting filament. In advanced red algae (class Florideophyceae), a filament transfers the zygote nucleus from the fertilized carpogonium to a nutritive or an auxiliary cell.
Contorted. With branches or axes that are irregular, twisted, or deformed (cf. torturous).
Convolute (convoluted). Rolled together (back) longitudinally.
Coquina. A hardened mixture of shell hash, limestone, and sand formed on beaches; also called beach rock.
Cordate. Heart-shaped or with a point and two equal rounded lobes at its base.
Corona inferior. In the green algal order Dasycladales (e.g., *Acetabularia*), a ring of short appendages on the underside of the cap at the insertion of the cap rays into the stalk.
Corona superior. In the green algal order Dasycladales (e.g., *Acetabularia*), a ring of short appendages on the upper side of a cap at the insertion of the cap rays into the stalk (main axis).
Corona superior protuberances. In the green algal order Dasycladales (e.g., *Acetabularia*), each corona superior segment produces a species-specific number of protuberances, which may remain undeveloped or grow out into branched deciduous hairs, leaving a scar.
Cornute. Horned, like a crown with horns; also, pointed, spiny, or having horns.
Corps en cerise (Fr. "body in cherry"). Bright cherry-red intracellular vesicles found in cortical cells of some red algae (see *Laurencia*) that have halogenated (terpenoid) secondary metabolites and appear as bright red spots in live cells.
Cortex. The outer tissue of a thallus that is external to the central medulla, axial cell, or pericentral cells and is usually made of compact cells or filaments and has plastids.
Corticated (cortication). Having an outer coating of cells, rhizoids, or filaments.
Corymbose. A branching pattern with most branch tips at the same level and forming a flat-topped cluster (corymb).
Costa (costae). A midrib or raised tissue on a blade that may extend its entire length.
Crenate. Blade margins with rounded, forward pointing teeth; also, having scalloped margins.
Crenulate. With minute rounded teeth, scallops, or notches.
Crystoliths. Cells containing calcium carbonate and usually having a distinct form (see *Peyssonnelia*); also refers to intracellular biogenic calcium carbonate of amorphous form.
Crisp. Curled up, especially near margins; also, strongly and finely undulate.
Crispate. Crinkled, irregularly waved or twisted; also, curled or roughened into small folds.
Crozier. A recurved hook-shaped tip of a branch, appearing like a shepherd's hooked staff.
Cruciate. Cross-shaped or divided by two successive planes perpendicular to each other; a type of meiotic division in tetrasporangia of red algae where all four tetraspores are visible.
Crustose. Crustlike and growing as a thin fleshy or calcified layer against the substratum.
Cryptostomata. A small indentation in the cortex of brown algae of the Fucales that often has sterile hairs extending to the surface.
Cyst. A resistant, usually thick-walled cell that is often a resting stage and produces reproductive cells; also, a sac, bladderlike cell, or cavity.
Cystocarp. The carposporophyte and accessory enveloping tissue (pericarp) of the gametophyte in advanced red algae.

Deciduous. Shedding, falling off, or not persistent.
Decumbent. Prostrate and curving upward.
Decussate. A type of branching in which lateral branches are in opposite pairs and each pair is at 90{dg} to those above and below to form a four-ranked vertical appearance on an axis. In tetrasporangia, the plane of secondary divisions that are perpendicular to each other.
Dentate. With margins toothed or having toothlike projections pointing outward.
Denuded. Axes that lose their lateral branches, hairs, or reproductive structures.
Determinate. Axes that have limited growth beyond a certain extent (e.g., branchlets).
Dichotomous. With two equal branches at each fork; also, paired branches (cf. bifurcate).
Dimerous. A type of thallus construction involving two sets of filaments (basal and erect) at more or less right angles to each other (cf. monomerous).
Dimorphous (dimorphic). Having two different forms or features.
Dioecious. With male and female organs on separate plants, or the male or female gametes produced by different individuals.
Diploid. Having twice the basic (i.e., haploid) number of chromosomes (2n).
Diplont (diplontic). A life history in which the plant is diploid with gametes being the only haploid cells.
Direct life history. Reproduction of the same phase of a potential life history by means of asexual spores, vegetative propagation, or propagules.
Discoid. A rounded, flat thallus; also, disclike or in the form of a disc.
Distal. Directed away from the base (e.g., near the tips); the opposite of proximal.
Distichous, distichous branching. With branches in two rows on opposite sides of an axis; branches may be alternate or opposite, with their arrangement suggesting a feather.
Distromatic. A blade composed of two layers of cells (also bistromatic).
Divaricate. Spreading widely; with wide branching patterns.
Dorsiventral. With distinct front (ventral) and back (dorsal) sides.
Ecorticate. Lacking a cortex or covering cells; the opposite of corticated.
Egg. A female gamete that is normally nonmotile and larger than a sperm.
Endogenous. Growth initiated from within.
Endophyte (endophytic). A plant growing in the tissue of another plant.
Endospores. An asexual reproductive unit formed inside a parent cell by division of the entire protoplast (e.g., autospores of some green algae).
Endozoic. An organism living within the cells or tissues of an animal.
Epilithic. Growing attached to an inorganic substratum, such as rocks or stones.
Epiphyte (epiphytic). A plant growing on the surface of another plant and not parasitic.
Epithallus, epithallium. The uppermost layer (surface cells) of crustose coralline algae
Epizoic. An organism living on the surface of an animal.
Eyespot. An orange (red) pigmented light spot found in some motile cells; it contains hematochrome dissolved in oil droplets and is usually associated with a plastid.
Exogenous. Formed at the surface or arising externally.
Fascicle (fasciculate). A cluster or tuft of branches (branchlets) that arise near the same area on an axis; also, a dense cluster or bundle.
Fastigiate. With many erect branches that are parallel to one another, producing an elongate habit; also, plants having upright branches.
Filamentous. A plant constructed of filaments or threadlike structures.

Filiform. Filamentous or threadlike; also, slender, cylindrical, or resembling a filament.
Flabellate. Fanlike; also, spreading upward from a narrow base; wedge-shaped.
Flaccid. Soft and flabby or limp; also, lacking rigidity and not able to hold up its own weight.
Flexuous. Flexed or bent alternately in opposite directions; also, curved, wavy, or zig-zag.
Foliose, foliaceous. Leaflike or broad and flat; also, with the shape of a leaf.
Floridean starch. The storage product of red algae, located in granules outside the chloroplast; an amylopectin (alpha-1,4-linked glucan, plus some alpha-1,6 linkages).
Fusiform. Spindle-shaped; elongated and tapered toward both ends.
Fusion cell. In red algae, a cell that has enlarged after fertilization, often joining with neighboring cells to form a larger multinucleated cell; common in the Corallinaceae.
Gametangium (gametangia). A cell or multicellular structure that produces gametes.
Gametophyte. A haploid generation that produces gametangia.
Gametophore. The portion of an algal filament that produces gametes; also, a structure in which gametangia are formed (e.g., *Batophora*).
Ganglia. Stellate or multi-branched cells in the outer medulla or inner cortex (cf. *Halymenia*); also, clusters of interconnected (fused), knotted cells in the medulla of certain red algae.
Geniculum (genicula). An uncalcified joint in the axis of articulated (i.e., geniculate) coralline red algae composed of a group of completely or partially noncalcified cells.
Gland cells. A dense or refractive cell that may contain halogenated compounds and stores or releases extracellular (anti-herbivorous) compounds.
Gonimoblast. In advanced red algae (class Florideophyceae), the carposporophyte tissue that develops from a fertilized carpogonium or auxiliary cell along with sterile cells and a fertile carposporangium.
Gonimoblastic filaments. In advanced red algae (class Florideophyceae), filaments that arise from fusion cells and produce carposporangia.
Gonimolobe. In advanced red algae (class Florideophyceae), carposporophyte tissue that is mostly fertile, from which carposporangia mature in sequential lobes (cf. gonimoblast).
Gregarious. Clustered or growing together in a dense aggregation but not united.
Haploid. Having one complete set (n) of chromosomes (cf. diploid).
Haplont (haplontic). An organism in which the vegetative stage is haploid and the zygote divides by meiosis.
Haplodiplont, haplodiplontic. A life history involving free-living diploid and haploid plants.
Hapteron (haptera). A rootlike multicellular structure that is more massive than a rhizoid.
Hapteral. Pertaining to a plant part adapted for attachment or support, as a hapteral filament.
Heteromorphic. An alternation of generations where haploid plants (gametophytic) produce gametes and are morphologically different from diploid ones (sporophytic) that produce spores.
Heterotrichous. A filamentous system composed of prostrate and erect filaments; in coralline red algae a type of thallus having two distinct types of filaments.
Heterotypic. A taxonomic synonym that is based on many nomenclatural types.
Holdfast. The basal attachment structure of a seaweed that may be unicellular or multicellular.
Homotypic. A taxonomic synonym that is based on the same nomenclatural type; also, a replacement name for a basionym.
Hyaline. Transparent, clear, or colorless; also, thin and transparent.
Hypogynous cell. The cell subtending a carpogonium and carrying the egg nucleus in the red algal order Ceramiales.

Hypothallium, hypothallus. The ventral-most layer(s) of filaments in a crustose thallus; also, the lowermost layer(s) of densely packed, fused filaments in a crustose red alga.
Indeterminate. Axes with unlimited growth potential and similar to the primary axis.
Indusium (indusia, indusiate). A sporangial covering or a membranous covering of superficial reproductive structures.
Intercalary. Lying between the apex and base or between nodes; also, inserted between.
Intergeniculum (intergenicula). The calcified segments between the joints of an articulated coralline alga, separated by noncalcified tissue.
Internode. An articulation or part of an axis intermediate between two nodes or joints; also, in a filament, the part of a cell between its end-walls.
Intertidal, intertidal zone. The region of shore lying between high and low tide levels.
Involucre. A group of sterile cells or filaments forming a protective envelope around a reproductive structure.
Isodiametric. Having equal dimensions or diameters (cf. parenchyma).
Isogamete, isogamous, isogamy. With gametes similar in size and shape (e.g., + and – strains).
Isomorphic. A life history in which diploid and haploid phases are morphologically similar.
Isomorphic alternation of generations. A life history in which the haploid gametophytic plants are structurally identical with diploid sporophytic plants.
Lamella (lamellae, lamellate). Thin layers or plates; to consist of thin layers (lamellate).
Lamina (laminae). The flat thin part of a leaf or blade; also, a leafy or bladelike expansion.
Lanceolate. An axis that is flat and narrow and two to three times as long as broad or widest in the middle or near the base and tapering to each end.
Lateral. To arise at or from or occur near the edge of another structure; also, a branch arising from the main axis.
Leucoplast. A colorless plastid, usually with starch grains and few thylakoids; its function is to produce and store polysaccharides.
Linnaean. In taxonomy, pertaining to names or works published after May 1, 1775.
Linear. Long, narrow, and with parallel sides; also, slightly broader than filiform.
Lithophyte (lithophytic). A plant growing on the surface of a rock or stone.
Littoral. A biological term that approximates the zone between the highest and lowest tides.
Locule. A compartment or space in a structure (see unilocular and plurilocular).
Mammillate. Having small protuberances, projections, or nipple-shaped lumps.
Mannitol. A sugar alcohol ($C_6H_{14}O_6$) product of photosynthesis in brown algae that functions as a balance to salinity and as a storage product.
Medulla. The central core of tissue in an axis that is surrounded by a cortex.
Mega-. Prefix meaning, big, large, or very large, from Greek *megas*.
Meiosporangium. A reproductive structure in which (meio-)spores are produced by meiosis.
Membranous, membranaceous. Flat, bladelike, and firm to the touch; also, having a gelatinous material that is scant and glossy.
Meristem (meristematic). The area in multicellular plants where cell division is located; also, a group or region of cells that rapidly divide and initiate growth.
Micro-. Prefix meaning, little, small, or microscopic, from Greek.
Microscopic. Not visible to the unaided eye or discernible only with a microscope.
Midrib. A thickened, central, longitudinal rib (costa) of a blade; also, a veinlike structure.
Moniliform. To resemble a string of spherical beads; also, regularly constricted filaments.

Monoecious. Having distinct male and female reproductive structures on the same thallus; one household; bisexual.
Monomerous. A thallus type composed of medulla and cortex (cf. dimerous)
Monopodial, monopodial branching. A type of growth in which secondary axes arise behind and remain subsidiary to a main axis that continues to grow indefinitely.
Monosiphonous. A single row of cells; also, a uniseriate structure or tube.
Monospore. A nonmotile spore produced singly from a parent cell (monosporangium).
Monosporangium (monosporangia). A sporangium with one spore and one to many nuclei.
Monostromatic. A thallus composed of a single layer of cells.
Monotypic. Axes or populations that are phenotypically consistent and not divisible into subunits; also, with only one component, as a genus with only one species.
Multiaxial. An axis composed of many elongate filaments and lacking a primary filament.
Multiaxial growth or organization. Having a central core of parallel longitudinal filaments, each derived from an apical cell; the primary axis is an aggregate of apposed filaments.
Multinucleate. With each cell having more than one nucleus (e.g., *Cladophora*).
Multiporate. A red algal conceptacle having a number of exit pores.
Multiseriate. With several columns or rows, more than two cells wide (cf. pluriseriate).
Multiseriate filament or thallus. A filament or plant body more than two cells in width.
Multistichous. An axis with branches on all sides.
Nemathecium (nemathecia). An elevated surface area of a thallus containing reproductive organs; also, a bulky, spore-containing tissue found on the surface.
Nodose, nodulose. Knotty, with local swellings, or minutely knobby in appearance.
Nom. Abbreviation for *nomen* (name).
Nom. cons. Abbreviation for *nomen conservadum*; a name that is conserved.
Nom. cons. prop. Abbreviation for *nomen conservandum propositum*; a Latin name proposed for conservation, but not yet formalized by the International Botanical Congress.
Nom. dub. Abbreviation for *nomen dubium* (doubtful name).
Nom. illegit. (or ***illeg.***). Abbreviation for *nomen illegitimum* (illegitimate name).
Nom. inval. Abbreviation for *nomen invalidum* (invalid name).
Nom. nov. Abbreviation for *nomen novum* (name proposed to replace another).
Nom. nud. Abbreviation for *nomen nudum* (invalid name published without a description, diagnosis, and/or reference to a previously validly published description or diagnosis).
Nutritive cell. A specialized cell in advanced red algae (class Florideophyceae) that receives the diploid nucleus or its division product from a fertilized carpogonium and provides nourishment.
Oblong. Elliptical or rectangular in shape with blunt ends and mostly parallel sides.
Obtuse. Blunt or rounded at the tip, not pointed or acute.
Ocellate. Shaped like an eye, having eyelike spots, or possessing circular patches or spots.
Ooblast. In advanced red algae (class Florideophyceae), a filament that affects the transfer of a zygote nucleus from the fertilized carpogonium to an auxiliary cell, directly or indirectly.
Oogamous. Sexual reproduction in which a small, motile sperm fuses with a larger nonmotile female cell.
Oogonium (oogonia). A single-celled female gametangium containing one or more eggs.
Orbicular. A flat disc that is circular or orb-shaped in outline; also, spherical or rounded.
Ostiole. A small pore, as in a conceptacle or cystocarp, for the escape of reproductive cells.

Oval, ovate, ovoid. Shaped like an egg; elongated with convex margins and equally rounded at the ends.
Palmelloid. A group of single nonmotile cells in a gelatinous mass (cf. *Pseudotetraspora*).
Palmate. Hand-shaped; flat and lobed to divided like the palm of a hand.
Papilla (papillae). A short, nipplelike outgrowth; also, a minute, rounded projection.
Paraphysis (paraphyses). Sterile hairs or filaments associated with reproductive cells and apparently acting as a spacer or for protection.
Paraspore. A reproductive cell similar to a tetraspore in red algae, but being tetraploid.
Parenchyma (parenchymatous). A three-dimensional tissue of thin-walled isodiametric cells.
Parietal. Outer, peripheral, or lying near a cell wall (e.g., chloroplasts along the edge of a cell).
Parthenogenesis. Development of gametes without fertilization.
Peltate. A flat circular blade attached by a central stalk on the ventral surface; also, shieldlike and fixed to the stalk at the center.
Pendent. Hanging or suspended from its support; also, growing downward.
Percurrent. Extending throughout the entire length or axis (e.g., a vein or midrib); also, a dominant axis visible throughout and not lost among lateral branches.
Perennial. Plants that live three or more years and reproduce after the second or third year.
Perforate. Having many pores; also, pierced or having translucent dots.
Pericarp. The haploid sterile tissue of a cystocarp that surrounds the female gametophyte as found in some advanced red algae (class Florideophyceae).
Pericentral cell. One of a number of cells cut off from and surrounding a central axial cell in some advanced red algae (class Florideophyceae).
Periclinal. Being parallel to a designated surface (cf. anticlinal).
Perithallium, perithallus. The central tissue of a crust that is produced by an intercalary meristem (e.g., coralline red algae). The term is often used to describe a tissue composed of filaments that arise from the hypothallus (cf. hypothallium).
Phycobilin, phycobiliprotein. A group of protein-bound, water soluble pigments in members of the Cyanophyta and Rhodophyta; the pigments reflect red and blue light.
Phycocolloid. A colloidal, mucilaginous substance in the cell walls of seaweeds (e.g., agar, carrageenan, laminarin) composed of high-molecular weight polymers of simple sugars. Used commercially for their stabilizing, thickening, and gelling properties.
Phycocyanin. A blue-light transmitting biliprotein; an accessory photosynthetic pigment.
Phycoerythrin. A red-light transmitting biliprotein; an accessory photosynthetic pigment.
Pinna (pinnae). Ultimate branches or branchlets in pinnately branched axes (cf. *Bryopsis*); also, the primary units of a pinnate leaf or one of a series of similar determinate laterals.
Pinnate, pinnate branching. With branches arranged in two rows on opposite sides of an axis; also, paired or alternate like a feather.
Pit connection. A connection between daughter cells of advanced red algae (class Florideophyceae) where adjacent cells have cytoplasmic continuity through a pit in their wall. The connections may be plugged with non-cell wall material.
Placenta, placental cell. In the class Florideophyceae, a multinuclate cell produced by the fusion of a carpogonium with an auxiliary cell and other nearby cells to produce carpospores.
Plethysmothallus (plethysmothallic). In the brown algae, a diploid filamentous stage that recycles itself until suitable conditions exist to form a macroscopic sporophyte (cf. protonema).

Plumose. Feathery, appearing featherlike or plumelike.
Plurilocular. Having several to many small chambers or compartments.
Plurilocular sporangium. In brown algae, multicellular reproductive structures on diploid or haploid plants that produce flagellated zooids. In diploid plants the zoospores are asexual and develop into diploid plants; in haploid plants they are gametes and may form haploid plants (i.e., via parthenogenesis) or sexually fuse.
Pluriseriate. A thallus more than two cells (or loculi) in thickness (cf. multiseriate).
Polychotomy (polychotomous). The branching of an axis into several lobes of equal status; also, a division into many parts or branches.
Polysiphonous. Characteristic of some advanced red algae (class Florideophyceae), where a ring of cells is produced by longitudinal divisions around a central uniseriate filament that forms pericentral cells (see *Polysiphonia*).
Polysporangium (polysporangia). A sporangium found in advanced red algae (class Florideophyceae) that produces many spores (multiples of fours) and is often meiotic in origin.
Polyspore. One of many spores formed in a polysporangia.
Polystromatic. A bladelike morphology with many cell layers.
Pre-Linnaean. Names or works that were published before May 1, 1775 (cf. Linnaean)
Prismatic. Having the shape of a prism and capable of refracting light.
Pro parte. A Latin phrase used in citations to show that a taxon in the sense of one author represents only part of the elements on which it was based by its original author.
Procarp. In the class Florideophyceae, a close association between a carpogonium and an auxiliary cell within a single branch system.
Proliferous. Production of outgrowths similar to and usually smaller than the main axis; also, production of many adventitious branches or branchlets as offshoots.
Propagula (propagulae, propagules). A multicellular, usually modified vegetative branch that detaches and propagates a new plant (e.g., *Sphacelaria*).
Protonema. In brown algae, a microscopic filamentous system that does not recycle itself but produces the macroscopic sporophyte directly (cf. plethysmothallus).
Proximal. Referring to the attached end; also, toward the base or near the point of origin (cf. distal).
Psammophyte (psammophytic). A plant that grows in unconsolidated sediments or on rocky substrata that is impacted by sand scouring; these plants show specialized morphological and/or reproductive adaptations.
Pseudodichotomous branching. Forming two unequal dichotomies or having two equal branches at branch points, with one being derived from a lateral branch.
Pseudofilament (pseudofilamentous). An incidental linear arrangement of cells that forms a somewhat filamentous thallus.
Pseudolateral. A lateral branch that arises through sympodial development by displacement of an apical shoot; also, a branched, pigmented, monosiphonous filament in red algae (cf. *Dasya*).
Pseudoparenchyma (pseudoparenchymatous). Tissue that resembles parenchyma with cells closely pressed together but arising from filaments.
Pseudoperennial. The equivalent of a terrestrial herbaceous perennial, with most of the plant dying back seasonally and later bearing new growth from a residual base (e.g., *Sargassum*).
Pulvinate. Cushionlike, cushion-shaped, having flattened pads, or padlike.

Pyramidal. Shaped like a pyramid with three or four sides and coming to a point.
Pyrenoid. A rounded organelle associated with starch synthesis; present in the chloroplasts of some green algae and distinguishable when stained with IKI; in red and brown algae it is not involved with starch synthesis.
Pyriform. Pear-shaped or being broader at the base (distally) than at the top (proximally).
Quadrate. Having four sides and four angles; also, squarish to rectangular.
Quadriflagellate. Having four flagella, as with a green algal zoospore.
Radial branching. A branching pattern with laterals arising from all sides of an axis.
Ramellus (ramelli). A minor or small branch (branchlet) in the final order of branching.
Receptacle. In the brown algae (e.g., Fucales) and some red algae (e.g., Corallinales), a fertile area, branch, or branchlet in which gametangia or sporangia are produced.
Refractive. Having optical qualities such that light is reflected, condensed, or refracted, making the object or body appear bright, especially under the microscope.
Repent. Bent over, prostrate, or growing along the substrata.
Reticulate (reticulated). Forming a network or lattice (e.g., *Dictyurus*).
Rhizine. A small-diameter, thick-walled rhizoidal filament of some red algae (see *Gelidium*).
Rhizoid. A hyphalike filament that is rootlike and functions in attachment.
Rhodolith. Nodular forms of unattached nongeniculate coralline red algae that form around grains, pebbles, or shell fragments; they roll around the bottom and have concentric layers.
Saccate. Saclike, appearing like a sack, pouched, or balloon-shaped.
Saxicolous. Growing attached to rock (cf. lithophytic).
Scutate. Shield-shaped or shield-bearing; also, a crustose, lobed to flattened plant.
Secund. Unilateral branching or branches arranged along one side of an axis.
Sedis incertae. Latin, "uncertain seat"; thus a category of uncertain taxonomic position.
Segregative cell divisions. Cell divisions where the protoplast cleaves into multinucleate packets inside the mother; these packets then develop walls that may divide the mother cell or grow outward to form a double-walled branch (e.g., *Valonia*); also, the cleavage of a multinucleate cell with individual nuclei to form new cells from the older ones.
Septum (septa). An in-growing cross wall of a filament; often found in multinucleated cells.
Septate. Transverse cell walls forming distinct cellular structures; also, compartmented.
Seriate. Cells or spores arranged in a series of transverse or longitudinal rows.
Serrate. With teeth on the margin and pointed toward the apex (e.g., like a saw).
Sessile. Lacking a stalk or supporting cell or borne directly on a branch.
Seta (setae). Thin, stiff, hairlike filaments; also, a bristle or bristle-shaped organ.
Siphon. A tubular structure that lacks cross walls. Some green algae have tubular structures without cross walls (e.g., *Codium)*. Some red algae have large elongated multinucleate cells (e.g., *Polysiphonia*).
Siphonous. Green algae that are tubular, multinucleate, and lack cross walls (see *Codium*).
Solution hole: A pool-like depression formed in limestone by dissolution in water.
Sorus (sori). A group of reproductive structures on the surface of a thallus that may be organized into specialized clusters and frequently with united bases.
Spermatangium (spermatangia). In red algae, cells that differentiate into spermatium.
Sperm. A flagellated male gamete.
Spermatium (spermatia). In red algae, a nonmotile or slightly amoeboid male gamete.
Spindle-shaped. Cylindrical, elongated axis or spore that tapers to both ends.
Spinelike. Branchlets or branches that taper to a point and are usually elongate.

Spinulose. A structure with small spines; also, to be covered with small spinelike hairs.
Sporangium (sporangia). Reproductive structures that produce spores via mitosis (mitospores) or meiosis (meiospores).
Sporophyte (sporophytic). The diploid phase in a life history that produces spores.
Stellate. Star-shaped, with multiple radiating structures, or provided with points like a star; also used to described some chloroplasts that are lobed and stellate in appearance
Sterile filaments. In the Delesseriaceae, cells produced during the formation of a procarp.
Sterile group. In the class Florideophyceae, cells that form during post-fertilization development of a carposporophyte and do not contribute to gonimoblast development.
Stichidium (stichidia). In the red algae, a specialized branchlet of distinctive shape that bears spermatangia or tetrasporangia.
Sub-. Prefix meaning under, below, beneath, close to, less than, or slightly, from Latin.
Subsidiary cell. In the class Florideophyceae, a vegetative cell produced on a supporting cell that becomes part of a fusion cell.
Supporting cell. In advanced red algae (class Florideophyceae), a cell that supports a carpogonial branch and other associated vegetative filaments that may function as an auxiliary cell.
Sympodial. Growth in which the apex of an axis is displaced by a subtending lateral branch and becomes a pseudolateral with the displacing lateral becoming the main axis; also, a structure in which the main axis consists of a succession of lateral shoots.
Synonym. Another name for the same taxon that was replaced by the correct one.
***Tabulae incertae*.** A Latin phrase meaning an uncertain record (tabulation).
***Tela arachnoidea*.** Latin phrase describing a network of elongated cells found on the inside of the pericarp and surrounding the carposporophyte in the red algal order Rhodymeniales.
Tenaculum (tenacular). The holdfast of an alga; also, minute cells at tips of filaments that attach to form netlike blades (cf. *Phyllodictyon*).
Terete. A slender cylinder that is more or less circular in cross section.
Tetrahedral. Four-sided, in reference to apical cells; also, a type of meiotic division in tetrasporangia of red algae that results in the four spores touching at the center and with only three walls visible.
Tetrasporangium (tetrasporangia). A unilocular sporangium in which four spores are produced by meiosis.
Tetraspore. A single spore in the tetrasporangium that was produced by meiosis.
Tetrasporophyte (tetrasporophytic). A diploid phase in a life history of a red alga that produces tetraspores via meiosis.
Thallus (thalli). A vegetative plant-body with simple internal differentiation of tissues and lacking vascular elements; also, an algal body that does not have true roots, stems, or leaves.
Tortuous. Having repeated bends or twists; also, winding or contorted.
Trabeculum (trabecula, trabeculae). Finger-{#}or strandlike wall ingrowths of *Caulerpa*.
Transverse, transverse section. A plane at right angles to the long axis (e.g., cross section).
Trichoblast. A simple or branched filament that is pigmented or colorless.
Trichocyte. A potential hair-producing cell in some nongeniculate coralline red algae. Cells are usually large, with thick walls, and may have intensely stained contents (also heterocyst, megacell).
Trichocyst. A large internal hyaline cell in crustose coralline red algae.
Trichogyne. In advanced red algae, the elongate distal part of a female gametangium (oogonium, carpogonium) that serves as a receptor site for male spermatia.
Trichotomous (trichotomously). Divided into three parts; also, a condition in which an axis branches into three lobes of equal status.

Trichothallic. In brown algae, growth in which the site of active cell division is located between the apex and base of a filament or group of filaments.
Trifurcate. Having three prongs or branches that diverge from the same point.
Triphasic. A life history in the red algal class Florideophyceae that has three morphological phases: gametophyte, carposporophyte, and tetrasporophyte.
Trimerous. Having three branches in each whorl; also, in groups of three or with three parts.
Triseriate. Consisting wholly or in part of three rows of cells in longitudinal series (cf. biseriate, multiseriate, pluriseriate, uniseriate).
Turbinate. Inversely conical; top-shaped.
Undulate, undulated. Wavy, with a wavy surface (edge), or corrugated.
Uniaxial, uniaxial construction. Having an axis with a single filament (cf. biaxial, multiaxial).
Unicellular. One-celled; an organism consisting of a single cell.
Unilateral, unilateral branching. One-sided or formed on one side of an axis; also, branches that arise on only one side of a main axis.
Unilocular, unilocular sporangium. A brown algal reproductive organ that contains a single locule (loculus) or cell that produces spores (biflagellated zoospores, aplanospores) via meiosis.
Uninucleate. A protoplast provided with a single nucleus.
Uniporate. A red algal conceptacle having a single exit pore.
Uniseriate, uniseriate filament. Occurring in a single longitudinal series; a transversely septate filament one cell wide (cf. biseriate, multiseriate, pluriseriate).
Urceolate. Urn-shaped, shaped like a pitcher or urn with a narrow extended opening.
Utricle. An enlarged tip of a sympodial branched coenocytic filament that forms the cortex of a plant and contains chloroplasts (e.g., *Codium*); a small sac, bottle, or vesicle.
Verticillate. Whorled or bearing successive whorls of branches (branchlets) along an axis.
Vesicle. A thin-walled, bladderlike sac (e.g., *Botryocladia*); a small float (e.g., *Sargassum*).
Whorl (whorled). Having parts arranged in a circle around an axis like spokes of a wheel.
Xanthophyll. Carotenoid pigments that are fat soluble, reflect yellow, orange, or red light, and contain oxygenated hydrocarbons.
Zonate. Divided by lines parallel to the surface as in zonate tetrasporangia in red algae with three parallel walls; also, divided by parallel planes or with bands or concentric layers.
Zooid. A motile reproductive cell with flagella (i.e., zoospore or gamete); term used when the sex, chromosome number, or function of a motile cell is unknown or uncertain.
Zoosporangium (zoosporangia). A sporangium that produces motile cells (zoospores).
Zoospore (zoosporic). A motile, flagellated, naked protoplast usually formed in a vegetative cell that grows into a new plant.
Zygospore. A thick-walled resting spore that develops from a zygote; also, a cell produced by the coalescence of two similar gametes.
Zygote. A diploid cell formed by the fusion of two gametes.
Zygotosporangium (zygotosporangia). In the genera *Bangia* and *Porphyra*, a sporangium formed internally by cell division in a fertilized carpogonium (zygote) and not externally from gonimoblast filaments as in most red algae.
Zygotospore. A spore produced from a zygotosporangium.

References

Abbott, I. A. 1945. The genus *Liagora* (Rhodophyceae) in Hawaii. Occasional Papers, Bernice P. Bishop Museum. 18: 145–169.

———. 1976. On the red algal genera *Grallatoria* Howe and *Callithamniella* Feldmann-Mazoyer (Ceramiales). *Br. Phycol. J.* 11: 143–149.

———. 1979. Some tropical species related to *Antithamnion* (Rhodophyta, Ceramiaceae). *Phycologia* 18: 213–227.

———. 1985. *Gracilaria* from Hawaii: key, list, and distribution of the species. In: I. A. Abbott and J. N. Norris (eds.), *Taxonomy of economic seaweeds with reference to some Pacific and Caribbean species*. Vol 1. Calif. Sea Grant College Program, Univ. Calif., La Jolla, pp. 85–87.

———. 1989. Marine algae of the northwest Hawaiian Islands. *Pac. Sci.* 42: 223–233.

———. 1990a. A taxonomic and nomenclatural assessment of the species of *Liagora* (Rhodophyta, Nemaliales) in the herbarium of Lamouroux. *Cryptogam. Algol.* 11: 111–136.

———. 1990b. A taxonomic assessment of the species of *Liagora* (Nemaliales, Rhodophyta) recognized by J. Agardh, based upon studies of type specimens. *Cryptogam. Bot.* 1: 308–322.

———. 1995. A decade of species of *Gracilaria* (*sensu latu*). In: I. A. Abbott (ed.), *Taxonomy of economic seaweeds with reference to some Pacific species*. Vol. 5. Calif. Sea Grant College Program, Univ. Calif., La Jolla, pp. 185–195.

———. 1999. *Marine red algae of the Hawaiian Islands*. Bishop Museum Press, Honolulu, xv + 477 pp.

Abbott, I. A., and M. S. Doty. 1960. Studies in the Helminthocladiaceae (Rhodophyta). II. *Trichogloeopsis*. *Am. J. Bot.* 47: 632–640.

Abbott, I. A., and G. J. Hollenberg. 1976. *Marine algae of California*. Stanford University Press, Stanford, xii + 827 pp.

Abbott, I. A., and J. M. Huisman. 2003. New species, observations, and a list of new records of brown algae (Phaeophyce) from the Hawaiian Islands. *Phycol. Res.* 51: 173–185.

———. 2004. Marine green and brown algae of the Hawaiian Islands. *Bishop Mus. Bull. Bot.* 4, Honolulu, xi + 259 pp.

Abbott, I. A., and M. M. Littler. 1969. Some Rhodymeniales from Hawaii. *Phycologia* 8: 165–169.

Abbott, I. A., J. Zhang, and B. M. Xia. 1991. *Gracilaria mixta, sp. nov.* and other western Pacific species of the genus (Rhodophyta: Gracilariaceae). *Pac. Sci.* 45: 12–27.

Adam, P. 1992. *Saltmarsh ecology*. Cambridge Univ. Press, Cambridge, xii + 461 pp.

Adams, D. A. 1963. Factors influencing vascular plant zonation in North Carolina salt marshes. *Ecology* 44: 445–456.

Adams, N. M. 1994. *Seaweeds of New Zealand*. Canterbury Univ. Press, Christchurch, New Zealand, 360 pp.

Adanson, M. 1763. *Familles des plantes*. II. partie. Paris. [24 +] 640 pp.

Adey, W. H. 1970. A revision of the Foslie crustose coralline herbarium. *K. Nor, Vidensk. Selsk., Skr. (Trondheim)* 1970 (1), 46 pp.

Adey, W. H., R. A. Townsend, and W. T. Boykins. 1982. The crustose coralline algae (Rhodophyta: Corallinaceae) of the Hawaiian Islands. *Smithson. Contrib. Mar. Sci.* 15. iv + 74 pp.

Afonso-Carrillo, J., M. Sansón, and C. Sangil. 2003. *Colaconema ophioglossum comb. nov.* and *Liagorophila endophytica*, two acrochaetioid algae (Rhodophyta) from the eastern Atlantic. *Cryptogam. Algol.* 24: 107–116.

Afonso-Carrillo, J., and C. Sobrino. 2004. Two amphi-Atlantic species of *Botryocladia* (Rhodymeniales, Rhodophyta) in the Canary Islands (Eastern Atlantic). *Cryptogam. Algol.* 25: 147–159.

Agardh, C. A. 1810–1812. Dispositio algarum Sueciae . . . Lundae [Lund]. Pp. [1]–16 (1810), 17–26 (1811), 27–45 (1812).

———. 1817. Synopsis algarum Scandinaviae . . . Lundae [Lund]. xl + 135 pp.

———. 1820–1828. Species algarum . . . Vol. 1, part 1. Lundae [Lund]. Pp. [i–iv +] [1]–168 (1820); Vol. 1, part 2. Lundae [Lund]. Pp. [i–viii +] 169–398 (1822), 399–531 (1823); Vol. 2, sect. 1. Gryphiae [Greifswald]. lxxvi + 189 pp. (1828).

———. 1821. Icones algarum ineditae. Fasc. 2. Holmiae [Stockholm]. [7] pp., pls. XI–XX.

———. 1822. Algae, Agardh. In: C. S. Kunth, Synopsis plantarum, quas, in itinere ad plagam aequinoctialem orbis novi, collegerunt Al. de Humboldt et Am. Bonpland. Vol. 1. Parisiis [Paris], pp. 1–6

———. 1824. Systema algarum. Lundae [Lund]. xxviii + 312 pp.

———. 1827. Aufzählung einiger in den östreichischen Ländern gefundenen neuen Gattungen und Arten von Algen, nebst ihrer Diagnostik und beigefügten Bermerkungen. Flora (Jena) 10: 625–646.

———. 1828–1835. Icones algarum europaearum. Leipsic [Leipzig].

Agardh, J. G. 1837. Novae species algarum, quas in itinere ad oras maris rubri collegit Eduardus Rüppell; cum observationibus nonnullis in species rariores antea cognitas. *Museum Senckenbergianum* 2: 169–174.

———. 1841. In historiam algarum symbolae. *Linnaea* 15: 1–50, 443–457.

———. 1842. *Algae maris Mediterranei et Adriatici, observationes in diagnosin specierum et dispositionem generum.* Apud Fortin, Masson et Cie. Parisiis (Paris). x + 164 pp.

———. 1846. *Caroli Ad. Agardh Icones algarum ineditae. Fasculi qui exstant duo.* Editio nova. Lundae [Lund]. [6] pp., 20 pls.

———. 1847('1848'). Nya alger från Mexico. *Öfvers. Förh. K. Sven. Vetensk.-Akad.* 4: 5–17.

———. 1848–1901. *Species genera et ordines algarum . . . Volumen primum: algas fucoideas complectens.* Lundae [Lund]. viii + 363 pp. (1848); . . . Volumen secundum: algas florideas complectens. Lundae [Lund]. xii + 1291 pp. Part 1, pp. [i]–xii + [1]–336 + 337–351 (Addenda and Index) (1851); part 2, fasc. 1, pp. 337–504 (1851); part 2, fasc. 2, pp. 505–700 + 701–720 (Addenda and Index) (1852); part 3, fasc. 1, pp. 701–786 (1852); part 3, fasc. 2, pp. 787–1291 (1139–1158 omitted) (1863); Volumen tertium: de Florideis curae posteriores. Part 1. Lipsiae [Leipzsig]. vii + 724 pp. (1876); Voluminis tertii, pars tertia: De dispositione Delesseriearum curae posteriores. Lundae [Lund]. [vii +] 239 pp. (1898); Voluminis terti, pars quarta. Supplementa ulteriora et indices sistens. Lundae [Lund]. [vi +] 149 pp. (1901).

———. 1849. Algologiska bidrag. *Öfvers. Förh. K. Sven. Vetensk.-Akad.* 6: 79–89.

———. 1854. Nya algformer. *Öfvers. Förh. K. Sven. Vetensk.-Akad.* 11: 107–111.

———. 1855. Till Algernes Systematik. VII. Florideae. *Acta Univ. Lund* 21: 1–117.

———. 1872('1871'). Bidrag till Florideernes systematik. *Lunds Universitets Års-Skrift, Afdelningen för Mathematik och Naturvetenskap*, 8(6), 60 pp.

———. 1873. Till algernes systematik. Nya bidrag. *Lunds Universitets Års-Skrift, Afdelningen för Mathematik och Naturvetenskap*, 9(8), 71 pp.

———. 1882. Till algernes systematik. Nya bidrag (Andra afdelningen). *Lunds Universitets Års-Skrift, Afdelningen för Mathematik och Naturvetenskap*, 17(4), 134 pp., 3 pls.

———. 1883. Till algernes systematik. Nya bidrag (Tredje afdelningen.). *Lunds Universitets Års-Skrift, Afdelningen för Mathematik och Naturvetenskap*, 19(2), 177 pp., 4 pls.

———. 1885. Till algernes systematik. Nya bidrag. (Fjerde afdelningen.) *Lunds Universitets Års-Skrift, Afdelningen för Mathematik och Naturvetenskap*, 21(8), 171 pp., [1] pl.

———. 1887. Till algernes systematik, Nya bidrag (Femte afdelningen). Pt. 5, (VIII. Siphoneae). *Lunds Universitets Års-Skrift, Afdelningen för Mathematik och Naturvetenskap*, 23(2), 174 pp., 5 pls.

———. 1889. Species Sargassorum australiae descriptae et dispositae. *K. [Svenska] Vetensk.-Akad. Handl.*, [ser. 4], 23(3), 133 pp., 31 pls.

———. 1890. Till algernes systematik. Nya bidrag. (Sjette afdelningen). *Lunds Universitets Års-Skrift, Andra Afdelningen, K. Fysiogr. Sällsk i Lund Handl.*, 26(3), 125 pp., 3 pls.

———. 1892. Analecta algologica. *Lunds Universitets Års-Skrift, Andra Afdelningen, K. Fysiogr. Sällsk i Lund Handl.*, 28(6), 182 pp., 3 pls.

———. 1894. Analecta algologica. Continuatio I. *Lunds Universitets Års-Skrift, Andra Afdelningen, K. Fysiogr. Sällsk i Lund Handl.*, 29(9), 144 pp., 2 pls.

———. 1896. Analecta algologica. Continuatio III. *Lunds Universitets Års-Skrift, Andra Afdelningen, Kongl. Fysiografiska Sällskapets i Lund Handlingar*, 32(2), 140 + 8 pp. [Index].

———. 1899. Analecta algologica. Continuatio V. *Lunds Universitets Års-Skrift, Andra Afdelningen, Kongl. Fysiogr. Sällsk i Lund Handl.*, 35(4), 160 pp., 3 pls.

Aguilar-Rosas, M. A., L. E. Aguilar-Rosas, and R. Aguilar-Rosas. 1998. Algas marinas de la región central de Quintana Roo, México. *Polibotánica* 7: 15–32.

Aguilar-Rosas, R., and L. E. Aguilar-Rosas. 2003. El género *Porphyra* (Bangiaceae, Rhodophyta) en la costa Pacifico de México. I. *Porphyra suborbiculata* Kjellman. *Hidrobiologia* 13: 51–56.

Allender, B. M., and G. T. Kraft 1983. The marine algae of Lord Howe Island (New South Wales): the Dictyotales and Cutleriales (Phaeophyta). *Brunonia* 6: 73–130.

Almodovar, L. R., and F. A. Pagan. 1971. Notes on a mangrove lagoon and mangrove channels at La Parguera, Puerto Rico. *Nova Hedwigia* 21: 241–259.

Ambronn, H. 1880. Ueber einige Fälle von Bilateralität bei den Florideen. *Bot. Zeitung (Berlin)* 38: 161–174, 177–185, 193–200, 209–216, 225–233, pls. III, IV.

Amsler, C. D. 1991. Family Ectocarpaceae. In: C. W. Schneider and R. B. Searles, *Seaweeds of the southeastern United States: Cape Hatteras to Cape Canaveral.* Duke Univ. Press, Durham, pp. 111–128.

Amsler, C. D., and D. F. Kapraun. 1985. *Giffordia onslowensis sp. nov.* (Phaeophyceae) from the North Carolina continental shelf and the relationship between *Giffordia* and *Acinetospora*. *J. Phycol.* 21: 94–99.

Andersen, R. A. 2004. Biology and systematics of heterokont and haptophyte algae. *Am. J. Bot.* 91: 1508–1522.

Andreakis, N., G. Procaccini, and W.H.C.F. Kooistra. 2004. *Asparagopsis taxiformis* and *Asparagopsis armata* (Bonnemaisoniales, Rhodophyta): genetic and morphological identification of Mediterranean populations. *Eur. J. Phycol.* 39: 273–283.

Anonymous. 1974. Florida coastal zone management atlas. Coastal Coord. Council, State of Florida, Tallahassee, 137 pp.

———. 1986. *Manasota 88 Inc. vs. Wilbour Boyd Corp.* Depositions of J. Carrol and G. Gilmore. State of Florida, Case 85–2904, Division of Administrative Hearings, Tallahassee.

Aponte, N. E., and D. L. Ballantine. 2001. Depth distribution of algal species on the deep insular fore reef at Lee Stocking Island, Bahamas. *Deep Sea Res.* 48: 2185–2194.

Aponte, N. E., D. L. Ballantine, and J. N. Norris. 1994. Culture studies on the morphology and life history of *Aglaothamnion herveyi* (Howe) *comb. nov.*, with notes on *A. felipponei* (Howe) *comb. nov.* (Ceramiaceae, Rhodophyta). *Phycologia* 33: 231–238.

———. 1997. *Aglaothamnion halliae comb. nov.* and *A. collinsii sp. nov.* (Ceramiales, Rhodophyta): resolution of nomenclatural and taxonomic confusion. *J. Phycol.* 33: 81–87.

Ardré, F., and P. Gayral. 1961. Quelques *Grateloupia* de l'Atlantique et du Pacifique. *Rev. Algol.*, N.S. 6: 38–48, pls. 2–4.

Ardissone, F. 1871. Revista dei Ceramii della flora italiana. *Nuovo Giornale Botanico Italiano* 3: 32–50.

———. 1878. Studi sulle alghe italiche della famiglia della Rhodomelacee. *Atti Soc. Crittog. Ital.* 1: 41–159.

Aregood, C. C. 1975. A study of the red alga *Calonitophyllum medium* (Hoyt) *comb. nov.* [= *Hymenena media* (Hoyt) Taylor]. *Br. Phycol. J.* 10: 347–362.

———. 1977. *Heterosiphonia hummii* (Dasyaceae-Rhodophyta), a new species from the Gulf of Mexico. *Bull. Mar. Sci.* 27: 393–399.

Areschoug, J. E. 1843. Algarum (phycearum) minus rite cognitarum pugillus secundus. *Linnaea* 17: 257–269.

———. 1847. Phycearum, quae in maribus Scandinaviae crescunt, enumeratio. Sectio prior Fucaceas continens. *Nova Acta Regiae Soc. Sci. Uppsaliensis* 13: 223–382, 9 pls.

———. 1850. Phycearum, quae in maribus Scandinaviae crescunt, enumeratio. Sectio posterior Ulvaceas continens. *Nova Acta Regiae Soc. Sci. Uppsaliensis* 14: 385–454, 3 pls. [also issued as Phyceae scandinavicae marinae, Uppsala with different pagination].

———. 1854. Phyceae novae et minus cognitae in maribus extra-Europaeis collectae, quas descriptionibus observationibusque illustravit. *Nova Acta Regiae Soc. Sci. Uppsaliensis*, ser. 3, 1: 329–372.

———. 1875. Observationes phycologicae III. *Nova Acta Regiae Soc. Sci. Uppsaliensis*, ser. 3, 10: 1–36.

Athanasiadis, A. 1985. North Aegean marine algae. I. New records and observations from the Sithonian Peninsula, Greece. *Bot. Mar.* 28: 453–468.

———. 1988. North Aegean marine algae II. Studies on the thallus structure and reproduction of *Nemastoma dichotomum* J. Agardh and *Predaea ollivierii* J. Feldmann (Rhodophyta, Gigartinales). *Bot. Mar.* 31: 23–32.

———. 1990. Evolutionary biogeography of the North Atlantic antithamnioid algae. In: D. J. Garbary and G. R. South (eds.), *Evolutionary biogeography of the marine algae of the North Atlantic*. Springer-Verlag, Berlin, pp. 219–240.

———. 1996a. Morphology and classification of the Ceramioideae (Rhodophyta) based on phylogenetic principles. *Opera Bot.* 128: 1–216.

———. 1996b. *Taxonomic litteratur och biogeografi av Skandinaviska rödalger oc brunalge*. Göteborg: Algologia, 280 pp.

———. 1997. Proposal to conserve the name *Antithamnion antillanum* against *Callithamnion lherminieri* (Rhodophyta, Ceramiales). *Taxon* 46: 765–767.

Athanasiadis, A., D. J. Garbary, and H. Vandermeulen. 1990. *Antithamnion makroklonion sp. nov.* (Rhodophyta, Ceramiales) from the Gulf of Aqaba (Red Sea). *Nord. J. Bot.* 10: 557–564.

Aziz, K. M. S. 1965. *Acrochaetium* and *Kylinia* in the southwestern North Atlantic. Ph.D. diss., Duke Univ., Durham, ix + 235 pp.

Babbini, L., and G. Bressan. 1997. Recensement de Corallinacées de la Mer Méditeranée et considérations phytogéographiques. *Bibl. Phycol.* 103: 1–421.

Bach, S. D., and M. N. Josselyn. 1978. Mass blooms of the alga *Cladophora* in Bermuda. *Mar. Pollut. Bull.* 9: 34–37.

Bailey, G. P., E. R. Cox, and R. Rezak. 1976. Morphological variability in the marine green alga *Acetabularia crenulata* Lamouroux (Dasycladaceae, Dasycladales). *Phycologia* 15: 19–23.

Bailey, J. C. 1999. Phylogenetic positions of *Lithophyllum incrustans* and *Titanoderma pustulatum* (Corallinaceae, Rhodophyta) based on 18S rDNA gene sequence analyses, with a revised classification of the Lithophylloideae. *Phycologia* 38: 208–216.

Bailey, J. C., J. E. Gabel, and D. W. Freshwater. 2004. Nuclear 18S rRNA gene sequence analyses indicate that the Mastophoroideae (Corallinaceae, Rhodophyta) is a polyphyletic taxon. *Phycologia* 43: 3–12.

Bailey, J. W. 1848. Continuation of the list of localities of algae in the United States. *Am. J. Sci. Arts* 6: 37–42.

Bakker, F. T., J. L. Olsen, and W. T. Stam. 1995. Global phylogeography in the cosmopolitan species *Cladophora vagabunda* (Chlorophyta) based on nuclear rDNA internal transcribed spacer sequences. *Eur. J. Phycol.* 30: 197–298.

Bakker, F. T., J. L. Olsen, W. T. Stam, and C. van den Hoek. 1994. The *Cladophora* complex (Chlorophyta): new views based on the 18S rRNA gene sequences. *Mol. Phylogenet. Evol.* 3: 365–382.

Balakrishnan, M. S. 1961 ('1960'). Reproduction in some Indian red algae and their taxonomy. In: P. Kachroo (ed.), *Proc. Symp. Algology*. Indian Council of Agricultural Research, New Delhi, pp. 85–98.

———. 1962. Studies on Indian Cryptonemiales. II. *Corynomorpha* J. Ag. *Phytomorphol.* 12: 77–86.

Balakrishnan, M. S., and V. N. Kinkar. 1981. A taxonomic account of Indian Ectocarpales and Ralfsiales. *Seaweed Res. Util.* 4(2): 1–57.

Baldauf, S. L. 2003. The deep roots of eukaryotes. *Science* 300: 1703–1706.

Ballantine, D. 1972. Algal epiphytes of four Florida seagrasses from Tarpon Springs, Florida. M.A. thesis, University of South Florida, Tampa, 87 pp.

———. 1982. *Halimeda hummii sp. nov., Halimeda cryptica* v. *acerifolia var. nov.* (Caulerpales, Chlorophyta), and additional records of *Halimeda* species from Puerto Rico. *J. Phycol.* 18: 86–91.

———. 1985. *Botryocladia wynnei sp. nov.* and *B. spinulifera* Taylor and Abbott (Rhodymeniales, Rhodophyta) from Puerto Rico. *Phycologia* 24: 199–204.

———. 1996. New records of benthic marine algal species records from Florida. *Gulf Mex. Sci.* 1: 11–15.

Ballantine, D., and N. E. Aponte. 1995. *Laurencia coelenterata* (Rhodomelaceae, Rhodophyta), a new diminutive species from the Dry Tortugas, Florida. *Bot. Mar.* 38: 417–421.

———. 1997a. A revised checklist of the benthic marine algae known to Puerto Rico. *Caribb. J. Sci.* 33: 150–179.

———. 1997b. Notes on the benthic marine algae of Puerto Rico VI. Additions to the flora. *Bot. Mar.* 40: 39–44.

———. 2002. *Ganonema farinosum* and *Ganonema dendroideum comb. nov.* (Liagoraceae, Rhodophyta) from Puerto Rico, Caribbean Sea. *Cryptogam. Algol.* 23: 211–222.

———. 2003. An annotated checklist of deep-reef benthic marine algae from Le Stocking Island, Bahamas (western Atlantic), I. Chlorophyta and Heterokontophyta. *Nova Hedwigia* 76: 113–127.

———. 2005. An annotated checklist of deep-reef benthic marine algae from Le Stocking Island, Bahamas (western Atlantic). II. Rhodophyta. *Nova Hedwigia* 80: 147–171.

Ballantine, D., and H. J. Humm. 1975. Benthic algae of the Anclote Estuary. I. Epiphytes of seagrass leaves. *Fla. Sci.* 38: 150–162.

Ballantine, D., and J. N. Norris. 1994. *Verdigellas*, a new deep-water genus (Tetrasporales, Chlorophyta) from the tropical western Atlantic. *Cryptogam. Bot.* 4: 368–372.

Ballantine, D., H. Ruiz, and N. E. Aponte. 2004. Notes on the benthic marine algae of Puerto Rico VIII. Additions to the flora. *Bot. Mar.* 47: 335–340.

Ballantine, D., and M. J. Wynne. 1988a. The life history and development of *Hypoglossum rhizophorum* (Delesseriaceae, Rhodophyta) in culture, a new deep-water species from the Caribbean. *J. Phycol.* 24: 8–12.

———. 1998b. *Ptilothamnion speluncarum* (Collins & Hervey) *comb. nov.* (Ceramiaceae, Rhodophyta) from Puerto Rico. *Cryptogam. Algol.* 19: 223–229.

Ballesteros S. E., and J. Romero. 1982. Catalogo de las algas bentonicas (con exclusion de las diatomeas) de la coast Catalina. *Collect. Bot. (Barcelona)*, 13: 723–765.

Barilotti, C. 1970. Non-genetic morphological variation in *Caulerpa prolifera*. M.A. thesis, Univ. South Florida, Tampa, 62 pp.

Barker-Webb, P., and S. Berthelot (eds.). 1839–1841. *Histoire naturelle des Iles Canaries.* Vol. 3, part 2, sect. 4. Paris. xv + 208 pp., 9 pls. [pp. 1–16 (1839), 17–160 (1840), 161–208, I–XV (1841)].

Barton, E. S. 1891. A systematic and structural account of the genus *Turbinaria*, Lamx. *Trans. Linn. Soc. Lond. Bot.* ser. 2, 3: 215–226, pls. 54, 55.

———. 1901. The genus *Halimeda*. *Siboga Exped. Monogr.* 60. Leiden, pp. 1–32 pp., 2 figs., 4 pls.

Basson, P. W. 1979. Marine algae of the Arabian Gulf Coast of Saudi Arabia (second half). *Bot. Mar.* 22: 65–82.

Batters, E. A. L. 1892. On *Conchocelis*, a new genus of perforating algae. *Phycol. Mem.* 1: 25–28.

———. 1893. On the necessity for removing *Ectocarpus secundus*, Kütz., to a new genus *Grevillea* 21: 85–86.

———. 1896. Some new British marine alge. *J. Bot., Lond.* 34: 6–11, 384–390.

———. 1902. A catalogue of the British marine algae. *J. Bot.* 40(suppl.): [2], 107 pp.

Beach, K., L. Walters, H. Borgeas, C. Smith, J. Coyer, and P. Vroom. 2003. The impact of *Dictyota* spp. on *Halimeda* populations of Conch Reef, Florida Keys. *J. Exp. Mar. Biol. Ecol.* 297: 141–159.

Beck, G. V., and A. Zahlbruckner. 1898. Schedae ad *kryptogamas exsiccatas* editae a Museo Palatino Vindobonensi, centuria IV. *Annalen des Kaiserlich Königlich Naturhistorischen Hofmuseums* 13: 443–472.

Bedinger, L. A., and S. S. Bell. 2006. Notes from underground: Bryopsidalean green algal holdfasts in psammophytic habitats. Meeting Program, Phycol. Soc. Am., Juneau, Alaska, pp. 42–43.

Bélanger, C. 1834. Voyagées aux Indies-Orientales . . . 1825, 1826, 1828 et 1829. Botanique. IIe Partie. Cryptogamie, pp. 81–192, 7 pls., Paris.

Bellorin, A. M., M. C. Oliveira, and E. C. Oliveira. 2002. Phylogeny and systematics of the marine algal family Gracilariaceae (Gracilariales, Rhodophyta) based on small subunit rDNA and ITS sequences of Atlantic and Pacific species. *J. Phycol.* 38: 551–563.

Ben Maiz, N., C.-F. Boudouresque, and F. Quahchi. 1987. Inventaire des algues et phanérogames marines benthiques de la Tunise. *Giomale Botanico Italiano* 121: 259–304.

Benhissoune, S., C.-F. Boudouresque, and M. Verlaque. 2001. A checklist of marine seaweeds of the Mediterranean and Atlantic coasts of Morocco. *Bot. Mar.* 44: 171–182.

———. 2002a. A checklist of the seaweeds of the Mediterranean and Atlantic coasts of Morocco. II. Phaeophyceae. *Bot. Mar.* 45: 217–230.

Benhissoune, S., C.-F. Boudouresque, M. Perret-Boudouresque, and M. Verlaque. 2002b. A checklist of the seaweeds of the Mediterranean and Atlantic coasts of Morocco. III. Rhodophyceae (excluding Ceramiales). *Bot. Mar.* 45: 391–412.

———. 2003. A checklist of the seaweeds of the Mediterranean and Atlantic coasts of Morocco. III. Rhodophyceae–Ceramiales. *Bot. Mar.* 46: 55–68.

Benz, M. C., N. J. Eiseman, and E. E. Gallagher. 1979. Seasonal occurrence and variation in standing crop of a drift algal community in the Indian River, Florida. *Bot. Mar.* 22: 413–420.

Berger, S., U. Fettweiss, S. Gleissberg, L. B. Liddle, U. Richter, H. Sawitzky, and G. C. Zuccarello. 2003. 18S rDNA phylogeny and evolution of cap development in Polyphysaceae (formerly Acetabulariaceae: Dasycladales, Chlorophyta). *Phycologia* 42: 506–561.

Berger, S., and M. J. Kaever. 1992. *Dasycladales. An illustrated monograph of a fascinating algal order.* G. Thieme, Stuttgart and New York, vii + 247 pp.

Berglund, H. 1969. On the cultivation of multicellular marine green algae in axenic culture. *Svensk. Bot. Tidskr.* 63: 251–264.

Berthold, G. 1882. Über die Vertheilung der Algen im Golf von Neapel nebst einem Verzeichnis der bisher daselbst beobachteten Arten. *Mitt. Zool. Sta. Neapel.* 3: 393–536.

Betancourt, L., and A. Herrera-Moreno. 2001. Algas marinas betónicas (Rhodophyta, Phaeophyta y Chloropyta) conocidas para la Hispaniola. *Moscosoa* 12: 105–134.

Bird, C. J., and McLachlan. 1992. *Seaweed flora of the maritimes 1. Rhodophyta: the red algae.* Biopress, Bristol, England, v + 177 pp.

Bird, C. J., M. A. Ragan, A. T. Critchley, E. L. Rice, and R. R. Gutell. 1994. Molecular relationships among the Gracilariaceae (Rhodophyta): further observations on some undetermined species. *Eur. J. Phycol.* 29: 195–202.

Bird, C. J., E. L. Rice, C. A. Murphy, and M. A. Ragan. 1992. Phylogenetic relationships in the Gracilariales (Rhodophyta) as determined by 18S rDNA sequences. *Phycologia* 31: 510–522.

Bitar, G., R. Dupuy de la Grandvire, M. Foulquie, and M. Verlaque. 2005. *Neomeris annulata* Dickie (Dasycladales, Chlorophyceae): a potential new invader from the Red Sea to the Mediterranean Sea. *Cryptogam. Bot.* 26: 309–317.

Bivona-Bernardi, A. de. 1822. *Scinaia* algarum marinarum novum genus. *L'Iride, Giornale di Scienze, Letteratura ed Arti per la Sicilia* 1: 232–234.

Blainville, H. M. D. de. 1834. *Manuel d'actinologie ou de zoophytologie.* F. G. Levrault, Paris, viii + 694 pp.

Blair, S. M. 1983. Taxonomic treatment of the *Chaetomorpha* and *Rhizoclonium* species (Cladophorales: Chlorophyta) in New England. *Rhodora* 85: 175–211.

Blair, S. M., A. C. Mathieson, and D. P. Cheney. 1982. Morphological and electrophoretic investigations of selected species of *Chaetomorpha* (Chlorophyta: Cladophorales). *Phycologia* 21: 164–172.

Bliding, C. 1963. A critical survey of European taxa in Ulvales. Part I. *Capsosiphon, Percursaria, Blidingia, Enteromorpha. Opera Bot.* 8(3), 160 pp.

———. 1968. A critical survey of European taxa in Ulvales. II. *Ulva, Ulvaria, Monostroma, Kormannia. Bot. Not.* 121(4): 535–629.

Blinks, L. R., and A. H. Blinks. 1931. Two genera of algae new to Bermuda. *Bull. Torrey Bot. Club* 57: 389–396.

Blomquist. H. L. 1954. A new species of *Myriotrichia* Harvey from the coast of North Carolina. *J. Elisha Mitchell Sci. Soc.* 70: 37–41.

———. 1958a. The taxonomy and chromatophores of *Pilayella antillarum* (Grunow) De Toni. *J. Elisha Mitchell Sci. Soc.* 74: 25–30.

———. 1958b. *Myriotrichia scutata* Blomquist conspecific with *Ectocarpus subcorymbosus* Holden in Collins. *J. Elisha Mitchell Sci. Soc.* 74: 24.

Blomster, J., E. M. Hovey, C. A. Maggs, and M. I. Stanhope. 2000. Species-specific oligonucleotide probes for macroalgae: molecular discrimination of two marine fouling species of *Enteromorpha* (Ulvophyceae). *Mol. Ecol.* 9: 177–186.

Blomster, J., C. A. Maggs, and M. J. Stanhope. 1998. Molecular and morphological analysis of *Enteromorpha intestinalis* and *E. compressa* (Chlorophyta) in the British Isles. *J. Phycol.* 34: 319–340.

———. 1999. Extensive intraspecific morphological variation in *Enteromorpha muscoides* (Chlorophyta) revealed by molecular analysis. *J. Phycol.* 35: 575–586.

Bodard, M. 1965. Le *Gracilaria occidentalis* (Borg.): une espèce de Rhodophycée pantropicale Atlantique. *Bull. Mus. Natl. Hist. Nat. (Paris)*, sér. 2, 36: 874–878.

Bodard, M., and J. Mollion. 1974. La végétation infralittorale de la petite côte sénégalaise. *Bull. Soc. Phycol. Fr.* 19: 193–221.

Boedeker, C., G. C. Zuccarello, S.G.A. Draisma, and W. F. Prud'homme van Reine. 2005. Phylogeny of the Cladophorales (Chlorophyta) inferred from 28S rDNA sequences, focusing on *Chaetomorpha* and *Rhizoclonium. Phycologia* 44(4, supplement): 10.

Bold, H. C., and M. W. Wynne. 1985. *Introduction to the algae*, 2nd ed. Prentice-Hall, Englewood Cliffs, N.J., xvi + 720 pp.

Bonneau, E. R. 1977. Polymorphic behavior of *Ulva lactuca* (Chlorophyta) in axenic culture. I. Occurrence of *Enteromorpha*-like plants in haploid clones. *J. Phycol.* 13: 133–140.

Bonnemaison, T. 1828. Essai sur les Hydrophytes loculées (ou articulées) de la famille des Épidermées et des Céramiées. *Mem. Mus. Natl. Hist. Nat. (France)*, sér. B, 16: 49–148, pls. 3–8.

Børgesen, F. 1905. Contribution à la connaissance du genre *Siphoncladus* Schmitz. Oversight Over Det *K. Dan. Vidensk. Selsk. Forhand.* 3: 259–291, 13 Figs.

———. 1907. An ecological and systematic account of the *Caulerpas* of the Danish West Indies. *K. Dan. Vidensk. Selsk. Skr., Biol. Medd.*, 7. Raekke, Naturvid. og Math. Afd. 4: 337–392.

———. 1909a. The species of *Avrainvilleas* hitherto found on the shores of the Danish West Indies. *Videnk. Medd. Naturh. Foren. Kjøbenhavn* 1908: 27–44.

———. 1909b. Some new or little known West Indian Florideae. *Bot. Tidsskr.* 30: 1–19.

———. 1910. Some new or little known West Indian Florideae. II. *Bot. Tidsskr.* 30: 177–207.

———. 1911. Some Chlorophyceae from the Danish West Indies. *Bot. Tidsskr.* 31: 127–152.

———. 1912. Some Chlorophyceae from the Danish West Indies. II. *Bot. Tidsskr.* 31: 241–273.
———. 1913. The marine algae of the Danish West Indies. Part 1. Chlorophyceae. *Dan. Bot. Ark.* 1(4). 158 [160] pp., folded map.
———. 1914a. The species of *Sargassum* found along the coast of the Danish West Indies with remarks upon the floating forms of the Sargasso Sea. In: H.F.E. Jungersen and E. Warming (eds.), *Mindeskrift i Anledning af Hundredaaret for Japetus Steenstrups Fødsel.*, Art. 32, pp. 1–20.
———. 1914b. The marine algae of the Danish West Indies. Part 2. Phaeophyceae. *Dan. Bot. Ark.* 2(2), 66[68] pp. [continuous pagination 159–226]
———. 1915. The marine algae of the Danish West Indies. Part 3. Rhodophyceae (1). *Dan. Bot. Ark.* 3: 1–80.
———. 1916. The marine algae of the Danish West Indies. Part 3. Rhodophyceae (2). *Dan. Bot. Ark.* 3: 81–144.
———. 1917. The marine algae of the Danish West Indies. Part 3. Rhodophyceae (3). *Dan. Bot. Ark.* 3: 145–240.
———. 1918. The marine algae of the Danish West Indies. Part 3. Rhodophyceae (4). *Dan. Bot. Ark.* 3: 241–304.
———. 1919. The marine algae of the Danish West Indies. Part 3. Rhodophyceae (5). *Dan. Bot. Ark.* 3: 305–368.
———. 1920. The marine algae of the Danish West Indies. Part 3. Rhodophyceae (6), with addenda to the Chlorophyceae, Phaeophyceae and Rhodophyceae. *Dan. Bot. Ark.* 3: 369–498 + [6].
———. 1924. Marine algae. In: Plants from Beata Island, St. Domingo, collected by C. H. Ostenfeld. (Botanical Results of the Dana-Expedition 1921–22, No. 1). *Dan. Bot. Ark.* 4(7): 14–35.
———. 1925. Marine algae from the Canary islands, especially from Teneriffe and Gran Canaria. I. Chlorophyceae. *K. Dan. Vidensk. Selsk., Biol. Medd.* 5(3): 1–123.
———. 1926. Marine algae from the Canary islands, especially from Teneriffe and Gran Canariad. II. Phaeophyceae. *K. Dan. Vidensk. Selsk., Biol. Medd.* 6(2): 1–112.
———. 1927. Marine algae from the Canary islands, especially from Teneriffe and Gran Canariad. III. Rhodophyceae. Part I, Bangiales and Nemalionales. *K. Dan. Vidensk. Selsk., Biol. Medd.* 6(6): 1–97.
———. 1929. Marine algae from the Canary islands, especially from Teneriffe and Gran Canariad. III. Rhodophyceae. Part 2, Cryptonemiales, Gigartinales and Rhodymeniales (Mélobésiées by Mme. P. Lemoine). *K. Dan. Vidensk. Selsk., Biol. Medd.* 8(1): 1–97.
———. 1930a. Marine algae from the Canary islands, especially from Teneriffe and Gran Canariad. III. Rhodophyceae. Part 3, Ceramiales. *K. Dan. Vidensk. Selsk., Biol. Medd.* 9(1): 1–159.
———. 1930b. Some Indian green and brown algae especially from the shore of the Presidency of Bombay. *J. Indian Bot. Soc.* 9: 151–174.
———. 1931. Some Indian Rhodophyceae especially from the shores of the Presidency of Bombay. *Bull. Misc. Inf. (Kew)* 1931: 1–24, pls. I, II.
———. 1932a. Some Indian Rhodophyceae, especially from the Presidency of Bombay. II. *Bull. Misc. Inf. (Kew)* 1932: 113–134, pls. I, II.
———. 1932b. A revision of Forsskàls algae mentioned in Flora Aegyptiaco-Arabica and found in his herbarium in the Botanical Museum of the University of Copenhagen. *Dan. Bot. Ark.* 8(2), 14 pp.
———. 1934. Some marine algae from the northern part of the Arabian Sea with remarks on their geographical distribution. *K. Dan. Vidensk. Selsk., Biol. Medd.* 11(6), 72 pp., 2 pls.
———. 1935. A list of marine algae from Bombay. *K. Dan. Vidensk. Selsk., Biol. Medd.* 12(2), 64 pp., 10 pls.
———. 1937. Contributions to a South Indian marine algal flora. I. *J. Indian Bot. Soc.* 16: 1–56, pl. I.
———. 1939. Marine algae from the Iranian Gulf especially from the innermost part near Bushire and the Island Kharg. In: K. Jessen and R. Spärck (eds.), *Danish scientific investigation in Iran*, Part I. Copenhagen, pp. 47–141, 43 Figs, map.
———. 1941. Some marine algae from Mauritius. II. Phaeophyceae. *K. Dan. Vidensk. Selsk., Biol. Medd.* 16(3), 81 pp., 8 pls.
———. 1942. Some marine algae from Mauritius. III. Rhodophyceae. Part 1. Porphyridiales, Bangiales, Nemalionales. *K. Dan. Vidensk. Selsk., Biol. Medd.* 17(5), 63 [64] pp., 27 figs., 2 pls.
———. 1943. Some marine algae from Mauritius. III. Rhodophyceae. Part 2. Gelidiales, Cryptonemiales, Gigartinales. *K. Dan. Vidensk. Selsk., Biol. Medd.* 19(1), 85 pp., I pl.
———. 1944. Some marine algae from Mauritius. III. Rhodophyceae. Part 3. Rhodymeniales. *K. Dan. Vidensk. Selsk., Biol. Medd.* 19(6): 1–32.

———. 1945. Some marine algae from Mauritius. III. Rhodophyceae. Part 4. Ceramiales. *K. Dan. Vidensk. Selsk., Biol. Medd.* 19(10), 68 pp.

———. 1949a. Some marine algae from Mauritius. Additions to the parts previously published. *K. Dan. Vidensk. Selsk., Biol. Medd.* 21(5), 48 pp.

———. 1949b. On the genus *Titanophora* (J. Ag.) Feldm. and description of a new species. *Dan. Bot. Ark.* 13(4), 8 pp., 2 pls.

———. 1950a. *Vaughaniella*, a new genus of Dictyotaceae. *K. Dan. Vidensk. Selsk., Biol. Medd.* 18(8), 10 pp, 8 figs.

———. 1950b. A new species of the genus *Predaea*. *Dan. Bot. Ark.* 14: 1–8.

———. 1952. Some marine algae from Mauritius. Additions to the parts previously published. IV. *K. Dan. Vidensk. Selsk., Biol. Medd.* 18(19), 72 pp., 5 pls.

Bornet, É. 1889. Note sur l'*Ectocarpus* (*Pylaiella*) *fulvescens* Thuret. *Rev. Gén. Bot.* 1: 5–10.

———. 1891. Note sur quelques *Ectocarpus*. *Bull. Soc. Bot. Fr.* 38: 353–372

———. 1892. Les algues de P.-K.-A. Schousboe récoltées au Maroc et dans le Mediterranées de 1815 à 1829. Mém. Soc. Natl. Sci. Math. Cherbourg 28: 165–376, 3 pls.

———. 1904. Deux *Chantransia corymbifera* Thuret. *Acrochaetium* et *Chantransia*. Bull. Soc. Bot. Fr. 51: XIV–XXIII (session jubilaire).

Bornet, É., and C. Flahault. 1888. Note sur deux nouveaux genres d'algues perforantes. *J. Bot. [Morot]* 2: 161–165.

———. 1889. Sur quelques plantes vivants dans le test calcaire des Mollusques. *Bull. Soc. Bot. Fr.* 36: CXLVII–CLXXVI, pls. VI–XII.

Bornet, É., and G. Thuret. 1876. Notes algologiques recueil d'observation sur les algues. Fasc. I. Paris xx + 70 [72], pls. I–XXV.

Borowitzka, M. A., and R. C. Lethbridge, R. C. 1989. Seagrass epiphytes. In: A.W.D. Larkum, A. J. McComb and S. A. Shepherd (eds.), *Biology of seagrasses*. Elsevier, Amsterdam, pp. 458–499.

Bory de St. Vincent, J.B.G.M. 1822–1831. *Dictionnaire Classique d'Histoire Naturelle*. Vols. 1–17. Rey and Gravier; Baudoin Frères, Paris. Vol. 1, pp. [i]–xvi, [1]–604, 1822a; Vol. 2, pp.[i–iii], [1]–621, 1822b; Vol. 3, pp. [i–iii], [1]–592, 1823a; Vol. 4, pp. [i–iii], [1]–628, 1823b; Vol. 5, pp. [i–iii], [1]–635, 1824a; Vol. 6, pp. [i–iii], [1]–593, [1]–4, 1824b; Vol. 7, pp. [i–iii], [1]–626, 1825a; Vol. 8, pp. [i–iii], [i]–iii, [1]–609, 1825b; Vol. 9, pp. [i–iii], [1]–596, 1825c; Vol. 10, pp. [i–iii], [1]–642, 1826; Vol. 11, pp. [i–iii], [1]–615, 1827a; Vol. 12, pp. [i–iii], [1]–634, 1827b; Vol. 13, pp. [i–iii], [1]–648, 1828a; Vol. 14, pp. [i–iii], [1]–710, 1828b; Vol. 15, pp. [i–iii], [1]–754, 1829; Vol. 16, pp. [i–iii], [1–4], [1]–748, 1830; Vol. 17 [atlas & illustrations], pp. [i–viii], [1]–141, pls. i–clx, 1831.

———. 1826–1829. Cryptogamie. In: L. I. Duperrey, Voyage autour du monde, exécuté par ordre du Roi, sur la corvette de sa majesté, "La Coquille," pendant les années 1822, 1823, 1824 et 1825. Paris. 301 pp. Atlas. Histoire Naturelle, Botanique. Pls. 1–13, 13 bis, 14–38. [Pp. 1–96 (1827); 97–200 (1828); 201–301 (1829); Atlas (1826).].

———. 1834. Hydrophytes, Hydrophytae. In: C. Bélanger (ed.), Voyage aux Indes-Orientales . . . pendant les années 1825, 1826, 1827, 1828 et 1829. Botanique. Cryptogamie. Paris, pp. 159–178, pls. XV, XVI.

Bot, P.V.M., W. T. Stam, and C. van den Hoek. 1990. Genotypic relations between geographic isolates of *Cladophora laetevirens* and *C. vagabunda*. *Bot. Mar.* 33: 441–446.

Boudouresque, C.-F., E. Coppejans, and J. Marcot. 1976. Un nouveau genre de Peyssonneliaceae, *Metapeyssonnelia* (Rhodophyta). *Phycologia* 15: 283–288.

Brauner, J. F. 1975. The seasonality of epiphytic algae on *Zostera marina* at Beaufort, North Carolina. *Nova Hedwigia* 26: 125–133.

Bravin, I. C., J. Torres, C.F.D. Gurgel, and Y. Yoneshigue-Valentin. 1999. Novas ocorrências de Clorofíceas marinhas de profundidade para o Brasil. *Hoehnea* 26: 121–133.

Bray, T. L., C. D. Neefus, and A. C. Mathieson. 2005. Two Asian species of *Porphyra* found on the New England coast. In Program and Abstracts, Annual Meeting, Northeast Algal Society, Samoset Resort, Rockport, Maine, 15–17 April, 2005, pp. 4.

Bressan, G., and L. Babbini. 2003. Biodiversità marina delle coste Italiane: Corallinales de Mar Mediterranea: guida all determinazione. *Biol. Mar. Mediterr.* 10(suppl. 2): 1–237.

Brodie, J., P. K. Hayes, G. L. Barker, L. M. Irvine and I. Bartsch. 1998. A reappraisal of *Porphyra* and *Bangia* (Bangophycidae, Rhodophyta) in the northeast Atlantic based on rbcL and rbcS intergeneric spacer. *J. Phycol.* 34: 1069–1074.

Brodie, J., and L. M. Irvine. 2003. *Seaweeds of the British Isles*, Vol. 1. Rhodophyta, Part 3B. Bangiophycidae.

Natural History Museum, London, xiii + 167 pp., 54 figs., 6 tables.
Brodie, J., and J. N. Norris. 1996. Life histories and morphology of some *Liagora* species (Liagoraceae, Rhodophyta) from the Florida Keys, USA. *Nova Hedwigia* 112: 121–134.
Broom, J. E., W. A. Nelson, W. A. Jones, C. Yarish, R. Aguilar Rosas, and L. E. Augilar Rosas. 2000. *Porphyra suborbiculata, Porphyra carolinensis* and *Porphyra lilliputiana*: three names for one small *Porphyra. J. Phycol.* 36(suppl.): 7–8.
Broom, J. E., W. A. Nelson, C. Yarish, W. A. Jones, R. Aguilar Rosas, and L. E. Aguilar Rosas. 2002. A reassessment of the taxonomic status of *Porphyra suborbiculata, Porphyra carolinensis* and *Porphyra lilliputiana* (Bangiales, Rhodophyta) based on molecular and morphological data. *Eur. J. Phycol.* 37: 227–235.
Brown, M. M. 2001. The drift algae of seagrass beds in Sunset Beach, Cockroach Bay, and Tarpon Bay on the west coast of Florida. M.S. thesis, Univ. South Florida, Tampa, 58 pp.
Browne, P. 1756 (1789). Algae, pp. 71–74. In: The civil and natural history of Jamaica. Part I, An accurate description of that island, 2 maps, pp. [4 +] i–viii, 1–27 [+ 1]. Part II, pp. [4 +] xxix–xxxiv; Book I, An account of the fossils, pp. 35–66; Book II, History of the vegetable productions, pp. lxvii–lxx, 71–374, pls. 1–38; Book III, Account of such animals, pp. ccclxxiv–ccclxxix [+ 1], pls. 39–49. [Part III not published.] 1756, London. Id., 2nd ed., pages as before, with: four additional indexes, 46 [unnumbered pages]. 1789. London.
Bucher, K. E., and J. N. Norris. 1992. A new deep-water red alga, *Titanophora submarina sp. nov.* (Gymnophloeaceae, Gigartinales), from the Caribbean Sea. *Phycologia* 31: 180–191.
Bucher, K. E., J. N. Norris, M. M. Littler, and D. S. Littler. 1990. Marine algae new to Florida, including *Trichosolen molassensis sp. nov.* (Chlorophyta) and *Diplothamnion jolyi* var. *ecellulare* var. nov. (Rhodophyta). *Cryptogam. Bot.* 1: 295–307.
Bula-Meyer, G. 1995. Macroalgas de la Isla de Gorgona (Pacífico Colombian) con nuevos registros y una explicación de la baja diversidad y biomasa. In: P. Pinto (ed.): *La isla de Gorgona*. Biblioteca J. J. Triana, pp. 23–45.
Bula-Meyer, G., and J. N. Norris. 2001. Notes on new records of red algae from the Columbian Caribbean. *Bot. Mar.* 44: 351–361.
Burkholder, P. R., and L. R. Almodovar. 1973. Studies on mangrove algal communities in Puerto Rico. *Fla. Sci.* 36: 66–74.
Burrows, E. M. 1991. *Seaweeds of the British Isles*. Vol. 2. *Chlorophyta*. Natural History Museum, London, Publ. xii + 238 pp., 9 pls.
Byrne, K., G. C. Zuccarello, J. West, M.-L. Liao, and G. T. Kraft. 2002. *Gracilaria* species (Gracilariaceae, Rhodophyta) from southeastern Australia, including a new species, *Gracilaria perplexa sp. nov.*: Morphology, molecular relationships and agar content. *Phycol. Res.* 50: 295–311.
Cabrera, R., and B. Martínez-Daranas. 2005. Variabildad morfológica de una población de *Avrainvillea asarifolia* f. *olivacea* D. S. Littler and M. M. Littler. *Rev. Investig. Mar., Univ. Habana* 26: 3–8.
Calvert, H. E. 1976. Culture studies on some Florida species of *Caulerpa*: morphological responses to reduced illumination. *Br. Phycol. J.* 11: 203–214.
Campbell, S. J., and W. J. Woelkerling. 1990. Are *Titanoderma* and *Lithophyllum* (Corallinaceae, Rhodophyta) distinct genera? *Phycologia* 29: 114–125.
Camus, C., A. P. Meynard, S. Faugeron, K. Kogame, and J. A. Correa. 2005. Differential life history phase expression in two coexisting species of *Scytosiphon* (Phaeophyceae) in northern Chile. *J. Phycol.* 41: 931–941.
Cardinal, A. 1964. Etude sur le Ectocarpacées de la Manche. Beih. *Nova Hedwigia* 15: 1–86.
Cassano, V., L.V.R. de Brito, and M.T.M. de Széchy. 2004. The occurrence of *Boodleosis vaucherioidea* Calderón-Sáenz et Schnetter (Udoteaceae, Chlorophyta) in Brazil. *Bot. Mar.* 47: 251–254.
Cassano, V., M.T.M. de Széchy, and M. T. Fujii. 2006. *Laurencia caduciramulosa* (Ceramiales, Rhodophyta) from Ilha Grande Bay, Rio De Janerio Brazil: a recent introduction into the Atlantic Ocean? *Cryptogamie, Algol.* 27: 265–277.
Cassano, V., and Y. Yoneshigue-Valentin. 2001. *Hincksia onslowensis* (Amsler & Kapraun) Silva (Phaeophyta, Ectocarpales): new occurrence in deep waters for the South American Atlantic coast. *Hoehnea* 28: 267–277.
Castagne, L. 1845. *Catalogue des plantes qui croissent naturellement aux environs de Marseille*. Aix, 263 pp., 7 pls.
———. 1851. *Supplément au catalogue des plantes qui croissent naturellement aux environs de Marseille.* Aix, 125 pp.

Cavalier-Smith, T. 1998. A revised six-kingdom system of life. *Biol. Rev.* 73: 203–266.

Cecere, E., M. Cormaci, G. Furnari, A. Petrocelli, O. Saracino, and D. Serio. 1996. Benthic algal flora of Cheradi Islands (Gulf of Taranto, Mediterranean Sea). *Nova Hedwigia* 62: 191–214.

Ceccherelli, G., and F. Cinelli.1997. Short-term effects of nutrient enrichment of the sediment and interactions between the seagrass *Cymodocea nodosa* and the introduced green alga *Caulerpa taxifolia* in a Mediterranean bay. *J. Exp. Mar. Biol. Ecol.* 217: 165–177.

———. 1999. Effects of *Posidonia oceanica* canopy on *Caulerpa taxifolia* size in a north-western Mediterranean bay. *J. Exp. Mar. Biol. Ecol.* 240: 19–36.

Ceccherelli, G., and L. Piazzi. 2001. Dispersal of *Caulerpa racemosa* fragments in the Mediterranean: lack of detachment time effect on establishment. *Bot. Mar.* 44: 209–213.

Chamberlain, Y. M. 1983. Studies in the Corallinaceae with special reference to *Fosliella* and *Pneophyllum* in the British Isles. *Bull. Br. Mus. (Nat. Hist.) Bot.* 11: 291–463.

———. 1994. Mastophoroideae Setchell. In: L. M. Irvine and Y. M. Chamberlain, *Seaweeds of the British Isles*. Vol. 1. *Rhodophyta. Part 2B. Corallinales, Hildenbrandiales*. Natural History Museum (London), HMSO, pp. 113–158.

Chamberlain, Y. M., L. M. Irvine, and R. Walker. 1991. A redescription of *Lithophyllum orbiculatum* (Rhodophyta, Corallinales) in the British Isles and a reassessment of generic delimitation in the Lithophylloideae. *Br. Phycol. J.* 26: 149–167.

Chang, C. F., and B. M. Xia. 1963. [*Polycavernosa*, a new genus of the Gracilariaceae.]. *Stud. Mar. Sin.* 3: 119–126, 2 pls. [in Chinese].

Chapman, R. L. 1971. The macroscopic marine algae on Sapelo Island and other sites on the Georgia coast. *Bull. Ga. Acad. Sci.* 29: 77–89.

Chapman, V. J. 1937. A revision of the marine algae of Norfolk. *J. Linn. Soc. Lond. Bot.* 51: 205–263.

———. 1952. New entities in the Chlorophyceae of New Zealand. *Trans. R. Soc. N. Z.* 80: 47–58.

———. 1956. The marine algae of New Zealand. Part 1. Myxophyceae and Chlorophyceae. *J. Linn. Soc. Lond. Bot.* 55: 333–501, pls. 24–50.

———. 1961. *The marine algae of Jamaica*. Part I. *Myxophyceae and Chlorophyceae*. Bull. Inst. Jamaica, Sci. Ser. 12, 159 pp.

———. 1963. *The marine algae of Jamaica*. Part 2. *Phaeophyceae and Rhodophyceae*. Bull. Inst. Jamaica, Sci. Ser. 12, 201 pp.

———. 1977a. *Ecosystems of the world*. Vol. 1. *Wet coastal ecosystems*. Elsevier, Amsterdam, xi + 428 pp.

———. 1977b. Marine algae of Norfolk Island and the Cook Islands. *Bot. Mar.* 20: 161–165

Chappell, D. F., C. J. O'Kelly, L. W. Wilcox, and G. L. Floyd. 1990. Zoospore apparatus architecture and the taxonomic position of *Phaeophila dendroides* (Ulvophyceae, Chlorophyta). *Phycologia* 29: 515–523.

Chen, E., and J. F. Gerber. 1990. Climate. In: Myers, R. L., and J. J. Ewel (eds.), *Ecosystems of Florida*. Univ. Central Florida Press, Orlando, pp. 11–34.

Cheney, D. P. 1977. R & C/P: a new and improved ratio for comparing seaweed floras *J. Phycol.* 13(suppl.): 12.

———. 1988. The genus *Eucheuma* J. Agardh in Florida and the Caribbean. In: I. A. Abbott (ed.), *Taxonomy of economic seaweeds with reference to some Pacific and Caribbean species*. Vol. 2. Calif. Sea Grant College Program, Univ. Calif., La Jolla, pp. 209–219.

Cheney, D. P., and J. P. Dyer III. 1974. Deep-water benthic algae of the Florida Middle Grounds. *Mar. Biol.* 27: 185–190.

Chiang, Y.-M. 1970. Morphological studies of red algae of the family Cryptonemiaceae. *Univ. Calif. Publ. Bot.* 58: 1–95.

Chihara, M. 1961. Life cycle of Bonnemaisoniaceous algae in Japan (1). *Sci. Rep. Tokyo Kyoiku Daigaku*, Sec. B. 10: 121–153.

———. 1967. *Collinsiellopsis*, a new genus of green algae. *Phycologia* 6: 87–95.

Chihara, M., M. Díaz-Piferrer, and G. F. Papenfuss. 1980. Proposal for the conservation of the green algal generic name *Pseudobryopsis* Berthold against *Trichosolen* Montagne. *Taxon* 29: 329–330.

Chisholm, J.R.M., J. M. Jaubert, and G. Giaccone. 1995. *Caulerpa taxifolia* in the northwest Mediterranean: introduced species or migrant from the Red Sea? *C. R. Hebd. Séances Acad. Sci. Paris* 318D: 1219–1226.

Chisholm, J.R.M., M. C. Dauga, E. Ageron, P.A.D. Grimont, and J. M. Jaubort. 1996. 'Roots' in mixotrophic algae. *Nature* 381: 382.

Cho, G. Y., S. H. Lee, and S. M. Boo. 2004. A new brown algal order, Ishigeales (Phaeophyceae), established on the basis of plastid protein-coding *rbc*L, *psa*A, and *psb*A region comparisons. *J. Phycol.* 40: 921–936.

Cho, T. O., G. Y. Cho, H. S. Yoon, S. M. Boo, and W. J. Lee. 2003. New records of *Myelophycus cavus* (Scytosiphonaceae, Phaeophyceae) in Korea and the taxonomic position of the genus on the basis of a plastid DNA phylogeny. *Nova Hedwigia* 76: 381–397.
Choi, H.-G. 1996. A systematic study of the Dasyaceae (Ceramiales, Rhodophyta). Ph.D. diss., Seoul National Univ., Seoul.
Choi, H.-G., I. Bárbara, and H.-S Kim. 2005. Phylogenetic relationships of the Compsothamnioideae (Ceramiaceae, Rhodophyta) based on nuclear SSU rDNA and plastid rbcL sequence data. *Phycologia* 44(4, supplement): 19.
Choi, H.-G., G. T. Kraft, I. K. Lee, and G. W. Saunders. 2002. Phylogenetic analyses of anatomical and nuclear SSU rDNA sequence data indicate that the Dasyaceae and Delesseriaceae (Ceramiales, Rhodophyta) are polyphyletic. *Eur. J. Phycol.* 37: 551–569.
Christensen, T. 1957. *Chaetomorpha linum* in the attached state. *Bot. Tidsskr.* 53: 311–316.
———. 1978. Annotations to a textbook of phycology. *Bot. Tidsskr.* 73: 65–70.
———. 1987a. *Seaweeds of the British Isles. 4. Tribophyceae (Xanthophyceae).* British Museum (Natural History), London, viii + 36 pp.
———. 1987b. Report of the committee for algae. *Taxon* 36: 66–69.
Chualáin, F. N. , C. A. Maggs, G. W. Saunders, and M. D. Guiry. 2004. The invasive genus *Asparagopsis* (Bonnemaisoniaceae, Rhodophyta): molecular systematics, morphology, and ecophysilogy of *Falkenbergia* isolates. *J. Phycol.* 40: 1112–1126.
Chynoweth, L. A. 1975. New primary production of *Spartina* and species diversity of associated macroinvertebrates of a semi-impounded salt marsh. Tech. Rep. 1, National Aeronautics and Space Administration, Kennedy Space Center, 123 pp.
Cintròn G. A., A. E. Lugo, and R. Martinez. 1985. Structural and functional properties of mangrove forests. In: W. G. D'Arcy and M. D. Correa (eds.), *The botany and natural history of Panama.* Monographs in Systematic Botany 10. Missouri Botanical Garden, St. Louis, pp. 53–66.
Clayton, M. N. 1974. Studies on the development, life history and taxonomy of the Ectocarpales (Phaeophyta) in Southern Australia. *Aust. J. Bot.* 22: 743–813.
———. 1980. Sexual reproduction—a rare occurrence in the life history of the complanate form of *Scytosiphon* (Scytosiphonaceae, Phaeophyta) from southern Australia. *Br. Phycol. J.* 15: 105–118.
Clemente y Rubio, S. de R. 1807. *Ensayo sorbe las variedades de la vid común que vegetan en Andalucía.* Madrid, xviii + 324 pp.
Clifton, K. E. 1997. Mass spawning by green algae on coral reefs. *Science* 275: 1113–1116.
Clifton, K. E., and Clifton, L. M. 1999. The phenology of sexual reproduction by tropical green algae. *J. Phycol.* 35: 24–34.
Cole, K. M., B. J. Hymes, and R. G. Sheath. 1983. Karyotypes and reproductive seasonality of the genus *Bangia* (Rhodophyta) in British Columbia. *Can. J. Bot.* 19: 136–145.
Coll, J., and J. Cox. 1977. The genus *Porphyra* C. Agardh (Rhodophyta, Bangiales) in the American North Atlantic. I. New species from North Carolina. *Bot. Mar.* 20: 155–159.
Coll, J., and E. C. Oliveira. 1976. The genus *Porphyra* C. Ag. (Rhodophyta-Bangiales) in the American South Atlantic. II. Uruguayan species. *Bot. Mar.* 19: 191–196.
Collado-Vides, L. 2002. Morphological plasticity of *Caulerpa prolifera* (Caulerpales, Chlorophyta) in relation to growth form in a coral reef lagoon. *Bot. Mar.* 45: 123–129.
Collins, F. S. 1900. Preliminary list of New England plants V. Marine algae. *Rhodora* 2: 41–52.
———. 1901. The algae of Jamaica. *Proc. Am. Acad. Arts Sci.* 37: 229–270.
———. 1902. The marine *Cladophoras* of New England. *Rhodora* 4: 111–127.
———. 1905. Phycological notes of the late Isaac Holden. II. *Rhodora* 7: 168–172, 222–243.
———. 1906a. New species in the Phycotheca. *Rhodora* 8: 104–113.
———. 1906b. *Acrochaetium* and *Chantransia* in North America. *Rhodora* 8: 189–196.
———. 1909a. New species of *Cladophora. Rhodora* 11: 17–26.
———. 1909b. The green algae of North America. *Tufts Coll. Stud. Sci. Ser.* 2: 79–480.
Collins, F. S., and A. B. Hervey. 1917. The algae of Bermuda. *Proc. Am. Acad. Arts Sci.* 53: 1–195.
Collins, F. S., I. Holden, and W. A. Setchell. 1895–1919. *Phycotheca boreali-americana.* A collection of dried specimens of the algae of North America. Malden, Massachusetts. Fasc. I–XLVI, A–E, nos. 1–2300, I–CXXV [Exsiccatae with printed labels].
Collins, F. S., and M. A. Howe. 1916. Notes on some species of *Halymenia. Bull. Torrey Bot. Club* 43: 169–182.

Compère, P. 1997. Report of the Committee for Algae: 3. *Taxon* 46: 319–321.

———2005. Report of the Committee for Algae: 9. *Taxon* 54: 523–524.

Conde, F., A. Flores-Moya, J. Soto, M. Altamirano and A. Sánchez. 1996. Check-list of Andalusia (S. Spain) seaweeds. III. Rhodophyceae. *Acta Bot. Malacit.* 21: 7–33.

Coppejans, E. 1977. Végétation marine de l'ile de Port-Cros (Parc National). XV. *Ceramium cingulatum* Weber van Bosse nouvelle pour la Méditerranée, et quelques populations d'un *Ceramium* sp. à parasporocystes. Biologisch Jaarboek Dodonaea 45: 51–61.

———. 1992. Marine algae of Papua New Guinea (Madang Prov.). 2. A revised and completed list of *Caulerpa* (Chlorophyta-Caulerpales). *Blumea* 36: 383–410.

Coppejans, E., and T. Beeckman. 1989. *Caulerpa* section Sedoideae (Chlorophyta, Caulerpales) from the Kenyan coast. *Nova Hedwigia* 49: 381–393.

———. 1990. *Caulerpa* (Chlorophyta, Caulerpales) from the Kenya Coast. *Nova Hedwigia* 50: 111–125.

Coppejans, E., O. De Clerck, and F. Leliaert. 2001a. Marine brown algae (Phaeophyta) from the north coast of Papua New Guinea, with a description of *Dictyota magneana sp. nov. Cryptogam. Algol.* 23: 15–40.

Coppejans, E., F. Leliaert, O. Dargent and O. De Clerck. 2001b. Marine green algae (Chlorophyta) from the north coast of Papua New Guinea. *Cryptogam. Algol.* 22: 375–443.

Coppejans, E., and W. F. Prud'homme van Reine. 1992. Seaweeds of the Snellius-II Expedition (E. Indonesia): the genus *Caulerpa* (Chlorophyta-Caulerpales). *Bull. Séances Acad. R. Sci. d'Outre-Mer* 37: 667–712.

Cordero, P. A., Jr. 1981. *Studies on Philippine marine red algae.* Nat. Mus. Philippines, Manila, 258 pp., XXVIII plates.

Cordeiro-Marino, M. 1978. Rhodofíceas bentônicas marinhas do Estado de Santa Catarina. *Rickia* 7: 1–243.

Cormaci, M., and G. Furnari. 1988. *Antithamnionella elegans* (Berthold) Cormaci *et* Furnari (Ceramiaceae, Rhodophyta) and related species, with the description of two new varieties. *Phycologia* 27: 340–346.

Cormaci, M., G. Furnari, and F. Pizzuto. 1994. Taxonomic and nomenclatural notes on *Anotrichium tenue* and related species (Ceramiaceae, Rhodophyta). *Taxon* 43: 633–637.

Correa, J. A., R. Nielsen, and D. W. Grund. 1988. Endophytic algae of *Chondrus crispus* (Rhodophyta). II. *Acrochaete heteroclada sp. nov., A. operculata sp. nov.* and *Phaeophila dendroides* (Chlorophyta). *J. Phycol.* 24: 528–539.

Cotton, A. D. 1907. New or little-known marine algae from the east. *Bull. Misc. Inf. (Kew)* 1907: 260–264, 1 pl.

———. 1910. On the growth of *Ulva latissima* in water polluted by sewage. *Bull. Misc. Inf. (Kew)*, pp. 15–19.

Cremades, J., and J. L. Pérez-Cirera. 1990. Nuevas combinaciones de algas bentónicas marinas, como resultado del estudio del Herbario de Simón de Rojas Clemente y Rubio (1777–1827). *An. Jard. Bot. Madr.* 47: 489–492.

Cribb, A. B. 1951. Invalidation of the genus *Vaughaniella. Nature* 168: 302.

———. 1958. Records of marine alge from southeastern Queensland. IV. *Caulerpa.* Univ. Queensland Press, Brisbane 3: 209–220.

———. 1966. The algae of Heron Island, Great Barrier Reef, Australia. Part 1. A general account. Univ. Queensland Papers, Heron Island Res. Station 1: 1–23.

———. 1983. *Marine algae of the southern Great Barrier Reef.* Part I. *Rhodophyta.* Handbook no. 2. Australian Coral Reef Society, Brisbane, 173 pp. + [2], pls. 1–71.

———. 1996. *Seaweeds of Queensland. A naturalist's guide.* The Queensland Naturalist's Club, Brisbane, 130 pp.

Croley, F. C., and C. J. Dawes. 1970. Ecology of the algae of a Florida Key. I. A preliminary checklist, zonation and seasonality. *Bull. Mar. Sci.* 20: 165–185.

Cronquist, A. 1960. The divisions and classes of plants. *Bot. Rev.* 26: 425–482.

Crouan, P. L., and H. M. Crouan. 1835. Observations microscopiques sur la genre *Mesogloia* Agardh. *Ann. Sci. Nat., Bot.*, sér. 2, 3: 98–99. pl. 2.

———. 1844. Observations sur le genre *Peyssonnelia* Dne. *Ann. Sci. Nat., Bot.*, sér. 3, 2: 367–368.

———. 1848. Sur l'organisation, la fructification et la classification du *Fucus wigghii* de Turner et de Smith, et de l' *Atractophora hypnoides. Ann. Sci. Nat., Bot.*, sér. 3, 12: 361–376.

———. 1852. Algues marines du Finistère . . . Brest. Vols. 1–3. Nos. 1–404. [Exsiccata with printed labels and indexes].

———. 1858. Note sur quelques algues marines nouvelles de la rade de Brest. *Ann. Sci. Nat., Bot.*, sér. 4, 9: 69–75.

———. 1859. Notice sur quelques espéce et genres nouveaux d'algues marines de la Rade de Brest. *Ann. Sci. Nat., Bot.*, sér. 4(12): 288–292, pl. 22.
———. 1860. Liste des algues marines découvertes dans le Finistère depuis la publication des algues de ce département en 1852. *Bull. Soc. Bot. Fr.* 7: 367–373 (pp. 1–7 in reprint).
———. 1867. Florule du Finistère . . . . Paris and Brest. x + 262 pp., 31 [+ 1] pls., frontispiece.
Cruz Adames, V. M., and D. L. Ballantine. 1996. Asexual reproduction in *Laurencia poiteaui* (Rhodomelaceae, Rhodophyta). *Bot. Mar.* 39: 75–77.
Dangeard, P. 1962. Sur le développement du *Petalonia fascia* (Müller) Kuntze et du *Scytosiphon lomentaria* (Lyngbye) Endlicher. *C. R. Hebd. Séances Acad. Sci. Paris* 254: 3290–3292.
D'Archiac, E. J. 1843. Description géologique du département de l'Aisne. *Mém. Soc. Géol. Fr.* 1: 129–418.
Davis, J. H. 1940. The ecology and geologic role of mangroves in Florida. Carnegie Institute, Washington, D.C., Publ. No. 517, pp. 303–412.
Davis, R. A., Jr., A. C. Hine, and E. A. Shinn. 1992. Holocene coastal development on the Florida Peninsula. Quaternary coasts of the United States: marine and lacustrine systems. SEPM Special Publ., 48 pp.
Dawes, C. J. 1967. *Marine algae of the vicinity of Tampa Bay*. Univ. South Florida, Tampa, 104 pp.
———. 1974. *The marine algae of the west coast of Florida*. Univ. Miami Press, Coral Gables, xvi + 201 pp.
———. 1980. A new variety of the calcareous coenocyte *Halimeda lacrimosa* Howe (Chlorophyta, Siphonales). *Bull. Mar. Sci.* 30: 142.
———. 1981. *Marine botany*. John Wiley and Sons, New York, viii + 628 pp.
———. 1985. Macroalgae of the Tampa Bay estuarine system. In: S. F. Treat, J. S. Simon, R. R. Lewis, E. D. Estevez, and S. K. Mahadevan (eds.), *Proceedings, Tampa Bay area scientific information symposium*. Burgess Publ., Minneapolis, pp. 184–209.
———. 1996. Macroalgal diversity, standing stock and productivity in a northern mangal on the west coast of Florida. *Nova Hedwigia* 112: 525–535.
———. 1998. *Marine botany*, 2nd ed. John Wiley and Sons, New York, xiv + 480 pp.
Dawes, C. J., S. Bell, R. A. Davis Jr., E. C. McCoy, H. R. Mushinbsky, and J. L. Simon. 1995. Initial effects of Hurricane Andrew on the shoreline habitats of southwestern Florida. *J. Coastal Res.* (special issue) 21: 103–110.
Dawes, C. J., S. A. Earle, and F. C. Croley. 1967. The offshore benthic flora of the southwest coast of Florida. *Bull. Mar. Sci.* 17: 211–231.
Dawes, C. J., D. Hanisak, and W. J. Kenworthy. 1995. Seagrass biodiversity in the Indian River Lagoon. *Bull. Mar. Sci.* 57: 59–66.
Dawes, C. J., and H. J. Humm. 1969. A new variety of the calcareous coenocyte, *Halimeda lacrimosa* Howe. *Bull. Mar. Sci.* 19: 428–431.
Dawes, C. J., and J. M. Lawrence. 1990. Seasonal changes in limestone and sand plant communities off the Florida west coast. P. S. Z. N. I. *Mar. Ecol.* 11: 97–104.
Dawes, C. J., and A. C. Mathieson. 1972. A new species of *Pseudocodium* (Chlorophyta, Siphonales) from the west coast of Florida. *Phycologia* 11: 273–277.
Dawes, C. J., A. C. Mathieson, and D. P. Cheney. 1974. Ecological studies of Floridian *Eucheuma* (Rhodophyta, Gigartinales). I. Seasonal growth and reproduction. *Bull. Mar. Sci.* 24: 235–273.
Dawes, C. J., E. D. McCoy, and K. L. Heck Jr. 1991. The tropical western Atlantic including the Caribbean Sea. In: A. C. Mathieson and P. H. Neinhuis (eds.), *Ecosystems of the world*. Vol. 24, *Intertidal and littoral ecosystems*. Elsevier, Amsterdam, pp. 215–234.
Dawes, C. J., R. C. Phillips, and G. Morrison. 2004. *Seagrass communities of the Gulf Coast of Florida: status and ecology*. Florida Fish and Wildlife Conservation Commission, Fish and Wildlife Research Institute and the Tampa Bay Estuary Program. St. Petersburg, iv + 84 pp.
Dawes, C. J., K. Siar, and D. Marlett. 1999a. Mangrove structure, litter, and macroalgal productivity in a northern-most forest of Florida. *Mangrove and Salt Marshes* 3: 259–267.
Dawes, C. J., C. Uranowski, J. Andorfer, and B. Teasdale. 1999b. Changes in the macroalgal taxa and zonation at the Content Keys, Florida. *Bull. Mar. Sci.* 64: 95–102.
Dawes, C. J., and J. F. van Breedveld. 1969. *U*. Vol. 1, Part 2. Florida Department of Conservation, Marine Laboratory, St. Petersburg, pp. 1–47.
Dawson, E. Y. 1941. A review of the genus *Rhodymenia* with descriptions of new species. *Allan Hancock Pac. Exped.* 3: 123–181, 13 pls.
———. 1944. The marine algae of the Gulf of California. *Allan Hancock Pac. Exped.* 3: v + 189–432, pl. 31–77

———. 1949a. *Studies of northeast Pacific Gracilariaceae.* Allan Hancock Found. Publ. Occas. Papers 7: 1–105, 25 pls.
———. 1949b, *Contributions toward a marine flora of the southern California Channel Islands.* I–III. Allan Hancock Found. Publ. Occas. Papers 8: 1–57.
———. 1950. A review of *Ceramium* along the Pacific coast of North America with special reference to its Mexican representatives. *Farlowia* 4: 113–138.
———. 1953. Marine red algae of Pacific Mexico. Part I. Bangiales to Corallinaceae subf. Corallinoideae. *Allan Hancock Pac. Exped.* 17: 1–239.
———. 1954. Marine plants in the vicinity of the Institute Océanographique de Nha Trang, Viêt Nam. *Pac. Sci.* 8: 372–469.
———. 1956. Some marine algae of the southern Marshall Islands. *Pac. Sci.* 10: 25–66.
———. 1957. An annotated list of marine algae from Eniwetok Atoll, Marshall Islands. *Pac. Sci.* 11: 92–132.
———. 1961. Marine red algae of Pacific Mexico. Part 4. Gigartinales. *Pac. Nat.* 2: 191–343.
———. 1962. Marine red algae of Pacific Mexico. Part 7: Ceramiales: Ceramiaceae, Delesseriaceae. *Allan Hancock Pacific Exped.* 26: 1–207.
———. 1963a. Marine red algae of Pacific Mexico. Part 6. Rhodymeniales. *Nova Hedwigia* 5: 437–476.
———. 1963b. Marine red algae of Pacific Mexico. Part 8. Ceramiales: Dasyaceae, Rhodomelaceae. *Nova Hedwigia* 6: 401–481.
Dawson, E. Y., C. Acleto, and N. Foldvik. 1964. The seaweeds of Peru. Beih. *Nova Hedwigia* 13: 1–111.
Day, J. W., Jr., C.A.S. Hall, W. M. Kemp, and Y. Yanez-Arancibia. 1989. *Estuarine ecology.* John Wiley and Sons, New York, xiv + 558 pp.
Decaisne, J. 1841. Plantes de l'Arabie Heureuse, recueillies par M. P.-E. Botta et décrites par M. J. Decaisne. *Arch. Mus. Hist. Nat. [Paris]* 2: 89–199, pls. 5–7.
———. 1842. Essais sur une classification des algues et des polypiers calcifères de Lamouroux. *Ann. Sci. Nat., Bot.*, sér. 2, 17: 297–380, pls. 14–17 (1842a); ibid. 18: 96–128 (1842b). [Second part, appearing in volume 18, bears subtitle, "Mémoire sur les corallines ou polypiers calcifères."].
De Candolle, A. P. 1801. Extrait d'un rapport sur les Conferves fait à la Société philomathique. *Bull. Soc. Philomath.* 3: 17–21.
De Clerck, O. 2003. The genus *Dictyota* in the Indian Ocean. *Opera Bot. Belg.* 13, 205 pp.
De Clerck, O., and E. Coppejans. 1997. The genus *Dictyota* J. V. Lamouroux (Dictyotaceae, Phaeophyta) from Indonesia in the herbarium of Weber-van Bosse, including the description of *Dictyota canaliculata spec. nov. Blumea* 42: 407–420.
De Clerck, O., B. Gavio, S. Fredericq, I. Bárbara, and E. Coppejans. 2005a. Systematics of *Grateloupia filicina* (Halymeniaceae, Rhodophyta), based on rbcL sequence analyses and morphological evidence, including the reinstatement of *G. minima* and the descrption of *G. capensis sp. nov. J. Phycol.* 41: 391–410.
De Clerck, O., C. Lane, F. Leliaert, D. A. Payo, and E. Coppejans. 2005b. The Dictyotales revisited: a morphological and molecular perspective. *Phycologia* 44(4, supplement): 26.
De Clerck, O., F. Leliaert, H. Verbruggen, C. E. Lane, J. C. De Paula, D. A. Payo, and E. Coppejans. 2006. A revised classification of the Dictyotaeae (Dictyotales, Phaeophyceae) based on *rbc*L and 26S ribosomal DNA sequence analyses. *J. Phycol.* 42: 1271–1288.
De Jong, Y. S. D. M.., A. W. G. van der Wurff, W. T. Stamm and J. L. Olsen. 1998. Studies on the Dasyaceae. 3. Towards a phylogeny of the Dasyaceae (Ceramiales, Rhodophyta), based on comparative rbcL gene sequences and morphology. *Eur. J. Phycol.* 33: 187–201.
De la Cruz, A. A. 1981. Differences between south Atlantic and Gulf Coast marshes. In: R. C. Carey, P. S. Markovits, and J. B. Kirkwood (eds.), Proc. U.S. Fish Wildl. Serv., *Workshop on coastal ecosystems of the southeastern United States.* Off. Biol. Serv. (Tech. Rep.) FWS/OBS 80–59, pp. 10–20.
Delépine, R., and A. Asensi. 1975. *Asteronema nov. gen*, nouveau genre de Phéophycée australe. *Bull. Soc. Bot. Fr.* 122: 295–304.
Delile, A. R. 1813–1826. Flore d' Égypte. Explication des planches. In: Description de l' Égypte . . . Histoire naturelle. Paris, Vol. 2, pp. 145–320 (1813). Atlas: 62 pls. (1826).
Demoulin, V., and L. Hoffmann. 1992. Proposal to conserve *Goniotrichum* Kützing with a conserved type specimen (Rhodophyta). *Taxon* 41: 759–761.
De Paula, E. J. 1988. The genus *Sargassum* C. Agardh (Phaeophyta-Fucales) from the coast of the State of São Paulo, Brazil. *Bol. Bot.* 10: 65–118.
De-Paula, J. C., A. D. Pedrini, M. D. Pinheiro, R. C. Pereira, and V. L. Teixeira. 2001. Chemical similarity between the brown algae *Dictyota cervicornis* and *D. pardalis* (Dictyotales, Phaeophyta). *Biochem. Syst. Ecol.* 29: 425–427.

Derbès, A., and A.J.J. Solier. 1850. Sur les organes reproducteurs des algues. *Ann. Sci. Nat., Bot.*, sér. 3, 14: 261–282.

———. 1851. Algues. In: J.L.M. Castagne, *Supplément au Catalogue des plantes qui croissent naturellement aux environs de Marseille.* Nicot and Pardigon: Aix, pp. 93–121.

———. 1856. Mémoire sur quelques points de la physiologie des Algues. *Suppl. C. R. Hebd. Séances Acad. Sci. Paris* 1, pp. 1–120, 23 pls.

De Reviers, B., and F. Rousseau. 1999. Towards a new classification of the brown algae. *Prog. Phycol. Res.* 13: 107–201.

Desfontaines, R. 1798–1799. Flora atlantica . . . Parisiis [Paris]. Vol. 2. 458 pp., pls. 121–261. [pp. 1–160, pls. 121–180 (1798); pp. 161–458, pls. 181–261 (1799)].

De Smedt, G., O. De Clerck, F. Leliaert, E. Coppejans, and L. M. Liao. 2001. Morphology and systematics of the genus *Halymenia* C. Agardh (Halymeniales, Rhodophyta) in the Philippines. *Nova Hedwigia* 73: 293–322.

Destombe, C., and S. E. Douglas. 1991. Rubisco spacer sequence divergence in the rhodophyte alga *Gracilaria verrucosa* and closely related species. *Curr. Genet.* 19: 395–398.

Desvaux, N. A. 1809. Neues Journal für die Botanik, herausgegeben, vom Professor Schrader: avec des observations sur le genre *Fluggia*, Rich. (*Slateria*, Sesv.); exgtrait par N. A. Desvaux. *J. Bot. [Desvaux]* 1: 240–246.

De Széchy, M.T.M., and E. J. De-Paula. 1998. Morphological types in *Sargassum* populations (Phaeophta, Fucales) from the coasts of Rio de Janeiro and São Paulo States, Brazil. *Leandra (Rio de Janeiro)* 13: 29–43.

De Toni, G. B. 1889. *Sylloge algarum omnium hucusque cognitarium.* Vol. I, *Chlorophyceae.* Patavii [Padova], pp. 12 + cxxxix + 1315.

———. 1895. *Sylloge algarum omnium hucusque cognitarium.* Vol. III, *Fucoideae.* Patavii [Padova], pp. xvi + 638.

———. 1897. *Sylloge algarum omnium hucusque cognitarium.* Vol. IV, *Florideae. Sectio 1.* Patavii [Padova], pp. [i]–xx + [i]–lxi + [1]–388.

———. 1900. *Sylloge algarum omnium hucusque cognitarium.* Vol. IV, *Florideae. Sectio II.* Patavii [Padova], pp. 387–776.

———. 1903. *Sylloge algarum omnium hucusque cognitarium.* Vol. IV, *Florideae. Sectio III.* Patavii [Padova], pp. 775–1525, 1 pl.

———. 1905. *Sylloge algarum omnium hucusque cognitarium.* Vol. IV, *Florideae. Sectio IV.* Patavii [Padova], pp. 1523–1973.

De Toni, G. [filius]. 1936. Noterelle di nomenclatura algologica. VII. Primo elenco di Floridee omonime. Brescia, [8] pp.

De Yoe, H. R., and D. L. Hockaday. 2001. Range extensions of *Codium taylorii* and *Caulerpa prolifera* into the Lower Laguna Madre, Texas. *Tex. J. Sci.* 53: 190–192.

Díaz-Martín, M. A., and J. Espinoza-Avalos. 2000. Distribution of brown seaweeds (Phaeophyta) in the Yucatán Peninsula, Mexico. *Bull. Mar. Sci.* 66: 279–289.

Diaz-Piferrer, M. 1965. A new species of *Pseudobryopsis* from Puerto Rico. *Bull. Mar. Sci.* 15: 463–474.

———. 1969. Distribution of the marine benthic flora of the Caribbean Sea. *Caribb. J. Sci.* 9: 151–178.

———. 1970. Adiciones a la flora marine de Venezuela. *Caribb. J. Sci.* 10: 159–198.

Díaz-Pulido, G., and G. Bula-Meyer. 1997. Marine algae from oceanic atolls in the southwestern Caribbean (Albuquerque Cays, Courtown Cays, Serrana Bank, and Roncador Bank). *Atoll Res. Bull.* 448, 8 pp.

Dickie, G. 1870. Notes on some algae found in the North-Atlantic Ocean. *J. Linn. Soc. Lond. Bot.* 11: 456–459.

———. 1874a. On the marine algae of Barbados. *J. Linn. Soc. Lond. Bot.* 14: 146–152.

———. 1874b. On the algae of Mauritius. *J. Linn. Soc. Lond. Bot.* 14: 190–202.

———. 1874c. [Enumeration of the algae collected at St. Thomas and Bermuda by H. N. Mosely, M. A., naturalist to H.M.S. *Challenger*]. *J. Linn. Soc. Lond. Bot.* 14: 312–316.

Dillwyn, L. W. 1802–1809. *British confervae.* London. 87 [+ 6] pp., pls. 1–109 + A-G. [Pls. 1–20 (1802), 21–38 (1803), 39–44 (1804), 45–56 (1805), 57–68, 70–81 (1806), 82–93 (1807), 94–99 (1808), 69, 100–109, A-G, pp. [1]–87, [1–6, Index and Errata] (1809).].

Dixon, P. S. 1959. Taxonomic and nomenclatural notes on the Florideae. I. *Bot. Not.* 112: 339–352.

———. 1960. Notes on important algal herbaria. II. The herbarium of John Ellis (?1710–1776). *Br. Phycol. Bull.* 2: 28–31.

———. 1973. *Biology of the Rhodophyta.* Hafner Press, New York, viii + 285 pp.

Dixon, P. S., and L. M. Irvine. 1970. Miscellaneous notes on algal taxonomy and nomenclature, III. *Bot. Not.* 123: 474–487.

———. 1977. *Seaweeds of the British Isles . . .* Vol. 1. *Rhodophyta. Part 1. Introduction, Nemaliales, Gigartinales.* British Museum (Natural History), London, xi + 252 pp.

Dolan, S. 2001. The use of medullary unit patterns of intergenicula and genicula in the taxonomy of *Amphiroa* (Corallinaceae, Rhodophyta). *Eur. J. Phycol.* 36: 397–407.

Dostál, R. 1929. *Caulerpa ollivieri n. sp.*, la seconde espèce européene des Cauerpacées. *Bull. Inst. Océanogr. Monaco* 531: 1–12.

Doty, M. S. 1947. The marine algae of Oregon. Part 1. Chlorophyta and Phaeophyta. *Farlowia* 3: 1–65, pls. 1–10.

———. 1948. The flora of Penikese, seventy-four years after. I. Penikese Island marine algae. *Rhodora* 50: 253–269.

Doty, M. S., and E. G. Meñez. 1960. *Tiffaniella*, a new genus in the Ceramiales. *Trans. Am. Microsc. Soc.* 79: 135–144.

Doweld, A. 2001. Prosyllabus tracheophytorum. GEOS, Moscow, USSR. *Draft highlights of reconnaissance level mapping and national trend analysis: results for the State of Florida.* 1984. Regional Wetlands Coordination. U.S. Fish Wildl. Serv., Atlanta.

Dragastan, O. N., D. S. Littler, and M. M. Littler. 2003. *Fossil siphonaceous green algal diversity of Key Largo and Miami limestone formations, South Florida* (part 1). Analele Universitatii Bucuresti, Geology Special Publication No. 1, University of Bucharest, pp. 5–56.

Draisma, S.G.A., J. L. Olsen, W. T. Stam, and W. F. Prud'homme van Reine. 2002. Phylogenetic relationships within the Spacelariales (Phaeophyceae): *rbc*L, RUBISCO spacer and morphology. *Eur. J. Phycol.* 37: 385–401.

Draisma, S.G.A., A. F. Peters, and R. L. Fletcher 2003. Evolution and taxonomy in the Phaeophyceae: effects of the molecular age on brown algal systematics. In: Norton, T. A. (ed). *Out of the past. Collected reviews to celebrate the Jubilee of the British Phycological Society.* British Phycological Society, Belfast, pp. 87–102.

Draisma, S.G.A., and W. F. Prud'homme van Reine. 2001. Onslowiaceae *fam. nov.* (Phaeophyceae). *J. Phycol.* 37: 647–649.

Draisma, S.G.A., W. F. Prud'homme van Reine, W. T. Stam, and J. L. Olsen. 2001. A reassessment of phylogenetic relationships within the Phaeophyceae based on rubisco large subunit and ribosomal DNA sequences. *J. Phycol.* 37: 586–603.

Drew, K. M. 1952. Studies in the Bangioideae 1. Observations on *Bangia fuscopurpurea* (Dillw.) Lyngb. in culture. *Phytomorphology* 2: 38–51.

———. 1956. *Conferva ceramicola* Lyngbye. *Bot. Tidsskr.* 53: 67–72.

Dring, M. J. 1982. *The biology of marine plants.* Edward Arnold, London, vi + 199 pp.

Duby, J. E. 1830. Aug. Pyrami de Candolle Botanicon gallicum seu Synopsis plantarum in Flora gallica descriptarum. Editio secunda. Ex herbariis et schedis Candollianis propriisque digestum. Pars secunda plantas cellulares continens. Paris. Pp. [i–vi +] 545–1068 + i–lviii.

Ducker, S. C. 1981. A history of Australian marine phycology. In: M. N. Clayton and R. J. King (eds.), *Marine botany: an Australasian perspective.* Longman Cheshire, Melbourne, pp. 1–14.

Ducluzeau, J. A. P. 1805. *Essai sur l'histoire naturelle des Conferves des environs de Montpellier.* Montpellier, 89 pp.

Durako, M. J. 1988. The seagrass bed. A community under assault. *Fla. Nat.* Fall: 6–8.

Durant, C. F. 1850. *Algae and corallines of the bay & harbor of New York, illustrated with natural types.* G. P. Putnam, New York, 43 pp.

Durieu de Maisonneuve, M. C. (ed.). 1846. *Exploration scientifique de l'Algérie pendant les années 1840, 1841, 1842 . . . Science physiques. Botanique. Cryptogamie.* Paris, pp. [1]–197, pls. 1–16.

Earle, S. A. 1969. The Phaeophyta of the eastern Gulf of Mexico. *Phycologia* 7: 71–254.

Eaton, D. C. 1875. List of the marine algae collected by Dr. Edward Palmer on the coast of Florida and at Nassau, Bahama Islands, March to August 1874. New Haven, Conn., 6 pp.

Edelstein, T., L.C.-M. Chen, and J. McLachlan. 1970. The life cycle of *Ralfsia clavata* and *R. bornetii*. *Can. J. Bot.* 48: 527–531.

Edwards, P. 1970. Illustrated guide to the seaweeds and sea grasses in the vicinity of Port Aransas, Texas. *Contrib. Mar. Sci.* 15 (supplement), 128 pp.

Egerod, L. E. 1952. An analysis of the siphonous Chlorophycophyta with special reference to the Siphonocladales, Siphonales, and Dasycladales of Hawaii. *Univ. Calif. Publ. Bot.* 25: 325–453, 14 pls.

———. 1975. Marine algae of the Andaman Sea Coast of Thailand: Chlorophyceae. *Bot. Mar.* 18: 41–66.

Eggert, A., E. M. Burger, and A. M. Breeman. 2003. Ecotypic differentiation in thermal traits in the tropical warm-temperate green macrophyte *Valonia utricularis*. *Bot. Mar.* 46: 69–81.

Eiseman, N. J. 1970. The green alga *Chalmasia antillana* from the Florida Keys. *Phycologia* 9: 45–47.
———. 1975. The marine algae of Point O' Rocks, Florida west coast. Ph.D. diss., Univ. South Florida, Tampa, 201 pp.
———. 1979. Marine algae of the east Florida continental shelf. I. Some new records of Rhodophyta, including *Scinaia incrassata sp. nov.* (Nemaliales: Chaetangiaceae). *Phycologia* 18: 355–361.
Eiseman, N. J., M. C. Benz, and D. E. Serbousek. 1978. A new record of *Acetabularia calyculus* from Florida. *Fla. Sci.* 41: 42–44.
Eiseman, N. J., and and S. M. Blair. 1982. New records and range extensions of deepwater algae from the East Flower Garden Bank, Northwestern Gulf of Mexico. *Contrib. Mar. Sci.* 25: 21–26.
Eiseman, N. J., and R. L. Moe. 1981. *Maripelta atlantica sp. nov.* (Rhodophyta, Rhodymeniales) a new deep-water alga from Florida. *J. Phycol.* 17: 299–308.
Eiseman, N. J., and J. N. Norris. 1981. *Dudresnaya patula sp. nov.*, an unusual deep-water red alga from Florida. *J. Phycol.* 17: 186–191.
Ellis, J. 1768. Extract of a letter from John Ellis, Esquire, F.R.S. to Dr. Linneaus, of Uppsal, F.R.S. on the animal nature of the genus of zoophytes, called *Corallina*. *Philos. Trans. [R. Soc. Lond.]* 57: 404–427, pls. XVII, XVIII.
Ellis, J. and D. Solander. 1786. *The natural history of many curious and uncommon zoophytes*, collected from various parts of the globe by the late John Ellis . . . Systematically arranged and described by the late Daniel Solander . . . B. White and Son, London, xii + 208 pp., 63 pls.
Endlicher, S. L. 1836–1840. *Genera plantarum secundum ordines naturales diposita*. Vindobonae [Wine]. [viii +] lx + 1334 pp. [pp. 1–160 (1836); pp. 161–400 (1837); pp. 401–640 (1838); pp. 641–960 (1839); pp. 961–1334 (1840)].
———. 1843. *Mantissa botanica altera. Sistens generum plantarum supplementum tertium*. Vindobonae [Wien]. [vi +] 111 pp.
Engler, A. and K. Prantl (eds.). 1890–1897. *Die natürlichen Pflanzenfamiliem*, I. Teil., Abt. 2. Leipzig, xii + 580 pp.
Esper, E. J. C. 1794–1797. *Fortsetzungen der Pflanzenthiere* . . . Erster Theil. Nürnberg. 230 pp. [114] pls. [numbered by genus]. [Part 1, pp. [1]–32, [18] pls. (1794); part 2, pp. 33–64, [15] pls. (1794); part 3, pp. 65–100, [18] pls. (1795); part 4, pp. 101–116, [12] pls. (1795); part 5, pp. 117–148, [16] pls. (1796); part 6, pp. 149–168, [11] pls. (1796); part 7, pp. 169–184, [14] pls. (1797); part 8, pp. 185–230, [10] pls. (1797).]
———. 1797–1800. *Icones fucorum* . . . Erster Theil. Nürnberg. 217 pp., pls. I–CXI. [Part 1, pp. [1]–54, pls. I–XXIV (1797); part 2, pp. [55–] 57–126, pls. XXV–LXIII (1798); part 3, pp. [127–] 129–166, pls. LXIV–LXXXVII (1799); part 4, pp. 167–217, pls. LXXXVIII–CXI (1800).]
———. 1805–1830. *Die Pflanzenthiere* . . . Dritter Theil. Nürnberg. 407 pp., [19] pls. [numbered by genus]. [Part 1 (13 of entire work), pp. [1]–24, [2] pls. (1805); part 2 (14 of entire work), pp. 25–76, [4] pls. (1806); part 3 (15 of entire work), pp. 77–144, [9] pls. (1809); part 4 (16 of entire work), by F. L. Hammer, pp. 145–283 [284], [2] pls. (1829); part 5 (17 of entire work), by F. L. Hammer, pp. 285–407, [2] pls. (1830).]
Etcheverry, D. H. 1986. *Algas Marinas Bentónicas de Chile*. UNESCO, Montevideo.
Eubank, L. 1946. Hawaiian representatives of the genus *Caulerpa*. *Univ. Calif. Publ. Bot.* 18: 409–432.
Ewel, J. J. 1990. Inshore marine habitats. In: R. L. Myers and J. J. Ewel (eds.), *Ecosystems of Florida*. Univ. Central Florida Press, Orlando, pp. 549–573.
Falkenberg, P. 1901. *Die Rhodomelaceen des Golfes von Neapel und der angrenzenden Meeres-Abschnitte*. Fauna und Flora des Golfes von Neapel, Monographie 26. Berlin, xvi + 754 pp., 24 pls.
Fama, P., B. Wysor, W.H.C.F. Kooistra, and G. C. Zuccarello. 2002. Molecular phylogeny of the genus *Caulerpa* (Caulerpales, Chlorophyta) inferred from chloroplast *tuf*A gene. *J. Phycol.* 38: 1040–1050.
Fan, K.-C. 1961. Morphological studies of the Gelidiales. *Univ. Calif. Publ. Bot.* 32: 315–368.
Fan, K.-C. and Y. C. Wang. 1974. [Studies on the marine algae of Hsisha Islands, China. I. *Ganonema gen. nov.*]. *Acta Phytotax. Sin.* 12: 489–495 [in Chinese with English summary].
Farghaly, M. S. 1980. Algues benthiques de la Mer Rouge et du bassin occidental de l'Océan Indien (étude taxinomique et essai de répartition, notamment des Udotéacées). 274 [+ 25] pp., 44 pls. [Diss., Université des Sciences et Techniques du Langedoc, Montpellier.]
Farlow, W. G. 1876. List of marine algae of the United States. Rep. U.S. Commission Fish and Fisheries for 1873–74 and 1874–75. Appendix E, Art. 32, pp. 691–718.
———. 1881. The marine algae of New England. In: Report of the U.S. Commission Fish and Fisheries for 1879. Appendix A-1, pp. 1–210, 15 pls. [completed volume issued 1882]
Farlow, W. G., C. L. Anderson, and D. C. Eaton. 1877–1889. *Algae exsiccatae Americae borealis*. Boston, Massachusetts. Fasc. I–V [Exsiccata with printed labels.]

Faye, E. J., H. W. Wang, S. Kawaguchi, S. Shimada and M. Masuda. 2004a. Reinstatement of *Grateloupia subpectinata* (Rhodophyta, Halymeniaceae) based on morphology and rbcL sequences. *Phycol. Res.* 52: 59–67.

Faye, E. J., S. Shimada, K. Kogame and M. Masuda. 2004b. A new red algal species *Meristotheca dakarensis* (Solieriaceae, Gigartinales) from Senegal, western Africa, with comments on the relegation of *Meristiella* Cheny (sic. Cheney) to synonymy with *Meristotheca* J. Agardh. *Cryptogam. Algol.* 25: 241–259.

Feldmann, J. 1931. Remarques sur les genres *Gelidium* Lamour., *Gelidiopsis* Schmitz *et Echinocaulon* (Kütz.) emend. In: *Travaux cryptogamiques dédiés à Louis Mangin*. Paris, pp. 151–166.

———. 1937. Recherches sur la végétation marine de la Mediterranée. La côte des Albères. *Rev. Algol.* 10: 1–339.

———. 1941. Les algues marines de la côte des Albères. IV. Rhodophycées (suite). *Rev. Algol.* 12: 77–100.

———. 1942. Remarques sur les Némastomacées. *Bull. Soc. Bot. Fr.* 89: 104–113.

———. 1955. Les plastes des *Caulerpa et* leur valeur systématique. *Rev. Gén. Bot.* 62: 422–431.

———. 1962. The Rhodophyta order Acrochaetiales and its classifcation. *Proc. 9th Pac. Sci. Congr.* 4: 219–222.

Feldmann, J., and G. Hamel. 1934. Observations sur quelques Gélidiacées. *Rev. Gén. Bot.* 46: 528–549.

———. 1936. Floridées de France VII: Gélidiales. *Rev. Algol.*, N.S. 9: 85–140, pls. 2–6 [or 9: 209–264, pls. 1–5].

Feldmann-Mazoyer, G. 1938a. Sur un nouveau genre de Céramiacée de la Méditerranée. *C. R. Hebd. Séances Acad. Sci. Paris* 207: 1119–1121.

———. 1938b. Les Céramiées de l'Afrique du Nord. *Extr. Bull. Soc. d'Hist. Nat. Afr. du Nord* 29: 317–331.

———. 1941. *Recherches sur les Céramiacées de la Méditerranée occidentale.* Alger. 510 pp. + 3 pp. [Errata graviora], 191 figs, 4 pls. [Printing finished June 6, 1940 but not used as a diss. until May 17, 1941].

Feussner, K.-D., P. A. Skelton, G. R. South, P. Alderslade, and W. Aalbersberg. 2004. *Ostreobium queketii* (Osteobiaceae: Chlorophyceae) invading the barnacle *Acastra* sp. (Pendunculata: Acastinae), endozoic in the octocoral *Rumphella suffruticosa* (Alcyonacea: Gorgoniidae) from Fiji, South Pacific. *N. Z. J. Mar. Freshw. Res.* 38: 87–90.

Fiore, J. 1975. A new generic name for *Farlowiella onusta* (Phaeophyta). *Taxon* 24: 497–498.

———. 1977. Life history and taxonomy of *Stictyosiphon subsimplex* Holden (Phaeophyta, Dictyosiphonales) and *Farlowia onusta* (Kützing) Kornmann in Kuckuck (Phaeophyta, Ectocarpales). *Phycologia* 16: 301–311.

Fletcher, R. L. 1978. Studies on the family Ralfsiaceae (Phaeophyceae) around the British Isles. In: D.E.G. Irvine and J. H. Price (eds.), *Modern approaches to the taxonomy of red and brown algae.* Systematics Association Special Volume 10. Academic Press, London and New York, pp. 371–398.

———. 1987. *Seaweeds of the British Isles . . .* Vol. 3, *Fucophyceae (Phaeophyceae), Part 1.* British Museum (Natural History), London, x + 359 pp.

Floyd, G. L., and C. J. O'Kelly . 1990. Phylum Chlorophyta, Class Ulvophyceae. In: L. Margulis, J. O. Corliss, M. Melkonian, and D. J. Chapman (eds.), *Handbook of Protoctista.* Jones and Bartlett, Boston, pp. 617–635.

Forsskål, P. 1775. *Flora aegyptiaco-arabica . . .* Post mortem auctoris edidit Carsten Niebuhr. Havniae [Copenhagen]. 32 + cxxvi + 219 [220] pp., frontispiece [map].

Foslie, M. 1897. On some Lithothamnia. *K. Nor. Vidensk. Selsk., Skr. (Trondheim)* 1897 (1): 1–20.

———. 1898a. Systematical survey of the Lithothamnia. *K. Nor. Vidensk. Selsk., Skr. (Trondheim)* 1898 (2), 7 pp.

———. 1898b. List of species of the Lithothamnia. *K. Nor. Vidensk. Selsk., Skr. (Trondheim)* 1898 (3), 11 pp.

———. 1899. Some new or critical *Lithothamnion. K. Nor. Vidensk. Selsk., Skr. (Trondheim)* 6: 1–19.

———. 1900. New or critical calcareous algae. *K. Nor. Vidensk. Selsk., Skr. (Trondheim)* 1899 (5), 19 pp.

———. 1901. New Melobesieae. *K. Nor. Vidensk. Selsk., Skr. (Trondheim)* 1900 (6), 34 pp.

———. 1904a. Lithothamnioneae, Melobesieae, Mastophoreae. In: A. Weber-van Bosse and M. Foslie (eds.), *The Corallinaceae of the Siboga-Expedition.* Siboga Exped. Monogr. 61. Leiden, pp. 10–77, pls. I–XIII.

———. 1904b. Algologiske notiser. *K. Nor. Vidensk. Selsk., Skr. (Trondheim)* 1904(2), 9 pp.

———. 1905. Remarks on northern Lithothamnia. *K. Nor. Vidensk. Selsk., Skr. (Trondheim)* 1905 (3): 1–138.

———. 1906. Den botaniske samling. *K. Nor. Vidensk. Selsk., Skr. (Trondheim)* 1905 (10), pp. 17–24.

———. 1907('1908'). Algologiske notiser IV. *K. Nor. Vidensk. Selsk., Skr. (Trondheim)* 1907(6), 30 pp.

———. 1908. Nye kalkalger. *K. Nor. Vidensk. Selsk., Skr. (Trondheim)* 1908 (12), 9 pp.
———. 1909. Algologiske notiser VI. *K. Nor. Vidensk. Selsk., Skr.* (Trondheim) 1909 (2), 63 pp.
———. 1929. *Contributions to a monograph of the* Lithothamnia. After the author's death collected and edited by Henrik Printz. Trondhjem [Trondheim], 60 pp., 75 pls.
Foslie, M., and M. A. Howe. 1906. New American coralline algae. *Bull. N. Y. Bot. Gard.* 4: 128–136.
Fotos, S. S. 1981. Observations on *Chnoospora minima* (Hering) Papenfuss (Phaeophyta, Scytosiphonales) in the field and in culture. Jpn. *J. Phycol.* 29: 101–108.
Fourqurean, J. M., and L. M. Rutten, 2004. The impact of Hurricane Georges on soft-bottom, back reef communities: site- and species-specific effects in south Florida seagrass beds. *Bull. Mar. Sci.* 75: 239–257.
Føyn, B. 1960. Sex-linked inheritance in *Ulva. Biol. Bull.* 118: 407–411.
———. 1961. Globose, a recessive mutant in *Ulva mutabilis. Bot. Mar.* 3: 60–64.
Fredericq, S., D. W. Freshwater, and M. H. Hommersand. 1999. Observations on the phylogenetic systematics and biogeography of the Solieriaceae (Gigartinales, Rhodophyta) inferred from rbcL sequences and morphological evidence. *Hydrobiologia* 398/399: 25–38.
Fredericq, S., and M. H. Hommersand. 1989a. Proposal of the Gracilariales *ord. nov.* (Rhodophyta) based on an analysis of the reproductive structures of *Gracilaria verrucosa. J. Phycol.* 25: 213–227.
———. 1989b. Comparative morphology and taxonomic status of *Gracilariopsis* (Gracilariales, Rhodophyta). *J. Phycol.* 25: 228–241.
———. 1990. Diagnoses and key to the genera of the Gracilariaceae (Gracilariales, Rhodophyta). *Hydrobiologia* 204/205: 172–178.
Fredericq, S., and J. N. Norris. 1985. Morphological studies on some tropical species of *Gracilaria* Grev. (Gracilariaceae, Rhodophyta): taxonomic concepts based on reproductive morphology. In: I. A. Abbott and J. N. Norris (eds.), *Taxonomy of economic seaweeds with reference to some Pacific and Caribbean species.* Vol. 1. Calif. Sea Grant Program, Univ. Calif., La Jolla, pp. 137–155.
———. 1992. Studies on cylindrical species of Western Atlantic *Gracilaria* (Gracilariales, Rhodophyta): *G. cylindrica* Børgesen and *G. blodgettii* Hervey. In: I. A. Abbott (ed.), *Taxonomy of economic seaweeds.* Vol. 3. Calif. Sea Grant Program, Univ. Calif., La Jolla, pp. 211–231.
———. 1995. A new order (Rhodogorgonales) and family (Rhodogorgonaceae) of red algae composed of two tropical calciferous genera, *Renouxia gen. nov.* and *Rhodogorgon. Cryptogam. Bot.* 4: 316–331.
Freshwater, D. W., and D. F. Kapraun. 1989. Spheroplast isolation and propagation of *Porphyra carolinensis* Coll *et* Cox for mariculture. *Aquaculture* 81: 101–110.
Friedmann, E. I., and W. C. Roth. 1977. Development of the siphonous green alga *Penicillus* and the *Espera* state. *Bot. J. Linn. Soc.* 74: 1898–214.
Fries, E. M. 1835–1837. *Corpus florarum provincialium Sueciae.* I. Floram scanicam scripsit Elias Fries. xxiv + 394 pp. Uppsalia [Uppsala] [pp.[1]–192 (1835), 193–346 (1836), 347–394 (1837), [i]–xxiv (1836)].
———. 1845. *Summa vegetatilium Scandinaviae* . . . Sectio prior. Uppsala, 258 pp.
Fries, L. 1975. Some observations of the morphology of *Enteromorpha linza* (L.) J. Ag. and *Enteromorpha compressa* (L.) Grev. in axenic culture. *Bot. Mar.* 18: 251–253.
Fritsch, F. E. 1945. *The structure and reproduction of the algae.* Vol. II. Foreword, Phaeophyceae, Rhodophyceae, Myxophyceae. Cambridge Univ. Press, xiv + 939 pp.
Furnari, G., M. Cormaci, and D. Serio. 1999. Catalogue of the benthic marine macroalgae of the Italian coast of the Adriatic Sea. *Bocconea* 12: 5–214.
———. 2001. The *Laurencia* complex (Rhodophyta, Rhodomelaceae) in the Mediterranean Sea: an overview. *Cryptogam. Algol.* 22: 331–373.
Furnari, G. M., M.-T. L'Hardy-Halos, J. Rueness, and D. Serio. 1998. On the conspecificity of *Aglaothamnion tenuissimum* and *A. byssoides* (Ceramiaceae, Rhodophyta). *Taxon* 47: 843–849.
Furnari, G. M., and D. Serio. 1993. The distinction of *Laurencia truncata* (Ceramiales, Rhodophyta) in the Mediterranean Sea from *Laurencia pinnatifida. Phycologia* 32: 367–372
Gabrielson, P. W. 1985. *Agardhiella* versus *Neoagardhiella* (Solieriaceae, Rhodophyta): another look at the lectotypification of *Gigartina tenera. Taxon* 34: 275–280.
Gabrielson, P. W., and D. P. Cheney. 1987. Morphology and taxonomy of *Meristiella gen. nov.* (Solieriaceae, Rhodophyta). *J. Phycol.* 23: 481–493.
Gabrielson, P. W., T. B. Widdowson, and S. C. Lindstrom. 2004. *Keys to the seaweeds and seagrasses of Oregon and California, north of Point Conception.* Phycological Contribution #6. PhycoID, Hillsborough, N.C., iv + 181 pp.
Gaillard, J. 1966. Un *Padina* nouveau des côtes d'Afrique: *Padina glabra sp. nova. Phycologia* 5: 222–226.

Gaillon, B. 1828. Thalassiophytes. *Dictionnaire des Sciences Naturelles* [Levrault] 53: 350–406.

Gallagher, S. B., and H. J. Humm. 1980. *Pilinia earleae n. sp.* (Chlorophyceae, Chroolepidaceae) from the Florida west coast. *J. Phycol.* 16: 532–536.

———. 1981. *Vaucheria* (Xanthophyceae, Vaucheriaceae) of the Central Florida Gulf Coast. *Bull. Mar. Sci.* 31: 184–190.

———. 1983. *Centroceras internitens n. sp.* (Rhodophyceae, Ceramiaceae) from the western tropical North Atlantic Ocean. *J. Phycol.* 19: 261–268.

Gallardo, T., A. Gómez Garreta, M. A. Ribera, M. A. Alvarez, and F. Conde. 1985. *A preliminary checklist of Iberian benthic marine algae.* Madrid, Real Jardín Botáánico, 83 pp.

Gallardo, T., A. Gómez Garreta, M. A. Ribera, M. Cormaci, G. Furnari, G. Giaccone, and C.-F. Boudouresque. 1993. Check-list of Mediterranean seaweeds. II. Chlorophyceae Wille s.l. *Bot. Mar.* 36: 399–421.

Ganesan, E. K. 1976. On *Kallymenia westii sp. nov.* (Rhodophyta, Cryptonemiales) from the Caribbean Sea. *Bol. Inst. Oceanogr.* (Cumaná). 15: 169–175.

———. 1990. *A catalog of benthic marine algae and seagrasses of Venezuela.* Fondo Editorial Conicit, Caracas, 237 pp., 15 maps.

Garbary, D. J. 1979. Numerical taxonomy and generic circumscription in the Acrochaetiaceae (Rhodophyta). *Bot. Mar.* 22: 477–492.

———. 1987. The Acrochaetiaceae (Rhodophyta): an annotated bibliography. *Bibl. Phycol.* 77, 267 pp.

Garbary, D. J., D. Grund, and J. McLachlan. 1978. The taxonomic status of *Ceramium rubrum* based on culture experiments. *Phycologia* 17: 85–94.

Garbary, D. J., G. I. Hansen, and R. F. Scagel. 1980. The marine algae of British Columbia and northern Washington: division Rhodophyta (red algae), class Bangiophyceae. *Syesis* 13: 137–195.

Garbary, D. J., and J. T. Harper. 1998. A phylogenetic analysis of the *Laurencia* complex (Rhodomelaceae) of the red algae. *Cryptogam. Algol.* 19: 185–200.

Garbary, D. J., and H. W. Johansen. 1982. Scanning electron microscopy of *Corallina* and *Haliptilon* (Corallinaceae, Rhodophyta): surface features and their taxonomic implications. *J. Phycol.* 18: 211–219.

García, M., and S. Gómez. 2001. Nuevos registros ficoflorísticos para el Estado Vargas, Litoral Central, Venezuela. *Acta Bot. Venez.* 24: 1–12.

Gargiulo, G. M., F. De Masi, and G. Tripodi. 1992. Morphology, reproduction and taxonomy of the Mediterranean species of *Gracilaria* (Gracilariales, Rhodophyta). *Phycologia* 3: 53–80.

Garreta, A. G. (ed.). 2001. *Flora Phycologica Iberica.* Vol. 1, Fucales. Universidad de Murcia, 192 pp.

Garreta, A. G., T. Gallardo, M. A. Ribera, M. Cormaci, G. Furnari, G. Giaccone, and C. F. Boudouresque. 2001. Checklist of Mediterranean seaweeds. III. Rhodophyceae Rabenh. 1. Ceramiales Oltm. *Bot. Mar.* 44: 425–460.

Gavio, B., and S. Fredericq. 2002. *Grateloupia turuturu* (Halymeniaceae, Rhodophyta) is the correct name of the non-native species in the Atlantic known as *Grateloupia doryphora. Eur. J. Phycol.* 37: 349–359.

———. 2003. *Botryocladia caraibica* (Rhodymeniales, Rhodophyta), a new species from the Caribbean. *Cryptogam. Algol.* 24: 93–106.

———. 2005. New species and new records of offshore members of the Rhodymeniales (Rhodophyta) in the northern Gulf of Mexico. *Gulf Mex. Sci.* 1: 58–83.

Gavio, B., E. Hickerson, and S. Fredericq. 2005. *Platoma chrysymenioides sp. nov.* (Schizymeniaceae), and *Sebdenia integra sp. nov.* (Sebdeniaceae), two new red algal species from the northwestern Gulf of Mexico, with a phylogenetic assessment of the Cryptonemiales complex (Rhodophyta). *Gulf Mex. Sci.*. 1: 38–57.

Gayral, P. 1959. Premières observations et réflexions sur des Ulvacées en culture. *Le Botaniste* 43: 85–100.

———. 1967. Mise au point sur le Ulvacées (Chlorophycées), particulièrement sur les résultats de leure étude en laboratoire. *Le Botaniste* 50: 205–251.

Geesink, R. 1973. Experimental investigations on marine and freshwater *Bangia* (Rhodophyta) from the Netherlands. *J. Exp. Mar. Biol. Ecol.* 11: 239–247.

Gepp, A., and E. S. Gepp. 1904. *Rhipidosiphon* and *Callipsygma. J. Bot.* 43: 363–366.

———. 1905. Notes on *Penicillus* and *Rhipocephalus. J. Bot., British and Foreign* 43: 1–5.

———. 1906. Some marine algae from New South Wales. *J. Bot., Lond.* 1906: 249–261, plate 481.

———. 1911. The Codiaceae of the Siboga Expedition including a monograph of Flabellarieae and Udoteae. *Siboga Exped. Monogr.* 62. Leiden, pp. 1–150, 22 pls.

Gerloff, J. 1959. *Bachelotia* (Bornet) Kuckuck ex Hamel oder *Bachelotia* (Bornet) Fox? *Nova Hedwigia* 1: 37–39.

Giaccone, G. 1968. New and interesting species of Rhodophyceae from the eastern Mediterranean. *G. Bot. Ital.* 102: 397–414.

———. 1969. Phytobenthos from Italian continental shelf. *G. Bot. Ital.* 103: 485–514.

Gilbert, W. J. 1962. Contribution to the marine Chlorophyta of Hawaii. I. *Pac. Sci.* 16: 134–144.

———. 1965. Contribution to the marine Chlorophyta of Hawaii. II. *Pac. Sci.* 19: 482–491.

Gmelin, S. G. 1768. *Historia fucorum.* Petropolis [St. Petersburg]. [xii +] 239 + 6 pp., 35 pl. [IA, IB, IIA, IIB, III–XXXIII].

Gobi, G. J. 1879. Bericht über die algologischen Forschungenn im Finnischen meerbusen, im sommer 1877 ausgefürt. Trudy Leningradskogoe [St. Petersburg] Obschchestvo Estestvpoispytatelei 10(2): 83–92.

Goff, L. J., D. A. Moon, and A. W. Coleman. 1994. Molecular delineation of species and species relationships in the red algal agarophytes *Gracilariopsis* and *Gracilaria* (Gracilariales). *J. Phycol.* 30: 521–537.

Golden, L., and D. Garbary. 1984. Studies on *Monostroma* (Monostromataceae) in British Columbia with emphasis on spore release. Jpn. *J. Phycol.* 32: 319–332, 41 figs.

Gómez Garreta A., T. Gallardo, M. A. Ribera, M. Cormaci, G. Furnari, G. Giaccone, and C. F. Boudouresque. 2001. Checklist of Mediterranean seaweeds. III. Rhodophyceae Rabenh. 1. Ceramiales Oltm. *Bot. Mar.* 44: 425–460.

Gómez-Poot, J. M., J. Espinoza-Avalos, and S. G. Jiménex-Flores. 2002. Vegetative and reproductive characteristics of two species of *Batophora* (Chlorophyta, Dasycladaceae) from Chetumal Bay, Quintant Roo, Mexico. *Bot. Mar.* 45: 189–195.

Gordon, E. M. 1972. Comparative morphology and taxonomy of the Wrangelieae, Sphondylothamnieae, and Spermothamnieae (Ceramiaceae, Rhodophyta). *Aust. J. Bot.*, suppl. 4, 1–180 pp., 3 tables.

Gordon-Mills, E. 1987. Morphology and taxonomy of *Chondria tenuissima* and *Chondria dasyphylla* (Rhodomelaceae, Rhodophyta) from European waters. *Br. Phycol. J.* 22: 237–255.

Graham, L. E., and L. W. Wilcox. 2000. *Algae.* Prentice-Hall, Upper Saddle River, N.J., xvi + 640 pp. [+ Glossary (G1–G14), *Taxonomic* Index (TI1–TI5), Subject Index (SI1–SI12), and Literature Cited (L1–L28)].

Grateloup, J.P.A.S. 1806. *Descriptiones aliquorum Ceramiorum novorum, cum iconum explicationibus.* (Appendix, with one unnumbered page and one unnumbered plate) Observations sur la constitution de l'été de 1806. Montpellier.

Gray, A. J. 1992. Saltmarsh plant ecology: zonation and succession revisited. In: J. R. L. Allen and K. Pye (eds.), *Saltmarshes: morphodynamics, conservation, and engineering significance.* Cambridge Univ. Press, Cambridge, pp. 63–79.

Gray, J. E. 1864. *Handbook of British water-weeds or algae: the Diatomaceae by W. Carruthers.* London, iv + 123 pp.

———. 1866. On *Anadyomene* and *Microdictyon*, with the description of three new allied genera, discovered by Menzies in the Gulf of Mexico. *J. Bot.* 4: 41–51, 65–72, pl. XLIV.

Gray, S. F. 1821. *A natural arrangement of British plants . . .* Vol. 1. London, xxviii + 824 pp., 21 pls.

Greer, S. P., and C. D. Amsler. 2004. Clonal variation in phototaxis and settlement behaviors of *Hincksia irregularis* (Phaeophyceae) spores. *J. Phycol.* 40: 44–53.

Greuter, W., J. McNeil, F. R. Barrie, H. M. Burdet, V. Demoulin, T. S. Filgueras, D. H. Nicolson, P. C. Silva, J. E. Skog, P. Trehane, N. J. Turland, and O. L. Hawksworth. 2000. *International Code of Botanical Nomenclature (St. Louis Code).* Koeltz Scientific Books, Königstein, Germany.

Greville, R. K. 1823–1824. *Scottish cryptogamic flora . . .* Vol. 2. Edinburgh, pls. 61–90 (1823), 91–120 (1824, with text).

———. 1824. *Flora Edinensis . . .* Edinburgh, lxxxi + 478 pp., 4 pl.

———. 1827a. *Scottish cryptogamic flora.* Vol. 5. Edinburg, pls. 241–300 (with text).

———. 1827b. Some account of a collection of cryptogamic plants from the Ionian Islands. *Trans. Linn. Soc. [Lond.]* 15: 335–348.

———. 1830. *Algae Brittanicae . . .* Edinburgh, lxxxviii + 218 pp., 19 pls.

Griselini, F. 1750. In: Pierre Bassaglia (ed.), *Observations de Francois Griselini de l'Academie des Sciences de Boulogne sur la scolopendre marine luisante et la baillouviana addresses a monsieur le chevalier de Baillou.* Venice, 32 pp., pl. 80.

Grossman, A. 2006. Regeneration of a cell from protoplasm. *J. Phycol.* 42: 1–5.

Grunow, A. 1867. Algae. In: *Reise seiner Majestät Fregatte Novara um die Erde, in den Jahren 1857, 1858, 1859.* Botanischer Theil. Erster Band. Sporenpflanzen. Wein, pp. 1–104, pls. I–XI. [Alternate title: Algae. In: E. Fenzl (ed.), *Reise der österreichischen Fregatte Novara un die Erde, in den Jahren 1857, 1858, 1859*].

———. 1874. Algen der Fidschi-, Tonga- und Samoa-Inseln, gesammelt von Dr. E. Graeffe. *J. Mus. Godeffroy [Hamburg]* 3: 23–50.

———. 1916. Additamenta ad cognitionem Sargassorum. *Verh. Zool.-Bot. Gesell. Wien* 66: 1–48, 136–185.

Guiler, E. R. 1952. The marine algae of Tasmania. Checklist with localities. *Pap. of the Proc. R. Soc. Tasman.* 86: 71–106.

Guimarães, S.M.P. de B. 1990. *Rodofíceas marinhas bentônicas do Estado do Espírito Santo: ordem Cryptonemiales.* Tese apresentada ao Instituto de Biociiencias da Universidade de São Paulo, 275 pp.

Guimarães, S.M.P. de B., and M. T. Fujii. 1998. Two species of foliose *Halymenia* (Halymeniaceae, Rhodophyta) from Brazil. *Bot. Mar.* 41: 495–504.

Guimarães, S.M.P. de B., M. T. Fujii, M. T. Pupo, and N. S. Yokoya. 2004. An assessment of the morphological characteristics and its taxonomical implications in the genus *Polysiphonia sensu lato* (Ceramiales, Rhodophyta) from the littoral of São Paulo and Espírito Santo States, Brazil. *Rev. Bras. Bot.*, São Paulo 27: 163–183.

Guimarães, S.M.P. de B., and E. C. Oliveria. 1996. *Taxon*omy of the flattened Solieriaceae (Rhodophyta) in Brazil: *Agardhiella* and *Meristiella*. *J. Phycol.* 32: 656–668

Guiry, M. D. 1977. Studies on marine algae of the British Isles. 10. The genus *Rhodymenia*. *Br. Phycol. J.* 12: 385–425.

———. 1990. Sporangia and spores. In: K. M. Cole and R. G. Sheath (eds.), *Biology of the red algae*. Cambridge University Press, Cambridge, pp. 347–376.

Guiry, M. D., and L. Irvine. 1989. Sporangial form and function in the Nemaliophycidae (Rhodophyta). In: H. D. Kumar (ed.), *Phykotalk*. Vol. I. Rastogi Publ., Meerut, India, pp. 155–184.

Guiry, M. D., and E. Nic Dhonncha. 2006. AlgaeBase version 3.0 Worldwide electronic publication, National University of Ireland, Galway (http://www.algaebase.org).

Güner, H.,V. Aysel, A. Sukatar, and M. Öztürk. 1983–1984. Check list of Izmir Bay marine algae: Phaeophyceae, Chlorophyceae. *Fac. Sci. J. Ege Univ.*, ser. B, 7: 57–65.

Gunnarsson, K., and S. Jónsson. 2002. Benthic marine algae of Iceland: revised checklist. *Cryptogam. Algol.* 23: 131–158.

Gurgel, C.F.D., and S. Fredericq. 2004. Systematics of the Gracilariaceae (Gracilariales, Rhodophyta): a critical assessment based on rbcL sequence analyses. *J. Phycol.* 40: 138–159.

Gurgel, C.F.D., S. Fredericq, and J. N. Norris. 2001. Phylogeny, taxonomy and biogeography of the Gracilariaceae (Gracilariales, Rhodophyta) with emphasis on the northwestern Atlantic. *Phycologia* 40(suppl.): 21.

———. 2004a. Molecular systematics and taxonomy of flattened species of *Gracilaria* Greville (Gracilariaceae, Gracilariales, Rhodophyta) from the western Atlantic. In: I. A. Abbott and K. McDermid. (eds.), *Taxonomy of economic seaweeds with reference to the Pacific and other locations*. Vol. 9. Hawaii Sea Grant College Program, Univ. Hawaii, Hilo, pp. 159–199.

———. 2004b. Phylogeography of *Gracilaria tikvahiae* (Gracilariaceae, Rhodophyta): a study of genetic discontinuity in a continuously distributed species based on molecular evidence. *J. Phycol.* 40: 748–758.

———. 2004c. *Gracilaria apiculata and G. flabelliformis* (Gracilariaceae, Rhodophyta): restoring old names for common tropical western Atlantic species, including the recognition of three new subspecies and a replacement name for "*G. lacinulata.*" *Cryptogam. Algol.* 25: 367–396.

Gurgel, C.F.D., L. M. Liao, S. Frederidcq, and M. H. Hommersand. 2003. Systematics of *Gracilariopsis* (Gracilariales, Rhodophyta) based on rbcL sequence analysis and morphological evidence. *J. Phycol.* 39: 154–171.

Güven, K. C., and F. Oztig. 1971. Über die marinen Algen a den Küsten der Turkei. *Bot. Mar.* 14: 121–128.

Hackett, H. E. 1974. *Laurencia brongniartii* (Rhodophyta-Rhodomelaceae) from Florida. *Fla. Sci.* 37: 50–51.

Hall, M. O. 1988. Dynamics and interaction of epiphytic macroalgae and meiofauna on the seagrass *Thalassia testudinum*. Ph.D. diss., Univ. South Florida, Tampa, 180 pp.

Hall, M. O., and N. J. Eiseman. 1981. The seagrass epiphytes of the Indian River, Florida. I. Species list with descriptions and seasonal occurrences. *Bot. Mar.* 24: 139–146.

Hamel, A., and G. Hamel. 1929. Sur l'heterogamie d'une Cladophoracée, *Lola* (*nov. gen.*) *lubrica* (Setch. and Gardn.). *C. R. Hebd. Séances Acad. Sci. Paris*. 189: 1094–1096.

Hamel, G. 1924. Floridées de France. II. *Rev. Algol.* 1: 427–457.

———. 1927. Recherches sur les genres *Acrochaetium* Naeg. et *Rhodochorton* Naeg. Saint-Lo., Paris, [v] + 117 pp.

———. 1930. Chlorophycées des côtes française. *Rev. Algol.* 5: 1–54.

———. 1931a. Chlorophycées des côtes française (fin). *Rev. Algol.* 6: 9–73.

———. 1931b. *Phéophycées de France*. Fasc. 1. Paris, pp. 1–80.

———. 1935. *Phéophycées de France*. Fasc. 2. Paris, pp. 81–176.

———. 1937. *Phéophycées de France*. Fasc. 3. Paris, pp. 177–240.

———. 1938. *Phéophycées de France*. Fasc. 4. Paris, pp. 241–336.
———. 1939a. Sur la classification des Ectocarpales. *Bot. Not.* 1939: 65–70.
———. 1939b. *Phéophycées de France*. Fasc. V. Paris, pp. [i]–xlvii + 337–431 [432], 10 pls.
Hamm, D., and H. J. Humm. 1976. Benthic algae of the Anclote Estuary. II. Bottom-dwelling species. *Fla. Sci.* 39: 209–229.
Hanisak, M. D. 2001. Macroalgal blooms in Florida's coastal waters: *Codium isthmocladum*. Seventh Int. Phycol. Congr. Thessaloniki, Greece. Program Abstracts, p. 82.
Hanisak, M. D., and S. M. Blair. 1988. The deep-water macroalgal community of the east Florida continental shelf (USA). *Helgol. Wiss. Meeresunters.* 42: 133–163.
Hansen 1977. *Cirrulicarpus carolinensis*, a new species in the Kallymeniaceae (Rhodophyta). Occas. Pap. Farlow Herb. Cryotogam. Bot. Harvard Univ. 12: 1–22.
Hansgirg, A. 1885. Ein Beitrag zur Kenntnis von der Verbreitung der Chromatophoren und Zellkerne bei den Schizophyceen (Phycochromaceen). *Ber. Deutsch. Bot. Ges.* 3: 14–22.
———. 1892. Prodomus der Algenflora von Böhmen. Teil 2. Prague.
Hardy, F. G., and M. D. Guiry. 2003. *A Check-list and Atlas of the Seaweeds of Britain and Ireland*. British Phycological Society, London, x + 435 pp.
Haroun, R. J., M. C. Gil-Rodríguez, J. Díaz de Castro, and W. F. Prud'homme van Reine. 2002. A checklist of the marine plants from the Canary Islands (central eastern Atlantic Ocean). *Bot. Mar.* 45: 139–169.
Harper, J. T., and G. W. Saunders. 2001. Molecular systematics of the florideophyte red algae (Florideophyceae, Rhodophyta) using nuclear large and small subunit rDNA sequence data. *J. Phycol.* 37: 1073–1082.
———. 2002. A re-classification of the Acrochaetiales based on molecular and morphological data, and establishent of the Colaconematales *ord. nov.* (Florideophyceae, Rhodophyta). Eur. *J. Phycol.* 37: 463–476.
Hartog, C. den. 1972. The effect of the salinity tolerance of algae on their distribution, as exemplified by *Bangia*. In: K. Nisizawa (ed.), *Proc. Seventh Int. Seaweed Symp.*, Sapporo, Japan. Univ. Tokyo Press, Tokyo, pp. 274–276.
Harvey, A. S., S. T. Broadwater, W. J. Woelkerling, and P. J. Mitrovski. 2003a. *Choreonema* (Corallinales, Rhodophyta): 18S rDNA phylogeny and resurrection of the Hapalidiaceae for the subfamilies Choreonematoideae, Austrolithoideae, and Melobesioideae. *J. Phycol.* 39: 988–998.
Harvey, A. S., W. J. Woelkerling, and A.J.K. Millar. 2003b. An account of the Hapalidiaceae (Corallinales, Rhodophyta) in south-eastern Australia. *Aust. Syst. Bot.* 16: 647–698.
Harvey, W. H. 1833. Confervoideae. In: W. J. Hooker (ed.), *The English flora of Sir James Edward Smith. Class XXIV. Cryptogamia*. London. Vol. 5 (or Vol. II of Dr. Hooker's British Flora), Part l, . . . x + 4 + 432 pp.
———. 1834. *Algological illustrations*. No. 1. Remarks on some British algae, and descriptions of new species recently added to our flora. J. Bot. [Hooker] 1: 296–305, pls. CXXXVIII, CXXXIX.
———. 1838. *The genera of South African plants arranged according to the natural system*. Cape Town, lxvi + 429 pp.
———. 1841. *A manual of the British Algae* . . . London, lvii + 229 pp.
———. 1846–1851. *Phycologia* britannica . . . Brothers: London. Vol. 1, pp. [i–iii] [title page ("1846") and dedication], v–viii [Advertisement], ix–xv [List], i–vi [Index], pls. 1–120 within unpaginated text. Vol. 2, pp. [i–ii] [title page ("1849")], i–vi [Index], pls. 121–240 with unpaginated text. Vol. 3, pp. [i–ii] [title page ("1851")], iii–iv [Preface], v–xxxix [Synopsis], xli–xlv [General index], pls. 241–360 with unpaginated text [pls. I–LXXII (1846); LXXIII–CXLIV (1847); CXLV–CCXVI (1848); CCXVII–CCCVI (1849); CCCVII–CCCLIV (1850); CCCLV–CCCLX (1851).]
———. 1847–1849. Nereis Australis . . . London. viii + 124 pp., 50 pls. [pp. [i]–viii + [1]–64, pls. I–XXV (1847); pp. 65–124, pls. XXVI–L (1849).]
———. 1849. *A manual of British marine algae* . . . London, lii + 252 pp., 27 pls.
———. 1852. Nereis boreali-americana . . . Part I. Melanospermae. Smithson. Contrib. Knowl. 3(4). 150 pp., pls. I–XII.
———. 1853. Nereis boreali-americana . . . Part II. Rhodospermae. Smithson. Contrib. Knowl. 5(5). 258 pp., pls. XIII–XXXVI.
———. 1855a. Some account of the marine botany of the colony of Western Australia. *Trans. R. Irish Acad.* 22 (Sci.): 525–566.
———. 1855b. Short characters of some new genera and species of algae discovered on the coast of the colony of Victoria, Australia. *Ann. Mag. Nat. Hist. (Lond.)* 5: 332–336.
———. 1858. Nereis boreali-americana . . . Part III. Chlorospermae, including supplements. Smithson. Contrib. Knowl. 10(2), 140 pp., pls. XXXVII–L.

———. 1859. *Phycologia* australica . . . Vol. 2. London, viii pp., pls. LXI–CXX. (with text).

———. 1860a. Characters of new algae, chiefly from Japan and adjacent regions, collected by Charles Wright in the North Pacific Exploring Expedition under Captain John Rodgers. *Proc. Am. Acad. Arts Sci.* 4: 327–335.

———. 1860b. *Phycologia* australica . . . Vol. 3. London, viii pp., pls. CXXI–CLXXX (with text).

Hatta, A. M., and W. F. Prud'homme van Reine. 1991. A taxonomic revision of Indonesian Gelidiales (Rhodophyta). *Blumea* 35: 347–380.

Hauck, F. 1876. Verzeichniss der im Golfe von Triest gesammelten Meeralgan (Fortsetzung). *Österr. Bot. Z.* 26: 24–26, 54–57, 91–93.

———. 1879. Beiträge zur Kenntniss der adriatischen Algen. XI. *Österr. Bot. Z.* 29: 242–245.

———. 1886–1889. Ueber einige von J. M. Hildebrandt im Rothen Meere und Indischen Ocean gesammelte Algen. *Hedwigia* 25: 165–168, 217–221 (1886); ibid. 26: 8–21, 41–45 (1887); ibid. 27: 86–93 (1888); ibid. 28: 188–190 (1889).

Hayden, H. S., and J. R. Waaland. 2002. Phylogenetic systematics of the Ulvaceae (Ulvales, Ulvophyceae) using chloroplast and nuclear DNA sequences. *J. Phycol.* 38: 1200–1212.

Hayden, H. S., J. Blomster, C. A. Maggs, P. C. Silva, M. J. Stanhope, and J. R. Waaland. 2003. Linnaeus was right all along: *Ulva* and *Enteromorpha* are not distinct genera. Eur. *J. Phycol.* 38: 277–294.

Haywood, J. 1974. Studies on the growth of *Stichococcus bacillaris* Naeg. in culture. *J. Mar. Biol. Assoc. UK* 54: 261–268.

Hazen, T. E. 1902. The Ulothricaceae and Chaetophoraceae of the United States. *Mem. Torrey Bot. Club* 11: 135–250.

Heerebout, G. R. 1968. Studies on the Erythropeltidaceae (Rhodophyceae-Bangiophycidae). *Blumea* 16: 139–157.

Hemminga, M., and C. M. Duarte. 2000. *Seagrass ecology*. Cambridge Univ. Press, Cambridge, 298 pp.

Henne, K.-D., and R. Schnetter. 1999. Revision of the *Pseudobryopsis/Trichosolen* complex (Bryopsidales, Chlorophyta) based on features of gametangial behavior and chloroplasts. *Phycologia* 38: 114–127.

Henry, E. C. 1984. Syringodermatales *ord. nov.* and *Syringoderma floridana sp. nov.* (Phaeophyceae). *Phycologia* 23: 419–426.

———. 1987a. The life history of *Onslowia endophytica* (Sphacelariales, Phaephyceae) in culture. *Phycologia* 26: 175–181.

———. 1987b. Morphology and life histories of *Onslowia bahamensis sp. nov.* and *Verosphacela ebrachia gen. et sp. nov.*, with a reassessment of the Choristocarpaceae (Sphacelariales, Phaeophyceae). *Phycologia* 26: 182–191.

Hering, C. 1841. Diagnoses algarum novarum a cl. Dre. Ferdinand Krauss in Africâ Australi lectarum. *Ann. Mag. Nat. Hist. (Lond.)* 8: 90–92.

Heydrich, F. 1897a. Corallinaceae, insbesondere Melobesieae. *Ber. Deutsch Bot. Ges.* 15: 34–70, 3 figs., pl. III.

———. 1897b. Melobesieae. *Ber. Deutsch Bot. Ges.* 15: 403–420, pl. 18.

———. 1901. Die Lithothamnien des Museum d'histoire naturelle in Paris. *Bot. Jahrb. Syst. Pflangesch. Pflanzengeogr.* 28: 529–545, plate XI.

Hillis, L. W. 1959. A revision of the genus *Halimeda* (order Siphonalels). *Publ. Inst. Mar. Sci.* [Univ. Texas] 6: 321–403.

———. 1997. Coral algal reefs from a calcareous green algal perspective, and a first carbonate budget. *Proc. 8th Int. Coral Reef Symp.* 1: 761–766.

Hillis, L. W., J. A. Engman, and W.H.C.F. Kooistra. 1998. Morphological and molecular phylogenies of *Halimeda* (Chlorophyta, Bryopsidales) identify three evolutionary lineages. *J. Phycol.* 34: 669–681.

Hillis-Colinvaux, L. W. 1980. Ecology and taxonomy of *Halimeda*: primary producer of coral reefs. *Adv. Mar. Biol.* 17: 1–327.

Hillson, C. J. 1977. *Seaweeds: A color-coded, illustrated guide to common marine plants of the east coast of the United States*. Keystone Books, Pennsylvania State University Press, University Park and London, 194 pp.

Hine, A., and H. J. Humm 1971. *Caulerpa ollivieri* in the Gulf of Mexico. *Bull. Mar. Sci.* 21: 552–555.

Hinson, T. D., and D. F. Kapraun. 1991. Karyology and nuclear DNA quantification of four species of *Chaetomorpha* (Cladophorales, Chlorophyta) from the western Atlantic. *Helgol. Wiss. Meeresunters.* 45: 273–285.

Hochstetter, C. F. 1841. *Nova genera plantarum Africae tum Australis tum tropicae borealis proponit et describit*. Flora, Regensburg 42: 657–672.

Hodgson, L. M., P. H. Tri, K. Lewmanomont, and K. J. McDermid. 2004. Annotated checklist of species of

Caulerpa and *Caulerpella* (Bryopsidales, Calerpaceae) from Vietnam, Thailand, and the Hawaiian Islands. In: I. A. Abbott and K. J. McDermid (eds.), *Taxonomy of economic seaweeds with reference to the Pacific and other locations*. Vol. 9. Hawaii Sea Grant College Program, Univ. Hawaii, Hilo. pp. 21–38.

Hoek, C. van den. 1963. *Revision of the European species of* Cladophora. E. J. Brill, Leiden, xi + 248 pp., 55 pls.

———. 1969a. Notes on *Cladophora* (Chlorophyceae) I. Two new Caribbean species: *Cladophora jongiorum nov. sp.* and *Cladophora longicellulata nov. sp. J. Phycol.* 5: 128–134.

———. 1969b. Notes on *Cladophora* (Chlorophyta) II. *Cladophora catenata* (Linnaeus) Kütz, 1843, *emend.* Van den Hoek, 1963. *J. Phycol.* 5: 134–135

———. 1978. Marine algae from the coral reef of Curaçao, Netherlands Antilles. I. Three new and one rarely observed species from the steep fore-reef slope. *Aquat. Bot.* 5: 47–61.

———. 1982. A taxonomic revision of the American species of *Cladophora* (Chlorophyceae) in the North Atlantic Ocean and their geographic distribution. *Verh. Kon. Ned. Akad. Wetensch., Afd. Natuurk., Tweede Sect.* 78: 1–236.

Hoek, C. van den, and M. Chihara. 2000. A taxonomic revision of the marine species of *Cladophora* (Chlorophyta) along the coasts of Japan and the Russian Far-east. *Natl. Sci. Mus. Monogr.* 19, 242 pp.

Hoek, C. van den, C. F. Colijn, A. M. Cortel-Breeman, and J. B. Wanders. 1972. Algal vegetation type along the shores and inner bays and lagoons of Curaçao and the lagoon Lac (Bonaire) Netherlands Antilles. *Ver. K. Ned. Acad. Wet.* 61: 1–72.

Hoek, C. van den., D. G. Mann, and H. M. Jahns. 1995. *Algae: an introduction to phycology*. Cambridge Univ. Press, Cambridge, xiv + 623 pp.

Hoek, C. van den, and H.B.S. Womersley. 1984. Genus *Cladophora* Kuetzing 1843: 262, *nom. cons.* In: H.B.S. Womersley, *The marine benthic flora of southern Australia*. Part I. [Adelaide], pp. 185–213.

Hoffman, A., and B. Santelices. 1999. *Flora Marina de Chile Central.* Pontificia Universidad Católica de Chile, Casilla Santiago, 434 pp.

Hoffman, B. A., and C. J. Dawes. 1997. Vegetational and abiotic analysis of the salterns of mangals and salt marshes of the west coast of Florida. *J. Coastal Res.* 13: 147–154

Hoffmann, L., C. Billard, M. Jannssens, M. Leruth, and V. Demoulin. 2000. Mass development of marine benthic Sarcinochrysidales (Chrysophyceae *s.l.*) in Corsica. *Bot. Mar.* 43: 223–231.

Höhnel, F. von. 1920. Mykologische Fragmente. *Ann. Mycol.* 18: 71–97.

Holden, I. 1899. Two new species of marine algae from Bridgeport, Connecticut. *Rhodora* 1: 197–198.

Hollenberg, G. J. 1958. Phycological notes. II. *Bull. Torrey Bot. Club* 85: 63–69.

———. 1968a. An account of the species of *Polysiphonia* of the central and western tropical Pacific Ocean. I. *Oligosiphonia*. *Pac. Sci.* 22: 56–98.

———. 1968b. An account of the species of the red alga *Herposiphonia* occurring in the central and western tropical Pacific Ocean. *Pac. Sci.* 22: 536–559.

Hommersand, M. H. 1963. The morphology and classification of some Ceramiaceae and Rhodomelaceae. *Univ. Calif. Publ. Bot.* 35: 165–366.

Hommersand, M. H., and S. Fredericq. 1988. An investigation of cystocarp development in *Gelidium pteridifolium* with a revised description of the Gelidiales (Rhodophyta). *Phycologia* 27: 254–272.

Hommersand, M. H., M. D. Guiry, S. Fredericq, and G. L. Leister. 1993. New perspectives in the taxonomy of the Gigartinaceae (Gigartinales, Rhodophyta). *Hydrobiologia* 260/261: 105–120.

Hooker, W. J. 1821. *Flora scotica; or a description of Scottish plants, arranged both according to the artificial and natural methods*. Richard and Arthur Taylor, Edinburgh. Part 1, pp. [i]–x, [1], [1]–292, [1–4, index]; Part 2, pp. [1]–297, [298–303, index].

———. 1833. Class XXIV. Cryptogamia. In: *The English flora of Sir James Edward Smith*. Vol. V (or Vol. II of Dr. Hooker's British flora). Part I. Comprising the mosses, Hepaticae, lichens, Characeae, and algae. London, pp. [i]–x + [1]–4 + [1]–432.

Hooker, W. J., and W. H. Harvey. 1845a. Algae antarcticae, being characters and descriptions of the hitherto unpublished species of algae, discovered in Lord Auckland's Group, Cambell's Island, Kerguelen's Land, Falklands Islands, Cape Horn and other southern circumporal regions, during the voyage of H. M. discovery ships "Erebus" and "Terror." *Lond. J. Bot.* 4: 239–276, 293–298.

———. 1845b. Algae Novae Zelandiae: being a catalogue of all the species of algae yet recorded as inhabiting the shores of New Zealand, with characters and brief descriptions of the new species discovered during the voyage of H. M. discovery ships 'Erebus" and "Terror" and of others communicated to Sir W. Hooker by Dr. Sinclair, the Rev. Colenso, and M. Raoul. *Lond. J. Bot.* 4: 521–551.

Hooper, J. 1850. *Introduction to algology, with a catalogue of American algae, or seaweeds, according to the latest classification of Prof. Harvey*. Brooklyn, 34 pp.

Hopkins, T. S., D. R. Blizzard, S. Brawley, S. A. Earle, D. E. Grimm, D. K. Gilbert, P. G. Johnson, E. H. Livingston, C. H. Lutz, J. K. Shaw, and B. Shaw. 1977. A preliminary characterization of the biotic components of composite strip transects on the Florida Middle Grounds, northeastern Gulf of Mexico. *Proc. Third Int. Coral Reef Symp*. 2: 31–37.

Hoppaugh, K. W. 1930. A taxonomic study of species of the genus *Vaucheria* collected in California. *Am. J. Bot.* 17: 329–347.

Hörnig, I and R. Schnetter. 1988. Notes on *Dictyota dichotoma, D. menstrualis, D. indica* and *D. pulchella spec. nov.* (Phaeophyta). *Phyton* 28: 277–291.

Hörnig, I., R. Schnetter and W. F. Prud'homme van Reine. 1993. Additional note to "The genus *Dictyota* (Phaeophyceae) in the North Atlantic. I. A new generic concept and new species." Correction and validation of new combinations in the genus *Dictyota*. *Nova Hedwigia* 56: 169–171.

———. 1992a. The genus *Dictyota* (Phaeophyta) in the North Atlantic. II. Key to the species. *Nova Hedwigia* 54: 397–402.

Hörnig, I., R. Schnetter, W. F. Prud'homme van Reine, E. Coppejans, K. Achenbach-Wege, and J. M. Over. 1992b. The genus *Dictyota* (Phaeophyceae) in the North Atlantic. I. A new generic concept and new species. *Nova Hedwigia* 54: 45–62.

Howe, M. A. 1901. Observations on the algae genera *Acicularia* and *Acetabulum*. *Bull. Torrey Bot. Club* 28: 321–334.

———. 1904. Collections of marine algae from Florida and the Bahamas. *J. N. Y. Bot. Gard.* 5: 164–166.

———. 1905a. Phycological studies. I. New Chlorophyceae from Florida and the Bahamas. *Bull. Torrey Bot. Club* 32: 241–252.

———. 1905b. Phycological studies. II. New Chlorophyceae, new Rhodophyceae, and miscellaneous notes. *Bull. Torrey Bot. Club* 32: 563–586.

———. 1907. Phycological studies. III. Further notes on *Halimeda* and *Avrainvillea*. *Bull. Torrey Bot. Club* 34: 491–516.

———. 1909. Phycological studies. IV. The genus *Neomeris* and notes on other Siphonales. *Bull. Torrey Bot. Club* 36: 75–104.

———. 1911. Phycological studies. V. Some marine algae of lower California, Mexico. *Bull. Torrey Bot. Club* 38: 489–514.

———. 1914. The marine algae of Peru. *Mem. Torrey Bot. Club* 15: 1–185.

———. 1915. Report on a visit to Porto Rico for collecting marine algae. *J. N. Y. Bot. Gard.* 16: 219–225.

———. 1917. A note on the structural dimorphism of sexual and tetrasporic plants of *Galaxaura obtusata*. *Bull. Torrey Bot. Club* 43: 621–624.

———. 1918. Class 3. Algae. In: N. L. Britton (ed.), *Flora of Bermuda* (illustrated). New York, pp. 489–540.

———. 1920a. Algae. In: N. L. Britton and C. F. Millspaugh (eds.), *The Bahama flora*. New York, pp. 553–618.

———. 1920b. Observations on monosporangial discs in the genus *Liagora*. *Bull. Torrey Bot. Club*. 47: 1–8.

Howe, M. A., and W. D. Hoyt. 1916. Notes on some marine algae from the vicinity of Beaufort, North Carolina. *Mem. N. Y. Bot. Gard.* 6: 105–123, pls. 11–15.

Howe, M. A., and W. R. Taylor 1931. Notes on new or little-known algae from Brazil. *Brittonia* 1: 7–33.

Hoyt, W. D. 1920. *Marine algae of Beaufort, N.C. and adjacent regions*. Bull. Bur. Fish. (U.S.) 36: 365–556, pls. 84–119.

———. 1927. The periodic fruiting of *Dictyota* and its relation to the environment. *Am. J. Bot.* 14: 592–619.

Huang, S.-F. 2000. *Seaweeds of northeastern Taiwan*. Vol. 12. National Taiwan Musem, Taipei, 233 pp.

Hudson, W. 1762. *Flora anglica* . . . Londini [London], viii [xvi] + 506 + [22] pp.

———. 1778. *Flora anglica* . . . Editio altera. Londini [London], [iii +] xxxviii [xxxix = Errata] + 690 pp.

Huisman, J. M. 1985. The *Scinaia* assemblage (Galaxauraceae, Rhodophyta): a re-appraisal. *Phycologia* 24: 403–418.

———. 1986. The red alga genus *Scinaia* (Nemaliales, Galaxauraceae) from Australia. *Phycologia* 25: 271–296.

———. 1987. The taxonomy and life history of *Gloiophloea* (Galaxauraceae, Rhodophyta). *Phycologia* 26: 167–174.

———. 1996. The red algal genus *Coelarthrum* Børgesen (Rhodymeniaceae, Rhodymeniales) in Australian seas, including the description of *Chamaebotrys gen. nov. Phycologia* 35: 95–112.

———. 2000. *Marine plants of Australia.* Univ. Western Australia Press, Nedlands, Western Australia, ix + 300 pp.

———. 2002. The type and Australian species of the red algal genera *Liagora* and *Ganonema* (Liagoraceae, Nemaliales). *Aust. Syst. Bot.* 15: 777–838.

Huisman, J. M., and M. A. Borowitzka. 1990. A revision of the Australian species of *Galaxaura* (Rhodophyta, Galaxauraceae), with a description of *Tricleocarpa gen. nov. Phycologia* 29: 150–172.

Huisman, J. M., and J. T. Harper. 1994. Studies of the Liagoraceae (Rhodophyta) of Western Australia: *Gloiotrichus fractalis gen. et sp. nov.* and *Ganonema helminthaxis sp. nov.* Eur. *J. Phycol.* 29: 73–85.

Huisman, J. M., J. T. Harper, and G. W. Saunders. 2004a. Phylogenetic study of the Nemaliales (Rhodophyta) based on large-subunit ribosomal DNA sequences supports segregation of the Scinaiaceae *fam. nov.* and resurrection of *Dichotomaria* Lamarck. *Phycol. Res.* 52: 224–234.

Huisman, J. M., and A.J.K. Millar. 1996. *Asteromenia* (Rhodymeniaceae, Rhodymeniales), a new red algal genus based on *Fauchea peltata. J. Phycol.* 32: 138–145.

Huisman, J. M., A. R. Sherwood, and I. A. Abbott. 2003. Morphology, reproduction, and the 18S gene sequence of *Pihiella liagoraciphila gen. et sp. nov.* (Rhodophyta), the so-called "monosporangial discs" associated with members of the Liagoraceae (Rhodophyta), and proposal of the Pihiellales *ord. nov. J. Phycol.* 39: 978–987.

———. 2004b. Studies of Hawaiian Galaxauraceae (Nemaliales, Rhodophyta): Large subunit rDNA gene sequences support conspecificity of *Galaxaura rugosa* and *G. subverticillata. Cryptogam. Algol.* 25: 337–352.

Huisman, J. M., and T. Schils. 2002. A re-assessment of the genus *Izziella* Doty (Liagoraceae, Rhodophyta). *Cryptogam. Algol.* 23: 237–249.

Huisman, J. M., and R. A. Townsend. 1993. An examination of Linnaean and pre-Linnaean taxa referable to *Galaxaura* and *Tricleocarpa* (Galaxauraceae, Rhodophyta). *J. Linn. Soc. Lond. Bot.* 113: 95–101.

Huisman, J. M., and D. I. Walker. 1990. A catalogue of the marine plants of Rottnest Island, Western Australia, with notes on their distribution and biogeography. *Kingia* 1: 349–459.

Huisman, J. M., and H.B.S. Womersley. 1992. Cystocarp development in the red alga *Nothogenia fastigiata* (Galaxauraceae, Rhodophyta). *Phycologia* 31: 359–364.

Huisman, J. M., and M. J. Wynne. 1999. *Liagora tsengii sp. nov.* (Liagoraceae, Rhodophyta) from Lesser Antilles, West Indies. *Bot. Mar.* 42: 219–225.

Humm, H. J. 1952a. Notes on the marine algae of Florida. I. The intertidal rocks at Marineland. *Fla. State Univ. Stud.* 7: 17–23.

———. 1952b. Notes on the marine algae of Florida. II. Flora of the rocky bottom off St. Marks Light, Wakulla County (Gulf of Mexico). *Phycol. Soc. Am., News Bull.* 6: 8.

———. 1963. Some new records and range extensions of Florida marine algae. *Bull. Mar. Sci.* 4: 305–308.

———. 1964. Epiphytes of the seagrass *Thalassia testudinum* in Florida. *Bull. Mar. Sci.* 14: 306–341.

———. 1979. *The marine algae of Virginia.* University Press of Virginia, Charlottesville, 263 pp.

Humm, H. J., and S. E. Taylor. 1961. Marine Chlorophyta of the upper West Coast of Florida. *Bull. Mar. Sci.* 11: 321–380.

Huvé, P., and H. Huvé. 1977. Notes de nomenclature algale. I. Le genre *Halichrysis* (J. Agardh 1851 emend. J. Agardh 1876) Schousboe mscr. In É. Bornet 1892 (Rhodyméniales, Rhodyméniacées). *Bull. Soc. Phycol. Fr.* 22: 99–107.

Irvine, D. E. G. 1956. Notes on British species of the genus *Sphacelaria* Lyngb. *Trans. Proc. Bot. Soc. Edinb.* 37: 24–45.

Irvine, L. M. 1976. In: M. W. Parke and P. S. Dixon: Checklist of British marine algae—third revision. *J. Mar. Biol. Assoc.* U.K. 56: 590.

———. 1983. *Seaweeds of the British Isles.* Vol. 1. Rhodophyta; Part 2A: Cryptonemiales (*sensu stricto*), Palmariales, Rhodymeniales. British Museum (Natural History), London, xii + 115 pp.

Irvine, L. M., and Y. M. Chamberlain. 1994. *Seaweeds of the British Isles.* Vol. 1. Rhodophyta; Part 2B: Corallinales, Hildenbrandiales. British Museum (Natural History), London, vii + 276 pp.

Irvine, L. M., and W. F. Farnham. 1983. Halymeniaceae. In: L. M. Irvine, *Seaweeds of the British Isles.* Vol. 1. Rhodophyta; Part 2A: Cryptonemiales (*sensu stricto*), Palmariales, Rhodymeniales. British Museum (Natural History), London, pp. 17–29.

Irvine, L. M., and M. D. Guiry. 1983. Rhodymeniales. In: L. M. Irvine, *Seaweeds of the British Isles.* Vol. 1. Rhodophyta; Part 2A: Cryptonemiales (*sensu stricto*), Palmariales, Rhodymeniales. British Museum (Natural History), London, pp. 77–98.

Irvine, L. M., and C. M. Pueschel. 1994. Hildenbrandiales. In: L. M. Irvine and Y. M. Chamberlain, *Seaweeds of the British Isles*. Vol. 1. Rhodophyta, Part 2B: Corallinales, Hildenbrandiales. British Museum (Natural History), London, pp. 235–241.

Isaac, W. E., and Y. M. Chamberlain. 1958. Marine algae of Inhaca Island and of the Inhaca Peninsula, II. *S. Afr. J. Bot.* 24: 123–158

Islam, A. K. M. N. 1976. Contribution to the study of the marine algae of Bangladesh. *Bibl. Phycol.* 19, [III +] 253 pp., including 73 pls., [2] maps.

Itono, H., and T. Tanaka. 1973. *Balliellla*, a new genus of Ceramiaceae (Rhodophyta). *Bot. Mag. (Tokyo)* 86: 241–252.

Iyer, R., E. M. Tronchin, and J. J. Bolton. 2005. Molecular systematics of the Gracilariaceae (Gracilariales, Rhodophyta) with emphasis on southern Africa. *J. Phycol.* 41: 672–684.

Jaap, W. C., and P. Hallock. 1990. Coral reefs. In: R. L. Myers and J. J. Ewel (eds.), *Ecosystems of Florida*. Univ. Central Florida Press, Orlando, pp. 574–616.

Jaasund, E. 1951. Marine algae from northern Norway. I. *Bot. Not.* 1951: 128–142.

———. 1976. *Intertidal seaweeds in Tanzania*. Univ. Tromsø [Tromsø, Norway], iii + 159 [160] pp., 12 pls.

———. 1977. Marine algae of Tanzania. 8. *Bot. Mar.* 20: 509–520.

Jacoby, C., B. Lapointe, and L. Creswell. 2004. Are native and non-indigenous seaweeds overgrowing Florida's east coast reefs? *Florida Seagrant Program SGEF-1256*. Gainesville.

Jacquin, N. J. 1787('1786'). *Collectanea ad botanicam, chemiam, et historiam naturalem spectantia, cum figuris*. Vindobonae [Wien], 1: 1–386 pp., 22 pls.

———. 1791('1789'). *Collectanea ad botanicam, chemiam, et historiam naturalem spectantia, cum figuris*. Vindobonae [Vienna], 3: 1–306, 23 pls.

Janzen, D. H. 1985. Mangroves: where's the understory? *J. Trop. Ecol.*: 1: 89–92.

Jao, C.-C. 1936. New Rhodophyceae from Woods Hole. *Bull. Torrey Bot. Club* 63: 237–258.

John, D. M. 2002. Orders Chaetophorales, Klebshormidiales, Microsporales, Ulothrichales. In: D. M. John, B. A. Whitton and A. J. Brook (eds.), *The freshwater algal flora of the British Isles. An identification guide to freshwater and terrestrial algae*. Cambridge Univ. Press, Cambridge, pp. 433–468.

John, D. M., G. W. Lawson, and G. K. Ameka. 2003. The marine macroalgae of the Tropical West Africa Subregion. *Nova Hedwigia* Beih. 125 pp.

John, D. M., J. H. Price, C. A. Maggs, and G. W. Lawson. 1979. Seaweeds of the western coast of tropical Africa and adjacent islands: a critical assessment. III. Rhodophyta (Bangiophyceae). *Bull. Br. Mus. (Nat. Hist.) Bot.* 7: 69–82.

John, D. M., W. F. Prud'homme van Reine, G. W. Lawson, T. B. Kostermans, and J. H. Price. 2004. A taxonomic and geographical catalogue of the seaweeds of the western coast of Africa and adjacent islands. *Nova Hedwigia* Beih. 127: 1–339, 1 fig.

Johansen, H. W. 1970. The diagnostic value of reproductive organs in some genera of articulated coralline red algae. *Br. Phycol. J.* 5: 79–86.

———. 1981. *Coralline algae, a first synthesis*. CRC Press, Boca Raton, Fla., 239 pp.

Johnson, J. H. 1961. *Limestone-building algae and algal limestones*. Johnson Publ., Boulder, Colo.

Johnson, J. H., and B. J. Ferris. 1948. Eocene algae from Florida. *J. Paleontol.* 22: 762–766.

Joly, A. B. 1957a. The genus *Bostrychia* Montagne, 1838 in southern Brazil. Taxonomic and ecological data. *Bol. Fac. Filos. Ciênc. Letras Univ. São Paulo, Ser. Bot.* 11: 53–74.

———. 1957b. Contribuição ao conhecimento da flora ficológica marinha da báia de Santos e arresdores. *Bol. Fac. Filos. Ciênc. Letras Univ. São Paulo, Ser. Bot.* 14: 1–196.

———. 1965. Flora marinha do litoral norte do Estado de São Paulo e regiões circunvizinhas. *Bol. Fac. Filos. Ciênc. Letras Univ. São Paulo, Ser. Bot.* 21: 1–393.

Joly, A. B., M. Cordeiro, and N. T. Yamaguishi. 1963. *Antithamnion tristicum* and *Ophidocladus herposiphonioides*, two new Rhodophyceae from southern Brazil. *Bol. Fac. Filos. Ciênc. Letras Univ. São Paulo, Ser. Bot.* 20: 57–64.

Joly, A. B., M. Cordeiro-Marino, Y. Ugadim, N. Yamaguishi-Tomita, and F. C. Pinheiro. 1965. New marine algae from Brazil. *Arquipel. Estac. Biol. Mar. Univ. Ceará* [Brazil] 5: 79–92.

Joly, A. B., and E. C. Oliveira. 1966. *Spyridiocolax* and *Heterodasya*, two new genera of the Rhodophyceae. *Sellowia* 18: 115–125.

Joly, A. B., and E. C. Oliveira. 1969('1968'). Notes on Brazilian algae II. A new *Anadyomene* of the deep water flora. *Phykos* 7: 27–31.

Joly, A. B., and Y. Ugadim. 1966('1965'). New marine algae from southern Brazil. *Rickia* 2: 159–181.

Jompa, J., and L. J. McCook. 2003. Contrasting effects of turf algae on corals: massive *Porites* spp. are unaffected by mixed-species turfs, but killed by the red alga *Anotrichiumn tenue*. *Mar. Ecol. Prog. Ser.* 258: 79–86.

Jones, A. C., S. A. Berkley, J. A. Bohnsack, S. A. Bortone, D. K. Camp, G. H. Darcey, J. C. Davis, K. D. Haddad, M. Y. Hedgpath, E. W. Irby Jr., W. C. Japp, F. S. Kennedy, W. G. Lyons, E. I. Nakamura, T. H. Perkins, J. K. Reed, K. A. Steidinger, J. T. Tilmant, and R. O. Williams. 1986. Oceanic habitat and fishery resources of Florida. In: W. Seaman Jr. (ed.), *Florida aquatic habitat and fishery resources*. Florida Chapter of the American Fisheries Society, Kissimmee, pp. 437–542.

Jones, P. L., and W. J. Woelkerling. 1984. An analysis of trichocyte and spore germination attributes as taxonomic characters in the *Pneophyllum-Fosliella* complex (Corallinaceae, Rhodophyta). *Phycologia* 23: 183–194.

Jordan, D. S. 1874. A key to the higher algae of the Atlantic coast between Newfoundland and Florida. *Am. Nat.* 8: 398–403.

Juett, L., C. J. Miller, S. J. Moore, and E. S. Ford. 1976. Summer marine algae at Vero Beach, Florida. *Fla. Sci.* 39: 76–80.

Kamiya, M. 2004. Speciation and biogeography of the *Caloglossa leprieurii* complex (Delesseriaceae, Rhodophyta). *J. Plant Res.* 117: 421–428.

Kamiya, M., J. Tanaka, and Y. Hara. 1997. Comparative morphology, crossability, and taxonomy within the *Caloglossa continua* (Delesseriaceae, Rhodophyta) complex from the western Pacific. *J. Phycol.* 33: 97–105.

Kamiya, M., and J. A. West. 2005. Asexual sporophytes produced from outcrossing of sexual plants. *Phycologia* 44(4, supplement): 50.

Kamiya, M., G. C. Zuccarello, and J. A. West. 2003. Evolutionary relationships of the genus *Caloglossa* (Delesseriaceae, Rhodophyta) inferred from large-subunit ribosomal RNA gene sequences, morphological evidence and reproductive compatibility with description of a new species from Guatemala. *Phycologia* 42: 478–497.

Kang, J. W. 1965. Marine algae of Ulrungdo Island. *Bull. Pusan Fish. Coll.* 6: 41–58.

Kapraun, D. F. 1970. Field and cultural studies of *Ulva* and *Enteromorpha* in the vicinity of Port Aransas, Texas. *Contrib. Mar. Sci.* 15: 205–285.

———. 1977. The genus *Polysiphonia* in North Carolina, USA. *Bot. Mar.* 20: 313–331.

———. 1979. The genus *Polysiphonia* (Ceramiales, Rhodophyta) in the vicinity of Port Aransas, Texas. *Contrib. Mar. Sci.* 22: 105–120.

Kapraun, D. F., and W. D. Freshwater. 1987. Karyological studies of five species in the genus *Porphyra* (Bangiales, Rhodophyta) from the North Atlantic and Mediterranean. *Phycologia* 26: 82–87.

Kapraun, D. F., A. J. Lemus, and G. Bula-Meyer. 1983. Genus *Polysiphonia* (Rhodophyta, Ceramiales) in the tropical western Atlantic: I. Colombia and Venezuela. *Bull. Mar. Sci.* 33: 881–898.

Kapraun, D. F., and J. N. Norris. 1982. The red alga *Polysiphonia* Greville (Rhodomelaceae) from Carrie Bow Cay and vicinity, Belize. In: K. Rützler and I. G. Macintyre (eds.), *The Atlantic Barrier Reef ecosystem at Carrie Bow Cay, Belize*. I. *Structure and communities*. Smithson. Contrib. Mar. Sci. 12: 225–238.

Kawaguchi, S. 2004. Morphological observations of the type and some authentic material of *Halymenia floresia* (Clemente y Rubio) C. Agardh, with notes on previous reports of this alga. In: I. A. Abbott and K. J. McDermid (eds.), *Taxonomy of economic seaweeds with reference to the Pacific and other locations*. Vol. 9. Hawaii Sea Grant College Program, Univ. Hawaii, Hilo, pp. 143–155.

Kawaguchi, S., H. W. Wang, T. Horiguchi, G. Sartoni, and M. Masuda. 2001. A comparative study of the red alga *Grateloupia filicina* (Halymeniaceae) from the northwestern Pacific and Mediterranean with the description of *Grateloupia asiatica, sp. nov*. *J. Phycol.* 37: 433–442.

Kawai, H., and H. Sasaki. 2004. Morphology, life history and molecular phylogeny of *Stschapovia flagellaris* (Tilopteridales, Phaeophyceae) and the erection of Stschapoviaceae *fam. nov*. *J. Phycol.* 40:1156–1169.

Kelly, B. O. 1995. Long term trends of macroalgae in Hillsborough Bay. *Fla. Sci.* 58: 179–191.

Kilar, J. A. 1988. Seasonal patterns of morphological variability in *Sargassum polyceratium* (Phaeophyta). *J. Phycol.* 24: 467–473.

———. 1992. Seasonal and between-plant variability in the morphology of *Sargassum mathiesonii sp. nov.* (Phaeophyta) from the Gulf of Mexico. *J. Phycol.* 24: 114–126.

Kim, G. H., T. A. Klotchkova, and Y. M. Kang. 2001. Life without a cell membrane: regeneration of protoplasts from disintegrated cells of the marine green alga *Bryopsis plumosa*. *J. Cell Sci.* 114: 2009–214.

Kim, G. H., T. A. Klotchkova, K.-S. Yoon, Y.-S Song, and K. P. Lee. 2006. Purification and characterization of a lectin, bryohealin, involved in the protoplast formation of a marine green alga *Bryopsis plumosa*. *J. Phycol.* 42: 86–95.

Kim, H. S., and I. K. Lee. 1994. Morphological differences among the populations of *Enteromorpha compressa* (L.) Greville (Chlorophyceae) due to environmental factors. *Korean J. Phycol.* 9: 29–35.

Kim, M. S., and I. K. Lee. 1994. Morphotaxonomic studies on the Korean Ectocarpaceae (Phaeophyta). II. Genus *Feldmannia* Hamel, specially referred to morphogenesis and phylogenetic relationship among related genera. *Korean J. Phycol.* 9: 153–168.

———. 1999. *Neosiphonia flavimarina gen. et sp. nov.* with a taxonomic reassessment of the genus *Polysiphonia* (Rhodomelaceae, Rhodophyta). *Phycol. Res.* 47: 271–281.

Kim, M. S., E. C. Yang, H.-G. Choi, and S. M. Boo. 2005. Phylogenetic relationships of *Polysiphonia* and *Neosiphonia* (Rhodophyta) based on rbcL and SSU rDNA sequence analyses. *Phycologia* 44(4, supplement): 54.

King, R. J., and C. F. Puttock. 1989. Morphology and taxonomy of *Bostrychia* and *Stictosiphonia* (Rhodomelaceae/Rhodophyta). *Aust. Syst. Bot.* 2: 1–73.

———. 1994. Morphology and taxonomy of *Caloglossa* (Delesseriaceae, Rhodophyta). *Aust. Syst. Bot.* 7: 89–124.

King, R. J., C. F. Puttock, and R. S. Vickery. 1988. A taxonomic study on the *Bostrychia tenella* complex (Rhodomelaceae, Rhodophyta). *Phycologia* 27: 10–19.

Kitayama, T., and T. Yoshida. 1992. First report of *Phaeophila dendroides* (P. Crouan *et* H. Crouan) Batters (Ulvophyceae, Chlorophyta) from Japan. Jpn. *J. Phycol.* 40: 47–50.

Kjellman, F. R. 1890. *Handbook I Skandinaviens hafsalgflora*. I. Fucoideae. Stockholm, iv + 103 pp, 17 figs.

———. 1897. Japanska arter af slägtet *Porphyra*. Bihang til Kongliga Svenska Vetenskaps-Akademiens Handlingar 23 (Afd. II, 11), 34 pp., 5 pls. (1–4 V).

———. 1900. Om Floridé-slägtet *Galaxaura*, dess organografi och systematik. Kongl. [Svenska] Vetensk.-Akad. Handl.[ser. 4,] 33(1). 109 pp., 20 pls.

Kjellman, F. R., and N. L. Svedelius. 1910('1911'). Lithodermataceae. In: A. Engler and K. Prantl (eds.), Die natürlichen Pflanzenfamilien, Nachträge zum I. Teil, Abt 2. Leipzig, pp. 173–176 (1911).

Klein, A. S., A. C. Mathieson, C. D. Neefus, D. F. Cain, H. A. Taylor, B. W. Teasdale, A. L. West, E. J. Hehre, J. Brodie, C. Yarish, and A. L. Wallace. 2003. Identification of north-western Atlantic *Porphyra* (Bangiaceae, Bangiales) based upon sequence variation in nuclear SSU and plastid rbcL genes. *Phycologia* 42: 109–122.

Knight, R. L., and W. F. Coggins. 1982. Record of estuarine and salt marsh metabolism at Crystal River, Florida, 1977–1981. Final summary report for Florida Power Corp. Department of Environmental and Engineering Sciences, Univ. Florida, Gainesville.

Knoepffler-Péguy, M. 1970. Quelques *Feldmannia* Hamel 1938 (Phaeophyceae, Ectocarpales) des Côtes d'Europe. *Vie et Milieu*, sér. A, 21: 137–188.

———. 1974. Le genre *Acinetospora* Bornet 1891 (Phaeophyceae- Ectocarpales). *Vie et Milieu*, sér. A, 24: 43–72.

Kobayashi, K., and Y. Kanaizuka. 1985. Reunification of sub-cellular fractions of *Bryopsis* into viable cells. *Plant Sci.* 40: 129–135.

Kogame, K. 1996. Morphology and life history of *Scytosiphon canaliculatus comb. nov.* (Scytosiphonales, Phaeophyceae) from Japan. *Phycol. Res.* 44: 85–94.

———. 1997. Life histories of *Colpomenia sinuosa* and *Hydroclathrus clathratus* (Scytosiphonaceae, Phaeophyceae) in culture. *Phycol. Res.* 45: 227–231.

———. 1998. A taxonomic study of Japanese *Scytosiphon* (Scytosiphonales, Phaeophyceae), including two new species. *Phycol. Res.* 46: 39–56.

Kogame, K., S. Uwai, S. Shimada, and M. Masuda. 2005. A study of sexual and asexual populations of *Scytosiphon lomentaria* (Scytosiphonaceae, Phaeophyceae) in Hokkaido, northern Japan, using molecular markers. *Eur. J. Phycol.* 40: 313–322.

Kooistra, W.H.C.F. 2002. Molecular phylogenies of Udoteaceae (Bryopsidales, Chlorophyta) reveal nonmonophyly for *Udotea*, *Penicillus*, and *Chlorodesmis*. *Phycologia* 41: 453–462.

Kooistra, W.H.C.F., E. C. Coppejans, and C. Payri. 2002. Molecular systematics, historical ecology, and phylogeography of *Halimeda* (Bryopsidales). *Mol. Phylogenet. Evol.* 24: 121–138.

Kooistra, W.H.C.F., and H. Verbruggen. 2005. Genetic patterns in the calcified tropical seaweeds *Halimeda opuntia*, *H. distorta*, *H. hederacea*, and *H. minima* (Bryopsidales, Chlorophyta) provide insights in species boundaries and interoceanic dispersal. *J. Phycol.* 41: 177–187.

Kornmann, P. 1953. Der Formenkreis von *Acinetospora crinita* (Carm.) *nov. comb. Helgol. Wiss. Meeresunters.* 4: 205–224.

———. 1962. Die Entwicklung von *Monostroma grevillei. Helgol. Wiss. Meeresunters.* 8: 195–202.

———. 1972. Ein Beitrag zur *Taxonomie* der Gattung *Chaetomorpha* (Cladophorales, Chlorophyta). *Helgol. Wiss. Meeresunters.* 23: 1–31.

———. 1989. *Sahlingia nov. gen.* based on *Erythrocladia subintegra* (Erythropeltidales [sic. 'Erythropelfidales'], Rhodophyta). *Br. Phycol. J.* 24: p. 223–228.

Kornmann, P., and P.-H. Sahling. 1977. Meeresalgen von Helgoland. Benthische Gruen-, Braun- und Rotalgen. *Helgol. Wiss. Meeresunters.* 29: 1–289.

———. 1980. *Ostreobium quekettii* (Codiales, Chlorophyta). *Helgol. Wiss. Meeresunters.* 34: 115–122.

———. 1985. Erythropeltidaceen (Bangiophyceae, Rhodophyta) von Helgoland. *Helgol. Wiss. Meeresunters.* 39: 213–236.

Koster, J. T. 1955. The genus *Rhizoclonium* Kütz. in the Netherlands. *Publ. Staz. Zool. Napoli* 27: 335–357.

Kraft, G. T. 1989. *Cylindraxis rotundatus gen. et sp. nov.* and its generic relationships within the Liagoraceae (Nemaliales, Rhodophyta). *Phycologia* 28: 275–304.

Kraft, G. T., and P. A. Robins. 1985. Is the order Cryptonemiales defensible? *Phycologia* 24: 67–77.

Kraft, G. T., and G. W. Saunders. 2000. Bringing order to red algal families: taxonomists ask the jurists 'Who's in charge here?' (Commentary). *Phycologia* 39: 358–361.

Kraft, G. T., and M. J. Wynne. 1979. An earlier name for the Atlantic North American red alga *Neoagardhiella baileyi* (Solieriaceae, Gigartinales). *Phycologia* 18: 325–329.

———. 1996. Delineation of the genera *Struvea* Sonder and *Phyllodictyon* J. E. Gray (Cladophorales, Chlorophyta). *Phycol. Res.* 44: 129–142.

Krellwitz, E. C., K. V. Kowallik, and P. S. Manos. 2001. Molecular and morphological analyses of *Bryopsis* (Bryopsidales, Chlorophyta) from the western North Atlantic and Caribbean. *Phycologia* 40: 330–339.

Krishnamurthy, V., and H. V. Joshi. 1970. A check-list of Indian marine algae. Central Salt and Marine Chemicals Research Institute, Bhavnagar, 36 pp. [37 = Errata].

Kristiansen, A., R. Nielsen, and P. M. Pedersen. 1993. An annotated list of marine algae collected on Lanzarote, Canary Islands, January 1986. *Cour. Forschungsinst. Senckenb.* 159: 93–102.

Kristiansen, J., and H. R. Preisig. 2001. Encyclopedia of Chrysophyte Genera. *Bibl. Phycol.* 110: 1–260.

Kruczynski, W. L. 1982. Salt marshes of the northeastern Gulf of Mexico. In: R. R. Lewis III (ed.), *Creation and restoration of coastal plant communities*. CRC Press, Boca Raton, Fla., pp. 71–87.

Kruczynski, W. L., C. B. Subrahmanyam, and S. H. Drake. 1978. Studies on the plant community of a north Florida salt marsh. Part I. Primary production. *Bull. Mar. Sci.* 28: 316–334.

Kuckuck, P. 1891. Beiträge zur Kenntniss einiger *Ectocarpus*-Arten der Kieler Föhrde. *Bot. Centralbl.* 48: 1–6, 33–41, 65–71, 97–104, 129–141.

———. 1894. Bemerkungen zur marinen Algenvegetation von Helgoland. *Helgol. Wiss. Meeresunters.* 1: 225–263.

———. 1929. Fragmente eniner Monographie der Phaeosporeen. *Helgol. Wiss. Meeresunters.*, Abt., N. F. 17: 1–93 (edited by W. Nienburg).

———. 1956. Ectocarpaceen-Studien. IV. *Herponema*, *Kützingiella nov. gen.*, *Farlowiella nov. gen. Helgol. Wiss. Meeresunters.* 5: 292–325 (edited by P. Kornman).

———. 1960. Ectocarpaceen-Studien VI. *Spongonema. Helgol. Wiss. Meeresunters.* 7: 93–113 (edited by P. Kornmann).

———. 1962. Zur Entwicklung von *Monostroma grevillei* und zur systematisch stellung von *Gomontia polyrhiza. Vortr. Gesamtbeb. Bot. N. F.* 1: 37–39.

Kumar, H. D. 1989. *Algal cell biology*. Affiliated East-West Press Private Limited, Delhi, vi + 186 pp.

Kuneida, H. 1934. On the life history of *Monostroma. Proc. Imp. Acad. Jpn.* 10: 103–106.

Kunth, C. S. 1822. *Synopsis plantarum.* Vol. 1. Paris, Levault.

Kuntze, O. 1891. *Revisio generum plantarum* . . . Part 2. Leipzig, pp. [375]–1011.

———. 1898. *Revisio generum plantarum* . . . Part 3 [3]. Arthur Felix: Leipzig, pp. i–iv + 1–576.

Kurihara, A., S. Arai, S. Shimada, and M. Masuda. 2005. The conspecificity of *Galaxaura apiculata* and *G. hystrix* (Nemaliales, Rhodophyta) inferred from comparative morphology and rbcL and ITS1 sequences. *Eur. J. Phycol.* 40: 39–52.

Kurz, H., and K. Wagner. 1957. Tidal marshes of the Gulf and Atlantic coasts of northern Florida and Charleston, South Carolina. Florida State Univ. Study no. 24, Tallahassee, 139 pp + 19 figs.

Kützing, F. T. 1833. Algologische Mitttheilugen. *Flora* 16: 513–521.

———. 1841('1842'). Ueber *Ceramium* Ag. *Linnaea* 15: 727–746.

———. 1843. *Phycologia* generalis (oder Anatomie, Physiologie und Systemkunde der Tange). Leipzig. xxxii + 458 [459= Verbesserungen] pp., 80 pls.

———. 1845. *Phycologia* germanica . . . Nordhausen, x + 340 pp.

———. 1847a. Diagnosen und Bemerkungen zu neuen oder kritischen Algen. *Bot. Zeitung (Berlin)* 5: 1–5, 22–25, 33–38, 52–55, 164–167, 177–180, 193–198, 219–223.
———. 1847b. *Tabulae phycologicae* . . . Vol. 1, fasc. 3–5. Nordhausen, pp. 17–36, pls. 21–50.
———. 1849. *Species algarum*. Lipsiae [Leipzig]. vi + 922 pp.
———. 1853. *Tabulae phycologicae* . . . Vol. 3. Nordhausen, 28 pp., 100 pls.
———. 1854. *Tabulae phycologicae* . . . Vol 4. Nordhausen, xvi + 23 pp., 100 pls.
———. 1855. *Tabulae phycologicae* . . . Vol 5. Nordhausen, ii + 30 pp., 100 pls.
———. 1856. *Tabulae phycologicae* . . . Vol 6. Nordhausen, iv + 35 pp., 100 pls.
———. 1857. *Tabulae phycologicae* . . . Vol. 7. Nordhausen, ii + 40 pp., 100 pls.
———. 1858. *Tabulae phycologicae* . . . Vol. 8. Nordhausen, ii + 48 pp., 100 pls.
———. 1859. *Tabulae phycologicae* . . . Vol. 9. Nordhausen, vii + 42 pp., 100 pls.
———. 1862. *Tabulae phycologicae* . . . Vol. 12. Nordhausen, iv + 30 pp., 100 pls.
———. 1863a. *Tabulae phycologicae* . . . Vol. 13. Nordhausen, iii + 31 pp., 100 pls.
———. 1863b. Diagnosen und Bemerkungen zu drei und siebenzig neuen Algenspecies. In: *Zu der öffentlichen Prüfung sämmtlicher Klassen der Realschule zur Nordhausen* . . . [1863]. Nordhausen, pp. 1–19. [Reprinted. Hedwigia 2: 86–95. 1863.]
———. 1864. *Tabulae phycologicae* . . . Vol. 14. Nordhausen, iii + 35 pp., 100 pls.
———. 1865. *Tabulae phycologicae* . . . Vol. 15. Nordhausen, iii + 36 pp., 100 pls.
———. 1866. *Tabulae phycologicae* . . . Vol. 16. Nordhausen, iii + 35 pp., 100 pls.
———. 1867. *Tabulae phycologicae* . . . Vol. 17. Nordhausen, iii + 30 pp., 100 pls.
———. 1868. *Tabulae phycologicae* . . . Vol. 18. Nordhausen, iii + 35 pp., 100 pls.
———. 1869. *Tabulae phycologicae* . . . Vol. 19. Nordhausen, iv + 36 pp., 100 pls.
Kylin, H. 1906. Zur Kenntnis einiger schwedischen *Chantransia*-Arten. In: Botaniska Studier tillägnade F. R. Kjellman den 4 November 1906. Uppsala, pp. 113–126.
———. 1907. Studien über die Algenflora der schwedischen Westküte. Akadem. Abhandl, iv + 287 pp., 7 pls., 6 tables, 41 text-figs. (Diss., Uppsala).
———. 1923. Studien über die Entwicklungsgeschichte der Florideen. *Kgl. Sven. Vetensk. Akad. Handl.* 63: 1–139, 82 Fig.
———. 1924. Studien über die Delesseriaceen. *Lunds Univ. Årsskr.*, N. F. Avd. 2, 20(6): 1–111.
———. 1925. The marine algae in the vicinity of the biological station at Friday Harbor, Wash. *Lunds Univ. Årsskr.*, N. F. Avd. 2, 21: 1–87.
———. 1931. Die Florideenordnung Rhodymeniales. *Lunds Univ. Årsskr.*, N. F. Avd. 2, 27(11): 1–48, 20 pls.
———. 1932. Die Florideenordnung Gigartinales. *Lunds Univ. Årsskr.*, N. F. Avd. 2, 28(8): 1–88 pp., 28 pls.
———. 1933. Über die Entwicklungsgeschichte der Phaeophyceen. *Lunds Univ. Årsskr.*, N. F. Avd. 2, 29(7): 1–102, 2 pls.
———. 1938a. Verzdichnis einiger Rhodophyceen von Südafrika. *Lunds Univ. Årsskr.*, N. F. Avd., 34(8). 25 [26] pp., 8 pls.
———. 1938b. Über die Chlorophyceengattungen *Entocladia* und *Ectochaete*. *Bot. Not.* 1938: 67–76.
———. 1940. Die Phaeophyceenordnung Chordariales. *Lunds Univ. Årsskr.*, N. F. Avd. 2, 36(9): 1–67, 8 pls.
———. 1941. Californische Rhodophyceen. *Acta Universitatis Lundensis* 37: 1–71, 7 Fig., 13 pls
———. 1944. Die Rhodophyceen der Schwedischen Westkueste. *Lunds Univ. Årsskr.*, N. F. Avd. 2, 40: 1–104, 32 pls.
———. 1947. Die Phaeophyceen der Schwedischen Westküste. *Lunds Univ. Årsskr.*, N.F. Avd. 2, 43: 1–99 pp., 18 pls.
———. 1949. Die Chlorophyceen der Schwedischen Westküste. *Lunds Univ. Årsskr.*, N. F. Avd. 2, 45(4): 1–79.
———. 1956. *Die Gattungen der Rhodophyceen*. C. W. K. Gleerups Förlag, Lund, xv + 673 pp.
Lagerheim, G. 1883. Bidrag till Sveriges Algflora. *Öfvers. K. Vetensk. Akad. Förh. Stockh.* 40: 37–78.
———. 1885. *Codiolum polyrhizum n. sp.* Ett bidrag till kännedomen om slägtet *Codiolum* A. Br. *Öfvers. Förh. Kongl. Svenska Vetensk.-Akad.* 42(8): 21–31, pl. 28.
Lam, D. W., and F. W. Zechman. 2006. Phylogenetic analyses of the Bryopsidales (Ulvophyceae, Chlorophyta) based on Rubisco large subunit gene sequences. *J. Phycol.* 42: 669–678.
Lamarck, J. B. de. 1813–1815. *Sur les polypiers empâtes*. Annalles du Muséum d'Histoire Naturelle [Paris] 20: 294–312, 370–386, 432–458 (1813); Mémoires du Muséum d' Histoire Naturelle [Paris] 1: 69–80, 162–168, 331–340 (1815).
———. 1816. *Histoire Naturelle des Animaux sans Vertèbres* . . . Vol. 2. Paris, 568 pp.

Lamarck, J. B. de, and A. P. DeCandolle. 1805. *Flore française*. Troisième édition. Vol. 2. Paris, xii + 600 pp.,[1] folded map.

Lamouroux, J. V. F. 1805. *Dissertations sur plusierus espèces* Fucus . . . Agen, xxiv + 83 [85 = Errata] pp., 36 pls.

———. 1809a. Exposition des caractères de genre *Dictyota*, et tableau de espèces qu'il renferme. *J. Bot.* [Desvaux] 2: 38–44.

———. 1809b. Mémoire sur trois nouveaux genres de la famille des algues marines. *J. Bot.* [Desvaux] 2: 129–135.

———. 1809c. Observations sur la physiologie des algues marines et description de cinq nouveaux genres de cette famille. *Nouv. Bull. Sci. Soc. Philom. Paris* 1: 330–333, pl. 6.

———. 1809d. Mémoire sur les *Caulerpes*, nouveau genre de la famille des Algues marines. *J. Bot.* [Desvaux] 2: 136–146.

———. 1812. Extrait d'un mémoire sur la classification des Polypiers coralligènes non entièrement pierreux. *Nouv. Bull. Sci. Soc. Philom. Paris* 3: 181–188.

———. 1813. Essai sur les genres de la famille des thalassiophytes non articulées. *Ann. Mus. Hist. Nat.* [Paris] 20: 21–47, 115–139, 267–293, plates 7–13 (reprint pp. 1–84).

———. 1816. *Histoire des polypiers coralligènes flexibles, vulgairement nommés zoophytes*. Caen., lxxxiv + 559 pp. [560 = errata] pp., 1 table, 19 pls.

———. 1821. *Exposition méthodique des genres de l'ordre des Polypiers* . . . des Zoophytes d'Ellis et Solander. Paris, viii + 115 pp., 84 pls.

———. 1824. *Delesseria*. Dictionnaire Classique d'Histoire Naturelle. [Paris] 5, pp. 386–388.

———. 1825. *Dictionnaire Classique d'Histoire Naturelle* [Paris] 7, pp. 1–626.

Lapointe, B. 1997. Nutrient thresholds for bottom-up control of macroalgal blooms on coral reefs in Jamaica and southeast Florida. *Limnol. Oceanogr.* 42: 1119–1131.

———. 1999. Simultaneous top-down and bottom-up forces control macroalgal blooms on coral reefs. (Reply to the comment by Hughes et al.). *Limnol. Oceanogr.* 44: 1586–1592.

Lapointe, B., and P. J. Barile. 1999. Seagrass die-off in Florida Bay: an alternative interpretion. *Estuaries* 22: 460–470.

———. 2001. Discrimination of nitrogen sources to harmful macroalgal blooms on coral reefs of southeast Florida. Final Report. Florida Institute of Oceanography, Subagreement #47 10 173 L3A, 38 pp.

———. 2004. "Comment on J. C. Zieman, J. W. Fourqurean, and T. A. Frankovich. 1999. Seagrass die-off in Florida Bay: long-term trends in abundance and growth of turtlegrass, *Thalassia testudinum* Estuaries 22: 460–470." *Estuaries* 27: 157–164.

Lapointe, B., P. J. Barile, and W. R. Matzie. 2004. Anthropogenic nutrient enrichment of seagrass and coral reef communities in the lower Florida Keys: discrimination of local versus regional nitrogen sources. *J. Exp. Mar. Biol. Ecol.* 308: 23–58.

Lapointe, B., P. J. Barile, M. M. Littler, and D. S. Littler. 2005a. Macroalgal blooms on southeast Florida coral reefs: II. Cross-shelf discrimination of nitrogen sources indicates widespread assimilation of sewage nitrogen. *Harmful Algae* 4: 1106–1122 (Online at *www.sciencedirect.com*).

Lapointe, B., P. J. Barile, M. J. Wynne, and C. S. Yentsch. 2005b. Reciprocal invasion: Mediterranean native *Caulerpa ollivieri* in the Bahamas supported by human nitrogen enrichment. *Aquatic Invaders* 16:1, 3–5.

Lapointe, B., and W. R. Matzie. 2002. Biotic phase-shifts in Florida Bay and fore reef communities of the Florida Keys: linkages with historical freshwater flows and nitrogen loading from Everglades runoff. In: J. W. Porter and K. G. Porter (eds.), *The Everglades, Florida Bay and coral reefs of the Florida Keys: an ecosystem sourcebook*. CRC Press, Boca Raton, Fla., pp. 629–648.

Lapointe, B., D. A. Tomasko, and W. R. Matzie. 1994. Eutrophication and tropic state classification of seagrass communities in the Florida Keys. *Bull. Mar. Sci.* 54: 696–717.

Larkum, A.W. D., A. J. McComb, and S. A. Shepherd (eds.). 1989. *Biology of seagrasses. A treatise on the biology of seagrasses with special reference to the Australian Region*. Elsevier, Amsterdam, 841 pp.

Larned, S. T. 1998. Nitrogen- versus phosphate-limited growth and sources of nutrients for coral reef macroalgae. *Mar. Biol.* 132: 269–287.

Lawson, G. W., and D. M. John. 1982. The marine algae and coastal environment of tropical West Africa. *Nova Hedwigia* Beih. 70: 455 pp.

———. 1987. The marine algae and coastal environment of tropical West Africa, 2nd ed. *Nova Hedwigia* Beih. 93: 415 pp.

Le Jolis, A. 1863. Liste des algues marines de Cherbourg. *Mém. Soc. Sci. Nat. Cherbourg* 10: 5–168, 6 pls.

Lee, H. B., and J. I. Kim. 1995. Notes on Gelidiales species from Korea. In: I. A. Abbott (ed.), *Taxonomy of economic seaweeds*. Vol. 5. California Sea Grant College Program, La Jolla, pp. 161–174.

Lee, I. K. 1978. Studies on Rhodymeniaceae from Hokkaido. *J. Fac. Sci. Hokkaido Imp. Univ., ser. 5, Bot.* 11: 1–194, 5 pls.

Lee, W. J., and K. S. Bae. 2002. Phylogenetic relationships among several genera of Dictyotaceae (Dictyotales, Phaeophyceae) based on 18 S rRNA and partial *rbc*L gene sequences. *Mar. Biol.* 140: 1107–1115.

Lee, Y.-P., and I. K. Lee. 1988. Contribution to the generic classification of the Rhodochortaceae (Rhodophyta, Nemaliales). *Bot. Mar.* 31: 119–131.

Lee, Y.-P., and T. Yoshida. 1997. The Acrochaetiaceae (Acrochaetiales, Rhodophyta) in Hokkaido. *Sci. Pap. Inst. Algol. Res. Fac. Sci. Hokkaido Imp. Univ.* 9: 155–229.

Leliaert, F., and E. Coppejans. 2003. The marine species of *Cladophora* (Chlorophyta) from the South African east coast. *Nova Hedwigia* 76: 45–82.

———. 2004. Crystalline cell inclusions: a new diagnostic character in the Cladophorophyceae (Chlorophyta). *Phycologia* 43: 189–203.

———. 2006. A revision of *Cladophoropsis* Børgesen (Siphonocladales, Chlorophyta). *Phycologia* 45: 657–679.

Leliaert, F., F. Rousseau, B. de Reviers, and E. Coppejans. 2003. Phylogeny of the Cladophorophyceae (Chlorophyta) inferred from partial LSU rRNA gene sequences: is the recognition of a separate order Siphonocladales justified? *Eur. J. Phycol.* 38: 233–246.

Lemoine, P. 1917. Corallinaceae. Subfam. 1 Melobesieae. In: F. Børgesen, The marine algae of the Danish West Indies. Part 3. Rhodophyceae (3). *Dan. Bot. Ark.* 3: 147–182.

———. 1928. Un nouveau genre de Mélobésiées: *Mesophyllum*. *Bull. Soc. Bot. Fr.* 75: 251–254.

Leonardi, P. I., E. J. Cáceres, and J. A. Correa. 2002. Ultrastructure and taxonomy of *Sporocladopsis novae-zelandiae* (Ulvophyceae, Chlorophyceae). *Bot. Mar.* 45: 324–330.

Levring, T. 1938. Verzeichnis einiger Chlorophyceen und Phaeophyceen von Südafrika. Univ. Lund. Ny Följd, Avd 2, 34(9): 1–25 pp., 10 figs., 4 pls.

———. 1940a. Die Phaeophyceengattung *Chlanidophora*, *Distromium* und *Syringoderma*. Kongl. Fysiogr. Sällsk i Lund Föhandl. 10: 217–227 [11 pp. as separate paper].

———. 1940b. Studien über die algenvegeation von Blekinge, Südschweden. Akademische Abhand. Lund Univ., 179 pp.

———. 1945. Marine algae from some Antarctic and sub-Antarctic islands. *Lunds Univ. Årsskr.*, N. F. Avd., 41(7): 1–36.

———. 1953. The marine algae of Australia. I. Rhodophyta: Goniotrichales, Bangiales and Nemalionales. *Arkiv für Botanik*, ser. 2, 2: 457–530.

Lewis, I. F., and H. F. Bryan. 1941. A new protophyte from the Dry Tortugas. *Am. J. Bot.* 28: 343–348.

Lewis, J. A., and H.B.S. Womersley. 1994. Family Phyllophoraceae Nägeli 1847: 248. In: H.B.S. Womersley, *The marine benthic flora of southern Australia. Rhodophyta*. Part IIIa, Bangiophyceae and Florideophyceae (Acrochaetales, Nemaliales, Gelidiales, Hildenbrandiales and Gigartinales *sensu lato*). Australian Biological Resources Study, Canberra, pp. 259–270.

Lewis, J. E., and J. N. Norris. 1987. A history and annotated account of the benthic marine algae of Taiwan. *Smithson. Contrib. Mar. Sci.* 29: 1–38.

Lewis, R. R., III, R. G. Gilmore Jr., D. W. Crewz, and W. E. Odum. 1985a. Mangrove habitat and fishery resources of Florida. In: W. Seaman Jr. (ed.), *Florida aquatic habitat and fishery resources*. Florida Chapter of American Fisheries Society, Kissimee, pp. 281–336.

Lewis, R. R., M. J. Durako., M. D. Moffler, and R. C. Phillips. 1985b. Seagrass meadows of Tampa Bay. In: S. F. Treat, J. L. Simon, R. R. Lewis, and R. L. Whitman Jr. (eds.): *Proceedings, Tampa Bay Area scientific information symposium*. Burgess Publ. Co., Minneapolis, pp. 210–246.

Lewmanomont, K., L. Wongrat, and C. Supanwanid. 1995. *Algae in Thailand*. Office of Environmental Policy and Planning, Bangkok, 334 pp.

L'Hardy-Halos, M.-T. 1965. Sur trois Callithamniées des environs de Roscoff. *Cah. Biol. Mar.* 6: 117–134.

L'Hardy-Halos, M.-T., and J. Rueness. 1990. Comparative morphology and crossability of related species of *Aglaothamnion* (Rhodophyta). *Phycologia* 29: 351–366, 2 tables.

Li, W. X., and Z. F. Ding. 1998. A new genus of Cryptonemiaceae—*Sinotubimorpha*. *J. Zhanjiang Ocean Univ.* 18(3): 1–5 (in Chinese).

Liao, L. M., and M. H. Hommersand. 2003. A morphological study and taxonomic reassessment of the generitype species in the Gracilariaceae. *J. Phycol.* 39: 1207–1232.

Lightfoot, J. 1777. *Flora scotica . . .* London. xli + 1151 [+ 24] pp., 35 pls.

Lin, S.-M., S. Fredericq, and M. H. Hommersand. 2001. Systematics of the Delesseriaceae (Ceramiales, Rhodophyta) based on large subunit rDNA and *rbcL* sequences, including the Phycodryoideae, *subfam. nov. J. Phycol.* 37: 881–899.

Lin, S-M., M. H. Hommersand, and S. Fredericq. 2005. Molecular systematics and biogeography of *Martensia* (Delesseriaceae, Rhodophyta). *Phycologia* 44(4, supplement): 62.

Lindley, J. 1846. *The vegetable kingdom.* Bradbury and Evans, London, lxviii + 908 pp.

Lindstrom, S. C. 1987. Acrosymphytaceae, a new family in the order Gigartinalies *sensu lato* (Rhodophyta). *Taxon* 36: 50–53.

Link, H. F. 1820. Epistola ad virum celeberrimum Nees ab Esenbeck . . . de algis aquaticis, in genera disponendis. In: C. G. Nees (ed.), *Horae Physicae Berolinensis . . .* Bonnae [Bonn.], pp. 1–8, pl. I.

———. 1833. *Handbuch zur Erkennung der nutzbarsten und am häufigsten vorkommenden Gewäche.* Part 3. Berlin, xviii + 536 pp.

Linnaeus, C. 1753. *Species plantarum . . .* Vol. 2. Holmiae [Stockholm], pp. 561–1200 + [1–31], 1 pl.

———. 1758. *Systema naturae per regna tria naturae . . .* Editio decima . . . Vol. I. Holmiae [Stockholm], pp. iv + 823 pp. [824 = Errata]

———. 1761. *Fauna Suecica*, 2nd ed., Stockholm, pp. i–xlvv + 1–578 pp., 2 pls.

———. 1766–1768. *Systema naturae per regna tria naturae, secundum classes, ordines, genera, species, cum characteribus, differentiis, synonyumis locis.* Ed. 12. Vols. 1–3. Salvii, Stockholm. Vol. 1, *Regnum animale*, part 1, 1766, pp. [2], [1]–532; part 2, 1767a, pp. [i], 533–1327. [1328], [1–36]. Vol. 2, *Regnum vegetabile*, 1767b, pp. [1]–736, [1–16]. Vol. 3, *Regnum Lapideum*, 1768, pp. [1]–236, [1–18], pls. 1–3.

Lipkin, Y., and P. C. Silva. 2002. Marine algae and seagrasses of the Dahlak Archipelago, southern Red Sea. *Nova Hedwigia* 75: 1–90.

Lirman, D., and P. Biber. 2000. Seasonal dynamics of macroalgal communities of the northern Florida reef tract. *Bot. Mar.* 43: 305–314.

Littler, D. S., and M. M. Littler. 1980. The evolution of thallus form and survival strategies in benthic marine macroalgae: field and laboratory test of a functional form model. *Am. Nat.* 116: 25–44.

———. 1990a. Systematics of *Udotea* species (Bryopsidales, Chlorophyta) in the tropical western Atlantic. *Phycologia* 29: 206–252.

———. 1990b. Re-establishment of the green algal genus *Rhipidosiphon* Montagne (Udoteaceae, Bryopsidales) with a description of *Rhipidosiphon floridensis sp. nov. Br. Phycol. J.* 25: 33–38.

———. 1991. Systematics of *Anadyomene* (Bryopsidales, Chlorophyta) in the tropical western Atlantic. *J. Phycol.* 27: 101–118.

———. 1992. Systematics of *Avrainvillea* species (Anadyomenaceae, Chlorophyta) in the tropical western Atlantic. *Phycologia* 31: 375–418.

———. 1997. An illustrated marine flora of the Pelican Cays, Belize. *Bull. Biol. Soc. Wash.* 9: 1–149.

———. 2000. *Caribbean reef plants.* Off Shore Graphics Inc., Washington, D.C., 542 pp.

———. 2003. *South Pacific reef plants.* Off Shore Graphics Inc., Washington, D.C., 331 pp.

———. 2004. *Taonia abbottiana sp. nov.* (Dictyotales, Phaeophyceae) from the tropical western Atlantic. *Cryptogam. Algol.* 25: 419–427.

Littler, D. S., M. M. Littler, K. E. Bucher, and J. N. Norris. 1989. *Marine plants of the Caribbean. A field guide from Florida to Brazil.* Smithsonian Institution Press, Washington, D.C., 263 pp.

Littler, M. M., and D. S. Littler. 2003. *Waterways and byways of the Indian River Lagoon: field guide for boaters, anglers and naturalists.* Offshore Graphics, Inc., Washington, D.C., 71 pp.

Littler, M. M., D. S. Littler, and B. L. Brooks. 2004. Extraordinary mound-building forms of *Avrainvillea* (Bryopsidales, Chlorophyta): their experimental taxonomy, comparative functional morphology and ecological strategies. *Atoll Res. Bull.* 515, National Museum of Natural History, Smithsonian Institution, Washington, D.C., 26 pp.

Livingston, R. J. 1990. Inshore marine habitats. In: R. L. Meyers and J. J. Ewel (eds.), *Ecosystems of Florida.* University of Central Florida Press, Orlando, pp. 549–573.

———. 2006. *Restoration of aquatic systems.* CRC Marine Science Series, Taylor and Francis Group, Boca Raton, Fla., 423 pp.

Loiseaux, S. 1967a. Morphologie et cytologie des Myrionémacées. Critères taxonomiques. *Rev. Gén. Bot.* 74: 329–347.

———. 1967b. Recherches sur les cycles de développement des Myrionématacées (Phéophycées). I and II. Hecatonématées et Myrionématées. *Rev. Gén Bot.* 74: 529–576.

———. 1969. Sur une espece de *Myriotrichia* obtenue en culture à partir zoïdes d' *Hecatonema maculans* Sauv. *Phycologia* 8: 11–15.
Lokhorst, G. M. 1978. *Taxon*omic studies on the marine and brackish-water species of *Ulothrix* (Ulotrichales, Chlorophyceae) in western Europe. *Blumea* 24: 191–299.
———. 1991. Synopsis of genera of Klebsormidiales and Ulotrichales. *Cryptogam. Bot.* 2: 274–288.
Lokhorst, G. M., and B. J. Trask. 1974. Taxonomic studies on the genus *Ulothrix* (Ulotrichales, Chlorophyceae) III. *Acta Bot. Neerl.* 23: 561–602.
López-Bautista, J., and D. F. Kapraun. 1995. Agar analysis, nuclear genome quantification and characteization of four agarophytes (*Gracilaria*) from the Mexican Gulf Coast. *J. Appl. Phycol.* 7: 351–357.
López-Piñero, I. Y., and D. L. Ballantine 2001. Ontogenesis and life history in culture of *Dasya caraibica* (Dasyaceae, Rhodophyta). *Phycologia* 40: 47–52.
Loureiro, J. de. 1790. *Flora Conchinchinensis: Sistens Plantas in regno* . . . Systema sexuale Linnaeanum. Ulyssipone, 744 pp.
Lucas, A.H.S., and F. Perrin. 1947. *The seaweeds of South Australia.* Part 2. The red seaweeds. Govt. Printer, Adelaide, pp. 107–458.
Lugo, A. E. 1986. Mangrove understory: an expensive luxury? *J. Trop. Ecol.* 2: 287–288.
Lugo, A. E., and S. C. Snedaker. 1974. The ecology of mangroves. *Ann. Rev. Ecol. Syst.* 5: 39–64.
Lukas, K. J. 1974. Two species of the chlorophyte genus *Ostreobium* from skeletons of Atlantic and Caribbean reef corals. *J. Phycol.* 10: 331–335.
Lund, S. 1959. The marine algae of East Greenland. I. Taxonomical part. *Medd. om Grønland* 156: 1–247.
Lyle, L. 1922. *Antithamnionella*, a new genus of algae. *J. Bot.* 60: 346–350.
Lyngbye, H. C. 1819. *Tentamen hydrophytologiae danicae* . . . Hafniae [Copenhagen], xxxii + 248 pp., 70 pls.
Mackay, H. T. 1836. *Flora Hibernica, comprising the Flowering Plants Ferns, Characeae, Musci Hepatiae Lichens and Algae of Ireland.* Fraser, Edinburgh, pp. xxxiv + 1–354 [part 1], 1–156 [part 2], 157–254 [part 3, algae], 255–279 [addenda + index].
Maggs, C. A. and M. D. Guiry. 1988('1987'). *Gelidiella calcicola sp. nov.* (Rhodophyta) from the British Isles and northern France. *Br. Phycol. J.* 22: 417–434.
Maggs, C. A. and M. H. Hommersand. 1993. *Seaweeds of the British Isles.* Volume 1. Rhodophyta. Part 3A. Ceramiales. British Museum (Natural History), London. xv + 444 pp.
Maggs, C. A. and C. M. Pueschel. 1989. Morphology and development of *Ahnfeltia plicata* (Rhodophyta): proposal of Ahnfeltiales *ord. nov. J. Phycol.* 25: 333–351.
Maggs, C. A., B. A. Ward, L. M. McIvor, C. M. Evans, J. Rueness, and M. J. Stanhope. 2002. Molecular analyses elucidate the taxonomy of fully corticated, nonspiny species of *Ceramium* (Ceramiaceae, Rhodophyta) in the British Isles. *Phycologia* 41: 409–420.
Magne, F. 1964. Recherches caryologiques chez les Floridées (Rhodophycées). *Cah. Biol. Mar.* 5: 461–617.
———. 1989. Classification et phylogénie des Rhodophycées. *Cryptogam. Algol.* 10: 101–115.
Malta, E.-J., S.G.A. Draisma, and P. Kamermans. 1999. Free-floating *Ulva* in the southwest Netherlands: species or morphotypes? A morphological, molecular and ecological comparison. *Eur. J. Phycol.* 34: 443–454.
Malta, E.-J., D. G. Ferreira, J. J. Vergara, and J. L. Perez. 2005. Nitrogen load and irradiance affect morphology, photosynthesis, and growth of *Caulerpa prolifera* (Bryopsidales, Chlorophyta). *Mar. Ecol. Prog. Ser.* 298: 101–114.
Marchewianka, M. 1925. Z flory glonów polskiego Baltyku. (Beiträge zu Algenflora der Ostsee). *Spraw. Komis. Fizogr.* 58/59: 33–45
Marcot-Coqueugniot, J. 1991. A preliminary list of marine algae from the Banc d'Arguin (Mauritania). *Bot. Mar.* 34: 195–199.
Mario, C., F. Giovanni, G. Giuseppe, and S. Donatella. 2004. Alien macrophytes in the Mediterranean Sea: a review. *Recent Res. Dev. Environ. Biol.* 1: 153–202.
Martens, G. von. 1868. Die Tange. In: *Die Preussische Expedition nach Ost-Asien. Nach amtlichen Quellen.* Botanischer Theil. Berlin, pp. 1–152, 8 pls. [Title page dated 1866, but published in 1868.].
———. 1869. Beiträge zur Algen-Flora Indiens. *Flora* (Jena) 52: 233–238.
Martinez-Daranas, B., M. Esquivel, and M. Hernández. 2001. Nuevos registros de algas marinas cubanas. *Avicennia* 14: 135–137.
Martius, C. F. P. 1828–1834. *Icones plantarum cryptogamicarum* . . . Monachii [München], 138 pp., 76 pls.
———. 1833. *Flora brasiliensis* . . . Vol. I. Pars prior. Algae, lichenes, hepaticae. Stuttgartiae [Stuttgart] et Tubingae [Tübingen], iv + 390 pp.
Masaki, T. 1968. Studies on the Melobesioideae of Japan. *Mem. Fac. Fish. Hokkaido Univ.* 16: 1–80.

Masuda, M. 1993. *Ahnfeltiopsis* (Gigartinales, Rhodophyta) in the western Pacific. Jpn. *J. Phycol.* 41: 1–6.

Masuda, M., and M. D. Guiry. 1995. Reproductive morphology of *Itonoa marginifera* (J. Agardh) *gen. et comb. nov.* (Nemastomatceae, Rhodophyta). *Eur. J. Phycol.* 30: 57–67.

Mateo-Cid, L. E., A. C. Mendoza-González, and R. B. Searles. 2002. New Mexican records of marine algae including *Crouania mayae sp. nov.* (Ceramiaceae, Rhodophyta). *Caribb. J. Sci.* 38: 205–221.

Mathieson, A. C. 1966. Morphological studies of the marine brown alga *Taonia lennebackerae* Farlow *ex* J. Agardh. I. Sporophytes, abnormal gametophytes and vegetative reproduction. *Nova Hedwigia* 12: 65–79.

Mathieson, A. C., and C. J. Dawes. 1975. Seasonal studies of Florida sublittoral marine algae. *Bull. Mar. Sci.* 25: 46–65.

———. 2002. *Chaetomorpha* balls foul New Hampshire, USA beaches. *Algae* 17: 283–292.

Mathieson, A. C., C. A. Penniman, and L. G. Harris. 1991. Northwest Atlantic rocky shore ecology. In: A. C. Mathieson and P. H. Nienhuis (eds.), *Ecosystems of the world*. Vol. 24, *Intertidal and littoral ecosystems*. Elsevier, Amsterdam, pp. 109–191.

May, V. 1965. A census and key to the species of Rhodophyceae (red algae) recorded from Australia. *Contrib. N. S. W. Natl. Herb.* 3: 349–429.

Mayhoub, H. 1974a. Observations et explication au sujet de l'ecologie et de la répartition des *Pseudobryopsis myura* (Ag.) Berthold (Codiales, Bryopsidacées). *Bull. Soc. Phycol. Fr.* 19: 164–167.

———. 1974b. Reproduction sexuée et cycle du développement de *Pseudobryopsis myura* (Ag.) Berthold (Chlorophycée, Codiale). *C. R. Hebd. Séances Acad. Sci. Paris* ser. D, 278: 867–870.

Mazé, H. 1868. *Hydrophytes de la Guyane Française* . . . Basse-Terre, Guadeloupe, iii + 41 pp.

Mazé, H., and A. Schramm. 1878 ['1870–1877']. *Essai de classification des algues de la Guadeloupe*, "Second edition." Government Printer, Basse-Terre (Guadelouope). xix + iii [= Errata] + 283 pp. (Although the second edition, the first edition of this work was originally issued with A. Schramm as the senior author).

McLachlan, J. 1979. *Gracilaria tikvahiae sp. nov.* (Rhodophyta, Gigartinales, Gracilariaceae) from the northwestern Atlantic. *Phycologia* 18: 19–23.

McLachlan, J., L. C-M. Chen, and T. Edelstein. 1970. The life history of *Microspongium* sp. *Phycologia* 10: 83–87.

Melvill, J. C. 1875. Notes on the marine algae of South Carolina and Florida. *J. Bot., British and Foreign* 12: 258–265.

Mendoza-González, A. C., L. E. Mateo-Cid, R. Aguilar-Rosas, and L. E. Aguilar-Rosas. 2001 ['2000']. La familia Sphacelariaceae (Sphacelariales, Phaeophyta) en las costas de México. *Polibotánica* 11: 21–48.

Mendoza-González, A. C., L. E. Mateo-Cid, and R. B. Searles. 2000. New records of benthic marine algae from Isla Cozumel, Mexico: Phaeophyta and Chlorophyta. *Bull. Mar. Sci.* 66: 119–130.

Meneghini, G. 1840a. Lettera del Prof. Giuseppe Meneghini al Dott. Iacob Corinaldi a Pisa. Pisa. Folded sheet without pagination, Bibliothèque Thuret-Bornet, Muséum National d'Histoire Naturelle, Paris.

———. 1840b. Botanische Notizen. *Flora* (Jena) 23: 510–512. [this appears to be the published form of Meneghini 1840a.

———. 1841. Memoria diretta a mostrare i rapporti di organizzazione fra le Alghe propriamente dette o Ficee, e le Alghe terrestri o Licheni. *Atti Riunione Sci. Ital.* 3: 417–431.

———. 1844. Algarum species novae vel minus notae a prof. J. Meneghini propositae. *G. Bot. Ital.*, Anno 1, 1(1): 296–306.

Menesguen, A. 1992. Modelling coastal eutrophication: The case of French *Ulva* mass blooms. In: R. A. Vollenweider, R. Marchetti and R. Viviani (eds.), *Symposium on marine coastal eutrophication*, Bologna, Italy, 21–24 March 1990. *Sci. Total Environ.*, suppl. 1992, pp. 979–992.

Millar, A.J.K. 1990. Marine red algae of Coffs Harbour region, northern New South Wales. *Aust. Syst. Bot.* 3: 293–593.

———. 1996. *Dasya roslyniae sp. nov.* (Dasyaceae, Rhodophyta), with a discussion on generic distinctions among *Dasya*, *Eupogodon*, *Rhodoptilum*, and *Pogonophorella*. *J. Phycol.* 32: 145–157.

———. 2000. *Veleroa magneana* (Brongniartelleae, Ceramiales), a new red alga species from the Coral Sea, South Pacific. *Cryptogam. Algol.* 21: 157–165.

Millar, A.J.K., and G. T. Kraft. 1993. Catalogue of marine and freshwater red algae (Rhodophyta) of New South Wales, including Lord Howe Island, south-western Pacific. *Aust. Syst. Bot.* 6: 1–90.

———. 2001. Monograph of the green macroalgal genus *Riphilia* (Udoteaceae, Halimedales), with a description of *R. crassa sp. nov.* from Australia and the Philippines. *Phycologia* 40: 21–34.

Min-Thein, U., and H.B.S. Womersley. 1976. Studies on the southern Australia taxa of Solieraceae, Rhabdoniaceae and Rhodophyllidaceae (Rhodophyta). *Aust. J. Bot.* 24: 1–166.

Misra, J. N. 1966. *The Phaeophyceae in India.* Indian Council of Agricultural Research, New Delhi, x + 203 pp., including 6 pls., 100 figs., 5 tables.

Mitsch, W. J., and J. G. Gosselink. 1993. *Wetlands*, 2nd ed. Van Nostrand Reinhold, New York, xiii + 722 pp.

Möbius, M. 1889. Bearbeitung der von H. Schenck in Brasilien gesammelten Algen. *Hedwigia* 18: 309–358, [1] fig., pls. X, XI.

Moe, M. A., Jr. 1963. *A survey of offshore fishing in Florida.* Florida State Board of Conservation Marine Laboratory. Professional Paper Series No. 4, 117 pp.

Moffler, M. D., and J. F. van Breedveld. 1979. *Nearshore marine ecology at Hutchinson Island, Florida: 1971–1974.* X. A species list of marine benthic macroalgae collected off Hutchinson Island Florida. Fla. Mar. Res. Publ. 34. St. Petersburg, iv + 84 pp

Monotilla, W. D., and M. Notoya. 2004. Morphological and physiological responses of *Porphyra suborbiculata* Kjellman (Bangiales, Rhodophyta) blades from five localities. *Bot. Mar.* 47: 323–334.

Montagne, C. 1837–1838. Centurie de plantes cellulaires exotiques nouvelles. *Ann. Sci. Nat., Bot.*, sér. 2, 8: 345–370.

———. 1839–1841. Plantae cellulares. In: P. Barker-Webb and S. Berthelot (ed.), *Histoire naturelle des Iles Canaries.* Vol. 3, part 2, sect. 4. Paris, xv + 208 pp., 9 pls. [pp. 1–16 (1839), 17–160 (1840), 161–208, I–XV (1841).].

———. 1840. Seconde centurie de plantes cellulaires exotiques nouvelles. Décades I et II. *Ann. Sci. Nat., Bot.*, sér. 2, 13: 193–207, pls. 5, 6.

———. 1842a. *Prodromus generum specierumque phycearum novarum, in itinere ad polum antarcticum . . .* collectarum . . . Parisiis [Paris], 16 pp.

———. 1842b. *Troisième centurie de plantes cellulaires exotiques nouvelles.* Décades V, VI, VII et VIII. Ann. Sci. Nat., Bot., Sér. 2, 18: 241–256, 7 pls.

———. 1842c. Botanique. Plantes cellulaires. In: R. de la Sagra (ed.), *Histoire physique, politique et naturelle de l'île de Cuba.* Paris [Vol. II.], x + 549 pp.; Atlas: pls. I–XX.

———. 1842d. *Bostrychia.* Dictionnaire Universel d'Histoire Naturelle. In A. d'Orbigny. Vol. 2: 660–661.

———. 1843. Quatrième centurie de plantes cellulaires exotiques nouvelles. Décades VIII, IX et X. *Ann. Sci. Nat., Bot.*, sér. 2, 20: 352–379, pls. 15, 16.

———. 1846. Ordo I. Phyceae Fries. In: M. C. Durieu de Maisonneuve (ed.), *Exploration scientifique de l'Algérie pendant les années 1840, 1841, 1842 . . . Science physiques. Botanique, Cryptogamie.* Paris, pp. [1]–197, pls. 1–16.

———. 1847. Phycologie. *Dictionnaire Universel d'Histoire Naturelle* [Orbigny] 10: 14–57.

———. 1849. Sixième centurie de plantes celulaires nouvelles, tant indigènes qu'éxotiques. Décades III à VI. *Ann. Sci. Nat., Bot.*, sér. 3, 11: 33–66.

———. 1850. Cryptogamia guayanensis, seu plantarum cellularium in Guyana gallica annis 1835–1849 a Cl. Leprieue collectarum enumeratio universalis. *Ann. Sci. Nat., Bot.*, sér. 3, 14: 283–309.

———. 1856. *Sylloge generum specierumque cryptogamarum* . . . Parisiis [Paris], xxiv + 498 pp.

———. 1857. Huitième centurie de plantes cellulaires nouvelles, tant indigègnes qu'exotiques. Décades VI et VII (1). *Ann. Sci. Nat., Bot.*, sér. 4, Vol. 8: p. 285–310.

———. 1861('1860'). Neuvième centurie de plantes cellulaires nouvelles tant indigènes qu' exotiques. Décades I et II. *Ann. Sci. Nat., Bot.*, sér. 4, 14: 167–185, pls. X, XI.

Montague, C. L., and R. C. Wiegert. 1990. Salt marshes. In: R. L. Myers and J. J. Ewel (eds.), *Ecosystems of Florida.* Univ. Central Florida Press, Orlando, pp. 481–516.

Montague, C. L., A. V. Zale, and H. F. Percival. 1987. Ecological effects of coastal marsh impoundments: A review. *Environ. Manage.* 11: 734–756.

Moon, R. E., and C. J. Dawes. 1976. Pigment changes and photosynthetic rates under selected wavelengths in the growing tips of *Eucheuma* during vegetative growth. *Br. Phycol. J.* 11: 165–174.

Morand, P., and X. Briand. 1996. Excessive growth of macroalgae: A symptom of environmental disturbance. *Bot. Mar.* 39: 491–516.

Moreira, L., and A. M. Suárez. 2002a. Estudio de género *Sargassum* C. Agardh, 1820 (Phaeophyta, Fucales, Sargassaceae) en aguas Cubanas. 1. *Sargassum furcatum* Kützing, nuevo reporte. *Rev. Investig. Mar., Univ. Habana* 23: 53–54.

———. 2002b. Estudio de género *Sargassum* C. Agardh, 1820 (Phaeophyta, Fucales, Sargassaceae) en aguas Cubanas. 2. Variaciones morfológicas en *Sargassum natans* (Linnaeus) Meyer en dos localidades. *Rev. Investig. Mar., Univ. Habana* 23: 55–58.

———. 2002c. Estudio de género *Sargassum* C. Agardh, 1820 (Phaeophyta, Fucales, Sargassaceae) en aguas Cubanas. 3. Variaciones morfológicas en *Sargassum filipendula* C. Agardh. *Rev. Investig. Mar., Univ. Habana* 23: 59–62.

———. 2002d. Estudio de género *Sargassum* C. Agardh, 1820 (Phaeophyta, Fucales, Sargassaceae) en aguas Cubanas. 4. Reproducción sexual en *Sargassum natans* (Linnaeus) Meyer y *S. fluitans* Borgesen. *Rev. Investig. Mar., Univ. Habana* 23: 63–65.

Moris, G., and G. De Notaris. 1839. Florula caprariae sive enumeratio plantarum in insula Capraria vel sponte nascentium vel ad utilitatem latius excultarum. *Mem. R. Accad. Sci. Torino*, ser. 2, 2 (Cl. Sc. Fis. e Mat.): 59–300, pls. I–VI.

Morrill, J. F. 1976. Comparative morphology and taxonomy of some dorsiventral and parasitic Rhodomelaceae. Ph.D. diss., Univ. North Carolina, Chapel Hill, ix + 224 pp., 90 pls.

Mshigeni, K. E., and D. Chapman. 1994. *Hypnea* (Gigartinales, Rhodophyta). In: I. Akatsuka (ed.), *Biology of economic algae*. SPB Academic Publishing, The Hague, pp. 245–281, 5 tables.

Müller, K. M., K. M. Cole, and R. G. Sheath. 2003. Systematics of *Bangia* (Bangiales, Rhodophyta) in North America. II. Biogeographical trends in karyology: chromosome numbers and linkage with gene sequence phylogenetic trees. *Phycologia* 42: 209–219.

Müller, K. M., R. G. Sheath, M. L. Vis, T. J. Crease, and K. M. Cole, 1998. Biogeography and systematics of *Bangia* (Bangiales, Rhodophyta) based on the Rubisco spacer, rbcL gene sequences and morphometric analysis. 1. North America. *Phycologia* 37: 195–207.

Müller, K. M., S. L. Thompson, J. J. Cannone, and R. G. Sheath. 2001. A molecular phylogenetic analysis of the Bangiales (Rhodophyta) and description of a new genus *Pseudobangia*. *Phycologia* 40(suppl.): 21.

Müller, O. F. 1778. *Icones plantarum . . . Florae danicae*. Vol. 5, fasc. 13. Havniae [Copenhagen], 8 pp., pls. 721–840.

Munier-Chalmas. 1879. Observations sur les algues calcaires confondues avec les Foramnifères et appartenant au groupe des Siphonées dichotomes. *Bull. Soc. Géol. Fr.*, III, 7: 661–670.

Murray, G. 1887. Catalogue of Ceylon algae in the Herbarium of the British Museum. *Ann. Mag. Nat. Hist. (Lond.)* ser. 5, 20: 21–44.

———. 1888. Catalogue of the marine algae of the West Indian region. *J. Bot.* [Lond.] 26: 193–196, 237–243, 303–307, 331–338, 358–363, pl. 284.

———. 1889. Catalogue of the marine algae of the West Indian region. *J. Bot.* [Lond.] 27: 237–242, 257–262, 298–305.

———. 1891. On new species of *Caulerpa*, with observations on the position of the genus. *Trans. Linn. Soc. Lond., Ser. 2, Bot.* 3: 207–213, pls. LII, LIII.

Murray, G., and L. A. Boodle. 1888. A structural and systematic account of the genus *Struvea*. *Ann. Bot.* 2: 265–282, pl. XVI.

———. 1889. A systematic and structural account of the genus *Avrainvillea* Decaisne. *J. Bot.* [Lond.] 27: 67–72, 97–101, pls. 288, 289.

Nägeli, C. 1846. *Herposiphonia*. *Zahrb. Wiss. Bot.* 1(3/4): 238–256, pl. 8.

———. 1847. Die neuern Algensysteme und Versuch zur Begründung eines eigenen Systems der Algen und Florideen. Neue Denkschriften der Allg. Schweizerischen Gesellschaft für die Gesammten Naturwissenschaften 9[2], 275 pp., 10 pls.

———. 1849. Gattungen einzelliger Algen physiologisch und systematisch bearbeitet. Neue Denkschr. Allg. Schweiz. Ges. Ges. Naturw. 10: 1–139.

———. 1858. Pflanzenphysiologische Untersuchungen (by C. Nägeli and C. Cramer). 2. Heft. Die Stärkekörner. Zürich, x + 623 pp., pls. XI–XXVI.

———. 1862. Beiträge zur Morphologie und Systematik der Ceramiaceae. Sitzungber. Königl. Bayer. Akad. Wiss. München 1861(2): 297–425.

Nägeli, C., and C. Cramer. 1855. Pflanzenphysiologische Untersuchungen. 1. Heft. Zürich, vi + 120 pp., pls. I–X.

———.1858. Pflanzenphysiologische Untersuchungen. 2. Heft. Die Stärkekörner. Zürich, x + 623 pp., pls. XI–XXVI.

Nakamura, Y. 1972. A proposal on the classification of the Phaeophyta. In: I. A. Abbott and M. Kurogi, (eds.), *Contributions to the systematics of benthic marine algae of the North Pacific*. Japanese Society of Phycology, Kobe, pp. 147–155.

Nam, K. W. 1999. Morphology of *Chondrophycus undulata* and *C. parvipapillata* and its implications for the taxonomy of the *Laurencia* (Ceramiales, Rhodophyta) complex. *Eur. J. Phycol.* 34: 455–468.

———. 2004. *Laurencia* complex: morphology of *Laurencia pinnata* (Rhodophyta), with a key to Korean species. In: I. A. Abbott and K. J. McDermid (eds.), *Taxonomy of economic seaweeds with reference to the Pacific and other locations*. Vol. 9. Hawaii Sea Grant College Program, Univ. Hawaii, Hilo, pp. 255–271.

——-. 2006. Phylogenetic re-evaluation of the *Laurencia* complex (Rhodophyta) with a description of *L. succulenta* sp. nov. from Korea. *J. Applied Phycol.* 18: 679–697.

———. 2007. Validation of the generic name *Palisada* (Rhodomelaceae, Rhodophyta). *ALGAE* 22: 52–53.

Nam, K. W., C. A. Maggs, and D. J. Garbary, 1994. Resurrection of the genus *Osmundea* with an emendation of the generic delineation of *Laurencia* (Ceramiales, Rhodophyta). *Phycologia* 33: 384–395.

Nardo, G. D. 1834. De novo genere algalrum cui nomen est *Hildenbrandtia prototypus. Isis* [Oken] 1834: 675–676.

Nasr, A. H. 1941. Some new and little known algae from the Red Sea. *Rev. Algol.* 12: 57–76, pl 2.

Neefus, C. D., A. C. Mathieson, C. Yarish, A. S. Klein, A. West, B. Teasdale, and E. J. Hehre. 2000. Five cryptic species of *Porphyra* from the Northwest Atlantic. *J. Phycol.* 36: (suppl.): 73.

Nees, C. G. 1820. *Horae physicae berolinensis* . . . Bonnae [Bonn], [xii] + 123 + [4] pp., 27 pls.

Nelson, W. A., T. J. Farr, and J.E.S. Broom. 2006. Phylogenetic relationships and generic concepts in the red order Bangiales: Challenges ahead. *Phycologia* 45: 249–259.

Nelson, W. A., G. A. Knight, and M. W. Hawkes. 1998. *Porphyra lilliputiana sp. nov.* (Bangiales, Rhodophyta): A diminutive New Zealand endemic with novel reproductive biology. *Phycol. Res.* 46: 57–61.

Neto, A. I., D. C. Cravo, and R. T. Haroun. 2001. Checklist of the benthic marine plants of the Madeira Archipelago. *Bot. Mar.* 44: 391–414.

Newton, L. 1931. *A handbook of the British seaweeds*. British Museum (Natural History), London, xiii + 478 pp.

Nielsen, R. 1972. A study of the shell-boring marine algae around the Danish Island Laesø. *Bot. Tidsskr.* 67: 245–269.

———. 1979. Culture studies on the type species of *Acrochaete, Bolbocloeon and Entocladia* (Chaetophoraceae, Chlorophyceae). *Bot. Not.* 132: 441–449.

———. 1983. Culture studies of *Acrochaete leptochaete comb. nov.* and *A. wittrockii comb. nov.* (Chaetophoraceae, Chlorophyceae). *Nord. J. Bot.* 3: 689–694.

———. 1994. New combinations within the genus *Colaconema* (Acrochaetiaceae, Bangiophyceae). *Nord. J. Bot.* 14: 715

Nielsen, R., and K. Gunnarsson. 2001. Seaweeds of the Faroe Islands: an annotated checklist. *Fródskaparrit* 49: 45–108.

Nielsen, R., and I. R. Price. 2001. Typification of *Caulerpa cupressoides* (Vahl) C. Agardh and *C. taxifolia* (Vahl) C. Agardh (Chlorophyta, Caulerpaceae). *Taxon* 50: 827–836.

Nishihara, G. N., Y. Mori, R. Terada, and T. Noro. 2003. Habitat characteristics and seasonal changes of *Laurencia brongniartii* (Ceramiales, Rhodophyta) in Kagoshima, southern Japan. *Phycol. Res.* 51: 30–37.

Nishihara, G. N., Y. Mori, and R. Terada. 2004. Photosynthesis and growth rate of *Laurencia brongniartii* J. Agardh (Rhodophyta, Ceramiales) in preparation for cultivation. *J. Appl. Phycol.* 16: 303–308.

Nishihara, G. N., R. Terada, and T. Noro. 2005. Effect of temperature and irradiance on the uptake of ammonium and nitrate by *Laurencia brongniartii* (Rhodophyta), Ceramiales). *J. Appl. Phycol.* 17: 371–377.

Niwa, K., N. Iijima, S. Kikuchi, T. Nagata, K. Ishihara, H.Saito, and M. Notoya. 2003. Molecular phylogenetic analysis of *Bangia* (Bangiales, Rhodophyta) in Japan. In: A.R.O. Chapman, R. J. Anderson, V. J. Vreeland, and I. R. Davison (eds.), *Proc. XVII Int. Seaweed Symp.*, Cape Town, South Africa, Oxford Univ. Press, Oxford, pp. 303–311.

Nizamuddin, M. 1964. Studies on the genus *Caulerpa* from Karachi. *Bot. Mar.* 6: 204–223.

Nizamuddin, M., and J. Gerloff 1980. New species and new combinations in the genus *Dilophus* J. Ag. *Nova Hedwigia* 31: 865–879.

Noda, M. 1969. The species of Phaeophyta from Sado Island in the Japan Sea. *Sci. Rep. Niigata Univ. Ser. D (Biol.)* 5: 47–69.

———. 1970. Some marine algae collected on the coast of Iwagasaki, Prov. Echigo facing the Japan Sea. *Sci. Rep. Niigata Univ. Ser. D (Biol.)* 7: 27–35.

———. 1987. Marine algae of the Japan Sea. Kazama-shobo, Tokyo.

Norris J. N., and K. E. Bucher. 1982. Marine algae and seagrasses from Carrie Bow Cay, Belize. In: K. Rützler and I. G. Macintyre (eds.), *The Atlantic Barrier Reef ecosystem at Carrie Bow Cay, Belize. I. Structure and communities*. Smithson. Contrib. Mar. Sci. 12: 167–223.

———. 1989. *Rhodogorgon*, an anamolous [sic] new red algal genus from the Caribbean Sea. *Proc. Biol. Soc. Wash.* 102: 1050–1066.

Norris J. N., and S. Fredericq. 1990. Studies on cylindrical species of western Atlantic species of *Gracilaria* (Gracilariales , Rhodophyta): *G. cylindrica* and *G. blodegetti* Harvey. *J. Phycol.* 26(suppl.): 20.

Norris J. N., and J. L. Olsen. 1991. Deep-water green algae from the Bahamas, including *Cladophora vandenhoekii sp. nov.* (Cladophorales). *Phycologia* 30: 315–328.

Norris, R. E. 1986 *Coelarthrum* (Rhodymeniaceae, Rhodophyceae), a genus new to South Africa. *S. Afr. J. Bot.* 52: 537–540.

———. 1987. The systematic position of *Gelidiopsis* and *Ceratodictyon* (Gigartinales, Rhodophyceae), genera new to South Africa. *S. Afr. J. Bot.* 53: 239–246.

———. 1991a. The structure, reproduction and taxonomy of *Vidalia* and *Osmundaria* (Rhodophyta, Rhodomelaceae). *J. Linn. Soc. Lond. Bot.* 106: 1–40.

———. 1991b. Some unusual marine algae (Rhodophyta) from South Africa. *Phycologia* 30: 582–596.

———. 1992a. Six marine macroalgal genera new to South Africa. *S. Afr. J. Bot.* 58: 2–12.

———. 1992b. A proposed phylogenetic scheme for the Gelidiales. In: I. A. Abbott (ed.), *Taxonomy of economic seaweeds with reference to some Pacific and western Atlantic species.* Vol. 3. Calif. Sea Grant College Program, Univ. Calif., La Jolla, pp. 151–171.

———. 1994. Some cumophytic Rhodomelaceae (Rhodophyta) occurring in Hawaiian surf. *Phycologia* 33: 434–443.

Norton, T. A., M. Melkonian, and R. A. Andersen. 1996. Algal biodiversity. *Phycologia* 35: 308–325.

Nugues, M. M., and R.P.M. Bak. 2006. Differential competitive abilities between Caribbean coral species and a brown alga: a year of experiments and a long-term perspective. *Mar. Ecol. Prog. Ser.* 315: 75–86.

Nugues, M. M., L. Delvoye, and R.P.M. Bak. 2004. Coral defense against macroalgae: differential effects of mesenterial filaments on the green alga *Halimeda opuntia. Mar. Ecol. Prog. Ser.* 278: 103–114.

N'Yeurt, A.D.R. 2001. Marine algae from the Suva Lagoon and reef, Fiji. *Aust. Syst. Bot.* 14: 689–869.

N'Yeurt, A.D.R., and C. E. Payri. 2006. Marine algal flora of French Polynesia. I. Phaeophyceae (Ochrophyta, brown algae). *Cryptogam. Algol.* 27: 111–152.

N'Yeurt, A.D.R., G. R. South, and D. W. Keats. 1996. A revised checklist of the benthic marine algae of the Fiji Islands, South Pacific (including the Island of Rotuma). *Micronesica* 29: 49–98.

O'Clair, R. M., and S. C. Lindstrom. 2000. *North Pacific seaweeds.* Plant Press, Auke Bay, Alaska, 159 pp.

Odum, W. E., and C. C. McIvor. 1990. Mangroves. In: R. L. Myers and J. J. Ewel (eds.), *Ecosystems of Florida.* Univ. Central Florida Press, Orlando, pp. 517–548.

Odum, W. E., C. C. McIvor, and T. J. Smith III. 1982. *The ecology of mangroves of South Florida: a community profile.* U.S. Fish Wildl. Serv., Office Biol. Serv. (Tech. Rep.) FWS/OBS 81–24, x + 144 pp.

Ogden, N. B. 1992. Morphology, reproduction and range extension of *Rhodogorgon* (?Nemaliales, Rhodophyta). *Phycologia* 31: 470–477.

Ohba, H., and S. Enomoto. 1987. Culture studies on *Caulerpa* (Caulerpales, Chlorophyta). II. Morphological variation of *C. racemosa* var. *laetevirens* under various culture conditions. *Jpn. J. Phycol.* 35: 178–188.

Ohba, H., H. Nashima, and S. Enomoto. 1992. Culture studies on *Caulerpa* (Caulerpales, Chlorophyceae). III. Reproduction, development and morphological variation of laboratory-cultured *C. racemosa* var. *peltata. Bot. Mag. (Tokyo)* 105: 589–600.

Okamura, K. 1900–1902. *Illustrations of the marine algae of Japan.* Tokyo. Vol. 1, 93 pp., 30 pls. [pp. 1–14, pls. I–V (1990); pp. 15–74, pls. VI–XXV (1901); pp. 75–93, pls. XXVI–XXX (1902)].

———. 1912–1915. *Icones of Japanese algae.* Tokyo. Vol. 2, 191 pp., 50 pls. (1912); Vol. 3, 218 pp., pls. CI–CL. [pp. 1–77, pls. CI–CXX (1913); pp. 79–121, pls. CXXI–CXXX (1914); pp. 123–218, pls. CXXXI–CL (1915)].

O'Kelly, C. J. 1989. The evolutionary origin of brown algae: information from studies of motile cell ultrastructure. In: J. C. Green, B.S.C. Leadbeater, and W. L. Diver (eds.), *The chromophyte algae: problems and perspectives.* Clarendon Press, Oxford, pp. 255–278.

O'Kelly, C. J., B. Wysor, and W. K. Bellows. 2004. *Collinsiella* (Ulvophyceae, Chlorophyta) and other ulotrichalean taxa with shell-boring sporophytes form a monophyletic clade. *Phycologia* 43: 41–49.

O'Kelly, C. J., and C. Yarish. 1981('1980'). Observations on marine Chaetophoraceae (Chlorophyta). I. Sporangial ontogeny in the type species of *Entocladia* and *Phaeophila. J. Phycol.* 16: 549–558.

Oliveira, E. C. 1969. Algas marinhas do Sul do Estado do Espírito Santo (Brasil). I. Ceramiales. *Bol. Fac. Filos. Ciênc. Letras Univ. São Paulo* 343 (Bot. 26): 1–277.

———. 1977. Algas Marinhas bentônicas do Brasil. São Paulo, Brazil: Universidade de São Paulo, Instituto de Biociências, [iv +] 407 pp. [Diss.]

Oliveira, E. C., C. J. Bird, and J. McLachlan. 1983. The genus *Gracilaria* (Rhodophyta, Gigartinales) in the western Atlantic, *Gracilaria domingensis*, *G. cervicornis*, and *G. ferox*. *Can. J. Bot.* 61: 2999–3008.

Oliveira, M. C., J. Kurniawan, C. J. Bird, E. L. Rice, C. A. Murphy, R. K. Singh, R. R. Gutell, and M. A. Ragan. 1995. A preliminary investigation of the order Bangiales (Bangiophycidae, Rhodophyta) based on the sequence of nuclear small-subunit ribosomal RNA gene. *Phycol. Res.* 43: 71–79.

Olsen, J. L., W. T. Stam, S. Berger, and D. Menzel. 1994. 18S rDNA and evolution in the Dasycladales (Chlorophyta): modern living fossils. *J. Phycol.* 30: 729–744.

Olsen, J. L., and J. A. West. 1988. *Ventricaria* (Siphonocladales-Cladophorales complex, Chlorophyta), a new genus for *Valonia ventricosa*. *Phycologia* 27: 103–108.

Olsen-Stojkovich, J. L. 1985. A phylogenetic look at selected genera in the Siphonocladales/Cladophorales complex using immunological data. *Proc. Fifth Int. Coral Reef Symp., Tahiti*, 1985, 5: 59–63.

Oltmanns, F. 1904. *Morphologie und Biologie der Algen*. Erster Band. Spezieller Teil. Jena, vi + 733 pp.

Ortega, M. M., J. L. Godínez, and G. G. Solórzano. 2001. *Catálogo de las Algas bénticas de las costas Mexicanas del Golfo de México y Mar Caribe*. Cuadernos 34, Instituto de Biologia, Universidad Nacional Autónoma de México, 594 pp.

Otero-Schmitt, J., and J. L. Pérez-Cirera, 2002. Infralittoral benthic biocoenoses from northern Ría de Muros, Atlantic coast of northwest Spain. *Bot. Mar.* 45: 93–122.

Otero-Schmitt, J., and A. Sanjuan. 1992. Epibiotic seaweeds of the Cape Verde Islands. *Bot. Mar.* 35: 379–390.

Ott, D. W., and M. H. Hommersand. 1974. Vaucheriae of North Carolina. I. Marine and brackish water species. *J. Phycol.* 10: 373–384.

Ouriques, L. C and Z. L. Bouzon. 2000. Stellate chloroplast organization in *Asteronema breviarticulatum comb. nov.* (Ectocarpales, Phaeophyta). *Phycologia* 39: 267–271.

Pacheco-Ruíz, I., and J. A. Zertuche-González. 2002. Red algae (Rhodophyta) from Bahía de los Ángeles, Gulf of California, Mexico. *Bot. Mar.* 45: 465–470.

Pacheco-Ruíz, I., J. A. Zertuche-González, and J. Espinoza-Avalos. 2005. The role of secondary attachment discs in the survival of *Chondracanthus squarrulosus* (Gigartinales, Rhodophyta). *Phycologia* 44: 629–631.

Page, J. Z. 1970. Existence of a *Derbesia* phase in the life history of *Halicystis osterhoutii* Blinks and Blinks. *J. Phycol.* 6: 375–380.

Pak, J. Y., C. Solorzano, M. Arai, and T. Nitta. 1991. Two distinct steps for spontaneous generation of subprotoplasts from a disintegrated *Bryopsis* cell. *Plant Physiol.* 96: 819–825.

Papenfuss, G. F. 1940. Notes on South African marine algae. I. *Bot. Not.* 1940: 200–226.

———. 1943a. Notes on algal nomenclature. II. *Gymnosorus* J. Agardh. *Am. J. Bot.* 30: 463–468.

———. 1943b. Notes on South African marine algae. II. *S. Afr. J. Bot.* 9: 79–92.

———. 1945. Review of the *Acrochaetium-Rhodochorton* complex of the red algae. *Univ. Calif. Publ. Bot.* 18: 299–334.

———. 1947. Further contributions toward an understanding of the *Acrochaetium-Rhodochorton* complex of the red algae. *Univ. Calif. Publ. Bot.* 18: 433–447.

———. 1950. Review of the genera of algae described by Stackhouse. *Hydrobiologia* 2: 181–208.

———. 1952. Notes on South African marine algae. III. *S. Afr. J. Bot.* 17: 167–188.

———. 1956. Notes on South African marine algae. IV. *S. Afr. J. Bot.* 22: 65–77.

———. 1960. On the genera of the Ulvales and the status of the order. *J. Linn. Soc. Lond. Bot.* 56: 303–318.

———. 1964. Catalogue and bibliography of Antarctic and sub-Antarctic benthic marine algae. In: M. O. Lee (ed.), *Bibliography of the Antarctic seas*. Vol. 1. American Geophysical Union, Washington, D.C., pp. 1–76

———. 1967. Notes on algal nomenclature. V. *Phykos* 5: 95–105.

———. 1968. A history, catalogue, and bibliography of Red Sea benthic algae. *Israel J. Bot.* 17: 1–118.

———. 1977. Review of the genera of Dictyotales (Phaeophycophyta). *Bull. Jpn. Soc. Phycol.* 25 (Suppl. Mem. Issue Yamada): 271–287.

Papenfuss, G. F., and Y.-M. Chiang. 1969. Remarks of the taxonomy of *Galaxaura* (Nemaliales, Chaetangiaceae). In: R. Margalef (ed.), *Proc. Sixth Int. Seaweed Symp.*, Santiago de Compostela, Subsecretaria de la Marina Mercante, Madrid, pp. 303–314.

Papenfuss, G. F., K. E. Mshigeni, and Y. M. Chiang. 1982. Revision of the red algal genus *Galaxaura* with special reference to the species occurring in the western Indian Ocean. *Bot. Mar.* 25: 401–444.

Park, C. H. 1980. The crustose coralline algae (Rhodophyta, Cryptonemiales) in Korea. *Bull. Natl. Fish. Univ. Pusan* 20: 1–30.

Parke, M. W., and P. S. Dixon. 1976. Check-list of British marine algae—third revision. *J. Mar. Biol. Assoc. U.K.* 56: 527–594.

Parkinson, P. G. 1980a. *Grateloupia*. The nomenclators' wreckage: an account of the chronic confusion attending the application of the name *Grateloupia* in the phycological literature, and explanation of their causes with reflections on a deficiency in the International Code of Botanical Nomenclature and a simple solution to the difficulties. Pettifogging Press, Auckland, 24 pp.

———. 1980b. *Halymenia*. Being a critical account of the confused nomenclature of *Halymenia* C. Agardh 1817 (Halymeniaceae, Cryptonemiales, Rhodophyta) with reflections on the International Code of Botanical Nomenclature and corrections to certain recent work in which it has been disregarded. Pettifogging Press, Auckland, 20 pp.

———. 1981. Remarks on some algal generic names recently proposed for conservation: *Halymenia*, *Grateloupia*, *Nemastoma* and *Schizmenia*. *Taxon* 32: 274–285.

Parsons, M. J. 1975. Morphology and taxonomy of the Dasyaceae and the Lophothalieae (Rhodomelaceae) of the Rhodophyta. *Aust. J. Bot.* 23: 549–713.

Parsons, M. J., and H.B.S. Womersley. 1998. Family Dasyaceae Kützing 1843: 413, 414. In: H.B.S. Womersley, *The marine benthic flora of Southern Australia*. Part IIIC, Ceramiales—Ceramiaceae, Dasyaceae. Australian Biological Resources Study, Canberra, pp. 422–510.

Paul, V. J., J. M. Cronan, Jr., and J. H. Cardellina II. 1993. Isolation of new brominated sesquiterpene feeding deterrents from the tropical green alga *Neomeris annulata* (Dasycladaceae: Chlorophyta). *J. Chem. Ecol.* 19: 1847–1860.

Payr, C., A.D.R. N'Yeurt, and J. Orempuller. 2000. *Algae of French Polynesia (Algues de Polynésie française)*. Au Vent des Iles Editions, Tahiti, 320 pp.

Pedersen, P. M. 1978. Culture studies on the pleomorphic brown alga *Myriotrichia clavaeformis* (Dictyosiphonales, Myriotrichiaceae). *Norw. J. Bot.* 25: 281–291.

———. 1984. Studies on primitive brown algae (Fucophyceae). *Opera Bot.* 74: 1–76.

Pedersen, P. M., and A. Kristiansen. 1994. (1111) Proposal to reject *Ulva simplicissima* Clemente, the basionym of *Scytosiphon simplicissimus* (Clemente) Cremades (Phaeophyceae). *Taxon* 43: 645.

———. 2001. On the enigmatic brown alga *Acinetospora crinita* (Ectocarpales, Fucophyceae). *Cryptogam. Algol.* 22: 209–218.

Pennings, S. C., M. P. Publisi, T. J. Pitlik, A. C. Himaya, and V. J. Paul. 1996. Effects of secondary metabolites and CaCO3 on feeding by surgeonfishes and parrotfishes: within-plant comparisons. *Mar. Ecol. Prog. Ser.* 134: 49–58.

Penrose, D. 1991. *Spongites fruticulosus* (Corallinaceae, Rhodophyta), the type species of *Spongites*, in southern Australia. *Phycologia* 30: 438–448.

———. 1992. *Neogoniolithon fosliei* (Corallinaceae, Rhodophyta), the type species of *Neogoniolithon*, in southern Australia. *Phycologia* 31: 338–350.

———. 1996a. Genus *Hydrolithon* (Foslie) Foslie 1909: 55. In: H.B.S. Womersley, *The marine benthic flora of southern Australia. Rhodophyta*. Part IIIB, Gracilariales, Rhodymeniales, Corallinales and Bonnemaisoniales. Flora of Australia Supplementary Series No. 5, Australian Biological Resources Study. Canberra, pp. 255–266.

———. 1996b. Genus *Pneophyllum* Kützing 1843: 385. In: H.B.S. Womersley: *The marine benthic flora of southern Australia. Rhodophyta*. Part IIIB, Gracilariales, Rhodymeniales, Corallinales and Bonnemaisoniales. Flora of Australia Supplementary Series No. 5, Australian Biological Resources Study, Canberra, pp. 266–272.

Penrose, D., and Y. M. Chamberlain. 1993. *Hydrolithon farinosum* (Lamouroux) *comb. nov.*: implications for generic concepts in the Mastophoroideae (Corallinaceae, Rhodophyta). *Phycologia* 32: 295–303.

Penrose, D., and W. J. Woelkerling. 1988. A taxonomic reassessment of *Hydrolithon* Foslie, *Porolithon* Foslie and *Pseudolithophylum* Lemoine (Corallinaceae, Rhodophyta) and their relationships to *Spongites*. *Phycologia* 27: 159–176.

———. 1991. *Pneophyllum fragile* in southern Australia: implications for generic concepts in the Mastophoroidea (Corallinaceae, Rhodophyta). *Phycologia* 30: 495–506.

———. 1992. A reappraisal of *Hydrolithon* and its relationship to *Spongites* (Corallinaceae, Rhodophyta). *Phycologia* 31: 81–88.

Pereira, R. C., A. R. Soares, V. L. Teixera, R. Villaça, and B. A. P. da Gama. 2004. Variation in chemical defenses against herbivory in southwestern Atlantic *Stypopodium zonale* (Phaeophyta). *Bot. Mar.* 47: 202–208.

Perestenko, L. P. 1994. *Red algae of the far-eastern seas of Russia*. St. Petersburg: Komarov Botanical Institute, Russian Academy of Science, 331 pp. (in Russian).

Peters, A. F., and M. E. Ramírez. 2001. Molecular phylogeny of small brown algae, with special reference to the systematic position of *Caepidium antarcticum* (Adenocystaceae, Ectocarpales). *Cryptogam. Algol.* 22: 187–200.

Peterson, R. D. 1972. Effects of light intensity on the morphology and productivity of *Caulerpa racemosa* (Forrskål) J. Agardh. *Micronesica* 8: 63–86.

Pfeiffer, W. J., and R. G. Wiegert. 1981. Grazers on *Spartina* and their predators. In: L. R. Pomeroy and R. G. Wiegert (eds.), *The ecology of a salt marsh*. Springer-Verlag, New York, pp. 87–112.

Pham-Hoang, Hô. 1969. *Rong biên Viêtnam* [Marine algae of South Vietnam]. Trung-tâm Hoc-Liêû Xuât-bán. [Ministry of Education and Youth, Saigon], [VI +] 558 pp., 67 + 224 + 88 + 119 + 2 figs. (in Vietnamese).

Philippi, R. A. 1837. Beweis, dass die Nulliporen Pflanzen sind. *Arch. Naturgesch.* 3: 387–393, pl. IX.

Phillips, J. A. 1997. Algae: In: R. J. F. Henderson (ed.), *Queensland plants: names and distribution*. Inodoroopily, Queensland, Queensland Herbarium, Department of Environment, pp. 223–240.

———. 2000. Systematics of the Australian species of *Dictyopteris* (Dictyotales, Phaeophyceae). *Aust. Syst. Bot.* 13: 283–323.

Phillips, R. C. 1960a. Ecology and distribution of marine algae found in Tampa Bay, Boca Ciega Bay and at Tarpon Springs, Florida. *Q. J. Fla. Acad. Sci.* 22: 155–162.

———. 1960b. The ecology of marine plants of Crystal Bay, Florida. *Q. J. Fla. Acad. Sci.* 22: 328–337.

———. 1961. Seasonal aspect of the marine algal flora of St. Lucie Inlet and adjacent Indian River, Florida. *J. Fla. Acad. Sci.* 24: 135–147.

Phillips, R. C., and V. C. Springer. 1960. Observations on the offshore benthic flora in the Gulf of Mexico off Pinellas County, Florida. *Am. Midl. Nat.* 64: 362–381.

Piazzi, L., G. Ceccherelli, and F. Cinelli. 2001. Threat to macroalgal diversity: effects of the introduced green alga *Caulerpa racemosa* in the Mediterranean. *Mar. Ecol. Prog. Ser.* 210: 149–159.

Piazzi, L., A. Meinesz, M. Verlaque, B. Akçali, B. Antolic,' M. Argyrou, D. Balata, E. Ballesteros, S. Calvo, F. Cinelli, S. Cirik, A. Cossu, R. D'Archino, S. Cirik, A. Cossu, R. D'Archino, A. S. Djellouli, F. Javel, E. Lanfranco, C. Mifsud, D. Papl, P. Panayotidis, A. Peirano, G. Pergent, A. Petrocelli, S. Ruitton, A. Zuljevic, and G. Ceccherelli. 2005. Invasion of *Caulerpa racemosa* var. *cylindracea* (Caulerpales, Chlorophyta) in the Mediterranean Sea: an assessment of the spread. *Cryptogam. Algol.* 26: 189–202.

Piccone, A. 1884. Crociera del Corsaro alle Isole Madera e Canarie del Capitano Enrico D'Albertis. Alghe. Genova [Genoa], 60 pp., 1 pl [= 5 Figs].

———. 1886. Alghe del viaggio di circumnavigazione della "Vettor Pisani." Genova, 97 pp., 2 pls.

Pilger, R. 1911. Die Meeresalgen von Kamerun. Nach der Sammlung von C. Ledermann. In: A. Engler, *Beiträge zur Flora von Africa*. XXXIX. Bot. Jahrbuch. 46: 293–323.

Plastino, E. M., and E. C. Oliveira. 1997. *Gracilaria caudata* J. Agardh (Gracilariales, Rhodophyta)—restoring an old name for a common western Atlantic alga. *Phycologia* 36: 225–232.

Pomeroy, L. R., W. M. Darley, E. L. Dunn, J. L. Gallagher, E. G. Haines, and D. M. Whitney. 1981. Primary production. In: L. R. Pomeroy and R. G. Wiegert (eds.), *The ecology of a salt marsh*. Ecological Studies 38. Springer-Verlag, New York, pp. 39–67.

Post, E. 1936. Systematische und pflanzengeographische Notizen zur *Bostrychia-Caloglossa*-Assoziation. *Rev. Algol.* 9: 1–84.

Price, I. R., and G. T. Kraft. 1991. Reproductive development and classification of the red algal genus *Ceratodictyon* (Rhodymeniales, Rhodophyta). *Phycologia* 30: 106–116.

Price, I. R., and F. J. Scott. 1992. *The turf algal flora of the Great Barrier Reef*. Pt. I: Rhodophyta. Botany Department, James Cook Univ., Townsville, Australia, xii + 266 pp.

Price, J. H., D. M. John, and G. W. Lawson. 1978. Seaweeds of the western coast of tropical Africa and adjacent islands: a critical reassessment. II. Phaeophyta. *Bull. Br. Mus. (Nat. Hist.) Bot.* 6: 87–182.

———. 1986. Seaweeds of the western coast of tropical Africa and adjacent islands: a critical reassessment. IV. Rhodophyta (Florideae). 1. Genera A–F. *Bull. Br. Mus. (Nat. Hist.) Bot.* 15: 1–122.

———. 1988. Seaweeds of the western coast of tropical Africa and adjacent islands: a critical reassessment. IV. Rhodophyta (Florideae). 2. Genera G. *Bull. Br. Mus. (Nat. Hist.) Bot.* 18: 195–273.

———. 1992. Seaweeds of the western coast of tropical Africa and adjacent islands: a critical reassessment. IV. Rhodophyta (Florideae). 3. Genera H–K. *Bull. Br. Mus. (Nat. Hist.) Bot.* 22: 123–146.

Price, W. M. 1967. Some aspects of the biology and taxonomy of the unbranched Cladophorales. Doctoral Diss., Univ. Liverpool, iii + 106 pp.

Prince, J. S., and S. Baker. 1984. *Batophora largoensis* new species (Chlorophyta, Dasycladaceae) from South Florida: morphological and ultrastructural evidence. *Bull. Mar. Sci.* 34: 321–329.

Prince, J. S., and C. D. Trowbridge. 2004. Reproduction in the green macroalga *Codium* (Chlorophyta): characterization of gametes. *Bot. Mar.* 47: 461–470.

Provasoli, L. 1965. Nutritional aspects of seaweed growth. *Proc. Can. Soc. Plant Physiol.* 6: 26–27.

Provasoli, L., and I. J. Pintner. 1980. Bacteria induced polymorphism in an axenic laboratory strain of *Ulva lactuca* (Chlorophyta). *J. Phycol.* 16: 196–201.

Provost, M. W. 1973. Mean high water mark and use of tidelands in Florida. *Fla. Sci.* 36: 558–563.

Prud'homme van Reine, W. F. 1982. *A taxonomic revision of the European Sphacelariaceae (Sphacelariales, Phaeophyceae).* Leiden Botanical Series, Vol. 6, Leiden, ix + 293 pp., 6 pls.

———. 1993. Sphacelariales (Phaeophyceae) of the world, a new synthesis. *Korean J. Phycol.* 8: 145–160.

Prud'homme van Reine, W. F., R. J. Haroun, and P. A. J. Audiffred. 1994. A reinvestigation of Macronesian seaweeds as studied by A. Piccone, with remarks on those studied by A. Grunow. *Nova Hedwigia* 58: 67–121.

Prud'homme van Reine, W. F., and G. M. Lokhorst. 1992. *Caulerpella gen. nov.* a non-holocarpic member of the Caulerpales (Chlorophyta). *Nova Hedwigia* 54: 113–126.

Pueschel, C. M., and K. M. Cole. 1982. Rhodophycean pit plugs: an ultrastructural survey with taxonomic implications. *Am. J. Bot.* 69: 703–720.

Pueschel, C. M., B. L. Judson, and S. Webegerg. 2005. Decalcification during epithallial cell turnover in *Jania adhaerens* (Corallinales, Rhodophyta). *Phycologia* 44: 156–162.

Quan-Young, L. I., M. A. Díaz-Martín, and J. Espinoza-Avalos. 2004. Floristics, cover, and phenology of marine macroalgae from Bajo Pepito, Isla Mujeres, Mexican Caribbean. *Bull. Mar. Sci.* 75: 11–26.

Quoy, J. R. C., and P. Gaimard. 1824. Zoologie, In: L. de Freycinet (ed.), *Voyage autour du monde . . . exécuté sur les corvettes de S. M. l'Uranie et la Physicienne, pendant les années 1817, 1818, 1819 et 1820.* Paris, [v +] 713 pp., 96 pls.

Rabenhorst, L. 1847. *Deutschlands Kryptogamen-Flora . . .* Zweiter Band. Zweite Abteilung: Algen. Leipzig, xix + 216 pp.

Raloff, J. 2000. Algal bloom is smothering Florida coral. *Sci. News* 157: 373.

Ramírez, M. E., and B. Santelices. 1991. *Catálago de las Algas Marinas Bentónicas de la Costa Temperada del Pacifico de Sudamerica.* Monogr. Biol. 57: 1–440.

Reedman, D. J., and H.B.S. Womersley. 1976. Southern Australian species of *Champia* and *Chylocladia* (Rhodymeniales: Rhodophyta). *Trans. R. Soc. S. Aust.* 100: 75–104.

Reinbold, T. 1893. Revision von Jürgens Algae aquaticae. I. Die Algen des Meeres- und des Brackwassers. *Nuova Notarisia* 4: 192–206.

Reinke, J. 1878. Ueber die Entwicklung von *Phyllitis*, *Scytosiphon* und *Asperococcus*. *Jahrb. Wiss. Bot.* 11: 262–273.

———. 1879. Zwei parasitische Algen. *Bot. Zeitung (Berlin)* 37: 473–478, pl. 6.

———. 1888. Einige neue braune und grüne Algen der Kieler Bucht. *Ber. Deutsch Bot. Ges.* 6: 240–241.

———. 1889a. *Algenflora der westlichen Ostee deutschen Anthiels. Eine systemaisch-pflanzengeographische Studie.* Komm. Wiss. Unters. Meere Kiel 6, xi + 101 pp., 1 map.

———. 1889b. *Atlas deutscher Meeresalgen*, 1. Berlin, iv + 34 pp., 25 pls.

———. 1900. Ueber *Caulerpa*. Ein Beitrag zur Biologie der Meeres-Organismen. *Wiss. Meeresunters.*, Abt. Kiel 5: 1–96.

Reinsch, P. F. 1875('1874/1875'). *Contributiones ad algologiam et fungologiam.* Norimbergae [Nürnberg]. xii + 103 pp., 131 pls. [I–III, IIIa, IV–VI, VIa, VII–XII, XIIa, XIII–XX, XXa, XXI–XXXV, XXXVa, XXXVI (Melanophyceae); I–XLII, XLIIa, XLIII–XLVII XLVIIa, XLVIII–LXI (Rhodophyceae); I–XVIII (Chlorophyllophyceae); I–IX (Fungi)].

Rezak, R., T. J. Bright, and M. McGrail. 1985. *Reefs and banks of the northwestern Gulf of Mexico: their geological, biological, and physical dynamics.* John Wiley and Sons, New York.

Rhodes, R. G., and M. V. Connell. 1973. The biology of brown algae on the Atlantic coast of Virginia. II. *Petalonia fascia* and *Scytosiphon lomentaria*. *Chesapeake Sci.* 14: 211–215.

Ribera, M. A., A. Gomez Garreta, T. Gallardo, M. Cormaci, G. Furnari, and G. Giaccone. 1992. Check-list of Mediterranean seaweeds. I. Fucophyceae (Warming, 1884). *Bot. Mar.* 35: 109–130.

Richards, H. M. 1901. *Ceramothamnion codii*, a new rhodophyceous alga. *Bull. Torrey Bot. Club* 28: 257–265, pls. 21, 22.

Richardson, W. D. 1975. The marine algae of Trinidad, West Indies. *Bull. Br. Mus. (Nat. Hist.) Bot.* 5(3): 1–143, 27 pls.

Ridd, P., M. W. Sandstrom, and E. Wolanski. 1988. Outwelling from tropical tidal salt flats. *Estuar. Coast. Shelf Sci.* 26: 243–253.

Rinidi, F., G. Sartoni, and F. Cinelli. 2002. A floristic account of the benthic marine algae of Tuscany (western Mediterranean Sea). *Nova Hedwigia* 74: 201–250.

Riosmena-Rodríguez and W. J. Woelkerling. 2000. Taxonomic biodiversity of Corallinales (Rhodophyta) in the Gulf of California, Mexico: towards an initial assessment. *Cryptogam. Algol.* 21: 315–354.

Robertson, A. I., and D. M. Alongi. (eds.) 1992. *Coastal and estuarine studies 41. Tropical mangrove ecosystems.* American Geophysical Union, Washington, D.C. 329 pp.

Rodriguez, C., and A. W. Stoner. 1990. The epiphyte community of mangrove roots in a tropical estuary: distribution and biomass. *Aquat. Bot.* 36: 117–126.

Rodríguez-Prieto, C., G. Michanek, and C. Ivon. 1999. Benthic marine algae from Martinique (Lesser Antilles). *Scientia Gerundensis* 24: 79–86.

Roeleveld, J. G., M. Duisterhof, and M. Vroman. 1974. On the year cycle of *Petalonia fascia* in the Netherlands. *Neth. J. Sea Res.* 8: 410–426.

Rojas, J. J., A. Lemus, and E. K. Ganesan. 1982. El ciclo vital "in vitro" del alga marina roja *Asparagopsis taxiformis* (Delile) Collins & Hervey (Bonnemaisoniales, Rhodophyta) del Mar Caribe. *Bol. Inst. Oceanogr.* (Cumaná) 21: 101–112.

Rosenvinge, L. K. 1893. Grønlands Havalger. *Medd. om Grønland* 3: 763–981, 2 pls.

———. 1909. *The marine algae of Denmark. Contributions to their natural history.* Part I. Introduction. Rhodophyceae I (Bangiales and Nemalionales). *K. Dan. Vidensk. Selsk. Skr.*, ser. 7, Naturv. Math. Afd., 7: 1–151, pls. I, II, 2 folded charts.

———. 1924. *The marine algae of Denmark. Contributions to their natural history.* Part III. Rhodophyceae III. (Ceramiales). *K. Dan. Vidensk. Selsk. Skr.*, ser. 7, Naturv. Math. Afd., 7: 285–487, pls. V–VII, [2] maps.

Rosenvinge, L. K., and S. Lund. 1941. *The marine algae of Denmark.* Vol. II. Phaeophyceae. *K. Dan. Vidensk. Selsk. Skr., Biol. Skr.* 1, 79 pp.

Ross, C., F. C. Küpper, V. Vreeland, J. H. Waite, and R. S. Jacobs. 2005a. Evidence of a latent oxidative burst in relation to wound repair in the giant unicellular chlorophyte *Dasycladus vermicularis. J. Phycol.* 41: 531–541.

Ross, C., V. Vreeland, J. H. Waite, and R. S. Jacobs. 2005b. Rapid assembly of a wound plug: stage one of a two stage wound repair mechanism in the giant unicellular chlorophyte *Dasycladus vermicularis. J. Phycol.* 41: 46–54.

Roth, A. W. 1797. *Catalecta botanica* . . . Fasc. 1. Lipisiae [Leipzig], viii + 244 + [10] pp., 8 pls.

———. 1800a. *Tentamen florae germanicae.* Vol. 3, part 1. Lipsiae [Leipzig], vii + 578 pp.

———. 1800b. *Catalecta botanica* . . . Fasc. 2. Lipsiae [Leipzig], [x +] 258 [+ 12] pp., 9 pls.

———. 1806. *Catalecta botanica* . . . Fasc. 3. Lipsiae [Leipzig], [viii +] 350 [+ 9] pp., 12 pls.

Rousseau, F., R. Burrowes, A. F. Peters, R. Kuhlenkamp, and B. de Reviers. 2001. A comprehensive phylogeny of the Phaeophyceae based on nrDNA sequences resolves the earliest divergences. *C. R. Hebd. Séances Acad. Sci. Paris* 324: 305–319.

Rousseau, F., and B. de Reviers. 1999a. Circumscription of the order Ectocarpales (Phaeophyceae): bibliographical synthesis and molecular evidence. *Cryptogam. Algol.* 20: 5–18.

———. 1999b. Phylogenetic relationships within the Fucales (Phaeophyceae) based on combined partial SSU + LSU rDNA sequence data. Eur. *J. Phycol.* 34: 53–64.

Rueness, J. 1977. *Norsk Algeflora.* Scandinavian University Books, Oslo, Bergen and Trondheim, 266 pp.

Rueness, J., and M.-T. L'Hardy-Halos. 1991. *Aglaothamnion westbrookiae sp. nov.* (Rhodophyta), a species previously confused under the name *Callithamnion byssoides. J. Phycol.* 27: 649–652.

Rull Lluch, J. 2002. Marine benthic algae of Namibia. *Sci. Mar.* 66 (suppl. 3): 5–256.

Rull Lluch, J., and A. Gómez Garreta. 2001. Contribution to the seaweed flora of Namibia: new records of Chlorophyta and Phaeophyceae. *Cryptogam. Algol.* 22: 297–304.

———. 2002. Contribution to the seaweed flora of Namibia: new records of Rhodophyta. *Bot. Mar.* 45: 518–529.

Ruprecht, F. J. 1851. *Algae ochotenses. Die ersten sicheren Nachrichten über die Tange des Ochotskischen Meeres.* St. Petersburg. 243 pp., 10 pls. [Preprint of: Tange des Ochotskischen Meeres. In: A. Th. V. Middendorff, *Reise in den äussersten Norden und Osten Sibiriens* . . . Band 1, Theil 2, Abth., 1, pp. 191–435, pls. 9–18. 1851.].

Russell, G. 1967. The ecology of some free-living Ectocarpaceae. *Helgol. Wiss. Meeresunters.* 15: 155–162.

Sahoo, D., Nivedita, and Debasish. 2001. *Seaweeds of Indian coast.* A.P.H. Publishing, New Delhi, xxi + 283 pp. (the two co-authors, apparently students, have no initials given)
Saito, Y. 1967. Studies on Japanese species of *Laurencia*, with respect to their comparative morphology. *Mem. Fac. Fish. Hokkaido Univ.* 15: 1–81.
Saito, Y., and H.B.S. Womersley. 1974. The southern Australian species of *Laurencia* (Ceramiales: Rhodophyta). *Aust. J. Bot.* 22: 815–874.
Sánchez, N., and C. Rodríguez-Prieto. 2005. Vegetative and reproductive morphology of the type species of *Gloiocladia*, *G. furcata* (Faucheaceae, Rhodophyta). *Phycologia* 44: 222–233.
Sansón, M., M. J. Martín, and J. Reyes. 2006. Vegetative and reproductive morphology of *Cladosiphon contortus*, *C. occidentalis* and *C. cymodoceae sp. nov.* (Ectocarpales, Phaeophyceae) from the Canary Islands. *Phycologia* 45: 529–545.
Santelices, B. 1976. Taxonomic and nomenclatural notes on some Gelidiales (Rhodophyta). *Phycologia* 15: 165–173.
Sartoni, G. 1978. Ricerche sulla flora algale della Somalia centromeridionale. 1. Il genere *Caulerpa*. *Webbia* 43: 397–416.
Sartoni, G., and U. Salghetti. 1987. Osservazioni morfologiche in *Taenioma perpusillum* (J. Agardh) J. Agardh e *T. nanum* (Kuetzing) Papenfuss (Delesseriaceae, Rhodophyta). *Inf. Bot. Ital.* 19: 144–154.
Saunders, G. W., and J. C. Bailey. 1997. Phylogenesis of pit-plug associated features in the Rhodophyta: inferences from molecular systematic data. *Can. J. Bot.* 75: 1436–1447.
———. 1999. Molecular systematic analyses indicate that the enigmatic *Apophlaea* is a member of the Hildenbrandiales (Rhodophyta, Florideophycidae). *J. Phycol.* 35: 171–175.
Saunders, G. W., A. Chiovitti, and G. T. Kraft. 2004. Small-subunit rDNA sequences from representatives of selected families of the Gigartinales and Rhodymeniales (Rhodophyta). 3. Delineating the Gigartinales *sensu stricto*. *Can. J. Bot.* 82: 43–74.
Saunders, G. W., and M. H. Hommersand. 2004. Assessing red algal supraordinal diversity and taxonomy in the context of contemporary systematic data. *Am. J. Bot.* 9: 1494–1507.
Saunders, G. W., and G. T. Kraft. 1994. Small-subunit rRNA gene sequences from representatives of selected families of the Gigartinales and Rhodymeniales (Rhodophta). I. Evidence for the Plocamiales *ord. nov. Can. J. Bot.* 72: 1250–1263.
———. 1996. Small-subunit rRNA gene sequences from representatives of selected families of the Gigartinales and Rhodymeniales (Rhodophta). 2 Recognition of the Halymeniales *ord. nov. Can. J. Bot.* 74: 694–707.
———. 1997. A molecular perspective on red algal evolution: focus on the Florideophycidae. In: D. Bhattacharya (ed.), *Origins of algae and their plastids.* Springer, New York, pp. 115–138.
———. 2002. Two new Australian species of *Predaea* (Nemastomataceae, Rhodophyta), with taxonomic recommendations for an emended Nemastomatales and expanded Halymeniales. *J. Phycol.* 38: 1245–1260.
Saunders, G. W., I. M. Strachan, and G. T. Kraft. 1999. The families of the order Rhodymeniales (Rhodophyta): a molecular-systematic investigation with a description of Faucheaceae *fam. nov. Phycologia* 38: 23–40.
Sauvageau, C. 1897. Sur quelques Myrionémacées. *Ann. Sci. Nat., Bot.*, sér. 8, 5: 161–288.
———. 1898. Ibid. title and journal (i.e. reprint): 1–130.
———. 1927. Sur les problèmes du *Giraudya*. *Bull. Stat. Biol. Arachnon* 24: 3–74.
———. 1929. Sur le dévelopment de quelques Phéosporées. *Bull. Stat. Biol. Arachnon* 26: 253–420.
Sawyer, C. M. 1965. The sea lettuce problem in Boston Harbor. *J. Water Pollut. Control Fed.* 37: 1122–1133.
Scagel, R. F. 1957. *An annotated list of the marine algae of British Columbia and Northern Washington.* Bull. 150, National Museum Canada, Ottawa, 289 pp., 1 fig.
———. 1966. *Marine algae of British Columbia and northern Washington.* Part 1, Chlorophyceae (green algae). Bull. Nat. Mus. Can. (Biol. Ser.) 207, viii + 257 pp.
Scagel, R. F., P. W. Gabrielson, D. J. Garbary, L. Golden, M. W. Hawkes, S. C. Lindstrom, J. C. Oliveira, and T. B. Widdowson. 1989. *A synopsis of the benthic marine algae of British Columbia, southeast Alaska, Washington and Oregon.* Phycological Contributions, Univ. British Columbia, Vancouver. 532 pp.
Schiffner, V. 1933. Meeresalgen aus Süd-Dalmatien, gessamelt von Franz Berger. *Öst. Bot. Z.* 82: 283–308.
Schils, T., O. De Clerck, and E. Coppejans. 2003. The red alga genus *Reticulocaulis* from the Arabian Sea, including *R. obpyriformis sp. nov.*, with comments on the family Naccariaceae. *Phycologia* 42: 44–55.
Schmitz, F. 1879. Ueber grüne Algen aus dem Golf von Athen. Sitzungsberichten der Naturf. Gesell. zu Halle, pp. 17–25 (1–7 on pages).

———. 1889. Systematische Übersicht der bisher bekannten Gattungen der Florideen. *Flora* (Jena) 72: 435–456, pl. XXI.

———. 1892. Klasse Rhodophyceae 2. Unterklasse Florideae. In: A. Engler (ed.), *Syllabus der Vorlesungen. Über Specielle und Medicinisch-Pharmaceutische Botanik Grosse Ausgabe.* Borntraeger, Berlin, pp. 16–23.

———. 1893. Die Gattung *Lophothalia* J. Ag. *Ber. Deutsch Bot. Ges.* 11: 212–232.

———. 1894. Kleinere Beiträge zur Kenntniss der Florideen. IV. *Nuova Notarisia* 5: 608–635.

———. 1895. Marine Florideen von Deutsch-Ostafrika. *Bot. Jahreb. für System.* 21: 137–177.

———. 1896. Bangiaceae. In: A. Engler and K. Prantl (eds.), *Die natürlichen Pflanzenfamilien* . . . I. Teil, Abt 2. Leipzig, pp. 307–316.

Schmitz, F., and P. Falkenberg. 1897. Rhodomelaceae. In: A. Engler and K. Prantl (eds.), *Die natürlichen Pflanzenfamilien* . . . I. Teil., Abt. 2. Leipzig, pp. 421–480.

Schmitz, F., and H. Hauptfleish. 1896('1897'). Rhodophyllidaceae. In: A. Engler and K. Prantl (eds.), *Die natürlichen Pflanzenfamilien* . . . I. Teil., Abt. 2. Leipzig, pp. 366–382.

———. 1897. Ceramiaceae . In: A. Engler and K. Prantl (eds.), *Die natürlichen Pflanzenfamilien* . . . I. Teil., Abt, 2. Leipzig, pp. 481–504.

Schneider, C. W. 1974. North Carolina marine algae. III. A community of Ceramiales (Rhodophyta) on a glass sponge from 60 meters. *Bull. Mar. Sci.* 24: 1093–1101.

———. 1975. North Carolina marine algae. V. Additions to the flora of Onslow Bay, including the reassignment of *Fauchea peltata* Taylor to *Weberella* Schmitz. *Br. Phycol. J.* 10: 129–138.

———. 1983. The red algal genus *Audouinella* Bory (Nemaliales: Acrochaetiaceae) from North Carolina. *Smithson. Contr. Mar. Sci.* 22. 25 pp.

———. 1984. Studies on *Antithamnionella*, *Callithamniella* and *Calloseris* (Rhodophyta, Ceramiales) from North Carolina, USA. *Phycologia* 23: 455–464.

———. 2000. Notes on the marine algae of the Bermudas. 5. Some Delesseriaceae (Rhodophyta), including the first record of *Hypoglossum barbatum* Okamura from the Atlantic Ocean. *Bot. Mar.* 43: 455–466.

———. 2003. An annotated checklist and bibliography of the marine macroalgae of the Bermuda Islands. *Nova Hedwigia* 76: 275–361.

———. 2004. Notes on the marine algae of the Bermudas. 6. Some rare or newly reported Ceramiales (Rhodophyta), including *Crouania elisiae sp. nov. Phycologia* 43: 563–578.

Schneider, C. W., and N. J. Eiseman. 1979. *Searlesia*, a new genus from the western Atlantic based on *Membranoptera subtropica* (Rhodophyta, Delesseriaceae). *Phycologia* 18: 319–324.

Schneider, C. W., and C. E. Lane. 2005. Notes on the marine algae of the Bermudas. 7. Additions to the flora including *Chondracanthus saundersii sp. nov.* (Rhodophyta, Gigartinaceae) based on rbcL sequence analysis. *Phycologia* 44: 72–83.

Schneider, C. W., and R. P. Reading. 1987. A revision of the genus *Peyssonnelia* (Rhodophyta, Cryptonemiales) from North Carolina, including *P. atlantica* new species. *Bull. Mar. Sci.* 40: 175–192.

Schneider, C. W., and R. B. Searles. 1975. North Carolina marine algae. IV. Further contributions from the continental shelf, including two new species of Rhodophyta. *Nova Hedwigia* 24: 83–103.

———. 1976. North Carolina marine algae. VII. New species of *Hypnea* and *Petroglossum* (Gigartinales) and additional records of other Rhodophyta. *Phycologia* 15: 51–60.

———. 1991. *Seaweeds of the southeastern United States, Cape Hatteras to Cape Canaveral.* Duke Univ. Press, Durham, xiv + 553 pp.

———. 1997. Notes on the marine algae of the Bermudas. I. New records of Antithamnieae and Dohrnielleae (Ceramiaceae, Rhodophyta), including *Antithamnionella bermudica sp. nov. Phycologia* 36: 12–23.

———. 1998. Notes on the marine algae of the Bermudas. 3. *Avrainvillea syllearleae*, *Discosporangium mesarthrocarpum* and *Peyssonnelia valentinii*. *J. Phycol.* 34: 180–188.

Schnetter, R. 1976. Marine Algen der karibischen Küsten von Kolumbien. I. Phaeophyceae. *Bibl. Phycol.*, J Cramer, Vol. 24, 125 pp.

Schnetter, R., I. Hörnig, and G. Weber-Peukert. 1987. Taxonomy of some North Atlantic *Dictyota* species (Phaeophyta). *Hydrobiologia* 151/152: 193–197.

Schramm. A., and H. Mazé. 1865. *Essai de classification des algues de la Guadeloupe.* [ed. 1]. Basse-Terre, Guadeloupe, 52 pp. [see comments under Mazé, H., and A. Schramm. 1878('1870–1877')].

———. 1866. *Essai de classification des algues de la Guadeloupe.* [ed. 1a]. Cayenne, Guyane Française, 144 pp.

Scopoli, J. A. 1772. *Flora carniolica*, editio secunda. Vindobona [Vienna], 496 pp.

Searles, R. B. 1972. North Carolina marine algae. I. Three new species from the continental shelf. *Phycologia* 11: 19–24.

———. 1984. North Carolina marine algae. XII. *Gloioderma rubrisporum sp. nov.* (Rhodophyta, Rhodymeniales). *Bull. Torrey Bot. Club* 111: 217–221.

———. 1988. *An illusrated field and laboratory guide to the seaweeds of Gray's Reef National Marine Sanctuary.* NOAA Technical Memorandum Series NOS MEMD 22, Washington, D.C., xv + 122 pp.

Searles, R. B., and G. L. Leister. 1980. North Carolina marine algae. IX. *Onslowia endophytica gen. et sp. nov.* (Phaeophyta, Sphacelariales) and notes on other new records for North Carolina. *J. Phycol.* 16: 35–40.

Sears, J. R. (ed.). 2002. *NEAS keys to benthic marine algae of the northeastern coast of North America from Long Island Sound to the Strait of Belle Isle.* Northeast Algal Society Contrib. no. 2, Dartmouth, Mass., xviii + 161 pp.

Sears, J. R., and S. H. Brawley. 1982. *Smithsoniella gen. nov.*, a possible evolutionary link between the multicellular and siphonous habits in the Ulvophyceae, Chlorophyta. *Am. J. Bot.* 69: 1450–1461.

———. 1983. Errata. Volume 69. *Am. J. Bot.* 70: 962.

Senerpont Domis, L. N. de, P. Famá, A. J. Bartlett, W. J. Prud'homme van Reine, C. A. Espinosa, and G. C. Trono, Jr. 2003. Defining taxon boundaries in members of the morphologically and genetically plastic genus *Caulerpa* (Caulerpales, Chlorophyta). *J. Phycol.* 39: 1019–1037.

Setchell, W. A. 1914. The *Scinaia* assemblage. *Univ. Calif. Publ. Bot.* 6: 79–152.

———. 1925. Notes on *Microdictyon. Univ. Calif. Publ. Bot.* 13: 101–107.

———. 1926. Tahitian algae and Tahitian spermatophytes collected by W. A. Setchell, C. B. Setchell, and H. E. Parks. *Univ. Calif. Publ. Bot.* 12: 61–240.

———. 1933. Some early algal confusions. II. *Univ. Calif. Publ. Bot.* 17: 187–254.

Setchell, W. A., and N. L. Gardner. 1903. Algae of northwestern America. *Univ. Calif. Publ. Bot.* 1: 165–418.

———. 1920. The marine algae of the Pacific coast of North America. Part II. Chlorophyceae. *Univ. Calif. Publ. Bot.* 8: 139–374.

———. 1924. New marine algae from the Gulf of California. *Proc. Calif. Acad. Sci.* ser. 4, 12: 695–949.

———. 1930. Marine algae of the Revillagigedo Islands Expedition in 1925. *Proc. Calif. Acad. Sci.* ser. 4, 19: 109–215.

———. 1937. A preliminary report on the algae. The Templeton Crocker Expedition of the California Academy of Science, 1932. *Proc. Calif. Acad. Sci.* ser. 4, 22: 65–98.

Setchell, W. A., and L. R. Mason. 1943. *Goniolithon* and *Neogoniolithon*: two genera of crustaceous coralline algae. *Proc. Natl. Acad. Sci. U.S.A.* 29: 87–92.

Sheath, R. G. 1984. The biology of freshwater red algae. In: F. E. Round and D. J. Chapman (eds.), *Prog. Phycol. Res.* 3: 89–157.

Sheath, R. G., K. M. Müller, and A. R. Sherwood. 2000. A proposal for a new algal order, the Thoreales. *J. Phycol.* 36 (suppl.): 62.

Sheridan, P. F., and R. J. Livingston. 1983. Abundance and seasonality of infauna and epifauna inhabiting a *Halodule wrightii* meadow in Apalachicola Bay, Florida. *Estuaries* 6: 407–419.

Sherwood, A. R., and R. G. Sheath. 2003. Systematics of the Hildenbrandiales (Rhodophyta): gene sequence and morphometric analyses of global collections. *J. Phycol.* 39: 409–422.

Shimada, S., M. Hiraoka, S. Nabata, M. Iima, and M. Masuda. 2003. Molecular phylogenetic analyses of the Japanese *Ulva* and *Enteromorpha* (Ulvales, Ulvophyceae), with special reference to the free-floating *Ulva. Phycol. Res.* 51: 99–108.

Silva, P. C. 1952. A review of nomenclatural conservation in the algae from the point of view of the type method. *Univ. Calif. Publ. Bot.* 25: 241–323.

———. 1960. *Codium* (Chlorophyta) in the tropical western Atlantic. *Nova Hedwigia* 1: 497–536.

———. 1972. Remarks on algal nomenclature. V. *Taxon* 21: 199–205.

———. 1980. Remarks on algal nomenclature. VI. *Taxon* 29: 121–145.

———. 1991. Nomenclatural notes on Clemente's *Ensay. An. Jard. Bot. Madr.* 49: 163–170.

———. 1996. Proposal to conserve the name *Asteronema* Delépine and Asensi (Phaeophyceae) against *Asteronema* Trevis. (Fungi). *Taxon* 45: 123–124.

———. 2002. Comments on the commentary by Kraft and Saunders (*Phycologia* 39: 258–261, 2000). *Phycologia* 41: 99–100.

———. 2003. Historical overview of the genus *Caulerpa. Cryptogam. Algol.* 24: 33–50.

———. 2004. Proposal to conserve the name *Caloglossa* against *Apiarium* (Rhodophyceae). *Taxon* 53: 555–556.
Silva, P. C., P. W. Basson, and R. L. Moe. 1996. *Catalogue of the benthic marine algae of the Indian Ocean.* Univ. Calif. Publ. Bot. 79: xiv + 1259 pp.
Silva, P. C., and T. C. DeCew. 1992. *Ahnfeltiopsis*, a new genus in the Phyllophoraceae (Gigartinales, Rhodophyceae). *Phycologia* 31: 576–580.
Silva, P. C., and H. W. Johansen. 1986. A reappraisal of the order Corallinales (Rhodophyceae). *Br. Phycol. J.* 21: 245–254.
Silva, P. C., and Y. Lipkin. 2003. Marine algae published by Savigny in the zoological portion of "Description de l'Egypte." *Cryptogam. Algol.* 24: 371–386.
Silva, P. C., E. G. Meñez, and R. L. Moe. 1987. *Catalog of the benthic marine algae of the Philippines.* Smithson. Contrib. Mar. Sci. 27: iv + 179 pp.
Sjöstedt, L. G. 1926. Floridean studies. *Acta Univ. Lunds. Årsskr.*, N. F. Avd. 2, 22(4): 95.
Skage, M., T. M. Gabrielsen, and J. Rueness. 2005. A molecular approach to investigate the phylogenetic basis of three widely used species groups in the red algal genus *Ceramium* (Ceramiales, Rhodophyta). *Phycologia* 44: 353–360.
Skelton, P. A., and G. R. South. 1999. A preliminary checklist of the benthic marine algae of the Samoan Archipelago. University of the South Pacific, Marine Studies Program Technical Report 99/1, 30 pp.
———. 2004. New records and notes on marine benthic algae of American Samoa—Chlorophyta & Phaeophyta. *Cryptogam. Algol.* 25: 291–312.
Skottsberg, C. 1923. Botanische ergebnisse der schwedischen expedition nach Patagonien und dem Feuerlande 1907–1909. IX. Marine Algae. 2. Rhodophyceae. *K. Sven. Vetensk.-Akad. Handl.* 63(8): 1–70.
Skuja, J. 1939. Versuch einer systematischen Einteilung der Bangioideen oder Protoflorideen. *Acta Horti bot. Univ. Latv.* 11/12: 23–40.
Sloane, H. 1696. Plantae submarinae. In: *Catalogus plantarum, quae in insula Jamaica sponte proveniunt vel bulgo coluntur cum earundem synonymis et locis natalibus; adjectis aliis quibusdam, quae in insulis Maderae, Barbados, Nieves, et St. Christophori nasuntur, seu prodromi historiae naturalis Jamaicae.* London, [pp. 1–7] 323 pp + 43 p.
———. 1707–1725. Submarine plants, Vol. 1. In: *A voyage to the islands Madera, Barbados, Nieves, St. Christophers and Jamaica, with the natural history of the herbs and trees, four-footed beasts, fishes, birds, insects, reptiles, etc., of the last of these islands.* Vol. 1, clvi + 264 pp., 156 + 4 pls., 1707. Vol. 2, xvii + 499 pp., 157–274 + 11 pls. Folio. London, [pp. 49–64].
Smith, G. M. 1944. *Marine algae of the Monterey Peninsula, California.* Stanford Univ. Press, Stanford, Calif., ix + 622 pp, 98 pls.
Smith, J. E., and J. Sowerby. 1790–1814. *English botany* . . . London. 36 vols., comprising 2,592 pls.
———. 1843. *Supplement to the English Botany of the late Sir J. E. Smith and Mr. Sowerby.* Vol. III. J. Sowerby; Longman and Co. Sherwood and Co., London, plates 2797–2867.
Smith III, T. J. 1992. Forest structure. In: A. I. Robertson and D. M. Alongi (eds.), Coastal and estuarine studies 41. *Tropical mangrove ecosystems.* American Geophysical Union, Washington, D.C., pp. 101–136.
Snedaker, S. C., and E. J. Lahmann. 1988. Mangrove understorey absence: a consequence of evolution? *J. Trop. Ecol.* 4: 311–314.
Solier, A.J.J. 1846. Sur deux Algues zoosporées formant le nouveau genre *Derbesia. Rev. Bot.* [Duchartre] 1: 452–454.
———. 1847. Mémoire sur deux algues zoosporées devant former un genre distinct, le genre *Derbesia. Ann. Sci. Nat., Bot.*, sér. 3, 7: 157–166.
Solms-Laubach, H. 1895. Monograph of the Acetabularieae. *Trans. Linn. Soc. Lond., Ser. 2., Bot.* 5: 1–39, pls. 1–4.
Sommerfelt, S. C. 1826. *Supplementum Florae lapponicae quam edidit Dr. Georgius Wahlenberg.* Christianiae [Oslo], xii + 331 pp., 3 pls.
Sonder, O. G. 1845. Nova algarum genera et species, quas in itinere ad oras occidentales Novae Hollandiae, collegit L. Preiss, Ph. Dr. Bot. Zeitung (Berlin) 3: 49–57.
———. 1846–1848. Algae L. Agardh. In: C. Lehmann, Plantae Preissianae sive enumeratio plantarum quas in Australasia occidentali et meridionali-occidentali annis 1838–1841 collegit Ludovicus Preiss . . . Vol. 2. Hamburgi [Hamburg], pp. 148–160 (1846), 161–195 (1848).
———. 1871. Die Algen des tropischen Australiens. Abh. Naturwiss. Verein Hamburg 5: 33–74, pls. 1–6.
Sonder, O. W. 1855. Algae annis 1852 et 1853 collectae. *Linnaea* 26: 506–528.

———. 1881. In: F. Mueller, Fragmenta Phytographiae Australiae. Supplementum ad volumen undecinum: Algae Austraianae hactenus cognitae, pp. 1–42, 105–107 (Melbourne).

Sosa, E. H. 1985. Adiciones a la flora marina de Cuba. Chlorophyta III. *Contributions, Symposium on Marine Sciences and 7th Scientific Meeting of the Institute of Oceanology, 20th Anniversary*, Academy of Sciences of Cuba, pp. 172–178.

———. 1996. Further observations on *Acetabularia farlowii* (Dasycladales). Additions to the marine flora of Cuba. Chlorophyta IV. Published by International Conference Center, City of Havana, Cuba, 23 pp.

South, G. R. 1984. A checklist of marine algae of eastern Canada, second revision. *Can. J. Bot.* 62: 680–704.

South, G. R., and A. Cardinal. 1970. A checklist of marine algae of eastern Canada. *Can. J. Bot.* 48: 2077–2095.

South, G. R., and R. G. Hooper. 1980. Catalogue and atlas of the benthic marine algae of the Island of Newfoundland. *Mem. Univ. Nfld. Occas. Pap. Biol.* 3, 136 pp.

South, G. R., and A.D.R. N'Yeurt. 1993. Contributions to a catalogue of benthic marine algae of Fiji. II. *Caulerpa* and *Caulerpella* (Chlorophyta-Caulerpales). *Micronesica* 26: 109–138.

South, G. R., and P. A. Skelton. 2003. Catalogue of the marine benthic macroalgae of the Fiji Islands, South Pacific. *Aust. Syst. Bot.* 16: 699–758.

South, G. R., and I. Tittley. 1986. *A checklist and distributional index of the benthic marine algae of the North Atlantic Ocean*. Huntsman Marine Laboratory and British Museum (Natural History), St. Andrews and London, 76 pp.

South, G. R., I. Tittley, and A. Yoshinga. 2001. Subtidal benthic marine algae of the Phoenix Islands, Republic of Kiribati, Central Pacific. *Bot. Mar.* 44: 559–570.

Sprengel, K. 1827. *Caroli Linnaei . . . Systema Vegetabilium*. Editio decima sexta. Voluminis IV. Pars I. Classis 24. Gottingae [Göttingen], iv + 592 pp.

Sreenivasa Rao, P., and M. K. Trivedi. 1980. Reproduction in *Gelidiella*. In: T. V. Desikachary and V. N. Raja Rao (eds.), *Taxonomy of algae*. Univ. Madras, pp. 255–260.

Stackhouse, J. 1795–1801. *Nereis britannica* . . . xl + 112 pp., 17 pls. Bathoniae [Bath]. [Fasc. 1, pp. i–viii + 1–30, pls. I–VIII (1795). Fasc. 2, pp. ix–xxiv + 31–70, pls. IX–XII (1797). Fasc. 3, pp. xxv–xl + 71–112, pls. XIII–XVII (1801).]

———. 1809. Tentamen marino-cryptogamicum, ordinem novum, in genera et species distributum, in Classe XXIV ta Linnaei sisten. *Mém. Soc. Imp. Nat. Moscou* 2: 50–97.

———. 1816. *Nereis Britannica*, Bath 1795–1801. Ed. 2. Oxford., xii + 68 pp., 20 pls.

Stafford, N. B., and S. S. Bell, 2006. Space competition between seagrass and *Caulerpa prolifera* (Forsskal) Lamouroux following simulated disturbances in Lassing Park, FL. *J. Exp. Mar. Biol. Ecol.* 333: 49–57.

Stam, W. T., J. L. Olsen, S. F. Zaleski, S. N. Murray, K. R. Brown, and L. J. Walters. 2006. A forensic and phylogenetic survey of *Caulerpa* species (Caulerpales, Chlorophyta) from the Florida coast, local aquarium shops, and e-commerce: establishing a proactive baseline for early detection. *J. Phycol.* 42: 1113–1124.

Steentoft, M. 1967. A revision of the marine algae of São Tomé and Princípe (Gulf of Guinea). *J. Linn. Soc. (Bot.)* 60: 99–146.

Stegenga, H. 1985. The marine Acrochaetiaceae (Rhodophyta) of southern Africa. *S. Afr. J. Bot.* 51: 291–330.

Stegenga, H., H. R. Engledow, J. J. Bolton, and R. J. Anderson. 2001. The distribution of the family Erythropeltidaceae (Rhodophyta) along the temperate Southern African coast, with the description of *Erythrotrichia platyphylla nov. spec. Nova Hedwiga* 73: 367–379.

Stegenga, H., and I. Mol. 1983. *Flora van de Nederlandse Zeewieren*. KNNV, Amsterdam, 263 pp.

———. 1988. An annotated checklist of the marine algae of the Caribbean islands of Aruba and Bonaire. *Nova Hedwigia* 46: 433–480.

Stegenga, H., I. Mol., W. F. Prud'homme van Reine, and G. M. Lokhorst. 1997. Checklist of the marine algae of the Netherlands. *Gorteria* Suppl. 4: 3–57.

Stegenga, H., and M. Vroman. 1987. Notes on some Ceramiaceae (Rhodophyta) from Curaçao, especially those from the exposed northeast coast. *Blumea* 32: 397–426.

———. 1988. Additions to the marine algal flora of Curaçao, Netherlands Antilles. *Blumea* 33: 299–311.

Steidinger, K. A., and J. F. van Breedveld. 1971. Benthic marine algae from the waters adjacent to the Crystal River electric power plant (1969 and 1970). Professional Papers Series 16, Florida Dept. Natural Resources, Marine Research Laboratory, St. Petersburg, 46 pp.

Stephenson, T. A. 1944. The constitution of the intertidal fauna and flora of South Africa. Part II. *Ann. Natal Mus.* 10: 261–358.

Stephenson, T. A., and A. Stephenson. 1972. *Life between tidemarks of rocky shores*. W. H. Freeman, San Francisco, xii + 425 pp.

Stewart, J., and J. Norris. 1981. Gelidiaceae (Rhodophyta) from the northern Gulf of California, Mexico. *Phycologia* 20: 273–284.

Stockmayer, S. 1890. Ueber die Algengattung *Rhizoclonium*. *Verh. Zool.-Bot. Gesell. Wien* 40 (Abh.): 571–586.

Stout, J. P. 1984. The ecology of irregularly flooded salt marshes of the northeastern Gulf of Mexico: a community profile. U.S. Fish Wild. Serv. Biol. Rep. 85(7.1), xiv + 98 pp.

Strate, H. J. van der, S. A. Boele-Bos, J. L. Olsen, L. van de Zande, and W. T. Stam. 2002. Phylogeographic studies in the tropical seaweed *Cladophoropsis membrancea* (Chlorophyta, Ulvophyceae) reveal a cryptic species complex. *J. Phycol.* 38: 572–582.

Strenth, N. E. 2001. *Caulerpa prolifera* (Chlorophyta: Caulerpaceae) from the Laguna Madre of South Texas. *Tex. J. Sci.* 53: 187–189.

Svedelius, N. 1906. Reports on the marine algae of Ceylon. No. 1. Ecological and systematic studies of Ceylon species of *Caulerpa*. *Rep. Ceylon Mar. Biol. Lab.* 2: 81–144.

Tai, V., S. C. Lindstrom, and G. W. Saunders. 2001. Phylogeny of the Dumontiaceae (Gigartinales, Rhodophyta) and associated families based on SSU rDNA and internal transcribed spacer sequence data. *J. Phycol.* 37: 184–196.

Takamatsu, M. 1943. The species of *Sphacelaria* from Japan, I. *J. Sigenkagaku Kenkyusyo* 1:153–187.

Tan, I. H., J. Blomster, G. Hansen, E. Leskinen, C. A. Maggs, D. G. Mann, H. J. Sluiman, and M. J. Stanhope. 1999. Molecular phylogenetic evidence for a reversible morphogenetic switch controlling the gross morphology of two common genera of green seaweeds, *Ulva* and *Enteromorpha*. *Mol. Biol. Evol.* 16: 1011–1018.

Tan, I. H., and L. D. Druehl. 1994. A molecular analysis of *Analipus* and *Ralfsia* (Phaeophyceae) suggests the order Ectocarpales is polyphyletic. *J. Phycol.* 30: 721–729.

Tanaka, J. 1990. Brown algae from the Amami Islands. *Mem. Nat. Sci. Mus., Tokyo* 23: 23–32.

Tanaka, J., and M. Chihara. 1984. *Taxonomic studies of Japanese mangrove macroalgae*. I. Genus *Bostrychia* (1) (Ceramiales, Rhodophyceae). *Bull. Nat. Sci. Mus. [Tokyo], Ser. B (Bot.)*, 10: 115–126.

Tanner, W. F. 1960. Florida coastal classification. *Trans. Gulf Coast Assoc. Geol. Soc.* 10: 259–260.

Tatewaki, M. 1966. Formation of a crustaceous sporophyte with unilocular sporangia in *Scytosiphon lomentaria*. *Phycologia* 6: 62–66.

Taylor W. R. 1928. *The marine algae of Florida, with special reference to the Dry Tortugas*. Publications of the Carnegie Institution 379, Washington, D.C., v + 219 pp.

———. 1929. Notes on the marine algae of Florida. *Bull. Torrey Bot. Club* 56: 199–210.

———. 1930. Algae collected by the "Hassler," "Albatross" and Schmitt expeditions. I. Marine algae from Brazil. *Am. J. Bot.* 17: 627–634.

———. 1933. Notes on algae from the tropical Atlantic Ocean, II. *Pap. Mich. Acad. Sci. Arts Lett.* 17: 395–407.

———. 1936. Notes on algae from the tropical Atlantic Ocean. III. *Pap. Mich. Acad. Sci. Arts Lett.* 21: 199–207.

———. 1937a. Marine Algae of the northeastern coast of North America. Univ. Michigan Press., Ann Arbor, vii + 427 pp.

———. 1937b. Notes on North Atlantic marine algae. I. *Pap. Mich. Acad. Sci. Arts Lett.* 22: 225–233.

———. 1939('1938'). Algae collected by the Hassler, Albatross and Schmitt expeditions, II. Marine algae from Uruguay, Argentina, the Falkland Islands and the Strait of Magellan. *Pap. Mich. Acad. Sci. Arts Lett.* 24: 127–164,7 pls.

———. 1939. Algae from the presidential cruise of 1938. *Smithson. Contrib. Knowl., Misc. Ser.* 98(9): 1–18, 2 pls.

———. 1942. *Caribbean marine algae of the Allan Hancock Expedition, 1939*. Allan Hancock Atlantic Exped. Rep. 2, 193 pp., including 20 pls.

———. 1943('1942'). Marine algae from Haiti collected by H. H. Bartlett in 1941. *Pap. Mich. Acad. Sci. Arts Lett.* 28: 143–163.

———. 1945. *Pacific marine algae of the Allan Hancock expeditions to the Galapagos Islands*. Allan Hancock Pacific Expedition. Rep. 12, 528 pp., 100 pls.

———. 1951. Structure and reproduction of *Chrysophaeum lewisii*. *Hydrobiologia* 3: 122–130.

———. 1952. The algal genus *Chrysophaeum*. *Bull. Torrey Bot. Club* 79: 79.

———. 1954a. Distribution of marine algae in the Gulf of Mexico. *Pap. Mich. Acad. Sci. Arts Lett.* 39: 85–109.
———. 1954b. Sketch of the character of the marine algal vegetation of the shores of the Gulf of Mexico. In: P. S. Galtstoff (ed.), *Gulf of Mexico, its origin, waters and marine life.* U.S. Fish. Bull. 89, pp. 177–192.
———. 1955. Notes on algae from the tropical Atlantic Ocean. IV. *Pap. Mich. Acad. Sci. Arts Lett.* 40: 67–76 [-78].
———. 1957. *Marine algae of the northeastern coast of North America.* Rev. ed. Univ. Michigan Press, Ann Arbor, ix + 509 pp.
———. 1960. *Marine algae of the eastern tropical and subtropical coasts of the Americas.* Univ. Michigan Press, Ann Arbor, xi + 870 pp.
———. 1962a. Marine algae from the tropical Atlantic Ocean: V. Algae from the Lesser Antilles. *Contrib. U.S. Natl. Herb.* 36: 43–62.
———. 1962b. Two undescribed species of *Halimeda. Bull. Torrey Bot. Club* 89: 172–177.
———. 1962c. Observations on *Pseudobryopsis* and *Trichosolen* (Chlorophyceae-Bryopsidaceae) in America. *Brittonia* 14: 58–65.
———. 1962d. A note on *Bryopsis* in the West Indies. *Phycologia* 2: 24–28.
———. 1966. Records of Asian and western Pacific marine algae, particularly algae from Indonesia and the Philippines. *Pac. Sci.* 20: 342–359.
———. 1967. Species of *Caulerpa* (Chlorophyceae) collected on the International Indian Expedition. *Blumea* 15: 45–53.
———. 1969. Notes on the distribution of West Indian marine algae particularly in the Lesser Antilles. *Contrib. Univ. Mich. Herb.* 9: 125–203.
———. 1975. A new species of *Halimeda* from Malaysia. *Contrib. Univ. Mich. Herb.* 11: 81–83.
Taylor W. R., and I. A. Abbott. 1973. A new species of *Botryocladia* from the West Indies. *Br. Phycol. J.* 8: 409–412.
Taylor W. R., A. B. Joly, and A. J. Bernatowicz. 1953. The relation of *Dichotomosiphon pusillus* to the algal genus *Boodleopsis. Pap. Mich. Acad. Sci. Arts Lett.* 38: 97–107 [–108].
Taylor, W. R., and C. F. Rhynne. 1970. Marine algae of Dominica. *Smithson. Contrib. Bot.* 3, 16 pp., 2 figs.
Teal, J. M. 1962. Energy flow in the salt marsh ecosystem of Georgia. *Ecology* 43: 614–624.
Thivy, F. 1959. On the morphology of the gametophytic generation of *Padina gymnospora* (Kuetz.) Vickers. *J. Mar. Biol. Assoc. India* 1: 69–76.
Thivy, F., and V. Visalakshmi. 1963. *Caulerpa charoides* (Harv. *ex* W.-v. Bosse) *nov. comb.*, a new record for the Indian Ocean region. *Bot. Mar.* 5: 101–104.
Thuret, G. 1854. Note sur la synonymie des *Ulva lactuca et latissima* L. suivie de quelques remarques sur la tribu des Ulvacees. *Mém. Soc. Sci. Nat. Cherbourg* 2: 17–32
Tittley, I., and A. I. Neto. 2005. The marine algal (seaweed) flora of the Azores: additions and amendments. *Bot. Mar.* 48: 248–255.
Tokida, J., and T. Masaki. 1956. Studies on the reproductive organs of red algae. II. On *Erythrophyllum gmelini* (Grun.) Yendo. *Bull. Fac. Fish. Hokkaido Univ.* 7: 63–71.
Tomasko, D. A., C. J. Dawes, and M. O. Hall. 1996. The effects of anthropogenic nutrient enrichment on turtle grass (*Thalassia testudinum*) in Sarasota Bay, Florida. *Estuaries* 19: 448–456.
Tomlinson, P. B. 1982. VII Helobiae (Alismatidae). In: C. R. Metchaf (ed.), *Anatomy of the Monocotylodens.* Caredon Press, Oxford, 539 pp.
Toste, M. F., M. I. Parente, A. I. Neto, and R. L. Fletcher. 2003a. Life history of *Colpomenia sinuosa* (Scytosiphoniacae, Phaeophyceae) in the Azores. *J. Phycol.* 39: 1268–1274.
———. 2003b. Life history of *Hydroclathrus clathratus* (Scytosiphonaceae, Phaeophyta) in the Azores. *Cryptogam. Algol.* 24: 209–218.
Trainor, F. R. 1978. *Introductory phycology.* John Wiley and Sons, New York, xvi + 525 pp.
Trevisan, V. B. A. 1845. *Nomenclator algarum . . .* Padoue [Padova], 80 pp.
Tseng, C. K. 1936. On marine algae new to China. *Bull. Fan Mem. Inst. Biol.* 7:169–196.
———. 1937. On marine algae new to China. *Bull. Fan Mem. Inst. Biol.* 7:231–255.
———. 1941. Studies on the Chinese species of *Liagora. Bull. Fan Mem. Inst. Biol.*, *Bot.* ser. 10: 265–282.
———. 1944. Marine algae of Hong Kong VI. The genus *Polysiphonia. Pap. Mich. Acad. Sci. Arts Lett.* 29: 67–82.
———. 1945. New and unrecorded marine algae of Hong Kong. *Pap. Mich. Acad. Sci. Arts Lett.* 30: 157–171.

———. 1946. Seaweed products and their uses in America. *J. N. Y. Bot. Gard.* 47: 1–10.
———. 1948. Marine algae of Hong Kong VII. The order Bangiales. *Lingnan Sci.* 22: 121–131.
———. 1983. *Common seaweeds of China.* Science Press, Beijing, x + 316 pp.
Tseng, C. K., and B. Lu. 2000. Fucales. Flora Algarum Marinarum Tomus III Phaeophyta, No. 11. Science Press, Beijing, xi + 238 pp.
Tsuda, R. T. 1972. Marine benthic algae of Guam. I. Phaeophyceae. *Micronesica* 8: 87–115.
———. 2004. *Dictyota* (Phaeophyceae) from Micronesia. In: I. A. Abbott and K. J. McDermid (eds.), *Taxonomy of economic seaweeds with reference to the Pacific and other locations.* Vol. 9. Hawaii Sea Grant College Program, Univ. Hawaii, Hilo, pp. 41–55.
Tsutsui, I., Q. N. Huybh, H. D. Nguyên, S. Arai, and T. Yoshida. 2005. *The common marine plants of southern Vietnam.* Japan Seaweed Association, 250 pp.
Turner, D. 1802. Descriptions of four new species of *Fucus. Trans. Linn. Soc. [Lond.]* 6: 125–136, pls. 8–10.
———. 1807–1808. Fuci sive Plantarum Fucorum Generi a Botanicis Ascriptarum Icones Descriptiones et Historia. Vol. 1. London, [iii +] 164 [+ 2] pp., pls. 1–71.
———. 1808–1809. Fuci sive Plantarum Fucorum Generi a Botanicis Ascriptarum Icones Descriptiones et Historia. Vol. 2. London, 162 [+ 2] pp., pls. 72–134.
———. 1809–1811. Fuci sive Plantarum Fucorum Generi a Botanicis Ascriptarum Icones Descriptiones et Historia. Vol. 3. London, 148 pp., pls. 135–196.
———. 1811–1819. Fuci sive Plantarum Fucorum Generi a Botanicis Ascriptarum Icones Descriptiones et Historia. Vol. 4. London, [iii +] 153 [+ 2] + 7 pp., pls. 197–258 [issued in parts, beginning at least as early as 1815].
Ugadim,Y., S.M.P. de B. Guimarães, and A. I. Kanagawa. 1986. Estudos em *Acrothamnion*, *Antithamnion* e *Antithamnionella* (Rhodophyta, Ceramiales) do Brasil. *Rickia* 13: 35–47.
Vahl, M. 1802. Endeel kryptogamiske Planter fra St.-Croix. *Skr. Naturhist.-Selsk.* 5: 29–47, pls. 133–162.
Valet, G. 1969. Contribution à l étude des Dasycladales. 2. Cytologie et reproduction. 3. Révision systématique. *Nova Hedwigia* 17: 551–664.
Verbruggen, H., O. De Clerck, E. Cocquyt, W.H.C.F. Kooistra, and E. Coppejans. 2005a. Morphometric taxonomy of siphonous green algae: a methodological study within the genus *Halimeda* (Bryopsidales). *J. Phycol.* 41: 126–139.
Verbruggen, H., O. De Clerck, and E. Coppejans. 2005b. Deviant segments hamper a mophometric approach towards *Halimeda* taxonomy. *Cryptogam. Bot.* 26: 259–274.
Verbruggen, H., O. De Clerck, A.D.R. N'Yeurt, H. Spalding, and P. S. Vroom. 2006. Phylogeny and taxonomy of *Halimeda incrassata*, including descriptions of *H. kanaloana* and *H. heteromorpha spp. nov.* (Bryopsidales, Chlorophyta). *Eur. J. Phycol.* 41: 337–362.
Verbruggen, H., O. De Clerck, W.H.C.F. Kooistra, and E. Coppejans. 2005c. Molecular and morphometric data pinpoint species boundaries in *Halimeda* section *Rhipsalis* (Bryopsidales, Chlorophyta). *J. Phycol.* 41: 606–621.
Verbruggen, H., O. De Clerck, W.H.C.F. Kooistra, P. S. Vroom, A.D.R. N'Yeurt, and E. Coppenjans. 2005d. Molecular and morphometric data pinpoint species boundaries in *Halimeda* section *Rhipsalis. Phycologia* 44(4, supplement): 105.
Verbruggen, H., and W.H.C.F. Kooistra. 2004. Morphological characterization of lineages within the calcified tropical genus *Halimeda* (Bryopsidales, Chlorophyta). *Eur. J. Phycol.* 39: 213–228.
Vergés, A., J. M. Utgé, and C. Rodríguez-Pieto. 2004. Life histories of *Predaea ollivieri* and *P. pusilla* (Nemastomatales, Rhodophyta). *Eur. J. Phycol.* 39: 411–421.
Verheij, E., and W. F. Prud'homme van Reine. 1993. Seaweeds of the Spermonde Archipelago, SW Sulawesi, Indonesia. *Blumea* 37: 385–510.
Verlaque, M., E. Ballesteros, and A. Antonius. 2000. *Metapeyssonnelia corallepida sp. nov.* (Peyssonneliaceae, Rhodophyta), an Atlantic encrusting red alga overgrowing corals. *Bot. Mar.* 43: 191–200.
Vickers, A. 1897('1896'). Contribution à la fore algologiques des Canaries. Ann. Sci. Nat., Bot., Sér. 8, 4: 293–306.
———. 1905. Liste des algues marines de la Barbade. *Ann. Sci. Nat., Bot.*, sér. 9, 1: 45–66.
———. 1908. *Phycologia* barbadensis. Iconographie des algues marines récoltées à l'île Barbade (Antilles) (Chlorophycées et Phéophycées). Paris, 44 pp., 87 pls. (avec texte explicatif par M. H. Shaw).
Villalard-Bohnsack, M. 1995. *Illustrated key to the seaweeds of New England.* Rhode Island Natural History Soc. Publ., Kingston, 144 pp.

Vinogradova, K. L. 1969. A contribution to the taxonomy of the order Ulvales (Chlorophyta). *Bot. Zh. SSSR* 54: 1347–1355.

Virnstein, R. W., and P. A. Carbonara. 1985. Seasonal abundance and distribution of drift algae and seagrasses in the mid-Indian River Lagoon, Florida. *Aquat. Bot.* 23: 67–82.

Vroom, P. S., and C. M. Smith. 2003. Reproductive features of Hawaiian *Halimeda velasquezii* (Bryopsidales, Chlorophyta), and an evolutionary assessment of reproductive characters in *Halimeda. Cryptogam. Algol.* 24: 355–370.

Vroom, P. S., C. M. Smith, and S. C. Keeley. 1998. Cladistics of the Bryopsidales: a preliminary analysis. *J. Phycol.* 34: 351–360.

Waisel, W. 1972. *Biology of halophytes.* Academic Press, New York, xiii + 395 pp.

Walters, L. J., and C. M. Smith, 1994. Rapid rhizoid production in *Halimeda discoidea* Decaisne (Chlorophyta, Cayulerpales) fragments: a mechanism for survival after separation from adult thalli. *J. Exp. Mar. Biol. Ecol.* 175: 105–120.

Walters, L. J., C. M. Smith, J. A. Coyer, C. L. Hunter, K. S. Beach, and P. S. Vroom. 2002. Asexual propagation in the coral reef macroalga *Halimeda* (Chlorophyta, Bryopsidales): production, dispersal and attachment of fragments. *J. Exp. Mar. Biol. Ecol.* 278: 47–65.

Wang, H. W., S. Kawaguchi, T. Horiguchi, and M. Masuda. 2000. Reinstatement of *Grateloupia catenata* (Rhodophyta, Halymeniaceae) on the basis of morphology and rbcL sequences. *Phycologia* 39: 228–237.

———. 2001. A morphological and molecular assessment of the genus *Prionitis* J. Agardh (Halymeniaceae, Rhodophyta). *Phycol. Res.* 49: 251–261.

Wang,W.-L., and Y.-M. Chiang. 2001. The reproductive development of the red alga *Actinotrichia fragilis* (Galaxauraceae, Nemaliales). *Eur. J. Phycol.* 36: 377–383.

Wang,W.-L., S.-L. Liu and S.-M. Lin. 2005. Systematics of the calcified genera of the Galaxauraceae (Nemaliales , Rhodophyta) with an emphasis on Taiwan species. *J. Phycol.* 41: 685–705.

Waters, D., M. M. Lopez-Bautista, S. Fredericq, and R. L. Chapman. 2001. Algae on offshore oil platforms in the Gulf of Mexico: biodiversity and biotechnology. *Phycologia* 40 (supplement): 115–116.

Webb, S. D. 1990. Historical biogeography. In: R. L. Myers and J. J. Ewel (eds.), *Ecosystems of Florida.* Univ. Central Florida Press, Orlando, pp. 481–516.

Weber, F., and D. M. H. Mohr. 1805. Einige Worte über unsre bishorigen, hauptsächlich carpologischen Zergliederungen von kryptogamischen Seegewächsen. *Beitr. Naturkd.* 1: 204–329.

Weber-van Bosse, A. 1896. On a genus of siphonean algae *Pseudocodium. J. Linn. Soc. Lond. Bot.* 32: 209–212, pl. I.

———. 1898. Monographie des *Caulerpes. Ann. Jard. Bot. Buitenzorg* 15: 243–401, pls. 20–34.

———. 1901. Ètudes sur les algues de l'Archipel Malaisien. [III.]. *Ann. Jard. Bot. Buitenzorg* 17: 126–141, pl. XVII–XIX.

———. 1904a. Note sur deux algues de l' Archipel Malaisien. *Recueil des Travaux Botaniques Néerlandais* 1: 96–105.

———. 1904b. Introduction. In: A. Weber-van Bosse and M. Foslie, *The Corallinaceae of the Siboga-Expedition.* Siboga Exped. Monogr. 61, Leiden, pp. 1–9, fig. 1, 2.

———. 1904c. Corallinaceae verae of the Malay Archipelago. In: A. Weber-van Bosse and M. Foslie, *The Corallinaceae of the Siboga-Expedition.* Siboga Exped. Monogr. 61, Leiden, pp. 79–110, pls. XIV–XVI.

———. 1910. Notice sur quelques genres nouveaux d'algues de l'Archipel Malaisien. *Ann. Jard. Bot. Buitenzorg*, Leiden. Sér. 2, 9: 1–25.

———. 1913. *Liste des algues du Siboga.* I. Myxophyceae, Chlorophyceae, Phaeophyceae avec le concours de M. Th. Reinbold. *Siboga Exped. Monogr.* 59a, Leiden, pp. 1–186, pls. 12–14.

———. 1914. The Percy Sladen Trust Expedition to the Indian Ocean in 1905, under the leadership of Mr. J. Stanley Gardiner, M.A. Reports: No. XIV. Marine algae, Rhodophyceae. *Trans. Linn. Soc. Lond., Ser. 2, Zoology* 16: 269–306, pls. 16–18.

———. 1921. *Liste de algues du Siboga.* II. Rhodophyceae. Première partie. Protoflorideae, Nemalionales, Cryptonemiales. *Siboga Exped. Monogr.* 59b, Leiden, pp. 187–310, pls. VI–VIII.

———. 1923. *Liste des algues du Siboga.* III. Rhodophyceae, Seconde partie. Ceramiales. *Siboga Exped. Monogr.* 59c, Leiden, pp. 311–392, pls. IX, X.

———. 1928. *Liste des algues du Siboga.* IV. Rhodophyceae. Troisième partie. Gigartinales et Rhodymeniales et tableau de la distribution des Chlorophycées, Phaeophycées et Rhodophycées de l'Archipel Malaisien. *Siboga Exped. Monogr.* 59d, Leiden, pp. 393–533, pls. XI–XVI.

Welten, M. C. M., P. A. J. Audiffred, and W. F. Prud'homme van Reine. 2002. Notes on marine algae collected in Guinea-Bissau, Tropical West Africa. *Bot. Mar.* 45: 380–384.

West, A. L., A. C. Mathieson, A. S. Klein, C. D. Neefus, and T. L. Bray. 2005. Molecular ecological studies of New England species of *Porphyra* (Rhodophyta, Bangiales). *Nova Hedwigia* 80: 1–24.

West, J. A., and G. C. Zuccarello. 1995. New records of *Bostrychia pinnata* and *Caloglossa ogasawaraensis* (Rhodophyta) from the Atlantic, U.S.A. *Bot. Mar.* 38: 303–306.

Wettstein, A. 1901. *Handbuch der systematischen Botanik*. Leipzig, Vienna, Austria.

Wilce, R. T., and A. N. Davis. 1984. Development of *Dumontia contorta* (Dumontiaceae, Cryptonemiales) compared with that of other red algae. *J. Phycol.* 20: 336–351.

Wilkes, R. J., L. M. McIvor, and M. D. Guiry. 2005a. Using rbcL sequence data to reassess the taxonomic position of some *Grateloupia* and *Dermocorynus* species (Halymeniaceae, Rhodophyta) from the north-eastern Atlantic. *Eur. J. Phycol.* 40: 53–60.

———. 2005b. The taxonomic position of *Maripelta rotata* (Rhodymeniaceae, Rhodophyta). *Cryptogam. Algol.* 26: 67–75.

———. 2006. Vegetative morphology and rbcL phylogeny of some members of the genera *Botryocladia* and *Irvinea* (Rhodymeiaceae, Rhodophyta). *Phycologia* 45: 481–494.

Wilkinson, M., and E. Burrows. 1972. An experimental taxonomic study of the algae confused under the name *Gomontia polyrhiza*. *J. Mar. Biol. Assoc. U.K.* 52: 49–57.

Willdenow, C. L. 1804. *Mertensia*, ett nytt slägte af Ormbunkarne. *Öfvers. Förh. K. Sven. Vetensk.-Akad. Handl. N. S.* 25: 165–170.

Wille, N. 1880. Om en ny endophytisk alge. Forh. Vidensk Selsk. Krist. 1880: 1–4

———. 1885. Alger fra Novaia-Zemlia og Kara-Havet. Dijmphna-Togtets zoöl.-bot. Udbytte, pp. 81–90, pl. 12, figs. 1, 7, 8.

———. 1890. Chlorophyceae. In: A. Engler and K. Prantl (eds.), *Die natürlichen Pflanzenfamilien* . . . I. Teil, Abt 2. Leipzig, pp. 24–161.

———. 1901. Studien über Chlorophyceen I–VII. Skr. Vidensk.-Selsk. Christiana, Math.-Naturvidensk. Kl. 1900 (6): 3–46.

———. 1906. Algologische untersuchungen an der biologischen station in Drontheim. I–VII. *K. Nor. Vidensk. Selsk., Skr. (Trondheim)* 1906 (3), 38 pp.

———. 1909–1910. Conjugatae und Chlorophyceae. In: A. Engler and K. Prantl (eds.), *Die natürlichen Pflanzenfamilien* . . . I. Teil, Abt 2. Leipzig, pp. 1–96 (1909); pp. 97–136 (1910).

Williams, L. G. 1948. Seasonal alternation of marine floras at Cape Lookout, North Carolina. *Am. J. Bot.* 35: 682–695.

———. 1951. Algae of the Black Rocks. *J. Elisha Mitchell Sci. Soc.* 67: 149–159.

Williams, R. B. 1962. The ecology of diatom populations in a Georgia salt marsh. Ph.D. diss., Harvard Univ., Cambridge, 146 pp.

Williams, S. L. 1990. Experimental studies of Caribbean seagrass bed development. *Ecol. Monogr.* 60: 449–469.

Williams, S. L., and T. R. Fischer. 1985. Kinetics of N^{15} labeled ammonium uptake by *Caulerpa cupressoides* (Chlorophyta). *J. Phycol.* 21: 287–296.

Wilson, H. L. 1910. *Gracilariophila*, a new parasite on *Gracilaria confervoides*. *Univ. Calif. Publ. Bot.* 4: 75–85.

Wiseman, D. R. 1966. A preliminary survey of the Rhodophyta of South Carolina. M.A. thesis, Duke Univ., Durham, N.C., v + 190 pp.

Withall, R. D., and G. W. Saunders. 2006. Combining small and large subunit ribosomal DNA genes to resolve relationships among orders of the Rhodymeniophycidae (Rhodophyta): recognition of the Acrosymphytales ord. nov. and Sebdeniales ord. nov. *Europ. J. Phycol.* 41: 379–394.

Withell, A. F., A.J.K. Millar, and G. T. Kraft. 1994. *Taxonomic* studies of the genus *Gracilaria* (Gracilariales, Rhodophyta) from Australia. *Aust. Syst. Bot.* 7: 281–352.

Withering, W. 1776. *A botanical arrangement of all the vegetables naturally growing in Great Britain* . . . Birmingham, xcvi + 838 pp., 12 pls.

———. 1796. *An arrangement of British plants* . . . Third edition . . . Vol. IV. Birmingham, 418 pp., pls. XVII, XVIII, XXXI.

Woelkerling, W. J. 1971. Morphology and taxonomy of the *Audouinella* complex (Rhodophyta) in southern Australia. *Aust. J. Bot.*, suppl. ser. 1, 91 pp.

———. 1972. Studies on the *Audouinella microscopica* (Naeg.) Woelk. complex (Rhodophyta). *Rhodora* 74: 85–96.

———. 1973a. The *Audouinella* complex (Rhodophyta) in the Western Sargasso Sea. *Rhodora* 75: 78–101.
———. 1973b. The morphology and systematics of the *Audouinella* complex (Acrochaetiaceae, Rhodophyta) in northeastern United States. *Rhodora* 75: 529–621
———. 1975. On the epibiotic and pelagic Chlorophyceae, Phaeophyceae, and Rhodophyceae of the western Sargasso Sea. *Rhodora* 77: 1–40.
———. 1985a. Proposal to conserve *Lithothamnion* against *Lithothamium* (Rhodophyta: Corallinaceae). *Taxon* 34: 302–303.
———. 1985b. A taxonomic reassessment of *Spongites* (Corallinaceae, Rhodophyta) based on studies of Kützing's original collections. *Br. Phycol. J.* 20: 123–153.
———. 1987. The status and disposition of *Paragoniolithon* (Corallinaceae, Rhodophyta). *Phycologia* 26: 144–148.
———. 1988. *The coralline red algae: an analysis of the genera and subfamilies of nongeniculate Corallinaceae.* British Museum (Natural History), London. Oxford Univ. Press, xi + 268 pp.
———. 1990. An introduction. In: K. M. Cole and R. G. Sheath (eds.), *Biology of the red algae.* Cambridge Univ. Press, Cambridge, pp. 1–6.
———. 1996a. Subfamily Melobesioideae Bizzozero 1885: 109 (as 'Melobesieae'). In: H.B.S. Womersley, *The marine benthic flora of southern Australia. Rhodophyta.* Part IIIB, Gracilariales, Rhodymeniales, Corallinales and Bonnemaisoniales. Flora of Australia Supplementary Series no. 5, Australian Biological Resources Study, Canberra, pp. 164–210.
———. 1996b. Subfamily Lithophylloideae Setchell 1943: 134 (as 'Lithophylleae). In: H.B.S. Womersley, *The marine benthic flora of southern Australia. Rhodophyta.* Part IIIB, Gracilariales, Rhodymeniales, Corallinales and Bonnemaisoniales. Flora of Australia Supplementary Series no. 5, Australian Biological Resources Study, Canberra, pp. 214–237.
Woelkerling, W. J., and S. J. Campbell. 1992. An account of southern Australian species of *Lithothamnion* (Corallinaceae, Rhodophyta). *Bull. Br. Mus. (Nat. Hist.) Bot.* 22(1): 1–107.
Woelkerling, W. J., Y. M. Chamberlain, and P. C. Silva. 1985. A taxonomic and nomenclatural reassessment of *Tenarea*, *Titanoderma* and *Dermatolithon* (Corallinaceae, Rhodophyta) based on studies of type and other critical specimens. *Phycologia* 24: 317–337.
Woelkerling, W. J., and W. A. Nelson. 2004. A baseline summary and analysis of the taxonomic biodiversity of coralline red algae (Corallinales, Rhodophya) recorded from the New Zealand region. *Cryptogam. Algol.* 25: 39–106.
Woelkerling, W. J., and D. Penrose. 1988. *Spongites* Kützing, 1841: 30. In: W. J. Woelkerling (ed.), *The coralline red algae: an analysis of the genera and subfamilies of nongeniculate Corallinaceae.* British Museum (Natural History), Oxford Univ. Press, London, pp. 150–157.
Woelkerling, W. J., and H.B.S. Womersley. 1994. Order Acrochaetales Feldmann 1953: 12. In: H.B.S. Womersley, *The marine benthic flora of southern Australia. Rhodophyta.* Part IIIa, Bangiophyceae and Florideophyceae (Acrochaetiales, Nemaliales, Gelidiales, Hildenbrandiales and Gigartinales *sensu lato*). Australian Biological Resources Study, Canberra, pp. 42–76.
Womersley, H.B.S. 1948. The marine algae of Kangaroo Island. II. The Pennington Bay Region. *Trans. R. Soc. Aust.* 72: 143–166.
———. 1950. The marine algae of Kangaroo Island III. List of species 1. *Trans. R. Soc. Aust.* 73: 137–197.
———. 1956. A critical survey of the marine algae of southern Australia. I. Chlorophyta. *Aust. J. Mar. Freshw. Res.* 7: 343–383.
———. 1967. A critical survey of the marine algae of southern Australia. II. Phaeophyta. *Aust. J. Bot.* 15: 189–270.
———. 1978. Southern Australian species of *Ceramium* Roth (Rhodophyta). *Aust. J. Mar. Freshw. Res.* 29: 205–257.
———. 1979. Southern Australian species of *Polysiphonia* Greville (Rhodophyta). *Aust. J. Bot.* 27:459–528.
———. 1984. *The marine benthic flora of southern Australia.* Part I. Government Printer, Adelaide, South Australia, 329 pp.
———. 1987. *The marine benthic flora of southern Australia.* Part II. South Australian Government Printing Division, Adelaide, 484 pp.
———. 1994. *The marine benthic flora of southern Australia. Rhodophyta.* Part IIIa, Bangiophyceae and Florideophyceae (Acrochaetiales, Nemaliales, Gelidiales, Hildenbrandiales and Gigartinales *sensu lato*). Australian Biological Resources Study, Canberra, 508 pp.
———. 1996. *The marine benthic flora of southern Australia. Rhodophyta.* Part IIIB, Gracilariales, Rhodymeniales, Corallinales and Bonnemaisoniales. Australian Biological Resources Study, Canberra, 392 pp.

———. 1998. *The marine benthic flora of southern Australia.* Part IIIC, Ceramiales—Ceramiaceae, Dasyaceae. Australian Biological Resources Study, Canberra, 535 pp.

———. 2003. *The marine benthic flora of southern Australia.* Part IIID, Ceramiales—Delesseriaceae, Sarcomeniaceae, Rhodomelaceae. Australian Biological Resources Study, Canberra, 533 pp.

Womersley, H.B.S., and A. Bailey. 1970. Marine algae of the Solomon Islands. *Philos. Trans. [R. Soc. Lond.], Ser. B, Biol. Sci.*, 259: 257–352.

Womersley, H.B.S., and S. A. Cartledge. 1975. The southern Australian species of *Spyridia* (Ceramiaceae: Rhodophyta). *Trans. R. Soc. S. Aust.* 99: 221–233.

Womersley, H.B.S., and M. D. Guiry 1994. Order Gelidiales Kylin 1923: 132. In: H.B.S. Womersley, *The marine benthic flora of southern Australia. Rhodophyta.* Part IIIa, Bangiophyceae and Florideophyceae (Acrochaetiales, Nemaliales, Gelidiales, Hildenbrandiales and Gigartinales *sensu lato*). Australian Biological Resources Study, Canberra, pp. 118–142.

Womersley, H.B.S., and H. W. Johansen. 1996. Subfamily Corallinoideae (Areschoug) Foslie 1908: 19. In: H.B.S. Womersley, *The marine benthic flora of southern Australia. Rhodophyta.* Part IIIB, Gracilariales, Rhodymeniales, Corallinales and Bonnemaisoniales. Flora of Australia Supplementary Series No. 5, Australian Biological Resources Study, Canberra, pp. 288–317.

Womersley, H.B.S., and G. T. Kraft. 1994. Family Nemastomataceae Schmitz 1892: 22, *nom. cons.* In: H.B.S. Womersley, *The marine benthic flora of southern Australia. Rhodophyta.* Part IIIa, Bangiophyceae and Florideophyceae (Acrochaetiales, Nemaliales, Gelidiales, Hildenbrandiales and Gigartinales *sensu lato*). Australian Biological Resources Study, Canberra, pp. 270–285.

Womersley, H.B.S., and J. A. Lewis. 1994. Family Halymeniaceae Bory 1828: 158. In: H.B.S. Womersley, *The marine benthic flora of southern Australia. Rhodophyta.* Part IIIa, Bangiophyceae and Florideophyceae (Acrochaetiales, Nemaliales, Gelidiales, Hildenbrandiales and Gigartinales *sensu lato*). Australian Biological Resources Study, Canberra, pp. 167–218.

Womersley, H.B.S., and E. M. Wollaston. 1998. Tribe Callithamnieae Schmitz 1889: 450. In: H.B.S. Womersley: *The marine benthic flora of Southern Australia.* Part IIIC, Ceramiales—Ceramiaceae, Dasyaceae. Australian Biological Resources Study, Canberra, pp. 231–269.

Won, B. Y., T. O. Cho, J. West, and S. Fredericq. 2005. *Spyridia filamentosa* (Ceramiales, Rhodophyta): a cosmopolitan red algal species? *Phycologia* 44(4, supplement): 112.

Woodward, T. J. 1794a. Description of *Fucus dasyphyllus. Trans. Linn. Soc. [Lond.]* 2: 239–241.

———. 1794b. Description of two new British Fuci. *Trans. Linn. Soc. [Lond.]* 2: 29–31.

———. 1797. Observation upon the generic character of *Ulva*, with descriptions of some new species. *Trans. Linn. Soc. [Lond.]* 3: 46–58.

Woolcott, G. W., K. Knöller, and R. J. King. 2001. Phylogeny of the Bryopsidaceae (Bryopsidales, Chlorophyta): cladistic analyses of morphological and molecular data. *Phycologia* 39: 471–481.

Woronin, M. 1869. Beitrag zur Kenntnis der Vaucherien. *Bot. Zeitung* 27: 137–144, 153–161.

Wulfen, F. X. 1803. Cryptogama aquatica. *Arch. Bot. (Leipzig)* 3: 1–64, pl. 1.

Wynne, M. J. 1969. Life history and systematic studies of some Pacific North American Phaeophyceae (brown algae). *Univ. Calif. Publ. Bot.* 50: 1–88.

———. 1984. The correct name for the type of *Hypoglossum* Kützing (Delesseriaceae, Rhodophyta). *Taxon* 33: 85–87.

———. 1985a. Notes on *Herposiphonia* (Rhodomelaceae, Rhodophyta) in South Africa, with a description of a new species. *Cryptogam. Algol.* 5: 167–177.

———. 1985b. Concerning the names *Scagelia corallina* and *Heterosiphonia wurdemannii* (Ceramiales, Rhodophyta). *Cryptogam. Algol.* 6: 81–90.

———. 1985c. Nomenclatural assessment of *Goniotrichum* Kützing, *Erythrotrichia* Areschoug, *Diconia* Harvey, and *Stylonema* Reinsch. *Taxon* 34: 502–505.

———. 1986. A checklist of benthic marine algae of the tropical and subtropical western Atlantic. *Can. J. Bot.* 64: 2239–2281.

———. 1989. The re-instatement of *Hydropuntia* Montagne (Gracilariaceae, Rhodophyta). *Taxon* 38: 476–479.

———. 1991. A change in the name of the type of *Chondria* C. Agardh (Rhodomelaceae, Rhodophyta). *Taxon* 40: 316–318.

———. 1994. The description of *Hypoglossum subsimplex sp. nov.* (Delesseriaceae, Rhodophyta) from the Florida Keys, Gulf of Mexico. *Cryptogam. Algol.* 15: 253–262.

———. 1997. *Rosenvingea antillarum* (P. Crouan and H. Crouan) *comb. nov.* to replace *R. floridana* (W. R. Taylor) W. R. Taylor (Scytosiphonales, Phaeophyta). *Cryptogam. Algol.* 18: 331–336.

———. 1998a. A checklist of benthic marine algae of the tropical and subtropical western Atlantic: first revision. *Nova Hedwigia Beih.* 116.155 pp.

———. 1998b. A study of *Padina antillarum* (Kützing) Piccone and a comparison with *P. tetrastromatica* Hauck (Dictyotales, Phaeophyta). *Cryptogam. Algol.* 4: 271–289.

———. 2001. New records of benthic marine algae from the Sultanate of Oman, northern Arabian Sea. II. *Nova Hedwigia* 72: 347–374.

———. 2002. *Turbinaria foliosa sp. nov.* (Fucales, Phaeophyceae) from the Sultanate of Oman, with a census of currently recognized species in the genus *Turbinaria*. *Phycol. Res.* 50: 283–293.

———. 2005a. A checklist of benthic marine algae of the tropical and subtropical western Atlantic: second revision. *Nova Hedwigia Beih.* 129, 152 pp.

———. 2005b. *Chrysymenia tigillis sp. nov.* (Rhodymeniales, Rhodophyta) from the Sultanate of Oman, with a census of currently recognized species in the genus *Chrysymenia*. *Phycol. Res.* 53: 215–223.

———. 2005c. Two new species of *Bryopsis* (Ulvophyceae, Chlorophyta) from the Sultanate of Oman, with a census of currently recognized species in the genus. *Contrib. Univ. Mich. Herb.* 24: 229–256.

———. 2006. Corrigenum. *Phycol. Res.* 54: 86.

Wynne, M. J., and D. L. Ballantine. 1985. Notes on the marine algae of Puerto Rico. IV. The taxonomic placement of *Grallatoria* (Ceramiaceae, Rhodophyta). *Crytotogam. Algol.* 6: 219–229.

———. 1986. The genus *Hypoglossum* Kützing (Delesseriaceae, Rhodophyta) in the tropical western Atlantic, including *H. anomalum sp. nov. J. Phycol.* 22: 185–193.

———. 1991. *Laurencia iridescens sp. nov.* (Rhodomelaceae, Ceramiales) from the Caribbean Sea. *Phycologia* 30: 394–401.

Wynne, M. J., and O. De Clerck. 1999a. First reports of *Padina antillarum* and *P. glabra* (Phaeophyta, Dictyotaceae) from Florida, with a key to the western Atlantic species of the genus. *Caribb. J. Sci.* 35: 286–295.

———. 1999b. *Taxonomic* notes on *Caloglossa monosticha* and *Caloglossa saigonensis* (Delesseriaceae, Rhodophyta). *Contrib. Univ. Mich. Herb.* 22: 209–213.

Wynne, M. J., and G. T. Kraft. 1985. *Hypoglossum caloglossoides sp. nov.* (Delesseriaceae, Rhodophyta) from Lord Howe Island, South Pacific. *Br. Phycol. J.* 20: 9–19.

Wynne, M. J., D. Serio, M. Cormaci, and G. Furnari. 2005. The species of *Chondrophycus* and *Laurencia* (Rhodomelaceae, Ceramiales) occurring in Dhofar, the Sultanate of Oman. *Phycologia* 44: 497–509.

Wynne, M. J., and W. R. Taylor. 1973. The status of *Agardhiella tenera* and *Agardiella baileyi* (Rhodophyta, Gigartinales). *Hydrobiologia* 43: 93–107.

Wysor, B. 2004. An annotated list of marine Chlorophyta from the Pacific Coast of the Republic of Panama with a comparison to Caribbean Panama species. *Nova Hedwigia* 78: 209–241.

Wysor, B., and O. De Clerck. 2003. An updated and annotated list of marine brown algae (Phaeophyceae) of the Caribbean Coast of the Republic of Panama. *Bot. Mar.* 46: 151–160.

Wysor, B., and W. H. C. Kooistra. 2003. An annotated list of marine Chlorophyta from the Caribbean coast of the Republic of Panama. *Nova Hedwigia* 77: 487–523.

Xia, B. M., and J. Zhang. 1999. *Flora algarum marinarum sinicatum*. Tomus II Rhodophyta No. V Ahnfeltiales, Gigartinales, Rhodymeniales. Science Press, Beijing, 201 pp., 11 pls.

Yamada, Y. 1931. Notes on *Laurencia*, with special reference to the Japanese species. *Univ. Calif. Publ. Bot.* 16: 185–310.

———. 1934. The marine Chlorophyceae from Ryukyu, especially from the vicinity of Nawa. *J. Fac. Sci., Hokkaido Imp. Univ.* Ser 5, 3: 33–88.

———. 1938a. The species of *Liagora* from Japan. *Sci. Pap. Inst. Algol. Res., Fac. Sci., Hokkaido Univ.* 2: 1–34.

———. 1938b. The marine algae from the island of Yonakuni. *Sci. Pap. Inst. Algol. Res., Fac. Sci., Hokkaido Univ.* 2: 53–86.

———. 1944a. List of marine algae from the atoll of Ant. *Sci. Pap. Inst. Algol. Res., Fac. Sci., Hokkaido Univ.* 3: 31–45.

———. 1944b. Notes on some Japanese algae X. *Sci. Pap. Inst. Algol. Res., Fac. Sci., Hokkaido Univ.* 3: 11–25.

———. 1950. A list of marine algae from Ryukyusho, Formosa, I. *Sci. Papers Inst. Algol. Res., Fac. Sci., Hokkaido Univ.* 3: 173–194.

Yamagishi, H., and M. Masuda. 2000. A taxonomic revision of a *Hypnea charoides-valentiae* complex (Rhodophyta, Gigartinales) in Japan, with a description of *H. flexicaulis sp. nov. Phycol. Res.* 48: 27–35.

Yamagishi, H., M. Masuda, T. Abe, S. Uwai, K. Kogame, S. Kawaguchi, and S. M. Phag. 2003. Taxonomic notes on marine algae from Malaysia. XI. Four species of Rhodophyceae. *Bot. Mar.* 46: 534–547.

Yamamoto, H. 1986. *Congracilaria babae gen et sp. nov.* (Gracilariaceae), an adelphoparasite growing on *Gracilaria salicornia* of Japan. *Bull. Fac. Fish. Hokkaido Univ.* 37: 281–290.

Yeh, W. J., and G. Y. Chen. 2004. Nuclear rDNA and internal transcribed spacer sequences clarify *Caulerpa racemosa* vars. from other *Caulerpa* species. *Aquat. Bot.* 80: 193–207.

Yoneshigue-Valentin, Y., and A. M. Loivos. 2005. Novas adições de macroalgas na plataforma continental do litoral brasileiro. 1. Ocorrência de *Carpomitra costata* e *Nereia tropica* (Sporochnales, Phaeophyta). In: Reunião Brasileira de Ficologia, 10., 2004, Salvador. *Formação de ficólogos: um compromisso com a sustentabilidade dos recursos aquáticos: anais. . .* Rio de Janeiro: Museu Nacional, pp. 325–338. Org. Sociedade Brasileira de Ficologia (Série Livros; 10).

York, P. H., D. J. Booth, T. M. Glasby, and B. C. Pease. 2006. Fish assemblages in habitats dominated by *Caulerpa taxifolia* and native seagrasses in south-eastern Australia. *Mar. Ecol. Prog. Ser.* 312: 223–234.

Yoshida, T. 1998. *Marine algae of Japan.* Uchida Rokakuho Publishing, Tokyo (in Japanese).

Yoshida, T., and H. Mikami. 1991. *Sorella pulchra* (Yamada) *comb. nov.*, based on *Erythroglossum pulchrum* Yamada (Delesseriaceae, Rhodophyta). *Jpn. J. Phycol.* 39: 123–129.

Yoshida, T., Y. Nakajima, and Y. Nakata. 1990. Check-list of marine algae of Japan (revised in 1990). *Jpn. J. Phycol.* 38: 269–320.

Yoshida, T., K. Yoshinga, and Y. Nakajima. 1995. Check list of marine algae of Japan (revised in 1995). *Jpn. J. Phycol.* 43: 115–171.

Young, D. N. 1981. Taxonomic observations on eastern Pacific *Antithamnion* species (Rhodophyta: Ceramiaceae) described by E. Y. Dawson. *Proc. Biol. Soc. Wash.* 95: 94–100.

Zaleski, S. F., and S. N. Murray. 2006. Taxonomic diversity and geographic distributions of aquarium-traded species of *Caulerpa* (Chlorophyta: Caulerpaceae) in southern California, USA. *Mar. Ecol. Prog. Ser.* 314: 97–108.

Zanardini, G. 1839. Species algarum novae vel minus cognitae. *Biblioteca Italiana* 96: 134–137.

———. 1840a. [Lettera] Alla Direzione della Biblioteca Italiana. *Biblioteca Italiana* [Milano] 96: 131–137.

———. 1840b. Sopra le Alghe del mare Adriatico. Lettera seconda di Giovanni Zanardini, medico fisico in Venezia, alla Direzione della Biblioteca Italiana. *Biblioteca Italiana* [Milano], 99: 195–229.

———. 1841('1842'). Synopsis algarum in mari Adriatico hucusque collectarum, cui accedunt monographia siphonearum nec non generales de algarum vita et structura disquisitiones cum tabulis auctoris manu ad vivum depictis. *Mem. Reale Accad. Sci. Torino*, ser. 2, 4: 1–255, pls. 1–8.

———. 1843. Saggio di classificazione naturale delle Ficee . . . Venezia. 64 pp., pl. [2] folded tables.

———. 1846. Memoria sulla *Desmarestia filiformis* di Giacobbe Agardh e sulle Chordariee in generale. Atti del VII Congresso degli scienziati italiani in Napoli, 1845, 2: 899–900.

Zechman, F. W. 2003. Phylogeny of the Dasycladales (Chlorophyta, Ulvophyceae) based on analyses of RUBISCO large subunit (rbcL) gene sequence. *J. Phycol.* 39: 819–836.

Zeh, W. 1912. Neue Arten der Gattung *Liagora.* Notizbl. des Köngil. Botan. Gart. und Mus. zu Berlin 5: 268–273.

Zieman, J. C. 1975. Tropical seagrass ecosystems and pollution. In: F.J.F. Wood and R. E. Johannes (eds.), *Tropical marine pollution*, Elsevier, Amsterdam, pp. 63–74.

———. 1982. *The ecology of the seagrasses of South Florida: a community profile.* U.S. Fish Wildl. Serv., Off. Biol. Serv. FWS/OBS-82/25, Washington, D.C., 185 pp.

Zieman, J. C., and R. T. Zieman. 1989. *The ecology of the seagrass meadows of the west coast of Florida: a community profile.* U.S. Fish Wildl. Serv., Biol. Rep. 85(7.25), Washington, D.C., ix + 155 pp.

Zimmerman, M. S., and R. J. Livingston. 1976. Seasonality and physico-chemical ranges of benthic macrophytes from a north Florida estuary (Apalachee Bay). *Contrib. Mar. Sci.* 20: 33–45.

Zinova, A. D. 1967. *A guide to the green, brown and red algae of the southern seas of the U.S.S.R.* V. L. Komarov Botanical Institute Academy of Science U S.S.R., Moscow and Lennigrad, 398 pp. (in Russian).

Zollinger, H. 1854. Systematische Verzeichniss der im indischen Archipel, in den Jahren 1842–1848 gesammelten sowie der aus Japan empfangenen Pflanzen. Zürich. Heft 1, xii + 80 pp.

Zuccarello, G., J. Buchanan, and J. A. West. 2006a. Increased sampling for inferring phylogeographic patterns in *Bostrychia radicans/B. moritziana* (Rhodomelaceae, Rhodophyta) in the eastern USA. *J. Phycol.* 42: 1349–1352.

Zuccarello, G., and J. A. West. 2002. Phylogeography of the *Bostrychia calliptera-B. pinnata* complex (Rhodomelaceae, Rhodophyta) and divergence rates based on nuclear, mitochondrial and plastid DNA markers. *Phycologia* 41: 49–60.

———. 2003. Multiple cryptic species: molecular diversity and reproductive isolation in the *Bostrychia radicans/B. moritziana* complex (Rhodomelaceae, Rhodophyta) with focus on North American isolates. *J. Phycol.* 39: 948–959.

———. 2006. Molecular phylogeny of the subfamily Bostrychioideae (Ceramiales, Rhodophyta): subsuming *Stictosiphonia* and highlighting polyphyly in species of *Bostrychia*. *Phycologia* 45: 24–36.

Zuccarello, G., J. A. West, A. Bitans, and G. Kraft. 2000. Molecular phylogeny of *Rhodochaete parvula* (Bangiophycidae, Rhodophyta). *Phycologia* 39: 75–81.

Zuccarello, G., J. A. West, and S. Loiseaux-de Goer. 2006b. Diversity of the *Bostrychia radicans/Bostrychia moritziana* species complex (Rhodomelaceae, Rhodophyta) in the mangroves of New Caledonia. *Cryptogamie Algol.* 27: 245–254.

Taxonomic Index

Taxa in **bold** occur in Florida. Pages numbers in **bold** refer to figure descriptions.

An internationally recognized expert on the ecology of seaweed taxa, Clinton J. Dawes is Distinguished Research Professor of biology at the University of South Florida. He is the author of eight books, including *Marine Botany*, second edition (1998).

Arthur C. Mathieson is professor of plant biology at the University of New Hampshire and the coeditor of *Ecosystems of the World, Vol. 14: Intertidal and Littoral Ecosystems* (1991).